A First Course in Linear Model Theory

CHAPMAN & HALL/CRC

Texts in Statistical Science Series

Recently Published Titles

Beyond Multiple Linear Regression
Applied Generalized Linear Models and Multilevel Models in R
Paul Roback, Julie Legler

Bayesian Thinking in Biostatistics
Gary L. Rosner, Purushottam W. Laud, and Wesley O. Johnson

Linear Models with Python
Julian J. Faraway

Modern Data Science with R, Second Edition
Benjamin S. Baumer, Daniel T. Kaplan, and Nicholas J. Horton

Probability and Statistical Inference
From Basic Principles to Advanced Models
Miltiadis Mavrakakis and Jeremy Penzer

Bayesian Networks
With Examples in R, Second Edition
Marco Scutari and Jean-Baptiste Denis

Time Series
Modeling, Computation, and Inference, Second Edition
Raquel Prado, Marco A. R. Ferreira and Mike West

A First Course in Linear Model Theory, Second Edition
Nalini Ravishanker, Zhiyi Chi, Dipak K. Dey

Foundations of Statistics for Data Scientists
With R and Python
Alan Agresti and Maria Kateri

Fundamentals of Causal Inference
With R
Babette A. Brumback

Sampling
Design and Analysis, Third Edition
Sharon L. Lohr

Theory of Statistical Inference
Anthony Almudevar

Probability, Statistics, and Data: A Fresh Approach Using R
Darrin Speegle and Brain Claire

For more information about this series, please visit: https://www.crcpress.com/Chapman--Hall/CRC-Texts-in-Statistical-Science/book-series/CHTEXSTASCI

A First Course in Linear Model Theory

Second Edition

Nalini Ravishanker
Zhiyi Chi
Dipak K. Dey

CRC Press
Taylor & Francis Group
Boca Raton London New York

CRC Press is an imprint of the
Taylor & Francis Group, an **informa** business
A CHAPMAN & HALL BOOK

Second edition published 2022
by CRC Press
6000 Broken Sound Parkway NW, Suite 300, Boca Raton, FL 33487-2742

and by CRC Press
2 Park Square, Milton Park, Abingdon, Oxon, OX14 4RN

First edition published by CRC Press 2001

CRC Press is an imprint of Taylor & Francis Group, LLC

Library of Congress Cataloging-in-Publication Data

Names: Ravishanker, Nalini, author. | Chi, Zhiyi, author. | Dey, Dipak, author.
Title: A first course in linear model theory / Nalini Ravishanker, Zhiyi Chi, Dipak K. Dey.
Description: Second edition. | Boca Raton : CRC Press, 2021. | Series: Chapman & Hall/CRC texts in statistical science | Includes bibliographical references and index.
Identifiers: LCCN 2021019402 (print) | LCCN 2021019403 (ebook) | ISBN 9781439858059 (hardback) | ISBN 9781032101392 (paperback) | ISBN 9781315156651 (ebook)
Subjects: LCSH: Linear models (Statistics)
Classification: LCC QA276 .R38 2021 (print) | LCC QA276 (ebook) | DDC 519.5/35--dc23
LC record available at https://lccn.loc.gov/2021019402
LC ebook record available at https://lccn.loc.gov/2021019403

ISBN: 978-1-439-85805-9 (hbk)
ISBN: 978-1-032-10139-2 (pbk)
ISBN: 978-1-315-15665-1 (ebk)

DOI: 10.1201/9781315156651

Typeset in CMR10 font
by KnowledgeWorks Global Ltd.

To my family and friends. N.R.

To my family. Z.C.

To Rita and Debosri. D.K.D.

Contents

Preface to the First Edition

Linear Model theory plays a fundamental role in the foundation of mathematical and applied statistics. It has a base in distribution theory and statistical inference, and finds application in many advanced areas in statistics including univariate and multivariate regression, analysis of designed experiments, longitudinal and time series analysis, spatial analysis, multivariate analysis, wavelet methods, etc. Most statistics departments offer at least one course on linear model theory at the graduate level. There are several excellent books on the subject, such as "Linear Statistical Inference and its Applications" by C.R. Rao, "Linear Models" by S.R. Searle, "Theory and Applications of the Linear Model" by F.A. Graybill, "Plane Answers to Complex Questions: The Theory of Linear Models" by R. Christiansen and "The Theory of Linear Models" by B. Jorgensen.

Our motivation has been to incorporate general principles of inference in linear models to the fundamental statistical education of students at the graduate level, while our treatment of contemporary topics in a systematic way will serve the needs of professionals in various industries. The three salient features of this book are: (1) developing standard theory of linear models with numerous applications in simple and multiple regression, as well as fixed, random and mixed-effects models, (2) introducing generalized linear models with examples, and (3) presenting some current topics including Bayesian linear models, general additive models, dynamic linear models and longitudinal models. The first two chapters introduce to the reader requisite linear and matrix algebra. This book is therefore a self-contained exposition of the theory of linear models, including motivational and practical aspects. We have tried to achieve a healthy compromise between theory and practice, by providing a sound theoretical basis, and indicating how the theory works in important special cases in practice. There are several examples throughout the text. In addition, we provide summaries of many numerical examples in different chapters, while a more comprehensive description of these is available in the first author's web site (http://www.stat.uconn.edu/∼nalini). There are several exercises at the end of each chapter that should serve to reinforce the methods.

Our entire book is intended for a two semester graduate course in linear models. For a one semester course, we recommend essentially the first eight chapters, omitting a few subsections, if necessary, and supplementing a few selected topics from chapters 9-11, if time permits. For instance, section 5.5, section 6.4, sections 7.5.2-7.5.4, and sections 8.5, 8.7 and 8.8 may be omitted in a one semester course. The first two chapters, which present a review on vectors and matrices specifically as they pertain to linear model theory, may also be assigned as background reading if the students had previous exposure to these topics. Our book requires some knowledge of statistics; in particular, a knowledge of elementary sampling distributions, basic estimation theory and hypothesis testing at an undergraduate level is definitely required. Occasionally, more advanced concepts of statistical inference are invoked in this book, for which suitable references are provided.

The plan of this book follows. The first two chapters develop basic concepts of linear and matrix algebra with a view towards application in linear models. Chapter 3 describes generalized inverses and solutions to systems of linear equations. We develop the notion of a general linear model in Chapter 4. An attractive feature of our book is that we unify full-rank and non full-rank models in the development of least squares inference and optimality via the Gauss-Markov theorem. Results for the full-rank (regression) case are provided as special cases. We also introduce via examples, balanced ANOVA models that are widely used in practice. Chapter 5 deals with multivariate normal and related distributions, as well as distributions of quadratic forms that are at the heart of inference. We also introduce the class of elliptical distributions that can serve as error distributions for linear models. Sampling from multivariate normal distributions is the topic of Chapter 6, together with assessment of and transformations to multivariate normality. This is followed by inference for the general linear model in Chapter 7. Inference under normal and elliptical errors is developed and illustrated on examples from regression and balanced ANOVA models. In Chapter 8, topics in multiple regression models such as model checking, variable selection, regression diagnostics, robust regression and nonparametric regression are presented. Chapter 9 is devoted to the study of unbalanced designs in fixed-effects ANOVA models, the analysis of covariance (ANACOVA) and some nonparametric test procedures. Random-effects models and mixed-effects models are discussed in detail in Chapter 10. Finally in Chapter 11, we introduce several special topics including Bayesian linear models, dynamic linear models, linear longitudinal models and generalized linear models (GLIM). The purpose of this chapter is to introduce to the reader some new frontiers of linear models theory; several references are provided so that the reader may explore further in these directions. Given the exploding nature of our subject area, it is impossible to be exhaustive in a text, and cover everything that should ideally be covered. We hope that our judgment in choice of material is appropriate and useful.

Most of our book was developed in the form of lecture notes for a sequence of two courses on linear models which both of us have taught for several years in the Department of Statistics at the University of Connecticut. The numerical examples in the text and in the web site were developed by NR over many years. In the text, we have acknowledged published work, wherever appropriate, for the use of data in our numerical examples, as well as for some of the exercise problems. We are indeed grateful for their use, and apologize for any inadvertent omission in this regard.

In writing this text, discussions with many colleagues were invaluable. In particular, we thank Malay Ghosh, for several suggestions that vastly improved the structure and content of this book. We deeply appreciate his time and goodwill. We thank Chris Chatfield and Jim Lindsey for their review and for the suggestion about including numerical examples in the text. We are also very grateful for the support and encouragement of our statistical colleagues, in particular Joe Glaz, Bani Mallick, Alan Gelfand and Yazhen Wang. We thank Ming-Hui Chen for all his technical help with Latex.

Many graduate students helped in proof reading the typed manuscript; we are especially grateful to Junfeng Liu, Madhuja Mallick and Prashni Paliwal. We also thank Karen Houle, a graduate student in Statistics, who helped with "polishing-up" the numerical examples in NR's web site. We appreciate all the help we received from people at Chapman & Hall/CRC – Bob Stern, Helena Redshaw, Gail Renard and Sean Davey.

Nalini Ravishanker and Dipak K. Dey
Department of Statistics
University of Connecticut
Storrs, CT

Preface to the Second Edition

Linear Model theory plays a fundamental role in the foundation of mathematical and applied statistics. Our motivation in writing the first edition of the book was to incorporate general principles of inference in linear models to the fundamental statistical education of students at the graduate level, while our treatment of contemporary topics in a systematic way will serve the needs of professionals in various industries.

The attractive features of the second edition are: (1) developing standard theory of linear models with numerous applications in simple and multiple regression, as well as fixed, random and mixed-effects models in the first few chapters, (2) devoting a chapter to the topic of generalized linear models with examples, (3) including two new chapters which contain detailed presentations on current and interesting topics such as multivariate linear models, Bayesian linear models, general additive models, dynamic linear models, longitudinal models, robust regression, regularized regression, etc., and (4) providing numerical examples in R for several methods, whose details are available here: https://github.com/nravishanker/FCILM-2. As in the first edition, here too, we have tried to achieve a healthy compromise between theory and practice, by providing a sound theoretical basis, and indicating how the theory works in important special cases in practice. There are several exercises at the end of each chapter that should serve to reinforce the methods.

Our entire book is intended for a two semester graduate course in linear models. For a one semester course, we recommend essentially the first eight chapters, omitting a few subsections, if necessary, and including a few selected topics from chapters 9-13, if time permits. For instance, section 5.5, section 6.4, sections 7.5.2-7.5.4, and sections 8.5, 8.7 and 8.8 may be omitted in a one semester course. The first two chapters, which present a review on vectors and matrices specifically as they pertain to linear model theory, may also be assigned as background reading if the students had previous exposure to these topics. Our book requires some knowledge of statistics; in particular, a knowledge of elementary sampling distributions, basic estimation theory and hypothesis testing at an undergraduate level is definitely required. When more advanced concepts of statistical inference are invoked in this book, suitable references are provided.

The plan of this book follows. The first two chapters develop basic concepts of linear and matrix algebra with a view towards application in linear models. Chapter 3 describes generalized inverses and solutions to systems of linear equations. We develop the notion of a general linear model in Chapter 4. An attractive feature of our book is that we unify full-rank and non full-rank models in the development of least squares inference and optimality via the Gauss–Markov theorem. Results for the full-rank (regression) case are provided as special cases. We also introduce via examples, balanced ANOVA models that are widely used in practice. Chapter 5 deals with multivariate normal and related distributions, as well as distributions of quadratic forms that are at the heart of inference. We also introduce the class of elliptical distributions that can serve as error distributions for linear models. Sampling from multivariate normal distributions is the topic of Chapter 6, together with assessment of and transformations to multivariate normality. In the second edition, we describe inference for the general linear model in Chapters and 8. Inference under normal and elliptical errors is developed and illustrated on examples from regression and balanced ANOVA models. Topics in multiple regression models such as model checking, variable selection, regression diagnostics, robust regression and nonparametric regression are presented in Chapter 9. Chapter 10 is devoted to the study of unbalanced designs in fixed-effects ANOVA models, the analysis of covariance (ANACOVA) and some nonparametric procedures. In this edition, we have included a new section on multiple comparisons. Random-effects models and mixed-effects models are discussed in detail in Chapter 11. Chapter 12 is devoted to generalized linear models, whereas in the first edition, we discussed this in just one section. In Chapter 13, we introduce several special topics including Bayesian linear models, dynamic linear models, linear longitudinal models and generalized linear models (GLIM). The purpose of this chapter is to introduce to the reader some new frontiers of linear models theory; several references are provided so that the reader may explore further in these directions. In Chapter 14, we have discussed the theory and applications for robust and regularized regressions, nonparametric methods for regression analysis and some ideas in missing data analysis in linear models. Given the exploding nature of our subject area, it is impossible to be exhaustive in a text, and cover everything that should ideally be covered. The second edition of the book considerably enhances the early parts of the book and further, also includes an in-depth treatment of a few useful and currently relevant additional topics. We hope that our judgment in choice of material for the second edition is useful. We have acknowledged published work, wherever appropriate, for the use of data and for some of the exercises. We are grateful for their use, and also grateful for the use of R and RStudio in the numerical examples. We apologize for any inadvertent omission in regard to acknowledgments.

We are very grateful to our colleagues, Haim Bar, Kun Chen, Min-Hui Chen, Yuwen Gu and Elizabeth Schifano, for useful discussions about the material. We also thank our PhD students, Sreeram Anantharaman, Surya Eada, Namitha Pais, and Patrick Toman for their help with the numerical examples and linking to github. We appreciate all the help we received from David Grubbs and Robin Lloyd Starkes at Chapman & Hall/CRC. Last, but not least, a very big thank you to our families and friends for their constant support.

Nalini Ravishanker, Zhiyi Chi, and Dipak K. Dey
Department of Statistics
University of Connecticut
Storrs, CT

1

Review of Vector and Matrix Algebra

In this chapter, we introduce basic results dealing with vector spaces and matrices, which are essential for an understanding of linear statistical methods. We provide several numerical and geometrical illustrations of these concepts. The material presented in this chapter is found in most textbooks that deal with matrix theory pertaining to linear models, including Graybill (1983), Harville (1997), Rao (1973), and Searle (1982). Unless stated otherwise, all vectors and matrices are assumed to be real, i.e., they have real numbers as elements.

1.1 Notation

An $m \times n$ matrix $\mathbf{A}$ is a rectangular array of real numbers of the form

$$\mathbf{A} = \begin{pmatrix} a_{11} & a_{12} & \cdots & a_{1n} \\ a_{21} & a_{22} & \cdots & a_{2n} \\ \vdots & \vdots & \vdots & \vdots \\ a_{m1} & a_{m2} & \cdots & a_{mn} \end{pmatrix} = \{a_{ij}\}$$

with row dimension m, column dimension n, and (i,j)th element a_{ij}. For example,

$$\mathbf{A} = \begin{pmatrix} 5 & 4 & 1 \\ -3 & 2 & 6 \end{pmatrix}$$

is a 2×3 matrix. We sometimes use $\mathbf{A} \in \mathcal{R}^{m \times n}$ to denote that $\mathbf{A}$ is an $m \times n$ matrix of real numbers. An n-dimensional column vector

$$\mathbf{a} = \begin{pmatrix} a_1 \\ \vdots \\ a_n \end{pmatrix}$$

can be thought of as a matrix with n rows and one column. For example,

$$\mathbf{a} = \begin{pmatrix} 1 \\ -1 \end{pmatrix}, \ \mathbf{b} = \begin{pmatrix} 3 \\ 1 \\ 5 \end{pmatrix}, \text{ and } \mathbf{c} = \begin{pmatrix} 0.25 \\ 0.50 \\ 0.75 \\ 1.00 \end{pmatrix}$$

are respectively 2-dimensional, 3-dimensional, and 4-dimensional vectors. An n-dimensional column vector with each of its n elements equal to unity is denoted by $\mathbf{1}_n$, while a column vector whose elements are all zero is called the null vector or the zero vector and is denoted by $\mathbf{0}_n$. When the dimension is obvious, we will drop the subscript. For any integer $n \geq 1$, we can write an n-dimensional column vector as $\mathbf{a} = (a_1, \cdots, a_n)'$, i.e., as the *transpose* of the

DOI: 10.1201/9781315156651-1

n-dimensional (row) vector with components $a_1, \cdots, a_n$. In this book, a vector denotes a column vector, unless stated otherwise. We use $\mathbf{a} \in \mathcal{R}^n$ to denote that $\mathbf{a}$ is an n-dimensional (column) vector.

An $m \times n$ matrix $\mathbf{A}$ with the same row and column dimensions, i.e., with $m = n$, is called a square matrix of order n. An $n \times n$ identity matrix is denoted by $\mathbf{I}_n$; each of its n diagonal elements is unity while each off-diagonal element is zero. An $m \times n$ matrix $\mathbf{J}_{mn}$ has each element equal to unity. An $n \times n$ unit matrix is denoted by $\mathbf{J}_n$. For example, we have

$$\mathbf{I}_3 = \begin{pmatrix} 1 & 0 & 0 \\ 0 & 1 & 0 \\ 0 & 0 & 1 \end{pmatrix}, \quad \mathbf{J}_{23} = \begin{pmatrix} 1 & 1 & 1 \\ 1 & 1 & 1 \end{pmatrix} \quad \text{and} \quad \mathbf{J}_3 = \begin{pmatrix} 1 & 1 & 1 \\ 1 & 1 & 1 \\ 1 & 1 & 1 \end{pmatrix}.$$

A $n \times n$ permutation matrix $\mathbf{R}$ is obtained by permuting the rows of $\mathbf{I}_n$. Each row and column of $\mathbf{R}$ contains exactly one 1 and has zeroes elsewhere. For example, when $n = 3$, $\mathbf{R} = \begin{pmatrix} 1 & 0 & 0 \\ 0 & 0 & 1 \\ 0 & 1 & 0 \end{pmatrix}$ is a permutation of $\mathbf{I}_3$.

An $n \times n$ matrix whose elements are zero except on the diagonal, where the elements are nonzero, is called a diagonal matrix. We will denote a diagonal matrix by $\mathbf{D} = \text{diag}(d_1, \cdots, d_n)$. Note that $\mathbf{I}_n$ is an $n \times n$ diagonal matrix, written as $\mathbf{I}_n = \text{diag}(1, \cdots, 1)$. An $m \times n$ matrix all of whose elements are equal to zero is called the null matrix or the zero matrix $\mathbf{O}$. An $n \times n$ matrix is said to be an upper triangular matrix if all the elements below and to the left of the main diagonal are zero. Similarly, if all the elements located above and to the right of the main diagonal are zero, then the $n \times n$ matrix is said to be lower triangular. For example,

$$\mathbf{U} = \begin{pmatrix} 5 & 4 & 3 \\ 0 & 2 & -6 \\ 0 & 0 & 5 \end{pmatrix} \quad \text{and} \quad \mathbf{L} = \begin{pmatrix} 5 & 0 & 0 \\ 4 & 2 & 0 \\ 3 & -6 & 5 \end{pmatrix}$$

are respectively upper triangular and lower triangular matrices. A square matrix is triangular if it is either upper triangular or lower triangular. A triangular matrix is said to be a unit triangular matrix if $a_{ij} = 1$ whenever $i = j$. Unless explicitly stated, we assume that vectors and matrices are non-null.

A submatrix of a matrix $\mathbf{A}$ is obtained by deleting certain rows and/or columns of $\mathbf{A}$. For example, let

$$\mathbf{A} = \begin{pmatrix} 1 & 3 & 5 & 7 \\ 5 & 4 & 1 & -9 \\ -3 & 2 & 6 & 4 \end{pmatrix} \quad \text{and} \quad \mathbf{B} = \begin{pmatrix} 5 & 4 & 1 \\ -3 & 2 & 6 \end{pmatrix}.$$

The 2×3 submatrix $\mathbf{B}$ is obtained by deleting row 1 and column 4 of the 3×4 matrix $\mathbf{A}$. Any matrix can be considered to be a submatrix of itself. We call a submatrix obtained by deleting the same rows and columns from $\mathbf{A}$ a principal submatrix of $\mathbf{A}$. For $r = 1, 2, \cdots, n$, the $r \times r$ leading principal submatrix of $\mathbf{A}$ is obtained by deleting the last $(n - r)$ rows and columns from $\mathbf{A}$. The 2×2 leading principal submatrix of the matrix $\mathbf{A}$ shown above is

$$\mathbf{C} = \begin{pmatrix} 1 & 3 \\ 5 & 4 \end{pmatrix}.$$

It may be easily verified that a principal submatrix of a diagonal, upper triangular or lower triangular matrix is respectively diagonal, upper triangular or lower triangular.

Some elementary properties of vectors and matrices are given in the following two sections. Familiarity with this material is recommended before a further study of properties of special matrices that are described in the following two chapters.

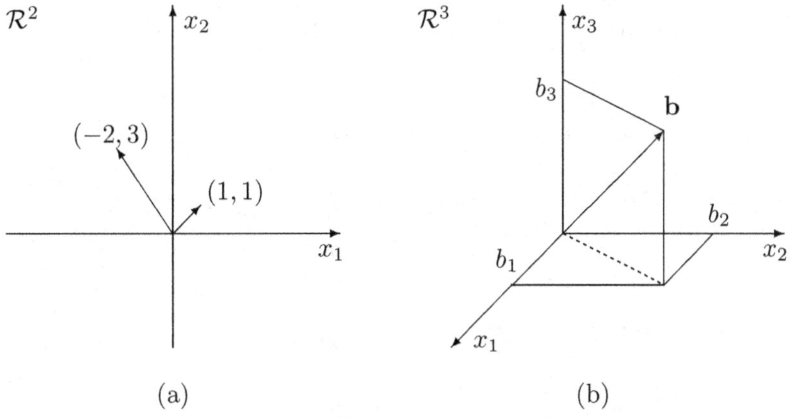

FIGURE 1.2.1. Geometric representation of 2- and 3-dimensional vectors.

1.2 Basic properties of vectors

An n-dimensional vector $\mathbf{a}$ is an ordered set of measurements, which can be represented geometrically as a directed line in n-dimensional space $\mathcal{R}^n$ with component a_1 along the first axis, component a_2 along the second axis, etc., and component a_n along the nth axis. We can represent 2-dimensional and 3-dimensional vectors respectively as points in the plane $\mathcal{R}^2$ and in 3-dimensional space $\mathcal{R}^3$. In this book, we always assume Cartesian coordinates for a Euclidean space, such as a plane or a 3-dimensional space.

Any 2-dimensional vector $\mathbf{a} = (a_1, a_2)'$ can be graphically represented by the point with coordinates (a_1, a_2) in the Cartesian coordinate plane, or as the arrow starting from the origin $(0, 0)$, whose tip is the point with coordinates (a_1, a_2). For $n = 2$, Figure 1.2.1 (a) shows the vectors $(1, 1)$ and $(-2, 3)$ as arrows starting from the origin. For $n = 3$, Figure 1.2.1 (b) shows a vector $\mathbf{b} = (b_1, b_2, b_3)'$ in $\mathcal{R}^3$.

Two vectors can be added (or subtracted) only if they have the same dimension, in which case the sum (or difference) of the two vectors is the vector of sums (or differences) of their elements, i.e.,

$$\mathbf{a} \pm \mathbf{b} = (a_1 \pm b_1, \cdots, a_n \pm b_n)'.$$

The sum of two vectors emanating from the origin is the diagonal of the parallelogram which has the vectors $\mathbf{a}$ and $\mathbf{b}$ as adjacent sides. Vector addition is commutative and associative, i.e., $\mathbf{a} + \mathbf{b} = \mathbf{b} + \mathbf{a}$, and $\mathbf{a} + (\mathbf{b} + \mathbf{c}) = (\mathbf{a} + \mathbf{b}) + \mathbf{c}$. The scalar multiple $c\mathbf{a}$ of a vector $\mathbf{a}$ is obtained by multiplying each element of $\mathbf{a}$ by the scalar c, i.e.,

$$c\mathbf{a} = (ca_1, \cdots, ca_n)'.$$

Definition 1.2.1. Inner product of vectors. The inner product of two n-dimensional vectors $\mathbf{a}$ and $\mathbf{b}$ is denoted by $\mathbf{a} \bullet \mathbf{b}$, $\mathbf{a}'\mathbf{b}$, or $\langle \mathbf{a}, \mathbf{b} \rangle$, and is the scalar

$$\mathbf{a}'\mathbf{b} = (a_1, \cdots, a_n) \begin{pmatrix} b_1 \\ \vdots \\ b_n \end{pmatrix} = a_1 b_1 + \cdots + a_n b_n = \sum_{i=1}^{n} a_i b_i.$$

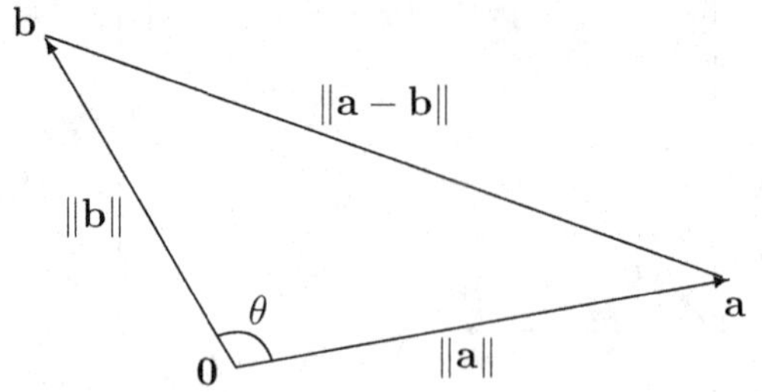

FIGURE 1.2.2. Distance and angle between two vectors.

The inner product of a vector $\mathbf{a}$ with itself is $\mathbf{a}'\mathbf{a}$. The positive square root of this quantity is called the *Euclidean norm*, or *length*, or *magnitude* of the vector, and is

$$\|\mathbf{a}\| = (a_1^2 + \cdots + a_n^2)^{1/2}.$$

Geometrically, the length of a vector $\mathbf{a} = (a_1, a_2)'$ in two dimensions may be viewed as the hypotenuse of a right triangle, whose other two sides are given by the vector components, a_1 and a_2. Scalar multiplication of a vector $\mathbf{a}$ changes its length,

$$\|c\mathbf{a}\| = (c^2a_1^2 + \cdots + c^2a_n^2)^{1/2} = |c|(a_1^2 + \cdots + a_n^2)^{1/2} = |c|\|\mathbf{a}\|.$$

If $|c| > 1$, $\mathbf{a}$ is expanded by scalar multiplication, while if $|c| < 1$, $\mathbf{a}$ is contracted. If $c = 1/\|\mathbf{a}\|$, the resulting vector is defined to be $\mathbf{b} = \mathbf{a}/\|\mathbf{a}\|$, the n-dimensional unit vector with length 1. A vector has both length and direction. If $c > 0$, scalar multiplication does not change the direction of a vector $\mathbf{a}$. However, if $c < 0$, the direction of the vector $c\mathbf{a}$ is in opposite direction to the vector $\mathbf{a}$. The unit vector $\mathbf{a}/\|\mathbf{a}\|$ has the same direction as $\mathbf{a}$.

The Euclidean distance between two vectors $\mathbf{a}$ and $\mathbf{b}$ is defined by

$$d(\mathbf{a}, \mathbf{b}) = d(\mathbf{b}, \mathbf{a}) = \|\mathbf{a} - \mathbf{b}\| = \sqrt{\sum_{i=1}^{n}(a_i - b_i)^2}.$$

The angle θ between two vectors $\mathbf{a}$ and $\mathbf{b}$ is defined in terms of their inner product as

$$\cos\theta = \mathbf{a}'\mathbf{b}/\|\mathbf{a}\|\|\mathbf{b}\| = \mathbf{a}'\mathbf{b}/(\sqrt{\mathbf{a}'\mathbf{a}}\sqrt{\mathbf{b}'\mathbf{b}})$$

(see Figure 1.2.2). Since $\cos\theta = 0$ only if $\mathbf{a}'\mathbf{b} = 0$, $\mathbf{a}$ and $\mathbf{b}$ are perpendicular (or orthogonal) when $\mathbf{a}'\mathbf{b} = 0$.

Result 1.2.1. Properties of inner product. Let $\mathbf{a}$, $\mathbf{b}$, and $\mathbf{c}$ be n-dimensional vectors and let d be a scalar. Then,

1. $\mathbf{a} \bullet \mathbf{b} = \mathbf{b} \bullet \mathbf{a}$.
2. $\mathbf{a} \bullet (\mathbf{b} + \mathbf{c}) = \mathbf{a} \bullet \mathbf{b} + \mathbf{a} \bullet \mathbf{c}$.
3. $d(\mathbf{a} \bullet \mathbf{b}) = (d\mathbf{a}) \bullet \mathbf{b} = \mathbf{a} \bullet (d\mathbf{b})$.
4. $\mathbf{a} \bullet \mathbf{a} \geq 0$, with equality if and only if $\mathbf{a} = \mathbf{0}$.
5. $\|\mathbf{a} \pm \mathbf{b}\|^2 = \|\mathbf{a}\|^2 + \|\mathbf{b}\|^2 \pm 2\mathbf{a} \bullet \mathbf{b}$.
6. $|\mathbf{a} \bullet \mathbf{b}| \leq \|\mathbf{a}\|\|\mathbf{b}\|$, with equality if and only if $\mathbf{a} = \mathbf{0}$ or $\mathbf{b} = c\mathbf{a}$ for some scalar c.

7. $\|\mathbf{a}+\mathbf{b}\| \le \|\mathbf{a}\| + \|\mathbf{b}\|$, with equality if and only if $\mathbf{a} = \mathbf{0}$ or $\mathbf{b} = c\mathbf{a}$ for some scalar $c \ge 0$.

The last two inequalities in Result 1.2.1 are respectively the Cauchy–Schwarz inequality and the triangle inequality, which we ask the reader to verify in Exercise 1.2. Geometrically, the triangle inequality states that the length of one side of a triangle does not exceed the sum of the lengths of the other two sides.

Definition 1.2.2. Outer product of vectors. The outer product of two vectors $\mathbf{a}$ and $\mathbf{b}$ is denoted by $\mathbf{a} \wedge \mathbf{b}$ or $\mathbf{ab}'$ and is obtained by post-multiplying the column vector $\mathbf{a}$ by the row vector $\mathbf{b}'$. There is no restriction on the dimensions of $\mathbf{a}$ and $\mathbf{b}$; if $\mathbf{a}$ is an $m \times 1$ vector and $\mathbf{b}$ is an $n \times 1$ vector, the outer product $\mathbf{ab}'$ is an $m \times n$ matrix.

Example 1.2.1. We illustrate all these vector operations by an example. Let

$$\mathbf{a} = \begin{pmatrix} 2 \\ 3 \\ 4 \end{pmatrix}, \quad \mathbf{b} = \begin{pmatrix} 6 \\ 7 \\ 9 \end{pmatrix}, \quad \text{and} \quad \mathbf{d} = \begin{pmatrix} 10 \\ 20 \end{pmatrix}.$$

Then,

$$\mathbf{a}+\mathbf{b} = \begin{pmatrix} 8 \\ 10 \\ 13 \end{pmatrix}, \quad \mathbf{a}-\mathbf{b} = \begin{pmatrix} -4 \\ -4 \\ -5 \end{pmatrix}, \quad 10\mathbf{b} = \begin{pmatrix} 60 \\ 70 \\ 90 \end{pmatrix},$$

$$\mathbf{a}'\mathbf{b} = 2 \times 6 + 3 \times 7 + 4 \times 9 = 69,$$

$$\mathbf{ab}' = \begin{pmatrix} 2 \\ 3 \\ 4 \end{pmatrix} \begin{pmatrix} 6 & 7 & 9 \end{pmatrix} = \begin{pmatrix} 12 & 14 & 18 \\ 18 & 21 & 27 \\ 24 & 28 & 36 \end{pmatrix}, \quad \text{and}$$

$$\mathbf{ad}' = \begin{pmatrix} 2 \\ 3 \\ 4 \end{pmatrix} \begin{pmatrix} 10 & 20 \end{pmatrix} = \begin{pmatrix} 20 & 40 \\ 30 & 60 \\ 40 & 80 \end{pmatrix}.$$

However, $\mathbf{a}+\mathbf{d}$, $\mathbf{a}'\mathbf{d}$, and $\mathbf{b}'\mathbf{d}$ are undefined. □

Definition 1.2.3. Vector space. A vector space is a set $\mathcal{V}$ of vectors $\mathbf{v} \in \mathcal{R}^n$, which is closed under addition and multiplication by a scalar, i.e., for any $\mathbf{u}$, $\mathbf{v} \in \mathcal{V}$ and $c \in \mathcal{R}$, $\mathbf{u}+\mathbf{v} \in \mathcal{V}$ and $c\mathbf{v} \in \mathcal{V}$.

For example, $\mathcal{R}^n$ is a vector space for any positive integer $n = 1, 2, \cdots$. As another example, consider k linear equations in n variables $x_1, \cdots, x_n$:

$$c_{i1}x_1 + \cdots + c_{in}x_n = 0, \quad i = 1, \cdots, k,$$

where c_{ij} are real constants. The totality of solutions $\mathbf{x} = (x_1, \cdots, x_n)'$ is a vector space. We discuss solutions of linear equations in Chapter 3.

Definition 1.2.4. Vector subspace and affine subspace. Let $\mathcal{S}$ be a subset of a vector space $\mathcal{V}$. If $\mathcal{S}$ is also a vector space, it is called a vector subspace, or simply, a subspace of $\mathcal{V}$. Further, for any $\mathbf{v} \in \mathcal{V}$, $\mathbf{v}+\mathcal{S} = \{\mathbf{v}+\mathbf{u}\text{: } \mathbf{u} \in \mathcal{S}\}$ is called an affine subspace of $\mathcal{V}$.

For example, $\{\mathbf{0}\}$ and $\mathcal{V}$ are (trivially) subspaces of $\mathcal{V}$. Any plane through the origin is a subspace of $\mathcal{R}^3$, and any plane is an affine subspace of $\mathcal{R}^3$. More generally, any vector space of n-dimensional vectors is a subspace of $\mathcal{R}^n$.

Definition 1.2.5. Linear span. Let $\mathcal{S}$ be a subset of a vector space $\mathcal{V}$. The set of all finite linear combinations of the vectors in $\mathcal{S}$, i.e., $\{\sum_{i=1}^{n} a_i\mathbf{v}_i\colon a_i \in \mathcal{R}, \mathbf{v}_i \in \mathcal{S}, n = 1, 2, \cdots\}$ is called the linear span, or simply, span of $\mathcal{S}$ and is denoted by Span($\mathcal{S}$).

The span of a set of vectors in a vector space is the smallest subspace that contains the set; it is also the intersection of all subspaces that contain the set. We are mostly interested in the case where $\mathcal{S}$ is a finite set of vectors $\{\mathbf{v}_1, \cdots, \mathbf{v}_l\}$. For example, the vectors $\mathbf{0}$, $\mathbf{v}_1$, $\mathbf{v}_2$, $\mathbf{v}_1 + \mathbf{v}_2$, $10\mathbf{v}_1$, $5\mathbf{v}_1 - 3\mathbf{v}_2$ all belong to $\text{Span}\{\mathbf{v}_1, \mathbf{v}_2\}$.

Definition 1.2.6. Linear dependence and independence of vectors. Let $\mathbf{v}_1, \cdots, \mathbf{v}_m$ be vectors in $\mathcal{V}$. The vectors are said to be linearly dependent if and only if there exist scalars $c_1, \cdots, c_m$, not all zero, such that $\sum_{i=1}^{m} c_i\mathbf{v}_i = \mathbf{0}$. If $\sum_{i=1}^{m} c_i\mathbf{v}_i \neq \mathbf{0}$ unless all c_i are zero, then $\mathbf{v}_1, \cdots, \mathbf{v}_m$ are said to be linearly independent (LIN) vectors.

For example, $\{\mathbf{0}\}$ is a linearly dependent set, as is any set of vectors containing $\mathbf{0}$.

Example 1.2.2. Let $\mathbf{v}_1 = (1, -1, 3)'$ and $\mathbf{v}_2 = (1, 1, 1)'$. Now, $\sum_{i=1}^{2} c_i\mathbf{v}_i = \mathbf{0} \implies c_1 + c_2 = 0$, $-c_1 + c_2 = 0$, and $3c_1 + c_2 = 0$, for which the only solution is $c_1 = c_2 = 0$. Hence, $\mathbf{v}_1$ and $\mathbf{v}_2$ are LIN vectors. □

Example 1.2.3. The vectors $\mathbf{v}_1 = (1, -1)'$, $\mathbf{v}_2 = (1, 2)'$, and $\mathbf{v}_3 = (2, 1)'$ are linearly dependent, which is verified by setting $c_1 = 1$, $c_2 = 1$, and $c_3 = -1$; we see that $\sum_{i=1}^{3} c_i\mathbf{v}_i = \mathbf{0}$. □

Result 1.2.2. For n-dimensional vectors $\mathbf{v}_1, \cdots, \mathbf{v}_m$, the following properties hold.

1. If the vectors are linearly dependent, we can express at least one of them as a linear combination of the others.
2. If $s \leq m$ of the vectors are linearly dependent, then all the vectors are linearly dependent.
3. If $m > n$, then the vectors are linearly dependent.

Proof. We show property 3 by induction on n. The proof of the other properties is left as Exercise 1.8. If $n = 1$, then we have scalars $v_1, \cdots, v_m$, $m > 1$. If $v_1 = 0$, then $1 \cdot v_1 + 0 \cdot v_2 + \cdots + 0 \cdot v_m = 0$. If $v_1 \neq 0$, then $c \cdot v_1 + 1 \cdot v_2 + \cdots + 1 \cdot v_m = 0$ with $c = -(v_2 + \cdots + v_m)/v_1$. In either case, $v_1, \cdots, v_m$ are linearly dependent. Suppose the property holds for $(n-1)$-dimensional vectors for some $n \geq 2$. We prove that the property must hold for n-dimensional vectors. Let $\mathbf{v}_1, \cdots, \mathbf{v}_m$ be n-dimensional vectors, where $m > n$. If $\mathbf{v}_1, \cdots, \mathbf{v}_n$ are linearly dependent, then by property 2, $\mathbf{v}_1, \cdots, \mathbf{v}_m$, $m > n$, are linearly dependent. Now suppose $\mathbf{v}_1, \cdots, \mathbf{v}_n$ are LIN. Let $\mathbf{v}_i = (v_{i1}, \ldots, v_{in})'$. Define the $(n-1)$-dimensional vector $\mathbf{u}_i$ by removing the first component of $\mathbf{v}_i$, i.e., $\mathbf{u}_i = (v_{i2}, \ldots, v_{in})'$. By the induction hypothesis, $\mathbf{u}_1, \cdots, \mathbf{u}_n$ are linearly dependent, so there are scalars $c_1, \cdots, c_n$ not all zero, such that $\sum_{j=1}^{n} c_j\mathbf{u}_j = \mathbf{0}$, i.e., $\sum_{j=1}^{n} v_{ij}c_j = 0$ for $i > 1$. Then, since $\mathbf{v}_1, \cdots, \mathbf{v}_n$ are LIN, $d = \sum_{j=1}^{n} v_{1j}c_j$ must be nonzero. Let $\tilde{c}_j = c_j/d$. Then $\sum_{j=1}^{n} v_{ij}\tilde{c}_j = 0$ for $i > 1$ and $\sum_{j=1}^{n} v_{1j}\tilde{c}_j = 1$. In other words, $\sum_{j=1}^{n} \tilde{c}_j\mathbf{v}_j = \mathbf{e}_1$, where $\mathbf{e}_i$ is the vector with 1 for its ith component and zeros elsewhere. Likewise, every $\mathbf{e}_i$ is a linear combination of $\mathbf{v}_1, \ldots, \mathbf{v}_n$. On the other hand, it is easy to see that every vector in $\mathcal{R}^n$ is a linear combination of $\mathbf{e}_1, \cdots, \mathbf{e}_n$. As a result, every vector $\mathbf{v}_i$, $i > n$, is a linear combination of $\mathbf{v}_1, \cdots, \mathbf{v}_n$, so $\mathbf{v}_1, \cdots, \mathbf{v}_m$ are linearly dependent. By induction, the proof of property 3 is complete. ■

Definition 1.2.7. Basis and dimension. Let $\mathcal{V}$ be a vector space. If $\{\mathbf{v}_1, \cdots, \mathbf{v}_m\} \subset \mathcal{V}$ is a set of LIN vectors that spans $\mathcal{V}$, it is called a basis (Hamel basis) for $\mathcal{V}$ and m is called the dimension of $\mathcal{V}$, denoted by $\dim(\mathcal{V}) = m$. If $\mathcal{S}$ is a subspace of $\mathcal{V}$, then the dimension of any affine subspace $\mathbf{v} + \mathcal{S}$, $\mathbf{v} \in \mathcal{V}$, is defined to be $\dim(\mathcal{S})$.

Every vector in $\mathcal{V}$ has a unique representation as a linear combination of vectors in a basis $\{\mathbf{v}_1, \cdots, \mathbf{v}_m\}$ of the space. First, by Definition 1.2.7, such a representation exists. Second, if a vector $\mathbf{x}$ is equal to both $\sum_{i=1}^m c_i\mathbf{v}_i$ and $\sum_{i=1}^m d_i\mathbf{v}_i$, then $\sum_{i=1}^m (c_i - d_i)\mathbf{v}_i = \mathbf{0}$, which, by the linear independence of $\mathbf{v}_1, \cdots, \mathbf{v}_m$, is possible only if $c_i = d_i$ for all i.

Every non-null vector space $\mathcal{V}$ of n-dimensional vectors has a basis, which is seen as follows. Choose any non-null $\mathbf{v}_1 \in \mathcal{V}$; $\mathbf{v}_1$ alone is LIN. Suppose we had chosen $\mathbf{v}_1, \cdots, \mathbf{v}_i \in \mathcal{V}$ that are LIN. If these vectors span $\mathcal{V}$, then stop. Otherwise, choose any $\mathbf{v}_{i+1} \in \mathcal{V}$ not in $\text{Span}\{\mathbf{v}_1, \cdots, \mathbf{v}_i\}$; it can be seen that $\mathbf{v}_1, \cdots, \mathbf{v}_{i+1}$ are LIN. By property 3 of Result 1.2.2, this process must stop after the choice of some $m \le n$ vectors $\mathbf{v}_1, \cdots, \mathbf{v}_m$, which form a basis of $\mathcal{V}$. Consequently, if $\mathcal{V}$ consists of n-dimensional vectors, $\dim(\mathcal{V})$ is strictly smaller than n unless $\mathcal{V} = \mathcal{R}^n$.

However, note that $\mathcal{V}$ does not have a unique basis. If $\{\mathbf{v}_1, \cdots, \mathbf{v}_m\}$ and $\{\mathbf{u}_1, \cdots, \mathbf{u}_k\}$ are two choices for a basis, then $m = k$, and so $\dim(\mathcal{V})$ is well-defined. To see this, suppose $k \ge m$. Since $\mathbf{u}_1 \in \text{Span}\{\mathbf{v}_1, \cdots, \mathbf{v}_m\}$, $\mathbf{u}_1 = \sum_{i=1}^m a_i\mathbf{v}_i$ for some $a_1, \ldots, a_m$ not all 0. Without loss of generality, suppose $a_1 \neq 0$. Since $\mathbf{v}_1 = \mathbf{u}_1/a_1 - \sum_{i=2}^m (a_i/a_1)\mathbf{v}_i$, $\mathcal{V} = \text{Span}\{\mathbf{u}_1, \cdots, \mathbf{u}_m\}$. Next, $\mathbf{u}_2 = b_1\mathbf{u}_1 + \sum_{i=2}^m b_i\mathbf{v}_i$ for some $b_1, \ldots, b_m$ not all 0. Since $\mathbf{u}_1$, $\mathbf{u}_2$ are LIN, not all $b_2, \ldots, b_m$ are 0. Without loss of generality, suppose $b_2 \neq 0$. Then as above, $\mathcal{V} = \text{Span}\{\mathbf{u}_1, \mathbf{u}_2, \mathbf{v}_3, \ldots, \mathbf{v}_m\}$. Continuing this process, it follows that $\mathcal{V} = \text{Span}\{\mathbf{u}_1, \cdots, \mathbf{u}_m\}$. This implies $k = m$. Since $\dim(\mathcal{V})$ does not depend on the choice of a basis, if $\mathcal{V} = \text{Span}\{\mathbf{w}_1, \cdots, \mathbf{w}_s\}$, then $\dim(\mathcal{V})$ is the maximum number of LIN vectors in $\{\mathbf{w}_1, \cdots, \mathbf{w}_s\}$.

Definition 1.2.8. Sum and direct sum of subspaces. Let $\mathcal{V}_1, \cdots, \mathcal{V}_k$ be subspaces of $\mathcal{V}$.

1. The sum of the subspaces is defined to be $\{\mathbf{v}_1 + \cdots + \mathbf{v}_k \colon \mathbf{v}_i \in \mathcal{V}_i, i = 1, \cdots, k\}$ and denoted by $\mathcal{V}_1 + \cdots + \mathcal{V}_k$ or $\sum_{i=1}^k \mathcal{V}_i$.

2. The sum is said to be a direct sum, denoted by $\mathcal{V}_1 \oplus \cdots \oplus \mathcal{V}_k$ or $\bigoplus_{i=1}^k \mathcal{V}_i$, if for any $\mathbf{v}_i, \mathbf{w}_i \in \mathcal{V}_i$, $i = 1, \ldots, k$, $\mathbf{v}_1 + \cdots + \mathbf{v}_k = \mathbf{w}_1 + \cdots + \mathbf{w}_k$ if and only if $\mathbf{v}_i = \mathbf{w}_i$ for all i; or alternately, if for every $\mathbf{v} \in \mathcal{V}_1 + \cdots + \mathcal{V}_k$, there are unique vectors $\mathbf{v}_i \in \mathcal{V}_i$, $i = 1, \cdots, k$, such that $\mathbf{v} = \mathbf{v}_1 + \cdots + \mathbf{v}_k$.

From Definition 1.2.8, $\mathcal{V}_1 + \cdots + \mathcal{V}_k$ is the smallest subspace of $\mathcal{V}$ that contains $\cup_{i=1}^k \mathcal{V}_i$. In the case of a direct sum, $\mathcal{V}_i \cap (\bigoplus_{j \neq i} \mathcal{V}_j) = \{\mathbf{0}\}$ for all i, $\dim(\mathcal{V}_1 \oplus \cdots \oplus \mathcal{V}_k) = \sum_{i=1}^k \dim(\mathcal{V}_i)$, and if $\{\mathbf{v}_{i1}, \ldots, \mathbf{v}_{im_i}\}$ is a basis of $\mathcal{V}_i$ for each i, where $m_i = \dim(\mathcal{V}_i)$, then the combined set $\{\mathbf{v}_{ij}, j = 1, \ldots, m_i,\ i = 1, \ldots, k\}$ is a basis of $\mathcal{V}_1 \oplus \cdots \oplus \mathcal{V}_k$.

The vectors $\mathbf{e}_1, \cdots, \mathbf{e}_n$ in the proof of property 3 in Result 1.2.2 are called the standard basis vectors of $\mathcal{R}^n$. For example, the standard basis vectors in $\mathcal{R}^2$ are $\mathbf{e}_1 = (1, 0)'$ and $\mathbf{e}_2 = (0, 1)'$, while those in $\mathcal{R}^3$ are $\mathbf{e}_1 = (1, 0, 0)'$, $\mathbf{e}_2 = (0, 1, 0)'$, and $\mathbf{e}_3 = (0, 0, 1)'$. Any $\mathbf{x} = (x_1, \cdots, x_n)' \in \mathcal{R}^n$ can be written as $\mathbf{x} = x_1\mathbf{e}_1 + \cdots + x_n\mathbf{e}_n$.

Definition 1.2.9. Orthogonal vectors and orthogonal subspaces. Two vectors $\mathbf{v}_1$ and $\mathbf{v}_2$ in $\mathcal{V}$ are orthogonal or perpendicular to each other if and only if $\mathbf{v}_1 \bullet \mathbf{v}_2 = \mathbf{v}_1'\mathbf{v}_2 = \mathbf{v}_2'\mathbf{v}_1 = 0$, denoted by $\mathbf{v}_1 \perp \mathbf{v}_2$. Two subspaces $\mathcal{V}_1$, $\mathcal{V}_2$ of $\mathcal{V}$ are orthogonal or perpendicular to each other if and only if every vector in $\mathcal{V}_1$ is orthogonal to every vector in $\mathcal{V}_2$, denoted by $\mathcal{V}_1 \perp \mathcal{V}_2$.

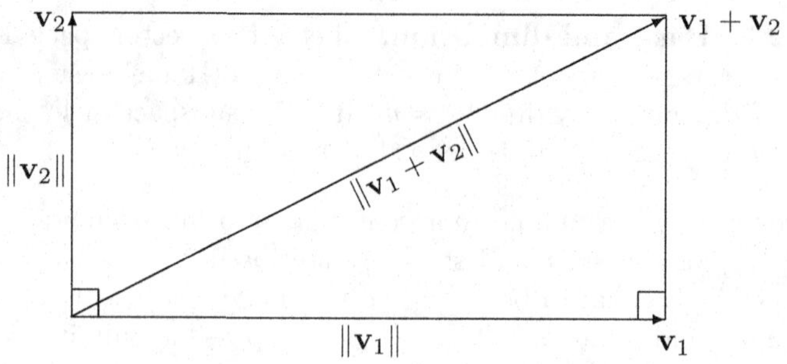

FIGURE 1.2.3. Pythagoras's theorem.

Pythagoras's Theorem states that for n-dimensional vectors $\mathbf{v}_1$ and $\mathbf{v}_2$,

$$\mathbf{v}_1 \perp \mathbf{v}_2 \text{ if and only if } \|\mathbf{v}_1 + \mathbf{v}_2\|^2 = \|\mathbf{v}_1\|^2 + \|\mathbf{v}_2\|^2; \tag{1.2.1}$$

this is illustrated in Figure 1.2.3.

Result 1.2.3.

1. If $\mathbf{v}_1, \cdots, \mathbf{v}_m$ are nonzero vectors which are mutually orthogonal, i.e., $\mathbf{v}_i'\mathbf{v}_j = 0$, $i \neq j$, then these vectors are LIN.

2. If $\mathcal{V}_1, \cdots, \mathcal{V}_m$ are subspaces of $\mathcal{V}$, and $\mathcal{V}_i \perp \mathcal{V}_j$ for any $i \neq j$, then their sum is a direct sum, i.e., every $\mathbf{v} \in \mathcal{V}_1 + \cdots + \mathcal{V}_m$ has a unique decomposition as $\mathbf{v} = \mathbf{v}_1 + \cdots + \mathbf{v}_m$ with $\mathbf{v}_i \in \mathcal{V}_i$, $i = 1, \cdots, m$.

The proof of the result is left as Exercise 1.9.

Definition 1.2.10. Orthonormal basis. A basis $\{\mathbf{v}_1, \cdots, \mathbf{v}_m\}$ of a vector space $\mathcal{V}$ such that $\mathbf{v}_i'\mathbf{v}_j = 0$, for all $i \neq j$ is called an orthogonal basis. If further, $\mathbf{v}_i'\mathbf{v}_i = 1$ for $i = 1, \cdots, m$, it is called an orthonormal basis of $\mathcal{V}$.

Result 1.2.4. Gram–Schmidt orthogonalization. Let $\{\mathbf{v}_1, \cdots, \mathbf{v}_m\}$ denote an arbitrary basis of $\mathcal{V}$. To construct an orthonormal basis of $\mathcal{V}$ starting from $\{\mathbf{v}_1, \cdots, \mathbf{v}_m\}$, we define

$$\mathbf{y}_1 = \mathbf{v}_1,$$
$$\mathbf{y}_k = \mathbf{v}_k - \sum_{i=1}^{k-1} \frac{\mathbf{y}_i'\mathbf{v}_k}{\|\mathbf{y}_i\|^2}\mathbf{y}_i, \quad k = 2, \cdots, m,$$
$$\mathbf{z}_k = \frac{\mathbf{y}_k}{\|\mathbf{y}_k\|}, \quad k = 1, \cdots, m.$$

Then $\{\mathbf{z}_1, \cdots, \mathbf{z}_m\}$ is an orthonormal basis of $\mathcal{V}$.

The proof of the result is left as Exercise 1.10. The stages in this process for a basis $\{\mathbf{v}_1, \mathbf{v}_2, \mathbf{v}_3\}$ are shown in Figure 1.2.4.

Example 1.2.4. We use Result 1.2.4 to find an orthonormal basis starting from the basis vectors $\mathbf{v}_1 = (1, -1, 1)'$, $\mathbf{v}_2 = (-2, 3, -1)'$, and $\mathbf{v}_3 = (1, 2, -4)'$. Let $\mathbf{y}_1 = \mathbf{v}_1$. We compute $\mathbf{y}_1'\mathbf{v}_2 = -6$ and $\mathbf{y}_1'\mathbf{y}_1 = 3$, so that $\mathbf{y}_2 = (0, 1, 1)'$. Next, $\mathbf{y}_1'\mathbf{v}_3 = -5$, $\mathbf{y}_2'\mathbf{v}_3 = -2$,

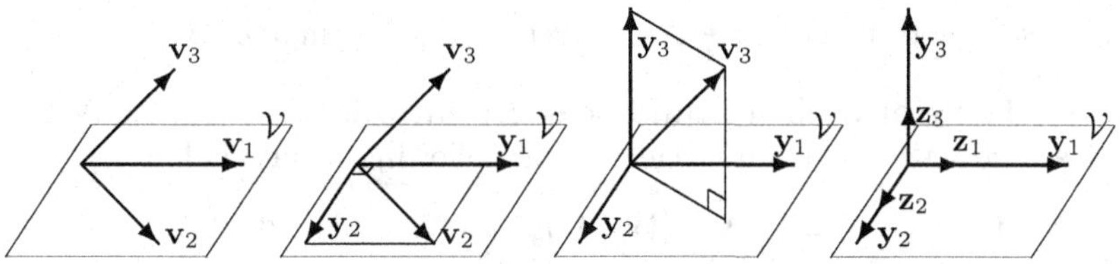

FIGURE 1.2.4. Gram–Schmidt orthogonalization in $\mathcal{R}^3$.

and $\mathbf{y}_2'\mathbf{y}_2 = 2$, so that $\mathbf{y}_3 = (8/3, 4/3, -4/3)'$. It is easily verified that $\{\mathbf{y}_1, \mathbf{y}_2, \mathbf{y}_3\}$ is an orthogonal basis and also that $\mathbf{z}_1 = (1/\sqrt{3}, -1/\sqrt{3}, 1/\sqrt{3})'$, $\mathbf{z}_2 = (0, 1/\sqrt{2}, 1/\sqrt{2})'$, and $\mathbf{z}_3 = (2/\sqrt{6}, 1/\sqrt{6}, -1/\sqrt{6})'$ form a set of orthonormal basis vectors. □

Definition 1.2.11. Orthogonal complement of a subspace. Let $\mathcal{W}$ be a vector subspace of a Euclidean space $\mathcal{R}^n$. Its orthogonal complement, written as $\mathcal{W}^\perp$, is the subspace consisting of all vectors in $\mathcal{R}^n$ that are orthogonal to every vector in $\mathcal{W}$.

It is easy to verify that $\mathcal{W}^\perp$ is indeed a vector subspace. If $\mathcal{V}$ is any subspace containing $\mathcal{W}$, then $\mathcal{W}^\perp \cap \mathcal{V}$, sometimes referred to as the orthogonal complement of $\mathcal{W}$ relative to $\mathcal{V}$, consists of all vectors in $\mathcal{V}$ that are orthogonal to every vector in $\mathcal{W}$. The proof of the next result is left as Exercise 1.11.

Result 1.2.5. If $\mathcal{W}$ and $\mathcal{V}$ are two subspaces of $\mathcal{R}^n$ and $\mathcal{W} \subset \mathcal{V}$, then

1. $\mathcal{V} = \mathcal{W} \oplus (\mathcal{W}^\perp \cap \mathcal{V})$ and $\dim(\mathcal{V}) = \dim(\mathcal{W}) + \dim(\mathcal{W}^\perp \cap \mathcal{V})$; and
2. $(\mathcal{W}^\perp \cap \mathcal{V})^\perp \cap \mathcal{V} = \mathcal{W}$.

1.3 Basic properties of matrices

We describe some elementary properties of matrices and provide illustrations. More detailed properties of special matrices that are relevant to linear model theory are given in Chapter 2.

Definition 1.3.1. Matrix addition and subtraction. For arbitrary $m \times n$ matrices $\mathbf{A}$ and $\mathbf{B}$, each of the same dimension, $\mathbf{C} = \mathbf{A} \pm \mathbf{B}$ is an $m \times n$ matrix whose (i, j)th element is $c_{ij} = a_{ij} \pm b_{ij}$. For example,

$$\begin{pmatrix} -5 & 4 & 1 \\ -3 & 2 & 6 \end{pmatrix} + \begin{pmatrix} 7 & -9 & 10 \\ 2 & 6 & -1 \end{pmatrix} = \begin{pmatrix} 2 & -5 & 11 \\ -1 & 8 & 5 \end{pmatrix}.$$

Definition 1.3.2. Multiplication of a matrix by a scalar. For an arbitrary $m \times n$ matrix $\mathbf{A}$, and an arbitrary real scalar c, $\mathbf{B} = c\mathbf{A} = \mathbf{A}c$ is an $m \times n$ matrix whose (i, j)th element is $b_{ij} = ca_{ij}$. For example,

$$5\begin{pmatrix} -5 & 4 & 1 \\ -3 & 2 & 6 \end{pmatrix} = \begin{pmatrix} -25 & 20 & 5 \\ -15 & 10 & 30 \end{pmatrix}.$$

When $c = -1$, we denote $(-1)\mathbf{A}$ as $-\mathbf{A}$, the negative of the matrix $\mathbf{A}$.

Result 1.3.1. Laws of addition and scalar multiplication. Let $\mathbf{A}, \mathbf{B}, \mathbf{C}$ be any $m \times n$ matrices and let a, b, c be any scalars. The following results hold:

1. $(\mathbf{A}+\mathbf{B})+\mathbf{C} = \mathbf{A}+(\mathbf{B}+\mathbf{C})$
2. $\mathbf{A}+\mathbf{B} = \mathbf{B}+\mathbf{A}$
3. $\mathbf{A}+(-\mathbf{A}) = (-\mathbf{A})+\mathbf{A} = \mathbf{O}$
4. $\mathbf{A}+\mathbf{O} = \mathbf{O}+\mathbf{A} = \mathbf{A}$
5. $c(\mathbf{A}+\mathbf{B}) = c\mathbf{A}+c\mathbf{B}$
6. $(a+b)\mathbf{C} = a\mathbf{C}+b\mathbf{C}$
7. $(ab)\mathbf{C} = a(b\mathbf{C}) = b(a\mathbf{C})$
8. $0\mathbf{A} = \mathbf{O}$
9. $1\mathbf{A} = \mathbf{A}$.

Definition 1.3.3. Matrix multiplication. For arbitrary matrices $\mathbf{A}$ and $\mathbf{B}$ of respective dimensions $m \times n$ and $n \times p$, $\mathbf{C} = \mathbf{AB}$ is an $m \times p$ matrix whose (i,j)th element is $c_{ij} = \sum_{l=1}^{n} a_{il} b_{lj}$. The product $\mathbf{AB}$ is undefined when the column dimension of $\mathbf{A}$ is not equal to the row dimension of $\mathbf{B}$. For example,

$$\begin{pmatrix} 5 & 4 & 1 \\ -3 & 2 & 6 \end{pmatrix} \begin{pmatrix} 7 \\ -3 \\ 2 \end{pmatrix} = \begin{pmatrix} 25 \\ -15 \end{pmatrix}.$$

In referring to the matrix product $\mathbf{AB}$, we say that $\mathbf{B}$ is pre-multiplied by $\mathbf{A}$, and $\mathbf{A}$ is post-multiplied by $\mathbf{B}$. Provided all the matrices are conformal under multiplication, the following properties hold:

Result 1.3.2. Laws of matrix multiplication. Let a be a scalar, let $\mathbf{A}$ be an $m \times n$ matrix and let the matrices $\mathbf{B}$ and $\mathbf{C}$ have appropriate dimensions so that the operations below are defined. Then,

1. $(\mathbf{AB})\mathbf{C} = \mathbf{A}(\mathbf{BC})$
2. $\mathbf{A}(\mathbf{B}+\mathbf{C}) = \mathbf{AB}+\mathbf{AC}$
3. $(\mathbf{A}+\mathbf{B})\mathbf{C} = \mathbf{AC}+\mathbf{BC}$
4. $a(\mathbf{BC}) = (a\mathbf{B})\mathbf{C} = \mathbf{B}(a\mathbf{C})$
5. $\mathbf{I}_m\mathbf{A} = \mathbf{A}\mathbf{I}_n = \mathbf{A}$
6. $0\mathbf{A} = \mathbf{O}$ and $\mathbf{A}0 = \mathbf{O}$.

In general, matrix multiplication is not commutative, i.e., $\mathbf{AB}$ is not necessarily equal to $\mathbf{BA}$. Note that depending on the row and column dimensions of $\mathbf{A}$ and $\mathbf{B}$, it is possible that (i) only $\mathbf{AB}$ is defined and $\mathbf{BA}$ is not, or (ii) both $\mathbf{AB}$ and $\mathbf{BA}$ are defined, but do not have the same dimensions, or (iii) $\mathbf{AB}$ and $\mathbf{BA}$ are defined and have the same dimensions, but $\mathbf{AB} \neq \mathbf{BA}$. Two $n \times n$ matrices $\mathbf{A}$ and $\mathbf{B}$ are said to commute under multiplication if $\mathbf{AB} = \mathbf{BA}$. A collection of $n \times n$ matrices $\mathbf{A}_1, \cdots, \mathbf{A}_k$ is said to be pairwise commutative if $\mathbf{A}_i\mathbf{A}_j = \mathbf{A}_j\mathbf{A}_i$ for $j > i$, $i, j = 1, \cdots, k$. Note that the product $\mathbf{A}^k = \mathbf{A} \cdots \mathbf{A}$ (k times) is defined only if $\mathbf{A}$ is a square matrix. It is easy to verify that $\mathbf{J}_{mn}\mathbf{J}_{np} = n\mathbf{J}_{mp}$.

Result 1.3.3. If $\mathbf{A}$ and $\mathbf{B}$ are $n \times n$ lower (resp. upper) triangular matrices with diagonal elements $a_1, \cdots, a_n$ and $b_1, \cdots, b_n$, respectively, then $\mathbf{AB}$ is also a lower (resp. upper) triangular matrix with diagonal elements $a_1b_1, \cdots, a_nb_n$.

The proof of Result 1.3.3 is left as Exercise 1.18.

Definition 1.3.4. Matrix transpose. The transpose of an $m \times n$ matrix $\mathbf{A}$ is an $n \times m$ matrix whose columns are the rows of $\mathbf{A}$ in the same order. The transpose of $\mathbf{A}$ is denoted by $\mathbf{A}'$. For example,

$$\mathbf{A} = \begin{pmatrix} 2 & 1 & 6 \\ 4 & 3 & 5 \end{pmatrix} \Longrightarrow \mathbf{A}' = \begin{pmatrix} 2 & 4 \\ 1 & 3 \\ 6 & 5 \end{pmatrix}, \quad \mathbf{B} = \begin{pmatrix} 6 & 7 \\ 8 & 9 \end{pmatrix} \Longrightarrow \mathbf{B}' = \begin{pmatrix} 6 & 8 \\ 7 & 9 \end{pmatrix}.$$

As we saw earlier, the transpose of an n-dimensional column vector with components $a_1, \cdots, a_n$ is the row vector $(a_1, \cdots, a_n)$. It is often convenient to write a column vector in this transposed form. The transpose of an upper (lower) triangular matrix is a lower (upper) triangular matrix. It may be easily verified that $\mathbf{J}_{mn} = \mathbf{1}_m \mathbf{1}_n'$.

Result 1.3.4. Laws of transposition. Let $\mathbf{A}$ and $\mathbf{B}$ conform under addition, and let $\mathbf{A}$ and $\mathbf{C}$ conform under multiplication. Let a, b, and c denote scalars and let $k \geq 2$ denote a positive integer. Then,

1. $(\mathbf{A}')' = \mathbf{A}$
2. $(a\mathbf{A} + b\mathbf{B})' = a\mathbf{A}' + b\mathbf{B}'$
3. $(c\mathbf{A})' = c\mathbf{A}'$
4. $\mathbf{A}' = \mathbf{B}'$ if and only if $\mathbf{A} = \mathbf{B}$
5. $(\mathbf{AC})' = \mathbf{C}'\mathbf{A}'$
6. $(\mathbf{A}_1 \cdots \mathbf{A}_k)' = \mathbf{A}_k' \cdots \mathbf{A}_1'$.

Definition 1.3.5. Symmetric matrix. A matrix $\mathbf{A}$ is said to be symmetric if $\mathbf{A}' = \mathbf{A}$. For example,

$$\mathbf{A} = \begin{pmatrix} 1 & 2 & -3 \\ 2 & 4 & 5 \\ -3 & 5 & 9 \end{pmatrix}$$

is a symmetric matrix. Note that a symmetric matrix is always a square matrix. Any diagonal matrix, written as $\mathbf{D} = \text{diag}(d_1, \cdots, d_n)$, is symmetric. Other examples of symmetric matrices include the variance-covariance matrix and the correlation matrix of any random vector, the identity matrix $\mathbf{I}_n$ and the unit matrix $\mathbf{J}_n$. A matrix $\mathbf{A}$ is said to be skew-symmetric if $\mathbf{A}' = -\mathbf{A}$.

Definition 1.3.6. Trace of a matrix. Let $\mathbf{A}$ be an $n \times n$ matrix. The trace of $\mathbf{A}$ is a scalar given by the sum of the diagonal elements of $\mathbf{A}$, i.e., $\text{tr}(\mathbf{A}) = \sum_{i=1}^n a_{ii}$. For example, if

$$\mathbf{A} = \begin{pmatrix} 2 & -4 & 5 \\ 6 & -7 & 0 \\ 3 & 9 & 7 \end{pmatrix},$$

then $\text{tr}(\mathbf{A}) = 2 - 7 + 7 = 2$.

Result 1.3.5. Properties of trace. Provided the matrices are conformable, and given scalars a and b,

1. $\text{tr}(\mathbf{I}_n) = n$
2. $\text{tr}(a\mathbf{A} \pm b\mathbf{B}) = a\,\text{tr}(\mathbf{A}) \pm b\,\text{tr}(\mathbf{B})$
3. $\text{tr}(\mathbf{AB}) = \text{tr}(\mathbf{BA})$
4. $\text{tr}(\mathbf{ABC}) = \text{tr}(\mathbf{CAB}) = \text{tr}(\mathbf{BCA})$
5. $\text{tr}(\mathbf{A}) = 0$ if $\mathbf{A} = \mathbf{O}$
6. $\text{tr}(\mathbf{A}') = \text{tr}(\mathbf{A})$
7. $\text{tr}(\mathbf{AA}') = \text{tr}(\mathbf{A}'\mathbf{A}) = \sum_{i,j=1}^n a_{ij}^2$
8. $\text{tr}(\mathbf{aa}') = \mathbf{a}'\mathbf{a} = \|\mathbf{a}\|^2 = \sum_{i=1}^n a_i^2$.

The trace operation in property 4 is valid under cyclic permutations only.

Definition 1.3.7. Determinant of a matrix. Let $\mathbf{A}$ be an $n \times n$ matrix. The determinant of $\mathbf{A}$ is a scalar given by

$$|\mathbf{A}| = \sum_{j=1}^n a_{ij}(-1)^{i+j}|\mathbf{M}_{ij}|, \text{ for any fixed } i, \quad \text{or}$$

$$|\mathbf{A}| = \sum_{i=1}^n a_{ij}(-1)^{i+j}|\mathbf{M}_{ij}|, \text{ for any fixed } j,$$

where $\mathbf{M}_{ij}$ is the $(n-1) \times (n-1)$ submatrix of $\mathbf{A}$ after deleting the ith row and the jth column from $\mathbf{A}$. We call $|\mathbf{M}_{ij}|$ the *minor* corresponding to a_{ij} and the signed minor, $F_{ij} = (-1)^{i+j}|\mathbf{M}_{ij}|$ the *cofactor* of a_{ij}. An alternative notation of the determinant is $\det(\mathbf{A})$, which is used in computing environments like R (R Core Team, 2018). We will use $|\mathbf{A}|$ and $\det(\mathbf{A})$ interchangeably in the text. We consider two special cases:

1. Suppose $n = 2$. Then $|\mathbf{A}| = a_{11}a_{22} - a_{12}a_{21}$.
2. Suppose $n = 3$. Fix $i = 1$ (row 1). Then

$$F_{11} = (-1)^{1+1}\begin{vmatrix} a_{22} & a_{23} \\ a_{32} & a_{33} \end{vmatrix}, \qquad F_{12} = (-1)^{1+2}\begin{vmatrix} a_{21} & a_{23} \\ a_{31} & a_{33} \end{vmatrix},$$
$$F_{13} = (-1)^{1+3}\begin{vmatrix} a_{21} & a_{22} \\ a_{31} & a_{32} \end{vmatrix},$$

and $|\mathbf{A}| = a_{11}F_{11} + a_{12}F_{12} + a_{13}F_{13}$. For example, if

$$\mathbf{A} = \begin{pmatrix} 2 & -4 & 5 \\ 6 & -7 & 0 \\ 3 & 9 & 7 \end{pmatrix}$$

then,

$$\begin{aligned} |\mathbf{A}| &= 2(-1)^{1+1}\begin{vmatrix} -7 & 0 \\ 9 & 7 \end{vmatrix} - 4(-1)^{1+2}\begin{vmatrix} 6 & 0 \\ 3 & 7 \end{vmatrix} + 5(-1)^{1+3}\begin{vmatrix} 6 & -7 \\ 3 & 9 \end{vmatrix} \\ &= 2(-49) + 4(42) + 5(75) = 445. \end{aligned}$$

Result 1.3.6. Properties of determinants. Let $\mathbf{A}$ and $\mathbf{B}$ be $n \times n$ matrices and let k be any integer. Then

1. $|\mathbf{A}| = |\mathbf{A}'|$.
2. $|c\mathbf{A}| = c^n|\mathbf{A}|$.
3. $|\mathbf{AB}| = |\mathbf{A}||\mathbf{B}|$.
4. If $\mathbf{A}$ is a diagonal matrix or an upper (or lower) triangular matrix, the determinant of $\mathbf{A}$ is equal to the product of its diagonal elements, i.e.,
$$|\mathbf{A}| = \prod_{i=1}^{n} a_{ii}.$$
5. If two rows (or columns) of a matrix $\mathbf{A}$ are equal, then $|\mathbf{A}| = 0$.
6. If $\mathbf{A}$ has a row (or column) of zeroes, then $|\mathbf{A}| = 0$.
7. If $\mathbf{A}$ has rows (or columns) that are multiples of each other, then $|\mathbf{A}| = 0$.
8. If a row (or column) of $\mathbf{A}$ is the sum of multiples of two other rows (or columns), then $|\mathbf{A}| = 0$.
9. Let $\mathbf{B}$ be obtained from $\mathbf{A}$ by multiplying one of its rows (or columns) by a nonzero constant c. Then, $|\mathbf{B}| = c|\mathbf{A}|$.
10. Let $\mathbf{B}$ be obtained from $\mathbf{A}$ by interchanging any two rows (or columns). Then, $|\mathbf{B}| = -|\mathbf{A}|$.

11. Let $\mathbf{B}$ be obtained from $\mathbf{A}$ by adding a multiple of one row (or column) to another row (or column). Then, $|\mathbf{B}| = |\mathbf{A}|$.

12. If $\mathbf{A}$ is an $m \times n$ matrix and $\mathbf{B}$ is an $n \times m$ matrix, then $|\mathbf{I}_m + \mathbf{AB}| = |\mathbf{I}_n + \mathbf{BA}|$.

Properties 1–11 in Result 1.3.6 are standard results. The proof of property 12 is deferred to Example 2.1.4.

Example 1.3.1. Vandermonde matrix. An $n \times n$ matrix $\mathbf{A}$ is a Vandermonde matrix if there are scalars $a_1, \cdots, a_n$, such that

$$\mathbf{A} = \begin{pmatrix} 1 & 1 & 1 & \cdots & 1 \\ a_1 & a_2 & a_3 & \cdots & a_n \\ a_1^2 & a_2^2 & a_3^2 & \cdots & a_n^2 \\ \vdots & \vdots & \vdots & \ddots & \vdots \\ a_1^{n-1} & a_2^{n-1} & a_3^{n-1} & \cdots & a_n^{n-1} \end{pmatrix}.$$

The determinant of $\mathbf{A}$ has a simple form:

$$\begin{aligned} |\mathbf{A}| &= \prod_{1 \le i < j \le n} (a_j - a_i) \\ &= (a_n - a_{n-1})(a_n - a_{n-2}) \cdots (a_n - a_2)(a_n - a_1) \\ &\quad \times (a_{n-1} - a_{n-2})(a_{n-1} - a_{n-3}) \cdots (a_{n-1} - a_1) \\ &\quad \times \cdots \times (a_2 - a_1). \end{aligned}$$

It is easily seen that $|\mathbf{A}| \neq 0$ if and only if $a_i \neq a_j$ for $i < j = 1, \cdots, n$, i.e., $a_1, \cdots, a_n$ are distinct. An example of a Vandermonde matrix is

$$\mathbf{A} = \begin{pmatrix} 1 & 1 & 1 \\ 1 & -1 & 2 \\ 1 & 1 & 4 \end{pmatrix},$$

with $a_1 = 1$, $a_2 = -1$, $a_3 = 2$, and $|\mathbf{A}| = -6$. □

Example 1.3.2. Intra-class correlation matrix. We define an $n \times n$ intra-class correlation matrix, which is also called an equicorrelation matrix, by

$$\mathbf{C} = \begin{pmatrix} 1 & \rho & \cdots & \rho \\ \rho & 1 & \cdots & \rho \\ \vdots & \vdots & \ddots & \vdots \\ \rho & \rho & \cdots & 1 \end{pmatrix} = (1 - \rho)\mathbf{I} + \rho\mathbf{J},$$

where $-1 < \rho < 1$. In the matrix $\mathbf{C}$, all the diagonal elements have the value 1, and all the off-diagonal elements have the same value ρ which is assumed to lie between -1 and 1. The determinant of **C** is easily computed by seeing that

$$\begin{vmatrix} 1 & \rho & \cdots & \rho \\ \rho & 1 & \cdots & \rho \\ \vdots & \vdots & \ddots & \vdots \\ \rho & \rho & \cdots & 1 \end{vmatrix} = \begin{vmatrix} 1+(n-1)\rho & \rho & \cdots & \rho \\ 1+(n-1)\rho & 1 & \cdots & \rho \\ \vdots & \vdots & \ddots & \vdots \\ 1+(n-1)\rho & \rho & \cdots & 1 \end{vmatrix} = [1+(n-1)\rho] \begin{vmatrix} 1 & \rho & \cdots & \rho \\ 1 & 1 & \cdots & \rho \\ \vdots & \vdots & \ddots & \vdots \\ 1 & \rho & \cdots & 1 \end{vmatrix}$$

$$= [1+(n-1)\rho] \begin{vmatrix} 1 & \rho & \cdots & \rho \\ 0 & 1-\rho & \cdots & 0 \\ \vdots & \vdots & \ddots & \vdots \\ 0 & 0 & \cdots & 1-\rho \end{vmatrix}$$

$$= [1+(n-1)\rho](1-\rho)^{n-1}$$

and $[1+(n-1)\rho](1-\rho)^{n-1} \geq 0$, which implies that $-\frac{1}{n-1} \leq \rho \leq 1$. So $|\mathbf{C}| = [1+(n-1)\rho](1-\rho)^{n-1}$; see Exercise 1.25 for another way to calculate $|\mathbf{C}|$. □

Definition 1.3.8. Nonsingular and singular matrices. If $|\mathbf{A}| \neq 0$, then **A** is said to be a nonsingular matrix. Otherwise, **A** is singular. For example, **A** is a nonsingular matrix and **B** is a singular matrix, where

$$\mathbf{A} = \begin{pmatrix} 1 & 6 \\ 0 & 3 \end{pmatrix} \quad \text{and} \quad \mathbf{B} = \begin{pmatrix} 1 & 6 \\ 1/2 & 3 \end{pmatrix}.$$

Definition 1.3.9. Inverse of a matrix. Let **A** be an $n \times n$ matrix. If there exists an $n \times n$ matrix **B** such that $\mathbf{AB} = \mathbf{I}_n$ (and $\mathbf{BA} = \mathbf{I}_n$), then **B** is called the (regular) inverse of **A**, and is denoted by $\mathbf{A}^{-1}$.

Result 1.3.7. A matrix **A** is invertible if and only if $|\mathbf{A}| \neq 0$.

Proof. To prove the "if" part, see that we compute the inverse of a matrix **A** using the formula

$$\mathbf{A}^{-1} = \frac{1}{|\mathbf{A}|}\text{Adj}(\mathbf{A}), \tag{1.3.1}$$

where $\text{Adj}(\mathbf{A})$ denotes the adjoint of **A**, and is defined to be the transpose of the matrix of cofactors of **A**, i.e., the (i,j)th element of $\text{Adj}(\mathbf{A})$ is the cofactor of a_{ji}. The proof of the "only if" part is left as Exercise 1.27. ■

Example 1.3.3. Suppose the matrix **A**, the matrix of cofactors **F** and the matrix $\text{Adj}(\mathbf{A})$ are given by

$$\mathbf{A} = \begin{pmatrix} -1 & 2 & 2 \\ 4 & 3 & -2 \\ -5 & 0 & 3 \end{pmatrix}, \quad \mathbf{F} = \begin{pmatrix} 9 & -2 & 15 \\ -6 & 7 & -10 \\ -10 & 6 & -11 \end{pmatrix} \quad \text{and}$$

$$\text{Adj}(\mathbf{A}) = \begin{pmatrix} 9 & -6 & -10 \\ -2 & 7 & 6 \\ 15 & -10 & -11 \end{pmatrix},$$

then, $|\mathbf{A}| = 17$, and

$$\mathbf{A}^{-1} = \frac{1}{|\mathbf{A}|}\text{Adj}(\mathbf{A}) = \begin{pmatrix} 9/17 & -6/17 & -10/17 \\ -2/17 & 7/17 & 6/17 \\ 15/17 & -10/17 & -11/17 \end{pmatrix}$$

is the inverse of **A**. □

Example 1.3.4. Let $\mathbf{A}$ be a $k \times k$ nonsingular matrix, and let $\mathbf{B}$ and $\mathbf{C}$ be any $k \times n$ and $n \times k$ matrices, respectively. Since we can write $\mathbf{A} + \mathbf{BC} = \mathbf{A}(\mathbf{I}_k + \mathbf{A}^{-1}\mathbf{BC})$, we see from property 3 of Result 1.3.6 that $|\mathbf{A} + \mathbf{BC}| = |\mathbf{A}||\mathbf{I}_k + \mathbf{A}^{-1}\mathbf{BC}|$. □

Definition 1.3.10. Reduced row echelon form (RREF). An $m \times n$ matrix $\mathbf{A}$ is said to be in RREF if the following conditions are met:

1. all zero rows are at the bottom of the matrix,
2. the leading nonzero entry of each nonzero row after the first occurs to the right of the leading nonzero entry of the previous row,
3. the leading nonzero entry in any nonzero row is 1, and
4. all entries in the column above and below a leading 1 are zero.

If only conditions 1 and 2 hold, the matrix has *row echelon form.* For example, among the following matrices,

$$\mathbf{A} = \begin{pmatrix} 1 & 0 & 2 & 3 & 0 \\ 0 & 1 & 4 & 5 & 0 \\ 0 & 0 & 0 & 0 & 0 \\ 0 & 0 & 0 & 0 & 0 \end{pmatrix}, \quad \mathbf{B} = \begin{pmatrix} 0 & 1 & 2 & 0 & 3 \\ 0 & 0 & 4 & 0 & 5 \\ 0 & 0 & 0 & 1 & 0 \\ 0 & 0 & 0 & 0 & 0 \end{pmatrix},$$

$$\mathbf{C} = \begin{pmatrix} 0 & 1 & 2 & 3 & 4 \\ 0 & 0 & 0 & 1 & 5 \\ 0 & 0 & 0 & 0 & 0 \end{pmatrix}, \quad \mathbf{D} = \begin{pmatrix} 0 & 0 & 0 & 0 & 0 \\ 0 & 1 & 2 & 0 & 4 \\ 0 & 0 & 0 & 0 & 0 \\ 0 & 0 & 0 & 1 & 5 \\ 0 & 0 & 0 & 0 & 0 \end{pmatrix}, \text{ and}$$

$$\mathbf{E} = \begin{pmatrix} 0 & 0 & 0 & 1 & 5 \\ 0 & 1 & 2 & 0 & 4 \end{pmatrix},$$

the matrix $\mathbf{A}$ is in RREF, whereas none of the other matrices is in RREF. In matrix $\mathbf{B}$, row 2 violates condition 3, matrix $\mathbf{C}$ violates condition 4, matrix $\mathbf{D}$ violates condition 1, while matrix $\mathbf{E}$ violates condition 2.

To verify invertibility and find the inverse (if it exists) of a square matrix $\mathbf{A}$,

(a) Perform elementary row operations on the augmented matrix $\begin{pmatrix}\mathbf{A} & \mathbf{I}\end{pmatrix}$ until $\mathbf{A}$ is in RREF.

(b) If RREF($\mathbf{A}$) $\neq \mathbf{I}$, then $\mathbf{A}$ is not invertible.

(c) If RREF($\mathbf{A}$) $= \mathbf{I}$, then the row operations that transformed $\mathbf{A}$ into $\mathbf{I}$ will have changed $\mathbf{I}$ into $\mathbf{A}^{-1}$.

Example 1.3.5. We describe an algorithm used to test whether an $n \times n$ matrix $\mathbf{A}$ is invertible, and if it is, to compute its inverse. The first step is to express the matrix $(\mathbf{A} : \mathbf{I})$ in reduced row echelon form, which we denote by RREF$(\mathbf{A} : \mathbf{I}) = (\mathbf{B} : \mathbf{C})$, say. If $\mathbf{B}$ has a row of zeroes, the matrix $\mathbf{A}$ is singular and is not invertible. Otherwise, the reduced matrix is now in the form $(\mathbf{I} : \mathbf{A}^{-1})$. We use this approach to find the inverse, if it exists, of the matrix

$$\mathbf{A} = \begin{pmatrix} 1 & 0 & -1 \\ 3 & 4 & -2 \\ 3 & 5 & -2 \end{pmatrix}.$$

We row reduce $(\mathbf{A} : \mathbf{I})$. In the display below, $\mathbf{B} \sim \mathbf{C}$ means $\mathbf{C}$ can be reduced from $\mathbf{B}$ by elementary row operations. For example, the second matrix can be reduced from $(\mathbf{A} : \mathbf{I})$ by subtracting 3 times the first row from the second and third rows, respectively:

$$\begin{pmatrix} 1 & 0 & -1 & : & 1 & 0 & 0 \\ 3 & 4 & -2 & : & 0 & 1 & 0 \\ 3 & 5 & -2 & : & 0 & 0 & 1 \end{pmatrix} \sim \begin{pmatrix} 1 & 0 & -1 & : & 1 & 0 & 0 \\ 0 & 4 & 1 & : & -3 & 1 & 0 \\ 0 & 5 & 1 & : & -3 & 0 & 1 \end{pmatrix} \sim$$

$$\begin{pmatrix} 1 & 0 & -1 & : & 1 & 0 & 0 \\ 0 & 4 & 1 & : & -3 & 1 & 0 \\ 0 & 0 & -1/4 & : & 3/4 & -5/4 & 1 \end{pmatrix} \sim \begin{pmatrix} 1 & 0 & -1 & : & 1 & 0 & 0 \\ 0 & 4 & 1 & : & -3 & 1 & 0 \\ 0 & 0 & 1 & : & -3 & 5 & -4 \end{pmatrix} \sim$$

$$\begin{pmatrix} 1 & 0 & 0 & : & -2 & 5 & -4 \\ 0 & 4 & 0 & : & 0 & -4 & 4 \\ 0 & 0 & 1 & : & -3 & 5 & -4 \end{pmatrix} \sim \begin{pmatrix} 1 & 0 & 0 & : & -2 & 5 & -4 \\ 0 & 1 & 0 & : & 0 & -1 & 1 \\ 0 & 0 & 1 & : & -3 & 5 & -4 \end{pmatrix},$$

so that

$$\mathbf{A}^{-1} = \begin{pmatrix} -2 & 5 & -4 \\ 0 & -1 & 1 \\ -3 & 5 & -4 \end{pmatrix}. \qquad \square$$

Result 1.3.8. Properties of inverse. Provided all the inverses exist,

1. $\mathbf{A}^{-1}$ is unique.
2. $(\mathbf{AB})^{-1} = \mathbf{B}^{-1}\mathbf{A}^{-1}$; $(\mathbf{A}_1 \cdots \mathbf{A}_k)^{-1} = \mathbf{A}_k^{-1} \cdots \mathbf{A}_1^{-1}$.
3. $(c\mathbf{A})^{-1} = (\mathbf{A}c)^{-1} = \frac{1}{c}\mathbf{A}^{-1}$.
4. If $|\mathbf{A}| \neq 0$, then $\mathbf{A}'$ and $\mathbf{A}^{-1}$ are nonsingular matrices and $(\mathbf{A}')^{-1} = (\mathbf{A}^{-1})'$.
5. *Sherman–Morrison–Woodbury formula* holds:
$$(\mathbf{A} + \mathbf{BCD})^{-1} = \mathbf{A}^{-1} - \mathbf{A}^{-1}\mathbf{B}(\mathbf{C}^{-1} + \mathbf{DA}^{-1}\mathbf{B})^{-1}\mathbf{DA}^{-1},$$
where $\mathbf{A}$, $\mathbf{B}$, $\mathbf{C}$, and $\mathbf{D}$ are respectively $m \times m$, $m \times n$, $n \times n$, and $n \times m$ matrices.
6. Provided $1 \pm \mathbf{b}'\mathbf{A}^{-1}\mathbf{a} \neq 0$, we have $(\mathbf{A} \pm \mathbf{ab}')^{-1} = \mathbf{A}^{-1} \mp \frac{(\mathbf{A}^{-1}\mathbf{a})(\mathbf{b}'\mathbf{A}^{-1})}{1 \pm \mathbf{b}'\mathbf{A}^{-1}\mathbf{a}}$.
7. $|\mathbf{A}^{-1}| = |\mathbf{A}|^{-1}$, i.e., the determinant of the inverse of $\mathbf{A}$ is equal to the reciprocal of $|\mathbf{A}|$.

Each of these properties is obtained by verifying that the product of the given matrix and its inverse is the identity matrix. Verification of property 6 is left for Exercise 1.28. See Harville (1997) for a proof of the Sherman–Morrison–Woodbury formula. It is useful to consider how one could come up with the formula in the first place. This will be discussed after Result 1.3.20. We consider the matrix inverse for several special matrices.

Result 1.3.9. If $\mathbf{A}$ is a lower (resp. upper) triangular matrix with diagonal elements $d_1, \cdots, d_n$, then $\mathbf{A}$ is invertible if and only if $d_i \neq 0$ for each i, in which case $\mathbf{A}^{-1}$ is also a lower (resp. upper) triangular matrix with diagonal elements $1/d_1, \cdots, 1/d_n$.

Proof. From property 4 of Result 1.3.6, $|\mathbf{A}| = d_1 \cdots d_n$, so by Result 1.3.7, $\mathbf{A}$ is invertible if and only if $d_i \neq 0$ for each i. If $\mathbf{A}$ is a lower triangular matrix, then it can shown that for $1 \leq j < i \leq n$, the submatrix $\mathbf{M}_{ij}$ of $\mathbf{A}$ obtained by deleting the ith row and jth column is a lower triangular matrix with at least one zero on the diagonal. Verification of this fact is a part of Exercise 1.20. By property 4, the cofactor $F_{ij} = (-1)^{i+j}|\mathbf{M}_{ij}|$ is zero. Then by (1.3.1), $\mathbf{A}^{-1}$ is lower triangular. Verification of the fact that the diagonal elements of $\mathbf{A}^{-1}$ are $1/d_1, \cdots, 1/d_n$ is also a part of Exercise 1.20. The case where $\mathbf{A}$ is upper triangular can be similarly proved. ■

Example 1.3.6. Inverse of an intra-class correlation matrix. We continue with Example 1.3.2. The cofactor of any diagonal element of $\mathbf{C}$ is based on the determinant of an $(n-1)\times(n-1)$ submatrix:

$$\begin{vmatrix} 1 & \rho & \cdots & \rho \\ \rho & 1 & \cdots & \rho \\ \vdots & \vdots & \ddots & \vdots \\ \rho & \rho & \cdots & 1 \end{vmatrix} = [1+(n-2)\rho](1-\rho)^{n-2},$$

while the cofactor of any off-diagonal element is

$$\begin{vmatrix} \rho & \rho & \cdots & \rho \\ \rho & 1 & \cdots & \rho \\ \vdots & \vdots & \ddots & \vdots \\ \rho & \rho & \cdots & 1 \end{vmatrix} = -\rho(1-\rho)^{n-2}.$$

Letting $D = (1-\rho)[1+(n-1)\rho]$,

$$\begin{aligned}\mathbf{A}^{-1} &= \frac{1}{D}\begin{pmatrix} 1+(n-2)\rho & -\rho & \cdots & -\rho \\ -\rho & 1+(n-2)\rho & \cdots & -\rho \\ \vdots & \vdots & \ddots & \vdots \\ -\rho & -\rho & \cdots & 1+(n-2)\rho \end{pmatrix} \\ &= \frac{1}{D}([1+(n-1)\rho]\mathbf{I} - \rho\mathbf{J}) \\ &= \frac{1}{(1-\rho)}\left(\mathbf{I} - \frac{\rho}{1+(n-1)\rho}\mathbf{J}\right).\end{aligned}$$

An alternate way to obtain $\mathbf{C}^{-1}$ is by using property 6 of Result 1.3.8. Suppose first that $\rho > 0$, and $\mathbf{C} = [(1-\rho)\mathbf{I} + \rho\mathbf{J}]$, so that $\mathbf{C}^{-1} = [(1-\rho)\mathbf{I} + \rho\mathbf{J}]^{-1}$. In property 6, set $\mathbf{A} = (1-\rho)\mathbf{I}$, $\mathbf{a} = \rho\mathbf{1}_n$, and $\mathbf{b} = \mathbf{1}_n$. Then $\mathbf{A}^{-1} = (1-\rho)^{-1}\mathbf{I}_n$, $\mathbf{A}^{-1}\mathbf{a} = (1-\rho)^{-1}\rho\mathbf{1}_n$, $\mathbf{b}'\mathbf{A}^{-1} = (1-\rho)^{-1}\mathbf{1}_n'$, and $\mathbf{b}'\mathbf{A}^{-1}\mathbf{a} = n\rho(1-\rho)^{-1}$, giving

$$\begin{aligned}\mathbf{C}^{-1} &= \left[\frac{1}{1-\rho}\mathbf{I} - \frac{(1-\rho)^{-2}\rho}{1+(1-\rho)^{-1}n\rho}\mathbf{J}\right] \\ &= \frac{1}{(1-\rho)}\left[\mathbf{I} - \frac{\rho}{1+(n-1)\rho}\mathbf{J}\right].\end{aligned}$$ □

Example 1.3.7. Toeplitz matrix. Consider the $n\times n$ Toeplitz matrix $\mathbf{A}$ which has the form

$$\mathbf{A} = \begin{pmatrix} 1 & \rho & \rho^2 & \cdots & \rho^{n-1} \\ \rho & 1 & \rho & \cdots & \rho^{n-2} \\ \vdots & \vdots & \vdots & \ddots & \vdots \\ \rho^{n-1} & \rho^{n-2} & \rho^{n-3} & \cdots & 1 \end{pmatrix}.$$

Note that all the elements on the jth subdiagonal and the jth superdiagonal coincide, for $j \geq 1$. It is easy to verify that, for $|\rho| < 1$, the inverse of $\mathbf{A}$ is

$$\mathbf{A}^{-1} = \frac{1}{1-\rho^2}\begin{pmatrix} 1 & -\rho & 0 & 0 & \cdots & 0 & 0 & 0 \\ -\rho & 1+\rho^2 & -\rho & 0 & \cdots & 0 & 0 & 0 \\ \vdots & \vdots & \vdots & \ddots & \cdots & 0 & 0 & 0 \\ 0 & 0 & 0 & 0 & \cdots & -\rho & 1+\rho^2 & -\rho \\ 0 & 0 & 0 & 0 & \cdots & 0 & -\rho & 1 \end{pmatrix},$$

which has a simpler Toeplitz form than $\mathbf{A}$. □

Definition 1.3.11. Orthogonal matrix. An $n \times n$ matrix $\mathbf{A}$ is orthogonal if $\mathbf{AA}' = \mathbf{A}'\mathbf{A} = \mathbf{I}_n$. For example,

$$\begin{pmatrix} \cos\theta & -\sin\theta \\ \sin\theta & \cos\theta \end{pmatrix}$$

is a 2×2 orthogonal matrix.

A direct consequence of Definition 1.3.11 is that, for an orthogonal matrix $\mathbf{A}$, $\mathbf{A}' = \mathbf{A}^{-1}$. Suppose $\mathbf{a}_i'$ denotes the ith row of $\mathbf{A}$, then, $\mathbf{AA}' = \mathbf{I}_n$ implies that $\mathbf{a}_i'\mathbf{a}_i = 1$, and $\mathbf{a}_i'\mathbf{a}_j = 0$ for $i \neq j$; so the rows of $\mathbf{A}$ have unit length and are mutually perpendicular (or orthogonal). Since $\mathbf{A}'\mathbf{A} = \mathbf{I}_n$, the columns of $\mathbf{A}$ have this property as well. If $\mathbf{A}$ is orthogonal, clearly, $|\mathbf{A}| = \pm 1$. It is also easy to show that the product of two orthogonal matrices $\mathbf{A}$ and $\mathbf{B}$ is itself orthogonal. Usually, orthogonal matrices are used to represent a change of basis or rotation.

Example 1.3.8. Helmert matrix. An $n \times n$ Helmert matrix $\mathbf{H}_n$ is an example of an orthogonal matrix and is defined by

$$\mathbf{H}_n = \begin{pmatrix} \frac{1}{\sqrt{n}}\mathbf{1}_n' \\ \mathbf{H}_0 \end{pmatrix},$$

where $\mathbf{H}_0$ is a $(n-1) \times n$ matrix such that for $i = 1, \ldots, n-1$, its ith row is $(\mathbf{1}_i', -i, 0, \cdots, 0)/\sqrt{\lambda_i}$ with $\lambda_i = i(i+1)$. For example, when $n = 4$, we have

$$\mathbf{H}_4 = \begin{pmatrix} 1/\sqrt{4} & 1/\sqrt{4} & 1/\sqrt{4} & 1/\sqrt{4} \\ 1/\sqrt{2} & -1/\sqrt{2} & 0 & 0 \\ 1/\sqrt{6} & 1/\sqrt{6} & -2/\sqrt{6} & 0 \\ 1/\sqrt{12} & 1/\sqrt{12} & 1/\sqrt{12} & -3/\sqrt{12} \end{pmatrix}.$$ □

We next define three important vector spaces associated with any matrix, viz., the *null space*, the *column space* and the *row space*. These concepts are closely related to properties of systems of linear equations, which are discussed in Chapter 3. A system of homogeneous linear equations is denoted by $\mathbf{Ax} = \mathbf{0}$, while $\mathbf{Ax} = \mathbf{b}$ denotes a system of nonhomogeneous linear equations.

Definition 1.3.12. Null space of a matrix. The null space $\mathcal{N}(\mathbf{A})$, of an $m \times n$ matrix $\mathbf{A}$ consists of all n-dimensional vectors $\mathbf{x}$ that are solutions to the homogeneous linear system $\mathbf{Ax} = \mathbf{0}$, i.e.,

$$\mathcal{N}(\mathbf{A}) = \{\mathbf{x} \in \mathcal{R}^n : \mathbf{Ax} = \mathbf{0}\}.$$

$\mathcal{N}(\mathbf{A})$ is a subspace of $\mathcal{R}^n$, and its dimension is called the nullity of $\mathbf{A}$. For example, the vector $\mathbf{x} = (1, 2)'$ belongs to the null space of the matrix $\mathbf{A} = \begin{pmatrix} 2 & -1 \\ -4 & 2 \end{pmatrix}$, since $\begin{pmatrix} 2 & -1 \\ -4 & 2 \end{pmatrix}\begin{pmatrix} 1 \\ 2 \end{pmatrix} = \begin{pmatrix} 0 \\ 0 \end{pmatrix}$.

We may use RREF($\mathbf{A}$) to find a basis of the null space of $\mathbf{A}$. We add or delete zero rows until RREF($\mathbf{A}$) is square. We then rearrange the rows to place the leading ones on the main diagonal to obtain $\widetilde{\mathbf{H}}$, which is the *Hermite form* of RREF($\mathbf{A}$). The nonzero columns of $\widetilde{\mathbf{H}} - \mathbf{I}$ are a basis for $\mathcal{N}(\mathbf{A})$. In general, an $n \times n$ matrix $\widetilde{\mathbf{H}}$ is in Hermite form if (i) each diagonal element is either 0 or 1; (ii) $\widetilde{h}_{ii} = 1$, the rest of column i has all zeroes; and (iii) $\widetilde{h}_{ii} = 0$, i.e., the ith row of $\widetilde{\mathbf{H}}$ is a vector of zeroes.

Example 1.3.9. We first find a basis for the null space of the matrix

$$\mathbf{A} = \begin{pmatrix} 1 & 0 & -5 & 1 \\ 0 & 1 & 2 & -3 \\ 0 & 0 & 0 & 0 \\ 0 & 0 & 0 & 0 \end{pmatrix},$$

which is in RREF, as we can verify. It is easy to see that the general solution to $\mathbf{Ax} = \mathbf{0}$ is the vector

$$\begin{pmatrix} x_1 \\ x_2 \\ x_3 \\ x_4 \end{pmatrix} = s\begin{pmatrix} 5 \\ -2 \\ 1 \\ 0 \end{pmatrix} + t\begin{pmatrix} -1 \\ 3 \\ 0 \\ 1 \end{pmatrix},$$

so that the vectors $(5, -2, 1, 0)'$ and $(-1, 3, 0, 1)'$ form a basis for $\mathcal{N}(\mathbf{A})$. This basis can also be obtained in an alternate way from RREF($\mathbf{A}$), which in this example coincides with $\mathbf{A}$. Computing

$$\mathbf{I} - \text{RREF}(\mathbf{A}) = \begin{pmatrix} 0 & 0 & 5 & -1 \\ 0 & 0 & -2 & 3 \\ 0 & 0 & 1 & 0 \\ 0 & 0 & 0 & 1 \end{pmatrix},$$

we see that the last two nonzero columns form a basis for $\mathcal{N}(\mathbf{A})$. □

Definition 1.3.13. Column space and row space of a matrix. Let $\mathbf{A}$ be an $m \times n$ matrix whose columns are the m-dimensional vectors $\mathbf{a}_1, \mathbf{a}_2, \cdots, \mathbf{a}_n$. The vector space Span$\{\mathbf{a}_1, \cdots, \mathbf{a}_n\}$ is called the column space (or range space) of $\mathbf{A}$, and is denoted by $\mathcal{C}(\mathbf{A})$. That is, the column space of $\mathbf{A}$ is the set consisting of all m-dimensional vectors that can be expressed as linear combinations of the n columns of $\mathbf{A}$ of the form

$$x_1\mathbf{a}_1 + x_2\mathbf{a}_2 + \cdots + x_n\mathbf{a}_n,$$

where $x_1, \cdots, x_n$ are scalars. The dimension of the column space of $\mathbf{A}$ is the number of LIN columns of $\mathbf{A}$, and it is called the column rank of $\mathbf{A}$. For example, given $\mathbf{A} = \begin{pmatrix} 1 & -2 \\ 2 & -4 \end{pmatrix}$, the vector $\mathbf{x}_1 = (-2, 2)'$ is not in $\mathcal{C}(\mathbf{A})$, whereas the vector $\mathbf{x}_2 = (3, 6)'$ is, because

$$\begin{pmatrix} 1 & -2 & -2 \\ 2 & -4 & 2 \end{pmatrix} \sim \begin{pmatrix} 1 & -2 & -2 \\ 0 & 0 & 6 \end{pmatrix}, \text{ while } \begin{pmatrix} 1 & -2 & 3 \\ 2 & -4 & 6 \end{pmatrix} \sim \begin{pmatrix} 1 & -2 & 3 \\ 0 & 0 & 0 \end{pmatrix}.$$

Likewise, if the rows of $\mathbf{A}$ are $\mathbf{b}_1', \cdots, \mathbf{b}_m'$, i.e., $\mathbf{A} = (\mathbf{b}_1, \ldots, \mathbf{b}_m)'$, then the vector space $\text{Span}\{\mathbf{b}_1, \cdots, \mathbf{b}_m\}$ is called the row space of $\mathbf{A}$, and is denoted by $\mathcal{R}(\mathbf{A})$. The row space of $\mathbf{A}$ is the set consisting of all n-dimensional vectors that can be expressed as linear combinations of the m rows of $\mathbf{A}$ of the form

$$x_1\mathbf{b}_1' + x_2\mathbf{b}_2' + \cdots + x_m\mathbf{b}_m'$$

where $x_1, \cdots, x_m$ are scalars. The dimension of the row space is called the row rank of $\mathbf{A}$. Concisely,

$$\mathcal{C}(\mathbf{A}) = \{\mathbf{Ax}: \mathbf{x} \in \mathcal{R}^n\}, \quad \text{and } \mathcal{R}(\mathbf{A}) = \mathcal{C}(\mathbf{A}') = \{\mathbf{A}'\mathbf{x}: \mathbf{x} \in \mathcal{R}^m\}.$$

The column space $\mathcal{C}(\mathbf{A})$ and the row space $\mathcal{R}(\mathbf{A})$ of any $m \times n$ matrix $\mathbf{A}$ are subspaces of $\mathcal{R}^m$ and $\mathcal{R}^n$, respectively. The symbol $\mathcal{C}^\perp(\mathbf{A})$ or $\{\mathcal{C}(\mathbf{A})\}^\perp$ represents the orthogonal complement of $\mathcal{C}(\mathbf{A})$ in $\mathcal{R}^m$. Likewise, $\mathcal{R}^\perp(\mathbf{A})$ or $\{\mathcal{R}(\mathbf{A})\}^\perp$ represents the orthogonal complement of $\mathcal{R}(\mathbf{A})$ in $\mathcal{R}^n$.

To find a basis of the column space of $\mathbf{A}$, we first find RREF($\mathbf{A}$). We select the columns of $\mathbf{A}$ which correspond to the columns of RREF($\mathbf{A}$) with leading ones. These are called the leading columns of $\mathbf{A}$ and form a basis for $\mathcal{C}(\mathbf{A})$. The nonzero rows of RREF($\mathbf{A}$) are a basis for $\mathcal{R}(\mathbf{A})$.

Example 1.3.10. We find a basis for $\mathcal{C}(\mathbf{A})$, where the matrix $\mathbf{A}$ and $\mathbf{B} = \text{RREF}(\mathbf{A})$ are shown below:

$$\mathbf{A} = \begin{pmatrix} 1 & -2 & 2 & 1 & 0 \\ -1 & 2 & -1 & 0 & 0 \\ 2 & -4 & 6 & 4 & 0 \\ 3 & -6 & 8 & 5 & 1 \end{pmatrix} \quad \text{and} \quad \mathbf{B} = \begin{pmatrix} 1 & -2 & 0 & -1 & 0 \\ 0 & 0 & 1 & 1 & 0 \\ 0 & 0 & 0 & 0 & 1 \\ 0 & 0 & 0 & 0 & 0 \end{pmatrix}.$$

We see that columns 1, 3 and 5 of $\mathbf{B}$ have leading ones. Then columns 1, 3 and 5 of $\mathbf{A}$ form a basis for $\mathcal{C}(\mathbf{A})$. □

Result 1.3.10. Let $\mathcal{C}(\mathbf{A})$ and $\mathcal{N}(\mathbf{A})$ respectively denote the column space and null space of an $m \times n$ matrix $\mathbf{A}$. Then,

1. $\dim[\mathcal{C}(\mathbf{A})] = n - \dim[\mathcal{N}(\mathbf{A})]$.
2. $\mathcal{N}(\mathbf{A}) = \{\mathcal{C}(\mathbf{A}')\}^\perp = \{\mathcal{R}(\mathbf{A})\}^\perp$.
3. $\dim[\mathcal{C}(\mathbf{A})] = \dim[\mathcal{R}(\mathbf{A})]$.
4. $\mathcal{C}(\mathbf{A}'\mathbf{A}) = \mathcal{C}(\mathbf{A}')$, and $\mathcal{R}(\mathbf{A}'\mathbf{A}) = \mathcal{R}(\mathbf{A})$.
5. $\mathcal{C}(\mathbf{A}) \subseteq \mathcal{C}(\mathbf{B})$ if and only if $\mathbf{A} = \mathbf{BC}$ for some matrix $\mathbf{C}$. Also, $\mathcal{R}(\mathbf{A}) \subseteq \mathcal{R}(\mathbf{B})$ if and only if $\mathbf{A} = \mathbf{CB}$ for some matrix $\mathbf{C}$.
6. $\mathcal{C}(\mathbf{ACB}) = \mathcal{C}(\mathbf{AC})$ if $\text{r}(\mathbf{CB}) = \text{r}(\mathbf{C})$.

Proof. 1. Suppose $\dim[\mathcal{C}(\mathbf{A})] = r$ and without loss of generality, let the first r columns of $\mathbf{A}$, i. e., $\mathbf{a}_1, \cdots, \mathbf{a}_r$ be LIN. Then for each $j = r+1, \ldots, n$, we can write $\mathbf{a}_j = \sum_{i=1}^r c_{ij}\mathbf{a}_i$, so for any $\mathbf{x} = (x_1, \cdots, x_n)'$, $\mathbf{Ax} = \sum_{i=1}^n x_i\mathbf{a}_i = \sum_{i=1}^r x_i\mathbf{a}_i + \sum_{j=r+1}^n x_j\big(\sum_{i=1}^r c_{ij}\mathbf{a}_i\big) = \sum_{i=1}^r \big(x_i + \sum_{j=r+1}^n c_{ij}x_j\big)\mathbf{a}_i$. Then $\mathbf{Ax} = \mathbf{0} \iff x_i = -\sum_{j=r+1}^n c_{ij}x_j$, $i = 1, \ldots, r$. In other words, any solution to $\mathbf{Ax} = \mathbf{0}$ is completely determined by $x_{r+1}, \cdots, x_n$, which can take any real values. As a result, $\dim[\mathcal{N}(\mathbf{A})] = n - r$.

2. $\mathbf{x} \in \mathcal{N}(\mathbf{A})$ if and only if $\mathbf{a}'\mathbf{x} = 0$, i.e., $\mathbf{x} \perp \mathbf{a}$, for every row $\mathbf{a}'$ of $\mathbf{A}$. The latter is equivalent to $\mathbf{x} \perp \mathcal{R}(\mathbf{A})$. Then $\mathcal{N}(\mathbf{A}) = \{\mathcal{R}(\mathbf{A})\}^{\perp}$.

3. By property 2, $\dim[\mathcal{N}(\mathbf{A})] = \dim[\{\mathcal{R}(\mathbf{A})\}^{\perp}] = n - \dim[\mathcal{R}(\mathbf{A})]$. By comparing to property 1, this gives $\dim[\mathcal{C}(\mathbf{A})] = \dim[\mathcal{R}(\mathbf{A})]$.

4. By property 2 and $(\mathbf{A}'\mathbf{A})' = \mathbf{A}'\mathbf{A}$, to show $\mathcal{C}(\mathbf{A}'\mathbf{A}) = \mathcal{C}(\mathbf{A}')$, it is enough to show that $\mathcal{N}(\mathbf{A}'\mathbf{A}) = \mathcal{N}(\mathbf{A})$, or $\mathbf{A}'\mathbf{A}\mathbf{x} = \mathbf{0} \iff \mathbf{A}\mathbf{x} = \mathbf{0}$. The $\Longleftarrow$ part is clear. On the other hand, $\mathbf{A}'\mathbf{A}\mathbf{x} = \mathbf{0} \implies \mathbf{x}'\mathbf{A}'\mathbf{A}\mathbf{x} = 0 \implies \|\mathbf{A}\mathbf{x}\|^2 = 0 \implies \mathbf{A}\mathbf{x} = \mathbf{0}$. The identity for the row spaces can be similarly proved.

5. Let the columns of $\mathbf{A}$ be $\mathbf{a}_1, \cdots, \mathbf{a}_n$. Suppose $\mathbf{A} = \mathbf{BC}$ for some $k \times n$ matrix $\mathbf{C}$ with entries c_{ij}. Let $\mathbf{c}_1, \cdots, \mathbf{c}_n$ be the columns of $\mathbf{C}$, and let $\mathbf{a}_j = \mathbf{B}\mathbf{c}_j \in \mathcal{C}(\mathbf{B})$ for each $j = 1, \cdots, n$. Thus $\mathcal{C}(\mathbf{A}) \subseteq \mathcal{C}(\mathbf{B})$. Conversely, if $\mathcal{C}(\mathbf{A}) \subseteq \mathcal{C}(\mathbf{B})$, then every column vector of $\mathbf{A}$ is a linear combination of the column vectors of $\mathbf{B}$, say $\mathbf{b}_1, \cdots, \mathbf{b}_k$, in other words, for each $j = 1, \cdots, n$, $\mathbf{a}_j = \mathbf{b}_1 c_{1j} + \cdots + \mathbf{b}_k c_{kj}$ for some c_{ij}. Then $\mathbf{A} = \mathbf{BC}$. The result on the row spaces can be proved similarly.

6. From property 5, $\mathcal{C}(\mathbf{ACB}) \subseteq \mathcal{C}(\mathbf{AC})$ and $\mathcal{C}(\mathbf{CB}) \subseteq \mathcal{C}(\mathbf{C})$. If $\mathrm{r}(\mathbf{CB}) = \mathrm{r}(\mathbf{C})$, then the dimensions of the two column spaces are equal, so with the first one being a subspace of the second one, they must be equal. Then by property 5 again, $\mathbf{C} = (\mathbf{CB})\mathbf{D}$ for some matrix $\mathbf{D}$. Then $\mathcal{C}(\mathbf{AC}) = \mathcal{C}(\mathbf{ACBD}) \subseteq \mathcal{C}(\mathbf{ACB})$. Thus $\mathcal{C}(\mathbf{ACB}) = \mathcal{C}(\mathbf{AC})$. ■

Definition 1.3.14. Rank of a matrix. Let $\mathbf{A}$ be an $m \times n$ matrix. From property 3 of Result 1.3.10, $\dim[\mathcal{C}(\mathbf{A})] = \dim[\mathcal{R}(\mathbf{A})]$. We call $\dim[\mathcal{C}(\mathbf{A})]$ the rank of $\mathbf{A}$, denoted by $\mathrm{r}(\mathbf{A})$. We say that $\mathbf{A}$ has full row rank if $\mathrm{r}(\mathbf{A}) = m$, which is possible only if $m \le n$, and has full column rank if $\mathrm{r}(\mathbf{A}) = n$, which is possible only if $n \le m$. A nonsingular matrix has full row rank and full column rank.

To find the rank of $\mathbf{A}$, we find $\mathrm{RREF}(\mathbf{A})$. We count the number of leading ones, which is then equal to $\mathrm{r}(\mathbf{A})$.

Example 1.3.11. Consider the matrices

$$\mathbf{A} = \begin{pmatrix} 1 & 2 & 2 & -1 \\ 1 & 3 & 1 & -2 \\ 1 & 1 & 3 & 0 \\ 0 & 1 & -1 & -1 \\ 1 & 2 & 2 & -1 \end{pmatrix} \quad \text{and} \quad \mathbf{B} = \begin{pmatrix} 1 & 2 & 2 & -1 \\ 0 & 1 & -1 & -1 \\ 0 & 0 & 0 & 0 \\ 0 & 0 & 0 & 0 \\ 0 & 0 & 0 & 0 \end{pmatrix},$$

where $\mathbf{B} = \mathrm{RREF}(\mathbf{A})$ has two nonzero rows. Hence, $\mathrm{r}(\mathbf{A}) = 2$. □

Result 1.3.11. Properties of rank.

1. An $m \times n$ matrix $\mathbf{A}$ has rank r if the largest nonsingular square submatrix of $\mathbf{A}$ has size r.
2. For an $m \times n$ matrix $\mathbf{A}$, $\mathrm{r}(\mathbf{A}) \le \min(m, n)$.
3. $\mathrm{r}(\mathbf{A} + \mathbf{B}) \le \mathrm{r}(\mathbf{A}) + \mathrm{r}(\mathbf{B})$.
4. $\mathrm{r}(\mathbf{AB}) \le \min\{\mathrm{r}(\mathbf{A}), \mathrm{r}(\mathbf{B})\}$, where $\mathbf{A}$ and $\mathbf{B}$ are conformal under multiplication.
5. For nonsingular matrices $\mathbf{A}$, $\mathbf{B}$, and an arbitrary matrix $\mathbf{C}$,

$$\mathrm{r}(\mathbf{C}) = \mathrm{r}(\mathbf{AC}) = \mathrm{r}(\mathbf{CB}) = \mathrm{r}(\mathbf{ACB}).$$

6. $\mathrm{r}(\mathbf{A}) = \mathrm{r}(\mathbf{A}') = \mathrm{r}(\mathbf{A}'\mathbf{A}) = \mathrm{r}(\mathbf{A}\mathbf{A}')$.

7. For any $n \times n$ matrix $\mathbf{A}$, $|\mathbf{A}| = 0$ if and only if $\mathrm{r}(\mathbf{A}) < n$.

8. $\mathrm{r}(\mathbf{A}, \mathbf{b}) \geq \mathrm{r}(\mathbf{A})$, i.e., inclusion of a column vector cannot decrease the rank of a matrix.

Result 1.3.12. Let $\mathbf{A}$ and $\mathbf{B}$ be $m \times n$ matrices. Let $\mathbf{C}$ be a $p \times m$ matrix with $\mathrm{r}(\mathbf{C}) = m$, and let $\mathbf{D}$ be an $n \times p$ matrix with $\mathrm{r}(\mathbf{D}) = n$.

1. If $\mathbf{CA} = \mathbf{CB}$, then $\mathbf{A} = \mathbf{B}$.

2. If $\mathbf{AD} = \mathbf{BD}$, then $\mathbf{A} = \mathbf{B}$.

3. If $\mathbf{CAD} = \mathbf{CBD}$, then $\mathbf{A} = \mathbf{B}$.

Proof. We prove only property 1 here; the proofs of the other two properties are similar. Let the column vectors of $\mathbf{C}$ be $\mathbf{c}_1, \cdots, \mathbf{c}_m$. Since $\mathrm{r}(\mathbf{C}) = m$, the column vectors are LIN. Let $\mathbf{A} = \{a_{ij}\}$ and $\mathbf{B} = \{b_{ij}\}$. The jth column vector of $\mathbf{CA}$ is $\mathbf{c}_1 a_{1j} + \cdots + \mathbf{c}_m a_{mj}$ and that of $\mathbf{CB}$ is $\mathbf{c}_1 b_{1j} + \cdots + \mathbf{c}_m b_{mj}$. Since these two column vectors are equal and $\mathbf{c}_i$ are LIN, $a_{ij} = b_{ij}$. Then $\mathbf{A} = \mathbf{B}$. ■

Result 1.3.13. Let $\mathbf{A}$ be an $m \times n$ matrix.

1. For $n \times p$ matrices $\mathbf{B}$ and $\mathbf{C}$, $\mathbf{AB} = \mathbf{AC}$ if and only if $\mathbf{A}'\mathbf{AB} = \mathbf{A}'\mathbf{AC}$.

2. For $p \times n$ matrices $\mathbf{E}$ and $\mathbf{F}$, $\mathbf{EA}' = \mathbf{FA}'$ if and only if $\mathbf{EA}'\mathbf{A} = \mathbf{FA}'\mathbf{A}$.

Proof. To prove property 1, the "only if" part is obvious. On the other hand, if $\mathbf{A}'\mathbf{AB} = \mathbf{A}'\mathbf{AC}$, then $\mathbf{O} = (\mathbf{B} - \mathbf{C})'(\mathbf{A}'\mathbf{AB} - \mathbf{A}'\mathbf{AC}) = (\mathbf{AB} - \mathbf{AC})'(\mathbf{AB} - \mathbf{AC})$, which implies that $\mathbf{AB} - \mathbf{AC} = \mathbf{O}$ (see Exercise 1.16). The proof of property 2 follows directly by transposing relevant matrices in property 1. ■

Definition 1.3.15. Equivalent matrices. Two matrices that have the same dimension and the same rank are said to be equivalent matrices.

Result 1.3.14. Equivalent canonical form of a matrix. An $m \times n$ matrix $\mathbf{A}$ with $\mathrm{r}(\mathbf{A}) = r$ is equivalent to $\mathbf{PAQ} = \begin{pmatrix} \mathbf{I}_r & \mathbf{O} \\ \mathbf{O} & \mathbf{O} \end{pmatrix}$, where $\mathbf{P}$ and $\mathbf{Q}$ are respectively $m \times m$ and $n \times n$ matrices, and are obtained as products of elementary matrices, i.e., matrices obtained from the identity matrix using elementary transformations. The matrices $\mathbf{P}$ and $\mathbf{Q}$ always exist, but need not be unique. Elementary transformations include

1. interchange of two rows (columns) of $\mathbf{I}$, or

2. multiplication of elements of a row (column) of $\mathbf{I}$ by a nonzero scalar c, or

3. adding to row j (column j) of $\mathbf{I}$, c times row i (column i).

Definition 1.3.16. Eigenvalue, eigenvector, and eigenspace of a matrix. A real or complex number λ is an eigenvalue (or characteristic root) of an $n \times n$ matrix $\mathbf{A}$ if $\mathbf{A} - \lambda\mathbf{I}_n$ is singular, i.e.,

$$|\mathbf{A} - \lambda\mathbf{I}_n| = 0.$$

The space $\mathcal{N}(\mathbf{A} - \lambda\mathbf{I}_n)$, containing vectors with possibly complex-valued components, is called the eigenspace corresponding to λ. Any non-null vector in the eigenspace is called an eigenvector of $\mathbf{A}$ corresponding to λ and satisfies

$$(\mathbf{A} - \lambda \mathbf{I}_n)\mathbf{v} = \mathbf{0}.$$

The dimension of the eigenspace, $g = \dim[\mathcal{N}(\mathbf{A} - \lambda \mathbf{I}_n)]$, is called the geometric multiplicity of λ.

The eigenvalues of $\mathbf{A}$ are solutions to the characteristic polynomial equation $P(\lambda) = |\mathbf{A} - \lambda \mathbf{I}| = 0$, which is a polynomial in λ of degree n. Note that the n eigenvalues of $\mathbf{A}$ are not necessarily all distinct or real-valued. Since $|\mathbf{A} - \lambda_j \mathbf{I}| = 0$, $\mathbf{A} - \lambda_j \mathbf{I}$ is a singular matrix, for $j = 1, \cdots, n$, and there exists a nonzero n-dimensional vector $\mathbf{v}_j$ which satisfies $(\mathbf{A} - \lambda_j \mathbf{I})\mathbf{v}_j = \mathbf{0}$, i.e., $\mathbf{A}\mathbf{v}_j = \lambda_j \mathbf{v}_j$. The eigenvectors of $\mathbf{A}$ are thus obtained by substituting each λ_j into $\mathbf{A}\mathbf{v}_j = \lambda_j \mathbf{v}_j$, $j = 1, \cdots, n$, and solving the resulting n equations. We say that an eigenvector $\mathbf{v}_j$ is normalized if its length is 1. If λ_j is complex-valued, then $\mathbf{v}_j$ may have complex elements. If some of the eigenvalues of the real matrix $\mathbf{A}$ are complex, then they must clearly be conjugate complex (a conjugate complex pair is defined as $a + \iota b$ and $a - \iota b$, where $\iota = \sqrt{-1}$).

If an eigenvalue λ is real, there is a corresponding real eigenvector $\mathbf{v}$. Also, if we multiply $\mathbf{v}$ by any complex scalar $c = a + \iota b$, then $c\mathbf{v}$ satisfies $\mathbf{A}c\mathbf{v} = \lambda c\mathbf{v}$, so that $c\mathbf{v}$ is an eigenvector of $\mathbf{A}$ corresponding to λ. Likewise, if λ is complex and $\mathbf{v}$ is a corresponding eigenvector while $\mathbf{u}$ is an eigenvector corresponding to the complex conjugate of λ, then $\mathbf{v}$ and $\mathbf{u}$ need not be conjugate, although there is a complex eigenvector $\mathbf{w}$, say, corresponding to λ, such that $\mathbf{v} = c_1\mathbf{w}$ and $\mathbf{u} = c_2\mathbf{w}^*$ for some scalars c_1 and c_2, and where $\mathbf{w}^*$ denotes the complex conjugate of $\mathbf{w}$.

Suppose $\mathbf{v}_{j1}$ and $\mathbf{v}_{j2}$ are nonzero eigenvectors of $\mathbf{A}$ corresponding to λ_j, it is easy to see that $\alpha_1\mathbf{v}_{j1} + \alpha_2\mathbf{v}_{j2}$ is also an eigenvector corresponding to λ_j, where α_1 and α_2 are real numbers. That is, we must have $\mathbf{A}(\alpha_1\mathbf{v}_{j1} + \alpha_2\mathbf{v}_{j2}) = \lambda_j(\alpha_1\mathbf{v}_{j1} + \alpha_2\mathbf{v}_{j2})$. The eigenvectors corresponding to any eigenvalue λ_j span a vector space, called the eigenspace of $\mathbf{A}$ for λ_j.

To find the eigenvectors of $\mathbf{A}$ corresponding to an eigenvalue λ, we find the basis of the null space of $\mathbf{A} - \lambda\mathbf{I}$. The nonzero columns of $\widetilde{\mathbf{H}} - \mathbf{I}$ are a basis for $\mathcal{N}(\mathbf{A} - \lambda\mathbf{I})$, where $\widetilde{\mathbf{H}}$ denotes the Hermite form of RREF$(\mathbf{A} - \lambda\mathbf{I})$ (see Example 1.3.14).

Result 1.3.15. Let $\mathbf{A}$ and $\mathbf{B}$ be $n \times n$ matrices such that $\mathbf{A} = \mathbf{M}^{-1}\mathbf{B}\mathbf{M}$, where $\mathbf{M}$ is nonsingular. Then the characteristic polynomials of $\mathbf{A}$ and $\mathbf{B}$ coincide.

Proof. $|\mathbf{A} - \lambda\mathbf{I}| = |\mathbf{M}^{-1}\mathbf{B}\mathbf{M} - \lambda\mathbf{I}| = |\mathbf{M}^{-1}||\mathbf{B}\mathbf{M} - \lambda\mathbf{M}| = |\mathbf{B}\mathbf{M}\mathbf{M}^{-1} - \lambda\mathbf{M}\mathbf{M}^{-1}| = |\mathbf{B} - \lambda\mathbf{I}|$. ■

The eigenspace $\mathcal{N}(\mathbf{A} - \lambda\mathbf{I})$ corresponding to an eigenvalue λ has the property that for any $\mathbf{x}$ in the eigenspace, $\mathbf{A}\mathbf{x} = \lambda\mathbf{x}$ is still in the space. This property is formalized below.

Definition 1.3.17. Let $\mathbf{A}$ be an $n \times n$ matrix. A subspace $\mathcal{V}$ of $\mathcal{R}^n$ is called an invariant subspace with respect to $\mathbf{A}$ if $\{\mathbf{A}\mathbf{x}\colon \mathbf{x} \in \mathcal{V}\} \subset \mathcal{V}$. That is, $\mathcal{V}$ gets mapped to itself under $\mathbf{A}$.

Result 1.3.16. If a non-null space $\mathcal{V}$ is an invariant subspace with respect to $\mathbf{A}$, then it contains at least one eigenvector of $\mathbf{A}$.

Proof. Let $\{\mathbf{v}_1, \cdots, \mathbf{v}_k\}$ be a basis of $\mathcal{V}$. Put $\mathbf{V} = (\mathbf{v}_1, \cdots, \mathbf{v}_k)$ so that $\mathcal{V} = \mathcal{C}(\mathbf{V})$. Since $\mathbf{A}\mathbf{v}_i \in \mathcal{V}$ for every i, then $\mathcal{C}(\mathbf{A}\mathbf{V}) \subset \mathcal{C}(\mathbf{V})$, so by property 5 of Result 1.3.10, $\mathbf{A}\mathbf{V} = \mathbf{V}\mathbf{B}$ for some $k \times k$ matrix $\mathbf{B}$. Now $\mathbf{B}$ has at least one eigenvalue, say λ, and a corresponding eigenvector $\mathbf{z}$. Let $\mathbf{x} = \mathbf{V}\mathbf{z}$. Then $\mathbf{x} \in \mathcal{C}(\mathbf{V}) = \mathcal{V}$. Since $\mathbf{V}$ has full column rank and $\mathbf{z} \neq \mathbf{0}$, we see that $\mathbf{x} \neq \mathbf{0}$. Then, $\mathbf{A}\mathbf{x} = \mathbf{A}\mathbf{V}\mathbf{z} = \mathbf{V}\mathbf{B}\mathbf{z} = \mathbf{V}\lambda\mathbf{z} = \lambda\mathbf{x}$, so $\mathbf{x}$ is an eigenvector in $\mathcal{V}$. ■

Definition 1.3.18. Spectrum of a matrix. The spectrum of an $n \times n$ matrix $\mathbf{A}$ is the set of its distinct (real or complex) eigenvalues $\{\lambda_1, \lambda_2, \cdots, \lambda_k\}$, so that the characteristic polynomial of $\mathbf{A}$ has factorization $P(\lambda) = (-1)^n(\lambda - \lambda_1)^{a_1} \cdots (\lambda - \lambda_k)^{a_k}$, where a_j are positive integers with $a_1 + \cdots + a_k = n$. The algebraic multiplicity of λ_j is defined to be a_j.

For each eigenvalue, its geometric multiplicity is at most as large as its algebraic multiplicity, i.e., $g_j \leq a_j$, the proof of which is left as Exercise 1.34. However, if $\mathbf{A}$ is symmetric, then $g_j = a_j$ (see Exercise 2.13).

Result 1.3.17. Suppose the spectrum of an $n \times n$ matrix $\mathbf{A}$ is $\{\lambda_1, \cdots, \lambda_k\}$. Then eigenvectors corresponding to different λ_j's are LIN.

Proof. Suppose $\mathbf{x}_j \in \mathcal{N}(\mathbf{A} - \lambda_j \mathbf{I}_n)$, $j = 1, \ldots, k$, and $\mathbf{x}_1 + \cdots + \mathbf{x}_k = \mathbf{0}$. We must show that all $\mathbf{x}_j = \mathbf{0}$. For any scalar c, $(\mathbf{A} - c\mathbf{I}_n)\mathbf{x}_j = (\lambda_j - c)\mathbf{x}_j$. Then for any scalars $c_1, \cdots, c_s$,

$$(\mathbf{A} - c_1\mathbf{I}_n) \cdots (\mathbf{A} - c_s\mathbf{I}_n)\mathbf{x}_j = (\lambda_j - c_1) \cdots (\lambda_j - c_s)\mathbf{x}_j.$$

For each $i = 1, \ldots, k$, $\mathbf{x}_i = -\sum_{j \neq i} \mathbf{x}_j$. From the display, pre-multiplying both sides by $\prod_{l \neq i}(\mathbf{A} - \lambda_l \mathbf{I}_n)$ gives $\prod_{j \neq i}(\lambda_j - \lambda_i)\mathbf{x}_i = \mathbf{0}$, and hence $\mathbf{x}_i = \mathbf{0}$. ■

Example 1.3.12. Let $\mathbf{A} = \begin{pmatrix} 1 & 1 \\ 0 & 1 \end{pmatrix}$. Then, $P(\lambda) = (\lambda - 1)^2$, with solutions $\lambda_1 = 1$ (repeated twice), so that $a_1 = 2$. Since $\mathbf{A} - \lambda_1 \mathbf{I}_2 = \begin{pmatrix} 0 & 1 \\ 0 & 0 \end{pmatrix}$, with rank 1, the geometric multiplicity of λ_1 is $g_1 = 2 - 1 = 1 < a_1$. □

Example 1.3.13. Let $\mathbf{A} = \begin{pmatrix} 0 & -1 \\ 1 & 0 \end{pmatrix}$. This matrix has no real eigenvalues, since, $P(\lambda) = \lambda^2 + 1$, with solutions $\lambda_1 = \iota$ and $\lambda_2 = -\iota$, where $\iota = \sqrt{-1}$. The corresponding eigenvectors are complex, and are $(\iota, 1)'$ and $(-\iota, 1)'$. □

Example 1.3.14. Let

$$\mathbf{A} = \begin{pmatrix} -1 & 2 & 0 \\ 1 & 2 & 1 \\ 0 & 2 & -1 \end{pmatrix}.$$

Then $|\mathbf{A} - \lambda \mathbf{I}| = -(\lambda + 1)(\lambda - 3)(\lambda + 2) = 0$, yielding solutions $\lambda_1 = -1$, $\lambda_2 = 3$, and $\lambda_3 = -2$, which are the distinct eigenvalues of $\mathbf{A}$. To obtain the eigenvectors corresponding to λ_i, we must solve the homogeneous linear system $(\mathbf{A} - \lambda_i \mathbf{I})\mathbf{v}_i = \mathbf{0}$, or in other words, identify the null space of the matrix $(\mathbf{A} - \lambda_i \mathbf{I})$, by completely reducing the augmented matrix $(\mathbf{A} - \lambda_i \mathbf{I} : \mathbf{0})$. Corresponding to $\lambda_1 = -1$, we see that

$$\mathbf{A} - (-1)\mathbf{I} = \begin{pmatrix} 0 & 2 & 0 \\ 1 & 3 & 1 \\ 0 & 2 & 0 \end{pmatrix}, \quad \text{RREF}(\mathbf{A} + \mathbf{I}) = \begin{pmatrix} 1 & 0 & 1 \\ 0 & 1 & 0 \\ 0 & 0 & 0 \end{pmatrix},$$

which is in Hermite form, i.e., $\text{RREF}(\mathbf{A} + \mathbf{I}) = \widetilde{\mathbf{H}}$. The nonzero column of

$$\widetilde{\mathbf{H}} - \mathbf{I} = \begin{pmatrix} 0 & 0 & 1 \\ 0 & 0 & 0 \\ 0 & 0 & -1 \end{pmatrix},$$

i.e., $(1, 0, -1)'$ is a basis of $\mathcal{N}(\mathbf{A} - (-1)\mathbf{I})$ and therefore of the eigenspace corresponding to

$\lambda_1 = -1$. Using a similar approach, we find that a basis of the eigenspace corresponding to $\lambda_2 = 3$ is $(-1, -2, -1)'$ and corresponding to $\lambda_3 = -2$ is $(-1, 1/2, -1)'$. These three vectors are the eigenvectors of the matrix $\mathbf{A}$. □

Result 1.3.18. Let $\mathbf{A}$ be an $n \times n$ matrix with n eigenvalues $\lambda_1, \cdots, \lambda_n$, counting multiplicities (i.e., the eigenvalues are not necessarily distinct), and let c be a real scalar. Then, counting multiplicities, the following properties hold for the eigenvalues of transformations of $\mathbf{A}$.

1. $\lambda_1^k, \cdots, \lambda_n^k$ are the eigenvalues of $\mathbf{A}^k$, for any positive integer k.
2. $c\lambda_1, \cdots, c\lambda_n$ are the eigenvalues of $c\mathbf{A}$.
3. $\lambda_1 + c, \cdots, \lambda_n + c$ are the eigenvalues of $\mathbf{A} + c\mathbf{I}$, while the eigenvectors of $\mathbf{A} + c\mathbf{I}$ coincide with the eigenvectors of $\mathbf{A}$.
4. If $\mathbf{A}$ is invertible, then $1/\lambda_1, \cdots, 1/\lambda_n$ are the eigenvalues of $\mathbf{A}^{-1}$, while the eigenvectors of $\mathbf{A}^{-1}$ coincide with the eigenvectors of $\mathbf{A}$.
5. $f(\lambda_1), \cdots, f(\lambda_n)$ are the eigenvalues of $f(\mathbf{A})$, where $f(.)$ is any polynomial.

The proof of Result 1.3.18 is quite simple by using the results in Chapter 2 and is left as Exercise 2.9. The result offers more than saying that if λ is an eigenvalue of $\mathbf{A}$, then $f(\lambda)$ is an eigenvalue of $f(\mathbf{A})$ (see Exercise 1.36), as the latter does not consider the multiplicity of the eigenvalue.

Result 1.3.19. Sum and product of eigenvalues. Let $\mathbf{A}$ be an $n \times n$ matrix with eigenvalues $\lambda_1, \cdots, \lambda_n$, counting multiplicities. Then,

1. $\text{tr}(\mathbf{A}) = \sum_{i=1}^{n} \lambda_i$.
2. $|\mathbf{A}| = \prod_{i=1}^{n} \lambda_i$.
3. $|\mathbf{I}_n \pm \mathbf{A}| = \prod_{i=1}^{n}(1 \pm \lambda_i)$.

The proof for the above result is left as Exercise 1.37.

Let $a_{ij}^{(k)}$ represent the (i, j)th element of $\mathbf{A}_k \in \mathcal{R}^{m\times n}$, $k = 1, 2, \cdots$. For every $i = 1, \cdots, m$ and $j = 1, \cdots, n$, suppose there exists a scalar a_{ij} which is the limit of the sequence of numbers $a_{ij}^{(1)}, a_{ij}^{(2)}, \cdots$, and suppose $\mathbf{A} = \{a_{ij}\}$. We say that the $m \times n$ matrix $\mathbf{A}$ is the limit of the sequence of matrices $\mathbf{A}_k$, $k = 1, 2, \cdots$, or that the sequence $\mathbf{A}_k$, $k = 1, 2, \cdots$ converges to the matrix $\mathbf{A}$, which we denote by $\lim_{k\to\infty} \mathbf{A}_k = \mathbf{A}$. If this limit exists, the sequence of matrices converges, otherwise it diverges. The following infinite series representation of the inverse of the matrix $\mathbf{I} - \mathbf{A}$ is used in Chapter 5.

Result 1.3.20. For a square matrix $\mathbf{A}$, the infinite series $\sum_{k=0}^{\infty} \mathbf{A}^k$, with $\mathbf{A}^0$ defined to be $\mathbf{I}$, converges if and only if $\lim_{k\to\infty} \mathbf{A}^k = \mathbf{O}$, in which case $\mathbf{I} - \mathbf{A}$ is nonsingular and

$$(\mathbf{I} - \mathbf{A})^{-1} = \sum_{k=0}^{\infty} \mathbf{A}^k.$$

As an application of the result, we derive the Sherman–Morrison–Woodbury formula in property 5 of Result 1.3.8. Put $\mathbf{M} = \mathbf{BCDA}^{-1}$. Then $(\mathbf{A} + \mathbf{BCD})^{-1} = (\mathbf{A} + \mathbf{MA})^{-1} = \mathbf{A}^{-1}(\mathbf{I} + \mathbf{M})^{-1}$. Provided $\mathbf{M}^k \to \mathbf{O}$ as $k \to \infty$,

$$(\mathbf{A} + \mathbf{BCD})^{-1} = \mathbf{A}^{-1} + \mathbf{A}^{-1}\sum_{k=1}^{\infty}(-1)^k\mathbf{M}^k.$$

Now, $\mathbf{A}^{-1}\mathbf{M} = \mathbf{A}^{-1}\mathbf{BCDA}^{-1} = (\mathbf{A}^{-1}\mathbf{BC})(\mathbf{DA}^{-1})$,

$$\mathbf{A}^{-1}\mathbf{M}^2 = \mathbf{A}^{-1}\mathbf{BCDA}^{-1}\mathbf{BCDA}^{-1} = (\mathbf{A}^{-1}\mathbf{BC})(\mathbf{DA}^{-1}\mathbf{BC})(\mathbf{DA}^{-1}),$$
$$\mathbf{A}^{-1}\mathbf{M}^3 = \mathbf{A}^{-1}\mathbf{BCDA}^{-1}\mathbf{BCDA}^{-1}\mathbf{BCDA}^{-1} = (\mathbf{A}^{-1}\mathbf{BC})(\mathbf{DA}^{-1}\mathbf{BC})^2(\mathbf{DA}^{-1}),$$

and so on. Then

$$\mathbf{A}^{-1}\sum_{k=1}^{\infty}(-1)^k\mathbf{M}^k = (\mathbf{A}^{-1}\mathbf{BC})\sum_{k=1}^{\infty}(-1)^k(\mathbf{DA}^{-1}\mathbf{BC})^{k-1}(\mathbf{DA}^{-1}) = -(\mathbf{A}^{-1}\mathbf{BC})(\mathbf{I}+\mathbf{DA}^{-1}\mathbf{BC})^{-1}(\mathbf{DA}^{-1}),$$

provided $(\mathbf{DA}^{-1}\mathbf{BC})^k \to \mathbf{O}$. Together, these formulas yield the Sherman–Morrison–Woodbury formula. A direct verification shows that it holds without assuming $\mathbf{M}^k \to \mathbf{O}$ or $(\mathbf{DA}^{-1}\mathbf{BC})^k \to \mathbf{O}$.

Definition 1.3.19. Exponential matrix. For any $n\times n$ matrix $\mathbf{A}$, we define the matrix $e^{\mathbf{A}}$ to be the $n \times n$ matrix given by:

$$e^{\mathbf{A}} = \sum_{i=0}^{\infty}\frac{\mathbf{A}^i}{i!},$$

when the expression on the right is a convergent series, i.e., all the $n \times n$ series $\sum_{i=0}^{\infty} a_{jk}^{(i)}$, $j = 1, \cdots, n$, $k = 1, \cdots, n$ are convergent, $a_{jk}^{(i)}$ being the (j,k)th element of $\mathbf{A}^i$ and $e^{\mathbf{0}} = \mathbf{I}$.

We conclude this section with some definitions of vector and matrix norms.

Definition 1.3.20. Vector norm. A vector norm on $\mathcal{R}^n$ is a function f: $\mathcal{R}^n \to \mathcal{R}$, denoted by $\|\mathbf{v}\|$, such that for every vector $\mathbf{v} \in \mathcal{R}^n$, and every $c \in \mathcal{R}$, we have

1. $f(\mathbf{v}) \geq 0$, with equality if and only if $\mathbf{v} = \mathbf{0}$,
2. $f(c\mathbf{v}) = |c|f(\mathbf{v})$, and
3. $f(\mathbf{u}+\mathbf{v}) \leq f(\mathbf{u}) + f(\mathbf{v})$ for every $\mathbf{u} \in \mathcal{R}^n$.

Specifically, the L_p-norm of a vector $\mathbf{v} = (v_1, \cdots, v_n)' \in \mathcal{R}^n$, which is also known in the literature as the *Minkowski metric* is defined by

$$\|\mathbf{v}\|_p = \{|v_1|^p + \cdots + |v_n|^p\}^{1/p}, \ p \geq 1.$$

We mention two special cases. The L_1-norm is defined by

$$\|\mathbf{v}\|_1 = \sum_{i=1}^{n}|v_i|$$

and forms the basis for the definition of *LAD* regression (see Chapter 9). The L_2-norm, which is also known as the Euclidean norm or the spectral norm is defined by

$$\|\mathbf{v}\|_2 = \{v_1^2 + \cdots + v_n^2\}^{1/2} = (\mathbf{v}'\mathbf{v})^{1/2},$$

and is the basis for least squares techniques (see the discussion below Definition 1.2.1). In this book, we will denote $\|\mathbf{v}\|_2$ simply as $\|\mathbf{v}\|$. An extension is to define the L_2-norm with respect to a nonsingular matrix $\mathbf{A}$ by

$$\|\mathbf{v}\|_{\mathbf{A}} = (\mathbf{v}'\mathbf{A}\mathbf{v})^{1/2}.$$

Definition 1.3.21. Matrix norm. A function f: $\mathcal{R}^{m\times n} \to \mathcal{R}$ is called a matrix norm on $\mathcal{R}^{m\times n}$, denoted by $\|\mathbf{A}\|$, if

1. $f(\mathbf{A}) \geq 0$ for all $m \times n$ real matrices $\mathbf{A}$, with equality if and only if $\mathbf{A} = \mathbf{O}$,
2. $f(c\mathbf{A}) = |c|f(\mathbf{A})$ for all $c \in \mathcal{R}$, and for all $m \times n$ matrices $\mathbf{A}$, and
3. $f(\mathbf{A}+\mathbf{B}) \leq f(\mathbf{A}) + f(\mathbf{B})$ for all $m \times n$ matrices $\mathbf{A}$ and $\mathbf{B}$.

The Frobenius norm of an $m \times n$ matrix $\mathbf{A} = \{a_{ij}\}$ with respect to the usual inner product is defined by

$$\|\mathbf{A}\| = [\text{tr}(\mathbf{A}'\mathbf{A})]^{1/2} = \left[\sum_{i=1}^{m}\sum_{j=1}^{n} a_{ij}^2\right]^{1/2}.$$

It is easy to verify that $\|\mathbf{A}\| \geq 0$, with equality holding only if $\mathbf{A} = \mathbf{O}$. Also, $\|c\mathbf{A}\| = |c|\|\mathbf{A}\|$.

1.4 R Code

There are R packages such as *pracma* and *Matrix* that enable computations on vectors and matrices. We show simple code that can be used for the notions defined in this chapter. Results from running the code are straightforward and we do not show them here.

```
library(pracma)
library(Matrix)

## Def. 1.3.1. Matrix addition and subtraction
# A and B must have the same order
(A <- matrix(c(-5, 4, 1, -3, 2, 6), nrow = 2, byrow = T))
(B <- matrix(c(7, -9, 10, 2, 6, -1), nrow = 2, byrow = T))
(C <- A + B)
(D <- A - B)

## Def. 1.3.2. Multiply a matrix by a scalar
c <- 5
(A2 <- c * A)

## Def. 1.3.3. Matrix multiplication
# ncol(A) must equal nrow(B)
(A <- matrix(c(5, 4, 1, -3, 2, 6), nrow = 2, byrow = T))
(B <- matrix(c(7, -3, 2), nrow = 3, byrow = T))
(M <- A %*% B)

## Result 1.3.3. Product of upper triangular matrices
(U <- matrix(c(1, 2, 3, 0, 4, 5, 0, 0, 6), nrow = 3, byrow = T))
(V <- matrix(c(2, 4, 6, 0, 8, 10, 0, 0, 12), nrow = 3,
               byrow = T))
(W <- U %*% V)

## Def. 1.3.4. Transpose of a matrix
(A <- matrix(c(2, 1, 6, 4, 3, 5), nrow = 2, byrow = T))
(tA <- t(A))
t(A)
```

```
## Def. 1.3.5. Symmetric matrix
S <- matrix(c(1, 2,-3, 2, 4, 5,-3, 5, 9), nrow = 3, byrow = T)
S
(table(S == t(S))) == (nrow(S) * ncol(S))

## Skew symmetric matrix
(SkS <- matrix(
  c(0,-1, 3, 6, 1, 0, 2,-5,-3,-2, 0, 4,-6, 5,-4, 0),
  nrow = 4,
  byrow = T
))
(table(-SkS == t(SkS))) == (nrow(SkS) * ncol(SkS))

## Def. 1.3.6. Trace of a square matrix
(T <- matrix(c(2,-4, 5, 6,-7, 0, 3, 9, 7), nrow = 3, byrow = T))
sum(diag(T))

## Def. 1.3.7. Determinant of a square matrix
(D <- matrix(c(2,-4, 5, 6,-7, 0, 3, 9, 7), nrow = 3, byrow = T))
det(D)

## Def. 1.3.8. Nonsingular matrix, |A| neq 0
(A <- matrix(c(1, 6, 0, 3), nrow = 2, byrow = T))
det(A) != 0

## Singular matrix , |A|= 0
(B <- matrix(c(1, 6, 1 / 2, 3), nrow = 2, byrow = T))
det(B)

## Def. 1.3.9. Inverse of a matrix
(I <- matrix(c(-1, 2, 2, 4, 3,-2,-5, 0, 3), nrow = 3, byrow = T))
solve(I)

## Def. 1.3.10. Reduced row echelon form
A <- matrix(c(2, 1,-1, 8,-3,-1, 2,-11,-2, 1, 2,-3),
            nrow = 3, byrow = T)
A
rref(A)

## Def. 1.3.12. Orthogonal matrix, t(A)=inv(A)
A <- randortho(4)
A %*% t(A)

## Diagonal matrices
Diagonal(4, 5)
diag(4)
Diag(c(1, 2, 3),-1)
Diag(c(1, 2, 3), 1)
diag(5, 3, 4)

## Identity and unit matrices
I <- diag(4)
one <- rep(1, 4)
J <- one %*% t(one)
```

```
## Intra-class correlation matrix
rho <- 0.5
I <- diag(4)
one <- rep(1, 4)
J <- one %*% t(one)
(C <-  (1 - rho) * I + rho * J)

## Toeplitz Matrix
rho <- 0.5
(A <- Toeplitz(c(1, rho, rho ^ 2, rho ^ 3, rho ^ 4)))
det(A)
solve(A)

## Example 1.3.9. Null space
A <-
  matrix(c(1, 0, 0, 0, 0, 1, 0, 0,-5, 2, 0, 0, 1,-3, 0, 0), nrow =
           4)
round(null(A), 4)
round(A %*% null(A), 4)

## Example 1.3.10. Row and column spaces
A <-
  matrix(c(1,-1, 2, 3,-2, 2,-4,-6, 2,-1, 6, 8, 1, 0, 4, 5, 0, 0, 0,
           1), nrow = 4)
round(orth(A), 4)  #Column space
round(orth(t(A)), 4)  #Row space

## Example 1.3.11. Rank of a matrix
A <-
  matrix(c(1, 1, 1, 0, 1, 2, 3, 1, 1, 2, 2, 1, 3,-1, 2,-1,-2,
           0,-1,-1), nrow = 5)
qr(A)$rank

## Example. 1.3.14. Eigenvalues and eigenvectors
A <- matrix(c(-1, 2, 0, 1, 2, 1, 0, 2,-1), nrow = 3, byrow = T)
eigen(A)
evalues <- eigen(A)$values
evectors <- eigen(A)$vectors    # V
#Is A = V L V^(-1)
table(round(evectors %*% diag(evalues) %*% solve(evectors), 1)==A)

## Def. 1.3.20. Vector norms
v <- c(-1, 2, 1)
Norm(v, 1)   #L-1 norm
Norm(v, 2)   #L-2 norm
sqrt(t(v) %*% v) #L-2 norm
max(abs(v))  #L-infinity norm

## Def. 1.3.21. Frobenius norm of a matrix
A <- matrix(c(2, 1, 6, 4, 3, 5), nrow = 2, byrow = T)
fnorm <- norm(A, type = ''F'')  #sqrt(sum(diag(t(A)%*%A)))
```

Exercises

1.1. Verify the Cauchy–Schwarz inequality for $\mathbf{a} = (-1, 2, 0, -1)'$ and $\mathbf{b} = (4, -2, -1, 1)'$.

1.2. Verify properties 6 and 7 in Result 1.2.1.

1.3. Suppose $\mathbf{x}$, $\mathbf{y}$ and $\mathbf{z}$ are orthonormal vectors. Let $\mathbf{u} = a\mathbf{x} + b\mathbf{y}$ and $\mathbf{v} = a\mathbf{x} + b\mathbf{z}$. Find a and b such that the vectors $\mathbf{u}$ and $\mathbf{v}$ are of unit length and the angle between them is $60°$.

1.4. Show that $\mathbf{v}_1 = (1, 1, 0, -1)'$, $\mathbf{v}_2 = (2, 0, 1, -1)'$ and $\mathbf{v}_3 = (0, -2, 1, 1)'$ are linearly dependent vectors. Find a set of two linearly independent vectors and express the third as a function of these two.

1.5. Show that the set of vectors $\mathbf{v}_1 = (2, 3, 2)'$, $\mathbf{v}_2 = (8, -6, 5)'$ and $\mathbf{v}_3 = (-4, 3, 1)'$ are linearly independent.

1.6. Verify whether the columns of $\mathbf{A}$ are linearly independent given

$$\mathbf{A} = \begin{pmatrix} -3 & 3 & 3 \\ 2 & 2 & 2 \\ 0 & 1 & 0 \end{pmatrix}.$$

1.7. Verify whether the vector $\mathbf{u} = (2, 3)'$ is in the span of the vectors $\mathbf{v}_1 = (1, 2)'$ and $\mathbf{v}_2 = (3, 5)'$.

1.8. Verify properties 1 and 2 in Result 1.2.2.

1.9. Verify Result 1.2.3.

1.10. Verify Result 1.2.4.

1.11. Verify Result 1.2.5.

1.12. Find all matrices that commute with the matrix

$$\mathbf{B} = \begin{pmatrix} b & 1 & 0 \\ 0 & b & 1 \\ 0 & 0 & b \end{pmatrix}.$$

1.13. Given $\mathbf{A} = \begin{pmatrix} a & b \\ 0 & 1 \end{pmatrix}$, find $\mathbf{A}^k$, for all $k \geq 2$.

1.14. If $\mathbf{AB} = \mathbf{BA}$, show that, for any given positive integer k, there exists a matrix $\mathbf{C}$ such that $\mathbf{A}^k - \mathbf{B}^k = (\mathbf{A} - \mathbf{B})\mathbf{C}$.

1.15. For any $n \times n$ matrix $\mathbf{A}$, show that the matrices $\mathbf{A}'\mathbf{A}$ and $\mathbf{A}\mathbf{A}'$ are symmetric.

1.16. For any $m \times n$ matrix $\mathbf{A}$, show that $\mathbf{A} = \mathbf{O}$ if and only if $\mathbf{A}'\mathbf{A} = \mathbf{O}$.

1.17. Let $\mathbf{A}$ be an $n \times n$ matrix and let $\mathbf{x}_i$ be an $n \times 1$ vector, $i = 1, \cdots, k$.

(a) Show that $\text{tr}(\mathbf{A}\sum_{i=1}^k \mathbf{x}_i\mathbf{x}_i') = \sum_{i=1}^k \mathbf{x}_i'\mathbf{A}\mathbf{x}_i$.

(b) Show that $\text{tr}(\mathbf{B}^{-1}\mathbf{A}\mathbf{B}) = \text{tr}(\mathbf{A})$.

1.18. Verify Result 1.3.3.

1.19. Verify Result 1.3.5.

1.20. Complete the proof of Result 1.3.9 by verifying the two facts mentioned in that proof

1.21. Find the determinant of the matrix

$$\mathbf{A} = \begin{pmatrix} 1 & 3 & -3 & 1 \\ 5 & 9 & -10 & 3 \\ 1 & 0 & 5 & -2 \\ 2 & 1 & -3 & 1 \end{pmatrix}.$$

1.22. Consider the determinant Δ_n of the $n \times n$ matrix

$$\begin{pmatrix} 1+a^2 & a & 0 & \cdots & 0 & 0 \\ a & 1+a^2 & a & \cdots & 0 & 0 \\ 0 & a & 1+a^2 & \cdots & 0 & 0 \\ \vdots & \vdots & \vdots & \ddots & \vdots & \vdots \\ 0 & 0 & 0 & \cdots & 1+a^2 & a \\ 0 & 0 & 0 & \cdots & a & 1+a^2 \end{pmatrix}.$$

Show that $\Delta_n - \Delta_{n-1} = a^2(\Delta_{n-1} - \Delta_{n-2})$, and hence find Δ_n.

1.23. If the row vectors of a square matrix are linearly dependent, show that the determinant of the matrix is zero.

1.24. Evaluate the determinant of

$$\begin{pmatrix} a_1+1 & a_2 & \cdots & a_n \\ a_1 & a_2+1 & \cdots & a_n \\ \vdots & \vdots & \ddots & \vdots \\ a_1 & a_2 & \cdots & a_n+1 \end{pmatrix}.$$

1.25. Apply property 12 of Result 1.3.6 to calculate the determinant of the intra-class correlation matrix in Example 1.3.2; see the alternative way to calculate the inverse of the matrix in Definition 1.3.7.

1.26. By reducing the matrix

$$\mathbf{A} = \begin{pmatrix} 1 & 2 & -1 \\ 1 & -2 & -1 \\ 1 & 6 & -1 \end{pmatrix},$$

show that it is singular.

1.27. In Result 1.3.7, show that a matrix $\mathbf{A}$ is invertible only if $|\mathbf{A}| \neq 0$.

1.28. Verify property 6 of Result 1.3.8.

1.29. (a) Show that $(\mathbf{I}+\mathbf{AB})^{-1} = \mathbf{I} - \mathbf{A}(\mathbf{I}+\mathbf{BA})^{-1}\mathbf{B}$, provided $\mathbf{AB}$ and $\mathbf{BA}$ exist.

(b) Using (a), show that $(a\mathbf{I}_k + b\mathbf{J}_k)^{-1} = (\mathbf{I}_k/a) - b\mathbf{J}_k/\{a(a+kb)\}$.

1.30. Let $\mathbf{A}$ be an $n \times n$ orthogonal matrix.

(a) Show that $|\mathbf{A}| = \pm 1$.

(b) Show that $\mathbf{r}_i\mathbf{r}_j' = \delta_{ij}$, and $\mathbf{c}_i'\mathbf{c}_j = \delta_{ij}$, where $\mathbf{r}_i$ is the ith row of $\mathbf{A}$, $\mathbf{c}_i$ is the ith column of $\mathbf{A}$ and δ_{ij} denotes the Kronecker delta, i.e., $\delta_{ij} = 1$ for $i = j$, and $\delta_{ij} = 0$ otherwise.

(c) Show that $\mathbf{AB}$ is orthogonal, where $\mathbf{B}$ is an $n \times n$ orthogonal matrix.

1.31. Let $\mathbf{A}$ be an $n \times n$ symmetric matrix and let $\text{r}(\mathbf{A}) = 1$. Show that $|\mathbf{I} + \mathbf{A}| = 1 + \text{tr}(\mathbf{A})$.

1.32. Find the dimension of the column space, null space and row space of the matrix

$$\mathbf{A} = \begin{pmatrix} 1 & 1 & 1 \\ 2 & 2 & 2 \\ -1 & 1 & -3 \\ 1 & 2 & 0 \end{pmatrix}.$$

1.33. Let $\mathbf{A} = \begin{pmatrix} 1 & 2 & 4 & 3 \\ 3 & -1 & 2 & -2 \\ 5 & -4 & 0 & -7 \end{pmatrix}$. Find the rank of $\mathbf{A}$.

1.34. Show that for any distinctive eigenvalue of a matrix, the geometric multiplicity is at most as large as the algebraic multiplicity.

1.35. Let $\lambda_1, \cdots, \lambda_k$ be all the distinct eigenvalues of a matrix $\mathbf{A}$ and $\mathcal{V}_1, \cdots, \mathcal{V}_k$ the corresponding eigenspaces. Show that the sum of the eigenspaces is a direct sum.

1.36. Let λ be an eigenvalue of $\mathbf{A}$.

(a) If $f(x)$ is a polynomial, then $f(\lambda)$ is an eigenvalue of $f(\mathbf{A})$.

(b) If $\mathbf{A}$ is invertible, then $1/\lambda$ is an eigenvalue of $\mathbf{A}^{-1}$.

1.37. Verify Result 1.3.19. Hint: Use Vieta's formulas relating coefficients of a polynomial of degree n to signed sums and products of its roots.

1.38. Verify Result 1.3.20.

1.39. Let $\mathbf{A} = \begin{pmatrix} a+1 & 1 & 1 \\ 1 & a+1 & 1 \\ 1 & 1 & a+1 \end{pmatrix}$. Show that $\mathbf{A} - a\mathbf{I}_3$ has a nonzero eigenvalue of 3. Find the corresponding eigenvector.

1.40. If $\mathbf{A}$ and $\mathbf{B}$ conform under multiplication, show that the nonzero eigenvalues of $\mathbf{AB}$ coincide with the nonzero eigenvalues of $\mathbf{BA}$.

1.41. Let $\mathbf{A}$ be a $k \times k$ matrix and $\mathbf{B}$ be a $k \times k$ nonsingular matrix. Show that $\mathbf{A}$ and $\mathbf{BAB}^{-1}$ have the same eigenvalues. If $\mathbf{A}\mathbf{v}_j = \lambda_j\mathbf{v}_j$, i.e., $\mathbf{v}_j$ is an eigenvector corresponding to the eigenvalue λ_j of $\mathbf{A}$, show that $\mathbf{B}\mathbf{v}_j$ is an eigenvector of $\mathbf{BAB}^{-1}$ for λ_j.

1.42. Let $\mathbf{A}$ be a symmetric matrix such that for any real vector $\mathbf{x} \neq \mathbf{0}$, $\mathbf{x}'\mathbf{A}\mathbf{x} > 0$. Verify the following results.

(a) $(\mathbf{x}'\mathbf{A}\mathbf{y})^2 \leq (\mathbf{x}'\mathbf{A}\mathbf{x})(\mathbf{y}'\mathbf{A}\mathbf{y})$ with equality holding if and only if $\mathbf{x} = \mathbf{0}$ or $\mathbf{y} = c\mathbf{x}$ for some scalar c.

(b) $(\mathbf{x}'\mathbf{y})^2 \leq (\mathbf{x}'\mathbf{A}\mathbf{x})(\mathbf{y}'\mathbf{A}^{-1}\mathbf{y})$ with equality holding if and only if $\mathbf{x} = \mathbf{0}$ or $\mathbf{y} = c\mathbf{A}\mathbf{x}$ for some scalar c.

2

Properties of Special Matrices

In this chapter, we define special matrices that find direct use in the theory of linear models, present some properties of such matrices, and illustrate these properties using several examples. The concepts and results discussed here will be used in subsequent chapters for the development of matrix results, distribution theory and statistical methods.

2.1 Partitioned matrices

Definition 2.1.1. An $m \times n$ partitioned matrix $\mathbf{A}$ is expressed as an array of submatrices (or blocks):

$$\mathbf{A} = \begin{pmatrix} \mathbf{A}_{11} & \mathbf{A}_{12} & \cdots & \mathbf{A}_{1c} \\ \mathbf{A}_{21} & \mathbf{A}_{22} & \cdots & \mathbf{A}_{2c} \\ \vdots & \vdots & \vdots & \vdots \\ \mathbf{A}_{r1} & \mathbf{A}_{r2} & \cdots & \mathbf{A}_{rc} \end{pmatrix}, \tag{2.1.1}$$

where $\mathbf{A}_{ij}$ is an $m_i \times n_j$ submatrix for $i = 1, \cdots, r$, $j = 1, \cdots, c$; $m_1, \cdots, m_r$ and $n_1, \cdots, n_c$ are positive integers such that $\sum_{i=1}^{r} m_i = m$, and $\sum_{j=1}^{c} n_j = n$. Note that each submatrix $\mathbf{A}_{ij}$, $j = 1, \cdots, c$ has the same number of rows for any i, and similarly each submatrix $\mathbf{A}_{ij}, i = 1, \cdots, r$ has the same number of columns for any j. For example,

$$\left(\begin{array}{ccc|c} 1 & 3 & 5 & 7 \\ \hline 5 & 4 & 1 & -9 \\ -3 & 2 & 6 & 4 \end{array}\right) \quad \text{and} \quad \left(\begin{array}{cc|cc} 1 & 3 & 5 & 7 \\ 5 & 4 & 1 & -9 \\ \hline -3 & 2 & 6 & 4 \end{array}\right)$$

are two different partitions of the same 3×4 matrix.

Definition 2.1.2. An $m \times n$ matrix $\mathbf{A}$ partitioned as

$$\mathbf{A} = \begin{pmatrix} \mathbf{A}_{11} & \mathbf{A}_{12} & \cdots & \mathbf{A}_{1r} \\ \mathbf{A}_{21} & \mathbf{A}_{22} & \cdots & \mathbf{A}_{2r} \\ \vdots & \vdots & \vdots & \vdots \\ \mathbf{A}_{r1} & \mathbf{A}_{r2} & \cdots & \mathbf{A}_{rr} \end{pmatrix} \tag{2.1.2}$$

is said to be a block-diagonal matrix if $\mathbf{A}_{ij} = \mathbf{O}$ for $i \neq j$, and is written as

$$\mathbf{A} = \text{diag}(\mathbf{A}_{11}, \mathbf{A}_{22}, \cdots, \mathbf{A}_{rr}).$$

In (2.1.2), if $\mathbf{A}_{ij} = \mathbf{O}$ for $j < i = 1, \cdots, r$, $\mathbf{A}$ is called an upper block-triangular matrix, while if $\mathbf{A}_{ij} = \mathbf{O}$ for $j > i = 1, \cdots, r$, then $\mathbf{A}$ is called a lower block-triangular matrix.

DOI: 10.1201/9781315156651-2

An $m \times n$ matrix $\mathbf{A}$ partitioned only by rows is written as

$$\mathbf{A} = \begin{pmatrix} \mathbf{A}_1 \\ \mathbf{A}_2 \\ \vdots \\ \mathbf{A}_r \end{pmatrix} \tag{2.1.3}$$

and if it is partitioned only by columns, we write

$$\mathbf{A}' = \begin{pmatrix} \mathbf{A}'_1 \\ \mathbf{A}'_2 \\ \vdots \\ \mathbf{A}'_c \end{pmatrix} = (\mathbf{A}_1, \mathbf{A}_2, \cdots, \mathbf{A}_c). \tag{2.1.4}$$

A partitioned n-dimensional column vector is denoted by

$$\mathbf{a} = \begin{pmatrix} \mathbf{a}_1 \\ \mathbf{a}_2 \\ \vdots \\ \mathbf{a}_r \end{pmatrix} \tag{2.1.5}$$

where $\mathbf{a}_i$ is an n_i-dimensional vector, and n_i, $i = 1, \cdots, r$ are positive integers such that $\sum_{i=1}^{r} n_i = n$. A partitioned n-dimensional row vector is of the form

$$\mathbf{a}' = (\mathbf{a}'_1, \mathbf{a}'_2, \cdots, \mathbf{a}'_r). \tag{2.1.6}$$

Consider a $p \times q$ matrix $\mathbf{B}$ which is partitioned as

$$\mathbf{B} = \begin{pmatrix} \mathbf{B}_{11} & \mathbf{B}_{12} & \cdots & \mathbf{B}_{1h} \\ \mathbf{B}_{21} & \mathbf{B}_{22} & \cdots & \mathbf{B}_{2h} \\ \vdots & \vdots & \vdots & \vdots \\ \mathbf{B}_{l1} & \mathbf{B}_{l2} & \cdots & \mathbf{B}_{lh} \end{pmatrix}, \tag{2.1.7}$$

where the dimension of $\mathbf{B}_{ij}$ is $p_i \times q_j$, for $i = 1, \cdots, l$, $j = 1, \cdots, h$. We define the following elementary operations on partitioned matrices.

Addition of partitioned matrices

The matrices $\mathbf{A}$ and $\mathbf{B}$ defined in (2.1.1) and (2.1.7) are conformal under addition if $p = m$, $q = n$, $l = r$, $h = c$, $p_i = m_i$, for $i = 1, \cdots, r$, and $q_j = n_j$, for $j = 1, \cdots, c$. Then, the (i, j)th submatrix of $\mathbf{C} = \mathbf{A} \pm \mathbf{B}$ is given by

$$\mathbf{C}_{ij} = \mathbf{A}_{ij} \pm \mathbf{B}_{ij} \quad \text{for } i = 1, \cdots, r \text{ and } j = 1, \cdots, c. \tag{2.1.8}$$

For example, when $r = c = l = h = 2$, and the submatrices have conformal dimensions for addition, (2.1.8) becomes

$$\mathbf{A} \pm \mathbf{B} = \begin{pmatrix} \mathbf{A}_{11} \pm \mathbf{B}_{11} & \mathbf{A}_{12} \pm \mathbf{B}_{12} \\ \mathbf{A}_{21} \pm \mathbf{B}_{21} & \mathbf{A}_{22} \pm \mathbf{B}_{22} \end{pmatrix}. \tag{2.1.9}$$

Multiplication of partitioned matrices

The product $\mathbf{AB}$ is defined if $n = p$, $l = c$, and $n_j = p_j$ for $j = 1, \cdots, c$, in which case the (i, j)th submatrix of $\mathbf{C} = \mathbf{AB}$ is given by

$$\mathbf{C}_{ij} = \sum_{k=1}^{c} \mathbf{A}_{ik}\mathbf{B}_{kj}. \tag{2.1.10}$$

When $r = c = l = h = 2$, (2.1.10) simplifies as

$$\mathbf{AB} = \begin{pmatrix} \mathbf{A}_{11}\mathbf{B}_{11} + \mathbf{A}_{12}\mathbf{B}_{21} & \mathbf{A}_{11}\mathbf{B}_{12} + \mathbf{A}_{12}\mathbf{B}_{22} \\ \mathbf{A}_{21}\mathbf{B}_{11} + \mathbf{A}_{22}\mathbf{B}_{21} & \mathbf{A}_{21}\mathbf{B}_{12} + \mathbf{A}_{22}\mathbf{B}_{22} \end{pmatrix}. \tag{2.1.11}$$

Example 2.1.1. Let $\mathbf{X}$ be a matrix with n rows $\mathbf{x}_1', \cdots, \mathbf{x}_n'$. Let $\mathbf{X}_{(i)}$ be the matrix obtained by deleting the ith row. By (2.1.10),

$$\mathbf{X}'\mathbf{X} = (\mathbf{x}_1, \cdots, \mathbf{x}_n) \begin{pmatrix} \mathbf{x}_1' \\ \vdots \\ \mathbf{x}_n' \end{pmatrix} = \sum_{j=1}^{n} \mathbf{x}_j\mathbf{x}_j'.$$

Likewise, $\mathbf{X}_{(i)}'\mathbf{X}_{(i)} = \sum_{j \neq i} \mathbf{x}_j\mathbf{x}_j' = \mathbf{X}'\mathbf{X} - \mathbf{x}_i\mathbf{x}_i'$. We can write the inverse of the symmetric matrix $\mathbf{X}'\mathbf{X}$ as

$$\begin{aligned}(\mathbf{X}'\mathbf{X})^{-1} &= (\mathbf{X}_{(i)}'\mathbf{X}_{(i)} + \mathbf{x}_i\mathbf{x}_i')^{-1} \\ &= (\mathbf{X}_{(i)}'\mathbf{X}_{(i)})^{-1} - \frac{(\mathbf{X}_{(i)}'\mathbf{X}_{(i)})^{-1}\mathbf{x}_i\mathbf{x}_i'(\mathbf{X}_{(i)}'\mathbf{X}_{(i)})^{-1}}{1 + \mathbf{x}_i'(\mathbf{X}_{(i)}'\mathbf{X}_{(i)})^{-1}\mathbf{x}_i},\end{aligned} \tag{2.1.12}$$

which follows directly by using Sherman–Morrison–Woodbury formula in Result 1.3.8, setting $\mathbf{A} = \mathbf{X}_{(i)}'\mathbf{X}_{(i)}$, $\mathbf{B} = \mathbf{x}_i$, $\mathbf{C} = 1$, and $\mathbf{D} = \mathbf{x}_i'$ therein. Similarly, setting $\mathbf{A} = \mathbf{X}'\mathbf{X}$, $\mathbf{B} = -\mathbf{x}_i$, $\mathbf{C} = 1$, and $\mathbf{D} = \mathbf{x}_i'$, we obtain

$$\begin{aligned}(\mathbf{X}_{(i)}'\mathbf{X}_{(i)})^{-1} &= (\mathbf{X}'\mathbf{X} - \mathbf{x}_i\mathbf{x}_i')^{-1} \\ &= (\mathbf{X}'\mathbf{X})^{-1} + \frac{(\mathbf{X}'\mathbf{X})^{-1}\mathbf{x}_i\mathbf{x}_i'(\mathbf{X}'\mathbf{X})^{-1}}{1 - \mathbf{x}_i'(\mathbf{X}'\mathbf{X})^{-1}\mathbf{x}_i}.\end{aligned} \tag{2.1.13}$$

These results are useful in studying the effect of deleting an observation in a linear model (see Chapter 9). □

Holing of a partitioned matrix

This refers to transforming a matrix into a block-triangular matrix. Suppose an $m \times n$ matrix $\mathbf{M}$ is partitioned as

$$\mathbf{M} = \begin{pmatrix} \mathbf{A} & \mathbf{B} \\ \mathbf{C} & \mathbf{D} \end{pmatrix}.$$

If $\mathbf{D}$ is nonsingular while $\mathbf{A}$ need not be a square matrix, then

$$\begin{aligned}\mathbf{M}\begin{pmatrix} \mathbf{I} & \mathbf{O} \\ -\mathbf{D}^{-1}\mathbf{C} & \mathbf{I} \end{pmatrix} &= \begin{pmatrix} \mathbf{A} - \mathbf{BD}^{-1}\mathbf{C} & \mathbf{B} \\ \mathbf{O} & \mathbf{D} \end{pmatrix}, \\ \begin{pmatrix} \mathbf{I} & -\mathbf{BD}^{-1} \\ \mathbf{O} & \mathbf{I} \end{pmatrix}\mathbf{M} &= \begin{pmatrix} \mathbf{A} - \mathbf{BD}^{-1}\mathbf{C} & \mathbf{O} \\ \mathbf{C} & \mathbf{D} \end{pmatrix}.\end{aligned} \tag{2.1.14}$$

The transformation (2.1.14) of $\mathbf{M}$ into an upper (resp. lower) block-triangular matrix by multiplying it by an invertible lower (resp. upper) block-triangular matrix is known as Hua's holing method (Bai and Silverstein, 2010). The matrix $\mathbf{A}-\mathbf{BD}^{-1}\mathbf{C}$ is known as the *Schur complement* of $\mathbf{D}$ (in $\mathbf{M}$).

Likewise, if $\mathbf{A}$ is nonsingular while $\mathbf{D}$ need not be a square matrix, then

$$\begin{aligned}\mathbf{M}\begin{pmatrix}\mathbf{I} & -\mathbf{A}^{-1}\mathbf{B}\\ \mathbf{O} & \mathbf{I}\end{pmatrix} &= \begin{pmatrix}\mathbf{A} & \mathbf{O}\\ \mathbf{C} & \mathbf{D}-\mathbf{CA}^{-1}\mathbf{B}\end{pmatrix},\\ \begin{pmatrix}\mathbf{I} & \mathbf{O}\\ -\mathbf{CA}^{-1} & \mathbf{I}\end{pmatrix}\mathbf{M} &= \begin{pmatrix}\mathbf{A} & \mathbf{B}\\ \mathbf{O} & \mathbf{D}-\mathbf{CA}^{-1}\mathbf{B}\end{pmatrix}.\end{aligned} \tag{2.1.15}$$

Now, the Schur complement of $\mathbf{A}$ (in $\mathbf{M}$) is $\mathbf{D}-\mathbf{CA}^{-1}\mathbf{B}$.

Result 2.1.1. Rank of a block-triangular matrix. Suppose $\mathbf{M}$ is a square lower block-triangular matrix and is partitioned as

$$\mathbf{M} = \begin{pmatrix}\mathbf{A} & \mathbf{O}\\ \mathbf{C} & \mathbf{D}\end{pmatrix},$$

where $\mathbf{A}$ is an $m\times n$ matrix and $\mathbf{D}$ is a $p\times q$ matrix (so $\mathbf{C}$ is a $p\times n$ matrix). If $\mathcal{R}(\mathbf{C})\subset\mathcal{R}(\mathbf{A})$ or $\mathcal{C}(\mathbf{C})\subset\mathcal{C}(\mathbf{D})$, such as when $\mathbf{A}$ or $\mathbf{D}$ is nonsingular, then $\mathrm{r}(\mathbf{M})=\mathrm{r}(\mathbf{A})+\mathrm{r}(\mathbf{D})$.

Proof. Let $\mathrm{r}(\mathbf{A}) = s$ and $\mathrm{r}(\mathbf{D}) = t$. Suppose $\mathcal{C}(\mathbf{C})\subset\mathcal{C}(\mathbf{D})$. Let $\mathbf{A}=(\mathbf{a}_1,\cdots,\mathbf{a}_n)$, $\mathbf{C}=(\mathbf{c}_1,\cdots,\mathbf{c}_n)$, and $\mathbf{D}=(\mathbf{d}_1,\cdots,\mathbf{d}_q)$. Without loss of generality, suppose $\{\mathbf{a}_1,\cdots,\mathbf{a}_s\}$ is a basis of $\mathcal{C}(\mathbf{A})$ and $\{\mathbf{d}_1,\cdots,\mathbf{d}_t\}$ a basis of $\mathcal{C}(\mathbf{D})$. Then, it is easy to verify that

$$\mathbf{u}_1=\begin{pmatrix}\mathbf{a}_1\\ \mathbf{0}_p\end{pmatrix},\cdots,\ \mathbf{u}_s=\begin{pmatrix}\mathbf{a}_s\\ \mathbf{0}_p\end{pmatrix},\ \mathbf{u}_{s+1}=\begin{pmatrix}\mathbf{0}_m\\ \mathbf{d}_1\end{pmatrix},\cdots,\ \mathbf{u}_{s+t}=\begin{pmatrix}\mathbf{0}_m\\ \mathbf{d}_t\end{pmatrix}$$

are LIN and $\mathcal{C}(\mathbf{M})\subset\mathrm{Span}\{\mathbf{u}_1,\cdots,\mathbf{u}_{s+t}\}$. Clearly $\mathbf{u}_j\in\mathcal{C}(\mathbf{M})$ for $j>s$. On the other hand, by assumption, for each $j\le s$, $\begin{pmatrix}\mathbf{0}_m\\ \mathbf{c}_j\end{pmatrix}\in\mathrm{Span}\{\mathbf{u}_{s+1},\cdots,\mathbf{u}_{s+t}\}\subset\mathcal{C}(\mathbf{M})$, so $\mathbf{u}_j=\begin{pmatrix}\mathbf{a}_j\\ \mathbf{c}_j\end{pmatrix}-\begin{pmatrix}\mathbf{0}_m\\ \mathbf{c}_j\end{pmatrix}\in\mathcal{C}(\mathbf{M})$. Thus $\{\mathbf{u}_1,\cdots,\mathbf{u}_{s+t}\}$ is a basis of $\mathcal{C}(\mathbf{M})$, so $\mathrm{r}(\mathbf{M})=s+t$. The case where $\mathcal{R}(\mathbf{C})\subset\mathcal{R}(\mathbf{A})$ can be proved similarly. ■

Result 2.1.2. Determinant and inverse of a block-triangular matrix. Suppose $\mathbf{M}$ is defined as in Result 2.1.1, where $\mathbf{A}$ is an $n\times n$ matrix and $\mathbf{D}$ is a $p\times p$ matrix. Then

$$|\mathbf{M}| = |\mathbf{A}|\,|\mathbf{D}|. \tag{2.1.16}$$

If both $\mathbf{A}$ and $\mathbf{D}$ are invertible, then $\mathbf{M}$ is invertible with

$$\mathbf{M}^{-1}=\begin{pmatrix}\mathbf{A}^{-1} & \mathbf{O}\\ -\mathbf{D}^{-1}\mathbf{CA}^{-1} & \mathbf{D}^{-1}\end{pmatrix}. \tag{2.1.17}$$

Similar results hold if $\mathbf{M}$ is an upper block-triangular square matrix, so for example, if $\mathbf{A}$ and $\mathbf{D}$ are invertible, then

$$\begin{pmatrix}\mathbf{A} & \mathbf{B}\\ \mathbf{O} & \mathbf{D}\end{pmatrix}^{-1}=\begin{pmatrix}\mathbf{A}^{-1} & -\mathbf{A}^{-1}\mathbf{BD}^{-1}\\ \mathbf{O} & \mathbf{D}^{-1}\end{pmatrix}. \tag{2.1.18}$$

Proof. We prove (2.1.16) by induction on n. Let the first row of $\mathbf{A}$ be $a_{11},\ldots,a_{1n}$. If $n=$

1, then $\mathbf{D}$ is exactly the submatrix of $\mathbf{M}$ obtained by deleting its first row and first column, so (2.1.16) follows from the cofactor expansion of the determinant (see Definition 1.3.7). Suppose (2.1.16) holds for $n \le k$. Now suppose $\mathbf{A}$ is a $(k+1)\times(k+1)$ matrix. Denote by $\mathbf{A}_i$ the submatrix of $\mathbf{A}$ obtained by deleting its first row and ith column. Then the corresponding submatrix of $\mathbf{M}$ is $\mathbf{M}_i = \begin{pmatrix} \mathbf{A}_i & \mathbf{O} \\ \mathbf{C}_i & \mathbf{D} \end{pmatrix}$, where $\mathbf{C}_i$ is the submatrix of $\mathbf{C}$ obtained by deleting its ith column. By induction hypothesis, $|\mathbf{M}_i| = |\mathbf{A}_i|\,|\mathbf{D}|$. By the cofactor expansion of the determinant, $|\mathbf{M}| = \sum_{i=1}^n a_{1i}(-1)^{1+i}|\mathbf{M}_i| = \sum_{i=1}^n a_{1i}(-1)^{1+i}|\mathbf{A}_i|\,|\mathbf{D}| = |\mathbf{A}|\,|\mathbf{D}|$. Hence by induction, (2.1.16) holds for any size of the square matrix $\mathbf{A}$, and the proof is complete. Note that if both $\mathbf{A}$ and $\mathbf{D}$ are nonsingular, then (2.1.17) can be verified by direct calculation. ■

As an application of Result 2.1.2, we reconsider Result 1.3.9. If $\mathbf{M}$ is an $n \times n$ lower triangular matrix, then it can be partitioned as

$$\mathbf{M} = \begin{pmatrix} d_1 & \mathbf{0} \\ \mathbf{c} & \mathbf{D} \end{pmatrix},$$

where $\mathbf{D}$ is an $(n-1)\times(n-1)$ lower triangular matrix. If $\mathbf{M}$ is invertible, then by Result 2.1.2,

$$\mathbf{M}^{-1} = \begin{pmatrix} 1/d_1 & \mathbf{0} \\ -\mathbf{D}^{-1}\mathbf{c}/d_1 & \mathbf{D}^{-1} \end{pmatrix}.$$

Result 2.1.3. Rank and determinant of a partitioned matrix. Suppose a matrix $\mathbf{M}$ is partitioned as

$$\mathbf{M} = \begin{pmatrix} \mathbf{A} & \mathbf{B} \\ \mathbf{C} & \mathbf{D} \end{pmatrix}, \tag{2.1.19}$$

where $\mathbf{D}$ is a square matrix. Suppose $|\mathbf{D}| \neq 0$. Then

$$\mathrm{r}(\mathbf{M}) = \mathrm{r}(\mathbf{D}) + \mathrm{r}(\mathbf{A} - \mathbf{B}\mathbf{D}^{-1}\mathbf{C}), \tag{2.1.20}$$

and furthermore, if $\mathbf{M}$ is a square matrix (so that $\mathbf{A}$ is also a square matrix), then

$$|\mathbf{M}| = |\mathbf{D}|\,|\mathbf{A} - \mathbf{B}\mathbf{D}^{-1}\mathbf{C}|. \tag{2.1.21}$$

Proof. Consider the first equality in (2.1.14). From Result 2.1.2, the determinant of $\begin{pmatrix} \mathbf{I} & \mathbf{O} \\ -\mathbf{D}^{-1}\mathbf{C} & \mathbf{I} \end{pmatrix}$ is 1, so the matrix is nonsingular. Then by property 5 of Result 1.3.11, $\mathbf{M}$ has the same rank as $\begin{pmatrix} \mathbf{A} - \mathbf{B}\mathbf{D}^{-1}\mathbf{C} & \mathbf{B} \\ \mathbf{O} & \mathbf{D} \end{pmatrix}$, which by Result 2.1.1 is $\mathrm{r}(\mathbf{D}) + \mathrm{r}(\mathbf{A} - \mathbf{B}\mathbf{D}^{-1}\mathbf{C})$, showing (2.1.20). Take the determinant on both sides of the first equality in (2.1.14). Then (2.1.21) follows by using property 3 of Result 1.3.6 and Result 2.1.2. ■

Example 2.1.2. (a) We will find $|\mathbf{M}|$, where

$$\mathbf{M} = \left(\begin{array}{cc|c} 1 & 2 & 0 \\ 2 & 5 & 0 \\ \hline 4 & 6 & 5 \end{array}\right)$$

has been partitioned into the form (2.1.19), with

$$\mathbf{A} = \begin{pmatrix} 1 & 2 \\ 2 & 5 \end{pmatrix}, \quad \mathbf{B} = \begin{pmatrix} 0 \\ 0 \end{pmatrix}, \quad \mathbf{C} = \begin{pmatrix} 4 & 6 \end{pmatrix}, \quad \text{and} \quad \mathbf{D} = (5).$$

Also, $|\mathbf{A}| = 1$, and $|\mathbf{D}| = 5$. From Result 2.1.2, we see that $|\mathbf{A}| = |\mathbf{A}|\,|\mathbf{D}| = 5$.

(b) We compute $|\mathbf{M}|$ where

$$\mathbf{M} = \left(\begin{array}{cc|c} 1 & 2 & 1 \\ 2 & 5 & 7 \\ \hline 4 & 6 & 5 \end{array}\right).$$

Here, $\mathbf{B} = \begin{pmatrix} 1 \\ 7 \end{pmatrix}$, while the other submatrices remain the same as in (a). Using Result 2.1.3, $|\mathbf{M}| = |\mathbf{D}|\,|\mathbf{A} - \mathbf{B}\mathbf{D}^{-1}\mathbf{C}| = 11$. □

We next show a result on the inverse of $\mathbf{M}$ partitioned as in (2.1.19).

Result 2.1.4. Inverse of a partitioned matrix. Suppose a nonsingular matrix $\mathbf{M}$ is partitioned as in (2.1.19), and $\mathbf{M}^{-1} = \begin{pmatrix} \mathbf{A}_* & \mathbf{B}_* \\ \mathbf{C}_* & \mathbf{D}_* \end{pmatrix}$ is partitioned the same way.

1. Suppose $|\mathbf{D}| \neq 0$. Then,

$$\begin{aligned} \mathbf{A}_* &= (\mathbf{A} - \mathbf{B}\mathbf{D}^{-1}\mathbf{C})^{-1}, \\ \mathbf{B}_* &= -\mathbf{A}_*\mathbf{B}\mathbf{D}^{-1}, \\ \mathbf{C}_* &= -\mathbf{D}^{-1}\mathbf{C}\mathbf{A}_* \\ \mathbf{D}_* &= \mathbf{D}^{-1} + \mathbf{D}^{-1}\mathbf{C}\mathbf{A}_*\mathbf{B}\mathbf{D}^{-1}. \end{aligned} \tag{2.1.22}$$

2. Suppose $|\mathbf{A}| \neq 0$. Then

$$\begin{aligned} \mathbf{A}_* &= \mathbf{A}^{-1} + \mathbf{A}^{-1}\mathbf{B}\mathbf{D}_*\mathbf{C}\mathbf{A}^{-1}, \\ \mathbf{B}_* &= -\mathbf{A}^{-1}\mathbf{B}\mathbf{D}_*, \\ \mathbf{C}_* &= -\mathbf{D}_*\mathbf{C}\mathbf{A}^{-1}, \\ \mathbf{D}_* &= (\mathbf{D} - \mathbf{C}\mathbf{A}^{-1}\mathbf{B})^{-1}. \end{aligned} \tag{2.1.23}$$

Proof. We prove property 1. The proof of property 2 is similar. Since $|\mathbf{M}| \neq 0$ and $|\mathbf{D}| \neq 0$, by Result 2.1.3, $|\mathbf{A} - \mathbf{B}\mathbf{D}^{-1}\mathbf{C}| \neq 0$, so $\mathbf{A} - \mathbf{B}\mathbf{D}^{-1}\mathbf{C}$ is invertible. Then by (2.1.14) followed by Result 2.1.2,

$$\begin{aligned} \mathbf{M}^{-1} &= \begin{pmatrix} \mathbf{I} & \mathbf{O} \\ -\mathbf{D}^{-1}\mathbf{C} & \mathbf{I} \end{pmatrix} \begin{pmatrix} \mathbf{A} - \mathbf{B}\mathbf{D}^{-1}\mathbf{C} & \mathbf{B} \\ \mathbf{O} & \mathbf{D} \end{pmatrix}^{-1} \\ &= \begin{pmatrix} \mathbf{I} & \mathbf{O} \\ -\mathbf{D}^{-1}\mathbf{C} & \mathbf{I} \end{pmatrix} \begin{pmatrix} \mathbf{A}_* & -\mathbf{A}_*\mathbf{B}\mathbf{D}^{-1} \\ \mathbf{O} & \mathbf{D}^{-1} \end{pmatrix}, \end{aligned}$$

yielding (2.1.22). ■

Example 2.1.2. Continued. We find the inverse of the matrix in (a). Since $|\mathbf{D}| = 5 \neq 0$, using (2.1.22) we see that $\mathbf{A}_* = \begin{pmatrix} 5 & -2 \\ -2 & 1 \end{pmatrix}$, $\mathbf{B}_* = \begin{pmatrix} 0 \\ 0 \end{pmatrix}$, $\mathbf{C}_* = (-8/5 \quad 2/5)$, and $\mathbf{D}_* = 1/5$, so that

$$\mathbf{M}^{-1} = \begin{pmatrix} 5 & -2 & 0 \\ -2 & 1 & 0 \\ -8/5 & 2/5 & 1/5 \end{pmatrix}.$$

Since $|\mathbf{M}| = 1$, we can also use (2.1.23) to get the same result. □

Example 2.1.3. Suppose $\mathbf{A}$ is a full column rank matrix and is partitioned into $\mathbf{A} = (\mathbf{A}_1, \mathbf{A}_2)$. Let

$$\begin{aligned}
&\mathbf{P} = \mathbf{A}(\mathbf{A}'\mathbf{A})^{-1}\mathbf{A}', \quad \mathbf{P}_1 = \mathbf{A}_1(\mathbf{A}_1'\mathbf{A}_1)^{-1}\mathbf{A}_1', \\
&\mathbf{B} = (\mathbf{I} - \mathbf{P}_1)\mathbf{A}_2, \quad \text{and} \\
&\mathbf{P}_2 = \mathbf{B}(\mathbf{B}'\mathbf{B})^{-1}\mathbf{B}' = (\mathbf{I} - \mathbf{P}_1)\mathbf{A}_2[\mathbf{A}_2'(\mathbf{I} - \mathbf{P}_1)\mathbf{A}_2]^{-1}\mathbf{A}_2'\mathbf{I} - \mathbf{P}_1).
\end{aligned}$$

We verify that $\mathbf{P} = \mathbf{P}_1 + \mathbf{P}_2$. Using (2.1.11), we first write

$$\mathbf{P} = \begin{pmatrix}\mathbf{A}_1 & \mathbf{A}_2\end{pmatrix}\begin{pmatrix}\mathbf{A}_1'\mathbf{A}_1 & \mathbf{A}_1'\mathbf{A}_2 \\ \mathbf{A}_2'\mathbf{A}_1 & \mathbf{A}_2'\mathbf{A}_2\end{pmatrix}^{-1}\begin{pmatrix}\mathbf{A}_1' \\ \mathbf{A}_2'\end{pmatrix}. \tag{2.1.24}$$

We use Result 2.1.4 to evaluate $(\mathbf{A}'\mathbf{A})^{-1}$ in partitioned form as

$$(\mathbf{A}'\mathbf{A})^{-1} = \begin{pmatrix}\mathbf{E} & \mathbf{F} \\ \mathbf{G} & \mathbf{N}\end{pmatrix}, \tag{2.1.25}$$

where

$$\begin{aligned}
\mathbf{E} &= (\mathbf{A}_1'\mathbf{A}_1)^{-1} + (\mathbf{A}_1'\mathbf{A}_1)^{-1}\mathbf{A}_1'\mathbf{A}_2\mathbf{N}\mathbf{A}_2'\mathbf{A}_1(\mathbf{A}_1'\mathbf{A}_1)^{-1}, \\
\mathbf{F} &= -(\mathbf{A}_1'\mathbf{A}_1)^{-1}\mathbf{A}_1'\mathbf{A}_2\mathbf{N}, \\
\mathbf{G} &= -\mathbf{N}\mathbf{A}_2'\mathbf{A}_1(\mathbf{A}_1'\mathbf{A}_1)^{-1}, \text{ and} \\
\mathbf{N} &= (\mathbf{A}_2'\mathbf{A}_2 - \mathbf{A}_2'\mathbf{A}_1(\mathbf{A}_1'\mathbf{A}_1)^{-1}\mathbf{A}_1'\mathbf{A}_2)^{-1} \\
&= (\mathbf{A}_2'[\mathbf{I} - \mathbf{A}_1(\mathbf{A}_1'\mathbf{A}_1)^{-1}\mathbf{A}_1']\mathbf{A}_2)^{-1} = [\mathbf{A}_2'(\mathbf{I} - \mathbf{P}_1)\mathbf{A}_2]^{-1}.
\end{aligned}$$

Substituting (2.1.25) into (2.1.24), we see that

$$\begin{aligned}
\mathbf{P} &= \mathbf{P}_1 + \mathbf{P}_1\mathbf{A}_2\mathbf{N}\mathbf{A}_2'\mathbf{P}_1 - \mathbf{P}_1\mathbf{A}_2\mathbf{N}\mathbf{A}_2' - \mathbf{A}_2\mathbf{N}\mathbf{A}_2'\mathbf{P}_1 + \mathbf{A}_2\mathbf{N}\mathbf{A}_2' \\
&= \mathbf{P}_1 + (\mathbf{I} - \mathbf{P}_1)\mathbf{A}_2[\mathbf{A}_2'(\mathbf{I} - \mathbf{P}_1)\mathbf{A}_2]^{-1}\mathbf{A}_2'(\mathbf{I} - \mathbf{P}_1) \\
&= \mathbf{P}_1 + \mathbf{P}_2.
\end{aligned}$$

This example is useful to show that an orthogonal projection matrix in a linear model can be decomposed into the sum of two (or more) orthogonal projection matrices. A geometric explanation of the example is given at the end of Result 2.6.7. □

Example 2.1.4. Let $\mathbf{A}$ be a $k \times k$ nonsingular matrix. We show that for any $k \times m$ matrix $\mathbf{B}$ and $m \times k$ matrix $\mathbf{C}$,

$$|\mathbf{A} - \mathbf{BC}| = |\mathbf{A}||\mathbf{I}_m - \mathbf{C}\mathbf{A}^{-1}\mathbf{B}|. \tag{2.1.26}$$

In particular, if $\mathbf{A} = \mathbf{I}$, this yields property 12 of Result 1.3.6. Setting $\mathbf{D} = \mathbf{I}$ in Result 2.1.3, we see that

$$\left|\begin{pmatrix}\mathbf{A} & \mathbf{B} \\ \mathbf{C} & \mathbf{I}_m\end{pmatrix}\right| = |\mathbf{A} - \mathbf{BC}|.$$

On the other hand, from Result 2.1.3, setting $\mathbf{D} = \mathbf{A}$ therein, we also see that

$$\left|\begin{pmatrix}\mathbf{A} & \mathbf{B} \\ \mathbf{C} & \mathbf{I}_m\end{pmatrix}\right| = |\mathbf{A}||\mathbf{I}_m - \mathbf{C}\mathbf{A}^{-1}\mathbf{B}|.$$

The required result follows by equating the two displays. □

2.2 Algorithms for matrix factorization

In this section, we present factorization methods that enable efficient computing in linear model theory and multivariate analysis. Although this is less a statistical problem than it is a problem in numerical computing, it is useful for a practicing statistician to have at least a rudimentary knowledge of such techniques. These methods are especially valuable for an appreciation of the development and computation of estimates and diagnostic measures in linear models and are employed by most statistical software in order to produce numerically stable results. The decomposition of a matrix $\mathbf{A}$ into a product of two or more matrices is useful in computing properties such as the rank, the determinant, or the inverse of $\mathbf{A}$. We do not give an exhaustive presentation, and the reader is referred to Golub and Van Loan (1989) or Stewart (1973) for details.

Result 2.2.1. Full-rank factorization. For any matrix $\mathbf{A} \in \mathcal{R}^{m\times n}$ with rank r, there are matrices $\mathbf{B} \in \mathcal{R}^{m\times r}$ and $\mathbf{C} \in \mathcal{R}^{r\times n}$ with rank r such that

$$\mathbf{A} = \mathbf{BC}. \tag{2.2.1}$$

Proof. Let $\mathbf{x}_1, \cdots, \mathbf{x}_r$ be a basis of $\mathcal{C}(\mathbf{A})$. Let $\mathbf{B} = (\mathbf{x}_1, \cdots, \mathbf{x}_r)$. Then $\mathrm{r}(\mathbf{B}) = r$ and every column vector $\mathbf{a}_j$ of $\mathbf{A}$, $j = 1, \ldots, n$, can be written as $\mathbf{a}_j = \sum_{i=1}^{r} c_{ij}\mathbf{x}_i = \mathbf{B}\mathbf{c}_j$, where $\mathbf{c}_j = (c_{1j}, \ldots, c_{rj})'$. Let $\mathbf{C} = (\mathbf{c}_1, \cdots, \mathbf{c}_n)$. Then $\mathbf{A} = \mathbf{BC}$. Since $\mathbf{C}$ has r rows, by property 2 of Result 1.3.11, $\mathrm{r}(\mathbf{C}) \le r$ while by property 4 of Result 1.3.11, $\mathrm{r}(\mathbf{C}) \ge \mathrm{r}(\mathbf{A}) = r$. Then, $\mathrm{r}(\mathbf{C}) = r$.

Alternatively, let $\mathbf{P}$ and $\mathbf{Q}$ be nonsingular $m \times m$ and $n \times n$ matrices, respectively, such that $\mathbf{A} = \mathbf{P}\begin{pmatrix}\mathbf{I}_r & \mathbf{O}\\ \mathbf{O} & \mathbf{O}\end{pmatrix}\mathbf{Q}$ (see Result 1.3.14). Let $\mathbf{B}$ be the $m \times r$ matrix consisting of the first r columns of $\mathbf{P}$ and $\mathbf{C}$ the $r \times n$ matrix consisting of the first r rows of $\mathbf{Q}$. Then $\mathbf{A} = \mathbf{BC}$. ■

Result 2.2.2. *QR* decomposition. Let $\mathbf{A}$ be an $m \times n$ matrix of full column rank. Then there exists an $m \times n$ matrix $\mathbf{Q}$ and an $n \times n$ nonsingular upper triangular matrix $\mathbf{R}$ such that

$$\mathbf{A} = \mathbf{QR}, \tag{2.2.2}$$

where the column vectors of $\mathbf{Q}$ form an orthogonal basis for $\mathcal{C}(\mathbf{A})$.

Proof. Let $\mathbf{a}_1, \cdots, \mathbf{a}_n$ denote the columns of $\mathbf{A}$, which are LIN. The Gram–Schmidt orthogonalization (see Result 1.2.4) can be used to construct an orthogonal set of vectors $\mathbf{b}_1, \cdots, \mathbf{b}_n$ which are defined recursively by

$$\mathbf{b}_1 = \mathbf{a}_1$$

$$\mathbf{b}_i = \mathbf{a}_i - \sum_{j=1}^{i-1} c_{ji}\mathbf{b}_j, \quad i = 2, \cdots, n,$$

where $c_{ij} = \mathbf{a}_j'\mathbf{b}_i/\mathbf{b}_i'\mathbf{b}_i$, $i < j = 1, \cdots, n$. By construction, we define $\mathbf{Q} = (\mathbf{b}_1, \cdots, \mathbf{b}_n)$ to be the required $m \times n$ orthogonal matrix, while $\mathbf{R}$ denotes the $n \times n$ upper-triangular matrix whose (i, j)th element is given by c_{ij}, $i < j = 1, \cdots, n$. Then, the QR decomposition of $\mathbf{A}$ has the form $\mathbf{A} = \mathbf{QR}$. ■

The QR decomposition is useful for computing numerically stable estimates of coefficients in a linear model. Various orthogonalization algorithms have been employed in the literature (see Golub and Van Loan (1989) or Stewart (1973)) which operate directly on the matrix of explanatory variables in a linear model. The QR decomposition also enables us to factor the orthogonal projection matrix in linear model theory into the product of two orthogonal matrices, which is useful in the study of regression diagnostics (see Chapter 9).

In the next three sections, we present results that are crucial for the development of linear model theory and are related to the diagonalization of general matrices, and in particular symmetric and positive definite (p.d.) matrices.

Definition 2.2.1. Diagonability of a matrix. An $n \times n$ matrix $\mathbf{A}$ is said to be diagonalizable (or diagonable) if there exists an $n \times n$ nonsingular matrix $\mathbf{Q}$ such that

$$\mathbf{Q}^{-1}\mathbf{A}\mathbf{Q} = \mathbf{D}, \tag{2.2.3}$$

where $\mathbf{D}$ is a diagonal matrix. The matrix $\mathbf{Q}$ diagonalizes $\mathbf{A}$ and further, $\mathbf{Q}^{-1}\mathbf{A}\mathbf{Q} = \mathbf{D}$ if and only if $\mathbf{A}\mathbf{Q} = \mathbf{Q}\mathbf{D}$, i.e., if and only if $\mathbf{A} = \mathbf{Q}\mathbf{D}\mathbf{Q}^{-1}$.

The process of constructing a matrix $\mathbf{Q}$ which diagonalizes $\mathbf{A}$ is referred to as the diagonalization of $\mathbf{A}$; in many cases, we can relate this to the eigensystem of $\mathbf{A}$. In Result 2.2.5, we show how to diagonalize an arbitrary $n \times n$ matrix $\mathbf{A}$. In Section 2.3, we show that a symmetric matrix $\mathbf{A}$ is *orthogonally diagonable.*

Definition 2.2.2. Orthogonal diagonability. An $n \times n$ matrix $\mathbf{A}$ is said to be orthogonally diagonable if and only if there exists an $n \times n$ orthogonal matrix $\mathbf{P}$ such that $\mathbf{P}'\mathbf{A}\mathbf{P}$ is a diagonal matrix.

Result 2.2.3. Let $\mathbf{A}$ be an $n \times n$ matrix. Suppose there exists an $n \times n$ nonsingular matrix $\mathbf{Q}$ such that $\mathbf{Q}^{-1}\mathbf{A}\mathbf{Q} = \mathbf{D} = \text{diag}(\lambda_1, \cdots, \lambda_n)$. Let $\mathbf{Q} = (\mathbf{q}_1, \cdots, \mathbf{q}_n)$. Then,

1. $\text{r}(\mathbf{A})$ is equal to the number of nonzero diagonal elements in $\mathbf{D}$.
2. $|\mathbf{A}| = \prod_{i=1}^n \lambda_i = |\mathbf{D}|$.
3. $\text{tr}(\mathbf{A}) = \sum_{i=1}^n \lambda_i = \text{tr}(\mathbf{D})$.
4. The characteristic polynomial of $\mathbf{A}$ is $P(\lambda) = (-1)^n \prod_{i=1}^n (\lambda - \lambda_i)$.
5. The eigenvalues of $\mathbf{A}$ are $\lambda_1, \cdots \lambda_n$, which are not necessarily all nonzero, nor are they necessarily distinct.
6. The columns of $\mathbf{Q}$ are the LIN eigenvectors of $\mathbf{A}$, where $\mathbf{q}_i$ corresponds to the eigenvalue λ_i.

Proof. Since $\mathbf{Q}$ is a nonsingular matrix, $\text{r}(\mathbf{A}) = \text{r}(\mathbf{Q}^{-1}\mathbf{A}\mathbf{Q}) = \text{r}(\mathbf{D})$, which is clearly equal to the number of nonzero diagonal elements λ_i, $i = 1, \cdots, n$, which proves property 1. The proof of property 2 follows from seeing that

$$|\mathbf{A}| = |\mathbf{Q}^{-1}\mathbf{Q}||\mathbf{A}| = |\mathbf{Q}^{-1}\mathbf{A}\mathbf{Q}| = |\mathbf{D}| = \prod_{i=1}^n \lambda_i.$$

Similarly, $\text{tr}(\mathbf{A}) = \text{tr}(\mathbf{Q}\mathbf{Q}^{-1}\mathbf{A}) = \text{tr}(\mathbf{Q}^{-1}\mathbf{A}\mathbf{Q}) = \text{tr}(\mathbf{D}) = \sum_{i=1}^n \lambda_i$, which proves property 3. By property 2, for any scalar λ, $|\mathbf{D} - \lambda\mathbf{I}| = |\mathbf{Q}^{-1}\mathbf{A}\mathbf{Q} - \lambda\mathbf{I}| = |\mathbf{Q}^{-1}(\mathbf{A} - \lambda\mathbf{I})\mathbf{Q}| = |\mathbf{A} - \lambda\mathbf{I}|$, so that the characteristic polynomials of $\mathbf{D}$ and $\mathbf{A}$ coincide, which proves property 4, of which property 5 is a direct consequence. Now, $\mathbf{Q}^{-1}\mathbf{A}\mathbf{Q} = \mathbf{D}$ implies that $\mathbf{A}\mathbf{Q} = \mathbf{Q}\mathbf{D}$, i.e., $\mathbf{A}\mathbf{q}_i = \lambda_i\mathbf{q}_i$, $i = 1, \cdots, n$. Since λ_i, $i = 1, \cdots, n$ are the eigenvalues of $\mathbf{A}$, property 6 follows. ■

The next result can be shown by following the argument used for property 6 of Result 2.2.3. Its proof is left as Exercise 2.14.

Result 2.2.4. Let $\mathbf{A}$ be a square matrix. Then the following properties hold.

1. For any nonsingular matrix $\mathbf{Q}$, $\mathbf{Q}^{-1}\mathbf{A}\mathbf{Q}$ is a diagonal matrix if and only if the columns of $\mathbf{Q}$ are n LIN eigenvectors of $\mathbf{A}$.

2. For any orthogonal matrix $\mathbf{P}$, $\mathbf{P}'\mathbf{A}\mathbf{P}$ is a diagonal matrix if and only if the columns of $\mathbf{P}$ are n orthonormal eigenvectors of $\mathbf{A}$.

For the next result, recall Definitions 1.3.16 and 1.3.18 on the geometric multiplicity and algebraic multiplicity of an eigenvalue, respectively.

Result 2.2.5. Diagonability theorem. Let $\mathbf{A}$ be an $n \times n$ matrix with k distinct eigenvalues $\lambda_1, \cdots, \lambda_k$. Let g_j and a_j respectively be the geometric multiplicity and algebraic multiplicity of λ_j, $j = 1, \ldots, k$. Then $\mathbf{A}$ is diagonable if and only if $g_j = a_j$ for all j, in which case $\mathbf{A}$ has n LIN eigenvectors. If $\mathbf{U}$ has the n LIN eigenvectors as columns, $\mathbf{U}^{-1}\mathbf{A}\mathbf{U}$ is a diagonal matrix with each λ_j appearing a_j times on the diagonal.

Proof. By Result 2.2.4, $\mathbf{A}$ is diagonable if and only if it has n LIN eigenvectors. We show that the latter holds if and only if $g_j = a_j$ for all j.

Sufficiency. Let $\mathcal{V}_j$ be the eigenspace corresponding to λ_j, $j = 1, \ldots, k$. Then $\dim(\mathcal{V}_j) = g_j$ and by Exercise 1.35, the sum of the eigenspaces is a direct sum. For each j, let $\{\mathbf{v}_{ij}, i = 1, \cdots, g_j\}$ be a basis of $\mathcal{V}_j$. Then $\mathbf{v}_{ij}, i = 1, \cdots, g_j$, $j = 1, \ldots, k$ are LIN eigenvectors. If $g_j = a_j$, then since $\sum_{i=1}^{k} a_j = n$, the $\mathbf{v}_{ij}$'s are n LIN eigenvectors.

Necessity. Let $\mathbf{u}_1, \cdots, \mathbf{u}_n$ be LIN eigenvectors of $\mathbf{A}$ with $\mathbf{A}\mathbf{u}_i = d_i\mathbf{u}_i$ for each i. Then $d_i \in \{\lambda_1, \cdots, \lambda_k\}$. If n_j is the number of columns of $\mathbf{U}$ that are eigenvectors corresponding to λ_j, then $n_j \le g_j \le a_j$. Since $\sum_{j=1}^{k} n_j = n = \sum_{j=1}^{k} a_j$, then $n_j = g_j = a_j$. This also shows that λ_j appears a_j times on the diagonal of $\mathbf{U}^{-1}\mathbf{A}\mathbf{U}$, where $\mathbf{U} = (\mathbf{u}_1, \cdots, \mathbf{u}_n)$. ■

Not every $n \times n$ matrix $\mathbf{A}$ is diagonable. For example, if $\mathbf{A} = \begin{pmatrix} 1 & 1 \\ 0 & 1 \end{pmatrix}$, then its unique eigenvalue is 1. Since the algebraic multiplicity of the eigenvalue is 2 while the geometric multiplicity is 1, by Result 2.2.5, $\mathbf{A}$ is not diagonable. On the other hand, we have the following general result, whose proof is beyond the scope of this book.

Result 2.2.6. Jordan canonical form. For every $n \times n$ matrix $\mathbf{A}$, there is a nonsingular matrix $\mathbf{U}$, such that $\mathbf{U}^{-1}\mathbf{A}\mathbf{U} = \mathbf{J} = \text{diag}(\mathbf{J}_1, \cdots, \mathbf{J}_s)$, where each $\mathbf{J}_i$ is either a scalar λ_i or an $n_i \times n_i$ matrix with $n_i > 1$ of the form

$$\mathbf{J}_i = \begin{pmatrix} \lambda_i & 1 & 0 & \cdots & 0 & 0 \\ 0 & \lambda_i & 1 & \cdots & 0 & 0 \\ 0 & 0 & \lambda_i & \cdots & 0 & 0 \\ \vdots & \vdots & \vdots & \ddots & \vdots & \vdots \\ 0 & 0 & 0 & \cdots & \lambda_i & 1 \\ 0 & 0 & 0 & \cdots & 0 & \lambda_i \end{pmatrix}.$$

Other than the order of appearance along the diagonal, the matrices $\mathbf{J}_1, \cdots, \mathbf{J}_s$ are unique. The matrix $\mathbf{J}$ is known as the Jordan canonical form of $\mathbf{A}$.

A general result on the decomposition of an $m \times n$ matrix $\mathbf{A}$ is given by the singular-value decomposition, which is shown in the next result. We leave its proof to the reader.

Result 2.2.7. Singular value decomposition. Any $m \times n$ matrix $\mathbf{A}$ of rank r has the decomposition

$$\mathbf{A} = \mathbf{P}\begin{pmatrix} \mathbf{D}_1 & \mathbf{O} \\ \mathbf{O} & \mathbf{O} \end{pmatrix}\mathbf{Q}' = \mathbf{P}_1\mathbf{D}_1\mathbf{Q}_1', \text{ or}$$

$$\mathbf{A} = \sum_{i=1}^{r} d_i \mathbf{p}_i \mathbf{q}_i',$$

where $\mathbf{P}$ is an $m \times m$ orthogonal matrix, $\mathbf{Q}$ is an $n \times n$ orthogonal matrix, $\mathbf{D}_1 = \text{diag}(d_1, \cdots, d_r)$ with $d_1 \geq d_2 \geq \cdots \geq d_r > 0$, $\mathbf{P}_1 = (\mathbf{p}_1, \cdots, \mathbf{p}_r)$ consists of the first r columns of $\mathbf{P}$, and $\mathbf{Q}_1 = (\mathbf{q}_1, \cdots, \mathbf{q}_r)$ consists of the first r columns of $\mathbf{Q}$.

The scalars $d_1, \cdots, d_r$, which are called the singular values of $\mathbf{A}$, are the positive square roots of the (not necessarily distinct) nonzero eigenvalues of $\mathbf{A}'\mathbf{A}$, which do not vary with the choice of $\mathbf{P}$ and $\mathbf{Q}$. The m columns of $\mathbf{P}$ are eigenvectors of $\mathbf{A}\mathbf{A}'$, with the first r columns corresponding to the nonzero eigenvalues $d_1^2, \cdots, d_r^2$, while the remaining $m - r$ columns correspond to the zero eigenvalues. Similarly, the n columns of $\mathbf{Q}$ are eigenvectors of $\mathbf{A}'\mathbf{A}$, with the first r columns corresponding to the nonzero eigenvalues $d_1^2, \cdots, d_r^2$, and the remaining $n-r$ columns corresponding to the zero eigenvalues. Once the first r columns of $\mathbf{P}$ are specified, the first r columns of $\mathbf{Q}$ are uniquely determined, and vice versa (see Harville, 1997, section 21.12 for more details).

2.3 Symmetric and idempotent matrices

Recall from Definition 1.3.5 that an $n \times n$ matrix $\mathbf{A}$ is symmetric if $\mathbf{A}' = \mathbf{A}$. We now give several results on symmetric matrices that are useful in the theory of linear models.

Result 2.3.1. The eigenvalues of every real symmetric matrix are real-valued.

Proof. For $z = a + \iota b$, where $\iota = \sqrt{-1}$, denote by z^* its complex conjugate $a - \iota b$, and for a matrix $\mathbf{M} = \{m_{ij}\}$, denote $\mathbf{M}^* = \{m_{ij}^*\}$. Let $\mathbf{A}$ be an $n \times n$ real symmetric matrix. Let λ be an eigenvalue of $\mathbf{A}$ and $\mathbf{x} = (x_1, \cdots, x_n)'$ a corresponding eigenvector; λ and x_i may be complex-valued. Then

$$\mathbf{x}^{*\prime}\mathbf{A}\mathbf{x} = \mathbf{x}^{*\prime}(\mathbf{A}\mathbf{x}) = \mathbf{x}^{*\prime}(\lambda\mathbf{x}) = \lambda\mathbf{x}^{*\prime}\mathbf{x}. \tag{2.3.1}$$

Since $\mathbf{A}$ is real-valued, we also have $\mathbf{A}\mathbf{x}^* = (\mathbf{A}\mathbf{x})^* = (\lambda\mathbf{x})^* = \lambda^*\mathbf{x}^*$, so that

$$\mathbf{x}^{*\prime}\mathbf{A}\mathbf{x} = (\mathbf{A}\mathbf{x}^*)'\mathbf{x} = (\lambda^*\mathbf{x}^*)'\mathbf{x} = \lambda^*\mathbf{x}^{*\prime}\mathbf{x}. \tag{2.3.2}$$

Equating (2.3.1) and (2.3.2), we get $\lambda\mathbf{x}^{*\prime}\mathbf{x} = \lambda^*\mathbf{x}^{*\prime}\mathbf{x}$. Since $\mathbf{x}^{*\prime}\mathbf{x} = |x_1|^2 + \cdots + |x_n|^2 > 0$, we must have that $\lambda = \lambda^*$, i.e., λ must be real-valued. ■

From Result 2.3.1, it follows that each eigenspace of a real symmetric matrix has a basis consisting of real-valued vectors. Henceforth, an eigenvector of a symmetric matrix will always be assumed to be real-valued.

Result 2.3.2. Let $\mathbf{x}_1$ and $\mathbf{x}_2$ be two eigenvectors corresponding to two distinct eigenvalues λ_1 and λ_2 of a symmetric matrix $\mathbf{A}$. Then, $\mathbf{x}_1$ and $\mathbf{x}_2$ are orthogonal.

Proof. We are given that $\mathbf{A} = \mathbf{A}'$, and $\mathbf{A}\mathbf{x}_j = \lambda_j \mathbf{x}_j$, $j = 1, 2$. Since $\mathbf{x}_j'\mathbf{A}\mathbf{x}_j$ is a scalar for $j = 1, 2$, we have

$$\lambda_1 \mathbf{x}_2'\mathbf{x}_1 = \mathbf{x}_2'\lambda_1\mathbf{x}_1 = \mathbf{x}_2'\mathbf{A}\mathbf{x}_1 = \mathbf{x}_1'\mathbf{A}\mathbf{x}_2 = \mathbf{x}_1'\lambda_2\mathbf{x}_2 = \lambda_2\mathbf{x}_1'\mathbf{x}_2 = \lambda_2\mathbf{x}_2'\mathbf{x}_1.$$

Since $\lambda_1 \neq \lambda_2$, then $\mathbf{x}_2'\mathbf{x}_1 = 0$. By Definition 1.2.9, $\mathbf{x}_1$ and $\mathbf{x}_2$ are orthogonal. For a more general alternate proof which is applicable when the eigenvalues are not necessarily distinct, see corollary 21.5.9 in Harville (1997). ■

Example 2.3.1. Let $\mathbf{A} = \begin{pmatrix} 2 & 2 \\ 2 & -1 \end{pmatrix}$. Then

$$|\mathbf{A} - \lambda\mathbf{I}| = \begin{vmatrix} 2-\lambda & 2 \\ 2 & -1-\lambda \end{vmatrix} = -(2-\lambda)(1+\lambda) - 4 = 0;$$

the solutions are $\lambda = 3$, and $\lambda = -2$, which are the eigenvalues of $\mathbf{A}$. It is easy to verify that the corresponding eigenvectors are $(2, 1)'$ and $(1, -2)'$, which are clearly orthogonal. □

Result 2.3.3. If $\mathbf{A}$ is an $n \times n$ symmetric matrix, any invariant subspace $\mathcal{V} \subset \mathcal{R}^n$ under $\mathbf{A}$ has k orthonormal eigenvectors of $\mathbf{A}$, where $k = \dim(\mathcal{V})$.

Proof. If $k = 0$, then $\mathcal{V} = \{\mathbf{0}\}$ and the result is trivial. Suppose the result has been shown for all k-dimensional invariant subspaces with respect to $\mathbf{A}$. Let $\mathcal{V}$ be a $(k+1)$-dimensional invariant subspace under $\mathbf{A}$. By Result 1.3.16, $\mathcal{V}$ has at least one eigenvector $\mathbf{q}$ of $\mathbf{A}$. Suppose $\mathbf{A}\mathbf{q} = \lambda\mathbf{q}$. Let $\mathcal{V}_1$ be the subspace of vectors in $\mathcal{V}$ that are orthogonal to $\mathbf{q}$. By property 1 of Result 1.2.5, $\mathcal{V}_1$ has dimension k. If $\mathbf{v} \in \mathcal{V}_1$, then $\mathbf{v} \in \mathcal{V}$, so by the invariance of $\mathcal{V}$, $\mathbf{A}\mathbf{v} \in \mathcal{V}$. On the other hand, since $\mathbf{v} \perp \mathbf{q}$, $(\mathbf{A}\mathbf{v})'\mathbf{q} = \mathbf{v}'\mathbf{A}\mathbf{q} = \mathbf{v}'\lambda\mathbf{q} = 0$, i.e., $\mathbf{A}\mathbf{v} \perp \mathbf{q}$. Thus $\mathbf{A}\mathbf{v} \in \mathcal{V}_1$. It follows that $\mathcal{V}_1$ is an invariant subspace under $\mathbf{A}$, so by the induction hypothesis, has k orthonormal eigenvectors of $\mathbf{A}$. These eigenvectors and $\mathbf{q}$ are $k+1$ orthonormal eigenvectors in $\mathcal{V}$. By induction, the result is proved. ■

Result 2.3.4. Spectral decomposition of symmetric matrices. An $n \times n$ matrix $\mathbf{A}$ with eigenvalues λ_k and corresponding eigenvectors $\mathbf{p}_k$, $k = 1, \cdots, n$, is diagonable by an orthogonal matrix $\mathbf{P} = (\mathbf{p}_1, \cdots, \mathbf{p}_n)$ such that

$$\mathbf{P}'\mathbf{A}\mathbf{P} = \mathbf{D} = \text{diag}(\lambda_1, \cdots, \lambda_n) \tag{2.3.3}$$

if and only if $\mathbf{A}$ is symmetric. In other words, every symmetric matrix is orthogonally diagonable. The spectral decomposition of $\mathbf{A}$ is

$$\mathbf{A} = \sum_{k=1}^{n} \lambda_k \mathbf{p}_k \mathbf{p}_k'. \tag{2.3.4}$$

Proof. Necessity. Let $\mathbf{A}$ be any $n \times n$ matrix, and suppose there exists an $n \times n$ orthogonal matrix $\mathbf{P}$ such that $\mathbf{P}'\mathbf{A}\mathbf{P} = \mathbf{D}$, where $\mathbf{D}$ is a diagonal matrix. It follows that $\mathbf{A} = \mathbf{P}\mathbf{D}\mathbf{P}'$. Note that $\mathbf{D}$ is symmetric. Then

$$\mathbf{A}' = (\mathbf{P}\mathbf{D}\mathbf{P}')' = \mathbf{P}\mathbf{D}'\mathbf{P}' = \mathbf{P}\mathbf{D}\mathbf{P}'$$

and so $\mathbf{A}$ is symmetric (see Definition 1.3.5).

Sufficiency. By Result 2.2.4, $\mathbf{A}$ is orthogonally diagonable if and only if it has n orthonormal eigenvectors. The latter is a consequence of Result 2.3.3, since $\mathcal{R}^n$ is an invariant space with respect to $\mathbf{A}$ (see Definition 1.3.17). ■

Result 2.3.5. If $\mathbf{A} \in \mathcal{R}^{n\times n}$ is symmetric with eigenvalues $\lambda_1, \cdots, \lambda_n$, then the following properties hold.

1. $\text{tr}(\mathbf{A}) = \sum_{i=1}^{n} \lambda_i$.
2. $\text{tr}(\mathbf{A}^s) = \sum_{i=1}^{n} \lambda_i^s$ for any nonnegative integer s.
3. $\text{tr}(\mathbf{A}^{-1}) = \sum_{i=1}^{n} 1/\lambda_i$, provided $\mathbf{A}$ is nonsingular.

The proof is an immediate consequence of Results 1.3.18 and 1.3.19.

Definition 2.3.1. An $n \times n$ matrix $\mathbf{A}$ is said to be idempotent if $\mathbf{A}^2 = \mathbf{A}$. $\mathbf{A}$ is symmetric and idempotent if $\mathbf{A}' = \mathbf{A}$ and $\mathbf{A}^2 = \mathbf{A}$.

Examples of symmetric and idempotent matrices include the identity matrix $\mathbf{I}_n$, the matrix $\overline{\mathbf{J}}_n = \frac{1}{n}\mathbf{J}_n$ and the centering matrix $\mathbf{C}_n = \mathbf{I}_n - \overline{\mathbf{J}}_n$. We complete this section with some properties of idempotent matrices.

Result 2.3.6. Properties of idempotent matrices.

1. $\mathbf{A}'$ is idempotent if and only if $\mathbf{A}$ is idempotent.
2. $\mathbf{I} - \mathbf{A}$ is idempotent if and only if $\mathbf{A}$ is idempotent.
3. If $\mathbf{A}$ is an $n \times n$ idempotent matrix, then it can be diagonalized as $\mathbf{Q}^{-1}\mathbf{A}\mathbf{Q} = \mathbf{D}$, where the diagonal elements of $\mathbf{D}$ are the eigenvalues of $\mathbf{A}$ which are equal to 1 or 0. Then, $\text{r}(\mathbf{A}) = \text{tr}(\mathbf{A})$ and $\text{r}(\mathbf{I}_n - \mathbf{A}) = n - \text{tr}(\mathbf{A})$.
4. If $\text{r}(\mathbf{A}) = n$ for an $n \times n$ idempotent matrix $\mathbf{A}$, then we must have $\mathbf{A} = \mathbf{I}_n$.

Proof. To prove property 1, assume first that $\mathbf{A}'\mathbf{A}' = \mathbf{A}'$. That $\mathbf{A}$ is idempotent follows by transposing both sides. The proof that $\mathbf{A}'$ is idempotent if $\mathbf{A}$ is idempotent is similar. To prove property 2, assume first that $\mathbf{I} - \mathbf{A}$ is idempotent, which implies that $(\mathbf{I} - \mathbf{A}) = (\mathbf{I} - \mathbf{A})(\mathbf{I} - \mathbf{A})$, from which idempotency of $\mathbf{A}$ follows immediately. The converse is similarly proved.

To show property 3, let $r = \text{r}(\mathbf{A})$. Let $\mathcal{V}_1 = \mathcal{C}(\mathbf{A})$ and $\mathcal{V}_0 = \mathcal{C}(\mathbf{I} - \mathbf{A})$. Each $\mathbf{v} \in \mathcal{R}^n$ is the sum of $\mathbf{A}\mathbf{v} \in \mathcal{V}_1$ and $\mathbf{v} - \mathbf{A}\mathbf{v} \in \mathcal{V}_0$, and $\mathcal{R}^n = \mathcal{V}_1 + \mathcal{V}_0$. If $\mathbf{v} \in \mathcal{V}_1$, then $\mathbf{v} = \mathbf{A}\mathbf{x}$ for some $\mathbf{x}$. Then $\mathbf{A}\mathbf{v} = \mathbf{A}^2\mathbf{x} = \mathbf{A}\mathbf{x} = \mathbf{v}$. As a result, $\mathcal{V}_1$ is the eigenspace corresponding to the eigenvalue 1. Likewise, $\mathcal{V}_0$ is the eigenspace corresponding to 0. By Result 1.3.17, $\mathcal{R}^n = \mathcal{V}_1 \oplus \mathcal{V}_0$. Note that $\dim(\mathcal{V}_1) = \text{r}(\mathbf{A}) = r$ and $\dim(\mathcal{V}_0) = n - \text{r}(\mathbf{A}) = n - r$. Let $\{\mathbf{q}_1, \cdots, \mathbf{q}_r\}$ be a basis of $\mathcal{V}_1$ and $\{\mathbf{q}_{r+1}, \ldots, \mathbf{q}_n\}$ be a basis of $\mathcal{V}_0$. Put $\mathbf{Q} = (\mathbf{q}_1, \cdots, \mathbf{q}_n)$. Then $\mathbf{Q}^{-1}\mathbf{A}\mathbf{Q} = \mathbf{D}$, where $\mathbf{D} = \text{diag}(\mathbf{I}_r, \mathbf{O})$. As a result, $\text{r}(\mathbf{A}) = \text{r}(\mathbf{D}) = r$ and from Result 1.3.15, the eigenvalues of $\mathbf{A}$ are 1 with multiplicity r and 0 with multiplicity $n - r$. From Result 1.3.19, $\text{tr}(\mathbf{A}) = r$. Since $\mathbf{I} - \mathbf{A}$ is also idempotent, then $\text{r}(\mathbf{I} - \mathbf{A}) = \text{tr}(\mathbf{I} - \mathbf{A}) = n - \text{tr}(\mathbf{A})$. To prove property 4, suppose that the $n \times n$ idempotent matrix $\mathbf{A}$ has rank n, so that $\mathbf{A}^{-1}$ exists. Then, $\mathbf{A} = \mathbf{I}_n\mathbf{A} = \mathbf{A}^{-1}\mathbf{A}\mathbf{A} = \mathbf{A}^{-1}\mathbf{A}^2 = \mathbf{A}^{-1}\mathbf{A} = \mathbf{I}_n$. ■

Result 2.3.7. A symmetric matrix $\mathbf{A} \in \mathcal{R}^{n\times n}$ is idempotent of rank m if and only if m of its eigenvalues are equal to 1 and the remaining $n - m$ eigenvalues are equal to 0.

Proof. Let $\mathbf{A}' = \mathbf{A}$, and let $\lambda_1, \cdots, \lambda_n$ denote the eigenvalues of $\mathbf{A}$, which are not necessarily all distinct. By Result 2.3.4, there exists an orthogonal matrix $\mathbf{P}$ such that $\mathbf{A} = \mathbf{P}\mathbf{D}\mathbf{P}'$, where $\mathbf{D} = \text{diag}(\lambda_1, \cdots, \lambda_n)$. Also,

$$\mathbf{A}^2 = \mathbf{P}\mathbf{D}\mathbf{P}'\mathbf{P}\mathbf{D}\mathbf{P}' = \mathbf{P}\mathbf{D}^2\mathbf{P}',$$

where $\mathbf{D}^2 = \text{diag}(\lambda_1^2, \cdots, \lambda_n^2)$. Suppose $\mathbf{A}^2 = \mathbf{A}$, i.e., $\mathbf{A}$ is idempotent. This must imply that $\mathbf{D}^2 = \mathbf{D}$, or $\lambda_j^2 - \lambda_j = 0$, which in turn implies that each eigenvalue is either 0 or 1. Conversely, let us suppose that each eigenvalue of $\mathbf{A}$ is either 0 or 1. This implies that $\lambda_j^2 = \lambda_j$ for all j, i.e., $\mathbf{A}^2 = \mathbf{A}$, so that $\mathbf{A}$ is idempotent. Clearly, $\text{r}(\mathbf{D})$ is equal to the number of nonzero eigenvalues in $\mathbf{A}$, which is also equal to $\text{r}(\mathbf{A})$ (since $\mathbf{P}$ is a nonsingular matrix). ■

We end this section by stating a result without proof of Cauchy's interlacing theorem (Horn and Johnson, 2013, theorem 4.3.17).

Result 2.3.8. If $\mathbf{A} \in \mathcal{R}^{n \times n}$ is symmetric with eigenvalues $\lambda_i(\mathbf{A})$ sorted in increasing order, and $\mathbf{A}_1$ is an $(n-1) \times (n-1)$ principal submatrix of $\mathbf{A}$ with eigenvalues $\lambda_i(\mathbf{A}_1)$ also sorted in increasing order, then $\lambda_i(\mathbf{A}) \leq \lambda_i(\mathbf{A}_1) \leq \lambda_{i+1}(\mathbf{A})$ for all $i = 1, \ldots, n-1$.

2.4 Nonnegative definite quadratic forms and matrices

We introduce quadratic forms and matrices of quadratic forms and describe their properties. First, we define a linear form in a vector $\mathbf{x}$, as well as a bilinear form in $\mathbf{x}$ and $\mathbf{y}$.

Definition 2.4.1. Linear form in x. Given an arbitrary vector $\mathbf{a} = (a_1, \cdots, a_n)'$, a linear form in $\mathbf{x} = (x_1, \cdots, x_n)'$ is a function that assigns to each vector $\mathbf{x} \in \mathcal{R}^n$ the value

$$\mathbf{a}'\mathbf{x} = \sum_{i=1}^{n} a_i x_i = a_1 x_1 + \cdots + a_n x_n. \tag{2.4.1}$$

Note that the linear form $\mathbf{a}'\mathbf{x}$ can also be written as $\mathbf{x}'\mathbf{a}$ and is a homogeneous polynomial of degree 1 with coefficient vector $\mathbf{a}$. For example, $4x_1 + 5x_2 - 3x_3$ is a linear form in $\mathbf{x} = (x_1, x_2, x_3)'$ with coefficient vector $\mathbf{a} = (4, 5, -3)'$. Two linear forms $\mathbf{a}'\mathbf{x}$ and $\mathbf{b}'\mathbf{x}$ are identically equal for all $\mathbf{x}$ if and only if $\mathbf{a}$=$\mathbf{b}$.

Definition 2.4.2. Bilinear form in x and y. Given an arbitrary $m \times n$ matrix $\mathbf{A} = \{a_{ij}\}$, a bilinear form is a function that assigns to each pair of vectors $\mathbf{x} = (x_1, \cdots, x_m)'$ and $\mathbf{y} = (y_1, \cdots, y_n)'$, the value

$$\mathbf{x}'\mathbf{A}\mathbf{y} = \sum_{i=1}^{m} \sum_{j=1}^{n} a_{ij} x_i y_j, \tag{2.4.2}$$

and $\mathbf{A}$ is the matrix of the bilinear form. The form in (2.4.2) can also be written as $\mathbf{y}'\mathbf{A}'\mathbf{x}$.

Two bilinear forms $\mathbf{x}'\mathbf{A}\mathbf{y}$ and $\mathbf{x}'\mathbf{B}\mathbf{y}$ are identically equal if and only if $\mathbf{A} = \mathbf{B}$. A bilinear form $\mathbf{x}'\mathbf{A}\mathbf{y}$ is symmetric if $\mathbf{x}'\mathbf{A}\mathbf{y} = \mathbf{y}'\mathbf{A}'\mathbf{x}$ for all $\mathbf{x}$ and $\mathbf{y}$, i.e., if and only if the matrix of the bilinear form is (square) symmetric, i.e., $\mathbf{A} = \mathbf{A}'$.

Example 2.4.1. The expression $x_1y_1 + 2x_1y_2 + 4x_2y_1 + 7x_2y_2 + 2x_3y_1 - 2x_3y_2$ is a bilinear form in $\mathbf{x} = (x_1, x_2, x_3)'$ and $\mathbf{y} = (y_1, y_2)'$, with the matrix of the bilinear form given by

$$\mathbf{A} = \begin{pmatrix} 1 & 2 \\ 4 & 7 \\ 2 & -2 \end{pmatrix}.$$

An example of a symmetric bilinear form in $\mathbf{x} = (x_1, x_2, x_3)'$ and $\mathbf{y} = (y_1, y_2, y_3)'$ is $x_1y_1 + 2x_1y_2 - 3x_1y_3 + 2x_2y_1 + 7x_2y_2 + 6x_2y_3 - 3x_3y_1 + 6x_3y_2 + 5x_3y_3$, the matrix of the bilinear form being

$$\mathbf{A} = \begin{pmatrix} 1 & 2 & -3 \\ 2 & 7 & 6 \\ -3 & 6 & 5 \end{pmatrix}.$$ □

Definition 2.4.3. Quadratic form in x. Given an arbitrary $n \times n$ matrix $\mathbf{A} = \{a_{ij}\}$, a quadratic form is a function that assigns to each vector $\mathbf{x} = (x_1, \cdots, x_n)' \in \mathcal{R}^n$, the value

$$\mathbf{x}'\mathbf{A}\mathbf{x} = \sum_{i=1}^{n} \sum_{j=1}^{n} a_{ij} x_i x_j, \tag{2.4.3}$$

which is a homogeneous polynomial of degree two.

Example 2.4.2. The expression $x_1^2 + 7x_2^2 + 4x_3^2 + 4x_1x_2 + 10x_1x_3 - 4x_2x_3$ is a quadratic form in $\mathbf{x} = (x_1, x_2, x_3)'$, the matrix of the quadratic form being $\mathbf{A} = \begin{pmatrix} 1 & 2 & 5 \\ 2 & 7 & -2 \\ 5 & -2 & 4 \end{pmatrix}$. When $\mathbf{x} = \mathbf{0}$, then $\mathbf{x}'\mathbf{A}\mathbf{x} = 0$ for all $\mathbf{A}$. □

Let $\mathbf{A} = \{a_{ij}\}$ and $\mathbf{B} = \{b_{ij}\}$ be two arbitrary $n \times n$ matrices. We say $\mathbf{x}'\mathbf{A}\mathbf{x}$ and $\mathbf{x}'\mathbf{B}\mathbf{x}$ are identically equal if and only if $\mathbf{A} + \mathbf{A}' = \mathbf{B} + \mathbf{B}'$. If $\mathbf{A}$ and $\mathbf{B}$ are symmetric matrices, then $\mathbf{x}'\mathbf{A}\mathbf{x}$ and $\mathbf{x}'\mathbf{B}\mathbf{x}$ are identically equal if and only if $\mathbf{A} = \mathbf{B}$. For any matrix $\mathbf{A}$, note that $\mathbf{C} = (\mathbf{A}+\mathbf{A}')/2$ is always symmetric and $\mathbf{x}'\mathbf{A}\mathbf{x} = \mathbf{x}'\mathbf{C}\mathbf{x}$. Hence, we may assume without loss of generality that corresponding to a given quadratic form, there exists a unique symmetric matrix $\mathbf{A}$ which is the matrix of that quadratic form. Let $\mathbf{x}'\mathbf{A}\mathbf{x}$ be a quadratic form in $\mathbf{x}$ and let $\mathbf{y} = \mathbf{C}^{-1}\mathbf{x}$, where $\mathbf{C}$ is an $n \times n$ nonsingular matrix. Then, $\mathbf{x}'\mathbf{A}\mathbf{x} = \mathbf{y}'\mathbf{C}'\mathbf{A}\mathbf{C}\mathbf{y} = \mathbf{y}'\mathbf{B}\mathbf{y}$, say. We refer to $\mathbf{A}$ and $\mathbf{B}$ as congruent matrices.

Definition 2.4.4. Nonnegative definite quadratic form. An arbitrary quadratic form $\mathbf{x}'\mathbf{A}\mathbf{x}$ is said to be nonnegative definite (n.n.d.) if $\mathbf{x}'\mathbf{A}\mathbf{x} \geq 0$ for every vector $\mathbf{x} \in \mathcal{R}^n$. The matrix $\mathbf{A}$ is called a nonnegative definite (n.n.d.) matrix.

Definition 2.4.5. Positive definite quadratic form. A nonnegative definite quadratic form $\mathbf{x}'\mathbf{A}\mathbf{x}$ is said to be positive definite (p.d.) if $\mathbf{x}'\mathbf{A}\mathbf{x} > 0$ for all non-null vectors $\mathbf{x} \in \mathcal{R}^n$ and $\mathbf{x}'\mathbf{A}\mathbf{x} = 0$ only when $\mathbf{x}$ is the null vector, i.e., when $\mathbf{x} = \mathbf{0}$. The matrix $\mathbf{A}$ is called a positive definite (p.d.) matrix.

Definition 2.4.6. Positive semidefinite quadratic form. A nonnegative definite quadratic form $\mathbf{x}'\mathbf{A}\mathbf{x}$ is said to be positive semidefinite (p.s.d.) if $\mathbf{x}'\mathbf{A}\mathbf{x} \geq 0$ for every $\mathbf{x} \in \mathcal{R}^n$ and $\mathbf{x}'\mathbf{A}\mathbf{x} = 0$ for some non-null $\mathbf{x}$. The matrix $\mathbf{A}$ is called a positive semidefinite (p.s.d.) matrix.

Example 2.4.3. The quadratic form $x_1^2 + \cdots + x_n^2 = \mathbf{x}'\mathbf{I}_n\mathbf{x} > 0$ for every non-null $\mathbf{x} \in \mathcal{R}^n$ and is p.d. The quadratic form $(x_1 + \cdots + x_n)^2 = \mathbf{x}'\mathbf{1}\mathbf{1}'\mathbf{x} = \mathbf{x}'\mathbf{J}_n\mathbf{x} \geq 0$ for every $\mathbf{x} \in \mathcal{R}^n$ and is equal to 0 when $x_1 + \cdots + x_n = 0$; it is a p.s.d. quadratic form. □

A quadratic form $\mathbf{x}'\mathbf{A}\mathbf{x}$ is respectively nonpositive definite, or negative definite or negative semidefinite if $-\mathbf{x}'\mathbf{A}\mathbf{x}$ is nonnegative definite, or positive definite or positive semidefinite. The only symmetric $n \times n$ matrix which is both nonnegative definite and nonpositive definite is the null matrix, $\mathbf{O}$. A quadratic form is said to be indefinite if $\mathbf{x}'\mathbf{A}\mathbf{x} > 0$ for some vectors $\mathbf{x}$ in $\mathcal{R}^n$ and $\mathbf{x}'\mathbf{A}\mathbf{x} \leq 0$ for some other vectors $\mathbf{x}$ in $\mathcal{R}^n$. The matrices of such quadratic forms have the corresponding names as well.

Result 2.4.1. Let $\mathbf{P}$ be an $n \times m$ matrix and let $\mathbf{A}$ be an $n \times n$ n.n.d. matrix. Then the matrix $\mathbf{P}'\mathbf{A}\mathbf{P}$ is n.n.d. If $\text{r}(\mathbf{P}) < m$, then $\mathbf{P}'\mathbf{A}\mathbf{P}$ is p.s.d. If $\mathbf{A}$ is p.d. and $\text{r}(\mathbf{P}) = m$, then $\mathbf{P}'\mathbf{A}\mathbf{P}$ is p.d.

Proof. Since $\mathbf{A}$ is n.n.d., by Definition 2.4.4, $\mathbf{x}'\mathbf{A}\mathbf{x} \geq 0$ for every $\mathbf{x} \in \mathcal{R}^n$. For any $\mathbf{y} \in \mathcal{R}^m$, $\mathbf{x} = \mathbf{P}\mathbf{y} \in \mathcal{R}^n$. Then

$$\mathbf{y}'(\mathbf{P}'\mathbf{A}\mathbf{P})\mathbf{y} = (\mathbf{P}\mathbf{y})'\mathbf{A}(\mathbf{P}\mathbf{y}) = \mathbf{x}'\mathbf{A}\mathbf{x} \geq 0, \tag{2.4.4}$$

which implies, by Definition 2.4.4 that $\mathbf{P}'\mathbf{A}\mathbf{P}$ is n.n.d. If $\text{r}(\mathbf{P}) < m$, then by property 4 of Result 1.3.11, we see that $\text{r}(\mathbf{P}'\mathbf{A}\mathbf{P}) \leq \text{r}(\mathbf{P}) < m$, so that $\mathbf{P}'\mathbf{A}\mathbf{P}$ is p.s.d. Further, if $\mathbf{A}$ is p.d., the quadratic form $(\mathbf{P}\mathbf{y})'\mathbf{A}(\mathbf{P}\mathbf{y}) = 0$ only when $\mathbf{P}\mathbf{y} = \mathbf{0}$, which implies that $\mathbf{y} = \mathbf{0}$ since $\text{r}(\mathbf{P}) = m$. Thus, in (2.4.4), $\mathbf{y}'(\mathbf{P}'\mathbf{A}\mathbf{P})\mathbf{y} = 0$ only when $\mathbf{y} = \mathbf{0}$, i.e., $\mathbf{P}'\mathbf{A}\mathbf{P}$ is p.d. ■

Result 2.4.2. Properties of n.n.d. matrices.

1. If an $n \times n$ matrix $\mathbf{A}$ is p.d. (or p.s.d.), and $c > 0$ is a positive scalar, then $c\mathbf{A}$ is also p.d. (or p.s.d.).

2. If two $n \times n$ matrices $\mathbf{A}$ and $\mathbf{B}$ are both n.n.d., then $\mathbf{A} + \mathbf{B}$ is n.n.d. If, in addition, either $\mathbf{A}$ or $\mathbf{B}$ is p.d., then $\mathbf{A} + \mathbf{B}$ is also p.d.

3. Any principal submatrix of a n.n.d. matrix is n.n.d. Any principal submatrix of a p.d. (or p.s.d.) matrix is p.d. (or p.s.d.).

Proof. To prove property 1, we see that by Definitions 2.4.5 and 2.4.6, the matrix $\mathbf{A}$ is p.d. (or p.s.d.) if the quadratic form $\mathbf{x}'\mathbf{A}\mathbf{x}$ is p.d. (or p.s.d.), or since $c > 0$, if $c\mathbf{x}'\mathbf{A}\mathbf{x} = \mathbf{x}'c\mathbf{A}\mathbf{x}$ is p.d. (or p.s.d.). This implies that $c\mathbf{A}$ is p.d. (or p.s.d.). Property 2 follows since $\mathbf{A}$ and $\mathbf{B}$ are both n.n.d., so that we have by Definition 2.4.4 that for every non-null vector $\mathbf{x} \in \mathcal{R}^n$, $\mathbf{x}'\mathbf{A}\mathbf{x} \geq 0$ and $\mathbf{x}'\mathbf{B}\mathbf{x} \geq 0$. Hence, $\mathbf{x}'\mathbf{A}\mathbf{x} + \mathbf{x}'\mathbf{B}\mathbf{x} = \mathbf{x}'(\mathbf{A} + \mathbf{B})\mathbf{x} \geq 0$, which implies that the matrix $\mathbf{A} + \mathbf{B}$ is n.n.d. In addition, suppose that $\mathbf{A}$ is p.d. Then, we must have by Definition 2.4.6 that $\mathbf{x}'\mathbf{A}\mathbf{x} > 0$, while $\mathbf{x}'\mathbf{B}\mathbf{x} \geq 0$ for every non-null $\mathbf{x} \in \mathcal{R}^n$. Hence, $\mathbf{x}'(\mathbf{A} + \mathbf{B})\mathbf{x} = \mathbf{x}'\mathbf{A}\mathbf{x} + \mathbf{x}'\mathbf{B}\mathbf{x} > 0$, so that $\mathbf{A} + \mathbf{B}$ is p.d. To prove property 3, consider the principal submatrix of an $n \times n$ matrix $\mathbf{A}$ obtained by deleting all its rows and columns except its $i_1, \cdots, i_m$th, where $i_1 < \cdots < i_m$. We can write the resulting submatrix as $\mathbf{P}'\mathbf{A}\mathbf{P}$, where $\mathbf{P}$ is the $n \times m$ matrix of rank m, whose columns are the $i_1, \cdots, i_m$th columns of $\mathbf{I}_n$. If $\mathbf{A}$ is n.n.d., it follows from Result 2.4.1 that $\mathbf{P}'\mathbf{A}\mathbf{P}$ is too. In particular, the principal minors of a p.d. matrix are all positive. ■

Result 2.4.3. An $n \times n$ p.d. matrix $\mathbf{A}$ is nonsingular and its inverse is also a p.d. matrix.

Proof. Suppose that, on the contrary, the p.d. matrix $\mathbf{A}$ is singular, with $\text{r}(\mathbf{A}) < n$. The columns of $\mathbf{A}$ are linearly dependent and hence there exists a vector $\mathbf{v} \neq \mathbf{0}$ such that $\mathbf{A}\mathbf{v} = \mathbf{0}$, which implies that $\mathbf{v}'\mathbf{A}\mathbf{v} = 0$, which is a contradiction to our assumption that $\mathbf{A}$ is p.d. Hence $\mathbf{A}$ must be nonsingular, and let $\mathbf{A}^{-1}$ denote the regular inverse of $\mathbf{A}$. Since $\mathbf{A}$ is p.d., by Result 2.4.1, $(\mathbf{A}^{-1})'\mathbf{A}\mathbf{A}^{-1}$ is p.d. But $(\mathbf{A}^{-1})' = (\mathbf{A}^{-1})'\mathbf{A}\mathbf{A}^{-1}$, implying that $(\mathbf{A}^{-1})'$ is p.d. and so is $\mathbf{A}^{-1}$. ■

Result 2.4.4. Let $\mathbf{A}$ be an $n \times n$ symmetric matrix of rank r with eigenvalues $\lambda_1, \cdots, \lambda_n$. Then

1. $\mathbf{A}$ is n.n.d. if and only if $\lambda_j \geq 0$, $j = 1, \cdots, n$, with exactly r of the eigenvalues being strictly positive.

2. $\mathbf{A}$ is p.d. if and only if $\lambda_j > 0$, $j = 1, \cdots, n$.

Proof. From Result 2.3.4, there exists an orthogonal matrix $\mathbf{Q}$ such that $\mathbf{A} = \mathbf{QDQ}'$, where $\mathbf{D} = \text{diag}(\lambda_1, \cdots, \lambda_n)$. Since $\text{r}(\mathbf{D}) = \text{r}(\mathbf{A}) = r$, exactly r of the eigenvalues are nonzero. If $\lambda_j \geq 0$, then $\mathbf{D}$ is n.n.d., and by Result 2.4.1, $\mathbf{A}$ is n.n.d. Conversely, since $\mathbf{D} = \mathbf{Q}'\mathbf{AQ}$, by Result 2.4.1, if $\mathbf{A}$ is n.n.d., then $\mathbf{D}$ is n.n.d., and hence $\lambda_j \geq 0$. This proves property 1. Property 2 can be proved similarly. ■

Result 2.4.5. Let $\mathbf{A}$ be an $n \times n$ symmetric matrix.

1. $\mathbf{A}$ can be factorized as $\mathbf{A} = \mathbf{PP}'$ for some $n \times r$ matrix $\mathbf{P}$ with rank r if and only if $\mathbf{A}$ is n.n.d. with rank r.

2. If $\mathbf{A}$ is n.n.d., then there is a unique $n \times n$ symmetric n.n.d. matrix $\mathbf{B}$ such that $\mathbf{A} = \mathbf{B}'\mathbf{B} = \mathbf{B}^2$; $\mathbf{B}$ is called the square root of $\mathbf{A}$, denoted by $\mathbf{B} = \mathbf{A}^{1/2}$.

3. If $\mathbf{A}$ is p.d., then $(\mathbf{A}^{1/2})^{-1} = (\mathbf{A}^{-1})^{1/2}$; we will denote both sides by $\mathbf{A}^{-1/2}$.

Proof. 1. Let $\mathbf{A} = \mathbf{PP}'$, where $\mathbf{P} \in \mathcal{R}^{n \times r}$ has rank r. Then $\mathbf{A}$ is symmetric and from property 6 of Result 1.3.11, $\text{r}(\mathbf{A}) = \text{r}(\mathbf{P}) = r$. For every $\mathbf{x} \in \mathcal{R}^n$, $\mathbf{x}'\mathbf{Ax} = \mathbf{x}'\mathbf{PP}'\mathbf{x} = (\mathbf{P}'\mathbf{x})'(\mathbf{P}'\mathbf{x}) \geq 0$. Then $\mathbf{A}$ is n.n.d. Conversely, let $\mathbf{A}$ be symmetric and n.n.d. with rank r. By Result 2.4.4 and, if necessary, by applying the same permutations to rows and columns, we have $\mathbf{A} = \mathbf{QDQ}'$, where $\mathbf{Q}$ is orthogonal and $\mathbf{D} = \text{diag}(\lambda_1, \cdots, \lambda_r, 0, \ldots, 0)$ with $\lambda_i > 0$. Define an $n \times r$ matrix $\mathbf{C} = \{c_{ij}\}$ such that $c_{ii} = \sqrt{\lambda_i}$, $i = 1, \ldots, r$ and all other $c_{ij} = 0$. Let $\mathbf{P} = \mathbf{QC}$. Then $\mathbf{P} \in \mathcal{R}^{n \times r}$ with rank r and since $\mathbf{CC}' = \mathbf{D}$, we have $\mathbf{PP}' = \mathbf{QCC}'\mathbf{Q}' = \mathbf{QDQ}' = \mathbf{A}$.

2. Let $\mathbf{Q}$ and $\mathbf{D}$ be defined as above. Define the $n \times n$ diagonal matrix $\mathbf{D}^{1/2} = \text{diag}(\lambda_1^{1/2}, \cdots, \lambda_r^{1/2}, 0, \ldots, 0)$ and let $\mathbf{B} = \mathbf{QD}^{1/2}\mathbf{Q}'$. Then we see that $\mathbf{B}$ is symmetric and n.n.d. and $\mathbf{B}^2 = \mathbf{B}'\mathbf{B} = \mathbf{A}$.

To show that the square root of $\mathbf{A}$ is unique, suppose $\mathbf{M}$ is symmetric and n.n.d. such that $\mathbf{QDQ}' = \mathbf{A} = \mathbf{M}^2$. Let $\mathbf{N} = \mathbf{Q}'\mathbf{MQ}$. Then $\mathbf{N}$ is symmetric and n.n.d. and $\mathbf{D} = \mathbf{N}^2$. All we need to show is that $\mathbf{N} = \mathbf{D}^{1/2}$, since then, $\mathbf{M} = \mathbf{QD}^{1/2}\mathbf{Q}' = \mathbf{B}$, as desired. Partition $\mathbf{N} = \begin{pmatrix} \mathbf{H} & \mathbf{K} \\ \mathbf{K}' & \mathbf{L} \end{pmatrix}$, where $\mathbf{H} \in \mathcal{R}^{r \times r}$ and $\mathbf{L} \in \mathcal{R}^{(n-r)\times(n-r)}$ are symmetric. Then the principal submatrix of $\mathbf{N}^2$ with the same entries as $\mathbf{L}$ is $\mathbf{K}'\mathbf{K} + \mathbf{L}^2$. However, since $\mathbf{N}^2 = \mathbf{D}$, this principal submatrix is $\mathbf{O}$. Then $\mathbf{K}'\mathbf{K} + \mathbf{L}^2 = \mathbf{O}$, and hence $\mathbf{K} = \mathbf{O}$ and $\mathbf{L} = \mathbf{O}$. As a result, $\mathbf{D} = \mathbf{N}^2 = \begin{pmatrix} \mathbf{H}^2 & \mathbf{O} \\ \mathbf{O} & \mathbf{O} \end{pmatrix}$, giving $\text{diag}(\lambda_1, \cdots, \lambda_r) = \mathbf{H}^2 = \mathbf{H}'\mathbf{H}$. Since $\mathbf{N}$ is n.n.d., from property 3 of Result 2.4.2, $\mathbf{H}$ is n.n.d., and we just showed that $\mathbf{H}$ has rank r, so it must be p.d. Partition $\mathbf{H}$ as $\begin{pmatrix} \mathbf{H}_1 & \mathbf{u} \\ \mathbf{u}' & c \end{pmatrix}$. By comparing $\text{diag}(\lambda_1, \cdots, \lambda_r)$ and the square of the partitioned $\mathbf{H}$, $\lambda_r = c^2 + \mathbf{u}'\mathbf{u}$, $\mathbf{H}_1\mathbf{u} + c\mathbf{u} = \mathbf{0}$, and $\mathbf{H}_1^2 + \mathbf{uu}' = \text{diag}(\lambda_1, \cdots, \lambda_{r-1})$. Since $\mathbf{H}$ is symmetric and p.d., $\mathbf{H}_1$ is symmetric and p.d., and $c > 0$. Then $\mathbf{H}_1 + c\mathbf{I}$ is p.d., so $\mathbf{H}_1\mathbf{u} + c\mathbf{u} = (\mathbf{H}_1 + c\mathbf{I})\mathbf{u} = \mathbf{0}$, which implies $\mathbf{u} = \mathbf{0}$. It follows that $c = \sqrt{\lambda_r}$ and $\mathbf{H}_1^2 = \text{diag}(\lambda_1, \cdots, \lambda_{r-1})$. By induction, we can show $\mathbf{H} = \text{diag}(\lambda_1^{1/2}, \cdots, \lambda_r^{1/2})$, yielding $\mathbf{N} = \mathbf{D}^{1/2}$.

3. Since $\mathbf{A}$ is p.d., all λ_j are positive. Let $\mathbf{D}^{-1/2} = \text{diag}(\lambda_1^{-1/2}, \cdots, \lambda_n^{-1/2})$. Then $(\mathbf{A}^{1/2})^{-1} = (\mathbf{QD}^{1/2}\mathbf{Q}')^{-1} = \mathbf{QD}^{-1/2}\mathbf{Q}'$ and $(\mathbf{A}^{-1})^{1/2} = (\mathbf{QD}^{-1}\mathbf{Q}')^{1/2} = \mathbf{QD}^{-1/2}\mathbf{Q}'$. Then $(\mathbf{A}^{1/2})^{-1} = (\mathbf{A}^{-1})^{1/2}$. ■

Example 2.4.4. For all symmetric p.d. $k \times k$ matrices $\mathbf{A}$ and $\mathbf{B}$, and $b > 0$, we show that

$$\frac{\exp\left\{-\frac{1}{2}\operatorname{tr}(\mathbf{A}^{-1}\mathbf{B})\right\}}{|\mathbf{A}|^b} \le \frac{(2b)^{kb}\exp(-kb)}{|\mathbf{B}|^b},$$

with equality holding only when $\mathbf{A} = \mathbf{B}/2b$. From property 3 of Result 1.3.5, $\operatorname{tr}(\mathbf{A}^{-1}\mathbf{B}) = \operatorname{tr}\{(\mathbf{A}^{-1}\mathbf{B}^{1/2})\mathbf{B}^{1/2}\} = \operatorname{tr}(\mathbf{B}^{1/2}\mathbf{A}^{-1}\mathbf{B}^{1/2})$ and from Result 2.4.3 followed by Result 2.4.1, $\mathbf{B}^{1/2}\mathbf{A}^{-1}\mathbf{B}^{1/2}$ is p.d. Let $\lambda_j > 0$, $j = 1, \cdots, k$, denote the eigenvalues of this matrix. Then

$$\operatorname{tr}(\mathbf{A}^{-1}\mathbf{B}) = \operatorname{tr}(\mathbf{B}^{1/2}\mathbf{A}^{-1}\mathbf{B}^{1/2}) = \sum_{j=1}^{k}\lambda_j, \quad \text{and}$$

$$|\mathbf{B}^{1/2}\mathbf{A}^{-1}\mathbf{B}^{1/2}| = |\mathbf{A}^{-1}||\mathbf{B}^{1/2}||\mathbf{B}^{1/2}| = |\mathbf{B}|/|\mathbf{A}| = \prod_{j=1}^{k}\lambda_j,$$

so that $|\mathbf{A}| = |\mathbf{B}|/\prod_{j=1}^{k}\lambda_j$. From these results, we see that

$$\begin{aligned}\frac{\exp\{-\frac{1}{2}\operatorname{tr}(\mathbf{A}^{-1}\mathbf{B})\}}{|\mathbf{A}|^b} &= \frac{\left(\prod_{j=1}^{k}\lambda_j\right)^b \exp(-\frac{1}{2}\sum_{j=1}^{k}\lambda_j)}{|\mathbf{B}|^b} \\ &= \frac{\prod_{j=1}^{k}\lambda_j^b \exp(-\frac{1}{2}\lambda_j)}{|\mathbf{B}|^b}.\end{aligned}$$

It can be verified that the function $\lambda_j^b \exp(-\frac{1}{2}\lambda_j)$ attains a maximum value of $(2b)^b\exp(-b)$ at $\lambda_j = 2b$, $j = 1, \cdots, k$, from which the result follows. □

Result 2.4.6. Let $\mathbf{P}$ be a symmetric p.d. matrix. For any vector $\mathbf{b}$,

$$\sup_{\mathbf{h}\neq\mathbf{0}} \frac{(\mathbf{h}'\mathbf{b})^2}{\mathbf{h}'\mathbf{P}\mathbf{h}} = \mathbf{b}'\mathbf{P}^{-1}\mathbf{b}. \tag{2.4.5}$$

Proof. For every constant $a \in \mathcal{R}$,

$$\begin{aligned}0 \le \|\mathbf{v} - a\mathbf{u}\|^2 &= a^2\|\mathbf{u}\|^2 - 2a\mathbf{u}'\mathbf{v} + \|\mathbf{v}\|^2 \\ &= \left\{a\|\mathbf{u}\| - \frac{\mathbf{u}'\mathbf{v}}{\|\mathbf{u}\|}\right\}^2 + \|\mathbf{v}\|^2 - \frac{(\mathbf{u}'\mathbf{v})^2}{\|\mathbf{u}\|^2}.\end{aligned}$$

For nonzero $\mathbf{u}$, the Cauchy–Schwarz inequality implies that

$$\sup_{\mathbf{v}\neq\mathbf{0}}\left\{\frac{(\mathbf{u}'\mathbf{v})^2}{\mathbf{v}'\mathbf{v}}\right\} = \mathbf{u}'\mathbf{u}. \tag{2.4.6}$$

Set $\mathbf{v} = \mathbf{P}^{1/2}\mathbf{h}$, and $\mathbf{u} = \mathbf{P}^{-1/2}\mathbf{b}$. Then, (2.4.6) yields (2.4.5) after simplification. ■

The next example shows a useful matrix inequality called the extended Cauchy–Schwarz inequality.

Example 2.4.5. Let $\mathbf{b}, \mathbf{d} \in \mathcal{R}^n$ and $\mathbf{B}$ be an $n \times n$ symmetric p.d. matrix. We show that

$$(\mathbf{b}'\mathbf{d})^2 \le (\mathbf{b}'\mathbf{B}\mathbf{b})(\mathbf{d}'\mathbf{B}^{-1}\mathbf{d}), \tag{2.4.7}$$

with equality if and only if $\mathbf{b} = a\mathbf{B}^{-1}\mathbf{d}$, or $\mathbf{d} = a\mathbf{B}\mathbf{b}$, for some constant a. Indeed, we see that $\mathbf{b}'\mathbf{d} = \mathbf{b}'\mathbf{B}^{1/2}\mathbf{B}^{-1/2}\mathbf{d} = (\mathbf{B}^{1/2}\mathbf{b})'(\mathbf{B}^{-1/2}\mathbf{d})$. Apply the Cauchy–Schwarz inequality (see Result 1.2.1) to the vectors $\mathbf{B}^{1/2}\mathbf{b}$ and $\mathbf{B}^{-1/2}\mathbf{d}$ to obtain the inequality in (2.4.7). □

We discuss triangular decomposition of a p.d. symmetric matrix. Let $\mathbf{A}$ be an $m \times m$ p.d. symmetric matrix. There exists a unique lower triangular matrix $\mathbf{L}$ with all diagonal elements equal to 1 and a unique diagonal matrix $\mathbf{D}$ with positive diagonal elements such that

$$\begin{aligned} \mathbf{A} &= \mathbf{L}^{-1}\mathbf{D}\mathbf{L}'^{-1}, && \text{or, equivalently,} \\ \mathbf{L}\mathbf{A}\mathbf{L}' &= \mathbf{D}, && \text{or, equivalently,} \\ \mathbf{A}^{-1} &= \mathbf{L}'\mathbf{D}^{-1}\mathbf{L}. && \end{aligned} \tag{2.4.8}$$

In the literature, there are three different, but mathematically equivalent, versions of the triangular decomposition that go under different names. The first involves $\mathbf{L}^{-1}$ and $\mathbf{D}$, and is known as the Crout decomposition: $\mathbf{A} = (\mathbf{L}^{-1}\mathbf{D})\mathbf{L}'^{-1} = \mathbf{U}\mathbf{L}'^{-1}$, say. The second form is called the Doolittle decomposition and combines matrices $\mathbf{D}$ and $\mathbf{L}^{-1\prime}$: $\mathbf{A} = \mathbf{L}^{-1}(\mathbf{D}\mathbf{L}'^{-1}) = \mathbf{L}^{-1}\mathbf{U}'$, say. The third version is the Cholesky decomposition of $\mathbf{A}$, which is discussed below.

Result 2.4.7. Cholesky decomposition of a p.d. symmetric matrix. A symmetric and p.d. matrix can be factored as

$$\begin{aligned} \mathbf{A} &= \mathbf{V}\mathbf{V}', \text{ where,} \\ \mathbf{V} &= \mathbf{L}^{-1}\mathbf{D}^{1/2}, \end{aligned} \tag{2.4.9}$$

where $\mathbf{D}$ is diagonal and $\mathbf{L}$ is a unit lower triangular matrix.

Proof. By Result 2.4.5, $\mathbf{A} = \mathbf{B}'\mathbf{B}$ for some nonsingular matrix $\mathbf{B}$. By the QR decomposition in Result 2.2.2, $\mathbf{B} = \mathbf{Q}\mathbf{R}$ for some orthogonal matrix $\mathbf{Q}$ and a nonsingular upper triangular matrix $\mathbf{R}$. Then $\mathbf{A} = (\mathbf{Q}\mathbf{R})'(\mathbf{Q}\mathbf{R}) = \mathbf{R}'\mathbf{R}$. Let the diagonal elements of $\mathbf{R}$ be $r_1, \cdots, r_n$. By property 4 of Result 1.3.6, $r_i \neq 0$. By Result 1.3.9, $\mathbf{R}^{-1}$ is upper triangular with diagonal elements $1/r_1, \cdots, 1/r_n$. Let $\mathbf{D}_0 = \text{diag}(r_1, \cdots, r_n)$ and $\mathbf{L} = \mathbf{D}_0\mathbf{R}'^{-1}$. Then $\mathbf{L}$ is a lower triangular matrix with all diagonal elements equal to 1. Since $\mathbf{R}' = \mathbf{L}^{-1}\mathbf{D}_0$, $\mathbf{A} = \mathbf{L}^{-1}\mathbf{D}_0^2\mathbf{L}'^{-1}$, and (2.4.9) follows by letting $\mathbf{D} = \mathbf{D}_0^2$.

To show the uniqueness of the decomposition, suppose there are two such decompositions, $\mathbf{A} = \mathbf{L}_1^{-1}\mathbf{D}_1\mathbf{L}_1'^{-1} = \mathbf{L}_2^{-1}\mathbf{D}_2\mathbf{L}_2'^{-1}$. Then, $\mathbf{D}_1 = \mathbf{M}\mathbf{D}_2\mathbf{M}'$, where by Results 1.3.3 and 1.3.9, $\mathbf{M} = \mathbf{L}_1\mathbf{L}_2^{-1}$ is lower triangular with diagonal elements equal to 1. Let the diagonal elements of $\mathbf{D}_i$ be $d_{i1}, \cdots, d_{in}$ and let $\mathbf{M} = \{m_{ij}\}$. By comparing the diagonal elements of $\mathbf{D}_1$ and $\mathbf{M}\mathbf{D}_2\mathbf{M}'$, we see that $d_{1i} = \sum_{j=1}^n m_{ij}^2 d_{2j}$. Since all d_{1i} and d_{2i} are positive and $m_{ii} = 1$, we have $d_{1i} \geq d_{2i}$. By a symmetric argument, $d_{2i} \geq d_{1i}$. Then, $d_{1i} = d_{2i}$, so that $\mathbf{D}_1 = \mathbf{D}_2$. It follows that $\sum_{j \neq i}^n m_{ij}^2 d_{2j} = 0$. Since $d_{2j} > 0$, $m_{ij}^2 = 0$ for $j \neq i$. As a result, $\mathbf{M} = \mathbf{I}$, and $\mathbf{L}_1 = \mathbf{L}_2$. ■

We end this section with a result on the spectral decomposition of a symmetric n.n.d. matrix.

Result 2.4.8. Let $\mathbf{A}$ be an $n \times n$ symmetric n.n.d. matrix. We can write $\mathbf{A}$ in the form

$$\mathbf{A} = \mathbf{Q}\begin{pmatrix} \mathbf{D}_1 & \mathbf{O} \\ \mathbf{O} & \mathbf{O} \end{pmatrix}\mathbf{Q}', \tag{2.4.10}$$

where $\mathbf{Q}$ is an $n \times n$ orthogonal matrix and $\mathbf{D}_1$ is a diagonal matrix with positive diagonal elements.

Proof. The proof follows directly from Result 2.3.4 and the nonnegativity of the eigenvalues of a n.n.d. matrix. ■

2.5 Simultaneous diagonalization of matrices

We present results that deal with finding a matrix $\mathbf{P}$ that will simultaneously diagonalize two $n \times n$ matrices with different properties in terms of symmetry and nonnegative definiteness.

Result 2.5.1. Let $\mathbf{A}$ and $\mathbf{B}$ be two $n \times n$ symmetric matrices. There exists an orthogonal matrix $\mathbf{P}$ such that $\mathbf{P'AP}$ and $\mathbf{P'BP}$ are both diagonal if and only if $\mathbf{AB} = \mathbf{BA}$.

Proof. **Sufficiency.** Assume $\mathbf{AB} = \mathbf{BA}$. Let $\lambda_1, \cdots, \lambda_s$ be the distinct eigenvalues of $\mathbf{A}$ and $\mathcal{V}_1, \cdots, \mathcal{V}_s$ be the corresponding eigenspaces. Suppose $\dim(\mathcal{V}_i) = g_i$. By Result 2.3.2, $\mathcal{V}_i \perp \mathcal{V}_j$, $i \neq j$. For $\mathbf{v} \in \mathcal{V}_i$, $\mathbf{Av} = \lambda_i \mathbf{v}$. Then,

$$\mathbf{A}(\mathbf{Bv}) = (\mathbf{AB})\mathbf{v} = \mathbf{BAv} = \mathbf{B}(\lambda_i \mathbf{v}) = \lambda_i(\mathbf{Bv}),$$

and $\mathbf{Bv} \in \mathcal{V}_i$. Therefore, $\mathcal{V}_i$ is an invariant space of $\mathcal{R}^n$ under $\mathbf{B}$. By Result 2.3.3, $\mathcal{V}_i$ has g_i orthonormal vectors $\mathbf{q}_{i1}, \cdots, \mathbf{q}_{ig_i}$ such that they are eigenvectors of $\mathbf{B}$. Obviously, $\mathbf{q}_{ij}$ are eigenvectors of $\mathbf{A}$. Let $\mathbf{Q}_i = (\mathbf{q}_{i1}, \cdots, \mathbf{q}_{ig_i})$ and $\mathbf{Q} = (\mathbf{Q}_1, \cdots, \mathbf{Q}_s)$. Then $\mathbf{Q}$ is orthogonal and diagonalizes both $\mathbf{A}$ and $\mathbf{B}$.

Necessity. Let $\mathbf{P'AP} = \mathbf{D}$ and $\mathbf{P'BP} = \boldsymbol{\Delta}$, where $\mathbf{D}$ and $\boldsymbol{\Delta}$ are diagonal matrices. Now, $\mathbf{D}\boldsymbol{\Delta} = \boldsymbol{\Delta}\mathbf{D}$, which implies that

$$\mathbf{AB} = \mathbf{PP'APP'BPP'} = \mathbf{PD\Delta P'} = \mathbf{P\Delta DP'} = \mathbf{PP'BPP'APP'} = \mathbf{BA}. \qquad \blacksquare$$

This result extends to $n \times n$ symmetric matrices $\mathbf{A}_1, \cdots, \mathbf{A}_k$, $k > 2$; these matrices are simultaneously diagonable by an orthogonal matrix $\mathbf{P}$ if and only if they commute under multiplication in pairs.

Result 2.5.2. Let $\mathbf{A}$ be an $n \times n$ symmetric p.d. matrix and let $\mathbf{B}$ be an $n \times n$ symmetric matrix. There exists a nonsingular matrix $\mathbf{P}$ such that $\mathbf{P'AP} = \mathbf{I}$ and $\mathbf{P'BP} = \boldsymbol{\lambda} = \text{diag}(\lambda_1, \cdots, \lambda_n)$, where λ_i are solutions to $|\mathbf{B} - \lambda\mathbf{A}| = 0$.

Proof. Since $\mathbf{A}$ is p.d., there is a nonsingular matrix $\mathbf{R}$ such that $\mathbf{R'AR} = \mathbf{I}$, so that $\mathbf{A} = (\mathbf{R'})^{-1}\mathbf{R}^{-1}$. Also, since $\mathbf{B}$ is symmetric, $\mathbf{R'BR}$ is symmetric. By Result 2.3.4, there exists an orthogonal matrix $\mathbf{Q}$ such that

$$\mathbf{Q'R'BRQ} = \mathbf{D} = \text{diag}(\lambda_1, \cdots, \lambda_n),$$

where λ_i's are solutions to the characteristic equation $|\mathbf{R'BR} - \lambda\mathbf{I}| = 0$. Note that

$$|\mathbf{R'BR} - \lambda\mathbf{I}| = |\mathbf{R'BR} - \lambda\mathbf{R'AR}| = |\mathbf{R'BR} - \mathbf{R'}\lambda\mathbf{AR}| = |\mathbf{R'}|\,|\mathbf{B} - \lambda\mathbf{A}|\,|\mathbf{R}| = 0.$$

Hence, the λ_i's are also solutions of $|\mathbf{B} - \lambda\mathbf{A}| = 0$. Let $\mathbf{P} = \mathbf{RQ}$. Then,

$$\mathbf{P'BP} = \mathbf{Q'R'BRQ} = \mathbf{D} \quad \text{and} \quad \mathbf{P'AP} = \mathbf{Q'R'ARQ} = \mathbf{Q'Q} = \mathbf{I},$$

which proves the result. $\blacksquare$

The problem of finding solutions to the equation $|\mathbf{B} - \lambda\mathbf{A}| = 0$ is called the *generalized eigenvalue problem*, and it reduces to the problem of finding the eigenvalues of $\mathbf{B}$ when $\mathbf{A} = \mathbf{I}$. Since

$$|\mathbf{B} - \lambda\mathbf{A}| = |\mathbf{R}|^2|\mathbf{R'BR} - \lambda\mathbf{I}| = |\mathbf{A}^{-1}\mathbf{B} - \lambda\mathbf{I}| = |\mathbf{BA}^{-1} - \lambda\mathbf{I}|,$$

the generalized eigenvalue problem is equivalent to that of finding the eigenvalues of $\mathbf{R'BR}$ or $\mathbf{A}^{-1}\mathbf{B}$ or $\mathbf{BA}^{-1}$.

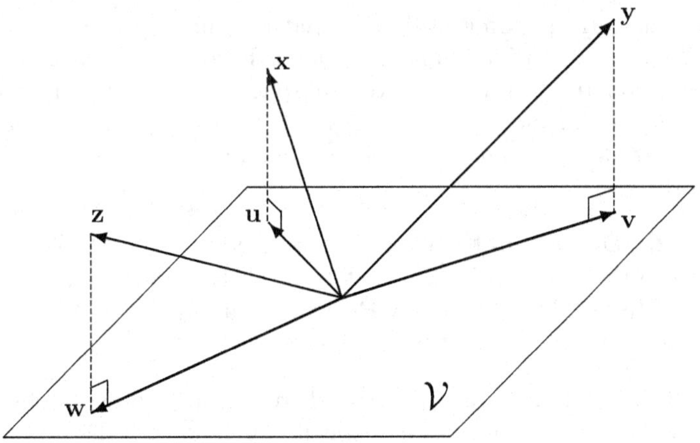

FIGURE 2.6.1. Orthogonal projection of three vectors onto a 2-dimensional subspace $\mathcal{V}$ of $\mathcal{R}^3$.

2.6 Geometrical perspectives

We discuss orthogonal projections and orthogonal projection matrices and their relevance to linear model theory. Recall that if $\mathbf{u}, \mathbf{v} \in \mathcal{R}^n$, we say that $\mathbf{u} \perp \mathbf{v}$ if $\mathbf{u}'\mathbf{v} = 0$ (under the Euclidean norm). If $\mathbf{u} \in \mathcal{R}^n$ and $\mathcal{V}$ is a subspace of $\mathcal{R}^n$, $\mathbf{u} \perp \mathcal{V}$ if $\mathbf{u}'\mathbf{v} = 0$ for every $\mathbf{v} \in \mathcal{V}$. Likewise, if $\mathcal{U}$ and $\mathcal{V}$ are two subspaces of $\mathcal{R}^n$, then $\mathcal{U} \perp \mathcal{V}$ if $\mathbf{u}'\mathbf{v} = 0$ for every $\mathbf{u} \in \mathcal{U}$ and every $\mathbf{v} \in \mathcal{V}$. The vector space $\mathcal{R}^n$ is said to be the direct sum of subspaces $\mathcal{U}$ and $\mathcal{V}$ if any vector $\mathbf{y} \in \mathcal{R}^n$ can be uniquely expressed as $\mathbf{y} = \mathbf{y}_1 + \mathbf{y}_2$, where $\mathbf{y}_1 \in \mathcal{U}$ and $\mathbf{y}_2 \in \mathcal{V}$. We denote this by

$$\mathcal{R}^n = \mathcal{U} \oplus \mathcal{V}$$

(see Definition 1.2.8). We now build upon the basic ideas that we introduced in Chapter 1. Recall the definition of the orthogonal complement of a linear subspace (Definition 1.2.11). We begin with the definition of the orthogonal projection of an n-dimensional vector.

Definition 2.6.1. The orthogonal projection of a vector $\mathbf{v}_1$ onto another vector $\mathbf{v}_2$ is given by

$$(\mathbf{v}_1'\mathbf{v}_2/\mathbf{v}_2'\mathbf{v}_2)\mathbf{v}_2 = (\mathbf{v}_1'\mathbf{v}_2/\|\mathbf{v}_2\|)(1/\|\mathbf{v}_2\|)\mathbf{v}_2.$$

Since the length of $\mathbf{v}_2/\|\mathbf{v}_2\|$ is unity, the length of the orthogonal projection of $\mathbf{v}_1$ on $\mathbf{v}_2$ is

$$|\mathbf{v}_1'\mathbf{v}_2|/\|\mathbf{v}_2\| = \|\mathbf{v}_1\||\mathbf{v}_1'\mathbf{v}_2|/(\|\mathbf{v}_1\|\|\mathbf{v}_2\|) = \|\mathbf{v}_1\||\cos(\theta)|,$$

where θ is the angle between $\mathbf{v}_1$ and $\mathbf{v}_2$.

Figure 2.6.1 illustrates the orthogonal projection of three vectors $\mathbf{x}, \mathbf{y}$, and $\mathbf{z}$ onto vectors $\mathbf{u}, \mathbf{v}$, and $\mathbf{w}$ in a 2-dimensional subspace $\mathcal{V}$ of a 3-dimensional space $\mathcal{R}^3$. We next discuss the notion of an orthogonal projection of a vector onto a subspace of $\mathcal{R}^n$. This concept is basic to an understanding of the geometry of the least squares approach which is a classical estimation tool in linear model theory. We show that such an orthogonal projection exists, it is unique and the corresponding projection matrix is unique as well. This is graphically represented in Figure 2.6.2, which illustrates the orthogonal projection of a 2-dimensional vector $\mathbf{y}$ onto a vector $\mathbf{u}$ which belongs to a subspace $\mathcal{V}$, and a vector $\mathbf{v}$ which belongs to $\mathcal{V}^\perp$, the orthogonal complement of $\mathcal{V}$ (see Definition 1.2.11). The null space of any $n \times k$ matrix $\mathbf{X}$ is the orthogonal complement of the column space of $\mathbf{X}'$, i.e., $\mathcal{N}(\mathbf{X}) = \mathcal{C}(\mathbf{X}')^\perp$ (see Result 1.3.10).

Definition 2.6.2. Orthogonal projection matrix. Let $\mathcal{V} \subset \mathcal{R}^n$. An $n \times n$ matrix $\mathbf{P}_\mathcal{V}$ is called the orthogonal projection matrix of $\mathcal{V}$ if for any $\mathbf{y} \in \mathcal{R}^n$, $\mathbf{P}_\mathcal{V}\mathbf{y} \in \mathcal{V}$ and $(\mathbf{I} - \mathbf{P}_\mathcal{V})\mathbf{y} = \mathbf{y} - \mathbf{P}_\mathcal{V}\mathbf{y} \in \mathcal{V}^\perp$. The matrix $\mathbf{P}_\mathcal{V}$ will be simply denoted by $\mathbf{P}$ when it is clear which subspace we are projecting onto.

Because $\mathcal{R}^n = \mathcal{V} \oplus \mathcal{V}^\perp$, for every $\mathbf{y} \in \mathcal{R}^n$, there is a unique decomposition $\mathbf{y} = \mathbf{u} + \mathbf{v}$, $\mathbf{u} \in \mathcal{V}$, $\mathbf{v} \in \mathcal{V}^\perp$. The next result gives an explicit expression for the orthogonal projection matrix, showing that it exists, and is the unique linear transformation that maps $\mathbf{y}$ to $\mathbf{u}$.

Result 2.6.1. Let $\mathbf{X}$ be any basis matrix of $\mathcal{V}$, i.e., the columns of $\mathbf{X}$ form a LIN basis of $\mathcal{V}$. Then $\mathbf{P} = \mathbf{X}(\mathbf{X}'\mathbf{X})^{-1}\mathbf{X}'$. Further, the orthogonal projection matrix onto $\mathcal{V}^\perp$ is $\mathbf{I}_n - \mathbf{P}$.

Proof. Let $\mathbf{Q} = \mathbf{X}(\mathbf{X}'\mathbf{X})^{-1}\mathbf{X}'$. By property 5 of Result 1.3.10, $\mathcal{C}(\mathbf{Q}) \subset \mathcal{C}(\mathbf{X}) = \mathcal{V}$, so for any $\mathbf{y}$, $\mathbf{v} = \mathbf{Q}\mathbf{y} \in \mathcal{V}$. To show that $\mathbf{P} = \mathbf{Q}$, it only remains to show that $\mathbf{y} - \mathbf{v} \in \mathcal{V}^\perp$. Let $\mathbf{u} \in \mathcal{V}$. We can write $\mathbf{u} = \mathbf{X}\mathbf{c}$, for some vector $\mathbf{c}$. Hence,

$$(\mathbf{y} - \mathbf{v})'\mathbf{u} = (\mathbf{y} - \mathbf{X}(\mathbf{X}'\mathbf{X})^{-1}\mathbf{X}'\mathbf{y})'\mathbf{X}\mathbf{c} = \mathbf{y}'\mathbf{X}\mathbf{c} - \mathbf{y}'\mathbf{X}(\mathbf{X}'\mathbf{X})^{-1}\mathbf{X}'\mathbf{X}\mathbf{c} = 0,$$

so that $\mathbf{y} - \mathbf{v} \perp \mathbf{u}$. Since $\mathbf{u} \in \mathcal{V}$ is arbitrary, this shows $\mathbf{y} - \mathbf{v} \perp \mathcal{V}$, as desired. Finally, since $(\mathbf{I}_n - \mathbf{P})\mathbf{y} \in \mathcal{V}^\perp$ and $\mathbf{P}\mathbf{y} \in \mathcal{V} = (\mathcal{V}^\perp)^\perp$, it follows that $\mathbf{I}_n - \mathbf{P}$ is the orthogonal projection matrix onto $\mathcal{V}^\perp$. ■

Result 2.6.2. The orthogonal projection matrices $\mathbf{P}$ and $\mathbf{I}_n - \mathbf{P}$ are symmetric and idempotent, and further $\mathbf{P}\mathbf{X} = \mathbf{X}$.

Proof. We see that

$$\mathbf{P}' = [\mathbf{X}(\mathbf{X}'\mathbf{X})^{-1}\mathbf{X}']' = \mathbf{X}(\mathbf{X}'\mathbf{X})^{-1}\mathbf{X}', \text{ and}$$
$$\mathbf{P}^2 = \mathbf{P}\mathbf{P} = \mathbf{X}(\mathbf{X}'\mathbf{X})^{-1}\mathbf{X}'\mathbf{X}(\mathbf{X}'\mathbf{X})^{-1}\mathbf{X}' = \mathbf{X}(\mathbf{X}'\mathbf{X})^{-1}\mathbf{X}',$$

so that symmetry and idempotency of $\mathbf{P}$ follow directly from Definition 1.3.5 and Definition 2.3.1. To show this in another way, observe that $\mathbf{P}\mathbf{c} \in \mathcal{V}$ and $(\mathbf{I}_n - \mathbf{P})\mathbf{d} \in \mathcal{V}^\perp$ for any vectors $\mathbf{c}$ and $\mathbf{d}$, so that by Definition 1.2.9, $\mathbf{c}'\mathbf{P}'(\mathbf{I}_n - \mathbf{P})\mathbf{d} = 0$, which in turn implies that $\mathbf{P}'(\mathbf{I}_n - \mathbf{P}) = \mathbf{O}$, that is, $\mathbf{P}' = \mathbf{P}'\mathbf{P}$. Then,

$$\mathbf{P} = (\mathbf{P}')' = (\mathbf{P}'\mathbf{P})' = \mathbf{P}'\mathbf{P} = \mathbf{P}',$$

which implies that $\mathbf{P}$ is symmetric. Since $\mathbf{P}^2 = \mathbf{P}$, it is also idempotent. That $\mathrm{r}(\mathbf{P}) = k$ follows directly from property 3 of Result 2.3.6. It is easy to verify that $\mathbf{P}\mathbf{X} = \mathbf{X}$, using properties of matrix inverse. The proof of the symmetry and idempotency of $\mathbf{I}_n - \mathbf{P}$ is obtained in a similar manner. ■

Result 2.6.3. The column space $\mathcal{C}(\mathbf{P})$ of $\mathbf{P}$ is $\mathcal{V}$, and the column space $\mathcal{C}(\mathbf{I}_n - \mathbf{P})$ is $\mathcal{V}^\perp$. If $\dim(\mathcal{V}) = k$, then $\mathrm{tr}(\mathbf{P}) = \mathrm{r}(\mathbf{P}) = k$ and $\mathrm{tr}(\mathbf{I}_n - \mathbf{P}) = \mathrm{r}(\mathbf{I}_n - \mathbf{P}) = n - k$.

Proof. Since $\mathbf{P}\mathbf{y} = \mathbf{u} \in \mathcal{V}$, it follows that $\mathcal{C}(\mathbf{P}) \subset \mathcal{V}$. Also, if $\mathbf{x} \in \mathcal{V}$, then by Result 2.6.1, the unique orthogonal decomposition of $\mathbf{x}$ is $\mathbf{x} = \mathbf{x} + \mathbf{0}$, which implies that $\mathbf{x} = \mathbf{P}\mathbf{x} \in \mathcal{C}(\mathbf{P})$.

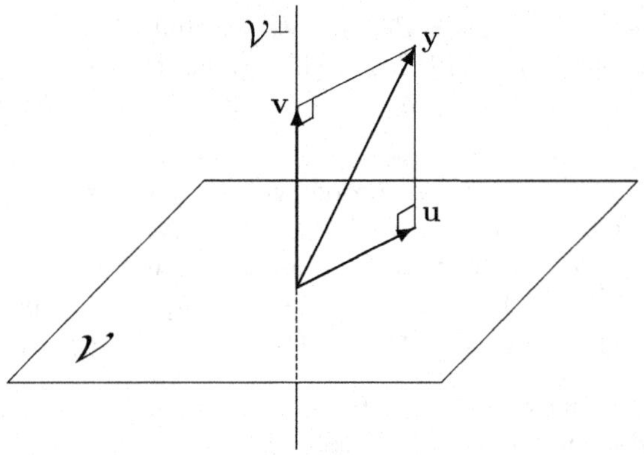

FIGURE 2.6.2. Orthogonal projection of a 3-dimensional vector $\mathbf{y}$ onto a 2-dimensional subspace $\mathcal{V}$ and its orthogonal complement $\mathcal{V}^\perp$.

The two spaces therefore coincide, and $\dim(\mathcal{V}) = \mathrm{r}(\mathbf{P})$. Since the orthogonal projection matrix $\mathbf{P}$ is symmetric and idempotent, it follows from Result 2.3.6 that

$$\mathrm{r}(\mathbf{P}) = \mathrm{tr}(\mathbf{P}) = \mathrm{tr}[\mathbf{X}(\mathbf{X}'\mathbf{X})^{-1}\mathbf{X}'] = \mathrm{tr}[\mathbf{X}'\mathbf{X}(\mathbf{X}'\mathbf{X})^{-1}] = \mathrm{tr}(\mathbf{I}_k) = k.$$

That $\mathrm{r}(\mathbf{I}_n - \mathbf{P}) = n - k$ follows immediately. ■

Result 2.6.4. If $\mathbf{v} \in \mathcal{V}$, then,

1. $\|\mathbf{y} - \mathbf{v}\|^2 = \|\mathbf{y} - \mathbf{P}\mathbf{y}\|^2 + \|\mathbf{P}\mathbf{y} - \mathbf{v}\|^2$.
2. $\|\mathbf{y} - \mathbf{P}\mathbf{y}\|^2 \le \|\mathbf{y} - \mathbf{v}\|^2$ for all $\mathbf{v} \in \mathcal{V}$ with equality holding if and only if $\mathbf{v} = \mathbf{P}\mathbf{y}$.

Proof. Note that $\mathbf{y} - \mathbf{P}\mathbf{y} \in \mathcal{V}^\perp$ and $\mathbf{P}\mathbf{y} - \mathbf{v} \in \mathcal{V}$. Hence, $\mathbf{y} - \mathbf{P}\mathbf{y} \perp \mathbf{P}\mathbf{y} - \mathbf{v}$, so by Pythagoras's Theorem in (1.2.1), $\|\mathbf{y} - \mathbf{v}\|^2 = \|\mathbf{y} - \mathbf{P}\mathbf{y}\|^2 + \|\mathbf{P}\mathbf{y} - \mathbf{v}\|^2$, showing property 1. Since $\|\mathbf{P}\mathbf{y} - \mathbf{v}\|^2 \ge 0$, we must have $\|\mathbf{y} - \mathbf{P}\mathbf{y}\|^2 \le \|\mathbf{y} - \mathbf{v}\|^2$, with equality if and only if $\|\mathbf{P}\mathbf{y} - \mathbf{v}\|^2 = 0$, i.e., $\mathbf{v} = \mathbf{P}\mathbf{y}$, and property 2 follows. ■

Result 2.6.5. Let $\mathbf{X}$ be an $n \times k$ matrix of rank k. There exists a matrix $\mathbf{Z} \in \mathcal{R}^{(n-k)\times n}$ such that $\mathbf{Z}\mathbf{X} = \mathbf{O}$ and $\mathcal{C}(\mathbf{X}) = \mathcal{N}(\mathbf{Z})$.

Proof. Let $\Omega = \mathcal{C}(\mathbf{X})$. By Definition 2.6.2 and Result 2.6.3, we see that $\mathbf{P}_\Omega = \mathbf{X}(\mathbf{X}'\mathbf{X})^{-1}\mathbf{X}'$ has rank k, while the $n \times n$ matrix $\mathbf{I}_n - \mathbf{P}_\Omega = \mathbf{I}_n - \mathbf{X}(\mathbf{X}'\mathbf{X})^{-1}\mathbf{X}'$ has rank $n-k$. Let $\mathbf{Z}$ consist of any $n-k$ LIN rows of $\mathbf{I}_n - \mathbf{P}_\Omega$. Since $(\mathbf{I}_n - \mathbf{P}_\Omega)\mathbf{P}_\Omega = \mathbf{P}_\Omega - \mathbf{P}_\Omega^2 = \mathbf{O}$ by Result 2.6.2, we have $\mathbf{Z}\mathbf{X} = \mathbf{O}$. Further, $\mathcal{N}(\mathbf{Z}) = \mathcal{C}(\mathbf{Z}')^\perp = \mathcal{C}(\mathbf{I}_n - \mathbf{P}_\Omega)^\perp = \Omega$. ■

Result 2.6.6. Let $\Omega \subset \mathcal{R}^n$, $\Omega_1 \subset \Omega$ and let $\mathbf{P}_\Omega$, $\mathbf{P}_{\Omega_1}$, and $\mathbf{P}_{\Omega_1^\perp}$ denote projection matrices (operators). Then,

1. $\mathbf{P}_\Omega\mathbf{P}_{\Omega_1} = \mathbf{P}_{\Omega_1}\mathbf{P}_\Omega = \mathbf{P}_{\Omega_1}$.
2. $\mathbf{P}_{\Omega_1}\mathbf{P}_{\Omega_1^\perp} = \mathbf{P}_{\Omega_1^\perp}\mathbf{P}_{\Omega_1} = \mathbf{O}$.
3. $\mathbf{P}_\Omega = \mathbf{P}_{\Omega_1} + \mathbf{P}_{\Omega_1^\perp\cap\Omega}$.
4. $\mathbf{P}_\Omega\mathbf{P}_{\Omega_1^\perp} = \mathbf{P}_{\Omega_1^\perp\cap\Omega}$.

Proof. Since by Result 2.6.3, $\Omega_1 = \mathcal{C}(\mathbf{P}_{\Omega_1})$, we see that $\mathbf{P}_\Omega \mathbf{P}_{\Omega_1} = \mathbf{P}_{\Omega_1}$. By symmetry of these projection matrices, property 1 follows. This implies that a vector $\mathbf{v}$ projected first onto Ω and then onto Ω_1 stays in Ω_1. To prove property 2, write $\mathbf{P}_{\Omega_1^\perp}\mathbf{P}_{\Omega_1} = \mathbf{P}_{\Omega_1^\perp}(\mathbf{P}_\Omega - \mathbf{P}_{\Omega_1^\perp}) = \mathbf{P}_{\Omega_1^\perp}\mathbf{P}_\Omega - \mathbf{P}^2_{\Omega_1^\perp} = \mathbf{P}_{\Omega_1^\perp} - \mathbf{P}_{\Omega_1^\perp} = \mathbf{O}$.

To prove property 3 (which states property 2 in a different way), consider $\mathbf{P}_\Omega \mathbf{y} = \mathbf{P}_{\Omega_1}\mathbf{y} + (\mathbf{P}_\Omega - \mathbf{P}_{\Omega_1})\mathbf{y}$, where $\mathbf{P}_\Omega \mathbf{y} \in \Omega$, and $\mathbf{P}_{\Omega_1}\mathbf{y} \in \Omega$. We also see that $(\mathbf{P}_\Omega - \mathbf{P}_{\Omega_1})\mathbf{y} \in \Omega$. Thus, $\mathbf{P}_\Omega \mathbf{y} = \mathbf{P}_{\Omega_1}\mathbf{y} + (\mathbf{P}_\Omega - \mathbf{P}_{\Omega_1})\mathbf{y}$ corresponds to the unique orthogonal decomposition described in Result 2.6.1, so that $\mathbf{P}_\Omega = \mathbf{P}_{\Omega_1} + \mathbf{P}_{\Omega_1^\perp \cap \Omega}$. For property 4, we see that by property 3, $\mathbf{P}_\Omega \mathbf{P}_{\Omega_1^\perp} = (\mathbf{P}_{\Omega_1} + \mathbf{P}_{\Omega_1^\perp \cap \Omega})\mathbf{P}_{\Omega_1^\perp} = \mathbf{P}_{\Omega_1^\perp \cap \Omega}\mathbf{P}_{\Omega_1^\perp} = \mathbf{P}_{\Omega_1^\perp \cap \Omega}$. ■

Result 2.6.6 considers the case where one subspace is contained in another subspace. For two arbitrary subspaces, the following result holds.

Result 2.6.7. Suppose Ω_1 and Ω_2 are two arbitrary subspaces of $\mathcal{R}^n$.

1. $(\Omega_1 \cap \Omega_2)^\perp = \Omega_1^\perp + \Omega_2^\perp$.
2. $(\Omega_1^\perp + \Omega_2) \cap \Omega_1 = \mathbf{P}_{\Omega_1}\Omega_2$.
3. $\Omega_1 + \Omega_2 = \Omega_1 \oplus \mathbf{P}_{\Omega_1^\perp}\Omega_2$.

Proof. 1. Since $(\Omega_1^\perp + \Omega_2^\perp)^\perp \subset (\Omega_i^\perp)^\perp = \Omega_i$, $i = 1, 2$, $(\Omega_1^\perp + \Omega_2^\perp)^\perp \subset \Omega_1 \cap \Omega_2$, so $\Omega_1^\perp + \Omega_2^\perp \supset (\Omega_1 \cap \Omega_2)^\perp$. On the other hand, if $\mathbf{x}_i \perp \Omega_i$, $i = 1, 2$, then $\mathbf{x}_i \perp \Omega_1 \cap \Omega_2$. As a result $\mathbf{x}_1 + \mathbf{x}_2 \perp \Omega_1 \cap \Omega_2$. This gives $\Omega_1^\perp + \Omega_2^\perp \subset (\Omega_1 \cap \Omega_2)^\perp$.

2. Let $\mathbf{x} = \mathbf{y} + \mathbf{z} \in (\Omega_1^\perp + \Omega_2) \cap \Omega_1$, where $\mathbf{y} \in \Omega_1^\perp$ and $\mathbf{z} \in \Omega_2$. Then, $\mathbf{x} - \mathbf{P}_{\Omega_1}\mathbf{z} = \mathbf{y} + \mathbf{P}_{\Omega_1^\perp}\mathbf{z}$. Since $\mathbf{x} - \mathbf{P}_{\Omega_1}\mathbf{z} \in \Omega_1$ while $\mathbf{y} + \mathbf{P}_{\Omega_1^\perp}\mathbf{z} \in \Omega_1^\perp$, they must be zero, so that $\mathbf{x} = \mathbf{P}_{\Omega_1}\mathbf{z} \in \mathbf{P}_{\Omega_1}\Omega_2$. Thus, $(\Omega_1^\perp + \Omega_2) \cap \Omega_1 \subset \mathbf{P}_{\Omega_1}\Omega_2$. On the other hand, for any $\mathbf{x} \in \Omega_2$, $\mathbf{P}_{\Omega_1}\mathbf{x}$ is clearly in Ω_1, while $\mathbf{P}_{\Omega_1}\mathbf{x} = -\mathbf{P}_{\Omega_1^\perp}\mathbf{x} + \mathbf{x} \in \Omega_1^\perp + \Omega_2$. Then, $\mathbf{P}_{\Omega_1}\mathbf{x} \in (\Omega_1^\perp + \Omega_2) \cap \Omega_1$, and hence $\mathbf{P}_{\Omega_1}\Omega_2 \subset (\Omega_1^\perp + \Omega_2) \cap \Omega_1$.

3. Since $\mathbf{P}_{\Omega_1^\perp}\Omega_2 \subset \Omega_1^\perp$, the sum of Ω_1 and $\mathbf{P}_{\Omega_1^\perp}\Omega_2$ is direct. For every $\mathbf{x} \in \Omega_1$ and $\mathbf{y} \in \Omega_2$, $\mathbf{x} + \mathbf{P}_{\Omega_1^\perp}\mathbf{y} = (\mathbf{x} - \mathbf{P}_{\Omega_1}\mathbf{y}) + \mathbf{y} \in \Omega_1 + \Omega_2$. Then, $\Omega_1 \oplus \mathbf{P}_{\Omega_1^\perp}\Omega_2 \subset \Omega_1 + \Omega_2$. On the other hand, $\mathbf{x} + \mathbf{y} = (\mathbf{x} + \mathbf{P}_{\Omega_1}\mathbf{y}) + \mathbf{P}_{\Omega_1^\perp}\mathbf{y} \in \Omega_1 \oplus \mathbf{P}_{\Omega_1^\perp}\Omega_2$, so that $\Omega_1 + \Omega_2 \subset \Omega_1 \oplus \mathbf{P}_{\Omega_1^\perp}\Omega_2$. Then, $\Omega_1 + \Omega_2 = \Omega_1 \oplus \mathbf{P}_{\Omega_1^\perp}\Omega_2$. ■

Property 3 of Result 2.6.7 is related to Example 2.1.3. Let $\mathbf{A} = (\mathbf{A}_1, \mathbf{A}_2)$ be a matrix with full column rank. If $\Omega_i = \mathcal{C}(\mathbf{A}_i)$, $i = 1, 2$, then $\mathcal{C}(\mathbf{A}) = \Omega_1 + \Omega_2$. Let $\tilde{\Omega} = \mathbf{P}_{\Omega_1^\perp}\Omega_2$. Since $\Omega_1 \perp \tilde{\Omega}$, property 3 implies $\mathbf{P}_{\mathcal{C}(\mathbf{A})} = \mathbf{P}_{\Omega_1} + \mathbf{P}_{\tilde{\Omega}}$. On the other hand, by Result 2.6.1, the matrices in Example 2.1.3 are $\mathbf{P} = \mathbf{P}_{\mathcal{C}(\mathbf{A})}$, $\mathbf{P}_1 = \mathbf{P}_{\mathcal{C}(\mathbf{A}_1)} = \mathbf{P}_{\Omega_1}$, and $\mathbf{P}_2 = \mathbf{P}_{\mathcal{C}(\mathbf{B})}$, with $\mathbf{B} = (\mathbf{I} - \mathbf{P}_1)\mathbf{A}_2 = \mathbf{P}_{\Omega_1^\perp}\mathbf{A}_2$. Since $\mathcal{C}(\mathbf{B})$ is exactly $\mathbf{P}_{\Omega_1^\perp}\Omega_2$, we see that property 3 leads to the same conclusion as Example 2.1.3, using geometric rather than algebraic reasoning.

2.7 Vector and matrix differentiation

We define derivatives of vectors, matrices, products of vectors and matrices, as well as scalar functions of vectors and matrices, results which commonly appear in linear model theory; see Magnus and Neudecker (1988) for more details. We assume throughout that all the derivatives exist and are continuous.

Definition 2.7.1. Let $\boldsymbol{\beta} = (\beta_1, \cdots, \beta_k)'$ be a k-dimensional vector and let $f(\boldsymbol{\beta})$ denote a scalar function of $\boldsymbol{\beta}$. The first partial derivative of f with respect to $\boldsymbol{\beta}$ is defined to be the k-dimensional vector of partial derivatives $\partial f/\partial \beta_i$:

$$\frac{\partial f(\boldsymbol{\beta})}{\partial \boldsymbol{\beta}} = \frac{\partial f}{\partial \boldsymbol{\beta}} = \begin{pmatrix} \partial f/\partial \beta_1 \\ \partial f/\partial \beta_2 \\ \vdots \\ \partial f/\partial \beta_k \end{pmatrix}. \tag{2.7.1}$$

Also,

$$\frac{\partial f}{\partial \boldsymbol{\beta}'} = (\partial f/\partial \beta_1, \quad \partial f/\partial \beta_2, \quad \cdots, \quad \partial f/\partial \beta_k). \tag{2.7.2}$$

Definition 2.7.2. The second partial derivative of f with respect to $\boldsymbol{\beta}$ is defined to be the $k \times k$ matrix of partial derivatives $\partial^2 f/\partial \beta_i \partial \beta_j = \partial^2 f/\partial \beta_j \partial \beta_i$:

$$\frac{\partial^2 f}{\partial \boldsymbol{\beta} \partial \boldsymbol{\beta}'} = \begin{pmatrix} \partial^2 f/\partial \beta_1^2 & \partial^2 f/\partial \beta_1 \partial \beta_2 & \cdots & \partial^2 f/\partial \beta_1 \partial \beta_k \\ \partial^2 f/\partial \beta_1 \partial \beta_2 & \partial^2 f/\partial \beta_2^2 & \cdots & \partial^2 f/\partial \beta_2 \partial \beta_k \\ \vdots & \vdots & \cdots & \vdots \\ \partial^2 f/\partial \beta_1 \partial \beta_k & \partial^2 f/\partial \beta_2 \partial \beta_k & \cdots & \partial^2 f/\partial \beta_k^2 \end{pmatrix}. \tag{2.7.3}$$

Example 2.7.1. Let $\boldsymbol{\beta} = (\beta_1, \beta_2)'$ and let $f(\boldsymbol{\beta}) = (\beta_1^2 - 2\beta_1\beta_2)$. Then,

$$\partial f/\partial \boldsymbol{\beta}' = (2\beta_1 - 2\beta_2, \quad -2\beta_1)$$

and

$$\partial^2 f/\partial \boldsymbol{\beta} \partial \boldsymbol{\beta}' = \begin{pmatrix} 2 & -2 \\ -2 & 0 \end{pmatrix}.$$ □

Result 2.7.1. Let f and g represent scalar functions of a k-dimensional vector $\boldsymbol{\beta}$, and let a and b be real constants. Then

$$\begin{aligned} \partial(af + bg)/\partial \beta_j &= a\partial f/\partial \beta_j + b\partial g/\partial \beta_j \\ \partial(fg)/\partial \beta_j &= f\partial g/\partial \beta_j + g\partial f/\partial \beta_j \\ \partial(f/g)/\partial \beta_j &= (1/g^2)(g\partial f/\partial \beta_j - f\partial g/\partial \beta_j). \end{aligned} \tag{2.7.4}$$

Definition 2.7.3. Let $\mathbf{A} = \{a_{ij}\}$ be an $m \times n$ matrix and let $f(\mathbf{A})$ be a real function of $\mathbf{A}$. The first partial derivative of f with respect to $\mathbf{A}$ is defined as the $m \times n$ matrix of partial derivatives $\partial f/\partial a_{ij}$:

$$\begin{aligned} \partial f(\mathbf{A})/\partial \mathbf{A} &= \{\partial f/\partial a_{ij}\}, i = 1, \cdots, m, j = 1, \cdots, n \\ &= \begin{pmatrix} \partial f/\partial a_{11} & \partial f/\partial a_{12} & \cdots & \partial f/\partial a_{1n} \\ \vdots & \vdots & & \vdots \\ \partial f/\partial a_{m1} & \partial f/\partial a_{m2} & \cdots & \partial f/\partial a_{mn} \end{pmatrix}. \end{aligned} \tag{2.7.5}$$

The results that follow give rules for finding partial derivatives of vector or matrix functions of matrices and vectors.

Result 2.7.2. Let $\boldsymbol{\beta} \in \mathcal{R}^n$ and let $\mathbf{A} \in \mathcal{R}^{m \times n}$. Then,

$$\partial \mathbf{A}\boldsymbol{\beta}/\partial \boldsymbol{\beta}' = \mathbf{A}, \text{ and } \partial \boldsymbol{\beta}'\mathbf{A}'/\partial \boldsymbol{\beta} = \mathbf{A}'. \tag{2.7.6}$$

Proof. We may write

$$\mathbf{A}\boldsymbol{\beta} = \begin{pmatrix} a_{11}\beta_1 + \cdots + a_{1n}\beta_n \\ a_{21}\beta_1 + \cdots + a_{2n}\beta_n \\ \vdots \\ a_{m1}\beta_1 + \cdots + a_{mn}\beta_n \end{pmatrix},$$

so that by Definition 2.7.2, $\partial \mathbf{A}\boldsymbol{\beta}/\partial \boldsymbol{\beta}'$ is given by

$$\begin{pmatrix} \partial(a_{11}\beta_1 + \cdots + a_{1n}\beta_n)/\partial\beta_1 & \cdots & \partial(a_{11}\beta_1 + \cdots + a_{1n}\beta_n)/\partial\beta_n \\ \partial(a_{21}\beta_1 + \cdots + a_{2n}\beta_n)/\partial\beta_1 & \cdots & \partial(a_{21}\beta_1 + \cdots + a_{2n}\beta_n)/\partial\beta_n \\ \vdots & & \vdots \\ \partial(a_{m1}\beta_1 + \cdots + a_{mn}\beta_n)/\partial\beta_1 & \cdots & \partial(a_{m1}\beta_1 + \cdots + a_{mn}\beta_n)/\partial\beta_n \end{pmatrix}$$

$$= \begin{pmatrix} a_{11} & \cdots & a_{1n} \\ \vdots & \vdots & \vdots \\ a_{m1} & & a_{mn} \end{pmatrix} = \mathbf{A}. \tag{2.7.7}$$

That $\partial \boldsymbol{\beta}'\mathbf{A}'/\partial \boldsymbol{\beta} = \mathbf{A}'$ follows by transposing both sides of (2.7.7). ■

Result 2.7.3. Let $\boldsymbol{\beta} \in \mathcal{R}^n$ and $\mathbf{A} \in \mathcal{R}^{n \times n}$. Then,

$$\begin{aligned} \partial \boldsymbol{\beta}'\mathbf{A}\boldsymbol{\beta}/\partial \boldsymbol{\beta} &= (\mathbf{A} + \mathbf{A}')\boldsymbol{\beta} \\ \partial \boldsymbol{\beta}'\mathbf{A}\boldsymbol{\beta}/\partial \boldsymbol{\beta}' &= \boldsymbol{\beta}'(\mathbf{A} + \mathbf{A}') \\ \partial^2 \boldsymbol{\beta}'\mathbf{A}\boldsymbol{\beta}/\partial \boldsymbol{\beta}\, \partial \boldsymbol{\beta}' &= \mathbf{A} + \mathbf{A}'. \end{aligned} \tag{2.7.8}$$

Further, if $\mathbf{A}$ is a symmetric matrix,

$$\begin{aligned} \partial \boldsymbol{\beta}'\mathbf{A}\boldsymbol{\beta}/\partial \boldsymbol{\beta} &= 2\mathbf{A}\boldsymbol{\beta} \\ \partial \boldsymbol{\beta}'\mathbf{A}\boldsymbol{\beta}/\partial \boldsymbol{\beta}' &= 2\boldsymbol{\beta}'\mathbf{A} \\ \partial^2 \boldsymbol{\beta}'\mathbf{A}\boldsymbol{\beta}/\partial \boldsymbol{\beta}\, \partial \boldsymbol{\beta}' &= 2\mathbf{A}. \end{aligned} \tag{2.7.9}$$

Proof. We prove the result for a symmetric matrix $\mathbf{A}$. Clearly,

$$\boldsymbol{\beta}'\mathbf{A}\boldsymbol{\beta} = \sum_{i,j=1}^{n} a_{ij}\beta_i\beta_j$$

so that

$$\begin{aligned} \frac{\partial \boldsymbol{\beta}'\mathbf{A}\boldsymbol{\beta}}{\partial \beta_r} &= \sum_{\substack{j=1 \\ r \neq j}}^{n} a_{rj}\beta_j + \sum_{\substack{i=1 \\ r \neq i}}^{n} a_{ir}\beta_i + 2a_{rr}\beta_r \\ &= \sum_{j=1}^{n} a_{rj}\beta_j \qquad \text{(by symmetry of } \mathbf{A}) \\ &= 2\mathbf{a}'_r\boldsymbol{\beta} \end{aligned}$$

where $\mathbf{a}'_r$ denotes the rth row vector of $\mathbf{A}$. By Definition 2.7.3, we get

$$\frac{\partial \boldsymbol{\beta}'\mathbf{A}\boldsymbol{\beta}}{\partial \boldsymbol{\beta}} = \begin{pmatrix} \partial \boldsymbol{\beta}'\mathbf{A}\boldsymbol{\beta}/\partial\beta_1 \\ \partial \boldsymbol{\beta}'\mathbf{A}\boldsymbol{\beta}/\partial\beta_2 \\ \vdots \\ \partial \boldsymbol{\beta}'\mathbf{A}\boldsymbol{\beta}/\partial\beta_n \end{pmatrix} = 2\begin{pmatrix} \mathbf{a}'_1 \\ \mathbf{a}'_2 \\ \vdots \\ \mathbf{a}'_n. \end{pmatrix}\boldsymbol{\beta} = 2\mathbf{A}\boldsymbol{\beta} \tag{2.7.10}$$

The second result follows by transposing both sides of (2.7.10). To show the last result, we again take the first partial derivative of $\partial \boldsymbol{\beta}'\mathbf{A}\boldsymbol{\beta}/\partial \boldsymbol{\beta}' = 2\boldsymbol{\beta}'\mathbf{A}$, and use Result 2.7.2. ■

Result 2.7.4. Let $\mathbf{C} \in \mathcal{R}^{m\times n}$, $\boldsymbol{\alpha} \in \mathcal{R}^m$, and $\boldsymbol{\beta} \in \mathcal{R}^n$. Then,

$$\frac{\partial \boldsymbol{\alpha}'\mathbf{C}\boldsymbol{\beta}}{\partial \mathbf{C}} = \boldsymbol{\alpha}\boldsymbol{\beta}'. \tag{2.7.11}$$

Proof. Since

$$\boldsymbol{\alpha}'\mathbf{C}\boldsymbol{\beta} = \sum_{i=1}^{m}\sum_{j=1}^{n} c_{ij}\alpha_i\beta_j,$$

we have

$$\partial(\boldsymbol{\alpha}'\mathbf{C}\boldsymbol{\beta})/\partial c_{kl} = \alpha_k\beta_l,$$

which is the (k,l)th element of $\boldsymbol{\alpha}\boldsymbol{\beta}'$, from which the result follows. ■

We next give a few useful results without proof.

Result 2.7.5. Let $\mathbf{A}$ be an $n \times n$ matrix. Then,

$$\frac{\partial\, \mathrm{tr}(\mathbf{A})}{\partial \mathbf{A}} = \mathbf{I}_n \quad \text{and} \quad \frac{\partial |\mathbf{A}|}{\partial \mathbf{A}} = [\mathrm{Adj}(\mathbf{A})], \tag{2.7.12}$$

where $\mathrm{Adj}(\mathbf{A})$ denotes the adjoint of $\mathbf{A}$.

Result 2.7.6. Suppose $\mathbf{A}$ is an $n \times n$ matrix with $|\mathbf{A}| > 0$. Then

$$\frac{\partial \ln |\mathbf{A}|}{\partial \mathbf{A}} = (\mathbf{A}')^{-1}. \tag{2.7.13}$$

Result 2.7.7. Let $\mathbf{A} \in \mathcal{R}^{m\times n}$ and $\mathbf{B} \in \mathcal{R}^{n\times m}$. Then,

$$\frac{\partial\, \mathrm{tr}(\mathbf{AB})}{\partial \mathbf{A}} = \mathbf{B}'. \tag{2.7.14}$$

Result 2.7.8. Let $\boldsymbol{\Omega}$ be a symmetric matrix, $\mathbf{y} \in \mathcal{R}^n$ $\boldsymbol{\beta} \in \mathcal{R}^k$, and $\mathbf{X} \in \mathcal{R}^{n\times k}$. Then,

$$\begin{aligned} \partial(\mathbf{y}-\mathbf{X}\boldsymbol{\beta})'\boldsymbol{\Omega}(\mathbf{y}-\mathbf{X}\boldsymbol{\beta})/\partial\boldsymbol{\beta} &= -2\mathbf{X}'\boldsymbol{\Omega}(\mathbf{y}-\mathbf{X}\boldsymbol{\beta}), \quad \text{and} \\ \partial^2(\mathbf{y}-\mathbf{X}\boldsymbol{\beta})'\boldsymbol{\Omega}(\mathbf{y}-\mathbf{X}\boldsymbol{\beta})/\partial\boldsymbol{\beta}\partial\boldsymbol{\beta}' &= 2\mathbf{X}'\boldsymbol{\Omega}\mathbf{X}. \end{aligned} \tag{2.7.15}$$

The next definition deals with partial derivatives of a matrix (or a vector) with respect to some scalar θ. We see that in this case, the partial derivative is itself a matrix or vector of the same dimension whose elements are the partial derivatives with respect to θ of each element of that matrix or vector.

Definition 2.7.4. Let $\mathbf{A} \in \mathcal{R}^{m\times n}$ be a function of a scalar θ, then

$$\begin{aligned} \partial\mathbf{A}/\partial\theta &= \{\partial a_{ij}/\partial\theta\}, \quad i=1,\cdots,m, \; j=1,\cdots,n \\ &= \begin{pmatrix} \partial a_{11}/\partial\theta & \partial a_{12}/\partial\theta & \cdots & \partial a_{1n}/\partial\theta \\ \vdots & \vdots & \ddots & \vdots \\ \partial a_{m1}/\partial\theta & \partial a_{m2}/\partial\theta & \cdots & \partial a_{mn}/\partial\theta \end{pmatrix}. \end{aligned} \tag{2.7.16}$$

2.8 Special operations on matrices

Definition 2.8.1. Kronecker product of matrices. Let $\mathbf{A} = \{a_{ij}\}$ be an $m \times n$ matrix and $\mathbf{B} = \{b_{ij}\}$ be a $p \times q$ matrix. The Kronecker product of $\mathbf{A}$ and $\mathbf{B}$ is denoted by $\mathbf{A} \otimes \mathbf{B}$ and is defined to be the $mp \times nq$ matrix

$$\mathbf{A} \otimes \mathbf{B} = \begin{pmatrix} a_{11}\mathbf{B} & a_{12}\mathbf{B} & \cdots & a_{1n}\mathbf{B} \\ a_{21}\mathbf{B} & a_{22}\mathbf{B} & \cdots & a_{2n}\mathbf{B} \\ \vdots & \vdots & \vdots & \vdots \\ a_{m1}\mathbf{B} & a_{m2}\mathbf{B} & \cdots & a_{mn}\mathbf{B} \end{pmatrix}. \tag{2.8.1}$$

The matrix in (2.8.1) is a partitioned matrix whose (i, j)th entry is a $p \times q$ submatrix $a_{ij}\mathbf{B}$. The Kronecker product $\mathbf{A} \otimes \mathbf{B}$ can be defined regardless of the dimensions of $\mathbf{A}$ and $\mathbf{B}$. The Kronecker product is also referred to in the literature as the *direct product* or the *tensor product.*

Example 2.8.1. Consider two matrices $\mathbf{A}$ and $\mathbf{B}$ where

$$\mathbf{A} = \begin{pmatrix} 3 & 4 & -1 \\ 2 & 0 & 0 \end{pmatrix}, \quad \text{and} \quad \mathbf{B} = \begin{pmatrix} 5 & -1 \\ 3 & 3 \end{pmatrix}.$$

Then,

$$\mathbf{A} \otimes \mathbf{B} = \begin{pmatrix} 15 & -3 & 20 & -4 & -5 & 1 \\ 9 & 9 & 12 & 12 & -3 & -3 \\ 10 & -2 & 0 & 0 & 0 & 0 \\ 6 & 6 & 0 & 0 & 0 & 0 \end{pmatrix},$$

and

$$\mathbf{B} \otimes \mathbf{A} = \begin{pmatrix} 15 & 20 & -5 & -3 & -4 & 1 \\ 10 & 0 & 0 & -2 & 0 & 0 \\ 9 & 12 & -3 & 9 & 12 & -3 \\ 6 & 0 & 0 & 6 & 0 & 0 \end{pmatrix}.$$ □

In general, $\mathbf{A} \otimes \mathbf{B}$ is not equal to $\mathbf{B} \otimes \mathbf{A}$. The elements in these two products are the same, except that they are in different positions. The definition $\mathbf{A} \otimes \mathbf{B}$ extends naturally to more than two matrices:

$$\mathbf{A} \otimes \mathbf{B} \otimes \mathbf{C} = \mathbf{A} \otimes (\mathbf{B} \otimes \mathbf{C}) \quad \text{and} \quad \bigotimes_{i=1}^{k} \mathbf{A}_i = \mathbf{A}_1 \otimes \mathbf{A}_2 \otimes \cdots \otimes \mathbf{A}_k. \tag{2.8.2}$$

Result 2.8.1. Properties of Kronecker product. Let $\mathbf{A}$ be an $m \times n$ matrix. Then

1. For a positive scalar c, we have $c \otimes \mathbf{A} = \mathbf{A} \otimes c = c\mathbf{A}$.

2. For any diagonal matrix $D = \text{diag}(d_1, \cdots, d_k)$,

$$\mathbf{D} \otimes \mathbf{A} = \text{diag}(d_1\mathbf{A}, \cdots, d_k\mathbf{A}).$$

3. $\mathbf{I} \otimes \mathbf{A} = \text{diag}(\mathbf{A}, \mathbf{A}, \cdots, \mathbf{A})$.

4. $\mathbf{I}_m \otimes \mathbf{I}_p = \mathbf{I}_{mp}$.

5. For a $p \times q$ matrix $\mathbf{B}$, we have $(\mathbf{A} \otimes \mathbf{B})' = \mathbf{A}' \otimes \mathbf{B}'$.

6. $(\mathbf{A} \otimes \mathbf{B})(\mathbf{C} \otimes \mathbf{D}) = (\mathbf{AC}) \otimes (\mathbf{BD})$, where we assume that relevant matrices are conformal for multiplication.

7. $\mathrm{r}(\mathbf{A} \otimes \mathbf{B}) = \mathrm{r}(\mathbf{A})\,\mathrm{r}(\mathbf{B})$.

8. $(\mathbf{A} + \mathbf{B}) \otimes (\mathbf{C} + \mathbf{D}) = (\mathbf{A} \otimes \mathbf{C}) + (\mathbf{A} \otimes \mathbf{D}) + (\mathbf{B} \otimes \mathbf{C}) + (\mathbf{B} \otimes \mathbf{D})$.

9. Suppose $\mathbf{A}$ is an $n \times n$ matrix, and $\mathbf{B}$ is an $m \times m$ matrix. The nm eigenvalues of $\mathbf{A} \otimes \mathbf{B}$ are products of the n eigenvalues $\lambda_i, i = 1, \cdots, n$ of $\mathbf{A}$ and the m eigenvalues $\gamma_j, j = 1, \cdots, m$ of $\mathbf{B}$.

10. $|\mathbf{A} \otimes \mathbf{B}| = |\mathbf{A}|^m |\mathbf{B}|^n = (\prod_{i=1}^n \lambda_i)^m (\prod_{j=1}^m \gamma_j)^n$.

11. Provided all the inverses exist, $(\mathbf{A} \otimes \mathbf{B})^{-1} = \mathbf{A}^{-1} \otimes \mathbf{B}^{-1}$.

Definition 2.8.2. Vectorization of matrices. Given an $m \times n$ matrix $\mathbf{A}$ with columns $\mathbf{a}_1, \cdots, \mathbf{a}_n$, we define $\mathrm{vec}(\mathbf{A}) = (\mathbf{a}_1', \cdots, \mathbf{a}_n')'$ to be an mn-dimensional column vector.

Result 2.8.2. Properties of the vec operator.

1. Given $m \times n$ matrices $\mathbf{A}$ and $\mathbf{B}$, $\mathrm{vec}(\mathbf{A} + \mathbf{B}) = \mathrm{vec}(\mathbf{A}) + \mathrm{vec}(\mathbf{B})$.

2. If $\mathbf{A}$, $\mathbf{B}$ and $\mathbf{C}$ are respectively $m \times n$, $n \times p$ and $p \times q$ matrices, then

   (i) $\mathrm{vec}(\mathbf{AB}) = (\mathbf{I}_p \otimes \mathbf{A})\,\mathrm{vec}(\mathbf{B}) = (\mathbf{B}' \otimes \mathbf{I}_m)\,\mathrm{vec}(\mathbf{A})$.

   (ii) $\mathrm{vec}(\mathbf{ABC}) = (\mathbf{C}' \otimes \mathbf{A})\,\mathrm{vec}(\mathbf{B})$.

   (iii) $\mathrm{vec}(\mathbf{ABC}) = (\mathbf{I}_q \otimes \mathbf{AB})\,\mathrm{vec}(\mathbf{C}) = (\mathbf{C}'\mathbf{B}' \otimes \mathbf{I}_n)\,\mathrm{vec}(\mathbf{A})$.

3. If $\mathbf{A}$ is $m \times n$ and $\mathbf{B}$ is $n \times m$,
$$\mathrm{vec}(\mathbf{B}')'\,\mathrm{vec}(\mathbf{A}) = \mathrm{vec}(\mathbf{A}')'\,\mathrm{vec}(\mathbf{B}) = \mathrm{tr}(\mathbf{AB}).$$

4. If $\mathbf{A}$, $\mathbf{B}$ and $\mathbf{C}$ are respectively $m \times n$, $n \times p$ and $p \times m$ matrices,
$$\begin{aligned}
\mathrm{tr}(\mathbf{ABC}) &= \mathrm{vec}(\mathbf{A}')'(\mathbf{C}' \otimes \mathbf{I}_n)\,\mathrm{vec}(\mathbf{B}) \\
&= \mathrm{vec}(\mathbf{A}')'(\mathbf{I}_m \otimes \mathbf{B})\,\mathrm{vec}(\mathbf{C}) = \mathrm{vec}(\mathbf{B}')'(\mathbf{A} \otimes \mathbf{I}_p)\,\mathrm{vec}(\mathbf{C}) \\
&= \mathrm{vec}(\mathbf{B}')'(\mathbf{I}_n \otimes \mathbf{C})\,\mathrm{vec}(\mathbf{A}) = \mathrm{vec}(\mathbf{C}')'(\mathbf{B}' \otimes \mathbf{I}_m)\,\mathrm{vec}(\mathbf{A}) \\
&= \mathrm{vec}(\mathbf{C}')'(\mathbf{I}_p \otimes \mathbf{A})\,\mathrm{vec}(\mathbf{B}).
\end{aligned}$$

Definition 2.8.3. Direct sum of matrices. The direct sum of two matrices $\mathbf{A}$ and $\mathbf{B}$ (which can be of any dimension) is defined as
$$\mathbf{A} \oplus \mathbf{B} = \begin{pmatrix} \mathbf{A} & \mathbf{O} \\ \mathbf{O} & \mathbf{B} \end{pmatrix}. \tag{2.8.3}$$

This operation extends naturally to more than two matrices:
$$\bigoplus_{i=1}^{k} \mathbf{A}_i = \mathbf{A}_1 \oplus \mathbf{A}_2 \oplus \cdots \oplus \mathbf{A}_k = \begin{pmatrix} \mathbf{A}_1 & \mathbf{O} & \cdots & \mathbf{O} \\ \mathbf{O} & \mathbf{A}_2 & \cdots & \mathbf{O} \\ \vdots & \vdots & \vdots & \vdots \\ \mathbf{O} & \mathbf{O} & \mathbf{O} & \mathbf{A}_k \end{pmatrix}. \tag{2.8.4}$$

This definition applies to vectors as well.

2.9 R Code

```
library(matlib)
library(matrixcalc)

## Result 2.4.7. Cholesky decomposition
A <- matrix(c(3, 4, 3, 4, 8, 6, 3, 6, 9), nrow = 3, byrow = 3)
A.chol <- chol(A)
A.chol
t(A.chol)
t(A.chol) %*% A.chol

## Result 2.2.2. QR decomposition
set.seed(1)
A <- matrix(rnorm(15), ncol = 3)
D <- qr(A)
qr.Q(D)
qr.R(D)
qr.X(D)
is.qr(D)

## Result 2.3.4. Spectral decomposition of a symmetric matrix
A <- matrix(c(13,-4, 2,-4, 11,-2, 2,-2, 8), 3, 3, byrow = TRUE)
ev <- eigen(A)
(L <- ev$values)
(V <- ev$vectors)
V %*% diag(L) %*% t(V)
diag(L)
zapsmall(t(V) %*% A %*% V)
A1 = L[1] * V[, 1] %*% t(V[, 1])
A2 = L[2] * V[, 2] %*% t(V[, 2])
A3 = L[3] * V[, 3] %*% t(V[, 3])
A1 + A2 + A3
all.equal(A, A1 + A2 + A3)

## Singular value decomposition; A=UDV'
A <- as.matrix(data.frame(c(4, 7, -1, 8), c(-5, -2, 4, 2), c(-1, 3, -3, 6)))
A
A.svd <- svd(A)
A.svd

## Example 2.8.1. Kronecker product of matrices
A <- matrix(c(3, 2, 4, 0, -1, 0), nrow = 2)
B <- matrix(c(5, 3, -1, 3), nrow = 2)
kronecker(A, B)
kronecker(B, A)

## Def. 2.8.2. Vectorization of matrices
A <- matrix(c(2, 9, 2, 6), nrow = 2)
c(A)

## Def. 2.8.3. Direct sum of matrices.
```

```
A <- matrix(c(3, 2, 4, 0,-1, 0), nrow = 2)
B <- matrix(c(5, 3,-1, 3), nrow = 2)
direct.sum(A, B)
```

Exercises

2.1. Let $\mathbf{P} = \begin{pmatrix}\mathbf{A} & \mathbf{B}\end{pmatrix}$, where $\mathbf{P}$ is an $n \times n$ orthogonal matrix and $\mathbf{A}$ is an $n \times m$ matrix. Show that $\mathbf{A}'\mathbf{A} = \mathbf{I}_m$ and $\mathbf{AA}'$ is an idempotent matrix.

2.2. Suppose an $m \times n$ matrix $\mathbf{M}$ is partitioned as $\mathbf{M} = \begin{pmatrix}\mathbf{A} & \mathbf{B} \\ \mathbf{C} & \mathbf{D}\end{pmatrix}$, where $\mathbf{A}$, $\mathbf{B}$, $\mathbf{C}$, and $\mathbf{D}$ are respectively $r \times r$, $r \times (n-r)$, $(n-r) \times r$, and $(n-r) \times (n-r)$ submatrices, such that $\mathrm{r}(\mathbf{A}) = \mathrm{r}(\mathbf{M}) = r > 0$. Show that $\mathbf{D} = \mathbf{CA}^{-1}\mathbf{B}$.

2.3. Suppose an $n \times n$ matrix $\mathbf{A}$ is partitioned as $\begin{pmatrix}\mathbf{P} & \mathbf{x} \\ \mathbf{x}' & 1\end{pmatrix}$, where $\mathbf{P}$ is an $(n-1) \times (n-1)$ dimensional nonsingular matrix, and $\mathbf{x}$ is an $(n-1)$-dimensional vector. Show that $\mathbf{x}'\mathbf{P}^{-1}\mathbf{x} = 1 - |\mathbf{P} - \mathbf{xx}'|/|\mathbf{P}|$.

2.4. Let $\mathbf{A}$, $\mathbf{B}$, $\mathbf{C}$, and $\mathbf{D}$ be respectively $m \times p$, $p \times q$, $n \times m$, and $n \times q$ matrices, and let $\mathbf{E} = \mathbf{D} - \mathbf{CAB}$.

(a) Show that $r\begin{pmatrix}\mathbf{O} & \mathbf{B} \\ \mathbf{A} & \mathbf{O}\end{pmatrix} = r\begin{pmatrix}\mathbf{B} & \mathbf{O} \\ \mathbf{O} & \mathbf{A}\end{pmatrix} = \mathrm{r}(\mathbf{A}) + \mathrm{r}(\mathbf{B})$.

(b) Show that $r\begin{pmatrix}\mathbf{A} & \mathbf{AB} \\ \mathbf{CA} & \mathbf{D}\end{pmatrix} = r\begin{pmatrix}\mathbf{D} & \mathbf{CA} \\ \mathbf{AB} & \mathbf{A}\end{pmatrix} = \mathrm{r}(\mathbf{A}) + \mathrm{r}(\mathbf{E})$.

2.5. Consider the partition in Example 2.1.3, and suppose that $\mathbf{A}_2$ contains only one column, i.e., $r = k - 1$. Show that

$$\mathbf{P} = \mathbf{P}_1 + \frac{(\mathbf{I} - \mathbf{P}_1)\mathbf{A}_2\mathbf{A}_2'(\mathbf{I} - \mathbf{P}_1)}{\mathbf{A}_2'(\mathbf{I} - \mathbf{P}_1)\mathbf{A}_2}.$$

2.6. (Rao and Toutenberg (1995), p. 295) Let $\mathbf{A}$ be an $n \times k$ matrix and $\mathbf{B}$ be a $k \times n$ matrix with $n \geq k$.

(a) Show that $\begin{vmatrix}-\lambda\mathbf{I}_n & -\mathbf{A} \\ \mathbf{B} & \mathbf{I}_k\end{vmatrix} = (-\lambda)^{n-k}|\mathbf{BA} - \lambda\mathbf{I}_k| = |\mathbf{AB} - \lambda\mathbf{I}_n|$.

(b) Hence, show that the n eigenvalues of $\mathbf{AB}$ are equal to the k eigenvalues of $\mathbf{BA}$, together with the eigenvalue 0, which has a multiplicity of $(n - k)$.

(c) Let $\mathbf{v}$ denote a nonzero eigenvector corresponding to a nonzero eigenvalue λ of $\mathbf{AB}$. Show that $\mathbf{u} = \mathbf{Bv}$ is a nonzero eigenvector of $\mathbf{BA}$ corresponding to this λ.

2.7. Let $\mathbf{A} = \mathbf{aa}'$, where $\mathbf{a}$ is a nonzero n-dimensional vector. Show that the only nonzero eigenvalue of $\mathbf{A}$ is $\lambda = \mathbf{a}'\mathbf{a}$ with corresponding eigenvector $\mathbf{a}$.

2.8. Let $\mathbf{A}$ be a $k \times k$ nonsingular matrix, and let $\mathbf{B}$ and $\mathbf{C}$ be respectively $k \times n$ and $n \times k$ matrices. Show that $|\mathbf{A} + \mathbf{BC}| = |\mathbf{A}||\mathbf{I}_n + \mathbf{CA}^{-1}\mathbf{B}|$.

2.9. Use the Jordan canonical form in Result 2.2.6 to verify Result 1.3.18.

2.10. Show that $|\mathbf{A}+\mathbf{a}\mathbf{a}'| = (1+\mathbf{a}'\mathbf{A}^{-1}\mathbf{a})|\mathbf{A}|$, where $\mathbf{A}$ is a $k \times k$ nonsingular matrix and $\mathbf{a}$ is a k-dimensional vector.

2.11. Let $\mathbf{A}$ be an $n \times n$ symmetric matrix. Show that there exist nonzero vectors $\mathbf{x}_1$ and $\mathbf{x}_2$ such that

$$\frac{\mathbf{x}_1'\mathbf{A}\mathbf{x}_1}{\mathbf{x}_1'\mathbf{x}_1} \le \frac{\mathbf{x}'\mathbf{A}\mathbf{x}}{\mathbf{x}'\mathbf{x}} \le \frac{\mathbf{x}_2'\mathbf{A}\mathbf{x}_2}{\mathbf{x}_2'\mathbf{x}_2}$$

for every nonzero vector $\mathbf{x} \in \mathcal{R}^n$, and that $\mathbf{x}_1$ and $\mathbf{x}_2$ are respectively eigenvectors corresponding to the smallest and largest eigenvalues of $\mathbf{A}$.

2.12. Let $\mathbf{A}$ be an $n \times n$ symmetric matrix with eigenvalues $\lambda_1, \cdots, \lambda_n$.

(a) Show that $\mathrm{r}(\mathbf{A})$ is equal to the number of nonzero eigenvalues of $\mathbf{A}$.

(b) Show that $\|\mathbf{A}\| = (\sum_{i=1}^{n} \lambda_i^2)^{1/2}$.

2.13. Let $\mathbf{A}$ be an $n \times n$ symmetric matrix which has k distinct eigenvalues $\lambda_1, \cdots, \lambda_k$ with geometric multiplicities $g_1, \cdots, g_k$ and algebraic multiplicities $a_1, \cdots, a_k$. Show that $\sum_{j=1}^{k} g_j = \sum_{j=1}^{k} a_j = n$, and that $g_j = a_j$ for $j = 1, \cdots, k$.

2.14. Verify Result 2.2.4.

2.15. Let $\mathbf{x}$ be an n-dimensional nonzero vector. Is

(a) $\frac{\mathbf{x}\mathbf{x}'}{\mathbf{x}'\mathbf{x}}$ symmetric and idempotent?

(b) $\mathbf{x}\mathbf{x}'$ symmetric and idempotent?

2.16. Let $\mathbf{A}$ be an $n \times n$ symmetric matrix, let $\mathbf{B}$ and $\mathbf{U}$ be $k \times k$ and $n \times k$ matrices such that $\mathbf{U}'\mathbf{U} = \mathbf{I}_k$, and $\mathbf{A}\mathbf{U} = \mathbf{U}\mathbf{B}$. Show that there exists an $n \times (n-k)$ matrix $\mathbf{V}$ such that the $n \times n$ matrix $(\mathbf{U}\ \mathbf{V})$ is orthogonal and $(\mathbf{U}\ \mathbf{V})'\mathbf{A}(\mathbf{U}\ \mathbf{V}) = \begin{pmatrix} \mathbf{B} & \mathbf{O} \\ \mathbf{O} & \mathbf{V}'\mathbf{A}\mathbf{V} \end{pmatrix}$.

2.17. Show that every symmetric idempotent matrix is nonnegative definite.

2.18. Let $\mathbf{A}$ be an $n \times n$ matrix. Show that the nonzero eigenvalues of $\mathbf{A}\mathbf{A}'$ coincide with the nonzero eigenvalues of $\mathbf{A}'\mathbf{A}$.

2.19. Let $\mathbf{A}$ and $\mathbf{B}$ be respectively $n \times k$ and $k \times n$ matrices, with $n \ge k$.

(a) Show that the n eigenvalues of $\mathbf{A}\mathbf{B}$ are equal to the p eigenvalues of $\mathbf{B}\mathbf{A}$, together with the eigenvalue 0 with multiplicity $(n-k)$.

(b) If $\mathbf{x}$ is a nonzero eigenvector of $\mathbf{A}\mathbf{B}$ corresponding to a nonzero eigenvalue λ, show that $\mathbf{y} = \mathbf{B}\mathbf{x}$ is a nonzero eigenvector of $\mathbf{B}\mathbf{A}$ corresponding to λ.

2.20. Prove the following min-max and max-min characterizations of the eigenvalues of a symmetric matrix $\mathbf{A} \in \mathcal{R}^{n\times n}$:

$$\lambda_k = \inf_{\mathcal{V}\subset\mathcal{R}^n:\ \dim(\mathcal{V})=n-k+1}\ \max_{\mathbf{v}\in\mathcal{V}:\ \|\mathbf{v}\|=1} \mathbf{v}'\mathbf{A}\mathbf{v} \qquad \text{(min-max)}$$

$$= \sup_{\mathcal{V}\subset\mathcal{R}^n:\ \dim(\mathcal{V})=k}\ \min_{\mathbf{v}\in\mathcal{V}:\ \|\mathbf{v}\|=1} \mathbf{v}'\mathbf{A}\mathbf{v}, \qquad \text{(max-min)}$$

where $\lambda_1 \ge \lambda_2 \ge \cdots \ge \lambda_n$ are the eigenvalues of $\mathbf{A}$.

2.21. Let $\mathbf{Q}$ be an $n \times n$ nonsingular matrix which diagonalizes a nonsingular matrix $\mathbf{A}$. Show that $\mathbf{A}^{-1}$ is also diagonalized by $\mathbf{Q}$.

2.22. Let $\mathbf{A}$ be an $n \times n$ symmetric matrix of rank r. Suppose $\text{tr}(\mathbf{A}) = \text{tr}(\mathbf{A}^2) = p$.

(a) Show that $0 \leq p \leq r$.

(b) If $\text{tr}(\mathbf{A}) = r$, show that $\mathbf{A}$ must be idempotent of rank r.

2.23. Suppose $\mathbf{A}_1$ and $\mathbf{A}_2$ are respectively $n \times p$ and $n \times q$ matrices of ranks p and q, and suppose the columns of $\mathbf{A}_1$ are LIN of the columns of $\mathbf{A}_2$. Let $\mathbf{B} = \mathbf{I} - \mathbf{A}_1(\mathbf{A}_1'\mathbf{A}_1)^{-1}\mathbf{A}_1'$. Show that $\mathbf{C} = \mathbf{A}_2'\mathbf{B}\mathbf{A}_2$ is nonsingular.

2.24. Let $\mathbf{P}$ be an $n \times n$ orthogonal matrix and let $\mathbf{A}$ be an $n \times n$ symmetric and idempotent matrix. Show that the matrix $\mathbf{P}'\mathbf{A}\mathbf{P}$ is symmetric and idempotent.

2.25. Let $\mathbf{A} = \begin{pmatrix} 2 & -1 & -1 \\ -1 & 1 & 1 \\ -1 & 1 & 4 \end{pmatrix}$. Verify that $\mathbf{A}$ is a p.d. matrix. Find a matrix $\mathbf{B}$ such that $\mathbf{A} = \mathbf{B}\mathbf{B}'$.

2.26. Let $\mathbf{A} = (1-a)\mathbf{I}_n + a\mathbf{J}_n$. For what values of a is the matrix $\mathbf{A}$ p.d.?

2.27. Let $\mathbf{B}$ be an $n \times n$ symmetric p.d. matrix and let $\mathbf{A}$ be an $n \times n$ symmetric matrix. Show that there exists a positive real number λ such that $\mathbf{B} - \lambda\mathbf{A}$ is p.d.

2.28. Let $\mathbf{A}$ be an $n \times n$ symmetric matrix and $\mathbf{x}$ be an n-dimensional vector.

(a) If $\mathbf{x}'\mathbf{A}\mathbf{x} = 0$ for all $\mathbf{x}$, show that $\mathbf{A} = \mathbf{O}$.

(b) Let $\mathbf{A}$ be p.s.d. Show that if $\mathbf{x}'\mathbf{A}\mathbf{x} = 0$, then $\mathbf{A}\mathbf{x} = \mathbf{0}$.

2.29. Let $\mathbf{A}$ and $\mathbf{B}$ be $n \times n$ symmetric matrices such that $\mathbf{A} + \mathbf{B}$ is nonsingular. Let $\mathbf{x}$, $\mathbf{a}$, and $\mathbf{b}$ be n-dimensional vectors. If $\mathbf{c} = (\mathbf{A}+\mathbf{B})^{-1}(\mathbf{A}\mathbf{a}+\mathbf{B}\mathbf{b})$, show that $(\mathbf{x}-\mathbf{a})'\mathbf{A}(\mathbf{x}-\mathbf{a}) + (\mathbf{x}-\mathbf{b})'\mathbf{B}(\mathbf{x}-\mathbf{b}) = (\mathbf{x}-\mathbf{c})'(\mathbf{A}+\mathbf{B})(\mathbf{x}-\mathbf{c}) + (\mathbf{a}-\mathbf{b})'\mathbf{A}(\mathbf{A}+\mathbf{B})^{-1}\mathbf{B}(\mathbf{a}-\mathbf{b})$.

2.30. Let $\mathbf{A}$, $\mathbf{B}$, and $\mathbf{C}$ be respectively $m \times m$, $n \times n$, and $n \times m$ matrices, and suppose $\mathbf{A}$ and $\mathbf{B}$ are p.d. Show that

(a) $(\mathbf{A}^{-1} + \mathbf{C}'\mathbf{B}^{-1}\mathbf{C})^{-1} = \mathbf{A} - \mathbf{A}\mathbf{C}'(\mathbf{C}\mathbf{A}\mathbf{C}' + \mathbf{B})^{-1}\mathbf{C}\mathbf{A}$,

(b) $(\mathbf{A}^{-1} + \mathbf{C}'\mathbf{B}^{-1}\mathbf{C})^{-1}\mathbf{C}'\mathbf{B}^{-1} = \mathbf{A}\mathbf{C}'(\mathbf{C}\mathbf{A}\mathbf{C}' + \mathbf{B})^{-1}$.

2.31. Let $\mathbf{A}_1, \cdots, \mathbf{A}_k$ be $n \times n$ matrices.

(a) Show that a necessary condition for $\mathbf{A}_1, \cdots, \mathbf{A}_k$ to be simultaneously diagonalizable (by an $n \times n$ nonsingular matrix $\mathbf{P}$) is that $\mathbf{A}_j\mathbf{A}_i = \mathbf{A}_i\mathbf{A}_j$, $j > i = 1, \cdots, k$, i.e., $\mathbf{A}_1, \cdots, \mathbf{A}_k$ commute in pairs.

(b) If $\mathbf{A}_1, \cdots, \mathbf{A}_k$ are symmetric, show that the condition in (a) is necessary and sufficient.

2.32. Let $\mathbf{A}$ and $\mathbf{B}$ be $n \times n$ symmetric n.n.d. matrices. Show that there exists a nonsingular matrix $\mathbf{P}$ such that $\mathbf{P}'\mathbf{A}\mathbf{P}$ and $\mathbf{P}'\mathbf{B}\mathbf{P}$ are both diagonal.

2.33. For square matrices $\mathbf{A}$ and $\mathbf{B}$, show that $\text{tr}(\mathbf{A} \otimes \mathbf{B}) = \text{tr}(\mathbf{A})\,\text{tr}(\mathbf{B})$.

2.34. If $\mathbf{u} = \mathbf{P}_{\mathcal{V}}\mathbf{y}$, $\mathbf{y} \in \mathcal{R}^n$, and $\mathcal{V} \subset \mathcal{R}^n$, show that $\mathbf{P}_{\mathcal{V}}$ must be unique.

2.35. Verify Result 2.7.8.

2.36. Verify properties 9–11 of Result 2.8.1.

3

Generalized Inverses and Solutions to Linear Systems

The notion of a generalized inverse of a matrix has its origin in the theory of simultaneous linear equations. Any matrix $\mathbf{A}$ has at least one generalized inverse, which is a matrix $\mathbf{G}$ such that $\mathbf{Gb}$ is a solution to a set of consistent linear equations $\mathbf{Ax} = \mathbf{b}$ (Rao and Mitra, 1971). We give the definition and properties of generalized inverses of matrices in Section 3.1, while in Section 3.2, we discuss solutions to systems of linear equations. Section 3.3 describes unconstrained and constrained optimization. These topics are fundamental in the development of linear model theory.

3.1 Generalized inverses

Definition 3.1.1. A generalized inverse (g-inverse) of an $m \times n$ matrix $\mathbf{A}$ is any $n \times m$ matrix $\mathbf{G}$ which satisfies the relation

$$\mathbf{AGA} = \mathbf{A}. \tag{3.1.1}$$

The matrix $\mathbf{G}$ is also referred to in the literature as "conditional inverse" or "pseudo-inverse". We will refer to $\mathbf{G}$ as g-inverse and also denote it by $\mathbf{A}^-$ (pronounced $\mathbf{A}$ minus).

Result 3.1.1. A g-inverse $\mathbf{G}$ of a real matrix $\mathbf{A}$ always exists. In general, $\mathbf{G}$ is not unique except in the special case where $\mathbf{A}$ is a square nonsingular matrix.

Proof. To show the existence of $\mathbf{G}$, we recall that by Result 2.2.1, the full-rank factorization of an $m \times n$ matrix $\mathbf{A}$ with $\mathrm{r}(\mathbf{A}) = r$ is given by $\mathbf{A} = \mathbf{BC}$, where $\mathbf{B}$ and $\mathbf{C}$ are respectively $m \times r$ and $r \times n$ matrices, each with rank r. By Result 1.3.11, $\mathbf{B}'\mathbf{B}$ and $\mathbf{CC}'$ are nonsingular. It is easy to verify that the $n \times m$ matrix defined by

$$\mathbf{A}_1^- = \mathbf{C}'(\mathbf{CC}')^{-1}(\mathbf{B}'\mathbf{B})^{-1}\mathbf{B}' \tag{3.1.2}$$

satisfies (3.1.1) and is a g-inverse of $\mathbf{A}$. For arbitrary $n \times m$ matrices $\mathbf{E}$ and $\mathbf{F}$, let

$$\mathbf{A}_2^- = \mathbf{A}_1^-\mathbf{A}\mathbf{A}_1^- + (\mathbf{I}_n - \mathbf{A}_1^-\mathbf{A})\mathbf{E} + \mathbf{F}(\mathbf{I}_m - \mathbf{A}\mathbf{A}_1^-). \tag{3.1.3}$$

By direct multiplication, it is easy to verify that $\mathbf{A}_2^-$ satisfies (3.1.1). Inserting different matrices $\mathbf{E}$ and $\mathbf{F}$ into (3.1.3) generates an infinite number of g-inverses of $\mathbf{A}$, starting from $\mathbf{A}_1^-$. The only matrix $\mathbf{A}$ which has a unique g-inverse is a square nonsingular matrix. Use the notation $\mathbf{A}^-$ for $\mathbf{G}$ in (3.1.1), and pre-multiply and post-multiply both sides by $\mathbf{A}^{-1}$ to get

$$\mathbf{A}^{-1}\mathbf{A}\mathbf{A}^-\mathbf{A}\mathbf{A}^{-1} = \mathbf{A}^{-1}\mathbf{A}\mathbf{A}^{-1}, \quad \text{i.e.,} \quad \mathbf{A}^- = \mathbf{A}^{-1},$$

and complete the proof. ■

DOI: 10.1201/9781315156651-3

Result 3.1.2. Algorithm to compute $\mathbf{A}^-$. Let $\mathbf{B}_{11}$ denote a submatrix of $\mathbf{A}$ with $\mathrm{r}(\mathbf{B}_{11}) = \mathrm{r}(\mathbf{A}) = r$. Let $\mathbf{R}$ and $\mathbf{S}$ denote elementary permutation matrices that bring $\mathbf{B}_{11}$ to the leading position, i.e.,

$$\mathbf{RAS} = \begin{pmatrix} \mathbf{B}_{11} & \mathbf{B}_{12} \\ \mathbf{B}_{21} & \mathbf{B}_{22} \end{pmatrix} = \mathbf{B}, \quad \text{say.}$$

Then,

$$\mathbf{A}^- = \mathbf{S}\mathbf{B}^-\mathbf{R} = \{\mathbf{R}'(\mathbf{B}^-)'\mathbf{S}'\}' \tag{3.1.4}$$

is a g-inverse of $\mathbf{A}$, where

$$\mathbf{B}^- = \begin{pmatrix} \mathbf{B}_{11}^{-1} & \mathbf{O} \\ \mathbf{O} & \mathbf{O} \end{pmatrix}. \tag{3.1.5}$$

Proof. Since $\mathbf{R}$ and $\mathbf{S}$ are orthogonal, $\mathbf{R}' = \mathbf{R}^{-1}$ and $\mathbf{S}' = \mathbf{S}^{-1}$, so that $\mathbf{A} = \mathbf{R}'\mathbf{B}\mathbf{S}'$. It is easy to verify that

$$\mathbf{B}\mathbf{B}^-\mathbf{B} = \begin{pmatrix} \mathbf{B}_{11} & \mathbf{B}_{12} \\ \mathbf{B}_{21} & \mathbf{B}_{21}\mathbf{B}_{11}^{-1}\mathbf{B}_{12} \end{pmatrix}.$$

From Result 2.1.3, $\mathrm{r}(\mathbf{B}) = \mathrm{r}(\mathbf{B}_{11}) + \mathrm{r}(\mathbf{B}_{22} - \mathbf{B}_{21}\mathbf{B}_{11}^{-1}\mathbf{B}_{12})$. Since $\mathrm{r}(\mathbf{B}) = \mathrm{r}(\mathbf{B}_{11})$, it follows that $\mathbf{B}_{22} = \mathbf{B}_{21}\mathbf{B}_{11}^{-1}\mathbf{B}_{12}$. Thus $\mathbf{B}^-$ satisfies (3.1.1), so that it is a g-inverse of $\mathbf{B}$. Now,

$$\mathbf{A}\mathbf{A}^-\mathbf{A} = \mathbf{R}'\mathbf{B}\mathbf{S}'\mathbf{S}\mathbf{B}^-\mathbf{R}\mathbf{R}'\mathbf{B}\mathbf{S}' = \mathbf{R}'\mathbf{B}\mathbf{B}^-\mathbf{B}\mathbf{S}' = \mathbf{R}'\mathbf{B}\mathbf{S}' = \mathbf{A},$$

since $\mathbf{R}\mathbf{R}' = \mathbf{I}_m$ and $\mathbf{S}'\mathbf{S} = \mathbf{I}_n$. ■

To summarize the algorithm, we find a nonsingular $r \times r$ submatrix $\mathbf{B}_{11}$ of $\mathbf{A}$, and compute $(\mathbf{B}_{11}^{-1})'$. In $\mathbf{A}$, we replace each element of $\mathbf{B}_{11}$ by the corresponding element of $(\mathbf{B}_{11}^{-1})'$, and replace all other elements of $\mathbf{A}$ by 0. We transpose the resulting matrix to obtain a g-inverse of $\mathbf{A}$.

Example 3.1.1. We find a g-inverse of the 3×4 matrix

$$\mathbf{A} = \begin{pmatrix} 4 & 1 & 2 & 0 \\ 1 & 1 & 5 & 15 \\ 3 & 1 & 3 & 5 \end{pmatrix}$$

using Result 3.1.2. We first verify that $\mathrm{r}(\mathbf{A}) = 2$. Next, suppose $\mathbf{B}_{11} = \begin{pmatrix} 4 & 1 \\ 1 & 1 \end{pmatrix}$; then $|\mathbf{B}_{11}| = 3$. Application of Result 3.1.2 gives the corresponding g-inverse as

$$\mathbf{A}^- = \begin{pmatrix} 1/3 & -1/3 & 0 \\ -1/3 & 4/3 & 0 \\ 0 & 0 & 0 \\ 0 & 0 & 0 \end{pmatrix}.$$

Suppose, we now choose another nonsingular submatrix of $\mathbf{A}$ of rank 2, i.e., we let $\mathbf{B}_{11} = \begin{pmatrix} 1 & 15 \\ 1 & 5 \end{pmatrix}$, with $|\mathbf{B}_{11}| = -10$. Using the algorithm, we find the corresponding g-inverse to be

$$\mathbf{A}^- = \begin{pmatrix} 0 & 0 & 0 \\ 0 & -1/2 & 3/2 \\ 0 & 0 & 0 \\ 0 & 1/10 & -1/10 \end{pmatrix}.$$

□

In Example 3.1.1, we see that **A** does not have a unique g-inverse. The entire set of g-inverses of a matrix **A** can be characterized as follows.

Result 3.1.3. Let **A** be an $m \times n$ matrix with rank r. By Result 1.3.14, there exists an $m \times m$ nonsingular matrix **B** and an $n \times n$ nonsingular matrix **C**, such that $\mathbf{A} = \mathbf{BDC}$, where **D** is the $m \times n$ partitioned matrix $\text{diag}(\mathbf{I}_r, \mathbf{O})$. Then the entire set of g-inverses of **A** is

$$\mathcal{G} = \left\{ \mathbf{C}^{-1} \begin{pmatrix} \mathbf{I}_r & \mathbf{K} \\ \mathbf{L} & \mathbf{M} \end{pmatrix} \mathbf{B}^{-1} \colon \mathbf{K} \in \mathcal{R}^{r \times (m-r)}, \mathbf{L} \in \mathcal{R}^{(n-r) \times r}, \mathbf{M} \in \mathcal{R}^{(n-r) \times (m-r)} \right\}.$$

In particular, $\mathcal{G}$ is an affine subspace of the linear space of $n \times m$ matrices, its dimension is $mn - r^2$, and the rank of a g-inverse of **A** can be any one of $r, r+1, \ldots, \min(m, n)$.

Proof. A matrix **G** is a g-inverse of **A** if and only if $\mathbf{BDCGBDC} = \mathbf{BDC}$. Since **B** and **C** are nonsingular, this is equivalent to $\mathbf{DCGBD} = \mathbf{D}$. Denote

$$\mathbf{CGB} = \begin{pmatrix} \mathbf{Q} & \mathbf{K} \\ \mathbf{L} & \mathbf{M} \end{pmatrix}.$$

Then

$$\mathbf{DCGBD} = \begin{pmatrix} \mathbf{I}_r & \mathbf{O} \\ \mathbf{O} & \mathbf{O} \end{pmatrix} \begin{pmatrix} \mathbf{Q} & \mathbf{K} \\ \mathbf{L} & \mathbf{M} \end{pmatrix} \begin{pmatrix} \mathbf{I}_r & \mathbf{O} \\ \mathbf{O} & \mathbf{O} \end{pmatrix} = \begin{pmatrix} \mathbf{Q} & \mathbf{O} \\ \mathbf{O} & \mathbf{O} \end{pmatrix},$$

so $\mathbf{DCGBD} = \mathbf{D}$ if and only if $\mathbf{Q} = \mathbf{I}_r$. This gives the form of $\mathcal{G}$. The rest of the result follows immediately. ■

The following result can be easily verified using (3.1.1).

Result 3.1.4. Let **A** be an $m \times n$ matrix and **G** be a g-inverse of **A**.

1. For a nonsingular $m \times m$ matrix **P** and an $n \times n$ matrix **Q**, $\mathbf{Q}^{-1}\mathbf{G}\mathbf{P}^{-1}$ is a g-inverse of **PAQ**.
2. **GA** is its own g-inverse.
3. If $c \neq 0$ is a scalar, then $\mathbf{G}/c$ is a g-inverse of $c\mathbf{A}$.
4. A g-inverse of $\mathbf{J}_n$ is $\mathbf{I}_n/n$.

Example 3.1.2. Let

$$\mathbf{A} = \begin{pmatrix} 2 & 2 & 6 \\ 2 & 3 & 8 \\ 6 & 8 & 22 \end{pmatrix}.$$

Then $|\mathbf{A}| = 0$, so $\text{r}(\mathbf{A}) \leq 2$. Let $\mathbf{B}_{11} = \begin{pmatrix} 2 & 6 \\ 2 & 8 \end{pmatrix}$. Then $|\mathbf{B}_{11}| = 4$, so $\mathbf{B}_{11}$ is nonsingular and $\text{r}(\mathbf{A}) = 2$. Using the algorithm in Result 3.1.2, we find the g-inverse corresponding to $\mathbf{B}_{11}$ to be

$$\mathbf{A}^- = \begin{pmatrix} 2 & -3/2 & 0 \\ 0 & 0 & 0 \\ -1/2 & 1/2 & 0 \end{pmatrix}.$$

On the other hand, if $\mathbf{B}_{11} = \begin{pmatrix} 2 & 2 \\ 2 & 3 \end{pmatrix}$, then $|\mathbf{B}_{11}| = 2$. Using the algorithm, we find the corresponding g-inverse

$$\mathbf{A}^{-} = \begin{pmatrix} 3/2 & -1 & 0 \\ -1 & 1 & 0 \\ 0 & 0 & 0 \end{pmatrix},$$

which is symmetric. □

The example demonstrates that even if $\mathbf{A}$ is a symmetric matrix, its g-inverse is not necessarily symmetric. However, we can always construct a symmetric g-inverse of a symmetric matrix. To see this, note that if $\mathbf{A}^{-}$ is any g-inverse of a symmetric matrix $\mathbf{A}$, so is $\mathbf{A}'$ (see Result 3.1.5), and therefore, the symmetric matrix $\frac{1}{2}(\mathbf{A}^{-} + \mathbf{A}^{-\prime})$ is a g-inverse of $\mathbf{A}$. In general, we may also obtain a symmetric g-inverse by applying the same permutation in the algorithm of Result 3.1.2 to the rows and columns. This would result in a symmetric $\mathbf{B}_{11}$ and therefore a symmetric g-inverse.

Result 3.1.5. Let $\mathbf{G}$ be a g-inverse of a symmetric matrix $\mathbf{A}$. Then $\mathbf{G}'$ is also a g-inverse of $\mathbf{A}$.

Proof. By (3.1.1), we have $\mathbf{AGA} = \mathbf{A}$. Transposing both sides, and using $\mathbf{A}' = \mathbf{A}$, we get the result. ■

Result 3.1.6. If $\mathbf{G}$ is a g-inverse of $\mathbf{A}$, then

$$\mathrm{r}(\mathbf{A}) \leq \mathrm{r}(\mathbf{G}) \leq \min(m, n). \tag{3.1.6}$$

Proof. The proof follows directly from $\mathbf{AGA} = \mathbf{A}$ and Result 1.3.11. ■

Result 3.1.7. Let $\mathbf{A}$ be a matrix of rank r and $\mathbf{G}$ be a g-inverse of $\mathbf{A}$. Then

1. $\mathbf{GA}$ and $\mathbf{AG}$ are idempotent.
2. $\mathbf{I} - \mathbf{GA}$ and $\mathbf{I} - \mathbf{AG}$ are idempotent.
3. $\mathrm{r}(\mathbf{GA}) = \mathrm{r}(\mathbf{AG}) = r$ and $\mathrm{r}(\mathbf{I} - \mathbf{GA}) = \mathrm{r}(\mathbf{I} - \mathbf{AG}) = n - r$.
4. $\mathrm{tr}(\mathbf{GA}) = \mathrm{tr}(\mathbf{AG}) = r$.

Proof. From $(\mathbf{GA})(\mathbf{GA}) = \mathbf{G}(\mathbf{AGA}) = \mathbf{GA}$ and $(\mathbf{AG})(\mathbf{AG}) = (\mathbf{AGA})\mathbf{G} = \mathbf{AG}$, property 1 follows. From property 2 of Result 2.3.6, property 2 follows. To prove property 3, we see that from Result 1.3.11, $\mathrm{r}(\mathbf{AA}^{-}) \leq \mathrm{r}(\mathbf{A})$ and since $(\mathbf{AA}^{-})\mathbf{A} = \mathbf{A}$, $\mathrm{r}(\mathbf{AA}^{-}) \geq \mathrm{r}(\mathbf{A})$. Then $\mathrm{r}(\mathbf{AA}^{-}) = \mathrm{r}(\mathbf{A}) = r$. Likewise, $\mathrm{r}(\mathbf{A}^{-}\mathbf{A}) = r$. The rest of property 3 as well as property 4 follow from property 1 and property 3 of Result 2.3.6. ■

Definition 3.1.2. If $\mathbf{A}^{-}\mathbf{A} = \mathbf{AA}^{-}$, then $\mathbf{A}^{-}$ is called a commuting g-inverse of $\mathbf{A}$, where $\mathbf{A}$ is a square matrix (Englefield, 1966).

Result 3.1.8. Let $\mathbf{D} = \mathrm{diag}(d_1, \cdots, d_n)$. Then the diagonal matrix with ith diagonal element equal to $1/d_i$ if $d_i \neq 0$ and 0 if $d_i = 0$ is a g-inverse of $\mathbf{D}$.

Proof. It is easy to verify that the diagonal matrix as described satisfies (3.1.1). ■

Result 3.1.9. Let $\mathbf{G}$ be a g-inverse of the symmetric matrix $\mathbf{A}'\mathbf{A}$, where $\mathbf{A}$ is any $m \times n$ matrix. Then,

1. $\mathbf{G}'$ is also a g-inverse of $\mathbf{A}'\mathbf{A}$.
2. $\mathbf{GA}'$ is a g-inverse of $\mathbf{A}$, so that
$$\mathbf{AGA}'\mathbf{A} = \mathbf{A}. \tag{3.1.7}$$
3. $\mathbf{AGA}'$ is invariant to $\mathbf{G}$, i.e.,
$$\mathbf{AG}_1\mathbf{A}' = \mathbf{AG}_2\mathbf{A}' \tag{3.1.8}$$
for any two g-inverses $\mathbf{G}_1$ and $\mathbf{G}_2$ of $\mathbf{A}'\mathbf{A}$.
4. Whether or not $\mathbf{G}$ is symmetric, $\mathbf{AGA}'$ is.

Because of properties 3 and 4, we can write $\mathbf{A}(\mathbf{A}'\mathbf{A})^-\mathbf{A}'$ without specifying which g-inverse of $\mathbf{A}'\mathbf{A}$ is used.

Proof. Since $\mathbf{A}'\mathbf{A}$ is symmetric, property 1 follows from Result 3.1.5. Using property 1 of Result 1.3.13, we see that $\mathbf{A}'\mathbf{AGA}'\mathbf{A} = \mathbf{A}'\mathbf{A}$ implies (3.1.7), proving property 2. To prove property 3, let $\mathbf{G}_1$ and $\mathbf{G}_2$ be two distinct g-inverses of $\mathbf{A}'\mathbf{A}$. From (3.1.7), $\mathbf{AG}_1\mathbf{A}'\mathbf{A} = \mathbf{AG}_2\mathbf{A}'\mathbf{A}$. Then by property 2 of Result 1.3.13, $\mathbf{AG}_1\mathbf{A}' = \mathbf{AG}_2\mathbf{A}'$, i.e., $\mathbf{AGA}'$ is invariant to the choice of a g-inverse $\mathbf{G}$. In particular, by choosing a symmetric g-inverse, which as mentioned earlier must exist for a symmetric matrix, it is seen that $\mathbf{AGA}$ is symmetric, proving property 4. ■

The next result is a useful application of Result 3.1.9.

Result 3.1.10. If $\mathbf{A}$ and $\mathbf{B}$ are n.n.d. $n \times n$ matrices, then $\mathbf{A}(\mathbf{A}+\mathbf{B})^-(\mathbf{A}+\mathbf{B}) = \mathbf{A}$.

Proof. From Result 2.4.5, $\mathbf{A} = \mathbf{K}'\mathbf{K}$ and $\mathbf{B} = \mathbf{L}'\mathbf{L}$ for some matrices $\mathbf{K}$ and $\mathbf{L}$, each having n columns. Let $\mathbf{M} = \begin{pmatrix}\mathbf{K}\\ \mathbf{L}\end{pmatrix}$. Then $\mathbf{A}+\mathbf{B} = \mathbf{M}'\mathbf{M}$. From property 2 of Result 3.1.9,
$$\begin{pmatrix}\mathbf{K}\\ \mathbf{L}\end{pmatrix}(\mathbf{A}+\mathbf{B})^-(\mathbf{A}+\mathbf{B}) = \mathbf{M}(\mathbf{M}'\mathbf{M})^-\mathbf{M}'\mathbf{M} = \mathbf{M} = \begin{pmatrix}\mathbf{K}\\ \mathbf{L}\end{pmatrix}.$$
As a result, $\mathbf{K}(\mathbf{A}+\mathbf{B})^-(\mathbf{A}+\mathbf{B}) = \mathbf{K}$. Pre-multiply both sides of the equation by $\mathbf{K}'$ to get the proof. ■

We will see later that Result 3.1.9, especially property 3, is very important for the discussion of inference for the less than full-rank linear model.

Example 3.1.3. Suppose
$$\mathbf{X} = \begin{pmatrix} 1 & 1 & 0 & 0\\ 1 & 1 & 0 & 0\\ 1 & 1 & 0 & 0\\ 1 & 0 & 1 & 0\\ 1 & 0 & 1 & 0\\ 1 & 0 & 0 & 1 \end{pmatrix}.$$
It is easy to verify that
$$\mathbf{A} = \mathbf{X}'\mathbf{X} = \begin{pmatrix} 6 & 3 & 2 & 1\\ 3 & 3 & 0 & 0\\ 2 & 0 & 2 & 0\\ 1 & 0 & 0 & 1 \end{pmatrix},$$

with rank 3. Two distinct g-inverses of $\mathbf{X}'\mathbf{X}$ are

$$\mathbf{G}_1 = \begin{pmatrix} 0 & 0 & 0 & 0 \\ 0 & 1/3 & 0 & 0 \\ 0 & 0 & 1/2 & 0 \\ 0 & 0 & 0 & 1 \end{pmatrix} \text{ and } \mathbf{G}_2 = \begin{pmatrix} 0 & 1/3 & 0 & 0 \\ 0 & 0 & 0 & 0 \\ 0 & -1/3 & 1/2 & 0 \\ 0 & -1/3 & 0 & 1 \end{pmatrix}$$

which respectively correspond to the full rank submatrices

$$\mathbf{A}_{11} = \begin{pmatrix} 3 & 0 & 0 \\ 0 & 2 & 0 \\ 0 & 0 & 1 \end{pmatrix} \text{ and } \mathbf{A}_{11} = \begin{pmatrix} 3 & 0 & 0 \\ 2 & 2 & 0 \\ 1 & 0 & 1 \end{pmatrix}.$$

It is easy to verify that

$$\mathbf{X}\mathbf{G}_1\mathbf{X}' = \mathbf{X}\mathbf{G}_2\mathbf{X}' = \begin{pmatrix} 1/3 & 1/3 & 1/3 & 0 & 0 & 0 \\ 1/3 & 1/3 & 1/3 & 0 & 0 & 0 \\ 1/3 & 1/3 & 1/3 & 0 & 0 & 0 \\ 0 & 0 & 0 & 1/2 & 1/2 & 0 \\ 0 & 0 & 0 & 1/2 & 1/2 & 0 \\ 0 & 0 & 0 & 0 & 0 & 1 \end{pmatrix}. \quad \square$$

Example 3.1.4. Let $\mathbf{B}$ be a $k \times k$ symmetric matrix of rank r. From Result 2.3.4, $\mathbf{B} = \mathbf{PDP}'$, where $\mathbf{D} = \text{diag}(\lambda_1, \cdots, \lambda_k)$ and $\mathbf{P} = (\mathbf{p}_1, \cdots, \mathbf{p}_k)$ is an orthogonal matrix. Suppose $\lambda_1, \cdots, \lambda_r$ are nonzero, while the last $\lambda_{r+1} = \cdots = \lambda_k = 0$. Then $\mathbf{B} = \sum_{i=1}^r \lambda_i \mathbf{p}_i \mathbf{p}_i'$. Let $\mathbf{D}^- = \text{diag}(1/\lambda_1, \cdots, 1/\lambda_r, 0, \cdots, 0)$ and $\mathbf{C} = \mathbf{PD}^-\mathbf{P}' = \sum_{i=1}^r \mathbf{p}_i\mathbf{p}_i'/\lambda_i$. By Result 3.1.8, $\mathbf{D}^-$ is a g-inverse of $\mathbf{D}$, and by Result 3.1.4, $\mathbf{C}$ is a g-inverse of $\mathbf{B}$. This result is useful in the discussion of estimable functions in the less than full-rank linear model. □

Example 3.1.5. Let $\mathbf{A}$ be an $m \times n$ matrix and let $\mathbf{P}$ and $\mathbf{Q}$ respectively denote nonsingular $m \times m$ and $n \times n$ matrices. We will show that $\mathbf{G}$ is a g-inverse of $\mathbf{PAQ}$ if and only if $\mathbf{G} = \mathbf{Q}^{-1}\mathbf{A}^-\mathbf{P}^{-1}$. The "if" part is just property 1 of Result 3.1.4. For the "only if" part, if $\mathbf{G}$ is a g-inverse of $\mathbf{PAQ}$, then $\mathbf{PAQGPAG} = \mathbf{PAG}$, and so $\mathbf{AQGPA} = \mathbf{A}$, i.e., $\mathbf{QGP}$ is a g-inverse of $\mathbf{A}$. Two special cases of this condition are given in Exercise 3.17. □

Result 3.1.11. Let $\mathbf{A}$ be an $m \times n$ matrix. For any $m \times p$ matrix $\mathbf{B}$, $\mathcal{C}(\mathbf{B}) \subset \mathcal{C}(\mathbf{A})$ if and only if $\mathbf{AA}^-\mathbf{B} = \mathbf{B}$, or equivalently, if and only if $(\mathbf{I} - \mathbf{AA}^-)\mathbf{B} = \mathbf{O}$. Similarly, for any $q \times n$ matrix $\mathbf{C}$, $\mathcal{R}(\mathbf{C}) \subset \mathcal{R}(\mathbf{A})$ if and only if $\mathbf{C} = \mathbf{CA}^-\mathbf{A}$, or equivalently, if and only if $\mathbf{C}(\mathbf{I} - \mathbf{A}^-\mathbf{A}) = \mathbf{O}$.

Proof. Sufficiency follows from property 5 of Result 1.3.10. To prove necessity, $\mathcal{C}(\mathbf{B}) \subset \mathcal{C}(\mathbf{A})$ implies that there exists a matrix $\mathbf{M}$ such that $\mathbf{B} = \mathbf{AM}$. Then $\mathbf{AA}^-\mathbf{B} = \mathbf{AA}^-\mathbf{AM} = \mathbf{AM} = \mathbf{B}$. The proof of the rest of the result is similar. ■

Result 3.1.12. G-inverse of a partitioned matrix. Let $\mathbf{A} = \begin{pmatrix} \mathbf{A}_{11} & \mathbf{A}_{12} \\ \mathbf{A}_{21} & \mathbf{A}_{22} \end{pmatrix}$ be an $m \times n$ partitioned matrix, $\mathbf{A}_{ij}$ having dimension $m_i \times n_j$, $i, j = 1, 2$, and let $\text{r}(\mathbf{A}) < m$. Let $\mathbf{E} = \mathbf{A}_{22} - \mathbf{A}_{21}\mathbf{A}_{11}^-\mathbf{A}_{12}$, $\mathcal{C}(\mathbf{A}_{12}) \subset \mathcal{C}(\mathbf{A}_{11})$, and $\mathcal{R}(\mathbf{A}_{21}) \subset \mathcal{R}(\mathbf{A}_{11})$. A g-inverse of $\mathbf{A}$ is

$$\begin{pmatrix} \mathbf{A}_{11}^- + \mathbf{A}_{11}^-\mathbf{A}_{12}\mathbf{E}^-\mathbf{A}_{21}\mathbf{A}_{11}^- & -\mathbf{A}_{11}^-\mathbf{A}_{12}\mathbf{E}^- \\ -\mathbf{E}^-\mathbf{A}_{21}\mathbf{A}_{11}^- & \mathbf{E}^- \end{pmatrix}$$
$$= \begin{pmatrix} \mathbf{A}_{11}^- & \mathbf{O} \\ \mathbf{O} & \mathbf{O} \end{pmatrix} + \begin{pmatrix} -\mathbf{A}_{11}^-\mathbf{A}_{12} \\ \mathbf{I}_{n_2} \end{pmatrix} \mathbf{E}^- \left(-\mathbf{A}_{21}\mathbf{A}_{11}^-, \;\; \mathbf{I}_{m_2} \right).$$

Proof. By Result 3.1.11, $\mathbf{A}_{12} = \mathbf{A}_{11}\mathbf{A}_{11}^{-}\mathbf{A}_{12}$ and $\mathbf{A}_{21} = \mathbf{A}_{21}\mathbf{A}_{11}^{-}\mathbf{A}_{11}$. Hence,

$$\begin{pmatrix} \mathbf{A}_{11} & \mathbf{A}_{12} \\ \mathbf{A}_{21} & \mathbf{A}_{22} \end{pmatrix} = \begin{pmatrix} \mathbf{A}_{11} & \mathbf{A}_{11}\mathbf{A}_{11}^{-}\mathbf{A}_{12} \\ \mathbf{A}_{21}\mathbf{A}_{11}^{-}\mathbf{A}_{11} & \mathbf{A}_{22} \end{pmatrix}.$$

The result follows by using Exercise 3.18, setting $\mathbf{A} = \mathbf{A}_{11}$, $\mathbf{D} = \mathbf{A}_{22}$, $\mathbf{B} = \mathbf{A}_{11}^{-}\mathbf{A}_{12}$, and $\mathbf{C} = \mathbf{A}_{21}\mathbf{A}_{11}^{-}$ therein, so that $\mathbf{C}\mathbf{A}_{11}\mathbf{B} = \mathbf{A}_{21}\mathbf{A}_{11}^{-}\mathbf{A}_{11}\mathbf{A}_{11}^{-}\mathbf{A}_{12} = \mathbf{A}_{21}\mathbf{A}_{11}^{-}\mathbf{A}_{12}$. ■

Example 3.1.6. Let

$$\mathbf{A} = \left(\begin{array}{cc|cc} 4 & 4 & 0 & 0 \\ 1 & 1 & 0 & 0 \\ \hline 0 & 0 & 3 & 5 \end{array}\right) = \left(\begin{array}{c|c} \mathbf{A}_{11} & \mathbf{O} \\ \hline \mathbf{O} & \mathbf{A}_{22} \end{array}\right).$$

Let

$$\mathbf{A}_{11}^{-} = \begin{pmatrix} 1/4 & 0 \\ 0 & 0 \end{pmatrix} \quad \text{and} \quad \mathbf{A}_{22}^{-} = \begin{pmatrix} 0 \\ 1/5 \end{pmatrix}.$$

Using Result 3.1.12,

$$\mathbf{A}^{-} = \begin{pmatrix} 1/4 & 0 & 0 \\ 0 & 0 & 0 \\ 0 & 0 & 0 \\ 0 & 0 & 1/5 \end{pmatrix} = \begin{pmatrix} \mathbf{A}_{11}^{-} & \mathbf{O} \\ \mathbf{O} & \mathbf{A}_{22}^{-} \end{pmatrix}$$

is a g-inverse of $\mathbf{A}$. □

Example 3.1.7. For an $m \times n$ matrix $\mathbf{A}$, and an $m \times p$ matrix $\mathbf{B}$, $\begin{pmatrix} \mathbf{A}^{-} \\ \mathbf{B}^{-} \end{pmatrix}$ is a g-inverse of $(\mathbf{A} \;\; \mathbf{B})$ if and only if

$$\begin{aligned} (\mathbf{A} \;\; \mathbf{B}) &= (\mathbf{A} \;\; \mathbf{B}) \begin{pmatrix} \mathbf{A}^{-} \\ \mathbf{B}^{-} \end{pmatrix} (\mathbf{A} \;\; \mathbf{B}) \\ &= (\mathbf{A} \;\; \mathbf{B}) \begin{pmatrix} \mathbf{A}^{-}\mathbf{A} & \mathbf{A}^{-}\mathbf{B} \\ \mathbf{B}^{-}\mathbf{A} & \mathbf{B}^{-}\mathbf{B} \end{pmatrix} \\ &= \left(\mathbf{A}\mathbf{A}^{-}\mathbf{A} + \mathbf{B}\mathbf{B}^{-}\mathbf{A} \quad \mathbf{A}\mathbf{A}^{-}\mathbf{B} + \mathbf{B}\mathbf{B}^{-}\mathbf{B}\right), \end{aligned}$$

i.e., if and only if $\mathbf{A}\mathbf{A}^{-}\mathbf{B} = \mathbf{O}$, and $\mathbf{B}\mathbf{B}^{-}\mathbf{A} = \mathbf{O}$. □

Result 3.1.13. Let $\mathbf{A}$ be an $m \times n$ matrix of rank r. Then $\mathbf{P} = \mathbf{A}(\mathbf{A}'\mathbf{A})^{-}\mathbf{A}'$ is the matrix of the orthogonal projection of m-dimensional vectors onto $\mathcal{C}(\mathbf{A})$. In particular, $\mathbf{P}\mathbf{A} = \mathbf{A}$, $\mathcal{C}(\mathbf{P}) = \mathcal{C}(\mathbf{A})$, and $\dim(\mathbf{P}) = r$.

Proof. For any $\mathbf{v} \in \mathcal{C}(\mathbf{A})$, we can write $\mathbf{v} = \mathbf{A}\mathbf{x}$ for some $\mathbf{x} \in \mathcal{R}^n$. From property 2 of Result 3.1.9, $\mathbf{A}\mathbf{G}\mathbf{A}'\mathbf{A} = \mathbf{A}$, so that $\mathbf{P}\mathbf{v} = \mathbf{A}\mathbf{G}\mathbf{A}'\mathbf{A}\mathbf{x} = \mathbf{A}\mathbf{x} = \mathbf{v}$. For any $\mathbf{u} \in \mathcal{C}(\mathbf{A})^{\perp}$, $\mathbf{A}'\mathbf{u} = \mathbf{0}$, and so $\mathbf{P}\mathbf{u} = \mathbf{A}\mathbf{G}\mathbf{A}'\mathbf{u} = \mathbf{0}$. Hence, $\mathbf{P}$ is the projection matrix onto $\mathcal{C}(\mathbf{A})$. The rest of the result follows immediately. ■

Note that if $\mathcal{V} = \mathcal{C}(\mathbf{A})$, then $\mathbf{P}$ is exactly the projection matrix onto $\mathcal{V}$ defined in Definition 2.6.2. It follows that for any matrix $\mathbf{B}$ with m rows, $\mathcal{C}(\mathbf{B}) = \mathcal{C}(\mathbf{A})$ if and only if $\mathbf{B}(\mathbf{B}'\mathbf{B})^{-}\mathbf{B}' = \mathbf{A}(\mathbf{A}'\mathbf{A})^{-}\mathbf{A}'$. From Result 2.6.2, $\mathbf{P}$ is symmetric and idempotent of rank r. If the columns of $\mathbf{A}$ are LIN, then the unique g-inverse of $\mathbf{A}'\mathbf{A}$ is $(\mathbf{A}'\mathbf{A})^{-1}$ and hence, $\mathbf{P} = \mathbf{A}(\mathbf{A}'\mathbf{A})^{-1}\mathbf{A}'$ (see Result 3.1.1).

Result 3.1.14. If $\mathbf{A}$ has m rows, then the matrix of the orthogonal projection onto $\mathcal{N}(\mathbf{A}')$ is $\mathbf{I}_m - \mathbf{A}(\mathbf{A}'\mathbf{A})^-\mathbf{A}'$.

The proof is left as an exercise.

The next result is a generalization of the Sherman–Morrison–Woodbury formula shown in Result 1.3.8. We use this result in Chapter 8.

Result 3.1.15. Let $\mathbf{A} \in \mathcal{R}^{n\times n}$ and $\mathbf{B} \in \mathcal{R}^{n\times k}$, with $\mathcal{C}(\mathbf{B}) \subset \mathcal{C}(\mathbf{A})$, and let $\mathbf{C} \in \mathcal{R}^{k\times n}$ with $\mathcal{R}(\mathbf{C}) \subset \mathcal{R}(\mathbf{A})$. Then

$$(\mathbf{A}+\mathbf{BC})^- = \mathbf{A}^- - \mathbf{A}^-\mathbf{B}(\mathbf{I}+\mathbf{CA}^-\mathbf{B})^-\mathbf{CA}^-, \tag{3.1.9}$$

where each $\mathbf{A}^-$ may represent a different g-inverse of $\mathbf{A}$.

Proof. Let $\mathbf{D} = (\mathbf{I}+\mathbf{CA}^-\mathbf{B})^-$. By the assumption, $\mathbf{B} = \mathbf{AR}$ and $\mathbf{C} = \mathbf{LA}$ for some matrices $\mathbf{R}$ and $\mathbf{L}$. Then $\mathbf{AA}^-\mathbf{B} = \mathbf{B}$, $\mathbf{CA}^-\mathbf{A} = \mathbf{C}$, and $\mathbf{D} = (\mathbf{I}+\mathbf{CR})^-$. Now

$$\begin{aligned}
&(\mathbf{A}+\mathbf{BC})(\mathbf{A}^- - \mathbf{A}^-\mathbf{BDCA}^-)(\mathbf{A}+\mathbf{BC})\\
&= (\mathbf{A}+\mathbf{BLA})(\mathbf{A}^- - \mathbf{A}^-\mathbf{BDCA}^-)(\mathbf{A}+\mathbf{ARC})\\
&= (\mathbf{I}+\mathbf{BL})(\mathbf{AA}^-\mathbf{A} - \mathbf{AA}^-\mathbf{BDCA}^-\mathbf{A})(\mathbf{I}+\mathbf{RC})\\
&= (\mathbf{I}+\mathbf{BL})(\mathbf{A} - \mathbf{BDC})(\mathbf{I}+\mathbf{RC})\\
&= (\mathbf{I}+\mathbf{BL})(\mathbf{A} - \mathbf{ARDC})(\mathbf{I}+\mathbf{RC})\\
&= (\mathbf{A}+\mathbf{BLA})(\mathbf{I} - \mathbf{RDC})(\mathbf{I}+\mathbf{RC})\\
&= (\mathbf{A}+\mathbf{BC})(\mathbf{I}+\mathbf{RC} - \mathbf{RDC} - \mathbf{RDCRC}),
\end{aligned}$$

so (3.1.9) follows if

$$(\mathbf{A}+\mathbf{BC})\mathbf{RC} = (\mathbf{A}+\mathbf{BC})(\mathbf{RDC}+\mathbf{RDCRC}).$$

Since the right side of the above display is $(\mathbf{AR}+\mathbf{ARCR})\mathbf{D}(\mathbf{I}+\mathbf{CR})\mathbf{C} = \mathbf{AR}(\mathbf{I}+\mathbf{CR})\mathbf{D}(\mathbf{I}+\mathbf{CR})\mathbf{C} = \mathbf{AR}(\mathbf{I}+\mathbf{CR})\mathbf{C} = (\mathbf{A}+\mathbf{BC})\mathbf{RC}$, the result follows. ■

Definition 3.1.3. Moore–Penrose inverse. The Moore–Penrose inverse $\mathbf{A}^+$ of an $m\times n$ matrix $\mathbf{A}$ satisfies the following conditions:

1. $\mathbf{AA}^+\mathbf{A} = \mathbf{A}$, i.e., $\mathbf{A}^+$ is a g-inverse of $\mathbf{A}$,
2. $\mathbf{A}^+\mathbf{AA}^+ = \mathbf{A}^+$, i.e., $\mathbf{A}$ is a g-inverse of $\mathbf{A}^+$,
3. $(\mathbf{AA}^+)' = \mathbf{AA}^+$, and
4. $(\mathbf{A}^+\mathbf{A})' = \mathbf{A}^+\mathbf{A}$.

If a g-inverse $\mathbf{G}$ of $\mathbf{A}$ satisfies only conditions 1 and 2, it is called a reflexive g-inverse of $\mathbf{A}$. We will not discuss these alternative inverses, except to mention that $\mathbf{A}^+$ is unique. The reader may refer to Rao and Mitra (1971) or to Pringle and Rayner (1971) for details.

3.2 Solutions to linear systems

Least squares estimation of the parameters in a linear statistical model starts with the mathematical problem of solving a system of linear equations involving those parameters and the data. These equations are called normal equations. In this section, we discuss systems of linear equations and some properties of solutions to such systems. A linear system of m equations in n unknown variables $\mathbf{x} = (x_1, \cdots, x_n)'$ is written as

$$\begin{array}{ccccc} a_{11}x_1 + a_{12}x_2 + \cdots + a_{1n}x_n &=& b_1 \\ a_{21}x_1 + a_{22}x_2 + \cdots + a_{2n}x_n &=& b_2 \\ \vdots \quad \vdots \quad \vdots \quad \vdots & & \vdots \\ a_{m1}x_1 + a_{m2}x_2 + \cdots + a_{mn}x_n &=& b_m \end{array}$$

or in matrix form as

$$\mathbf{A}\mathbf{x} = \mathbf{b}, \tag{3.2.1}$$

where $\mathbf{A} = \{a_{ij}\}$ is an $m \times n$ coefficient matrix, $\mathbf{b} = (b_1, \cdots, b_m)'$ is called the right side and a_{ij}'s and b_i's are fixed scalars. Solving (3.2.1) is the process of finding a solution, provided it exists, i.e., a value of $\mathbf{x}$ which satisfies (3.2.1). The matrix $\mathbf{B_A} = (\mathbf{A} \quad \mathbf{b})$ is called the augmented matrix.

The linear system $\mathbf{Ax} = \mathbf{b}$ is said to be homogeneous if $\mathbf{b} = \mathbf{0}$, i.e., the system is

$$\mathbf{A}\mathbf{x} = \mathbf{0}, \tag{3.2.2}$$

and it is said to be nonhomogeneous if $\mathbf{b} \neq \mathbf{0}$. The collection of solutions to (3.2.1) is the set of all n-dimensional vectors $\mathbf{x}$ that satisfy $\mathbf{Ax} = \mathbf{b}$ and is called the solution set of the system.

Definition 3.2.1. Consistent linear system. A linear system $\mathbf{Ax} = \mathbf{b}$ is said to be consistent if it has at least one solution; otherwise, if no solution exists, the system is inconsistent.

Example 3.2.1. The system

$$\begin{pmatrix} 1 & 3 \\ 2 & 6 \end{pmatrix} \begin{pmatrix} x_1 \\ x_2 \end{pmatrix} = \begin{pmatrix} 5 \\ 10 \end{pmatrix},$$

which may be written out as

$$\begin{array}{ccccc} x_1 & + & 3x_2 & = & 5 \\ 2x_1 & + & 6x_2 & = & 10, \end{array}$$

has a solution given by $x_1 = 2, x_2 = 1$, and is consistent. Note that row 2 in the coefficient matrix is twice row 1 and the same relationship exists between the corresponding elements of the right side. This linear system is said to be compatible. It is easy to see that the system

$$\begin{pmatrix} 1 & 3 \\ 2 & 6 \end{pmatrix} \begin{pmatrix} x_1 \\ x_2 \end{pmatrix} = \begin{pmatrix} 5 \\ 19 \end{pmatrix}$$

is not consistent. Note that $|\mathbf{A}| = 0$. □

Definition 3.2.2. A linear system $\mathbf{Ax} = \mathbf{b}$ is said to be compatible if every linear relationship that exists among the rows of $\mathbf{A}$ also exists among the corresponding rows of $\mathbf{b}$, i.e., $\mathbf{c}'\mathbf{b} = 0$ for every $\mathbf{c}'$ such that $\mathbf{c}'\mathbf{A} = \mathbf{0}$.

We state two useful results without proof.

Result 3.2.1. A linear system $\mathbf{Ax} = \mathbf{b}$ is consistent if and only if it is compatible.

For consistency of the system $\mathbf{Ax} = \mathbf{b}$, we do not require existence of linear relationships among the rows of $\mathbf{A}$. For instance, when the rows of $\mathbf{A}$ are LIN, so that $\mathbf{A}^{-1}$ exists, (3.2.1) is consistent. However, Result 3.2.1 states that should there exist linear relationships among the rows of $\mathbf{A}$, the same relationships should exist among the rows of $\mathbf{b}$ as well. Solutions to linear equations exist if and only if the equations are consistent. Henceforth, we will assume that the equations are consistent.

Result 3.2.2. The linear system $\mathbf{Ax} = \mathbf{b}$ is consistent

1. if and only if $\mathcal{C}(\mathbf{A}, \mathbf{b}) = \mathcal{C}(\mathbf{A})$; or
2. if and only if $\mathrm{r}(\mathbf{A}, \mathbf{b}) = \mathrm{r}(\mathbf{A})$; or
3. if $\mathbf{A}$ has full row rank m.

A homogeneous linear system $\mathbf{Ax} = \mathbf{0}$ is always consistent; its (nonempty) solution set is a linear space called the solution space. For an $n \times n$ matrix $\mathbf{A}$, the solution space corresponding to $\mathbf{Ax} = \mathbf{0}$ is

$$\mathcal{N}(\mathbf{A}) = \{\mathbf{x}\text{: } (\mathbf{I} - \mathbf{A})\mathbf{x} = \mathbf{x}\} \subset \mathcal{C}(\mathbf{I} - \mathbf{A}). \tag{3.2.3}$$

Also, $\mathcal{N}(\mathbf{I} - \mathbf{A}) = \{\mathbf{x}\text{: } \mathbf{Ax} = \mathbf{x}\} \subset \mathcal{C}(\mathbf{A})$. The solution set of a nonhomogeneous system (3.2.1) is *not* a linear space. This is seen immediately by noting that the null vector $\mathbf{0}$ is not a solution to the system.

Result 3.2.3. For every m-dimensional vector $\mathbf{b}$ for which $\mathbf{Ax} = \mathbf{b}$ is consistent, $\mathbf{Gb}$ is a solution to the system if and only if the $n \times m$ matrix $\mathbf{G}$ is a g-inverse of $\mathbf{A}$.

Proof. Let $\mathbf{G}$ be a g-inverse of $\mathbf{A}$, let $\mathbf{b}$ be any m-dimensional vector for which $\mathbf{Ax} = \mathbf{b}$ is consistent, and suppose $\mathbf{x}^0$ is any solution to $\mathbf{Ax} = \mathbf{b}$. Then $\mathbf{A}(\mathbf{Gb}) = (\mathbf{AG})\mathbf{b} = \mathbf{AGAx}^0 = \mathbf{Ax}^0 = \mathbf{b}$, so $\mathbf{Gb}$ is a solution. To show the converse, suppose that $\mathbf{Gb}$ is a solution to $\mathbf{Ax} = \mathbf{b}$ for every m-dimensional vector $\mathbf{b}$ for which the system is consistent, so that $\mathbf{AGb} = \mathbf{b}$. Note that $\mathbf{b} \in \mathcal{C}(\mathbf{A})$. In particular, let $\mathbf{b} = \mathbf{a}_j$, the jth column of $\mathbf{A}$, be one choice of $\mathbf{b}$. By property 1 of Result 3.2.2, the system $\mathbf{Ax} = \mathbf{b}$ is consistent for this choice of $\mathbf{b}$, and $\mathbf{Ga}_j$ is a solution, i.e., $\mathbf{AGa}_j = \mathbf{a}_j$. Since this holds for every column of $\mathbf{A}$, we see that $\mathbf{AGA} = \mathbf{A}$, i.e., $\mathbf{G}$ is a g-inverse of $\mathbf{A}$. ■

Let $\mathbf{A}$ be an $n \times n$ nonsingular matrix, and $\mathbf{G}$ be any $n \times n$ matrix. $\mathbf{Gb}$ is a solution to $\mathbf{Ax} = \mathbf{b}$ for every n-dimensional vector $\mathbf{b}$ if and only if $\mathbf{G} = \mathbf{A}^{-1}$.

Example 3.2.2. A g-inverse of the coefficient matrix $\mathbf{A} = \begin{pmatrix} 1 & 3 \\ 2 & 6 \end{pmatrix}$ corresponding to the consistent linear system in Example 3.2.1 is $\mathbf{A}^{-} = \begin{pmatrix} 0 & 0 \\ 1/3 & 0 \end{pmatrix}$ and $\mathbf{A}^{-}\mathbf{b} = (0 \quad 5/3)'$ is a solution. □

For convenience of notation, we use $\mathbf{z}^0$ to denote a solution to the homogeneous system $\mathbf{Az} = \mathbf{0}$ (in $\mathbf{z}$) and $\mathbf{x}^0$ to denote a solution to the nonhomogeneous system $\mathbf{Ax} = \mathbf{b}$ (in $\mathbf{x}$).

Result 3.2.4. Let $\mathbf{A}$ be a matrix with n columns. The following results hold.

1. An n-dimensional vector $\mathbf{z}^0$ is a solution to the homogeneous system $\mathbf{Az} = \mathbf{0}$ (in $\mathbf{z}$) if and only if, for some column vector $\mathbf{y}$,

$$\mathbf{z}^0 = (\mathbf{I} - \mathbf{A}^{-}\mathbf{A})\mathbf{y}. \tag{3.2.4}$$

2. $\mathcal{N}(\mathbf{A}) = \mathcal{C}(\mathbf{I} - \mathbf{A}^{-}\mathbf{A})$ and $\dim[\mathcal{N}(\mathbf{A})] = n - \mathrm{r}(\mathbf{A})$.

Proof. Let $\mathbf{z}^0 = (\mathbf{I} - \mathbf{A}^{-}\mathbf{A})\mathbf{y}$ for some $\mathbf{y} \in \mathcal{R}^n$; using (3.1.1),

$$\mathbf{A}\mathbf{z}^0 = (\mathbf{A} - \mathbf{A}\mathbf{A}^{-}\mathbf{A})\mathbf{y} = (\mathbf{A} - \mathbf{A})\mathbf{y} = \mathbf{0},$$

so that $\mathbf{z}^0$ solves $\mathbf{A}\mathbf{z} = \mathbf{0}$. Conversely, if $\mathbf{z}^0$ is a solution to $\mathbf{A}\mathbf{z} = \mathbf{0}$, then

$$\mathbf{z}^0 = \mathbf{z}^0 - \mathbf{A}^{-}(\mathbf{A}\mathbf{z}^0) = (\mathbf{I} - \mathbf{A}^{-}\mathbf{A})\mathbf{z}^0,$$

i.e., $\mathbf{y} = \mathbf{z}^0$, satisfying (3.2.4). This proves property 1, which together with property 1 of Result 1.3.10, yields property 2. ■

If $\mathrm{r}(\mathbf{A}) = n$, i.e., if $\mathbf{A}$ has full column rank, then $\dim[\mathcal{N}(\mathbf{A})] = 0$, and the system $\mathbf{A}\mathbf{z} = \mathbf{0}$ has a unique solution, viz., $\mathbf{0}$. If $\mathrm{r}(\mathbf{A}) < n$, then the homogeneous system $\mathbf{A}\mathbf{z} = \mathbf{0}$ has an infinite number of solutions.

Result 3.2.5. Let $\mathbf{x}^*$ be a particular solution to a consistent linear system $\mathbf{A}\mathbf{x} = \mathbf{b}$. A vector $\mathbf{x}^0$ is a solution to this system if and only if

$$\mathbf{x}^0 = \mathbf{x}^* + \mathbf{z}^0, \tag{3.2.5}$$

for some $\mathbf{z}^0$ which is a solution to the homogeneous system $\mathbf{A}\mathbf{z} = \mathbf{0}$.

Proof. If $\mathbf{x}^0 = \mathbf{x}^* + \mathbf{z}^0$, then $\mathbf{A}\mathbf{x}^0 = \mathbf{A}\mathbf{x}^* + \mathbf{A}\mathbf{z}^0 = \mathbf{b} + \mathbf{0} = \mathbf{b}$, so that $\mathbf{x}^0$ is a solution to (3.2.1). To prove the converse, let $\mathbf{x}^0$ solve the system $\mathbf{A}\mathbf{x} = \mathbf{b}$. Defining $\mathbf{z}^0 = \mathbf{x}^0 - \mathbf{x}^*$ and since $\mathbf{A}\mathbf{z}^0 = \mathbf{A}\mathbf{x}^0 - \mathbf{A}\mathbf{x}^* = \mathbf{b} - \mathbf{b} = \mathbf{0}$, we see that $\mathbf{z}^0$ solves $\mathbf{A}\mathbf{z} = \mathbf{0}$. ■

All vectors in the solution set $\mathbf{A}\mathbf{x} = \mathbf{b}$ can be generated from

$$\mathbf{x} = \mathbf{x}^* + \mathbf{z},$$

where $\mathbf{x}^*$ is a particular solution of (3.2.1) and $\mathbf{z}$ ranges over all vectors in the solution space of $\mathbf{A}\mathbf{z} = \mathbf{0}$.

Result 3.2.6. A vector $\mathbf{x}^0$ is a solution to the system $\mathbf{A}\mathbf{x} = \mathbf{b}$ if and only if, for some vector $\mathbf{y}$,

$$\mathbf{x}^0 = \mathbf{A}^{-}\mathbf{b} + (\mathbf{I} - \mathbf{A}^{-}\mathbf{A})\mathbf{y}. \tag{3.2.6}$$

Proof. The proof follows directly from Results 3.2.3–3.2.5. ■

Example 3.2.3. Let $\mathbf{y} = (y_1, y_2)'$. We see that

$$\mathbf{A}^{-}\mathbf{b} + (\mathbf{I} - \mathbf{A}^{-}\mathbf{A})\mathbf{y} = \begin{pmatrix} y_1 \\ 5/3 - 1/3y_1 \end{pmatrix}. \tag{3.2.7}$$

The solution set then consists of all vectors of the general form (3.2.7). □

Example 3.2.4. Consider the system of equations $\mathbf{A}\mathbf{x} = \mathbf{b}$, where

$$\mathbf{A} = \begin{pmatrix} 1 & -2 & 3 & 2 \\ 1 & 0 & 1 & -3 \\ 1 & 2 & -3 & 0 \end{pmatrix}, \quad \mathbf{x} = \begin{pmatrix} x_1 \\ x_2 \\ x_3 \\ x_4 \end{pmatrix}, \quad \text{and } \mathbf{b} = \begin{pmatrix} 2 \\ -4 \\ -4 \end{pmatrix}.$$

This system is consistent (see Exercise 3.19). A g-inverse of $\mathbf{A}$ is

$$\mathbf{A}^{-} = \begin{pmatrix} 1/2 & 0 & 1/2 \\ -1 & 3/2 & -1/2 \\ -1/2 & 1 & -1/2 \\ 0 & 0 & 0 \end{pmatrix},$$

and a solution to the system of equations is $\mathbf{x} = \mathbf{A}^{-}\mathbf{b} = (-1, -6, -3, 0)'$. Another solution is $\mathbf{x}_1 = \mathbf{A}^{-}\mathbf{b} + (\mathbf{I} - \mathbf{A}^{-}\mathbf{A})\mathbf{y}$, for an arbitrary $\mathbf{y} \in \mathcal{R}^4$. □

Result 3.2.7. If $\mathbf{A}$ is nonsingular, then $\mathbf{Ax} = \mathbf{b}$ has a unique solution given by $\mathbf{A}^{-1}\mathbf{b}$.

Proof. By property 3 of Result 3.2.2, the system $\mathbf{AX} = \mathbf{b}$ is consistent. $\mathbf{Ax} = \mathbf{b}$ implies $\mathbf{A}^{-1}\mathbf{Ax} = \mathbf{A}^{-1}\mathbf{b}$, i.e., $\mathbf{Ix} = \mathbf{A}^{-1}\mathbf{b}$, or $\mathbf{x} = \mathbf{A}^{-1}\mathbf{b}$. ■

Result 3.2.8. Let $\mathbf{Ax} = \mathbf{b}$ denote a consistent linear system and let $\mathbf{c}$ be an n-dimensional vector. The value of $\mathbf{c}'\mathbf{x}$ is the same for every solution to $\mathbf{Ax} = \mathbf{b}$ if and only if $\mathbf{c}' \in \mathcal{R}(\mathbf{A})$ (also see Exercise 3.21).

Proof. If $\mathbf{c}' \in \mathcal{R}(\mathbf{A})$, there exists an m-dimensional vector $\mathbf{t}$ such that $\mathbf{c}' = \mathbf{t}'\mathbf{A}$. Let $\mathbf{x}^*$ and $\mathbf{x}^0$ respectively denote any solution and a particular solution to $\mathbf{Ax} = \mathbf{b}$. Then,

$$\mathbf{c}'\mathbf{x}^* = \mathbf{t}'\mathbf{A}\mathbf{x}^* = \mathbf{t}'\mathbf{b} = \mathbf{t}'\mathbf{A}\mathbf{x}^0 = \mathbf{c}'\mathbf{x}^0,$$

i.e., $\mathbf{c}'\mathbf{x}$ is the same for each solution to $\mathbf{Ax} = \mathbf{b}$. To prove the converse, it follows from Result 3.2.6 that for every $\mathbf{y}$, the value of $\mathbf{c}'[\mathbf{A}^{-}\mathbf{b} + (\mathbf{I} - \mathbf{A}^{-}\mathbf{A})\mathbf{y}]$ remains the same. Then, $\mathbf{c}'(\mathbf{I} - \mathbf{A}^{-}\mathbf{A})\mathbf{y} = \mathbf{0}$ for every $\mathbf{y} \in \mathcal{R}^n$, i.e., $\mathbf{c}'\mathbf{A}^{-}\mathbf{A} = \mathbf{c}'$, i.e., $\mathbf{c}' \in \mathcal{R}(\mathbf{A})$. We will recall this result in the discussion of estimability in the less than full-rank linear model in Chapter 4. ■

We next state without proof (see Harville, 1997, p.155) a result on *absorption*, which enables us to solve a linear system with an arbitrary number of equations in an arbitrary number of unknown variables. Consider the linear system $\mathbf{Ax} = \mathbf{b}$, where $\mathbf{A}$ is an $m \times n$ matrix, while $\mathbf{b}$ and $\mathbf{x}$ are respectively m-dimensional and n-dimensional vectors. Suppose these are partitioned as

$$\mathbf{A} = \begin{pmatrix} \mathbf{A}_{11} & \mathbf{A}_{12} \\ \mathbf{A}_{21} & \mathbf{A}_{22} \end{pmatrix}, \quad \mathbf{b} = \begin{pmatrix} \mathbf{b}_1 \\ \mathbf{b}_2 \end{pmatrix}, \quad \text{and } \mathbf{x} = \begin{pmatrix} \mathbf{x}_1 \\ \mathbf{x}_2 \end{pmatrix},$$

where $\mathbf{A}_{11}$ is $m_1 \times n_1$, $\mathbf{A}_{12}$ is $m_1 \times n_2$, $\mathbf{A}_{21}$ is $m_2 \times n_1$, $\mathbf{A}_{22}$ is $m_2 \times n_2$, $\mathbf{x}_1$ is $n_1 \times 1$, $\mathbf{x}_2$ is $n_2 \times 1$, $\mathbf{b}_1$ is $m_1 \times 1$, and $\mathbf{b}_2$ is $m_2 \times 1$. We can express $\mathbf{Ax} = \mathbf{b}$ as

$$\mathbf{A}_{11}\mathbf{x}_1 + \mathbf{A}_{12}\mathbf{x}_2 = \mathbf{b}_1,$$
$$\mathbf{A}_{21}\mathbf{x}_1 + \mathbf{A}_{22}\mathbf{x}_2 = \mathbf{b}_2.$$

The following result implicitly requires solving the first m_1 equations in $\mathbf{Ax} = \mathbf{b}$ for $\mathbf{x}_1$ in terms of $\mathbf{x}_2$, substituting this solution for $\mathbf{x}_1$ into the last m_2 equations, thereby absorbing the first m_1 equations into the last m_2 equations, solving the resulting *reduced* linear system, and finally back-solving for $\mathbf{x}_1$. This procedure is called *absorption*, and is useful, for example, in the context of solving the system of normal equations in a two-way fixed-effects ANOVA model without interaction (see Example 4.2.6).

Result 3.2.9. Absorption. Consider the linear system $\mathbf{Ax} = \mathbf{b}$, where $\mathbf{A}$, $\mathbf{b}$ and $\mathbf{x}$ are partitioned as above. Suppose $\mathcal{C}(\mathbf{A}_{12}) \subset \mathcal{C}(\mathbf{A}_{11})$, $\mathcal{R}(\mathbf{A}_{21}) \subset \mathcal{R}(\mathbf{A}_{11})$ and $\mathbf{b}_1 \in \mathcal{C}(\mathbf{A}_{11})$.

The vector $\mathbf{x}^* = \begin{pmatrix}\mathbf{x}_1^* \\ \mathbf{x}_2^*\end{pmatrix}$ is a solution to $\mathbf{Ax} = \mathbf{b}$ if and only if $\mathbf{x}_2^*$ is a solution to the linear system (in $\mathbf{x}_2$)

$$(\mathbf{A}_{22} - \mathbf{A}_{21}\mathbf{D})\mathbf{x}_2 = \mathbf{b}_2 - \mathbf{A}_{21}\mathbf{c},$$

where $\mathbf{c} = \mathbf{A}_{11}^- \mathbf{b}_1$, $\mathbf{D} = \mathbf{A}_{11}^- \mathbf{A}_{12}$, and $\mathbf{x}_1^*$ and $\mathbf{x}_2^*$ are solutions to the system (in $\mathbf{x}_1$ and $\mathbf{x}_2$)

$$\mathbf{A}_{11}\mathbf{x}_1 + \mathbf{A}_{12}\mathbf{x}_2 = \mathbf{b}_1.$$

3.3 Linear optimization

Both unconstrained minimization and constrained minimization of a quadratic function of a real vector $\mathbf{x}$ are important topics in the least squares approach to linear model theory and will be discussed in Chapter 4. For more details, see Harville (1997), chapter 19.

3.3.1 Unconstrained minimization

Consider the problem of finding the minimum of a function $f(x_1, x_2)$. The usual approach is to create a set of equations in x_1 and x_2 by setting the partial derivatives of $f(x_1, x_2)$ with respect to x_1 and x_2 to zero, and solving the equations for the minimizing values x_1^* and x_2^*.

For a setup useful for linear models, suppose we wish to minimize the (objective) function of an n-dimensional vector $\mathbf{x}$,

$$f(\mathbf{x}) = \mathbf{x}'\mathbf{V}\mathbf{x} - 2\mathbf{c}'\mathbf{x}, \tag{3.3.1}$$

where $\mathbf{V}$ is a $n \times n$ symmetric n.n.d. matrix and $\mathbf{c} \in \mathcal{C}(\mathbf{V})$. The nature of the unconstrained minimization of $f(\mathbf{x})$ in $\mathbf{x}$ is stated in Result 3.3.1, while Example 3.3.1 shows an example when $n = 2$.

Result 3.3.1. The function $f(\mathbf{x}) = \mathbf{x}'\mathbf{V}\mathbf{x} - 2\mathbf{c}'\mathbf{x}$ has a minimum at $\mathbf{x}^*$ if and only if $\mathbf{x}^*$ solves the system of (consistent) equations $\mathbf{Vx} = \mathbf{c}$. Then, $f(\mathbf{x}^*) = -\mathbf{c}'\mathbf{x}^*$.

Proof. Since $\mathbf{c} \in \mathcal{C}(\mathbf{V})$, consistency of the system $\mathbf{Vx} = \mathbf{c}$ follows from Result 3.2.2.

Sufficiency: Let $\mathbf{x}^*$ be any solution to $\mathbf{Vx} = \mathbf{c}$. For any arbitrary $\mathbf{x} \in \mathcal{R}^n$, we can write

$$\begin{aligned}\mathbf{x}'\mathbf{V}\mathbf{x} &= [\mathbf{x}^* + (\mathbf{x} - \mathbf{x}^*)]'\mathbf{V}[\mathbf{x}^* + (\mathbf{x} - \mathbf{x}^*)] \\ &= \mathbf{x}^{*\prime}\mathbf{V}\mathbf{x}^* + 2\mathbf{c}'(\mathbf{x} - \mathbf{x}^*) + (\mathbf{x} - \mathbf{x}^*)'\mathbf{V}(\mathbf{x} - \mathbf{x}^*),\end{aligned}$$

from which it follows that for any $\mathbf{x}$,

$$f(\mathbf{x}) - f(\mathbf{x}^*) = (\mathbf{x} - \mathbf{x}^*)'\mathbf{V}(\mathbf{x} - \mathbf{x}^*) \geq 0.$$

Necessity: Assume that $f(\mathbf{x}) - f(\mathbf{x}^*) \geq 0$. Let $\mathbf{x}^0$ be any n-dimensional vector such that $f(\mathbf{x}) - f(\mathbf{x}^0) \geq 0$ for all $\mathbf{x}$; in particular, $f(\mathbf{x}^*) - f(\mathbf{x}^0) \geq 0$. On the other hand, by assumption, we had $f(\mathbf{x}^0) - f(\mathbf{x}^*) \geq 0$, which implies $\mathbf{V}(\mathbf{x}^0 - \mathbf{x}^*) = \mathbf{0}$, i.e., $\mathbf{x}^0$ solves $\mathbf{Vx} = \mathbf{c}$. ■

Example 3.3.1. Let $\mathbf{x} = (x_1, x_2)'$, let $f(\mathbf{x}) = x_1^2 + \frac{1}{2}x_1 + 3x_1x_2 + 5x_2^2 = \mathbf{x}'\mathbf{V}\mathbf{x} - 2\mathbf{c}'\mathbf{x}$, where $\mathbf{V} = \begin{pmatrix} 1 & 3/2 \\ 3/2 & 5 \end{pmatrix}$ and $\mathbf{c} = (-1/4, 0)'$. Since $\mathbf{V}$ is p.d., the vector $\mathbf{x}^*$ which solves $\mathbf{Vx} = \mathbf{c}$ is unique and is $\mathbf{x}^* = (-0.455, 0.136)$. □

3.3.2 Constrained minimization

Consider the problem of minimizing a function $f(x_1, x_2)$ subject to a constraint relating x_1 and x_2 which is written as

$$g(x_1, x_2) = 0. \tag{3.3.2}$$

A "substituting and solving" approach to minimization consists of (a) expressing x_2 as a function $h(x_1)$ of x_1 by solving (3.3.2), (b) substituting $x_2 = h(x_1)$ into $f(x_1, x_2)$ to obtain $f(x_1, h(x_1))$, and (c) minimizing the function $f(x_1, h(x_1))$ of a single variable x_1 in the "usual" way using differential calculus.

Example 3.3.2. Let $\mathbf{x} = (x_1, x_2)'$, let $f(\mathbf{x}) = x_1^2 + \frac{1}{2}x_1 + 3x_1x_2 + 5x_2^2$ and let $g(\mathbf{x}) = 3x_1 + 2x_2 + 2 = 0$. In step (a), we write $x_2 = h(x_1) = -\frac{3}{2}x_1 - 1$. In step (b), we obtain

$$f(x_1, h(x_1)) = x_1^2 + \frac{1}{2}x_1 + 3x_1(-\frac{3}{2}x_1 - 1) + 5(-\frac{3}{2}x_1 - 1)^2.$$

Step (c) consists of setting $\frac{d}{dx_1} f(x_1, h(x_1))$ to zero and solving for $x_1^* = -0.8065$. Then, $x_2^* = 0.2097$. □

It may be difficult to employ this approach when n is large or when we have several constraints. A simpler, and more elegant method for constrained minimization of a function $f(\mathbf{x})$ is the Lagrange multiplier approach. The technique of Lagrange multipliers (sometimes called undetermined multipliers) can be used to find the stationary points of a function of several variables subject to one or more constraints. Suppose $\mathbf{x} = (x_1, \cdots, x_n)' \in \mathcal{D} \subset \mathcal{R}^n$ and we wish to minimize $f(\mathbf{x})$ subject to a single constraint equation $g(\mathbf{x}) = 0$.

We denote the Lagrangian function by

$$\mathcal{L}(\mathbf{x}, \lambda) = f(\mathbf{x}) + \lambda g(\mathbf{x}), \tag{3.3.3}$$

where $\lambda \in \mathcal{R}$ is called the Lagrangian multiplier. Note that $\partial\mathcal{L}/\partial\lambda = 0$ leads to the constraint $g(\mathbf{x}) = 0$, while $\partial\mathcal{L}/\partial\mathbf{x} = \mathbf{0}$ involves $\mathbf{x}$ and λ. We find the stationary point of $\mathcal{L}(\mathbf{x}, \lambda)$ with respect to both $\mathbf{x}$ and λ by solving these $n+1$ equations, leading to stationary solutions $\mathbf{x}^*$ and λ^*.

Example 3.3.3. We continue with Example 3.3.2. The Lagrangian function is

$$\mathcal{L}(x_1, x_2, \lambda) = x_1^2 + \frac{1}{2}x_1 + 3x_1x_2 + 5x_2^2 + \lambda(3x_1 + 2x_2 + 2).$$

Differentiating $\mathcal{L}(x_1, x_2, \lambda)$ with respect to x_1, x_2 and λ and setting each partial derivative to zero yields the system of consistent linear equations

$$\begin{pmatrix} 2 & 3 & 3 \\ 3 & 10 & 2 \\ 3 & 2 & 0 \end{pmatrix} \begin{pmatrix} x_1 \\ x_2 \\ \lambda \end{pmatrix} = \begin{pmatrix} -\frac{1}{2} \\ 0 \\ -2 \end{pmatrix}$$

whose solution is $\mathbf{x}^* = (-0.8065, 0.2097)$ which is the constrained minimizer of $f(\mathbf{x})$ subject to $g(\mathbf{x})$; we get $\lambda^* = 0.1613$. □

Note that if we are not interested in λ, we can eliminate it from the equations using the substitution method without the necessity of finding its value (hence λ is called the "undetermined multiplier").

The Lagrangian multiplier technique for constrained minimization can be extended to the situation where there are K constraints $g_j(\mathbf{x})$, $j = 1, \cdots, K$. In this case, the Lagrangian function becomes

$$\mathcal{L}(\mathbf{x}, \boldsymbol{\lambda}) = f(\mathbf{x}) + \sum_{j=1}^{K} \lambda_j g_j(\mathbf{x}), \tag{3.3.4}$$

where $\boldsymbol{\lambda} = (\lambda_1, \cdots, \lambda_K)'$ denotes an arbitrary K-dimensional vector of Lagrangian multipliers. Minimize (3.3.4) with respect to $\mathbf{x}$ and $\boldsymbol{\lambda}$ to obtain $\mathbf{x}^*$ and $\boldsymbol{\lambda}^*$ (Dixon, 1972).

For a setup useful in the theory of linear models (as we will see in Chapters 4 and 7), we show a result (Harville, 1997, section 19.2) to describe the Lagrange multiplier approach for constrained minimization of an objective function $f(\mathbf{x})$ in $\mathbf{x} \in \mathcal{R}^n$ subject to a set of $K \geq 1$ linear constraints given by $\mathbf{A}'\mathbf{x} = \mathbf{b}$, where $\mathbf{A}$ is an $n \times K$ matrix and $\mathbf{b} \in \mathcal{C}(\mathbf{A}')$. These constraints can also be written as $\mathbf{a}_j'\mathbf{x} = b_j$, $j = 1, \cdots, K$. Let $f(\mathbf{x}) = \mathbf{x}'\mathbf{V}\mathbf{x} - 2\mathbf{c}'\mathbf{x}$ where $\mathbf{V}$ is a $n \times n$ symmetric n.n.d. matrix and $\mathbf{c} \in \mathcal{C}(\mathbf{V}, \mathbf{A})$. To minimize $f(\mathbf{x})$ subject to $\mathbf{A}'\mathbf{x} = \mathbf{b}$, we incorporate an arbitrary K-dimensional vector of constants $\boldsymbol{\lambda} = (\lambda_1, \cdots, \lambda_K)'$ and write the Lagrangian function as

$$\mathcal{L}(\mathbf{x}, \boldsymbol{\lambda}) = (\mathbf{x}'\mathbf{V}\mathbf{x} - 2\mathbf{c}'\mathbf{x}) + \boldsymbol{\lambda}'(\mathbf{A}'\mathbf{x} - \mathbf{b}). \tag{3.3.5}$$

The constrained minimizer of $f(\mathbf{x})$ is discussed in Result 3.3.2.

Result 3.3.2. The linear system

$$\begin{pmatrix} 2\mathbf{V} & \mathbf{A} \\ \mathbf{A}' & \mathbf{O} \end{pmatrix} \begin{pmatrix} \mathbf{x} \\ \boldsymbol{\lambda} \end{pmatrix} = \begin{pmatrix} 2\mathbf{c} \\ \mathbf{b} \end{pmatrix} \tag{3.3.6}$$

is consistent. The vector $\mathbf{x}^*$ is the constrained minimizer of $f(\mathbf{x}) = \mathbf{x}'\mathbf{V}\mathbf{x} - 2\mathbf{c}'\mathbf{x}$ subject to the constraint $g(\mathbf{x})$, i.e., $\mathbf{A}'\mathbf{x} = \mathbf{b}$ if and only if there exists a vector $\boldsymbol{\lambda}^*$ so that $(\mathbf{x}^*, \boldsymbol{\lambda}^*)$ solves (3.3.6). Then

$$f(\mathbf{x}^*) = -\mathbf{c}'\mathbf{x}^* - \mathbf{b}'\boldsymbol{\lambda}^*.$$

Proof. The proof is left as Exercise 3.23. ■

Example 3.3.4. We use Result 3.3.2 on the objective function $f(\mathbf{x})$ and linear constraint $g(\mathbf{x})$ from Example 3.3.3 written respectively as $f(\mathbf{x}) = \mathbf{x}'\mathbf{V}\mathbf{x} - 2\mathbf{c}'\mathbf{x}$ and $g(\mathbf{x}) = \mathbf{A}'\mathbf{x} = \mathbf{b}$, with $\mathbf{V} = \begin{pmatrix} 1 & 3/2 \\ 3/2 & 5 \end{pmatrix}$, $\mathbf{c} = (-1/4, 0)'$, $\mathbf{A} = (3, 2)'$ and $b = -2$. Solving the set of equations (3.3.6) yields $x_1^* = -0.8065$ and $x_2^* = 0.2097$ as the constrained minimizing values of x_1 and x_2, while $\lambda^* = 0.1613$. □

The following example shows an application of Result 3.3.2 to a situation with two constraints. R codes for all these examples are shown in Section 3.4.

Example 3.3.5. We minimize the objective function $f(\mathbf{x}) = x_1^2 + \frac{1}{2}x_1 + 3x_1x_2 + 5x_2^2$ subject to two constraints, $g_1(\mathbf{x}) = 3x_1 + 2x_2 + 2 = 0$ and $g_2(\mathbf{x}) = 15x_1 - 3x_2 - 1 = 0$. Let $\boldsymbol{\lambda} = (\lambda_1, \lambda_2)'$. While $\mathbf{V}$ and $\mathbf{c}$ are the same in Example 3.3.4, $\mathbf{A}' = \begin{pmatrix} 3 & 2 \\ 15 & -3 \end{pmatrix}$ and $\mathbf{b} = (-2, 1)'$. We see that $\mathbf{x}^* = (-0.1026, -0.8462)$, and $\boldsymbol{\lambda}^* = (3.5454, -0.5595)'$. □

3.4 R Code

```
## Example 3.1.1. Generalized inverse
library(pracma)
A <- matrix(c(4, 1, 3, 1, 1, 1, 2, 5, 3, 0, 15, 5), nrow = 3)
round(pinv(A), 4) # Moore-Penrose generalized inverse

        [,1]    [,2]    [,3]
[1,]  0.1411 -0.0351  0.0823
[2,]  0.0332 -0.0056  0.0203
[3,]  0.0584  0.0016  0.0395
[4,] -0.0406  0.0641 -0.0057

## Example 3.2.1. Solving nonhomogeneous linear equations, Ax=b
library(matlib)
A <- matrix(c(2, -1, 6, 1, -2, 0, -6, 2, 0, 3, 1, -7, 4, 1,
              8, -16), 4, 4)
A
     [,1] [,2] [,3] [,4]
[1,]    2   -2    0    4
[2,]   -1    0    3    1
[3,]    6   -6    1    8
[4,]    1    2   -7  -16

b <- c(2, 6, 12, -7)
showEqn(A, b)

 2*x1 - 2*x2 + 0*x3  + 4*x4  =   2
-1*x1 + 0*x2 + 3*x3  + 1*x4  =   6
 6*x1 - 6*x2 + 1*x3  + 8*x4  =  12
 1*x1 + 2*x2 - 7*x3 - 16*x4  =  -7

c(R(A), R(cbind(A, b)))

[1] 4 4

all.equal(R(A), R(cbind(A, b)))

[1] TRUE

Solve(A, b, fractions = TRUE)

x1         =  -1
  x2       =  -4
    x3     =   2
      x4   =  -1

## Example 3.2.2. Solve Ax=b, plotting equations
A <- matrix(c(1, 3, 2, 6), 2, 2)
b <- c(5, 10)
```

```
showEqn(A, b)

1*x1 + 2*x2  =   5
3*x1 + 6*x2  =  10

plotEqn(A, b, solution = T) #graph not shown here
Solve(A, b)

x1 + 2*x2  =  3.33333333
        0  =  1.66666667

## Example 3.2.4 (modified). Solving homogeneous linear
## equations; Az=0
A <- matrix(c(1,-2, 3, 2, 1, 0, 1,-3, 1, 2 ,-3, 0),
            nrow = 3,
            byrow = T)
b <- c(0, 0, 0)
showEqn(A, b)
# Check: homogeneous equations must always be consistent
all.equal(R(A), R(cbind(A, b)))
# Solution space is N(A) with dimension L
(n <- ncol(A))
(L <- n - R(A))
# Check whether there is a unique solution, i.e., is L=0?
(L == 0)
Solve(A, b, fractions = TRUE)    # use Solve
# Basis of the non-unique solution space:
# Z0 = (I- GA)y for some y
G <- MASS::ginv(A)    # generalized-inverse
# four y vectors = columns of I_4 to get basis of solution space
Z0 <- (diag(4) - (G %*% A)) %*% diag(4)
echelon(t(Z0))
```

Exercises

3.1. Find a g-inverse of

$$\text{(a) } \mathbf{A} = \begin{pmatrix} 1 & 2 \\ 1 & 1 \\ -1 & 0 \end{pmatrix}, \text{ and (b) } \mathbf{A} = \begin{pmatrix} 1 & 2 & 4 & 3 \\ 3 & -1 & 2 & -2 \\ 5 & -4 & 0 & -7 \end{pmatrix}.$$

3.2. Let $\mathbf{A} = \{a_{ij}\}$ be an $n \times n$ upper diagonal matrix such that $a_{i,i+1} = 1$ while all other $a_{ij} = 0$. Find all the g-inverses of $\mathbf{A}$.

3.3. Let $\mathbf{K}$ be an idempotent matrix, let $\mathbf{B} = \mathbf{KAK}$, and let $\mathbf{B}^-$ denote a g-inverse of $\mathbf{B}$. Show that $\mathbf{KB}^-\mathbf{K}$ is also a g-inverse of $\mathbf{B}$.

3.4. If $\mathbf{G} = \mathbf{A}^-$ and $\mathbf{AG} = \mathbf{GA}$, then for any positive integer k, show that $\mathbf{G}^k$ is a g-inverse of $\mathbf{A}^k$.

3.5. Let $\mathbf{G}$ be a g-inverse of a matrix $\mathbf{A}'\mathbf{A}$. Show that

(a) $\mathbf{A}\mathbf{G}'\mathbf{A}'\mathbf{A} = \mathbf{A}$.

(b) $\mathbf{A}'\mathbf{A}\mathbf{G}\mathbf{A}' = \mathbf{A}'$.

(c) $\mathbf{A}'\mathbf{A}\mathbf{G}'\mathbf{A}' = \mathbf{A}'$.

(d) $\mathbf{A}\mathbf{G}'\mathbf{A}' = \mathbf{A}\mathbf{G}\mathbf{A}'$.

(e) $\mathbf{A}\mathbf{G}'\mathbf{A}'$ is symmetric.

3.6. Let $\mathbf{B}^-\mathbf{A}^-$ be a g-inverse of $\mathbf{AB}$. Show that $\mathbf{A}^-\mathbf{A}\mathbf{B}\mathbf{B}^-$ is idempotent.

3.7. Let $\mathbf{A}$ be an $n \times n$ symmetric matrix of rank $r < n$, and let $\mathbf{A} = \mathbf{C}\mathbf{C}'$, where $\mathrm{r}(\mathbf{C}) = r$. If $\mathbf{A}^-$ denotes a g-inverse of $\mathbf{A}$, show that $(\mathbf{A}^-\mathbf{C})(\mathbf{A}^-\mathbf{C})'$ is also a g-inverse of $\mathbf{A}$.

3.8. Let $\mathbf{G}$ be a g-inverse of $\mathbf{A}$. Show that $\mathbf{A}$ is also a g-inverse of $\mathbf{G}$ if and only if $\mathrm{r}(\mathbf{G}) = \mathrm{r}(\mathbf{A})$. In particular, this holds for the g-inverses obtained in Result 3.1.2.

3.9. Let $\mathbf{G}$ be a g-inverse of $\mathbf{A}$. Show that $\mathbf{AG}$ is symmetric if and only if any of the following conditions holds:

(a) $\mathbf{AG}$ is the matrix of orthogonal projection onto $\mathcal{C}(\mathbf{A})$.

(b) $\mathbf{G}$ has the form $\mathbf{G} = (\mathbf{A}'\mathbf{A})^-\mathbf{A}' + (\mathbf{I} - \mathbf{A}^-\mathbf{A})\mathbf{U}$ for some matrix $\mathbf{U}$.

3.10. [Englefield (1966)]. Let $\mathbf{A}$ be an $n \times n$ matrix. If $\mathrm{r}(\mathbf{A}) = r \leq n$, show that any one of the following conditions is necessary and sufficient for the existence of a commuting g-inverse of $\mathbf{A}$:

(a) $\mathrm{r}(\mathbf{A}) = \mathrm{r}(\mathbf{A}^2)$.

(b) there exists a nonsingular matrix $\mathbf{B}$ such that $\mathbf{B}\mathbf{A}\mathbf{B}^{-1} = \begin{pmatrix} \mathbf{A}_1 & \mathbf{O} \\ \mathbf{O} & \mathbf{O} \end{pmatrix}$, where $\mathbf{A}_1$ is an $r \times r$ nonsingular matrix.

3.11. Show that a symmetric matrix $\mathbf{A}$ always has a commuting g-inverse and that it is also possible to construct a symmetric commuting g-inverse of $\mathbf{A}$.

3.12. Let $\mathbf{G}_1$ and $\mathbf{G}_2$ be two g-inverses of an $m \times n$ matrix $\mathbf{A}$. Let $\mathbf{x}$ be any vector such that $\mathbf{A}\mathbf{G}_1\mathbf{x} = \mathbf{x}$. Show that $\mathbf{A}\mathbf{G}_2\mathbf{x} = \mathbf{x}$.

3.13. In Example 3.1.4, let $\mathbf{A} = \mathbf{CB}$. Show that $\mathbf{p}_i'\mathbf{A} = \mathbf{p}_i'$, $i = 1, \cdots, r$, while $\mathbf{p}_i'\mathbf{A} = \mathbf{0}$, $i = r+1, \cdots, k$.

3.14. Prove Result 3.1.14.

3.15. Show that $\mathbf{PX} = \mathbf{X}$, where $\mathbf{P} = \mathbf{X}(\mathbf{X}'\mathbf{X})^-\mathbf{X}'$ denotes the projection matrix onto $\mathcal{C}(\mathbf{X})$.

3.16. Suppose a square matrix $\mathbf{M}$ is partitioned as $\mathbf{M} = \begin{pmatrix} \mathbf{A}'\mathbf{A} & \mathbf{A}'\mathbf{B} \\ \mathbf{B}'\mathbf{A} & \mathbf{B}'\mathbf{B} \end{pmatrix}$. Show that a g-inverse of $\mathbf{M}$ is given by

$$\mathbf{M}^- = \begin{pmatrix} (\mathbf{A}'\mathbf{A})^- & \mathbf{O} \\ \mathbf{O} & \mathbf{O} \end{pmatrix} + \begin{pmatrix} -(\mathbf{A}'\mathbf{A})^-\mathbf{A}'\mathbf{B} \\ \mathbf{I} \end{pmatrix} \mathbf{S}^- \left(-\mathbf{B}'\mathbf{A}(\mathbf{A}'\mathbf{A})^- \quad \mathbf{I}\right),$$

where $\mathbf{S} = \mathbf{B}'\mathbf{B} - \mathbf{B}'\mathbf{A}(\mathbf{A}'\mathbf{A})^-\mathbf{A}'\mathbf{B}$.

3.17. Let $\mathbf{B}$ be an $m \times n$ matrix and let $\mathbf{A}$ and $\mathbf{C}$ respectively denote nonsingular $m \times m$ and $n \times n$ matrices. Show that

(a) $\mathbf{G}$ is a g-inverse of $\mathbf{AB}$ if and only if $\mathbf{G} = \mathbf{B}^{-}\mathbf{A}^{-1}$; and

(b) $\mathbf{G}$ is a g-inverse of $\mathbf{BC}$ if and only if $\mathbf{G} = \mathbf{C}^{-1}\mathbf{B}^{-}$.

3.18. Given the matrices defined in Exercise 2.4, show that a g-inverse of the matrix $\begin{pmatrix} \mathbf{A} & \mathbf{AB} \\ \mathbf{CA} & \mathbf{D} \end{pmatrix}$ is

$$\begin{pmatrix} \mathbf{A}^{-} + \mathbf{BE}^{-}\mathbf{C} & -\mathbf{BE}^{-} \\ -\mathbf{E}^{-}\mathbf{C} & \mathbf{E}^{-} \end{pmatrix} = \begin{pmatrix} \mathbf{A}^{-} & \mathbf{O} \\ \mathbf{O} & \mathbf{O} \end{pmatrix} + \begin{pmatrix} -\mathbf{B} \\ \mathbf{I}_r \end{pmatrix} \mathbf{E}^{-} \left(-\mathbf{C}, \;\; \mathbf{I}_{n-r}\right).$$

3.19. Show that the system of equations $\mathbf{Ax} = \mathbf{b}$ in Example 3.2.4 is consistent.

3.20. Solve the system of linear equations $\mathbf{Ax} = \mathbf{b}$ where

$$\mathbf{A} = \begin{pmatrix} 2 & -2 & 0 & 4 \\ -1 & 0 & 3 & 1 \\ 6 & -6 & 1 & 8 \\ 1 & 2 & -7 & -16 \end{pmatrix}, \text{ and } \mathbf{b} = \begin{pmatrix} 2 \\ 6 \\ 12 \\ -7 \end{pmatrix}.$$

3.21. Show that $\mathbf{c}'\mathbf{x}$ has a unique value for each solution to $\mathbf{Ax} = \mathbf{b}$ if and only if $\mathbf{c}'\mathbf{A}^{-}\mathbf{A} = \mathbf{c}'$, where $\mathbf{A}^{-}$ is any g-inverse of the matrix $\mathbf{A}$.

3.22. Consider the matrix $\begin{pmatrix} \mathbf{A} & \mathbf{b} \\ \mathbf{c}' & 0 \end{pmatrix}$, where $\mathbf{A}$ is an $m \times n$ matrix, $\mathbf{b}$ is an m-dimensional vector which belongs to the column space of $\mathbf{A}$, and $\mathbf{c}$ is an n-dimensional vector which belongs to the row space of $\mathbf{A}$. Show that the value of $\mathbf{c}'\mathbf{x}$ is the same at all solutions of $\mathbf{Ax} = \mathbf{b}$ and is given by $\mathbf{c}'\mathbf{A}^{-}\mathbf{b}$ where $\mathbf{A}^{-}$ is a g-inverse of $\mathbf{A}$.

3.23. Prove Result 3.3.2.

4

General Linear Model

The theory of linear statistical models underlies several important and widely used procedures such as univariate and multivariate regression analysis, analysis of variance, analysis of covariance, random-effects modeling, time series analysis, spatial analysis, etc. In this chapter, we introduce the general linear model (GLM) to explain an unknown response vector as a linear function of known predictors and an unknown vector of model parameters. We discuss general results for the situation where the matrix of predictors need not have full rank. We define the notion of estimability of linear functions of the vector of model parameters, and derive the least squares estimates of such functions. The Gauss–Markov theorem guarantees an important optimality property for these estimates. We discuss optimality under the least squares (LS) framework as well as under the generalized least squares (GLS) framework which assumes that the error vector has a p.d. covariance matrix which is not the identity matrix. We also construct LS estimates for the GLM subject to linear constraints on the model parameters. The results in this chapter are distribution free, so that we do not need to specify a probability distribution for the model errors.

4.1 Model definition and examples

Given data $(Y_i, X_{i1}, X_{i2}, \cdots, X_{ik})$, $i = 1, \cdots, N$, the general linear model has the form

$$\mathbf{y} = \mathbf{X}\boldsymbol{\beta} + \boldsymbol{\varepsilon}, \tag{4.1.1}$$

where $\mathbf{y} = (Y_1, \cdots, Y_N)'$ is an N-dimensional vector of observed responses, $\boldsymbol{\beta} = (\beta_0, \beta_1, \cdots, \beta_k)'$ is a $(k+1)$-dimensional vector of unknown parameters, $\mathbf{X}$ is an $N \times (k+1)$ matrix of known predictors of rank r, and $\boldsymbol{\varepsilon} = (\varepsilon_1, \cdots, \varepsilon_N)'$ is an N-dimensional random vector of unobserved errors. The matrix $\mathbf{X}$ is written as

$$\mathbf{X} = \begin{pmatrix} 1 & X_{11} & \cdots & X_{1k} \\ 1 & X_{21} & \cdots & X_{2k} \\ \vdots & \vdots & \vdots & \vdots \\ 1 & X_{N1} & \cdots & X_{Nk} \end{pmatrix}.$$

For convenience of notation, we let $p = k+1$. Note that unless specified otherwise, we assume that the first column of $\mathbf{X}$ is the vector $\mathbf{1}_N = (1, \cdots, 1)'$ and the other columns are vectors of known predictors, so that the first coefficient in $\boldsymbol{\beta}$ is the intercept. We can write $\mathbf{X} = (\mathbf{1}_N, \widetilde{\mathbf{X}})$, where $\widetilde{\mathbf{X}}$ is an $N \times k$ matrix whose columns correspond to the k predictors. Note that we may also write the model (4.1.1) as

$$Y_i = \beta_0 + \beta_1 X_{i1} + \cdots + \beta_k X_{ik} + \varepsilon_i, \quad i = 1, \cdots, N, \tag{4.1.2}$$

or as

$$Y_i = \mathbf{x}_i'\boldsymbol{\beta} + \varepsilon_i, \quad i = 1, \cdots, N, \tag{4.1.3}$$

DOI: 10.1201/9781315156651-4

where $\mathbf{x}_i = (1, X_{i1}, \cdots, X_{ik})'$ denotes a $p \times 1$ vector corresponding to the ith subject. Occasionally, we will consider a model without intercept. In this case $\mathbf{X} = \widetilde{\mathbf{X}}$ and β_0 is not present in (4.1.2). To keep the discussion meaningful, we shall always assume that $\mathbf{X} \neq \mathbf{O}$. Whether or not the general linear model (4.1.1) has an intercept, it will be approached the same way in the book.

Suppose

$$\mathrm{E}(\boldsymbol{\varepsilon}) = \mathbf{0} \quad \text{and} \quad \mathrm{Cov}(\boldsymbol{\varepsilon}) = \sigma^2 \mathbf{I}_N, \tag{4.1.4}$$

i.e., the errors are uncorrelated, each with zero mean and the same variance σ^2. In Chapter 7, where we discuss detailed inference for the linear model, we will assume some probability distribution for the error vector, usually the normal distribution. Here, we see that it follows from (4.1.4) and the properties of the expectation and covariance operators that (see Appendix A)

$$\begin{aligned} \mathrm{E}(\mathbf{y}) &= \mathrm{E}(\mathbf{X}\boldsymbol{\beta} + \boldsymbol{\varepsilon}) = \mathbf{X}\boldsymbol{\beta} + \mathrm{E}(\boldsymbol{\varepsilon}) = \mathbf{X}\boldsymbol{\beta} \quad \text{and} \\ \mathrm{Cov}(\mathbf{y}) &= \mathrm{Cov}(\mathbf{X}\boldsymbol{\beta} + \boldsymbol{\varepsilon}) = \mathrm{Cov}(\boldsymbol{\varepsilon}) = \sigma^2 \mathbf{I}_N. \end{aligned} \tag{4.1.5}$$

Multiple regression models, fixed-effects and random-effects analysis of variance (ANOVA) models, and analysis of covariance (ANACOVA) models that are frequently encountered in applied statistics all fall under the umbrella of the general linear model. In regression models, the X_j's may be continuous or categorical observed variables, whereas in the family of ANOVA models, the explanatory variables generally correspond to levels of different factors of interest in designed experiments. The symmetric matrix $\mathbf{X}'\mathbf{X}$, and the symmetric, idempotent matrices $\mathbf{P} = \mathbf{X}(\mathbf{X}'\mathbf{X})^{-}\mathbf{X}'$ (the orthogonal projection matrix, or hat matrix, or prediction matrix), and $\mathbf{I} - \mathbf{P}$ play a central role in the development of statistical theory for linear models. If $\mathrm{r}(\mathbf{X}'\mathbf{X}) = p$, we have the *full-rank model*, while $\mathrm{r}(\mathbf{X}'\mathbf{X}) = r < p$ corresponds to the *less than full-rank model*, or the *design model*. The multiple regression model is a "full-rank linear model", whereas the ANOVA models are examples of "less than full-rank linear models". In this chapter, we describe the form of the general linear model, derive the least squares estimates of relevant parameters, and prove an important theorem called the Gauss–Markov Theorem. This theorem states a useful optimality result for the least squares estimators of certain linear parametric functions of the parameter vector $\boldsymbol{\beta}$. We begin with some examples of the linear model.

Example 4.1.1. Simple linear regression. Suppose there is a single continuous-valued response (dependent variable) Y, and a single continuous-valued predictor (independent variable) X, and we wish to explain the variability in Y due to X. The simple linear regression model has the form (4.1.2) with $k = 1$:

$$Y_i = \beta_0 + \beta_1 X_i + \varepsilon_i, \quad i = 1, \cdots, N. \tag{4.1.6}$$

We have postulated a straight line relationship between X and Y, with intercept β_0 and slope β_1. Both β_0 and β_1 are unknown model parameters, which must be estimated together with the error variance σ^2, based on data pairs (X_i, Y_i), $i = 1, \cdots, N$. The usual interpretation for β_0 is that it is the value assumed by Y when $X = 0$. When $\beta_0 = 0$, (4.1.6) reduces to a simple linear regression model without intercept, represented by a straight line through the origin. The coefficient β_1 is the change in Y for a unit increase in X. When $\beta_1 > 0$, there is a positive association between X and Y, so that Y is expected to increase as X increases. When $\beta_1 < 0$, there is an inverse or negative relationship between the two variables, while if $\beta_1 = 0$, Y is unaffected by changes in X. When β_1 is very close to zero, we interpret it as a weak, or virtually nonexistent linear regression relationship. A scatterplot of Y versus X

is one of the first exploratory data analysis (EDA) steps that enables a quick assessment of the validity of a linear model fit to the data.

Suppose we transform both the response and predictor variables as follows: $Y^* = c_1 + c_2Y$, and $X^* = d_1 + d_2X$. We investigate the effect of this data transformation on the regression coefficients. Suppose the regression model on the transformed variables is

$$Y^* = \beta_0^* + \beta_1^* X^*.$$

After some algebra, we can see that

$$\beta_1^* = c_2\beta_1/d_2, \text{ and } \beta_0^* = c_1 - \{c_2d_1\beta_1/d_2\} + c_2\beta_0.$$

The slope is unaffected by a shift in the location but is affected by a scale change, while the intercept is affected by both a location shift and a scale change. □

Example 4.1.2. Multiple linear regression. We relate a single continuous-valued response Y to multiple predictors $X_1, \cdots, X_k$, $k > 1$, using a linear model of the form (4.1.2), which represents a hyperplane in $\mathcal{R}^p$. The response surface has a linear functional form whose parameters are the coefficients β_j, $j = 1, \cdots, k$, which are called partial regression coefficients. We interpret β_j as the amount by which Y is expected to change when X_j is increased by one unit, while all other predictor variables are held at fixed values. The coefficient β_0 corresponds to the constant term. If a particular β_j is zero, the interpretation is that the corresponding predictor X_j is unrelated to Y (in a linear model), and may be dropped from the regression. A multivariate scatterplot is recommended as a preliminary step which enables us to assess the usefulness of a linear model fit to such data. Let $\overline{Y} = \sum_{i=1}^N Y_i/N$ and $\overline{X}_j = \sum_{i=1}^N X_{ij}/N$, $j = 1, \cdots, k$. By expressing each observation on the response and the predictors in terms of deviations from their respective means, we may write (4.1.2) in *deviations form* as

$$y_i = \beta_1 x_{i1} + \cdots + \beta_k x_{ik} + \varepsilon_i,$$

where $y_i = Y_i - \overline{Y}$, and $x_{ij} = X_{ij} - \overline{X}_j$, $j = 1, \cdots, k$.

The *centered and scaled form* of the multiple linear regression model is

$$Y_i^* = \beta_1^* X_{i1}^* + \beta_2^* X_{i2}^* + \cdots + \beta_k^* X_{ik}^* + \varepsilon_i,$$

where $\beta_j^* = \beta_j(S_{jj}/S_{yy})^{1/2}$, $j = 1, \cdots, k$ and

$$X_{ij}^* = (X_{ij} - \overline{X}_j)/\sqrt{S_{jj}}, \text{ and } Y_i^* = (Y_i - \overline{Y})/\sqrt{S_{yy}},$$

for $i = 1, \cdots, N$, $j = 1, \cdots, k$. Let $\mathbf{X}^*$ denote the regression matrix of centered and scaled variables, i.e.,

$$\mathbf{X}^* = \begin{pmatrix} X_{11}^* & X_{12}^* & \cdots & X_{1k}^* \\ X_{21}^* & X_{22}^* & \cdots & X_{2k}^* \\ \vdots & \vdots & \vdots & \vdots \\ X_{N1}^* & X_{N2}^* & \cdots & X_{Nk}^* \end{pmatrix},$$

so that $\mathbf{X}^{*\prime}\mathbf{X}^*$ denotes the $k \times k$ correlation matrix of the explanatory variables, excluding the intercept column, which becomes zero by the centering.

Standardized coefficients describe the relative importance of the explanatory variables in the model. They are the partial regression coefficients in the model where each variable is normalized by subtracting its sample mean and dividing by its sample standard deviation.

A comparison of the different forms yields the following relationship between the partial regression coefficients and the corresponding standardized coefficients:

$$\beta_j^* = \beta_j s_{X_j}/s_Y, \quad j = 1, \cdots, k.$$

We see that the standardized coefficients have the same sign as the β_j's. A standardized coefficient of 0.6 means that an increase of one standard deviation in the independent variable is expected to cause a change of 0.6 standard deviations in the response variable. An elasticity measures the percent change in the response variable Y corresponding to an increase of 1% in the explanatory variable. We end this example with the notion of column-equilibrating the matrix $\mathbf{X}$. This consists of dividing each column of $\mathbf{X}$ by the sum of squares of its elements. The sum of squares of the elements in each column of the resulting matrix $\mathbf{X}_E$, say, should be 1 since $X_{E,ij} = X_{ij}/\sum_{l=1}^N X_{lj}^2$. □

Example 4.1.3. One-way fixed-effects ANOVA model. Consider the model

$$Y_{ij} = \mu + \tau_i + \varepsilon_{ij}, \quad j = 1, \cdots, n_i, \ i = 1, \cdots, a, \tag{4.1.7}$$

which can be written in the form (4.1.1) with

$$\begin{aligned}
\mathbf{y} &= (Y_{11}, \cdots, Y_{1,n_1}, \cdots, Y_{a1}, \cdots, Y_{a,n_a})', \\
\boldsymbol{\varepsilon} &= (\varepsilon_{11}, \cdots, \varepsilon_{1,n_1}, \cdots, \varepsilon_{a1}, \cdots, \varepsilon_{a,n_a})', \\
\boldsymbol{\beta} &= (\mu, \tau_1, \cdots, \tau_a)', \\
\mathbf{X} &= [\mathbf{1}_N \quad \mathbf{1}_{n_1} \oplus \cdots \oplus \mathbf{1}_{n_a}]
\end{aligned}$$

where $N = \sum_{i=1}^a n_i$, $\mathrm{E}(\boldsymbol{\varepsilon}) = \mathbf{0}$, $\mathrm{Cov}(\boldsymbol{\varepsilon}) = \sigma^2 \mathbf{I}_N$, and the direct sum of vectors was introduced in Definition 2.8.3. In general, Y_{ij} represents the observed response from the jth subject in the ith group, where $j = 1, \cdots, n_i$, and $i = 1, \cdots, a$. The design or incidence matrix $\mathbf{X}$ consists of 1's and 0's and $p = a+1$. The ith column of $\mathbf{X}$ has 1's in its $\sum_{k=1}^{i-1}(n_k+1)$th row to its $\sum_{k=1}^{i} n_k$th row, and zeroes elsewhere. It is easy to verify that $\mathrm{r}(\mathbf{X}) = r = a$, since the last a columns of the design matrix add up to the first column, imposing one dependence. For example, when $a = 3$, $n_1 = 3$, $n_2 = 2$, and $n_3 = 1$, we write

$$\begin{aligned}
Y_{1j} &= \mu + \tau_1 + \varepsilon_{1j}, \quad j = 1, 2, 3, \\
Y_{2h} &= \mu + \tau_2 + \varepsilon_{2h}, \quad h = 1, 2, \\
Y_{31} &= \mu + \tau_3 + \varepsilon_{31},
\end{aligned}$$

so that $\mathrm{E}(Y_{1j}) = \mu + \tau_1$, $j = 1, 2, 3$, $\mathrm{E}(Y_{2h}) = \mu + \tau_2$, $h = 1, 2$ and $\mathrm{E}(Y_{31}) = \mu + \tau_3$. In this case, the response vector is $\mathbf{y} = (Y_1, \cdots, Y_6)'$, the parameter vector is $\boldsymbol{\beta} = (\mu, \tau_1, \tau_2, \tau_3)'$, the error vector is $\boldsymbol{\varepsilon} = (\varepsilon_1, \cdots, \varepsilon_6)'$ and the design matrix is

$$\mathbf{X} = \begin{pmatrix} 1 & 1 & 0 & 0 \\ 1 & 1 & 0 & 0 \\ 1 & 1 & 0 & 0 \\ 1 & 0 & 1 & 0 \\ 1 & 0 & 1 & 0 \\ 1 & 0 & 0 & 1 \end{pmatrix},$$

with $\mathrm{r}(\mathbf{X}) = 3$. When $n_i = n$ for $i = 1, \cdots, a$, we say the model is *balanced*; otherwise, it is an *unbalanced* model. Using Kronecker product notation (see Definition 2.8.1), it is easy to verify that the design matrix in a balanced model can be written as

$$\mathbf{X} = [(\mathbf{1}_a, \mathbf{I}_a) \otimes \mathbf{1}_n]$$ □

Example 4.1.4. One-way random-effects ANOVA model. Consider the balanced version of the model in (4.1.8)

$$Y_{ij} = \mu + \gamma_i + \varepsilon_{ij}, \quad j = 1, \cdots, n, \ i = 1, \cdots, a, \tag{4.1.8}$$

where the errors are independently distributed with $\mathrm{E}(\varepsilon_{ij}) = 0$, and $\mathrm{Var}(\varepsilon_{ij}) = \sigma_\varepsilon^2$. Unlike the fixed-effects model, the γ_i's are no longer fixed unknown constants, but are themselves assumed to be independent random variables, with $\mathrm{E}(\gamma_i) = 0$, and $\mathrm{Var}(\gamma_i) = \sigma_\gamma^2$. We also assume that the γ_i's and ε_{ij}'s are independently distributed. In this case, it is straightforward to verify that

$$\mathrm{Cov}(Y_{ij}, Y_{i',j'}) = \begin{cases} \sigma_\gamma^2 + \sigma_\varepsilon^2, & i = i', j = j', \\ \sigma_\gamma^2, & i = i', j \neq j', \\ 0, & i \neq i'. \end{cases} \tag{4.1.9}$$

Concisely, we can write

$$\begin{aligned} \mathbf{y} &= (\mathbf{1}_a \otimes \mathbf{1}_n)\mu + (\mathbf{I}_a \otimes \mathbf{1}_n)\boldsymbol{\gamma} + \boldsymbol{\varepsilon} \text{ and} \\ \mathrm{Cov}(\mathbf{y}) &= \mathbf{I}_a \otimes (\sigma_\gamma^2 \mathbf{J}_n + \sigma_\varepsilon^2 \mathbf{I}_n). \end{aligned} \tag{4.1.10}$$

The quantity

$$\rho = \frac{\sigma_\gamma^2}{\sigma_\gamma^2 + \sigma_\varepsilon^2} \tag{4.1.11}$$

is called the intra-class correlation. We will discuss inference for random-effects models in Chapter 11. We will also study mixed-effects models which are useful in designed experiments where some factors are fixed, while others are random. □

In the general linear model (4.1.1), the parameter vector $\boldsymbol{\beta}$ and the error variance σ^2 are usually unknown. The least squares approach, which is described in the next section, is a simple and elegant procedure that enables estimation of functions of the parameters in a linear model.

4.2 Least squares approach

Given data on the response variable and predictors, either from an observational study or from a designed experiment, the objective is inference on the model parameters, or functions of the model parameters, as well as predictions for the response variable based on the general linear model (4.1.1). The method of least squares, which was introduced in the early 19th century, enables such inference under minimal assumptions. In particular, we need not specify any parametric form for the probability distribution of the errors ε_i. In the full-rank linear model, i.e., when $\mathrm{r}(\mathbf{X}) = p = k+1$, the least squares approach enables us to construct the best linear unbiased estimator of the parameter vector $\boldsymbol{\beta}$, "best" in the sense of having minimum variance in the class of all linear unbiased estimators. When $\mathrm{r}(\mathbf{X}) = r < p$, we will obtain least squares estimates of certain linear functions of $\boldsymbol{\beta}$; although, as we shall see, there does not exist a unique estimator for $\boldsymbol{\beta}$ itself. In order to proceed with inference beyond point estimation and prediction for the linear model, i.e., in order to construct confidence interval estimates or to do hypothesis tests, it is usual to assume some parametric form for the error

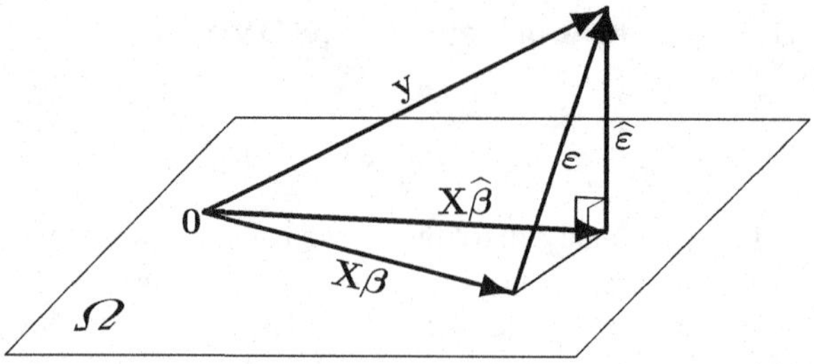

FIGURE 4.2.1. Geometry of least squares.

distribution. The simplest and most popular distributional assumption is the assumption of normality of the linear model errors. In Chapter 5, we introduce suitable families of multivariate probability distributions, including the multivariate normal distribution, and return to classical inference for linear models in Chapter 7. We now describe the least squares principle.

Geometrically, the response $\mathbf{y} = (Y_1, \cdots, Y_N)$ represents a point (or a vector from the origin $\mathbf{0}$) in the N-dimensional Euclidean space $\mathcal{R}^N$. Let $\widetilde{\mathbf{x}}_j = (X_{1,j}, \cdots, X_{N,j})'$, $j = 1, \cdots, k$, so that $\mathbf{X} = (\mathbf{1}_N, \widetilde{\mathbf{X}}) = (\mathbf{1}_N, \widetilde{\mathbf{x}}_1, \cdots, \widetilde{\mathbf{x}}_k)$, and let $\mathcal{C}(\mathbf{X})$ denote the vector subspace of $\mathcal{R}^N$ defined by these columns of $\mathbf{X}$ (see Definition 1.3.13). The vector $\mathbf{X}\boldsymbol{\beta} = \beta_0\mathbf{1}_N + \beta_1\widetilde{\mathbf{x}}_1 + \cdots + \beta_k\widetilde{\mathbf{x}}_k$ is in $\mathcal{C}(\mathbf{X})$. When $\mathrm{r}(\mathbf{X}) = p$, the p LIN columns of $\mathbf{X}$ span the p-dimensional *estimation space* $\Omega = \mathcal{C}(\mathbf{X})$. In Figure 4.2.1, the points $\mathbf{0}, \mathbf{y}$, and $\mathbf{X}\boldsymbol{\beta}$ form a triangle in $\mathcal{R}^N$, whose sides are the vectors $\mathbf{y}, \mathbf{X}\boldsymbol{\beta}$, and $\boldsymbol{\varepsilon} = \mathbf{y} - \mathbf{X}\boldsymbol{\beta}$. The method of least squares consists of minimizing $\boldsymbol{\varepsilon}'\boldsymbol{\varepsilon} = \|\mathbf{y} - \mathbf{X}\boldsymbol{\beta}\|^2$ with respect to $\boldsymbol{\beta}$, or equivalently minimizing $\|\mathbf{y} - \boldsymbol{\theta}\|^2$ where $\boldsymbol{\theta} = \mathbf{X}\boldsymbol{\beta} \in \mathcal{C}(\mathbf{X})$. As $\boldsymbol{\theta}$ varies in $\mathcal{C}(\mathbf{X})$, the square of the length of the vector $\mathbf{y} - \boldsymbol{\theta}$ will be minimum when $\boldsymbol{\theta} = \widehat{\boldsymbol{\theta}}$, for $\widehat{\boldsymbol{\theta}}$ in $\mathcal{C}(\mathbf{X})$. Then, $\mathbf{X}'(\mathbf{y} - \widehat{\boldsymbol{\theta}}) = \mathbf{0}$ or $\mathbf{X}'\widehat{\boldsymbol{\theta}} = \mathbf{X}'\mathbf{y}$. The vector $\widehat{\boldsymbol{\theta}}$ denotes the unique orthogonal projection of $\mathbf{y}$ onto $\mathcal{C}(\mathbf{X})$, so $\mathbf{X}\widehat{\boldsymbol{\beta}}$ (which is equal to $\widehat{\boldsymbol{\theta}}$) is unique. In other words, when $\mathbf{X}$ has full rank p, the least squares estimate of $\boldsymbol{\beta}$ is the vector $\widehat{\boldsymbol{\beta}}$ which uniquely minimizes the quadratic form $(\mathbf{y} - \mathbf{X}\boldsymbol{\beta})'(\mathbf{y} - \mathbf{X}\boldsymbol{\beta})$, which denotes the squared length of the vector joining $\mathbf{y}$ and $\mathbf{X}\boldsymbol{\beta}$ in Figure 4.2.1. From the figure, we also see that this quadratic form is minimum when $\mathbf{X}\widehat{\boldsymbol{\beta}}$ joins the origin $\mathbf{0}$ to the foot of the perpendicular (denoted by $\widehat{\mathbf{y}}$) from $\mathbf{y}$ onto the estimation space Ω. The fitted vector $\widehat{\mathbf{y}} = \mathbf{X}\widehat{\boldsymbol{\beta}}$ is orthogonal to $\widehat{\boldsymbol{\varepsilon}} = \mathbf{y} - \mathbf{X}\widehat{\boldsymbol{\beta}}$. The residual vector $\widehat{\boldsymbol{\varepsilon}}$ lies in the subspace $\Omega^{\perp} = \mathcal{C}^{\perp}(\mathbf{X})$ of $\mathcal{R}^N$, called the *error space*. Any vector in the estimation space is orthogonal to any vector in the error space, i.e., the estimation space is orthogonal to the estimation space. The dimensions of the estimation and error spaces are respectively r and $N - r$.

Starting from the general linear model (4.1.1), consider the function of $\boldsymbol{\beta}$

$$
\begin{aligned}
S(\boldsymbol{\beta}) = \boldsymbol{\varepsilon}'\boldsymbol{\varepsilon} = \sum_{i=1}^{N}(Y_i - \mathbf{x}_i'\boldsymbol{\beta})^2 &= (\mathbf{y} - \mathbf{X}\boldsymbol{\beta})'(\mathbf{y} - \mathbf{X}\boldsymbol{\beta}) \\
&= \mathbf{y}'\mathbf{y} - 2\boldsymbol{\beta}'\mathbf{X}'\mathbf{y} + \boldsymbol{\beta}'\mathbf{X}'\mathbf{X}\boldsymbol{\beta}. \qquad (4.2.1)
\end{aligned}
$$

The least squares solution of $\boldsymbol{\beta}$ is chosen to minimize (4.2.1). Differentiating (4.2.1) with respect to the vector $\boldsymbol{\beta}$ (see Result 2.7.8) yields the set of p *normal equations*

$$\mathbf{X}'\mathbf{X}\boldsymbol{\beta}^0 = \mathbf{X}'\mathbf{y}, \qquad (4.2.2)$$

which admits a solution if $\mathrm{r}(\mathbf{X}'\mathbf{X}, \mathbf{X}'\mathbf{y}) = \mathrm{r}(\mathbf{X}'\mathbf{X})$ (see Result 3.2.2). We ask the reader to verify this (Exercise 4.1). The set of normal equations has a unique solution if and only if the matrix $\mathbf{X}'\mathbf{X}$ has full rank p, i.e., $(\mathbf{X}'\mathbf{X})^{-1}$ exists. We denote the unique solution by

$$\widehat{\boldsymbol{\beta}} = (\mathbf{X}'\mathbf{X})^{-1}\mathbf{X}'\mathbf{y} \tag{4.2.3}$$

and refer to it as the least squares (LS) or ordinary least squares (OLS) estimate of $\boldsymbol{\beta}$. From Result 1.3.11, $\text{r}(\mathbf{X}'\mathbf{X}) = \text{r}(\mathbf{X})$. Therefore, $\mathbf{X}'\mathbf{X}$ is invertible if and only if $\mathbf{X}$ has full column rank p. When $\text{r}(\mathbf{X}'\mathbf{X}) = \text{r}(\mathbf{X}) = r < p$, from Result 3.2.3, there are infinitely many solutions to (4.2.2), one of which is

$$\boldsymbol{\beta}^0 = \mathbf{G}\mathbf{X}'\mathbf{y}, \tag{4.2.4}$$

where $\mathbf{G}$ is any g-inverse of the symmetric $p \times p$ matrix $\mathbf{X}'\mathbf{X}$. Note that according to Result 3.2.5, for any vector $\mathbf{u}$ and g-inverse $\mathbf{G}_1$ of $\mathbf{X}'\mathbf{X}$, not necessarily the same as $\mathbf{G}$, $\boldsymbol{\beta}^0 + (\mathbf{I} - \mathbf{G}_1\mathbf{X}\mathbf{X}')\mathbf{u}$ is also a solution to (4.2.2). However, as far as inference is concerned, the extension does not provide any extra information, so it will not be considered further. We refer to $\boldsymbol{\beta}^0$ in (4.2.4) as an LS or OLS solution. We saw in Result 3.1.1 that $\mathbf{G}$ is not unique. What a solution $\boldsymbol{\beta}^0$ "estimates" depends on which g-inverse of $\mathbf{X}'\mathbf{X}$ we use! It is clear that $\boldsymbol{\beta}^0$ cannot be regarded as an estimator of $\boldsymbol{\beta}$, but merely as one possible solution to the set of normal equations (4.2.2).

The solution $\boldsymbol{\beta}^0$ indeed minimizes the sum of squares function $S(\boldsymbol{\beta})$ since

$$\begin{aligned} S(\boldsymbol{\beta}) &= (\mathbf{y} - \mathbf{X}\boldsymbol{\beta}^0 + \mathbf{X}\boldsymbol{\beta}^0 - \mathbf{X}\boldsymbol{\beta})'(\mathbf{y} - \mathbf{X}\boldsymbol{\beta}^0 + \mathbf{X}\boldsymbol{\beta}^0 - \mathbf{X}\boldsymbol{\beta}) \\ &= (\mathbf{y} - \mathbf{X}\boldsymbol{\beta}^0)'(\mathbf{y} - \mathbf{X}\boldsymbol{\beta}^0) + (\mathbf{y} - \mathbf{X}\boldsymbol{\beta}^0)'\mathbf{X}(\boldsymbol{\beta}^0 - \boldsymbol{\beta}) \\ &\quad + (\boldsymbol{\beta}^0 - \boldsymbol{\beta})'\mathbf{X}'(\mathbf{y} - \mathbf{X}\boldsymbol{\beta}^0) + (\boldsymbol{\beta}^0 - \boldsymbol{\beta})'\mathbf{X}'\mathbf{X}(\boldsymbol{\beta}^0 - \boldsymbol{\beta}) \\ &= S(\boldsymbol{\beta}^0) + (\boldsymbol{\beta}^0 - \boldsymbol{\beta})'\mathbf{X}'\mathbf{X}(\boldsymbol{\beta}^0 - \boldsymbol{\beta}), \end{aligned} \tag{4.2.5}$$

since $(\boldsymbol{\beta}^0 - \boldsymbol{\beta})'\mathbf{X}'(\mathbf{y} - \mathbf{X}\boldsymbol{\beta}^0) = (\mathbf{y} - \mathbf{X}\boldsymbol{\beta}^0)'\mathbf{X}(\boldsymbol{\beta}^0 - \boldsymbol{\beta}) = 0$ by (4.2.2). The last term on the right side of (4.2.5) is non-negative, so that $S(\boldsymbol{\beta}) - S(\boldsymbol{\beta}^0) \geq 0$ for all $\boldsymbol{\beta} \in \mathcal{R}^p$. The minimum value $S(\boldsymbol{\beta}^0)$ is denoted by SSE, and provides the basis for an estimate of the error variance σ^2.

Letting $\widehat{\mathbf{y}} = \mathbf{X}\boldsymbol{\beta}^0$, the least squares estimate of the error variance is

$$\widehat{\sigma}^2 = \frac{1}{N-r}(\mathbf{y} - \widehat{\mathbf{y}})'(\mathbf{y} - \widehat{\mathbf{y}}), \tag{4.2.6}$$

for which a computationally simple form is obtained after some algebraic simplification as

$$\widehat{\sigma}^2 = \frac{1}{N-r}(\mathbf{y}'\mathbf{y} - \boldsymbol{\beta}^{0\prime}\mathbf{X}'\mathbf{y}). \tag{4.2.7}$$

Example 4.2.1. Let $Y_1, \cdots, Y_N$ be a random sample from a normal population with mean θ and variance σ^2. In the framework of (4.1.1),

$$Y_i = \theta + \varepsilon_i, \quad i = 1, \cdots, N,$$

where ε_i are independently distributed with $\text{E}(\varepsilon_i) = 0$, and $\text{Var}(\varepsilon_i) = \sigma^2$. In matrix notation, let $\mathbf{y} = (Y_1, \cdots, Y_N)'$, $\boldsymbol{\varepsilon} = (\varepsilon_1, \cdots, \varepsilon_N)'$, $\mathbf{X} = (1, \cdots, 1)' = \mathbf{1}_N$, and $\boldsymbol{\beta} = \theta$. Since $\mathbf{X}'\mathbf{X} = N$, and $\mathbf{X}'\mathbf{y} = \sum_{i=1}^N Y_i$, the least squares estimate of the scalar parameter θ and its variance are

$$\widehat{\theta} = \sum_{i=1}^N Y_i/N = \overline{Y} \quad \text{and} \quad \text{Var}(\widehat{\theta}) = \sigma^2(\mathbf{1}_N'\mathbf{1}_N)^{-1} = \sigma^2/N. \qquad \square$$

Definition 4.2.1. Fitted vector. The vector of fitted values $\widehat{\mathbf{y}}$ is uniquely defined as a function of the solution vector $\boldsymbol{\beta}^0$:

$$\widehat{\mathbf{y}} = \mathbf{X}\boldsymbol{\beta}^0 = \mathbf{XGX}'\mathbf{y}. \tag{4.2.8}$$

For the full-rank model, the fitted vector has the form $\mathbf{X}(\mathbf{X}'\mathbf{X})^{-1}\mathbf{X}'\mathbf{y}$.

Definition 4.2.2. Residual vector. The vector of least squares residuals $\widehat{\boldsymbol{\varepsilon}}$ is also uniquely defined as a function of the solution vector $\boldsymbol{\beta}^0$:

$$\widehat{\boldsymbol{\varepsilon}} = \mathbf{y} - \widehat{\mathbf{y}} = (\mathbf{I}_N - \mathbf{XGX}')\mathbf{y}. \tag{4.2.9}$$

For the full-rank model, the residual vector has the form $[\mathbf{I}_N - \mathbf{X}(\mathbf{X}'\mathbf{X})^{-1}\mathbf{X}']\mathbf{y}$.

Note that when $\mathrm{r}(\mathbf{X}) = r < p$, since there are infinitely many LS solutions $\boldsymbol{\beta}^0$, we might suspect that there are infinitely many $\widehat{\mathbf{y}}$. However, by Result 3.1.13, $\mathbf{P} = \mathbf{XGX}'$ is invariant to the choice of $\mathbf{G}$ and represents the linear transformation of the orthogonal projection from $\mathcal{R}^N$ onto the estimation space $\mathcal{C}(\mathbf{X})$, while $\mathbf{I} - \mathbf{P} = \mathbf{I} - \mathbf{XGX}'$ represents the orthogonal projection from $\mathcal{R}^N$ onto the error space $\mathcal{C}^{\perp}(\mathbf{X})$. As a result the fitted vector $\widehat{\mathbf{y}}$ and the residual vector $\widehat{\boldsymbol{\varepsilon}}$ are unique, being the projections of $\mathbf{y}$ onto $\mathcal{C}(\mathbf{X})$ and $\mathcal{C}(\mathbf{X})^{\perp}$, respectively.

Result 4.2.1. Properties of P, I − P and H = GX′X. Suppose $\mathrm{r}(\mathbf{X}) = r$. Then,

1. $\mathbf{P}$ and $\mathbf{I} - \mathbf{P}$ are symmetric and idempotent matrices.
2. $\mathrm{r}(\mathbf{P}) = r$ and $\mathrm{r}(\mathbf{I} - \mathbf{P}) = N - r$.
3. $\mathbf{P}(\mathbf{I} - \mathbf{P}) = \mathbf{O}$.
4. $\mathbf{H}$ is idempotent, with $\mathrm{r}(\mathbf{H}) = r$. Further, $\mathbf{X}'\mathbf{XH} = \mathbf{X}'\mathbf{X}$.

Proof. All the properties except the last one on $\mathbf{H}$ have already appeared in Results 2.6.2 and 3.1.13. Idempotency of $\mathbf{H}$ in property 4 is a direct consequence of the fact that $\mathbf{X}'\mathbf{XGX}' = \mathbf{X}'$ (see Exercise 3.5), so that $\mathbf{H}^2 = \mathbf{GX}'\mathbf{XGX}'\mathbf{X} = \mathbf{GX}'\mathbf{X} = \mathbf{H}$. Hence, $\mathrm{r}(\mathbf{H}) = \mathrm{r}(\mathbf{X}'\mathbf{X}) = \mathrm{r}(\mathbf{X}) = r$. Now, $\mathbf{X}'\mathbf{XH} = \mathbf{X}'\mathbf{XGX}'\mathbf{X} = \mathbf{X}'\mathbf{X}$, which also follows immediately. ■

Result 4.2.2. For the general linear model (4.1.1),

$$\widehat{\mathbf{y}} = \mathbf{Py} = \mathbf{X}\boldsymbol{\beta} + \mathbf{P}\boldsymbol{\varepsilon}, \quad \widehat{\boldsymbol{\varepsilon}} = (\mathbf{I} - \mathbf{P})\mathbf{y} = (\mathbf{I} - \mathbf{P})\boldsymbol{\varepsilon}, \quad \widehat{\mathbf{y}} \perp \widehat{\boldsymbol{\varepsilon}},$$

where $\widehat{\mathbf{y}} \perp \widehat{\boldsymbol{\varepsilon}}$ can be also written as

$$\widehat{\mathbf{y}}'(\mathbf{y} - \widehat{\mathbf{y}}) = \sum_{i=1}^{N} \widehat{Y}_i(Y_i - \widehat{Y}_i) = 0.$$

The result immediately follows from Definitions 4.2.1 and 4.2.2 and the projection interpretation of the least squares approach.

In Figure 4.2.1 (a), $\mathbf{y}$, $\widehat{\mathbf{y}}$ and $\widehat{\boldsymbol{\varepsilon}}$ are the sides of a right triangle. An application of Pythagoras's Theorem gives

$$\mathbf{y}'\mathbf{y} = \widehat{\mathbf{y}}'\widehat{\mathbf{y}} + \widehat{\boldsymbol{\varepsilon}}'\widehat{\boldsymbol{\varepsilon}}.$$

We next characterize these squared distances in terms of quadratic forms in $\mathbf{y}$.

Definition 4.2.3. Sums of squares. Let SST, SSR and SSE respectively denote the total variation in Y, the variation explained by the fitted model, and the unexplained (residual) variation. We define these sums of squares by

$$\begin{aligned} SST &= \mathbf{y}'\mathbf{y} = \sum_{i=1}^{N} Y_i^2, \\ SSR &= \widehat{\mathbf{y}}'\widehat{\mathbf{y}} = \mathbf{y}'\mathbf{P}\mathbf{y} = \boldsymbol{\beta}^{0\prime}\mathbf{X}'\mathbf{X}\boldsymbol{\beta}^0, \quad \text{and} \\ SSE &= \widehat{\boldsymbol{\varepsilon}}'\widehat{\boldsymbol{\varepsilon}} = \mathbf{y}'(\mathbf{I}_N - \mathbf{P})\mathbf{y} = \mathbf{y}'\mathbf{y} - \boldsymbol{\beta}^{0\prime}\mathbf{X}'\mathbf{X}\boldsymbol{\beta}^0. \end{aligned} \tag{4.2.10}$$

SST refers to the total sum of squares, and the ANOVA decomposition represents a partition of SST as

$$SST = SSR + SSE, \tag{4.2.11}$$

where SSR is the model sum of squares and SSE is the error sum of squares.

The ANOVA decomposition in (4.2.11) can also be written as

$$\mathbf{y}'\mathbf{y} = \mathbf{y}'\mathbf{X}\mathbf{G}\mathbf{X}'\mathbf{y} + \mathbf{y}'(\mathbf{I}_N - \mathbf{X}\mathbf{G}\mathbf{X}')\mathbf{y}.$$

Orthogonality of the fitted vector $\widehat{\mathbf{y}}$ and the residual vector $\widehat{\boldsymbol{\varepsilon}}$ is necessary for an unambiguous partition of SST as the sum of SSR and SSE. The dimensions of the linear spaces $\mathcal{R}^N$, $\mathcal{C}(\mathbf{X})$ and $\mathcal{C}^{\perp}(\mathbf{X})$ also split as $N = r + (N - r)$. In some cases, it is useful to express the ANOVA decomposition in terms of mean corrected sums of squares:

$$SST_c = SSR_c + SSE, \tag{4.2.12}$$

where

$$\begin{aligned} SST_c &= \sum_{i=1}^{N} (Y_i - \overline{Y})^2 = \mathbf{y}'\mathbf{y} - N\overline{Y}^2 \quad \text{and} \\ SSR_c &= \boldsymbol{\beta}^{0\prime}\mathbf{X}'\mathbf{y} - N\overline{Y}^2. \end{aligned} \tag{4.2.13}$$

We can also express the ANOVA decomposition by including the sum of squares due to the mean

$$SSM = N\overline{Y}^2 \tag{4.2.14}$$

to write SST as

$$SST = SSM + SSR_c + SSE. \tag{4.2.15}$$

Note that the error sum of squares can also be written as

$$SSE = \widehat{\boldsymbol{\varepsilon}}'\widehat{\boldsymbol{\varepsilon}} = (\mathbf{y} - \widehat{\mathbf{y}})'(\mathbf{y} - \widehat{\mathbf{y}}), \tag{4.2.16}$$

so that the least squares estimate of the error variance is

$$\widehat{\sigma}^2 = \frac{SSE}{N - r}. \tag{4.2.17}$$

Example 4.2.2. We continue with Example 4.1.1. In the simple linear regression model, $\mathbf{y} = (Y_1, \cdots, Y_N)'$ and $\mathbf{x} = (X_1, \cdots, X_N)'$ are vectors in $\mathcal{R}^N$. Let

$$S(\beta_0, \beta_1) = \boldsymbol{\varepsilon}'\boldsymbol{\varepsilon} = \sum_{i=1}^{N} \varepsilon_i^2 = \sum_{i=1}^{N} (Y_i - \beta_0 - \beta_1 X_i)^2.$$

Minimizing $S(\beta_0, \beta_1)$ with respect to β_0 and β_1, and setting the normal equations equal to zero,

$$\begin{aligned}
\frac{\partial}{\partial \beta_0} S(\beta_0, \beta_1) &= -2 \sum_{i=1}^{N} (Y_i - \beta_0 - \beta_1 X_i) = 0 \\
&\implies N\widehat{\beta}_0 + \widehat{\beta}_1 \sum_{i=1}^{N} X_i = \sum_{i=1}^{N} Y_i, \\
\frac{\partial}{\partial \beta_1} S(\beta_0, \beta_1) &= -2 \sum_{i=1}^{N} X_i (Y_i - \beta_0 - \beta_1 X_i) = 0 \\
&\implies \widehat{\beta}_0 \sum_{i=1}^{N} X_i + \widehat{\beta}_1 \sum_{i=1}^{N} X_i^2 = \sum_{i=1}^{N} X_i Y_i.
\end{aligned}$$

Solving these equations simultaneously for the parameters, we obtain

$$\begin{aligned}
\widehat{\beta}_1 &= \frac{\sum_{i=1}^{N} (X_i - \overline{X})(Y_i - \overline{Y})}{\sum_{i=1}^{N} (X_i - \overline{X})^2} = \frac{\sum_{i=1}^{N} X_i Y_i - N\overline{X}\,\overline{Y}}{\sum_{i=1}^{N} X_i^2 - N\overline{X}^2}, \\
\widehat{\beta}_0 &= \overline{Y} - \widehat{\beta}_1 \overline{X}
\end{aligned} \tag{4.2.18}$$

and

$$\widehat{\sigma}^2 = \sum_{i=1}^{N} (Y_i - \widehat{\beta}_0 - \widehat{\beta}_1 X_i)^2 / (N-2). \tag{4.2.19}$$

We give a geometric interpretation. Consider the vector $\boldsymbol{\varepsilon} = (Y_1 - \beta_0 - \beta_1 X_1, \cdots, Y_N - \beta_0 - \beta_1 X_N)'$ for arbitrary coefficients β_0 and β_1. The least squares criterion chooses β_0 and β_1 such that the squared length

$$\|\boldsymbol{\varepsilon}\|^2 = \|\mathbf{y} - \beta_0 \mathbf{1}_N - \beta_1 \mathbf{x}\|^2 = \sum_{i=1}^{N} (Y_i - \beta_0 - \beta_1 X_i)^2$$

is a minimum (see Figure 4.2.2). Varying (β_0, β_1) varies $\beta_0 \mathbf{1}_N + \beta_1 \mathbf{x}$ through all the vectors in the space spanned by $\mathbf{1}_N$ and $\mathbf{x}$ (point A). The choice of (β_0, β_1) minimizing $\|\boldsymbol{\varepsilon}\|^2$ corresponds to the projection of the response vector $\mathbf{y}$ onto this plane, such that the vector $\widehat{\boldsymbol{\varepsilon}}$ is orthogonal to the plane (point B). The least squares estimates $\widehat{\beta}_0$ and $\widehat{\beta}_1$ are these minimizing values. The vector of "fitted" or "predicted" values is $\widehat{\mathbf{y}} = \widehat{\beta}_0 \mathbf{1}_N + \widehat{\beta}_1 \mathbf{x}$, i.e., the ith predicted value is $\widehat{Y}_i = \widehat{\beta}_0 + X_i \widehat{\beta}_1$, $i = 1, \cdots, N$.

The least squares residual vector is $\widehat{\boldsymbol{\varepsilon}} = \mathbf{y} - \widehat{\beta}_0 \mathbf{1}_N - \widehat{\beta}_1 \mathbf{x}$, i.e., the ith residual is $\widehat{\varepsilon}_i = Y_i - \widehat{Y}_i$, $i = 1, \cdots, N$. The least squares predictor $\widehat{\mathbf{y}}$ is the vector with the smallest error, i.e., the vector corresponding to the shortest error vector, $\widehat{\boldsymbol{\varepsilon}}$. This happens when $\widehat{\mathbf{y}} \perp \widehat{\boldsymbol{\varepsilon}}$. □

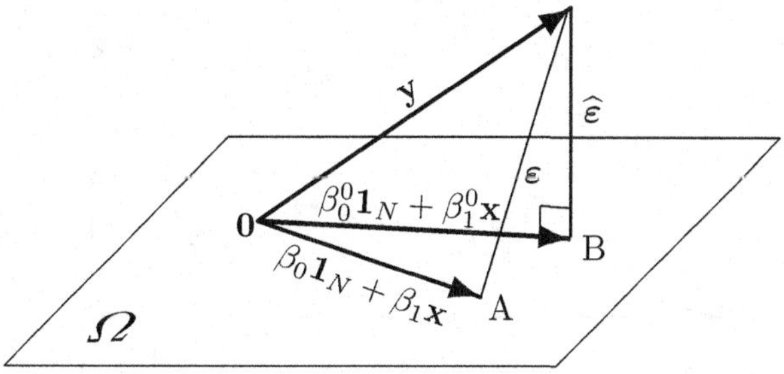

FIGURE 4.2.2. Geometry of simple linear regression. Ω is the estimation space spanned by $\mathbf{1}_N$ and $\mathbf{x}$.

Example 4.2.3. Consider the linear model

$$\begin{pmatrix} Y_1 \\ Y_2 \\ Y_3 \end{pmatrix} = \begin{pmatrix} 1 & -1 & 1 \\ 1 & 0 & -2 \\ 1 & 1 & 1 \end{pmatrix} \begin{pmatrix} \beta_0 \\ \beta_1 \\ \beta_2 \end{pmatrix} + \begin{pmatrix} \varepsilon_1 \\ \varepsilon_2 \\ \varepsilon_3 \end{pmatrix}.$$

We can verify that the columns of $\mathbf{X}$ are mutually orthogonal. The least squares estimate of $\boldsymbol{\beta} = (\beta_0, \beta_1, \beta_2)'$ is

$$\widehat{\boldsymbol{\beta}} = \begin{pmatrix} \frac{1}{3}(Y_1 + Y_2 + Y_3) \\ -\frac{1}{2}(Y_1 - Y_3) \\ \frac{1}{6}(Y_1 - 2Y_2 + Y_3) \end{pmatrix}.$$

Suppose we set $\beta_2 = 0$ in the above model. The resulting model is

$$\begin{pmatrix} Y_1 \\ Y_2 \\ Y_3 \end{pmatrix} = \begin{pmatrix} 1 & -1 \\ 1 & 0 \\ 1 & 1 \end{pmatrix} \begin{pmatrix} \beta_0 \\ \beta_1 \end{pmatrix} + \begin{pmatrix} \varepsilon_1 \\ \varepsilon_2 \\ \varepsilon_3 \end{pmatrix},$$

and the least squares estimate of β_0 and β_1 are $\widehat{\beta}_0 = \frac{1}{3}(Y_1+Y_2+Y_3)$ and $\widehat{\beta}_1 = -\frac{1}{2}(Y_1 - Y_3)$, which are unchanged by omitting the parameter β_2. It is not difficult to see that this is a consequence of the orthogonality of $\mathbf{X}$, which leads to a diagonal $\mathbf{X}'\mathbf{X}$ matrix. □

Measuring the adequacy of the least squares fit is an important practical problem. Recall that the vectors $\mathbf{y}$, $\widehat{\mathbf{y}}$ and $\widehat{\boldsymbol{\varepsilon}}$ form a right triangle, with the latter two vectors containing the right angle. Either the angle between the vectors $\mathbf{y}$ and $\widehat{\mathbf{y}}$ (which lies between zero and 90°) or their relative lengths can be used to assess adequacy of fit (see Figure 4.2.3). When the model provides a good fit (as in (a)), the angle between $\mathbf{y}$ and $\widehat{\mathbf{y}}$ is small, and the two vectors have nearly the same length. On the contrary, the angle is large and $\widehat{\mathbf{y}}$ is much shorter than $\mathbf{y}$ if there is a poor fit (as in (b)). If $\mathbf{y} \perp \mathcal{C}(\mathbf{X})$, the projection of $\mathbf{y}$ onto $\mathcal{C}(\mathbf{X})$ is the null vector, i.e., $\widehat{\mathbf{y}} = \mathbf{0}$. The square of the cosine of the angle between $\mathbf{y}$ and $\widehat{\mathbf{y}}$, which is also equal to the square of the relative lengths of $\widehat{\mathbf{y}}$ and $\mathbf{y}$ is called the coefficient of determination, i.e., $R^2 = \|\widehat{\mathbf{y}}\|^2/\|\mathbf{y}\|^2$. In other words, we can interpret R^2 as the proportion by which the fitted vector $\widehat{\mathbf{y}}$ is shorter than the observed response vector $\mathbf{y}$. We give a formal definition.

Definition 4.2.4. R-square. We define the coefficient of determination of the linear model as the proportion of the total variation in Y explained by the explanatory variables,

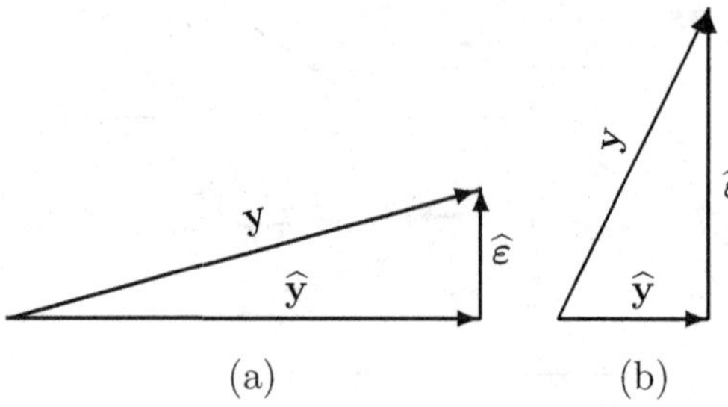

FIGURE 4.2.3. Adequacy of least squares.

i.e.,

$$R^2 = \frac{SSR}{SST} = 1 - \frac{SSE}{SST}. \tag{4.2.20}$$

Clearly, $0 \le R^2 \le 1$.

This measure, which is also the multiple correlation coefficient between Y and the set of predictors $(X_1, \cdots, X_k)$ (see Definition 5.2.8), is widely used to assess goodness of fit of the linear regression. A value of R^2 close to 0 indicates a poor linear regression fit, while a value close to 1 indicates a good fit. An R^2 of 0 says that the linear model does not explain any variation in the response Y. In a simple regression model, it is possible that (a) the values of Y lie randomly around the horizontal line $Y = \overline{Y}$, or (b) the observations lie on a circle. An R^2 of 1 can occur only when all the Y values lie on the estimated regression line. In this case, it is easily verified that R^2 is the square of the simple product-moment correlation between X and Y. One disadvantage with using R^2 as a measure of goodness of fit is that it does not account for the number of degrees of freedom associated with SSR. We define a measure that makes this d.f. correction in the context of a full-rank model.

Definition 4.2.5. Adjusted R-square. The adjusted R^2, or the corrected R^2 is defined as

$$\begin{aligned} R^2_{adj.} &= 1 - \frac{\widehat{\text{Var}}(\widehat{\boldsymbol{\varepsilon}})}{\widehat{\text{Var}}(\mathbf{y})} = 1 - \frac{\sum_{i=1}^N \widehat{\varepsilon}_i^2/(N-p)}{\sum_{i=1}^N (Y_i - \overline{Y})^2/(N-1)} \\ &= 1 - \frac{SSE/(N-p)}{SST_c/(N-1)}, \end{aligned} \tag{4.2.21}$$

where p is the total number of parameters in the fitted model including the intercept, SST_c is the total mean corrected sum of squares and SSE denotes the error sum of squares with $(N-p)$ degrees of freedom.

Since an "adjustment" has been made for the corresponding degrees of freedom in the relevant sums of squares, $R^2_{adj.}$ is useful for comparing different regression fits to the same data as well as for comparing different datasets, although in the latter situation, its usefulness is rather limited. It is a gross initial indicator, and one is better off using other model comparison criteria. The adjusted R^2 is closely related to Mallows' C_p statistic, which we describe in Chapter 9. We next derive a general formula for the mean of a quadratic form which is useful for obtaining the covariance matrix of $\boldsymbol{\beta}^0$.

Result 4.2.3. Suppose $\mathrm{E}(\mathbf{x}) = \boldsymbol{\mu}$, and $\mathrm{Var}(\mathbf{x}) = \boldsymbol{\Sigma}$, then,

$$\mathrm{E}(\mathbf{x}'\mathbf{A}\mathbf{x}) = \mathrm{tr}(\mathbf{A}\boldsymbol{\Sigma}) + \boldsymbol{\mu}'\mathbf{A}\boldsymbol{\mu}. \tag{4.2.22}$$

Proof. Similar to the well-known univariate expression $\mathrm{E}(X^2) = \mathrm{Var}(X) + \{\mathrm{E}(X)\}^2$, we can write in the vector case $\mathrm{E}(\mathbf{x}\mathbf{x}') = \boldsymbol{\Sigma} + \boldsymbol{\mu}\boldsymbol{\mu}'$, i.e., $\mathrm{E}(\mathbf{x}-\boldsymbol{\mu})(\mathbf{x}-\boldsymbol{\mu})' = \boldsymbol{\Sigma}$. Since $\mathrm{tr}(\mathbf{x}'\mathbf{A}\mathbf{x}) = \mathrm{tr}(\mathbf{A}\mathbf{x}\mathbf{x}')$ from property 4 of Result 1.3.5,

$$\begin{aligned}\mathrm{E}(\mathbf{x}'\mathbf{A}\mathbf{x}) &= \mathrm{E}[\mathrm{tr}(\mathbf{A}\mathbf{x}\mathbf{x}')] = \mathrm{tr}[\mathbf{A}\,\mathrm{E}(\mathbf{x}\mathbf{x}')] = \mathrm{tr}[\mathbf{A}\boldsymbol{\Sigma} + \mathbf{A}\boldsymbol{\mu}\boldsymbol{\mu}'] \\ &= \mathrm{tr}(\mathbf{A}\boldsymbol{\Sigma}) + \mathrm{tr}(\boldsymbol{\mu}'\mathbf{A}\boldsymbol{\mu}) = \mathrm{tr}(\mathbf{A}\boldsymbol{\Sigma}) + \boldsymbol{\mu}'\mathbf{A}\boldsymbol{\mu},\end{aligned}$$

which completes the proof. ■

Result 4.2.4. Mean and variance.

1. Letting $\mathbf{H} = \mathbf{G}\mathbf{X}'\mathbf{X} = (\mathbf{X}'\mathbf{X})^-\mathbf{X}'\mathbf{X}$, we have

$$\mathrm{E}(\boldsymbol{\beta}^0) = \mathbf{H}\boldsymbol{\beta} \quad \text{and} \quad \mathrm{Cov}(\boldsymbol{\beta}^0) = \sigma^2\mathbf{G}\mathbf{X}'\mathbf{X}\mathbf{G}'. \tag{4.2.23}$$

2. The mean and covariance of the fitted vector and residual vector are

$$\begin{aligned}\mathrm{E}(\widehat{\mathbf{y}}) &= \mathbf{X}\boldsymbol{\beta} \quad \text{and} \quad \mathrm{Cov}(\widehat{\mathbf{y}}) = \sigma^2\mathbf{P}, \\ \mathrm{E}(\widehat{\boldsymbol{\varepsilon}}) &= \mathbf{0} \quad \text{and} \quad \mathrm{Cov}(\widehat{\boldsymbol{\varepsilon}}) = \sigma^2(\mathbf{I}-\mathbf{P}).\end{aligned} \tag{4.2.24}$$

3. $\boldsymbol{\beta}^0$ and $\widehat{\boldsymbol{\varepsilon}}$ are uncorrelated, i.e., $\mathrm{Cov}(\boldsymbol{\beta}^0, \widehat{\boldsymbol{\varepsilon}}) = \mathbf{O}$, so in particular, $\widehat{\mathbf{y}}$ and $\widehat{\boldsymbol{\varepsilon}}$ are uncorrelated.

4. The expectation of the mean square error is

$$\mathrm{E}(\widehat{\sigma}^2) = \sigma^2. \tag{4.2.25}$$

Proof. Since by (4.2.4), $\boldsymbol{\beta}^0 = \mathbf{G}\mathbf{X}'\mathbf{y}$, then $\mathrm{E}(\boldsymbol{\beta}^0) = \mathbf{G}\mathbf{X}'\,\mathrm{E}(\mathbf{y}) = \mathbf{G}\mathbf{X}'\mathbf{X}\boldsymbol{\beta} = \mathbf{H}\boldsymbol{\beta}$ and $\mathrm{Cov}(\boldsymbol{\beta}^0) = \mathbf{G}\mathbf{X}'\,\mathrm{Cov}(\mathbf{y})\mathbf{X}\mathbf{G}' = \mathbf{G}\mathbf{X}'(\sigma^2\mathbf{I}_N)\mathbf{X}\mathbf{G}' = \sigma^2\mathbf{G}\mathbf{X}'\mathbf{X}\mathbf{G}'$, which shows property 1. To see property 2, see that from $\widehat{\mathbf{y}} = \mathbf{P}\mathbf{y}$, $\mathrm{E}(\widehat{\mathbf{y}}) = \mathbf{P}\,\mathrm{E}(\mathbf{y}) = \mathbf{P}\mathbf{X}\boldsymbol{\beta} = \mathbf{X}\boldsymbol{\beta}$ and $\mathrm{Cov}(\widehat{\mathbf{y}}) = \mathbf{P}\,\mathrm{Cov}(\mathbf{y})\mathbf{P}' = \mathbf{P}(\sigma^2\mathbf{I})\mathbf{P}' = \sigma^2\mathbf{P}$. The formulas for $\widehat{\boldsymbol{\varepsilon}}$ can be similarly proved.

$$\begin{aligned}\mathrm{Cov}(\boldsymbol{\beta}^0, \widehat{\boldsymbol{\varepsilon}}) &= \mathrm{Cov}(\mathbf{G}\mathbf{X}'\mathbf{y}, (\mathbf{I}-\mathbf{P})\boldsymbol{\varepsilon}) = \mathbf{G}\mathbf{X}'\,\mathrm{Cov}(\mathbf{y}, \boldsymbol{\varepsilon})(\mathbf{I}-\mathbf{P}) \\ &= \sigma^2\mathbf{G}\mathbf{X}'(\mathbf{I}-\mathbf{P}) = \sigma^2\mathbf{G}(\mathbf{X}' - \mathbf{X}'\mathbf{P}) = \mathbf{O}.\end{aligned}$$

Since $\widehat{\mathbf{y}} = \mathbf{X}\boldsymbol{\beta}^0$ is a linear function of $\boldsymbol{\beta}^0$, it is also uncorrelated with $\widehat{\boldsymbol{\varepsilon}}$, showing property 3. To show property 4, since $SSE = \widehat{\boldsymbol{\varepsilon}}'\widehat{\boldsymbol{\varepsilon}} = [(\mathbf{I}_N - \mathbf{P})\mathbf{y}]'[(\mathbf{I}-\mathbf{p})\mathbf{y}] = \mathbf{y}'(\mathbf{I}_N - \mathbf{P})\mathbf{y}$, by Result 4.2.3,

$$\begin{aligned}\mathrm{E}(SSE) &= \mathrm{E}[\mathbf{y}'(\mathbf{I}_N - \mathbf{X}\mathbf{G}\mathbf{X}')\mathbf{y}] \\ &= \mathrm{tr}[\sigma^2(\mathbf{I}_N - \mathbf{X}\mathbf{G}\mathbf{X}')] + \boldsymbol{\beta}'\mathbf{X}'(\mathbf{I}_N - \mathbf{X}\mathbf{G}\mathbf{X}')\mathbf{X}\boldsymbol{\beta} \\ &= \sigma^2(N-r).\end{aligned} \tag{4.2.26}$$

Therefore, the least squares estimator $\widehat{\sigma}^2$ is an unbiased estimate of σ^2 and is invariant to the choice of $\mathbf{G}$. ■

Corollary 4.2.1. In the full-rank model, the properties of the least squares estimator of $\boldsymbol{\beta}$ follow as a special case of Result 4.2.4 and are given below:

$$\mathrm{E}(\widehat{\boldsymbol{\beta}}) = \boldsymbol{\beta} \quad \text{and} \quad \mathrm{Cov}(\widehat{\boldsymbol{\beta}}) = \sigma^2(\mathbf{X}'\mathbf{X})^{-1}. \tag{4.2.27}$$

Example 4.2.4. Let Y_1, Y_2, and Y_3 be independently distributed with $E(Y_i) = i\theta$, and $Var(Y_i) = \sigma^2$, for $i = 1, 2, 3$. We first set this up as a full-rank linear model of the form (4.1.1), with

$$\mathbf{y} = \begin{pmatrix} Y_1 \\ Y_2 \\ Y_3 \end{pmatrix}, \quad \boldsymbol{\varepsilon} = \begin{pmatrix} \varepsilon_1 \\ \varepsilon_2 \\ \varepsilon_3 \end{pmatrix}, \quad \mathbf{X} = \begin{pmatrix} 1 \\ 2 \\ 3 \end{pmatrix} \quad \text{and } \boldsymbol{\beta} = \theta$$

The least squares estimator of θ is $\widehat{\theta} = (\mathbf{X}'\mathbf{X})^{-1}\mathbf{X}'\mathbf{y} = \frac{1}{14}(Y_1 + 2Y_2 + 3Y_3)$, with variance $Var(\widehat{\theta}) = \sigma^2/14$. The estimator is unbiased for θ since $E(\widehat{\theta}) = \frac{1}{14}\{\theta + 2(2\theta) + 3(3\theta)\} = \theta$. The fitted vector $\widehat{\mathbf{y}}$ and the residual vector $\widehat{\boldsymbol{\varepsilon}}$ are respectively given by

$$\widehat{\mathbf{y}} = \begin{pmatrix} \widehat{\theta} \\ 2\widehat{\theta} \\ 3\widehat{\theta} \end{pmatrix} = \begin{pmatrix} \frac{1}{14}(Y_1 + 2Y_2 + 3Y_3) \\ \frac{1}{7}(Y_1 + 2Y_2 + 3Y_3) \\ \frac{3}{14}(Y_1 + 2Y_2 + 3Y_3) \end{pmatrix} \quad \text{and} \quad \widehat{\boldsymbol{\varepsilon}} = \mathbf{y} - \widehat{\mathbf{y}} = \begin{pmatrix} Y_1 - \widehat{\theta} \\ Y_2 - 2\widehat{\theta} \\ Y_3 - 3\widehat{\theta} \end{pmatrix}.$$

The error sum of squares is

$$\begin{aligned} SSE = \widehat{\boldsymbol{\varepsilon}}'\widehat{\boldsymbol{\varepsilon}} &= (Y_1^2 + Y_2^2 + Y_3^2) - \widehat{\theta}(Y_1 + 2Y_2 + 3Y_3) \\ &= (Y_1^2 + Y_2^2 + Y_3^2) - \frac{1}{14}(Y_1 + 2Y_2 + 3Y_3)^2 \end{aligned}$$

and the mean square error (MSE) is $\widehat{\sigma}^2 = SSE/(N - p) = SSE/2$. □

Numerical Example 4.1. Simple linear regression. We consider the dataset "cars" in the R package *car*, and model the speed of cars Y as a function of the stopping distances X. The data was recorded in the 1920's.

```
library(car); data("cars"); attach(cars); plot(dist, speed) # see Figure 4.2.4

# simple correlation
corr.test <- cor(dist, speed)
cor.test


  Pearson's product-moment correlation


data:  cars$speed and cars$dist
t = 9.464, df = 48, p-value = 1.49e-12
alternative hypothesis: true correlation is not equal to 0
95 percent confidence interval:
 0.6816422 0.8862036
sample estimates:
      cor
0.8068949
```

The estimated simple correlation between Y and X is $\widehat{\rho} = 0.807$, which indicates a moderately high, positive linear relationship between the variables.

```
# OLS estimation in an SLR model
mod.slr <- lm(speed ~ dist, data = cars)
summary(mod.slr)


Call:
lm(formula = speed ~ dist, data = cars)
```

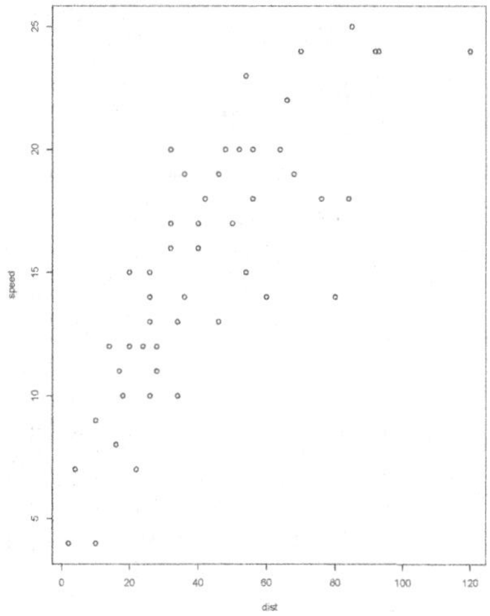

FIGURE 4.2.4. Scatterplot of speed versus distance.

```
Residuals:
    Min      1Q  Median      3Q     Max
-7.5293 -2.1550  0.3615  2.4377  6.4179

Coefficients:
            Estimate Std. Error t value Pr(>|t|)
(Intercept)  8.28391    0.87438   9.474 1.44e-12 ***
dist         0.16557    0.01749   9.464 1.49e-12 ***
---
Signif. codes:  0 '***' 0.001 '**' 0.01 '*' 0.05 '.' 0.1 ' ' 1

Residual standard error: 3.156 on 48 degrees of freedom
Multiple R-squared:  0.6511,    Adjusted R-squared:  0.6438
F-statistic: 89.57 on 1 and 48 DF,  p-value: 1.49e-12
```

The fitted simple linear regression model for speed is

$$\widehat{Y} = 8.284 + 0.166X.$$

The output shows the estimated coefficients together with their standard errors. The coefficient of determination is $R^2 = 0.651$, which is the same as $\widehat{\rho}^2$. Inference for this model is discussed in Chapter 7. ▲

Numerical Example 4.2. Multiple linear regression. We fit an MLR model to operational data from an industrial plant related to the oxidation of ammonia to nitric acid. The dataset "stackloss" is available in the R package *datasets*. The response variable is Y (stack loss). There are $k = 3$ predictor variables, X_1 (cooling water inlet temperature), X_2 (concentration of acid) and X_3 (flow of cooling air).

```
library(datasets); library(car)
data(stackloss)
attach(stackloss)
data <- cbind(stack.loss, Air.Flow, Water.Temp, Acid.Conc.)
```

```
scatterplotMatrix(stackloss) # see Figure 4.2.5

# matrix of pairwise correlations
(cor.mat <- cor(data))

             stack.loss Air.Flow Water.Temp Acid.Conc.
stack.loss        1.000    0.920      0.876      0.400
Air.Flow          0.920    1.000      0.782      0.500
Water.Temp        0.876    0.782      1.000      0.391
Acid.Conc.        0.400    0.500      0.391      1.000
```

The correlation matrix shows pairwise correlations between the variables (Y, X_1, X_2, X_3). The fitted multiple regression model is

$$\widehat{Y} = -39.920 + 0.716X_1 + 1.295X_2 - 0.152X_3,$$

with an R^2 value of 0.914.

```
# OLS estimation in an MLR model
mod <- lm(stack.loss ~ Air.Flow + Water.Temp
    + Acid.Conc., data = stackloss)
(smod <- summary(mod))

Call:
lm(formula = stack.loss ~ Air.Flow + Water.Temp + Acid.Conc.,
    data = stackloss)

Residuals:
   Min     1Q Median     3Q    Max
-7.238 -1.712 -0.455  2.361  5.698

Coefficients:
            Estimate Std. Error t value Pr(>|t|)
(Intercept)  -39.920     11.896   -3.36   0.0038 **
Air.Flow       0.716      0.135    5.31  5.8e-05 ***
Water.Temp     1.295      0.368    3.52   0.0026 **
Acid.Conc.    -0.152      0.156   -0.97   0.3440
---
Signif. codes:  0 '***' 0.001 '**' 0.01 '*' 0.05 '.' 0.1 ' ' 1

Residual standard error: 3.24 on 17 degrees of freedom
Multiple R-squared:  0.914,     Adjusted R-squared:  0.898
F-statistic: 59.9 on 3 and 17 DF,  p-value: 3.02e-09
```

Entries in the last two columns of the table of coefficients pertain to inference based on an assumption of normal errors, and they will be discussed in Chapter 7. ▲

Although the normal equations in (4.2.2) presumably consist of p equations in p unknowns, there are only r LIN equations, since the remaining $(p-r)$ equations are linear combinations of these. To obtain a solution of $\boldsymbol{\beta}$, $(p-r)$ additional *consistent* equations, say, $\mathbf{a}_j'\boldsymbol{\beta} = b_j$, $j = 1, \cdots, p-r$, are required, satisfying the condition that $\mathbf{a}_j$ is not a linear combination of the rows of $\mathbf{X}'\mathbf{X}$. Suppose that, on the contrary, $\mathbf{a}_j'$ is such a linear combination. Then, we can either obtain the corresponding additional equation from (4.2.2), or we will face inconsistency, i.e., we will obtain two different values for $\mathbf{a}_j'\boldsymbol{\beta}^0$. In other words,

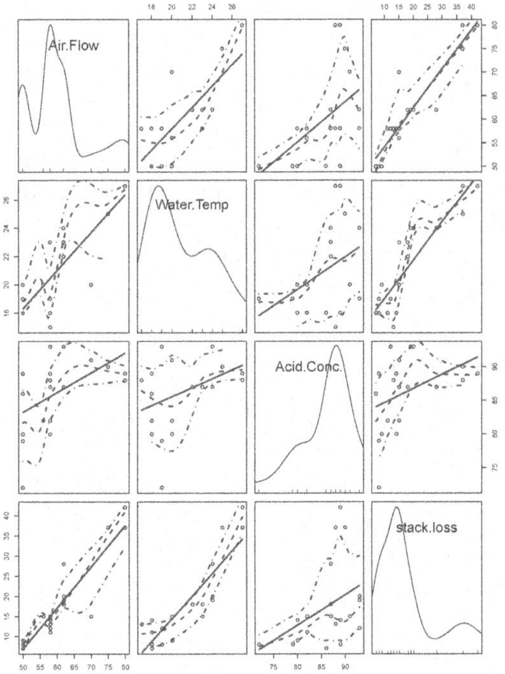

FIGURE 4.2.5. Matrix scatterplot for the stackloss data.

this condition requires that $\mathbf{a}_j'\boldsymbol{\beta}$ is a *nonestimable* function (see Section 4.3). Suppose we represent the $p - r$ additional equations in the form $\mathbf{A}\boldsymbol{\beta} = \mathbf{b}$; each equation (row) must correspond to a nonestimable function. We will refer to these additional equations as *constraints*; they are generally chosen after inspecting the meaning of the parameters in the model, and are included for the purpose of solving the normal equations for interpretable solutions.

Example 4.2.5. We continue with Example 4.1.3. In the one-way fixed-effects ANOVA model, we have seen that $\mathrm{r}(\mathbf{X}) = a < p$, so that the least squares solution $\boldsymbol{\beta}^0$ is not unique. Let $Y_{i\cdot}$ and $\overline{Y}_{i\cdot}$ respectively denote the total and average of the sample observations under the ith treatment. Let $Y_{\cdot\cdot}$ and $\overline{Y}_{\cdot\cdot}$ respectively denote the grand total and average of all the sample observations. We have used the "dot" subscript notation which implies summation over the subscript that it replaces. Symbolically, $Y_{i\cdot} = \sum_{j=1}^{n_i} Y_{ij}$, $\overline{Y}_{i\cdot} = Y_{i\cdot}/n_i$, $i = 1, \cdots, a$, $Y_{\cdot\cdot} = \sum_{i=1}^{a}\sum_{j=1}^{n_i} Y_{ij}$, and $\overline{Y}_{\cdot\cdot} = Y_{\cdot\cdot}/N$, where $N = \sum_{i=1}^{a} n_i$ is the total number of observations. The method of least squares consists of minimizing the sum of squares

$$S(\mu, \tau_1, \cdots, \tau_a) = \sum_{i=1}^{a}\sum_{j=1}^{n_i} \varepsilon_{ij}^2 = \sum_{i=1}^{a}\sum_{j=1}^{n_i} (Y_{ij} - \mu - \tau_i)^2;$$

the $(a+1)$ normal equations are

$$\left.\frac{\partial S}{\partial \mu}\right|_{\mu^0, \tau_i^0} = 0, \quad \text{and} \quad \left.\frac{\partial S}{\partial \tau_i}\right|_{\mu^0, \tau_i^0} = 0, \quad i = 1, \cdots, a,$$

which have the form

$$N\mu^0 + \sum_{i=1}^{a} n_i\tau_i^0 = Y_{\cdot\cdot} \quad \text{and} \quad n_i(\mu^0 + \tau_i^0) = Y_{i\cdot}, \quad i = 1, \cdots, a.$$

Note that the sum of the last a normal equations is equal to the first; the $a+1$ normal equations are linearly dependent and hence no unique solution exists. In other words, there are $p = a+1$ parameters, and $\mathrm{r}(\mathbf{X}) = a$, so we can set one element of $\boldsymbol{\beta}^0$ equal to zero, and solve for the remaining a parameters uniquely. In general, we may obtain a solution to the normal equations subject to a "model constraint" such as $\mu^0 = 0$, or $\tau_a^0 = 0$, or $\sum_{i=1}^{a} n_i\tau_i = 0$.

1. If we set $\mu^0 = 0$, the model can be simplified as

   $$Y_{ij} = \tau_i + \varepsilon_{ij}, \; j = 1, \cdots, n_i, \; i = 1, \cdots, a.$$

   Minimizing

   $$S(\tau_1, \cdots, \tau_a) = \sum_{i=1}^{a}\sum_{j=1}^{n_i}(Y_{ij} - \tau_i)^2$$

   with respect to $\tau_1, \cdots, \tau_a$ and solving, yields $\tau_i^0 = \overline{Y}_{i\cdot}$, $i = 1, \cdots, a$, so that

   $$\boldsymbol{\beta}^0 = (0, \overline{Y}_{1\cdot}, \cdots, \overline{Y}_{a\cdot})'.$$

   The g-inverse corresponding to this solution is obtained by deleting the first row and column from $\mathbf{X}'\mathbf{X}$ in the algorithm in Result 3.1.2 as

   $$\mathbf{G} = \begin{pmatrix} 0 & \mathbf{0} \\ \mathbf{0} & \mathbf{D} \end{pmatrix},$$

   where $\mathbf{D} = \mathrm{diag}(1/n_1, \cdots, 1/n_a)$ is an $a \times a$ diagonal matrix. The orthogonal projection matrix $\mathbf{P}$ has the block diagonal form

   $$\mathbf{P} = \mathrm{diag}(\mathbf{J}_{n_1}/n_1, \cdots, \mathbf{J}_{n_a}/n_a)$$

   and the fitted vector $\widehat{\mathbf{y}}$ is

   $$\widehat{\mathbf{y}} = \mathbf{P}\mathbf{y} = (\overline{Y}_{1\cdot}\mathbf{1}'_{n_1}, \cdots, \overline{Y}_{a\cdot}\mathbf{1}'_{n_a})'.$$

   The ANOVA decomposition can be written as

   $$\sum_{i=1}^{a}\sum_{j=1}^{n_i} Y_{ij}^2 = \sum_{i=1}^{a} n_i\overline{Y}_{i\cdot}^2 + \sum_{i=1}^{a}\sum_{j=1}^{n_i}(Y_{ij} - \overline{Y}_{i\cdot})^2.$$

2. Suppose we impose an alternate constraint $\sum_{i=1}^{a} n_i\tau_i^0 = 0$. We minimize the Lagrangian function

   $$\mathcal{L}(\mu, \tau_1, \cdots, \tau_a, \lambda) = \sum_{i=1}^{a}\sum_{j=1}^{n_i}(Y_{ij} - \mu - \tau_i)^2 + 2\lambda\sum_{i=1}^{a} n_i\tau_i,$$

   leading to the normal equations

   $$N\mu + \sum_{i=1}^{a} n_i\tau_i = Y_{\cdot\cdot}$$

   $$n_i\mu + n_i\tau_i + \lambda n_i = Y_{i\cdot}, \; i = 1, \cdots, a$$

   $$\sum_{i=1}^{a} n_i\tau_i = 0.$$

We solve these simultaneously to get $\lambda^0 = 0$ and the solution vector is

$$\boldsymbol{\beta}^0 = (\overline{Y}_{..}, \overline{Y}_{1\cdot} - \overline{Y}_{..}, \cdots, \overline{Y}_{a\cdot} - \overline{Y}_{..}).$$

Subject to this constraint, the estimate of the overall mean μ is the grand sample average of all the N observations, while the estimate of the ith treatment effect τ_i is the difference between the average of the sample observations under the ith treatment and the grand average. The estimate of the error variance in both cases is

$$\widehat{\sigma}^2 = \frac{SSE}{(N-a)}, \tag{4.2.28}$$

where

$$\begin{aligned} SSE &= \sum_{i=1}^{a}\sum_{j=1}^{n_i}(Y_{ij} - \overline{Y}_{..} - \overline{Y}_{i\cdot} + \overline{Y}_{..})^2 \\ &= \sum_{i=1}^{a}\sum_{j=1}^{n_i} Y_{ij}^2 - \sum_{i=1}^{a} n_i \overline{Y}_{i\cdot}^2 \end{aligned} \tag{4.2.29}$$

A solution for $\boldsymbol{\beta}$ depends on the constraint under which we choose to solve the normal equations. □

Although this might seem like an unfortunate event, since it might lead different data analysts to draw completely different inferences, we show that estimation and inference is indeed unique for certain functions of the parameters called "estimable functions", which are described in the next section.

Numerical Example 4.3. One-way fixed-effects ANOVA model. The R dataset "PlantGrowth" describes an experiment to compare yields obtained under a control and two different treatment conditions. The response variable Y is weight and the factor variable X has $a = 3$ levels, *ctrl, trt1, trt2.* This is a balanced design with $n_1 = n_2 = n_3 = 10$.

```
data(PlantGrowth)
fit.lm <- lm(weight ~ group, data = PlantGrowth)
summary(fit.lm)

Call:
lm(formula = weight ~ group, data = PlantGrowth)

Residuals:
   Min     1Q Median     3Q    Max
-1.071 -0.418 -0.006  0.263  1.369

Coefficients:
            Estimate Std. Error t value Pr(>|t|)
(Intercept)    5.032      0.197   25.53   <2e-16 ***
grouptrt1     -0.371      0.279   -1.33    0.194
grouptrt2      0.494      0.279    1.77    0.088 .
---
Signif. codes:  0 ‘***’ 0.001 ‘**’ 0.01 ‘*’ 0.05 ‘.’ 0.1 ‘ ’ 1

Residual standard error: 0.623 on 27 degrees of freedom
Multiple R-squared:  0.264,     Adjusted R-squared:  0.21
```

From the output, we see that $\mu^0 = 5.032$, $\tau_1^0 = -0.371$ and $\tau_2^0 = 0.494$. The estimated error variance is $\widehat{\sigma}^2 = 0.388$.

```
fit <- aov(weight ~ group, data = PlantGrowth)
summary(fit)

   Df Sum Sq Mean Sq F value Pr(>F)
group        2   3.77   1.883    4.85  0.016 *
Residuals   27  10.49   0.389
---
Signif. codes:  0 '***' 0.001 '**' 0.01 '*' 0.05 '.' 0.1 ' ' 1

coef(fit)

(Intercept)   grouptrt1   grouptrt2
      5.032      -0.371       0.494
```

As shown above, we can use the aov() function to obtain the ANOVA table for carrying out inference, which is discussed in Chapter 7. ▲

We conclude this section with an example of another fixed-effects linear model involving two factors.

Example 4.2.6. Two-way cross-classified model. Suppose an experiment has two factors, Factor A at a levels and Factor B at b levels. In a cross-classified model, every level of Factor A can be studied in combination with every level of Factor B. We introduce two relevant models.

Additive model.

This model involves only main effects due to each factor and has the form

$$Y_{ijk} = \mu + \tau_i + \theta_j + \varepsilon_{ijk}, \quad 1 \le k \le n_{ij},\ 1 \le i \le a,\ 1 \le j \le b, \tag{4.2.30}$$

where Y_{ijk} denotes the kth replicate in the (i,j)th cell, μ is the overall mean, τ_i denotes the effect due to the ith level of Factor A, θ_j denotes the effect due to the jth level of Factor B, and the errors ε_{ijk} are i.i.d. with zero mean and variance σ^2. Consider the balanced design where $n_{ij} = n$ for all $1 \le i \le a$ and $1 \le j \le b$. Write $\boldsymbol{\beta} = (\mu, \tau_1, \cdots, \tau_a, \theta_1, \cdots, \theta_b)'$ and

$$\mathbf{X} = \begin{pmatrix} \mathbf{1}_{nb} & \mathbf{1}_{nb} & \mathbf{0} & \cdots & \mathbf{0} & \mathbf{L}_b \\ \mathbf{1}_{nb} & \mathbf{0} & \mathbf{1}_{nb} & \cdots & \mathbf{0} & \mathbf{L}_b \\ \vdots & \vdots & \vdots & \ddots & \vdots & \vdots \\ \mathbf{1}_{nb} & \mathbf{0} & \mathbf{0} & \cdots & \mathbf{1}_{nb} & \mathbf{L}_b \end{pmatrix}, \tag{4.2.31}$$

where $\mathbf{L}_b = \text{diag}(\mathbf{1}_n, \cdots, \mathbf{1}_n)$ is an $(nb \times b)$ matrix and is repeated a times in $\mathbf{X}$. Next, write $\mathbf{y} = (\mathbf{y}'_{11}, \cdots, \mathbf{y}'_{1b}, \cdots, \mathbf{y}'_{a1}, \cdots, \mathbf{y}'_{ab})'$, where $\mathbf{y}_{ij} = (Y_{ij1}, \cdots, Y_{ijn})'$, and write $\boldsymbol{\varepsilon}$ in a similar way. Then we can write (4.2.30) in the form (4.1.1). In this balanced model, $\mathbf{X}$ is an $N \times p$ design matrix with $N = nab$ and $p = (a + b + 1)$, while $\text{r}(\mathbf{X}) = (a + b - 1)$. The normal equations $\mathbf{X}'\mathbf{X}\boldsymbol{\beta}^0 = \mathbf{X}'\mathbf{y}$ have the form

$$nab\mu^0 + nb\sum_{i=1}^{a}\tau_i^0 + na\sum_{j=1}^{b}\theta_j^0 = Y_{...}$$

$$nb\mu^0 + nb\tau_i^0 + n\sum_{j=1}^{b}\theta_j^0 = Y_{i..}, \quad i = 1, \cdots, a$$

$$na\mu^0 + n\sum_{i=1}^{a}\tau_i^0 + na\theta_j^0 = Y_{.j.}, \quad j = 1, \cdots, b.$$

We impose two constraints $\sum_{i=1}^{a}\tau_i^0 = 0$, and $\sum_{j=1}^{b}\theta_j^0 = 0$, i.e., $\mathbf{A}'\boldsymbol{\beta} = \mathbf{0}$ with $\mathbf{A}' = \begin{pmatrix} 0 & \mathbf{1}_a' & \mathbf{0} \\ 0 & \mathbf{0} & \mathbf{1}_b' \end{pmatrix}$, and solve the system of normal equations to obtain

$$\mu^0 = \overline{Y}_{...}, \ \tau_i^0 = \overline{Y}_{i..} - \overline{Y}_{...}, \ \theta_j^0 = \overline{Y}_{.j.} - \overline{Y}_{...}, \quad 1 \leq i \leq a, \ 1 \leq j \leq b.$$

Searle (1971), p. 266, has described the use of *absorbing equations* for solving the normal equations (see Result 3.2.9). The fitted values are

$$\widehat{Y}_{ijk} = \mu^0 + \tau_i^0 + \theta_j^0 = \overline{Y}_{i..} + \overline{Y}_{.j.} - \overline{Y}_{...}.$$

Model with interaction.

In some experiments, we find that the difference in response between the levels of Factor A is not the same at all levels of Factor B. In this case, there is *interaction* between the two factors. A balanced model which includes main effects due to each factor as well as interactions between them is

$$Y_{ijk} = \mu + \tau_i + \theta_j + (\tau\theta)_{ij} + \varepsilon_{ijk}, \ 1 \leq k \leq n, \ 1 \leq i \leq a, \ 1 \leq j \leq b, \tag{4.2.32}$$

where $(\tau\theta)_{ij}$ denotes the interaction effect between the ith level of Factor A and jth level of Factor B, and the other terms are as before. The interaction term is sometimes denoted in the literature by γ_{ij}. Write $\mathbf{y}$ and $\boldsymbol{\varepsilon}$ as before, while $\boldsymbol{\beta} = (\mu, \tau_1, \cdots, \tau_a, \theta_1, \cdots, \theta_b, (\tau\theta)_{11}, \cdots, (\tau\theta)_{1b}, \cdots, (\tau\theta)_{a1}, \cdots, (\tau\theta)_{ab})'$ and

$$\mathbf{X} = \begin{pmatrix} \mathbf{1}_{nb} & \mathbf{1}_{nb} & \mathbf{0} & \cdots & \mathbf{0} & \mathbf{L}_b & \mathbf{L}_b & \mathbf{0} & \cdots & \mathbf{0} \\ \mathbf{1}_{nb} & \mathbf{0} & \mathbf{1}_{nb} & \cdots & \mathbf{0} & \mathbf{L}_b & \mathbf{0} & \mathbf{L}_b & \cdots & \mathbf{0} \\ \vdots & \vdots & \vdots & \ddots & \vdots & \vdots & \vdots & \vdots & \ddots & \vdots \\ \mathbf{1}_{nb} & \mathbf{0} & \mathbf{0} & \cdots & \mathbf{1}_{nb} & \mathbf{L}_b & \mathbf{0} & \mathbf{0} & \cdots & \mathbf{L}_b \end{pmatrix}. \tag{4.2.33}$$

Then we can write (4.2.32) in the form (4.1.1). The design matrix $\mathbf{X}$ has dimension $N \times p$ with $N = nab$ and $p = 1 + a + b + ab$, but its rank is only ab. If $n = 1$, i.e., there is no replication, we can show that we cannot separate interaction from the error (see Exercise 4.12). Let $Y_{...} = \sum_{k=1}^{n}\sum_{j=1}^{b}\sum_{i=1}^{a} Y_{ijk}$, $Y_{i..} = \sum_{k=1}^{n}\sum_{j=1}^{b} Y_{ijk}$, $Y_{.j.} = \sum_{k=1}^{n}\sum_{i=1}^{a} Y_{ijk}$ and $Y_{ij.} = \sum_{k=1}^{n} Y_{ijk}$. The normal equations are obtained by minimizing

$$\sum_{k=1}^{n}\sum_{j=1}^{b}\sum_{i=1}^{a}(Y_{ijk} - \mu - \tau_i - \theta_j - (\tau\theta)_{ij})^2$$

and are given by

$$Y_{\cdot\cdot\cdot} = abn\mu^0 + bn\sum_{i=1}^{a}\tau_i^0 + an\sum_{j=1}^{b}\theta_j^0 + n\sum_{i=1}^{a}\sum_{j=1}^{b}(\tau\theta)_{ij}^0$$

$$Y_{i\cdot\cdot} = bn\mu^0 + bn\tau_i^0 + n\sum_{j=1}^{b}\theta_j^0 + n\sum_{j=1}^{b}(\tau\theta)_{ij}^0$$

$$Y_{\cdot j\cdot} = an\mu^0 + n\sum_{i=1}^{a}\tau_i^0 + an\theta_j^0 + n\sum_{i=1}^{a}(\tau\theta)_{ij}^0$$

$$Y_{ij\cdot} = n\mu^0 + n\tau_i^0 + n\theta_j^0 + n(\tau\theta)_{ij}^0, \quad 1 \le i \le a,\, 1 \le j \le b.$$

Only the last set of ab equations is LIN, and the remaining equations can be derived from these, by adding over i, or over j, or both. Of the constraints $\sum_{i=1}^{a}\tau_i^0 = 0$, $\sum_{j=1}^{b}\theta_j^0 = 0$, $\sum_{i=1}^{a}(\tau\theta)_{ij}^0 = 0$, $j = 1, \cdots, b$, and $\sum_{j=1}^{b}(\tau\theta)_{ij}^0 = 0$, $i = 1, \cdots, a$, only $(a+b+1)$ are LIN, since any $(a+b-1)$ of the constraints on $(\tau\theta)_{ij}$ imply $\sum_{i=1}^{a}\sum_{j=1}^{b}(\tau\theta)_{ij}^0 = 0$, which in turn implies the one remaining constraint. Imposing these constraints, the normal equations yield the following solutions:

$$\mu^0 = \overline{Y}_{\cdots}, \quad \tau_i^0 = \overline{Y}_{i\cdot\cdot} - \overline{Y}_{\cdots}, \quad \theta_j^0 = \overline{Y}_{\cdot j\cdot} - \overline{Y}_{\cdots},$$
$$(\tau\theta)_{ij}^0 = \overline{Y}_{ij\cdot} - \overline{Y}_{i\cdot\cdot} - \overline{Y}_{\cdot j\cdot} + \overline{Y}_{\cdots}, \quad 1 \le i \le a,\; 1 \le j \le b. \tag{4.2.34}$$

A numerical example illustrating these and further inference are described in Chapter 7. □

Example 4.2.7. Two-factor nested model. In some multi-factor experiments, the levels of one factor, say Factor B, are similar, but not identical for different levels of another factor, Factor A. Then, the levels of Factor B may be considered to be nested within (under) the levels of Factor A, and we have a nested or hierarchical model. The two-way nested model has the form

$$Y_{ijk} = \mu + \tau_i + \theta_{j(i)} + \varepsilon_{(ij)k}, \quad 1 \le k \le n_{ij},\, 1 \le j \le b_i,\, 1 \le i \le a. \tag{4.2.35}$$

There are a levels of Factor A, and b_i levels of Factor B nested within each of the levels of Factor A in this unbalanced model. For example, suppose that a clinical trial involves a counties, and there are b_i hospitals in the ith county, and that data is collected on a sample of n_{ij} patients from the jth hospital in the ith county. The analysis of such data requires a nested or hierarchical model. Notice that we cannot include an interaction term between Factor A and Factor B since every level of Factor B does not appear with every level of Factor A. If $n_{ij} = n$, and $b_i = b$ for all i, j, we have a balanced model with $N = nab$ observations. The total number of model parameters in this case is $p = 1 + a + ab$ corresponding to μ, τ_i, and $\theta_{j(i)}$, $1 \le j \le b$, $1 \le i \le a$.

In the unbalanced case, the normal equations are obtained by minimizing the function

$$\sum_{i=1}^{a}\sum_{j=1}^{b_i}\sum_{k=1}^{n_{ij}}(Y_{ijk} - \mu^0 - \tau_i^0 - \theta_{j(i)}^0)^2,$$

and have the form

$$Y_{\cdot\cdot\cdot} = N\mu^0 + \sum_{i=1}^{a} n_{i\cdot}\tau_i^0 + \sum_{i=1}^{a}\sum_{j=1}^{b_i} n_{ij}\theta_{j(i)}^0,$$

$$Y_{i\cdot\cdot} = n_{i\cdot}\mu^0 + n_{i\cdot}\tau_i^0 + \sum_{j=1}^{b_i} n_{ij}\theta_{j(i)}^0,$$

$$Y_{ij\cdot} = n_{ij}(\mu^0 + \tau_i^0 + \theta_{j(i)}^0), \quad 1 \le j \le b_i, \, 1 \le i \le a,$$

where $n_{i\cdot} = \sum_{j=1}^{b_i} n_{ij}$, and $N = \sum_{i=1}^{a}\sum_{j=1}^{b_i} n_{ij}$. We may verify that the first equation is obtained by summing the last set of equations over both i and j, while the next set of a equations are obtained by summing the last set over j. Although the total number of parameters in the unbalanced model is $p = 1 + a + \sum_{i=1}^{a} b_i$, the number of LIN normal equations is only $\sum_{i=1}^{a} b_i$. We impose the following $(a+1)$ constraints, $\sum_{i=1}^{a} n_{i\cdot}\tau_i^0 = 0$, and $\sum_{j=1}^{b_i} n_{ij}\theta_{j(i)}^0 = 0$, $1 \le i \le a$. The least squares solutions of the parameters under these constraints are

$$\mu^0 = \overline{Y}_{\cdots}, \quad \tau_i^0 = \overline{Y}_{i\cdot\cdot} - \overline{Y}_{\cdots}, \quad \theta_{j(i)}^0 = \overline{Y}_{ij\cdot} - \overline{Y}_{i\cdot\cdot}, \qquad 1 \le i \le a, \, 1 \le j \le b_i.$$

The fitted values are $\widehat{Y}_{ijk} = \overline{Y}_{ij\cdot}$, and the residuals are $\widehat{\varepsilon}_{ijk} = Y_{ijk} - \overline{Y}_{ij\cdot}$. For the balanced model, the least squares solutions are obtained by setting $n_{ij} = n$, and $b_i = b$ in the above formulas. □

Numerical examples for these ANOVA models will be described in detail in Chapter 7, where we discuss inference for the balanced models. In Chapter 10, we discuss unbalanced ANOVA models. The setup in the next result enables us to discuss consequences of model mis-specification on solution vectors and predictions.

Result 4.2.5. Consider the linear model

$$\mathbf{y} = \mathbf{X}_1\boldsymbol{\beta}_1 + \mathbf{X}_2\boldsymbol{\beta}_2 + \boldsymbol{\varepsilon},$$

where $\text{E}(\boldsymbol{\varepsilon}) = \mathbf{0}$ and $\text{Cov}(\boldsymbol{\varepsilon}) = \sigma^2\mathbf{I}_N$. Let $\mathbf{P}_i = \mathbf{P}_{\mathcal{C}(\mathbf{X}_i)}$, $i = 1, 2$. Then the LS solutions of $\boldsymbol{\beta}_1^0$ and $\boldsymbol{\beta}_2^0$ are

$$\begin{aligned} \boldsymbol{\beta}_1^0 &= (\mathbf{X}_1'\mathbf{X}_1)^-\mathbf{X}_1'(\mathbf{y} - \mathbf{X}_2\boldsymbol{\beta}_2^0), \\ \boldsymbol{\beta}_2^0 &= [\mathbf{X}_2'(\mathbf{I} - \mathbf{P}_1)\mathbf{X}_2]^-\mathbf{X}_2'(\mathbf{I} - \mathbf{P}_1)\mathbf{y}. \end{aligned} \tag{4.2.36}$$

Proof. The model can be written as $\mathbf{y} = \mathbf{X}\boldsymbol{\beta} + \boldsymbol{\varepsilon}$, where $\mathbf{X} = (\mathbf{X}_1, \mathbf{X}_2)$ and $\boldsymbol{\beta} = (\boldsymbol{\beta}_1', \boldsymbol{\beta}_2')'$. The LS solutions of $\boldsymbol{\beta}_1$ and $\boldsymbol{\beta}_2$ are obtained by solving the normal equations $\mathbf{X}'\mathbf{X}\boldsymbol{\beta} = \mathbf{X}\mathbf{y}$, i.e., by simultaneously solving

$$\begin{aligned} \mathbf{X}_1'\mathbf{X}_1\boldsymbol{\beta}_1^0 + \mathbf{X}_1'\mathbf{X}_2\boldsymbol{\beta}_2^0 &= \mathbf{X}_1'\mathbf{y}, \\ \mathbf{X}_2'\mathbf{X}_1\boldsymbol{\beta}_1^0 + \mathbf{X}_2'\mathbf{X}_2\boldsymbol{\beta}_2^0 &= \mathbf{X}_2'\mathbf{y}. \end{aligned}$$

For each possible $\boldsymbol{\beta}_2^0$, from the first equation, $\boldsymbol{\beta}_1^0 = (\mathbf{X}_1'\mathbf{X}_1)^-\mathbf{X}_1'(\mathbf{y} - \mathbf{X}_2\boldsymbol{\beta}_2^0)$. Plugging this into the second equation,

$$\mathbf{X}_2'\mathbf{X}_1(\mathbf{X}_1'\mathbf{X}_1)^-\mathbf{X}_1'(\mathbf{y} - \mathbf{X}_2\boldsymbol{\beta}_2^0) + \mathbf{X}_2'\mathbf{X}_2\boldsymbol{\beta}_2^0 = \mathbf{X}_2'\mathbf{y}.$$

Since $\mathbf{P}_1 = \mathbf{X}_1(\mathbf{X}_1'\mathbf{X}_1)^-\mathbf{X}_1'$, we have $\mathbf{X}_2'(\mathbf{I} - \mathbf{P}_1)\mathbf{X}_2\boldsymbol{\beta}_2^0 = \mathbf{X}_2'(\mathbf{I} - \mathbf{P}_1)\mathbf{y}$, proving (4.2.36). ■

4.3 Estimable functions

In the previous section, we saw that unless $\text{r}(\mathbf{X}) = p$, $\boldsymbol{\beta}^0$ is not unique. Although in the full-rank model, we can estimate any function of $\boldsymbol{\beta}$, we must restrict ourselves to estimating only certain linear functions of $\boldsymbol{\beta}$ when $\mathrm{r}(\mathbf{X}) < p$. Such a linear function of $\boldsymbol{\beta}$ is called an *estimable function.* In other words, a linear function of $\boldsymbol{\beta}$ for which a (unique) estimator based on $\boldsymbol{\beta}^0$ exists, which is invariant to the solution $\boldsymbol{\beta}^0$, is called an estimable function. A more precise definition, which also provides an approach for identification of an estimable function of $\boldsymbol{\beta}$, is given below.

Definition 4.3.1. A linear parametric function $\mathbf{c}'\boldsymbol{\beta}$ is said to be an estimable function of $\boldsymbol{\beta}$ if there exists an N-dimensional vector $\mathbf{t} = (t_1, \cdots, t_N)'$ such that the expectation (with respect to the distribution of $\mathbf{y}$) of the linear combination $\mathbf{t}'\mathbf{y} = t_1Y_1 + \cdots + t_NY_N$ is equal to $\mathbf{c}'\boldsymbol{\beta}$, i.e.,

$$\mathrm{E}(\mathbf{t}'\mathbf{y}) = \mathbf{c}'\boldsymbol{\beta}. \tag{4.3.1}$$

In other words, $\mathbf{c}'\boldsymbol{\beta}$ is estimable if there exists a linear function of $\mathbf{y}$ whose expected value is identically equal to $\mathbf{c}'\boldsymbol{\beta}$. Note that for a particular linear function $\mathbf{c}'\boldsymbol{\beta}$, the vector $\mathbf{t}$ is not required to be uniquely defined in any sense; the existence of such a vector $\mathbf{t}$ suffices. Usually, there may exist several linear functions of $\mathbf{y}$, each of whose expectations is equal to $\mathbf{c}'\boldsymbol{\beta}$. For establishing estimability, it is sufficient to verify the existence of any one of these functions. If $\mathrm{r}(\mathbf{X}) = p$, i.e., if we have a full-rank linear model, then any linear function of $\boldsymbol{\beta}$ is a linear estimable function, which implies that we may estimate and carry out inference for any linear function of $\boldsymbol{\beta}$. This, however, is not the case for the less than full-rank model, and we must first check whether a function of interest is an estimable function. Only then, we can estimate it and proceed with further inference. Note also that the definition of estimability does not depend on the error variance specification. We state and prove some properties of estimable functions of $\boldsymbol{\beta}$; these results are trivially true for the full-rank case.

Result 4.3.1.

1. The expected value of any observation Y_i is estimable.

2. Any linear combination of estimable functions is estimable.

3. The function $\mathbf{c}'\boldsymbol{\beta}$ is estimable if and only if

$$\mathbf{c}' = \mathbf{t}'\mathbf{X} \tag{4.3.2}$$

   for some vector $\mathbf{t}$, i.e., $\mathbf{c} \in \mathcal{R}(\mathbf{X})$.

4. The function $\mathbf{c}'\boldsymbol{\beta}$ is estimable if and only if

$$\mathbf{c}' = \mathbf{c}'\mathbf{H}, \tag{4.3.3}$$

   where $\mathbf{H} = \mathbf{G}\mathbf{X}'\mathbf{X}$.

5. The function $\mathbf{c}'\boldsymbol{\beta}$ is estimable if and only if the quantity $\mathbf{c}'\boldsymbol{\beta}^0$ is invariant to the LS solution $\boldsymbol{\beta}^0$.

Property 3 and property 4 each gives a *necessary and sufficient condition* for estimability.

Proof. 1. Since $\mathrm{E}(Y_i) = \mathrm{E}(\mathbf{t}'\mathbf{y})$ with $\mathbf{t} = \mathbf{e}_i$, by Definition 4.3.1, $\mathrm{E}(Y_i)$ is estimable.

2. If $\mathbf{c}_1'\boldsymbol{\beta}$ and $\mathbf{c}_2'\boldsymbol{\beta}$ are estimable functions of $\boldsymbol{\beta}$, then for some N-dimensional vectors $\mathbf{t}_1$ and $\mathbf{t}_2$, $\mathbf{c}_i'\boldsymbol{\beta} = \mathrm{E}(\mathbf{t}_i'\mathbf{y})$, $i = 1, 2$. Then for constants a_1 and a_2, $a_1\mathbf{c}_1'\boldsymbol{\beta} + a_2\mathbf{c}_2'\boldsymbol{\beta} = \mathrm{E}(\mathbf{t}'\mathbf{y})$ with $\mathbf{t} = a_1\mathbf{t}_1 + a_2\mathbf{t}_2$, which proves property 2.

3. Since $\mathrm{E}(\mathbf{y}) = \mathbf{X}\boldsymbol{\beta}$, we see from Definition 4.3.1 that $\mathbf{c}'\boldsymbol{\beta}$ is estimable if and only if $\mathbf{c}'\boldsymbol{\beta} = \mathbf{t}'\mathbf{X}\boldsymbol{\beta}$ holds for *all* values of $\boldsymbol{\beta}$. The latter holds if and only if $\mathbf{c}' = \mathbf{t}'\mathbf{X}$ for some N-dimensional vector $\mathbf{t}$, proving property 3.

4. First assume that $\mathbf{c}'\boldsymbol{\beta}$ is estimable. From (4.3.2), $\mathbf{c}' = \mathbf{t}'\mathbf{X}$ for some $\mathbf{t}$. Then, by property 2 of Result 3.1.9, $\mathbf{c}'\mathbf{H} = \mathbf{t}'\mathbf{X}\mathbf{H} = \mathbf{t}'\mathbf{X}\mathbf{G}\mathbf{X}'\mathbf{X} = \mathbf{t}'\mathbf{X} = \mathbf{c}'$. Conversely, if $\mathbf{c}'\mathbf{H} = \mathbf{c}'$, then $\mathbf{c}' = \mathbf{c}'\mathbf{G}\mathbf{X}'\mathbf{X} = \mathbf{t}'\mathbf{X}$, say, where $\mathbf{t}' = \mathbf{c}'\mathbf{G}\mathbf{X}'$, so that by (4.3.2), $\mathbf{c}'\boldsymbol{\beta}$ is estimable.

5. If $\mathbf{c}'\boldsymbol{\beta}$ is estimable, then from property 3, $\mathbf{c}' = \mathbf{t}'\mathbf{X}$ for some $\mathbf{t}$, so that

$$\mathbf{c}'\boldsymbol{\beta}^0 = \mathbf{t}'\mathbf{X}\boldsymbol{\beta}^0 = \mathbf{t}'\mathbf{X}\mathbf{G}\mathbf{X}'\mathbf{y} = \mathbf{t}'\mathbf{P}\mathbf{y},$$

where $\mathbf{P}$ is the (unique) orthogonal projection matrix onto $\mathcal{C}(\mathbf{X})$. Then $\mathbf{c}'\boldsymbol{\beta}^0$ is invariant to $\boldsymbol{\beta}^0$. Conversely, let $\mathbf{c}'\boldsymbol{\beta}^0$ be invariant to $\boldsymbol{\beta}^0 = \mathbf{G}\mathbf{X}'\mathbf{y}$. We have $\mathbf{c} = \mathbf{c}_0 + \mathbf{d}$, where $\mathbf{c}_0 \in \mathcal{R}(\mathbf{X})$ and $\mathbf{d} \in \mathcal{R}(\mathbf{X})^\perp = \mathcal{N}(\mathbf{X})$. As already shown, $\mathbf{c}_0'\boldsymbol{\beta}^0$ is invariant to $\boldsymbol{\beta}^0$. Then so is $\mathbf{d}'\boldsymbol{\beta}^0 = \mathbf{c}'\boldsymbol{\beta}^0 - \mathbf{c}_0'\boldsymbol{\beta}_0$, yielding invariance of $\mathrm{E}(\mathbf{d}'\boldsymbol{\beta}^0) = \mathbf{d}'\mathbf{G}\mathbf{X}'\mathbf{X}\boldsymbol{\beta}$ to $\mathbf{G}$. With $\boldsymbol{\beta}$ being arbitrary, $\mathbf{d}'\mathbf{G}\mathbf{X}'\mathbf{X}$ has the same invariance. For any $\mathbf{u} \in \mathcal{R}^p$, since $\mathbf{X}\mathbf{d} = \mathbf{0}$, it is readily seen that $\mathbf{G}+\mathbf{d}\mathbf{u}'$ is a g-inverse of $\mathbf{X}'\mathbf{X}$. Then by the invariance, $\mathbf{d}'(\mathbf{d}\mathbf{u}')\mathbf{X}'\mathbf{X} = \|\mathbf{d}\|^2\mathbf{u}'\mathbf{X}'\mathbf{X} = \mathbf{0}$. Because $\mathbf{X}'\mathbf{X} \neq \mathbf{O}$ and $\mathbf{u}$ is arbitrary, this yields $\mathbf{d} = \mathbf{0}$. Then $\mathbf{c} = \mathbf{c}_0 \in \mathcal{R}(\mathbf{X})$, so by property 3, $\mathbf{c}'\boldsymbol{\beta}$ is estimable. ■

Example 4.3.1. In the model

$$\begin{aligned}
\mathrm{E}(Y_1) &= \beta_1 + \beta_2 \\
\mathrm{E}(Y_2) &= \beta_1 + \beta_3 \\
\mathrm{E}(Y_3) &= \beta_1 + \beta_2,
\end{aligned}$$

we find a necessary and sufficient condition for estimability of $c_1\beta_1 + c_2\beta_2 + c_3\beta_3$, for real constants c_i, $i = 1, \cdots, 3$. The function $\sum_{i=1}^3 c_i\beta_i$ is estimable if and only if there exist t_i, $i = 1, \cdots, 3$ such that $\mathrm{E}(\sum_{i=1}^3 t_iY_i) = \sum_{i=1}^3 c_i\beta_i$, i.e., if and only if c_1, c_2 and c_3 satisfy the conditions $c_1 = \sum_{i=1}^3 t_i$, $c_2 = t_1 + t_3$, and $c_3 = t_2$, i.e., if and only if $c_1 = c_2 + c_3$. For example, while $\beta_1 + \beta_2/2 + \beta_3/2$ is estimable, the function $\beta_1 - \beta_2/2 - \beta_3/2$ is not estimable. □

From properties 2 and 3 of Result 4.3.1, the following characterization of the entire set of estimable functions is obvious.

Result 4.3.2. The space of vectors $\mathbf{c}$ such that $\mathbf{c}'\boldsymbol{\beta}$ is estimable is $\mathcal{R}(\mathbf{X})$, so there exist exactly r LIN estimable functions of $\boldsymbol{\beta}$, where $r = \mathrm{r}(\mathbf{X})$.

If $\mathbf{C} = (\mathbf{c}_1, \cdots, \mathbf{c}_k)$ is a $p \times k$ matrix, then $\mathbf{C}'\boldsymbol{\beta}$ is said to be estimable if every $\mathbf{c}_i'\boldsymbol{\beta}$ is estimable. From property 3 of Result 4.3.1, $\mathbf{C}' = \mathbf{T}'\mathbf{X}$ for some $N \times k$ matrix $\mathbf{T}$. Although $\mathbf{T}$ need not be unique, its orthogonal projection onto $\mathcal{C}(\mathbf{X})$ is unique, i.e., if $\mathbf{T}_1$ and $\mathbf{T}_2$ are such that $\mathbf{C}' = \mathbf{T}_1'\mathbf{X} = \mathbf{T}_2'\mathbf{X}$, then

$$\mathbf{P}\mathbf{T}_1 = \mathbf{X}(\mathbf{X}'\mathbf{X})^-\mathbf{X}'\mathbf{T}_1 = \mathbf{X}(\mathbf{X}'\mathbf{X})^-\mathbf{C} = \mathbf{X}(\mathbf{X}'\mathbf{X})^-\mathbf{X}'\mathbf{T}_2 = \mathbf{P}\mathbf{T}_2.$$

The following result is an immediate consequence of the previous result, and we leave its proof as an exercise.

Result 4.3.3.

1. $\mathbf{X}\boldsymbol{\beta}$ is estimable, and any linear function of $\mathbf{X}\boldsymbol{\beta}$ is also estimable.

2. $\mathbf{X}'\mathbf{X}\boldsymbol{\beta}$ is estimable, and any linear function of $\mathbf{X}'\mathbf{X}\boldsymbol{\beta}$ is also estimable.

As an application of the above results, we have the following.

Result 4.3.4. Estimable functions in the one-way ANOVA model. Let $\boldsymbol{\beta} = (\mu, \tau_1, \cdots, \tau_a)'$ in the one-way fixed-effects ANOVA model in Example 4.1.3. The following properties hold.

1. For $\mathbf{c} = (c_0, c_1, \cdots, c_a)'$, $\mathbf{c}'\boldsymbol{\beta}$ is estimable if and only if $c_0 = \sum_{i=1}^a c_i$.

2. If $\mathbf{c}'\boldsymbol{\beta}$ is estimable, then $\mathbf{c}'\boldsymbol{\beta}^0 = c_1\overline{Y}_{1\cdot} + \cdots + c_a\overline{Y}_{a\cdot}$.

Proof. From property 3 of Result 4.3.1, $\mathbf{c}'\boldsymbol{\beta}$ is estimable if and only if $\mathbf{c} \in \mathcal{R}(\mathbf{X})$ with $\mathbf{X}$ given in Example 4.1.3. The set of vectors $\{\mathbf{v}_i = (1, \mathbf{e}_i')', \ i = 1, \cdots, a\}$ is a basis for $\mathcal{R}(\mathbf{X})$, where $\mathbf{e}_i$ is the ith standard basis vector of $\mathcal{R}^a$. Hence $\mathbf{c} \in \mathcal{R}(\mathbf{X})$ implies $\mathbf{c} = \sum_{i=1}^a b_i\mathbf{v}_i$ for some $b_1, \cdots, b_a$, giving $c_0 = \sum_{i=1}^a b_i$ and $c_i = a_i$, $i = 1, \ldots, a$, so $c_0 = \sum_{i=1}^a c_i$. Conversely, $c_0 = \sum_{i=1}^a c_i$ implies $\mathbf{c} = \sum_{i=1}^a c_i\mathbf{v}_i$, so $\mathbf{c} \in \mathcal{R}(\mathbf{X})$. Property 1 follows. From property 5 of Result 4.3.1, if $\mathbf{c}'\boldsymbol{\beta}$ is estimable, then $\mathbf{c}'\boldsymbol{\beta}^0$ is invariant to the choice of $\boldsymbol{\beta}^0$. Property 2 follows by choosing $\boldsymbol{\beta}^0 = (0, \overline{Y}_{1\cdot}, \cdots, \overline{Y}_{a\cdot})'$ (see Example 4.2.5). ■

Definition 4.3.2. Contrast. A contrast in the p-dimensional parameter vector $\boldsymbol{\beta}$ is a linear function $\mathbf{c}'\boldsymbol{\beta}$ such that $\mathbf{c}'\mathbf{1}_p = \sum_{i=1}^p c_i = 0$. Such a vector $\mathbf{c} = (c_1, \cdots, c_p)'$ is called a contrast vector.

Definition 4.3.3. Orthogonal contrasts. Two contrasts $\mathbf{c}_j'\boldsymbol{\beta}$ and $\mathbf{c}_l'\boldsymbol{\beta}$ are orthogonal if $\mathbf{c}_j'\mathbf{G}\mathbf{c}_l = 0$.

Example 4.3.2. Contrasts in the one-way ANOVA model. We discuss useful properties about contrasts in the one-way ANOVA model in Example 4.1.3. Again, the parameter vector is $\boldsymbol{\beta} = (\mu, \tau_1, \cdots, \tau_a)'$.

1. The function $\gamma = \sum_{i=1}^a c_i\tau_i$ is estimable if and only if $\sum_{i=1}^a c_i = 0$. It is easy to see that this important result is just a special case of property 1 of Result 4.3.4. Note that γ is a contrast in the treatment effects τ_i, $i = 1, \cdots, a$, and includes functions such as $\tau_1 - \tau_2$, $2\tau_1 - \tau_2 - \tau_4$, etc.

2. Functions such as $\tau_j - \tau_l$, for $j \neq l$ are elementary contrasts since they represent simple comparisons of two effects or groups.

3. Other contrasts are called general contrasts and can be expressed as linear combinations of elementary contrasts. For instance, we can write $\tau_1 + \tau_2 - 2\tau_3 = (\tau_1 - \tau_3) + (\tau_2 - \tau_3)$, and $c_1\tau_1 + c_2\tau_2 + \cdots + c_a\tau_a = c_1(\tau_1 - \tau_a) + c_2(\tau_2 - \tau_a) + \cdots + c_{a-1}(\tau_{a-1} - \tau_a)$, since $\sum_{i=1}^a c_i = 0$.

4. There exist at most $a - 1$ LIN contrasts and $a - 1$ corresponding contrast vectors. One such set of contrast vectors is $(0, 1, -1, 0, \cdots, 0)'$, $(0, 1, 0, -1, \cdots, 0)'$, $\cdots$, $(0, 1, 0, 0, \cdots, -1)'$, which corresponds to contrasts $\tau_1 - \tau_2, \tau_1 - \tau_3, \cdots, \tau_1 - \tau_a$.

5. Consider the contrast vectors $\mathbf{c}_l' = (0, \mathbf{1}_l', -l, 0, \cdots, 0)/\sqrt{l(l+1)}$, $l = 1, \cdots, a-1$, which are orthonormal (see the Helmert matrix in Example 1.3.8). The corresponding contrasts are $(\tau_1 + \cdots + \tau_l - l\tau_{l+1})/\sqrt{l(l+1)}$. From Definition 4.3.3, note that two contrasts $\sum_{i=1}^a c_i\tau_i$ and $\sum_{i=1}^a d_i\tau_i$ are orthogonal if $\sum_{i=1}^a c_id_i/n_i = 0$. In the balanced design, this reduces to the condition $\sum_{i=1}^a c_id_i = 0$. □

Example 4.3.3. Nonestimable functions in the one-way ANOVA model. Neither the overall mean μ nor the level effects $\tau_1, \cdots, \tau_a$ are estimable.

1. We have $\mu = \mathbf{c}'\boldsymbol{\beta}$, where $c = (c_0, c_1, \cdots, c_a)'$, with $c_0 = 1$, and $c_1 = \cdots = c_a = 0$. Then from property 1 of Result 4.3.4, μ is not estimable.

2. The level effect τ_i is not estimable for any i. We leave verification of this as Exercise 4.16. □

We present a result for a balanced two-way cross-classified additive model. We discuss this further in Chapter 7.

Result 4.3.5. Consider the model (4.2.30) with $n_{ij} = 1$, $i = 1, \cdots, a$, $j = 1, \cdots, b$.

1. The function $\sum_{i=1}^{a} c_i\tau_i$ is estimable if and only if $\sum_{i=1}^{a} c_i = 0$, i.e., the function is a contrast in τ_i.

2. Contrasts in $\{\tau_i\}$ are orthogonal to contrasts in $\{\theta_j\}$.

Proof. Following the discussion on the additive model (4.2.30) in Example 4.2.6, let $\boldsymbol{\beta} = (\mu, \tau_1, \cdots, \tau_a, \theta_1, \cdots, \theta_b)$ and let $\mathbf{X}$ be as in (4.2.31). By direct calculation,

$$\mathbf{X}'\mathbf{X} = \begin{pmatrix} nab & nb\mathbf{1}_a' & na\mathbf{1}_b' \\ nb\mathbf{1}_a & nb\mathbf{I}_a & n\mathbf{J}_{ab} \\ na\mathbf{1}_b & n\mathbf{J}_{ba} & na\mathbf{I}_b \end{pmatrix}$$

and, letting $d = b - 1$,

$$\mathbf{G} = \frac{1}{n}\begin{pmatrix} 0 & \mathbf{0}_a' & \mathbf{0}_d' & 0 \\ \mathbf{0}_a & b^{-1}(\mathbf{I}_a + \frac{d}{a}\mathbf{J}_a) & -a^{-1}\mathbf{J}_{a,d} & \mathbf{0}_a \\ \mathbf{0}_d & -a^{-1}\mathbf{J}_{d,a} & a^{-1}(\mathbf{I}_d + \mathbf{J}_d) & \mathbf{0}_d \\ 0 & \mathbf{0}_a' & \mathbf{0}_d' & 0 \end{pmatrix}$$

is a g-inverse of $\mathbf{X}'\mathbf{X}$. Let $\mathbf{c} = (c_1, \cdots, c_a)'$ and $\mathbf{f} = (0, \mathbf{c}', \mathbf{0}_b')'$. Then $\sum_{i=1}^{a} c_i\tau_i = \mathbf{f}'\boldsymbol{\beta}$, so that from Result 1.3.10 and Result 4.3.1, $\sum_{i=1}^{a} c_i\tau_i$ is estimable if and only if $\mathbf{f} = \mathbf{X}'\mathbf{X}\mathbf{t}$ for some $\mathbf{t} = (t_0, \mathbf{t}_a', \mathbf{t}_b')'$, where $\mathbf{t}_a \in \mathcal{R}^a$ and $\mathbf{t}_b \in \mathcal{R}^b$. This is equivalent to the existence of a solution to the system of equations

$$\begin{cases} abt_0 + b\mathbf{1}_a'\mathbf{t}_a + a\mathbf{1}_b'\mathbf{t}_b = 0 \\ bt_0\mathbf{1}_a + b\mathbf{t}_a + (\mathbf{1}_b'\mathbf{t}_b)\mathbf{1}_a = \mathbf{c} \\ at_0\mathbf{1}_b + (\mathbf{1}_a'\mathbf{t}_a)\mathbf{1}_b + a\mathbf{t}_b = \mathbf{0}. \end{cases}$$

If a solution exists, then $\sum_{i=1}^{a} c_i = \mathbf{c}'\mathbf{1}_a = [bt_0\mathbf{1}_a + b\mathbf{t}_a + (\mathbf{1}_b'\mathbf{t}_b)\mathbf{1}_a]'\mathbf{1}_a = abt_0 + b\mathbf{1}_a'\mathbf{t}_a + a\mathbf{1}_b'\mathbf{t}_b = 0$. Conversely, if $\sum_{i=1}^{a} c_i = 0$, then $t_0 = 0$, $\mathbf{t}_a = \mathbf{c}/b$, and $\mathbf{t}_b = \mathbf{0}$ constitutes a solution. This proves property 1.

Next, let $\sum_{i=1}^{a} c_i\tau_i$ be a contrast and let $\mathbf{f}$ be defined as before. It can be directly checked that with the above g-inverse $\mathbf{G}$, $\mathbf{G}\mathbf{f} = \mathbf{f}/b$; in particular, the last b components of $\mathbf{G}\mathbf{f}$ are all 0. On the other hand, any contrast in $\{\theta_j\}$ can be written as $\mathbf{g}'\boldsymbol{\beta}$, where only the last b components of $\mathbf{g}$ can be non-zero. As a result, $\mathbf{f}'\mathbf{g} = \mathbf{0}$, so that $\mathbf{f}'\mathbf{G}\mathbf{g} = (\mathbf{f}'\mathbf{g})/b = 0$, proving property 2. ■

4.4 Gauss–Markov theorem

We now state one of the most fundamental results in linear model theory, which gives an optimality result for linear estimates of estimable functions of $\boldsymbol{\beta}$ without any distributional assumption.

Result 4.4.1. Suppose in the general linear model (4.1.1), $\boldsymbol{\varepsilon}$ satisfies (4.1.4), i.e., $\mathrm{E}(\boldsymbol{\varepsilon}) = \mathbf{0}$ and $\mathrm{Cov}(\boldsymbol{\varepsilon}) = \sigma^2\mathbf{I}_N$. Let $\mathbf{c}'\boldsymbol{\beta}$ be an estimable function of $\boldsymbol{\beta}$, and let $\boldsymbol{\beta}^0$ denote any solution to the normal equations (4.2.2). Then, $\mathbf{c}'\boldsymbol{\beta}^0$ is the unique best linear unbiased estimator (b.l.u.e.) of $\mathbf{c}'\boldsymbol{\beta}$ with $\mathrm{Var}(\mathbf{c}'\boldsymbol{\beta}^0) = \sigma^2\mathbf{c}'\mathbf{G}\mathbf{c}$, in the sense that if $d_0+\mathbf{d}'\mathbf{y}$ is any linear unbiased estimator of $\mathbf{c}'\boldsymbol{\beta}$, then $\mathrm{Var}(d_0+\mathbf{d}'\mathbf{y}) \geq \mathrm{Var}(\mathbf{c}'\boldsymbol{\beta}^0)$, with equality if and only if $d_0+\mathbf{d}'\mathbf{y} = \mathbf{c}'\boldsymbol{\beta}^0$.

Proof. Clearly, $\mathbf{c}'\boldsymbol{\beta}^0 = \mathbf{c}'\mathbf{G}\mathbf{X}'\mathbf{y}$ is a linear function of $\mathbf{y}$. As we mentioned earlier, $\mathbf{c}'\boldsymbol{\beta}^0$ is invariant to $\mathbf{G}$ and hence to $\boldsymbol{\beta}^0$, and can therefore be regarded as a unique estimator of $\mathbf{c}'\boldsymbol{\beta}$. Now, from property 4 of Result 4.3.1,

$$\mathrm{E}(\mathbf{c}'\boldsymbol{\beta}^0) = \mathbf{c}'\,\mathrm{E}(\boldsymbol{\beta}^0) = \mathbf{c}'\mathbf{G}\mathbf{X}'\mathbf{X}\boldsymbol{\beta} = \mathbf{c}'\boldsymbol{\beta}, \tag{4.4.1}$$

showing that $\mathbf{c}'\boldsymbol{\beta}^0$ is an unbiased estimator of $\mathbf{c}'\boldsymbol{\beta}$. Further,

$$\mathrm{Var}(\mathbf{c}'\boldsymbol{\beta}^0) = \sigma^2\mathbf{c}'\mathbf{G}\mathbf{X}'\mathbf{X}\mathbf{G}'\mathbf{c} = \sigma^2(\mathbf{c}'\mathbf{G}\mathbf{X}'\mathbf{X})\mathbf{G}\mathbf{c} = \sigma^2\mathbf{c}'\mathbf{G}\mathbf{c}. \tag{4.4.2}$$

We now show that $\mathbf{c}'\boldsymbol{\beta}^0$ has minimum variance among the class of linear unbiased estimators of $\mathbf{c}'\boldsymbol{\beta}$. Let $d_0 + \mathbf{d}'\mathbf{y}$ denote any linear unbiased estimator of $\mathbf{c}'\boldsymbol{\beta}$. Then, we must have $\mathrm{E}(d_0 + \mathbf{d}'\mathbf{y}) = d_0 + \mathbf{d}'\mathbf{X}\boldsymbol{\beta} = \mathbf{c}'\boldsymbol{\beta}$, which implies that $d_0 = 0$ and $\mathbf{d}'\mathbf{X} = \mathbf{c}'$. Then $\mathbf{d}'\mathbf{y} = \mathbf{d}'(\widehat{\mathbf{y}} + \widehat{\boldsymbol{\varepsilon}}) = \mathbf{d}'\widehat{\mathbf{y}} + \mathbf{d}'\widehat{\boldsymbol{\varepsilon}} = \mathbf{d}'\mathbf{X}\boldsymbol{\beta}^0 + \mathbf{d}'\widehat{\boldsymbol{\varepsilon}} = \mathbf{c}'\boldsymbol{\beta}^0 + \mathbf{d}'\widehat{\boldsymbol{\varepsilon}}$. Since $\widehat{\mathbf{y}}$ and $\widehat{\boldsymbol{\varepsilon}}$ have covariance $\mathbf{0}$ according to property 3 in Result 4.2.4, $\mathbf{c}'\boldsymbol{\beta}^0$ and $\mathbf{d}'\widehat{\boldsymbol{\varepsilon}}$ are uncorrelated, giving

$$\mathrm{Var}(\mathbf{d}'\mathbf{y}) = \mathrm{Var}(\mathbf{c}'\boldsymbol{\beta}^0) + \mathrm{Var}(\mathbf{d}'\widehat{\boldsymbol{\varepsilon}}) \geq \mathrm{Var}(\mathbf{c}'\boldsymbol{\beta}^0),$$

with equality if and only if $\mathbf{d}'\widehat{\boldsymbol{\varepsilon}} = 0$, i.e., $\mathbf{d}'\mathbf{y} = \mathbf{c}'\boldsymbol{\beta}^0$. ■

Consider a projection approach for constructing the b.l.u.e. of an estimable function $\mathbf{c}'\boldsymbol{\beta}$. Let $\mathbf{t}$ be any vector such that $\mathbf{c}' = \mathbf{t}'\mathbf{X}$. Then

$$\mathbf{t}'\widehat{\mathbf{y}} = \mathbf{t}'\mathbf{X}\boldsymbol{\beta}^0 = \mathbf{c}'\boldsymbol{\beta}^0.$$

So, $\mathbf{t}'\widehat{\mathbf{y}} = \mathbf{t}'\mathbf{P}\mathbf{y}$ is the b.l.u.e. of $\mathbf{c}'\boldsymbol{\beta}$. On the other hand, $\mathbf{t}'\mathbf{y}$ is also an unbiased estimator of $\mathbf{c}'\boldsymbol{\beta}$ since $\mathrm{E}(\mathbf{t}'\mathbf{y}) = \mathbf{t}'\,\mathrm{E}(\mathbf{y}) = \mathbf{t}'\mathbf{X}\boldsymbol{\beta} = \mathbf{c}'\boldsymbol{\beta}$, so that $\mathbf{t}'\widehat{\boldsymbol{\varepsilon}} = \mathbf{t}'(\mathbf{I} - \mathbf{P})\mathbf{y}$ belongs to the error space. Thus, $\mathbf{t}'\mathbf{P}$ is the projection of $\mathbf{t}'$ onto the estimation space.

Corollary 4.4.1. In the full-rank model, the Gauss–Markov theorem states that the b.l.u.e. of $\mathbf{c}'\boldsymbol{\beta}$ is $\mathbf{c}'\widehat{\boldsymbol{\beta}} = \mathbf{c}'(\mathbf{X}'\mathbf{X})^{-1}\mathbf{X}'\mathbf{y}$, with variance $\sigma^2\mathbf{c}'(\mathbf{X}'\mathbf{X})^{-1}\mathbf{c}$.

Result 4.4.2. Let $\mathbf{c}_1'\boldsymbol{\beta}$ and $\mathbf{c}_2'\boldsymbol{\beta}$ be two estimable functions of $\boldsymbol{\beta}$. Then

$$\mathrm{Cov}(\mathbf{c}_1'\boldsymbol{\beta}^0, \mathbf{c}_2'\boldsymbol{\beta}^0) = \sigma^2\mathbf{c}_1'\mathbf{G}\mathbf{c}_2.$$

Proof. By definition, $\mathrm{Cov}(\mathbf{c}_1'\boldsymbol{\beta}^0, \mathbf{c}_2'\boldsymbol{\beta}^0) = \sigma^2\mathbf{c}_1'\,\mathrm{Cov}(\boldsymbol{\beta}^0)\mathbf{c}_2 = \sigma^2\mathbf{c}_1'\mathbf{G}\mathbf{c}_2$. ■

From Definition 4.3.3 and Result 4.4.2, two estimable contrasts $\mathbf{c}_i'\boldsymbol{\beta}$, $i = 1, 2$, are orthogonal if and only if $\mathbf{c}_1'\boldsymbol{\beta}^0$ and $\mathbf{c}_2'\boldsymbol{\beta}^0$ are uncorrelated.

Example 4.4.1. For the simple linear regression in Example 4.1.1, we have

$$\begin{aligned}\mathrm{E}(\widehat{\beta}_0) = \beta_0 \quad &\text{and} \quad \mathrm{Var}(\widehat{\beta}_0) = \sigma^2 \textstyle\sum_{i=1}^N X_i^2 / N \sum_{i=1}^N (X_i - \overline{X})^2, \\ \mathrm{E}(\widehat{\beta}_1) = \beta_1 \quad &\text{and} \quad \mathrm{Var}(\widehat{\beta}_1) = \sigma^2 / \textstyle\sum_{i=1}^N (X_i - \overline{X})^2.\end{aligned}$$

These results follow directly from Corollary 4.2.1. Further,

$$\mathrm{Cov}(\widehat{\beta}_0, \widehat{\beta}_1) = -\sigma^2 \overline{X} / \textstyle\sum_{i=1}^N (X_i - \overline{X})^2.$$

The estimators $\widehat{\beta}_0$ and $\widehat{\beta}_1$ are clearly unbiased. When $\overline{X} = 0$, $\widehat{\beta}_0$ and $\widehat{\beta}_1$ are uncorrelated. Intuitively, we may see that this is so because when $\overline{X} = 0$, $\widehat{\beta}_0 = \overline{Y}$, while $\widehat{\beta}_1$ is unaffected by a shift in location of the Y variable, i.e., it is uncorrelated with $\overline{Y}$. That $\mathrm{E}(\widehat{Y}_i) = \beta_0 + \beta_1 X_i$, and

$$\begin{aligned}\mathrm{Var}(\widehat{Y}_i) &= \mathrm{Var}(\widehat{\beta}_0) + X_i^2 \, \mathrm{Var}(\widehat{\beta}_1) + 2X_i \, \mathrm{Cov}(\widehat{\beta}_0, \widehat{\beta}_1) \\ &= \sigma^2/N + \sigma^2 (X_i - \overline{X})^2 / \textstyle\sum_{i=1}^N (X_i - \overline{X})^2\end{aligned}$$

follow directly from property 2 of Result 4.2.4. □

Example 4.4.2. Let $\boldsymbol{\beta} = (\mu, \tau_1, \cdots, \tau_a)'$ in the one-way fixed-effects ANOVA model in Example 4.1.3 and $\mathbf{c} = (c_0, c_1, \cdots, c_a)'$. From Result 4.3.4, $\mathbf{c}'\boldsymbol{\beta}$ is estimable if and only if $c_0 = \sum_{i=1}^a c_i$, in which case $\mathbf{c}'\boldsymbol{\beta}^0 = \sum_{i=1}^a c_i \overline{Y}_{i\cdot}$. If $\mathrm{Var}(\varepsilon_{ij}) = \sigma^2$, $j = 1, \cdots, n_i$, $i = 1, \cdots, a$, then from Result 4.4.1, $\mathbf{c}'\boldsymbol{\beta}^0$ is the b.l.u.e. of $\mathbf{c}'\boldsymbol{\beta}$. Since $\mathrm{Var}(\overline{Y}_{i\cdot}) = \sigma^2/n_i$, we see that $\mathrm{Var}(\mathbf{c}'\boldsymbol{\beta}^0) = \sigma^2 \sum_{i=1}^a (1/n_i)$. □

Example 4.4.3. In the two-way fixed-effects cross-classified additive model with $n = 1$, the following functions are estimable:

1. $\mu + \tau_i + \beta_j$ is estimable with b.l.u.e. $\overline{Y}_{i\cdot} + \overline{Y}_{\cdot j} - \overline{Y}_{\cdot\cdot}$;
2. $\tau_i - \tau_h$ is estimable, for $i \neq h$, with b.l.u.e. $\overline{Y}_{i\cdot} - \overline{Y}_{h\cdot}$;
3. $\beta_j - \beta_l$ is estimable, for $j \neq l$, with b.l.u.e. $\overline{Y}_{\cdot j} - \overline{Y}_{\cdot l}$;

estimability of these functions is directly verified since they are in turn $\mathrm{E}(Y_{ij})$, $\mathrm{E}(Y_{ij}) - \mathrm{E}(Y_{hj})$, and $\mathrm{E}(Y_{ij}) - \mathrm{E}(Y_{il})$. The b.l.u.e. of $\mu + \tau_i + \beta_j$ is $\mu^0 + \tau_i^0 + \beta_j^0 = \overline{Y}_{\cdot\cdot} + \overline{Y}_{i\cdot} - \overline{Y}_{\cdot\cdot} + \overline{Y}_{\cdot j} - \overline{Y}_{\cdot\cdot} = \overline{Y}_{i\cdot} + \overline{Y}_{\cdot j} - \overline{Y}_{\cdot\cdot}$. The b.l.u.e. of $\tau_i - \tau_h$ is $\tau_i^0 - \tau_h^0 = (\overline{Y}_{i\cdot} - \overline{Y}_{\cdot\cdot}) - (\overline{Y}_{h\cdot} - \overline{Y}_{\cdot\cdot}) = \overline{Y}_{i\cdot} - \overline{Y}_{h\cdot}$, with estimated variance $2\widehat{\sigma}^2/b$. To verify the variance estimate, we may see that $\mathrm{Var}(\overline{Y}_{i\cdot}) = \mathrm{Var}(\sum_{j=1}^b Y_{ij}/b) = \widehat{\sigma}^2/b$. Similarly, the b.l.u.e. of $\beta_j - \beta_l$ is $\beta_j^0 - \beta_l^0 = \overline{Y}_{\cdot j} - \overline{Y}_{\cdot l}$, with estimated variance $2\widehat{\sigma}^2/a$. □

Note: Rao (1952) proposed a class of nonnegative quadratic unbiased estimates $\mathbf{y}'\mathbf{A}\mathbf{y}$ of $(N - r)\sigma^2$, where $\mathbf{A}$ is symmetric and n.n.d. We discuss this in Chapter 11.

4.5 Generalized least squares

Consider the linear model

$$\mathbf{y} = \mathbf{X}\boldsymbol{\beta} + \boldsymbol{\varepsilon}, \quad \mathrm{Var}(\boldsymbol{\varepsilon}) = \sigma^2 \mathbf{V}, \tag{4.5.1}$$

where $\text{r}(\mathbf{X}) = r \le p$ and $\mathbf{V}$ is a known $N \times N$ symmetric, p.d. matrix. The other assumptions of the general linear model still hold. We show that the least squares procedure produces optimal estimates (in the sense of the Gauss–Markov theorem). Since $\mathbf{V}$ is p.d., from Result 2.4.5, we have that $\mathbf{K} = \mathbf{V}^{1/2}$ is a symmetric and n.n.d. matrix. Let $\mathbf{L} = \mathbf{K}^{-1}$. Then $\mathbf{V}^{-1} = \mathbf{L}'\mathbf{L}$. Let

$$\mathbf{z} = \mathbf{L}\mathbf{y}, \quad \mathbf{B} = \mathbf{L}\mathbf{X}, \quad \text{and} \quad \boldsymbol{\eta} = \mathbf{L}\boldsymbol{\varepsilon}. \tag{4.5.2}$$

Since $\text{r}(\mathbf{X}) = r$ and $\mathbf{L}$ is nonsingular, it follows from property 5 of Result 1.3.11 that $\text{r}(\mathbf{B}) = r$. It may also be verified that

$$\text{E}(\boldsymbol{\eta}) = \mathbf{0}, \quad \text{and} \quad \text{Var}(\boldsymbol{\eta}) = \mathbf{L}(\sigma^2\mathbf{V})\mathbf{L}' = \sigma^2\mathbf{L}\mathbf{K}\mathbf{K}'\mathbf{L}' = \sigma^2\mathbf{I}_N.$$

Consider the “transformed” linear model

$$\mathbf{z} = \mathbf{B}\boldsymbol{\beta} + \boldsymbol{\eta}, \quad \text{Var}(\boldsymbol{\eta}) = \sigma^2\mathbf{I}_N, \tag{4.5.3}$$

which resembles the general linear model in (4.1.1) with $\text{r}(\mathbf{B}) = r$. Minimizing $\boldsymbol{\eta}'\boldsymbol{\eta} = (\mathbf{z} - \mathbf{B}\boldsymbol{\beta})'(\mathbf{z} - \mathbf{B}\boldsymbol{\beta})$ with respect to $\boldsymbol{\beta}$, we obtain the generalized least squares (GLS) solution to the normal equations as

$$\begin{aligned}\boldsymbol{\beta}^0_{GLS} &= (\mathbf{B}'\mathbf{B})^-\mathbf{B}'\mathbf{z} = (\mathbf{X}'\mathbf{L}'\mathbf{L}\mathbf{X})^-\mathbf{X}'\mathbf{L}'\mathbf{L}\mathbf{y} \\ &= (\mathbf{X}'\mathbf{V}^{-1}\mathbf{X})^-\mathbf{X}'\mathbf{V}^{-1}\mathbf{y},\end{aligned} \tag{4.5.4}$$

where $(\mathbf{X}'\mathbf{V}^{-1}\mathbf{X})^-$ is a g-inverse of the $p \times p$ matrix $\mathbf{X}'\mathbf{V}^{-1}\mathbf{X}$. There is an alternate way to obtain the solution $\boldsymbol{\beta}^0_{GLS}$. Write

$$\begin{aligned}\boldsymbol{\eta}'\boldsymbol{\eta} &= \boldsymbol{\varepsilon}'\mathbf{V}^{-1}\boldsymbol{\varepsilon} = (\mathbf{y} - \mathbf{X}\boldsymbol{\beta})'\mathbf{V}^{-1}(\mathbf{y} - \mathbf{X}\boldsymbol{\beta}) \\ &= \mathbf{y}'\mathbf{y} - 2\boldsymbol{\beta}'\mathbf{X}'\mathbf{V}^{-1}\mathbf{y} + \boldsymbol{\beta}'\mathbf{X}'\mathbf{V}^{-1}\mathbf{X}\boldsymbol{\beta}.\end{aligned}$$

Differentiating with respect to $\boldsymbol{\beta}$ and setting equal to zero, i.e.,

$$\partial(\boldsymbol{\eta}'\boldsymbol{\eta})/\partial\boldsymbol{\beta} = -2\mathbf{X}'\mathbf{V}^{-1}\mathbf{y} + 2\mathbf{X}'\mathbf{V}^{-1}\mathbf{X}\boldsymbol{\beta} = \mathbf{0}$$

gives $\boldsymbol{\beta}^0_{GLS}$ in (4.5.4). If $r = p$, i.e., if we have a full-rank model, then $(\mathbf{X}'\mathbf{V}^{-1}\mathbf{X})^{-1}$ exists and the solution vector (4.5.4) is the unique GLS estimator of $\boldsymbol{\beta}$ given by

$$\widehat{\boldsymbol{\beta}}_{GLS} = (\mathbf{X}'\mathbf{V}^{-1}\mathbf{X})^{-1}\mathbf{X}'\mathbf{V}^{-1}\mathbf{y}. \tag{4.5.5}$$

The fitted vector and residual vector by the GLS estimation are

$$\begin{aligned}\widehat{\mathbf{y}}_{GLS} &= \mathbf{X}\boldsymbol{\beta}^0_{GLS} = \mathbf{K}\widehat{\mathbf{z}} = \mathbf{W}\mathbf{y}, \text{ and} \\ \widehat{\boldsymbol{\varepsilon}}_{GLS} &= \mathbf{y} - \widehat{\mathbf{y}}_{GLS} = \mathbf{K}\widehat{\boldsymbol{\eta}} = (\mathbf{I} - \mathbf{W})\mathbf{y},\end{aligned} \tag{4.5.6}$$

where $\widehat{\mathbf{z}}$ and $\widehat{\boldsymbol{\eta}}$ are respectively the fitted vector and residual vector by the OLS estimation for the model (4.5.3), and

$$\mathbf{W} = \mathbf{X}(\mathbf{X}'\mathbf{V}^{-1}\mathbf{X})^-\mathbf{X}'\mathbf{V}^{-1} = \mathbf{K}\mathbf{P}_{\mathcal{C}(\mathbf{L}\mathbf{X})}\mathbf{L}. \tag{4.5.7}$$

It is straightforward to check (see Exercise 4.23) that

$$\mathcal{C}(\mathbf{W}) = \mathcal{C}(\mathbf{X}), \quad \mathbf{W}\mathbf{X} = \mathbf{X}, \quad \text{and} \quad \mathbf{W}^2 = \mathbf{W}. \tag{4.5.8}$$

In general, $\mathbf{W}$ is not an orthogonal projection onto $\mathcal{C}(\mathbf{X})$ (under the Euclidean inner product) and $\widehat{\mathbf{y}}_{GLS}$ is not equal to the fitted vector obtained by OLS estimation. On the other hand, if the inner product between two vectors $\mathbf{u}$ and $\mathbf{v}$ is defined as $\mathbf{u}'\mathbf{V}^{-1}\mathbf{v}$, then $\mathbf{W}$ is the orthogonal projection onto $\mathcal{C}(\mathbf{X})$, as it can be directly verified that $(\mathbf{y}-\mathbf{W}\mathbf{y})'\mathbf{V}^{-1}\mathbf{z}=0$ for all $\mathbf{z}$ of the form $\mathbf{X}\mathbf{v}$.

We can define the following sums of squares in the GLS setup:

$$\begin{aligned} SST_{GLS} &= \mathbf{z}'\mathbf{z} = \mathbf{y}'\mathbf{V}^{-1}\mathbf{y}, \\ SSR_{GLS} &= \widehat{\mathbf{z}}'\widehat{\mathbf{z}} = \widehat{\mathbf{y}}_{GLS}\mathbf{V}^{-1}\widehat{\mathbf{y}}_{GLS}, \\ SSE_{GLS} &= \widehat{\boldsymbol{\eta}}'\widehat{\boldsymbol{\eta}} = \widehat{\boldsymbol{\varepsilon}}'_{GLS}\mathbf{V}^{-1}\widehat{\boldsymbol{\varepsilon}}_{GLS}, \end{aligned} \tag{4.5.9}$$

and write the ANOVA decomposition as

$$SST_{GLS} = SSR_{GLS} + SSE_{GLS}.$$

Corresponding to the mean and covariance under the LS estimation (see Result 4.2.4), we have the following result.

Result 4.5.1. Denote $\mathbf{S}=\mathbf{X}'\mathbf{V}^{-1}\mathbf{X}$. Then the following properties hold.

1. $\mathrm{E}(\boldsymbol{\beta}^0_{GLS}) = \mathbf{S}^-\mathbf{S}\boldsymbol{\beta}$ and $\mathrm{Cov}(\boldsymbol{\beta}^0) = \sigma^2\mathbf{S}^-\mathbf{S}\mathbf{S}^-$.
2. The mean and covariance of the fitted vector and residual vector under the GLS estimation are

$$\begin{aligned} \mathrm{E}(\widehat{\mathbf{y}}_{GLS}) = \mathbf{X}\boldsymbol{\beta} \quad &\text{and} \quad \mathrm{Cov}(\widehat{\mathbf{y}}_{GLS}) = \sigma^2\mathbf{W}\mathbf{V}, \\ \mathrm{E}(\widehat{\boldsymbol{\varepsilon}}_{GLS}) = \mathbf{0} \quad &\text{and} \quad \mathrm{Cov}(\widehat{\boldsymbol{\varepsilon}}_{GLS}) = \sigma^2(\mathbf{I}-\mathbf{W})\mathbf{V}. \end{aligned} \tag{4.5.10}$$

3. $\boldsymbol{\beta}^0_{GLS}$ and $\widehat{\boldsymbol{\varepsilon}}_{GLS}$ are uncorrelated, so that $\widehat{\mathbf{y}}_{GLS}$ and $\widehat{\boldsymbol{\varepsilon}}_{GLS}$ are uncorrelated.
4. $\widehat{\sigma}^2_{GLS} = SSE_{GLS}/(N-r)$ is an unbiased estimator of σ^2.

Proof. Property 1 follows from (4.5.4) and $\mathrm{Cov}(\mathbf{y}) = \mathrm{Cov}(\boldsymbol{\varepsilon}) = \sigma^2\mathbf{V}$. From (4.5.6) and Result 4.2.4 applied to (4.5.3), $\mathrm{E}(\widehat{\mathbf{y}}_{GLS}) = \mathrm{E}(\mathbf{K}\widehat{\mathbf{z}}) = \mathbf{K}(\mathbf{B}\boldsymbol{\beta}) = \mathbf{X}\boldsymbol{\beta}$ and $\mathrm{Cov}(\widehat{\mathbf{y}}_{GLS}) = \mathbf{K}\,\mathrm{Cov}(\widehat{\mathbf{z}})\mathbf{K}' = \sigma^2\mathbf{K}\mathbf{P}_{\mathcal{C}(\mathbf{B})}\mathbf{K}' = \sigma^2\mathbf{X}(\mathbf{X}'\mathbf{V}^{-1}\mathbf{X})^-\mathbf{X}' = \mathbf{W}\mathbf{V}$. The mean and covariance of $\widehat{\boldsymbol{\varepsilon}}_{GLS}$ can be derived similarly, proving property 2. To prove property 3, note that $\boldsymbol{\beta}^0_{GLS}$ is in fact the OLS solution under (4.5.3), so by Result 4.2.4, it is uncorrelated with $\widehat{\boldsymbol{\eta}}$, and so is uncorrelated with $\widehat{\varepsilon}_{GLS} = K\widehat{\boldsymbol{\eta}}$. Finally, by Result 4.2.4, $SSE_{GLS} = \mathrm{E}(\widehat{\boldsymbol{\eta}}'\widehat{\boldsymbol{\eta}}) = (N-1)\sigma^2$, yielding property 4. ■

We note that a function $\mathbf{c}'\boldsymbol{\beta}$ is estimable in model (4.5.1) if and only if it is estimable in the transformed model (4.5.3); see Exercise 4.20 and Exercise 4.21.

Result 4.5.2. If $\mathbf{c}'\boldsymbol{\beta}$ is an estimable function of $\boldsymbol{\beta}$, then the statistic $\mathbf{c}'\boldsymbol{\beta}^0_{GLS}$ is the unique b.l.u.e. of $\mathbf{c}'\boldsymbol{\beta}$. For two estimable functions $\mathbf{c}_1'\boldsymbol{\beta}$ and $\mathbf{c}_2'\boldsymbol{\beta}$,

$$\mathrm{Cov}(\mathbf{c}_1'\boldsymbol{\beta}^0_{GLS}, \mathbf{c}_2'\boldsymbol{\beta}^0_{GLS}) = \sigma^2\mathbf{c}_1'(\mathbf{X}'\mathbf{V}^{-1}\mathbf{X})^-\mathbf{c}_2.$$

The proof mimics the proof of Result 4.4.1 for the transformed model (4.5.3). Its detail is left as Exercise 4.22.

Corollary 4.5.1. For the full-rank model with $\mathrm{r}(\mathbf{X}) = p$, the GLS estimator $\widehat{\boldsymbol{\beta}}_{GLS}$ has the following properties:

1. $\widehat{\boldsymbol{\beta}}_{GLS}$ is an unbiased estimator of $\boldsymbol{\beta}$.

2. The variance-covariance matrix of $\widehat{\boldsymbol{\beta}}_{GLS}$ is $\text{Var}(\widehat{\boldsymbol{\beta}}_{GLS}) = \sigma^2(\mathbf{X}'\mathbf{V}^{-1}\mathbf{X})^{-1}$.

3. The error sum of squares is $SSE_{GLS} = (\mathbf{y} - \mathbf{X}\widehat{\boldsymbol{\beta}}_{GLS})'\mathbf{V}^{-1}(\mathbf{y} - \mathbf{X}\widehat{\boldsymbol{\beta}}_{GLS})$.

4. $\mathbf{c}'\widehat{\boldsymbol{\beta}}_{GLS}$ is the best linear unbiased estimator (b.l.u.e.) of $\mathbf{c}'\boldsymbol{\beta}$, with variance $\sigma^2\mathbf{c}'(\mathbf{X}'\mathbf{V}^{-1}\mathbf{X})^{-1}\mathbf{c}$.

Proof. We have $\text{E}(\widehat{\boldsymbol{\beta}}_{GLS}) = (\mathbf{X}'\mathbf{V}^{-1}\mathbf{X})^{-1}\mathbf{X}'\mathbf{V}^{-1}\mathbf{X}\boldsymbol{\beta} = \boldsymbol{\beta}$, which proves property 1. The other properties directly follow from Result 4.5.1. ■

Example 4.5.1. Consider the model

$$Y_i = \beta X_i + \varepsilon_i, \quad i = 1, \cdots, N,$$

where ε_i are normal random variables with $\text{E}(\varepsilon_i) = 0$ and $\text{Cov}(\varepsilon_i, \varepsilon_j) = \sigma^2\rho^{|j-i|}$, $i, j = 1, \cdots, N$, $|\rho| < 1$. We can write this in the form (4.5.1) with $\mathbf{y} = (Y_1, \cdots, Y_N)'$, $\boldsymbol{\varepsilon} = (\varepsilon_1, \cdots, \varepsilon_N)'$, $\mathbf{x} = (X_1, \cdots, X_N)'$, β a scalar, and

$$\mathbf{V} = \begin{pmatrix} 1 & \rho & \rho^2 & \cdots & \rho^{N-1} \\ \rho & 1 & \rho & \cdots & \rho^{N-2} \\ \vdots & \vdots & \vdots & \ddots & \vdots \\ \rho^{N-1} & \rho^{N-2} & \rho^{N-3} & \cdots & 1 \end{pmatrix}. \tag{4.5.11}$$

This is a linear regression model with autoregressive order 1 (AR(1)) errors , i.e., a serially correlated simple linear regression model. From Example 1.3.7,

$$\{\mathbf{V}^{-1}\}_{i,j} = \begin{cases} 1/(1-\rho^2), & \text{if } i = j = 1, N \\ (1+\rho^2)/(1-\rho^2), & \text{if } i = j = 2, \cdots, N-1 \\ -\rho/(1-\rho^2), & \text{if } |j-i| = 1 \\ 0, & \text{otherwise ,} \end{cases}$$

so that

$$\mathbf{x}'\mathbf{V}^{-1}\mathbf{x} = \frac{1}{1-\rho^2}\left\{\sum_{i=1}^{N} X_i^2 + \rho^2 \sum_{i=2}^{N-1} X_i^2 - 2\rho \sum_{i=2}^{N} X_i X_{i-1}\right\}, \text{ and}$$

$$\mathbf{x}'\mathbf{V}^{-1}\mathbf{y} = \frac{1}{1-\rho^2}\left\{\sum_{i=1}^{N} X_i Y_i + \rho^2 \sum_{i=2}^{N-1} X_i Y_i - 2\rho \sum_{i=2}^{N}(X_i Y_{i-1} + Y_i X_{i-1})\right\}.$$

Then, $\widehat{\beta}_{GLS} = (\mathbf{x}'\mathbf{V}^{-1}\mathbf{x})^{-1}\mathbf{x}'\mathbf{V}^{-1}\mathbf{y}$ is the b.l.u.e. of β with variance given by $\sigma^2(\mathbf{x}'\mathbf{V}^{-1}\mathbf{x})^{-1}$ (see Corollary 4.5.1). By substituting an estimate of σ^2, i.e., $\widehat{\sigma}^2_{GLS} = (\mathbf{y} - \mathbf{x}\widehat{\beta}_{GLS})'\mathbf{V}^{-1}(\mathbf{y} - \mathbf{x}\widehat{\beta}_{GLS})$, we obtain the estimated variance of the GLS estimator of β. The variance of the OLS estimator of β is $\text{Var}(\widehat{\beta}) = \sigma^2(\mathbf{x}'\mathbf{x})^{-1}\mathbf{x}'\mathbf{V}\mathbf{x}(\mathbf{x}'\mathbf{x})^{-1}$. Therefore,

$$\begin{aligned} \text{Var}(\widehat{\beta}) - \text{Var}(\widehat{\beta}_{GLS}) &= \sigma^2(\mathbf{x}'\mathbf{x})^{-1}\mathbf{x}'\mathbf{V}\mathbf{x}(\mathbf{x}'\mathbf{x})^{-1} \\ &\quad - \sigma^2(\mathbf{x}'\mathbf{V}^{-1}\mathbf{x})^{-1}\mathbf{x}'\mathbf{V}^{-1}\mathbf{V}\mathbf{V}^{-1}\mathbf{x}(\mathbf{x}'\mathbf{V}^{-1}\mathbf{x})^{-1} \\ &= \sigma^2\mathbf{s}\mathbf{V}\mathbf{s}', \quad \text{say,} \end{aligned}$$

where $\mathbf{s} = (\mathbf{x}'\mathbf{x})^{-1}\mathbf{x}' - (\mathbf{x}'\mathbf{V}^{-1}\mathbf{x})^{-1}\mathbf{x}'\mathbf{V}^{-1}$. This expression is p.s.d. since $\mathbf{V}$ is p.d., verifying that $\text{Var}(\widehat{\beta}) \geq \text{Var}(\widehat{\beta}_{GLS})$; equality holds when $\mathbf{V} = \mathbf{I}_N$. □

Under certain conditions, the OLS estimator and the GLS estimator of $\boldsymbol{\beta}$ coincide, as shown in the next example. We then present a result due to McElroy (1967) which gives a necessary and sufficient condition for this to happen in the full-rank linear model.

Example 4.5.2. Let $\mathbf{y} = (Y_1, \cdots, Y_N)'$ be a random vector with mean $\theta \mathbf{1}_N$ and covariance matrix $\sigma^2 \mathbf{V}$, where $\mathbf{V}$ is the equicorrelation matrix with $V_{ii} = 1$, $V_{ij} = \rho$, $i \neq j$, $i, j = 1, \cdots, N$ (see Example 1.3.2). We write this in the form (4.5.1) with $\mathbf{x} = \mathbf{1}_N$, $\boldsymbol{\beta} = \theta$, and $\boldsymbol{\varepsilon} = (\varepsilon_1, \cdots, \varepsilon_N)'$. Using the form of the inverse of $\sigma^2 \mathbf{V}$ derived in Example 1.3.6, we obtain $(\mathbf{x}'\mathbf{V}^{-1}\mathbf{x})^{-1} = \{1 + (N-1)\rho\}/N$, and $\mathbf{x}'\mathbf{V}^{-1}\mathbf{y} = N\overline{Y}/\{1 + (N-1)\rho\}$. From (4.5.5), the GLS estimate of θ is $\widehat{\theta}_{GLS} = \overline{Y}$, which coincides with its OLS estimate. □

Result 4.5.3. For the model (4.5.1), let $\widehat{\mathbf{y}} = \mathbf{X}\boldsymbol{\beta}^0$ denote the fitted vector under the OLS estimation. Then $\widehat{\mathbf{y}}$ and $\widehat{\mathbf{y}}_{GLS}$ are identical for all $\mathbf{y}$ if and only if $\mathcal{C}(\mathbf{V}^{-1}\mathbf{X}) = \mathcal{C}(\mathbf{X})$.

Proof. From (4.5.6) and (4.5.7), $\widehat{\mathbf{y}}_{GLS} = \widehat{\mathbf{y}}$ if and only if $\mathbf{L}^{-1}\mathbf{P}_{\mathcal{C}(\mathbf{LX})}\mathbf{L}\mathbf{y} = \mathbf{P}\mathbf{y}$, where $\mathbf{P}$ is the orthogonal projection onto $\mathcal{C}(\mathbf{X})$. It is already seen that the equality holds if $\mathbf{y} \in \mathcal{C}(\mathbf{X})$, so in order for it to hold for all $\mathbf{y}$, it is necessary and sufficient that for all $\mathbf{y} \perp \mathcal{C}(\mathbf{X})$, $\mathbf{L}^{-1}\mathbf{P}_{\mathcal{C}(\mathbf{LX})}\mathbf{L}\mathbf{y} = \mathbf{0}$, i.e., $\mathbf{L}\mathbf{y} \perp \mathcal{C}(\mathbf{LX})$, or $(\mathbf{L}\mathbf{y})'\mathbf{L}\mathbf{X} = \mathbf{y}'\mathbf{V}^{-1}\mathbf{X} = \mathbf{0}$. It follows that $\widehat{\mathbf{y}}_{GLS}$ is identical to $\widehat{\mathbf{y}}$ if and only if for all $\mathbf{y} \perp \mathcal{C}(\mathbf{X})$, $\mathbf{y} \perp \mathcal{C}(\mathbf{V}^{-1}\mathbf{X})$, which is equivalent to $\mathcal{C}(\mathbf{V}^{-1}\mathbf{X}) \subset \mathcal{C}(\mathbf{X})$. Since $\mathbf{V}$ is nonsingular, the last condition is equivalent to $\mathcal{C}(\mathbf{V}^{-1}\mathbf{X}) = \mathcal{C}(\mathbf{X})$, completing the proof. ■

First, note that since $\mathbf{V}$ is nonsingular, $\mathcal{C}(\mathbf{V}^{-1}\mathbf{X}) = \mathcal{C}(\mathbf{X})$ if and only $\mathcal{C}(\mathbf{X}) = \mathcal{C}(\mathbf{VX})$. Also, note that $\mathcal{C}(\mathbf{V}^{-1}\mathbf{X}) = \mathcal{C}(\mathbf{X})$ is a necessary and sufficient condition for $\boldsymbol{\beta}^0$ and $\boldsymbol{\beta}^0_{GLS}$ to coincide for all $\mathbf{y}$. Further, in the case of the full-rank model, since $\widehat{\mathbf{y}} = \mathbf{X}\widehat{\boldsymbol{\beta}}$ implies $\widehat{\boldsymbol{\beta}} = (\mathbf{X}'\mathbf{X})^{-1}\mathbf{X}'\widehat{\mathbf{y}}$, and likewise, $\widehat{\boldsymbol{\beta}}_{GLS} = (\mathbf{X}'\mathbf{X})^{-1}\mathbf{X}'\widehat{\mathbf{y}}_{GLS}$, from Result 4.5.3, $\widehat{\boldsymbol{\beta}}$ and $\widehat{\boldsymbol{\beta}}_{GLS}$ are identical for all $\mathbf{y}$ if and only if $\mathcal{C}(\mathbf{V}^{-1}\mathbf{X}) = \mathcal{C}(\mathbf{X})$. There are other cases where the GLS estimate of $\boldsymbol{\beta}$ coincides with its OLS estimate, when $\mathbf{y}$ satisfies certain conditions; see Exercise 4.19.

Corollary 4.5.2. Weighted least squares. Consider the linear model (4.5.1) where $\mathbf{V} = \text{diag}(\sigma_1^2, \cdots, \sigma_N^2)$ is an $N \times N$ diagonal matrix. Let $\mathbf{W} = \mathbf{V}^{-1} = \text{diag}(\sigma_1^{-2}, \cdots, \sigma_N^{-2})$

1. When $\text{r}(\mathbf{X}) < p$, we refer to $\boldsymbol{\beta}^0_{GLS}$ as the weighted least squares solution vector and denote it as $\boldsymbol{\beta}^0_{WLS} = (\mathbf{X}'\mathbf{W}\mathbf{X})^{-}\mathbf{X}'\mathbf{W}\mathbf{y}$.

2. When $\text{r}(\mathbf{X}) = p$, $\widehat{\boldsymbol{\beta}}_{GLS}$ is the WLS estimator of $\boldsymbol{\beta}$ and is denoted by $\widehat{\boldsymbol{\beta}}_{WLS} = (\mathbf{X}'\mathbf{W}\mathbf{X})^{-1}\mathbf{X}'\mathbf{W}\mathbf{y}$.

When $\mathbf{V}$ is diagonal, the effect of the generalized least squares analysis is to "weight" the original observations by the corresponding entries in the diagonal matrix $\mathbf{W}$.

Example 4.5.3. Let $Y_1, \cdots, Y_N$ denote independent observations, where $Y_i \sim N(i\theta, i^2\sigma^2)$, $i = 1, \cdots, N$. We estimate θ by the method of least squares and obtain the variance of this estimate. First, write this in the form (4.5.1) with $\mathbf{y} = (Y_1, \cdots, Y_N)'$, $\mathbf{x} = (1, \cdots, N)'$, $\boldsymbol{\beta} = \theta$, $\boldsymbol{\varepsilon} = (\varepsilon_1, \cdots, \varepsilon_N)'$, and $\mathbf{V} = \text{Var}(\boldsymbol{\varepsilon}) = \sigma^2 \,\text{diag}(1^2, 2^2, \cdots, N^2)$. It follows that the weighted least squares (WLS) estimate of θ and its variance are $\widehat{\theta}_{WLS} = (\mathbf{x}'\mathbf{V}^{-1}\mathbf{x})^{-1}\mathbf{x}'\mathbf{V}^{-1}\mathbf{y} = \frac{1}{N}\sum_{i=1}^{N} Y_i/i$, and $\text{Var}(\widehat{\theta}_{WLS}) = \sigma^2/N$. □

4.6 Estimation subject to linear constraints

Suppose we wish to obtain estimates of parameters in the linear model (4.1.1) subject to the linear constraints denoted by $\mathbf{A}'\boldsymbol{\beta} = \mathbf{b}$, where $\mathbf{A}$ is a known $p \times q$ matrix of full column rank, and $\mathbf{b} \in \mathcal{R}(\mathbf{A})$ is a known vector. The constrained estimation problem arises in two situations. First, there are many examples where a constraint on the parameters is an integral part of the model specification. In this case, it only makes sense to estimate under these constraints on the model. For example, recall the constraints that we imposed in order to solve the normal equations in the one-way ANOVA example in Section 4.2, when $\mathrm{r}(\mathbf{X}) < p$. The linear model without the model constraints is referred to as the *unrestricted model.* Second, in general linear hypothesis testing, it is required to obtain estimates under constraints imposed by hypotheses on certain estimable functions of the parameters (see Section 7.2). Here, we must search for the constrained least squares solutions under the hypotheses from among parameter values satisfying the constraints under the hypotheses, even if the unknown true parameter values do not necessarily satisfy them, i.e., the constraint is false.

We discuss two approaches, i.e., the method of Lagrangian multipliers, which is an algebraic approach, and the method of orthogonal projections, a geometrical approach.

4.6.1 Method of Lagrangian multipliers

We obtain the least squares solution of $\boldsymbol{\beta}$ under the constraint $\mathbf{A}'\boldsymbol{\beta} = \mathbf{b}$ using the method of Lagrangian multipliers (see Section 3.3). Let $2\boldsymbol{\lambda}$ denote a q-dimensional vector of Lagrange multipliers (for algebraic convenience, we have used $2\boldsymbol{\lambda}$ instead of $\boldsymbol{\lambda}$). We minimize

$$S_r(\boldsymbol{\beta}, \boldsymbol{\lambda}) = (\mathbf{y} - \mathbf{X}\boldsymbol{\beta})'(\mathbf{y} - \mathbf{X}\boldsymbol{\beta}) + 2\boldsymbol{\lambda}'(\mathbf{A}'\boldsymbol{\beta} - \mathbf{b}) \tag{4.6.1}$$

with respect to $\boldsymbol{\beta}$ and $\boldsymbol{\lambda}$. Setting equal to zero the first partial derivatives of $S_r(\boldsymbol{\beta}, \boldsymbol{\lambda})$ with respect to $\boldsymbol{\beta}$ and $\boldsymbol{\lambda}$ results in the following set of normal equations:

$$\frac{\partial}{\partial \boldsymbol{\beta}} S_r(\boldsymbol{\beta}, \boldsymbol{\lambda}) = \mathbf{0} \Rightarrow \mathbf{X}'\mathbf{X}\boldsymbol{\beta}_r^0 + \mathbf{A}\boldsymbol{\lambda}_r^0 = \mathbf{X}'\mathbf{y}, \tag{4.6.2}$$

$$\frac{\partial}{\partial \boldsymbol{\lambda}} S_r(\boldsymbol{\beta}, \boldsymbol{\lambda}) = \mathbf{0} \Rightarrow \mathbf{A}'\boldsymbol{\beta}_r^0 = \mathbf{b}. \tag{4.6.3}$$

We will distinguish between the cases when $\mathbf{A}'\boldsymbol{\beta}$ is estimable and when it is not. We first show the following useful result.

Result 4.6.1. For all the solutions $(\boldsymbol{\beta}_r^0, \boldsymbol{\lambda}_r^0)$ to the set of normal equations (4.6.2) and (4.6.3), $(\mathbf{X}\boldsymbol{\beta}_r^0, \mathbf{A}\boldsymbol{\lambda}_r^0)$ is the same.

Proof. Let $(\boldsymbol{\beta}_r^0, \boldsymbol{\lambda}_r^0)$ and $(\boldsymbol{\beta}_*^0, \boldsymbol{\lambda}_*^0)$ be any two solutions. Let $\mathbf{u} = \boldsymbol{\beta}_r^0 - \boldsymbol{\beta}_*^0$ and $\mathbf{v} = \boldsymbol{\lambda}_r^0 - \boldsymbol{\lambda}_*^0$. Then $\mathbf{X}'\mathbf{X}\mathbf{u} + \mathbf{A}\mathbf{v} = \mathbf{0}$ and $\mathbf{A}'\mathbf{u} = \mathbf{0}$. Pre-multiply the first equality by $\mathbf{u}'$ and then apply the second equality to get $\mathbf{u}'\mathbf{X}'\mathbf{X}\mathbf{u} = 0$. Then $\mathbf{X}\mathbf{u} = \mathbf{0}$, which in turn yields $\mathbf{A}\mathbf{v} = \mathbf{0}$. Then $\mathbf{X}\boldsymbol{\beta}_r^0 = \mathbf{X}\boldsymbol{\beta}_*^0$ and $\mathbf{A}\boldsymbol{\lambda}_r^0 = \mathbf{A}\boldsymbol{\lambda}_*^0$, yielding the proof. ■

4.6.1.1 Case I: $\mathbf{A}'\boldsymbol{\beta}$ is estimable

We start by noting that $\mathbf{A}' = \mathbf{T}'\mathbf{X}$ for some full column rank matrix $\mathbf{T}$.

Result 4.6.2. $\mathbf{A}'\mathbf{G}\mathbf{A}$ is nonsingular.

Proof. It suffices to show that for any vector $\mathbf{v}$, $\mathbf{A}'\mathbf{G}\mathbf{A}\mathbf{v} = \mathbf{T}'\mathbf{X}\mathbf{G}\mathbf{X}'\mathbf{T}\mathbf{v} = \mathbf{0}$ only if $\mathbf{v} = \mathbf{0}$. Since $\mathbf{P} = \mathbf{X}\mathbf{G}\mathbf{X}'$ is the orthogonal projection matrix onto $\mathcal{C}(\mathbf{X})$, $(\mathbf{P}\mathbf{T})'(\mathbf{P}\mathbf{T})\mathbf{v} = \mathbf{0}$, property 1 of Result 1.3.13 yields $\mathbf{P}\mathbf{T}\mathbf{v} = \mathbf{0}$. Then $\mathbf{T}\mathbf{v} \perp \mathcal{C}(\mathbf{X})$, so $\mathbf{X}'\mathbf{T}\mathbf{v} = \mathbf{A}\mathbf{v} = \mathbf{0}$. Since $\mathbf{A}$ has full column rank, then $\mathbf{v} = \mathbf{0}$. ■

We now state and prove the main result.

Result 4.6.3. The following properties hold.

1. $\mathbf{X}\boldsymbol{\beta}_r^0$ is invariant to the solutions of the normal equations (4.6.2) and (4.6.3).
2. One choice for $\boldsymbol{\beta}_r^0$ is

$$\boldsymbol{\beta}_r^0 = \boldsymbol{\beta}^0 - \mathbf{G}\mathbf{A}(\mathbf{A}'\mathbf{G}\mathbf{A})^{-1}(\mathbf{A}'\boldsymbol{\beta}^0 - \mathbf{b}). \tag{4.6.4}$$

Proof. Let $(\boldsymbol{\beta}_r^0, \boldsymbol{\lambda}_r^0)$ be any solution to (4.6.2) and (4.6.3). Pre-multiplying both sides of (4.6.2) by $\mathbf{A}'\mathbf{G}$ gives $\mathbf{A}'\mathbf{G}\mathbf{X}'\mathbf{X}\boldsymbol{\beta}_r^0 + \mathbf{A}'\mathbf{G}\mathbf{A}\boldsymbol{\lambda}_r^0 = \mathbf{A}'\mathbf{G}\mathbf{X}'\mathbf{y}$. By property 4 of Result 4.3.1 and (4.6.3), $\mathbf{A}'\mathbf{G}\mathbf{X}'\mathbf{X}\boldsymbol{\beta}_r^0 = \mathbf{A}'\boldsymbol{\beta}_r^0 = \mathbf{b}$. On the other hand,

$$\mathbf{A}'\mathbf{G}\mathbf{X}'\mathbf{y} = \mathbf{T}'\mathbf{X}\mathbf{G}\mathbf{X}'\mathbf{y} = \mathbf{T}'\mathbf{X}\boldsymbol{\beta}^0 = \mathbf{A}'\boldsymbol{\beta}^0.$$

Hence, $\mathbf{b} + \mathbf{A}'\mathbf{G}\mathbf{A}\boldsymbol{\lambda}_r^0 = \mathbf{A}'\boldsymbol{\beta}^0$. From Result 4.6.2, $\boldsymbol{\lambda}_r^0 = (\mathbf{A}'\mathbf{G}\mathbf{A})^{-1}(\mathbf{A}'\boldsymbol{\beta}^0 - \mathbf{b})$. Since $\boldsymbol{\lambda}_r^0$ is unique, this shows that $\mathbf{X}'\mathbf{X}\boldsymbol{\beta}_r^0 = \mathbf{X}'\mathbf{y} - \mathbf{A}\boldsymbol{\lambda}_r^0$ is invariant to the solutions to (4.6.2) and (4.6.3). Property 1 follows from property 2 of Result 1.3.13. Since (4.6.2) can be written as $\mathbf{X}'\mathbf{X}\boldsymbol{\beta}_r^0 = \mathbf{X}'(\mathbf{y} - \mathbf{T}\boldsymbol{\lambda}_r^0)$,

$$\boldsymbol{\beta}_r^0 = \mathbf{G}\mathbf{X}'(\mathbf{y} - \mathbf{T}\boldsymbol{\lambda}_r^0) = \mathbf{G}\mathbf{X}'\mathbf{y} - \mathbf{G}\mathbf{A}\boldsymbol{\lambda}_r^0 = \boldsymbol{\beta}^0 - \mathbf{G}\mathbf{A}\boldsymbol{\lambda}_r^0$$

is a solution to (4.6.2), completing the proof of property 2. ■

We refer to $\boldsymbol{\beta}_r^0$ in (4.6.4) as the constrained LS solution of $\boldsymbol{\beta}$ in the linear model $\mathbf{y} = \mathbf{X}\boldsymbol{\beta} + \boldsymbol{\varepsilon}$ subject to $\mathbf{A}'\boldsymbol{\beta} = \mathbf{b}$. Let

$$\widehat{\mathbf{y}}_r = \mathbf{X}\boldsymbol{\beta}_r^0, \quad \widehat{\boldsymbol{\varepsilon}}_r = \mathbf{y} - \widehat{\mathbf{y}}_r, \quad \text{and} \quad SSE_r = \widehat{\boldsymbol{\varepsilon}}_r'\widehat{\boldsymbol{\varepsilon}}_r \tag{4.6.5}$$

be the fitted vector, residual vector and error sum of squares in the restricted model.

Result 4.6.4. The constrained LS solution $\boldsymbol{\beta}_r^0$ has the following properties.

1. $\mathrm{E}(\boldsymbol{\beta}_r^0) = \mathbf{H}\boldsymbol{\beta} - \mathbf{G}\mathbf{A}(\mathbf{A}'\mathbf{G}\mathbf{A})^{-1}(\mathbf{A}'\boldsymbol{\beta} - \mathbf{b})$.
2. $\mathrm{Cov}(\boldsymbol{\beta}_r^0) = \sigma^2\mathbf{G}[\mathbf{X}'\mathbf{X} - \mathbf{A}(\mathbf{A}'\mathbf{G}\mathbf{A})^{-1}\mathbf{A}']\mathbf{G}'$.
3. $\mathrm{E}(SSE_r) = (N - r + q)\sigma^2 + (\mathbf{A}'\boldsymbol{\beta} - \mathbf{b})'(\mathbf{A}'\mathbf{G}\mathbf{A})^{-1}(\mathbf{A}'\boldsymbol{\beta} - \mathbf{b})$.

Thus, if the linear constraints are true, i.e., $\mathbf{A}'\boldsymbol{\beta} = \mathbf{b}$, then $\mathrm{E}(\boldsymbol{\beta}_r^0) = \mathbf{H}\boldsymbol{\beta}$, $\mathrm{E}(\widehat{\mathbf{y}}_r) = \mathbf{X}\boldsymbol{\beta}$, and $\mathrm{E}(SSE_r) = (N - r + q)\sigma^2$.

Proof. Take expectations on both sides of (4.6.4). Property 1 follows by using (4.2.23) and (4.3.3), i.e., $\mathrm{E}(\boldsymbol{\beta}^0) = \mathbf{H}\boldsymbol{\beta}$ and $\mathbf{A}'\mathbf{H} = \mathbf{A}'$. Next,

$$\mathrm{Cov}(\boldsymbol{\beta}_r^0) = \mathrm{Cov}\{[\mathbf{I} - \mathbf{G}\mathbf{A}(\mathbf{A}'\mathbf{G}\mathbf{A})^{-1}\mathbf{A}']\boldsymbol{\beta}^0\},$$

which, after some simplification, and using (4.2.23) gives property 2. Next, since $\widehat{\boldsymbol{\varepsilon}}_r = \widehat{\boldsymbol{\varepsilon}} + \mathbf{X}(\boldsymbol{\beta}^0 - \boldsymbol{\beta}_r^0)$ and $\widehat{\boldsymbol{\varepsilon}} \perp \mathbf{X}(\boldsymbol{\beta}^0 - \boldsymbol{\beta}_r^0)$, where $\widehat{\boldsymbol{\varepsilon}} = \mathbf{y} - \mathbf{X}\boldsymbol{\beta}^0$,

$$\begin{aligned} SSE_r &= \widehat{\boldsymbol{\varepsilon}}'\widehat{\boldsymbol{\varepsilon}} + (\boldsymbol{\beta}^0 - \boldsymbol{\beta}_r^0)'\mathbf{X}'\mathbf{X}(\boldsymbol{\beta}^0 - \boldsymbol{\beta}_r^0) \\ &= SSE + (\mathbf{A}'\boldsymbol{\beta}^0 - \mathbf{b})'(\mathbf{A}'\mathbf{G}\mathbf{A})^{-1}(\mathbf{A}'\boldsymbol{\beta}^0 - \mathbf{b}). \end{aligned}$$

Then, by taking expectations and using Result 4.2.3, property 3 can be proved. Details are left as Exercise 4.29. ■

Corollary 4.6.1. When $\mathrm{r}(\mathbf{X}) = p$, the constrained least squares estimate $\widehat{\boldsymbol{\beta}}_r$ is

$$\widehat{\boldsymbol{\beta}}_r = \widehat{\boldsymbol{\beta}} - (\mathbf{X}'\mathbf{X})^{-1}\mathbf{A}[\mathbf{A}'(\mathbf{X}'\mathbf{X})^{-1}\mathbf{A}]^{-1}(\mathbf{A}'\widehat{\boldsymbol{\beta}} - \mathbf{b}). \tag{4.6.6}$$

Example 4.6.1. Suppose $Y_i = \theta_i + \varepsilon_i$, $i = 1, \cdots, 5$, ε_i are i.i.d. $N(0, \sigma^2)$ variables, and θ_i are unknown real values subject to the model constraint $\sum_{i=1}^{5} \theta_i = 100$. We write $\mathbf{y} = \mathbf{X}\boldsymbol{\beta} + \boldsymbol{\varepsilon}$, where, $\mathbf{y} = (Y_1, \cdots, Y_5)'$, $\mathbf{X} = \mathbf{I}_5$, and $\boldsymbol{\beta} = (\theta_1, \cdots, \theta_5)'$, subject to the constraint $\mathbf{A}'\boldsymbol{\beta} = \mathbf{b}$, with $\mathbf{A} = \mathbf{1}_5$, and $b = 100$. The unconstrained least squares estimate of $\boldsymbol{\beta}$ is $\mathbf{y}$, while the constrained estimate (see (4.6.4)) is $\widehat{\boldsymbol{\beta}}_r' = (Y_1 - \overline{Y} + 20, \cdots, Y_5 - \overline{Y} + 20)$ with $\mathrm{Cov}(\widehat{\boldsymbol{\beta}}_r) = \sigma^2(\mathbf{I} - \mathbf{J}/5)$, and $SSE_r = (\sum_{i=1}^{5} Y_i - 100)^2/5$. □

4.6.1.2 Case II: $\mathbf{A}'\boldsymbol{\beta}$ is not estimable

This case is possible only when the model has less than full rank, i.e., $\mathrm{r}(\mathbf{X}) < p$, since otherwise $\mathcal{R}(\mathbf{X}) = \mathcal{R}^p \supseteq \mathcal{C}(\mathbf{A})$. Let $\Omega_0 = \mathcal{C}(\mathbf{A}) \cap \mathcal{R}(\mathbf{X})$, so that

$$s = \dim(\Omega_0) < \dim(\mathcal{C}(\mathbf{A})) = q.$$

We find a basis $\{\mathbf{v}_1, \cdots, \mathbf{v}_q\}$ of $\mathcal{C}(\mathbf{A})$, such that $\{\mathbf{v}_1, \cdots, \mathbf{v}_s\}$ is a basis of Ω_0 and $\mathrm{Span}\{\mathbf{v}_{s+1}, \cdots, \mathbf{v}_q\} \cap \mathcal{R}(\mathbf{X}) = \{\mathbf{0}\}$; we will discuss how to do this later. Let $\mathbf{A}_1 = (\mathbf{v}_1, \cdots, \mathbf{v}_s)$ and $\mathbf{A}_2 = (\mathbf{v}_{s+1}, \ldots, \mathbf{v}_q)$.

Result 4.6.5. The following properties hold.

1. $\mathbf{X}\boldsymbol{\beta}_r^0$ is invariant to the solutions to the normal equations (4.6.2) and (4.6.3).
2. One choice for $\boldsymbol{\beta}_r^0$ is

$$\boldsymbol{\beta}_r^0 = \boldsymbol{\beta}_*^0 - (\mathbf{H} - \mathbf{I})[\mathbf{A}_2'(\mathbf{H} - \mathbf{I})]^-(\mathbf{A}_2'\boldsymbol{\beta}_*^0 - \mathbf{b}_2), \tag{4.6.7}$$

where $\mathbf{H} = \mathbf{G}\mathbf{X}'\mathbf{X}$, $\mathbf{b}_i = \mathbf{R}_i'\mathbf{b}$ with $\mathbf{R}_i = (\mathbf{A}'\mathbf{A})^{-1}\mathbf{A}'\mathbf{A}_i$, $i = 1, 2$, and

$$\boldsymbol{\beta}_*^0 = \boldsymbol{\beta}^0 - \mathbf{G}\mathbf{A}_1(\mathbf{A}_1'\mathbf{G}\mathbf{A}_1)^{-1}(\mathbf{A}_1'\boldsymbol{\beta}^0 - \mathbf{b}_1). \tag{4.6.8}$$

3. The vectors of fitted values and residuals and the SSE of the model $\mathbf{y} = \mathbf{X}\boldsymbol{\beta} + \boldsymbol{\varepsilon}$ under the constraint $\mathbf{A}'\boldsymbol{\beta} = \mathbf{b}$ are equal to those of the same model under the constraint $\mathbf{A}_1'\boldsymbol{\beta} = \mathbf{b}_1$.

Proof. Since $\mathcal{C}(\mathbf{A}_i) \subset \mathcal{C}(\mathbf{A})$, $\mathbf{A}_i = \mathbf{A}\mathbf{R}_i$. Note that $\mathbf{R} = (\mathbf{R}_1, \mathbf{R}_2) \in \mathcal{R}^q$ is nonsingular. Let $\boldsymbol{\theta}^0 = \mathbf{R}^{-1}\boldsymbol{\lambda}_r^0$. Partition $\boldsymbol{\theta}^0$ as $\begin{pmatrix}\boldsymbol{\theta}_1^0 \\ \boldsymbol{\theta}_2^0\end{pmatrix}$, where $\dim(\boldsymbol{\theta}_1^0) = s$. Then (4.6.2) and (4.6.3) can be written as

$$\mathbf{X}'\mathbf{X}\boldsymbol{\beta}_r^0 + \mathbf{A}_1\boldsymbol{\theta}_1^0 + \mathbf{A}_2\boldsymbol{\theta}_2^0 = \mathbf{X}'\mathbf{y}, \tag{4.6.9}$$

$$\mathbf{A}_1'\boldsymbol{\beta}_r^0 = \mathbf{b}_1, \tag{4.6.10}$$

$$\mathbf{A}_2'\boldsymbol{\beta}_r^0 = \mathbf{b}_2. \tag{4.6.11}$$

Note that $\mathbf{A}_1$ has full column rank s and $\mathcal{C}(\mathbf{A}_1) \subset \mathcal{R}(\mathbf{X})$, while $\mathbf{A}_2$ has full column rank $q - s$ and $\mathcal{C}(\mathbf{A}_2) \cap \mathcal{R}(\mathbf{X}) = \{\mathbf{0}\}$. Then, (4.6.9) yields

$$\mathbf{X}'\mathbf{X}\boldsymbol{\beta}_r^0 + \mathbf{A}_1\boldsymbol{\theta}_1^0 = \mathbf{X}'\mathbf{y}. \tag{4.6.12}$$

From property 1 of Result 4.6.3, $\mathbf{X}\boldsymbol{\beta}_r^0$ is invariant to the solutions to (4.6.10) and (4.6.12), which implies property 1.

Next, from property 2 of Result 4.6.3, a solution to (4.6.10) and (4.6.12) is $\boldsymbol{\beta}^0_*$ in (4.6.8). To solve (4.6.10)–(4.6.12) simultaneously, note that for any $\mathbf{z} \in \mathcal{R}^p$, $\boldsymbol{\beta}^0_* + (\mathbf{H} - \mathbf{I})\mathbf{z}$ solves (4.6.10) and (4.6.12), and in order for it to solve (4.6.11), it is necessary and sufficient that

$$\mathbf{A}_2'(\mathbf{H} - \mathbf{I})\mathbf{z} = -\mathbf{A}_2'\boldsymbol{\beta}^0_* + \mathbf{b}_2. \tag{4.6.13}$$

Since $\mathbf{b}_2 \in \mathcal{C}(\mathbf{A}_2')$, to show that the equation has a solution, it suffices to show that $\mathcal{C}(\mathbf{A}_2') \subset \mathcal{C}(\mathbf{A}_2'(\mathbf{H} - \mathbf{I}))$. Let $\mathbf{v} \in \mathcal{C}(\mathbf{A}_2'(\mathbf{H} - \mathbf{I}))^\perp$. Then $\mathbf{v}'\mathbf{A}_2'(\mathbf{H} - \mathbf{I}) = \mathbf{0}$ (see Result 1.3.18), giving $\mathbf{H}'\mathbf{A}_2\mathbf{v} = \mathbf{A}_2\mathbf{v}$. Since $\mathbf{H}'\mathbf{A}_2\mathbf{v} = \mathbf{X}'\mathbf{X}\mathbf{G}'\mathbf{A}_2\mathbf{v} \in \mathcal{R}(\mathbf{X})$, then $\mathbf{A}_2\mathbf{v} \in \mathcal{R}(\mathbf{X})$, so $\mathbf{A}_2\mathbf{v} \in \mathcal{R}(\mathbf{X}) \cap \mathcal{C}(\mathbf{A}_2)$. However, from the definition of $\mathbf{A}_2$, the intersection of the two spaces is $\{\mathbf{0}\}$. Then $\mathbf{A}_2\mathbf{v} = \mathbf{0}$, i.e., $\mathbf{v} \perp \mathcal{C}(\mathbf{A}_2')$. Then $\mathcal{C}(\mathbf{A}_2'(\mathbf{H} - \mathbf{I}))^\perp \subset \mathbf{C}(\mathbf{A}_2')^\perp$, and hence $\mathcal{C}(\mathbf{A}_2') \subset \mathcal{C}(\mathbf{A}_2'(\mathbf{H} - \mathbf{I}))$. As a result, a solution to (4.6.13) is $\mathbf{z} = -[\mathbf{A}_2'(\mathbf{H} - \mathbf{I})]^-(\mathbf{A}_2'\boldsymbol{\beta}^0_* - \mathbf{b}_2)$. Plugging this into $\boldsymbol{\beta}^0_* + (\mathbf{H} - \mathbf{I})\mathbf{z}$ proves (4.6.7). Finally, $\mathbf{X}\boldsymbol{\beta}^0_r = \mathbf{X}\boldsymbol{\beta}^0_*$ and $\boldsymbol{\beta}^0_*$ is a LS solution under the constraint $\mathbf{A}_1'\boldsymbol{\beta} = \mathbf{d}_1$. This shows property 3. ■

Note that $\mathbf{A}_2$ has no role in Result 4.6.5, although it appears in the constrained LS solution $\boldsymbol{\beta}^0_r$. We briefly describe how to find $\mathbf{A}_1$ and $\mathbf{A}_2$. For any $\mathbf{A}\mathbf{z} \in \mathcal{C}(\mathbf{A})$, from Result 4.3.1, $\mathbf{A}\mathbf{z} \in \mathcal{R}(\mathbf{X})$ if and only if $\mathbf{z} \in \mathcal{N}(\mathbf{K}) = \mathcal{C}(\mathbf{I} - \mathbf{K}^-\mathbf{K})$, where $\mathbf{K} = (\mathbf{I} - \mathbf{H}')\mathbf{A}$. As a result, $\mathcal{C}(\mathbf{A}) \cap \mathcal{R}(\mathbf{X}) = \mathbf{A}\mathcal{N}(\mathbf{K}) = \mathcal{C}(\mathbf{A} - \mathbf{A}\mathbf{K}^-\mathbf{K})$. Any s LIN column vectors of $\mathbf{A} - \mathbf{A}\mathbf{K}^-\mathbf{K}$ can be used as the column vectors of $\mathbf{A}_1$. Then $\mathbf{A} - \mathbf{P}_{\mathcal{C}(\mathbf{A}_1)}\mathbf{A}$ consists of column vectors in $\mathcal{C}(\mathbf{A}_1)^\perp \cap \mathcal{C}(\mathbf{A})$. Any $q - s$ LIN column vectors of the matrix $\mathbf{A} - \mathbf{A}\mathbf{K}^-\mathbf{K}$ can be used as the column vectors of $\mathbf{A}_2$.

In summary, as far as fitted values, residuals, and SSE are concerned, constraints on a nonestimable function of $\boldsymbol{\beta}$ can be reduced in an explicit way to constraints on an estimable function of $\boldsymbol{\beta}$.

Example 4.6.2. Let $Y_{ij} = \mu + \tau_i + \varepsilon_{ij}$, with $\mathrm{E}(\varepsilon_{ij}) = 0$, $i = 1, \cdots, a$, $j = 1, \cdots, n$, and suppose that we impose the model constraint $\sum_{i=1}^a c_i\tau_i = 0$, with $\sum_{i=1}^a c_i \neq 0$. We obtain the LS solutions for μ and $\boldsymbol{\tau} = (\tau_1, \cdots, \tau_a)'$ using the method of Lagrangian multipliers. Let

$$S_r(\mu, \boldsymbol{\tau}, \lambda) = \sum_{i=1}^a \sum_{j=1}^n (Y_{ij} - \mu - \tau_i)^2 + 2\lambda \sum_{i=1}^a c_i\tau_i.$$

Then we need to solve

$$\frac{\partial S}{\partial \mu} = -2\sum_{i=1}^a \sum_{j=1}^n (Y_{ij} - \mu - \tau_i) = 0,$$

$$\frac{\partial S}{\partial \tau_i} = -2\sum_{j=1}^n (Y_{ij} - \mu - \tau_i) + 2\lambda c_i = 0, \quad 1 \le i \le a.$$

Adding the last a equations and subtracting the first one gives $2\lambda \sum_{i=1}^a c_i = 0$. Since $\sum_{i=1}^a c_i \neq 0$, then $\lambda = 0$. Thus $(\mu^0, \tau_1^0, \cdots, \tau_a^0)$ is a solution under the constraint if and only if $\tau_i^0 = \mu^0 - \overline{Y}_{i\cdot}$ and $\sum_{i=1}^a c_i\tau_i = 0$. It then follows that $\mu^0 = \sum_{i=1}^a c_i\overline{Y}_{i\cdot} / \sum_{i=1}^a c_i$ and $\tau_i^0 = \overline{Y}_{i\cdot} - \mu^0$, $i = 1, \cdots, a$, form the unique solution. The fitted values are $\widehat{Y}_{ij} = \mu^0 + \tau_i^0 = \overline{Y}_{i\cdot}$, which are the same as in the unconstrained model. Indeed, since $\sum_{i=1}^a c_i\tau_i = \mathbf{A}'\boldsymbol{\beta}$, where $\mathbf{A} = (0, c_1, \cdots, c_a)'$ and $\boldsymbol{\beta} = (\mu, \tau_1, \cdots, \tau_a)'$, it is easy to see that $\mathcal{C}(\mathbf{A}) \cap \mathcal{R}(\mathbf{X}) = \{\mathbf{0}\}$. Then by Result 4.6.5, the fitted values are the same under the model constraint as those in the unconstrained case. □

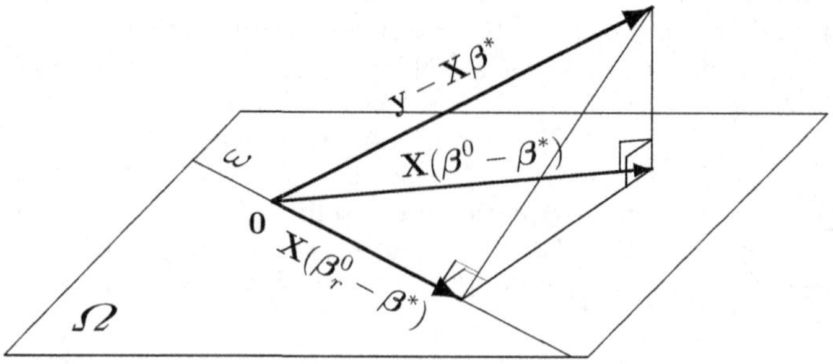

FIGURE 4.6.1. Geometry of constrained least squares. $\boldsymbol{\beta}^*$ is any solution to $\mathbf{A}'\boldsymbol{\beta} = \mathbf{b}$.

4.6.2 Method of orthogonal projections

We describe the method of orthogonal projections for the case where $\mathbf{A}'\boldsymbol{\beta}$ is estimable. Let $\Omega = \mathcal{C}(\mathbf{X})$ and $\mathbf{P}_\Omega = \mathbf{XGX}'$ denote the orthogonal projection matrix so that $\mathbf{P}_\Omega\mathbf{y} = \mathbf{X}\boldsymbol{\beta}^0$. Suppose $\boldsymbol{\beta}^*$ satisfies $\mathbf{A}'\boldsymbol{\beta} = \mathbf{b}$ and let $\boldsymbol{\theta} = \boldsymbol{\beta} - \boldsymbol{\beta}^*$. We define $\omega = \{\mathbf{X}\boldsymbol{\theta}\colon \mathbf{A}'\boldsymbol{\theta} = \mathbf{0}\}$. Again, $\mathbf{A}' = \mathbf{T}'\mathbf{X}$. We can write $\omega = \{\mathbf{X}\boldsymbol{\theta}\colon \mathbf{T}'\mathbf{X}\boldsymbol{\theta} = \mathbf{0}\} = \Omega \cap \mathcal{C}(\mathbf{T})^\perp$. Let $\omega^\perp$ be the orthogonal complement of ω in $\mathcal{R}^N$. Then the orthogonal complement of ω relative to Ω is $\omega^\perp \cap \Omega$. Subtracting $\mathbf{X}\boldsymbol{\beta}^*$ from both sides of (4.1.1), we can write

$$\mathbf{y} - \mathbf{X}\boldsymbol{\beta}^* = \mathbf{X}(\boldsymbol{\beta} - \boldsymbol{\beta}^*) + \boldsymbol{\varepsilon} = \mathbf{X}\boldsymbol{\theta} + \boldsymbol{\varepsilon}.$$

Thus, $\boldsymbol{\beta}_r^0$ is a least squares solution of $\boldsymbol{\beta}$ under the constraint $\mathbf{A}'\boldsymbol{\beta}_r^0 = \mathbf{b}$ if and only if

$$\mathbf{P}_\omega(\mathbf{y} - \mathbf{X}\boldsymbol{\beta}^*) = \mathbf{X}(\boldsymbol{\beta}_r^0 - \boldsymbol{\beta}^*). \tag{4.6.14}$$

From property 3 of Result 2.6.6, $\mathbf{P}_\omega = \mathbf{P}_\Omega - \mathbf{P}_{\omega^\perp\cap\Omega}$. Since $\mathbf{P}_\Omega(\mathbf{y} - \mathbf{X}\boldsymbol{\beta}^*) = \mathbf{X}(\boldsymbol{\beta}^0 - \boldsymbol{\beta}^*)$, (4.6.14) is equivalent to

$$\mathbf{X}\boldsymbol{\beta}_r^0 = \mathbf{X}\boldsymbol{\beta}^0 - \mathbf{P}_{\omega^\perp\cap\Omega}(\mathbf{y} - \mathbf{X}\boldsymbol{\beta}^*). \tag{4.6.15}$$

From Result 2.6.7,

$$\omega^\perp \cap \Omega = (\Omega \cap \mathcal{C}(\mathbf{T})^\perp)^\perp \cap \Omega = (\Omega^\perp + \mathcal{C}(\mathbf{T})) \cap \Omega = \mathbf{P}_\Omega\mathcal{C}(\mathbf{T}) = \mathcal{C}(\mathbf{P}_\Omega\mathbf{T}).$$

From Result 4.6.4, $\mathbf{A}'\mathbf{GA} = \mathbf{T}'\mathbf{XGX}'\mathbf{T} = (\mathbf{P}_\Omega\mathbf{T})'(\mathbf{P}_\Omega\mathbf{T})$ is nonsingular, so that $\mathbf{P}_{\omega^\perp\cap\Omega} = \mathbf{P}_\Omega\mathbf{T}[(\mathbf{P}_\Omega\mathbf{T})'\mathbf{P}_\Omega\mathbf{T}]^{-1}\mathbf{T}'\mathbf{P}_\Omega$. Then,

$$\begin{aligned}\mathbf{P}_{\omega^\perp\cap\Omega}(\mathbf{y} - \mathbf{X}\boldsymbol{\beta}^*) &= \mathbf{P}_\Omega\mathbf{T}[(\mathbf{P}_\Omega\mathbf{T})'\mathbf{P}_\Omega\mathbf{T}]^{-1}\mathbf{T}'\mathbf{P}_\Omega(\mathbf{y} - \mathbf{X}\boldsymbol{\beta}^*)\\ &= \mathbf{XGA}(\mathbf{A}'\mathbf{GA})^{-1}(\mathbf{A}'\boldsymbol{\beta}^0 - \mathbf{b}).\end{aligned} \tag{4.6.16}$$

From (4.6.15) and (4.6.16), we see that the constrained least squares solution of $\boldsymbol{\beta}$ adjusts $\boldsymbol{\beta}^0$ by an amount that depends on $\mathbf{X}$, $\mathbf{A}$, and $\mathbf{A}'\boldsymbol{\beta}^0 - \mathbf{b}$ and has the form

$$\boldsymbol{\beta}_r^0 = \boldsymbol{\beta}^0 - \mathbf{GA}(\mathbf{A}'\mathbf{GA})^{-1}(\mathbf{A}'\boldsymbol{\beta}^0 - \mathbf{b}),$$

which is the same as (4.6.4). This is illustrated geometrically in Figure 4.6.1 for the full-rank case when $\mathrm{r}(\mathbf{X}) = p$.

Exercises

4.1. Verify the consistency condition for existence of a least squares solution, i.e., $\text{r}(\mathbf{X}'\mathbf{X}, \mathbf{X}'\mathbf{y}) = \text{r}(\mathbf{X}'\mathbf{X})$.

4.2. Show that

(a) $\mathbf{X}'\mathbf{X}\mathbf{H} = \mathbf{X}'\mathbf{X}$, where $\mathbf{H} = \mathbf{G}\mathbf{X}'\mathbf{X}$, and

(b) $\mathbf{I}_N - \mathbf{X}\mathbf{G}\mathbf{X}'$ is symmetric and idempotent.

4.3. When an object of unit mass is subjected to an unknown force θ for a length of time t, its position changes to $t^2\theta/2$. Observed positions $Y_1, \cdots, Y_N$ are measured at known times $t_1, \cdots, t_N$. Assume the linear model $Y_i = t_i^2\theta/2 + \varepsilon_i$, $i = 1, \cdots, N$ and that the measurement errors ε_i have $\text{E}(\varepsilon_i) = 0$ and $\text{Var}(\varepsilon_i) = \sigma^2$. Obtain the least squares estimator of the unknown force θ, and derive the variance of the estimator.

4.4. The displacement S_i of the ith object at time t_i is given by the formula

$$S_i = vt_i + \varepsilon_i, \quad i = 1, \cdots, N,$$

where all the N objects are subjected to an equal velocity. Assume that $\text{E}(\varepsilon_i) = 0$, and $\text{Var}(\varepsilon_i) = \sigma^2$. Find the least squares estimate of the unknown velocity v and estimate the variance of this estimate.

4.5. Let $\mathbf{y} = \mathbf{X}\boldsymbol{\beta} + \boldsymbol{\varepsilon}$, where $\boldsymbol{\varepsilon} \sim N(\mathbf{0}, \sigma^2\mathbf{I}_N)$, and $\boldsymbol{\beta}$ is a p-dimensional vector, with $N = 10$ and $p = 3$. Given $\mathbf{y}'\mathbf{y} = 58$, and given the following normal equations,

$$\begin{aligned} 4\widehat{\beta}_1 + 2\widehat{\beta}_2 - 2\widehat{\beta}_3 &= 4 \\ 2\widehat{\beta}_1 + 2\widehat{\beta}_2 + \widehat{\beta}_3 &= 7 \\ -2\widehat{\beta}_1 + \widehat{\beta}_2 + 6\widehat{\beta}_3 &= 9, \end{aligned}$$

find the least squares estimates of $\boldsymbol{\beta}$ and σ^2. What are the estimates of β_1, β_2, β_3, $\beta_1 - \beta_2$, and $\beta_1 + \beta_3$?

4.6. Suppose we have $Y_{1j} = \beta_1 + \varepsilon_{1j}$, $j = 1, \cdots, n_1$, $Y_{2j} = \beta_1 + \beta_2 + \varepsilon_{2j}$, $j = 1, \cdots, n_1$, and $Y_{3j} = \beta_1 - 2\beta_2 + \varepsilon_{3j}$, $j = 1, \cdots, n_2$, where ε_{ij} are i.i.d. $N(0, \sigma^2)$ variables. Under what condition on n_1 and n_2 are the least squares estimates of β_1 and β_2 uncorrelated?

4.7. (Rao (1973), p. 236). Let $\mathbf{X} = [\mathbf{x}_0, \mathbf{x}_1, \cdots, \mathbf{x}_{k-1}, \mathbf{x}_k] = [\mathbf{W}, \mathbf{x}_k]$, and let $\text{r}(\mathbf{X}) = k + 1$.

(a) Show that $|\mathbf{X}'\mathbf{X}| = |\mathbf{W}'\mathbf{W}|(\mathbf{x}_k'\mathbf{x}_k - \mathbf{x}_k'\mathbf{W}(\mathbf{W}'\mathbf{W})^{-1}\mathbf{W}'\mathbf{x}_k)$.

(b) From (a), deduce that $|\mathbf{W}'\mathbf{W}|/|\mathbf{X}'\mathbf{X}| \geq 1/(\mathbf{x}_k'\mathbf{x}_k)$. Use this to show that in the usual linear model $\mathbf{y} = \mathbf{X}\boldsymbol{\beta} + \boldsymbol{\varepsilon}$, $Var(\widehat{\beta}_k) \geq \sigma^2(\mathbf{x}_k'\mathbf{x}_k)^{-1}$, with equality holding if and only if $\mathbf{x}_k'\mathbf{x}_j = 0$, $j = 0, 1, \cdots, k-1$.

4.8. Consider LS estimation in the simple linear regression model (4.1.6).

(a) Prove that $\text{Var}(\widehat{\beta}_0)$ is a minimum if the observations on the predictor variable X_i, $i = 1, \cdots, N$ are chosen such that $\overline{X} = 0$.

(b) If the X_i's can be selected anywhere in the interval $[a, b]$, and if N is an even integer, show that $\text{Var}(\widehat{\beta}_1)$ is minimized if $N/2$ values of the predictor are selected equal to a and the remaining $N/2$ values are chosen to equal b.

4.9. Let $Y_1, \cdots, Y_6$ denote the yield of a production process on six consecutive days. Machine A was used on days 1, 3, and 5, while Machine B was used on days 2, 4, and 6. Consider the models

(i) $Y_i = \beta_0 + (-1)^i \beta_1 + \varepsilon_i$,

(ii) $Y_i = \beta_0 + (-1)^i \beta_1 + i\beta_2 + \varepsilon_i$.

In each case, assume that $E(\varepsilon_i) = 0$, and $Var(\varepsilon_i) = \sigma^2$. Obtain the least squares estimate of β_1 under each model, and show that the ratio of $Var(\widehat{\beta}_1)$ under Model (i) to $Var(\widehat{\beta}_1)$ under Model (ii) is 32/35.

4.10. Consider the simple linear regression model (4.1.6).

(a) Verify that the residuals $\widehat{\varepsilon}_1, \cdots, \widehat{\varepsilon}_N$ are uncorrelated with the fitted values $\widehat{Y}_1, \cdots, \widehat{Y}_N$, whereas they are correlated with the observed responses $Y_1, \cdots, Y_N$.

(b) Follow Example 4.2.2 to show that the slope of the straight line fit of $\widehat{\varepsilon}_i$ on $\widehat{Y}_i$ is zero, while that of the fit of $\widehat{\varepsilon}_i$ on Y_i is equal to $1 - r^2$, where r denotes the sample correlation coefficient between X and Y defined by

$$r^2 = [\textstyle\sum_{i=1}^N (X_i - \overline{X})(Y_i - \overline{Y})]^2 / \sum_{i=1}^N (X_i - \overline{X})^2 \sum_{i=1}^N (Y_i - \overline{Y})^2.$$

4.11. When $\beta_0 = 0$ in (4.1.6), corresponding to a straight line through the origin, i.e., a simple linear regression without intercept, show that

$$\widehat{\beta}_1 = \textstyle\sum_{i=1}^N X_i Y_i / \sum_{i=1}^N X_i^2 \quad \text{and} \quad Var(\widehat{\beta}_1) = \sigma^2 / \sum_{i=1}^N X_i{}^2.$$

4.12. In (4.2.32) with $n = 1$, show that it is not possible to carry out inference on the effects, although it is possible to obtain point estimates.

4.13. Prove Result 4.3.3.

4.14. Consider the model $\mathbf{y} = \mathbf{X}_1\boldsymbol{\beta}_1 + \mathbf{X}_2\boldsymbol{\beta}_2 + \boldsymbol{\varepsilon}$. Show that $\mathbf{X}_1\boldsymbol{\beta}_1$ is estimable if and only if $\mathcal{C}(\mathbf{X}_1) \cap \mathcal{C}(\mathbf{X}_2) = \{\mathbf{0}\}$, i.e., the columns of $\mathbf{X}_1$ are LIN of the columns of $\mathbf{X}_2$.

4.15. In the general linear model (4.1.1), prove that all linear functions $\mathbf{c}'\boldsymbol{\beta}$ are estimable if and only if the columns of $\mathbf{X}$ are LIN.

4.16. In the model (4.1.7), show that τ_i is not estimable for any i.

4.17. In the model (4.2.30), show that for constants $c_1, \cdots, c_a$, $\sum_{i=1}^a c_i\tau_i$ is estimable if and only if $\sum_{i=1}^a c_i = 0$.

4.18. Let Y_{i1} denote a pre-treatment score and Y_{i2} a post-treatment score on the ith individual, $i = 1, \cdots, N$. For $i, l = 1, \cdots, N$ and $j = 1, 2$, let $E(Y_{ij}) = \tau_j$, $Var(Y_{ij}) = \sigma^2$ and $Corr(Y_{i1}, Y_{l2}) = \rho$ if $i = l$, and 0 if $i \neq l$.

(a) Estimate the parameters τ_1 and τ_2 in this linear model.

(b) How will you obtain the estimated standard errors of the estimates in (a)?

4.19. Consider the model (4.5.1). For the full-rank case, show that $\widehat{\boldsymbol{\beta}}_{GLS}$ and $\widehat{\boldsymbol{\beta}}$ coincide when

(a) $\mathbf{y} \in \mathcal{C}(\mathbf{X})$, or

(b) $\mathbf{V} = (1 - \rho)\mathbf{I} + \rho\mathbf{J}$ and $\mathbf{y}$ is orthogonal to $\mathbf{X}$, the first column of $\mathbf{X}$ being $\mathbf{1}_N$.

4.20. Show that $\mathbf{c}'\boldsymbol{\beta}$ is estimable in the model (4.5.1) if and only if it is estimable in the model (4.5.3).

4.21. In the model (4.5.1), show that

(a) $(\mathbf{X}'\mathbf{V}^{-1}\mathbf{X})\boldsymbol{\beta}$ is estimable.

(b) $\mathbf{c}'\boldsymbol{\beta}$ is estimable if $\mathbf{c}'(\mathbf{X}'\mathbf{V}^{-1}\mathbf{X})^{-}(\mathbf{X}'\mathbf{V}^{-1}\mathbf{X}) = \mathbf{c}'$.

4.22. Verify Result 4.5.2.

4.23. Verify equation (4.5.8).

4.24. Suppose Y_1, Y_2, and Y_3 are measurements of the angles of a triangle subject to error. The information is given as a linear model $Y_i = \theta_i + \varepsilon_i$, where θ_i are the true angles, $i = 1, 2, 3$. Assume that $\text{E}(\varepsilon_i) = 0$, and $\text{Var}(\varepsilon_i) = \sigma^2$. Obtain the least squares estimates of θ_i (subject to the constraint $\theta_1 + \theta_2 + \theta_3 = 180°$).

4.25. In the model $Y_i = \beta_0 + \beta_1 X_i + \varepsilon_i$, $i = 1, \cdots, N$, suppose the X_i's are constrained to lie in the interval $[-10, 15]$. Choose the X_i's to minimize (a) $\text{Var}(\widehat{\beta_0})$, and (b) $\text{Var}(\widehat{\beta_1})$. Repeat the exercise assuming that the X_i's are constrained to lie in the interval $[-10, 10]$.

4.26. Consider the model $\mathbf{y} = \boldsymbol{\beta} + \boldsymbol{\varepsilon}$, $\boldsymbol{\varepsilon} \sim N(\mathbf{0}, \sigma^2\mathbf{V})$, where $\mathbf{V}$ is an $N \times N$ p.d. matrix. If $\mathbf{z} = \mathbf{B}\mathbf{y}$, and $\boldsymbol{\eta} = \mathbf{B}\boldsymbol{\varepsilon}$, where $\mathbf{B} = \mathbf{V}^{-1/2}$, show that the linear model can be expressed as $\mathbf{z} = \mathbf{B}\boldsymbol{\beta} + \boldsymbol{\eta}$, with $\boldsymbol{\eta} \sim N(\mathbf{0}, \sigma^2\mathbf{I}_N)$. Obtain the least squares estimate of $\boldsymbol{\beta}$. Also obtain the estimate of $\boldsymbol{\beta}$ under the constraint $\mathbf{C}'\boldsymbol{\beta} = \mathbf{0}$, where $\mathbf{C}$ is an $N \times q$ matrix of rank $q < N$.

4.27. Consider the model $\mathbf{y} = \mathbf{X}\boldsymbol{\beta} + \boldsymbol{\varepsilon}$, where

$$\mathbf{y} = \begin{pmatrix} Y_1 \\ Y_2 \\ Y_3 \\ Y_4 \end{pmatrix}, \quad \mathbf{X} = \begin{pmatrix} 0 & 1 & 0 \\ 1 & 0 & 1 \\ -1 & 1 & -1 \\ 1 & -1 & 1 \end{pmatrix}, \quad \boldsymbol{\beta} = \begin{pmatrix} \beta_1 \\ \beta_2 \\ \beta_3 \end{pmatrix},$$

and $\boldsymbol{\varepsilon} \sim N(0, \sigma^2\mathbf{I})$. If possible, obtain the b.l.u.e.'s of (i) $\beta_1 + \beta_3$, and (ii) β_2, and compute their variances.

4.28. For the constrained estimates (4.6.5), show that $SSE_r = SSE + \|\widehat{\mathbf{y}} - \widehat{\mathbf{y}}_r\|^2$.

4.29. Complete the proof of property 3 of Result 4.6.4.

4.30. Continuing Example 4.6.2, suppose we impose the constraints $\mu = s$ and $\tau_1 + \tau_2 = t$, where s and t are constants.

(a) Show that μ and $\tau_1 + \tau_2$ are not estimable while $2\mu + \tau_1 + \tau_2$ is.

(b) Show that the vector of fitted values under the above constraints is identical to the one under the constraint $2\mu + \tau_1 + \tau_2 = 2s + t$.

5

Multivariate Normal and Related Distributions

This chapter describes distributions that are at the heart of linear model theory. We begin with the definition and properties of the multivariate normal distribution, and then define various distributions such as the chi-square, F, and t distributions that are derived from the normal distribution. Using these, we are able to describe the distributions of functions of quadratic forms in normal random vectors, which form the backbone of statistical inference in linear model theory. We also highlight some distributions that are useful as alternatives to the multivariate normal distribution in a contemporary treatment of linear models.

5.1 Intergral evaluation theorems

In this section, we present two results that are useful for evaluating integrals that arise in the context of multivariate normal distributions (see Section 5.2). From integral calculus, we recall that for integer $k \geq 0$,

$$\int_{-\infty}^{\infty} x^k \exp\{-x^2/2\}\, dx = \begin{cases} \dfrac{k!\sqrt{2\pi}}{2^{k/2}(k/2)!} & \text{if } k \text{ is even} \\ 0 & \text{if } k \text{ is odd.} \end{cases} \tag{5.1.1}$$

Result 5.1.1. Aitken's integral. Let $\mathbf{A}$ be an $n \times n$ p.d. symmetric matrix. Then

$$\int_{\mathcal{R}^n} \exp\{-\mathbf{x}'\mathbf{A}\mathbf{x}/2\}\, d\mathbf{x} = (2\pi)^{n/2}|\mathbf{A}|^{-1/2}, \tag{5.1.2}$$

where $\int_{\mathcal{R}^n}$ denotes an n-dimensional integral.

Proof. By Result 2.4.5, there exists a nonsingular matrix $\mathbf{P}$ satisfying $\mathbf{A} = \mathbf{P}'\mathbf{P}$. Letting $\mathbf{Q} = \mathbf{P}^{-1}$, we see that $\mathbf{Q}'\mathbf{A}\mathbf{Q} = \mathbf{I}_n$, so that $|\mathbf{Q}'\mathbf{A}\mathbf{Q}| = |\mathbf{Q}|^2|\mathbf{A}| = 1$, and hence $|\mathbf{Q}| = |\mathbf{A}|^{-1/2}$. Under the transformation $\mathbf{y} = \mathbf{P}\mathbf{x}$, we have $\mathbf{x} = \mathbf{Q}\mathbf{y}$, $\mathbf{x}'\mathbf{A}\mathbf{x} = \mathbf{y}'\mathbf{Q}'\mathbf{A}\mathbf{Q}\mathbf{y} = \mathbf{y}'\mathbf{y}$, and using (5.1.1),

$$\begin{aligned}\int_{\mathcal{R}^n} \exp\{-\mathbf{x}'\mathbf{A}\mathbf{x}/2\}\, d\mathbf{x} &= \int_{\mathcal{R}^n} |\mathbf{Q}| \exp\{-\mathbf{y}'\mathbf{y}/2\} d\mathbf{y} \\ &= |\mathbf{A}|^{-1/2} \int_{-\infty}^{\infty} \cdots \int_{-\infty}^{\infty} \exp\left\{-\frac{1}{2}\sum_{i=1}^{n} y_i^2\right\} dy_1 \cdots dy_n \\ &= |\mathbf{A}|^{-1/2} \prod_{i=1}^{n} \int_{-\infty}^{\infty} \exp\{-y_i^2/2\} dy_i \\ &= (2\pi)^{n/2}|\mathbf{A}|^{-1/2}.\end{aligned}$$

■

DOI: 10.1201/9781315156651-5

Result 5.1.2. General integral evaluation theorem. Let a_0 denote a scalar, $\mathbf{a}$ and $\mathbf{b}$ be n-dimensional vectors of constants, $\mathbf{A}$ be an $n \times n$ symmetric matrix of constants, and $\mathbf{B}$ be a p.d. symmetric matrix of constants. Then

$$I = \int_{\mathcal{R}^n} (\mathbf{x}'\mathbf{A}\mathbf{x} + \mathbf{x}'\mathbf{a} + a_0) \exp\{-(\mathbf{x}'\mathbf{B}\mathbf{x} + \mathbf{x}'\mathbf{b} + b_0)\}\, d\mathbf{x} \tag{5.1.3}$$

is evaluated as

$$\begin{aligned} I &= \frac{1}{2}\pi^{n/2}|\mathbf{B}|^{-1/2}\left[\text{tr}(\mathbf{A}\mathbf{B}^{-1}) - \mathbf{b}'\mathbf{B}^{-1}\mathbf{a} + \frac{1}{2}\mathbf{b}'\mathbf{B}^{-1}\mathbf{A}\mathbf{B}^{-1}\mathbf{b} + 2a_0\right] \\ &\quad \times \exp\left\{\frac{1}{4}\mathbf{b}'\mathbf{B}^{-1}\mathbf{b} - b_0\right\}. \end{aligned} \tag{5.1.4}$$

Proof. Let $\mathbf{c} = -\mathbf{B}^{-1}\mathbf{b}/2$ and $\mathbf{y} = \mathbf{x} - \mathbf{c}$. The exponent in (5.1.3) can be written as

$$\mathbf{x}'\mathbf{B}\mathbf{x} + \mathbf{x}'\mathbf{b} + b_0 = (\mathbf{y}+\mathbf{c})'\mathbf{B}(\mathbf{y}+\mathbf{c}) + (\mathbf{y}+\mathbf{c})'\mathbf{b} + b_0 = \mathbf{y}'\mathbf{B}\mathbf{y} + \frac{1}{2}\mathbf{b}'\mathbf{c} + b_0, \tag{5.1.5}$$

while the nonexponent can be written as

$$\begin{aligned} \mathbf{x}'\mathbf{A}\mathbf{x} + \mathbf{x}'\mathbf{a} + a_0 &= (\mathbf{y}+\mathbf{c})'\mathbf{A}(\mathbf{y}+\mathbf{c}) + (\mathbf{y}+\mathbf{c})'\mathbf{a} + a_0 \\ &= \mathbf{y}'\mathbf{A}\mathbf{y} + \mathbf{y}'(\mathbf{a} + 2\mathbf{A}\mathbf{c}) + \mathbf{c}'\mathbf{A}\mathbf{c} + \mathbf{c}'\mathbf{a} + a_0. \end{aligned} \tag{5.1.6}$$

Denote

$$\begin{aligned} a_1 &= \mathbf{c}'\mathbf{A}\mathbf{c} + \mathbf{c}'\mathbf{a} + a_0 = \mathbf{b}'\mathbf{B}^{-1}\mathbf{A}\mathbf{B}^{-1}\mathbf{b}/4 - \mathbf{b}'\mathbf{B}^{-1}\mathbf{a}/2 + a_0, \\ b_1 &= \mathbf{b}'\mathbf{c}/2 + b_0 = -\mathbf{b}'\mathbf{B}^{-1}\mathbf{b}/4 + b_0. \end{aligned}$$

Let $\mathbf{d} = \mathbf{a} + 2\mathbf{A}\mathbf{c}$ and $\mathbf{R} = 2\mathbf{B}$. Substituting (5.1.5) and (5.1.6) into (5.1.3), we see that

$$\begin{aligned} I &= e^{-b_1}\left[\int_{\mathcal{R}^n} \mathbf{y}'\mathbf{A}\mathbf{y} \exp\{-\mathbf{y}'\mathbf{R}\mathbf{y}/2\}\, d\mathbf{y} + \int_{\mathcal{R}^n} \mathbf{y}'\mathbf{d} \exp\{-\mathbf{y}'\mathbf{R}\mathbf{y}/2\}\, d\mathbf{y} \right. \\ &\quad \left. + \int_{\mathcal{R}^n} a_1 \exp\{-\mathbf{y}'\mathbf{R}\mathbf{y}/2\}\, d\mathbf{y}\right] \\ &= e^{-b_1}(I_1 + I_2 + I_3), \text{ say.} \end{aligned} \tag{5.1.7}$$

Since $\mathbf{R}$ is p.d., there exists an $n \times n$ nonsingular matrix $\mathbf{P}$ such that $\mathbf{R} = \mathbf{P}'\mathbf{P}$. Let $\mathbf{z} = \mathbf{P}\mathbf{y}$, so that $\mathbf{y} = \mathbf{P}^{-1}\mathbf{z}$, and the Jacobian is

$$J = |\mathbf{P}^{-1}| = |\mathbf{P}|^{-1} = |\mathbf{R}|^{-1/2}.$$

Let $\mathbf{W} = \{w_{ij}\} = (\mathbf{P}')^{-1}\mathbf{A}\mathbf{P}^{-1}$. We see that $\mathbf{y}'\mathbf{A}\mathbf{y} = \mathbf{z}'\mathbf{W}\mathbf{z}$ and so

$$I_1 = \int_{\mathcal{R}^n} \mathbf{z}'\mathbf{W}\mathbf{z} \exp\{-\mathbf{z}'\mathbf{z}/2\}\, J\, d\mathbf{z} = J \sum_{i,j=1}^{n} w_{ij} \int_{\mathcal{R}^n} z_i z_j \exp\{-\mathbf{z}'\mathbf{z}/2\}\, d\mathbf{z}.$$

For $i = j$,

$$\int_{\mathcal{R}^n} z_i^2 \exp\{-\mathbf{z}'\mathbf{z}/2\}\, d\mathbf{z} = \int_{-\infty}^{\infty} z_i^2 e^{-z_i^2/2}\, dz_i \prod_{s \neq i} \int_{-\infty}^{\infty} e^{-z_s^2/2}\, dz_s = (2\pi)^{n/2},$$

while for $i \neq j$, since $z_i z_j \exp\{-\mathbf{z}'\mathbf{z}/2\}$ is an odd function in z_i,

$$\int_{\mathcal{R}^n} z_i z_j \exp\{-\mathbf{z}'\mathbf{z}/2\}\, d\mathbf{z} = 0.$$

Finally, we get

$$\begin{aligned} I_1 &= J \sum_{i=1}^{n} w_{ii} (2\pi)^{n/2} = (2\pi)^{n/2} J \operatorname{tr}\{(\mathbf{P}')^{-1}\mathbf{A}\mathbf{P}^{-1}\} \\ &= (2\pi)^{n/2} J \operatorname{tr}\{\mathbf{A}(\mathbf{P}')^{-1}\mathbf{P}^{-1}\} = (2\pi)^{n/2} |\mathbf{R}|^{-1/2} \operatorname{tr}(\mathbf{A}\mathbf{R}^{-1}) \\ &= \frac{1}{2}\pi^{n/2} |\mathbf{B}|^{-1/2} \operatorname{tr}(\mathbf{A}\mathbf{B}^{-1}). \end{aligned} \tag{5.1.8}$$

Next, since $\mathbf{y}'\mathbf{d} \exp\{-\mathbf{y}'\mathbf{R}\mathbf{y}/2\}$ is an odd function of $\mathbf{y}$, $I_2 = 0$, and from Result 5.1.1, $I_3 = a_1(2\pi)^{n/2}|\mathbf{R}|^{-1/2} = a_1\pi^{n/2}|\mathbf{B}|^{-1/2}$. Substituting these expressions for I_1, I_2, and I_3 into (5.1.7), we get the required expression for I. ■

5.2 Multivariate normal distribution and properties

The multivariate normal distribution is the most widely used distribution to characterize the probabilistic behavior of a k-dimensional random vector. In this section, we define this distribution, and describe some important properties associated with this distribution. The normal distribution forms the basis for the development of the classical theory of linear models and multivariate analysis. We begin with the definition of the standard multivariate normal distribution.

Definition 5.2.1. $N_k(\mathbf{0}, \mathbf{I})$. A random vector $(Z_1, \cdots, Z_k)'$ has a multivariate standard normal distribution, denoted by $(Z_1, \cdots, Z_k)' \sim N_k(\mathbf{0}, \mathbf{I})$, if and only if it has p.d.f.

$$f(\mathbf{z}) = \frac{1}{(2\pi)^{k/2}} \exp(-\mathbf{z}'\mathbf{z}/2). \tag{5.2.1}$$

The mean of this distribution is $\mathrm{E}(\mathbf{z}) = \mathbf{0}$, and its variance-covariance matrix is $\mathrm{Cov}(\mathbf{z}) = \mathbf{I}_k$. It is easily verified that $f(\mathbf{z}) \geq 0$ for all $\mathbf{z} \in \mathcal{R}^k$, and using Aitken's integral in Result 5.1.1, we can easily see that $\int_{\mathcal{R}^k} f(\mathbf{z}) d\mathbf{z} = 1$, so (5.2.1) is a proper p.d.f.

Result 5.2.1. If $(Z_1, \cdots, Z_k)' \sim N_k(\mathbf{0}, \mathbf{I})$, then its m.g.f. is

$$M_{Z_1,\cdots,Z_k}(\mathbf{t}) = \exp(\mathbf{t}'\mathbf{t}/2) \tag{5.2.2}$$

for $\mathbf{t} = (t_1, \cdots, t_k)'$ in $\mathcal{R}^k$.

Proof. From (A.5),

$$\begin{aligned} M_{Z_1,\cdots,Z_k}(\mathbf{t}) &= \int_{\mathcal{R}^k} \exp(\mathbf{t}'\mathbf{z}) f(\mathbf{z})\, d\mathbf{z} \\ &= \frac{1}{(2\pi)^{k/2}} \int_{\mathcal{R}^k} \exp\{-(\mathbf{z}'\mathbf{z}/2 - \mathbf{z}'\mathbf{t})\}\, d\mathbf{z}. \end{aligned}$$

An application of Result 5.1.1 or Result 5.1.2 then yields the result. ■

It can be shown that if $X_1, \cdots, X_k$ are mutually independent normal random variables with respective means $\mu_1, \cdots, \mu_k$, and a common variance σ^2, then $(X_1, \cdots, X_k) \sim N_k(\boldsymbol{\mu}, \sigma^2\mathbf{I})$, where $\boldsymbol{\mu} = (\mu_1, \cdots, \mu_k)'$.

Result 5.2.2. Let $(Z_1, \cdots, Z_k)' \sim N_k(\mathbf{0}, \mathbf{I})$. Let $\mathbf{c} = (c_1, \cdots, c_k)'$ be a nonzero vector of constants and c_0 be a scalar constant. Then the univariate linear function of $(Z_1, \cdots, Z_k)'$ defined by $X = \sum_{i=1}^k c_i Z_i + c_0$ has a $N(c_0, \mathbf{c}'\mathbf{c})$ distribution. That is, any linear function of a standard multivariate normal vector itself has a normal distribution.

Proof. The m.g.f. of X is $M_X(t) = \mathrm{E}(e^{tX}) = \mathrm{E}[\exp(t\sum_{i=1}^k c_i Z_i + c_0)]$, which we evaluate using Aitken's integral as

$$M_X(t) = \frac{\exp(c_0 t)}{(2\pi)^{k/2}} \int_{\mathcal{R}^k} \exp(t\mathbf{c}'\mathbf{z} - \mathbf{z}'\mathbf{z}/2)\, d\mathbf{z} = e^{c_0 t + \mathbf{c}'\mathbf{c}t^2/2}. \tag{5.2.3}$$

From (A4), this is the m.g.f. of a $N(c_0, \mathbf{c}'\mathbf{c})$ random variable. ■

There are several equivalent definitions of a general multivariate normal distribution that may have a singular variance-covariance matrix. We give one definition below, and present other equivalent definitions in Result 5.2.4.

Definition 5.2.2. $N_k(\boldsymbol{\mu}, \boldsymbol{\Sigma})$. Let $\boldsymbol{\mu} \in \mathcal{R}^k$ be a vector of constants and $\boldsymbol{\Sigma}$ be a $k \times k$ constant symmetric n.n.d. matrix. A random vector $\mathbf{x} = (X_1, \cdots, X_k)'$ is said to have a $N_k(\boldsymbol{\mu}, \boldsymbol{\Sigma})$ distribution if and only if $\mathbf{x}$ has the same distribution as $\boldsymbol{\mu} + \mathbf{B}\mathbf{z}$ for some $k \times s$ matrix $\mathbf{B}$ satisfying $\mathbf{B}\mathbf{B}' = \boldsymbol{\Sigma}$ and $\mathbf{z} \sim N_s(\mathbf{0}, \mathbf{I})$. If $\mathrm{r}(\boldsymbol{\Sigma}) < k$, then $N_k(\boldsymbol{\mu}, \boldsymbol{\Sigma})$ is called a singular multivariate normal distribution. If $\boldsymbol{\Sigma} = \mathbf{O}$, then $N_k(\boldsymbol{\mu}, \mathbf{O})$ is the degenerate distribution concentrated at $\boldsymbol{\mu}$.

The next result follows directly from Definition 5.2.2.

Result 5.2.3. If $\mathbf{x} \sim N_k(\boldsymbol{\mu}, \boldsymbol{\Sigma})$, then $\mathrm{E}(\mathbf{x}) = \boldsymbol{\mu}$ and $\mathrm{Cov}(\mathbf{x}) = \boldsymbol{\Sigma}$.

Proof. From Definition 5.2.2, $\mathbf{x}$ has the same distribution as $\boldsymbol{\mu} + \mathbf{B}\mathbf{z}$, where $\mathbf{z} \sim N_s(\mathbf{0}, \mathbf{I})$ and $\mathbf{B}$ satisfies $\mathbf{B}\mathbf{B}' = \boldsymbol{\Sigma}$. Then $\mathrm{E}(\mathbf{x}) = \boldsymbol{\mu} + \mathrm{E}(\mathbf{B}\mathbf{z}) = \boldsymbol{\mu} + \mathbf{B}\,\mathrm{E}(\mathbf{z}) = \boldsymbol{\mu}$ and by (A.12), $\mathrm{Cov}(\mathbf{x}) = \mathrm{Cov}(\boldsymbol{\mu} + \mathbf{B}\mathbf{z}) = \mathbf{B}\,\mathrm{Cov}(\mathbf{z})\mathbf{B}' = \mathbf{B}\mathbf{B}' = \boldsymbol{\Sigma}$. ■

In the proofs of many of the following results, we will use the m.g.f. of the normal vector $\mathbf{x}$. There are two reasons for this, the first being that if $\boldsymbol{\Sigma}$ is not p.d., then there is no density function defined on $\mathcal{R}^k$, while the m.g.f. exists. The second reason is that the m.g.f. is a natural tool to use for deriving the distributional results that we require.

Result 5.2.4. Equivalent conditions on normality. Fix a k-dimensional vector $\boldsymbol{\mu}$ and a $k \times k$ symmetric n.n.d. matrix $\boldsymbol{\Sigma}$. Then $\mathbf{x} \sim N_k(\boldsymbol{\mu}, \boldsymbol{\Sigma})$ if and only if any one of the following properties holds.

1. The m.g.f. of $\mathbf{x}$ is $M_{\mathbf{x}}(\mathbf{t}) = \exp\{\mathbf{t}'\boldsymbol{\mu} + \mathbf{t}'\boldsymbol{\Sigma}\mathbf{t}/2\}$.

2. For every $\mathbf{c} \in \mathcal{R}^k$, $\mathbf{c}'\mathbf{x}$ has a univariate $N(\mathbf{c}'\boldsymbol{\mu}, \mathbf{c}'\boldsymbol{\Sigma}\mathbf{c})$ distribution.

Proof. We first show the equivalence between $\mathbf{x}$ having a $N_k(\boldsymbol{\mu}, \boldsymbol{\Sigma})$ distribution and property 1. From Definition 5.2.2 and properties of m.g.f., $\mathbf{x} \sim N_k(\boldsymbol{\mu}, \boldsymbol{\Sigma})$ if and only if $M_{\mathbf{x}}(t) = M_{\boldsymbol{\mu}+\mathbf{B}\mathbf{z}}(\mathbf{t})$ for some $k \times s$ matrix $\mathbf{B}$ satisfying $\mathbf{B}\mathbf{B}' = \boldsymbol{\Sigma}$ and $\mathbf{z} \sim N_s(\mathbf{0}, \mathbf{I})$. Using

Result 5.2.1,

$$M_{\boldsymbol{\mu}+\mathbf{Bz}}(\mathbf{t}) = \mathrm{E}[\exp\{\mathbf{t}'(\boldsymbol{\mu}+\mathbf{Bz})\}] = \exp(\mathbf{t}'\boldsymbol{\mu})\,\mathrm{E}[\exp\{(\mathbf{B}'\mathbf{t})'\mathbf{z}\}]$$
$$= \exp\{\mathbf{t}'\boldsymbol{\mu} + (\mathbf{B}'\mathbf{t})'(\mathbf{B}'\mathbf{t})/2\} = \exp(\mathbf{t}'\boldsymbol{\mu} + \mathbf{t}'\boldsymbol{\Sigma}\mathbf{t}/2).$$

The equivalence follows. We next show that property 1 implies property 2. Given $\mathbf{c} \in \mathcal{R}^k$, let $Y = \mathbf{c}'\mathbf{x}$. Then for any $t \in \mathcal{R}$, letting $\mathbf{d} = t\mathbf{c}$,

$$\mathrm{E}[e^{tY}] = \mathrm{E}[\exp(\mathbf{d}'\mathbf{x})] = \exp(\mathbf{d}'\boldsymbol{\mu} + \mathbf{d}'\boldsymbol{\Sigma}\mathbf{d}/2)$$
$$= \exp\{t(\mathbf{c}'\boldsymbol{\mu}) + (\mathbf{c}'\boldsymbol{\Sigma}\mathbf{c})t^2/2\},$$

which implies that $Y = \mathbf{c}'\mathbf{x}$ has a $N(\mathbf{c}'\boldsymbol{\mu}, \mathbf{c}'\boldsymbol{\Sigma}\mathbf{c})$ distribution. Finally, if property 2 holds, then for any $\mathbf{t} \in \mathcal{R}^k$, $\mathbf{t}'\mathbf{x} \sim N(\mathbf{t}'\boldsymbol{\mu}, \mathbf{t}'\boldsymbol{\Sigma}\mathbf{t})$, so that

$$\mathrm{E}[\exp(\mathbf{t}'\mathbf{x})] = \mathrm{E}[\exp\{1\cdot(\mathbf{t}'\mathbf{x})\}] = \exp(\mathbf{t}'\boldsymbol{\mu} + \mathbf{t}'\boldsymbol{\Sigma}\mathbf{t}/2),$$

yielding property 1. ■

From Result 5.2.4, we see that in order to show that a random vector follows a certain normal distribution, it is enough to show that its one-dimensional projections follow the corresponding univariate distributions. This technique of reducing a proof on a multivariate distribution to one on univariate distributions is known as the Cramér-Wold technique. From Result 5.2.4, the following holds.

Corollary 5.2.1. Let $\mathbf{x}$ be a k-variate random vector. If for any $\mathbf{c} \in \mathcal{R}^k$, $\mathbf{c}'\mathbf{x}$ has a normal distribution, possibly a degenerate one, then $\mathbf{x} \sim N_k(\boldsymbol{\mu}, \boldsymbol{\Sigma})$.

The proof of the corollary is left as Exercise 5.1. Note that unlike property 2 of Result 5.2.4, the result makes no assumption on the parameters of the normal distribution for each $\mathbf{c}'\mathbf{x}$.

The next result is fundamental and is widely used in linear model theory. It states that if a k-variate normal random vector is subject to a linear transformation, then the resulting vector also has a multivariate normal distribution of appropriate dimension. The usefulness of the result may be appreciated by recalling that in linear model theory, the least squares estimates (which correspond to the maximum likelihood estimates under normality) of the model parameters are linear functions of the response vector $\mathbf{y}$. The distributions of such estimators will be derived in Chapter 7 using this result.

Result 5.2.5. Let $\mathbf{x} \sim N_k(\boldsymbol{\mu}, \boldsymbol{\Sigma})$. For a given $q \times k$ matrix $\mathbf{B}$ of constants and a q-dimensional vector $\mathbf{b}$ of constants,

$$\mathbf{y} = \mathbf{Bx} + \mathbf{b} \sim N_q(\mathbf{B}\boldsymbol{\mu} + \mathbf{b}, \mathbf{B}\boldsymbol{\Sigma}\mathbf{B}'). \tag{5.2.4}$$

Proof. For any $\mathbf{t} = (t_1, \cdots, t_q)' \in \mathcal{R}^q$,

$$M_{\mathbf{y}}(\mathbf{t}) = \mathrm{E}[\exp(\mathbf{t}'\mathbf{y})] = \mathrm{E}[\exp\{\mathbf{t}'(\mathbf{Bx}+\mathbf{b})\}]$$
$$= \exp(\mathbf{t}'\mathbf{b})\,\mathrm{E}[\exp(\mathbf{t}'\mathbf{Bx})] = \exp(\mathbf{t}'\mathbf{b})M_{\mathbf{x}}(\mathbf{B}'\mathbf{t})$$
$$= \exp\{\mathbf{t}'(\mathbf{B}\boldsymbol{\mu}+\mathbf{b}) + \mathbf{t}'\mathbf{B}\boldsymbol{\Sigma}\mathbf{B}'\mathbf{t}/2\}.$$

The proof follows from Result 5.2.4. ■

Example 5.2.1. Let $\mathbf{x} = (X_1, X_2, X_3)' \sim N_3(\boldsymbol{\mu}, \boldsymbol{\Sigma})$. We use Result 5.2.5 to find the distribution of

$$\mathbf{B}\mathbf{x} = \begin{pmatrix} X_1 - X_2 \\ X_2 - X_3 \end{pmatrix} = \begin{pmatrix} 1 & -1 & 0 \\ 0 & 1 & -1 \end{pmatrix} \begin{pmatrix} X_1 \\ X_2 \\ X_3 \end{pmatrix}.$$

This vector clearly has a bivariate normal distribution with mean

$$\mathbf{B}\boldsymbol{\mu} = \begin{pmatrix} 1 & -1 & 0 \\ 0 & 1 & -1 \end{pmatrix} \begin{pmatrix} \mu_1 \\ \mu_2 \\ \mu_3 \end{pmatrix} = \begin{pmatrix} \mu_1 - \mu_2 \\ \mu_2 - \mu_3 \end{pmatrix}$$

and covariance matrix

$$\mathbf{B}\boldsymbol{\Sigma}\mathbf{B}' = \begin{pmatrix} 1 & -1 & 0 \\ 0 & 1 & -1 \end{pmatrix} \begin{pmatrix} \sigma_{11} & \sigma_{12} & \sigma_{13} \\ \sigma_{12} & \sigma_{22} & \sigma_{23} \\ \sigma_{13} & \sigma_{23} & \sigma_{33} \end{pmatrix} \begin{pmatrix} 1 & 0 \\ -1 & 1 \\ 0 & -1 \end{pmatrix}. \qquad \square$$

Example 5.2.2. Let $\mathbf{x} = (X_1, \cdots, X_n)' \sim N_n(\mu \mathbf{1}_n, \sigma^2 \mathbf{I}_n)$. Here, $\boldsymbol{\mu} = \mu \mathbf{1}_n$ and $\boldsymbol{\Sigma} = \sigma^2 \mathbf{I}_n$. Note that $\mathbf{x}$ represents a "random sample" of size n from a normal population with mean μ and variance σ^2. Then $\overline{X} = \sum_{i=1}^n X_i/n = \mathbf{c}'\mathbf{x}$, where $\mathbf{c}' = \mathbf{1}'/n$. It is easily verified that $\mathbf{c}'\boldsymbol{\mu} = \sum_{i=1}^n \mu/n = \mu$, and $\mathbf{c}'\boldsymbol{\Sigma}\mathbf{c} = \sigma^2/n$. By Result 5.2.5, $\overline{X}$ is a linear function of $\mathbf{x}$, and has a $N(\mu, \sigma^2/n)$ distribution. $\square$

When a normal distribution has a nonsingular variance-covariance matrix, it has a p.d.f. given below.

Result 5.2.6. Normal p.d.f. A normal distribution $N_k(\boldsymbol{\mu}, \boldsymbol{\Sigma})$ with $\boldsymbol{\Sigma}$ being p.d. has a p.d.f. (with respect to Lebesgue measure) given by

$$f(\mathbf{x}; \boldsymbol{\mu}, \boldsymbol{\Sigma}) = \frac{1}{(2\pi)^{k/2} |\boldsymbol{\Sigma}|^{1/2}} \exp\left\{ -\frac{1}{2} (\mathbf{x} - \boldsymbol{\mu})' \boldsymbol{\Sigma}^{-1} (\mathbf{x} - \boldsymbol{\mu}) \right\}. \tag{5.2.5}$$

Proof. Since $\boldsymbol{\Sigma}$ is symmetric and p.d., from Result 2.4.5, there is a nonsingular matrix $\boldsymbol{\Gamma}$ such that $\boldsymbol{\Sigma} = \boldsymbol{\Gamma}\boldsymbol{\Gamma}'$. Define $\mathbf{x} = \mathbf{g}(\mathbf{z}) = \boldsymbol{\Gamma}\mathbf{z} + \boldsymbol{\mu}$. The inverse transformation is $\mathbf{g}^*(\mathbf{x}) = \boldsymbol{\Gamma}^{-1}(\mathbf{x} - \boldsymbol{\mu})$ with Jacobian $J(\mathbf{x}) \equiv |\boldsymbol{\Gamma}^{-1}| = |\boldsymbol{\Sigma}|^{-1/2}$. Let $(Z_1, \cdots, Z_k)' \sim N_k(\mathbf{0}, \mathbf{I})$. By Definition 5.2.2,

$$\begin{pmatrix} X_1 \\ \vdots \\ X_k \end{pmatrix} = \boldsymbol{\Gamma} \begin{pmatrix} Z_1 \\ \vdots \\ Z_k \end{pmatrix} + \boldsymbol{\mu}$$

has a $N_k(\boldsymbol{\mu}, \boldsymbol{\Sigma})$ distribution and from Result A.2, its p.d.f. is given by

$$f(\mathbf{x}; \boldsymbol{\mu}, \boldsymbol{\Sigma}) = h(\mathbf{g}^*(\mathbf{x})) |\boldsymbol{\Sigma}|^{-1/2},$$

where $h(\mathbf{z})$ denotes a $N_k(\mathbf{0}, \mathbf{I})$ p.d.f. Since $h(\mathbf{z}) = (2\pi)^{-k/2} \exp\{-\mathbf{z}'\mathbf{z}/2\}$, plugging in the expression for $\mathbf{g}^*(\mathbf{x})$ then yields the formula for $f(\mathbf{x}; \boldsymbol{\mu}, \boldsymbol{\Sigma})$. ∎

Example 5.2.3. Bivariate normal distribution. Let $\boldsymbol{\mu} = (\mu_1, \mu_2)'$ and $\boldsymbol{\Sigma} = \begin{pmatrix} \sigma_1^2 & \rho\sigma_1\sigma_2 \\ \rho\sigma_1\sigma_2 & \sigma_2^2 \end{pmatrix}$, where $-\infty < \mu_i < \infty$, $\sigma_i > 0$, for $i = 1, 2$, and $|\rho| < 1$. Then, $|\boldsymbol{\Sigma}| = \sigma_1^2\sigma_2^2(1 - \rho^2)$ and

$$\boldsymbol{\Sigma}^{-1} = \frac{1}{1 - \rho^2} \begin{pmatrix} 1/\sigma_1^2 & -\rho/\sigma_1\sigma_2 \\ -\rho/\sigma_1\sigma_2 & 1/\sigma_2^2 \end{pmatrix}.$$

From Result 5.2.6, the $N_2(\boldsymbol{\mu}, \boldsymbol{\Sigma})$ p.d.f. is

$$f(x_1, x_2) = \frac{1}{2\pi|\boldsymbol{\Sigma}|^{1/2}} \exp\left\{-\frac{1}{2}\begin{pmatrix}x_1 - \mu_1 & x_2 - \mu_2\end{pmatrix} \boldsymbol{\Sigma}^{-1} \begin{pmatrix}x_1 - \mu_1 \\ x_2 - \mu_2\end{pmatrix}\right\}$$

$$= \frac{1}{2\pi\sigma_1\sigma_2(1-\rho^2)^{1/2}} \exp\left\{-\frac{z_1^2 + z_2^2 - 2\rho z_1 z_2}{2(1-\rho^2)}\right\}, \tag{5.2.6}$$

where $z_i = (x_i - \mu_i)/\sigma_i$ is the "standardized" value of x_i. We sometimes denote this bivariate normal distribution by $N_2(\mu_1, \mu_2, \sigma_1^2, \sigma_2^2, \rho)$. □

Example 5.2.4. Suppose $\mathbf{x} = (X_1, X_2)' \in \mathcal{R}^2$ has p.d.f.

$$f(x_1, x_2) = C \exp\left\{-\frac{1}{2}(x_1^2 + 2x_2^2 - 2x_1x_2 - 2x_1 + 2x_2)\right\},$$

where C is the normalizing constant. We show that $\mathbf{x}$ follows a bivariate normal distribution and identify its mean $\boldsymbol{\mu}$ and variance $\boldsymbol{\Sigma}$. It is easy to see that

$$x_1^2 + 2x_2^2 - 2x_1x_2 - 2x_1 + 2x_2 = (x_1 - 1)^2 + 2x_2^2 - 2(x_1 - 1)x_2 - 1.$$

Then,

$$f(x_1, x_2) \propto \exp\left\{-\frac{1}{2}(x_1 - 1, x_2)' \begin{pmatrix}1 & -1 \\ -1 & 2\end{pmatrix} \begin{pmatrix}x_1 - 1 \\ x_2\end{pmatrix}\right\},$$

so that, $\mathbf{x} \sim N_2(\boldsymbol{\mu}, \boldsymbol{\Sigma})$ with

$$\boldsymbol{\mu} = (1, 0)', \text{ and } \boldsymbol{\Sigma} = \begin{pmatrix}1 & -1 \\ -1 & 2\end{pmatrix}^{-1} = \begin{pmatrix}2 & 1 \\ 1 & 1\end{pmatrix}.$$ □

From (5.2.6), it is also evident that the paths of $\mathbf{x}$ values yielding a constant value for the bivariate normal p.d.f. (called contours) are ellipsoids. In general, the multivariate normal p.d.f. assumes a constant value on surfaces where the squared distance $(\mathbf{x} - \boldsymbol{\mu})'\boldsymbol{\Sigma}^{-1}(\mathbf{x} - \boldsymbol{\mu})$ is constant, i.e.,

$$(\mathbf{x} - \boldsymbol{\mu})'\boldsymbol{\Sigma}^{-1}(\mathbf{x} - \boldsymbol{\mu}) = c^2. \tag{5.2.7}$$

The center of this ellipsoid is $\boldsymbol{\mu}$, its size is determined by c, and its shape and orientation are characterized in terms of the eigenvalues and eigenvectors of $\boldsymbol{\Sigma}$. Let $\lambda_1 \geq \cdots \geq \lambda_k > 0$ denote the eigenvalues of $\boldsymbol{\Sigma}$ and let $\mathbf{v}_1, \cdots, \mathbf{v}_k$ denote the corresponding orthonormal eigenvectors. If the multiplicity m_j of λ_j is 1, then $\mathbf{v}_j$ determines the direction of the jth principal axis, i.e., the jth principal axis is in the direction of the jth eigenvector of $\boldsymbol{\Sigma}$, and its length is $2c\sqrt{\lambda_j}$, $j = 1, \cdots, k$. However, if $m_j > 1$, then the corresponding eigenvectors and hence the corresponding principal axes are not uniquely defined. The volume of this ellipsoid is in proportion to $c^k|\boldsymbol{\Sigma}^{-1}|$. If $\boldsymbol{\Sigma}$ is a constant diagonal matrix, i.e., $\boldsymbol{\Sigma} = \sigma^2\mathbf{I}$, then (5.2.7) represents a k-dimensional sphere with radius $c\sigma$. If we choose $c^2 = \chi^2_{k,\alpha}$, which is the upper (100α)th percentile from a chi-square distribution with k degrees of freedom (d.f.), we obtain contours that contain $100(1 - \alpha)\%$ of the probability.

Example 5.2.5. We discuss contours of the bivariate normal density defined in (5.2.6), assuming that $\sigma_1^2 = \sigma_2^2$ (Johnson and Wichern, 2007). It is easy to verify that the eigenvalues and corresponding eigenvectors of $\boldsymbol{\Sigma}$ are

$$\lambda_1 = \sigma_1^2(1 + \rho), \quad \lambda_2 = \sigma_1^2(1 - \rho), \text{ and}$$

$$\mathbf{v}_1 = \begin{pmatrix}1/\sqrt{2} \\ 1/\sqrt{2}\end{pmatrix}, \quad \mathbf{v}_2 = \begin{pmatrix}1/\sqrt{2} \\ -1/\sqrt{2}\end{pmatrix}.$$

When $\rho > 0$, λ_1 is the largest eigenvalue and $\mathbf{v}_1$ lies along the 45° line through the point (μ_1, μ_2); the major axis is associated with λ_1. When $\rho < 0$, then λ_2 is the largest eigenvalue and the major axis of the constant density ellipse lies along a straight line at right angles to the 45° line through (μ_1, μ_2). □

Example 5.2.6. Suppose $(X_1, X_2)'$ has a bivariate normal distribution with mean and covariance respectively given by

$$\boldsymbol{\mu} = \begin{pmatrix} 2 \\ 5 \end{pmatrix} \quad \text{and} \quad \boldsymbol{\Sigma} = \begin{pmatrix} 2 & 1 \\ 1 & 1 \end{pmatrix},$$

so that we identify $\mu_1 = 2$, $\mu_2 = 5$, $\sigma_1 = \sqrt{2}$, $\sigma_2 = 1$, and $\rho = 1/\sqrt{2}$. We will find an ellipse which contains $(X_1, X_2)'$ with probability α, $0 < \alpha < 1$. From Example 5.2.3, contours of equal density have the form

$$(X_1 - 2)^2 - 2(X_1 - 2)(X_2 - 5) + 2(X_2 - 5)^2 = c^2.$$

The value of c corresponds to the upper (100α)th percentile of the chi-square distribution with 2 degrees of freedom (see B.5) which is given as the value of u such that the chi-square c.d.f. $F(u)$ is equal to α. It follows that $c = -2\ln(1 - \alpha)$. By computing the contour for different α values, we generate the elliptical contours of the bivariate normal p.d.f. □

The p.d.f. in (5.2.5) involves k parameters of location, and $k(k+1)/2$ distinct parameters that characterize $\boldsymbol{\Sigma}$. From Result 5.2.6, if $\mathbf{x} \sim N_k(\boldsymbol{\mu}, \boldsymbol{\Sigma})$ is nonsingular, then for any $k \times k$ matrix $\boldsymbol{\Gamma}$ with $\boldsymbol{\Gamma}\boldsymbol{\Gamma}' = \boldsymbol{\Sigma}$, $\mathbf{z} = \boldsymbol{\Gamma}^{-1}(\mathbf{x} - \boldsymbol{\mu})$ has a $N_k(\mathbf{0}, \mathbf{I})$ distribution. If, on the other hand, $\boldsymbol{\Sigma}$ is p.s.d., so that $\mathrm{r}(\boldsymbol{\Sigma}) = r < k$, then the inverse transformation does not exist and no explicit determination of the p.d.f. of $\mathbf{x}$ with respect to Lebesgue measure in $\mathcal{R}^k$ exists. However, $\mathbf{x}$ is concentrated on an r-dimensional affine space. We describe this in the following result and then give some examples of singular multivariate normal distributions.

Result 5.2.7. Let $\mathbf{x} \sim N_k(\boldsymbol{\mu}, \boldsymbol{\Sigma})$. Suppose $\boldsymbol{\Sigma} \neq \mathbf{O}$ and $\mathrm{r}(\boldsymbol{\Sigma}) = r \leq k$.

1. The support of $\mathbf{x}$ is the entire affine space $\boldsymbol{\mu} + \mathcal{C}(\boldsymbol{\Sigma})$.
2. Let $\boldsymbol{\Gamma}$ be any $k \times r$ matrix with $\boldsymbol{\Sigma} = \boldsymbol{\Gamma}\boldsymbol{\Gamma}'$, so it has full column rank. Let $\mathbf{z} = (\boldsymbol{\Gamma}'\boldsymbol{\Gamma})^{-1}\boldsymbol{\Gamma}'(\mathbf{x} - \boldsymbol{\mu})$. Then $\mathbf{z} \sim N_r(\mathbf{0}, \mathbf{I})$ and $\mathbf{x} = \boldsymbol{\Gamma}\mathbf{z} + \boldsymbol{\mu}$.

Proof. From Result A.1, the support of $\mathbf{x}$ is contained in $\boldsymbol{\mu} + \mathcal{C}(\boldsymbol{\Sigma})$. This only partially proves property 1 because we have not yet shown that $\mathbf{x}$ can assume any values in $\boldsymbol{\mu} + \mathcal{C}(\boldsymbol{\Gamma})$. To do this, we show property 2. It is easily seen that $\mathrm{E}(\mathbf{z}) = \mathbf{0}$. Since $\mathrm{Cov}(\mathbf{z}) = \mathrm{Cov}((\boldsymbol{\Gamma}'\boldsymbol{\Gamma})^{-1}\boldsymbol{\Gamma}'\mathbf{x}) = (\boldsymbol{\Gamma}'\boldsymbol{\Gamma})^{-1}\boldsymbol{\Gamma}'\boldsymbol{\Sigma}\boldsymbol{\Gamma}(\boldsymbol{\Gamma}'\boldsymbol{\Gamma})^{-1} = (\boldsymbol{\Gamma}'\boldsymbol{\Gamma})^{-1}(\boldsymbol{\Gamma}'\boldsymbol{\Gamma})(\boldsymbol{\Gamma}'\boldsymbol{\Gamma})(\boldsymbol{\Gamma}'\boldsymbol{\Gamma})^{-1} = \mathbf{I}$, we see that $\mathbf{z} \sim N_r(\mathbf{0}, \mathbf{I})$. Then, since $\boldsymbol{\Gamma}\mathbf{z} = \boldsymbol{\Gamma}(\boldsymbol{\Gamma}\boldsymbol{\Gamma}')^{-1}\boldsymbol{\Gamma}'(\mathbf{x} - \boldsymbol{\mu}) = \mathbf{P}(\mathbf{x} - \boldsymbol{\mu}) = \mathbf{x} - \boldsymbol{\mu}$, where the last equality is due to property 1, property 2 follows. Finally, since $\mathbf{z} \sim N_r(\mathbf{0}, \mathbf{I})$, it can assume any values in $\mathcal{R}^k$. Then $\mathbf{x}$ can take values anywhere in $\boldsymbol{\mu} + \mathcal{C}(\boldsymbol{\Gamma})$, completing the proof of property 1. ■

Example 5.2.7. When $k = 1$, $\boldsymbol{\Sigma}$ corresponds to a scalar, and a singular (univariate) normal distribution is obtained only if the variance is zero. The m.g.f. of the singular normal random variable X is $M_X(t) = \exp(\mu t)$, which is the m.g.f. of a random variable X degenerate at μ, i.e., $\mathrm{P}(X = \mu) = 1$. Hence, the singular normal distribution when $k = 1$ is a discrete distribution degenerate at a single value. When $k = 2$, since $\boldsymbol{\Sigma} = \begin{pmatrix} \sigma_1^2 & \sigma_{12} \\ \sigma_{12} & \sigma_2^2 \end{pmatrix}$ is n.n.d., $\sigma_1^2 \geq 0$, $\sigma_2^2 \geq 0$, and $\sigma_1^2\sigma_2^2 - \sigma_{12}^2 \geq 0$, and $\boldsymbol{\Sigma}$ is p.d. if and only if $\sigma_1^2 > 0$, $\sigma_2^2 > 0$, and $\sigma_1^2\sigma_2^2 - \sigma_{12}^2 > 0$, bivariate singular normal distributions occur under the following four scenarios:

(i) $\sigma_1 = \sigma_2 = 0$, so that $\sigma_{12} = 0$. Then, $\mathbf{x} \sim N_2(\boldsymbol{\mu}, \mathbf{O})$, so that $\mathbf{x}$ is degenerate at $\boldsymbol{\mu} = (\mu_1, \mu_2)'$.

(ii) $\sigma_1 = 0$, $\sigma_2 > 0$, so that $\sigma_{12} = 0$. Then, $X_1 \sim N(\mu_1, 0)$ and is degenerate, while $X_2 \sim N(\mu_2, \sigma_2^2)$ and is nondegenerate. Since a degenerate random variable is independent of any other random variable, X_1 and X_2 are independent.

(iii) $\sigma_1 > 0$, $\sigma_2 = 0$, so that $\sigma_{12} = 0$. This case is similar to (ii), with the roles of X_1 and X_2 reversed.

(iv) $\sigma_1 > 0$, $\sigma_2 > 0$, and $\sigma_1^2\sigma_2^2 = \sigma_{12}^2$. Then both X_1 and X_2 have nondegenerate distributions. However, we can verify that $\text{Var}(\sigma_2^2 X_1 - \sigma_{12}X_2) = 0$, so $(X_1, X_2)'$ lies on the line $\sigma_2^2 x_1 - \sigma_{12}x_2 = \sigma_2^2\mu_1 - \sigma_{12}\mu_2$ with probability one. □

So far we have focused on single normal random vectors. We now consider properties of jointly distributed normal random vectors.

Definition 5.2.3. Let $\mathbf{x}_1, \cdots, \mathbf{x}_n$ be jointly distributed random vectors that may have different dimensions. The random vectors are said to be jointly normally distributed if the random vector $\mathbf{x} = (\mathbf{x}_1', \cdots, \mathbf{x}_n')'$ has a multivariate normal distribution.

Note that normality of all the individual random vectors does not guarantee that they are jointly normally distributed. For example, let $X \sim N(0,1)$. Let

$$Y = \begin{cases} X & \text{if } |X| \leq 1 \\ -X & \text{if } |X| > 1 \end{cases}$$

be a nonlinear function of X. Then $Y \sim N(0,1)$ as well. However, $X + Y = 2X\, I\{|X| \leq 1\}$, where $I\{.\}$ is an indicator function, has support $[-2, 2]$ and is not normally distributed. Hence, (X, Y) is not normally distributed. The next result provides two sufficient conditions for joint normality.

Result 5.2.8. Joint normality of random vectors. Let $\mathbf{x}_1, \cdots, \mathbf{x}_n$ be jointly distributed random vectors, $\mathbf{x}_i$ having dimension p_i. Let $\mathbf{x} = (\mathbf{x}_1', \cdots, \mathbf{x}_n')'$ and $p = p_1 + \cdots + p_n$.

1. If for some $\mathbf{y} \sim N_k(\mathbf{a}, \mathbf{V})$, $\mathbf{x}_i = \mathbf{B}_i\mathbf{y} + \mathbf{b}_i$, $i = 1, \ldots, n$, where $\mathbf{B}_i$ is a $p_i \times k$ matrix of constants and $\mathbf{b}_i$ is a p_i-dimensional vector of constants, then $\mathbf{x} \sim N_p(\boldsymbol{\mu}, \boldsymbol{\Sigma})$, where $\boldsymbol{\mu} = (\boldsymbol{\mu}_1', \cdots, \boldsymbol{\mu}_n')'$ with $\boldsymbol{\mu}_i = \mathbf{B}_i\mathbf{a} + \mathbf{b}_i$ and $\boldsymbol{\Sigma} \in \mathcal{R}^{n \times n}$ can be partitioned as $\{\boldsymbol{\Sigma}_{ij}\}$ with $\boldsymbol{\Sigma}_{ij} = \text{Cov}(\mathbf{x}_i, \mathbf{x}_j) = \mathbf{B}_i\mathbf{V}\mathbf{B}_j'$.

2. If $\mathbf{x}_i \sim N_{p_i}(\boldsymbol{\mu}_i, \boldsymbol{\Sigma}_i)$, $i = 1, \ldots, n$, are independent, then $\mathbf{x} \sim N_p(\boldsymbol{\mu}, \boldsymbol{\Sigma})$, where $\boldsymbol{\mu} = (\boldsymbol{\mu}_1', \cdots, \boldsymbol{\mu}_n')'$ and $\boldsymbol{\Sigma} = \text{diag}(\boldsymbol{\Sigma}_1, \cdots, \boldsymbol{\Sigma}_n)$.

Proof. Given $\mathbf{c} = (\mathbf{c}_1', \cdots, \mathbf{c}_n')'$, where $\mathbf{c}_i \in \mathcal{R}^{p_i}$, $\mathbf{c}'\mathbf{x} = \sum_{i=1}^n \mathbf{c}_i'\mathbf{x}_i$. Under the condition of property 1, $\mathbf{c}'\mathbf{x} = \sum_{i=1}^n \mathbf{c}_i'(\mathbf{B}_i\mathbf{y} + \mathbf{b}_i) = (\sum_{i=1}^n \mathbf{B}_i'\mathbf{c}_i)'\mathbf{y} + \sum_{i=1}^n \mathbf{c}_i'\mathbf{b}_i$, so by Result 5.2.4, it has a univariate normal distribution with mean $(\sum_{i=1}^n \mathbf{B}_i'\mathbf{c}_i)'\boldsymbol{\mu} + \sum_{i=1}^n \mathbf{c}_i'\mathbf{b}_i = \sum_{i=1}^n \mathbf{c}_i'(\mathbf{B}_i\boldsymbol{\mu} + \mathbf{b}_i) = \sum_{i=1}^n \mathbf{c}_i'\boldsymbol{\mu}_i = \mathbf{c}'\boldsymbol{\mu}$ and variance

$$\left(\sum_{i=1}^n \mathbf{B}_i'\mathbf{c}_i\right)' \mathbf{V} \left(\sum_{i=1}^n \mathbf{B}_i'\mathbf{c}_i\right) = \sum_{i,j=1}^n \mathbf{c}_i'(\mathbf{B}_i\mathbf{V}\mathbf{B}_j')\mathbf{c}_j = \sum_{i,j=1}^n \mathbf{c}_i'\boldsymbol{\Sigma}_{ij}\mathbf{c}_j = \mathbf{c}'\boldsymbol{\Sigma}\mathbf{c},$$

where $\boldsymbol{\mu}$ and $\boldsymbol{\Sigma}$ are described in property 1. By property 2 of Result 5.2.4, $\mathbf{x} \sim N_p(\boldsymbol{\mu}, \boldsymbol{\Sigma})$. Furthermore, by (A.12), $\text{Cov}(\mathbf{x}_i, \mathbf{x}_j) = \text{Cov}(\mathbf{B}_i\mathbf{y}, \mathbf{B}_j\mathbf{y}) = \mathbf{B}_i \text{Cov}(\mathbf{y})\mathbf{B}_j' = \mathbf{B}_i\mathbf{V}\mathbf{B}_j$. This completes the proof of property 1.

Next, if $\mathbf{c}_i'\mathbf{x}_i \sim N(\mathbf{c}_i'\boldsymbol{\mu}_i, \mathbf{c}_i'\boldsymbol{\Sigma}_i\mathbf{c}_i)$ are independent univariate random variables, then their sum $\mathbf{c}'\mathbf{x}$ is also normally distributed with mean $\sum_{i=1}^n \mathbf{c}_i'\boldsymbol{\mu}_i = \mathbf{c}'\boldsymbol{\mu}$ and variance $\sum_{i=1}^n \mathbf{c}_i'\boldsymbol{\Sigma}_i\mathbf{c}_i = \mathbf{c}'\,\mathrm{diag}(\boldsymbol{\Sigma}_1, \cdots, \boldsymbol{\Sigma}_n)\mathbf{c}$, and property 2 follows. ■

Result 5.2.9. Reproductive property. Suppose $\mathbf{x}_j$, $j = 1, \cdots, p$ are independently distributed as $N_k(\boldsymbol{\mu}_j, \boldsymbol{\Sigma}_j)$. For any constants $a_1, \cdots, a_p$,

$$\sum_{j=1}^p a_j\mathbf{x}_j \sim N_k\left(\sum_{j=1}^p a_j\boldsymbol{\mu}_j, \sum_{j=1}^p a_j^2\boldsymbol{\Sigma}_j\right).$$

Proof. This result is a direct consequence of Result 5.2.8. Alternately, it follows from Result 5.2.4 and the fact that the m.g.f. of $\mathbf{y} = \sum_{j=1}^p a_j\mathbf{x}_j$ is

$$\begin{aligned} M_{\mathbf{y}}(\mathbf{t}) &= \mathrm{E}\left[\exp\left\{\mathbf{t}'\sum_{j=1}^p a_j\mathbf{x}_j\right\}\right] = \prod_{j=1}^p \mathrm{E}[\exp\{a_j\mathbf{t}'\mathbf{x}_j\}] \\ &= \prod_{j=1}^p \exp\left\{a_j\mathbf{t}'\boldsymbol{\mu}_j + \frac{1}{2}a_j^2\mathbf{t}'\boldsymbol{\Sigma}_j\mathbf{t}\right\} \\ &= \exp\left\{\mathbf{t}'\left(\sum_{j=1}^p a_j\boldsymbol{\mu}_j\right) + \frac{1}{2}\mathbf{t}'\left(\sum_{j=1}^p a_j^2\boldsymbol{\Sigma}_j\right)\mathbf{t}\right\}. \end{aligned}$$

■

It is well known that independent random variables are uncorrelated, but uncorrelated random variables are not necessarily independent. Importantly, for jointly normally distributed random vectors, independence and zero correlation are equivalent. More specifically, we have the following.

Result 5.2.10. Let $\mathbf{x}_1, \cdots, \mathbf{x}_m$ be jointly normally distributed. Then they are mutually independent if and only if $\mathrm{Cov}(\mathbf{x}_i, \mathbf{x}_j) = \mathbf{O}$ for all $i \neq j$.

Proof. Suppose $\mathbf{x}_1, \cdots, \mathbf{x}_m$ are jointly independently distributed. Then, letting $\boldsymbol{\mu}_i = \mathrm{E}(\mathbf{x}_i)$,

$$\boldsymbol{\Sigma}_{ij} = \mathrm{E}[(\mathbf{x}_i - \boldsymbol{\mu}_i)(\mathbf{x}_j - \boldsymbol{\mu}_j)'] = [\mathrm{E}(\mathbf{x}_i - \boldsymbol{\mu}_i)][\mathrm{E}(\mathbf{x}_j - \boldsymbol{\mu}_j)'] = \mathbf{O}.$$

Conversely, suppose $\boldsymbol{\Sigma}_{ij} = \mathbf{O}$ for $i \neq j$, $i, j = 1, \cdots, m$. Let $\mathbf{x} = (\mathbf{x}_1', \cdots, \mathbf{x}_m')'$. Then for any $\mathbf{t} = (\mathbf{t}_1', \cdots, \mathbf{t}_m')'$,

$$\begin{aligned} M_{\mathbf{x}}(\mathbf{t}) &= \exp\{\mathbf{t}'\boldsymbol{\mu} + \mathbf{t}'\boldsymbol{\Sigma}\mathbf{t}/2\} = \exp\left\{\sum_{j=1}^m \mathbf{t}_j'\boldsymbol{\mu}_j + \frac{1}{2}\sum_{i,j=1}^m \mathbf{t}_i'\boldsymbol{\Sigma}_{ij}\mathbf{t}_j\right\} \\ &= \prod_{j=1}^m \exp\{\mathbf{t}_j'\boldsymbol{\mu}_j + \mathbf{t}_j'\boldsymbol{\Sigma}_{jj}\mathbf{t}_j/2\} = \prod_{j=1}^m M_{\mathbf{x}_j}(\mathbf{t}_j), \end{aligned}$$

so that, by (A.14), $\mathbf{x}_1, \cdots, \mathbf{x}_m$ are jointly independently distributed. ■

Example 5.2.8. Let $\mathbf{x} \sim N_k(\mathbf{0}, \sigma^2\mathbf{I})$. Let $\mathbf{A}$ be an arbitrary $q \times k$ matrix and $\mathbf{G}$ be a g-inverse of $\mathbf{A}$. Let $\mathbf{y}_1 = \mathbf{A}\mathbf{x}$ and $\mathbf{y}_2 = (\mathbf{I} - \mathbf{G}\mathbf{A})'\mathbf{x}$. From Result 5.2.8, $\mathbf{y}_1$ and $\mathbf{y}_2$ are jointly normally distributed. We see that $\mathbf{y}_1 \sim N_q(\mathbf{0}, \sigma^2\mathbf{A}\mathbf{A}')$, while $\mathbf{y}_2 \sim N_k(\mathbf{0}, \sigma^2(\mathbf{I} - \mathbf{G}\mathbf{A})(\mathbf{I} - \mathbf{G}\mathbf{A})')$. Since

$$\mathrm{Cov}\{\mathbf{A}\mathbf{x}, (\mathbf{I} - \mathbf{G}\mathbf{A})'\mathbf{x}\} = \sigma^2\mathbf{A}(\mathbf{I} - \mathbf{G}\mathbf{A}) = \mathbf{O},$$

from Result 5.2.10, $\mathbf{y}_1$ and $\mathbf{y}_2$ are independent. □

Result 5.2.11. Let $\mathbf{x} \sim N_k(\boldsymbol{\mu}, \boldsymbol{\Sigma})$. Suppose we partition $\mathbf{x}$ as $(\mathbf{x}_1', \mathbf{x}_2')'$, where $\mathbf{x}_1$ is q-dimensional and $\mathbf{x}_2$ is $(k-q)$-dimensional. We conformably partition $\boldsymbol{\mu} = (\boldsymbol{\mu}_1', \boldsymbol{\mu}_2')'$ and $\boldsymbol{\Sigma} = \{\boldsymbol{\Sigma}_{ij}\}_{i,j=1,2}$. Then the subvectors $\mathbf{x}_1$ and $\mathbf{x}_2$ are jointly normally distributed, the marginal distribution of $\mathbf{x}_1$ is $N_q(\boldsymbol{\mu}_1, \boldsymbol{\Sigma}_{11})$, the marginal distribution of $\mathbf{x}_2$ is $N_{k-q}(\boldsymbol{\mu}_2, \boldsymbol{\Sigma}_{22})$, and $\text{Cov}(\mathbf{x}_1, \mathbf{x}_2) = \boldsymbol{\Sigma}_{12}$.

Proof. Since $\mathbf{x}_1 = (\mathbf{I}_q \ \mathbf{O})\mathbf{x}$ and $\mathbf{x}_2 = (\mathbf{O} \ \mathbf{I}_{k-q})\mathbf{x}$, the result is a direct consequence of property 1 of Result 5.2.8. ■

It is clear from Result 5.2.11 that all subsets of a multivariate normal random vector themselves have normal distributions. The mean and variance of the normal distribution which corresponds to particular subvector of $\mathbf{x}$ is, in fact, obtained by simply selecting appropriate elements from $\boldsymbol{\mu}$ and $\boldsymbol{\Sigma}$.

Example 5.2.9. Let $\mathbf{x} \sim N_4(\boldsymbol{\mu}, \boldsymbol{\Sigma})$, where

$$\boldsymbol{\mu} = (2, 3, 0, -1)' \text{ and } \boldsymbol{\Sigma} = \begin{pmatrix} 4 & 1 & 0 & -1 \\ 1 & 3 & 0 & 2 \\ 0 & 0 & 2 & 0 \\ -1 & 2 & 0 & 5 \end{pmatrix}.$$

We first note that $\boldsymbol{\Sigma}$ is p.d., since all its principal minors are positive. We find the distribution of $\mathbf{x}_1 = (X_2, X_4)'$. By Result 5.2.11, $\mathbf{x}_1$ has a bivariate normal distribution with mean $\boldsymbol{\mu}_1 = (3, -1)'$ and variance-covariance matrix $\boldsymbol{\Sigma}_{11} = \begin{pmatrix} 3 & 2 \\ 2 & 5 \end{pmatrix}$. Since the off-diagonal elements of $\boldsymbol{\Sigma}_{11}$ are nonzero, X_2 and X_4 are dependent.

Next, we verify that $\mathbf{x}_2 = (X_1, X_2)'$ and $\mathbf{x}_3 = X_3$ are independently distributed. Once again, from Result 5.2.11, the distribution of $\mathbf{x}_{(1)} = (X_1, X_2, X_3)'$ is a trivariate normal with mean and variance-covariance given by

$$\boldsymbol{\mu}_{(1)} = \begin{pmatrix} 2 \\ 3 \\ 0 \end{pmatrix} \text{ and } \boldsymbol{\Sigma}_{(1)} = \begin{pmatrix} 4 & 1 & 0 \\ 1 & 3 & 0 \\ 0 & 0 & 2 \end{pmatrix}.$$

In the notation of Result 5.2.10, this gives $\text{Cov}(\mathbf{x}_2, \mathbf{x}_3) = \begin{pmatrix} 0 \\ 0 \end{pmatrix}$, so that $\mathbf{x}_2$ and $\mathbf{x}_3$ are independent. □

Result 5.2.12. Conditional distribution. Let $\mathbf{x} \sim N_k(\boldsymbol{\mu}, \boldsymbol{\Sigma})$. Suppose we partition $\mathbf{x}$ as $(\mathbf{x}_1', \mathbf{x}_2')'$, where $\mathbf{x}_1$ is a q-dimensional subvector, $0 < q < k$, $\mathbf{x}_2$ is $(k-q)$-dimensional, and suppose that $\boldsymbol{\mu}$ and $\boldsymbol{\Sigma}$ are partitioned conformably. For any $\mathbf{c}_2$ in the support of $\mathbf{x}_2$, the conditional distribution of $\mathbf{x}_1$ given that $\mathbf{x}_2 = \mathbf{c}_2$ is multivariate normal with mean vector

$$\boldsymbol{\mu}_{1.2} = \boldsymbol{\mu}_1 + \boldsymbol{\Sigma}_{12}\boldsymbol{\Sigma}_{22}^{-}(\mathbf{c}_2 - \boldsymbol{\mu}_2) \tag{5.2.8}$$

and variance-covariance matrix equal to the Schur complement of $\boldsymbol{\Sigma}_{11}$, i.e.,

$$\boldsymbol{\Sigma}_{11.2} = \boldsymbol{\Sigma}_{11} - \boldsymbol{\Sigma}_{12}\boldsymbol{\Sigma}_{22}^{-}\boldsymbol{\Sigma}_{21}. \tag{5.2.9}$$

Both $\mu_{1.2}$ and $\boldsymbol{\Sigma}_{11.2}$ are invariant to the choice of $\boldsymbol{\Sigma}_{22}^{-}$.

Proof. We first give an analytic proof of this important result under the stronger assumption that $\boldsymbol{\Sigma}$ is p.d. and hence $\mathbf{x}$ has a p.d.f. $f(\mathbf{u})$; we use $\mathbf{u}$ to denote the argument of the p.d.f. to distinguish it from $\mathbf{x}$, which is a random variable. Corresponding to the partition $\mathbf{x} = (\mathbf{x}_1', \mathbf{x}_2')'$, we partition $\mathbf{u} = (\mathbf{u}_1', \mathbf{u}_2')'$ and write $f(\mathbf{u}_1, \mathbf{u}_2)$ for the p.d.f. of $\mathbf{x}$, and $f_i(\mathbf{u}_i)$ for the marginal p.d.f. of $\mathbf{x}_i$, $i = 1, 2$. The conditional p.d.f. of $\mathbf{x}_1$ given $\mathbf{x}_2 = \mathbf{c}_2$, is then

$$g(\mathbf{u}_1 \,|\, \mathbf{x}_2 = \mathbf{c}_2) = \frac{f(\mathbf{u}_1, \mathbf{c}_2)}{f_2(\mathbf{c}_2)} = \frac{(2\pi)^{-k/2}|\boldsymbol{\Sigma}|^{-1/2}\exp\{-Q/2\}}{(2\pi)^{-(k-q)/2}|\boldsymbol{\Sigma}_{22}|^{-1/2}\exp\{-Q_2/2\}}$$
$$= (2\pi)^{-q/2}|\boldsymbol{\Sigma}_{22}|^{1/2}|\boldsymbol{\Sigma}|^{-1/2}\exp\{-(Q - Q_2)/2\},$$

where, letting $\tilde{\mathbf{u}}_i = \mathbf{u}_i - \boldsymbol{\mu}_i$ and $\tilde{\mathbf{c}}_i = \mathbf{c}_i - \boldsymbol{\mu}_i$, $i = 1, 2$,

$$Q = (\tilde{\mathbf{u}}_1' \;\tilde{\mathbf{c}}_2') \begin{pmatrix} \boldsymbol{\Sigma}_{11} & \boldsymbol{\Sigma}_{12} \\ \boldsymbol{\Sigma}_{21} & \boldsymbol{\Sigma}_{22} \end{pmatrix}^{-1} \begin{pmatrix} \tilde{\mathbf{u}}_1 \\ \tilde{\mathbf{c}}_2 \end{pmatrix}, \text{ and } Q_1 = \tilde{\mathbf{c}}_2'\boldsymbol{\Sigma}_{22}^{-1}\tilde{\mathbf{c}}_2.$$

From Result 2.1.4,

$$\begin{pmatrix} \boldsymbol{\Sigma}_{11} & \boldsymbol{\Sigma}_{12} \\ \boldsymbol{\Sigma}_{21} & \boldsymbol{\Sigma}_{22} \end{pmatrix}^{-1} = \begin{pmatrix} \boldsymbol{\Sigma}_{11.2}^{-1} & -\boldsymbol{\Sigma}_{11.2}^{-1}\boldsymbol{\Sigma}_{12}\boldsymbol{\Sigma}_{22}^{-1} \\ -\boldsymbol{\Sigma}_{22}^{-1}\boldsymbol{\Sigma}_{21}\boldsymbol{\Sigma}_{11.2}^{-1} & \boldsymbol{\Sigma}_{22.1}^{-1} \end{pmatrix} \tag{5.2.10}$$

or

$$\begin{pmatrix} \boldsymbol{\Sigma}_{11} & \boldsymbol{\Sigma}_{12} \\ \boldsymbol{\Sigma}_{21} & \boldsymbol{\Sigma}_{22} \end{pmatrix}^{-1} = \begin{pmatrix} \boldsymbol{\Sigma}_{11.2}^{-1} & -\boldsymbol{\Sigma}_{11}^{-1}\boldsymbol{\Sigma}_{12}\boldsymbol{\Sigma}_{22.1}^{-1} \\ -\boldsymbol{\Sigma}_{22.1}^{-1}\boldsymbol{\Sigma}_{21}\boldsymbol{\Sigma}_{11}^{-1} & \boldsymbol{\Sigma}_{22.1}^{-1} \end{pmatrix}, \tag{5.2.11}$$

where $\boldsymbol{\Sigma}_{11.2}$ is shown in (5.2.9) and $\boldsymbol{\Sigma}_{22.1} = \boldsymbol{\Sigma}_{22} - \boldsymbol{\Sigma}_{21}\boldsymbol{\Sigma}_{11}^{-1}\boldsymbol{\Sigma}_{12}$ is the Schur compliment of $\boldsymbol{\Sigma}_{22}$. Plugging (5.2.10) into the expression for Q, we get

$$Q - Q_2 = \tilde{\mathbf{u}}_1'\boldsymbol{\Sigma}_{11.2}^{-1}(\tilde{\mathbf{u}}_1 - \boldsymbol{\Sigma}_{12}\boldsymbol{\Sigma}_{22}^{-1}\tilde{\mathbf{c}}_2) - \tilde{\mathbf{c}}_2'\boldsymbol{\Sigma}_{22}^{-1}\boldsymbol{\Sigma}_{21}\boldsymbol{\Sigma}_{11.2}^{-1}\tilde{\mathbf{u}}_1$$
$$+ \tilde{\mathbf{c}}_2'(\boldsymbol{\Sigma}_{22.1}^{-1} - \boldsymbol{\Sigma}_{22}^{-1})\tilde{\mathbf{c}}_2.$$

From the expression for $\boldsymbol{\Sigma}_{22.1}$,

$$\boldsymbol{\Sigma}_{22.1}^{-1} - \boldsymbol{\Sigma}_{22}^{-1} = \boldsymbol{\Sigma}_{22.1}^{-1}(\boldsymbol{\Sigma}_{22} - \boldsymbol{\Sigma}_{22.1})\boldsymbol{\Sigma}_{22}^{-1} = \boldsymbol{\Sigma}_{22.1}^{-1}\boldsymbol{\Sigma}_{21}\boldsymbol{\Sigma}_{11}^{-1}\boldsymbol{\Sigma}_{12}\boldsymbol{\Sigma}_{22}^{-1}.$$

By comparing (5.2.10) and (5.2.11), $\boldsymbol{\Sigma}_{22.1}^{-1}\boldsymbol{\Sigma}_{21}\boldsymbol{\Sigma}_{11}^{-1} = \boldsymbol{\Sigma}_{22}^{-1}\boldsymbol{\Sigma}_{21}\boldsymbol{\Sigma}_{11.2}^{-1}$. Then

$$Q - Q_2 = \tilde{\mathbf{u}}_1'\boldsymbol{\Sigma}_{11.2}^{-1}(\tilde{\mathbf{u}}_1 - \boldsymbol{\Sigma}_{12}\boldsymbol{\Sigma}_{22}^{-1}\tilde{\mathbf{c}}_2) - \tilde{\mathbf{c}}_2'\boldsymbol{\Sigma}_{22}^{-1}\boldsymbol{\Sigma}_{21}\boldsymbol{\Sigma}_{11.2}^{-1}\tilde{\mathbf{u}}_1$$
$$+ \tilde{\mathbf{c}}_2'\boldsymbol{\Sigma}_{22}^{-1}\boldsymbol{\Sigma}_{21}\boldsymbol{\Sigma}_{11.2}^{-1}\boldsymbol{\Sigma}_{12}\boldsymbol{\Sigma}_{22}^{-1}\tilde{\mathbf{c}}_2$$
$$- (\tilde{\mathbf{u}}_1 - \boldsymbol{\Sigma}_{12}\boldsymbol{\Sigma}_{22}^{-1}\tilde{\mathbf{c}}_2)'\boldsymbol{\Sigma}_{11.2}^{-1}(\tilde{\mathbf{u}}_1 - \boldsymbol{\Sigma}_{12}\boldsymbol{\Sigma}_{22}^{-1}\tilde{\mathbf{c}}_2).$$

From Result 2.1.3, $|\boldsymbol{\Sigma}| = |\boldsymbol{\Sigma}_{22}|\,|\boldsymbol{\Sigma}_{11.2}|$. Since $\tilde{\mathbf{u}}_1 - \boldsymbol{\Sigma}_{12}\boldsymbol{\Sigma}_{22}^{-1}\tilde{\mathbf{c}}_2 = \mathbf{u}_1 - \boldsymbol{\mu}_{1.2}$,

$$g(\mathbf{u}_1 \,|\, \mathbf{x}_2 = \mathbf{c}_2) = (2\pi)^{-q/2}|\boldsymbol{\Sigma}_{11.2}|^{-1/2}\exp\left\{-(\mathbf{u}_1 - \boldsymbol{\mu}_{1.2})'\boldsymbol{\Sigma}_{11.2}^{-1}(\mathbf{u}_1 - \boldsymbol{\mu}_{1.2})/2\right\}.$$

By (5.2.5), this is a normal p.d.f. with mean $\boldsymbol{\mu}_{1.2}$ and covariance $\boldsymbol{\Sigma}_{11.2}$, which completes the proof.

We now give a proof that strictly follows the condition of the result, i.e., $\boldsymbol{\Sigma}$ is not assumed to be p.d. First, we verify the invariance of the right sides of (5.2.8) and (5.2.9) to the choice of g-inverse of $\boldsymbol{\Sigma}_{22}$. From Result 5.2.7, the support of $\mathbf{x}_2$ is $\boldsymbol{\mu}_2 + \mathcal{C}(\boldsymbol{\Sigma}_{22}) = \boldsymbol{\mu}_2 + \mathcal{R}(\boldsymbol{\Sigma}_{22})$. Then, the rows of $(\mathbf{x}_1 - \boldsymbol{\mu}_1)(\mathbf{x}_2 - \boldsymbol{\mu}_2)'$ are in $\mathcal{R}(\boldsymbol{\Sigma}_{22})$ with probability one. It follows that the rows of $\boldsymbol{\Sigma}_{12} = \mathrm{E}[(\mathbf{x}_1 - \boldsymbol{\mu}_1)(\mathbf{x}_2 - \boldsymbol{\mu}_2)']$ are in $\mathcal{R}(\boldsymbol{\Sigma}_{22})$, so that

$$\boldsymbol{\Sigma}_{12} = \mathbf{L}\boldsymbol{\Sigma}_{22} \text{ for some matrix } \mathbf{L}. \tag{5.2.12}$$

Also, given $\mathbf{c}_2$ in the support of $\mathbf{x}_2$, $\mathbf{c}_2 - \boldsymbol{\mu}_2 = \boldsymbol{\Sigma}_{22}\mathbf{d}$ for some vector $\mathbf{d}$. As a result,

$\mathbf{\Sigma}_{12}\mathbf{\Sigma}_{22}^{-}(\mathbf{c}_2-\boldsymbol{\mu}_2)=\mathbf{L}\mathbf{\Sigma}_{22}\mathbf{\Sigma}_{22}^{-}\mathbf{\Sigma}_{22}\mathbf{d}=\mathbf{L}\mathbf{\Sigma}_{22}\mathbf{d}$ and $\mathbf{\Sigma}_{12}\mathbf{\Sigma}_{22}^{-}\mathbf{\Sigma}_{21}=\mathbf{L}\mathbf{\Sigma}_{22}\mathbf{\Sigma}_{22}^{-}\mathbf{\Sigma}_{22}\mathbf{L}'=\mathbf{L}\mathbf{\Sigma}_{22}\mathbf{L}'$, showing the required invariance.

Let $\mathbf{G}$ be any fixed g-inverse of $\mathbf{\Sigma}_{22}$ and write

$$\mathbf{x}_1-\boldsymbol{\mu}_1=\mathbf{\Sigma}_{12}\mathbf{G}(\mathbf{x}_2-\boldsymbol{\mu}_2)+\mathbf{z}, \tag{5.2.13}$$

where $\mathbf{z}=(\mathbf{x}_1-\boldsymbol{\mu}_1)-\mathbf{\Sigma}_{12}\mathbf{G}(\mathbf{x}_2-\boldsymbol{\mu}_2)$. By (5.2.12),

$$\text{Cov}(\mathbf{z},\mathbf{x}_2)=\mathbf{\Sigma}_{12}-\mathbf{\Sigma}_{12}\mathbf{G}\mathbf{\Sigma}_{22}=\mathbf{\Sigma}_{12}-\mathbf{L}\mathbf{\Sigma}_{22}\mathbf{G}\mathbf{\Sigma}_{22}=\mathbf{\Sigma}_{12}-\mathbf{L}\mathbf{\Sigma}_{22}=\mathbf{O},$$

so from Result 5.2.10, $\mathbf{z}$ and $\mathbf{x}_2$ are independent. Hence, given $\mathbf{x}_2=\mathbf{c}_2$, the conditional distribution of $\mathbf{x}_1$ is the same as the distribution of $\boldsymbol{\mu}_1+\mathbf{\Sigma}_{12}\mathbf{G}(\mathbf{c}_2-\boldsymbol{\mu}_2)+\mathbf{z}=\boldsymbol{\mu}_{1.2}+\mathbf{z}$. We see that $\text{E}(\mathbf{z})=\mathbf{0}$ while $\text{Cov}(\mathbf{z})=\text{Cov}(\mathbf{x}_1-\mathbf{\Sigma}_{12}\mathbf{G}\mathbf{x}_2)=\mathbf{\Sigma}_{11}-2\mathbf{\Sigma}_{12}\mathbf{G}\mathbf{\Sigma}_{21}+\mathbf{\Sigma}_{12}\mathbf{G}\mathbf{\Sigma}_{22}\mathbf{G}'\mathbf{\Sigma}_{21}$. Now $\mathbf{\Sigma}_{12}\mathbf{G}\mathbf{\Sigma}_{22}\mathbf{G}'\mathbf{\Sigma}_{21}=\mathbf{L}\mathbf{\Sigma}_{22}\mathbf{G}\mathbf{\Sigma}_{22}\mathbf{G}'\mathbf{\Sigma}_{22}\mathbf{L}'=\mathbf{L}\mathbf{\Sigma}_{22}\mathbf{G}\mathbf{\Sigma}_{22}\mathbf{L}'=\mathbf{\Sigma}_{12}\mathbf{G}\mathbf{\Sigma}_{21}$, so that $\text{Cov}(\mathbf{z})=\mathbf{\Sigma}_{11.2}$. This completes the proof. ■

Example 5.2.10. Let $\mathbf{x}=(X_1,X_2,X_3)'\sim N_3(\mathbf{0},\mathbf{\Sigma})$, where

$$\mathbf{\Sigma}=\begin{pmatrix}4&1&0\\1&2&1\\0&1&3\end{pmatrix}.$$

We illustrate the computation of marginal and conditional distributions as well as distributions of a linear combination of the components of $\mathbf{x}$. By Result 5.2.11, we see that $X_1\sim N(0,4)$, $X_2\sim N(0,2)$, and $X_3\sim N(0,3)$. The marginal distribution of the vector $(X_2,X_3)'$ is bivariate normal with mean vector and covariance matrix given respectively by

$$\begin{pmatrix}0\\0\end{pmatrix}\text{ and }\begin{pmatrix}2&1\\1&3\end{pmatrix}.$$

From Result 5.2.12, the conditional distribution of X_1 given $(X_2,X_3)=(x_2,x_3)$ is univariate normal with mean and variance given respectively by

$$\text{E}(X_1\,|\,X_2=x_2,X_3=x_3)=(3x_2-x_3)/5,\ \text{and}$$
$$\text{Var}(X_1\,|\,X_2=x_2,X_3=x_3)=17/5.$$

To find the distribution of a linear function of $\mathbf{x}$, viz., $Y=4X_1-6X_2+X_3-18$, we use Result 5.2.5 with $q=1$, $\mathbf{B}=(4,-6,1)'$, and $b=-18$. Then, $Y\sim N(-18,79)$. □

The conditional variance-covariance matrix $\mathbf{\Sigma}_{11.2}$ has the following useful property. This result is a simple consequence of Result 2.1.4 and its proof is left as Exercise 5.12.

Result 5.2.13. Suppose $\mathbf{\Sigma}=\{\mathbf{\Sigma}_{ij}\}_{i,j=1,2}$ is nonsingular and $\mathbf{\Sigma}^{-1}$ is partitioned conformably as $\{\mathbf{\Delta}_{ij}\}$. Then $\mathbf{\Sigma}_{11.2}=\mathbf{\Delta}_{11}^{-1}$. Furthermore, the components of $\mathbf{x}_1$ are conditionally mutually independent given $\mathbf{x}_2=\mathbf{c}_2$ if and only if $\mathbf{\Delta}_{11}^{-1}$ is a diagonal matrix.

We give a characterization of $\boldsymbol{\mu}_{1.2}$ analogous to the b.l.u.e. in linear regression. Recall that for symmetric matrices $\mathbf{Q}$ and $\mathbf{S}$, we denote $\mathbf{Q}\leq\mathbf{S}$ if $\mathbf{S}-\mathbf{Q}$ is n.n.d. If $\mathbf{x}_1$ and $\mathbf{x}_2$ are jointly distributed, then a linear function of $\mathbf{x}_2$ of the form $\mathbf{A}\mathbf{x}_2+\mathbf{a}$ is said to be an unbiased predictor of $\mathbf{x}_1$ if $\text{E}(\mathbf{x}_1)=\text{E}(\mathbf{A}\mathbf{x}_2+\mathbf{a})$.

Result 5.2.14. Best linear unbiased predictor. Let $\mathbf{x}=(\mathbf{x}_1',\mathbf{x}_2')'$ have mean $\boldsymbol{\mu}=(\boldsymbol{\mu}_1',\boldsymbol{\mu}_2')'$ and variance-covariance matrix $\mathbf{\Sigma}=\{\mathbf{\Sigma}_{ij}\}$ with $\boldsymbol{\mu}_i=\text{E}(\mathbf{x}_i)$ and $\mathbf{\Sigma}_{ij}=\text{Cov}(\mathbf{x}_i,\mathbf{x}_j)$. Let $\boldsymbol{\mu}_{1.2}(\mathbf{x}_2)=\boldsymbol{\mu}_1+\mathbf{\Sigma}_{12}\mathbf{\Sigma}_{22}^{-}(\mathbf{x}_2-\boldsymbol{\mu}_2)$. Then $\boldsymbol{\mu}_{1.2}(\mathbf{x}_2)$ is an unbiased predictor of $\mathbf{x}_1$, and when $\mathbf{x}_2 = \mathbf{c}_2$, we say that $\boldsymbol{\mu}_{1.2}(\mathbf{c}_2)$ is a realized value of the unbiased predictor of $\mathbf{x}_1$. The prediction error $\boldsymbol{\zeta} = \mathbf{x}_1 - \boldsymbol{\mu}_{1.2}(\mathbf{x}_2)$ is uncorrelated with $\mathbf{x}_2$ such that

$$\text{Cov}(\mathbf{x}_1) = \text{Cov}(\boldsymbol{\mu}_{1.2}(\mathbf{x}_2)) + \text{Cov}(\boldsymbol{\zeta}) \text{ and}$$
$$\text{Cov}(\boldsymbol{\zeta}) = \boldsymbol{\Sigma}_{11.2} \leq \text{Cov}(\mathbf{x}_1 - \mathbf{f}(\mathbf{x}_2))$$

for any unbiased linear predictor $\mathbf{f}(\mathbf{x}_2) = \mathbf{A}\mathbf{x}_2 + \mathbf{a}$, with equality holding if and only if $\mathbf{f}(\mathbf{x}_2) = \boldsymbol{\mu}_{1.2}(\mathbf{x}_2)$. Hence $\boldsymbol{\mu}_{1.2}(\mathbf{x}_2)$ is known as the best linear predictor (b.l.u.p.) of $\mathbf{x}_1$. Moreover, for normally distributed $\mathbf{x}$, $\boldsymbol{\mu}_{1.2}(\mathbf{x}_2)$ has the smallest variance-covariance matrix among *all* unbiased (linear or nonlinear) predictors $\mathbf{f}(\mathbf{x}_2)$ of $\mathbf{x}_1$.

Proof. It is straightforward to check that $\boldsymbol{\mu}_{1.2}(\mathbf{x}_2)$ is an unbiased predictor of $\mathbf{x}_1$. By exactly the same argument following (5.2.13), $\text{Cov}(\mathbf{x}_2, \boldsymbol{\zeta}) = \mathbf{O}$ and $\text{Cov}(\boldsymbol{\zeta}) = \text{Cov}(\mathbf{x}_1 - \boldsymbol{\mu}_{1.2}(\mathbf{x}_2)) = \boldsymbol{\Sigma}_{11.2}$. Let $\mathbf{f}(\mathbf{x}_2)$ be an unbiased linear predictor of $\mathbf{x}_1$. Then $\text{Var}(\mathbf{x}_1 - \mathbf{f}(\mathbf{x}_2)) = \text{Var}((\boldsymbol{\mu}_{1.2}(\mathbf{x}_2) - \mathbf{f}(\mathbf{x}_2)) + \boldsymbol{\zeta})$. Since $\boldsymbol{\mu}_{1.2}(\mathbf{x}_2) - \mathbf{f}(\mathbf{x}_2)$ is a linear function of $\mathbf{x}_2$, it is uncorrelated with $\boldsymbol{\zeta}$. Then $\text{Var}(\mathbf{x}_1 - \mathbf{f}(\mathbf{x}_2)) = \text{Var}(\boldsymbol{\mu}_{1.2}(\mathbf{x}_2) - \mathbf{f}(\mathbf{x}_2)) + \text{Var}(\boldsymbol{\zeta}) \geq \text{Var}(\boldsymbol{\zeta})$ with equality holding if and only if $\text{Var}(\boldsymbol{\mu}_{1.2}(\mathbf{x}_2) - \mathbf{f}(\mathbf{x}_2)) = \mathbf{O}$. Since $\boldsymbol{\mu}_{1.2}(\mathbf{x}_2) - \mathbf{f}(\mathbf{x}_2)$ has mean $\mathbf{0}$, $\text{Var}(\boldsymbol{\mu}_{1.2}(\mathbf{x}_2) - \mathbf{f}(\mathbf{x}_2)) = \mathbf{O}$ if and only if $\boldsymbol{\mu}_{1.2}(\mathbf{x}_2) = \mathbf{f}(\mathbf{x}_2)$. This completes the proof that $\boldsymbol{\mu}_{1.2}(\mathbf{x}_2)$ is the unique b.l.u.p. of $\mathbf{x}_1$.

Finally, if $\mathbf{x} = (\mathbf{x}_1', \mathbf{x}_2')'$ is normally distributed, then for any unbiased estimator $\mathbf{f}(\mathbf{x}_2)$ with finite variance, $\boldsymbol{\varepsilon}$ is independent of $\mathbf{f}(\mathbf{x}_2)$, so the two are uncorrelated. The same argument as above then shows that $\text{Var}(\mathbf{x}_1 - \boldsymbol{\mu}_{1.2}(\mathbf{x}_2)) \leq \text{Var}(\mathbf{x}_1 - \mathbf{f}(\mathbf{x}_2))$, with equality holding if and only if $\mathbf{f}(\mathbf{x}_2) = \boldsymbol{\mu}_{1.2}(\mathbf{x}_2)$. ■

Suppose $\mathbf{x}_1 = (X_1, \cdots, X_q)'$ and $\mathbf{x}_2 = (X_{q+1}, \cdots, X_k)'$. For each $1 \leq i \leq q$, let $\boldsymbol{\sigma}_{i2} = \text{Cov}(X_i, \mathbf{x}_2)$. Then $\mu_{i.2}(\mathbf{x}_2) = \text{E}(X_i) + \boldsymbol{\sigma}_{i2}\boldsymbol{\Sigma}_{22}^{-}(\mathbf{x}_2 - \boldsymbol{\mu}_2)$ is the b.l.u.p. of X_i based on $\mathbf{x}_2$. Since $\boldsymbol{\sigma}_{12}, \ldots, \boldsymbol{\sigma}_{q2}$ form the rows of $\boldsymbol{\Sigma}_{12}$, the b.l.u.p. of $\mathbf{x}_1$ is the vector of b.l.u.p.s of the X_i's.

In many cases, in order to determine whether a linear function of $\mathbf{x}_2$ is the b.l.u.p. of $\mathbf{x}_1$, it is unnecessary to check if it has the form of the b.l.u.p. in Result 5.2.14. Instead, the following criterion can be used.

Result 5.2.15. Suppose $\mathbf{f}(\mathbf{x}_2) = \mathbf{A}\mathbf{x}_2 + \mathbf{a}$ has the same mean value as $\mathbf{x}_1$. Then $\mathbf{f}(\mathbf{x}_2)$ is the b.l.u.p. of $\mathbf{x}_1$ if and only if the error $\boldsymbol{\zeta} = \mathbf{x}_1 - \mathbf{f}(\mathbf{x}_2)$ is uncorrelated with $\mathbf{x}_2$, i.e., $\text{Cov}(\boldsymbol{\zeta}, \mathbf{x}_2) = \mathbf{O}$.

Proof. If $\mathbf{f}(\mathbf{x}_2)$ is the b.l.u.p. of $\mathbf{x}_1$, then the proof of Result 5.2.14 already shows that $\text{Cov}(\boldsymbol{\zeta}, \mathbf{x}_2) = \mathbf{O}$. Conversely, if $\text{Cov}(\boldsymbol{\zeta}, \mathbf{x}_2) = \mathbf{O}$, then for any unbiased linear predictor $\mathbf{g}(\mathbf{x}_2)$, $\text{Cov}(\boldsymbol{\zeta}, \mathbf{g}(\mathbf{x}_2)) = \mathbf{O}$. Similar to the proof of Result 5.2.14, $\text{Var}[\mathbf{x}_1 - \mathbf{g}(\mathbf{x}_2)] = \text{Var}[\mathbf{f}(\mathbf{x}_2) - \mathbf{g}(\mathbf{x}_2) + \boldsymbol{\zeta}] = \text{Var}[\mathbf{f}(\mathbf{x}_2) - \mathbf{g}(\mathbf{x}_2)] + \text{Var}(\boldsymbol{\zeta}) \geq \text{Var}(\boldsymbol{\zeta})$. Therefore, $\mathbf{f}(\mathbf{x}_2)$ is the b.l.u.p. of $\mathbf{x}_1$. ■

Definition 5.2.4. Simple correlation coefficient. Suppose the random vector $\mathbf{x} = (X_1, \cdots, X_k)'$ has a $N_k(\boldsymbol{\mu}, \boldsymbol{\Sigma})$ distribution. The simple correlation between any two random variables X_i and X_j is defined by

$$\rho_{ij} = \frac{\text{Cov}(X_i, X_j)}{[\text{Var}(X_i)\,\text{Var}(X_j)]^{1/2}} = \frac{\sigma_{ij}}{\sigma_i \sigma_j}, \tag{5.2.14}$$

provided $\sigma_{ii} = \sigma_i^2 > 0$ and $\sigma_{jj} = \sigma_j^2 > 0$. If either term in the denominator is zero, then ρ_{ij} is undefined.

For a collection of algebraic, geometric, and trigonometric interpretations of the simple correlation coefficient, see Lee Rodgers and Nicewander (1988). The connection between b.l.u.p. and the correlation coefficient in the bivariate normal case is particularly simple.

Result 5.2.16. Suppose $\sigma_j > 0$. Let $\tilde{X}_i$ denote the b.l.u.p. of X_i based on X_j. Then the error of the b.l.u.p. has variance

$$\mathrm{Var}(X_i - \tilde{X}_i) = \mathrm{Var}(X_i) - \mathrm{Var}(\tilde{X}_i) = (1 - \rho_{ij}^2)\sigma_i^2.$$

Furthermore, $(\tilde{X}_i - \mu_i)/\sigma_i = \rho_{ij}(X_j - \mu_j)/\sigma_j$. That is, the standardized b.l.u.p. of X_i is the standardized X_j multiplied by the correlation between X_i and X_j.

The proof of the result is left as Exercise 5.16.

Result 5.2.17. The simple correlation coefficient satisfies the inequality $-1 \leq \rho_{ij} \leq 1$, with $|\rho_{ij}| = 1$ if and only if there exist constants $a \neq 0$ and b such that $\mathrm{P}(X_j = aX_i + b) = 1$. If $\rho_{ij} = 1$, then $a > 0$, and if $\rho_{ij} = -1$, then $a < 0$.

Proof. For every real constant c, we have $\mathrm{E}[(X_j - \mu_j) - c(X_i - \mu_i)]^2 \geq 0$, which implies that the discriminant of the nonnegative quadratic function $\sigma_j^2 - 2c\sigma_{ij} + c^2\sigma_i^2$ is nonpositive, i.e., $4\sigma_{ij}^2 - 4\sigma_i^2\sigma_j^2 \leq 0$. This implies that $\sigma_{ij}^2/(\sigma_i^2\sigma_j^2) \leq 1$, or $\rho_{ij}^2 \leq 1$, from which it follows that $-1 \leq \rho_{ij} \leq 1$. Also, $|\rho_{ij}| = 1$ if and only if the discriminant is equal to zero, i.e., $\mathrm{E}[(X_j - \mu_j) - c(X_i - \mu_i)]^2 = 0$ for some (unique) c. But the latter is equivalent to $X_j - \mu_j = c(X_i - \mu_i)$ with probability one, i.e., $X_j = aX_i + b$ with probability one with $a = c$, and $b = \mu_j - c\mu_i$. The single root of the quadratic function is $c = \mathrm{Cov}(X_i, X_j)/\mathrm{Var}(X_i)$, so that if $\rho_{ij} = 1$, then $a > 0$, while if $\rho_{ij} = -1$, then $a < 0$. ■

Result 5.2.18. If X_i and X_j are independent random variables, then $\rho_{ij} = 0$.

Proof. If X_i and X_j are independent, $\mathrm{E}(X_iX_j) = \mathrm{E}(X_i)\,\mathrm{E}(X_j)$. Hence,

$$\mathrm{Cov}(X_i, X_j) = \mathrm{E}(X_iX_j) - \mathrm{E}(X_i)\,\mathrm{E}(X_j) = 0,$$

and using (5.2.14), it follows that $\rho_{ij} = 0$. ■

Note that if X_i and X_j are uncorrelated, they need not necessarily be independent. The covariance and correlation measure only a *linear* relationship between X_i and X_j. It is possible for two dependent random variables to have zero correlation, for example, X_1 and X_2 uniform on the square with vertices $(-1, 0)$, $(0, 1)$, $(1, 0)$, and $(0, -1)$. In Example 5.2.11, we show that for the bivariate normal distribution discussed in Example 5.2.3, $\rho = 0$ does indeed imply independence of X_1 and X_2.

Example 5.2.11.

(a) In the definition of the bivariate normal distribution for $\mathbf{x} = (X_1, X_2)'$ given in Example 5.2.3, let $\rho = 0$. Then, $\boldsymbol{\Sigma} = \mathrm{diag}(\sigma_1^2, \sigma_2^2)$, $|\boldsymbol{\Sigma}| = \sigma_1^2\sigma_2^2$, and $\boldsymbol{\Sigma}^{-1} = \mathrm{diag}(1/\sigma_1^2, 1/\sigma_2^2)$, so that (5.2.6) becomes

$$f(x_1, x_2) = (2\pi\sigma_1\sigma_2)^{-1}\exp\{-(x_1 - \mu_1)^2/2\sigma_1^2\}\exp\{-(x_2 - \mu_2)^2/2\sigma_2^2\}$$

for $-\infty < x_1, x_2 < \infty$, which is the product of two univariate normal p.d.f.s, indicating independence of X_1 and X_2.

(b) Let $\mathbf{x}$ have a k-variate normal distribution with mean vector $\boldsymbol{\mu}$ and covariance matrix $\mathbf{I}_k$. Consider $Y_1 = \mathbf{a}'\mathbf{x}$, $Y_2 = \mathbf{b}'\mathbf{x}$, with $\mathbf{a}'\mathbf{b} = 0$. By Result 5.2.5, $\mathbf{y} = (Y_1, Y_2)'$ has a bivariate normal distribution with mean vector $(\mathbf{a}'\boldsymbol{\mu}, \mathbf{b}'\boldsymbol{\mu})'$, and covariance matrix $\boldsymbol{\Sigma} = \mathrm{diag}(\mathbf{a}'\mathbf{a}, \mathbf{b}'\mathbf{b})$, since $\mathbf{a}'\mathbf{b} = 0$. The joint distribution of $\mathbf{y}$ factors into the marginal distributions of Y_1 and Y_2, so that Y_1 and Y_2 are independent. □

Result 5.2.19. For any two random variables X_i and X_j with respective means μ_i and μ_j, respective variances σ_i^2 and σ_j^2, and correlation ρ_{ij},

$$\text{Var}(aX_i + bX_j) = a^2\sigma_i^2 + b^2\sigma_j^2 + 2ab\rho_{ij}\sigma_i\sigma_j. \tag{5.2.15}$$

If X_i and X_j are independent, then

$$\text{Var}(aX_i + bX_j) = a^2\sigma_i^2 + b^2\sigma_j^2. \tag{5.2.16}$$

Proof. Let $\mathbf{x} = (X_i, X_j)'$ and $\mathbf{c} = (a, b)'$. Then by (A.12),

$$\begin{aligned}\text{Var}(aX_i + bX_j) &= \text{Var}(\mathbf{c}'\mathbf{x}) = \mathbf{c}'\,\text{Cov}(\mathbf{x})\mathbf{c} \\ &= (a,b)'\begin{pmatrix}\sigma_i^2 & \sigma_{ij} \\ \sigma_{ij} & \sigma_j^2\end{pmatrix}\begin{pmatrix}a \\ b\end{pmatrix} = a^2\sigma_i^2 + b^2\sigma_j^2 + 2ab\sigma_{ij},\end{aligned}$$

and we get (5.2.15) using the definition in (5.2.14); (5.2.16) follows directly from Result 5.2.18. ■

Definition 5.2.5. Let $\mathbf{x} = (X_1, \cdots, X_k)'$ be a random vector with componentwise means μ_i, $i = 1, \cdots, k$, and pairwise correlations ρ_{ij}, $i, j = 1, \cdots, k$. The $k \times k$ symmetric matrix $\mathbf{R} = \{\rho_{ij}\}$ is called the correlation matrix of $\mathbf{x}$. Each diagonal element of $\mathbf{R}$ is equal to 1, while each off-diagonal element lies between -1 and 1.

Result 5.2.20. The correlation matrix $\mathbf{R}$ is nonnegative definite.

Proof. We have seen that $\boldsymbol{\Sigma} = \text{Cov}(\mathbf{x})$ is n.n.d. Let $\mathbf{D} = \text{diag}(\sigma_1, \cdots, \sigma_k)$, where $\sigma_i = \sqrt{\text{Var}(X_i)}$. We can write $\mathbf{R} = \mathbf{D}^{-1}\boldsymbol{\Sigma}\mathbf{D}^{-1}$, from which it follows that $\mathbf{R}$ too is n.n.d. ■

Example 5.2.12. Let $\mathbf{x} = (X_1, X_2, X_3, X_4)' \sim N_4(\boldsymbol{\mu}, \boldsymbol{\Sigma})$ where

$$\boldsymbol{\mu} = \begin{pmatrix}1 \\ 2 \\ -1 \\ 3\end{pmatrix} \text{ and } \boldsymbol{\Sigma} = \begin{pmatrix}2 & 0 & 1 & -1 \\ 0 & 3 & 0 & 2 \\ 1 & 0 & 5 & 0 \\ -1 & 2 & 0 & 3\end{pmatrix}.$$

It is easy to verify that $\rho_{11} = \rho_{22} = \rho_{33} = \rho_{44} = 1$, $\rho_{12} = \rho_{23} = \rho_{34} = 0$, $\rho_{13} = 1/\sqrt{10}$, $\rho_{14} = -1/\sqrt{6}$, and $\rho_{24} = 2/3$. The correlation matrix is

$$\mathbf{R} = \begin{pmatrix}1 & 0 & 1/\sqrt{10} & -1/\sqrt{6} \\ 0 & 1 & 0 & 2/3 \\ 1/\sqrt{10} & 0 & 1 & 0 \\ -1/\sqrt{6} & 2/3 & 0 & 1\end{pmatrix}.$$ □

Definition 5.2.6. Fisher's z-transformation of ρ_{ij}. The z-transformation or the arctanh transformation of ρ_{ij} is

$$\delta_{ij} = \frac{1}{2}\ln[(1+\rho_{ij})/(1-\rho_{ij})] = \text{arctanh}(\rho_{ij}).$$

In Chapter 6, we describe sampling from a multivariate normal distribution and properties of various sample statistics that arise in this context. In particular, we will consider Fisher's z-transformation of the sample correlation coefficient and derive its distribution. This finds use in linear model theory for conducting inference on ρ_{ij}.

Definition 5.2.7. Partial correlation coefficient. Suppose the random vector $\mathbf{x} = (X_1, \cdots, X_k)'$ has a $N_k(\boldsymbol{\mu}, \boldsymbol{\Sigma})$ distribution, and suppose we partition it as $\mathbf{x} = (\mathbf{x}_1', \mathbf{x}_2')'$, where $\mathbf{x}_1 = (X_1, \cdots, X_q)'$ is a q-dimensional vector, and $\mathbf{x}_2 = (X_{q+1}, \cdots, X_k)'$ is a $(k-q)$-dimensional vector. We partition $\boldsymbol{\mu}$ and $\boldsymbol{\Sigma}$ conformably, as in Result 5.2.11. Let X_j and X_l be components of the subvector $\mathbf{x}_1$. The partial correlation coefficient of X_j and X_l given $\mathbf{x}_2 = \mathbf{c}_2$ is defined by

$$\rho_{jl|(q+1,\cdots,k)} = \frac{\sigma_{jl|(q+1,\cdots,k)}}{\{\sigma_{jj|(q+1,\cdots,k)}\sigma_{ll|(q+1,\cdots,k)}\}^{1/2}}, \tag{5.2.17}$$

provided the expressions in the denominator are nonzero, and where $\sigma_{jl|(q+1,\cdots,k)}$ is the (j,l)th element in $\boldsymbol{\Sigma}_{11.2} = \boldsymbol{\Sigma}_{11} - \boldsymbol{\Sigma}_{12}\boldsymbol{\Sigma}_{22}^{-}\boldsymbol{\Sigma}_{21}$.

Result 5.2.21. A partial correlation coefficient defined by (5.2.17) satisfies

$$-1 \le \rho_{ij|(q+1,\cdots,k)} \le 1. \tag{5.2.18}$$

Let ζ_i be the error of the b.l.u.p. of X_i based in $\mathbf{x}_2$, i.e., $\zeta_i = X_i - \mu_i - \boldsymbol{\sigma}_{i2}\boldsymbol{\Sigma}_{22}^{-}(\mathbf{x}_2 - \mathbf{c}_2)$, where $\boldsymbol{\sigma}_{i2} = \text{Cov}(X_i, \mathbf{x}_2)$. Then $\rho_{jl|(q+1,\cdots,k)} = \text{Cov}(\zeta_j, \zeta_l)$.

Proof. Since $\boldsymbol{\Sigma}_{11.2} = \text{Cov}(\boldsymbol{\zeta})$, where $\boldsymbol{\zeta} = (\zeta_1, \cdots, \zeta_q)'$, then the result follows directly from Result 5.2.17. ■

Example 5.2.13. Continuing with Example 5.2.12, we compute the partial correlations between components of $\mathbf{x}$. To compute $\rho_{12|(3,4)}$, we first compute the conditional variance-covariance matrix of (X_1, X_2) given (X_3, X_4) as

$$\begin{pmatrix} 2 & 0 \\ 0 & 3 \end{pmatrix} - \begin{pmatrix} 1 & -1 \\ 0 & 2 \end{pmatrix} \begin{pmatrix} 5 & 0 \\ 0 & 3 \end{pmatrix}^{-1} \begin{pmatrix} 1 & 0 \\ -1 & 2 \end{pmatrix} = \begin{pmatrix} 22/15 & 2/3 \\ 2/3 & 5/3 \end{pmatrix}.$$

Hence, $\rho_{12|(3,4)} = \frac{2/3}{\sqrt{22/15}\sqrt{5/3}} = \sqrt{2/11}$. The other partial correlations may be computed similarly. □

Definition 5.2.8. Multiple correlation coefficient. Suppose the random vector $(X_0, X_1, \cdots, X_k)'$ has a $N_{k+1}(\boldsymbol{\mu}, \boldsymbol{\Sigma})$ distribution, and suppose

$$\mathbf{x} = \begin{pmatrix} X_0 \\ \mathbf{x}^{(1)} \end{pmatrix}, \ \boldsymbol{\mu} = \begin{pmatrix} \mu_0 \\ \boldsymbol{\mu}^{(1)} \end{pmatrix}, \ \text{and } \boldsymbol{\Sigma} = \begin{pmatrix} \sigma_{00} & \boldsymbol{\sigma}_{01} \\ \boldsymbol{\sigma}_{10} & \boldsymbol{\Sigma}^{(1)} \end{pmatrix},$$

where $\mathbf{x}^{(1)} = (X_1, \cdots, X_k)'$, $\boldsymbol{\mu}^{(1)} = (\mu_1, \cdots, \mu_k)'$, $\boldsymbol{\Sigma}^{(1)}$ is the lower $k \times k$ submatrix of $\boldsymbol{\Sigma}$, $\boldsymbol{\sigma}_{10} = (\sigma_{10}, \cdots \sigma_{k0})'$ is a k-dimensional vector consisting of covariances between X_0 and $X_1, \cdots, X_k$, and $\boldsymbol{\sigma}_{01} = \boldsymbol{\sigma}_{10}'$. The multiple correlation coefficient of X_0 and $\mathbf{x}^{(1)}$ is

$$\rho_{0(1,\cdots,k)} = \frac{\text{Cov}(W, X_0)}{\{\text{Var}(W)\,\text{Var}(X_0)\}^{1/2}}, \tag{5.2.19}$$

provided the denominator is nonzero, where $W = \mu_0 + \boldsymbol{\sigma}_{01}[\boldsymbol{\Sigma}^{(1)}]^{-}(\mathbf{x}^{(1)} - \boldsymbol{\mu}^{(1)})$ is the b.l.u.p. of X_0 based on $\mathbf{x}^{(1)}$. That is, the multiple correlation coefficient of X_0 and $\mathbf{x}^{(1)}$ is the correlation coefficient of X_0 and the b.l.u.p. of X_0 based on $\mathbf{x}^{(1)}$.

Example 5.2.14. We continue with Example 5.2.12, and compute $\rho_{1(2,3)}$, the multiple correlation coefficient of X_1 and (X_2, X_3) using (5.2.20). First, we find the marginal distribution of (X_1, X_2, X_3). From Result 5.2.11, this distribution is normal with mean and variance given by

$$\begin{pmatrix} 1 \\ 2 \\ -1 \end{pmatrix} \text{ and } \begin{pmatrix} 2 & 0 & 1 \\ 0 & 3 & 0 \\ 1 & 0 & 5 \end{pmatrix}.$$

Further, $\sigma_{11} = \text{Var}(X_1) = 2$, $\boldsymbol{\sigma}_{12} = (\sigma_{12}, \sigma_{13})' = (0, 1)$, and $\boldsymbol{\Sigma}^{(1)} = \begin{pmatrix} 3 & 0 \\ 0 & 5 \end{pmatrix}$. Therefore, $\rho_{1(2,3)} = 0$. Next, using (5.2.20), we compute $\rho_{1(2,3,4)} = \sqrt{3/5}$. □

Result 5.2.22. Suppose $(X_0, X_1, \cdots, X_k)'$ has a $N_{k+1}(\boldsymbol{\mu}, \boldsymbol{\Sigma})$ distribution, and suppose it is partitioned as in Definition 5.2.8. Let $\sigma^2 = \sigma_{00} - \boldsymbol{\sigma}_{01}[\boldsymbol{\Sigma}^{(1)}]^{-}\boldsymbol{\sigma}_{10}$ denote the conditional variance $\text{Var}(X_0 \mid X_1 = x_1, \cdots, X_k = x_k)$. Then, the multiple correlation coefficient can be computed as

$$\rho_{0(1,\cdots,k)} = \left\{ \frac{\sigma_{00} - \sigma^2}{\sigma_{00}} \right\}^{1/2} = \left\{ \frac{\boldsymbol{\sigma}_{01}[\boldsymbol{\Sigma}^{(1)}]^{-}\boldsymbol{\sigma}_{10}}{\sigma_{00}} \right\}^{1/2}. \tag{5.2.20}$$

Proof. From Result 5.2.14, $X_0 = W + \zeta$ with $\text{Var}(\zeta) = \sigma_{00} - \boldsymbol{\sigma}_{01}[\boldsymbol{\Sigma}^{(1)}]^{-}\boldsymbol{\sigma}_{10} = \sigma^2$ and $\text{Cov}(W, \zeta) = 0$. Then

$$\begin{aligned} \text{Var}(W) &= \text{Var}(X_0) - \text{Var}(\zeta) = \sigma_{00} - \sigma^2, \\ \text{Cov}(X_0, W) &= \text{Cov}(W + \zeta, W) = \text{Var}(W). \end{aligned}$$

The proof follows from substituting into (5.2.19) and simplifying. ■

Result 5.2.23. The multiple correlation coefficient in Definition 5.2.8 satisfies $0 \le \rho_{0(1,\cdots,k)} \le 1$.

Proof. The expression $\text{Cov}(X_0, W)$ in the proof of Result 5.2.22 is nonnegative, from which it follows that $\rho_{0(1,2,\cdots,k)} \ge 0$. That $\rho_{0(1,2,\cdots,k)} \le 1$ follows from observing that $\sigma^2 \le \sigma_{00}$ (since $\sigma_{00} - \sigma^2 = \boldsymbol{\sigma}_{01}[\boldsymbol{\Sigma}^{(1)}]^{-1}\boldsymbol{\sigma}_{10} \ge 0$, being equal to $\text{Var}(W)$). ■

Result 5.2.24. Let $\mathbf{y} = (X_0, X_1, \cdots, X_k)' \sim N_{k+1}(\boldsymbol{\mu}_y, \boldsymbol{\Sigma}_y)$. The following relationship between the multiple and partial correlation coefficients holds for $q = 2, 3, \cdots, k$:

$$\rho^2_{0(1,2,\cdots,q)} = \rho^2_{0(1,2,\cdots,q-1)} + [1 - \rho^2_{0(1,2,\cdots,q-1)}]\rho^2_{0q|(1,2,\cdots,q-1)}. \tag{5.2.21}$$

Proof. First, we give a proof assuming $\boldsymbol{\Sigma}_y$ is p.d. Given q, write $\mathbf{y}^* = (X_0, X_q, X_1, \cdots, X_{q-1})'$. Then by Result 5.2.11, $\mathbf{y}^*$ has a $(q+1)$-variate normal distribution with mean and covariance given by

$$\boldsymbol{\mu}^* = (\mu_0, \mu_q, \mu_1, \cdots, \mu_{q-1})' \text{ and } \boldsymbol{\Sigma}^* = \begin{pmatrix} \sigma_{00} & \sigma_{0q} & \boldsymbol{\sigma}_0' \\ \sigma_{q0} & \sigma_{qq} & \boldsymbol{\sigma}_q' \\ \boldsymbol{\sigma}_0 & \boldsymbol{\sigma}_q & \boldsymbol{\Sigma} \end{pmatrix},$$

where $\boldsymbol{\sigma}_0 = (\sigma_{10}, \cdots \sigma_{q-1,0})'$, $\boldsymbol{\sigma}_q = (\sigma_{1q}, \cdots \sigma_{q-1,q})'$, and $\boldsymbol{\Sigma}$ is the variance-covariance matrix of $(X_1, \cdots, X_{q-1})'$. Partition $\boldsymbol{\Sigma}^{*-1}$ in the same way to get

$$\boldsymbol{\Sigma}^{*-1} = \begin{pmatrix} \delta_{00} & \delta_{0q} & \boldsymbol{\delta}_0' \\ \delta_{q0} & \delta_{qq} & \boldsymbol{\delta}_q' \\ \boldsymbol{\delta}_0 & \boldsymbol{\delta}_q & \boldsymbol{\Delta} \end{pmatrix}.$$

We consider the conditional distribution of $(X_0, X_q)'$ given $X_1 = x_1, \cdots, X_{q-1} = x_{q-1}$; by Result 5.2.13, this distribution is bivariate normal with covariance

$$\begin{pmatrix} \sigma_{00|(1,2,\cdots,q-1)} & \sigma_{0q|(1,2,\cdots,q-1)} \\ \sigma_{q0|(1,2,\cdots,q-1)} & \sigma_{qq|(1,2,\cdots,q-1)} \end{pmatrix}$$
$$= \begin{pmatrix} \delta_{00} & \delta_{0q} \\ \delta_{q0} & \delta_{qq} \end{pmatrix}^{-1} = \frac{1}{(\delta_{00}\delta_{qq} - \delta_{0q}^2)} \begin{pmatrix} \delta_{qq} & -\delta_{q0} \\ -\delta_{0q} & \delta_{00} \end{pmatrix}.$$

By Definition 5.2.7,

$$\rho^2_{0q|(1,2,\cdots,q-1)} = \sigma^2_{0q|(1,2,\cdots,q-1)}/\{\sigma_{00|(1,2,\cdots,q-1)}\sigma_{qq|(1,2,\cdots,q-1)}\}$$
$$= (-\delta_{0q})^2/(\delta_{00}\delta_{qq}).$$

Meanwhile, from Result 5.2.22,

$$\rho^2_{0(1,2,\cdots,q-1)} = [\sigma_{00} - \sigma_{00|(1,\cdots,q-1)}]/\sigma_{00}$$
$$= [\sigma_{00} - \delta_{qq}(\delta_{00}\delta_{qq} - \delta_{0q}^2)^{-1}]/\sigma_{00}.$$

On the other hand, from Result 5.2.13, δ_{00}^{-1} is the conditional variance of X_0 given $X_q, X_1, \cdots, X_{q-1}$. Then from Result 5.2.22, we see that

$$\rho^2_{0(1,2,\cdots,q)} = (\sigma_{00} - \delta_{00}^{-1})/\sigma_{00},$$

Hence,

$$[1 - \rho^2_{0(1,2,\cdots,q-1)}][1 - \rho^2_{0q|(1,2,\cdots,q-1)}]$$
$$= [\delta_{qq}(\delta_{00}\delta_{qq} - \delta_{0q}^2)^{-1}\sigma_{00}^{-1}]\left(1 - \frac{\delta_{0q}^2}{\delta_{00}\delta_{qq}}\right) \tag{5.2.22}$$
$$= \frac{1}{\sigma_{00}\delta_{00}} = 1 - \rho^2_{0(1,2,\cdots,q)}.$$

After simplification of (5.2.22), we get the required result.

We next give another proof based on the perspective of linear prediction. Let $\mathbf{x}_2 = (X_1, \cdots, X_{q-1})'$. Let $f_0(\mathbf{x}_2)$ be the b.l.u.p. of X_0 and $\zeta_0 = X_0 - f_0(\mathbf{x}_2)$. How does the addition of X_q improve the prediction of X_0? Intuitively, the improvement should come from the part of X_q uncorrelated with $\mathbf{x}_2$ that can be used to reduce ζ_0. Thus, let $\zeta_q = X_q - f_q(\mathbf{x}_2)$, where $f_q(\mathbf{x}_2)$ is the b.l.u.p. of X_q, and let $g(\zeta_q)$ be the b.l.u.p. of ζ_0 based on ζ_q. Now $f_0(\mathbf{x}_2) + g(\zeta_q)$ is an unbiased linear predictor of X_0 based on $(X_1, \cdots, X_q)'$ with error $\tilde{\zeta}_0 = X_0 - f_0(\mathbf{x}_2) - g(\zeta_q) = \zeta_0 - g(\zeta_q)$. Then $\text{Cov}(\tilde{\zeta}_0, \zeta_q) = \text{Cov}(\tilde{\zeta}_0, X_q - f_q(\mathbf{x}_2)) = 0$. Meanwhile, since both ζ_0 and ζ_q are uncorrelated with $\mathbf{x}_2$, $\text{Cov}(\tilde{\zeta}_0, \mathbf{x}_2) = \mathbf{0}$. As a result, $\tilde{\zeta}_0$ is uncorrelated with $\mathbf{y}$, so by Result 5.2.15, $f_0(\mathbf{x}_2) + g(\zeta_q)$ is the b.l.u.p. of X_0 based on $(X_1, \cdots, X_q)'$.

Now from Result 5.2.22,

$$1 - \rho^2_{0(1,2,\cdots,q-1)} = \frac{\text{Var}(\zeta_0)}{\sigma_{00}},$$

and since $\rho_{0q|(1,\cdots,q-1)}$ is the correlation between ζ_0 and ζ_q, from Result 5.2.16,

$$[1 - \rho^2_{0q|(1,\cdots,q-1)}]\,\text{Var}(\zeta_0) = \text{Var}(\zeta_0 - g(\zeta_q)) = \text{Var}(\tilde{\zeta}_0),$$

which is the variance of the error of the b.l.u.p. based on $(X_1, \cdots, X_q)$. Combining the above two displays

$$[1 - \rho^2_{0(1,2,\cdots,q-1)}][1 - \rho^2_{0q|(1,\cdots,q-1)}] = \frac{\text{Var}(\tilde{\zeta}_0)}{\sigma_{00}}$$

which according to Result 5.2.22 is (5.2.22). Then the proof is complete. ■

Result 5.2.25. The following inequality holds:

$$\rho^2_{0(1)} \leq \rho^2_{0(1,2)} \leq \cdots \leq \rho^2_{0(1,2,\cdots,k)}$$

Proof. This is clear from (5.2.21). ■

5.3 Some noncentral distributions

The noncentral chi-square, noncentral F, and noncentral t-distributions are derived from the multivariate normal distribution and are useful for a discussion of inference for linear models. In general, these distributions arise as the sampling distributions of statistics when a null hypothesis of interest is not true. The p.d.f. of the central chi-square and the noncentral chi-square distributions have been variously derived in the literature. For instance, section 11.2 of Kendall and Stuart (1958) gave a geometrical derivation of the central chi-square p.d.f. using spherical (or polar) coordinates while Guenther (1964) extended this approach to the noncentral chi-square distribution. In this section, we present derivations that are based on the moment generating function method. We begin with the derivation of the central chi-square distribution, starting from the $N_k(\mathbf{0}, \mathbf{I})$ distribution.

Result 5.3.1. Let $\mathbf{z} \sim N_k(\mathbf{0}, \mathbf{I})$, and let $U = \mathbf{z}'\mathbf{z} = \sum_{i=1}^k Z_i^2$. Then, $U \sim \chi^2_k$, i.e., U has a (central) chi-square distribution with k degrees of freedom (d.f.) and p.d.f. given by

$$f(u) = \frac{1}{2^{k/2}\Gamma(k/2)} u^{(k-2)/2} \exp(-u/2), \quad u > 0. \tag{5.3.1}$$

Proof. Since U is a function of a multivariate standard normal random vector $\mathbf{z}$, we have

$$\begin{aligned} M_U(t) = \text{E}[\exp\{tU\}] &= \int_{\mathcal{R}^k} \frac{1}{(2\pi)^{k/2}} \exp\left\{t\sum_{i=1}^k z_i^2 - \frac{1}{2}\sum_{i=1}^k z_i^2\right\} d\mathbf{z} \\ &= \frac{1}{(2\pi)^{k/2}} \int_{\mathcal{R}^k} \exp\left\{-\frac{1}{2}(1-2t)\sum_{i=1}^k z_i^2\right\} d\mathbf{z} = (1-2t)^{-k/2}, \end{aligned} \tag{5.3.2}$$

which follows from the use of an integral evaluation theorem, either Result 5.1.1 or Result 5.1.2. Comparing (5.3.2) with (B.6), we see that (5.3.2) is the m.g.f. of a χ^2_k random variable. The p.d.f. of U follows from (B.5). ■

In particular, if $Z \sim N(0,1)$, then $U = Z_1^2 \sim \chi^2_1$. However, $U \sim \chi^2_k$ does not imply that $Y = \sqrt{U}$ has a standard normal distribution. Properties of the central chi-square distribution are given in Appendix B. It is useful to recall that the χ^2_k distribution, where k is an integer, is a special case of the Gamma(α, β) distribution with $\alpha = k/2$ and $\beta = 2$. The chi-square distribution plays an important role in statistical inference, especially when sampling from a normal population. In tests for the variance of a single normal population using information from a random sample of size n, it is well known that under the null hypothesis, the test statistic follows a central chi-square distribution with $(n-1)$ degrees of freedom. Critical values from the corresponding central chi-square distribution enables us to construct the rejection region for the test. The next example shows a property of the central chi-square distribution.

Example 5.3.1. If $m > n$, both being integers, then $P(\chi^2_m > c) > P(\chi^2_n > c)$, where c is a known constant. This result is obvious since clearly $P(\sum_{j=1}^m X_j^2 > c) > P(\sum_{j=1}^n X_j^2 > c)$ where X_j's are i.i.d. $N(0,1)$ variables. □

When the mean of a normal random vector $\mathbf{x}$ is nonzero, the distribution of the quadratic form $\mathbf{x}'\mathbf{x}$ no longer has a central chi-square distribution, as the next result shows.

Result 5.3.2. Let $\mathbf{x} \sim N_k(\boldsymbol{\mu}, \mathbf{I})$, where $\boldsymbol{\mu} = (\mu_1, \cdots, \mu_k)' \neq \mathbf{0}$, and let $U = \mathbf{x}'\mathbf{x}$. The p.d.f. of U is

$$f(u) = \sum_{j=0}^{\infty} \frac{\exp(-\lambda)\lambda^j}{j!} \frac{u^{(k+2j-2)/2}\exp\{-u/2\}}{2^{j+\frac{k}{2}}\Gamma(\frac{1}{2}(k+2j))}, \quad u > 0, \tag{5.3.3}$$

where $\lambda = \frac{1}{2}\boldsymbol{\mu}'\boldsymbol{\mu} = \frac{1}{2}\sum_{j=1}^k \mu_j^2$. We say $U \sim \chi^2(k, \lambda)$, i.e., U has a noncentral chi-square distribution with k degrees of freedom and noncentrality parameter equal to λ.

Proof. We derive the m.g.f. of $U = \mathbf{x}'\mathbf{x}$, where $\mathbf{x} \sim N_k(\boldsymbol{\mu}, \mathbf{I})$, and show that this coincides with the m.g.f. of a random variable with p.d.f. (5.3.3). Since $\mathbf{x} \sim N_k(\boldsymbol{\mu}, \mathbf{I})$, we have

$$\begin{aligned} M_{\mathbf{x}'\mathbf{x}}(t) &= \int_{\mathcal{R}^k} \frac{1}{(2\pi)^{k/2}} \exp\left\{t\mathbf{x}'\mathbf{x} - \frac{1}{2}(\mathbf{x}-\boldsymbol{\mu})'(\mathbf{x}-\boldsymbol{\mu})\right\} d\mathbf{x} \\ &= (1-2t)^{-k/2}\exp\{2t\lambda/(1-2t)\}, \quad t < 1/2, \end{aligned} \tag{5.3.4}$$

which follows directly from Result 5.1.2, with $\mathbf{B} = \text{diag}\{(1-2t)/2\}$, $\mathbf{b} = -\boldsymbol{\mu}$, $\mathbf{A} = \mathbf{0}$, $\mathbf{a} = \mathbf{0}$, and $a_0 = 1$. Alternately, we can write the m.g.f. in the form $M_{\mathbf{x}'\mathbf{x}}(t) = (1-2t)^{-k/2}\exp\{-\lambda[1-(1-2t)^{-1}]\}$, for $t < 1/2$. Next, we evaluate the m.g.f. of a random variable U with p.d.f. (5.3.3) as $M_U(t) = E[\exp(Ut)]$, i.e.,

$$\begin{aligned} M_U(t) &= \int_0^{\infty} \exp\{ut\} \sum_{j=0}^{\infty} \frac{e^{-\lambda}\lambda^j}{j!} \frac{u^{(k+2j-2)/2}\exp\{-u/2\}}{2^{j+\frac{k}{2}}\Gamma(\frac{1}{2}(k+2j))} du \\ &= (1-2t)^{-k/2} \sum_{j=0}^{\infty} \frac{e^{-\lambda}\lambda^j}{j!(1-2t)^j} \\ &= (1-2t)^{-k/2}\exp\{2\lambda t/(1-2t)\}, \quad t < 1/2 \\ &= (1-2t)^{-k/2}\exp\{-\lambda[1-(1-2t)^{-1}]\}, \quad t < 1/2. \end{aligned} \tag{5.3.5}$$

Therefore, $U = \mathbf{x}'\mathbf{x} \sim \chi^2(k, \lambda)$ distribution. ■

Note that $\lambda = 0$ if and only if $\boldsymbol{\mu} = \mathbf{0}$, in which case, the distribution of U is a central chi-square. The noncentral chi-square distribution is the distribution of a quadratic form $\mathbf{x}'\mathbf{x}$ when $\mathbf{x}$ has a nonzero mean. The jth term in (5.3.3) is the product of a Poisson p.m.f. with mean λ at count j and a central chi-square density with $k + 2j$ degrees of freedom. In other words, the noncentral chi-square distribution defined in Result 5.3.2 is an example of a *mixture distribution* involving central chi-square and Poisson distributions. The hierarchy is (see Casella and Berger (1990), chapter 4)

$$U \mid Y \sim \chi^2_{k+2Y} \text{ and } Y \sim \text{Poisson}(\lambda). \tag{5.3.6}$$

Example 5.3.2. Suppose $X_1, \cdots, X_k$ are independently distributed random variables, with $X_j \sim N(\mu_j, \sigma_j^2)$, $j = 1, \cdots, k$. We obtain the distribution of $\sum_{j=1}^k X_j^2/\sigma_j^2$. Note that if we set $Z_j = X_j/\sigma_j$, $j = 1, \cdots, k$, then $Z_1, \cdots, Z_k$ are independently distributed, with $Z_j \sim N(\mu_j/\sigma_j, 1)$, $j = 1, \cdots, k$. Hence, by Result 5.3.2, $\sum_{j=1}^k Z_j^2 = \sum_{j=1}^k X_j^2/\sigma_j^2$ has a $\chi^2(k, \lambda)$ distribution, where $\lambda = \sum_{j=1}^k \mu_j^2/2\sigma_j^2$. □

Example 5.3.3. Suppose $U \sim \chi^2(k, \lambda)$ and $V \sim \chi^2_k$. We show that $\text{P}(U > c) \geq \text{P}(V > c)$, for any constant $c > 0$. We have

$$\begin{aligned}
\text{P}(U > c) &= \sum_{j=0}^{\infty} \text{P}(U > c \mid J = j)\, \text{P}(J = j), \quad \text{where } J \sim \text{Poisson}(\lambda/2) \\
&= \sum_{j=0}^{\infty} \text{P}(\chi^2_{2j+k} > c)\, \text{P}(J = j) \\
&\geq \sum_{j=0}^{\infty} \text{P}(\chi^2_k > c)\, \text{P}(J = j) \quad \text{(see Example 5.3.1)} \\
&= \text{P}(\chi^2_k > c) \sum_{j=0}^{\infty} \text{P}(J = j) \\
&= \text{P}(\chi^2_k > c).
\end{aligned}$$

□

Example 5.3.4. Let $U \sim \chi^2(k, \lambda)$. For a given constant c, we show that a property of the noncentral chi-square distribution is that $\text{P}(U > c)$ is an increasing function of λ. Again,

$$h(\lambda) = \text{P}(U > c) = \sum_{j=0}^{\infty} \text{P}(\chi^2_{2j+k} > c) \frac{1}{j!} e^{-\lambda/2} (\lambda/2)^j,$$

so that the first derivative of $h(\lambda)$ is

$$\begin{aligned}
&h'(\lambda) \\
&= \sum_{j=1}^{\infty} \text{P}(\chi^2_{2j+k} > c) \left[-\frac{1}{2j!} e^{-\lambda/2} (\lambda/2)^j + \frac{1}{2(j-1)!} e^{-\lambda/2} (\lambda/2)^{j-1} \right] \\
&\quad - \frac{1}{2} e^{-\lambda/2}\, \text{P}(\chi^2_k > c) \\
&= \frac{1}{2} \sum_{j=0}^{\infty} \text{P}(\chi^2_{2j+2+k} > c) \frac{1}{j!} e^{-\lambda/2} (\lambda/2)^j - \frac{1}{2} \sum_{j=0}^{\infty} \text{P}(\chi^2_{2j+k} > c) \frac{1}{j!} e^{-\lambda/2} (\lambda/2)^j \\
&\geq 0,
\end{aligned}$$

which shows that $\text{P}(U > c)$ is an increasing function of λ. □

Result 5.3.3. Let $\mathbf{x} \sim N_k(\boldsymbol{\mu}, \boldsymbol{\Sigma})$, with $\text{r}(\boldsymbol{\Sigma}) = k$. Then

$$\begin{aligned}
U_1 &= (\mathbf{x} - \boldsymbol{\mu})' \boldsymbol{\Sigma}^{-1} (\mathbf{x} - \boldsymbol{\mu}) \sim \chi^2_k, \text{ and} \\
U_2 &= \mathbf{x}' \boldsymbol{\Sigma}^{-1} \mathbf{x} \sim \chi^2(k, \lambda),
\end{aligned}$$

where $\lambda = \frac{1}{2} \boldsymbol{\mu}' \boldsymbol{\Sigma}^{-1} \boldsymbol{\mu}$.

Proof. Let $\boldsymbol{\Sigma} = \boldsymbol{\Gamma}\boldsymbol{\Gamma}'$, with $\text{r}(\boldsymbol{\Gamma}) = k$, and let $\mathbf{z}_1 = \boldsymbol{\Gamma}^{-1}(\mathbf{x} - \boldsymbol{\mu})$. From Result 5.2.5, $\mathbf{z}_1 \sim N_k(\mathbf{0}, \mathbf{I})$. Using Result 5.3.1, we see that $\mathbf{z}_1'\mathbf{z}_1 \sim \chi^2_k$. But

$$\mathbf{z}_1'\mathbf{z}_1 = (\mathbf{x} - \boldsymbol{\mu})'\boldsymbol{\Gamma}'^{-1}\boldsymbol{\Gamma}^{-1}(\mathbf{x} - \boldsymbol{\mu}) = (\mathbf{x} - \boldsymbol{\mu})'\boldsymbol{\Sigma}^{-1}(\mathbf{x} - \boldsymbol{\mu}) = U_1,$$

which proves the first result. To derive the distribution of U_2, we see again that $\mathbf{z}_2 = \boldsymbol{\Gamma}^{-1}\mathbf{x} \sim N_k(\boldsymbol{\Gamma}^{-1}\boldsymbol{\mu}, \mathbf{I})$, so that using Result 5.3.2, we see that $\mathbf{z}_2'\mathbf{z}_2 \sim \chi^2(k, \lambda)$. Again,

$$\mathbf{z}_2'\mathbf{z}_2 = \mathbf{x}'\boldsymbol{\Gamma}^{-1}\boldsymbol{\Gamma}'^{-1}\mathbf{x} = \mathbf{x}'\boldsymbol{\Sigma}^{-1}\mathbf{x} = U_2,$$

which proves the second result. ■

Result 5.3.4. Let $U \sim \chi^2(k, \lambda)$. Then $\text{E}(U) = k + 2\lambda$ and $\text{Var}(U) = 2(k + 4\lambda)$.

Proof. Using the hierarchical setup given in (5.3.6), we have

$$\begin{aligned}
\text{E}(U) &= \text{E}[\text{E}(U \mid Y)] = \text{E}(k + 2Y) = k + 2\lambda, \text{ and} && (5.3.7)\\
\text{Var}(U) &= \text{E}[\text{Var}(U \mid Y)] + \text{Var}[\text{E}(U \mid Y)] \\
&= \text{E}(2k + 4Y) + \text{Var}(k + 2Y) \\
&= 2k + 4\lambda + 4\lambda = 2(k + 4\lambda). && (5.3.8)
\end{aligned}$$

Alternatively, we can derive these expressions using

$$\text{E}(U^j) = \frac{\partial^j}{\partial t^j} M_U(t) \mid_{t=0}, \quad j = 1, 2, \tag{5.3.9}$$

and $\text{Var}(U) = \text{E}(U^2) - [\text{E}(U)]^2$. Note that all moments of the p.d.f. (5.3.3) exist and we may interchange summation and integration in order to compute these moments. ■

Result 5.3.5. Let $U_i \sim \chi^2(k_i, \lambda_i)$, $i = 1, \cdots, K$, and suppose the U_i's are all independent. Then

$$U = \sum_{i=1}^{K} U_i \sim \chi^2\left(\sum_{i=1}^{K} k_i, \sum_{i=1}^{K} \lambda_i\right). \tag{5.3.10}$$

Proof. For $t < 1/2$, the m.g.f. of U is

$$\begin{aligned}
M_U(t) &= \prod_{i=1}^{K} M_{U_i}(t) \quad \text{(by independence)} \\
&= \prod_{i=1}^{K} (1 - 2t)^{-k_i/2} \exp\{-\lambda_i[1 - (1 - 2t)^{-1}]\}, \\
&= (1 - 2t)^{-\sum_{i=1}^{K} k_i/2} \exp\left\{-[1 - (1 - 2t)^{-1}] \sum_{i=1}^{K} \lambda_i\right\},
\end{aligned}$$

which is the m.g.f. of a $\chi^2(\sum_{i=1}^{K} k_i, \sum_{i=1}^{K} \lambda_i)$ distribution, thus establishing the additive property of the noncentral chi-square distribution. ■

Result 5.3.6. Let $U_1 \sim \chi^2(k_1, \lambda)$, let $U_2 \sim \chi^2_{k_2}$, and let U_1 and U_2 be independently distributed. Then,

$$F = \frac{U_1/k_1}{U_2/k_2} \sim F(k_1, k_2, \lambda), \tag{5.3.11}$$

i.e., a noncentral F-distribution with numerator and denominator degrees of freedom equal to k_1 and k_2 respectively, noncentrality parameter λ, and with p.d.f.

$$f(v;k_1,k_2,\lambda)=\sum_{j=0}^{\infty}\frac{e^{-\lambda}\lambda^j}{j!}\frac{k_1^{(k_1+2j)/2}k_2^{k_2/2}\Gamma(k_1/2+k_2/2+j)v^{k_1/2+j-1}}{\Gamma(k_1/2+j)\Gamma(k_2/2)(k_2+k_1v)^{k_1/2+k_2/2+j}},\quad v>0.$$

Proof. Since U_1 and U_2 are independently distributed, we have from (5.3.1) and (5.3.3) that for $u_1>0$, $u_2>0$,

$$\begin{aligned}f(u_1,u_2)&=f(u_1)f(u_2)\\&=\sum_{j=0}^{\infty}\frac{\exp(-\lambda)\lambda^j}{j!}\frac{u_1^{(k_1+2j-2)/2}\exp\{-u_1/2\}}{2^{(k_1+2j)/2}\Gamma((k_1+2j)/2)}\frac{u_2^{(k_2-2)/2}\exp(-u_2/2)}{2^{k_2/2}\Gamma(k_2/2)}.\end{aligned}\tag{5.3.12}$$

Let

$$\alpha_j=\exp(-\lambda)\lambda^j/\{j!2^{(k_1+k_2+2j)/2}\Gamma((k_1+2j)/2)\Gamma(k_2/2)\}.$$

Consider the following transformation of variables:

$$f=k_2u_1/k_1u_2,\quad z=u_1+u_2,$$

with Jacobian

$$|J|=k_2(u_1+u_2)/(k_1u_2^2),$$

so that

$$\begin{aligned}f(v;k_1,k_2,\lambda)&=\int_0^{\infty}\frac{1}{|J|}f(v,z)dz\\&=\int_0^{\infty}\sum_{j=0}^{\infty}\alpha_j\left(\frac{k_1vz}{k_1v+k_2}\right)^{\frac{1}{2}k_1+j-1}\left(\frac{k_2z}{k_1v+k_2}\right)^{\frac{1}{2}k_2-1}\frac{k_1k_2z}{(k_1v+k_2)^2}e^{-z/2}\,dz\\&=\sum_{j=0}^{\infty}\alpha_jk_1^{\frac{1}{2}k_1+j}k_2^{\frac{1}{2}k_2}\frac{v^{\frac{1}{2}k_1+j-1}}{(k_1v+k_2)^{\frac{1}{2}k_1+\frac{1}{2}k_2+j}}\int_0^{\infty}z^{\frac{1}{2}k_1+\frac{1}{2}k_2+j-1}e^{-z/2}\,dz.\end{aligned}$$

Since

$$\int_0^{\infty}z^{\frac{1}{2}k_1+\frac{1}{2}k_2+j-1}\exp(-z/2)dz=2^{\frac{1}{2}k_1+\frac{1}{2}k_2+j}\Gamma(\frac{1}{2}k_1+\frac{1}{2}k_2+j),$$

the result follows directly. ■

The noncentral F-distribution is an extension of the central F-distribution (see Appendix B) to the case when the normal random vector $\mathbf{x}$ has nonzero mean, i.e., the quadratic form in the numerator has a noncentral chi-square distribution. The noncentral F-distribution is useful for evaluating the power of tests of hypotheses, which requires the evaluation of

$$G(\lambda)=\int_g^{\infty}f(v;k_1,k_2,\lambda)dv,\tag{5.3.13}$$

where $g=F_{k_1,k_2,\alpha}(\lambda)$ denotes the 100αth percentile point of the noncentral $F(k_1,k_2,\lambda)$-distribution, and α denotes the level of significance of the test. The noncentral beta distribution, which is discussed in Result 5.3.7 is also related to a noncentral chi-square and an independent central chi-square random variable.

Result 5.3.7. Let $U_1 \sim \chi^2(k_1, \lambda)$, let $U_2 \sim \chi^2_{k_2}$, and let U_1 and U_2 be independently distributed. Then

$$G = U_1/(U_1 + U_2) \sim \text{Beta}(k_1, k_2, \lambda),$$

i.e., a noncentral Beta distribution with numerator and denominator degrees of freedom equal to k_1 and k_2 respectively, noncentrality parameter λ, and p.d.f.

$$f(w; k_1, k_2, \lambda) = \exp(-\frac{\lambda^2}{2}) \sum_{j=0}^{\infty} \left(\frac{\lambda^2}{2}\right)^j \frac{1}{j!} B\left(w; \frac{k_1}{2} + j, \frac{k_2}{2}\right), \quad 0 < w < 1,$$

where $B(w; \alpha, \beta)$ corresponds to the (central) Beta p.d.f. with shape parameter α and scale parameter β (see Appendix B).

Proof. We begin with the joint p.d.f. of U_1 and U_2 which was shown in (5.3.12) and use the transformation

$$G = U_1/(U_1 + U_2).$$

The derivation of the p.d.f. of the noncentral beta distribution is then similar to that of the noncentral F-distribution. ■

We define two other noncentral distributions, the noncentral t-distribution and the doubly noncentral F-distribution. Scheffé (1959) presented an application of the doubly noncentral F-distribution and also showed a procedure for approximating it via the noncentral F-distribution.

Result 5.3.8. If $X \sim N(\mu, \sigma^2)$, $U/\sigma^2 \sim \chi^2_k$, and X is distributed independently of U,

$$T = \frac{X}{\sqrt{U/k}} \sim t(k, \delta) \tag{5.3.14}$$

has a noncentral t-distribution with p.d.f. (see Rao (1973), p. 172)

$$f(t; k, \delta) = \frac{k^{k/2}}{\Gamma(k/2)} \frac{\exp(-\delta^2/2)}{(k + t^2)^{(k+1)/2}} \sum_{s=0}^{\infty} \Gamma\left(\frac{k + s + 1}{2}\right) \left(\frac{\delta^s}{s!}\right) \left(\frac{2t^2}{k + t^2}\right)^{s/2} \tag{5.3.15}$$

for $-\infty < t < \infty$, where $\delta = \mu/\sigma$.

Proof. The joint distribution of X and U is

$$c_1 \exp\left\{-\frac{(x - \mu)^2}{2\sigma^2}\right\} \exp\left\{-\frac{u}{\sigma^2}\right\} u^{k/2-1},$$

where $c_1^{-1} = \sqrt{2\pi}\sigma^{k+1} 2^{k/2} \Gamma(k/2)$. Consider the transformation

$$X = R \sin\theta, \quad U = R^2 \cos^2\theta;$$

the joint density of R and θ is equal to

$$c_2 r^k (\cos\theta)^{k-1} \exp\{-(r^2 - 2\mu r \sin\theta)/2\sigma^2\},$$

where $c_2 = c_1 \exp\{-\mu^2/2\sigma^2\} = c_1 \exp\{-\delta^2/2\}$. The joint p.d.f. does not factor into two expressions, one involving r alone, and the other involving only θ, so R and θ are not independent. By expanding $\exp\{\mu r \sin\theta/\sigma^2\}$, we write the joint density as an infinite series

$$c_2 \exp\{-r^2/2\sigma^2\}(\cos\theta)^{k-1} \sum_j \frac{(\mu \sin\theta)^j r^{k+j}}{j!\sigma^{2j}}. \tag{5.3.16}$$

We obtain the marginal density of θ by integrating out term by term with respect to r in (5.3.16), to get

$$(c_2/2)(\cos\theta)^{k-1} \sum_j \Gamma\left(\frac{k+j+1}{2}\right) 2^{(j+k+1)/2} \frac{(\delta \sin\theta)^j}{j!}.$$

Now, transforming from θ to $T = \sqrt{k}\tan\theta$, we obtain the p.d.f. in (5.3.15). ■

Note also that if $Z \sim N(0,1)$, $U \sim \chi^2_k$, δ is a constant, and Z and U are independently distributed, then

$$T = (Z+\delta)/\sqrt{U/k} \sim t(k,\delta)$$

has a noncentral t-distribution. The noncentral t-distribution plays the following role in statistical inference. Suppose we consider a test H_0: $\mu = 0$ versus H_1: $\mu \neq 0$ on the basis of observations $\mathbf{x} = (X_1, \cdots, X_n)' \sim N(\mu \mathbf{1}_n, \sigma^2 \mathbf{I}_n)$, where both μ and σ^2 are unknown. The test statistic is $\sqrt{n}\overline{X}/S$ where $\overline{X} = \sum_{i=1}^n X_i/n$, and $S^2 = \sum_{i=1}^n (X_i - \overline{X})^2/(n-1)$. Under H_0, $\overline{X} \sim N(0, \sigma^2/n)$, $(n-1)S^2/\sigma^2 \sim \chi^2_{n-1}$, and they are distributed independently. Hence, $\sqrt{n}\overline{X}/S \sim t_{n-1}$, which is the central t-distribution with $(n-1)$ degrees of freedom. Under the alternative hypothesis H_1, however, $\overline{X} \sim N(\mu, \sigma^2/n)$, and $(n-1)S^2/\sigma^2$ is still distributed independently as a χ^2_{n-1} variable, so that $\sqrt{n}\overline{X}/S$ has a noncentral t-distribution with $(n-1)$ degrees of freedom and noncentrality parameter $\lambda = \mu^2/2\sigma^2$. The noncentral t-distribution is useful for power calculations.

Result 5.3.9. Let U_1 and U_2 have independent noncentral chi-square distributions, i.e., $U_1 \sim \chi^2(k_1, \lambda_1)$, and $U_2 \sim \chi^2(k_2, \lambda_2)$. Then

$$F^* = \frac{U_1/k_1}{U_2/k_2} \sim F''(k_1, k_2, \lambda_1, \lambda_2),$$

where $F''(k_1, k_2, \lambda_1, \lambda_2)$ refers to the doubly noncentral F-distribution with degrees of freedom k_1 and k_2, and noncentrality parameters λ_1 and λ_2. The doubly noncentral F-distribution is based on the ratio of two independent noncentral chi-square random variables, and for $v > 0$ has p.d.f.

$$f(v) = \frac{k_1}{k_2} \exp\left(-\frac{\lambda_1+\lambda_2}{2}\right) \sum_{r=0}^{\infty}\sum_{s=0}^{\infty} N(r,s,k_1,k_2)/D(r,s,k_1,k_2),$$

where

$$N(r,s,k_1,k_2) = \left(\frac{\lambda_1}{2}\right)^r \left(\frac{\lambda_2}{2}\right)^s \left(\frac{k_1 v}{k_2}\right)^{k_1/2+r-1} \quad \text{and}$$

$$D(r,s,k_1,k_2) = r!s!B(k_1/2+r, k_2/2+s)\left(1+\frac{k_1 v}{k_2}\right)^{k_1/2+k_2/2+r+s}.$$

Proof. Recall that we can write $U_1 \sim \chi^2_{2J_1+k_1}$, and $U_2 \sim \chi^2_{2J_2+k_2}$, where J_1 and J_2 are independent Poisson random variables with respective means $\lambda_1/2$ and $\lambda_2/2$. Given $J_1 = j_1$ and $J_2 = j_2$, U_1 and U_2 are independently distributed and

$$U = U_1/U_2 \sim \{(2j_1 + k_1)/(2j_2 + k_2)\}F_{2j_1+k_1, 2j_2+k_2}$$

i.e., conditional on J_1 and J_2, the ratio of U_1 and U_2 is proportional to a central F-distribution. The p.d.f. of U is then

$$\begin{aligned} f(u) = & \sum_{j_2=0}^{\infty} \sum_{=0}^{\infty} j_1 \frac{1}{j_1!} \exp\left(-\frac{\lambda_1}{2}\right)\left(\frac{\lambda_1}{2}\right)^{j_1} \frac{1}{j_2!} \exp\left(-\frac{\lambda_2}{2}\right)\left(\frac{\lambda_2}{2}\right)^{j_2} \\ & \times \left\{B\left(j_1 + \frac{k_1}{2}, j_2 + \frac{k_2}{2}\right)\right\}^{-1} (1+u)^{-(2j_1+2j_2+k_1+k_2)/2} \\ & \times u^{(2j_1+k_1)/2-1} u^{(2j_2+k_2)/2-1}, \end{aligned}$$

which after some simplification yields the required result. ■

5.4 Distributions of quadratic forms

There is a rich literature on the distribution of quadratic forms (see Graybill, 1961; Rao, 1973). The basic theorems dealing with distributions of quadratic forms in normal random vectors are due to Cochran (1934), Craig (1943), and Rao (1973), while Shanbhag (1966) dealt with independence of quadratic forms. Although the discussion that we present is by no means complete, it gives the basic results that are needed for the development of linear model theory. Other useful references are Driscoll (1999), Hayes and Haslett (1999), Khuri (1999), and Seely et al. (1997). As before, we assume that the matrix of a quadratic form is symmetric.

Result 5.4.1. Let $\mathbf{x} \sim N_k(\mathbf{0}, \mathbf{I})$. Then the quadratic form $\mathbf{x}'\mathbf{A}\mathbf{x}$ has a chi-square distribution if and only if $\mathbf{A}$ is idempotent, in which case $\mathbf{x}'\mathbf{A}\mathbf{x} \sim \chi^2_m$ with $m = \text{r}(\mathbf{A})$.

Proof. Let $\mathbf{A}$ be a symmetric and idempotent matrix of rank m. We will show that $\mathbf{x}'\mathbf{A}\mathbf{x} \sim \chi^2_m$ distribution. By Result 2.3.4, there exists a $k \times k$ orthogonal matrix $\mathbf{P}$ such that $\mathbf{P}'\mathbf{A}\mathbf{P} = \begin{pmatrix} \mathbf{I}_m & \mathbf{O} \\ \mathbf{O} & \mathbf{O} \end{pmatrix}$. Define $\mathbf{z} = \mathbf{P}'\mathbf{x}$. Partition $\mathbf{P}$ as $(\mathbf{P}_1 \ \mathbf{P}_2)$, where $\mathbf{P}_1'\mathbf{P}_1 = \mathbf{I}_m$, and partition $\mathbf{z}' = (\mathbf{z}_1' \ \mathbf{z}_2')$. Then, $\mathbf{z}_1 = \mathbf{P}_1'\mathbf{x} \sim N_m(\mathbf{0}, \mathbf{I})$ and

$$\mathbf{x}'\mathbf{A}\mathbf{x} = (\mathbf{P}\mathbf{z})'\mathbf{A}(\mathbf{P}\mathbf{z}) = \mathbf{z}'\mathbf{P}'\mathbf{A}\mathbf{P}\mathbf{z} = (\mathbf{z}_1' \ \mathbf{z}_2') \begin{pmatrix} \mathbf{I}_m & \mathbf{O} \\ \mathbf{O} & \mathbf{O} \end{pmatrix} \begin{pmatrix} \mathbf{z}_1 \\ \mathbf{z}_2 \end{pmatrix} = \mathbf{z}_1'\mathbf{z}_1. \tag{5.4.1}$$

Since $\mathbf{z}_1 \sim N_m(\mathbf{0}, \mathbf{I})$, the proof follows from Result 5.3.1. To prove the converse, assume that $\mathbf{x}'\mathbf{A}\mathbf{x} \sim \chi^2_m$. We must show that this implies that the symmetric matrix $\mathbf{A}$ is idempotent of rank m. Since $\mathbf{x}'\mathbf{A}\mathbf{x} \sim \chi^2_m$, we have by (B.6),

$$M_{\mathbf{x}'\mathbf{A}\mathbf{x}}(t) = (1-2t)^{-m/2}, \quad t < 1/2. \tag{5.4.2}$$

Since $\mathbf{x} \sim N_k(\mathbf{0}, \mathbf{I})$, we can also obtain the m.g.f. of $\mathbf{x}'\mathbf{A}\mathbf{x}$ using an integral evaluation theorem, either Result 5.1.1 or Result 5.1.2, as

$$M_{\mathbf{x}'\mathbf{A}\mathbf{x}}(t) = \int_{\mathcal{R}^k} \frac{1}{(2\pi)^{k/2}} \exp\{t(\mathbf{x}'\mathbf{A}\mathbf{x}) - \mathbf{x}'\mathbf{x}/2\}\, d\mathbf{x}$$
$$= |\mathbf{I}_k - 2t\mathbf{A}|^{-1/2} = \prod_{i=1}^{k} (1 - 2t\lambda_i)^{-1/2}, \tag{5.4.3}$$

where λ_i, $i = 1, \cdots, k$ are the eigenvalues of the symmetric matrix $\mathbf{A}$. Comparing the two expressions in (5.4.2) and (5.4.3) which must be equal to one another, we should have

$$\prod_{i=1}^{k} (1 - 2t\lambda_i)^{-1/2} = (1 - 2t)^{-m/2} \tag{5.4.4}$$

for every t in some neighborhood of zero. This will be true if m of the eigenvalues of $\mathbf{A}$ are equal to 1, and the remaining $k - m$ eigenvalues are equal to 0. By Result 2.3.7, this implies that $\mathbf{A}$ is an idempotent matrix of rank m. ■

Result 5.4.2. Let $\mathbf{x} \sim N_k(\boldsymbol{\mu}, \boldsymbol{\Sigma})$. The rth cumulant of the quadratic form $\mathbf{x}'\mathbf{A}\mathbf{x} + \mathbf{a}'\mathbf{x}$ is

$$\kappa_r = \kappa_r(\mathbf{x}'\mathbf{A}\mathbf{x} + \mathbf{a}'\mathbf{x})$$
$$= \begin{cases} \text{tr}(\mathbf{A}\boldsymbol{\Sigma}) + \boldsymbol{\mu}'\mathbf{A}\boldsymbol{\mu} + \mathbf{a}'\boldsymbol{\mu} & \text{if } r = 1 \\ 2^{r-1}(r-1)! \left[\text{tr}\{(\mathbf{A}\boldsymbol{\Sigma})^r\} + r\mathbf{b}'\boldsymbol{\Sigma}(\mathbf{A}\boldsymbol{\Sigma})^{r-2}\mathbf{b}\right] & \text{if } r \geq 2, \end{cases} \tag{5.4.5}$$

where $\mathbf{b} = \mathbf{A}\boldsymbol{\mu} + \frac{1}{2}\mathbf{a}$.

Proof. We first consider the special case where $\mathbf{x} \sim N_k(\mathbf{0}, \mathbf{I})$. From Result 5.1.2, the m.g.f. of $\mathbf{x}'\mathbf{A}\mathbf{x} + \mathbf{a}'\mathbf{x}$ is finite in a neighborhood of the origin and for $|t| > 0$ small enough,

$$M_{\mathbf{x}'\mathbf{A}\mathbf{x}+\mathbf{a}'\mathbf{x}}(t) = \int_{\mathcal{R}^k} \frac{1}{(2\pi)^{k/2}} \exp\left\{t(\mathbf{z}'\mathbf{A}\mathbf{z} + \mathbf{a}'\mathbf{z}) - \frac{1}{2}\mathbf{z}'\mathbf{z}\right\} d\mathbf{z}$$
$$= |\mathbf{I}_k - 2t\mathbf{A}|^{-1/2} \exp\left\{\frac{1}{2}t^2\mathbf{a}'(\mathbf{I} - 2t\mathbf{A})^{-1}\mathbf{a}\right\}. \tag{5.4.6}$$

The cumulant generating function of $\mathbf{x}'\mathbf{A}\mathbf{x} + \mathbf{a}'\mathbf{x}$ is $\mathcal{K}(t) = \sum_{r=1}^{\infty} \kappa_r t^r / r! = \log[M_{\mathbf{x}'\mathbf{A}\mathbf{x}+\mathbf{a}'\mathbf{x}}(t)]$, so that

$$\sum_{r=1}^{\infty} \frac{1}{r!}\kappa_r t^r = -\frac{1}{2}\log|\mathbf{I} - 2t\mathbf{A}| + \frac{1}{2}t^2\mathbf{a}'(\mathbf{I} - 2t\mathbf{A})^{-1}\mathbf{a}. \tag{5.4.7}$$

Let δ_j, $j = 1, \cdots, k$, denote the eigenvalues of $\mathbf{A}$. Then $1 - 2t\delta_j$ are the eigenvalues of $\mathbf{I} - 2t\mathbf{A}$. For $|t| < \min(1/|2\delta_1|, \cdots, 1/|2\delta_k|)$,

$$-\frac{1}{2}\log|\mathbf{I} - 2t\mathbf{A}| = -\frac{1}{2}\sum_{j=1}^{k} \log\{1 - 2t\delta_j\} = -\frac{1}{2}\sum_{j=1}^{k}\sum_{r=1}^{\infty} \frac{-(2t\delta_j)^r}{r}$$
$$= \sum_{r=1}^{\infty} \frac{2^{r-1}t^r}{r} \sum_{j=1}^{k} \delta_j^r = \sum_{r=1}^{\infty} \frac{2^{r-1}t^r}{r} \text{tr}(\mathbf{A}^r). \tag{5.4.8}$$

Also, for $|t|$ small enough, $(2t)^k\mathbf{A}^k \to \mathbf{0}$ as $t \to \infty$. By Result 1.3.20, $(\mathbf{I}-2t\mathbf{A})$ is nonsingular and

$$(\mathbf{I}-2t\mathbf{A})^{-1} = \sum_{r=0}^{\infty} 2^r t^r \mathbf{A}^r. \tag{5.4.9}$$

Substituting from (5.4.8) and (5.4.9) into the right side of (5.4.7), and equating coefficients of like powers of t^r on both sides gives the required result for the cumulants for the standard normal case.

For the general case, from Section 5.1, let $\mathbf{x} = \mathbf{\Gamma z} + \boldsymbol{\mu}$, where $\mathbf{z} \sim N_r(\mathbf{0}, \mathbf{I})$ and $\mathbf{\Gamma}$ is a matrix with $\mathbf{\Gamma\Gamma}' = \mathbf{\Sigma}$. Then

$$\begin{aligned}
\mathbf{x}'\mathbf{A}\mathbf{x} + \mathbf{a}'\mathbf{x} &= (\mathbf{\Gamma z} + \boldsymbol{\mu})'\mathbf{A}(\mathbf{\Gamma z} + \boldsymbol{\mu}) + \mathbf{a}'(\mathbf{\Gamma z} + \boldsymbol{\mu}) \\
&= \mathbf{z}'(\mathbf{\Gamma}'\mathbf{A}\mathbf{\Gamma})\mathbf{z} + (2\boldsymbol{\mu}'\mathbf{A}\mathbf{\Gamma} + \mathbf{a}'\mathbf{\Gamma})\mathbf{z} + \boldsymbol{\mu}'\mathbf{A}\boldsymbol{\mu} + \mathbf{a}'\boldsymbol{\mu} \\
&= \mathbf{z}'\tilde{\mathbf{A}}\mathbf{z} + \tilde{\mathbf{a}}'\mathbf{z} + a_0,
\end{aligned}$$

where $\tilde{\mathbf{A}} = \mathbf{\Gamma}'\mathbf{A}\mathbf{\Gamma}$, $\tilde{\mathbf{a}} = 2\mathbf{\Gamma}'\mathbf{A}\boldsymbol{\mu} + \mathbf{\Gamma}'\mathbf{a}$, and $a_0 = \boldsymbol{\mu}'\mathbf{A}\boldsymbol{\mu} + \mathbf{a}'\boldsymbol{\mu}$. Then the cumulant generating function of $\mathbf{x}'\mathbf{A}\mathbf{x} + \mathbf{a}'\mathbf{x}$ is $\log[M_{\mathbf{z}'\tilde{\mathbf{A}}\mathbf{z}+\tilde{\mathbf{a}}\mathbf{z}}(t)] + a_0 t$ so that from the result for the standard normal case,

$$\kappa_r = \begin{cases} \operatorname{tr}(\tilde{\mathbf{A}}) + a_0 & \text{if } r = 1 \\ 2^{r-1}(r-1)!\left[\operatorname{tr}(\tilde{\mathbf{A}}^r) + r\tilde{\mathbf{a}}'\tilde{\mathbf{A}}^{r-2}\tilde{\mathbf{a}}/4\right] & \text{if } r \geq 2. \end{cases}$$

Since for $r \geq 0$, $\operatorname{tr}(\tilde{\mathbf{A}}^r) = \operatorname{tr}\{(\mathbf{\Gamma}'\mathbf{A}\mathbf{\Gamma})^r\} = \operatorname{tr}\{\mathbf{\Gamma}'(\mathbf{A\Sigma})^{r-1}\mathbf{A}\mathbf{\Gamma}\} = \operatorname{tr}\{(\mathbf{A\Sigma})^r\}$, and $\tilde{\mathbf{a}}'\tilde{\mathbf{A}}^r\tilde{\mathbf{a}}/4 = (2\mathbf{A}\boldsymbol{\mu} + \mathbf{a})'\mathbf{\Gamma}(\mathbf{\Gamma}'\mathbf{A}\mathbf{\Gamma})^r\mathbf{\Gamma}'(2\mathbf{A}\boldsymbol{\mu} + \mathbf{a})/4 = (\mathbf{A}\boldsymbol{\mu} + \mathbf{a}/2)'\mathbf{\Sigma}(\mathbf{A\Sigma})^r(\mathbf{A}\boldsymbol{\mu} + \mathbf{a}/2)$, the proof for the general case follows. ■

From Result 5.4.2, we can show directly (see Exercise 5.26) that

$$\begin{aligned}
\mathrm{E}(\mathbf{x}'\mathbf{A}\mathbf{x}) &= \operatorname{tr}(\mathbf{A\Sigma}) + \boldsymbol{\mu}'\mathbf{A}\boldsymbol{\mu}, \\
\mathrm{Var}(\mathbf{x}'\mathbf{A}\mathbf{x}) &= 2\operatorname{tr}(\mathbf{A\Sigma A\Sigma}) + 4\boldsymbol{\mu}'\mathbf{A\Sigma A}\boldsymbol{\mu}, \text{ and} \\
\mathrm{Cov}(\mathbf{x}, \mathbf{x}'\mathbf{A}\mathbf{x}) &= 2\mathbf{\Sigma A}\boldsymbol{\mu}.
\end{aligned}$$

These properties are useful to characterize the first two moments of the distribution of $\mathbf{x}'\mathbf{A}\mathbf{x}$. We sometimes encounter the need to employ the following result on the moments of quadratic forms, which we state without proof (see Magnus and Neudecker, 1988).

Result 5.4.3. Let $\mathbf{A}$, $\mathbf{B}$, and $\mathbf{C}$ be symmetric $k \times k$ matrices and let $\mathbf{x} \sim N_k(\mathbf{0}, \mathbf{\Sigma})$. Then, letting $\mathbf{A}_1 = \mathbf{A\Sigma}$, $\mathbf{B}_1 = \mathbf{B\Sigma}$, and $\mathbf{C}_1 = \mathbf{C\Sigma}$,

$$\begin{aligned}
\mathrm{E}[(\mathbf{x}'\mathbf{A}\mathbf{x})(\mathbf{x}'\mathbf{B}\mathbf{x})] &= \operatorname{tr}(\mathbf{A}_1)\operatorname{tr}(\mathbf{B}_1) + 2\operatorname{tr}(\mathbf{A}_1\mathbf{B}_1), \\
\mathrm{E}[(\mathbf{x}'\mathbf{A}\mathbf{x})(\mathbf{x}'\mathbf{B}\mathbf{x})(\mathbf{x}'\mathbf{C}\mathbf{x})] &= \operatorname{tr}(\mathbf{A}_1)\operatorname{tr}(\mathbf{B}_1)\operatorname{tr}(\mathbf{C}_1) + 2[\operatorname{tr}(\mathbf{A}_1)][\operatorname{tr}(\mathbf{B}_1\mathbf{C}_1)] \\
&\quad + 2[\operatorname{tr}(\mathbf{B}_1)][\operatorname{tr}(\mathbf{A}_1\mathbf{C}_1)] + 2[\operatorname{tr}(\mathbf{C}_1)][\operatorname{tr}(\mathbf{A}_1\mathbf{B}_1)] \\
&\quad + 8\operatorname{tr}(\mathbf{A}_1\mathbf{B}_1\mathbf{C}_1).
\end{aligned}$$

We next state and prove an important result which gives a condition under which a quadratic form in a normal random vector has a noncentral chi-square distribution. This result is fundamental for a discussion of linear model inference under normality.

Result 5.4.4. Let $\mathbf{x} \sim N_k(\boldsymbol{\mu}, \mathbf{\Sigma})$, $\mathbf{\Sigma}$ being p.d., and let $\mathbf{A} \neq \mathbf{O}$ be a symmetric matrix of rank r. Then the quadratic form $U = \mathbf{x}'\mathbf{A}\mathbf{x}$ has a noncentral chi-square distribution if and only if $\mathbf{A\Sigma}$ is idempotent, in which case $U \sim \chi^2(r, \lambda)$ with $\lambda = \boldsymbol{\mu}'\mathbf{A}\boldsymbol{\mu}/2$.

Proof. We present a proof which only uses simple calculus and matrix algebra (see Khuri, 1999; Driscoll, 1999).

Necessity. We assume that $U = \mathbf{x}'\mathbf{A}\mathbf{x} \sim \chi^2(r, \lambda)$, and must show that this implies $\mathbf{A\Sigma}$ is idempotent of rank r. By Result 2.4.5, we can write $\mathbf{\Sigma} = \mathbf{PP}'$, for nonsingular $\mathbf{P}$. Let $\mathbf{y} = \mathbf{P}^{-1}\mathbf{x}$. Clearly, $U = \mathbf{y}'\mathbf{P}'\mathbf{APy}$. Since $\mathbf{P}'\mathbf{AP}$ is symmetric, by Result 2.3.4, there exists an orthogonal matrix $\mathbf{Q}$ such that $\mathbf{P}'\mathbf{AP} = \mathbf{QDQ}'$, where $\mathbf{D} = \text{diag}(\mathbf{O}, d_1\mathbf{I}_{m_1}, \cdots, d_p\mathbf{I}_{m_p})$, with $d_1 < \cdots < d_p$ denoting the distinct *nonzero* eigenvalues of $\mathbf{P}'\mathbf{AP}$, with multiplicities $m_1, \ldots, m_p$. Then $\mathbf{z} = \mathbf{Q}'\mathbf{y} \sim N_k(\boldsymbol{\nu}, \mathbf{I})$ with $\boldsymbol{\nu} = \mathbf{Q}'\mathbf{P}^{-1}\boldsymbol{\mu}$. Partition $\mathbf{z}$ as $(\mathbf{z}_0', \mathbf{z}_1', \cdots, \mathbf{z}_p')'$, so that $\mathbf{z}_i$ has dimension m_i, $i \geq 1$. Partition $\boldsymbol{\nu}$ conformably as $(\boldsymbol{\nu}_0', \boldsymbol{\nu}_1', \cdots, \boldsymbol{\nu}_p')'$. Then

$$U = \mathbf{z}'\mathbf{Dz} = \sum_{i=1}^{p} d_i \mathbf{z}_i'\mathbf{z}_i. \tag{5.4.10}$$

Since $\mathbf{z}_i'\mathbf{z}_i \sim \chi^2(m_i, \theta_i)$ are independent, with $\theta_i = \boldsymbol{\nu}_i'\boldsymbol{\nu}_i/2 \geq 0$,

$$M_U(t) = \prod_{i=1}^{p} M_{\mathbf{z}_i'\mathbf{z}_i}(d_i t), \tag{5.4.11}$$

where, each m.g.f. on the right side has the form in (5.3.4) and is positive, i.e., greater than zero. From (5.3.4), we also see that $M_U(t) < \infty$ if and only if $t < 1/2$, and $M_{\mathbf{z}_i'\mathbf{z}_i}(d_i t) < \infty$ if and only if $d_i t < 1/2$. If possible, let $d_1 < 0$. Then, for all $t < 1/(2d_1) < 0$, the left side of (5.4.11) is finite, while the right side is ∞, which is a contradiction. This implies that each d_i must be positive. Next, let us consider three possible cases, i.e., $d_p > 1$, $d_p = 1$ and $d_p < 1$.

Case (i). If possible, let $d_p > 1$. Then, for any $t \in (1/(2d_p), 1/2)$, the left side of (5.4.11) is finite, while the right side is ∞, which is again a contradiction. Thus, we must have $d_i \in (0, 1]$ for $i = 1, \ldots, p$.

Case (ii). Suppose $d_p = 1$. From (5.4.11),

$$\frac{M_U(t)}{M_{\mathbf{z}_p'\mathbf{z}_p}(t)} = \prod_{i=1}^{p-1} M_{\mathbf{z}_i'\mathbf{z}_i}(d_i t). \tag{5.4.12}$$

From (5.3.4), for $t < 1/2$,

$$\begin{aligned}\frac{M_U(t)}{M_{\mathbf{z}_p'\mathbf{z}_p}(t)} &= \frac{(1-2t)^{-r/2}\exp\{2\lambda t/(1-2t)\}}{(1-2t)^{-m_p/2}\exp\{2\theta_p t/(1-2t)\}} \\ &= (1-2t)^{-(r-m_p)/2}\exp\{2(\lambda-\theta_p)t/(1-2t)\}.\end{aligned}$$

If $r - m_p \neq 0$ or $\lambda - \theta_p \neq 0$, then, depending on their signs, as $t \uparrow 1/2$ (i.e., increases from 0 to 1/2), the limit of the left side of (5.4.12) is either 0 or ∞. On the other hand, since each d_i on the right side of (5.4.12) (if there are any), lies in $(0, 1)$, the limit of the right side is strictly positive and finite. This contradiction implies that $r - m_p = \lambda - \theta_p = 0$, so that both sides of (5.4.12) are constant and equal to 1, implying that $p = 1$. As a result, the only nonzero eigenvalue of $\mathbf{P}'\mathbf{AP}$ is 1. That is, $\mathbf{P}'\mathbf{AP}$, and hence $\mathbf{A\Sigma}$ is idempotent with rank equal to the multiplicity of 1, which is r.

Case (iii). Finally, suppose $d_p < 1$. A similar argument as above shows that as $t \uparrow 1/2$, the left side of (5.4.11) tends to either 0 or ∞, while the right side of (5.4.11) tends to a positive finite number. This contradiction then rules out the possibility that $d_p < 1$ and completes the proof of necessity.

Sufficiency. We must show that if $\mathbf{A\Sigma}$ is idempotent of rank r, then U has a $\chi^2(r,\lambda)$ distribution. Since $\mathbf{A\Sigma}$ is idempotent, in (5.4.10), $p = 1$, $d_1 = 1$, and $m_1 = r$. Hence, $\mathbf{x}'\mathbf{A}\mathbf{x}$ has a $\chi^2(r,\lambda)$ distribution, where $\lambda = \theta_1$. An alternate proof of sufficiency amounts to showing that the m.g.f. of $\mathbf{x}'\mathbf{A}\mathbf{x}$ coincides with the m.g.f. of a $\chi^2(r,\lambda)$ random variable, which is obvious from (5.3.5) and (5.4.6). ■

We leave the following consequence of Result 5.4.4 as Exercise 5.24.

Result 5.4.5. Let $\mathbf{x} \sim N_k(\boldsymbol{\mu}, \boldsymbol{\Sigma})$ with $\mathrm{r}(\boldsymbol{\Sigma}) = k$. Then $\mathbf{x}'\mathbf{A}\mathbf{x}$ follows a noncentral chi-square distribution if and only if any one of the following three conditions is met:

1. $\mathbf{A\Sigma}$ is an idempotent matrix of rank m,
2. $\mathbf{\Sigma A}$ is an idempotent matrix of rank m,
3. $\boldsymbol{\Sigma}$ is a g-inverse of $\mathbf{A}$ with $\mathrm{r}(\mathbf{A}) = m$.

Under any of the conditions, $\mathbf{x}'\mathbf{A}\mathbf{x} \sim \chi^2(m,\lambda)$ with $\lambda = \boldsymbol{\mu}'\mathbf{A}\boldsymbol{\mu}/2$.

Example 5.4.1. Suppose $\mathbf{x} \sim N_k(\boldsymbol{\mu}, \boldsymbol{\Sigma})$, where $\boldsymbol{\Sigma}$ is p.d. We show that $\mathbf{x}'\mathbf{A}\mathbf{x}$ is distributed as a linear combination of independent noncentral chi-square variables. First, since $\boldsymbol{\Sigma}$ is p.d., there is a nonsingular matrix $\mathbf{P}$ such that $\boldsymbol{\Sigma} = \mathbf{PP}'$. Since $\mathbf{P}'\mathbf{AP}$ is symmetric, there exists an orthogonal matrix $\mathbf{Q}$ such that $\mathbf{Q}'\mathbf{P}'\mathbf{APQ} = \mathbf{D} = \mathrm{diag}(\lambda_1, \cdots, \lambda_r, 0, \cdots, 0)$, these being the distinct eigenvalues of $\mathbf{P}'\mathbf{AP}$. Let $\mathbf{x} = \mathbf{PQz}$. From Result 5.2.5, $\mathbf{z} \sim N(\mathbf{Q}'\mathbf{P}^{-1}\boldsymbol{\mu}, \mathbf{I})$; also $\mathbf{x}'\mathbf{A}\mathbf{x} = \mathbf{z}'\mathbf{D}\mathbf{z} = \sum_{i=1}^{k} \lambda_i Z_i^2$. The required result follows from Result 5.3.2. □

When $\mathbf{x}$ may have a singular normal distribution, the following result on a quadratic form holds (Ogasawara and Takahashi, 1951). Its proof is left as Exercise 5.28.

Result 5.4.6. Let $\mathbf{x} \sim N_k(\mathbf{0}, \boldsymbol{\Sigma})$. The quadratic form $\mathbf{x}'\mathbf{A}\mathbf{x}$ has a chi-square distribution if and only if $\mathbf{\Sigma A \Sigma A \Sigma} = \mathbf{\Sigma A \Sigma}$, in which case $\mathbf{x}'\mathbf{A}\mathbf{x} \sim \chi^2_p$ with $p = \mathrm{r}(\mathbf{\Sigma A \Sigma})$.

The next result is often referred to as Laha's theorem (Laha, 1956). It is a generalization of Craig's theorem which deals with the independence of $\mathbf{x}'\mathbf{A}\mathbf{x}$ and $\mathbf{x}'\mathbf{B}\mathbf{x}$ when $\mathbf{x} \sim N_k(\mathbf{0}, \mathbf{I})$ (Craig, 1943; Shanbhag, 1966).

Result 5.4.7. Let $\mathbf{x} \sim N_k(\boldsymbol{\mu}, \boldsymbol{\Sigma})$, where $\boldsymbol{\Sigma}$ is p.d. For $\mathbf{x}'\mathbf{A}\mathbf{x} + \mathbf{a}'\mathbf{x}$ and $\mathbf{x}'\mathbf{B}\mathbf{x} + \mathbf{b}'\mathbf{x}$ to be independent, it is necessary and sufficient that (i) $\mathbf{A\Sigma B} = \mathbf{O}$, (ii) $\mathbf{a}'\mathbf{\Sigma B} = \mathbf{0}$, (iii) $\mathbf{b}'\mathbf{\Sigma A} = \mathbf{0}$, and (iv) $\mathbf{a}'\mathbf{\Sigma b} = 0$.

Proof. The proof presented below uses the fact that two random variables W_1 and W_2 with bounded c.g.f.'s in a neighborhood of the origin are independent if and only if their cumulants satisfy

$$\kappa_r(yW_1 + zW_2) = \kappa_r(yW_1) + \kappa_r(zW_2)$$

for all real y and z and for all positive integers r.

Sufficiency. Since $\mathbf{A}' = \mathbf{A}$, $\mathbf{x}'\mathbf{A}\mathbf{x} = (\mathbf{Ax})'(\mathbf{A}'\mathbf{A})^-\mathbf{A}'(\mathbf{Ax})$ is a function of $\mathbf{Ax}$. Likewise $\mathbf{x}'\mathbf{B}\mathbf{x}$ is a function of $\mathbf{Bx}$. If (i)–(iv) are all satisfied, then by Result 5.2.10, each of $\mathbf{a}'\mathbf{x}$ and $\mathbf{Ax}$ is independent of both $\mathbf{b}'\mathbf{x}$ and $\mathbf{Bx}$, and hence $\mathbf{x}'\mathbf{A}\mathbf{x}+\mathbf{a}'\mathbf{x}$ is independent of $\mathbf{x}'\mathbf{B}\mathbf{x}+\mathbf{b}'\mathbf{x}$.

Necessity. We first consider the case where $\mathbf{x} \sim N_k(\mathbf{0}, \mathbf{I})$. For any y, z, $\xi_{yz} = y(\mathbf{x}'\mathbf{A}\mathbf{x} + \mathbf{a}'\mathbf{x}) + z(\mathbf{x}'\mathbf{B}\mathbf{x} + \mathbf{b}'\mathbf{x}) = \mathbf{x}'(y\mathbf{A} + z\mathbf{B})\mathbf{x} + (y\mathbf{a} + z\mathbf{b})'\mathbf{x}$. By Result 5.4.2, for $r \geq 2$,

$$\frac{\kappa_r(\xi_{yz})}{2^{r-1}(r-1)!} = \mathrm{tr}\{(y\mathbf{A} + z\mathbf{B})^r\} + r(y\mathbf{a} + z\mathbf{b})'(y\mathbf{A} + z\mathbf{B})^{r-2}(y\mathbf{a} + z\mathbf{b})/4.$$

On the other hand,

$$\frac{\kappa_r(y(\mathbf{x}'\mathbf{A}\mathbf{x}+\mathbf{a}'\mathbf{x}))}{2^{r-1}(r-1)!} = y^r \operatorname{tr}(\mathbf{A}^r) + ry^r\mathbf{a}'\mathbf{A}^{r-2}\mathbf{a}/4,$$
$$\frac{\kappa_r(z(\mathbf{x}'\mathbf{B}\mathbf{x}+\mathbf{b}'\mathbf{x}))}{2^{r-1}(r-1)!} = z^r \operatorname{tr}(\mathbf{B}^r) + rz^r\mathbf{b}'\mathbf{B}^{r-2}\mathbf{b}/4,$$

By independence,

$$\begin{aligned}&\operatorname{tr}\{(y\mathbf{A}+z\mathbf{B})^r\} + \mathrm{r}(y\mathbf{a}+z\mathbf{b})'(y\mathbf{A}+z\mathbf{B})^{r-2}(y\mathbf{a}+z\mathbf{b})/4\\ &= y^r \operatorname{tr}(\mathbf{A}^r) + ry^r\mathbf{a}'\mathbf{A}^{r-2}\mathbf{a}/4 + z^r\operatorname{tr}(\mathbf{B}^r) + rz^r\mathbf{b}'\mathbf{B}^{r-2}\mathbf{b}/4.\end{aligned} \tag{5.4.13}$$

Let $\lambda_1, \cdots, \lambda_p$ denote all the distinct nonzero elements in the spectra (set of eigenvalues) of $y\mathbf{A}$, $z\mathbf{B}$, and $y\mathbf{A}+z\mathbf{B}$ (so that the λ_i are distinct). Let $\mathbf{C}$ represent one of the three matrices. From the spectral decomposition of a symmetric matrix (see Result 2.3.4), $\mathbf{C} = \sum_{i=1}^p \lambda_i \mathbf{P}_{i,\mathbf{C}}\mathbf{P}'_{i,\mathbf{C}}$, where $\mathbf{P}_{i,\mathbf{C}}$ consists of $m_{i,\mathbf{C}}$ orthonormal eigenvectors of $\mathbf{C}$ corresponding to λ_i, with $m_{i,\mathbf{C}}$ being the multiplicity of λ_i as an eigenvalue of $\mathbf{C}$; if λ_i is not an eigenvalue of $a\mathbf{A}$, take $m_{i,\mathbf{C}} = 0$, and $\mathbf{P}_{i,\mathbf{C}} = \mathbf{O}$. Let $\mathbf{u}_\mathbf{C}$ be one of the vectors $y\mathbf{a}$, $z\mathbf{b}$, and $y\mathbf{a}+z\mathbf{b}$ corresponding to $\mathbf{C}$. Let $\mu_{i,\mathbf{C}} = \mathbf{u}'_\mathbf{C}\mathbf{P}_{i,\mathbf{C}}\mathbf{P}'_{i,\mathbf{C}}\mathbf{u}_\mathbf{C}$. Then for any positive integer $r \geq 2$,

$$\operatorname{tr}(\mathbf{C}^r) = \sum_{i=1}^p \lambda_i^r m_{i,\mathbf{C}}, \quad r\mathbf{u}'_\mathbf{C}\mathbf{C}^{r-2}\mathbf{u}_\mathbf{C} = r\sum_{i=1}^p \lambda_i^{r-2}\mu_{i,\mathbf{C}}, \tag{5.4.14}$$

so from (5.4.13)

$$\begin{aligned}&\sum_{i=1}^p \lambda_i^r \underbrace{(m_{i,y\mathbf{A}+z\mathbf{B}} - m_{i,y\mathbf{A}} - m_{i,z\mathbf{B}})}_{\nu_i}\\ &+\sum_{i=1}^p r\lambda_i^{r-2}\underbrace{(\mu_{i,y\mathbf{A}+z\mathbf{B}} - \mu_{i,y\mathbf{A}} - \mu_{i,z\mathbf{B}})}_{\nu_{p+i}} = 0.\end{aligned} \tag{5.4.15}$$

Recall that all $\lambda_1, \cdots, \lambda_p$ are distinct and nonzero. Suppose $\lambda_1 < \cdots < \lambda_p$ and let $\rho = \max(|\lambda_1|, |\lambda_p|)$. Divide both sides of (5.4.15) by $r\rho^{r-2}$ and let $r \to \infty$.

There are three cases, (a) $|\lambda_1| = |\lambda_p|$, so that $-\lambda_1 = \lambda_p = \rho$, (b) $|\lambda_1| < \lambda_p = \rho$, and (c) $|\lambda_p| < -\lambda_1 = \rho$. Consider case (a). If r stays even as $r \to \infty$, then $\nu_1 + \nu_p = 0$, while if r stays odd as $r \to \infty$, then $-\nu_1 + \nu_p = 0$. As a result, $\nu_1 = \nu_p = 0$. Now divide both sides of (5.4.15) by ρ^r. By a similar argument, $\mu_1 = \mu_p = 0$. The other two cases can be treated similarly. Thus, the coefficients corresponding to the eigenvalue(s) with the largest absolute value are all zero. We then can reduce the number of terms of the summations in (5.4.15) by two in case (a), and by one in cases (b) and (c). Then, by the same argument, the coefficients corresponding to the eigenvalue(s) with the second largest absolute value are all zero. Repeating the argument, it follows that all the ν_i's in (5.4.15) are zero, giving

$$m_{i,y\mathbf{A}+z\mathbf{B}} - m_{i,y\mathbf{A}} - m_{i,z\mathbf{B}} = 0 \quad \text{and} \quad \mu_{i,y\mathbf{A}+z\mathbf{B}} - \mu_{i,y\mathbf{A}} - \mu_{i,z\mathbf{B}} = 0.$$

Then from (5.4.14),

$$\operatorname{tr}\{(y\mathbf{A}+z\mathbf{B})^r\} = y^r\operatorname{tr}(\mathbf{A}^r) + b^r\operatorname{tr}(\mathbf{B}^r), \tag{5.4.16}$$
$$(y\mathbf{a}+z\mathbf{b})'(y\mathbf{A}+z\mathbf{B})^{r-2}(y\mathbf{a}+z\mathbf{b}) = y^r\mathbf{a}'\mathbf{A}^{r-2}\mathbf{a} + z^r\mathbf{b}'\mathbf{B}^{r-2}\mathbf{b}. \tag{5.4.17}$$

Since the equation holds for all y and z, for any integers $r \geq 2$, $m \geq 0$, and $n \geq 0$, the coefficients of $y^m z^n$ on both sides are equal. In particular, for $r = 4$ and $m = n = 2$, comparing the coefficients of y^2z^2 in (5.4.16) yields

$$\begin{aligned}&\mathrm{tr}(\mathbf{A}^2\mathbf{B}^2 + \mathbf{B}^2\mathbf{A}^2 + (\mathbf{AB} + \mathbf{BA})^2)\\ &= \mathrm{tr}((\mathbf{AB})'(\mathbf{AB}) + (\mathbf{BA})'(\mathbf{BA}) + (\mathbf{AB} + \mathbf{BA})'(\mathbf{AB} + \mathbf{BA})) = 0.\end{aligned}$$

It then follows that $\mathbf{AB} = \mathbf{BA} = \mathbf{O}$. Plugging this into (5.4.17) and comparing the coefficients of y^2z^2 therein when $r = 4$ yields

$$\mathbf{b}'\mathbf{A}^2\mathbf{b} + \mathbf{a}'\mathbf{B}^2\mathbf{a} = (\mathbf{Ab})'(\mathbf{Ab}) + (\mathbf{Ba})'(\mathbf{Ba}) = 0,$$

so $\mathbf{Ab} = \mathbf{Ba} = \mathbf{0}$. Finally, letting $r = 2$ in (5.4.17) and comparing the coefficients of yz on both sides gives $\mathbf{a}'\mathbf{b} = 0$. Then the necessity is proved for the case $\mathbf{x} \sim N_k(\mathbf{0}, \mathbf{I})$.

For the general case, $\mathbf{x} = \mathbf{\Gamma z} + \boldsymbol{\mu}$, where $\mathbf{\Gamma}$ is a nonsingular matrix with $\mathbf{\Gamma\Gamma}' = \mathbf{\Sigma}$. Then

$$\begin{aligned}\mathbf{x}'\mathbf{Ax} + \mathbf{a}'\mathbf{x} &= (\mathbf{\Gamma z} + \boldsymbol{\mu})'\mathbf{A}(\mathbf{\Gamma z} + \boldsymbol{\mu}) + \mathbf{a}'(\mathbf{\Gamma z} + \boldsymbol{\mu})\\ &= \mathbf{z}'(\mathbf{\Gamma}'\mathbf{A\Gamma})\mathbf{z} + (2\mathbf{\Gamma}'\mathbf{A}\boldsymbol{\mu} + \mathbf{\Gamma}'\mathbf{a})'\mathbf{z} + (\boldsymbol{\mu}'\mathbf{\Gamma}\boldsymbol{\mu} + \mathbf{a}'\boldsymbol{\mu}). \qquad (5.4.18)\end{aligned}$$

Likewise $\mathbf{x}'\mathbf{Bx} + \mathbf{b}'\mathbf{x}$ can be written in terms of $\mathbf{z}$. Then the necessity of (i)–(iv) in the general case follows from the standard normal case. Further details of the proof are left as a part of Exercise 5.34. ■

Result 5.4.7 has the following consequence.

Result 5.4.8. Let $\mathbf{x} \sim N_k(\boldsymbol{\mu}, \mathbf{\Sigma})$ with $\mathbf{\Sigma}$ being p.d. Let $\mathbf{B}$ be an $m \times k$ matrix. For $\mathbf{Bx}$ and $\mathbf{x}'\mathbf{Ax} + \mathbf{a}'\mathbf{x}$ to be independent, it is necessary and sufficient that (i) $\mathbf{B\Sigma A} = \mathbf{O}$ and (ii) $\mathbf{B\Sigma a} = \mathbf{0}$.

Proof. Denote the rows of $\mathbf{B}$ by $\mathbf{b}_1', \cdots, \mathbf{b}_m'$. Then $\mathbf{Bx}$ and $\mathbf{x}'\mathbf{Ax}+\mathbf{a}'\mathbf{x}$ are independent if and only if each $\mathbf{b}_i'\mathbf{x}$ is independent of $\mathbf{x}'\mathbf{Ax}+\mathbf{a}'\mathbf{x}$. The result then follows from Result 5.4.7. A less technical proof of the necessity of (i) and (ii), which does not rely on the cumulant generating function argument used in Result 5.4.7, is sketched in Exercise 5.29. ■

For more details, see Ogawa (1950, 1993), Laha (1956), and Reid and Driscoll (1988). Using only linear algebra and calculus, Driscoll and Krasnicka (1995) proved the general case where $\mathbf{x} \sim N_k(\boldsymbol{\mu}, \mathbf{\Sigma})$ with $\mathbf{\Sigma}$ possibly being singular. In this case, they showed that a necessary and sufficient condition for $\mathbf{x}'\mathbf{Ax} + \mathbf{a}'\mathbf{x}$ and $\mathbf{x}'\mathbf{Bx} + \mathbf{b}'\mathbf{x}$ to be independently distributed is that

$$\begin{aligned}\mathbf{\Sigma A\Sigma B\Sigma} = \mathbf{O}, \quad & \mathbf{\Sigma A\Sigma}(\mathbf{B}\boldsymbol{\mu} + \mathbf{b}) = \mathbf{0},\\ \mathbf{\Sigma B\Sigma}(\mathbf{A}\boldsymbol{\mu} + \mathbf{a}) = \mathbf{0}, \quad \text{and} \quad & (\mathbf{A}\boldsymbol{\mu} + \mathbf{a}/2)'\mathbf{\Sigma}(\mathbf{B}\boldsymbol{\mu} + \mathbf{b}/2) = 0.\end{aligned} \qquad (5.4.19)$$

The proof is left as a part of Exercise 5.34.

Example 5.4.2. We show independence between the mean and sum of squares. Let $\mathbf{x} = (X_1, \cdots, X_n)' \sim N_n(\mathbf{0}, \mathbf{I})$ so that we can think of $X_1, \cdots, X_n$ as a random sample from a $N(0, 1)$ population. The sample mean $\overline{X} = \mathbf{1}'\mathbf{x}/n$ has a $N(0, 1/n)$ distribution (using Result 5.2.9), and the sample sum of squares $\sum_{i=1}^n (X_i - \overline{X})^2$ has a central chi-square distribution with $n - 1$ degrees of freedom (by Result 5.4.1). It is easily verified by expressing $\overline{X}^2$ and $\sum_{i=1}^n (X_i - \overline{X})^2$ as quadratic forms in $\mathbf{x}$ that $\overline{X}^2$, and hence $\overline{X}$ is independent of $\sum_{i=1}^n (X_i - \overline{X})^2$. □

Example 5.4.3. Let $\mathbf{x} \sim N_k(\boldsymbol{\mu}, \mathbf{I})$ and suppose a $k \times k$ orthogonal matrix $\mathbf{T}$ is partitioned as $\mathbf{T} = (\mathbf{T}_1', \mathbf{T}_2')'$, where $\mathbf{T}_i$ is a $k \times k_i$ matrix, $i = 1, 2$, such that $k_1 + k_2 = k$. It is easy to verify that $\mathbf{T}_1\mathbf{T}_1' = \mathbf{I}_{k_1}$, $\mathbf{T}_2\mathbf{T}_2' = \mathbf{I}_{k_2}$, $\mathbf{T}_1\mathbf{T}_2' = \mathbf{O}$, $\mathbf{T}_2\mathbf{T}_1' = \mathbf{O}$, and $\mathbf{T}'\mathbf{T} = \mathbf{I}_k$. Also, $\mathbf{T}_i'\mathbf{T}_i$ is idempotent of rank k_i, $i = 1, 2$. By Result 5.4.5, $\mathbf{x}'\mathbf{T}_i'\mathbf{T}_i\mathbf{x} \sim \chi^2(k_i, \boldsymbol{\mu}'\mathbf{T}_i'\mathbf{T}_i\boldsymbol{\mu})$, $i = 1, 2$. By Result 5.4.7, these two quadratic forms are independently distributed. □

Note that Result 5.4.7 applies whether or not the quadratic forms $\mathbf{x}'\mathbf{A}\mathbf{x}$ and $\mathbf{x}'\mathbf{B}\mathbf{x}$ have chi-square distributions. We next discuss without proof a more general theorem dealing with quadratic forms in normal random vectors. The basic result is due to Cochran (1934), which was later modified by James (1952). A proof is sketched in Exercises 5.37 and 5.38.

Result 5.4.9. Cochran's theorem. Let $\mathbf{x} \sim N_k(\boldsymbol{\mu}, \mathbf{I})$. Let $Q_j = \mathbf{x}'\mathbf{A}_j\mathbf{x}$, $j = 1, \cdots, L$, where $\mathbf{A}_j$ are symmetric matrices with $\text{r}(\mathbf{A}_j) = r_j$ and $\sum_{j=1}^L \mathbf{A}_j = \mathbf{I}$. Then, in order for Q_j's to be independent noncentral chi-square variables, it is necessary and sufficient that $\sum_{j=1}^L r_j = k$, in which case, $Q_j \sim \chi^2(r_j, \boldsymbol{\mu}'\mathbf{A}_j\boldsymbol{\mu}/2)$.

5.5 Remedies for non-normality

Section 5.5.1 discusses transformations to normality, while Section 5.5.2 describes a few alternate distributions for the responses in the GLM.

5.5.1 Transformations to normality

A general family of univariate and multivariate transformations to normality is introduced in this section.

5.5.1.1 Univariate transformations

A transformation of X is a function T which replaces X by a new "transformed" variable $T(X)$. The simplest and most widely used transformations belong to the family of power transformations, which are defined below.

Definition 5.5.1. Power transformation. The family of power transformations have the form

$$T_P(X) = \begin{cases} aX^p + b & \text{if } p \neq 0, \\ c \log X + d & \text{if } p = 0, \end{cases} \tag{5.5.1}$$

where a, b, c, d, and p are arbitrary real scalars.

The power transformation is useful for bringing skewed distributions of random variables closer to symmetry, and thence, to normality. The square root transformation or the logarithmic transformation, for instance, have the effect of "pulling in " one tail of the distribution. Any power transformation is either concave or convex throughout its domain of positive numbers, i.e., there is no point of inflection. This implies that a power transformation either compresses the scale for larger X values more than it does for smaller X values (for example, $T_P(X) = \log X$), or it does the reverse (for example, $T_P(X) = X^2$). We cannot, however, use a power transformation to expand the scale of X for large and small values, while compressing it for values in between! Tukey (1957) considered the family of transformations

$$X^{(\lambda)} = \begin{cases} X^\lambda & \text{if } \lambda \neq 0, \\ \log X & \text{if } \lambda = 0 \text{ and } X > 0, \end{cases} \tag{5.5.2}$$

which is a special case of (5.5.1). This family of transformations, indexed by λ, includes the well-known square root (when $\lambda = 1/2$), logarithmic (when $\lambda = 0$), and reciprocal (when $\lambda = -1$) transformations. Notice that (5.5.2) has a discontinuity at $\lambda = 0$. Box and Cox (1964) offered a remedy to this problem which is defined below.

Definition 5.5.2. Box–Cox transformation.

$$X^{(\lambda)} = \begin{cases} (X^\lambda - 1)/\lambda & \text{if } \lambda \neq 0, \\ \log X & \text{if } \lambda = 0 \text{ and } X > 0, \end{cases} \tag{5.5.3}$$

which has been widely used in practice to achieve transformation to normality.

If some of the X_i's assume negative values, a positive constant may be added to all the variables to make them positive. Box and Cox (1964) also proposed the shifted power transformation defined as follows.

Definition 5.5.3. Box–Cox shifted power transformation.

$$X^{(\lambda)} = \begin{cases} [(X + \delta)^\lambda - 1]/\lambda & \text{if } \lambda \neq 0, \\ \log[X + \delta] & \text{if } \lambda = 0, \end{cases} \tag{5.5.4}$$

where the parameter δ is chosen such that $X > -\delta$.

Several modifications of the Box–Cox transformation exist in the literature, of which a few are mentioned here. Manly (1976) proposed a modification which allows the incorporation of negative observations and is an effective tool to transform skewed unimodal distributions to approximate normality. This is given by

$$X^{(\lambda)} = \begin{cases} [\exp(\lambda X) - 1]/\lambda & \text{if } \lambda \neq 0 \\ X & \text{if } \lambda = 0. \end{cases} \tag{5.5.5}$$

Bickel and Doksum (1981) proposed a modification which incorporates unbounded support for $X^{(\lambda)}$:

$$X^{(\lambda)} = \{|X|^\lambda \operatorname{sign}(X) - 1\}/\lambda, \tag{5.5.6}$$

where

$$\operatorname{sign}(u) = \begin{cases} -1 & u < 0, \\ 0 & u = 0, \\ 1 & u > 0. \end{cases} \tag{5.5.7}$$

With any of these transformations, it is important to note that very often the range of the transformed variable $X^{(\lambda)}$ is restricted based on the sign of λ, in turn implying that the transformed values may not cover the entire real line. Consequently, only approximate normality may result from the transformation.

Definition 5.5.4. Modulus transformation. The following transformation was suggested by John and Draper (1980):

$$X^{(\lambda)} = \begin{cases} \text{sign}(X)[(|X|+1)^{\lambda} - 1]/\lambda & \text{if } \lambda \neq 0 \\ \text{sign}(X)[\log(|X|+1)] & \text{if } \lambda = 0. \end{cases} \tag{5.5.8}$$

The modulus transformation works best to achieve approximate normality when the distribution of X is already approximately symmetric about some location. It alters each half of the distribution about this central value via the same power transformation in order to bring the distribution closer to a normal distribution. It is not difficult to see that when $X > 0$, (5.5.8) is equivalent to the power transformation. Given a random sample $X_1, \cdots, X_N$, estimation of λ using the maximum likelihood approach and the Bayesian framework was proposed by Box and Cox (1964), while Carroll and Ruppert (1988) discussed several robust adaptations. The maximum likelihood approach for estimating λ in the context of linear regression is described in Section 8.2.1. A generalization of the Box–Cox transformation to symmetric distributions was considered by Hinkley (1975), while Solomon (1985) extended it to random-effects models.

In a general linear model, if normality is suspect, a possible remedy is a transformation of the data. In general, the parameter of the transformation, i.e., λ, is unknown and must be estimated from the data. We give a brief description of the maximum likelihood approach for estimating λ (Box and Cox, 1964). Suppose we fit the model (4.1.2) to the data $(\mathbf{x}_i, Y_i)$, $i = 1, \cdots, N$. The estimation procedure consists of the following steps. We first choose a set of λ values in a pre-selected real interval, such as $(-5, 5)$. For each chosen λ, we compute the vector of transformed variables $\mathbf{y}^{(\lambda)} = (Y_1^{(\lambda)}, \cdots, Y_N^{(\lambda)})$ using (5.5.3), say. We then fit the normal linear model (4.1.2) to $(\mathbf{x}_i, Y_i^{(\lambda)})$, and compute $SSE(\lambda)$ based on the maximum likelihood estimates (which coincide with the OLS estimates under normality). In the plot of $SSE(\lambda)$ versus λ, we locate the value of λ which corresponds to the minimum value of $SSE(\lambda)$. This is the MLE of λ.

5.5.1.2 Multivariate transformations

Andrews et al. (1971) proposed a multivariate generalization of the Box–Cox transformation. A transformation of a k-dimensional random vector $\mathbf{x}$ may be defined either with the objective of transforming each component X_j, $j = 1, \cdots, k$, marginally to normality, or to achieve joint normality. They defined the simple family of "marginal" transformations as follows.

Definition 5.5.5. Transformation to marginal normality. The transformation for the jth component X_j is defined for $j = 1, \cdots, k$ by

$$X_j^{(\lambda_j)} = \begin{cases} (X_j^{\lambda_j} - 1)/\lambda_j & \text{if } \lambda_j \neq 0, \\ \log X_j & \text{if } \lambda_j = 0 \text{ and } X_j > 0, \end{cases} \tag{5.5.9}$$

where the λ_j are chosen to improve the marginal normality of $X_j^{(\lambda_j)}$ for $j = 1, \cdots, k$, via maximum likelihood estimation.

It is well known that marginal normality of the components does not imply multivariate normality; in using the marginal transformations in the previous section, it is hoped that the marginal normality of $X_j^{(\lambda_j)}$ might lead to a transformed $\mathbf{x}$ vector which is more amenable to procedures assuming multivariate normality. The following set of marginal transformations was proposed by Andrews et al. (1971) in order to achieve *joint normality* of the transformed data.

Definition 5.5.6. Transformation to joint normality. Suppose $X_j^{(\lambda_j)}$, $j = 1, \cdots, k$ denote the marginal transformations, and suppose the vector $\boldsymbol{\lambda} = (\lambda_1, \cdots, \lambda_k)'$ denotes the set of parameters that yields joint normality of the vector $\mathbf{x}^{(\boldsymbol{\lambda})} = (X_1^{(\lambda_1)}, \cdots, X_k^{(\lambda_k)})'$. Suppose the mean and covariance of this multivariate normal distribution are $\boldsymbol{\mu}$ and $\boldsymbol{\Sigma}$. The joint p.d.f. of $\mathbf{x}$ is

$$f(\mathbf{x}; \boldsymbol{\mu}, \boldsymbol{\Sigma}, \boldsymbol{\lambda}) = \frac{|J|}{(2\pi)^{k/2}|\boldsymbol{\Sigma}|^{1/2}} \exp\left\{-\frac{1}{2}(\mathbf{x}^{(\boldsymbol{\lambda})} - \boldsymbol{\mu})'\boldsymbol{\Sigma}^{-1}(\mathbf{x}^{(\boldsymbol{\lambda})} - \boldsymbol{\mu})\right\}, \tag{5.5.10}$$

where the Jacobian of the transformation is given by $J = \prod_{j=1}^k X_j^{(\lambda_j)}$.

Given N independent observations $\mathbf{x}_1, \cdots, \mathbf{x}_N$, the estimate of $\boldsymbol{\lambda}$, along with $\boldsymbol{\mu}$ and $\boldsymbol{\Sigma}$ is obtained by numerically maximizing the likelihood function which has the form in (5.5.10). There are some situations where nonnormality is manifest only in some directions in the k-dimensional space. Andrews et al. (1971) suggested an approach (a) to identify these directions, and (b) to then estimate a power transformation of the projections of $\mathbf{x}_1, \cdots, \mathbf{x}_N$ onto these selected directions in order to improve the normal approximation.

5.5.2 Alternatives to multivariate normal distribution

The multivariate normal distributions constitute a very useful family of symmetric distributions that have found widespread use in the classical theory of linear models and multivariate analysis. However, the normal distribution is not the only choice to characterize the response variable in many situations. Particularly, in robustness studies, where interest lies in assessing sensitivity of procedures to the assumption of normality, interest has centered on a more general class of multivariate distributions (see Kotz et al. (2000)).

Of special interest are distributions whose contours of equal density have elliptical shapes (see Section 5.2), and whose tail behavior differs from that of the normal. We begin the discussion by introducing a finite mixture distribution of multivariate normal distributions, as well as scale mixtures of multivariate normals. We then extend these to a more general class of spherically symmetric distributions, and finally to the class of elliptically symmetric distributions.

5.5.2.1 Mixture of normals

A finite parametric mixture of normal distributions is useful in several practical applications. We give a definition and some examples.

Definition 5.5.7. We say that $\mathbf{x}$ has an L-component mixture of k-variate normal distributions if its p.d.f. is

$$\begin{aligned} &f(\mathbf{x}; \boldsymbol{\mu}_1, \boldsymbol{\Sigma}_1, p_1, \cdots, \boldsymbol{\mu}_L, \boldsymbol{\Sigma}_L, p_L) \\ &= \sum_{j=1}^{L} p_j (2\pi)^{-k/2} |\boldsymbol{\Sigma}_j|^{-1/2} \exp\left\{-\frac{1}{2}\left(\mathbf{x} - \boldsymbol{\mu}_j\right)' \boldsymbol{\Sigma}_j^{-1} \left(\mathbf{x} - \boldsymbol{\mu}_j\right)\right\}, \quad \mathbf{x} \in \mathcal{R}^k. \end{aligned}$$

This p.d.f. exhibits multi-modality with up to L distinct peaks.

Example 5.5.1. We saw the form of the bivariate normal p.d.f. in Example 5.2.3. Mixtures of L bivariate normal distributions enable us to generate a rich class of bivariate densities which have up to L distinct peaks. Consider two mixands. Let $\boldsymbol{\mu}_1 = \boldsymbol{\mu}_2 = \mathbf{0}$, $\sigma_{1,j}^2 = \sigma_{2,j}^2 = 1$, for $j = 1, 2$, $\rho_1 = 1/2$, and $\rho_2 = -1/2$. With mixing proportions $p_1 = p_2 = 1/2$, the mixture p.d.f. of $(X_1, X_2)'$ is

$$f(\mathbf{x}, \boldsymbol{\mu}_1, \boldsymbol{\Sigma}_1, p_1, \boldsymbol{\mu}_2, \boldsymbol{\Sigma}_2, p_2)$$
$$= \frac{1}{2\pi\sqrt{3}} e^{-2(x_1^2+x_2^2)/3} \left(e^{-2x_1x_2/3} + e^{2x_1x_2/3}\right), \quad \mathbf{x} = (x_1, x_2)' \in \mathcal{R}^2.$$

A plot of this p.d.f. reveals regions where one of the two components dominates, and there are also regions of transition where the p.d.f. does not appear to be "normal". A well-known property of a bivariate normal mixture is that all conditional and marginal distributions are univariate normal mixtures. □

Example 5.5.2. Consider the following ε-contaminated normal distribution which is a mixture of a $N(\mathbf{0}, \mathbf{I})$ distribution and a $N(\mathbf{0}, \sigma^2\mathbf{I})$ distribution, with $0 \le \varepsilon \le 1$. Its p.d.f. is

$$f(\mathbf{x}; \sigma^2, \varepsilon) = \frac{(1-\varepsilon)}{(2\pi)^{k/2}} \exp\left\{-\frac{\mathbf{x}'\mathbf{x}}{2}\right\} + \frac{\varepsilon}{(2\pi)^{k/2}\sigma^k} \exp\left\{-\frac{\mathbf{x}'\mathbf{x}}{2\sigma^2}\right\}. \quad \square$$

The mixture of normals accommodates modeling in situations where the data exhibits multi-modality. Suppose λ denotes a discrete random variable, assuming two distinct positive values λ_1 and λ_2, with respective probabilities p_1 and p_2, where $p_1 + p_2 = 1$. Let $\mathbf{x}$ be a k-dimensional random vector which is defined as follows: conditionally on $\lambda = \lambda_j$, $\mathbf{x} \sim N_k(\mathbf{0}, \lambda_j \mathbf{I})$, $j = 1, 2$. The "conditional" p.d.f. of $\mathbf{x}$ (conditional on λ) is

$$f(\mathbf{x} \mid \lambda_j) = (2\pi)^{-k/2} \lambda_j^{-k/2} \exp\{-\mathbf{x}'\mathbf{x}/2\lambda_j\}, \quad \mathbf{x} \in \mathcal{R}^k.$$

The unconditional distribution of $\mathbf{x}$ has the mixture p.d.f.

$$f(\mathbf{x}) = (2\pi)^{-k/2}\{p_1\lambda_1^{-k/2}\exp(-\mathbf{x}'\mathbf{x}/2\lambda_1) + p_2\lambda_2^{-k/2}\exp(-\mathbf{x}'\mathbf{x}/2\lambda_2)\}.$$

This distribution is called a scale mixture of multivariate normals. In general, we can include L mixands, $L \ge 2$. By varying the mixing proportions p_j and the values λ_j, we can generate a flexible class of distributions that are useful in modeling a variety of multivariate data. It can be shown that all marginal distributions of this scale distribution mixture are themselves scale mixtures of normals of appropriate dimensions, a property which this distribution shares with the multivariate normal (see Result 5.2.11). Suppose we wish to maintain unimodality while allowing for heavy-tailed behavior, we would assume that the mixing random variable λ has a continuous distribution with p.d.f. $\pi(\lambda)$. We define the resulting flexible class of distributions and show several examples which have useful applications in modeling multivariate data.

Definition 5.5.8. Multivariate scale mixture of normals (SMN distribution). A k-dimensional random vector $\mathbf{x}$ has a multivariate SMN distribution with mean vector $\boldsymbol{\theta}$ and covariance matrix $\boldsymbol{\Sigma}$ if its p.d.f. has a "mixture" form

$$f(\mathbf{x}; \boldsymbol{\theta}, \boldsymbol{\Sigma}) = \int_{\mathcal{R}^+} N_k(\mathbf{x}; \boldsymbol{\theta}, \kappa(\lambda)\boldsymbol{\Sigma})\, dF(\lambda), \tag{5.5.11}$$

where $\kappa(.)$ is a positive function defined on $\mathcal{R}^+$, and $F(.)$ is a c.d.f., which may be either discrete or continuous. The scalar λ is called the mixing parameter, and $F(.)$ is the mixing distribution.

Example 5.5.3. Suppose we set $\kappa(\lambda) = 1/\lambda$ in (5.5.11), and assume that the parameter $\lambda \sim \text{Gamma}(\nu/2, \nu/2)$, i.e.,

$$\pi(\lambda) = \frac{1}{\Gamma(\nu/2)}(\nu/2)^{\nu/2}\lambda^{\nu/2-1}\exp\{-\nu\lambda/2\}, \quad -\infty < \lambda < \infty.$$

The resulting multivariate t-distribution is a special example of the scale mixtures of normals family with ν degrees of freedom and p.d.f.

$$f(\mathbf{z};\boldsymbol{\theta},\boldsymbol{\Sigma},\nu)=\frac{\Gamma\{\frac{1}{2}(k+\nu)\}}{\Gamma(\nu/2)(\nu\pi)^{k/2}|\boldsymbol{\Sigma}|^{1/2}}\left[1+\frac{1}{\nu}(\mathbf{z}-\boldsymbol{\theta})'\boldsymbol{\Sigma}^{-1}(\mathbf{z}-\boldsymbol{\theta})\right]^{-(\nu+k)/2} \tag{5.5.12}$$

for $\mathbf{z}\in\mathcal{R}^k$. It can be shown that if

$$\mathbf{z}=\frac{\mathbf{x}}{\sqrt{\xi/\nu}}+\boldsymbol{\theta}\quad\text{with } \mathbf{x}\sim N_k(\mathbf{0},\boldsymbol{\Sigma}),\ \xi\sim\chi^2_\nu \text{ independent},$$

then $\mathbf{z}\sim f(\mathbf{z};\boldsymbol{\theta},\boldsymbol{\Sigma},\nu)$ and, furthermore, $(\mathbf{z}-\boldsymbol{\theta})'\boldsymbol{\Sigma}^{-1}(\mathbf{z}-\boldsymbol{\theta})/k\sim F_{k,\nu}$. When $\nu\to\infty$, the multivariate t-distribution approaches the multivariate normal distribution. By setting $\boldsymbol{\theta}=\mathbf{0}$, and $\boldsymbol{\Sigma}=\mathbf{I}_k$ in (5.5.12), we get the standard distribution, usually denoted by $f(\mathbf{z})$. When $\nu=1$, the distribution corresponds to a k-variate Cauchy distribution. In particular, when $k=2$, let $\mathbf{z}=(Z_1,Z_2)$ denote a random vector with a Cauchy distribution. The p.d.f. of $\mathbf{z}$ is

$$f(\mathbf{z})=\frac{1}{2\pi}(1+\mathbf{z}'\mathbf{z})^{-3/2},\quad \mathbf{z}\in\mathcal{R}^2,$$

and corresponds to a (standard) bivariate Cauchy distribution, which is a simple example of a bivariate scale mixture of normal distributions. □

Example 5.5.4. If we assume $\kappa(\lambda)=4\lambda^2$, where λ follows an asymptotic Kolmogorov distribution with p.d.f.

$$\pi(\lambda)=8\sum_{j=1}^{\infty}(-1)^{j+1}j^2\lambda\exp\{-2j^2\lambda^2\},$$

the resulting multivariate logistic distribution is a special case of the scale mixture of normals family. This distribution finds use in modeling multivariate binary data. □

Example 5.5.5. If we set $\kappa(\lambda)=2\lambda$, and assume that $\pi(\lambda)$ is a positive stable p.d.f. $S^P(\alpha,1)$ (see item 12 in Appendix B) whose polar form of the p.d.f. is given by (Samorodnitsky and Taqqu, 1994)

$$\pi_{S^P}(\lambda;\alpha,1)=\{\alpha/(1-\alpha)\}\lambda^{-\{\alpha/(1-\alpha)+1\}}\int_0^1 s(u)\exp\{-s(u)/\lambda^{\alpha/(1-\alpha)}\}du,$$

for $0<\alpha<1$, and

$$s(u)=\{\sin(\alpha\pi u)/\sin(\pi u)\}^{\alpha/(1-\alpha)}\{\sin[(1-\alpha)\pi u]/\sin(\pi u)\}.$$

The resulting scale mixture of normals distribution is called the multivariate symmetric stable distribution. □

5.5.2.2 Spherical distributions

In this section, we define the class of spherical (or radial) distributions.

Definition 5.5.9. A k-dimensional random vector $\mathbf{z}=(Z_1,\cdots,Z_k)'$ is said to have a spherical (or spherically symmetric) distribution if its distribution does not change under rotations of the coordinate system, i.e., if the distribution of the vector $\mathbf{Az}$ is the same as the distribution of $\mathbf{z}$ for any orthogonal $k\times k$ matrix $\mathbf{A}$. If the p.d.f. of $\mathbf{z}$ exists in $\mathcal{R}^k$, it depends on $\mathbf{z}$ only through $\mathbf{z}'\mathbf{z} = \sum_{i=1}^{k} Z_i^2$; for any function h (called the density generator function),

$$f(\mathbf{z}) \propto h(\mathbf{z}'\mathbf{z}) = c_k h(\mathbf{z}'\mathbf{z}), \tag{5.5.13}$$

where c_k is a constant. The mean and covariance of $\mathbf{z}$, provided they exist, are

$$\mathrm{E}(\mathbf{z}) = \mathbf{0}, \quad \text{and} \quad \mathrm{Cov}(\mathbf{z}) = c\mathbf{I}_k,$$

where $c \geq 0$ is some constant.

Different choices of the function h give rise to different examples of the spherical distributions (Muirhead, 1982; Fang et al., 1990). Contours of constant density of a spherical random vector $\mathbf{z}$ are circles when $k = 2$, or spheres for $k > 2$, which are centered at the origin. The spherical normal distribution shown in the following example is a popular member of this class.

Example 5.5.6. Let $\mathbf{z}$ have a k-variate normal distribution with mean $\mathbf{0}$ and covariance $\sigma^2\mathbf{I}_k$. We say $\mathbf{z}$ has a spherical normal distribution with p.d.f.

$$f(\mathbf{z};\sigma^2) = \frac{1}{(2\pi)^{k/2}\sigma^k} \exp\left\{-\frac{1}{2\sigma^2}\mathbf{z}'\mathbf{z}\right\}, \quad \mathbf{z} \in \mathcal{R}^k.$$

The density generator function is clearly $h(u) = c\exp\{-u/2\}, u \geq 0$. □

The ϵ-contaminated normal distribution shown in Example 5.5.2 is also an example of a spherical distribution, as is the standard multivariate t-distribution defined in Example 5.5.3. The following example generalizes the well-known double-exponential (Laplace) distribution to the multivariate case; this distribution is useful for modeling data with outliers.

Example 5.5.7. Consider the bivariate generalization of the standard double-exponential distribution to a vector $\mathbf{z} = (Z_1, Z_2)$ with p.d.f.

$$f(\mathbf{z}) = \frac{1}{2\pi}\exp\{-(\mathbf{z}'\mathbf{z})^{1/2}\}, \quad \mathbf{z} \in \mathcal{R}^2.$$

This is an example of a spherical distribution; notice the similarity of this p.d.f. to that of the bivariate standard normal vector. □

Definition 5.5.10. The squared radial random variable $T = \|\mathbf{z}\|$ has p.d.f.

$$\frac{\pi^{k/2}}{\Gamma(k/2)} t^{(k/2)-1} h(t), \quad t > 0. \tag{5.5.14}$$

We say T has a radial-squared distribution with k d.f. and density generator h, i.e., $T \sim R_k^2(h)$.

The main appeal of spherical distributions lies in the fact that many results that we have seen for the multivariate normal hold for the general class of spherical distributions. For example, if $\mathbf{z} = (Z_1, Z_2)'$ is a bivariate spherically distributed random vector, the ratio $V = Z_1/Z_2$ has a Cauchy distribution provided $\mathrm{P}(Z_2 = 0) = 0$. If $\mathbf{z} = (Z_1, \cdots, Z_k)'$ has a k-variate spherical distribution, $k \geq 2$, with $\mathrm{P}(\mathbf{z} = \mathbf{0}) = 0$, we can show that

$$V = Z_1/\{(Z_2^2 + \cdots + Z_k^2)^{1/2}/(k-1)\} \sim t_{k-1}.$$

In many cases, we wish to extend the definition of a spherical distribution to include random vectors with a nonzero mean $\boldsymbol{\mu}$ and a general covariance matrix $\boldsymbol{\Sigma}$. This generalization leads us from spherical distributions to elliptical (or elliptically contoured distributions), which form the topic of the next subsection.

5.5.2.3 Elliptical distributions

The family of elliptical or elliptically contoured distributions is the most general family that we will consider as alternatives to the multivariate normal distribution. We derive results on the forms of the corresponding marginal and conditional distributions. There is a vast literature on spherical and elliptical distributions (Kelker, 1970; Devlin et al., 1976; Chmielewski, 1981; Fang et al., 1990; Fang and Anderson, 1990), and the reader is referred to these for more details on this interesting and useful class of distributions.

Definition 5.5.11. Let $\mathbf{z} \in \mathcal{R}^k$ follow a spherical distribution, $\boldsymbol{\mu} \in \mathcal{R}^k$ be a fixed vector, and $\boldsymbol{\Gamma}$ be a $k \times k$ matrix. The random vector $\mathbf{x} = \boldsymbol{\mu} + \boldsymbol{\Gamma}\mathbf{z}$ is said to have an elliptical, or elliptically contoured, or elliptically symmetric distribution. Provided they exist, the mean and covariance of $\mathbf{x}$ are

$$\mathrm{E}(\mathbf{x}) = \boldsymbol{\mu} \quad \text{and} \quad \mathrm{Cov}(\mathbf{x}) = c\boldsymbol{\Gamma}\boldsymbol{\Gamma}' = c\mathbf{V},$$

where $c \geq 0$ is a constant. The m.g.f. of the distribution, if it exists, has the form

$$M_{\mathbf{x}}(\mathbf{t}) = \psi(\mathbf{t}'\mathbf{V}\mathbf{t}) \exp\{\mathbf{t}'\boldsymbol{\mu}\} \tag{5.5.15}$$

for some function ψ. In case the m.g.f. does not exist, we invoke the characteristic function of the distribution for proof of distributional properties.

In order for an elliptically contoured random vector to admit a density (with respect to Lebesgue measure), the matrix $\mathbf{V}$ must be p.d. and the density generator function $h(.)$ in (5.5.13) must satisfy the condition

$$\int_0^\infty \frac{\pi^{k/2}}{\Gamma(k/2)} t^{(k/2)-1} h(t)\, dt = 1.$$

If the p.d.f. of $\mathbf{x}$ exists, it will be a function only of the norm $\|\mathbf{x}\| = (\mathbf{x}'\mathbf{x})^{1/2}$. We denote the class of elliptical distributions by $E_k(\boldsymbol{\mu}, \mathbf{V}, h)$. If a random vector $\mathbf{x} \sim E_k(\mathbf{0}, \mathbf{I}_k, h)$, then $\mathbf{x}$ has a spherical distribution. Suppose $\boldsymbol{\mu}$ is a fixed k-dimensional vector, and $\mathbf{y} = \boldsymbol{\mu} + \mathbf{P}\mathbf{x}$, where $\mathbf{P}$ is a nonsingular $k \times k$ matrix. Then, $\mathbf{y} \sim E_k(\boldsymbol{\mu}, \mathbf{V}, h)$, with $\mathbf{V} = \mathbf{P}\mathbf{P}'$.

Result 5.5.1. Let $\mathbf{z}$ denote a spherically distributed random vector with p.d.f. $f(\mathbf{z})$, and let $\mathbf{x} = \boldsymbol{\mu} + \boldsymbol{\Gamma}\mathbf{z}$ have an elliptical distribution, where $\boldsymbol{\Gamma}$ is a $k \times k$ nonsingular matrix. Let $\mathbf{V} = \boldsymbol{\Gamma}\boldsymbol{\Gamma}'$, and note that $\mathbf{z} = \boldsymbol{\Gamma}^{-1}(\mathbf{x} - \boldsymbol{\mu})$. Then the p.d.f. of $\mathbf{x}$ has the form

$$f(\mathbf{x}) = c_k |\mathbf{V}|^{-1/2} h[(\mathbf{x} - \boldsymbol{\mu})'\mathbf{V}^{-1}(\mathbf{x} - \boldsymbol{\mu})], \quad \mathbf{x} \in \mathcal{R}^k \tag{5.5.16}$$

for some function $h(.)$ which can be independent of k, and such that $r^{k-1}h(r^2)$ is integrable over $[0, \infty)$.

Proof. The transformation from $\mathbf{z}$ to $\mathbf{x} = \boldsymbol{\mu} + \boldsymbol{\Gamma}\mathbf{z}$ has Jacobian $J = |\boldsymbol{\Gamma}^{-1}|$. By Result A.2, we have for $\mathbf{x} \in \mathcal{R}^k$,

$$\begin{aligned} f(\mathbf{x}) &= c_k |\boldsymbol{\Gamma}|^{-1} f\{\boldsymbol{\Gamma}^{-1}(\mathbf{x} - \boldsymbol{\mu})\} \\ &= c_k [|\boldsymbol{\Gamma}|^{-1} |\boldsymbol{\Gamma}|^{-1}]^{1/2} h[(\mathbf{x} - \boldsymbol{\mu})'\boldsymbol{\Gamma}'^{-1}\boldsymbol{\Gamma}^{-1}(\mathbf{x} - \boldsymbol{\mu})] \\ &= c_k |\mathbf{V}|^{-1/2} h[(\mathbf{x} - \boldsymbol{\mu})'\mathbf{V}^{-1}(\mathbf{x} - \boldsymbol{\mu})]. \end{aligned}$$

Note that the same steps are used in the derivation of the multivariate normal p.d.f. in Section 5.2. The relation between the spherical distribution and the (corresponding) elliptical distribution is the same as the relationship between the multivariate standard normal distribution (Definition 5.2.1) and the corresponding normal distribution with nonzero mean and covariance $\boldsymbol{\Sigma}$ (Definition 5.2.2). ■

The distribution of $\mathbf{x}$ will have m moments provided the function $r^{m+k-1}h(r^2)$ is integrable on $[0, \infty)$. We show two examples.

Example 5.5.8. Let $\mathbf{x}$ have a k-variate normal distribution with mean $\boldsymbol{\mu}$ and covariance $\sigma^2\mathbf{I}$. Then $\mathbf{x}$ has an elliptical distribution. A rotation about $\boldsymbol{\mu}$ is given by $\mathbf{y} = \mathbf{P}(\mathbf{x}-\boldsymbol{\mu})+\boldsymbol{\mu}$, where $\mathbf{P}$ is an orthogonal matrix. We see that $\mathbf{y} \sim N_k(\boldsymbol{\mu}, \sigma^2\mathbf{I})$ (see Exercise 5.9), so that the distribution is unchanged under rotations about $\boldsymbol{\mu}$. We say that the distribution is spherically symmetric about $\boldsymbol{\mu}$. In fact, the normal distribution is the only multivariate distribution with independent components X_j, $j = 1, \cdots, k$, that is spherically symmetric. □

Example 5.5.9. Suppose $\mathbf{x} = \boldsymbol{\mu}+\boldsymbol{\Gamma}\mathbf{z}$, where $\mathbf{z}$ was defined in Example 5.5.3, $\boldsymbol{\mu} = (\mu_1, \mu_2)'$ is a fixed vector, and $\boldsymbol{\Gamma}$ is a nonsingular 2×2 matrix. Let $\mathbf{A} = \boldsymbol{\Gamma}\boldsymbol{\Gamma}'$; the p.d.f. of $(X_1, X_2)'$ is

$$f(\mathbf{x}; \boldsymbol{\mu}, \mathbf{A}) = (2\pi)^{-1}|\mathbf{A}|^{-1/2}[1 + (\mathbf{x} - \boldsymbol{\mu})'\mathbf{A}^{-1}(\mathbf{x} - \boldsymbol{\mu})]^{-3/2}, \quad \mathbf{x} \in \mathcal{R}^2.$$

This is the multivariate Cauchy distribution, which is a special case of the multivariate t-distribution. The density generator for the k-variate Cauchy distribution is $h(u) = c\{1 + u\}^{-(k+1)/2}$, while the density generator for the k-variate t-distribution with ν degrees of freedom is $h(u) = c\{1 + u/\nu\}^{-(k+\nu)/2}$. In terms of its use in linear model theory, the Cauchy distribution and the Student's t-distribution with small ν are considered useful as robust alternatives to the multivariate normal distribution in terms of error distribution specification. □

Example 5.5.10. Let $\mathbf{z}$ be the standard double exponential variable specified in Example 5.5.7, and suppose we define $\mathbf{x} = \boldsymbol{\mu} + \boldsymbol{\Gamma}\mathbf{z}$ where $\boldsymbol{\mu} = (\mu_1, \mu_2)'$ is a fixed vector, and $\boldsymbol{\Gamma}$ is a nonsingular 2×2 matrix. Let $\mathbf{A} = \boldsymbol{\Gamma}\boldsymbol{\Gamma}'$; the p.d.f. of $(X_1, X_2)'$ is

$$f(\mathbf{x}; \boldsymbol{\mu}, \mathbf{A}) = (2\pi)^{-1}|\mathbf{A}|^{-1/2} \exp\{-[(\mathbf{x} - \boldsymbol{\mu})'\mathbf{A}^{-1}(\mathbf{x} - \boldsymbol{\mu})]^{1/2}\}, \quad \mathbf{x} \in \mathcal{R}^2.$$

A comparison of the contours of this distribution with those of a bivariate normal distribution having the same location and spread shows that this distribution is more peaked at the center and has heavier tails. □

The next result specifies the marginal distributions and the conditional distributions. Result 5.5.3 characterizes the class of normal distributions within the family of elliptically symmetric distributions. Let $\mathbf{x} = (X_1, \cdots, X_k)' \sim E_k(\boldsymbol{\mu}, \mathbf{V}, h)$.

Result 5.5.2. Suppose we partition $\mathbf{x}$ as $\mathbf{x} = (\mathbf{x}_1', \mathbf{x}_2')'$, where $\mathbf{x}_1$ and $\mathbf{x}_2$ are respectively q-dimensional and $(k-q)$-dimensional vectors. Suppose $\boldsymbol{\mu}$ and $\mathbf{V}$ are partitioned conformably (similar to Result 5.2.11).

1. The marginal distribution of $\mathbf{x}_i$ is elliptical, i.e., $\mathbf{x}_1 \sim E_q(\boldsymbol{\mu}_1, \mathbf{V}_{11})$ and $\mathbf{x}_2 \sim E_{k-q}(\boldsymbol{\mu}_2, \mathbf{V}_{22})$. Unless $f(\mathbf{x})$ has an atom of weight at the origin, the p.d.f. of each marginal distribution exists.

2. The conditional distribution of $\mathbf{x}_1$ given $\mathbf{x}_2 = \mathbf{c}_2$ is q-variate elliptical with mean

$$E(\mathbf{x}_1 \,|\, \mathbf{x}_2 = \mathbf{c}_2) = \boldsymbol{\mu}_1 + \mathbf{V}_{12}\mathbf{V}_{22}^{-1}(\mathbf{c}_2 - \boldsymbol{\mu}_2), \tag{5.5.17}$$

while the conditional covariance of $\mathbf{x}_1$ given $\mathbf{x}_2 = \mathbf{c}_2$ only depends on $\mathbf{c}_2$ through the quadratic form $(\mathbf{x}_2 - \mathbf{c}_2)'\mathbf{V}_{22}^{-1}(\mathbf{x}_2 - \mathbf{c}_2)$. The distribution of $\mathbf{x}_2$ given $\mathbf{x}_1 = \mathbf{c}_1$ is derived similarly.

Proof. The m.g.f. of $\mathbf{x}_i$ is $\psi(\mathbf{t}_i'\mathbf{V}_{ii}\mathbf{t}_i)\exp\{\mathbf{t}_i'\boldsymbol{\mu}_i\}$, $i = 1, 2$. As a result, $\mathbf{x}_1 \sim E_q(\boldsymbol{\mu}_1, \mathbf{V}_{11})$ and $\mathbf{x}_2 \sim E_{k-q}(\boldsymbol{\mu}_2, \mathbf{V}_{22})$. The p.d.f. of $\mathbf{x}_1$, if it exists, has the form

$$f_1(\mathbf{x}_1) = c_q|\mathbf{V}_{11}|^{-1/2}h_q[(\mathbf{x}_1 - \boldsymbol{\mu}_1)'\mathbf{V}_{11}^{-1}(\mathbf{x}_1 - \boldsymbol{\mu}_1)],$$

where the function h_q depends only on h and q, and is independent of $\boldsymbol{\mu}$ and $\mathbf{V}$. This completes the proof of property 1. To show property 2, we see that by definition, the conditional mean is

$$\mathrm{E}(\mathbf{x}_1 \,|\, \mathbf{x}_2 = \mathbf{c}_2) = \int \mathbf{x}_1\, dF_{\mathbf{x}_1 \,|\, \mathbf{c}_2}(\mathbf{x}_1).$$

Substituting $\mathbf{y} = \mathbf{x}_1 - \boldsymbol{\mu}_1 - \mathbf{V}_{12}\mathbf{V}_{22}^{-1}(\mathbf{c}_2 - \boldsymbol{\mu}_2)$ and simplifying, we get

$$\mathrm{E}(\mathbf{x}_1 \,|\, \mathbf{x}_2 = \mathbf{c}_2) = \int \mathbf{y}dF_{\mathbf{y} \,|\, \mathbf{c}_2}(\mathbf{y}) + \boldsymbol{\mu}_1 + \mathbf{V}_{12}\mathbf{V}_{22}^{-1}(\mathbf{c}_2 - \boldsymbol{\mu}_2).$$

Since it can be verified that the joint m.g.f. of $\mathbf{y}$ and $\mathbf{x}_2$, when it exists, satisfies $M_{\mathbf{y},\mathbf{x}_2}(-\mathbf{t}_1, \mathbf{t}_2) = M_{\mathbf{y},\mathbf{x}_2}(\mathbf{t}_1, \mathbf{t}_2)$, we see that $\int \mathbf{y}dF_{\mathbf{y} \,|\, \mathbf{c}_2}(\mathbf{y}) = 0$, proving (5.5.17). The conditional covariance is

$$\begin{aligned}
&\mathrm{Cov}(\mathbf{x}_1 \,|\, \mathbf{x}_2 = \mathbf{c}_2) \\
&= \int [\mathbf{x}_1 - \mathrm{E}(\mathbf{x}_1 \,|\, \mathbf{x}_2 = \mathbf{c}_2)][\mathbf{x}_1 - \mathrm{E}(\mathbf{x}_1 \,|\, \mathbf{x}_2 = \mathbf{c}_2)]' f_{\mathbf{x}_1 \,|\, \mathbf{x}_2}(\mathbf{x}_1)d\mathbf{x}_1 \\
&= \frac{c_k}{|\mathbf{V}|^{1/2}} \int \mathbf{z}\mathbf{z}' \frac{h[\mathbf{z}'\mathbf{V}_{11.2}\mathbf{z} + (\mathbf{c}_2 - \boldsymbol{\mu}_2)'\mathbf{V}_{22}^{-1}(\mathbf{c}_2 - \boldsymbol{\mu}_2)]}{f_{\mathbf{x}_2}(\mathbf{c}_2)}\, d\mathbf{z},
\end{aligned}$$

where $\mathbf{V}_{11.2} = \mathbf{V}_{11} - \mathbf{V}_{12}\mathbf{V}_{22}^{-1}\mathbf{V}_{21}$. The result follows since $f_{\mathbf{x}_2}(\mathbf{c}_2)$ is a function of the quadratic form $(\mathbf{x}_2 - \mathbf{c}_2)'\mathbf{V}_{22}^{-1}(\mathbf{x}_2 - \mathbf{c}_2)$. ■

Result 5.5.3. Suppose $\mathbf{x}$, $\boldsymbol{\mu}$, and $\mathbf{V}$ are partitioned as in Result 5.5.2.

1. If any marginal p.d.f. of a random vector $\mathbf{x}$ which has an $E_k(\boldsymbol{\mu}, \mathbf{V}, h)$ distribution is normal, then $\mathbf{x}$ must have a normal distribution.

2. If the conditional distribution of $\mathbf{x}_1$ given $\mathbf{x}_2 = \mathbf{c}_2$ is normal for any q, $q = 1, \cdots, k-1$, then $\mathbf{x}$ has a normal distribution.

3. Let $k > 2$, and assume that the p.d.f. of $\mathbf{x}$ exists. The conditional covariance of $\mathbf{x}_1$ given $\mathbf{x}_2 = \mathbf{c}_2$ is independent of $\mathbf{x}_2$ only if $\mathbf{x}$ has a normal distribution.

4. If $\mathbf{x} \sim E_k(\boldsymbol{\mu}, \mathbf{V}, h)$, and $\mathbf{V}$ is diagonal, then the components of $\mathbf{x}$ are independent only if the distribution of $\mathbf{x}$ is normal.

Proof. By Result 5.5.2, the m.g.f. (or characteristic function) of $\mathbf{x}$ has the same form as the m.g.f. (or characteristic function) of $\mathbf{x}_1$, from which property 1 follows. Without loss of generality, let $\boldsymbol{\mu} = \mathbf{0}$ and $\mathbf{V} = \mathbf{I}$, so that by Result 5.5.2, the conditional mean of $\mathbf{x}_1$ given $\mathbf{x}_2 = \mathbf{c}_2 = \mathbf{0}$, and its conditional covariance has the form $\phi(\mathbf{c}_2)\mathbf{I}_{k-q}$. Also, $g(\mathbf{x}_1 \,|\, \mathbf{x}_2 = \mathbf{c}_2) = c_k h(\mathbf{x}_1'\mathbf{x}_1 + \mathbf{x}_2'\mathbf{x}_2)/f_2(\mathbf{c}_2)$ is a function of $\mathbf{x}_1'\mathbf{x}_1$. If the conditional distribution is normal,

$$c_k h(\mathbf{x}_1'\mathbf{x}_1 + \mathbf{x}_2'\mathbf{x}_2) = \{2\pi\sigma(\mathbf{c}_2)\}^{-(k-q)/2} f_2(\mathbf{c}_2) \exp\left\{-\frac{\mathbf{x}_1'\mathbf{x}_1}{2\sigma(\mathbf{c}_2)}\right\},$$

so that

$$c_k h(\mathbf{x}_1'\mathbf{x}_1) = \{2\pi\sigma(\mathbf{c}_2)\}^{-(k-q)/2} f_2(\mathbf{c}_2) \exp\left\{\frac{\mathbf{x}_2'\mathbf{x}_2}{2\sigma(\mathbf{c}_2)}\right\} \exp\left\{-\frac{\mathbf{x}_1'\mathbf{x}_1}{2\sigma(\mathbf{c}_2)}\right\}.$$

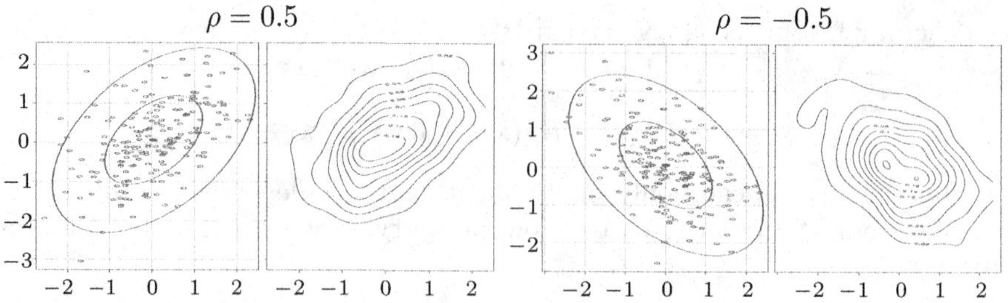

FIGURE 5.5.1. Contours of a bivariate normal density versus contours of kernel density estimates using a sample from the density; $\rho = 0.5$ in the left two panels and $\rho = -0.5$ in the right two panels.

This implies that $f(\mathbf{x}) = c_k h(\mathbf{x}'\mathbf{x})$, i.e., $\mathbf{x}$ has a normal distribution, proving property 2. The proof of properties 3 and 4 is left as an exercise. ■

We defer discussion of results on distributions of quadratic forms in elliptical random vectors (some of which are analogous to the theory for normal distributions) to Chapter 13, where we describe elliptically contoured linear models.

5.6 R Code

```
library(car)
library(mvtnorm)
library(MASS)

## Example 5.2.5. Bivariate normal
# simulate bivariate normal with rho=0.5
set.seed(1);n <- 200;rho <- 0.5
mu <- c(0,0);Sigma <- matrix(c(1,0.5,0.5,1),2)
bvn1 <- mvtnorm::rmvnorm(n,mu,Sigma, method="svd")

# simulate bivariate normal with rho=-0.5
set.seed(1); n <- 200; rho <- -0.5
mu <- c(0,0);Sigma <- matrix(c(1,-0.5,-0.5,1),2)
bvn2 <- mvtnorm::rmvnorm(n,mu,Sigma, method="svd")
par(mfrow=c(1,2))
dataEllipse(bvn1[,1],bvn1[,2],levels=c(0.5,0.95),xlab="x1",
ylab="x2",sub="rho=0.5")
dataEllipse(bvn2[,1],bvn2[,2],levels=c(0.5,0.95),xlab="x1",
ylab="x2",sub="rho=-0.5")

# contours based on kde
par(mfrow=c(1,2))
kde1 <- kde2d(bvn1[,1],bvn1[,2],n=n)
contour(kde1,xlab="x1",ylab="x2",sub="rho=0.5")
kde2 <- kde2d(bvn2[,1],bvn2[,2],n=n)
contour(kde2,xlab="x1",ylab="x2",sub="rho=-0.5")
```

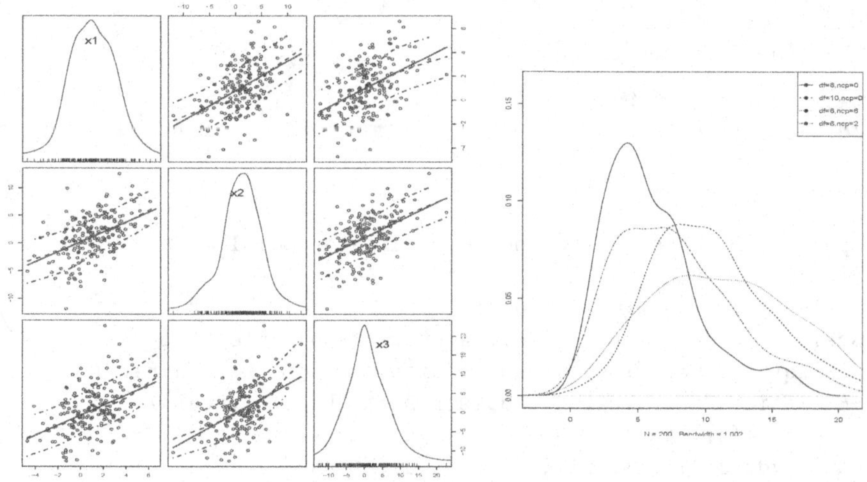

FIGURE 5.6.1. Scatterplot matrix of trivariate normal samples (left) and empirical densities of samples from central and noncentral chi-square distributions (right).

The contour plots of two normal densities together with scatter plots of random samples from the densities are shown in Figure 5.5.1. For comparison, each is accompanied by a contour plot of the kernel density estimates using the sample.

```
## Sample a trivariate normal distribution; see Figure 5.6.1.
set.seed(1); n <- 200; rho <- 0.5
mu1 <- 1;mu2 <- 1;mu3 <- 1
s1 <- 2;s2 <- 4;s3 <- 6
mu <- c(mu1,mu2,mu3); mu
Sigma <- matrix(c(s1^2,s1*s2*rho,s1*s3*rho,s2*s1*rho,s2^2,
s2*s3*rho,s3*s1*rho,s3*s2*rho,s3^2),3);Sigma
tvn <- mvtnorm::rmvnorm(n,mu,Sigma, method="svd")
colnames(tvn) <- c("x1","x2","x3")
scatterplotMatrix(tvn[,1:3])

## Section 5.3. Chi-square distributions: see Figure 5.6.1
n<-200
df1<-6; df2<-10 #d.f.
ncp1<-6; ncp2<-2 #noncentrality parameter
#Central chi-square
sim.1<-rchisq(n,df1);sim.2<-rchisq(n,df2)
#Noncentral chi-square
sim.3<-rchisq(n,df1,ncp=ncp1);sim.4<-rchisq(n,df1,ncp=ncp2)
plot(density(sim.1),col="red",main="Chi-Square",ylim=c(0,0.16))
lines(density(sim.2),col="blue")
lines(density(sim.3),col="green")
lines(density(sim.4),col="black")
legend("topright",c("df=6,ncp=0","df=10,ncp=0","df=6,ncp=6",
 "df=6,ncp=2"),col=c("red","blue","green","black"),pch=19)

## Section 5.3. F-distributions
n<-100;df1<-4;df2<-9;ncp<-7
```

```
Sim.cf<-rf(n,df1,df2)
Sim.ncf<-rf(n,df1,df2,ncp=ncp)

## Section 5.3. t-distributions
n<-100;df<-8;ncp<-4
Sim.ct<-rt(n,df)
Sim.nct<-rt(n,df,ncp=ncp)

## Example 5.5.6. Spherical standard bivariate normal dist.
library("distrEllipse")
k<-2; n<-1000; loc <- c(0,0) # origin
c<-4; scale <- c*diag(length(loc)) #constant multiple of I
RL1 <- Chisq(df=length(loc))# radial dist.sqrt of chi-square(k)
z1_f <-EllipticalDistribution(radDistr=sqrt(RL1),loc,scale)
z1 <- t(r(z1_f)(n)) # samples
k1 <- MASS::kde2d(z1[,1], z1[,2]);contour(k1,asp=1)
MVN::mvn(z1,multivariatePlot="qq") #Check MVN

## Example 5.5.7. Spherical standard bivariate Laplace dist.
k<-2; n<-1000; loc <- c(0,0)
RL2 <-(DExp()^2+DExp()^2) #Radial dist.r=sqrt(x'x)
z2_f <- EllipticalDistribution(radDistr=sqrt(RL2),loc,scale)
z2 <- t(r(z2_f)(n))
k2 <- MASS::kde2d(z2[,1],z2[,2]);contour(k2,asp=1,xlim=c(-3,3),
ylim=c(-3,3))
MVN::mvn(z2, multivariatePlot="qq") #Check MVN

## Example  5.5.8. Elliptical bivariate normal dist.
k<-2;n<-1000;loc<-c(1,2)
scale<-matrix(c(2,0,0,3),nrow=2,byrow=T) # P where PP'=V
x1_f
  <-EllipticalDistribution(radDistr=sqrt(Chisq(df=length(loc))),
    loc,scale)
x1 <- t(r(x1_f)(n))
k1 <- MASS::kde2d(x1[,1],x1[,2]);contour(k1,asp = 1)
MVN::mvn(x1, multivariatePlot="qq")

## Example  5.5.10. Elliptical bivariate Laplace distribution
k<-2;n<-1000;loc<-c(1,2)
scale<-matrix(c(2,0,0,3),nrow=2,byrow=T) # P where PP'=V
RL <- (DExp()^2 + DExp()^2)
z2_f <- EllipticalDistribution(radDistr=sqrt(RL))
x2 <- t(r(z2_f)(n))%*% t(scale)+loc
k2 <- MASS::kde2d(x2[,1],x2[,2])
contour(k2,asp=1,xlim=c(-10, 10),ylim=c(-10, 10))
MVN::mvn(x2, multivariatePlot="qq")
```

The contours and the chi-squared (or gamma) plot to check multivariate normality are shown in Figure 5.6.2.

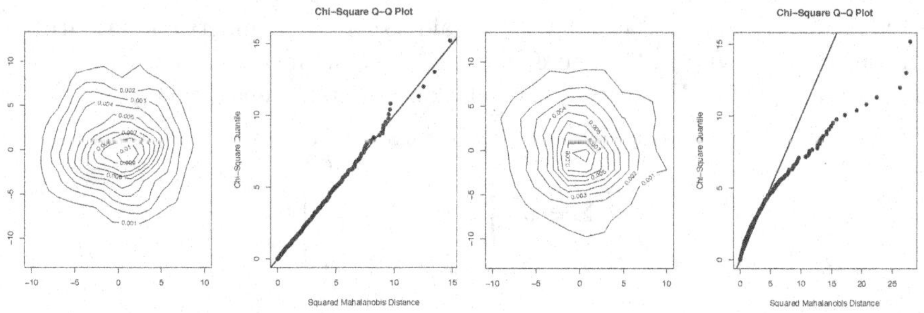

FIGURE 5.6.2. Contours and gamma plot for the spherical standard bivariate normal distribution (left two panels) and bivariate Laplace distribution (right two panels).

Exercises

5.1. Prove Corollary 5.2.1.

5.2. Let $\mathbf{x} = (X_1, \cdots, X_k)' \sim N_k(\boldsymbol{\mu}, \boldsymbol{\Sigma})$, with $\text{r}(\boldsymbol{\Sigma}) = k$.

(a) Show that

$$\int_{\mathcal{R}^k} \exp\{-(\mathbf{x} - \boldsymbol{\mu})'\boldsymbol{\Sigma}^{-1}(\mathbf{x} - \boldsymbol{\mu})/2\}\, d\mathbf{x} = (2\pi)^{k/2}|\boldsymbol{\Sigma}|^{1/2}.$$

(b) Evaluate $\int_{-\infty}^{\infty}\int_{-\infty}^{\infty} \exp\{-(x_1^2 + 2x_1x_2 + 4x_2^2)\}\, dx_1\, dx_2$.

5.3. (Graybill, 1961). Let $(X_1, X_2)'$ have a bivariate normal distribution with p.d.f.

$$f(\mathbf{x}; \boldsymbol{\mu}, \boldsymbol{\Sigma}) = \frac{1}{k}\exp(-Q/2)$$

where $Q = 2x_1^2 - x_1x_2 + 4x_2^2 - 11x_1 - 5x_2 + 19$, and k is a constant. Find a constant a such that $\text{P}(3X_1 - X_2 < a) = 0.01$.

5.4. Let X_1 and X_2 be random variables such that $X_1 + X_2$ and $X_1 - X_2$ have independent standard normal distributions. Show that $(X_1, X_2)'$ has a bivariate normal distribution.

5.5. The c.g.f. of a trivariate random vector $\mathbf{x} = (X_1, X_2, X_3)'$ is given by

$$\mathcal{K}_{\mathbf{x}}(\mathbf{t}) = 5t_1^2 + 3t_2^2 + 6t_3^2 - 2t_1t_2 + 4t_1t_3 + 2t_2t_3 + 4t_1 - 2t_2 + t_3.$$

Show that $\mathbf{x}$ has a trivariate normal distribution. Identify the mean and variance-covariance matrix of the distribution.

5.6. (a) Show that $(X_1, X_2)'$ has a bivariate normal distribution with means μ_1, μ_2, variances σ_1^2 and σ_2^2, and correlation coefficient ρ if and only if every linear combination $c_1X_1 + c_2X_2$ has a univariate normal distribution with mean $c_1\mu_1 + c_2\mu_2$, and variance $c_1^2\sigma_1^2 + c_2^2\sigma_2^2 + 2c_1c_2\rho\sigma_{12}$, where c_1 and c_2 are real constants, not both equal to zero.

(b) Let $Y_i = X_i/\sigma_i$, $i = 1, 2$. Show that $\text{Var}(Y_1 - Y_2) = 2(1 - \rho)$.

5.7. The random vector $\mathbf{x} = (X_1, \cdots, X_k)'$ is said to have a symmetric multivariate normal distribution if $\mathbf{x} \sim N_k(\boldsymbol{\mu}, \boldsymbol{\Sigma})$ where $\boldsymbol{\mu} = \mu \mathbf{1}_k$, i.e., the mean of each X_j is equal to the same constant μ, and $\boldsymbol{\Sigma}$ is the equicorrelation dispersion matrix, i.e.,

$$\boldsymbol{\Sigma} = \sigma^2 \begin{pmatrix} 1 & \rho & \cdots & \rho \\ \rho & 1 & \cdots & \rho \\ \vdots & \vdots & \ddots & \vdots \\ \rho & \rho & \cdots & 1 \end{pmatrix}.$$

When $k = 3$, $\mu = 0$, $\sigma^2 = 2$, and $\rho = 1/2$, find the probability that $X_3 = \min(X_1, X_2, X_3)$. [Hint: Recall that if $(X_1, \cdots, X_k)'$ has a continuous symmetric distribution, then all possible permutations of $X_1, \cdots, X_k$ are equally likely, each having probability $\mathrm{P}(X_{i_1} < \cdots < X_{i_k}) = 1/k!$ for any permutation $(i_1, \cdots, i_k)$ of the first k positive integers.]

5.8. Let $(X_1, \cdots, X_k)' \sim N_k(\mathbf{0}, \boldsymbol{\Sigma})$ with p.d.f. $f(\mathbf{x})$, where $\boldsymbol{\Sigma} = \{\sigma_{ij}\}$. The entropy of the random vector is defined as

$$h(X_1, \cdots, X_k) = -\int f(\mathbf{x}) \ln f(\mathbf{x})\, d\mathbf{x}.$$

(a) Show that $h(X_1, \cdots, X_k) = \frac{1}{2} \ln\{(2\pi e)^k |\boldsymbol{\Sigma}|\}$.

(b) Hence, or otherwise, show that $|\boldsymbol{\Sigma}| \le \prod_{i=1}^k \sigma_{ii}$, with equality holding if and only if $\sigma_{ij} = 0$, for $i \ne j$. The inequality is known as Hadamard's inequality.

5.9. Let $\mathbf{x} \sim N_k(\mathbf{0}, \sigma^2 \mathbf{I})$ and let $\mathbf{y} = \mathbf{P}\mathbf{x}$, where $\mathbf{P}$ is a $k \times k$ orthogonal matrix. Show that $\mathbf{y}$ has a $N_k(\mathbf{0}, \sigma^2 \mathbf{I})$ distribution.

5.10. Let $\mathbf{x} \sim N_k(\boldsymbol{\mu}, \boldsymbol{\Sigma})$, where $\mathrm{r}(\boldsymbol{\Sigma}) = r \le k$. Show that there exists an $r \times k$ matrix $\mathbf{B}$ and an r-dimensional vector $\mathbf{b}$ such that the vector $\mathbf{z} = \mathbf{B}\mathbf{x} + \mathbf{b}$ has a $N_r(\mathbf{0}, \mathbf{I})$ distribution.

5.11. Let $\mathbf{x} \sim N_k(\boldsymbol{\mu}, \boldsymbol{\Sigma})$ and suppose that $\mathbf{x} = (\mathbf{x}_1', \mathbf{x}_2', \mathbf{x}_3')'$, where $\mathbf{x}_i$ is a k_i-dimensional vector, and $\sum_{i=1}^3 k_i = k$. Assume that $\boldsymbol{\mu}$ and $\boldsymbol{\Sigma}$ are partitioned conformably. Derive the conditional distribution of $\mathbf{x}_3$ given $\mathbf{x}_1 = \mathbf{c}_1$, and $\mathbf{x}_2 = \mathbf{c}_2$.

5.12. Verify Result 5.2.13.

5.13. For $\mathbf{x}_1$ and $\mathbf{x}_2$ as in Result 5.2.14, show that the b.l.u.p. of $\mathbf{x}_1$ based on $\mathbf{x}_2$ is exactly $\mathbf{x}_1$ if and only if $\mathbf{x}_1 = \mathbf{A}\mathbf{x}_2 + \mathbf{a}$ for some matrix $\mathbf{A}$ and vector $\mathbf{a}$ of constants.

5.14. Let $\mathbf{x} = (X_1, X_2, X_3)' \sim N_3(\boldsymbol{\mu}, \boldsymbol{\Sigma})$, where $\boldsymbol{\mu} = (\mu_1, \mu_2, \mu_3)'$ and $\boldsymbol{\Sigma} = \sigma^2 \begin{pmatrix} 1 & \rho & 0 \\ \rho & 1 & \rho \\ 0 & \rho & 1 \end{pmatrix}$.

(a) What are the marginal distributions of X_2 and X_3?

(b) Write down the conditional distribution of X_1 given X_2 and X_3. Under what condition does this distribution coincide with the marginal distribution of X_1?

(c) For what value of ρ are the two random variables $X_1 + X_2 + X_3$ and $X_1 - X_2 - X_3$ independently distributed?

5.15. Let $\mathbf{x} = (X_1, X_2, X_3)' \sim N_3(\mathbf{0}, \boldsymbol{\Sigma})$, where $\boldsymbol{\Sigma} = \begin{pmatrix} 1 & \rho & \rho \\ \rho & 1 & \rho \\ \rho & \rho & 1 \end{pmatrix}$. Show that

(a) $\text{Corr}(X_1^2, X_2^2) = \rho^2$;

(b) $\rho_{12|3} = \rho/(1+\rho)$;

(c) $\rho_{1(2,3)} = [2\rho^2/(1+\rho)]^{1/2}$.

5.16. Verify Result 5.2.16.

5.17. Suppose that under the setting of Definition 5.2.7, the partial correlation coefficient $\rho_{jk|(q+1,\cdots,k)}$ is well-defined. Show that $|\rho_{jk|(q+1,\cdots,k)}| = 1$ if and only if $X_j = \mathbf{a}'\mathbf{x}_2 + bX_k + c$ for some vector $\mathbf{a}$, $b \neq 0$, and c.

5.18. Let X_1 and X_2 be jointly distributed random variables with the property that $\text{E}(X_2 \mid X_1) = \alpha + \beta X_1$. Show that $\text{Corr}(X_1, X_2) = \beta\sigma_1/\sigma_2$, where σ_1 and σ_2 denote the respective standard deviations of X_1 and X_2.

5.19. Under the setting of Definition 5.2.8, show that $\rho_{0(1,\cdots,k)} = 1$ if and only if $X_0 = \mathbf{a}'\mathbf{x}^{(1)} + a_0$ for some constant vector $\mathbf{a}$ and scalar a_0.

5.20. Show that $\chi^2(k,\lambda) = \chi^2(1,\lambda) + \chi^2_{k-1}$.

5.21. Let $U \sim \chi^2(k,\lambda)$, where $k > 0, \lambda > 0$. For $u > 0$, show that $\text{P}(U \leq u) = \text{P}(X_1 - X_2 \geq k/2)$ where X_1 and X_2 are independent Poisson random variables with respective means $u/2$ and λ.

5.22. Let $(X_1, X_2)' \sim N_2(\mu_1, \mu_2, \sigma_1^2, \sigma_2^2, \rho)$, $|\rho| \neq 1$. Show that

$$T = \frac{Z_1^2 - 2\rho Z_1 Z_2 + Z_2^2}{1-\rho^2}$$

has a χ^2 distribution, where $Z_i = (X_i - \mu_i)/\sigma_i$. What are its parameters?

5.23. (a) Let $\mathbf{x} \sim N_k(\boldsymbol{\mu}, \mathbf{D})$, where $\mathbf{D} = \text{diag}(\sigma_1^2, \cdots, \sigma_k^2)$, $\text{r}(\mathbf{D}) = k$. Find the mean and variance of the random variable $U = \mathbf{x}'\mathbf{D}^{-1}\mathbf{x}$.

(b) Let $\mathbf{x} \sim N_k(\boldsymbol{\mu}, \boldsymbol{\Sigma})$, with $\text{r}(\boldsymbol{\Sigma}) = k$. What is the distribution of $U = (\mathbf{x}-\boldsymbol{\mu})'\boldsymbol{\Sigma}^{-1}(\mathbf{x}-\boldsymbol{\mu})$?

(c) Let $\mathbf{x} \sim N_k(\boldsymbol{\mu}, \mathbf{D})$. Suppose $\mathbf{A} = \mathbf{D}^{-1} - (\mathbf{D}^{-1}\mathbf{1}\mathbf{1}'\mathbf{D}^{-1})/\mathbf{1}'\mathbf{D}^{-1}\mathbf{1}$. Find the distribution of $\mathbf{x}'\mathbf{A}\mathbf{x}$.

5.24. Verify Result 5.4.5.

5.25. Let $V \sim F(k_1, k_2, \lambda)$. Show that

$$\text{E}(V) = \frac{k_2}{k_2 - 2}\left(1 + \frac{2\lambda}{k_1}\right)$$

and

$$\text{Var}(V) = \frac{2k_2^2}{k_1^2(k_2-2)}\left[\frac{(k_1+2\lambda)^2}{(k_2-2)(k_2-4)} + \frac{k_1+4\lambda}{k_2-4}\right].$$

5.26. If $\mathbf{x} \sim N_k(\boldsymbol{\mu}, \boldsymbol{\Sigma})$, show that

(a) $\text{E}(\mathbf{x}'\mathbf{A}\mathbf{x}) = \text{tr}(\mathbf{A}\boldsymbol{\Sigma}) + \boldsymbol{\mu}'\mathbf{A}\boldsymbol{\mu}$

(b) $\text{Var}(\mathbf{x}'\mathbf{A}\mathbf{x}) = 2\,\text{tr}(\mathbf{A}\boldsymbol{\Sigma})^2 + 4\boldsymbol{\mu}'\mathbf{A}\boldsymbol{\Sigma}\mathbf{A}\boldsymbol{\mu}$.

(c) $\text{Cov}(\mathbf{x}, \mathbf{x}'\mathbf{A}\mathbf{x}) = 2\boldsymbol{\Sigma}\mathbf{A}\boldsymbol{\mu}$.

5.27. Let $\mathbf{x} \sim N_k(\boldsymbol{\mu}, \mathbf{I})$. Show that $\mathbf{x}'\mathbf{A}\mathbf{x} \sim \chi^2(k, \boldsymbol{\mu}'\mathbf{A}\boldsymbol{\mu}/2)$ if and only if $\mathbf{A}$ is idempotent of rank m.

5.28. Verify Result 5.4.6. [Hint: one possible way is to use the fact that if $\mathbf{x} \sim N_k(\mathbf{0}, \boldsymbol{\Sigma})$, then $\mathbf{x} = \boldsymbol{\Gamma}\mathbf{z}$ where $\mathbf{z} \sim N_k(\mathbf{0}, \mathbf{I})$ and $\boldsymbol{\Gamma}\boldsymbol{\Gamma}' = \boldsymbol{\Sigma}$. Then write $\mathbf{x}'\mathbf{A}\mathbf{x}$ in terms of $\mathbf{z}$, so that Result 5.4.4 is applicable].

5.29. Verify the necessity part of Result 5.4.8 by the following steps.

(a) Suppose $\mathbf{x} \sim N_k(\mathbf{0}, \mathbf{I})$. Show that the b.l.u.p. of $\mathbf{x}$ based on $\mathbf{B}\mathbf{x}$ is $\widehat{\mathbf{x}} = \mathbf{P}\mathbf{x}$ and $\mathbf{y} = \mathbf{x} - \widehat{\mathbf{x}} \sim N_k(\mathbf{0}, \mathbf{I} - \mathbf{P})$, where $\mathbf{P}$ is the matrix of orthogonal projection on $\mathcal{C}(\mathbf{B}')$.

(b) Verify $\mathbf{x}'\mathbf{A}\mathbf{x} + \mathbf{a}'\mathbf{x} = \mathbf{y}'\mathbf{A}\mathbf{y} + (2\mathbf{A}\widehat{\mathbf{x}} + \mathbf{a})'\mathbf{y} + (\widehat{\mathbf{x}}'\mathbf{A}\widehat{\mathbf{x}} + \mathbf{a}'\widehat{\mathbf{x}})$.

(c) Show that independence of $\mathbf{B}\mathbf{x}$ and $\mathbf{x}'\mathbf{A}\mathbf{x} + \mathbf{a}'\mathbf{x}$ implies that for any fixed $\mathbf{x}_0$, $U(\mathbf{x}_0) = \mathbf{y}'\mathbf{A}\mathbf{y} + (2\mathbf{A}\mathbf{P}\mathbf{x}_0 + \mathbf{a})'\mathbf{y} + (\mathbf{x}_0'\mathbf{P}\mathbf{A}\mathbf{P}\mathbf{x}_0 + \mathbf{a}'\mathbf{P}\mathbf{x}_0)$ has the same distribution.

(d) Use the invariance of $E[U(\mathbf{x}_0)]$ to show $\mathbf{P}\mathbf{A}\mathbf{P} = \mathbf{O}$ and $\mathbf{a}'\mathbf{P} = \mathbf{0}$, and the latter is equivalent to $\mathbf{B}\mathbf{a} = \mathbf{0}$.

(e) Use the invariance of $\text{Var}[U(\mathbf{x}_0)]$ to show that $\mathbf{B}\mathbf{A} = \mathbf{O}$.

(f) Complete the proof by reducing the case where $\mathbf{x} \sim N_k(\boldsymbol{\mu}, \boldsymbol{\Sigma})$ to the standard normal case.

5.30. Let $\mathbf{x} \sim N_k(\boldsymbol{\mu}, \mathbf{Q}\mathbf{A})$, where $\mathbf{A}$ is a positive definite matrix, and $\mathbf{Q}$ is symmetric and idempotent with $\text{tr}(\mathbf{Q}) = m$. What is the distribution of $U = \mathbf{x}'\mathbf{A}^{-1}\mathbf{x}$?

5.31. Let $(X_1, X_2, X_3)' \sim N_3(\mathbf{0}, \mathbf{I})$. Let $Q = \frac{1}{6}(X_1^2 + 4X_2^2 + X_3^2 - 4X_1X_2 + 2X_1X_3 - 4X_2X_3)$. Find the distribution of Q.

5.32. Let $\mathbf{x} \sim N_k(\mathbf{0}, \boldsymbol{\Sigma})$, where $\boldsymbol{\Sigma} = \sigma^2[(1-\rho)\mathbf{I}_k + \rho\mathbf{J}_k]$, $0 \le \rho < 1$.

(a) Show that the distinct eigenvalues of $\boldsymbol{\Sigma}$ are $\lambda_1 = \sigma^2(1-\rho)$, with multiplicity $g_1 = k-1$, and $\lambda_2 = \sigma^2[1 + (k-1)\rho]$ with multiplicity $g_2 = 1$.

(b) Define $\mathbf{A}_1 = \mathbf{I}_k - \mathbf{J}_k/k$, and $\mathbf{A}_2 = \mathbf{J}_k/k$. Show that $\mathbf{A}_1$ and $\mathbf{A}_2$ are idempotent, $\mathbf{A}_1\mathbf{A}_2 = \mathbf{0}$, and that $\boldsymbol{\Sigma} = \lambda_1\mathbf{A}_1 + \lambda_2\mathbf{A}_2$.

(c) Let $Q_i = \mathbf{x}'\mathbf{A}_i\mathbf{x}/\lambda_i$, $i = 1, 2$. Show that Q_1 and Q_2 have independent chi-square distributions. Find the parameters of these distributions.

5.33. Let $\mathbf{x} \sim N_k(\boldsymbol{\Sigma}\boldsymbol{\mu}, \sigma^2\boldsymbol{\Sigma})$, where $\boldsymbol{\Sigma}$ is symmetric and p.d. and $\sigma^2 > 0$. Let $\mathbf{B} = \boldsymbol{\Sigma}^{-1} - \boldsymbol{\Sigma}^{-1}\mathbf{1}_k(\mathbf{1}_k'\boldsymbol{\Sigma}^{-1}\mathbf{1}_k)^{-1}\mathbf{1}_k'\boldsymbol{\Sigma}^{-1}$.

(a) Derive the distribution of $\mathbf{B}\mathbf{x}$.

(b) Derive the distribution of $\mathbf{y}'\boldsymbol{\Sigma}\mathbf{y}$ when (i) $\boldsymbol{\mu} = \mathbf{0}$, and (ii) $\boldsymbol{\mu} \neq \mathbf{0}$.

[Hint: Show that $\mathbf{B}$ is symmetric and $\mathbf{B}\boldsymbol{\Sigma}$ is idempotent.]

5.34. (a) Complete the proof of Result 5.4.7 for the general case where $\mathbf{x} \sim N_k(\mu, \boldsymbol{\Sigma})$ with $\boldsymbol{\Sigma}$ being p.d.

(b) Use (5.4.18) and Result 5.4.7 for the case where $\mathbf{x} \sim N_k(\mathbf{0}, \mathbf{I})$ to prove (5.4.19).

5.35. Let $(X_1, X_2)' \sim N_2(\mu\mathbf{1}, \boldsymbol{\Sigma})$, where $\boldsymbol{\Sigma} = (1-\rho)\mathbf{I}_2 + \rho\mathbf{J}_2$, $-1 < \rho < 1$. Let $Q_1 = (X_1 - X_2)^2$ and $Q_2 = (X_1 + X_2)^2$. Show that $Q_1/2(1-\rho)$ has a chi-square distribution and that Q_1 and Q_2 are distributed independently.

5.36. Let $\mathbf{x} \sim N_k(\boldsymbol{\mu}, \boldsymbol{\Sigma})$ with $\boldsymbol{\Sigma}$ being p.d. Show that a necessary and sufficient condition for $(\mathbf{x} - \boldsymbol{\mu})'\mathbf{A}(\mathbf{x} - \boldsymbol{\mu})$ and $(\mathbf{x} - \boldsymbol{\mu})'\mathbf{B}(\mathbf{x} - \boldsymbol{\mu})$ to be independently distributed is $\boldsymbol{\Sigma}\mathbf{A}\boldsymbol{\Sigma}\mathbf{B}\boldsymbol{\Sigma} = \mathbf{O}$.

5.37. Verify the necessity part of Result 5.4.9.

5.38. Verify the sufficiency part of Result 5.4.9 by the following steps. By assumption, $\sum_{j=1}^{L} \mathbf{A}_j = \mathbf{I}_k$ and $\sum_{j=1}^{L} \mathrm{r}(\mathbf{A}_j) = k$. Recall the rank inequality $\mathrm{r}(\mathbf{A}+\mathbf{B}) \le \mathrm{r}(\mathbf{A})+\mathrm{r}(\mathbf{B})$, see Result 1.3.11.

(a) Show that for any $\mathbf{A}_j$, $\mathrm{r}(\mathbf{I} - \mathbf{A}_j) + \mathrm{r}(\mathbf{A}_j) = k$.

(b) Let m_a denote the multiplicity of a as an eigenvalue of $\mathbf{A}_j$. Show that $\mathrm{r}(\mathbf{I} - \mathbf{A}_j) = k - m_1$ and $\mathrm{r}(\mathbf{A}_j) = k - m_0$.

(c) Use (a) and (b) to show that $\mathbf{A}_j$ is idempotent.

(d) Show that $\sum_{i \neq j} \mathbf{A}_j \mathbf{A}_i \mathbf{A}_j = \mathbf{O}$.

(e) From (d) derive $\mathbf{A}_i \mathbf{A}_j = \mathbf{O}$ for $i \neq j$.

(f) Use (c) and (e) to complete the proof on the joint distribution of $Q_j = \mathbf{x}'\mathbf{A}\mathbf{x}$.

5.39. Prove properties 3 and 4 in Result 5.5.3.

5.40. Let $\mathbf{y} \sim E_k(\boldsymbol{\mu}, \mathbf{V}, h)$, and let $\mathbf{B}$ be a $q \times k$ matrix of constants, with $\mathrm{r}(\mathbf{B}) = q \le k$. Show that the distribution of $\mathbf{B}\mathbf{y} \sim E_q(\mathbf{B}\boldsymbol{\mu}, \mathbf{B}\mathbf{V}\mathbf{B}')$. [Hint: Derive the m.g.f. of $\mathbf{B}\mathbf{y}$.]

6

Sampling from the Multivariate Normal Distribution

We construct the joint density of a random sample from a multivariate normal distribution and describe estimation of the parameters of the distribution along with properties of these estimates. While most of these results, such as results related to the Wishart distribution, are only related to the multivariate linear model that we discuss in Section 13.1, we include them here for the sake of completeness and since knowledge of these ideas is becoming increasingly more important in multivariate and high-dimensional linear models. Based on random samples, we describe inference for simple, multiple, and partial correlation coefficients, as well as some notions related to assessing multivariate normality. Readers may defer Chapter 6 for later, and go directly to Chapter 7 to continue with inference for the general linear model that we discussed in Chapter 4.

6.1 Distribution of sample mean and covariance

The mean of a random sample $\mathbf{x}_1, \cdots, \mathbf{x}_N$ is defined to be the k-dimensional vector

$$\overline{\mathbf{x}}_N = \frac{1}{N}\sum_{i=1}^{N} \mathbf{x}_i, \tag{6.1.1}$$

and the sample covariance matrix is the $k \times k$ matrix $\mathbf{S}_N = \mathbf{W}_N/(N-1)$, where

$$\mathbf{W}_N = \sum_{i=1}^{N}(\mathbf{x}_i - \overline{\mathbf{x}}_N)(\mathbf{x}_i - \overline{\mathbf{x}}_N)' = \sum_{i=1}^{N} \mathbf{x}_i\mathbf{x}_i' - N\overline{\mathbf{x}}_N\overline{\mathbf{x}}_N'. \tag{6.1.2}$$

If $\mathbf{x}_i = (X_{i,1}, \ldots, X_{i,k})'$ and $\overline{\mathbf{x}}_N = (\overline{X}_1, \cdots, \overline{X}_k)'$, we can write $\mathbf{S}_N = \{S_{jl}\}$, $j, l = 1, \cdots, k$, with

$$S_{jl} = \sum_{i=1}^{N}(X_{i,j} - \overline{X}_j)(X_{i,l} - \overline{X}_l)/(N-1).$$

The sample covariance matrix has k variances S_{jj}, $j = 1, \cdots, k$ and $k(k-1)/2$ possibly distinct covariances, S_{jl}, $j < l$, $j, l = 1, \cdots, k$. The generalized sample variance is the scalar quantity $|\mathbf{S}_N|$, which is a natural summary measure of variability in the sample $\mathbf{x}_1, \cdots, \mathbf{x}_N$. Another summary is $\text{tr}(\mathbf{S}_N)$, the total variance.

Given a univariate random sample $X_1, \cdots, X_N$ from a $N(\mu, \sigma^2)$ population, we know that the sample mean $\overline{X}$ has a normal distribution with mean μ and variance σ^2/N, the statistic $(N-1)S^2/\sigma^2$ is distributed as a χ^2_{N-1} variable, and the two distributions are independent (Casella and Berger, 1990; Mukhopadhyay, 2000). The corresponding distributional result for the k-variate situation is given in Result 6.1.2. We first define a multivariate distribution called the Wishart distribution, which is derived from the multivariate normal

DOI: 10.1201/9781315156651-6

distribution as the sampling distribution of the sample statistic $\sum_{i=1}^N (\mathbf{x}_i - \overline{\mathbf{x}}_N)(\mathbf{x}_i - \overline{\mathbf{x}}_N)'$. The Wishart distribution is a multivariate extension of the chi-square distribution.

Definition 6.1.1. Wishart distribution. Let $\mathbf{x}_1, \cdots, \mathbf{x}_m$ be i.i.d. $N_k(\mathbf{0}, \mathbf{\Sigma})$ with $\mathbf{\Sigma}$ being p.d. The distribution of the random $k \times k$ matrix $\mathbf{W} = \sum_{j=1}^m \mathbf{x}_j \mathbf{x}_j'$ is called the k-dimensional Wishart distribution with m d.f. and scale matrix $\mathbf{\Sigma}$, and is denoted by $W_k(\mathbf{\Sigma}, m)$.

Provided $m \geq k$, the p.d.f. of $\mathbf{W}$ is

$$f(\mathbf{W}; \mathbf{\Sigma}) = \frac{|\mathbf{W}|^{(m-k-1)/2} \exp\{-\frac{1}{2} \operatorname{tr}(\mathbf{\Sigma}^{-1}\mathbf{W})\}}{2^{km/2} |\mathbf{\Sigma}|^{m/2} \Gamma_k(m/2)}, \tag{6.1.3}$$

where $\Gamma_k(m/2) = \pi^{k(k-1)/4} \prod_{j=1}^k \Gamma(\frac{1}{2}(m+1-j))$. The additivity property holds for Wishart distributions. That is, if $\mathbf{W}_j \sim W_k(\mathbf{\Sigma}, m_j)$, $j = 1, \cdots, J$, are independent, then $\sum_{j=1}^J \mathbf{W}_j \sim W_k(\mathbf{\Sigma}, \sum_{j=1}^J m_j)$.

The following result, which is a multivariate version of Cochran's theorem (see Result 5.4.9), is useful not only in this chapter, but also in the study on multivariate general linear models in Chapter 13.

Result 6.1.1. Let $\mathbf{X} = (\mathbf{x}_1, \cdots, \mathbf{x}_m)'$ with $\mathbf{x}_j$ being i.i.d. $N_k(\mathbf{0}, \mathbf{\Sigma})$. Let $\mathbf{A}$ be an $m \times m$ symmetric and idempotent matrix, i.e., a projection matrix. If $\mathrm{r}(\mathbf{A}) = r$, then $\mathbf{X}'\mathbf{A}\mathbf{X} \sim W_k(\mathbf{\Sigma}, r)$.

Proof. Let $\mathbf{A} = \mathbf{P}' \operatorname{diag}(\mathbf{I}_r, \mathbf{O}_{m-r})\mathbf{P}$, where $\mathbf{P}$ is an orthogonal matrix. Let the *columns* of $\mathbf{X}$ be $\mathbf{x}_{(1)}, \ldots, \mathbf{x}_{(k)}$, so that $\mathbf{x}_{(i)} = (X_{1,i}, X_{2,i}, \cdots, X_{m,i})'$. Note that $\operatorname{Cov}(\mathbf{x}_{(i)}, \mathbf{x}_{(j)}) = \sigma_{ij}\mathbf{I}_m$, $i, j = 1, \ldots, k$, where σ_{ij} is the (i,j)th entry of $\mathbf{\Sigma}$. Then $\operatorname{Cov}(\mathbf{P}\mathbf{x}_{(i)}, \mathbf{P}\mathbf{x}_{(j)}) = \mathbf{P} \operatorname{Cov}(\mathbf{x}_{(i)}, \mathbf{x}_{(j)})\mathbf{P}' = \sigma_{ij}\mathbf{I}_m$. Since the components of $\mathbf{X}$ are jointly normally distributed with mean $\mathbf{0}$, we see that $\mathbf{P}\mathbf{X} \sim \mathbf{X}$, where the symbol $\sim$ means "has the same distribution as". Then, $\mathbf{X}'\mathbf{A}\mathbf{X} = (\mathbf{P}\mathbf{X})' \operatorname{diag}(\mathbf{I}_r, \mathbf{O}_{m-r})\mathbf{P}\mathbf{X} \sim \mathbf{X}' \operatorname{diag}(\mathbf{I}_r, \mathbf{O}_{m-r})\mathbf{X} = \sum_{i=1}^r \mathbf{x}_i\mathbf{x}_i' \sim W_k(\mathbf{\Sigma}, r)$. ■

Result 6.1.2. Distribution of $\overline{\mathbf{x}}_N$ and $\mathbf{S}_N$. Let $\mathbf{x}_1, \cdots, \mathbf{x}_N$ denote a random sample from a $N_k(\boldsymbol{\mu}, \mathbf{\Sigma})$ population. Then

1. the distribution of $\overline{\mathbf{x}}_N$ is $N_k(\boldsymbol{\mu}, \mathbf{\Sigma}/N)$,
2. for $N \geq 2$, $(N-1)\mathbf{S}_N$ follows a Wishart $W_k(\mathbf{\Sigma}, N-1)$ distribution, and
3. $\overline{\mathbf{x}}_N$ and $\mathbf{S}_N$ are independently distributed.

Proof. Property 1 is a direct consequence of the reproductive property of normal distributions; see Result 5.2.9. To prove property 2, since $\mathbf{x}_i - \overline{\mathbf{x}}_N$ is invariant to $\boldsymbol{\mu}$, assume without loss of generality that $\boldsymbol{\mu} = \mathbf{0}$. Let $\mathbf{X} = (\mathbf{x}_1, \cdots, \mathbf{x}_N)'$. From (6.1.2), $(N-1)\mathbf{S}_N = \mathbf{X}'\mathbf{X} - N\overline{\mathbf{x}}_N\overline{\mathbf{x}}_N'$. Since $\overline{\mathbf{x}}_N = \mathbf{X}'\mathbf{1}_N/N$, then $\mathbf{S}_N = \mathbf{X}'\mathbf{A}\mathbf{X}$, where $\mathbf{A} = \mathbf{I}_N - \mathbf{J}_N/N$ is symmetric and idempotent with $\mathrm{r}(\mathbf{A}) = \operatorname{tr}(\mathbf{A}) = N-1$. Then property 2 follows from Result 6.1.1. To prove property 3, note that $\mathbf{x}_1, \cdots, \mathbf{x}_N, \overline{\mathbf{x}}_N$ are jointly normally distributed, and $\operatorname{Cov}(\mathbf{x}_i - \overline{\mathbf{x}}_N, \overline{\mathbf{x}}_N) = \mathbf{O}$, as can be easily verified. Since $\mathbf{x}_i - \overline{\mathbf{x}}_N$ is independent of $\overline{\mathbf{x}}_N$, it follows directly that $\mathbf{S}_N$ is independent of $\overline{\mathbf{x}}_N$. ■

Ghosh (1996) gives an alternate proof of Result 6.1.2 using induction (see Exercise 6.5), while Anderson (2003) uses the Helmert transformation.

Example 6.1.1. For any nonzero vector $\mathbf{a} \in \mathcal{R}^k$, we show that the statistic $\mathbf{a}'\mathbf{S}_N\mathbf{a}/\mathbf{a}'\boldsymbol{\Sigma}\mathbf{a} \sim \chi^2_{N-1}$. Decomposing $\mathbf{S}_N$ as in (6.1.2), $\mathbf{a}'\mathbf{S}_N\mathbf{a} = \sum_{i=1}^N (z_i - \overline{z}_N)^2$, where $z_i = \mathbf{a}'\mathbf{x}_i$ are i.i.d. $N(\mathbf{a}'\boldsymbol{\mu}, \mathbf{a}'\boldsymbol{\Sigma}\mathbf{a})$. That $\mathbf{a}'\mathbf{S}_N\mathbf{a}$ is a scaled chi-square is an immediate consequence of property 2 of Result 6.1.2. □

The factorization theorem (Halmos and Savage, 1949) enables us to identify a sufficient statistic by inspecting the form of the p.d.f. or p.m.f. of the random sample.

Result 6.1.3. The sample mean and sample variance are sufficient statistics for the mean $\boldsymbol{\mu}$ and covariance $\boldsymbol{\Sigma}$ in a k-variate normal random sample $\mathbf{x}_1, \cdots, \mathbf{x}_N$.

Proof. The joint density function of N random vectors $\mathbf{x}_1, \cdots, \mathbf{x}_N$ is a product of the marginal k-variate normal densities:

$$\begin{aligned} f(\mathbf{x}_1, \cdots, \mathbf{x}_N; \boldsymbol{\mu}, \boldsymbol{\Sigma}) &= \prod_{i=1}^N \frac{\exp\{-\frac{1}{2}(\mathbf{x}_i - \boldsymbol{\mu})'\boldsymbol{\Sigma}^{-1}(\mathbf{x}_i - \boldsymbol{\mu})\}}{(2\pi)^{k/2}|\boldsymbol{\Sigma}|^{1/2}} \\ &= \frac{\exp\{-\frac{1}{2}\sum_{i=1}^N(\mathbf{x}_i - \boldsymbol{\mu})'\boldsymbol{\Sigma}^{-1}(\mathbf{x}_i - \boldsymbol{\mu})\}}{(2\pi)^{Nk/2}|\boldsymbol{\Sigma}|^{N/2}}. \end{aligned} \tag{6.1.4}$$

Since $\sum_{i=1}^N(\mathbf{x}_i - \overline{\mathbf{x}}_N)(\overline{\mathbf{x}}_N - \boldsymbol{\mu})' = \mathbf{O}$, and $\sum_{i=1}^N(\overline{\mathbf{x}}_N - \boldsymbol{\mu})(\mathbf{x}_i - \overline{\mathbf{x}}_N)' = \mathbf{O}$, using Result 1.3.5, we can write

$$\begin{aligned} \sum_{i=1}^N(\mathbf{x}_i - \boldsymbol{\mu})'\boldsymbol{\Sigma}^{-1}(\mathbf{x}_i - \boldsymbol{\mu}) &= \sum_{i=1}^N \mathrm{tr}[(\mathbf{x}_i - \boldsymbol{\mu})'\boldsymbol{\Sigma}^{-1}(\mathbf{x}_i - \boldsymbol{\mu})] \\ &= \sum_{i=1}^N \mathrm{tr}\left[\boldsymbol{\Sigma}^{-1}(\mathbf{x}_i - \boldsymbol{\mu})(\mathbf{x}_i - \boldsymbol{\mu})'\right] = \mathrm{tr}\left[\boldsymbol{\Sigma}^{-1}\sum_{i=1}^N\{(\mathbf{x}_i - \boldsymbol{\mu})(\mathbf{x}_i - \boldsymbol{\mu})'\}\right] \\ &= \mathrm{tr}[\boldsymbol{\Sigma}^{-1}\sum_{i=1}^N\{(\mathbf{x}_i - \overline{\mathbf{x}}_N + \overline{\mathbf{x}}_N - \boldsymbol{\mu})(\mathbf{x}_i - \overline{\mathbf{x}}_N + \overline{\mathbf{x}}_N - \boldsymbol{\mu})'\}] \\ &= \mathrm{tr}\left[\boldsymbol{\Sigma}^{-1}\sum_{i=1}^N\{(\mathbf{x}_i - \overline{\mathbf{x}}_N)(\mathbf{x}_i - \overline{\mathbf{x}}_N)'\}\right] + N(\overline{\mathbf{x}}_N - \boldsymbol{\mu})'\boldsymbol{\Sigma}^{-1}(\overline{\mathbf{x}}_N - \boldsymbol{\mu}). \end{aligned}$$

Hence the joint density function can be written as

$$\begin{aligned} &f(\mathbf{x}_1, \cdots, \mathbf{x}_N; \boldsymbol{\mu}, \boldsymbol{\Sigma}) \\ &= \frac{\exp\left\{-\frac{1}{2}\mathrm{tr}[(N-1)\boldsymbol{\Sigma}^{-1}\mathbf{S}_N] - \frac{N}{2}(\overline{\mathbf{x}}_N - \boldsymbol{\mu})'\boldsymbol{\Sigma}^{-1}(\overline{\mathbf{x}}_N - \boldsymbol{\mu})\right\}}{(2\pi)^{Nk/2}|\boldsymbol{\Sigma}|^{N/2}}, \end{aligned} \tag{6.1.5}$$

and depends on the sample observations $\mathbf{x}_1, \cdots, \mathbf{x}_N$ only through $\overline{\mathbf{x}}_N$ and $\mathbf{S}_N$. By the factorization theorem, these two sample statistics are *sufficient statistics* for $\boldsymbol{\mu}$ and $\boldsymbol{\Sigma}$. ■

Result 6.1.3 states a special property of a multivariate normal sample. Further, by Basu's theorem (Basu, 1964), $\overline{\mathbf{x}}_N$ and $\mathbf{S}_N$ are independently distributed. In general, the sample mean and variance are not sufficient statistics in nonnormal samples. Given observed values of $\mathbf{x}_1, \cdots, \mathbf{x}_N$, (6.1.5) viewed as a function of $\boldsymbol{\mu}$ and $\boldsymbol{\Sigma}$ is the *likelihood function*, maximizing which yields the MLE's of these parameters. The MLE's of the mean vector and covariance matrix of a k-variate normal population are derived in the following result.

Result 6.1.4. Maximum likelihood estimation of $\boldsymbol{\mu}$ and $\boldsymbol{\Sigma}$. Based on a random sample $\mathbf{x}_1, \cdots, \mathbf{x}_N$ from a $N_k(\boldsymbol{\mu}, \boldsymbol{\Sigma})$ distribution, the maximum likelihood estimates of $\boldsymbol{\mu}$ and $\boldsymbol{\Sigma}$ are

$$\widehat{\boldsymbol{\mu}}_{ML} = \frac{1}{N}\sum_{i=1}^{N} \mathbf{x}_i = \overline{\mathbf{x}}_N, \text{ and} \tag{6.1.6}$$

$$\widehat{\boldsymbol{\Sigma}}_{ML} = \frac{1}{N}\sum_{i=1}^{N} (\mathbf{x}_i - \overline{\mathbf{x}}_N)(\mathbf{x}_i - \overline{\mathbf{x}}_N)' = \frac{N-1}{N}\mathbf{S}_N. \tag{6.1.7}$$

Proof. The likelihood function $L(\boldsymbol{\mu}, \boldsymbol{\Sigma}; \mathbf{x}_1, \cdots, \mathbf{x}_N)$ has the form shown on the right side of (6.1.5). The MLE's of $\boldsymbol{\mu}$ and $\boldsymbol{\Sigma}$ are denoted by $\widehat{\boldsymbol{\mu}}_{ML}$ and $\widehat{\boldsymbol{\Sigma}}_{ML}$, and are the values that maximize $L(\boldsymbol{\mu}, \boldsymbol{\Sigma}; \mathbf{x}_1, \cdots, \mathbf{x}_N)$. Since $\boldsymbol{\Sigma}^{-1}$ is p.d. (see Definition 2.4.5), the distance $(\overline{\mathbf{x}}_N - \boldsymbol{\mu})'\boldsymbol{\Sigma}^{-1}(\overline{\mathbf{x}}_N - \boldsymbol{\mu})$ in the exponent of the likelihood function is positive unless $\boldsymbol{\mu} = \overline{\mathbf{x}}_N$. The MLE of $\boldsymbol{\mu}$ is then $\overline{\mathbf{x}}_N$, since it is the value of $\boldsymbol{\mu}$ that maximizes the likelihood function. We next maximize the following function with respect to $\boldsymbol{\Sigma}$:

$$L(\boldsymbol{\Sigma}; \widehat{\boldsymbol{\mu}}_{ML}, \mathbf{x}_1, \cdots, \mathbf{x}_N) = \frac{\exp\left\{-\frac{1}{2}\operatorname{tr}\left[\boldsymbol{\Sigma}^{-1}\sum_{i=1}^{N}(\mathbf{x}_i - \overline{\mathbf{x}}_N)(\mathbf{x}_i - \overline{\mathbf{x}}_N)'\right]\right\}}{(2\pi)^{Nk/2}|\boldsymbol{\Sigma}|^{N/2}}. \tag{6.1.8}$$

Using Example 2.4.4 with $\mathbf{A} = \boldsymbol{\Sigma}$, $\mathbf{B} = \sum_{i=1}^{N}(\mathbf{x}_i - \overline{\mathbf{x}}_N)(\mathbf{x}_i - \overline{\mathbf{x}}_N)'$, and $b = N/2$, we can show that $\widehat{\boldsymbol{\Sigma}}_{ML} = \frac{1}{N}\sum_{i=1}^{N}(\mathbf{x}_i - \overline{\mathbf{x}}_N)(\mathbf{x}_i - \overline{\mathbf{x}}_N)'$ maximizes the likelihood in (6.1.8). The maximized likelihood is

$$L(\widehat{\boldsymbol{\mu}}_{ML}, \widehat{\boldsymbol{\Sigma}}_{ML}; \mathbf{x}_1, \cdots, \mathbf{x}_N) = \frac{\exp(-\frac{Nk}{2})}{(2\pi)^{Nk/2}|\widehat{\boldsymbol{\Sigma}}_{ML}|^{N/2}}, \tag{6.1.9}$$

which completes the proof. ■

Corollary 6.1.1. Using the invariance property of MLE's, which states that the MLE of a function $g(\boldsymbol{\theta})$ of a parameter $\boldsymbol{\theta}$ is $g(\widehat{\boldsymbol{\theta}}_{ML})$, we see that

1. The MLE of the function $\boldsymbol{\mu}'\boldsymbol{\Sigma}^{-1}\boldsymbol{\mu}$ is $\widehat{\boldsymbol{\mu}}'_{ML}\widehat{\boldsymbol{\Sigma}}^{-1}_{ML}\widehat{\boldsymbol{\mu}}_{ML}$.
2. The MLE of $\sqrt{\sigma_{lj}}$, the (l, j)th element of $\boldsymbol{\Sigma}$ is given by $\sqrt{\widehat{\sigma}_{lj}}$, the square-root of the MLE of the (l, j)th entry in $\widehat{\boldsymbol{\Sigma}}_{ML}$.

Although the estimator $\widehat{\boldsymbol{\mu}}_{ML}$ is a unbiased estimator of $\boldsymbol{\mu}$, $\widehat{\boldsymbol{\Sigma}}_{ML}$ is a biased estimator of $\boldsymbol{\Sigma}$. Analogous to the univariate case, an unbiased estimator of $\boldsymbol{\Sigma}$ is $\widehat{\boldsymbol{\Sigma}} = \frac{1}{N-1}\mathbf{S}_N$. The unbiased estimators of $\boldsymbol{\mu}$ and $\boldsymbol{\Sigma}$ denoted by $\widehat{\boldsymbol{\mu}}$ and $\widehat{\boldsymbol{\Sigma}}$ respectively are complete, sufficient statistics for these parameters (Seber, 1984).

6.2 Distributions related to correlation coefficients

In Chapter 5, we defined theoretical versions of simple, partial, and multiple correlations; we described properties of each type of correlation, and discussed relationships between them. In this section, we discuss estimation of these correlation coefficients, as well as distributional properties of these estimators that enable inference. We begin with the simple correlation based on i.i.d. bivariate normal samples $(X_{i,j}, X_{i,l})$, $i = 1, \cdots, N$, where X_j and X_l denote the jth and lth components of the k-dimensional vector $\mathbf{x}$. Suppose $\mathrm{E}(X_{i,j}) = \mu_j$, $\mathrm{E}(X_{i,l}) = \mu_l$, $\mathrm{Var}(X_{i,j}) = \sigma_j^2$, $\mathrm{Var}(X_{i,l}) = \sigma_l^2$, and $\mathrm{Corr}(X_{i,j}, X_{i,l}) = \rho_{jl}$.

Result 6.2.1. Estimation of simple correlations.

1. The MLE of the simple correlation between X_j and X_l is

$$\widehat{\rho}_{jl} = \frac{\widehat{\sigma}_{jl}}{\sqrt{\widehat{\sigma}_{jj}\widehat{\sigma}_{ll}}}, \tag{6.2.1}$$

   where the quantities on the right side have been defined in property 2 of Corollary 6.1.1. The estimator $\widehat{\rho}_{jl}$ can be expressed as a function of the complete, sufficient statistics $\overline{\mathbf{x}}_N$ and $\mathbf{S}_N$.

2. When $\rho_{jl} = 0$, the p.d.f. of $\widehat{\rho}_{jl}$ is

$$f(r) = \frac{1}{B(1/2, (N-2)/2)}(1-r^2)^{(N-4)/2}, \quad -1 \leq r \leq 1. \tag{6.2.2}$$

3. When $\rho_{jl} = 0$, the random variable

$$T_{jl} = \widehat{\rho}_{jl}\sqrt{\frac{N-2}{1-\widehat{\rho}_{jl}^2}} \tag{6.2.3}$$

   has a Student's t-distribution with $(N-2)$ degrees of freedom.

Proof. Property 1 follows from the invariance of the MLE (see Corollary 6.1.1). To prove property 2, note that the sample correlation coefficient is invariant under change of location and scale. Hence we can assume without loss of generality that $\mu_j = \mu_l = 0$, and $\sigma_{jj}^2 = \sigma_{ll}^2 = 1$. If $\rho_{jl} = 0$, then $\mathbf{y} = (X_{1,j}, \cdots, X_{N,j})'$ and $\mathbf{z} = (X_{1,l}, \cdots, X_{N,l})'$ are i.i.d. $\sim N(\mathbf{0}, \mathbf{I}_N)$. From the proof of property 2 of Result 6.1.2,

$$\widehat{\rho}_{jl} = \frac{\mathbf{y}'(\mathbf{I}_N - \mathbf{J}_N/N)\mathbf{z}}{\sqrt{\mathbf{y}'(\mathbf{I}_N - \mathbf{J}_N/N)\mathbf{y}}\sqrt{\mathbf{z}'(\mathbf{I}_N - \mathbf{J}_N/N)\mathbf{z}}} \sim \frac{\mathbf{u}'\mathbf{v}}{\|\mathbf{u}\|\,\|\mathbf{v}\|} = \boldsymbol{\eta}'\boldsymbol{\xi},$$

where $\mathbf{u}$ and $\mathbf{v}$ are i.i.d. $\sim N(\mathbf{0}, \mathbf{I}_{N-1})$, $\boldsymbol{\eta} = \mathbf{u}/\|\mathbf{u}\|$, and $\boldsymbol{\xi} = \mathbf{v}/\|\mathbf{v}\|$. Note that $\boldsymbol{\eta}$ and $\boldsymbol{\xi}$ are independent and uniformly distributed on the unit sphere in $\mathcal{R}^{N-1}$. It follows that $\widehat{\rho}_{jl} = \boldsymbol{\eta}'\boldsymbol{\xi} \sim \mathbf{e}_1'\boldsymbol{\xi} = \xi_1$. Let $\mathbf{v} = (V_1, \cdots, V_{N-1})'$. Since $V_1^2 \sim \chi_1^2$ and $V_2^2 + \cdots + V_{N-1}^2 \sim \chi_{N-2}^2$ are independent,

$$\xi_1^2 = \frac{V_1^2}{V_1^2 + V_2^2 + \cdots + V_{N-1}^2} \sim \text{Beta}(1/2, N/2-1).$$

Since

$$\xi_1^2 \sim s^{-1/2}(1-s)^{N/2-2}/B(1/2, N/2-1), \; 0 < s < 1,$$

the p.d.f. of $\widehat{\rho}_{jl} \sim \xi_1$ follows immediately, proving property 2. Finally, property 3 follows from

$$T_{jl} \sim \xi_1\sqrt{\frac{N-2}{1-\xi_1^2}} = V_1\sqrt{\frac{N-2}{V_2^2 + \cdots + V_{N-1}^2}}. \qquad \blacksquare$$

For an alternate proof of this result, as well as for a proof of the sampling distribution of $\widehat{\rho}_{jl}$ when $\rho_{jl} \neq 0$, see Rao (1973). Here, we present two alternate forms for the non-null distribution of $\widehat{\rho}_{jl}$. Starting from the joint distribution of S_{jj}, S_{ll} and $\widehat{\rho}_{jl}$ (see Exercise 6.8), it can be shown that this has the form

$$f(r) = \frac{2^{N-3}(1-\rho_{jl}^2)^{(N-1)/2}}{\pi\Gamma(N-2)}(1-r^2)^{(N-4)/2}\sum_{t=0}^{\infty}\frac{(2r\rho_{jl})^t}{t!}[\Gamma(\frac{N+t-1}{2})]^2. \qquad (6.2.4)$$

Another form of the distribution of $\widehat{\rho}_{jl}$ is

$$\begin{aligned} f(r) = & \frac{(1-\rho_{jl}^2)^{(N-1)/2}}{\pi}(N-2)(1-r^2)^{(N-4)/2} \\ & \times \int_0^1 \frac{u^{N-2}}{(1-u^2)^{1/2}}\frac{1}{(1-\rho_{jl}ru)^{N-1}}du, \end{aligned} \qquad (6.2.5)$$

which is obtained by verifying that

$$\begin{aligned} & \int_0^1 u^{N-2}(1-u^2)^{-1/2}(1-ru\rho_{jl})^{-(N-1)}du \\ & = \frac{2^{N-3}}{\Gamma(N-1)}\sum_{t=0}^{\infty}\frac{(2r\rho_{jl})^t}{t!}[\Gamma(\frac{N+t-1}{2})]^2. \end{aligned}$$

It would be natural to test H_0: $\rho_{jl} = 0$ versus H_1: $\rho_{jl} \neq 0$ using property 2 in Result 6.2.1. Unfortunately, the critical values under this null distribution of $\widehat{\rho}_{jl}$ have not been widely tabulated. The null distribution of the transformed variable in property 3 is generally used instead. The two-sided test procedure rejects H_0 at level of significance α if $|(N-2)^{1/2}\widehat{\rho}_{jl}/\sqrt{1-\widehat{\rho}_{jl}}| \geq t_{N-2,\alpha/2}$. In practice, Fisher's z-transformation of $\widehat{\rho}_{jl}$ leads to simpler inference, as shown in the next result. For a proof of this result, see Kendall and Stuart (1958), section 16.33.

Result 6.2.2. Let

$$\begin{aligned} Z_{jl} &= \frac{1}{2}\ln\left(\frac{1+\widehat{\rho}_{jl}}{1-\widehat{\rho}_{jl}}\right) = \tanh^{-1}(\widehat{\rho}_{jl}), \\ \delta_{jl} &= \frac{1}{2}\ln\left(\frac{1+\rho_{jl}}{1-\rho_{jl}}\right) = \tanh^{-1}(\rho_{jl}), \quad \text{and} \\ \upsilon^2 &= \frac{1}{N-3}; \end{aligned} \qquad (6.2.6)$$

then, Z_{jl} has an approximate $N(\delta_{jl}, \upsilon^2)$ distribution (Fisher, 1921).

We now consider estimation of the partial correlations (see Definition 5.2.7) based on a random sample $\mathbf{x}_1, \cdots, \mathbf{x}_N \sim N_k(\boldsymbol{\mu}, \boldsymbol{\Sigma})$. Supose we partition the random vector $\mathbf{x}_i = (X_{i,1}, \cdots, X_{i,k})'$ as $\mathbf{x}_i = (\mathbf{y}_i', \mathbf{z}_i')'$, where $\mathbf{y}_i' = (X_{i,1}, \cdots, X_{i,q})$ is a q-dimensional vector, and $\mathbf{z}_i' = (X_{i,q+1}, \cdots, X_{i,k})$ is a $(k-q)$-dimensional vector. We partition $\boldsymbol{\mu}$, $\boldsymbol{\Sigma}$, $\overline{\mathbf{x}}_N$, and $\mathbf{S}_N$ conformably (as in Result 5.2.11). In particular, suppose

$$\mathbf{S}_N = \begin{pmatrix} \mathbf{S}_{11} & \mathbf{S}_{12} \\ \mathbf{S}_{21} & \mathbf{S}_{22} \end{pmatrix},$$

where $\mathbf{S}_{11}$ is a $q \times q$ matrix. If $N-1 \geq k-q$, then by (6.1.2), $\mathbf{S}_{22}$ is nonsingular with probability one. Define

$$\mathbf{S}_{11.2} = \mathbf{S}_{11} - \mathbf{S}_{12}\mathbf{S}_{22}^{-1}\mathbf{S}_{21}.$$

Result 6.2.3. Let $N > k - q$. Then, $\mathbf{S}_{11.2}$ has a $W_q(\boldsymbol{\Sigma}_{11.2}, N-1-k+q)$ distribution. For more results on partitioned Wishart random matrices, see Mardia et al. (1979), section 3.4.3.

Proof. Let $\mathbf{Y} = (\mathbf{y}_1, \cdots, \mathbf{y}_N)'$, $\mathbf{Z} = (\mathbf{z}_1, \cdots, \mathbf{z}_N)'$, and $\mathbf{A} = \mathbf{I}_N - \mathbf{J}_N/N$. Then $\mathbf{S}_{11} = \mathbf{Y}'\mathbf{A}\mathbf{Y}$, $\mathbf{S}_{22} = \mathbf{Z}'\mathbf{A}\mathbf{Z}$, and $\mathbf{S}_{12} = \mathbf{S}_{21}' = \mathbf{Y}'\mathbf{A}\mathbf{Z}$. We show the following stronger result. Given any value of $\mathbf{Z}$ such that $\mathbf{S}_{22}$ is nonsingular, the conditional distribution of $\mathbf{S}_{11.2}$ is $W_k(\boldsymbol{\Sigma}_{11.2}, N-1-k+q)$.

Without loss of generality, assume $\boldsymbol{\mu} = \mathbf{0}$. By Result 5.2.12, $\mathbf{y}_i = \mathbf{B}\mathbf{z}_i + \mathbf{u}_i$, where $\mathbf{B} = \boldsymbol{\Sigma}_{12}\boldsymbol{\Sigma}_{22}^{-1}$ and $\mathbf{u}_i \sim N(\mathbf{0}, \boldsymbol{\Sigma}_{11.2})$ is independent of $\mathbf{z}_i$. Let $\mathbf{U} = (\mathbf{u}_1, \cdots, \mathbf{u}_N)'$, so that $\mathbf{Y} = \mathbf{Z}\mathbf{B}' + \mathbf{U}$. Then,

$$\begin{aligned}
\mathbf{S}_{11.2} &= \mathbf{Y}'\mathbf{A}\mathbf{Y} - \mathbf{Y}'\mathbf{A}\mathbf{Z}(\mathbf{Z}'\mathbf{A}\mathbf{Z})^{-1}(\mathbf{Z}'\mathbf{A}\mathbf{Y}) \\
&= (\mathbf{A}\mathbf{Y})'[\mathbf{I}_N - \mathbf{A}\mathbf{Z}(\mathbf{Z}'\mathbf{A}\mathbf{Z})^{-1}(\mathbf{A}\mathbf{Z})'](\mathbf{A}\mathbf{Y}) \\
&= (\mathbf{A}\mathbf{Z}\mathbf{B}' + \mathbf{A}\mathbf{U})'[\mathbf{I}_N - \mathbf{P}_{\mathcal{C}(\mathbf{AZ})}](\mathbf{A}\mathbf{Z}\mathbf{B}' + \mathbf{A}\mathbf{U}).
\end{aligned}$$

Since $(\mathbf{I}_N - \mathbf{P}_{\mathcal{C}(AZ)})\mathbf{A}\mathbf{Z}\mathbf{B}' = \mathbf{O}$ and, as can be easily checked, $\mathbf{A}\mathbf{P}_{\mathcal{C}(\mathbf{AZ})} = \mathbf{P}_{\mathcal{C}(\mathbf{AZ})}\mathbf{A} = \mathbf{P}_{\mathcal{C}(\mathbf{AZ})}$, we have $\mathbf{S}_{11.2} = \mathbf{U}'[\mathbf{A} - \mathbf{P}_{\mathcal{C}(\mathbf{AZ})}]\mathbf{U}$. Now $\mathbf{A} - \mathbf{P}_{\mathcal{C}(\mathbf{AZ})}$ is symmetric and idempotent, and its rank is $\text{tr}(\mathbf{A}) - \text{tr}(\mathbf{P}_{\mathcal{C}(\mathbf{AZ})}) = N-1-\text{r}(\mathbf{AZ}) = N-1-\text{r}((\mathbf{AZ})'\mathbf{AZ}) = N-1-\text{r}(\mathbf{S}_{22})$. Given any $\mathbf{Z}$ with $\mathbf{S}_{22}$ being nonsingular, the rank is $N-1-(k-q) = N-1-k+q$ and $\mathbf{u}_i$ are i.i.d. $\sim N(\mathbf{0}, \boldsymbol{\Sigma}_{11.2})$. The proof follows from Result 6.1.1. ■

Result 6.2.4. Estimation of partial correlation coefficients.

1. The maximum likelihood estimator of $\rho_{jl|(q+1,\cdots,k)}$ is

$$\widehat{\rho}_{jl|(q+1,\cdots,k)} = \frac{[\mathbf{S}_{11.2}]_{j,l}}{[\mathbf{S}_{11.2}]_{j,j}^{1/2}[\mathbf{S}_{11.2}]_{l,l}^{1/2}}, \tag{6.2.7}$$

where $[\mathbf{S}_{11.2}]_{m,n}$ denotes the (m,n)th element of the $q \times q$ matrix $\mathbf{S}_{11.2}$.

2. If $\rho_{jl|(q+1,\cdots,k)} = 0$, then

$$\frac{[N-(k-q)-2]^{1/2}\widehat{\rho}_{jl|(q+1,\cdots,k)}}{\{1-\widehat{\rho}^2_{jl|(q+1,\cdots,k)}\}^{1/2}} \sim t_{N-(k-q)-2}. \tag{6.2.8}$$

Proof. The form of the estimator follows from the invariance property of the MLE. When $\rho_{jl|(q+1,\cdots,k)} = 0$, the p.d.f. of $\widehat{\rho}_{jl|(q+1,\cdots,k)}$ has the form in (6.2.2), replacing the empirical and theoretical simple correlations by corresponding partial correlations, and replacing N by $N-(k-q)$. ■

We next discuss estimation of the multiple correlation coefficient based on N samples from a $(k+1)$-variate normal distribution. Using notation similar to Definition 5.2.8, let

$$\overline{\mathbf{x}}_N = \begin{pmatrix} \overline{X}_0 \\ \overline{\mathbf{x}}^{(1)} \end{pmatrix} \text{ and } \mathbf{S} = \begin{pmatrix} S_{00} & \mathbf{s}_{01} \\ \mathbf{s}_{10} & \mathbf{S}^{(1)} \end{pmatrix} \tag{6.2.9}$$

denote the sample mean and sample covariance matrix of the random sample $(X_{i,0}, X_{i,1}, \cdots, X_{i,k})$, $i = 1, \cdots, N$.

Result 6.2.5. Estimation of the multiple correlation coefficient.

1. The maximum likelihood estimator of $\rho_{0(1,\cdots,k)}$ is

$$\widehat{\rho}_{0(1,\cdots,k)} = \frac{\{\mathbf{s}_{01}[\mathbf{S}^{(1)}]^{-1}\mathbf{s}_{10}\}^{1/2}}{S_{00}^{1/2}}. \tag{6.2.10}$$

2. When $\rho_{0(1,\cdots,k)} = 0$,

$$\frac{(N-k-1)\widehat{\rho}^2_{0(1,\cdots,k)}}{k[1-\widehat{\rho}^2_{0(1,\cdots,k)}]} \sim F_{k,N-k-1}. \tag{6.2.11}$$

Proof. The proof of property 1 follows directly from the form of the multiple correlation coefficient and invariance of the MLE. Let $S_{00.1} = S_{00} - \mathbf{s}_{01}[\mathbf{S}^{(1)}]^{-1}\mathbf{s}_{10}$. Then, $1-\widehat{\rho}^2_{0(1,\cdots,k)} = S_{00.1}/S_{00}$, so that

$$\widehat{\rho}^2_{0(1,\cdots,k)}/(1-\widehat{\rho}^2_{0(1,\cdots,k)}) = \{\mathbf{s}_{01}[\mathbf{S}^{(1)}]^{-1}\mathbf{s}_{10}\}/S_{00.1}.$$

When $\rho_{0(1,\cdots,k)} = 0$, $\sigma_{01} = \mathbf{0}$ (see Definition 5.2.8), so that $S_{00.1} \sim W_1(\Sigma_{00}, N-k-1)$, $\mathbf{s}_{01}[\mathbf{S}^{(1)}]^{-1}\mathbf{s}_{10} \sim W_1(\Sigma_{00}, k)$, and they are independently distributed. The proof of property 2 follows from the definition of the F-distribution (see (B.8)). ■

When $\rho_{0(1,\cdots,k)} = 0$, the statistic $B = \widehat{\rho}^2_{0(1,\cdots,k)}/[1-\widehat{\rho}^2_{0(1,\cdots,k)}]$ follows a Beta$(k-1, N-k)$ distribution, as we ask the reader to show in Exercise 6.12. The non-null distribution of B is a noncentral beta (see Result 5.3.7).

6.3 Assessing the normality assumption

In practice, we must often assess whether a random sample does indeed come from a normal distribution. We discuss a few procedures that enable us to detect situations where the observed data depart, to a small or large extent, from the assumption of a normal parent population. There exist several procedures in the literature for assessing the hypothesis of univariate normality, such as (a) skewness and kurtosis tests, (b) omnibus tests such as Shapiro and Wilk's test, (c) likelihood ratio tests with specific nonnormal alternatives, (d) goodness of fit tests such as the χ^2-test and the Kolmogorov–Smirnov test, and (e) graphical methods such as normal probability plots. Assume that we have a random sample $X_1, \cdots, X_N$ from some continuous distribution with c.d.f. $F(.)$. We denote the empirical c.d.f. by

$$F_N(x) = \frac{1}{N}(\text{Number of observations} \leq x).$$

The Kolmogorov–Smirnoff test is useful for testing H_0: $F(x) = F_0(x)$ for all x versus H_1: $F(x) \neq F_0(x)$ for some x, where F_0 corresponds to a $N(\mu, \sigma^2)$ distribution. The Kolmogorov–Smirnoff test statistic is

$$D_N = \sup_x |F_N(x) - F_0(x)|,$$

which will be large if the data are not consistent with H_0. The p-values can be computed based on the asymptotic null distribution of D_N given by

$$\lim_{N\to\infty} \mathrm{P}\{\sqrt{N}D_N \le z\} = 1 - 2\sum_{k=1}^{\infty}(-1)^{k-1}\exp(-2k^2z^2) := Q(z)$$

for every $z > 0$. The function $Q(z)$ is the c.d.f. of a continuous distribution called the *Kolmogorov distribution.* In general, the parameters μ and σ^2 are unknown, and may be replaced by their sample counterparts.

Another test for normality is due to Shapiro and Wilk (1965) and is based on a comparison of ordered sample values with their expected locations under the null hypothesis of normality. Let $Z_{(1)} \le \cdots \le Z_{(N)}$ be an ordered sample from a standard normal distribution, and let $m_i = \mathrm{E}(Z_{(i)})$, $i = 1, \cdots, N$ be the normal scores. Under the hypothesis of normality,

$$\mathrm{E}(X_{(i)}) = \mu + \sigma\,\mathrm{E}(Z_{(i)}) = \mu + \sigma m_i,$$

i.e., we expect that the ordered observations $X_{(i)}$'s are linearly related to the m_i's. Shapiro and Wilk (1965) proposed the statistic

$$SW = \frac{(\mathbf{m}'\boldsymbol{\Omega}^{-1}\mathbf{u})^2}{(\mathbf{m}'\boldsymbol{\Omega}^{-2}\mathbf{m})\sum_{i=1}^{N}(X_i - \overline{X})^2},$$

where $\mathbf{m} = (m_1, \cdots, m_N)'$, $\boldsymbol{\Omega} = \{\omega_{ij}\}$ with $\omega_{ij} = \mathrm{Cov}(Z_{(i)}, Z_{(j)})$, and $\mathbf{u} = (X_{(1)}, \cdots, X_{(N)})'$. The statistic SW may be interpreted as the ratio of an estimate of variability based on ordered statistics and normality to the usual residual mean square. The authors provided extensive tables of percentage points of SW.

The well-known *normal probability plot,* which is a plot of the empirical quantiles versus the theoretical quantiles from a standard normal distribution, is available with most software packages. When the points in the normal probability plot lie very nearly along a straight line, the normality assumption is reasonable. The pattern of possible deviations of the scatter points from a straight line will indicate the nature of departure from normality, such as skewness, kurtosis, outliers or multi-modality.

Next, we discuss some approaches for assessing multivariate normality, including tests based on multivariate measures of skewness and kurtosis, as well as graphical procedures (Gnanadesikan, 1997). In Chapter 5, we have seen that (a) contours of the multivariate normal distribution are ellipsoids, and (b) linear combinations of normal random variables also have a normal distribution. As a first step in assessing multivariate normality, we may check to see whether the univariate marginal empirical distributions are approximately univariate normal using any of the approaches described earlier.

To assess bivariate normality, scatterplots of pairs of variables, X_j and X_l are useful. By Result 5.2.5, the vector $\mathbf{x}_2 = (X_j, X_l)$ has a bivariate normal distribution, i.e., $N_2(\boldsymbol{\mu}_2, \boldsymbol{\Sigma}_2)$, say. In Example 5.2.6, we showed that contours of constant density would be ellipses defined by $(\mathbf{x}_2 - \boldsymbol{\mu}_2)'\boldsymbol{\Sigma}_2^{-1}(\mathbf{x}_2 - \boldsymbol{\mu}_2) = \chi^2_{2,\alpha}$, where $\chi^2_{2,\alpha}$ is the upper (100α)th percentile from a chi-square distribution with $k = 2$ degrees of freedom. If the data follows a multivariate normal distribution, the scatterplot of $X_{i,j}$ versus $X_{i,l}$, $i = 1, \cdots, N$ should exhibit a pattern that is (nearly) elliptical and roughly $100\alpha\%$ of the points should lie inside the estimated ellipse $(\mathbf{x}_2 - \overline{\mathbf{x}}_2)'\mathbf{S}_2^{-1}(\mathbf{x}_2 - \overline{\mathbf{x}}_2) \le \chi^2_{2,\alpha}$, where $\overline{\mathbf{x}}_2$ is the mean of the bivariate sample, and $\mathbf{S}_2$ is the sample covariance matrix. This comparison of the empirical proportion of points within the estimated ellipse to the theoretical probability gives a rough assessment of bivariate normality. Alternately, for any $k \ge 2$, one can compute the squared generalized distances

$$d_j^2 = (N-1)(\mathbf{x}_j - \overline{\mathbf{x}}_N)'\mathbf{S}_N^{-1}(\mathbf{x}_j - \overline{\mathbf{x}}_N), \quad j = 1, \cdots, N.$$

Provided the data follows a k-variate normal distribution, and both $N \geq 30$ and $N-k \geq 30$, $\overline{\mathbf{x}}_N$ converges in probability to $\boldsymbol{\mu}$, while $\mathbf{S}_N$ converges in probability to $\boldsymbol{\Sigma}$; then, d_j^2 behaves like a chi-square random variable. The following plot, called the *chi-square plot*, or the *gamma plot* enables us to assess multivariate normality. First, order the squared generalized distances, $d_{(1)}^2 \leq \cdots \leq d_{(N)}^2$. Plot the pairs of points $(d_{(j)}^2, \chi^2_{k,\, j^*})$, where $j^* = \frac{1}{N}(j - 1/2)$, and $\chi^2_{k,\, j^*}$ denotes the upper $(100j^*)$th percentile of the chi-square distribution with k degrees of freedom. If the points lie approximately on a straight line, normality may be assumed, but not otherwise. Once again, the pattern of points may suggest the nature of departure from normality (Andrews et al., 1971; Gnanadesikan, 1997). In Figure 6.4.1, we show a gamma plot to assess multivariate normality for data simulated from a 4-variate normal distribution.

Mardia (1970) proposed a test for normality based on multivariate skewness and kurtosis, arguing that these quantities provide direct measures of departure from normality. Let $\mathbf{x}$ follow a k-variate distribution with mean $\boldsymbol{\mu}$ and covariance $\boldsymbol{\Sigma}$, and suppose $\boldsymbol{\Sigma}^{-1} = \{\sigma^{rr^*}\}$.

Definition 6.3.1. A multivariate measure of skewness which is invariant under nonsingular transformations (such as $\mathbf{x} = \mathbf{A}\mathbf{y} + \mathbf{b}$) is defined by

$$\beta_{1,k} = \sum_{r,s,t} \sum_{r^*s^*t^*} \sigma^{rr^*} \sigma^{ss^*} \sigma^{tt^*} \mu_{111}^{(rst)} \mu_{111}^{(r^*s^*t^*)},$$

where $\mu_{111}^{(rst)} = \mathrm{E}\{(X_r - \mu_r)(X_s - \mu_s)(X_t - \mu_t)\}$. For any symmetric distribution about $\boldsymbol{\mu}$, $\beta_{1,k} = 0$.

When $k = 1$, $\beta_{1,1} = \beta_1$, the usual univariate measure of skewness. When $k = 2$, with (r, s)th central moment $\mu_{r,s}$, $\mathrm{Var}(X_1) = 1$, $\mathrm{Var}(X_2) = 1$, and $\mathrm{Corr}(X_1, X_2) = \rho = 0$, the bivariate skewness measure is

$$\beta_{1,2} = \mu_{3,0}^2 + \mu_{0,3}^2 + 3\mu_{1,2}^2 + 3\mu_{2,1}^2,$$

which is identically zero if and only if μ_{30}, $\mu_{0,3}$, $\mu_{1,2}$, and $\mu_{2,1}$ all vanish.

Definition 6.3.2. A measure of multivariate kurtosis invariant under nonsingular transformations is

$$\beta_{2,k} = \mathrm{E}\{(\mathbf{x} - \boldsymbol{\mu})'\boldsymbol{\Sigma}^{-1}(\mathbf{x} - \boldsymbol{\mu})\}^2.$$

When $\boldsymbol{\mu} = \mathbf{0}$ and $\boldsymbol{\Sigma} = \mathbf{I}_k$, the measure of kurtosis is invariant under orthogonal transformations and is $\beta_{2,k} = \mathrm{E}\{(\mathbf{x}'\mathbf{x})^2\}$.

Let $\mathbf{x}_1, \cdots, \mathbf{x}_N$ denote a random sample from a k-dimensional population with mean $\boldsymbol{\mu}$ and covariance matrix $\boldsymbol{\Sigma}$. Let $\overline{\mathbf{x}}_N$ and $\mathbf{S}_N = \{S_{ij}\}$ respectively denote the sample mean and sample covariance matrix, and let $\{S^{ij}\}$ denote the inverse of $\mathbf{S}_N$. A sample measure of skewness corresponding to $\beta_{1,k}$ is

$$b_{1,k} = (N-1)^3 \sum_{r,s,t} \sum_{r^*s^*t^*} S^{rr^*} S^{ss^*} S^{tt^*} M_{111}^{(rst)} M_{111}^{(r^*s^*t^*)}, \tag{6.3.1}$$

where $M_{111}^{(rst)} = \sum_{i=1}^N (X_{i,r} - \overline{X}_r)(X_{i,s} - \overline{X}_s)(X_{i,t} - \overline{X}_t)/N$. An alternate expression for $b_{1,k}$ is (Mardia, 1975)

$$b_{1,k} = \frac{1}{N^2} \sum_{i=1}^N \sum_{j=1}^N r_{ij}^3, \tag{6.3.2}$$

where we define

$$r_{ij} = (N-1)(\mathbf{x}_i - \overline{\mathbf{x}}_N)'\mathbf{S}_N^{-1}(\mathbf{x}_j - \overline{\mathbf{x}}_N).$$

The sample skewness $b_{1,k}$ is also invariant under nonsingular transformations of the data. A sample measure of kurtosis is

$$b_{2,k} = \frac{N-1}{N}\sum_{i=1}^{N}\{(\mathbf{x}_i - \overline{\mathbf{x}}_N)'\mathbf{S}_N^{-1}(\mathbf{x}_i - \overline{\mathbf{x}}_N)\}^2, \tag{6.3.3}$$

which is also invariant under nonsingular transformations such as $\mathbf{x} = \mathbf{A}\mathbf{y} + \mathbf{b}$. For large samples, we can test for k-variate normality based on these measures as shown in the following result. For a proof of the result, the reader is referred to Mardia (1970).

Result 6.3.1. The following tests for multivariate normality hold in large samples:

1. The statistic $A = \frac{1}{6}Nb_{1,k}$ has a χ^2 distribution with $k(k+1)(k+2)/6$ d.f. under the null hypothesis H_0: $\beta_{1,k} = 0$. We reject the null hypothesis of normality for large values of A. We can also use the approximation that for $k > 7$, $\sqrt{2A} \sim N(\frac{1}{3}[k(k+1)(k+2)-3], 1)$.

2. Under the null hypothesis H_0: $\beta_{2,k} = k(k+2)$, the statistic

$$B = \sqrt{N}(b_{2,k} - \beta_{2,k})/[8k(k+2)]^{1/2}$$

has a standard normal distribution.

Geometrically, the *Mahalanobis distance* between $\mathbf{x}_i$ and $\mathbf{x}_j$ is

$$d_{ij}^2 = (N-1)(\mathbf{x}_i - \mathbf{x}_j)'\mathbf{S}_N^{-1}(\mathbf{x}_i - \mathbf{x}_j).$$

Let r_{ii} be the square of the Mahalanobis distance between $\mathbf{x}_i$ and $\overline{\mathbf{x}}_N$. It may be verified that $r_{ij} = \frac{1}{2}(r_{ii} + r_{jj} - d_{ij}^2)$. The Mahalanobis angle between the vectors $\mathbf{x}_i - \overline{\mathbf{x}}_N$ and $\mathbf{x}_j - \overline{\mathbf{x}}_N$ is defined by

$$\cos\theta_{ij} = r_{ij}/[r_{ii}r_{jj}]^{1/2}.$$

We can express the multivariate skewness and kurtosis in terms of these quantities:

$$b_{1,k} = \frac{1}{N^2}\sum_{i=1}^{N}\sum_{j=1}^{N}\{r_{ii}r_{jj}\}^{3/2}\{\cos\theta_{ij}\}^3,$$

$$b_{2,k} = \frac{1}{N}\sum_{i=1}^{N} r_{ii}^2.$$

If the sample points $\mathbf{x}_i$, $i = 1, \cdots, N$ are uniformly distributed on a k-dimensional hypersphere or ellipsoid, then $b_{1,k} \simeq 0$. If, however, the data departs from spherical symmetry, then $b_{1,k}$ will be large; this might occur for instance when there is an abnormal clustering of points. In such cases, the value of $b_{2,k}$ will also be abnormally large. Figures 6.3.1 (a) and (b) represent the symmetric and abnormal clustering cases, respectively.

6.4 R Code

```
## Def. 6.1.1. Wishart Distributions
```

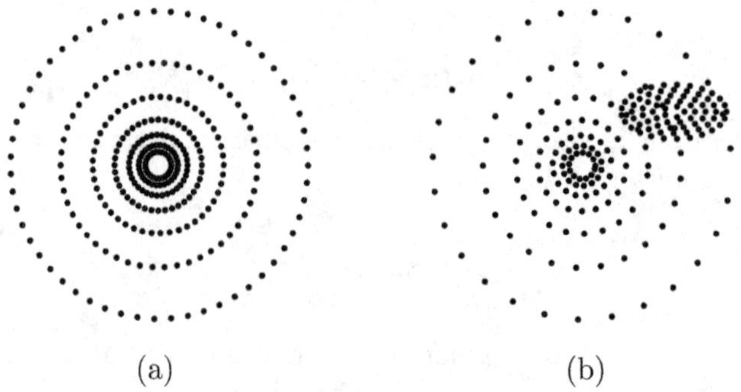

FIGURE 6.3.1. Graphical representation of skewness.

```
library(matrixsampling)
n <- 200
df <- 7
sig <- matrix(c(4,-1, 0,-1, 4, 2, 0, 2, 9), nrow = 3)

# Central Wishart distribution
sample <- rwishart(n = n, nu = df, Sigma = sig)
# Noncentral Wishart distribution
ncp <- diag(rep(1, 3))
sample <- rwishart(
  n = n,
  nu = df,
  Sigma = sig,
  Theta = ncp
)
# Inverse noncentral Wishart distribution
sig <- matrix(c(4, -1, 0, -1, 4, 2, 0, 2, 9), nrow = 3)
sample <- rinvwishart(n = n, nu = df, Omega = sig)
```

```
## Section 6.2. Estimate correlations
X <- iris
```

Note that X_5 is a factor variable showing the species; we therefore exclude X_5 from the computations.

```
# corr(X1,X3)
cor(X[, 1], X[, 3])
[1] 0.8717538

cor.test(X[, 1], X[, 3])

Pearson's product-moment correlation
data:  X[, 1] and X[, 3]
t = 21.646, df = 148, p-value < 2.2e-16
alternative hypothesis: true correlation is not equal to 0
95 percent confidence interval:
 0.8270363 0.9055080
sample estimates:
      cor
```

```
0.8717538

# correlation matrix
cor(X[,-5])

Sepal.Length Sepal.Width Petal.Length Petal.Width
Sepal.Length    1.0000000  -0.1175698    0.8717538   0.8179411
Sepal.Width    -0.1175698   1.0000000   -0.4284401  -0.3661259
Petal.Length    0.8717538  -0.4284401    1.0000000   0.9628654
Petal.Width     0.8179411  -0.3661259    0.9628654   1.0000000
```

We look at the partial correlations defined in Result 6.2.4. Note that the matrix of partial correlations shown below is quite different from the matrix of simple correlations shown above.

```
# Partial correlations
library(ppcor)
# Partial correlation of X1 and X2 given X3,X4
pcor.test(X[, 1], X[, 2], X[, c(3, 4)])

   estimate      p.value statistic   n gp  Method
1 0.6285707 1.199846e-17   9.76538 150  2 pearson

# all partial correlations
pcor(X[-5])$estimate

 Sepal.Length Sepal.Width Petal.Length Petal.Width
Sepal.Length    1.0000000   0.6285707    0.7190656  -0.3396174
Sepal.Width     0.6285707   1.0000000   -0.6152919   0.3526260
Petal.Length    0.7190656  -0.6152919    1.0000000   0.8707698
Petal.Width    -0.3396174   0.3526260    0.8707698   1.0000000
```

For illustration, we show the multiple correlation of X_1 with X_2, X_3, X_4 defined in Result 6.2.5. This correlation is useful when one of the variables is a response, and we seek to see how it is correlated with a set of predictors.

```
# Multiple correlation
library(mro)
mcr.test(X[,-5], 1, c(2, 3, 4))

Testing Multiple Correlation Coefficient is Zero
data:  X[, -5]
Statistic = 295.54, df1 = 4.00000, df2 = 146.00000, sample.MC
= 0.92661, p-value < 2.2e-16
alternative hypothesis: true population multiple correlation
greater than 0

## Section 6.2. Assess MVN
library(mvtnorm)
k <- 4
n <- 200
mu <- c(1, 2, 3, 4)
V  <- matrix(
  c(1, 0.2, 0.3, 0.1, 0.2, 4, 0.6,
```

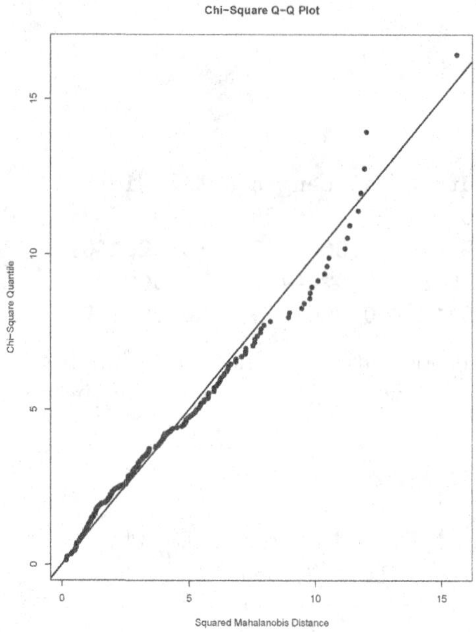

FIGURE 6.4.1. Gamma plot to assess multivariate normality. Based on $n = 200$ samples from a 4-variate normal distribution.

```
    0.2, 0.3, 0.6, 9, 0.3, 0.1, 0.2, 0.3, 1),
  nrow = 4,
  byrow = T
)
x <- rmvnorm(n, mu, V)  # samples
library(MVN)
mvn(x, multivariatePlot = "qq")$multivariateNormality

             Test        Statistic               p value Result
1 Mardia Skewness  39.06431274613 0.00654567925423297     NO
2 Mardia Kurtosis 1.0381452404758   0.299202441963671    YES
3             MVN            <NA>                <NA>     NO

$univariateNormality
          Test  Variable Statistic   p value Normality
1 Shapiro-Wilk  Column1     0.9866    0.0565    YES
2 Shapiro-Wilk  Column2     0.9879    0.0880    YES
3 Shapiro-Wilk  Column3     0.9935    0.5222    YES
4 Shapiro-Wilk  Column4     0.9902    0.1928    YES

$Descriptives
    n     Mean   Std.Dev   Median       Min       Max      25th
1 200 1.081625 1.0056085 1.106687 -2.030094  3.672031 0.5309078
2 200 1.836980 1.8492980 2.021865 -2.980825  6.449778 0.6897461
3 200 2.877046 2.9404112 3.269656 -4.363457 12.625767 0.8527974
4 200 3.960598 0.9594657 3.965766  1.237546  5.994668 3.2863629
```

```
        75th        Skew    Kurtosis
1 1.600990 -0.10689588  0.48675713
2 3.171851 -0.31238448 -0.21365832
3 4.763341  0.02168587 -0.05018561
4 4.632018 -0.24934379 -0.44506528

# Skewness and kurtosis measures based on MVN sample
library(psych)
M <- mardia(x, na.rm = TRUE, plot = TRUE)
M$b1p # sample skewness
[1] 1.154438
M$b2p # Sample kurtosis
[1] 24.76762

## Result 6.3.1.
# Property 1
M$skew  # test statistic
[1] 38.48127
M$p.skew # p-value
[1] 0.007729752

## Result 6.3.1.
# Property 2
M$kurtosis # test statistic
[1] 0.7834531
M$p.kurt   # p-value
[1] 0.4333611
```

Exercises

6.1. Show that the generalized variance $|\mathbf{S}_N|$ of a random sample $\mathbf{x}_1, \cdots, \mathbf{x}_N$ is zero if and only if the columns of the matrix of deviations $\mathbf{X} - \mathbf{1}_N\mathbf{x}'$ are linearly dependent, where $\mathbf{X} = (\mathbf{x}_1, \cdots, \mathbf{x}_N)'$.

6.2. Let $\mathbf{W} \sim W_k(\mathbf{\Sigma}, m)$. Show that for any $k \times q$ matrix $\mathbf{B}$, $\mathbf{B}'\mathbf{W}\mathbf{B} \sim W_q(\mathbf{B}'\mathbf{\Sigma}\mathbf{B}, m)$.

6.3. Let $\mathbf{x}_1, \cdots, \mathbf{x}_N$ be i.i.d. $\sim N(\boldsymbol{\mu}, \mathbf{\Sigma})$ and $\mathbf{W} = \sum_{i=1}^N \mathbf{x}_i\mathbf{x}_i'$, where $\mathbf{\Sigma}$ is p.d.

(a) Show that $\text{E}(\mathbf{W}) = N(\mathbf{\Sigma} + \boldsymbol{\mu}\boldsymbol{\mu}')$.

(b) Show that given $\mathbf{\Sigma}$, other than changing $\boldsymbol{\mu}$ to $-\boldsymbol{\mu}$, any change to $\boldsymbol{\mu}$ gives a different distribution of $\mathbf{W}$. When $\boldsymbol{\mu} \neq \mathbf{0}$, the distribution is known as a noncentral Wishart distribution.

6.4. Express the maximized likelihood function in (6.1.9) as a function of the generalized variance $|\mathbf{S}_N|$ of the random sample $\mathbf{x}_1, \cdots, \mathbf{x}_N$.

6.5. Let $\overline{\mathbf{x}}_m$ and $\mathbf{S}_m$ be the sample mean and covariance matrix of $\mathbf{x}_1, \cdots, \mathbf{x}_m$. Use the following identities (Ghosh, 1996) to give an alternate proof of Result 6.1.2:

$$\overline{\mathbf{x}}_{m+1} = \frac{1}{m+1}(m\overline{\mathbf{x}}_m + \mathbf{x}_{m+1}), \text{ and}$$

$$\mathbf{S}_{m+1} = \mathbf{S}_m + \frac{m}{m+1}(\mathbf{x}_{m+1} - \overline{\mathbf{x}}_m)(\mathbf{x}_{m+1} - \overline{\mathbf{x}}_m)'.$$

6.6. If $\mathbf{S}_N \sim W_k(\mathbf{\Sigma}, N-1)$, and $\mathbf{a} = (a_1, \cdots, a_k)'$ is an arbitrary vector, show that $\max_{\mathbf{a}}(\mathbf{a}'\mathbf{S}_N\mathbf{a}/\mathbf{a}'\mathbf{\Sigma}\mathbf{a})$ and $\min_{\mathbf{a}}(\mathbf{a}'\mathbf{S}_N\mathbf{a}/\mathbf{a}'\mathbf{\Sigma}\mathbf{a})$ are respectively the largest and smallest roots of the determinantal equation $|\mathbf{S}_N - c\mathbf{\Sigma}| = 0$.

6.7. Suppose $\mathbf{S}_{N_1} \sim W_k(\Sigma, N_1 - 1)$, which is independent of $\mathbf{S}_{N_2} \sim W_k(\Sigma, N_2 - 1)$. For any $\mathbf{a} = (a_1, \cdots, a_k)'$, find the distribution of $\mathbf{a}'\mathbf{S}_{N_1}\mathbf{a}/\mathbf{a}'\mathbf{S}_{N_2}\mathbf{a}$.

6.8. Derive the joint distribution of S_{jj}, S_{ll} and $\widehat{\rho}_{jl}$.

6.9. Let $\mathbf{S}_{ij}$, $i, j = 1, 2$, be as in Result 6.2.3. Show that $\mathbf{S}_{11.2}$ is independent of $\mathbf{S}_{12}$ and $\mathbf{S}_{22}$.

6.10. The following table gives the mean vector and the variance-covariance matrix of four variables X_0, X_1, X_2, and X_3:

	X_0	X_1	X_2	X_3	Means
X_0	60.516	0.998	3.511	21.122	18.3
X_1		15.129	23.860	1.793	14.9
X_2			54.756	3.633	30.5
X_3				18.225	7.8

(a) Compute the multiple correlation coefficient of X_0 on X_1, X_2, and X_3.

(b) Compute the partial correlation coefficient between X_0 and X_3, eliminating the effects of X_1 and X_2.

6.11. Suppose $\mathbf{S}$ is partitioned as in (6.2.9). Evaluate $\widehat{\rho}^2_{0(1,\cdots,k)}$ in terms of $|\mathbf{S}|$, $|\mathbf{S}^{(1)}|$ and S_{00}.

6.12. When $\rho_{0(1,\cdots,k)} = 0$, show that the statistic $\widehat{\rho}^2_{0(1,\cdots,k)}$ follows a beta distribution. What are the parameters of this distribution?

6.13. Find $\mathrm{E}(d_j^2)$, where d_j^2 are the squared generalized distances.

6.14. Denote by $Z_{(1)} \leq \cdots \leq Z_{(N)}$ the order statistics of an i.i.d. sample of size N from a $N(0,1)$ distribution. Let $\mathbf{m}$ and $\mathbf{\Omega}$ be the mean and covariance matrix of $\mathbf{z} = (Z_{(1)}, \cdots, Z_{(N)})'$.

(a) Show that

$$\textstyle\sum_{j=1}^N \mathrm{Cov}(Z_{(i)}, Z_{(j)}) = \mathrm{Cov}(Z_{(i)}, N\overline{Z}_N) = \mathrm{Cov}(\overline{Z}_N, N\overline{Z}_N) = 1,$$

$i = 1, \ldots, N$, so that $\mathbf{\Omega}\mathbf{1}_N = \mathbf{1}_N$.

(b) Show that if $X_{(1)} \leq \cdots \leq X_{(N)}$ is an ordered sample, then its Shapiro–Wilk test statistic SW is the square of the sample correlation between $\mathbf{\Omega}^{-1}\mathbf{m}$ and $(X_{(1)} - \overline{X}_N, \cdots, X_{(N)} - \overline{X}_N)'$.

6.15. Let $\mathbf{x} \sim N(\boldsymbol{\mu}, \mathbf{\Sigma})$. Show that $b_{1,k}$ and $b_{2,k}$ defined in (6.3.2) and (6.3.3) are invariant under a nonsingular linear transformation $\mathbf{y} = \mathbf{P}\mathbf{x} + \mathbf{q}$ (i.e., $\mathbf{P}$ is nonsingular). Discuss whether these coefficients depend on $\boldsymbol{\mu}$ and $\mathbf{\Sigma}$.

7

Inference for the General Linear Model-I

In this chapter, we describe classical inference for the general linear model (GLM) that we introduced in Chapter 4. The least squares approach did not require any distributional assumptions on the model errors. However, in many cases, one is interested in constructing interval estimates for the parameters of interest, as well as in testing hypotheses about these parameters or functions of these parameters. Parametric inference proceeds by assuming that the errors are generated by some probability model. The most widely used assumption is that of normality, and the multivariate normal distribution was defined in Chapter 5. In Chapter 6, we looked at methods for assessing multivariate normality and studied some transformations to normality. The properties of the multivariate normal distribution ensures that the least squares solutions of the general linear model parameters are also multivariate normal, as Result 7.1.1 shows. Using results from distributions of quadratic forms that were discussed in Section 5.4, we derive tests of hypotheses in Section 7.2, followed by a discussion of a nested sequence of hypotheses in Section 7.3. Section 7.4 describes construction of marginal confidence intervals and joint confidence regions for suitable functions of model parameters.

7.1 Properties of least squares solutions

In Chapter 4, we derived the least squares solution $\boldsymbol{\beta}^0$ for the general linear model (4.1.1),

$$\mathbf{y} = \mathbf{X}\boldsymbol{\beta} + \boldsymbol{\varepsilon},$$

where $\mathbf{X} \in \mathcal{R}^{N\times p}$, $\mathrm{E}(\boldsymbol{\varepsilon}) = \mathbf{0}$ and $\mathrm{Var}(\boldsymbol{\varepsilon}) = \sigma^2\mathbf{I}_N$, and defined the estimate of the error variance $\widehat{\sigma}^2$. We also defined the fitted vector $\widehat{\mathbf{y}} = \mathbf{X}\boldsymbol{\beta}^0$ and residual vector $\widehat{\boldsymbol{\varepsilon}} = \mathbf{y} - \widehat{\mathbf{y}}$ (Definitions 4.2.1 and 4.2.2) as

$$\widehat{\mathbf{y}} = \mathbf{P}\mathbf{y} = \mathbf{X}\boldsymbol{\beta} + \mathbf{P}\boldsymbol{\varepsilon} \quad \text{and} \quad \widehat{\boldsymbol{\varepsilon}} = \mathbf{y} - \widehat{\mathbf{y}} = (\mathbf{I}-\mathbf{P})\mathbf{y} = (\mathbf{I}-\mathbf{P})\boldsymbol{\varepsilon},$$

where $\mathbf{P} = \mathbf{X}\mathbf{G}\mathbf{X}'$ is the projection matrix onto $\mathcal{C}(\mathbf{X})$ (Result 4.2.2).

In this section, we assume that $\boldsymbol{\varepsilon}$ follows a multivariate normal distribution, $N(\mathbf{0}, \sigma^2\mathbf{I}_N)$, with p.d.f.

$$\begin{aligned} f(\boldsymbol{\varepsilon};\sigma^2\mathbf{I}_N) &= \frac{1}{(2\pi\sigma^2)^{N/2}}\exp\left\{-\frac{1}{2\sigma^2}\boldsymbol{\varepsilon}'\boldsymbol{\varepsilon}\right\} \\ &= \frac{1}{(2\pi\sigma^2)^{N/2}}\exp\left\{-\frac{1}{2\sigma^2}\sum_{i=1}^{N}\varepsilon_i^2\right\}, \quad \boldsymbol{\varepsilon} \in \mathcal{R}^N. \end{aligned}$$

We show that the least squares estimator of any estimable function of $\boldsymbol{\beta}$ has a multivariate normal distribution. We derive the mean and variance of this estimator, and discuss

DOI: 10.1201/9781315156651-7

the full-rank model as a special case. Based on the point estimates and precision estimates for estimable functions of $\boldsymbol{\beta}$, we construct a joint confidence region and marginal confidence intervals. We develop test procedures for various hypotheses of interest about estimable functions of $\boldsymbol{\beta}$. We also derive the distribution of the fitted vector and the residual vector. In (4.2.11), we introduced the ANOVA decomposition of the total sum of squares $SST = \mathbf{y}'\mathbf{y}$ into the model sum of squares $SSR = \widehat{\mathbf{y}}'\widehat{\mathbf{y}}$ and the error sum of squares $SSE = \widehat{\boldsymbol{\varepsilon}}'\widehat{\boldsymbol{\varepsilon}}$. Using results from Chapter 5, we now derive the distributions of these quadratic forms in normal random vectors.

Result 7.1.1. Let $\boldsymbol{\varepsilon} \sim N(\mathbf{0}, \sigma^2\mathbf{I}_N)$ and $\text{r}(\mathbf{X}) = r$. Then,

1. $\boldsymbol{\beta}^0$ has a singular normal distribution
$$\boldsymbol{\beta}^0 \sim N(\mathbf{H}\boldsymbol{\beta}, \sigma^2\mathbf{G}\mathbf{X}'\mathbf{X}\mathbf{G}'), \tag{7.1.1}$$
where $\mathbf{H} = \mathbf{G}\mathbf{X}'\mathbf{X}$.
2. $\widehat{\mathbf{y}} \sim N(\mathbf{X}\boldsymbol{\beta}, \sigma^2\mathbf{P})$ and $\widehat{\boldsymbol{\varepsilon}} \sim N(\mathbf{0}, \sigma^2(\mathbf{I} - \mathbf{P}))$.
3. SSE has a scaled chi-square distribution:
$$\frac{SSE}{\sigma^2} \sim \chi^2_{N-r}. \tag{7.1.2}$$
4. $\boldsymbol{\beta}^0$ and $\widehat{\boldsymbol{\varepsilon}}$ are distributed independently.
5. The model sum of squares has a scaled noncentral chi-square distribution:
$$\frac{SSR}{\sigma^2} \sim \chi^2(r, \lambda), \quad \text{where} \quad \lambda = \frac{1}{2\sigma^2}\boldsymbol{\beta}'\mathbf{X}'\mathbf{X}\boldsymbol{\beta}. \tag{7.1.3}$$
6. SSR and SSE are independently distributed.
7. Denote $SSM = N(\overline{Y})^2$ (see (4.2.14)). Then
$$\frac{SSM}{\sigma^2} \sim \chi^2(1, \lambda), \quad \text{where} \quad \lambda = \frac{(\mathbf{1}_N'\mathbf{X}\boldsymbol{\beta})^2}{2N\sigma^2}. \tag{7.1.4}$$
8. SSM is distributed independently of SSE if and only if $\mathbf{1}_N \in \mathcal{C}(\mathbf{X})$.

Proof. First, $\mathbf{y} \sim N(\mathbf{X}\boldsymbol{\beta}, \sigma^2\mathbf{I}_N)$. Since $\boldsymbol{\beta}^0 = \mathbf{G}\mathbf{X}'\mathbf{y}$, property 1 follows from Result 5.2.5. Property 2 also follows from Result 5.2.5. Since $SSE = \boldsymbol{\varepsilon}'(\mathbf{I} - \mathbf{P})\boldsymbol{\varepsilon}$ and $\mathbf{I} - \mathbf{P}$ is idempotent with rank $N - r$, property 3 follows from Result 5.4.4. Since $\mathbf{X}'\mathbf{y}$ and $\widehat{\boldsymbol{\varepsilon}}$ are jointly normally distributed and $\text{Cov}(\mathbf{X}'\mathbf{y}, \widehat{\boldsymbol{\varepsilon}}) = \sigma^2\mathbf{X}'(\mathbf{I} - \mathbf{P}) = \mathbf{O}$, they are independent, proving property 4. Property 5 follows from Result 5.4.4. Independence of $SSR = \boldsymbol{\beta}^{0\prime}\mathbf{X}'\mathbf{X}\boldsymbol{\beta}^0$ and $SSE = \widehat{\boldsymbol{\varepsilon}}'\widehat{\boldsymbol{\varepsilon}}$ follows from the independence of $\boldsymbol{\beta}^0$ and $\widehat{\boldsymbol{\varepsilon}}$ proved above. Recall that $\overline{\mathbf{J}}_N = \mathbf{1}_N\mathbf{1}_N'/N$ is symmetric and idempotent of rank 1. We can write $SSM/\sigma^2 = \mathbf{y}'\overline{\mathbf{J}}_N\mathbf{y}/\sigma^2$, so that property 7 is a direct consequence of Result 5.4.4. From Result 5.4.7, SSM and SSE are independent if and only if $\mathbf{1}_N\mathbf{1}_N'(\mathbf{I} - \mathbf{P}) = \mathbf{0}$. The latter holds if and only if $(\mathbf{I} - \mathbf{P})\mathbf{1}_N = \mathbf{1}_N - \mathbf{P}\mathbf{1}_N = \mathbf{0}$, which in turn is equivalent to $\mathbf{1}_N \in \mathcal{C}(\mathbf{X})$. Then property 8 follows. ■

Result 7.1.2. Let $\mathbf{c}_j'\boldsymbol{\beta}$, $j = 1, \cdots, s$, denote s estimable functions of $\boldsymbol{\beta}$ (see Definition 4.3.1). Let $\mathbf{C} = (\mathbf{c}_1, \cdots, \mathbf{c}_s)$.

1. $\mathbf{C}'\boldsymbol{\beta}^0$ is invariant to the choice of the least squares solution $\boldsymbol{\beta}^0$ and
$$\mathbf{C}'\boldsymbol{\beta}^0 \sim N(\mathbf{C}'\boldsymbol{\beta}, \sigma^2\mathbf{C}'\mathbf{G}\mathbf{C}).$$

2. If $\mathbf{c}_1, \cdots, \mathbf{c}_s$ are LIN, then $\mathbf{C}'\mathbf{G}\mathbf{C}$ is nonsingular and given a vector of constants $\mathbf{d} \in \mathcal{R}^s$, if

$$\begin{aligned} Q &= (\mathbf{C}'\boldsymbol{\beta}^0 - \mathbf{d})'[\text{Cov}(\mathbf{C}'\boldsymbol{\beta}^0)/\sigma^2]^{-1}(\mathbf{C}'\boldsymbol{\beta}^0 - \mathbf{d}) \\ &= \sigma^2(\mathbf{C}'\boldsymbol{\beta}^0 - \mathbf{d})'(\mathbf{C}'\mathbf{G}\mathbf{C})^{-1}(\mathbf{C}'\boldsymbol{\beta}^0 - \mathbf{d}), \end{aligned}$$

then $Q/\sigma^2 \sim \chi^2(s, \lambda)$ with $\lambda = (\mathbf{C}'\boldsymbol{\beta} - \mathbf{d})'(\mathbf{C}'\mathbf{G}\mathbf{C})^{-1}(\mathbf{C}'\boldsymbol{\beta} - \mathbf{d})/2\sigma^2$.

Proof. From property 5 of Result 4.3.1, $\mathbf{C}'\boldsymbol{\beta}^0$ is invariant to the choice of $\boldsymbol{\beta}^0$. From property 1 of Result 7.1.1 and Result 5.2.5, $\mathbf{C}'\boldsymbol{\beta}^0$ is normally distributed, so we only need to check its mean and variance. From Result 4.4.1, $\text{E}(\mathbf{C}'\boldsymbol{\beta}^0) = \mathbf{C}'\boldsymbol{\beta}$ and for any $\mathbf{a} \in \mathcal{R}^s$, $\mathbf{a}'\,\text{Var}(\mathbf{C}'\boldsymbol{\beta}^0)\mathbf{a} = \text{Var}((\mathbf{C}\mathbf{a})'\boldsymbol{\beta}^0) = \sigma^2(\mathbf{C}\mathbf{a})'(\mathbf{C}\mathbf{a}) = \sigma^2\mathbf{a}'\mathbf{C}'\mathbf{G}\mathbf{C}\mathbf{a}$. Since $\mathbf{a}$ is arbitrary, then $\text{Var}(\mathbf{C}'\boldsymbol{\beta}^0) = \sigma^2\mathbf{C}'\mathbf{G}\mathbf{C}$. This proves property 1.

Estimability of $\mathbf{C}'\boldsymbol{\beta}$ implies that $\mathbf{C}' = \mathbf{T}'\mathbf{X}$ for some matrix $\mathbf{T}$. Then $\mathbf{C}'\mathbf{G}\mathbf{C} = \mathbf{T}'\mathbf{X}\mathbf{G}\mathbf{X}'\mathbf{T} = \mathbf{T}'\mathbf{P}\mathbf{T} = (\mathbf{P}\mathbf{T})'(\mathbf{P}\mathbf{T})$. From property 6 of Result 1.3.11, $\mathbf{C}'\mathbf{G}\mathbf{C}$ is nonsingular if and only if $\mathbf{P}\mathbf{T}$ has full column rank. Suppose $\mathbf{P}\mathbf{T}\mathbf{v} = \mathbf{0}$. Then $\mathbf{T}\mathbf{v} \perp \mathcal{C}(\mathbf{X})$, so $\mathbf{X}'\mathbf{T}\mathbf{v} = \mathbf{0}$, giving $\mathbf{C}\mathbf{v} = \mathbf{0}$. Since $\mathbf{C}$ has full column rank, then $\mathbf{v} = \mathbf{0}$, showing $\mathbf{P}\mathbf{T}$ has full column rank. The distribution of Q/σ^2 directly follows from Result 5.4.4, proving property 2. ■

In the following corollary, we state some results for the full-rank case, i.e., when $\mathbf{X}$ has full column rank. In this case, $\boldsymbol{\beta}$ is estimable, as is any linear function of $\boldsymbol{\beta}$. Further, the least squares estimate of $\boldsymbol{\beta}$, in addition to being the b.l.u.e., is also the minimum variance unbiased estimate (MVUE).

Corollary 7.1.1. When $\text{r}(\mathbf{X}) = p$, i.e., for the full-rank model, the following properties hold for $\widehat{\boldsymbol{\beta}}$:

1. $\widehat{\boldsymbol{\beta}} \sim N_p(\boldsymbol{\beta}, \sigma^2(\mathbf{X}'\mathbf{X})^{-1})$.

2. For any given vector $\mathbf{a} = (a_1, \cdots, a_p)'$, $\mathbf{a}'\widehat{\boldsymbol{\beta}} \sim N(\mathbf{a}'\boldsymbol{\beta}, \sigma^2\mathbf{a}'(\mathbf{X}'\mathbf{X})^{-1}\mathbf{a})$.

3. $\text{Cov}(\widehat{\boldsymbol{\beta}}, \widehat{\boldsymbol{\varepsilon}}) = \mathbf{O}$, so that $\widehat{\boldsymbol{\beta}}$ is independent of SSE.

The proof is left as Exercise 7.1.

Example 7.1.1. We continue with Example 4.1.1. For the simple linear regression model, it may be shown that

$$\begin{aligned} &\widehat{\beta}_0 \sim N(\beta_0, \sigma^2 \textstyle\sum_{i=1}^N X_i^2/N\sum_{i=1}^N (X_i - \overline{X})^2), \\ &\widehat{\beta}_1 \sim N(\beta_1, \sigma^2/\textstyle\sum_{i=1}^N (X_i - \overline{X})^2). \end{aligned}$$

The distributions of $\widehat{\beta}_0$ and $\widehat{\beta}_1$ involve the unknown error variance σ^2. The estimate of σ^2 is the sample variance of the residuals (residual mean square) defined in (4.2.19) as

$$\widehat{\sigma}^2 = \sum_{i=1}^N \widehat{\varepsilon}_i^2/(N-2) = \sum_{i=1}^N (Y_i - \widehat{\beta}_0 - \widehat{\beta}_1 X_i)^2/(N-2),$$

where $\widehat{\varepsilon}_i = Y_i - \widehat{Y}_i$ is the ith residual. Provided the model assumptions are met, $\widehat{\sigma}^2$ is an unbiased and consistent estimator of σ^2. It may be shown that $(N-2)\widehat{\sigma}^2/\sigma^2$ has a chi-square distribution with $N-2$ d.f. Substituting the value of $\widehat{\sigma}^2$ for the unknown quantity σ^2, we obtain estimated variances of $\widehat{\beta}_0$ and $\widehat{\beta}_1$ as well as their estimated covariance. Let us denote the estimated standard errors of $\widehat{\beta}_0$ and $\widehat{\beta}_1$ respectively by $s_{\widehat{\beta}_0}$ and $s_{\widehat{\beta}_1}$. The estimates of β_0 and β_1 and their associated standard errors will be used in the derivation of confidence intervals and hypotheses tests. □

Example 7.1.2. Suppose $\mathbf{y} = \mathbf{X}\boldsymbol{\beta} + \boldsymbol{\varepsilon}$, $\mathbf{X}$ is an $N \times p$ matrix with $\mathrm{r}(\mathbf{X}) = p$, and $\boldsymbol{\varepsilon} \sim N(\mathbf{0}, \sigma^2\mathbf{V})$, where $\mathbf{V}$ is a known p.d. matrix. Let $\widehat{\boldsymbol{\beta}}_{GLS}$ denote the GLS estimator of $\boldsymbol{\beta}$ (see Section 4.5). By Result 5.2.5, $\widehat{\boldsymbol{\beta}}_{GLS} \sim N(\boldsymbol{\beta}, \sigma^2(\mathbf{X}'\mathbf{V}^{-1}\mathbf{X})^{-1})$. Defining

$$Q = (\mathbf{y} - \mathbf{X}\widehat{\boldsymbol{\beta}}_{GLS})'\mathbf{V}^{-1}(\mathbf{y} - \mathbf{X}\widehat{\boldsymbol{\beta}}_{GLS}),$$

we show below that Q/σ^2 has a χ^2_{N-p} distribution. Now, from Section 4.5, if $\mathbf{z} = \mathbf{V}^{-1/2}\mathbf{y}$, $\mathbf{B} = \mathbf{V}^{-1/2}\mathbf{X}$, and $\boldsymbol{\eta} = \mathbf{V}^{-1/2}\boldsymbol{\varepsilon}$, then $\boldsymbol{\eta} \sim N(\mathbf{0}, \sigma^2\mathbf{I})$ and $\mathbf{z} - \mathbf{B}\widehat{\boldsymbol{\beta}}_{GLS} = (\mathbf{I}-\mathbf{P})\mathbf{z} = (\mathbf{I}-\mathbf{P})\boldsymbol{\eta}$, where $\mathbf{P}$ is the orthogonal projection onto $\mathcal{C}(\mathbf{B})$. Then $Q = \boldsymbol{\eta}'(\mathbf{I} - \mathbf{P})\boldsymbol{\eta}$. Since $\mathbf{I} - \mathbf{P}$ is idempotent of rank $N - \mathrm{r}(\mathbf{B}) = N - p$, from Result 5.4.5, $Q/\sigma^2 \sim \chi^2_{N-p}$. □

We remark that $Q/(N-p)$ is optimal in the sense that it is the minimum norm quadratic unbiased estimator (MINQUE) of σ^2. For more details on MINQUE, see Chapter 11.

7.2 General linear hypotheses

In this section, we derive tests for hypotheses about certain parametric linear functions of $\boldsymbol{\beta}$. The hypothesis

$$H\colon \mathbf{C}'\boldsymbol{\beta} = \mathbf{d} \tag{7.2.1}$$

is called a general linear hypothesis, where $\mathbf{C}$ is a $p \times s$ matrix of rank s with known coefficients and $\mathbf{d} = (d_1, \cdots, d_s)'$ is a vector of known constants. This hypothesis can also be written as

$$H\colon \mathbf{c}_1'\boldsymbol{\beta} = d_1, \ \mathbf{c}_2'\boldsymbol{\beta} = d_2, \ \cdots, \ \mathbf{c}_s'\boldsymbol{\beta} = d_s. \tag{7.2.2}$$

We assume that $\mathrm{r}(\mathbf{C}) = s$, since otherwise, some of the relations in H are redundant, and may be obtained from the others. In practice, we expect that such redundant hypotheses about parameters have been eliminated.

Unless $\mathrm{r}(\mathbf{X}) = p$, not all such linear hypotheses are in general, testable. A hypothesis is said to be *testable* if $\mathbf{C}'\boldsymbol{\beta}$ can be written in terms of estimable functions of $\boldsymbol{\beta}$; otherwise, it is a non-testable hypothesis. The matrix $\mathbf{C}$ must satisfy the condition $\mathbf{T}'\mathbf{X} = \mathbf{C}'$ or $\mathbf{C}'\mathbf{H} = \mathbf{C}'$ in order that the hypothesis H is testable (see Section 4.3).

Definition 7.2.1. Consider the normal general linear model $\mathbf{y} = \mathbf{X}\boldsymbol{\beta} + \boldsymbol{\varepsilon}$, where $\mathrm{r}(\mathbf{X}) = r < p$. A hypothesis $H\colon \mathbf{C}'\boldsymbol{\beta} = \mathbf{d}$ is testable if the s rows of $\mathbf{C}'$ are linearly dependent on the rows of $\mathbf{X}$; i.e., $\mathbf{c}_i' = \mathbf{t}_i'\mathbf{X}$, or equivalently, if there exists an $N \times s$ matrix $\mathbf{T}$ such that $\mathbf{C}' = \mathbf{T}'\mathbf{X}$. Equivalently, a hypothesis $H\colon \mathbf{C}'\boldsymbol{\beta} = \mathbf{d}$ is testable if and only if $\mathbf{C}'\mathbf{H} = \mathbf{C}'$.

To obtain a solution for $\boldsymbol{\beta}$ under the constraint imposed by H, we use the approach in Section 4.6, and minimize the sum of squares function $S(\boldsymbol{\beta})$ subject to $\mathbf{C}'\boldsymbol{\beta} = \mathbf{d}$. The equations corresponding to (4.6.2) and (4.6.3) are

$$\mathbf{X}'\mathbf{X}\boldsymbol{\beta}^0_H + \mathbf{C}\boldsymbol{\lambda} = \mathbf{X}'\mathbf{y}, \quad \mathbf{C}'\boldsymbol{\beta}^0_H = \mathbf{d}, \tag{7.2.3}$$

where $2\boldsymbol{\lambda}$ denotes the vector of Lagrangian multipliers. Provided $\mathbf{C}'\boldsymbol{\beta} = \mathbf{d}$ (as a system of equations in $\boldsymbol{\beta}$) is consistent, the least squares solution of $\boldsymbol{\beta}$ constrained by H has the form (see (4.6.4))

$$\boldsymbol{\beta}^0_H = \boldsymbol{\beta}^0 - \mathbf{G}\mathbf{C}(\mathbf{C}'\mathbf{G}\mathbf{C})^{-1}(\mathbf{C}'\boldsymbol{\beta}^0 - \mathbf{d}). \tag{7.2.4}$$

7.2.1 Derivation of and motivation for the F-test

We develop a statistic based on the F-distribution to test the (testable) null hypothesis H: $\mathbf{C}'\boldsymbol{\beta} = \mathbf{d}$ (see Definition 7.2.1). In Section 4.6, we discussed estimation of a linear model subject to linear constraints. It is useful to distinguish between

(a) a *constraint on a model*, which we denote by $\mathbf{A}'\boldsymbol{\beta} = \mathbf{b}$, versus

(b) a *constraint by a hypothesis*, which we denote by H: $\mathbf{C}'\boldsymbol{\beta} = \mathbf{d}$.

A model with constraints on its parameters is called a *restricted model.* A submodel with "constraints by a hypothesis" is called a *reduced model.* Using simple examples, we illustrate below (i) restricted models, (ii) reduced models, and (iii) reduced, restricted models.

(i) **Restricted model.** A model which specifies certain (natural) constraints on its parameters is called a restricted model, i.e., $\mathbf{y} = \mathbf{X}\boldsymbol{\beta} + \boldsymbol{\varepsilon}$, subject to $\mathbf{A}'\boldsymbol{\beta} = \mathbf{b}$. An example is a balanced one-way ANOVA model, $Y_{ij} = \mu + \tau_i + \varepsilon_{ij}$, subject to a natural model constraint $\sum_{i=1}^{a} \tau_i = 0$.

(ii) **Reduced model.** A model with a constraint imposed by a hypothesis is called a reduced model, i.e., $\mathbf{y} = \mathbf{X}\boldsymbol{\beta} + \boldsymbol{\varepsilon}$, subject to H: $\mathbf{C}'\boldsymbol{\beta} = \mathbf{d}$. An example is a multiple linear regression model $Y_i = \beta_0 + \sum_{j=1}^{k} \beta_j X_{ij} + \varepsilon_i$ subject to the hypothesis H: $\beta_j = 0$, $j = q+1, \cdots, k$. The reduced model is then $Y_i = \beta_0 + \sum_{j=1}^{q} \beta_j X_{ij} + \varepsilon_i$.

(iii) **Reduced, restricted model.** A constrained model with an additional constraint by a hypothesis is called a reduced, restricted model, i.e., $\mathbf{y} = \mathbf{X}\boldsymbol{\beta}+\boldsymbol{\varepsilon}$, subject to a constraint on the model $\mathbf{A}'\boldsymbol{\beta} = \mathbf{b}$ and then a constraint by a hypothesis $\mathbf{C}'\boldsymbol{\beta} = \mathbf{d}$. An example is a balanced one-way ANOVA model $Y_{ij} = \mu + \tau_i + \varepsilon_{ij}$ subject to a model constraint $\sum_{i=1}^{a} \tau_i = 0$ and a constraint by a hypothesis H: $\mu = 0$.

Result 7.2.1. Suppose the hypothesis H: $\mathbf{C}'\boldsymbol{\beta} = \mathbf{d}$ is testable and $\mathbf{C}'\boldsymbol{\beta} = \mathbf{d}$ is a consistent system of equations in $\boldsymbol{\beta}$. When $\mathbf{y} \sim N(\mathbf{X}\boldsymbol{\beta}, \sigma^2\mathbf{I}_N)$, the test statistic

$$F(H) = \frac{Q/s}{SSE/(N-r)} \sim F(s, N-r, \lambda), \tag{7.2.5}$$

i.e., a noncentral F-distribution with numerator d.f. s, denominator d.f. $N - r$, and noncentrality parameter λ, where

$$\begin{aligned} Q &= (\mathbf{C}'\boldsymbol{\beta}^0 - \mathbf{d})'[\text{Cov}(\mathbf{C}'\boldsymbol{\beta}^0)/\sigma^2]^{-1}(\mathbf{C}'\boldsymbol{\beta}^0 - \mathbf{d}) \\ &= (\mathbf{C}'\boldsymbol{\beta}^0 - \mathbf{d})'(\mathbf{C}'\mathbf{G}\mathbf{C})^{-1}(\mathbf{C}'\boldsymbol{\beta}^0 - \mathbf{d}), \text{ and} \end{aligned} \tag{7.2.6}$$

$$\lambda = \frac{1}{2\sigma^2}(\mathbf{C}'\boldsymbol{\beta} - \mathbf{d})'(\mathbf{C}'\mathbf{G}\mathbf{C})^{-1}(\mathbf{C}'\boldsymbol{\beta} - \mathbf{d}). \tag{7.2.7}$$

In particular, under H, $F(H) \sim F_{s,N-r}$.

Proof. From Result 5.3.3, $Q/\sigma^2 \sim \chi^2(s, \lambda)$, a noncentral chi-square distribution with d.f. s and noncentrality parameter λ. From property 3 of Result 7.1.1, $SSE/\sigma^2 = \widehat{\boldsymbol{\varepsilon}}'\widehat{\boldsymbol{\varepsilon}}/\sigma^2 \sim \chi^2_{N-r}$. Since Q is a function of $\boldsymbol{\beta}^0$ only and SSE a function of $\widehat{\boldsymbol{\varepsilon}}$ only, by property 4, Q and SSE are independent. By Result 5.3.6,

$$F(H) = \frac{Q/s\sigma^2}{SSE/(N-r)\sigma^2} \sim F(s, N-r, \lambda).$$

Under H: $\mathbf{C}'\boldsymbol{\beta} = \mathbf{d}$, so that $\lambda = 0$, and then, $F(H) \sim F_{s,N-r}$. ■

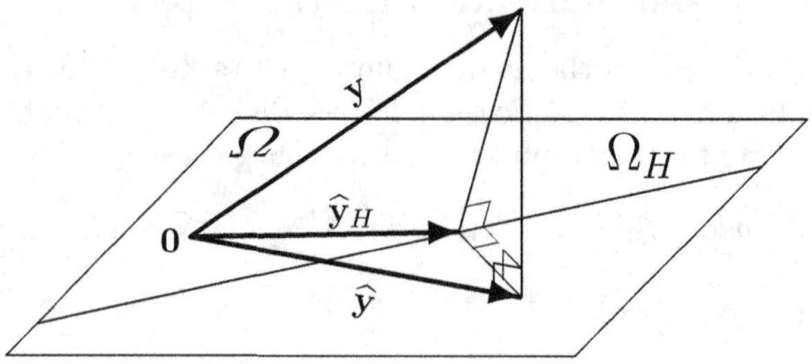

FIGURE 7.2.1. Geometry of the F-test. $\Omega_H = \{\mathbf{X}\boldsymbol{\beta}\colon \mathbf{C}'\boldsymbol{\beta} = \mathbf{d}\}$. Note that for $\mathbf{d} \neq \mathbf{0}$, Ω_H need not contain $\mathbf{0}$.

Let Ω_H denote the estimation space Ω reduced by the hypothesis H. Figure 7.2.1 illustrates the F-test geometrically. Also, the F-test amounts to checking whether $\mathbf{d}$ belongs to the $100(1-\alpha)\%$ confidence ellipsoid for $\mathbf{C}'\boldsymbol{\beta}$, which is centered at $\mathbf{C}'\boldsymbol{\beta}^0$, i.e., the ellipsoid

$$\{\mathbf{C}'\boldsymbol{\beta}\colon (\mathbf{C}'\boldsymbol{\beta} - \mathbf{C}'\boldsymbol{\beta}^0)'(\mathbf{C}'\mathbf{G}\mathbf{C})^{-1}(\mathbf{C}'\boldsymbol{\beta} - \mathbf{C}'\boldsymbol{\beta}^0) \leq s\widehat{\sigma}^2 F_{s,N-r}\}. \tag{7.2.8}$$

Example 7.2.1. In the simple linear regression model, we have seen that the estimation space Ω is a plane spanned by $\mathbf{1}_N$ and $\mathbf{x} = (X_1, \cdots, X_N)'$. The estimation space Ω_H reduced by the null hypothesis $H\colon \mathbf{C}'\boldsymbol{\beta} = \mathbf{d}$, i.e., $H\colon c_0\beta_0 + c_1\beta_1 = d$ is a straight line formed by combining the vectors $d\mathbf{x}/c_1$ (of constant length), and $\beta_0\{\mathbf{1}_N - c_0\mathbf{x}/c_1\}$ (of variable length). This straight line is parallel to $\mathbf{1}_N - c_0\mathbf{x}/c_1$, but displaced by a distance of $d\mathbf{x}/c_1$. Clearly, Ω_H is a subset of Ω, while $\Omega \cap \Omega_H^{\perp}$ is the space spanned by all vectors in Ω that are orthogonal to Ω_H. When $p = 2$, and $s = 1$, two vectors span Ω, while Ω_H and $\Omega \cap \Omega_H^{\perp}$ are each spanned by a single vector denoted respectively by $\mathbf{v} = \beta_0\mathbf{1}_N + (d - c_0\beta_0)\mathbf{x}/c_1$, and $(\mathbf{v}'\mathbf{x})\mathbf{1}_N - (\mathbf{v}'\mathbf{1}_N)\mathbf{x}$. □

Example 7.2.2. Consider the model (4.1.2), the multiple linear regression model with k predictors and an intercept. We derive the F-test for each $H_j\colon \beta_j = d_j$, where d_j is a fixed constant, $1 \leq j \leq k$. We can write the hypotheses as $H\colon \mathbf{C}'\boldsymbol{\beta} = \mathbf{d}$, where $\mathbf{C}'$ is a p-dimensional vector with 1 in the $(j+1)$th position and zero elsewhere. For $j = 1, \cdots, k$, straightforward calculations yield

$$Q_j = (\widehat{\beta}_j - d_j)^2/\{(\mathbf{X}'\mathbf{X})^{-1}\}_{j+1,j+1},$$

so that under H_j,

$$F(H_j) = Q_j/\{SSE/(N-p)\} \sim F_{1,N-p}. \qquad \square$$

Example 7.2.3. Let A_1 and A_2 denote objects of unknown weights β_1 and β_2 respectively. The weights of A_1 and A_2 are measured on a balance using the following scheme, all of these actions being repeated twice:

(1) both objects on the balance, resulting in weights $Y_{1,1}$ and $Y_{1,2}$,

(2) A_1 on the balance, resulting in weights $Y_{2,1}$ and $Y_{2,2}$, and

(3) A_2 on the balance, resulting in weights $Y_{3,1}$ and $Y_{3,2}$.

We assume that $Y_{i,j}$'s are independent, normally distributed variables, with common variance σ^2. We also assume that the balance has an unknown systematic error β_0. We wish to test the hypothesis that both objects have the same weight. The model can be written as

$$\begin{aligned} Y_{1,j} &= \beta_0 + \beta_1 + \beta_2 + \varepsilon_{1,j} \\ Y_{2,j} &= \beta_0 + \beta_1 + \varepsilon_{2,j} \\ Y_{3,j} &= \beta_0 + \beta_2 + \varepsilon_{3,j}, \qquad j = 1, 2, \end{aligned}$$

where $\varepsilon_{i,j}$ are i.i.d. $N(0, \sigma^2)$ variables, $i = 1, 2, 3$, $j = 1, 2$. Let $\overline{Y}_{i\cdot} = (Y_{i,1} + Y_{i,2})/2$. It is easy to verify that

$$\widehat{\boldsymbol{\beta}} = \begin{pmatrix} -\overline{Y}_{1\cdot} + \overline{Y}_{2\cdot} + \overline{Y}_{3\cdot} \\ \overline{Y}_{1\cdot} - \overline{Y}_{3\cdot} \\ \overline{Y}_{1\cdot} - \overline{Y}_{2\cdot} \end{pmatrix}$$

with

$$\text{Cov}(\widehat{\boldsymbol{\beta}}) = \sigma^2 \begin{pmatrix} 3/2 & -1 & -1 \\ -1 & 1 & 1/2 \\ -1 & 1/2 & 1 \end{pmatrix},$$

and that $\widehat{\boldsymbol{\beta}}$ has a normal distribution. Also,

$$\widehat{\sigma}^2 = \frac{1}{3} \sum_{i=1}^{3} \sum_{j=1}^{2} (Y_{i,j} - \overline{Y}_{i\cdot})^2 \sim \chi^2_3.$$

The test of H: $\beta_1 = \beta_2$ can be written as H: $\mathbf{c}'\boldsymbol{\beta} = \mathbf{d}$, with $\mathbf{c} = (0, 1, -1)'$ and $d = 0$. Then $\mathbf{c}'\widehat{\boldsymbol{\beta}} - d = \overline{Y}_{2\cdot} - \overline{Y}_{3\cdot}$ and $\text{Var}(\mathbf{c}'\widehat{\boldsymbol{\beta}}) = \text{Var}(\overline{Y}_{2\cdot} - \overline{Y}_{3\cdot}) = \sigma^2$. By (7.2.6), the test statistic is

$$F(H) = (\overline{Y}_{2\cdot} - \overline{Y}_{3\cdot})^2 / \widehat{\sigma}^2 \sim F_{1,3}$$

under H. □

An alternative derivation of the F-statistic is shown below. The least squares solution for $\boldsymbol{\beta}$ in the linear model $\mathbf{y} = \mathbf{X}\boldsymbol{\beta} + \boldsymbol{\varepsilon}$ subject to the linear constraint by the hypothesis H: $\mathbf{C}'\boldsymbol{\beta} = \mathbf{d}$ was given in (7.2.4) as $\boldsymbol{\beta}^0_H$. The corresponding fitted vector and residual vector are

$$\widehat{\mathbf{y}}_H = \mathbf{X}\boldsymbol{\beta}^0_H, \text{ and } \widehat{\boldsymbol{\varepsilon}}_H = \mathbf{y} - \widehat{\mathbf{y}}_H.$$

Then $SSE_H = \widehat{\boldsymbol{\varepsilon}}'_H \widehat{\boldsymbol{\varepsilon}}_H$, where $\widehat{\boldsymbol{\varepsilon}}_H = (\mathbf{y} - \widehat{\mathbf{y}}) + (\widehat{\mathbf{y}} - \widehat{\mathbf{y}}_H) = \widehat{\boldsymbol{\varepsilon}} + (\widehat{\mathbf{y}} - \widehat{\mathbf{y}}_H)$.

Result 7.2.2. The statistic for testing H: $\mathbf{C}'\boldsymbol{\beta} = \mathbf{d}$ may also be written as

$$F(H) = \frac{(SSE_H - SSE)/\text{r}(\mathbf{C})}{SSE/(N - r)}, \tag{7.2.9}$$

which has an $F_{\text{r}(\mathbf{C}), N-r}$ distribution when H is true.

Proof. Note that $\widehat{\boldsymbol{\varepsilon}} \perp \mathcal{C}(\mathbf{X})$, while both $\widehat{\mathbf{y}}$ and $\widehat{\mathbf{y}}_H$ belong to $\mathcal{C}(\mathbf{X})$. As a result, $\widehat{\boldsymbol{\varepsilon}} \perp (\widehat{\mathbf{y}} - \widehat{\mathbf{y}}_H)$, so by Pythagoras's Theorem (see (1.2.1)), $\widehat{\boldsymbol{\varepsilon}}'_H \widehat{\boldsymbol{\varepsilon}}_H = \widehat{\boldsymbol{\varepsilon}}'\widehat{\boldsymbol{\varepsilon}} + \|\widehat{\mathbf{y}} - \widehat{\mathbf{y}}_H\|^2$. On the other hand, from (7.2.4),

$$\begin{aligned} \|\widehat{\mathbf{y}} - \widehat{\mathbf{y}}_H\|^2 &= (\boldsymbol{\beta}^0 - \boldsymbol{\beta}^0_H)\mathbf{X}'\mathbf{X}(\boldsymbol{\beta}^0 - \boldsymbol{\beta}^0_H) \\ &= (\mathbf{C}'\boldsymbol{\beta}^0 - \mathbf{d})'(\mathbf{C}'\mathbf{G}\mathbf{C})^{-1}\mathbf{C}'\mathbf{G}'\mathbf{X}'\mathbf{X}\mathbf{G}\mathbf{C}(\mathbf{C}'\mathbf{G}\mathbf{C})^{-1}(\mathbf{C}'\boldsymbol{\beta}^0 - \mathbf{d}). \end{aligned}$$

Estimability of $\mathbf{C}'\boldsymbol{\beta}$ implies that $\mathbf{X}'\mathbf{X}\mathbf{G}\mathbf{C} = \mathbf{H}'\mathbf{C} = \mathbf{C}$. Plugging this into the above display, it follows that

$$\|\widehat{\mathbf{y}} - \widehat{\mathbf{y}}_H\|^2 = Q,$$

where Q is defined in (7.2.6). As a result, $SSE_H - SSE = Q$. ■

The F-statistic in (7.2.9) is useful when $\mathbf{C}$ is not of full column rank, or when it is easier to compute SSE_H and SSE than to compute Q according to (7.2.6); see Example 7.2.5. Golub and Styan (1973) have suggested a numerically efficient procedure for computing Q.

Here is a motivation for the F-test statistic. Let us consider the quantity Q/s. We have seen that Q/σ^2 has a noncentral $\chi^2(s,\lambda)$ distribution, with $\lambda = (\mathbf{C}'\boldsymbol{\beta}-\mathbf{d})'(\mathbf{C}'\mathbf{G}\mathbf{C})^{-1}(\mathbf{C}'\boldsymbol{\beta}-\mathbf{d})/2\sigma^2$. By Result 5.3.4, we have $\mathrm{E}(Q/\sigma^2) = s + 2\lambda$, so that

$$\mathrm{E}(Q/s) = \sigma^2 + (\mathbf{C}'\boldsymbol{\beta} - \mathbf{d})'(\mathbf{C}'\mathbf{G}\mathbf{C})^{-1}(\mathbf{C}'\boldsymbol{\beta} - \mathbf{d})/s.$$

We know that $\mathrm{E}(\widehat{\sigma}^2) = \sigma^2$. Under H: $\mathbf{C}'\boldsymbol{\beta} - \mathbf{d} = \mathbf{0}$, the second term in $\mathrm{E}(Q/s)$ becomes zero, and $\mathrm{E}(Q/s) = \sigma^2$. We therefore expect that under H, the statistic $F(H)$ is approximately equal to 1. When the null hypothesis is false, $\mathrm{E}(Q/s) > \sigma^2 = \mathrm{E}(\widehat{\sigma}^2)$. Now,

$$\mathrm{E}\{F(H)\} = \mathrm{E}(Q/s)\,\mathrm{E}(1/\widehat{\sigma}^2)$$

and we know from the Markov inequality (see (C.3)) that $\mathrm{E}(1/\widehat{\sigma}^2) > 1/\,\mathrm{E}(\widehat{\sigma}^2)$. Therefore,

$$\mathrm{E}\{F(H)\} > \mathrm{E}(Q/s)/\,\mathrm{E}(\widehat{\sigma}^2) > 1,$$

and H is rejected if $F(H)$ is significantly large.

Example 7.2.4. Sets of regression lines. Consider L regression lines

$$Y_{l,i} = \beta_{l,0} + \beta_{l,1}X_{l,i} + \varepsilon_{l,i}, \quad i = 1, \cdots, n_l, \; l = 1, \cdots, L, \tag{7.2.10}$$

where $\varepsilon_{l,i}$'s are uncorrelated with $\mathrm{E}(\varepsilon_{l,i}) = 0$ and $\mathrm{Var}(\varepsilon_{l,i}) = \sigma^2$. The model can be written in the form (4.1.1) with

$$\mathbf{y} = \begin{pmatrix} \mathbf{y}_1 \\ \mathbf{y}_2 \\ \vdots \\ \mathbf{y}_L \end{pmatrix}, \; \mathbf{X} = \begin{pmatrix} \mathbf{X}_1 & \mathbf{O} & \cdots & \mathbf{O} \\ \mathbf{O} & \mathbf{X}_2 & \cdots & \mathbf{O} \\ \vdots & \vdots & \ddots & \vdots \\ \mathbf{O} & \mathbf{O} & \cdots & \mathbf{X}_L \end{pmatrix}, \; \boldsymbol{\beta} = \begin{pmatrix} \boldsymbol{\beta}_1 \\ \boldsymbol{\beta}_2 \\ \vdots \\ \boldsymbol{\beta}_L \end{pmatrix}, \text{ and } \boldsymbol{\varepsilon} = \begin{pmatrix} \boldsymbol{\varepsilon}_1 \\ \boldsymbol{\varepsilon}_2 \\ \vdots \\ \boldsymbol{\varepsilon}_L \end{pmatrix}$$

where $\mathbf{y}_l = (Y_{l,1}, \cdots, Y_{l,n_l})'$, $\boldsymbol{\beta}_l = (\beta_{l,0}, \beta_{l,1})'$, $\boldsymbol{\varepsilon}_l = (\varepsilon_{l,1}, \cdots, \varepsilon_{l,n_l})'$, and

$$\mathbf{X}_l = \begin{pmatrix} 1 & X_{l,1} \\ 1 & X_{l,2} \\ \vdots & \vdots \\ 1 & X_{l,n_l} \end{pmatrix}.$$

Here, $N = \sum_{l=1}^{L} n_l$, $p = 2L$, and $\mathrm{r}(\mathbf{X}) = 2L$. The least squares estimate $\widehat{\boldsymbol{\beta}}$ is obtained by minimizing $S(\boldsymbol{\beta}) = (\mathbf{y} - \mathbf{X}\boldsymbol{\beta})'(\mathbf{y} - \mathbf{X}\boldsymbol{\beta}) = \sum_{l=1}^{L}(\mathbf{y}_l - \mathbf{X}_l\boldsymbol{\beta}_l)'(\mathbf{y}_l - \mathbf{X}_l\boldsymbol{\beta}_l)$. This can be done by minimizing each $(\mathbf{y}_l - \mathbf{X}_l\boldsymbol{\beta}_l)'(\mathbf{y}_l - \mathbf{X}_l\boldsymbol{\beta}_l)$, which corresponds to a simple linear regression. From results similar to Example 7.1.1,

$$\widehat{\beta}_{l,1} = \frac{\sum_{i=1}^{n_l}(Y_{l,i} - \overline{Y}_{l\cdot})(X_{l,i} - \overline{X}_{l\cdot})}{\sum_{i=1}^{n_l}(X_{l,i} - \overline{X}_{l\cdot})^2}, \tag{7.2.11}$$

$$\widehat{\beta}_{l,0} = \overline{Y}_{l\cdot} - \widehat{\beta}_{l,1}\overline{X}_{l\cdot}. \tag{7.2.12}$$

where $\overline{Y}_{l\cdot} = \sum_{i=1}^{n_l} Y_{l,i}/n_l$, and $\overline{X}_{l\cdot} = \sum_{i=1}^{n_l} X_{l,i}/n_l$, for $l = 1, \cdots, L$. Then,

$$SSE = \sum_{l=1}^{L}\left[\sum_{i=1}^{n_l}(Y_{l,i} - \overline{Y}_{l\cdot})^2 - \widehat{\beta}_{l,1}^2 \sum_{i=1}^{n_l}(X_{l,i} - \overline{X}_{l\cdot})^2\right]$$
$$= \sum_{l=1}^{L}\sum_{i=1}^{n_l}(Y_{l,i} - \overline{Y}_{l\cdot})^2 - \sum_{l=1}^{L}\widehat{\beta}_{l,1}^2 \sum_{i=1}^{n_l}(X_{l,i} - \overline{X}_{l\cdot})^2. \tag{7.2.13}$$

Suppose we impose the constraint by the hypothesis that the slopes of the L lines are the same, i.e., we hypothesize that

$$H\colon \beta_{1,1} = \beta_{2,1} = \cdots = \beta_{L,1} = \beta.$$

The least squares estimates of $\beta_{l,0}$, $l = 1, \cdots, L$ and β are obtained by minimizing the function

$$S(\beta_{1,0}, \cdots, \beta_{L,0}, \beta) = \sum_{l=1}^{L}\sum_{i=1}^{n_l}(Y_{l,i} - \beta_{l,0} - \beta X_{l,i})^2$$

with respect to $\beta_{l,0}$, $l = 1, \cdots, L$, and β. The normal equations are

$$\sum_{i=1}^{n_l}(Y_{l,i} - \widehat{\beta}_{l,0,H} - \widehat{\beta}_H X_{l,i}) = 0, \quad l = 1, \cdots, L,$$
$$\sum_{l=1}^{L}\sum_{i=1}^{n_l}(Y_{l,i} - \widehat{\beta}_{l,0,H} - \widehat{\beta}_H X_{l,i})X_{l,i} = 0;$$

solving these simultaneously, we obtain the least squares estimates of the parameters under the reduction imposed by H as

$$\widehat{\beta}_{l,0,H} = \overline{Y}_{l\cdot} - \widehat{\beta}_H \overline{X}_{l\cdot}, \tag{7.2.14}$$
$$\widehat{\beta}_H = \frac{\sum_{l=1}^{L}\sum_{i=1}^{n_l} X_{l,i}(Y_{l,i} - \overline{Y}_{l\cdot})}{\sum_{l=1}^{L}\sum_{i=1}^{n_l} X_{l,i}(X_{l,i} - \overline{X}_{l\cdot})}. \tag{7.2.15}$$

Clearly, SSE is the error sum of squares in (7.2.13), while

$$SSE_H = \sum_{l=1}^{L}\sum_{i=1}^{n_l}\{Y_{l,i} - \overline{Y}_{l\cdot} - \widehat{\beta}_H(X_{l,i} - \overline{X}_{l\cdot})\}^2$$
$$= \sum_{l=1}^{L}\sum_{i=1}^{n_l}(Y_{l,i} - \overline{Y}_{l\cdot})^2 - \widehat{\beta}_H^2 \sum_{l=1}^{L}\sum_{i=1}^{n_l}(X_{l,i} - \overline{X}_{l\cdot})^2, \tag{7.2.16}$$

so that

$$SSE_H - SSE = \sum_{l=1}^{L}\widehat{\beta}_l^2 \sum_{i=1}^{n_l}(X_{l,i} - \overline{X}_{l\cdot})^2 - \widehat{\beta}_H^2 \sum_{l=1}^{L}\sum_{i=1}^{n_l}(X_{l,i} - \overline{X}_{l\cdot})^2.$$

The test statistic is

$$F(H) = \frac{(SSE_H - SSE)/(L-1)}{SSE/(N-2L)}, \tag{7.2.17}$$

which follows an $F_{L-1,N-2L}$ distribution under H. In Chapter 9, we further explore inference for sets of regression lines. □

Source	d.f.	SS	MS
Model	r	$SSR = \boldsymbol{\beta}^{0\prime}\mathbf{X}'\mathbf{y}$	$MSR = SSR/r$
Residual	$N-r$	$SSE = \mathbf{y}'\mathbf{y} - \boldsymbol{\beta}^{0\prime}\mathbf{X}'\mathbf{y}$	$MSE = SSE/(N-r)$
Total	N	$SST = \mathbf{y}'\mathbf{y}$	

TABLE 7.2.1. ANOVA table for Example 7.2.5.

We next discuss the relation between the F-test statistic and the coefficient of determination R^2 in the full-rank model.

Result 7.2.3. In the model (4.1.2) where ε_i are i.i.d. $N(0, \sigma^2)$ variables, consider the test of H: $\beta_1 = \cdots = \beta_k = 0$. The F-test statistic can be written in terms of the coefficient of determination as

$$F(H) = (N-k-1)R^2/[k(1-R^2)] \sim F_{k,N-k-1}$$

under H. The null distribution of R^2 is Beta$(k/2, (N-k-1)/2)$.

Proof. Under H, the model is $Y_i = \beta_0 + \varepsilon_i$, with $\widehat{\beta}_H = \overline{Y}$ and $SSE_H = SST = \sum_{i=1}^{N}(Y_i - \overline{Y})^2$. From (7.2.9) and Definition 4.2.4, the form of $F(H)$ in terms of R^2 follows. Since $F(H) \sim F_{k,N-k-1}$ under H, the null distribution of R^2 follows directly from (B.13). ■

In Chapter 4, we introduced sums of squares associated with the general linear model. In particular, we distinguished two ways of writing the ANOVA decomposition, $SST = SSR + SSE$, or $SST_c = SSR_c + SSE$, where SST, and SSR denote the total and model sums of squares, respectively, SSE denotes the residual sum of squares, while SST_c and SSR_c denote the mean corrected total and model sums of squares. SSM is the sum of squares due to fitting the mean. We now present the ANOVA table in special cases.

Example 7.2.5. Suppose we wish to test the hypothesis H: $\mathbf{X}\boldsymbol{\beta} = \mathbf{0}$ in the general linear model (4.1.1). Since $\mathbf{X}\boldsymbol{\beta}$ includes all estimable parametric functions of $\boldsymbol{\beta}$, this hypothesis requires that all estimable parametric functions are null. Clearly this is a testable hypothesis. However, in this case, since $\mathbf{C} = \mathbf{X}'$ is not of full column rank as $N > p$, we cannot compute Q in (7.2.6). Instead, we use (7.2.9) to derive the F-statistic. Under the constraint by the hypothesis H: $\mathbf{X}\boldsymbol{\beta} = \mathbf{0}$, SSE_H is exactly $\mathbf{y}'\mathbf{y}$. Then $SSE_H - SSE = SSR$, so

$$F(H) = \{SSR/r\}/\{SSE/(N-r)\} \sim F(r, N-r, \boldsymbol{\beta}'\mathbf{X}'\mathbf{X}\boldsymbol{\beta}/2\sigma^2).$$

Under the null hypothesis, the noncentrality parameter is zero since $\mathbf{X}\boldsymbol{\beta} = \mathbf{0}$, so that the null distribution of the test statistic $F(H)$ is an $F_{r,N-r}$ distribution. It is not surprising that the numerator d.f. is r, since there are exactly r LIN estimable functions in $\mathbf{X}\boldsymbol{\beta}$ (see Result 4.3.2). It is usual to represent this information via an ANOVA table as shown in Table 7.2.1.

It is important to keep in mind that this statistic does not provide a test of $\boldsymbol{\beta} = \mathbf{0}$; in fact, unless $\mathrm{r}(\mathbf{X}) = p$, $\boldsymbol{\beta}$ is not even estimable, and H: $\boldsymbol{\beta} = \mathbf{0}$ is *not a testable hypothesis.* In the full-rank case when $(\mathbf{X}'\mathbf{X})^{-1}$ exists, the hypotheses $\mathbf{X}\boldsymbol{\beta} = \mathbf{0}$ and $\boldsymbol{\beta} = \mathbf{0}$ are equivalent, since $\mathbf{X}\boldsymbol{\beta} = \mathbf{0}$ implies that $\mathbf{X}'\mathbf{X}\boldsymbol{\beta} = \mathbf{0}$. It is left as an exercise to the reader to verify that SSR, and hence the F-test is unchanged if we replace $\mathbf{X}\boldsymbol{\beta} = \mathbf{0}$ by $\mathbf{L}\boldsymbol{\beta} = \mathbf{0}$, where $\mathbf{L}\boldsymbol{\beta}$ is a different set of r LIN estimable parametric functions of $\boldsymbol{\beta}$. The ANOVA table separating out the mean fitting is shown in Table 7.2.2.

Under the null hypothesis H: $\mathbf{X}\boldsymbol{\beta} = \mathbf{0}$, $F_M = MSM/MSE \sim F_{1,N-r}$, while $F_c =$

Source	d.f.	SS	MS
Mean	1	$SSM = N\overline{Y}^2$	$MSM = SSM/1$
Model*	$r-1$	$SSR_c = \beta^{0\prime}\mathbf{X}'\mathbf{y} - N\overline{Y}^2$	$MSR_c = SSR_c/(r-1)$
Residual	$N-r$	$SSE = \mathbf{y}'\mathbf{y} - \beta^{0\prime}\mathbf{X}'\mathbf{y}$	$MSE = SSE/(N-r)$
Total	N	$SST = \mathbf{y}'\mathbf{y}$	

$*$ Mean corrected

TABLE 7.2.2. ANOVA table separating out the mean.

$MSR_c/MSE \sim F_{r-1,N-r}$. While F_M tests whether $\text{E}(\overline{Y}) = 0$, F_c tests for effects apart from the mean. □

Example 7.2.6. Consider the linear model

$$Y_1 = \beta_1 + 3\beta_2 + \varepsilon_1$$
$$Y_2 = 2\beta_1 - \beta_2 + \varepsilon_2$$
$$Y_3 = 3\beta_1 - 4\beta_2 + \varepsilon_3,$$

where ε_i are i.i.d. $N(0, \sigma^2)$ variables, $i = 1, 2, 3$. We derive the F-statistic for testing H: $\beta_1 = \beta_2$, which can be written as $\mathbf{c}'\boldsymbol{\beta} = \mathbf{d}$, with $\mathbf{c} = (1, -1)'$, $\boldsymbol{\beta} = (\beta_1, \beta_2)'$, and $d = 0$. We can verify that $\widehat{\boldsymbol{\beta}}$ and $\mathbf{I} - \mathbf{P}$ are

$$\widehat{\boldsymbol{\beta}} = \frac{1}{243}\begin{pmatrix} 59Y_1 + 41Y_2 + 34Y_3 \\ 53Y_1 + 8Y_2 - 23Y_3 \end{pmatrix}, \quad \mathbf{I} - \mathbf{P} = \frac{1}{243}\begin{pmatrix} 25 & -65 & 35 \\ -65 & 169 & -91 \\ 35 & -91 & 49 \end{pmatrix},$$

$\mathbf{c}'\widehat{\boldsymbol{\beta}} - d = (6Y_1 + 33Y_2 + 57Y_3)/243$ and $\mathbf{c}'(\mathbf{X}'\mathbf{X})^{-1}\mathbf{c} = 2/27$. Also, $\widehat{\sigma}^2 = \mathbf{y}'(\mathbf{I} - \mathbf{P})\mathbf{y}$. The F-test statistic has the form in (7.2.5), and follows an $F_{1,1}$ distribution under H. □

Example 7.2.7. We continue with Example 4.1.3. In the one-way fixed-effects ANOVA model, a primary interest is in testing equality of the a treatment means, i.e.,

$$H_0\colon \mu_1 = \mu_2 = \cdots = \mu_a, \text{ versus}$$
$$H_1\colon \mu_i \neq \mu_j \text{ for at least one pair } (i, j),$$

or equivalently, the hypothesis that the a treatment effects are equal, i.e.,

$$H_0\colon \tau_1 = \tau_2 = \cdots = \tau_a, \text{ versus}$$
$$H_1\colon \tau_i \neq \tau_j \text{ for at least one pair } (i, j).$$

The ANOVA identity leads to an orthogonal partitioning of the total variability in the response variable Y into its component parts:

$$\sum_{i=1}^{a}\sum_{j=1}^{n_i}(Y_{ij} - \overline{Y}_{\cdot\cdot})^2 = \sum_{i=1}^{a} n_i(\overline{Y}_{i\cdot} - \overline{Y}_{\cdot\cdot})^2 + \sum_{i=1}^{a}\sum_{j=1}^{n_i}(Y_{ij} - \overline{Y}_{i\cdot})^2.$$

The term on the left is the total corrected sum of squares SST_c. The first term on the right side is a sum of squares of the differences *between* treatment means and the grand mean and is called the treatment sum of squares, $SSTr$, while the second term on the right is the sum of squares of the differences of observations *within* a treatment from the treatment

Source	d.f.	SS	MS
Treatment	$a-1$	$SSTr = \sum_{i=1}^{a} n_i(\overline{Y}_{i\cdot} - \overline{Y}_{\cdot\cdot})^2$	$MSTr = SSTr/(a-1)$
Residual*	$N-a$	$SSE = \sum_{i=1}^{a}\sum_{j=1}^{n_i}(Y_{ij} - \overline{Y}_{i\cdot})^2$	$MSE = SSE/(N-a)$
Total	$N-1$	$SST_c = \sum_{i=1}^{a}\sum_{j=1}^{n_i}(Y_{ij} - \overline{Y}_{\cdot\cdot})^2$	

$*$ Mean corrected

TABLE 7.2.3. ANOVA table for the one-factor model.

mean and is the error sum of squares, SSE. Symbolically, we can write the decomposition as

$$SST_c = SSTr + SSE.$$

This algebraic identity is easily verified by seeing that $N\overline{Y}_{\cdot\cdot} = Y_{\cdot\cdot} = \sum_{i=1}^{a} n_i \overline{Y}_{i\cdot}$. Each of these sums of squares is a quadratic form in

$$\mathbf{y} = (Y_{11}, Y_{12,}, \cdots, Y_{1n_1}, \cdots, Y_{a1}, Y_{a2}, \cdots, Y_{a,n_a})',$$

an N-dimensional normal random vector. It may be shown using results from Chapter 5 that SST/σ^2 and $SSTr/\sigma^2$ have chi-square distributions with $N-1$ d.f. and $a-1$ d.f., respectively, when H_0 is true, while SSE/σ^2 has a chi-square distribution with $N-a$ d.f. Moreover, $SSTr$ and SSE are independently distributed. We define the mean squares, $MSTr = SSTr/(a-1)$ and $MSE = SSE/(N-a)$; it is easily seen that $\text{E}(MSE) = \sigma^2$, while $\text{E}(MSTr) = \sigma^2 + \sum_{i=1}^{a} n_i\tau_i^2/(a-1)$.

The ANOVA identity provides us with two alternate estimates of σ^2, one based on the variability within treatments and the other based on the variability between treatments. Specifically, MSE is an estimate of σ^2 and $MSTr$ estimates σ^2 under H_0. If there are significant differences in the treatment means, then $\text{E}(MSTr)$ exceeds σ^2. The F-statistic for testing H_0 compares $MSTr$ with MSE. From the independent chi-square distributions of $SSTr/\sigma^2$ and SSE/σ^2 under H_0, it follows that the ratio for testing H_0 is

$$F(H_0) = \{SSTr/(a-1)\}/\{SSE/(N-a)\} = MSTr/MSE,$$

which is distributed as an $F(a-1, N-a)$-distribution. These details are summarized in Table 7.2.3.

We saw that under H_0, both $MSTr$ and MSE are unbiased estimates of σ^2. Under the alternative hypothesis, however, $\text{E}(MSTr) > \sigma^2$. Hence, if the null hypothesis were false, the expected value of the numerator would be significantly greater than the expectation of the denominator, and consequently, we would reject H_0 for a large value of the test statistic $F(H_0)$. That is, we reject H_0 at $\alpha\%$ level of significance if $F(H_0) > F_{a-1,N-a,\alpha}$, where, $F_{a-1,N-a,\alpha}$ denotes the upper $\alpha\%$ critical point from an $F_{a-1,N-a}$ distribution. Alternatively, in practice, we may compute the p-value of the test as $\text{P}\{F > F(H_0)\}$. Most statistical software give the p-value of this upper-tailed test. If the p-value is less than α, we reject H_0. □

Numerical Example 7.1. One-way fixed-effects ANOVA model. We continue with Numerical Example 4.3 and study the effect of the factor variable group (with three levels, "ctrl", "trt1", "trt2") on the response variable weight for the data set "PlantGrowth". We use the R function `aov()`.

```
fit <- aov(weight ~ group, data = PlantGrowth)
summary(fit)
            Df Sum Sq Mean Sq F value Pr(>F)
group        2  3.766  1.8832   4.846 0.0159 *
Residuals   27 10.492  0.3886
---
Signif. codes:  0 '***' 0.001 '**' 0.01 '*' 0.05 '.' 0.1 ' ' 1

coef(fit)  #OLS coefficients

(Intercept)   grouptrt1   grouptrt2
      5.032      -0.371       0.494
```

The factor group is significant in explaining weight as indicated by the small p-value of 0.016 corresponding to an observed F-stat of 4.85. ▲

7.2.2 Power of the F-test

By definition, the power of the F-test for the general linear hypothesis H at significance level α is

$$P(F(H) > F_{s,N-r,\alpha} \,|\, H \text{ is false }) = \int_{F_{s,N-r,\alpha}}^{\infty} g(x)\,dx,$$

where $g(x)$ is the p.d.f. of the F statistic

$$F(H) = \frac{(SSE_H - SSE)/s}{SSE/(N-r)}.$$

From Result 7.2.1, when the errors are i.i.d. normally distributed with mean 0, $F(H)$ has a noncentral F-distribution. An explicit expression for the p.d.f. of a noncentral F-distribution is given in Result 5.3.6.

It is clear that an evaluation of the power function requires an estimate of the noncentrality parameter λ. If $U \sim \chi^2(k, \lambda)$, it is easy to see that $(U-k)/2$ is an unbiased estimator of λ. However, since it may assume negative values, it is not a useful estimator, and we prefer to use $(U-k)^+/2$. The MLE of λ does not have a closed form and must be computed numerically (Saxena and Alam, 1982).

7.2.3 Testing independent and orthogonal contrasts

Let $\mathbf{c}_i'$ and $\mathbf{c}_j'$ denote two distinct rows of the $p \times s$ matrix $\mathbf{C}$. The quadratic forms

$$Q_i = \boldsymbol{\beta}^{0\prime}\mathbf{c}_i(\mathbf{c}_i'\mathbf{G}\mathbf{c}_i)^{-1}\mathbf{c}_i'\boldsymbol{\beta}^0 = \mathbf{y}'\mathbf{X}\mathbf{G}'\mathbf{c}_i(\mathbf{c}_i'\mathbf{G}\mathbf{c}_i)^{-1}\mathbf{c}_i'\mathbf{G}\mathbf{X}'\mathbf{y}, \text{ and}$$
$$Q_j = \boldsymbol{\beta}^{0\prime}\mathbf{c}_j(\mathbf{c}_j'\mathbf{G}\mathbf{c}_j)^{-1}\mathbf{c}_j'\boldsymbol{\beta}^0 = \mathbf{y}'\mathbf{X}\mathbf{G}'\mathbf{c}_j(\mathbf{c}_j'\mathbf{G}\mathbf{c}_j)^{-1}\mathbf{c}_j'\mathbf{G}\mathbf{X}'\mathbf{y}$$

appear as the numerator sums of squares in the F-statistics for testing H_i: $\mathbf{c}_i'\boldsymbol{\beta} = 0$, and H_j: $\mathbf{c}_j'\boldsymbol{\beta} = 0$. That is,

$$F(H_i) = Q_i \,/\{SSE/(N-r)\} \sim F_{1,N-r}$$

under H_i, so that we reject H_i at level of significance α if $F(H_i) > F_{1,N-r,\alpha}$. A similar result holds for testing H_j. It can be verified that Q_i and Q_j are independently distributed if and only if

$$\mathbf{X}\mathbf{G}'\mathbf{c}_i(\mathbf{c}_i'\mathbf{G}\mathbf{c}_i)^{-1}\mathbf{c}_i'\mathbf{G}\mathbf{X}'\mathbf{X}\mathbf{G}'\mathbf{c}_j(\mathbf{c}_j'\mathbf{G}\mathbf{c}_j)^{-1}\mathbf{c}_j'\mathbf{G}\mathbf{X}' = \mathbf{O},$$

which in turn is true if and only if $\mathbf{c}_i'\mathbf{G}\mathbf{X}'\mathbf{X}\mathbf{G}'\mathbf{c}_j = 0$. Since $\mathbf{c}_j' = \mathbf{t}_j'\mathbf{X}$, the condition is equivalent to $\mathbf{c}_i'\mathbf{G}\mathbf{X}'\mathbf{X}\mathbf{G}'\mathbf{X}'\mathbf{t}_j = 0$. Since $\mathbf{X}'\mathbf{X}\mathbf{G}'\mathbf{X}' = \mathbf{X}'$, this in turn is equivalent to $\mathbf{c}_i'\mathbf{G}\mathbf{X}'\mathbf{t}_j = 0$, i.e.,

$$\mathbf{c}_i'\mathbf{G}\mathbf{c}_j = 0. \tag{7.2.18}$$

Equation (7.2.18) gives a necessary and sufficient condition for independence of Q_i and Q_j. If (7.2.18) holds, we refer to $\mathbf{c}_i'\boldsymbol{\beta}$ and $\mathbf{c}_j'\boldsymbol{\beta}$ as orthogonal contrasts. It can be verified that when $\mathbf{c}_i'\mathbf{G}\mathbf{X}'\mathbf{t}_j = 0$, for $i, j = 1, \cdots, s$, $i \neq j$, then $(\mathbf{C}'\mathbf{G}\mathbf{C})^{-1}$ reduces to a diagonal matrix. The statistic for testing the general linear hypothesis, H: $\mathbf{C}'\boldsymbol{\beta} = \mathbf{d}$, can then be written as

$$\begin{aligned} Q &= \sum_{i=1}^{s} (\mathbf{c}_i'\boldsymbol{\beta}^0 - d_i)'(\mathbf{c}_i'\mathbf{G}\mathbf{c}_i)^{-1}(\mathbf{c}_i'\boldsymbol{\beta}^0 - d_i) \\ &= \sum_{i=1}^{s} \frac{(\mathbf{c}_i'\boldsymbol{\beta}^0 - d_i)^2}{\mathbf{c}_i'\mathbf{G}\mathbf{c}_i} = \sum_{i=1}^{s} Q_i. \end{aligned}$$

7.3 Restricted and reduced models

In Chapter 4, we discussed estimation subject to linear constraints on the parameters. We now describe partitioning of the sums of squares under such constraints. We also discuss reduced models, and reduced, restricted models in detail. These ideas are extremely useful for building a general theory of hypothesis testing in a linear model framework.

We describe the situation where the linear model (4.1.1), i.e., $\mathbf{y} = \mathbf{X}\boldsymbol{\beta} + \boldsymbol{\varepsilon}$, is subject to a "natural" restriction or constraint on the model, i.e., $\mathbf{A}'\boldsymbol{\beta} = \mathbf{b}$, and we wish to test the hypothesis H: $\mathbf{C}'\boldsymbol{\beta} = \mathbf{d}$, i.e., we impose a constraint by hypothesis. We develop inference in the framework of a nested sequence of hypotheses. A few remarks are in order.

Remark 1. In all the discussion up to this point, we have only considered estimable/testable constraints. It is useful to note that in general, a constraint on the model, $\mathbf{A}'\boldsymbol{\beta} = \mathbf{b}$, may not be estimable/testable under the unrestricted model (4.1.1).

Remark 2. From the theory on solutions to linear systems (see Section 3.2), recall that $\mathbf{A}'\boldsymbol{\beta} = \mathbf{b}$ is satisfied if and only if $\boldsymbol{\beta} = \boldsymbol{\beta}^* + [\mathbf{I} - (\mathbf{A}')^-\mathbf{A}']\mathbf{z}$, where $\boldsymbol{\beta}^*$ is a solution to the equation and $\mathbf{z}$ is any vector of the same dimension as $\boldsymbol{\beta}^*$. By substituting this expression for $\boldsymbol{\beta}$, the restricted model can be reformulated as an equivalent, unrestricted model and $\boldsymbol{\beta}$ can be estimated (see Section 3.3). Nevertheless, in many situations, the original parameter $\boldsymbol{\beta}$ has interpretations rooted in the problem under consideration, and the constraint on the model also encodes useful relationships between the components of $\boldsymbol{\beta}$. This makes it desirable or even necessary to use the restricted form of the model, thus giving rise to the issue of hypothesis testing under the restricted model and leading to consideration of a reduced, restricted model.

Remark 3. In many examples, we deal with the *constraints* additionally imposed by a hypothesis H: $\mathbf{C}'\boldsymbol{\beta} = \mathbf{d}$. Although both a constraint on the model and a constraint by the hypothesis are expressed as linear equations in $\boldsymbol{\beta}$, the former is an integral part of a restricted model, while the latter is specifically associated with a hypothesis under which a smaller, or *reduced* model becomes true.

7.3.1 Estimation space and estimability under constraints

Consider a restricted model

$$\mathbf{y} = \mathbf{X}\boldsymbol{\beta} + \boldsymbol{\varepsilon}, \quad \mathbf{A}'\boldsymbol{\beta} = \mathbf{b}, \tag{7.3.1}$$

where $\boldsymbol{\varepsilon}$ satisfies the same conditions as under the regression model (4.1.1). We shall always assume that $\mathbf{A}'\boldsymbol{\beta} = \mathbf{b}$ is consistent, i.e., $\mathbf{b} \in \mathcal{R}(\mathbf{A})$. In particular, if $\mathbf{A}$ has full column rank, then $\boldsymbol{\beta}^* = \mathbf{A}(\mathbf{A}'\mathbf{A})^{-1}\mathbf{b}$ is a special solution.

Estimation of $\boldsymbol{\beta}$ under the restricted model (7.3.1) was discussed in Section 4.6. Here, we focus on the estimation space of the model. This is useful, since a linear model is characterized by its estimation space, as we have seen in Section 4.2. For the model in (7.3.1), the estimation space is

$$\mathcal{S} = \{\mathbf{X}\boldsymbol{\beta}\colon \mathbf{A}'\boldsymbol{\beta} = \mathbf{b}\}, \tag{7.3.2}$$

and the least squares approach provides the vector of fitted values from the restricted model to be

$$\widehat{\mathbf{y}}_r = \arg\min_{\mathbf{z}\in\mathcal{S}} (\mathbf{y} - \mathbf{z})'(\mathbf{y} - \mathbf{z}).$$

From Result 4.6.1, $\widehat{\mathbf{y}}_r$ is unique and

$$\widehat{\mathbf{y}}_r = \mathbf{X}\boldsymbol{\beta}_r^0, \tag{7.3.3}$$

where $\boldsymbol{\beta}_r^0$ is any vector that together with some $\boldsymbol{\lambda}_r^0$ solves the set of normal equations (4.6.2) and (4.6.3). To characterize $\widehat{\mathbf{y}}_r$ from a projection point of view, fix any solution $\boldsymbol{\beta}^*$ to $\mathbf{A}'\boldsymbol{\beta} = \mathbf{b}$ and let $\boldsymbol{\theta} = \boldsymbol{\beta} - \boldsymbol{\beta}^*$. Then, we have

$$\mathcal{S} = \mathbf{X}\boldsymbol{\beta}^* + \omega \quad \text{with} \quad \omega = \{\mathbf{X}\boldsymbol{\theta}\colon \mathbf{A}'\boldsymbol{\theta} = \mathbf{0}\}; \tag{7.3.4}$$

see Section 4.6.2. Since ω is a linear subspace of $\mathcal{R}^N$, we have $\widehat{\mathbf{y}}_r - \mathbf{X}\boldsymbol{\beta}^* = \mathbf{P}_\omega(\mathbf{y} - \mathbf{X}\boldsymbol{\beta}^*)$. Therefore,

$$\widehat{\mathbf{y}}_r = \mathbf{P}_\omega\mathbf{y} + (\mathbf{I} - \mathbf{P}_\omega)\mathbf{X}\boldsymbol{\beta}^*. \tag{7.3.5}$$

Note that, in some situations, despite the constraint on the model in (7.3.1), the estimation space $\mathcal{S}$ under the restricted model may be the same as that of the unrestricted model, i.e., $\mathcal{C}(\mathbf{X})$, so that the LS fit of $\mathbf{y}$ is the same under both models. Such a situation arises when the unrestricted model is not of full rank and a model constraint is specified to uniquely identify the parameter; see Section 7.3.2.

Result 7.3.1. In order for the estimation space $\mathcal{S}$ to be the same as that of the unrestricted model, i.e., $\mathcal{S} = \mathcal{C}(\mathbf{X})$, it is necessary and sufficient that $\mathcal{C}(\mathbf{A}) \cap \mathcal{R}(\mathbf{X}) = \{\mathbf{0}\}$.

Proof. From (7.3.4), $\omega = \mathcal{C}(\mathbf{X}(\mathbf{I} - \mathbf{P}_{\mathcal{C}(\mathbf{A})}))$. Since $\mathcal{S} = \mathbf{X}\boldsymbol{\beta}^* + \omega \subseteq \mathcal{C}(\mathbf{X})$, we see that $\mathcal{S} = \mathcal{C}(\mathbf{X})$ if and only if $\omega = \mathcal{C}(\mathbf{X})$. First, let $\omega = \mathcal{C}(\mathbf{X})$. If $\mathbf{v} \in \mathcal{C}(\mathbf{A}) \cap \mathcal{R}(\mathbf{X})$, then $\mathbf{v} = \mathbf{X}'\mathbf{t}$ for some $\mathbf{t}$ and $\mathbf{P}_{\mathcal{C}(\mathbf{A})}\mathbf{v} = \mathbf{v}$, so that $\mathbf{t}'\mathbf{X}(\mathbf{I} - \mathbf{P}_{\mathcal{C}(\mathbf{A})}) = \mathbf{v}'(\mathbf{I} - \mathbf{P}_{\mathcal{C}(\mathbf{A})}) = \mathbf{0}$. It follows that $\mathbf{t} \perp \omega = \mathcal{C}(\mathbf{X})$, and so $\mathbf{v} = \mathbf{X}'\mathbf{t} = \mathbf{0}$, showing $\mathcal{C}(\mathbf{A}) \cap \mathcal{R}(\mathbf{X}) = \{\mathbf{0}\}$.

Conversely, let $\mathcal{C}(\mathbf{A}) \cap \mathcal{R}(\mathbf{X}) = \{\mathbf{0}\}$. If $\mathbf{v} \perp \omega$, then $\mathbf{v}'\mathbf{X}(\mathbf{I} - \mathbf{P}_{\mathcal{C}(\mathbf{A})}) = \mathbf{0}$, so that $\mathbf{X}'\mathbf{v} = \mathbf{P}_{\mathcal{C}(\mathbf{A})}\mathbf{X}'\mathbf{v}$. Since the left side of the above expression is in $\mathcal{R}(\mathbf{X})$, while the right side is in $\mathcal{C}(\mathbf{A})$, we must have $\mathbf{X}'\mathbf{v} = \mathbf{0}$, so that $\mathbf{v} \perp \mathcal{C}(\mathbf{X})$. Hence $\omega^\perp \subset [\mathcal{C}(\mathbf{X})]^\perp$ and since $\omega \subset \mathcal{C}(\mathbf{X})$, $\omega = \mathcal{C}(\mathbf{X})$. ■

For the general case, let $\mathcal{V}_1 = \mathcal{C}(\mathbf{A}) \cap \mathcal{R}(\mathbf{X})$ and $\mathcal{V}_2$ be a subspace of $\mathcal{C}(\mathbf{A})$ such that $\mathcal{C}(\mathbf{A}) = \mathcal{V}_1 \oplus \mathcal{V}_2$. Note that $\mathcal{V}_2 \cap \mathcal{R}(\mathbf{X}) = \{\mathbf{0}\}$. Each column $\mathbf{a}_i$ of $\mathbf{A}$ has a unique decomposition $\mathbf{a}_{i1} + \mathbf{a}_{i2}$, with $\mathbf{a}_{ij} \in \mathcal{V}_j$, $j = 1, 2$. Let $\mathbf{A}_j$ be the matrix consisting of column vectors $\mathbf{a}_{ij}$, $j = 1, 2$, so that $\mathbf{A} = \mathbf{A}_1 + \mathbf{A}_2$. Then, $\mathcal{C}(\mathbf{A}) = \mathcal{C}(\mathbf{A}_1) \oplus \mathcal{C}(\mathbf{A}_2)$, since the sum of $\mathcal{C}(\mathbf{A}_j)$ contains $\mathcal{C}(\mathbf{A})$, while each is contained in the corresponding $\mathcal{V}_j$. Since $\mathcal{C}(\mathbf{A}_j) \subset \mathcal{C}(\mathbf{A})$, $\mathbf{A}_j = \mathbf{A}\mathbf{R}_j$ for some matrix $\mathbf{R}_j$. As a result, the model constraint $\mathbf{A}'\boldsymbol{\beta} = \mathbf{b}$ can be written as a pair of constraints

$$\mathbf{A}_1'\boldsymbol{\beta} = \mathbf{R}_1'\mathbf{b} = \mathbf{b}_1, \quad \mathbf{A}_2'\boldsymbol{\beta} = \mathbf{R}_2'\mathbf{b} = \mathbf{b}_2;$$

see Result 4.6.5. The point is that $\mathbf{A}_1'\boldsymbol{\beta}$ is estimable while $\mathcal{C}(\mathbf{A}_2) \cap \mathcal{C}(\mathbf{X}) = \{\mathbf{0}\}$. Result 4.6.5 implies that the estimation space of the restricted model (7.3.1) is the same as the one under the restriction $\mathbf{A}_1'\boldsymbol{\beta} = \mathbf{b}_1$. The result was proved by looking into the LS solution of the parameter. Here we restate it as a part of the result below and prove it by only considering the estimation space.

Result 7.3.2. The estimation space $\mathcal{S}$ has the following properties:

1. Define $\mathbf{A}_j$ and $\mathbf{b}_j$ as above. Then $\mathcal{S} = \{\mathbf{X}\boldsymbol{\beta}\colon \mathbf{A}_1'\boldsymbol{\beta} = \mathbf{b}_1\}$ and $\mathcal{C}(\mathbf{A}_1) = \mathcal{C}(\mathbf{A}) \cap \mathcal{R}(\mathbf{X})$.
2. For ω in (7.3.4), $\dim(\omega) = \mathrm{r}(\mathbf{X}) - \dim[\mathcal{C}(\mathbf{A}) \cap \mathcal{R}(\mathbf{X})] = \mathrm{r}(\mathbf{A}, \mathbf{X}') - \mathrm{r}(\mathbf{A})$.

Proof. The discussion above already shows $\mathcal{C}(\mathbf{A}_1) = \mathcal{C}(\mathbf{A}) \cap \mathcal{R}(\mathbf{X})$. Write $\mathbf{A}_1' = \mathbf{T}'\mathbf{X}$ for some matrix $\mathbf{T}$. Clearly, $\mathcal{S} = \{\mathbf{X}\boldsymbol{\beta}\colon \mathbf{A}_1'\boldsymbol{\beta} = \mathbf{b}_1, \mathbf{A}_2'\boldsymbol{\beta} = \mathbf{b}_2\} \subset \{\mathbf{X}\boldsymbol{\beta}\colon \mathbf{A}_1'\boldsymbol{\beta} = \mathbf{b}_1\}$. Write $\mathcal{V} = \mathcal{C}(\mathbf{X})$. Then $\{\mathbf{X}\boldsymbol{\beta}\colon \mathbf{A}_1'\boldsymbol{\beta} = \mathbf{b}_1\} = \{\mathbf{X}\boldsymbol{\beta}\colon \mathbf{T}'\mathbf{X}\boldsymbol{\beta} = \mathbf{b}_1\} = \{\mathbf{y} \in \mathcal{V}\colon \mathbf{T}'\mathbf{y} = \mathbf{b}_1\}$. From Result 7.3.1, $\mathcal{V} = \{\mathbf{X}\boldsymbol{\beta}\colon \mathbf{A}_2'\boldsymbol{\beta} = \mathbf{b}_2\}$. Then $\{\mathbf{y} \in \mathcal{V}\colon \mathbf{T}'\mathbf{y} = \mathbf{b}_1\} = \{\mathbf{y} \in \{\mathbf{X}\boldsymbol{\beta}\colon \mathbf{A}_2'\boldsymbol{\beta} = \mathbf{b}_2\}\colon \mathbf{T}'\mathbf{y} = \mathbf{b}_1\} = \{\mathbf{X}\boldsymbol{\beta}\colon \mathbf{A}_2'\boldsymbol{\beta} = \mathbf{b}_2, \mathbf{T}'\mathbf{X}\boldsymbol{\beta} = \mathbf{b}_1\} = \{\mathbf{X}\boldsymbol{\beta}\colon \mathbf{A}_1'\boldsymbol{\beta} = \mathbf{b}_1, \mathbf{A}_2'\boldsymbol{\beta} = \mathbf{b}_2\} = \mathcal{S}$. This completes the proof of property 1. Next, by applying property 1 to $\mathbf{b} = \mathbf{0}$, $\omega = \{\mathbf{X}\boldsymbol{\theta}\colon \mathbf{T}'\mathbf{X}\boldsymbol{\theta} = \mathbf{0}\} = \{\mathbf{X}\boldsymbol{\theta}\colon \mathbf{T}'\mathbf{P}_{\mathcal{C}(\mathbf{X})}'\mathbf{X}\boldsymbol{\theta} = \mathbf{0}\} = \{\mathbf{y} \in \mathcal{C}(\mathbf{X})\colon \mathbf{y} \perp \mathcal{C}(\mathbf{P}_{\mathcal{C}(\mathbf{X})}\mathbf{T})\}$. Since $\mathcal{C}(\mathbf{P}_{\mathcal{C}(\mathbf{X})}\mathbf{T})\} \subset \mathcal{C}(\mathbf{X})$, then

$$\dim(\omega) = \dim[\mathcal{C}(\mathbf{X})] - \dim[\mathcal{C}(\mathbf{P}_{\mathcal{C}(\mathbf{X})}\mathbf{T})\}] = \mathrm{r}(\mathbf{X}) - \mathrm{r}(\mathbf{P}_{\mathcal{C}(\mathbf{X})}\mathbf{T}).$$

We see that $\mathcal{R}(\mathbf{P}_{\mathcal{C}(\mathbf{X})}\mathbf{T}) = \mathcal{R}(\mathbf{X}'\mathbf{T})$. Then $\mathrm{r}(\mathbf{P}_{\mathcal{C}(\mathbf{X})}\mathbf{T}) = \mathrm{r}(\mathbf{X}'\mathbf{T}) = \mathrm{r}(\mathbf{A}_1)$, and from property 1, the first equality in property 2 holds. Since $\mathcal{C}(\mathbf{A}, \mathbf{X}') = \mathcal{C}(\mathbf{A}) + \mathcal{R}(\mathbf{X}) = \mathcal{C}(\mathbf{A}_2) \oplus \mathcal{R}(\mathbf{X})$, $\mathrm{r}(\mathbf{A}, \mathbf{X}') = \mathrm{r}(\mathbf{X}) + \mathrm{r}(\mathbf{A}_2)$. Similarly, $\mathrm{r}(\mathbf{A}) = \mathrm{r}(\mathbf{A}_1) + \mathrm{r}(\mathbf{A}_2)$, showing the second equality in property 2. ■

We now consider estimability and testability under the restricted model (7.3.1).

Definition 7.3.1. A linear function $\mathbf{c}'\boldsymbol{\beta}$ is said to be estimable under the restricted model (7.3.1) if there are constant vectors $\mathbf{s}$ and $\mathbf{t}$, such that for any $\boldsymbol{\beta}$ satisfying the model constraint in (7.3.1), $\mathbf{s}'\mathbf{b} + \mathrm{E}(\mathbf{t}'\mathbf{y}) = \mathbf{c}'\boldsymbol{\beta}$.

Note the extra $\mathbf{s}'\mathbf{b}$ in comparison to Definition 4.3.1 for an estimable function under an unrestricted model. This is because, with $\mathbf{t}$ taking values in $\mathcal{R}^N$, $\mathrm{E}(\mathbf{t}'\mathbf{y})$ may be zero; however, the set of solutions to $\mathbf{A}'\boldsymbol{\beta} = \mathbf{b}$ is an affine space that may not contain $\mathbf{0}$, so that $\mathbf{c}'\boldsymbol{\beta}$ may never be zero on the space. The term $\mathbf{s}'\mathbf{b}$ is needed to compensate for this.

Result 7.3.3. The function $\mathbf{c}'\boldsymbol{\beta}$ is estimable under the restricted model (7.3.1) if and only if any one of the following properties holds:

1. $\mathbf{c} \in \mathcal{R}(\mathbf{X}) + \mathcal{C}(\mathbf{A}) = \mathcal{C}(\mathbf{A}, \mathbf{X}')$. In particular, if $\mathbf{c}'\boldsymbol{\beta}$ is estimable under the unrestricted model (4.1.1), then it is estimable under the restricted model.
2. $\mathbf{c}' = \mathbf{c}'(\mathbf{X}'\mathbf{X} + \mathbf{A}\mathbf{A}')^-(\mathbf{X}'\mathbf{X} + \mathbf{A}\mathbf{A}')$.

3. $\mathbf{c}'\boldsymbol{\beta}_r^0$ is invariant to the solutions to the set of normal equations (4.6.2) and (4.6.3).

Proof. If $\mathbf{c}'\boldsymbol{\beta}$ is estimable under the model (7.3.1), then from Definition 7.3.1, there are constant vectors $\mathbf{s}$ and $\mathbf{t}$, such that $\mathbf{s}'\mathbf{b} + \mathbf{t}'\mathbf{X}\boldsymbol{\beta} = \mathbf{c}'\boldsymbol{\beta}$ for any $\boldsymbol{\beta}$ satisfying $\mathbf{A}'\boldsymbol{\beta} = \mathbf{b}$. Fix any $\boldsymbol{\beta}^*$ satisfying the model constraint. Then $\mathbf{t}'\mathbf{X}(\boldsymbol{\beta} - \boldsymbol{\beta}^*) = \mathbf{c}'(\boldsymbol{\beta} - \boldsymbol{\beta}^*)$ for any $\boldsymbol{\beta}$ satisfying $\mathbf{A}'(\boldsymbol{\beta} - \boldsymbol{\beta}^*) = \mathbf{0}$. As a result, for any vector $\mathbf{u} \perp \mathcal{C}(\mathbf{A})$, $\mathbf{u} \perp (\mathbf{c} - \mathbf{X}'\mathbf{t})$, and hence $\mathbf{c} - \mathbf{X}'\mathbf{t} \in \mathcal{C}(\mathbf{A})$. This shows $\mathbf{c} \in \mathcal{R}(\mathbf{X}) + \mathcal{C}(\mathbf{A})$. Conversely, suppose $\mathbf{c} = \mathbf{X}'\mathbf{t} + \mathbf{A}\mathbf{s}$. Then under the restricted model, $\mathrm{E}(\mathbf{t}'\mathbf{y}) = \mathbf{t}'\mathbf{X}\boldsymbol{\beta} = (\mathbf{t}'\mathbf{X} + \mathbf{s}'\mathbf{A}')\boldsymbol{\beta} - \mathbf{s}'\mathbf{A}'\boldsymbol{\beta} = \mathbf{c}'\boldsymbol{\beta} - \mathbf{s}'\mathbf{b}$, so by definition, $\mathbf{c}'\boldsymbol{\beta}$ is estimable under the model (7.3.1). This shows that the estimability of $\mathbf{c}'\boldsymbol{\beta}$ is equivalent to property 1. Since the property can be written as $\mathbf{c} \in \mathcal{R}(\mathbf{M})$, where $\mathbf{M} = (\mathbf{X}', \mathbf{A})'$, from property 4 of Result 4.3.1, it is equivalent to $\mathbf{c}' = \mathbf{c}'(\mathbf{M}'\mathbf{M})^-(\mathbf{M}'\mathbf{M})$, which is property 2. If $\mathbf{c} \in \mathcal{R}(\mathbf{X}) + \mathcal{C}(\mathbf{A})$, then $\mathbf{c}' = \mathbf{t}'\mathbf{X} + \mathbf{u}'\mathbf{A}'$ for some vectors $\mathbf{t}$ and $\mathbf{u}$. Suppose $(\boldsymbol{\beta}_r^0, \boldsymbol{\lambda}_r^0)$ is a solution to (4.6.2) and (4.6.3). Then $\mathbf{c}'\boldsymbol{\beta}_r^0 = (\mathbf{t}'\mathbf{X} + \mathbf{u}'\mathbf{A}')\boldsymbol{\beta}_r^0 = \mathbf{t}'\mathbf{X}\boldsymbol{\beta}_r^0$, which is invariant to $(\boldsymbol{\beta}_r^0, \boldsymbol{\lambda}_r^0)$ by Result 4.6.1. Conversely, let $\mathbf{c}'\boldsymbol{\beta}_r^0$ have property 3. For any $\mathbf{u} \perp \mathcal{R}(\mathbf{X}) + \mathcal{C}(\mathbf{A})$, by $\mathbf{X}\mathbf{u} = \mathbf{0}$ and $\mathbf{A}'\mathbf{u} = \mathbf{0}$, it is easy to see that $(\boldsymbol{\beta}_r^0 + \mathbf{u}, \boldsymbol{\lambda}_r^0)$ is also a solution. Then the invariance implies $\mathbf{c}'\mathbf{u} = 0$, i.e., $\mathbf{c} \perp \mathbf{u}$. Since this holds for all $\mathbf{u} \perp \mathcal{R}(\mathbf{X}) + \mathcal{C}(\mathbf{A})$, then $\mathbf{c} \in \mathcal{R}(\mathbf{X}) + \mathcal{C}(\mathbf{A})$. Hence property 3 is equivalent to property 1. ■

Consider a hypothesis H: $\mathbf{C}'\boldsymbol{\beta} = \mathbf{d}$. It is trivially false if it cannot be satisfied by any $\boldsymbol{\beta}$ that satisfies the model constraint in (7.3.1). In other words, the hypothesis is not trivially false if $\mathbf{D}'\boldsymbol{\beta} = \mathbf{f}$ is consistent, where $\mathbf{D} = (\mathbf{A}, \mathbf{C})$ and $\mathbf{f}' = (\mathbf{b}', \mathbf{d}')$. We will only consider hypotheses that are not trivially false.

Definition 7.3.2. Reduced, restricted model. A hypothesis H: $\mathbf{C}'\boldsymbol{\beta} = \mathbf{d}$ is said to be testable under the restricted model (7.3.1) if, under the model, $\mathbf{C}'\boldsymbol{\beta}$ is estimable and H is not trivially false. As we saw before, the restricted model that is further subject to the constraints imposed by H will be called a reduced, restricted model.

For many inference problems, a hypothesis is worth considering only if it reduces the estimation space. Denote by $A \subsetneq B$ or $B \supsetneq A$ if A is a strict subset of B, or equivalently, if B is a strict superset of A.

Result 7.3.4. Let H: $\mathbf{C}'\boldsymbol{\beta} = \mathbf{d}$ be a testable hypothesis under the restricted model (7.3.1). Let $\mathcal{S}_1$ denote the estimation space of the reduced model under H, i.e., $\mathcal{S}_1 = \{\mathbf{X}\boldsymbol{\beta}\text{: } \mathbf{A}'\boldsymbol{\beta} = \mathbf{b}, \mathbf{C}'\boldsymbol{\beta} = \mathbf{d}\}$. Then $\mathcal{S}_1 \subsetneq \mathcal{S}$ if and only if $\mathcal{C}(\mathbf{A}, \mathbf{C}) \cap \mathcal{R}(\mathbf{X}) \supsetneq \mathcal{C}(\mathbf{A}) \cap \mathcal{R}(\mathbf{X})$.

Proof. Fix $\boldsymbol{\beta}^* \in \mathcal{S}_1$. Then $\mathcal{S}_1 = \mathbf{X}\boldsymbol{\beta}^* + \omega_1$ and $\mathcal{S} = \mathbf{X}\boldsymbol{\beta}^* + \omega$, where $\omega_1 = \{\mathbf{X}\boldsymbol{\theta}\text{: } \mathbf{A}'\boldsymbol{\theta} = \mathbf{0}, \mathbf{C}'\boldsymbol{\theta} = \mathbf{0}\}$ and $\omega = \{\mathbf{X}\boldsymbol{\theta}\text{: } \mathbf{A}'\boldsymbol{\theta} = \mathbf{0}\}$. Clearly, $\omega_1 \subseteq \omega$ and $\mathcal{C}(\mathbf{A}, \mathbf{C}) \cap \mathcal{R}(\mathbf{X}) \supseteq \mathcal{C}(\mathbf{A}) \cap \mathcal{R}(\mathbf{X})$. From Result 7.3.2, $\dim(\omega_1) = \mathrm{r}(\mathbf{X}) - \dim[\mathcal{C}(\mathbf{A}, \mathbf{C}) \cap \mathcal{R}(\mathbf{X})]$ and $\dim(\omega) = \mathrm{r}(\mathbf{X}) - \dim[\mathcal{C}(\mathbf{A}) \cap \mathcal{R}(\mathbf{X})]$. Then $\omega_1 \neq \omega$ if and only if $\dim[\mathcal{C}(\mathbf{A}, \mathbf{C}) \cap \mathcal{R}(\mathbf{X})] > \dim[\mathcal{C}(\mathbf{A}) \cap \mathcal{R}(\mathbf{X})]$, which is equivalent to $\mathcal{C}(\mathbf{A}, \mathbf{C}) \cap \mathcal{R}(\mathbf{X}) \supsetneq \mathcal{C}(\mathbf{A}) \cap \mathcal{R}(\mathbf{X})$. ■

The next result will be useful.

Result 7.3.5. Suppose in the restricted model (7.3.1), $\mathcal{C}(\mathbf{A}) \cap \mathcal{R}(\mathbf{X}) = \{\mathbf{0}\}$ and H: $\mathbf{C}'\boldsymbol{\beta} = \mathbf{d}$ is testable so that $\mathbf{C}' = \mathbf{A}\mathbf{S} + \mathbf{X}'\mathbf{T}$ for some unique matrices $\mathbf{S}$ and $\mathbf{T}$. Then the estimation space of the reduced, restricted model under H is $\{\mathbf{X}\boldsymbol{\beta}\text{: } \mathbf{T}'\mathbf{X}\boldsymbol{\beta} = \mathbf{d} - \mathbf{S}'\mathbf{b}\}$.

Proof. The reduced, restricted model is subject to $\mathbf{A}'\boldsymbol{\beta} = \mathbf{b}$ and $\mathbf{C}'\boldsymbol{\beta} = \mathbf{S}'\mathbf{A}'\boldsymbol{\beta} + \mathbf{T}'\mathbf{X}\boldsymbol{\beta} = \mathbf{d}$, which can be combined into a single form $\mathbf{B}'\boldsymbol{\beta} = \mathbf{f}$, where $\mathbf{B} = (\mathbf{A}, \mathbf{T}'\mathbf{X})$ and $\mathbf{f} = (\mathbf{b}', \mathbf{d}' - \mathbf{b}'\mathbf{S})'$. Then $\mathbf{B} = \mathbf{B}_1 + \mathbf{B}_2$, where $\mathbf{B}_1 = (\mathbf{O}, \mathbf{T}'\mathbf{X}) = \mathbf{B} \cdot \mathrm{diag}(\mathbf{O}, \mathbf{I})$ and $\mathbf{B}_2 = (\mathbf{A}, \mathbf{O})$. From $\mathcal{C}(\mathbf{A}) \cap \mathcal{R}(\mathbf{X}) = \{\mathbf{0}\}$ and Result 7.3.1, the estimation space of the reduced model is $\{\mathbf{X}\boldsymbol{\beta}: \mathbf{B}_1'\boldsymbol{\beta} = \text{diag}(\mathbf{O}, \mathbf{I})\mathbf{f}\}$, which is the same as the one claimed in the result. ■

Example 7.3.1. Consider the model $\mathbf{y} = \boldsymbol{\beta} + \boldsymbol{\varepsilon}$ with $\boldsymbol{\beta} = (\beta_1, \beta_2, \beta_3, \beta_4)'$ satisfying the restriction H_0: $\sum_{i=1}^4 \beta_i = \mathbf{1}'\boldsymbol{\beta} = 0$, where $\boldsymbol{\varepsilon} \sim N_4(\mathbf{0}, \sigma^2\mathbf{I})$. We derive the test statistic for H: $\beta_1 = \beta_2$ under the restricted model. The LS estimates subject to H_0 are

$$\widehat{\mathbf{y}}_{H_0} = \widehat{\boldsymbol{\beta}}_{H_0} = (Y_1 - \overline{Y}, Y_2 - \overline{Y}, Y_3 - \overline{Y}, Y_4 - \overline{Y})',$$

where $\overline{Y} = \sum_{i=1}^4 Y_i/4$. On the other hand, the LS estimates subject to both H_0 and the hypothesis H are

$$\widehat{\mathbf{y}}_{H_1} = \widehat{\boldsymbol{\beta}}_{H_1} = ((Y_1 + Y_2)/2 - \overline{Y}, (Y_1 + Y_2)/2 - \overline{Y}, Y_3 - \overline{Y}, Y_4 - \overline{Y})'.$$

Then $SSE_{H_0} = (\mathbf{y} - \widehat{\mathbf{y}}_{H_1})'(\mathbf{y} - \widehat{\mathbf{y}}_{H_1}) = (\sum Y_i)^2/4$ and $Q = (\widehat{\mathbf{y}}_{H_0} - \widehat{\mathbf{y}}_{H_1})'(\widehat{\mathbf{y}}_{H_0} - \widehat{\mathbf{y}}_{H_1}) = (Y_1 - Y_2)^2/2$, so that $F = \{2(Y_1 - Y_2)^2\}/\{(\sum Y_i)^2\} \sim F_{1,1}$ under H. □

The reduced model setup enables us to discuss consequences of model misspecification on solution vectors and predictions. We start with an example that shows a general result.

Example 7.3.2. Consider the linear model

$$\mathbf{y} = \mathbf{X}_1\boldsymbol{\beta}_1 + \mathbf{X}_2\boldsymbol{\beta}_2 + \boldsymbol{\varepsilon}, \tag{7.3.6}$$

where $\text{E}(\boldsymbol{\varepsilon}) = \mathbf{0}$ and $\text{Cov}(\boldsymbol{\varepsilon}) = \sigma^2\mathbf{I}_N$. Let $r = \text{r}(\mathbf{X}_1, \mathbf{X}_2)$ and $r_i = \text{r}(\mathbf{X}_i)$, $i = 1, 2$. It can be shown that for the null hypothesis H: $\boldsymbol{\beta}_1 = \mathbf{0}$,

$$\text{E}(Q) = \text{E}(SSE_H - SSE) = \sigma^2(r - r_2) + \boldsymbol{\beta}_1'\mathbf{X}_1'(\mathbf{I}_N - \mathbf{P}_2)\mathbf{X}_1\boldsymbol{\beta}_1.$$

The formula holds even when H is not testable under the full model; the proof of this is left as Exercise 7.10. □

Example 7.3.3. We describe hypothesis tests in three situations that correspond to imposing a model constraint on the (unrestricted) one-way ANOVA model (4.1.7). We use the solution vector $\boldsymbol{\beta}^0 = (0, \overline{Y}_{1\cdot}, \cdots, \overline{Y}_{a\cdot})'$, and corresponding g-inverse $\mathbf{G} = \begin{pmatrix} 0 & \mathbf{0} \\ \mathbf{0} & \mathbf{D} \end{pmatrix}$, where $\mathbf{D} = \text{diag}(1/n_1, \cdots, 1/n_a)$.

1. The test for H: $\mu = 0$ in the model (4.1.7) subject to the model constraint $\sum_{i=1}^a n_i\tau_i = 0$ is equivalent to testing H: $\mu + N^{-1}\sum_{i=1}^a n_i\tau_i = 0$ in the unrestricted model, and the test statistic $SSM/\widehat{\sigma}^2$ follows an $F_{1,N-a}$ distribution under H.

2. The test for H: $\mu = 0$ in the model (4.1.7) subject to the model constraint $\sum_{i=1}^a \tau_i = 0$ is equivalent to testing H: $\mu + \sum_{i=1}^a \tau_i = 0$ in the unrestricted model. The test statistic $[\sum_{i=1}^a \overline{Y}_{i\cdot}]^2/[\widehat{\sigma}^2 \sum_{i=1}^a (1/n_i)]$ follows an $F_{1,N-a}$ distribution under H.

3. The test for H: $\mu = 0$ in the model (4.1.7) subject to the model constraint $\sum_{i=1}^a c_i\tau_i = 0$, where $c_1, \cdots, c_a$ are constants satisfying $\sum_{i=1}^a c_i = 1$, is equivalent to testing H: $\mu + \sum_{i=1}^a c_i\tau_i = 0$ in the unrestricted model, and the test statistic $[\sum_{i=1}^a c_i\overline{Y}_{i\cdot}]^2/[\widehat{\sigma}^2 \sum_{i=1}^a (c_i^2/n_i)]$ follows an $F_{1,N-a}$ distribution under H.

These results follow from Result 7.3.5. They can also be easily verified directly. We discuss situation 1 in detail. Recall that $\mu + \tau_1, \cdots, \mu + \tau_a$ are all estimable. The function $\mu + N^{-1}\sum_{i=1}^a n_i\tau_i$ is therefore estimable under the unrestricted model, and yields μ upon imposing the model constraint $\sum_{i=1}^a n_i\tau_i = 0$. Therefore, testing H: $\mu + N^{-1}\sum_{i=1}^a n_i\tau_i = 0$ in the unrestricted model is the same as testing H: $\mu = 0$ in the model with the model constraint $\sum_{i=1}^a n_i\tau_i = 0$. We show two approaches for computing the F-statistic in case 1.

(a) In the notation of this section, the hypothesis H: $\mu + N^{-1}\sum_{i=1}^{a} n_i\tau_i = 0$ can be written as H: $\mathbf{c}'\boldsymbol{\beta} = 0$, where $\mathbf{c} = (N, n_1, \cdots, n_a)'$ is an $(a+1)$-dimensional vector. By Result 7.2.1, $F = Q/\widehat{\sigma}^2 \sim F_{1,N-a}$ under H, where $Q = \boldsymbol{\beta}^{0\prime}\mathbf{c}(\mathbf{c}'\mathbf{G}\mathbf{c})^{-1}\mathbf{c}'\boldsymbol{\beta}^0 = N\overline{Y}_{..}^2 = SSM$.

(b) From Example 4.2.5, for the model (4.1.7) subject to the model constraint $\sum_{i=1}^{a}\tau_i = 0$, the solution vector is $\boldsymbol{\beta}^0 = (\overline{Y}_{..}, \overline{Y}_{1\cdot} - \overline{Y}_{..}, \cdots, \overline{Y}_{a\cdot} - \overline{Y}_{..})$ and

$$SSE = \sum_{i=1}^{a}\sum_{j=1}^{n_i}(Y_{ij} - \overline{Y}_{..} - \overline{Y}_{i\cdot} + \overline{Y}_{..})^2 = \sum_{i=1}^{a}\sum_{j=1}^{n_i} Y_{ij}^2 - \sum_{i=1}^{a} n_i\overline{Y}_{i\cdot}^2.$$

To compute SSE_H, note that under H: $\mu = 0$, the model (4.1.7) becomes

$$Y_{ij} = \tau_i + \varepsilon_{ij},\ j = 1, \ldots, n_i,\ i = 1, \ldots, a \text{ s.t. } \sum_{i=1}^{a} n_i\tau_i = 0.$$

We minimize

$$\mathcal{L}(\tau_1, \cdots, \tau_a, \lambda) = \sum_{i=1}^{a}\sum_{j=1}^{n_i}(Y_{ij} - \tau_i)^2 + 2\lambda\sum_{i=1}^{a} n_i\tau_i$$

with respect to $\tau_1, \cdots, \tau_a$ and λ to get $\boldsymbol{\beta}^0 = (0, \overline{Y}_{1\cdot} - \overline{Y}_{..}, \cdots, \overline{Y}_{a\cdot} - \overline{Y}_{..})'$ and $\lambda^0 = \overline{Y}_{..}$. Then,

$$SSE_H = \sum_{i=1}^{a}\sum_{j=1}^{n_i}(Y_{ij} - \overline{Y}_{i\cdot} + \overline{Y}_{..})^2,$$

and, $Q = SSE_H - SSE = N\overline{Y}_{..}^2$, leading to the F-statistic given above.

To show cases 2 and 3, we again write each hypothesis under the unrestricted model as H: $\mathbf{c}'\boldsymbol{\beta} = 0$, where $\mathbf{c}'$ is the vector $(1, 1/a, \cdots, 1/a)$ for case 2, and the vector $(1, c_1, \cdots, c_a)$ for case 3. In each case, the F-test statistic follows from Result 7.2.1 or using (7.2.9).

This example shows that when the constraint on the model is not an estimable function, an interesting situation develops. Under different restrictions, the same parameter (here, μ) is estimable, although it is non-estimable in the unrestricted model. However, its b.l.u.e. is different under the different restrictions; in fact, its b.l.u.e. under a particular restriction is equal to the b.l.u.e. of that estimable function in the unrestricted model from which we obtain under the restriction. In each case, we obtain a different F-statistic for testing H: $\mu = 0$. □

7.3.2 Nested sequence of models or hypotheses

Consider a set of constraints $\mathbf{C}_j'\boldsymbol{\beta} = \mathbf{d}_j$, $j = 1, \ldots, L$, on the parameter $\boldsymbol{\beta}$ in the restricted model (7.3.1). Assume that all $\mathbf{C}_j'\boldsymbol{\beta}$ are estimable under the restricted model, so that $\mathcal{C}(\mathbf{C}_j) \subset \mathcal{C}(\mathbf{A}, \mathbf{X}')$, and there is at least one $\boldsymbol{\beta}^*$ satisfying all these constraints as well as the model constraint in (7.3.1). Furthermore, assume that

$$\mathcal{C}(\mathbf{A}, \mathbf{C}_1, \cdots, \mathbf{C}_L) = \mathcal{C}(\mathbf{A}, \mathbf{X}') \tag{7.3.7}$$

and for $l = 1, \ldots, L$

$$\mathcal{C}(\mathbf{A}, \mathbf{C}_1, \cdots, \mathbf{C}_{l-1}) \cap \mathcal{R}(\mathbf{X}) \subsetneq \mathcal{C}(\mathbf{A}, \mathbf{C}_1, \cdots, \mathbf{C}_l) \cap \mathcal{R}(\mathbf{X}). \tag{7.3.8}$$

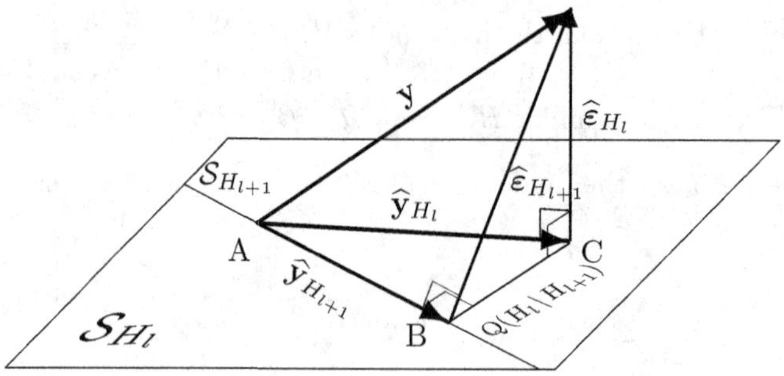

FIGURE 7.3.1. Nested sequence of hypotheses.

Since (4.1.1) is equivalent to $\tilde{\mathbf{y}} = \mathbf{X}\tilde{\boldsymbol{\beta}} + \boldsymbol{\varepsilon}$ with $\tilde{\mathbf{y}} = \mathbf{y} - \boldsymbol{\beta}^*$ and $\tilde{\boldsymbol{\beta}} = \boldsymbol{\beta} - \boldsymbol{\beta}^*$ and the model constraint and constraints by hypotheses on $\boldsymbol{\beta}$ are equivalent to $\mathbf{A}\tilde{\boldsymbol{\beta}} = \mathbf{0}$ and $\mathbf{C}_j'\tilde{\boldsymbol{\beta}} = \mathbf{0}$, respectively, we can and will assume without loss of generality that $\mathbf{b} = \mathbf{0}$ and $\mathbf{d}_j = \mathbf{0}$. With a slight abuse of notation, let H_0 denote the model constraint $\mathbf{A}'\boldsymbol{\beta} = \mathbf{0}$ and $\mathcal{S}_{H_0} = \{\mathbf{X}\boldsymbol{\beta}\colon \mathbf{A}'\boldsymbol{\beta} = \mathbf{0}\}$ denote the estimation space of the restricted model (7.3.1).

For $l = 1, 2, \cdots, L$, let H_l denote the constraints by hypotheses H_l: $\mathbf{C}_j'\boldsymbol{\beta} = \mathbf{0}$, $j = 1, \ldots, l$, and $\mathcal{S}_{H_l}$ denote the estimation space of the reduced, restricted model under H_l, i.e., $\mathcal{S}_{H_l} = \{\mathbf{X}\boldsymbol{\beta}\colon \mathbf{A}'\boldsymbol{\beta} = \mathbf{0}, \mathbf{C}_j'\boldsymbol{\beta} = \mathbf{0}, j = 1, \cdots, l\}$. From (7.3.7), $\mathcal{S}_{H_L} = \{\mathbf{0}\}$. From (7.3.8) and Result 7.3.4,

$$\mathcal{S}_{H_0} \supsetneq \mathcal{S}_{H_1} \supsetneq \cdots \supsetneq \mathcal{S}_{H_l} \supsetneq \cdots \supsetneq \mathcal{S}_{H_L} = \{\mathbf{0}\}.$$

Correspondingly, we obtain a nested sequence of reduced, restricted models $\mathbf{y} = \mathbf{X}\boldsymbol{\beta} + \boldsymbol{\varepsilon}$, $\mathbf{A}'\boldsymbol{\beta} = \mathbf{0}$, $\mathbf{C}_l'\boldsymbol{\beta} = \mathbf{0}$, $l = 1, \ldots, L$, going from the model (7.3.1) down to a model where $\mathbf{y} = \boldsymbol{\varepsilon}$. We will also use $\mathcal{S}_{H_l}$ to denote these nested models.

Denote by $\mathbf{P}_{H_l}$ the projection of $\mathbf{y}$ onto $\mathcal{S}_{H_l}$. For each l, the restricted LS fit under H_l is $\widehat{\mathbf{y}}_{H_l} = \mathbf{P}_{H_l}\mathbf{y}$, and the model SSR and SSE are, respectively,

$$Q(H_l) = \widehat{\mathbf{y}}_{H_l}'\widehat{\mathbf{y}}_{H_l} = \mathbf{y}'\mathbf{P}_{H_l}\mathbf{y}, \text{ and}$$
$$SSE_{H_l} = (\mathbf{y} - \widehat{\mathbf{y}}_{H_l})'(\mathbf{y} - \widehat{\mathbf{y}}_{H_l}) = \mathbf{y}'(\mathbf{I} - \mathbf{P}_{H_l})\mathbf{y}.$$

As l increases, SSR decreases, with $Q(H_0) = \mathbf{y}'\mathbf{P}_{H_0}\mathbf{y}$ being the full model SSR and $Q(H_L) = 0$. Consider the lth and $(l+1)$th stages of the nesting. We denote by $Q(H_l \mid H_{l+1})$ the model SSR due to the fit under H_l beyond that obtained by fitting (4.1.1) under H_{l+1}. In other words, $Q(H_l \mid H_{l+1})$ is the SSR of H_l adjusted for H_{l+1}, and is defined to be

$$Q(H_l \mid H_{l+1}) = Q(H_l) - Q(H_{l+1}) = SSE_{H_{l+1}} - SSE_{H_l}. \tag{7.3.9}$$

The following partition of SST clearly holds,

$$\begin{aligned}
SST &= Q(H_0) + SSE_{H_0} \\
&= Q(H_1) + Q(H_0 \mid H_1) + SSE_{H_0} \\
&\cdots\cdots\cdots \\
&= \sum_{l=0}^{L-1} Q(H_l \mid H_{l+1}) + SSE_{H_0}.
\end{aligned}$$

The quantity $Q(H_l \mid H_{l+1})$ tests whether $\mathbf{C}_{l+1}'\boldsymbol{\beta} = \mathbf{0}$, given that $\mathbf{A}'\boldsymbol{\beta} = \mathbf{0}$ and $\mathbf{C}_j'\boldsymbol{\beta} = \mathbf{0}$ for all $j \leq l$. Note that the fitted model has more parameters under H_l than under H_{l+1}. In the triangle in Figure 7.3.1 with vertices A, B, C, we see that $\|AB\|$ denotes the length of the vector $\widehat{\mathbf{y}}_{H_l}$, so that $\|AB\|^2 = \widehat{\mathbf{y}}'_{H_l}\widehat{\mathbf{y}}_{H_l} = \mathbf{y}'\mathbf{P}_{H_l}\mathbf{y} = Q(H_l)$. Similarly, $\|AC\|$ denotes the length of $\widehat{\mathbf{y}}_{H_{l+1}}$, so that $\|AC\|^2 = \widehat{\mathbf{y}}'_{H_{l+1}}\widehat{\mathbf{y}}_{H_{l+1}} = \mathbf{y}'\mathbf{P}_{H_{l+1}}\mathbf{y} = Q(H_{l+1})$. From Result 2.6.6, $\|CB\|$ is the length of

$$\widehat{\mathbf{y}}_{H_l} - \widehat{\mathbf{y}}_{H_{l+1}} = (\mathbf{P}_{H_l} - \mathbf{P}_{H_{l+1}})\mathbf{y} = \mathbf{R}_{l+1}\mathbf{y}, \tag{7.3.10}$$

where $\mathbf{R}_{l+1}$ is the orthogonal projection onto $\mathcal{S}_{H_l} \cap \mathcal{S}^{\perp}_{H_{l+1}}$, so that $\widehat{\mathbf{y}}_{H_l} - \widehat{\mathbf{y}}_{H_{l+1}} \perp \mathbf{P}_{H_{l+1}}\mathbf{y} = \widehat{\mathbf{y}}_{H_{l+1}}$. Hence,

$$\|CB\|^2 = \|AB\|^2 - \|AC\|^2 = Q(H_l) - Q(H_{l+1}) = Q(H_l \mid H_{l+1})$$

and we can write

$$Q(H_l \mid H_{l+1}) = (\widehat{\mathbf{y}}_{H_l} - \widehat{\mathbf{y}}_{H_{l+1}})'(\widehat{\mathbf{y}}_{H_l} - \widehat{\mathbf{y}}_{H_{l+1}}). \tag{7.3.11}$$

From (7.3.10),

$$\widehat{\mathbf{y}}_{H_l} - \widehat{\mathbf{y}}_{H_{l+1}} = (\boldsymbol{\mu}_{H_l} - \boldsymbol{\mu}_{H_{l+1}}) + \mathbf{R}_{l+1}\boldsymbol{\varepsilon}, \quad \text{where} \quad \boldsymbol{\mu}_{H_l} = \mathbf{P}_{H_l}\mathbf{X}\boldsymbol{\beta}.$$

Let $\boldsymbol{\varepsilon} \sim N(\mathbf{0}, \sigma^2\mathbf{I}_N)$. From Result 5.4.4,

$$\begin{aligned} &Q(H_l \mid H_{l+1})/\sigma^2 \sim \chi^2(s_{l+1}, \lambda_{l+1}) \quad \text{with} \\ &s_{l+1} = \dim(\mathcal{S}_{H_l}) - \dim(\mathcal{S}_{H_{l+1}}), \\ &\lambda_{l+1} = (\boldsymbol{\mu}_{H_l} - \boldsymbol{\mu}_{H_l+1})'(\boldsymbol{\mu}_{H_l} - \boldsymbol{\mu}_{H_l+1})/2\sigma^2. \end{aligned} \tag{7.3.12}$$

When H_{l+1} holds for $\boldsymbol{\beta}$, the noncentrality parameter is zero and

$$\{Q(H_l \mid H_{l+1})/s_{l+1}\}/\{SSE_{H_0}/(N-r)\} \sim F_{s_{l+1}, N-r} \tag{7.3.13}$$

which is the usual extra sum of squares F-test in linear model applications. When the extra d.f. is equal to 1, we call it a "partial F-test".

Example 7.3.4. We describe in detail an application to the fixed-effects two-way model (4.2.32) with interaction effects for the balanced case. As in Example 4.2.6, $\mathbf{y} = \mathbf{X}\boldsymbol{\beta} + \boldsymbol{\varepsilon}$, where $\mathbf{y} = (\mathbf{y}'_{11}, \cdots, \mathbf{y}'_{1b}, \cdots, \mathbf{y}'_{a1}, \cdots, \mathbf{y}'_{ab})'$ with $\mathbf{y}_{ij} = (Y_{ij1}, \cdots, Y_{ijn})'$, $\boldsymbol{\varepsilon}$ is written in a similar way, $\mathbf{X}$ is given in (4.2.33), and $\boldsymbol{\beta} = (\mu, \boldsymbol{\tau}', \boldsymbol{\theta}', (\boldsymbol{\tau\theta})')'$, with $\boldsymbol{\tau} = (\tau_1, \cdots, \tau_a)'$, $\boldsymbol{\theta} = (\theta_1, \cdots, \theta_b)'$, and $(\boldsymbol{\tau\theta}) = ((\tau\theta)_{11}, \cdots (\tau\theta)_{1b}, \cdots, (\tau\theta)_{a1}, \cdots, (\tau\theta)_{ab})'$. The matrix $\mathbf{X}$ is $N \times p$ with rank ab, where $N = nab$ and $p = 1 + a + b + ab$. Its row space is the same as that of

$$\mathbf{M} = \begin{pmatrix} \mathbf{1}_b & \mathbf{1}_b & \mathbf{0} & \cdots & \mathbf{0} & \mathbf{I}_b & \mathbf{I}_b & \mathbf{0} & \cdots & \mathbf{0} \\ \mathbf{1}_b & \mathbf{0} & \mathbf{1}_b & \cdots & \mathbf{0} & \mathbf{I}_b & \mathbf{0} & \mathbf{I}_b & \cdots & \mathbf{0} \\ \vdots & \vdots & \vdots & \ddots & \vdots & \vdots & \vdots & \vdots & \ddots & \vdots \\ \mathbf{1}_b & \mathbf{0} & \mathbf{0} & \cdots & \mathbf{1}_b & \mathbf{I}_b & \mathbf{0} & \mathbf{0} & \cdots & \mathbf{I}_b \end{pmatrix},$$

because $\mathbf{M}$ consists of all the different rows of $\mathbf{X}$ without repetition. Consider the restricted model under the model constraints

$$\sum_{i=1}^{a} \tau_i = \sum_{j=1}^{b} \theta_j = 0, \ \sum_{i=1}^{a} (\tau\theta)_{ij} = 0, \ 1 \le j \le b, \ \sum_{j=1}^{b} (\tau\theta)_{ij} = 0, \ 1 \le i \le a. \tag{7.3.14}$$

In matrix form, these constraints can be written as $\mathbf{A}'\boldsymbol{\beta} = \mathbf{0}$, where the column vectors of $\mathbf{A}$ are

$$(0, \mathbf{1}'_a, \mathbf{0}'_b, \underbrace{\mathbf{0}'_b, \cdots, \mathbf{0}'_b}_{a \text{ repeats}})' = (0, \mathbf{1}'_a, \mathbf{0}'_b, \mathbf{1}'_a \otimes \mathbf{0}'_b)',$$
$$(0, \mathbf{0}'_a, \mathbf{1}'_b, \mathbf{0}'_b, \cdots, \mathbf{0}'_b)' = (0, \mathbf{0}'_a, \mathbf{1}'_b, \mathbf{1}'_a \otimes \mathbf{0}'_b)',$$
$$(0, \mathbf{0}'_a, \mathbf{0}'_b, \mathbf{e}'_j, \cdots, \mathbf{e}'_j)' = (0, \mathbf{0}'_a, \mathbf{0}'_b, \mathbf{1}'_a \otimes \mathbf{e}'_j)', \quad 1 \le j \le b,$$
$$(0, \mathbf{0}'_a, \mathbf{0}'_b, \underbrace{\mathbf{0}'_b, \cdots, \mathbf{0}'_b}_{i-1 \text{ repeats}}, \mathbf{1}'_b, \mathbf{0}'_b, \cdots, \mathbf{0}'_b)', \quad 1 \le i \le a,$$

with $\mathbf{e}_j$ denoting the jth standard basis vector of $\mathcal{R}^b$. We see that $\mathcal{R}(\mathbf{X}) = \mathcal{R}(\mathbf{M}) = \{(\sum_{i=1}^a \mathbf{1}'_b\mathbf{u}_i, \mathbf{1}'_b\mathbf{u}_1, \cdots, \mathbf{1}'_b\mathbf{u}_a, \sum_{i=1}^a \mathbf{u}'_i, \mathbf{u}'_1, \cdots, \mathbf{u}'_a)',\ \mathbf{u}_i \in \mathcal{R}^b\}$ and $\mathcal{C}(\mathbf{A}) = \{(0, s\mathbf{1}'_a, t\mathbf{1}'_b, \mathbf{v}' + d_1\mathbf{1}'_b, \cdots, \mathbf{v}' + d_a\mathbf{1}'_b)',\ s, t, d_i \in \mathcal{R}, \mathbf{v} \in \mathcal{R}^b\}$, so that $\mathcal{R}(\mathbf{X}) \cap \mathcal{C}(\mathbf{A}) = \{\mathbf{0}\}$. From Result 7.3.1, the estimation space of the restricted model is the same as that of the unrestricted model. Next, the column vectors have one linear dependency, so that $\mathrm{r}(\mathbf{A}) = a + b + 1$. Hence $\mathcal{R}(\mathbf{X}) \oplus \mathcal{C}(\mathbf{A})$ has dimension $ab + (a+b+1) = p$, so is equal to $\mathcal{R}^p$. Then, from Result 7.3.3 and Definition 7.3.2, all linear hypotheses under the restricted model are testable.

Consider the following hypotheses, H_1: $(\tau\theta)_{ij}$ are all equal, H_2: $\theta_1 = \cdots = \theta_b$, H_3: $\tau_1 = \cdots = \tau_a$, and H_4: $\mu = 0$. As a passing remark, under the unrestricted model, none of the hypotheses is testable because each can be written as $\mathbf{C}'\boldsymbol{\beta} = \mathbf{0}$ with $\mathcal{C}(\mathbf{C}) \cap \mathcal{R}(\mathbf{X}) = \{\mathbf{0}\}$.

As the constraints imposed by H_l, $1 \le l \le 4$, are added on in sequel, the restricted two-way model with interaction is reduced step-by-step to the restricted two-way additive model, the restricted one-way ANOVA model, the mean only model, and finally, the noise only model $\mathbf{y} = \boldsymbol{\varepsilon}$. Starting with the full model, i.e., the restricted two-way model with interaction, this sequence of nested models can be written as follows:

(i) $\mathcal{S}_{H_0} = \{Y_{ijk} = \mu + \tau_i + \theta_j + (\tau\theta)_{ij} + \varepsilon_{ijk},\ \sum_{i=1}^a \tau_i = 0,\ \sum_{j=1}^b \theta_j = 0,\ \sum_{i=1}^a (\tau\theta)_{ij} = 0,\ 1 \le j \le b,\ \sum_{j=1}^b (\tau\theta)_{ij} = 0,\ 1 \le i \le a\}$; $\dim(\mathcal{S}_{H_0}) = ab$;

(ii) $\mathcal{S}_{H_1} = \{Y_{ijk} = \mu + \tau_i + \theta_j + \varepsilon_{ijk},\ \sum_{i=1}^a \tau_i = 0,\ \sum_{j=1}^b \theta_j = 0\}$; $\dim(\mathcal{S}_{H_1}) = a + b - 1$;

(iii) $\mathcal{S}_{H_2} = \{Y_{ijk} = \mu + \tau_i + \varepsilon_{ijk},\ \sum_{i=1}^a \tau_i = 0\}$; $\dim(\mathcal{S}_{H_2}) = a$;

(iv) $\mathcal{S}_{H_3} = \{Y_{ijk} = \mu + \varepsilon_{ijk}\}$; $\dim(\mathcal{S}_{H_3}) = 1$;

(v) $\mathcal{S}_{H_4} = \{Y_{ijk} = \varepsilon_{ijk}\}$; $\dim(\mathcal{S}_{H_4}) = 0$.

Denote the vectors of fitted values under the models by $\widehat{\mathbf{y}}_{H_l} = \{\widehat{Y}_{ijk,H_l}\}$. Generally speaking, vectors of fitted values under nested models do not have explicit relations among them and have to be derived individually. However, the restricted two-way model has an orthogonality property that simplifies the derivation. Let $\mathbf{X}_A = \mathrm{diag}(\mathbf{1}_{nb}, \cdots, \mathbf{1}_{nb})$, $\mathbf{X}_B = (\mathbf{L}_b, \cdots, \mathbf{L}_b)'$, $\mathbf{X}_{AB} = \mathrm{diag}(\mathbf{L}_b, \cdots, \mathbf{L}_b)$, where in each case, the number of repeats is a. Then $\mathbf{X} = (\mathbf{1}_N, \mathbf{X}_A, \mathbf{X}_B, \mathbf{X}_{AB})$ and $\mathbf{X}\boldsymbol{\beta} = \mu\mathbf{1}_N + \mathbf{X}_A\boldsymbol{\tau} + \mathbf{X}_B\boldsymbol{\theta} + \mathbf{X}_{AB}(\boldsymbol{\tau\theta})$. Let

$$\mathcal{V}_M = \mathrm{Span}(\mathbf{1}_N),$$
$$\mathcal{V}_A = \{\mathbf{X}_A\boldsymbol{\tau}\colon \boldsymbol{\tau}'\mathbf{1}_a = 0\},$$
$$\mathcal{V}_B = \{\mathbf{X}_B\boldsymbol{\theta}\colon \boldsymbol{\theta}'\mathbf{1}_b = 0\},$$
$$\mathcal{V}_{AB} = \{\mathbf{X}_{AB}(\boldsymbol{\tau\theta})\colon \textstyle\sum_{i=1}^a (\tau\theta)_{ij} = 0, 1 \le j \le b, \sum_{j=1}^b (\tau\theta)_{ij} = 0, 1 \le i \le a\}.$$

Then the estimation space of each nested model is either the sum or a partial sum of the above spaces. The critical observation is that $\mathcal{V}_M$, $\mathcal{V}_A$, $\mathcal{V}_B$, $\mathcal{V}_{AB}$ are orthogonal to each other, which is not hard to verify. For example, to show $\mathcal{V}_A \perp \mathcal{V}_{AB}$, any $\mathbf{X}_A\boldsymbol{\tau} \in \mathcal{V}_A$ and $\mathbf{X}_{AB}(\boldsymbol{\tau\theta}) \in \mathcal{V}_{AB}$ have inner product $\boldsymbol{\tau}'\mathbf{X}_A'\mathbf{X}_{AB}(\boldsymbol{\tau\theta})$. From $\mathbf{1}_{nb}'\mathbf{L}_b = n^2\mathbf{1}_b$, $\mathbf{X}_A'\mathbf{X}_{AB} = n^2\operatorname{diag}(\mathbf{1}_b, \cdots, \mathbf{1}_b)$, where $\mathbf{1}_b$ is repeated a times. Write $(\boldsymbol{\tau\theta}) = ((\boldsymbol{\tau\theta})_1', \cdots, (\boldsymbol{\tau\theta})_a')'$ with $(\boldsymbol{\tau\theta})_i = ((\tau\theta)_{i1}, \cdots, (\tau\theta)_{ib})'$. Then $\mathbf{1}_b'(\boldsymbol{\tau\theta})_i = 0$, giving $\mathbf{X}_A'\mathbf{X}_{AB}(\boldsymbol{\tau\theta}) = n^2(\mathbf{1}_b'(\boldsymbol{\tau\theta})_1, \cdots, \mathbf{1}_b'(\boldsymbol{\tau\theta})_a)' = \mathbf{0}$. Thus $(\mathbf{X}_A\boldsymbol{\tau})'(\mathbf{X}_{AB}(\boldsymbol{\tau\theta})) = 0$, giving $\mathbf{X}_A\boldsymbol{\tau} \perp \mathbf{X}_{AB}(\boldsymbol{\tau\theta})$. Because both vectors are arbitrary, this gives $\mathcal{V}_A \perp \mathcal{V}_{AB}$.

Thus, the estimation spaces of the nested models, still denoted by $\mathcal{S}_{H_l}$, can be decomposed as

$$\mathcal{S}_{H_0} = \mathcal{V}_M \oplus \mathcal{V}_A \oplus \mathcal{V}_B \oplus \mathcal{V}_{AB}, \quad \mathcal{S}_{H_1} = \mathcal{V}_M \oplus \mathcal{V}_A \oplus \mathcal{V}_B,$$
$$\mathcal{S}_{H_2} = \mathcal{V}_M \oplus \mathcal{V}_A, \quad \mathcal{S}_{H_3} = \mathcal{V}_M, \quad \mathcal{S}_{H_4} = \{\mathbf{0}\}. \tag{7.3.15}$$

and the orthogonal projection of $\mathbf{y}$ onto an estimation space is equal to the sum of the orthogonal projections of $\mathbf{y}$ onto its component $\mathcal{V}$'s. From Example 4.2.6, $\mathbf{X}\beta^0 = \mu^0\mathbf{1}_N + \mathbf{X}_A\boldsymbol{\tau}^0 + \mathbf{X}_B\boldsymbol{\theta}^0 + \mathbf{X}_{AB}(\boldsymbol{\tau\theta})^0$ is the orthogonal projection of $\mathbf{y}$ onto $\mathcal{S}_{H_0}$. It follows that the terms in the sum are projections of $\mathbf{y}$ onto $\mathcal{V}_M$, $\mathcal{V}_A$, $\mathcal{V}_B$, and $\mathcal{V}_{AB}$, respectively. As a result, $\widehat{\mathbf{y}}_{H_1} = \mu^0\mathbf{1}_N + \mathbf{X}_A\boldsymbol{\tau}^0 + \mathbf{X}_B\boldsymbol{\theta}^0$, $\widehat{\mathbf{y}}_{H_2} = \mu^0\mathbf{1}_N + \mathbf{X}_A\boldsymbol{\tau}^0$, and $\widehat{\mathbf{y}}_{H_3} = \mu^0\mathbf{1}_N$. Clearly, $\widehat{\mathbf{y}}_{H_4} = \mathbf{0}$. Component-wise,

$$\widehat{Y}_{ijk,H_0} = \mu^0 + \tau_i^0 + \theta_j^0 + (\tau\theta)_{ij}^0 = \overline{Y}_{ij\cdot},$$
$$\widehat{Y}_{ijk,H_1} = \mu^0 + \tau_i^0 + \theta_j^0 = \overline{Y}_{i\cdot\cdot} + \overline{Y}_{\cdot j\cdot} - \overline{Y}_{\cdots},$$
$$\widehat{Y}_{ijk,H_2} = \mu^0 + \tau_i^0 = \overline{Y}_{i\cdot\cdot}, \quad \widehat{Y}_{ijk,H_3} = \mu^0 = \overline{Y}_{\cdots}.$$

We now consider the test statistics of the hypothesis. To start, the total sum of squares (SST), the SSR of the full model $\mathcal{S}_{H_0}$ and its SSE are

$$SST = \mathbf{y}'\mathbf{y} = \sum_{i=1}^a \sum_{j=1}^b \sum_{k=1}^n Y_{ijk}^2,$$
$$Q(H_0) = \widehat{\mathbf{y}}_{H_0}'\widehat{\mathbf{y}}_{H_0} = n\sum_{i=1}^a \sum_{j=1}^b \overline{Y}_{ij\cdot}^2, \text{ and}$$
$$SSE = SST - Q(H_0) = \sum_{i=1}^a \sum_{j=1}^b \sum_{k=1}^n (Y_{ijk} - \overline{Y}_{ij\cdot})^2.$$

Let ε_{ijk} be i.i.d. $\sim N(0, \sigma^2)$. Then under the full model, $Q(H_0)/\sigma^2 \sim \chi^2_{ab}$, while $SSE/\sigma^2 \sim \chi^2_{ab(n-1)}$. Next,

$$Q(H_1) = \widehat{\mathbf{y}}_{H_1}'\widehat{\mathbf{y}}_{H_1} = n\sum_{i=1}^a \sum_{j=1}^b (\overline{Y}_{i\cdot\cdot} + \overline{Y}_{\cdot j\cdot} - \overline{Y}_{\cdots})^2,$$

and it follows that the *interaction sum of squares* is

$$\begin{aligned} Q(H_0 \mid H_1) &= (\widehat{\mathbf{y}}_{H_1} - \widehat{\mathbf{y}}_{H_0})'(\widehat{\mathbf{y}}_{H_1} - \widehat{\mathbf{y}}_{H_0}) \\ &= n\sum_{i=1}^a \sum_{j=1}^b (\overline{Y}_{ij\cdot} - \overline{Y}_{i\cdot\cdot} - \overline{Y}_{\cdot j\cdot} + \overline{Y}_{\cdots})^2, \end{aligned}$$

Since $\dim(\mathcal{S}_0) - \dim(\mathcal{S}_1) = (a-1)(b-1)$, from (7.3.12) and (7.3.13), under $\mathcal{S}_{H_1}$,

$Q(H_0 \,|\, H_1)/\sigma^2 \sim \chi^2_{(a-1)(b-1)}$ and

$$F(H_1) = \frac{Q(H_0 \,|\, H_1)/(a-1)(b-1)}{SSE/ab(n-1)} \sim F_{(a-1)(b-1),ab(n-1)}.$$

We see that H_2 corresponds to a hypothesis of no difference among b levels of Factor B. Although H_2 is testable under the full model, from the point of view of model interpretation, it makes more sense to test H_2 under a model without interaction, i.e., under $\mathcal{S}_{H_1}$. We have

$$Q(H_1 \,|\, H_2) := (\widehat{\mathbf{y}}_{H_1} - \widehat{\mathbf{y}}_{H_2})'(\widehat{\mathbf{y}}_{H_1} - \widehat{\mathbf{y}}_{H_2})$$
$$= \sum_{i=1}^{a}\sum_{j=1}^{b}\sum_{k=1}^{n}(\overline{Y}_{\cdot j\cdot} - \overline{Y}_{\dots})^2 = an\sum_{j=1}^{b}(\overline{Y}_{\cdot j\cdot} - \overline{Y}_{\dots})^2.$$

Since $\dim(\mathcal{S}_{H_1}) - \dim(\mathcal{S}_{H_2}) = b-1$, from (7.3.12) and (7.3.13), under $\mathcal{S}_{H_2}$, $Q(H_1 \,|\, H_2)/\sigma^2 \sim \chi^2_{b-1}$ and

$$F(H_2) = \frac{Q(H_1 \,|\, H_2)/(b-1)}{SSE/ab(n-1)} \sim F_{(b-1),ab(n-1)}.$$

Third, H_3 corresponds to the hypothesis of no difference among a levels of Factor A. Suppose we wish to test H_3 after H_1 and H_2 are accepted. Then

$$Q(H_2 \,|\, H_3) := (\widehat{\mathbf{y}}_{H_2} - \widehat{\mathbf{y}}_{H_3})'(\widehat{\mathbf{y}}_{H_2} - \widehat{\mathbf{y}}_{H_3})$$
$$= \sum_{i=1}^{a}\sum_{j=1}^{b}\sum_{k=1}^{n}(\overline{Y}_{i\cdot\cdot} - \overline{Y}_{\dots})^2 = bn\sum_{i=1}^{a}(\overline{Y}_{i\cdot\cdot} - \overline{Y}_{\dots})^2.$$

Since $\dim(\mathcal{S}_{H_2}) - \dim(\mathcal{S}_{H_3}) = a-1$, from (7.3.12) and (7.3.13), under $\mathcal{S}_{H_3}$, $Q(H_2 \,|\, H_3)/\sigma^2 \sim \chi^2_{a-1}$ and

$$F(H_3) = \frac{Q(H_2 \,|\, H_3)/(a-1)}{SSE/ab(n-1)} \sim F_{a-1,ab(n-1)}.$$

Finally, from $\widehat{\mathbf{y}}_{H_4} = \mathbf{0}$,

$$Q(H_3 \,|\, H_4) = Q(H_3) = \widehat{\mathbf{y}}'_{H_3}\widehat{\mathbf{y}}_{H_3} = \sum_{i=1}^{a}\sum_{j=1}^{b}\sum_{k=1}^{n}\overline{Y}^2_{\dots} = nab\overline{Y}^2_{\dots},$$

so that under $\mathcal{S}_{H_4}$, $Q(H_3 \,|\, H_4) \sim \chi^2_1$ and

$$F(H_4) = nab\overline{Y}^2_{\dots}/\{SSE/ab(n-1)\} \sim F_{1,ab(n-1)}.$$

Since $SST - nab\overline{Y}^2_{\dots} = SST_c = \sum_{ijk}(Y_{ijk} - \overline{Y}_{\dots})^2$,

$$SST_c = n\sum_{i=1}^{a}\sum_{j=1}^{b}(\overline{Y}_{ij\cdot} - \overline{Y}_{i\cdot\cdot} - \overline{Y}_{\cdot j\cdot} + \overline{Y}_{\dots})^2$$
$$+ an\sum_{j=1}^{b}(\overline{Y}_{\cdot j\cdot} - \overline{Y}_{\dots})^2. + bn\sum_{i=1}^{a}(\overline{Y}_{i\cdot\cdot} - \overline{Y}_{\dots})^2.$$

From the interpretation of the tests underlying the decomposition, the sums of squares on

Source	d.f.	SS
Mean μ	1	$SS_M = abn\overline{Y}_{\dots}^2$
τ after μ	$a-1$	$SS_A = nb\sum_i(\overline{Y}_{i\cdot\cdot} - \overline{Y}_{\dots})^2$
θ after (μ, τ)	$b-1$	$SS_B = na\sum_j(\overline{Y}_{\cdot j\cdot} - \overline{Y}_{\dots})^2$
$(\tau\theta)$ after (μ, τ, θ)	$(a-1)(b-1)$	$SS_{AB} = n\sum_{ij}(\overline{Y}_{ij\cdot} - \overline{Y}_{i\cdot\cdot} - \overline{Y}_{\cdot j\cdot} + \overline{Y}_{\dots})^2$
Error	$ab(n-1)$	$SSE = \sum_{ijk}(Y_{ijk} - \overline{Y}_{ij\cdot})^2$
Total	nab	$SST = \sum_{ijk} Y_{ijk}^2$

TABLE 7.3.1. ANOVA table for two-factor model with interaction.

the right side are referred to as the SS due to interaction, additive effect due to Factor B, and additive effect due to Factor A, respectively denoted by SS_{AB}, SS_B, and SS_A. Thus we have the following ANOVA decomposition

$$SST_c = SS_A + SS_B + SS_{AB} + SSE,$$

which is shown in Table 7.3.1.

Remark 1. The forms of $Q(H_\ell \,|\, H_{\ell+1})$ and their interpretations critically depend on the model constraints (7.3.14), and may not hold under other constraints. For example, if the model constraints are changed to $\tau_1 = \cdots = \tau_a = \theta_1 = \cdots = \theta_b = \sum_{i=1}^a \sum_{j=1}^b (\tau\theta)_{ij} = 0$, the restricted model still has the same estimation space as the unrestricted one. However, if the hypotheses H_1, H_2, H_3, and H_4 are tested in sequel, since now $\mathcal{S}_{H_0} = \{Y_{ijk} = \mu + (\tau\theta)_{ij} + \varepsilon_{ijk}, \sum_{i=1}^a \sum_{j=1}^b (\tau\theta)_{ij} = 0\}$ and $\mathcal{S}_{H_1} = \mathcal{S}_{H_2} = \mathcal{S}_{H_3} = \{Y_{ijk} = \mu + \varepsilon_{ijk}\}$, it is clear that $Q(H_0 \,|\, H_1) = SST_c > SS_{AB}$ and $Q(H_1 \,|\, H_2) = Q(H_2 \,|\, H_3) = 0$, so they cannot be reasonably interpreted as the SS due to the interaction between Factors A and B, or as additive effects due to the individual factors.

Remark 2. For the balanced restricted model in the example, because the nested estimation spaces have the orthogonal decomposition (7.3.15), SS_A, SS_B, and SS_{AB} are sums of squares of the projections of $\mathbf{y}$ onto $\mathcal{V}_A$, $\mathcal{V}_B$, and $\mathcal{V}_{AB}$, respectively. We say that the τ effects, θ effects, and $(\tau\theta)$ effects are *orthogonal*. A consequence is that the decomposition of SST_c is independent of the order in which the hypotheses are tested. That is, if, say H_2 is tested first, H_1 the second, H_4 the third, and H_3 the last, then $Q(H_0 \,|\, H_2)$, $Q(H_2 \,|\, H_1)$, and $Q(H_4 \,|\, H_3)$ are still SS_B, SS_{AB}, and SS_A, respectively. The stability against the order of tests makes these sums of squares rather attractive in application besides being intuitive. Unfortunately, this property does not hold in the unbalanced case, as we discuss in Chapter 10. Moreover, for most (restricted) models and hypotheses, the decomposition of SST_c depends on the order of tests. □

Example 7.3.5. We now consider the fixed-effects additive two-way model (4.2.30). Consider two hypotheses, H_A: $\tau_1 = \cdots = \tau_a$ and H_B: $\theta_1 = \cdots = \theta_b$. It is easy to see that, in contrast to the two-way model with interaction in Example 7.3.4, both hypotheses are testable under the unrestricted model. On the other hand, as in the two-way model with interaction, in the balanced design, the τ effects and θ effects are orthogonal, but in the unbalanced design, this property does not hold, as we discuss in Chapter 10.

Consider the balanced case where $n_{ij} = n$ for all $1 \le i \le a$ and $1 \le j \le b$. By testing H_A and H_B in sequel, the ANOVA decomposition for the model is

$$SST_c = SS_A + SS_B + SSE,$$

Source	d.f.	SS
Mean μ	1	$SS_M = abn\overline{Y}^2_{...}$
τ after μ	$a-1$	$SS_A = nb\sum_i(\overline{Y}_{i..} - \overline{Y}_{...})^2$
θ after (μ,τ)	$b-1$	$SS_B = na\sum_j(\overline{Y}_{.j.} - \overline{Y}_{...})^2$
Error	$nab-a-b+1$	$SSE = \sum_{ijk}(Y_{ijk} - \overline{Y}_{i..} - \overline{Y}_{.j.} + \overline{Y}_{...})^2$
Total	nab	$SST = \sum_{ijk} Y^2_{ijk}$

TABLE 7.3.2. ANOVA table for the two-factor main effects model.

where

$$SST_c = \sum_{k=1}^{n}\sum_{i=1}^{a}\sum_{j=1}^{b}(Y_{ijk} - \overline{Y}_{...})^2 \text{ with } (nab-1) \text{ d.f.}$$

$$SS_A = Q(H_0 \mid H_A) = nb\sum_{i=1}^{a}(\overline{Y}_{i..} - \overline{Y}_{...})^2 \text{ with } (a-1) \text{ d.f.}$$

$$SS_B = Q(H_A \mid H_B) = na\sum_{j=1}^{b}(\overline{Y}_{.j.} - \overline{Y}_{...})^2 \text{ with } (b-1) \text{ d.f.}$$

$$SSE = \sum_{k=1}^{n}\sum_{i=1}^{a}\sum_{j=1}^{b}(Y_{ijk} - \overline{Y}_{i..} - \overline{Y}_{.j.} + \overline{Y}_{...})^2 \text{ with } (nab-a-b+1) \text{ d.f.}$$

From (7.3.13), the F-statistic for H_A is

$$F(H_A) = \frac{MS_A}{MSE} = \frac{SS_A/(a-1)}{SSE/(nab-a-b-1)}$$

which has an $F_{a-1,nab-a-b+1}$ distribution under H_A, and the F-statistic for H_B is

$$F(H_B) = \frac{MS_B}{MSE} = \frac{SS_A/(b-1)}{SSE/(nab-a-b-1)}$$

which has an $F_{b-1,nab-a-b+1}$ distribution under H_B. □

Numerical Example 7.2. Two-way additive model. The "ToothGrowth" data is from a study evaluating the effect of vitamin C on tooth growth in Guinea pigs. The experiment was performed on 60 pigs, where each animal received one of three dose levels of vitamin C by one of two delivery methods, (orange juice or ascorbic acid, a form of vitamin C). The dataset is available in the R package *datasets*. The response variable Y is the tooth length. The two factor variables are X_1 (delivery method with two levels) and X_2 (dose, with three levels). We show code and results for fitting a two-way model without interaction between the factors.

```
attach(ToothGrowth)
dose <-
  factor(dose,
         levels = c(0.5, 1, 2),
         labels = c("D0.5", "D1", "D2"))
fit <- aov(len ~ supp + dose, data = ToothGrowth)
summary(fit)
```

```
            Df Sum Sq Mean Sq F value    Pr(>F)
supp         1  205.4   205.4   11.45   0.0013 **
dose         1 2224.3  2224.3  123.99 6.31e-16 ***
Residuals   57 1022.6    17.9
---
Signif. codes:  0 ‘***’ 0.001 ‘**’ 0.01 ‘*’ 0.05 ‘.’ 0.1 ‘ ’ 1
```

Both the main effects are significant, as seen from large values of the F-statistics and small p-values. We can obtain concise and detailed least squares solutions of the coefficients of the fit as follows.

```
coef(fit)   # coefficients
(Intercept)      suppVC        dose
   9.272500   -3.700000    9.763571

summary(lm(len ~ supp + dose, data = ToothGrowth))

Call:
lm(formula = len ~ supp + dose, data = ToothGrowth)

Residuals:
   Min      1Q Median      3Q     Max
-6.600 -3.700  0.373  2.116  8.800

Coefficients:
            Estimate Std. Error t value Pr(>|t|)
(Intercept)   9.2725     1.2824   7.231 1.31e-09 ***
suppVC       -3.7000     1.0936  -3.383   0.0013 **
dose          9.7636     0.8768  11.135 6.31e-16 ***
---
Signif. codes:  0 ‘***’ 0.001 ‘**’ 0.01 ‘*’ 0.05 ‘.’ 0.1 ‘ ’ 1

Residual standard error: 4.236 on 57 degrees of freedom
Multiple R-squared:  0.7038,Adjusted R-squared:  0.6934
F-statistic: 67.72 on 2 and 57 DF,  p-value: 8.716e-16
```

The overall F-stat value is 67.7 and the p-value is 8.72e-16. ▲

Numerical Example 7.3. Two-way model with interaction. The data "weightgain" in the R package *HSAUR* arise from an experiment to study the gain in weight (in gms) of rats (Y) fed on four different diets, distinguished by the "source" of protein (source, $X1$) with two levels (beef, or cereal) and the "amount" of protein (type, $X2$) at two levels, (low, or high).

```
library(HSAUR)
data(weightgain)
fit <- aov(weightgain ~ source*type,data=weightgain)
summary(fit)

 Df Sum Sq Mean Sq F value Pr(>F)
source       1    221   220.9   0.988 0.3269
type         1   1300  1299.6   5.812 0.0211 *
source:type  1    884   883.6   3.952 0.0545 .
```

```
Residuals   36   8049   223.6
---
Signif. codes:  0 '***' 0.001 '**' 0.01 '*' 0.05 '.' 0.1 ' ' 1
```

We notice that while the interaction effect is (barely) significant at the 5% level and the type of feed is significant, the source of feed is not. The least squares coefficients are shown below.

```
summary(lm(fit))$coefficients

                    Estimate Std. Error   t value     Pr(>|t|)
(Intercept)            100.0   4.728577 21.148009 6.842420e-22
sourceCereal           -14.1   6.687218 -2.108500 4.201233e-02
typeLow                -20.8   6.687218 -3.110411 3.644273e-03
sourceCereal:typeLow    18.8   9.457155  1.987913 5.446757e-02
```

▲

Example 7.3.6. Two-way nested or hierarchical model. In Example 4.2.7, we derived the least squares solutions of the parameters in the balanced model. Following the same method in Example 7.3.4, the ANOVA decomposition is

$$SST_c = SS_A + SS_{B(A)} + SSE,$$

where

$$\begin{aligned} SST_c &= \sum_{i=1}^{a}\sum_{j=1}^{b}\sum_{k=1}^{n}(Y_{ijk} - \overline{Y}_{...})^2 \text{ with } (abn-1) \text{ d.f.} \\ SS_A &= nb\sum_{i=1}^{a}(\overline{Y}_{i..} - \overline{Y}_{...})^2 \text{ with } (a-1) \text{ d.f.} \\ SS_{B(A)} &= n\sum_{i=1}^{a}\sum_{j=1}^{b}(\overline{Y}_{ij.} - \overline{Y}_{i..})^2 \text{ with } a(b-1) \text{ d.f.} \\ SSE &= \sum_{i=1}^{a}\sum_{j=1}^{b}\sum_{k=1}^{n}(Y_{ijk} - \overline{Y}_{ij.})^2 \text{ with } ab(n-1) \text{ d.f.} \end{aligned}$$

The ANOVA table is shown in Table 7.3.3. The expected mean squares are

$$\begin{aligned} E(MS_A) &= \sigma^2 + \frac{bn\sum_{i=1}^{a}\tau_i^2}{a-1}, \\ E(MS_{B(A)}) &= \sigma^2 + \frac{n\sum_{i=1}^{a}\sum_{j=1}^{b}\theta_{j(i)}^2}{a(b-1)}, \\ E(MSE) &= \sigma^2. \end{aligned}$$

The F-test statistic for H_A: $\tau_1 = \cdots = \tau_a$ is $F(H_A) = MS_A/MSE$, which has an $F_{a-1,ab(n-1)}$ distribution under H_A. The test statistic for $H_{B(A)}$: all $\theta_{j(i)}$ are equal, $1 \le j \le b$, $1 \le i \le a$, is $F(H_{B(A)}) = MS_{B(A)}/MSE$, which has an $F_{a(b-1),ab(n-1)}$ under $H_{B(A)}$. This setup may be extended to higher order nested models in a similar way. □

Example 7.3.7. Consider the balanced model with nested and crossed factors:

$$Y_{ijkl} = \mu + \tau_i + \theta_j + \eta_{k(j)} + (\tau\theta)_{ij} + (\tau\eta)_{ik(j)} + \varepsilon_{(ijk)l}, \tag{7.3.16}$$

Source	SS	d.f.	MS
Factor A	SS_A	$a-1$	MS_A
Factor B (within Factor A)	$SS_{B(A)}$	$a(b-1)$	$MS_{B(A)}$
Error	SSE	$ab(n-1)$	MSE
Total	SST	$abn-1$	

TABLE 7.3.3. ANOVA table for Example 7.3.6.

$1 \le i \le a$, $1 \le j \le b$, $1 \le k \le c$, $1 \le l \le n$. It may be verified that the ANOVA decomposition is

$$SST = SS_A + SS_B + SS_{C(B)} + SS_{AB} + SS_{AC(B)} + SSE, \tag{7.3.17}$$

where

$$SST = \sum_{i=1}^{a}\sum_{j=1}^{b}\sum_{k=1}^{c}\sum_{l=1}^{n}(Y_{ijkl} - \overline{Y}_{....})^2 \text{ with } (abcn-1) \text{ d.f.}$$

$$SS_A = \sum_{i=1}^{a}\frac{Y_{i...}^2}{bcn} - \frac{Y_{....}^2}{abcn} \text{ with } (a-1) \text{ d.f.}$$

$$SS_B = \sum_{j=1}^{b}\frac{Y_{\cdot j..}^2}{acn} - \frac{Y_{....}^2}{abcn} \text{ with } (b-1) \text{ d.f.}$$

$$SS_{C(B)} = \sum_{j=1}^{b}\sum_{k=1}^{c}\frac{Y_{\cdot jk\cdot}^2}{an} - \frac{Y_{\cdot j..}^2}{acn} \text{ with } b(c-1)$$

$$SS_{AB} = \sum_{i=1}^{a}\sum_{j=1}^{b}\frac{Y_{ij..}^2}{cn} - \frac{Y_{....}^2}{abcn} - SS_A - SS_B \text{ with } (a-1)(b-1) \text{ d.f.}$$

$$SS_{AC(B)} = \sum_{i=1}^{a}\sum_{j=1}^{b}\sum_{k=1}^{c}\frac{Y_{ijk\cdot}^2}{n} - \sum_{j=1}^{b}\sum_{k=1}^{c}\frac{Y_{\cdot jk\cdot}^2}{an} - \sum_{i=1}^{a}\sum_{j=1}^{b}\sum_{j=1}^{b}\frac{Y_{\cdot j..}^2}{acn} \text{ with } b(a-1)(c-1) \text{ d.f.}$$

As we did in Example 7.3.4, it is straightforward to explain this in the framework of a nested sequence of hypothesis leading to suitable F-tests. □

Example 7.3.8. Higher-order crossed models. It is possible to extend the methods of the two-factor models to deal with models involving several factors. We illustrate the ideas using three factors, Factor A, Factor B and Factor C with levels a, b, and c, respectively. We only consider the balanced case. The three-way cross-classified model with all possible interactions is

$$Y_{ijkl} = \mu + \tau_i + \theta_j + \eta_k + (\tau\theta)_{ij} + (\tau\eta)_{ik} + (\theta\eta)_{jk} + (\tau\theta\eta)_{ijk} + \varepsilon_{ijkl},$$

for $i = 1, \cdots, a$, $j = 1, \cdots, b$, $k = 1, \cdots, c$, and $l = 1, \cdots, n$, with $n > 1$. We assume that $\varepsilon_{ijkl} \sim N(0, \sigma^2)$. The effects $(\tau\theta)_{ij}$, $(\tau\eta)_{ik}$ and $(\theta\eta)_{jk}$ are called two-factor interactions, while $(\tau\theta\eta)_{ijk}$ is a three-factor interaction. Suppose all three factors are fixed effects. We obtain least squares solutions of the main effects, two-way and three-way interactions by imposing the following model constraints on the normal equations:

$$\sum_{i=1}^{a}\tau_i = \sum_{j=1}^{b}\theta_j = \sum_{k=1}^{c}\eta_k = 0$$

$$\sum_{i=1}^{a}(\tau\theta)_{ij} = \sum_{j=1}^{b}(\tau\theta)_{ij} = \sum_{i=1}^{a}(\tau\eta)_{ik} = \sum_{k=1}^{c}(\tau\eta)_{ik} = 0,$$
$$\sum_{j=1}^{b}(\theta\eta)_{jk} = \sum_{k=1}^{c}(\theta\eta)_{ik} = 0,$$
$$\sum_{i=1}^{a}(\tau\theta\eta)_{ijk} = \sum_{j=1}^{b}(\tau\theta\eta)_{ijk} = \sum_{k=1}^{c}(\tau\theta\eta)_{ijk} = 0.$$

For $1 \le i \le a$, $1 \le j \le b$, $1 \le k \le c$, and$1 \le l \le n$, the solutions of the parameters are

$$\begin{aligned}
\mu^0 &= \overline{Y}_{....}\\
\tau_i^0 &= \overline{Y}_{i...} - \overline{Y}_{....},\ i = 1, \cdots, a,\\
\theta_j^0 &= \overline{Y}_{\cdot j\cdot\cdot} - \overline{Y}_{....},\ j = 1, \cdots, b,\\
\eta_k^0 &= \overline{Y}_{\cdot\cdot k\cdot} - \overline{Y}_{....},\ k = 1, \cdots, c,\\
(\tau\theta)_{ij}^0 &= \overline{Y}_{ij\cdot\cdot} - \overline{Y}_{i...} - \overline{Y}_{\cdot j\cdot\cdot} + \overline{Y}_{....},\ i = 1, \cdots, a;\ j = 1, \cdots, b,\\
(\tau\eta)_{ik}^0 &= \overline{Y}_{i\cdot k\cdot} - \overline{Y}_{i...} - \overline{Y}_{\cdot\cdot k\cdot} + \overline{Y}_{....},\ i = 1, \cdots, a;\ k = 1, \cdots, c,\\
(\theta\eta)_{jk}^0 &= \overline{Y}_{\cdot jk\cdot} - \overline{Y}_{\cdot j\cdot\cdot} - \overline{Y}_{\cdot\cdot k\cdot} + \overline{Y}_{....},\ j = 1, \cdots, b;\ k = 1, \cdots, c,\\
(\tau\theta\eta)_{ijk}^0 &= \overline{Y}_{ijk\cdot} - \overline{Y}_{ij\cdot\cdot} - \overline{Y}_{i\cdot k\cdot} - \overline{Y}_{\cdot jk\cdot} + \overline{Y}_{i...} + \overline{Y}_{\cdot j\cdot\cdot} + \overline{Y}_{\cdot\cdot k\cdot} - \overline{Y}_{.....}
\end{aligned}$$

The ANOVA decomposition is

$$SST_c = SS_A + SS_B + SS_C + SS_{AB} + SS_{AC} + SS_{BC} + SS_{ABC} + SSE,$$

where

$$SST_c = \sum_{i=1}^{a}\sum_{j=1}^{b}\sum_{k=1}^{c}\sum_{l=1}^{n}(Y_{ijkl} - \overline{Y}_{....})^2 \text{ with } (abcn - 1) \text{ d.f.}$$

$$SS_A = \sum_{i=1}^{a}\frac{Y_{i...}^2}{bcn} - \frac{Y_{....}^2}{abcn} \text{ with } (a-1) \text{ d.f.}$$

$$SS_B = \sum_{j=1}^{b}\frac{Y_{\cdot j\cdot\cdot}^2}{acn} - \frac{Y_{....}^2}{abcn} \text{ with } (b-1) \text{ d.f.}$$

$$SS_C = \sum_{k=1}^{c}\frac{Y_{\cdot\cdot k\cdot}^2}{abn} - \frac{Y_{....}^2}{abcn} \text{ with } (c-1) \text{ d.f.}$$

$$SS_{AB} = \sum_{i=1}^{a}\sum_{j=1}^{b}\frac{Y_{ij\cdot\cdot}^2}{cn} - \frac{Y_{....}^2}{abcn} - SS_A - SS_B \text{ with } (a-1)(b-1) \text{ d.f.}$$

$$SS_{AC} = \sum_{i=1}^{a}\sum_{k=1}^{c}\frac{Y_{i\cdot k\cdot}^2}{bn} - \frac{Y_{....}^2}{abcn} - SS_A - SS_C \text{ with } (a-1)(c-1) \text{ d.f.}$$

$$SS_{BC} = \sum_{j=1}^{b}\sum_{k=1}^{c}\frac{Y_{\cdot jk\cdot}^2}{an} - \frac{Y_{....}^2}{abcn} - SS_B - SS_C \text{ with } (b-1)(c-1) \text{ d.f.}$$

$$SS_{ABC} = \sum_{i=1}^{a}\sum_{j=1}^{b}\sum_{k=1}^{c}\frac{Y_{ijk\cdot}^2}{n} - \frac{Y_{....}^2}{abcn} - SS_A - SS_B - SS_C - SS_{AB}$$

$$- SS_{AC} - SS_{BC} \text{ with } (a-1)(b-1)(c-1) \text{ d.f.},$$

while SSE is obtained by subtracting these sums of squares from SST_c and has a χ^2 distribution with $abc(n-1)$ d.f. These may be used to construct confidence interval estimates of estimable functions of the parameter vector as well as for carrying out various tests of hypotheses.

The model simplifies considerably when there are no interactions:

$$Y_{ijkl} = \mu + \tau_i + \theta_j + \eta_k + \varepsilon_{ijkl},$$

for $i = 1, \cdots, a$, $j = 1, \cdots, b$, $k = 1, \cdots, c$, and $l = 1, \cdots, n$. We assume that $\varepsilon_{ijkl} \sim N(0, \sigma^2)$. Suppose all three factors are fixed effects. We obtain least squares solutions of the main effects by imposing the following model constraints on the normal equations:

$$\sum_{i=1}^{a} \tau_i = \sum_{j=1}^{b} \theta_j = \sum_{k=1}^{c} \eta_k = 0.$$

The ANOVA decomposition in this case is

$$SST_c = SS_A + SS_B + SS_C + SSE,$$

where

$$SST_c = \sum_{i=1}^{a}\sum_{j=1}^{b}\sum_{k=1}^{c}\sum_{l=1}^{n}(Y_{ijkl} - \overline{Y}_{....})^2 \text{ with } (abcn-1) \text{ d.f.}$$

$$SS_A = \sum_{i=1}^{a} \frac{Y_{i...}^2}{bcn} - \frac{Y_{....}^2}{abcn} \text{ with } (a-1) \text{ d.f.}$$

$$SS_B = \sum_{j=1}^{b} \frac{Y_{\cdot j\cdot\cdot}^2}{acn} - \frac{Y_{....}^2}{abcn} \text{ with } (b-1) \text{ d.f.}$$

$$SS_C = \sum_{k=1}^{c} \frac{Y_{\cdot\cdot k\cdot}^2}{abn} - \frac{Y_{....}^2}{abcn} \text{ with } (c-1) \text{ d.f.},$$

while SSE is obtained by subtracting these sums of squares from SST_c and has a χ^2 distribution with $abcn - a - b - c + 1$ d.f. Based on this decomposition, we can test for the significance of these main effect differences. Exercise 7.21 looks at an $a \times a$ Latin Square Design (LSD) in which all three factors have a levels each, and $n = 1$. □

The following result summarizes the consequences of underfitting and suggests that deleting variables corresponding to small coefficients (relative to their standard errors) will lead to higher precision in the estimates of coefficients corresponding to the retained variables.

Result 7.3.6. Underfitting. Suppose the true model is given by (7.3.6), but the model fitted to the data is

$$\mathbf{y} = \mathbf{X}_1\boldsymbol{\beta}_1 + \boldsymbol{\varepsilon}. \tag{7.3.18}$$

Let $\boldsymbol{\beta}_{1,H}^0 = (\mathbf{X}_1'\mathbf{X}_1)^- \mathbf{X}_1'\mathbf{y}$ be a LS solution vector for $\boldsymbol{\beta}_1$, and let $\widehat{\sigma}_{1,H}^2 = \{\mathbf{y}'(\mathbf{I}-\mathbf{P}_1)\mathbf{y}\}/(N - r_1)$ denote the estimate of σ^2. Then,

1. $E(\boldsymbol{\beta}^0_{1,H}) = \mathbf{H}_1\boldsymbol{\beta}_1 + (\mathbf{X}_1'\mathbf{X}_1)^-\mathbf{X}_1'\mathbf{X}_2\boldsymbol{\beta}_2$, where $\mathbf{H}_1 = (\mathbf{X}_1'\mathbf{X}_1)^-\mathbf{X}_1'\mathbf{X}_1$. In the full rank case, $E(\widehat{\boldsymbol{\beta}}_{1,H}) = \boldsymbol{\beta}_1 + (\mathbf{X}_1'\mathbf{X}_1)^{-1}\mathbf{X}_1'\mathbf{X}_2\boldsymbol{\beta}_2$.

2. $E(\widehat{\sigma}^2_{1,H}) = \sigma^2 + \frac{1}{N-r_1}\boldsymbol{\beta}_2'\mathbf{X}_2'(\mathbf{I}-\mathbf{P}_1)\mathbf{X}_2\boldsymbol{\beta}_2$ and $\widehat{\boldsymbol{\beta}}_{1,H}$ is an unbiased estimator of $\boldsymbol{\beta}_1$ regardless of $\boldsymbol{\beta}_2$ if and only if the columns vectors of $\mathbf{X}_1$ are orthogonal to those of $\mathbf{X}_2$.

3. Let $\widehat{\mathbf{y}}_{0,H} = \mathbf{X}_{1,0}\boldsymbol{\beta}^0_{1,H}$ denote the prediction corresponding to a new set of values $\mathbf{X}_{1,0}$. Then,

$$E(\widehat{\mathbf{y}}_{0,H}) = \mathbf{X}_{1,0}[\boldsymbol{\beta}_1 + (\mathbf{X}_1'\mathbf{X}_1)^-\mathbf{X}_1'\mathbf{X}_2\boldsymbol{\beta}_2].$$

Proof. The model (7.3.18) corresponds to imposing the reduction H: $\mathbf{X}_2\boldsymbol{\beta}_2 = \mathbf{0}$ on the model (7.3.6). Under the reduced fitted model, we have $\text{Cov}(\boldsymbol{\beta}^0_1) = \sigma^2(\mathbf{X}_1'\mathbf{X}_1)^-$. Now,

$$\begin{aligned} E(\boldsymbol{\beta}^0_{1,H}) &= (\mathbf{X}_1'\mathbf{X}_1)^-\mathbf{X}_1'\,E(\mathbf{y}) \\ &= (\mathbf{X}_1'\mathbf{X}_1)^-\mathbf{X}_1'(\mathbf{X}_1\boldsymbol{\beta}_1 + \mathbf{X}_2\boldsymbol{\beta}_2) \\ &= (\mathbf{X}_1'\mathbf{X}_1)^-\mathbf{X}_1'\mathbf{X}_1\boldsymbol{\beta}_1 + (\mathbf{X}_1'\mathbf{X}_1)^-\mathbf{X}_1'\mathbf{X}_2\boldsymbol{\beta}_2 \\ &= \mathbf{H}_1\boldsymbol{\beta}_1 + (\mathbf{X}_1'\mathbf{X}_1)^-\mathbf{X}_1'\mathbf{X}_2\boldsymbol{\beta}_2. \end{aligned}$$

In the full rank case, $\mathbf{H}_1 = \mathbf{I}$ so $E(\widehat{\boldsymbol{\beta}}_{1,H}) = \boldsymbol{\beta}_1 + (\mathbf{X}_1'\mathbf{X}_1)^{-1}\mathbf{X}_1'\mathbf{X}_2\boldsymbol{\beta}_2$. Thus, $\widehat{\boldsymbol{\beta}}_{1,H}$ is an unbiased estimate for $\boldsymbol{\beta}_1$ regardless of $\boldsymbol{\beta}_2$ if and only if $\mathbf{X}_1'\mathbf{X}_2 = \mathbf{O}$. This proves property 1. To prove property 2, we use Result 4.2.3, setting $\mathbf{x} = \mathbf{y}$ and $\mathbf{A} = \mathbf{I} - \mathbf{P}_1$ therein. Since $\mathbf{X}_1'(\mathbf{I}-\mathbf{P}_1) = \mathbf{O}$ and $\text{tr}(\mathbf{I}-\mathbf{P}_1) = N - r_1$,

$$\begin{aligned} E(\mathbf{y}'(\mathbf{I}-\mathbf{P}_1)\mathbf{y}) &= \sigma^2\,\text{tr}(\mathbf{I}-\mathbf{P}_1) + (\mathbf{X}_1\boldsymbol{\beta}_1 + \mathbf{X}_2\boldsymbol{\beta}_2)'(\mathbf{I}-\mathbf{P}_1)(\mathbf{X}_1\boldsymbol{\beta}_1 + \mathbf{X}_2\boldsymbol{\beta}_2) \\ &= (N-r_1)\sigma^2 + \boldsymbol{\beta}_2'\mathbf{X}_2'(\mathbf{I}-\mathbf{P}_1)\mathbf{X}_2\boldsymbol{\beta}_2, \end{aligned}$$

so that

$$E(\widehat{\sigma}^2_{1,H}) = \sigma^2 + \frac{1}{N-r_1}\boldsymbol{\beta}_2'\mathbf{X}_2'(\mathbf{I}-\mathbf{P}_1)\mathbf{X}_2\boldsymbol{\beta}_2 \neq \sigma^2$$

unless $\boldsymbol{\beta}_2 = \mathbf{0}$. Property 3 follows directly since $E(\widehat{\mathbf{y}}_{0,H}) = E(\mathbf{X}_{1,0}\boldsymbol{\beta}^0_{1,H})$. ■

Example 7.3.9. While the *true* model is $E(Y) = \beta_0 + \beta_1 X_1 + \beta_2 X_1^2$, suppose we *fit* the model $Y_i = \beta_0 + \beta_1 X_1 + \varepsilon_i$, $i = 1, 2, 3$, using data $(X_1, Y_1) = (-2, 15)$, $(X_2, Y_2) = (0, 21)$, and $(X_3, Y_3) = (2, 55)$. Using Result 7.3.6, we can verify that $\widehat{\beta}_0$ has bias $8\beta_2/3$, while $\widehat{\beta}_1$ is unbiased. □

Result 7.3.7. Overfitting. Suppose the true model is (7.3.18), but we fit the full model (7.3.6) to the data. Then for the solutions $\boldsymbol{\beta}^0_1$ and $\boldsymbol{\beta}^0_2$ in (4.2.36),

1. $E(\boldsymbol{\beta}^0_1) = \mathbf{H}_1\boldsymbol{\beta}_1$.

2. MSE is an unbiased estimate of σ^2.

Proof. From (4.2.36), $E(\boldsymbol{\beta}^0_2) = [\mathbf{X}_2'(\mathbf{I}-\mathbf{P}_1)\mathbf{X}_2]^-\mathbf{X}_2'(\mathbf{I}-\mathbf{P}_1)\mathbf{X}_1\boldsymbol{\beta}_1 = \mathbf{0}$, and so $E(\boldsymbol{\beta}^0_1) = E\{(\mathbf{X}_1'\mathbf{X}_1)^-\mathbf{X}_1'(\mathbf{y} - \mathbf{X}_2\boldsymbol{\beta}^0_2)\} = \mathbf{H}_1\boldsymbol{\beta}_1$. This proves property 1. The proof of property 2 is left as Exercise 7.11. ■

For more results on the effects of underfitting and overfitting on estimation, see Exercises 7.12 and 7.13.

7.4 Confidence intervals

The least squares approach and the method of maximum likelihood (see Section 8.1.1) provide point estimates of components of $\boldsymbol{\beta}$. Under the assumption of normal errors, we constructed the F-test for a single hypothesis on each of the estimable functions $\mathbf{c}_j'\boldsymbol{\beta}$, $j = 1, \cdots, s$. We next construct the joint confidence region for the estimable functions as well as marginal confidence intervals for each of them.

7.4.1 Joint and marginal confidence intervals

Let $\mathbf{C}'\boldsymbol{\beta}$ denote a vector of estimable functions of $\boldsymbol{\beta}$, where $\mathbf{C}$ is a $p \times s$ matrix of known coefficients. We have seen that the least squares estimator of $\mathbf{C}'\boldsymbol{\beta}$ is $\mathbf{C}'\boldsymbol{\beta}^0$, with covariance $\sigma^2\mathbf{C}'\mathbf{G}\mathbf{C}$.

Result 7.4.1. The $100(1-\alpha)\%$ joint confidence region for $\mathbf{C}'\boldsymbol{\beta}$ is

$$(\mathbf{C}'\boldsymbol{\beta}^0 - \mathbf{C}'\boldsymbol{\beta})'(\mathbf{C}'\mathbf{G}\mathbf{C})^{-1}(\mathbf{C}'\boldsymbol{\beta}^0 - \mathbf{C}'\boldsymbol{\beta}) \le s\widehat{\sigma}^2 F_{s,N-r,\alpha}, \tag{7.4.1}$$

where $F_{s,N-r,\alpha}$ denotes the upper $\alpha\%$ critical point from an $F_{s,N-r}$ distribution.

Proof. The proof follows directly from inverting the F-statistic in (7.2.5). ■

Corollary 7.4.1. If $\mathbf{X}$ has full column rank, then the $100(1-\alpha)\%$ joint confidence region for $\boldsymbol{\beta}$ is given by

$$(\widehat{\boldsymbol{\beta}} - \boldsymbol{\beta})'(\mathbf{X}'\mathbf{X})(\widehat{\boldsymbol{\beta}} - \boldsymbol{\beta}) \le p\widehat{\sigma}^2 F_{p,N-p,\alpha}. \tag{7.4.2}$$

In Figure 7.4.1, for $p = 2$, the ellipsoid represents the joint confidence region for $\boldsymbol{\beta}$, while the marginals intervals for β_1 and β_2 form the rectangle ABCD.

The marginal $100(1-\alpha)\%$ confidence intervals for s estimable functions $\mathbf{c}_i'\boldsymbol{\beta}$, $i = 1, \cdots, s$ of $\boldsymbol{\beta}$ are given by

$$\mathbf{c}_i'\boldsymbol{\beta}^0 \pm t_{N-r,\alpha/2}\widehat{\sigma}(\mathbf{c}_i'\mathbf{G}\mathbf{c}_i)^{1/2}, \quad i = 1, \cdots, s. \tag{7.4.3}$$

While each marginal interval contains the true parameter $\mathbf{c}_i'\boldsymbol{\beta}$ with probability $(1-\alpha)$, the probability that "simultaneously" all s functions $\mathbf{c}_i'\boldsymbol{\beta}$, $i = 1, \cdots, s$ are contained in the intervals in (7.4.3) is certainly much less than $(1-\alpha)$. We may, of course, construct simultaneous confidence intervals for estimable functions of $\boldsymbol{\beta}$, as described in Section 7.4.2.

Example 7.4.1. Continuing with Example 4.1.1, we construct confidence intervals for the model parameters and functions of the parameters in a simple linear regression model by inverting t-tests (or, equivalently, F-tests having one numerator d.f.). We can test the null hypothesis H_0: $\beta_1 = \beta_{1,0}$ using the t-statistic $t = (\widehat{\beta}_1 - \beta_{1,0})/s_{\widehat{\beta}_1}$, which follows a t_{N-2} distribution under H_0. For a two-sided alternative H_1: $\beta_1 \ne \beta_{1,0}$, the test rejects H_0 at level of significance α if $|t_{obs}| > t_{N-2,\alpha/2}$, i.e., the observed test statistic is greater in absolute value than the upper $100(\alpha/2)\%$ critical value from a t-distribution with $N-2$ d.f. The $100(1-\alpha)\%$ confidence interval for β_1 is given by $\widehat{\beta}_1 \pm s_{\widehat{\beta}_1} t_{N-2,\alpha/2}$. Using a similar procedure, the $100(1-\alpha)\%$ confidence interval for β_0 is given by $\widehat{\beta}_0 \pm s_{\widehat{\beta}_0} t_{N-2,\alpha/2}$. □

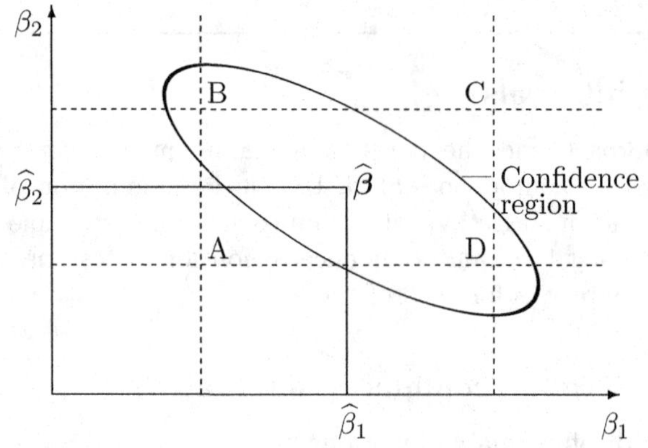

FIGURE 7.4.1. Joint confidence region for $\boldsymbol{\beta}$.

Example 7.4.2. We continue with Example 4.1.2. From the distributional properties of the multiple linear regression model, it follows that $\widehat{\sigma}^2(\mathbf{X}'\mathbf{X})^{-1}$ is an unbiased estimator of $\text{Cov}(\widehat{\boldsymbol{\beta}})$. Since (i) $\widehat{\beta}_j - \beta_j$, $j = 0, \cdots, k$, has a normal distribution with mean 0 and variance $\sigma^2(\mathbf{X}'\mathbf{X})^{-1}_{j+1,j+1}$ (which is the $(j+1)$th diagonal element of $\text{Cov}(\widehat{\boldsymbol{\beta}})$), (ii) $(N-k-1)\widehat{\sigma}^2/\sigma^2 \sim \chi^2_{N-k-1}$, and (iii) $(N-k-1)\widehat{\sigma}^2/\sigma^2$ is distributed independently of $\widehat{\beta}_j - \beta_j$, $j = 0, \cdots, k$, it follows that the statistic

$$t_j = (\widehat{\beta}_j - \beta_j)/s_{\widehat{\beta}_j} \sim t_{N-k-1}. \tag{7.4.4}$$

This framework enables us to carry out hypothesis tests on the components of $\boldsymbol{\beta}$ as well as to construct confidence intervals for β_j. To test the null hypothesis H_0: $\beta_j = \beta_{j,0}$, substitute $\beta_{j,0}$ for β_j in (7.4.4); if $|t_{obs}| > t_{N-k-1.\alpha/2}$, we reject H_0 at level of significance α. The $100(1-\alpha)\%$ confidence interval for β_j is

$$\widehat{\beta}_j \pm s_{\widehat{\beta}_j} t_{N-k-1,\alpha/2}. \qquad \square$$

Example 7.4.3. We continue with Example 4.5.1, where we looked at the GLS estimation of β in a model with AR(1) errors. We see that

$$\widehat{\beta}_{GLS} \sim N(\beta, \sigma^2(\mathbf{x}'\mathbf{V}^{-1}\mathbf{x})^{-1}) \quad \text{and} \quad (N-1)\widehat{\sigma}^2_{GLS} \sim \chi^2_{N-1}$$

(see Example 7.1.2). Hence,

$$(\widehat{\beta}_{GLS} - \beta)/\{\widehat{\sigma}^2_{GLS}(\mathbf{x}'\mathbf{V}^{-1}\mathbf{x})^{-1}\}^{1/2} \sim t_{N-1}.$$

The $100(1-\alpha)\%$ confidence interval for β is

$$\widehat{\beta}_{GLS} \pm t_{N-1,\alpha/2}\frac{\widehat{\sigma}_{GLS}}{\sqrt{\mathbf{x}'\mathbf{V}^{-1}\mathbf{x}}}. \qquad \square$$

Example 7.4.4. Suppose

$$\begin{pmatrix} Y_1 \\ Y_2 \\ Y_3 \end{pmatrix} = \begin{pmatrix} 1 & 0 \\ 2 & -1 \\ 1 & 2 \end{pmatrix} \begin{pmatrix} \beta_1 \\ \beta_2 \end{pmatrix} + \begin{pmatrix} \varepsilon_1 \\ \varepsilon_2 \\ \varepsilon_3 \end{pmatrix},$$

where ε_i are i.i.d. $N(0, \sigma^2)$ variables, $i = 1, 2, 3$. We construct a 95% confidence interval for $\theta = \beta_1 + 3\beta_2$. We have $\theta = \mathbf{c}'\boldsymbol{\beta}$, where $\mathbf{c}' = (1, 3)$, and with $\mathbf{X}$ having full column rank 2, $\mathbf{G} = (\mathbf{X}'\mathbf{X})^{-1} = \text{diag}(1/6, 1/5)$. Then,

$$\widehat{\boldsymbol{\beta}} = \mathbf{G}\mathbf{X}'\mathbf{y} = \begin{pmatrix} (Y_1 + 2Y_2 + Y_3)/6 \\ -Y_2/5 + 2Y_3/5 \end{pmatrix}, \text{ and}$$

$$\widehat{\boldsymbol{\varepsilon}} = \mathbf{y} - \mathbf{X}\widehat{\boldsymbol{\beta}} = \begin{pmatrix} (5Y_1 - 2Y_2 - Y_3)/6 \\ -Y_1/3 + 2Y_2/15 + Y_3/15 \\ -Y_1/6 + Y_2/15 + Y_3/30 \end{pmatrix}.$$

By the Gauss–Markov theorem (Result 4.4.1), the b.l.u.e. of θ is $\widehat{\theta} = \widehat{\beta}_1 + 3\widehat{\beta}_2 = (5Y_1 - 8Y_2 + 41Y_3)/30$ and $\text{Var}(\widehat{\theta}) = \mathbf{c}'\mathbf{G}\mathbf{c}\sigma^2 = 59\sigma^2/30$. By (4.2.17) and since $N - r = 1$, we have $\widehat{\sigma}^2 = \widehat{\boldsymbol{\varepsilon}}'\widehat{\boldsymbol{\varepsilon}}$. The 95% confidence interval for θ is the t-interval

$$\widehat{\theta} \pm t_{1,.025}(59\widehat{\sigma}^2/30)^{1/2}.$$

□

Example 7.4.5. Consider the linear model

$$\begin{pmatrix} Y_1 \\ Y_2 \\ Y_3 \end{pmatrix} = \begin{pmatrix} 1 & 1 & 0 \\ 1 & 0 & 1 \\ 1 & 1 & 0 \end{pmatrix} \begin{pmatrix} \beta_1 \\ \beta_2 \\ \beta_3 \end{pmatrix} + \begin{pmatrix} \varepsilon_1 \\ \varepsilon_2 \\ \varepsilon_3 \end{pmatrix},$$

where $\boldsymbol{\varepsilon} \sim N(\mathbf{0}, \sigma^2\mathbf{I})$. We construct a 95% confidence interval for $\theta = \beta_1 + \beta_2/3 + 2\beta_3/3$. First, since $\text{r}(\mathbf{X}) = 2 < 3$, we must verify that the linear function θ is estimable. Given $\mathbf{c} = (c_1, c_2, c_3)'$, it is easy to verify using (4.3.3) that $\mathbf{c}'\boldsymbol{\beta}$ is estimable if and only if $c_1 = c_2 + c_3$ (see Example 4.3.1). Based on

$$\mathbf{G} = (\mathbf{X}'\mathbf{X})^{-} = \begin{pmatrix} 0 & 1/2 & 0 \\ 0 & 0 & 0 \\ 1 & -3/2 & 0 \end{pmatrix} \text{ and } \boldsymbol{\beta}^0 = \begin{pmatrix} (Y_1 + Y_3)/2 \\ 0 \\ Y_2 - (Y_1 + Y_3)/2 \end{pmatrix},$$

$\widehat{\boldsymbol{\varepsilon}} = ((Y_1 - Y_3)/2, 0, -(Y_1 - Y_3)/2)'$, the b.l.u.e. of θ is $\widehat{\theta} = \mathbf{c}'\boldsymbol{\beta}^0 = (Y_1 + 4Y_2 + Y_3)/6$, and $\text{Var}(\widehat{\theta}) = \mathbf{c}'\mathbf{G}\mathbf{c}\sigma^2 = \sigma^2/2$. Since $N - r = 1$, $\widehat{\sigma}^2 = (Y_1 - Y_3)^2/2$. Using (7.4.3), the 95% confidence interval of θ is

$$\mathbf{c}'\boldsymbol{\beta}^0 \pm t_{2,.025}(\mathbf{c}'\mathbf{G}\mathbf{c}\widehat{\sigma}^2)^{1/2} = (Y_1 + 4Y_2 + Y_3)/6 \pm t_{1,.025}|Y_1 - Y_3|/2.$$

□

7.4.2 Simultaneous confidence intervals

The marginal $100(1-\alpha)\%$ confidence intervals for s contrasts $\mathbf{c}_i'\boldsymbol{\beta}$, $i = 1, \cdots, s$ given in (7.4.3) provide coverage probability of $(1-\alpha)$ for *each* $\mathbf{c}_i'\boldsymbol{\beta}$. For simultaneous inference, we wish to claim in general that $\text{P}[\mathbf{c}'\boldsymbol{\beta} \in I_{\mathbf{c}}, \mathbf{c} \in \mathcal{S}] = 1 - \alpha$, where $\mathcal{S}$ is a subset of $\mathcal{C}(\mathbf{X})$; then, $\{I_{\mathbf{c}}, \mathbf{c} \in \mathcal{S}\}$ is called a "family" of confidence intervals with confidence coefficient $(1-\alpha)$.

We describe two types of simultaneous confidence intervals: Scheffé intervals and Bonferroni t-intervals. Although Scheffé intervals can achieve simultaneous coverage for a large or even infinite number of estimable functions, their construction heavily relies on the normality of $\boldsymbol{\varepsilon}$. On the other hand, Bonferroni t-intervals are only useful for a small number of estimable functions. However, they are easier to calculate and can provide approximate simultaneous coverage even when normality does not hold.

7.4.2.1 Scheffé intervals

Scheffé (1953) proposed a method for comparing *all possible contrasts* in $\boldsymbol{\beta}$ such that the probability of Type I Error is at most α. In other words, the probability is at least $1-\alpha$ that the associated confidence intervals simultaneously contain the true values of *all* the contrasts in $\boldsymbol{\beta}$. While confidence intervals on contrasts are often of interest, for example, in the fixed-effects one-way ANOVA model, in simple linear regression, we might be interested in confidence intervals on all linear combinations $c_0\beta_0 + c_1\beta_1$. Scheffé's method applies to estimable functions of $\boldsymbol{\beta}$ that are not contrasts as well.

Suppose the objective is to construct simultaneous confidence intervals for $\mathbf{c}'\boldsymbol{\beta}$, $\mathbf{c} \in \mathcal{S} \subset \mathcal{R}(\mathbf{X})$. Let $\mathcal{L} = \text{Span}(\mathcal{S})$ and $\dim(\mathcal{L}) = s$. Scheffé's method constructs confidence intervals for $\mathbf{c}'\boldsymbol{\beta} = \sum_{i=1}^{p} c_i\beta_i$ for all $\mathbf{c} \in \mathcal{L}$ with simultaneous $100(1-\alpha)\%$ coverage, hence achieving the objective. Let $\mathbf{c}_i = (c_{i1}, \cdots, c_{ip})'$, $i = 1, \cdots, s$, be a basis of $\mathcal{L}$ and set

$$\mathbf{C} = (\mathbf{c}_1, \cdots, \mathbf{c}_s) = \begin{pmatrix} c_{11} & \cdots & c_{s1} \\ \vdots & \ddots & \vdots \\ c_{1p} & \cdots & c_{sp} \end{pmatrix}.$$

From Result 5.2.5, $\mathbf{C}'\boldsymbol{\beta}^0 \sim N_s(\mathbf{C}'\boldsymbol{\beta}, \sigma^2\mathbf{C}'\mathbf{G}\mathbf{C})$. A test of H: $\mathbf{C}'\boldsymbol{\beta} = \mathbf{d}$, say, is based on

$$(\mathbf{C}'\boldsymbol{\beta}^0 - \mathbf{d})'(\mathbf{C}'\mathbf{G}\mathbf{C})^{-1}(\mathbf{C}'\boldsymbol{\beta}^0 - \mathbf{d}) \,/ s\widehat{\sigma}^2 \sim F_{s,N-r} \tag{7.4.5}$$

under the null hypothesis.

Result 7.4.2. The $100(1-\alpha)\%$ simultaneous confidence intervals via the "F-projections" approach for estimable functions $\mathbf{c}'\boldsymbol{\beta}$ for all $\mathbf{c} \in \mathcal{L}$ are given by

$$\mathbf{c}'\boldsymbol{\beta}^0 \pm \widehat{\sigma}\{s(\mathbf{c}'\mathbf{G}\mathbf{c})F_{s,N-r,\alpha}\}^{1/2}. \tag{7.4.6}$$

The simultaneous confidence interval for any $\mathbf{c}'\boldsymbol{\beta}$, $\mathbf{c} \in \mathcal{L}$ is the projection of the confidence ellipsoid (7.2.8) onto the one-dimensional space spanned by $\mathbf{c}$.

Proof. Let $\mathbf{P} = \mathbf{C}'\mathbf{G}\mathbf{C}$, $\gamma = s\widehat{\sigma}^2 F_{s,N-r,\alpha}$, and $\mathbf{C}'\boldsymbol{\beta}^0 - \mathbf{C}'\boldsymbol{\beta} = \mathbf{b}$. Using Result 2.4.6,

$$\begin{aligned} 1-\alpha &= \text{P}(F_{s,N-r} \le F_{s,N-r,\alpha}) \\ &= \{(\mathbf{C}'\boldsymbol{\beta}^0 - \mathbf{C}'\boldsymbol{\beta})'(\mathbf{C}'\mathbf{G}\mathbf{C})^{-1}(\mathbf{C}'\boldsymbol{\beta}^0 - \mathbf{C}'\boldsymbol{\beta}) \le s\widehat{\sigma}^2 F_{s,N-r,\alpha}\} \\ &= \text{P}\{\mathbf{b}'\mathbf{P}^{-1}\mathbf{b} \le \gamma\} = \text{P}\left\{\sup_{\mathbf{h}\neq\mathbf{0}} \frac{(\mathbf{h}'\mathbf{b})^2}{\mathbf{h}'\mathbf{P}\mathbf{h}} \le \gamma\right\} \\ &= \text{P}\left\{\frac{(\mathbf{h}'\mathbf{b})^2}{\mathbf{h}'\mathbf{P}\mathbf{h}} \le \gamma \text{ for all } \mathbf{h} \neq \mathbf{0}\right\} \\ &= \text{P}\left\{\frac{|\mathbf{h}'\mathbf{C}'\boldsymbol{\beta}^0 - \mathbf{h}'\mathbf{C}'\boldsymbol{\beta}|}{\widehat{\sigma}(\mathbf{h}'\mathbf{P}\mathbf{h})^{1/2}} \le [sF_{s,N-r,\alpha}]^{1/2} \text{ for all } \mathbf{h} \neq \mathbf{0}\right\}. \end{aligned}$$

Since $\mathcal{L} = \{\mathbf{C}\mathbf{h}, \mathbf{h} \in \mathcal{R}^s\}$, setting $\mathbf{c} = \mathbf{C}\mathbf{h}$ gives $1-\alpha = \text{P}\{|\mathbf{c}'\boldsymbol{\beta}^0 - \mathbf{c}'\boldsymbol{\beta}| \le \widehat{\sigma}\{s(\mathbf{c}'\mathbf{G}\mathbf{c})F_{s,N-r,\alpha}\}^{1/2}$ for all $\mathbf{c} \in \mathcal{L}\}$, which proves the result. ■

In Exercise 7.25, we ask the reader to derive the Scheffé F-intervals for the linear model with a p.d. covariance structure. For ANOVA models, Scheffé's method is generally recommended when (a) the sample sizes n_i, $i = 1, \cdots, a$ are unequal, and (b) we are interested in more complicated comparisons between treatment means than simple pairwise comparisons.

Note that by Result 7.1.2, for any estimable $\mathbf{c}'\boldsymbol{\beta}$, $\text{Var}(\mathbf{c}'\boldsymbol{\beta}^0) = \sigma^2\mathbf{c}'\mathbf{G}\mathbf{c}$. Therefore, the confidence interval (7.4.6) can also be written as

$$\mathbf{c}'\boldsymbol{\beta}^0 \pm \{s\widehat{\text{Var}}(\mathbf{c}'\boldsymbol{\beta}^0)F_{s,N-r,\alpha}\}^{1/2}.$$

Example 7.4.6. We apply Scheffé's simultaneous procedure to simple linear regression to construct simultaneous confidence intervals for $c_0\beta_0 + c_1\beta_1$, which are given by

$$\{(c_0\widehat{\beta}_0 + c_1\widehat{\beta}_1) \pm [2\widehat{\text{Var}}(c_0\widehat{\beta}_0 + c_1\widehat{\beta}_1)F_{2,N-2,\alpha}]^{1/2}\},$$

where, $\text{Var}(c_0\widehat{\beta}_0+c_1\widehat{\beta}_1) = c_0^2\,\text{Var}(\widehat{\beta}_0)+c_1^2\,\text{Var}(\widehat{\beta}_1)+2c_0c_1\,\text{Cov}(\widehat{\beta}_0,\widehat{\beta}_1)$ can be computed from (4.2.27). Letting $c_0 = 1$, and $c_1 = X$, we can obtain the simultaneous confidence intervals for $\beta_0 + \beta_1 X$. □

Suppose $\text{r}(\mathbf{X}) = p$. Then $\mathbf{x}'\boldsymbol{\beta}$ is an estimable function of $\boldsymbol{\beta}$ for all $\mathbf{x} \in \mathcal{R}^p$. From Result 7.4.2, with probability $1-\alpha$, the entire graph of $\mathbf{x}'\boldsymbol{\beta}$ as a function of $\mathbf{x}$ lies between those of

$$\mathbf{x}'\widehat{\boldsymbol{\beta}} - \widehat{\sigma}\{pF_{p,N-p,\alpha}[\mathbf{x}(\mathbf{X}'\mathbf{X})^{-1}\mathbf{x}]\}^{1/2} \text{ and } \mathbf{x}'\widehat{\boldsymbol{\beta}} + \widehat{\sigma}\{pF_{p,N-p,\alpha}[\mathbf{x}(\mathbf{X}'\mathbf{X})^{-1}\mathbf{x}]\}^{1/2}. \quad (7.4.7)$$

The two functions are sometimes referred to as simultaneous confidence surfaces of $\mathbf{x}'\boldsymbol{\beta}$.

7.4.2.2 Bonferroni *t*-intervals

Let us consider the problem of constructing simultaneous $100(1-\alpha)\%$ confidence intervals for L pairwise treatment differences $\tau_i - \tau_l$, $i \neq l$, $i, l = 1, \cdots, a$ in ANOVA models, such that the probability that the intervals jointly contain the true differences is at least $1-\alpha$. This goal can be achieved by using the Bonferroni inequality.

For any events $A_1, \cdots, A_L$, we have

$$\text{P}(A_1 \cap \cdots \cap A_L) = 1 - \text{P}(\bar{A}_1 \cup \cdots \cup \bar{A}_L),$$

where $\bar{A}_j$ denotes the complement of the event A_j, and

$$\text{P}(\bar{A}_1 \cup \cdots \cup \bar{A}_L) \leq \sum_{i=1}^{L} \text{P}(\bar{A}_i).$$

The first-order Bonferroni inequality gives

$$\text{P}(A_1 \cap \cdots \cap A_L) \geq 1 - \sum_{i=1}^{L} \text{P}(\bar{A}_i).$$

Suppose we denote the confidence intervals by $I_1, \cdots, I_L$, and suppose $\text{P}(\theta_i \in I_i) = 1 - \alpha_i$, $i = 1, \cdots, L$, where θ_i denotes a difference between any two treatment means. Let A_i denote the event $\{\theta_i \in I_i)$. Then

$$\text{P}(\theta_1 \in I_1, \cdots, \theta_L \in I_L) \geq 1 - \sum_{i=1}^{L} \text{P}(\theta_i \notin I_i) = 1 - \sum_{i=1}^{L} \alpha_i.$$

If $\alpha_i = \alpha$ for all $i = 1, \cdots, L$, then the simultaneous confidence coefficient could be as small as $(1-L\alpha) < (1-\alpha)$ when $L > 1$. For example, if $L = 10$ and $\alpha = 0.05$, the joint confidence coefficient is 0.50. We achieve an overall confidence of $1-\alpha$ by setting $\sum_{i=1}^{L}\alpha_i = \alpha$. One way to do this is to have each individual confidence coefficient to be $1-(\alpha/L)$. This is known as the Bonferroni procedure. It can be conservative. The resulting confidence intervals are known as Bonferroni intervals.

To construct Bonferroni intervals for estimable functions, we see that

$$T_i = (\mathbf{c}_i'\boldsymbol{\beta}^0 - \mathbf{c}_i'\boldsymbol{\beta}) \Big/ \{\widehat{\sigma}(\mathbf{c}_i'\mathbf{G}\mathbf{c}_i)^{1/2}\}, \quad i = 1, \cdots, s$$

has a Student's t-distribution with $N-r$ d.f. Let $t_{\alpha/2s}$ denote upper $\alpha/2s$th percentile points from this distribution. Suppose E_i denotes the event that the interval $\mathbf{c}_i'\boldsymbol{\beta}^0 \pm t_{\alpha/2}\widehat{\sigma}[\mathbf{c}_i'\mathbf{G}\mathbf{c}_i]^{1/2}$ contains $\mathbf{c}_i'\boldsymbol{\beta}$, for $i = 1, \cdots s$, so that $\text{P}(E_i) = 1 - \alpha/s$. Hence, the probability that $\mathbf{c}_i'\boldsymbol{\beta}$ lies in the interval $\mathbf{c}_i'\boldsymbol{\beta}^0 \pm t_{\alpha/2s}\widehat{\sigma}[\mathbf{c}_i'\mathbf{G}\mathbf{c}_i]^{1/2}$, for all $i = 1, \cdots, s$ is at least $(1 - \alpha)$.

Exercises

7.1. Prove Corollary 7.1.1.

7.2. Consider the model in Example 8.1.1.

(a) How will you test H: $\beta_0 = 0$?

(b) Derive an F-statistic to test H: $\beta_1 = \beta_2$.

(c) For $N = 12$, how would you test H_0: $X^* = 0$ versus H_1: $X^* \neq 0$ at the 5% level of significance?

7.3. Let $Y_i = \beta_0 + \beta_1 X_{i1} + \cdots + \beta_5 X_{i5} + \varepsilon_i$, $i = 1, \cdots, N$, $\varepsilon_i \sim N(0, \sigma^2)$. Derive the F-statistic for testing H: $\beta_3 = \beta_4 = \beta_5 = 0$. By what amount is this test statistic changed if mean subtracted responses $Y_i - \overline{Y}$ are used instead of the original responses?

7.4. In the model (4.1.7) with $a = 3$, assume that ε_{ij} are i.i.d. $N(0, \sigma^2)$ variables. Derive a suitable test for H: $\mu + \tau_1 = \mu + \tau_2 = \mu + \tau_3$.

7.5. In an experiment where several treatments are compared with a control, it may be desirable to replicate the control more than the experimental treatments, since the control enters into every difference investigated. Suppose each of m experimental treatments is replicated t times while the control is replicated c times. Let Y_{ij} denote the jth observation on the ith experimental treatment, $j = 1, \cdots, t$, $i = 1, \cdots, m$, and let Y_{0j} denote the jth observation on the control, $j = 1, \cdots, c$. Assume that $Y_{ij} = \tau_i + \varepsilon_{ij}$, $i = 0, \cdots, m$, where ε_{ij} are i.i.d. $N(0, \sigma^2)$ variables. Find the distribution of the least squares estimates of $\theta_i = \tau_i - \tau_0$, $i = 1, \cdots, m$.

7.6. Derive the F-statistic to test the hypothesis that two straight lines $Y_{l,i} = \beta_{l,0} + \beta_{l,1} X_{l,i} + \varepsilon_{l,i}$, $i = 1, \cdots, n_l$, $l = 1, 2$, $\varepsilon_{l,i} \sim N(0, \sigma^2)$ intersect at a point (x_0, y_0). Hint. By looking at a shifted model $Y_{l,i} = \beta^*_{l,0} + \beta^*_{l,1} X^*_{l,i} + \varepsilon_{l,i}$, where $\beta^*_{l,0} = \beta_{l,0} + \beta_{l,1} x_0$, and $X^*_{l,i} = (X_{l,i} - x_0)$, show that the two lines intersect at (x_0, y_0) if and only if they intersect at a point $(0, y_0^*)$.

7.7. Obtain $\text{E}(R^2)$ in the multiple linear regression model 4.1.2 with normal errors when $\beta_1 = \cdots = \beta_k = 0$.

7.8. (Wu et al., 1993) For $i = 1, \cdots, N$, let

$$\begin{aligned}
Y_i &= \varepsilon_i, \quad i \neq k, k+1, k+2 \\
Y_k &= \lambda_1 + \varepsilon_k \\
Y_{k+1} &= -c\lambda_1 + \lambda_2 + \varepsilon_{k+1} \\
Y_{k+2} &= -c\lambda_2 + \varepsilon_{k+2},
\end{aligned}$$

where k is a fixed integer, $1 \leq k \leq N-2$, $|c| < 1$ is a known constant, and ε_i's are i.i.d. $N(0, \sigma^2)$ variables. Let $\boldsymbol{\beta} = (\lambda_1, \lambda_2)'$, and suppose σ^2 is known.

(a) Derive the least squares estimate of $\boldsymbol{\beta}$, and the variance of the estimate.

(b) Derive the least squares estimate of $\boldsymbol{\beta}$ subject to the restriction $\lambda_1 + \lambda_2 = 0$. What is the variance of this estimate?

(c) Derive a statistic for testing H: $\lambda_1 + \lambda_2 = 0$ versus the alternative hypothesis that λ_1 and λ_2 are unrestricted.

7.9. Consider the model $\mathbf{y} = \boldsymbol{\beta} + \boldsymbol{\varepsilon}$, where $\boldsymbol{\varepsilon} \sim N_4(\mathbf{0}, \sigma^2\mathbf{I})$, and $\sum_{i=1}^4 \beta_i = 0$. Derive the F-statistic for testing H: $\beta_1 = \beta_2$.

7.10. Apply Result 4.2.3 to prove the last formula for $\mathrm{E}(Q)$ in Example 7.3.2.

7.11. Prove property 2 in Result 7.3.7.

7.12. Suppose the true model is $\mathbf{y} = \mathbf{X}_1\boldsymbol{\beta}_1 + \boldsymbol{\varepsilon}$, where $\mathrm{Cov}(\boldsymbol{\varepsilon}) = \sigma^2\mathbf{I}_N$. Let $\widehat{\mathbf{y}}_H$ be the LS estimate of $\mathbf{X}_1\boldsymbol{\beta}_1$ under the model and $\widehat{\sigma}^2_H$ the LS estimate of σ^2. The data can also be fitted to the model $\mathbf{y} = \mathbf{X}_1\boldsymbol{\beta}_1 + \mathbf{X}_2\boldsymbol{\beta}_2 + \boldsymbol{\varepsilon}$, yielding LS estimate $\widehat{\mathbf{y}}$ and LS estimate $\widehat{\sigma}^2$.

(a) Show that

$$\mathrm{Cov}(\widehat{\mathbf{y}}) = \mathrm{Cov}(\widehat{\mathbf{y}}_H) + \sigma^2(\mathbf{I} - \mathbf{P}_1)\mathbf{X}_2[\mathbf{X}_2'(\mathbf{I} - \mathbf{P}_1)\mathbf{X}_2]^-\mathbf{X}_2'(\mathbf{I} - \mathbf{P}_1),$$

where $\mathbf{P}_1$ is the projection onto $\mathcal{C}(\mathbf{X}_1)$. In other words, the variance of $\widehat{\mathbf{y}}$ is larger than $\widehat{\mathbf{y}}_H$, where for symmetric matrix $\mathbf{A}$ is said to be larger than symmetric matrix $\mathbf{B}$ if $\mathbf{A} - \mathbf{B}$ is n.n.d.

(b) Show that $\mathrm{Var}(\widehat{\sigma}^2)/\mathrm{Var}(\widehat{\sigma}^2_H) \geq 1$ and identify the ratio.

7.13. Suppose the true model is $\mathbf{y} = \mathbf{X}_1\boldsymbol{\beta}_1 + \mathbf{X}_2\boldsymbol{\beta}_2 + \boldsymbol{\varepsilon}$ with $\mathbf{X}_2\boldsymbol{\beta}_2 \neq \mathbf{0}$, where $\mathrm{Cov}(\boldsymbol{\varepsilon}) = \sigma^2\mathbf{I}_N$ and $\mathcal{C}(\mathbf{X}_1) \cap \mathcal{C}(\mathbf{X}_2) = \{\mathbf{0}\}$. Let $r_i = \mathrm{r}(\mathbf{X}_i)$. Let $\widehat{\mathbf{y}}$ be the LS fit under the model. If the data is fitted to the model $\mathbf{y} = \mathbf{X}_1\boldsymbol{\beta}_1 + \boldsymbol{\varepsilon}$, then the LS fit $\widehat{\mathbf{y}}_H$ is a biased estimator of $\boldsymbol{\mu} = \mathbf{X}_1\boldsymbol{\beta}_1 + \mathbf{X}_2\boldsymbol{\beta}_2$. However, $\widehat{\mathbf{y}}_H$ can be better than $\widehat{\mathbf{y}}$ in the sense that $\mathrm{E}(\|\widehat{\mathbf{y}}_H - \boldsymbol{\mu}\|^2) < \mathrm{E}(\|\widehat{\mathbf{y}} - \boldsymbol{\mu}\|^2)$. Find a necessary and sufficient condition on $\boldsymbol{\beta}_2$ for this to hold.

7.14. In Example 8.1.1, test H_0: $X^* = 0$ versus H_1: $X^* \neq 0$ at the 5% level of significance when $N = 40$.

7.15. Consider the balanced incomplete block design (BIBD) with a levels of treatment, each replicated r times, b blocks, and k treatments per block. The total number of observations is $N = ar = bk$. The statistical model is

$$Y_{ij} = \mu + \tau_i + \theta_j + \varepsilon_{ij},$$

where Y_{ij} is the ith observation in block j, μ is the overall mean, τ_i is the effect of the ith treatment, and θ_j is the effect of block j. Assume $\varepsilon_{ij} \sim N(0, \sigma^2)$. Let $\lambda = \mathrm{r}(k-1)/(a-1)$ denote the number of times each pair of treatments appears in the same block. Test the equality of treatment effects τ_i, $i = 1, \cdots, a$.

7.16. In the regression model

$$Y_i = \beta_0 + \beta_1 X_i + \beta_2(3X_i^2 - 2) + \varepsilon_i, \ i = 1, 2, 3,$$

with $X_1 = -1$, $X_2 = 0$, and $X_3 = 1$, what happens to the least squares estimates of β_0 and β_1 when $\beta_2 = 0$? Why?

7.17. In the model $Y_{ij} = \mu + \tau_i + \varepsilon_{ij}$, $j = 1, \cdots, n_i$, $i = 1, \cdots, 3$, $\varepsilon_{ij} \sim N(0, \sigma^2)$, derive a test statistic for H: $(\mu + \tau_1)/5 = (\mu + \tau_2)/10 = (\mu + \tau_3)/15$.

7.18. In the one-way ANOVA model $Y_{ij} = \mu + \tau_i + \varepsilon_{ij}$, $j = 1, \cdots, n$, $i = 1, \cdots, a$, with $N(0, \sigma^2)$ errors, derive a test of H: $\mu + \tau_1 = 2(\mu + \tau_2) = \cdots = a(\mu + \tau_a)$.

7.19. Let $Y_{ij} = \mu + \tau_i + \varepsilon_{ij}$, $j = 1, \cdots, n$, $i = 1, \cdots, 3$, and $\varepsilon_{ij} \sim N(0, \sigma^2)$. Derive a test for H: $\tau_2 = (\tau_1 + \tau_3)/2$.

7.20. Consider the three factor model with normal errors, i.e., $Y_{ijk} = \mu + \tau_i + \theta_j + (\tau\theta)_{ij} + \eta_k + (\theta\eta)_{jk} + \varepsilon_{ijk}$, $i = 1, \cdots, a$, $j = 1, \cdots, b$, $k = 1, \cdots, c$, with the constraints $\sum_i \tau_i = 0$, $\sum_j \theta_j = 0$, $\sum_k \eta_k = 0$, $\sum_j (\tau\theta)_{ij} = 0$, $\sum_i (\tau\theta)_{ij} = 0$, $\sum_j (\theta\eta)_{jk} = 0$, and $\sum_k (\theta\eta)_{jk} = 0$. Develop a suitable sequence of nested hypotheses, giving the relevant sums of squares. Complete the ANOVA table.

7.21. Consider an ANOVA model with three factors A, B, C as in Example 7.3.8. Suppose all three factors have a levels each, and $n = 1$, we can consider the $a \times a$ Latin Square Design (LSD):

$$Y_{ijk} = \mu + \tau_i + \theta_j + \eta_k + \varepsilon_{ijk},$$

for $i, j, k = 1, \cdots, a$. Develop a sequence of F-statistics for testing the effects due to the three factors and the overall mean effect.

7.22. Consider the measurement error model in Exercise 4.3. Assuming that ε_i are i.i.d. $N(0, \sigma^2)$ variables, derive the distribution of the least squares estimator of the unknown force θ, and the level $(1 - \alpha)$ confidence interval for θ.

7.23. In Exercise 4.5, assume that the errors have a normal distribution. Construct 95% confidence intervals for β_1, β_2, β_3, $\beta_1 - \beta_2$, and $\beta_1 + \beta_3$.

7.24. Consider two parallel regression lines $Y_{l,i} = \beta_{l,0} + \beta X_{l,i} + \varepsilon_{l,i}$, $i = 1, \cdots, n_l$, $l = 1, 2$, $\varepsilon_{l,i} \sim N(0, \sigma^2)$. Obtain a point estimate and an interval estimate for the horizontal distance $\delta = (\beta_{2,0} - \beta_{1,0})/\beta$ between the two lines.

7.25. Consider the regression model $\mathbf{y} = \mathbf{X}\boldsymbol{\beta} + \boldsymbol{\varepsilon}$, $\boldsymbol{\varepsilon} \sim N(\mathbf{0}, \sigma^2\mathbf{V})$, where $\mathbf{V}$ is a known p.d. matrix. Obtain a simultaneous confidence set for $\mathbf{c}'\boldsymbol{\beta}$ for all $\mathbf{c}$ in $\mathcal{L}$, using Scheffé's projection method.

8

Inference for the General Linear Model-II

This chapter continues with topics on inference for the general linear model (GLM). We begin with a description of likelihood based approaches for parameter estimation and model selection using information criteria. This is followed by a brief section on departures from model assumptions in Section 8.2. We then describe a topic that is usually referred to as "regression diagnostics", but that may be applied more widely to the GLM. We end this chapter with a discussion of prediction and calibration in Section 8.4.

8.1 Likelihood-based approaches

Least squares estimation does not require specification of a probability distribution for the errors $\boldsymbol{\varepsilon}$. In this section, we assume that $\boldsymbol{\varepsilon}$ follows a normal probability distribution, derive the maximum likelihood estimate of $\boldsymbol{\beta}$ and study its properties. In general, if $\mathbf{z} = (Z_1, \cdots, Z_N)'$ is a random sample from a population with p.d.f. or p.m.f. $f(z; \boldsymbol{\theta})$, where $\boldsymbol{\theta} = (\theta_1, \cdots, \theta_q)'$, the likelihood function is defined by

$$L(\boldsymbol{\theta}; \mathbf{z}) = \prod_{i=1}^{N} f(z_i; \boldsymbol{\theta}).$$

The maximum likelihood estimator of $\boldsymbol{\theta}$ is a point estimator, denoted by $\widehat{\boldsymbol{\theta}}_{ML}$, and is the value in the parameter space Θ that maximizes $L(\boldsymbol{\theta}; \mathbf{z})$, i.e., the value in Θ for which the observed sample $\mathbf{z}$ is most likely. In practice, it is convenient to obtain $\widehat{\boldsymbol{\theta}}_{ML}$ by maximizing $\log L(\boldsymbol{\theta}; \mathbf{z})$, which is a monotonic function of $L(\boldsymbol{\theta}; \mathbf{z})$. That is, we find $\widehat{\boldsymbol{\theta}}_{ML}$ such that

$$\log L(\widehat{\boldsymbol{\theta}}_{ML}; \mathbf{z}) = \sup_{\boldsymbol{\theta} \in \Theta} \log L(\boldsymbol{\theta}; \mathbf{z}).$$

There are situations where this supremum is attained at an interior point of Θ, and other situations where the supremum is attained on the boundary of Θ. In the former case, if $\log L(\boldsymbol{\theta}; \mathbf{z})$ is a differentiable function of $\boldsymbol{\theta}$, $\widehat{\boldsymbol{\theta}}_{ML}$ may be obtained as a solution to the equations

$$\partial \log L(\boldsymbol{\theta}; \mathbf{z}) / \partial \theta_i = 0, \quad i = 1, \cdots, q,$$

which satisfies $\log L(\boldsymbol{\theta}; \mathbf{z}) \geq \log L(\widehat{\boldsymbol{\theta}}_{ML}; \mathbf{z})$ for all $\boldsymbol{\theta} \in \Theta$. The MLE of $\boldsymbol{\theta}$ is unique if $f(z; \boldsymbol{\theta})$ is unimodal, which can be verified in practice by the log concavity of $f(z; \boldsymbol{\theta})$. For instance, as discussed in Chapter 6, given a random sample $\mathbf{x}_1, \cdots, \mathbf{x}_N$ from a $N_k(\boldsymbol{\mu}, \boldsymbol{\Sigma})$ population, the MLEs of $\boldsymbol{\mu}$ and $\boldsymbol{\Sigma}$ are unique.

Consider a situation where we are interested in estimating some function $h(\boldsymbol{\theta})$, although $f(z; \boldsymbol{\theta})$ is itself indexed by $\boldsymbol{\theta}$. By the *invariance property* of the maximum likelihood estimator, $h(\widehat{\boldsymbol{\theta}}_{ML})$ is the MLE of $h(\boldsymbol{\theta})$. This property is useful in linear model theory, as the following result shows. Let $\boldsymbol{\theta} = \mathbf{C}'\boldsymbol{\beta}$ denote an estimable function of $\boldsymbol{\beta}$.

DOI: 10.1201/9781315156651-8

8.1.1 Maximum likelihood estimation under normality

Assuming that $\boldsymbol{\varepsilon} \sim N(\mathbf{0}, \sigma^2\mathbf{I}_N)$ in (4.1.1), we obtain the maximum likelihood estimates of parameters.

Result 8.1.1. Consider the general linear model $\mathbf{y} = \mathbf{X}\boldsymbol{\beta} + \boldsymbol{\varepsilon}$ with normal errors. The maximum likelihood estimate of an estimable function $\boldsymbol{\theta} = \mathbf{C}'\boldsymbol{\beta}$ of $\boldsymbol{\beta}$ coincides with its least squares estimate, while $\widehat{\sigma}^2_{ML}$ is a scalar multiple of $\widehat{\sigma}^2$. In the full-rank case, by setting $\mathbf{C} = \mathbf{I}_N$, we obtain the MLEs of $\boldsymbol{\beta}$ and $\widehat{\sigma}^2_{ML}$ as

$$\begin{aligned} \widehat{\boldsymbol{\beta}}_{ML} &= (\mathbf{X}'\mathbf{X})^{-1}\mathbf{X}'\mathbf{y}, \\ \widehat{\sigma}^2_{ML} &= (\mathbf{y} - \mathbf{X}\widehat{\boldsymbol{\beta}}_{ML})'(\mathbf{y} - \mathbf{X}\widehat{\boldsymbol{\beta}}_{ML})/N. \end{aligned} \tag{8.1.1}$$

Proof. When $\mathbf{y} \sim N(\mathbf{X}\boldsymbol{\beta}, \sigma^2\mathbf{I}_N)$, the logarithm of the likelihood function is (see (B.1))

$$\ell(\boldsymbol{\beta}, \sigma^2; \mathbf{y}, \mathbf{X}) = -\frac{N}{2}\log(2\pi\sigma^2) - \frac{1}{2\sigma^2}(\mathbf{y} - \mathbf{X}\boldsymbol{\beta})'(\mathbf{y} - \mathbf{X}\boldsymbol{\beta}). \tag{8.1.2}$$

We set the first partial derivatives of $\ell(\boldsymbol{\beta}, \sigma^2; \mathbf{y}, \mathbf{X})$ with respect to $\boldsymbol{\beta}$ and σ^2 equal to zero, and solve the resulting equations simultaneously. Specifically, using Result 2.7.2 and Result 2.7.3, we obtain the equations

$$\begin{aligned} \frac{\partial}{\partial \boldsymbol{\beta}}\ell(\boldsymbol{\beta}, \sigma^2; \mathbf{y}, \mathbf{X}) &= \frac{1}{2\sigma^2}(2\mathbf{X}'\mathbf{y} - 2\mathbf{X}'\mathbf{X}\boldsymbol{\beta}^0) \equiv \mathbf{0}, \\ \frac{\partial}{\partial \sigma^2}\ell(\boldsymbol{\beta}, \sigma^2; \mathbf{y}, \mathbf{X}) &= -\frac{N}{2\widehat{\sigma}^2_{ML}} + \frac{1}{2\widehat{\sigma}^4_{ML}}(\mathbf{y} - \mathbf{X}\boldsymbol{\beta}^0)'(\mathbf{y} - \mathbf{X}\boldsymbol{\beta}^0) \equiv 0, \end{aligned}$$

which we solve to obtain $\boldsymbol{\beta}^0 = (\mathbf{X}'\mathbf{X})^{-}\mathbf{X}'\mathbf{y} = \mathbf{G}\mathbf{X}'\mathbf{y}$ and $\widehat{\sigma}^2_{ML} = (\mathbf{y} - \mathbf{X}\boldsymbol{\beta}^0)'(\mathbf{y} - \mathbf{X}\boldsymbol{\beta}^0)/N$. When $\mathbf{C}'\boldsymbol{\beta}$ is estimable,

$$\mathbf{C}'\boldsymbol{\beta}^0 = \mathbf{C}'\mathbf{G}\mathbf{X}'\mathbf{y} = \mathbf{T}'\mathbf{X}\mathbf{G}\mathbf{X}'\mathbf{y}$$

using the definition of estimability. Since $\mathbf{X}\mathbf{G}\mathbf{X}'$ is invariant to the choice of $\mathbf{G}$, it follows that $\widehat{\mathbf{C}'\boldsymbol{\beta}} = \mathbf{C}'\boldsymbol{\beta}^0$ is unique, and is the MLE of $\mathbf{C}'\boldsymbol{\beta}$. In the full-rank case, by setting $\mathbf{C} = \mathbf{I}_N$, we can obtain the MLEs of $\boldsymbol{\beta}$ and σ^2 shown in (8.1.1). ■

In the full-rank case, $\widehat{\boldsymbol{\beta}}_{ML}$ is an unbiased estimate of $\boldsymbol{\beta}$, with covariance $\sigma^2(\mathbf{X}'\mathbf{X})^{-1}$ (obtained as the inverse of the Fisher information matrix). The minimal sufficient statistics for $\boldsymbol{\beta}$ and σ^2 are respectively $\widehat{\boldsymbol{\beta}}$ and $\widehat{\sigma}^2$, and $\widehat{\boldsymbol{\beta}}$ is the *MVUE* of $\boldsymbol{\beta}$ (Rao, 1973). Also, $E(\widehat{\sigma}^2_{ML}) = (N-p)\sigma^2/N$, so that $\widehat{\sigma}^2_{ML}$ is a biased estimator of σ^2, with the bias decreasing as N increases (relative to p).

Example 8.1.1. Let $Y_i = \beta_0 + \beta_1 X_i + \beta_2 X_i^2 + \varepsilon_i$, $i = 1, \cdots, N$, $(N > 3)$, where $\sum_{i=1}^N X_i = 0$, $\sum_{i=1}^N X_i^3 = 0$, and ε_i are i.i.d. $N(0, \sigma^2)$ variables. We assume that $\beta_2 \neq 0$. We wish to find the value X^*, $(-\infty < X^* < \infty)$, for which the quadratic model function attains an extremum (maximum for a profit function with $\beta_2 < 0$, and minimum for a cost function with $\beta_2 > 0$). Setting the first derivative of the function $f(x) = \beta_0 + \beta_1 x + \beta_2 x^2$ equal to zero, and solving, we get $x^* = -\beta_1/2\beta_2$. Under normality, the maximum likelihood estimator of $\boldsymbol{\beta} = (\beta_0, \beta_1, \beta_2)'$ coincides with its LS estimator $\widehat{\boldsymbol{\beta}} = (\mathbf{X}'\mathbf{X})^{-1}\mathbf{X}'\mathbf{y}$, with

$$\mathbf{X}'\mathbf{X} = \begin{pmatrix} N & 0 & \sum X_i^2 \\ 0 & \sum X_i^2 & 0 \\ \sum X_i^2 & 0 & \sum X_i^4 \end{pmatrix} \quad \text{and} \quad \mathbf{X}'\mathbf{y} = \begin{pmatrix} \sum Y_i \\ \sum X_i Y_i \\ \sum X_i^2 Y_i \end{pmatrix},$$

i.e.,

$$\widehat{\beta}_0 = \{\textstyle\sum_{i=1}^N Y_i - \widehat{\beta}_2 \sum_{i=1}^N X_i^2\}/N,$$
$$\widehat{\beta}_1 = \{\textstyle\sum_{i=1}^N Y_i X_i\}/\{\sum_{i=1}^N X_i^2\}, \quad \text{and}$$
$$\widehat{\beta}_2 = \{\textstyle\sum_{i=1}^N Y_i X_i^2 - \widehat{\beta}_0 \sum_{i=1}^N X_i^2\}/\sum_{i=1}^N X_i^4.$$

By the invariance property of the MLE, $\widehat{X}_{ML}^* = -\widehat{\beta}_1/2\widehat{\beta}_2$. □

Example 8.1.2. In model (4.5.1), suppose $\boldsymbol{\varepsilon} \sim N(\mathbf{0}, \sigma^2\mathbf{V})$, then $\mathbf{c}'\boldsymbol{\beta}_{GLS}^0$ is the maximum likelihood estimator of the function $\mathbf{c}'\boldsymbol{\beta}$ provided the function is estimable. □

We can also motivate the F-statistic for testing H_0: $\boldsymbol{\beta} = \mathbf{0}$ in the full-rank linear model by using the likelihood ratio test, as the following result shows. It is straightforward to show the result for testing an estimable hypothesis H_0: $\mathbf{C}'\boldsymbol{\beta} = \mathbf{d}$ when $\mathrm{r}(\mathbf{X}) < p$.

Result 8.1.2. The statistic

$$F_0 = \frac{(N-p)}{s}(\Lambda^{-N/2} - 1) \tag{8.1.3}$$

is a monotonically decreasing function of the likelihood ratio test statistic Λ.

Proof. For the full-rank case, the maximum likelihood estimates of $\boldsymbol{\beta}$ and σ^2 obtained by maximizing the likelihood function over the entire (unrestricted) parameter space $\mathcal{R}^p \times \mathcal{R}^+$ were shown in (8.1.1) as

$$\widehat{\boldsymbol{\beta}}_{ML} = (\mathbf{X}'\mathbf{X})^{-1}\mathbf{X}'\mathbf{y} \quad \text{and} \quad \widehat{\sigma}_{ML}^2 = (\mathbf{y} - \mathbf{X}\widehat{\boldsymbol{\beta}}_{ML})'(\mathbf{y} - \mathbf{X}\widehat{\boldsymbol{\beta}}_{ML})/N.$$

The corresponding (unrestricted) maximized likelihood function is

$$L(\widehat{\boldsymbol{\beta}}_{ML}, \widehat{\sigma}_{ML}^2; \mathbf{y}, \mathbf{X}) = (2\pi\widehat{\sigma}_{ML}^2)^{-N/2}\exp\{-N/2\}.$$

The maximum likelihood estimates of $\boldsymbol{\beta}$ and σ^2 subject to the constraint $\mathbf{C}'\boldsymbol{\beta} = \mathbf{d}$ under H_0 are obtained by using the method of Lagrangian multipliers for maximizing the log-likelihood function subject to the linear constraint. It may be verified that $\widetilde{\boldsymbol{\beta}}_{ML} = \widetilde{\boldsymbol{\beta}}$, the restricted least squares estimate of $\boldsymbol{\beta}$ shown in Corollary 4.6.1, while

$$\widetilde{\sigma}_{ML}^2 = (\mathbf{y} - \mathbf{X}\widetilde{\boldsymbol{\beta}}_{ML})'(\mathbf{y} - \mathbf{X}\widetilde{\boldsymbol{\beta}}_{ML})/N.$$

The corresponding (restricted under H_0) maximized likelihood function is

$$L(\widetilde{\boldsymbol{\beta}}_{ML}, \widetilde{\sigma}_{ML}^2; \mathbf{y}, \mathbf{X}) = (2\pi\widetilde{\sigma}_{ML}^2)^{-N/2}\exp\{-N/2\}.$$

The likelihood ratio test statistic is

$$\Lambda = L(\widetilde{\boldsymbol{\beta}}_{ML}, \widetilde{\sigma}_{ML}^2; \mathbf{y}, \mathbf{X})/L(\widehat{\boldsymbol{\beta}}_{ML}, \widehat{\sigma}_{ML}^2; \mathbf{y}, \mathbf{X}) = (\widehat{\sigma}_{ML}^2/\widetilde{\sigma}_{ML}^2)^{N/2}, \tag{8.1.4}$$

and we reject H_0 if Λ is too small. We can see that F_0 defined in (8.1.3) is a monotonically decreasing function of Λ, and we reject H_0 if the statistic F_0 is too large. ■

8.1.2 Model selection criteria

Suppose Y is the observed response with unknown p.d.f. $g(.)$. In model selection, a set of models is available, each corresponding to a parametric family of p.d.f.'s $\{f(y;\theta); \theta \in \Theta\}$.

The functional form of $f(y;\theta)$, or the parameter space Θ, or both, are usually different for different models. The question is which model is most suitable for modeling Y. This section describes two criteria based on the likelihood function that are widely used in model selection: Akaike Information Criterion (AIC, Akaike, 1974) and Bayesian Information Criterion (BIC, Schwarz, 1978).

The AIC starts with the point of view that, given the sample Y, a suitable model should have small Kullback–Leibler (KL) divergence (see (C.1)) $\mathrm{KL}(g \,\|\, \widehat{f})$, where $\widehat{f} = f(y;\widehat{\theta}_{ML})$ with $\widehat{\theta}_{ML} = \widehat{\theta}_{ML}(Y)$, the MLE of θ under the model. Let $\ell(\theta;y) = \log f(y;\theta)$. Let $Y^* \sim g(.)$ be independent of Y. Then,

$$\mathrm{KL}(g \,\|\, \widehat{f}) = \int g(y) \log \frac{g(y)}{f(y;\widehat{\theta}_{ML})}\, dy = \mathrm{E}\left[\log g(Y^*)\right] - \mathrm{E}[\ell(\widehat{\theta}_{ML};Y^*) \,|\, Y]$$
$$= \mathrm{E}\left[\log g(Y^*)\right] - \ell(\widehat{\theta}_{ML};Y) + \Delta/2,$$

where

$$\Delta = 2\ell(\widehat{\theta}_{ML};Y) - 2\,\mathrm{E}[\ell(\widehat{\theta}_{ML};Y^*) \,|\, Y].$$

Since $\widehat{\theta}_{ML}$ is a function of Y only, it is independent of Y^* and is a constant conditional on Y. Also, if $D(Y,Y^*) = 2[\ell(\widehat{\theta}_{ML};Y) - \ell(\widehat{\theta}_{ML};Y^*)]$, then $\Delta = \mathrm{E}[D(Y,Y^*) \,|\, Y]$ and $\mathrm{E}\,\Delta = \mathrm{E}[D(Y,Y^*)]$.

Since $-\,\mathrm{E}[\log g(Y^*)]$, known as the entropy of g, does not depend on the model, minimizing $\mathrm{KL}(g \,\|\, \widehat{f})$ is equivalent to minimizing $\Delta - 2\ell(\widehat{\theta}_{ML};Y)$. However, while $\ell(\widehat{\theta}_{ML};Y)$ is available from Y, Δ is less tractable. The AIC of the model is obtained by replacing Δ by an approximation of $\mathrm{E}\,\Delta$. We then choose a model with the smallest AIC as the best model. The approximation relies on the asymptotic normality of the MLE, which is beyond the scope of this book; see chapter 7 of Hastie et al. (2009). The AIC thus evaluates the model from the standpoint of making a prediction of independent *future* (unobserved) data Y^* generated from the same unknown true distribution as the observed data Y.

The BIC arises from a very different point of view, i.e., a suitable model should have high posterior probability (likelihood) given the data. Assume a uniform prior on the set of models. By Bayes' rule, the posterior probability of a model M given data Y is

$$\pi(\mathcal{M} \,|\, Y) \propto f(Y \,|\, \mathcal{M}) = \int e^{\ell(\theta;Y)} \pi(\theta \,|\, \mathcal{M})\, d\theta,$$

where $\pi(\theta \,|\, \mathcal{M})$ is the prior on the parameter θ of $\mathcal{M}$. The BIC is obtained as an approximation of $-2\log \int e^{\ell(\theta;Y)} \pi(\theta \,|\, \mathcal{M})\, d\theta$ and a model with the smallest BIC is considered optimal. According to the asymptotic normality of the MLE, to a certain extent, $\ell(\theta;Y) \approx \ell(\widehat{\theta}_{ML};Y) - (\theta - \widehat{\theta}_{ML})^2/(2I)$, where I is the Fisher information. This enables the application of the so-called Laplace approximation (see item 4 in Appendix C) to get the specific form of BIC.

We show that these measures have simple forms under the normal linear model. The derivations do not need the asymptotic normality of the MLE. Suppose the observed response $\mathbf{y} = (Y_1, \cdots, Y_N)' \sim N(\mathbf{X}\boldsymbol{\beta}, \sigma^2 \mathbf{I}_N)$, where $\mathbf{X}$ is an $N \times p$ matrix of full column rank p (assumed to be the dimension of the model) and $\boldsymbol{\beta}$ and σ^2 are unknown parameters. Note that we can assume that $\mathbf{X}$ has full column rank without loss of generality because, if it does not, we can always construct the model $\mathbf{y} = \mathbf{X}_1\boldsymbol{\beta}_1 + \boldsymbol{\varepsilon}$, where $\mathbf{X}_1$ is a submatrix of $\mathbf{X}$ with LIN columns; and the two models will have the same AIC and BIC. Let $\ell(\boldsymbol{\beta}, \sigma^2; \mathbf{y}, \mathbf{X})$ denote the log-likelihood function. Let $\widehat{\boldsymbol{\beta}}_{ML}$ and $\widehat{\sigma}^2_{ML}$ denote the maximum likelihood estimators of $\boldsymbol{\beta}$ and σ^2, respectively.

Result 8.1.3. The AIC for a model with dimension p is

$$\text{AIC} = 2p - 2\ell(\widehat{\boldsymbol{\beta}}_{ML}, \widehat{\sigma}^2_{ML}; \mathbf{y}, \mathbf{X}) = N\log(2\widehat{\sigma}^2_{ML}) + N + 2p. \tag{8.1.5}$$

Proof. The task is to calculate

$$\begin{aligned} \text{E}(\Delta) &= \text{E}[D(\mathbf{y}, \mathbf{y}^*)], \text{ where,} \\ D(\mathbf{y}, \mathbf{y}^*) &= 2\ell(\widehat{\boldsymbol{\beta}}_{ML}, \widehat{\sigma}^2_{ML}; \mathbf{y}, \mathbf{X}) - 2\ell(\widehat{\boldsymbol{\beta}}_{ML}, \widehat{\sigma}^2_{ML}; \mathbf{y}^*, \mathbf{X}), \end{aligned}$$

where $\mathbf{y}^* = (Y_1^*, \cdots, Y_N^*)'$ is an independent random vector with the same distribution as $\mathbf{y}$. For any $\mathbf{b} \in \mathcal{R}^p$ and $s^2 > 0$,

$$\begin{aligned} &\ell(\mathbf{b}, s^2; \mathbf{y}, \mathbf{X}) - \ell(\mathbf{b}, s^2; \mathbf{y}^*, \mathbf{X}) = \frac{\|\mathbf{y}^* - \mathbf{X}\mathbf{b}\|^2 - \|\mathbf{y} - \mathbf{X}\mathbf{b}\|^2}{2s^2} \\ &= \frac{\|\mathbf{y}^* - \mathbf{X}\widetilde{\boldsymbol{\beta}}_{ML}\|^2 + \|\mathbf{X}(\mathbf{b} - \widetilde{\boldsymbol{\beta}}_{ML})\|^2 - \|\mathbf{y} - \mathbf{X}\widehat{\boldsymbol{\beta}}_{ML}\|^2 - \|\mathbf{X}(\mathbf{b} - \widehat{\boldsymbol{\beta}}_{ML})\|^2}{2s^2}, \end{aligned}$$

where $\widetilde{\boldsymbol{\beta}}_{ML}$ and $\widetilde{\sigma}^2_{ML}$ are the MLE's of $\boldsymbol{\beta}$ and σ^2 based on $\mathbf{y}^*$. Then

$$D(\mathbf{y}, \mathbf{y}^*) = (N\widetilde{\sigma}^2_{ML} + \|\mathbf{X}(\widetilde{\boldsymbol{\beta}}_{ML} - \widehat{\boldsymbol{\beta}}_{ML})\|^2)/\widehat{\sigma}^2_{ML} - N,$$

yielding

$$\text{E}(\Delta) = \text{E}(N\widetilde{\sigma}^2_{ML}/\widehat{\sigma}^2_{ML}) + \text{E}\{\|\mathbf{X}(\widetilde{\boldsymbol{\beta}}_{ML} - \widehat{\boldsymbol{\beta}}_{ML})\|^2/\widehat{\sigma}^2_{ML}\} - N. \tag{8.1.6}$$

Throughout the rest of the proof, keep in mind that $(\widehat{\boldsymbol{\beta}}_{ML}, \widehat{\sigma}^2_{ML})$ is a function of $\mathbf{y}$, while $(\widetilde{\boldsymbol{\beta}}_{ML}, \widetilde{\sigma}^2_{ML})$ is a function of $\mathbf{y}^*$, and because $\mathbf{y}$ and $\mathbf{y}^*$ are independent, $(\widehat{\boldsymbol{\beta}}_{ML}, \widehat{\sigma}^2_{ML})$ and $(\widetilde{\boldsymbol{\beta}}_{ML}, \widetilde{\sigma}^2_{ML}$ are independent. Since $N\widetilde{\sigma}^2_{ML}/\sigma^2$ and $N\widehat{\sigma}^2_{ML}/\sigma^2$ are independent $\sim \chi^2_{N-p}$, $\widetilde{\sigma}^2_{ML}/\widehat{\sigma}^2_{ML} \sim F_{N-p,N-p}$, we see that

$$\text{E}(\widetilde{\sigma}^2_{ML}/\widehat{\sigma}^2_{ML}) = (N-p)/(N-p-2). \tag{8.1.7}$$

Next,

$$\text{E}(\|\mathbf{X}(\widetilde{\boldsymbol{\beta}}_{ML} - \widehat{\boldsymbol{\beta}}_{ML})\|^2/\widehat{\sigma}^2_{ML}) = \text{E}\{\text{E}(\|\mathbf{X}(\widetilde{\boldsymbol{\beta}}_{ML} - \widehat{\boldsymbol{\beta}}_{ML})\|^2 \,|\, \mathbf{y})/\widehat{\sigma}^2_{ML}\}.$$

Since

$$\begin{aligned} &\|\mathbf{X}(\widetilde{\boldsymbol{\beta}}_{ML} - \widehat{\boldsymbol{\beta}}_{ML})\|^2 \\ &= \|\mathbf{X}(\widetilde{\boldsymbol{\beta}}_{ML} - \boldsymbol{\beta})\|^2 + \|\mathbf{X}(\widehat{\boldsymbol{\beta}}_{ML} - \boldsymbol{\beta})\|^2 + 2(\mathbf{X}\widetilde{\boldsymbol{\beta}}_{ML} - \mathbf{X}\boldsymbol{\beta})'(\mathbf{X}\widehat{\boldsymbol{\beta}}_{ML} - \mathbf{X}\boldsymbol{\beta}), \end{aligned}$$

and $\text{E}(\mathbf{X}\widetilde{\boldsymbol{\beta}}_{ML}) = \mathbf{X}\boldsymbol{\beta}$ (see property 2 in Result 7.1.1),

$$\text{E}\{\|\mathbf{X}(\widetilde{\boldsymbol{\beta}}_{ML} - \widehat{\boldsymbol{\beta}}_{ML})\|^2 \,|\, \mathbf{y}\} = \text{E}\{\|\mathbf{X}(\widetilde{\boldsymbol{\beta}}_{ML} - \boldsymbol{\beta})\|^2\} + \|\mathbf{X}(\widehat{\boldsymbol{\beta}}_{ML} - \boldsymbol{\beta})\|^2.$$

As a result,

$$\begin{aligned} &\text{E}\{\|\mathbf{X}(\widetilde{\boldsymbol{\beta}}_{ML} - \widehat{\boldsymbol{\beta}}_{ML})\|^2/\widehat{\sigma}^2_{ML}\} \\ &= \text{E}\{\text{E}(\|\mathbf{X}(\widetilde{\boldsymbol{\beta}}_{ML} - \boldsymbol{\beta})\|^2)/\widehat{\sigma}^2_{ML}\} + \text{E}\{\|\mathbf{X}(\widehat{\boldsymbol{\beta}}_{ML} - \boldsymbol{\beta})\|^2/\widehat{\sigma}^2_{ML}\}. \end{aligned}$$

From property 5 of Result 7.1.1, $\mathrm{E}(\|\mathbf{X}(\widetilde{\boldsymbol{\beta}}_{ML}-\boldsymbol{\beta})\|^2)=\sigma^2 p$ and $\|\mathbf{X}(\widehat{\boldsymbol{\beta}}_{ML}-\boldsymbol{\beta})\|^2/\sigma^2\sim\chi^2_p$. Then

$$\begin{aligned}
&\mathrm{E}\{\|\mathbf{X}(\widetilde{\boldsymbol{\beta}}_{ML}-\widehat{\boldsymbol{\beta}}_{ML})\|^2/\widehat{\sigma}^2_{ML}\}\\
&= N\,\mathrm{E}\left(\frac{\sigma^2 p}{N\widehat{\sigma}^2_{ML}}\right)+\frac{Np}{N-p}\,\mathrm{E}\left[\frac{\|\mathbf{X}(\widehat{\boldsymbol{\beta}}_{ML}-\boldsymbol{\beta})\|^2/p}{N\widehat{\sigma}^2_{ML}/(N-p)}\right]\\
&= Np\,\mathrm{E}(1/\chi^2_{N-p})+\frac{Np}{N-p}\,\mathrm{E}(F_{p,N-p})=\frac{2Np}{N-p-2}.
\end{aligned} \tag{8.1.8}$$

Combining (8.1.6)–(8.1.8) yields $\mathrm{E}\,\Delta=2N(p+1)/(N-p-2)$. Since p is fixed, for $N\gg p$, $\mathrm{E}\,\Delta\approx 2p+2$. Ignoring the constant summands, this leads to (8.1.5). ■

Result 8.1.4. The BIC is

$$\mathrm{BIC}=p\log N-2\ell(\widehat{\boldsymbol{\beta}}_{ML},\widehat{\sigma}^2_{ML};\mathbf{y},\mathbf{X})=N\log(2\pi\widehat{\sigma}^2_{ML})+N+p\log N. \tag{8.1.9}$$

Proof. Let $\boldsymbol{\beta}$ and σ^2 have uniform priors on $\mathcal{R}^p$ and $(0,\infty)$, respectively. Suppose $\mathbf{X}$ has full column rank. Then

$$\begin{aligned}
f(\mathbf{y}\,|\,\mathcal{M}) &= \int_0^\infty\left(\int_{\mathcal{R}^p}\frac{1}{(2\pi v)^{N/2}}e^{-\|\mathbf{y}-\mathbf{X}\mathbf{b}\|^2/2v}\,d\mathbf{b}\right)dv\\
&= \int_0^\infty\frac{1}{(2\pi v)^{N/2}}\left(\int_{\mathcal{R}^p}e^{-\|\mathbf{y}-\mathbf{X}\widehat{\boldsymbol{\beta}}_{ML}\|^2/2v}\times e^{-\|\mathbf{X}(\mathbf{b}-\widehat{\boldsymbol{\beta}}_{ML})\|^2/2v}\,d\mathbf{b}\right)dv\\
&= \int_0^\infty\frac{1}{(2\pi v)^{N/2}}\left(\int_{\mathcal{R}^p}e^{-N\widehat{\sigma}^2_{ML}/2v}\times e^{-\|\mathbf{X}\mathbf{b}\|^2/2v}\,d\mathbf{b}\right)dv\\
&= \frac{1}{\sqrt{\det(\mathbf{X}'\mathbf{X})}}\int_0^\infty\frac{1}{(2\pi v)^{(N-p)/2}}e^{-N\widehat{\sigma}^2_{ML}/2v}\,dv.
\end{aligned}$$

Put $m=(N-p)/2$. By change of variable $t=1/v$,

$$\begin{aligned}
f(\mathbf{y}\,|\,\mathcal{M}) &= \frac{1}{\sqrt{\det(\mathbf{X}'\mathbf{X})}}\int_0^\infty\frac{s^{m-2}}{(2\pi)^m}e^{-N\widehat{\sigma}^2_{ML}s/2}\,dv\\
&= \frac{1}{\sqrt{\det(\mathbf{X}'\mathbf{X})}}\frac{\Gamma(m-1)}{2\pi^m}\frac{1}{(N\widehat{\sigma}^2_{ML})^{m-1}}.
\end{aligned}$$

By Stirling's formula (see item 5 in Appendix C), $\Gamma(t)\sim(t/e)^t\sqrt{2\pi t}$ as $t\to\infty$. For fixed p and as $N\to\infty$,

$$\begin{aligned}
f(\mathbf{y}\,|\,\mathcal{M}) &\sim \frac{1}{\sqrt{\det(\mathbf{X}'\mathbf{X})}}\left(\frac{m-1}{e}\right)^{m-1}\sqrt{2\pi m}\cdot\frac{1}{2\pi^m}\frac{1}{(N\widehat{\sigma}^2_{ML})^{m-1}}\\
&= \frac{1}{\sqrt{\det(\mathbf{X}'\mathbf{X})}}\left(\frac{N-p-2}{N}\right)^{(N-p-2)/2}\frac{\sqrt{m/2\pi}}{(2e\pi\widehat{\sigma}^2_{ML})^{m-1}}\\
&\sim \frac{1}{\sqrt{\det(\mathbf{X}'\mathbf{X})}}\frac{\sqrt{m/2\pi}}{(2e\pi\widehat{\sigma}^2_{ML})^{m-1}}e^{-(p+2)/2},
\end{aligned}$$

so that

$$\begin{aligned}
\log f(\mathbf{y}\,|\,\mathcal{M}) &= -\log\sqrt{\det(\mathbf{X}'\mathbf{X})}+(1/2)\log N-(m-1)\log(2e\pi\widehat{\sigma}^2_{ML})\\
&\quad -p/2+O(1).
\end{aligned}$$

Since $\ell(\widehat{\boldsymbol\beta}_{ML}, \widehat{\sigma}^2_{ML}; \mathbf{y}, \mathbf{X}) = -(N/2)\log(2\pi e \widehat{\sigma}^2_{ML})$, we have

$$\log f(\mathbf{y} \mid \mathcal{M}) = -\log\sqrt{\det(\mathbf{X}'\mathbf{X})} + (1/2)\log N \\ + (1 - p/N)\ell(\widehat{\boldsymbol\beta}_{ML}, \widehat{\sigma}^2_{ML}; \mathbf{y}, \mathbf{X}) - p/2 + O(1).$$

Assume that the rows of $\mathbf{X}$ are randomly sampled, but fixed. In many cases, for example, when the rows of $\mathbf{X}$ are i.i.d., for $N \gg p$, $(1/N)(\mathbf{X}'\mathbf{X}) \approx \mathbf{V}$, where $\mathbf{V} \in \mathbf{R}^{p\times p}$ is nonrandom and p.d. Then, $\det(\mathbf{X}'\mathbf{X}) \approx \det(N\mathbf{V}) = N^p \det(\mathbf{V})$ and hence,

$$\log f(\mathbf{y} \mid \mathcal{M}) = -[(p-1)/2]\log N + (1 - p/N)\ell(\widehat{\boldsymbol\beta}_{ML}, \widehat{\sigma}^2_{ML}; \mathbf{y}, \mathbf{X}) \\ - p/2 + O(1).$$

As mentioned earlier, BIC is defined as an approximation of $-2\log f(\mathbf{y} \mid \mathcal{M})$. This leads to (8.1.9) and completes the proof. ■

8.2 Departures from model assumptions

Residual analysis is a crucial step in assessing the adequacy of a fitted regression model. The analysis is mostly developed for the GLM subject to the assumption (4.1.4) on the errors, i.e., $\mathbf{y} = \mathbf{X}\boldsymbol\beta + \boldsymbol\varepsilon$, $E(\boldsymbol\varepsilon) = \mathbf{0}$ and $\text{Cov}(\boldsymbol\varepsilon) = \sigma^2\mathbf{I}_N$. Under the assumption, the vector of least squares (LS) fitted values and the vector of LS residuals are uncorrelated, i.e., $\text{Cov}(\widehat{\mathbf{y}}, \widehat{\boldsymbol\varepsilon}) = \mathbf{O}$, where $\widehat{\mathbf{y}} = \mathbf{P}\mathbf{y}$ and $\widehat{\boldsymbol\varepsilon} = \mathbf{y} - \widehat{\mathbf{y}}$, $\mathbf{P}$ being the orthogonal projection matrix onto $\mathcal{C}(\mathbf{X})$. If the fitted model is accurate, the behavior of the residuals, which may be viewed as "estimates" of the errors, should be similar to the underlying errors. A careful perusal of the residuals should therefore enable us to conclude either that the fitting procedure has not violated any assumptions, or that some or all of the assumptions may have been violated and there is merit in revising the model fit. The usual assumptions in a linear model include (a) linearity of the model, (b) homoscedasticity, (c) independence, and (d) normality. Residuals provide graphical and nongraphical summaries that enable us to verify departures from these assumptions.

8.2.1 Graphical procedures

Three basic residual plots that enable verification of the model assumptions are included routinely in almost all statistical software packages: (i) a plot of residuals versus each predictor, (ii) a plot of residuals versus fitted values $\widehat{Y}_i$, and (iii) a normal probability plot of residuals.

When a model gives a good fit, we expect that $\text{Cov}(\widehat{\boldsymbol\varepsilon}, \widehat{\mathbf{y}}) \approx \mathbf{O}$, i.e., the residuals are uncorrelated with the fitted values, so that an ideal plot of residuals versus fitted values should contain a random scatter of points in an approximate horizontal band centered at zero. Likewise, since $\widehat{\boldsymbol\varepsilon} \perp \mathcal{C}(\mathbf{X})$, an ideal plot of residuals versus predictors X_j, $j = 1, \cdots, k$ is also expected to show a random scatter of points. On the other hand, since $\text{Cov}(\widehat{\boldsymbol\varepsilon}, \mathbf{y})$ is non-zero, i.e., residuals are always correlated with the actual responses even when the model gives a perfect fit, a plot of residuals versus observed responses would be useless.

Departures from the "ideal" plot can occur in several ways, and correspond to violation of specific assumptions.

(a) A departure from the linearity assumption will be indicated by a tendency of the scatter to exhibit curvature. A remedy could be to include polynomial powers of suitable orders of the X_j's as additional predictors in the model.

(b) A funnel-shaped plot, which widens (becomes narrow) to the right indicates that the error variance increases (decreases) with increasing values of the predictor (or fit), i.e., heteroscedasticity. This departure from homoscedasticity of errors may be corrected by an appropriate (variance stabilizing) transformation or by using weighted (instead of ordinary) least squares (see Section 4.5).

(c) In some cases, especially when the data is observed over time, the independence assumption on the errors is violated. A plot of residuals versus order (or time) indicates this by a pattern of runs, i.e., high (low) values followed by high (low) values. A remedy would be to use GLS instead of LS (again, see Section 4.5).

(d) The normality assumption is checked via a normal probability plot of the residuals, which is a plot of the empirical quantiles of the residuals versus corresponding quantiles from a standard normal distribution. Departure from a straight line indicates violation of the normality assumption. Remedies may include a transformation of Y_i to normality (see Section 5.5.1) or the use of alternative distributions to model the response (see Section 5.5.2).

8.2.2 Heteroscedasticity

A GLM is heteroscedastic if $\text{Var}(\boldsymbol{\varepsilon}) = \text{diag}(\sigma_1^2, \cdots, \sigma_N^2)$, where not all the σ_i^2 are the same. That is, the errors are still uncorrelated, but do not have identical variances. Heteroscedasticity is common in many applications. For example, in a cross-sectional study of firms within an industry, revenues of large firms might be more variable than revenues of smaller firms. A detailed discussion of modeling and diagnostics under heterogeneity is given in Carroll and Ruppert (1988).

Example 8.2.1. Consider the simple linear regression model

$$\mathbf{y} = \beta\mathbf{x} + \boldsymbol{\varepsilon},$$

where $\text{Var}(\boldsymbol{\varepsilon}) = \text{diag}(\sigma_1^2, \cdots, \sigma_N^2)$. The LS estimate of β is $\widehat{\beta} = \mathbf{x}'\mathbf{y}/(\mathbf{x}'\mathbf{x})$ and we can see that

$$\text{Var}(\widehat{\beta}) = V^{-1}\sum_{i=1}^{N} a_i\sigma_i^2,$$

where $V = \mathbf{x}'\mathbf{x}$ and $a_i = X_i^2/V$. Under the assumption that σ_i^2 are all equal to some unknown $\sigma^2 > 0$, the LS estimators of σ^2 and $\text{Var}(\widehat{\beta})$ are given by (4.2.17) and Corollary 4.2.1, so that $\widehat{\sigma}^2 = (\mathbf{y}-\widehat{\beta}\mathbf{x})'(\mathbf{y}-\widehat{\beta}\mathbf{x})/(N-1)$ and $\widehat{\text{Var}}(\widehat{\beta}) = \widehat{\sigma}^2/(\mathbf{x}'\mathbf{x})$. Since $\mathbf{P} = \mathbf{x}\mathbf{x}'/(\mathbf{x}'\mathbf{x})$,

$$\text{E}(\widehat{\sigma}^2) = (N-1)^{-1}\,\text{E}\{tr[(\mathbf{I}-\mathbf{P})\boldsymbol{\varepsilon}\boldsymbol{\varepsilon}']\} = (N-1)^{-1}\sum_{i=1}^{N}(1-a_i)\sigma_i^2,$$

so that

$$\text{E}(\widehat{\text{Var}}(\widehat{\beta})) = \frac{\sum_{i=1}^{N}(1-a_i)\sigma_i^2}{(N-1)V}.$$

Note that $a_1 + \cdots + a_N = 1$ and $a_i \geq 0$. If the model is actually heteroscedastic, then from

Chapter 4, $\widehat{\beta}$ is unbiased but not efficient, i.e, it is no longer the b.l.u.e. for β. Furthermore, from the above displays, the LS estimator of $\text{Var}(\widehat{\beta})$ is biased, which can result in misleading conclusions from hypothesis tests about the regression parameter β. □

Example 8.2.2. For the simple regression model in Example 8.2.1, if $\text{Var}(\varepsilon) = \text{diag}(\sigma^2 s_1, \cdots, \sigma^2 s_N)$, where $s_i > 0$ are known and $\sigma^2 > 0$ is unknown, then from Chapter 4, the b.l.u.e. of β is the weighted least squares (WLS) estimate given by Corollary 4.5.2. We have

$$\widehat{\beta}_{WLS} = \frac{\sum_{i=1}^N X_i Y_i / s_i}{\sum_{i=1}^N X_i^2 / s_i}, \quad \widehat{\sigma}^2_{WLS} = \frac{1}{N-1} \sum_{i=1}^N \frac{(Y_i - \widehat{\beta}_{WLS} X_i)^2}{s_i},$$

and $\text{Var}(\widehat{\beta}_{WLS}) = \sigma^2 / \sum_{i=1}^N (X_i^2 / s_i)$.

If $\sigma_i^2 = \sigma^2 X_i^2$, where $X_i \neq 0$, then

$$\widehat{\beta}_{WLS} = \frac{1}{N} \sum_{i=1}^N \frac{Y_i}{X_i}, \quad \widehat{\sigma}^2_{WLS} = \frac{1}{N-1} \sum_{i=1}^N \frac{(Y_i - \widehat{\beta}_{WLS} X_i)^2}{X_i^2},$$

and $\text{Var}(\widehat{\beta}_{WLS}) = \sigma^2 / N$.

On the other hand, if $Y_i = \sum_{j=1}^{n_i} U_{ij}/n_i$ and $X_i = \sum_{j=1}^{n_i} V_{ij}/n_i$ are aggregated variables, and the U's are uncorrelated with the same variance σ^2, then $\sigma_i^2 = \sigma^2 / n_i$, giving

$$\widehat{\beta}_{WLS} = \frac{\sum_{i=1}^N n_i X_i Y_i}{\sum_{i=1}^N n_i X_i^2}, \quad \widehat{\sigma}^2_{WLS} = \frac{1}{N-1} \sum_{i=1}^N n_i (Y_i - \widehat{\beta}_{WLS} X_i)^2,$$

and $\text{Var}(\widehat{\beta}_{WLS}) = \sigma^2 / (\sum_{i=1}^N n_i X_i^2)$. □

In addition to graphical methods that help us to diagnose heteroscedasticity, we can also carry out formal tests for departures from homoscedasticity. Recall the two-sample F-test of H_0: $\sigma_1^2 = \sigma_2^2$ versus H_1: $\sigma_1^2 > \sigma_2^2$, based on normal random samples of sizes n_1 and n_2. If the observed test statistic $F_0 = S_1^2 / S_2^2$ is greater than the critical $F_{n_1-1, n_2-1, \alpha}$, we reject the null hypothesis at level α. This approach can be extended for testing the assumption of homogeneity of variances in multiple populations; see Judge et al. (1985), section 11.2.

The following results describe some tests for heteroscedasticity without actually specifying its functional form. Let Y_{ij} $1 \le j \le n_i$, $1 \le i \le a$, be independent observations, such that for each i, $Y_{i1}, \cdots, Y_{in_i}$ are i.i.d. with the same unknown variance σ_i^2, i.e., $\text{Var}(\varepsilon_{ij})$ are known to be the same within each of the a groups. The total sample size is $N = n_1 + \cdots + n_a$. Suppose we wish to test the homogeneity of the variances, i.e.,

$$H_0: \sigma_1^2 = \cdots = \sigma_a^2 \quad \text{versus} \quad H_1: \sigma_i^2 \text{ are not all equal.}$$

Result 8.2.1. Bartlett's test for homogeneity of variances. Let

$$S_i^2 = \sum_{j=1}^{n_i} \frac{(Y_{ij} - \overline{Y}_{i\cdot})^2}{n_i - 1}, \quad i = 1, \cdots, a, \quad \overline{S}^2 = \frac{1}{N-a} \sum_{i=1}^a (n_i - 1) S_i^2.$$

Define

$$M = (N-a) \log \overline{S}^2 - \sum_{i=1}^a (n_i - 1) \log S_i^2, \text{ and}$$

$$C = 1 + \frac{1}{3(a-1)} \left\{ \sum_{i=1}^a \frac{1}{n_i - 1} - \frac{1}{N-a} \right\}.$$

If the Y_{ij}'s are normally distributed, then under H_0, M/C is approximately χ^2_{a-1} distributed. Bartlett's test rejects H_0 at significance level α if $M/C > \chi^2_{a-1,\alpha}$.

Proof. For $i = 1, \cdots, a$, letting $\nu_i = n_i - 1$, $\nu_i S_i^2 \sim \chi^2_{\nu_i}$. Likewise, letting $\nu = N - a$, $\nu \overline{S}^2 \sim \chi^2_\nu$. Since $-M + \nu \log \overline{S}^2 = \sum_{i=1}^a \nu_i \log S_i^2$, and since the S_i's are independent, it follows that $\mathrm{E}[e^{t(-M+\nu \log \overline{S}^2)}] = \prod_{i=1}^a \mathrm{E}(e^{t\nu_i \log S_i^2})$ for any t. On the other hand, since $(S_1^2/\overline{S}^2, \cdots, S_a^2/\overline{S}^2)'$ is independent of $\overline{S}^2$, then so is $M = \sum_{i=1}^a \nu_i \log(S_i^2/\overline{S}^2)$. Hence, $\mathrm{E}(e^{-tM})\,\mathrm{E}[(\overline{S}^2)^{t\nu}] = \prod_{i=1}^a \mathrm{E}[(S_i^2)^{t\nu_i}]$, or

$$\mathcal{K}(-t) = \sum_{i=1}^a \log \mathrm{E}[(S_i^2)^{t\nu_i}] - \log \mathrm{E}[(\overline{S}^2)^{t\nu}], \tag{8.2.1}$$

where $\mathcal{K}(.)$ is the cumulant generating function (c.g.f.) of M, and each term on the right side can be explicitly evaluated under a chi-square distribution. When all $n_i \to \infty$ such that $n_i/n_{i'}$ stay bounded for $i \neq i'$, we can compare the asymptotic expansion of the right side of (8.2.1) and that of the c.g.f. of a χ^2_{a-1} variable to see that M/C is more closely approximated by a χ^2_{a-1} distribution than M is; for details, see Bartlett (1937). ■

Bartlett's test is sensitive to the normality assumption. The test proposed by Levene (1960) is an approximate test for the homogeneity of variances from a populations, and is somewhat less sensitive to departures from normality than Bartlett's test since it uses the average of absolute deviations instead of mean square deviations as a measure of variation within a group.

Result 8.2.2. Levene's test for homogeneity of variances. Define the quantity

$$Z_{ij} = |Y_{ij} - \overline{Y}_{i\cdot}|/n_i;$$

compute

$$W = \frac{(N-a)\sum_{i=1}^a n_i(\overline{Z}_{i\cdot} - \overline{Z}_{\cdot\cdot})^2}{(a-1)\sum_{i=1}^a \sum_{j=1}^{n_i}(Z_{ij} - \overline{Z}_{i\cdot})^2}.$$

Then, Levene's test rejects H_0 at signficance level α if $W > F_{a-1,N-a,\alpha}$.

Now, consider the GLM $\mathbf{y} = \mathbf{X}\boldsymbol{\beta} + \boldsymbol{\varepsilon}$, where $\mathbf{X}$ has full column rank p and $\boldsymbol{\varepsilon} \sim N(\mathbf{0}, \boldsymbol{\Sigma})$ with $\boldsymbol{\Sigma} = \mathrm{diag}(\sigma_1^2, \cdots, \sigma_N^2)$. Suppose we wish to test the homogeneity of the variances, i.e.,

$$H_0\colon \sigma_1^2 = \cdots = \sigma_N^2. \tag{8.2.2}$$

We describe without proof three widely used procedures for testing H_0 versus certain alternatives. These procedures are available in many statistical software for checking homoscedasticity and are respectively due to Goldfeld and Quandt (1965), Breusch and Pagan (1979), and White (1980).

Result 8.2.3. Goldfeld–Quandt test. This test is used when it is suspected that σ_i^2 is an increasing function of a scalar function $f(\mathbf{z}_i) \geq 0$, where f is known and $\mathbf{z}_i$ may include some or all predictor variables in $\mathbf{x}_i$ as well as other predictor variables. To test H_0 in (8.2.2) versus H_1: σ_i^2 is increasing in $f(\mathbf{z}_i)$, sort the observations by the value of $f(\mathbf{z}_i)$ in ascending order and fit two regressions, one for the first $(N-m)/2$ observations and one for the last $(N-m)/2$ observations, where $m \approx N/5$. Let MSE_S and MSE_L be the MSEs of the two regressions. Then, asymptotically, i.e., as $N \to \infty$

$$MSE_L/MSE_S \sim F_{(N-m-2p)/2,(N-m-2p)/2} \text{ under } H_0.$$

The test rejects H_0 in favor of H_1 at significance level α if $MSE_L/MSE_S > F_{(N-m-2p)/2,(N-m-2p)/2,\alpha}$.

The Goldfeld–Quandt test is a popular test for structural breaks in variance. However, when it is suspected that more than one predictor variable is causing heteroscedasticity, the test may not be useful.

Result 8.2.4. Breusch–Pagan test. The Breusch–Pagan test is used when an assumption that the error variance is an increasing function of a given scalar function of the predictor variables is untenable. Suppose

$$\sigma_i^2 = h(\mathbf{z}_i'\boldsymbol{\gamma}), \tag{8.2.3}$$

where h is a twice differentiable function, $\mathbf{z}_i = (1, Z_{i1}, \cdots, Z_{iq})'$ is a vector of predictors which can include predictor variables other than the X_i's, and $\boldsymbol{\gamma} = (\gamma_0, \gamma_1, \cdots, \gamma_q)'$. Then (8.2.2) is equivalent to H_0: $\gamma_1 = \cdots = \gamma_q = 0$. To test H_0 versus H_1: not all $\gamma_1, \cdots, \gamma_q$ are 0, first, obtain the vector of LS residuals $\widehat{\boldsymbol{\varepsilon}}$ from the linear model $\mathbf{y} = \mathbf{X}\boldsymbol{\beta} + \boldsymbol{\varepsilon}$. Letting $\mathbf{u} = (U_1, \cdots, U_N)'$ with $U_i = \widehat{\varepsilon}_i^2/(\widehat{\boldsymbol{\varepsilon}}'\widehat{\boldsymbol{\varepsilon}})$, the Breusch–Pagan test statistic is defined to be half of the SSR of the LS fit of the linear model $\mathbf{u} = \mathbf{Z}\boldsymbol{\gamma} + \boldsymbol{\eta}$. Under H_0, the Breusch–Pagan test statistic asymptotically follows a χ_q^2 distribution.

The assumption of normal errors is critical for the Breusch–Pagan test. The test proposed by White (1980) is closely related to the Breusch–Pagan test, but does not depend critically on normality of errors.

Result 8.2.5. White's test. To test H_0 in (8.2.2) versus H_1: not all σ_i^2 are equal, let $\mathbf{z}_i$ be a vector with 1 and all products of the form $X_{ij}X_{il}$, $j \le l$ as components. Let $\widehat{\boldsymbol{\varepsilon}}$ be the vector of LS residuals of the linear model $\mathbf{y} = \mathbf{X}\boldsymbol{\beta} + \boldsymbol{\varepsilon}$. Obtain LS fit for the model

$$\widehat{\varepsilon}_i^2 = \mathbf{z}_i'\boldsymbol{\eta} + u_i, \quad i = 1, \cdots, N,$$

and calculate the coefficient of determination R^2. The statistic for White's test is defined to be NR^2. As $N \to \infty$, $NR^2 \sim \chi^2_{p(p+1)/2}$ under H_0.

Numerical Example 8.1. We consider the data set "compasst" from the R package *HoRM* which is based on a study of computer-assisted learning by 12 students in an effort to assess the cost of computer time. The response variable is cost (in cents), while the predictor variable is num.responses (total number of responses in completing a lesson).

```
library(HoRM)
data(compasst)
ols <- lm(cost ~ num.responses, data = compasst)
plot(
  ols$fitted.values,
  ols$residuals,
  xlab = ''OLS fits'',
  ylab = ''OLS residuals''
)
```

The plot of OLS residuals versus fitted values shows a funnel-shaped pattern. We show results from White's test using the *lmtest* package. The data rejects the homoscedasticity assumption. Other tests may be done in a similar way.

```
lmtest::bptest(ols, ~ I(num.responses ^ 2), data = compasst)

studentized Breusch-Pagan test

data:  ols
BP = 6.7651, df = 1, p-value = 0.009296
```

We handle the heteroscedasticity by WLS estimation. To get the weights, we carry out an OLS regression of the square (or absolute values) of the OLS residuals on the OLS fits. We square the fitted values from this regression and employ these values as the weights for the WLS fit. The plot of WLS residuals versus fits does not have the pronounced funnel shaped pattern.

```
new.fits = lm(abs(ols$residuals) ~ ols$fitted)$fitted^2
wls <- lm(cost ~ num.responses, weights = 1 / new.fits,
          data = compasst)
lmtest::bptest(wls, ~ I(num.responses ^ 2), data = compasst)
summary(wls)$coefficients # WLS Estimation

 Estimate Std. Error  t value     Pr(>|t|)
(Intercept)    17.300637   4.827736 3.583592 4.981868e-03
num.responses  3.421106   0.370310 9.238492 3.268919e-06

summary(wls)$r.squared     # WLS R-square
[1] 0.8951229

summary(wls)$sigma^2       # Estimate of sigma^2 WLS
[1] 1.344104

plot(wls$fitted.values, wls$residuals,xlab=''WLS fits'',
ylab=''WLS residuals''))
```

▲

8.2.3 Serial correlation

When data is observed over time, the assumption of independent errors is often suspect, as indicated by a plot of residuals versus order (time). The linear model with serially correlated errors has the form

$$Y_t = \mathbf{x}_t'\boldsymbol{\beta} + \varepsilon_t = \beta_0 + \sum_{j=1}^{k} \beta_j X_{tj} + \varepsilon_t, \quad t = 1, \cdots, N, \tag{8.2.4}$$

where the subscript t is used to indicate time, and error terms from different time periods are correlated, i.e., $\text{Corr}(\varepsilon_t, \varepsilon_s)$ may be nonzero for $t \neq s$. This model can be expressed in the form (4.5.1) with $\text{E}(\boldsymbol{\varepsilon}\boldsymbol{\varepsilon}') = \sigma^2\mathbf{V}$, where $\mathbf{V}$ is an $N \times N$ p.d. matrix whose form is specified under an assumption that $\{\varepsilon_t\}$ is a stationary stochastic process. This assumption implies that the first two moments of the distribution of $\boldsymbol{\varepsilon}$ do not depend on t, and we can write

$$\mathbf{V} = \begin{pmatrix} 1 & \rho_1 & \rho_2 & \cdots & \rho_{N-1} \\ \rho_1 & 1 & \rho_1 & \cdots & \rho_{N-2} \\ \rho_2 & \rho_1 & 1 & \cdots & \rho_{N-3} \\ \vdots & \vdots & \vdots & \ddots & \vdots \\ \rho_{N-1} & \rho_{N-2} & \rho_{N-3} & \cdots & 1 \end{pmatrix},$$

where for $j \geq 1$, $\rho_j = \text{E}(\varepsilon_t \varepsilon_{t-j})/\sigma^2 = \text{E}(\varepsilon_t \varepsilon_{t+j})/\sigma^2$ is the autocorrelation between two random errors that are j time periods apart. This specification is still rather general, and the most commonly assumed specification for the stationary stochastic process is the first-order autoregressive ($AR(1)$) process, which we introduced in Example 4.5.1, in which case, $\mathbf{V}$ has the form in (4.5.11) which is a function of a single parameter $\rho \in (-1, 1)$. Let $\mathbf{X} = (\mathbf{x}_1, \cdots, \mathbf{x}_N)'$ and $\widehat{\boldsymbol{\varepsilon}} = (\mathbf{I} - \mathbf{P})\mathbf{y} = (\mathbf{I} - \mathbf{P})\boldsymbol{\varepsilon}$ the vector of LS residuals, where $\mathbf{P}$ is the matrix of orthogonal projection onto $\mathcal{C}(\mathbf{X})$.

Example 8.2.3. Consider the multiple regression model with serially correlated errors, i.e., (8.2.4) where $\text{E}(\varepsilon_t) = 0$ and $\text{Cov}(\varepsilon_t, \varepsilon_s) = \sigma^2 \rho^{|t-s|}$, $s, t = 1, \cdots, N$, with $|\rho| < 1$. Then

$$\widehat{\rho} = \frac{\sum_{t=2}^{N} \widehat{\varepsilon}_t \widehat{\varepsilon}_{t-1}}{\sum_{t=1}^{N} \widehat{\varepsilon}_t^2} \tag{8.2.5}$$

is an estimator of the serial correlation ρ. □

In addition to the residual versus time plot to detect serial correlation, we may use significance tests and enhanced graphical procedures to diagnose correlation in errors. The most popular test for serial correlation is the Durbin–Watson test (Durbin and Watson, 1950, 1951). The null hypothesis is H_0: $\rho = 0$, while the alternative hypothesis is either H_1: $\rho \neq 0$, or H_1: $\rho > 0$, or H_1: $\rho < 0$. The Durbin–Watson test statistic is defined as

$$DW = \frac{\sum_{t=2}^{N} (\widehat{\varepsilon}_t - \widehat{\varepsilon}_{t-1})^2}{\sum_{t=1}^{N} \widehat{\varepsilon}_t^2}. \tag{8.2.6}$$

Note that the summation in the numerator of the DW statistic runs from $t = 2$ to N since $\widehat{\varepsilon}_0$ is not available; this is referred to as an "end effect". The statistic DW lies in the range of 0 to 4, with a value of 2 corresponding to $\rho = 0$. When successive values of $\widehat{\varepsilon}_t$ are close together, DW is small, indicating the presence of positive serial correlation.

Now, DW can be written as a ratio of quadratic forms in the LS residuals, i.e.,

$$DW = \frac{\widehat{\boldsymbol{\varepsilon}}' \mathbf{A} \widehat{\boldsymbol{\varepsilon}}}{\widehat{\boldsymbol{\varepsilon}}' \widehat{\boldsymbol{\varepsilon}}}, \tag{8.2.7}$$

where $\widehat{\boldsymbol{\varepsilon}} \sim N(\mathbf{0}, \sigma^2(\mathbf{I} - \mathbf{P}))$ under H_0 (see property 2 of Result 7.1.1) and

$$\mathbf{A} = \begin{pmatrix} 1 & -1 & 0 & \cdots & 0 & 0 \\ -1 & 2 & -1 & \cdots & 0 & 0 \\ \vdots & \vdots & \vdots & \ddots & \vdots & \vdots \\ 0 & 0 & 0 & \cdots & 2 & -1 \\ 0 & 0 & 0 & \cdots & -1 & 1 \end{pmatrix}. \tag{8.2.8}$$

Since $\mathbf{x}'\mathbf{A}\mathbf{x} = \sum_{t=2}^{N} (x_t - x_{t-1})^2 \geq 0$ for any $\mathbf{x} = (x_1, \cdots, x_N)'$, with equality holding if and only if all the x's are equal, the symmetric matrix $\mathbf{A}$ is n.n.d. of rank $N - 1$, with eigenvalues $0 = \lambda_1 < \lambda_2 \leq \cdots \leq \lambda_N$.

Since $\widehat{\boldsymbol{\varepsilon}} = (\mathbf{I} - \mathbf{P})\boldsymbol{\varepsilon}$, we can write DW as the ratio of quadratic forms in $\boldsymbol{\varepsilon}$. Denote by $\mathbf{K}$ the symmetric and n.n.d. matrix $(\mathbf{I} - \mathbf{P})\mathbf{A}(\mathbf{I} - \mathbf{P})$. Then

$$DW = \frac{\boldsymbol{\varepsilon}' \mathbf{K} \boldsymbol{\varepsilon}}{\boldsymbol{\varepsilon}' (\mathbf{I} - \mathbf{P}) \boldsymbol{\varepsilon}}. \tag{8.2.9}$$

In Exercise 8.5, we ask the reader to show an approximate relation between DW and $\widehat{\rho}$ and to obtain approximate formulas for $\text{E}(DW)$ and $\text{Var}(DW)$.

The exact distribution of the Durbin–Watson test statistic DW depends on the matrix $\mathbf{X}$, which can be unfortunate since the data $\mathbf{X}$ is different in each situation. For this reason, the bounds test for H_0: $\rho = 0$, where the decision rule does not depend on $\mathbf{X}$, is useful.

Result 8.2.6. Let $r = \text{r}(\mathbf{X})$. The following properties hold.

1. $\text{r}(\mathbf{K}) = N - r - 1$ if $\mathbf{1}_N'\mathbf{X} = \mathbf{0}$, i.e., $\sum_{t=1}^N X_{tj} = 0$ for each $1 \le j \le k$, otherwise $\text{r}(\mathbf{K}) = N - r$.

2. Let $\nu_1 \le \cdots \le \nu_{N-r}$ be the $N-r$ largest eigenvalues of $\mathbf{K}$; the remaining r eigenvalues are 0. Then for $i = 1, \cdots, N-r$, $\lambda_i \le \nu_i \le \lambda_{i+r}$, with $\nu_1 > 0$ if and only if $\mathbf{1}_N'\mathbf{X} \ne \mathbf{0}$.

3. $\mathbf{K}$ and $\mathbf{I}-\mathbf{P}$ can be simultaneously diagonalized by an orthogonal matrix $\mathbf{Q}$, such that $\mathbf{Q}'\mathbf{K}\mathbf{Q}$ has on its diagonal $\nu_1, \cdots, \nu_{N-r}$ followed by r 0's, and $\mathbf{Q}'(\mathbf{I}-\mathbf{P})\mathbf{Q}$ has on its diagonal $N-r$ 1's followed by r 0's. Let $\boldsymbol{\xi} = (\xi_1, \cdots, \xi_N)' = \mathbf{Q}'\boldsymbol{\varepsilon}$. Then

$$DW = \frac{\sum_{i=1}^{N-r} \nu_i \xi_i^2}{\sum_{i=1}^{N-r} \xi_i^2}, \tag{8.2.10}$$

   and under H_0, ξ_i are i.i.d. $N(0, \sigma^2)$ variables.

4. Let

$$d_L = \frac{\sum_{i=1}^{N-r} \lambda_i \xi_i^2}{\sum_{i=1}^{N-r} \xi_i^2} \quad \text{and} \quad d_U = \frac{\sum_{i=1}^{N-r} \lambda_{i+r} \xi_i^2}{\sum_{i=1}^{N-r} \xi_i^2}.$$

   Then, $d_L \le DW \le d_U$.

Proof. Consider the null space of $\mathbf{K}$. If $\mathbf{K}\mathbf{x} = \mathbf{0}$, then $[(\mathbf{I}-\mathbf{P})\mathbf{x}]'\mathbf{A}[(\mathbf{I}-\mathbf{P})\mathbf{x}] = 0$, so either $(\mathbf{I}-\mathbf{P})\mathbf{x} = \mathbf{0}$ or $(\mathbf{I}-\mathbf{P})\mathbf{x} = c\mathbf{1}_N$ for some $c \ne 0$. The two cases are respectively equivalent to $\mathbf{x} \in \mathcal{C}(\mathbf{X})$ and $\mathbf{1}_N \in \mathcal{C}(\mathbf{I}-\mathbf{P}) = \mathcal{C}(\mathbf{X})^\perp$ with $\mathbf{x} = (\mathbf{I}-\mathbf{P})^- \mathbf{1}_N$. As a result, $\mathcal{N}(\mathbf{K})$ is $\mathcal{C}(\mathbf{X})$ if $\mathbf{1}_N'\mathbf{X} \ne \mathbf{0}$, and $\mathcal{C}(\mathbf{X}) \oplus \text{Span}(\mathbf{1}_N)$ otherwise. Then from property 1 of Result 1.3.10, property 1 follows. Let $\mathbf{L}$ be an orthogonal matrix such that $\mathbf{L}'(\mathbf{I}-\mathbf{P})\mathbf{L} = \text{diag}(\mathbf{I}_{N-r}, \mathbf{O})$. Let $\mathbf{L}'\mathbf{A}\mathbf{L} = \mathbf{B} = \begin{pmatrix} \mathbf{B}_1 & \mathbf{B}_2 \\ \mathbf{B}_2' & \mathbf{B}_3 \end{pmatrix}$, where $\mathbf{B}_1$ is $(N-r) \times (N-r)$. Then

$$\begin{aligned} \mathbf{L}'\mathbf{K}\mathbf{L} &= \mathbf{L}'(\mathbf{I}-\mathbf{P})\mathbf{L}\mathbf{B}\mathbf{L}'(\mathbf{I}-\mathbf{P})\mathbf{L} \\ &= \begin{pmatrix} \mathbf{I}_{N-p} & \mathbf{O} \\ \mathbf{O} & \mathbf{O} \end{pmatrix} \begin{pmatrix} \mathbf{B}_1 & \mathbf{B}_2 \\ \mathbf{B}_2' & \mathbf{B}_3 \end{pmatrix} \begin{pmatrix} \mathbf{I}_{N-p} & \mathbf{O} \\ \mathbf{O} & \mathbf{O} \end{pmatrix} = \begin{pmatrix} \mathbf{B}_1 & \mathbf{O} \\ \mathbf{O} & \mathbf{O} \end{pmatrix}. \end{aligned}$$

Since $\mathbf{L}'\mathbf{K}\mathbf{L}$ has the same spectrum (i.e., the set of eigenvalues) as $\mathbf{K}$, $\nu_1, \cdots, \nu_{N-r}$ are precisely the eigenvalues of $\mathbf{B}_1$. Note that $\mathbf{B}_1$ is an $(N-r) \times (N-r)$ principal submatrix of $\mathbf{L}'\mathbf{A}\mathbf{L}$, whose eigenvalues are $\lambda_1 \le \cdots \le \lambda_N$. Now, let $\mathbf{M}_s$ denote the principal submatrix of $\mathbf{B}$ obtained by deleting the last r rows and columns of $\mathbf{B}$. Then, $\mathbf{B}_1 = \mathbf{M}_r$ and $\mathbf{B} = \mathbf{M}_0$. From the interlacing theorem (Result 2.3.8), for all $i = 1, \cdots, N-r$, $\lambda_i = \lambda_i(\mathbf{M}_0) \le \lambda_i(\mathbf{M}_1) \cdots \le \lambda_i(\mathbf{M}_r) = \nu_r \le \lambda_{i+1}(M_{r-1}) \le \cdots \le \lambda_{i+r}(\mathbf{M}_0) = \lambda_{i+r}$. From property 1, $\nu_1 > 0$ if and only if $\mathbf{1}_N'\mathbf{X} \ne \mathbf{0}$. This proves property 2. Since $\mathbf{K}(\mathbf{I}-\mathbf{P}) = (\mathbf{I}-\mathbf{P})\mathbf{K} = \mathbf{K}$, from Result 2.5.1, $\mathbf{K}$ and $\mathbf{I}-\mathbf{P}$ can be simultaneously diagonalized by an orthogonal matrix $\mathbf{Q} = (\mathbf{q}_1, \cdots, \mathbf{q}_N)$. Arrange the $\mathbf{q}$'s such that $(\mathbf{I}-\mathbf{P})\mathbf{q}_i = \mathbf{q}_i$ for $i \le N-r$ and $(\mathbf{I}-\mathbf{P})\mathbf{q}_i = \mathbf{0}$ for $i > N-r$. It follows that $\mathbf{K}\mathbf{q}_i = (\mathbf{I}-\mathbf{P})\mathbf{A}(\mathbf{I}-\mathbf{P})\mathbf{q}_i = \mathbf{0}$ for $i > N-r$ as well. By rearranging $\mathbf{q}_1, \cdots, \mathbf{q}_{N-r}$ if necessary, we can also have $\mathbf{K}\mathbf{q}_i = \nu_i\mathbf{q}_i$ for $i \le N-r$. Under H_0, $\boldsymbol{\varepsilon} \sim N(\mathbf{0}, \sigma^2\mathbf{I}_N)$. Then, $\boldsymbol{\xi} = \mathbf{Q}'\boldsymbol{\varepsilon} \sim N(\mathbf{0}, \sigma^2\mathbf{I}_N)$ by Result 5.2.4, completing the proof of property 3. Property 4 follows immediately from the last two properties. ■

Note that the distributions of d_L and d_U are determined by the spectrum of $\mathbf{A}$ and the rank of $\mathbf{X}$, and do not depend on the particular predictor matrix $\mathbf{X}$ or the unknown variance σ^2. This leads to the Durbin–Watson bounds test. Details on the derivation of these distributions are beyond the scope of this book, and the reader is referred to Durbin and Watson (1950). Critical values $d_{L,N,r}$ and $d_{U,N,r}$ are available in most software, which provide p-values for the Durbin–Watson test. For a given level of significance α, the bounds test has the following decision rules based on lower and upper critical values (at level α) $d_{L,N,r}$ and $d_{U,N,r}$.

1. If H_1 is $\rho > 0$, reject H_0 at significance level α if $DW < d_{L,N,r}$; accept H_0 if $DW > d_{U,N,r}$; and the test is inconclusive if $d_{L,N,r} \le DW \le d_{U,N,r}$.

2. If H_1 is $\rho < 0$, the decision rule has the form of the rule for H_1: $\rho > 0$, replacing DW by $(4 - DW)$.

3. If H_1 is $\rho \neq 0$, reject H_0 at significance level 2α if $DW < d_{L,N,r}$ or $4 - DW < d_{L,N,r}$; accept H_0 if $DW > d_{U,N,r}$ or $4 - DW > d_{U,N,r}$; and the test is inconclusive otherwise.

The test is illustrated in the numerical example at the end of this section. It is important to realize that not rejecting the null hypothesis does not necessarily mean that the errors are uncorrelated; it simply means that there is no significant first-order autocorrelation. More complex linear stationary time series models may be employed in order to model autocorrelation in regression errors. An example is the autoregressive moving-average process of order (p, q) defined as

$$\varepsilon_t = \phi_1 \varepsilon_{t-1} + \phi_2 \varepsilon_{t-2} + \cdots + \phi_p \varepsilon_{t-p} + u_t + \theta_1 u_{t-1} + \theta_2 u_{t-2} + \cdots \theta_q u_{t-q},$$

where u_t's are i.i.d. $N(0, \sigma^2)$ variables, and the $(p+q)$ parameters $(\phi_1, \cdots, \phi_p)$, and $(\theta_1, \cdots, \theta_q)$ must satisfy certain restrictions. Plots of the sample autocorrelation function and partial autocorrelation function enable us to identify the model order in these cases. For details, see Brockwell and Davis (2006) or Shumway and Stoffer (2017).

Example 8.2.4. Cochrane–Orcutt procedure. If the errors in a linear model are serially correlated, an iterative estimation procedure called the Cochrane–Orcutt procedure based on a transformation of the data may be used. Let $\widehat{\boldsymbol{\beta}}$ denote the LS estimate of $\boldsymbol{\beta}$, and let $\widehat{\boldsymbol{\varepsilon}}$ be the LS residual vector. In this initial step, the first-order serial correlation is estimated by (8.2.5). Denote these quantities by $\widehat{\boldsymbol{\beta}}^{(0)}$, $\widehat{\boldsymbol{\varepsilon}}^{(0)}$, and $\widehat{\rho}^{(0)}$ respectively. Using $\widehat{\rho}^{(0)}$, we transform the original model and obtain LS estimates from the transformed model

$$Y_t^* = \beta_0(1 - \widehat{\rho}^{(0)}) + \beta_1 X_{t1}^* + \cdots + \beta_k X_{tk}^* + u_t, \quad t = 1, \cdots, N,$$

where $Y_t^* = Y_t - \widehat{\rho}^{(0)} Y_{t-1}$, and $X_{tj}^* = X_{tj} - \widehat{\rho}^{(0)} X_{t-1,j}$, $j = 1, \cdots, k$. Let $\widehat{\boldsymbol{\beta}}^{(1)}$ and $\widehat{\boldsymbol{\varepsilon}}^{(1)}$ respectively denote the vector of LS estimates, and residual vector from this first iteration. We repeat this procedure several times until convergence. This procedure works well in practice, although there is always the danger that the procedure may tend to a local rather than a global solution. □

Numerical Example 8.2. The data set "Air.df" in the R package *EnvStats* consists of daily measurements of ozone concentration, wind speed, temperature, and solar radiation in New York City for 153 consecutive days between May 1 and September 30, 1973.

```
library(EnvStats)
data(Air.df)
mod <-
  lm(ozone ~ radiation + temperature + wind + I(temperature^2) +
    I(wind^2), data = Air.df, na.action = na.exclude
  )
summary(mod)$coef

                     Estimate   Std. Error   t value     Pr(>|t|)
(Intercept)      6.1291962159 2.6079277182  2.350217 2.063124e-02
radiation        0.0023149992 0.0005182987  4.466535 2.010275e-05
temperature     -0.0932427175 0.0702224722 -1.327819 1.871174e-01
wind            -0.2660591907 0.0593631547 -4.481891 1.892616e-05
I(temperature^2) 0.0009137791 0.0004586062  1.992514 4.891001e-02
I(wind^2)        0.0087710259 0.0026007280  3.372527 1.043803e-03

# Test for serial correlation
lmtest::dwtest(mod)

Durbin-Watson test

data:  mod
DW = 1.7459, p-value = 0.06956
alternative hypothesis: true autocorrelation is greater than 0
```

We fit the model using GLS using the gls{nlme} function (see Section 4.5) and compare these coefficients with the OLS coefficients.

```
library(nlme)
mod.gls <- gls(
  ozone ~ radiation + temperature + wind + I(temperature ^ 2)
  + I(wind ^ 2),
  data = Air.df,
  na.action = na.exclude,
  correlation = corARMA(p = 1),
  method = "ML"
)
summary(mod.gls)

Generalized least squares fit by maximum likelihood
  Model: ozone ~ radiation + temperature + wind
       + I(temperature^2) + I(wind^2)
  Data: Air.df
       AIC      BIC   logLik
  156.2408 177.917 -70.1204

Correlation Structure: AR(1)
 Formula: ~1
 Parameter estimate(s):
      Phi
0.1316824
```

```
Coefficients:
                       Value Std.Error   t-value p-value
(Intercept)         7.017139 2.7603111  2.542155  0.0125
radiation           0.002395 0.0005164  4.638260  0.0000
temperature        -0.116755 0.0739558 -1.578709  0.1174
wind               -0.268966 0.0575356 -4.674766  0.0000
I(temperature^2)    0.001067 0.0004826  2.210333  0.0293
I(wind^2)           0.008922 0.0025251  3.533248  0.0006

Residual standard error: 0.4590685
Degrees of freedom: 111 total; 105 residual
```
▲

8.3 Diagnostics for the GLM

In this section, we continue to explore aspects of fitting a GLM. In particular, we look more closely at properties of the residuals and the projection matrix that enable us to diagnose abnormal features in the linear model. *Diagnostic measures* are statistical measures used for detecting problems with the models and/or observations or variables in the data set, including problems of model misspecification, outliers, influential observations, and collinearity. Belsley et al. (1980) gave an extensive description of these topics, while a more theoretical discussion is found in Chatterjee and Hadi (1988). Since residual analysis can be carried out in general without assuming full column rank of $\mathbf{X}$, except for situations that directly look at the LS estimator of $\boldsymbol{\beta}$, we present results for the general case when relevant. These methods are usually referred to as *regression diagnostics* in the literature.

8.3.1 Further properties of the projection matrix

As seen earlier, the symmetric and idempotent hat matrix or projection matrix $\mathbf{P} = \{p_{ij}\}$ onto $\mathcal{C}(\mathbf{X})$ plays an important role in the analysis of a linear model. The matrix $(\mathbf{I}-\mathbf{P})$ is also symmetric and idempotent and is often called the *residuals matrix.* Since $\mathbf{P}$ is symmetric and idempotent, it is easily seen that

$$p_{ii} = \sum_{j=1}^{N} p_{ij}^2, \quad i = 1, \cdots, N. \tag{8.3.1}$$

We have seen in Section 4.2 that $\mathbf{PX} = \mathbf{X}$, $(\mathbf{I}-\mathbf{P})\mathbf{X} = \mathbf{O}$, $\widehat{\mathbf{y}} = \mathbf{Py}$, and $\widehat{\boldsymbol{\varepsilon}} = (\mathbf{I}-\mathbf{P})\mathbf{Y} = (\mathbf{I}-\mathbf{P})\boldsymbol{\varepsilon}$. That is, $\widehat{\varepsilon}_i = \varepsilon_i - \sum_{j=1}^{N} p_{ij}\varepsilon_j$, and the relation between the ith error and the ith residual only depends on $\mathbf{P}$. If p_{ij}'s are sufficiently small, then $\widehat{\boldsymbol{\varepsilon}}$ is a reasonable substitute for $\boldsymbol{\varepsilon}$. Further, $\text{Var}(\widehat{Y}_i) = \sigma^2 p_{ii}$, $\text{Var}(\widehat{\varepsilon}_i) = \sigma^2(1-p_{ii})$, and $\text{Cov}(\widehat{\varepsilon}_i, \widehat{\varepsilon}_j) = -\sigma^2 p_{ij}$, $i \neq j$. The ith predicted response can be written as

$$\widehat{Y}_i = \sum_{i=1}^{N} p_{ij} Y_j = p_{ii} Y_i + \sum_{i \neq j} p_{ij} Y_j, \tag{8.3.2}$$

from which we see that

$$\partial \widehat{Y}_i / \partial Y_j = p_{ij}, \quad i, j = 1, \cdots, N. \tag{8.3.3}$$

Result 8.3.1. Let $\mathbf{P} = \{p_{ij}\}$ denote the $N \times N$ projection matrix. Then,

1. $\sum_{i=1}^{N}\sum_{j=1}^{N} p_{ij}^2 = \text{r}(\mathbf{X})$.
2. $0 \le p_{ii} \le 1$, $i = 1, \cdots, N$.
3. For $i \neq j$, $p_{ij}^2 \le p_{ii}(1-p_{ii})$ and $-0.5 \le p_{ij} \le 0.5$, so that if p_{ii} is large (i.e., near 1) or small (i.e., near 0), then $|p_{ij}|$ is small.
4. In a linear model with intercept, $p_{ii} \ge 1/N$, $i = 1, \cdots, N$.

Proof. Since $\mathbf{P}$ is symmetric and idempotent, $\sum_{i,j=1}^{N} p_{ij}^2 = \text{tr}(\mathbf{P}^2) = \text{r}(\mathbf{P})$, proving property 1. From (8.3.1), for $j \neq i$, $p_{ii} \ge p_{ii}^2 + p_{ij}^2$, so that $0 \le p_{ij}^2 \le p_{ii}(1-p_{ii}) \le 1/4$, yielding both properties 2 and 3. In a linear model with intercept, $\mathbf{1}_N \in \mathcal{C}(\mathbf{X})$, so from Result 2.6.6, $\mathbf{P}-\mathbf{P}_1$ is the projection matrix onto $\mathcal{C}(\mathbf{1}_N)^{\perp} \cap \mathcal{C}(\mathbf{X})$, where $\mathbf{P}_1$ is the projection matrix onto $\mathcal{C}(\mathbf{1}_N)$. In particular, all the diagonal elements of $\mathbf{P}-\mathbf{P}_1$ are nonnegative. From $\mathbf{P}_1 = \mathbf{1}_N\mathbf{1}_N'/N$, it follows that $p_{ii} \ge 1/N$, showing property 4. ■

We next consider (1) cases when p_{ii} is 0 or 1, and (2) a relationship between the elements of $\mathbf{P}$ and the residuals $\widehat{\varepsilon}_i$, which will be useful for a later discussion on the relative usefulness of different transformed residuals for diagnostic purposes. Henceforth, a parenthesis-enclosed subscript (i) will denote omission of the ith observation, e.g., $\mathbf{X}_{(i)}$ is the submatrix of $\mathbf{X}$ with its ith row deleted, and $\mathbf{y}_{(i)}$ is the subvector of $\mathbf{y}$ with the ith component deleted.

For the next result, we need the generalized Sherman–Morrison–Woodbury formula shown in Result 3.1.15.

Result 8.3.2. Let $\mathbf{X}$ be a matrix with N rows $\mathbf{x}_1', \cdots, \mathbf{x}_N'$ (as discussed in Example 2.1.1). The following properties hold for the diagonal elements p_{ii} of the projection matrix:

1. $p_{ii} = 1$ if and only if $\mathbf{x}_i \notin \mathcal{R}(\mathbf{X}_{(i)})$;
2. If $\mathbf{x}_i \in \mathcal{R}(\mathbf{X}_{(i)})$, then
$$p_{ii} = \frac{\mathbf{x}_i'(\mathbf{X}_{(i)}'\mathbf{X}_{(i)})^{-}\mathbf{x}_i}{1+\mathbf{x}_i'(\mathbf{X}_{(i)}'\mathbf{X}_{(i)})^{-}\mathbf{x}_i},$$
and $p_{ii} = 0$ if and only if $\mathbf{x}_i = \mathbf{0}$.
3. If $\widehat{\boldsymbol{\varepsilon}} \neq \mathbf{0}$, then $p_{ii} + \widehat{\varepsilon}_i^2/\widehat{\boldsymbol{\varepsilon}}'\widehat{\boldsymbol{\varepsilon}} \le 1$. As a result, observations with larger p_{ii} will tend to have smaller residuals $\widehat{\varepsilon}_i$, and the usual residual plots will not identify these observations as anomalies.

Proof. Let $i = 1$ without loss of generality. If $p_{11} = 1$, then from property 3 of Result 8.3.1, $p_{1j} = p_{j1} = 0$ for all $j > 1$, so $\mathbf{P} = \begin{pmatrix} 1 & \mathbf{0} \\ \mathbf{0} & \mathbf{P}_{(1)} \end{pmatrix}$. If $a\mathbf{x}_1' + \mathbf{b}'\mathbf{X}_{(1)} = \mathbf{0}$, then $(a, \mathbf{b}')\mathbf{X} = \mathbf{0}$, so $(a, \mathbf{b}')\mathbf{P} = (a, \mathbf{b}'\mathbf{P}_{(1)}) = \mathbf{0}$. Hence $a = 0$, showing $\mathbf{x}_1 \notin \mathcal{R}(\mathbf{X}_{(1)})$. Conversely, if $\mathbf{x}_1 \notin \mathcal{R}(\mathbf{X}_{(1)})$, then let $\mathbf{P}_{(1)}$ be the projection matrix for $\mathbf{X}_{(1)}$ and $\mathbf{P}$ be the block diagonal matrix as above. On the one hand, $\mathbf{PX} = \begin{pmatrix} 1 & \mathbf{0} \\ \mathbf{0} & \mathbf{P}_{(1)} \end{pmatrix}\begin{pmatrix} \mathbf{x}_1' \\ \mathbf{X}_{(1)} \end{pmatrix} = \mathbf{X}$. On the other hand, $\mathbf{v} \perp \mathcal{C}(\mathbf{X})$; then writing $\mathbf{v}' = (a, \mathbf{b}')$, $a\mathbf{x}_1' + \mathbf{b}'\mathbf{X}_{(1)} = \mathbf{0}$, and since $\mathbf{x}_1 \notin \mathcal{R}(\mathbf{X}_{(1)})$, $a = 0$ and $\mathbf{b} \perp \mathcal{C}(\mathbf{X}_{(1)})$, giving $\mathbf{v}'\mathbf{P} = (0, \mathbf{b}'\mathbf{P}_{(1)}) = \mathbf{0}$. Therefore, $\mathbf{P}$ is the projection matrix onto $\mathcal{C}(\mathbf{X})$, in particular, $p_{11} = 1$. This proves property 1. Denote $\mathbf{G}_{(1)} = (\mathbf{X}'_{(1)}\mathbf{X}_{(1)})^{-}$. If $\mathbf{x}_1 \in \mathcal{R}(\mathbf{X}_{(1)})$, then $\mathbf{x}_1 \in \mathcal{C}(\mathbf{X}'_{(1)}\mathbf{X}_{(1)}) = \mathcal{R}(\mathbf{X}'_{(1)}\mathbf{X}_{(1)})$, so from $\mathbf{X}'\mathbf{X} = \mathbf{X}'_{(1)}\mathbf{X}_{(1)} + \mathbf{x}_1\mathbf{x}'_1$ and Result 3.1.15,

$$p_{11} = \mathbf{x}'_1(\mathbf{X}'\mathbf{X})^{-}\mathbf{x}_1 = \mathbf{x}'_1\left[\mathbf{G}_{(1)} - \frac{\mathbf{G}_{(1)}\mathbf{x}_1\mathbf{x}'_1\mathbf{G}_{(1)}}{1+\mathbf{x}'_1\mathbf{G}_{(1)}\mathbf{x}_1}\right]\mathbf{x}_1,$$

yielding the formula in property 2. Clearly $p_{11} = 0$ if and only if $\mathbf{x}'_1(\mathbf{X}'_{(1)}\mathbf{X}_{(1)})^{-}\mathbf{x}_1 = 0$. Since $\mathbf{x}'_1 = \mathbf{t}'\mathbf{X}_{(1)}$ for some $\mathbf{t}$, $p_{11} = \mathbf{x}'_1(\mathbf{X}'_{(1)}\mathbf{X}_{(1)})^{-}\mathbf{x}_1 = \mathbf{t}'\mathbf{P}_{(1)}\mathbf{t} = \|\mathbf{P}_{(1)}\mathbf{t}\|^2$, so $p_{11} = 0$ if and only if $\mathbf{P}_{(1)}\mathbf{t} = \mathbf{0}$, which is equivalent to $\mathbf{x}_1 = \mathbf{X}'_{(1)}\mathbf{t} = \mathbf{0}$. This proves property 2.

Let $\widehat{\boldsymbol{\varepsilon}} \neq \mathbf{0}$ and $\mathbf{u} = \widehat{\boldsymbol{\varepsilon}}/\|\widehat{\boldsymbol{\varepsilon}}\|$. Then $\mathbf{u} \perp \mathcal{C}(\mathbf{X})$ is a unit vector. Let $\mathbf{v} = (\mathbf{I}-\mathbf{P})\mathbf{e}_1$. Then, $1 - p_{11} - \widehat{\varepsilon}_1^2/\widehat{\boldsymbol{\varepsilon}}'\widehat{\boldsymbol{\varepsilon}} = \mathbf{e}'_1(\mathbf{I}-\mathbf{P}-\mathbf{u}\mathbf{u}')\mathbf{e}_1 = \|\mathbf{v}\|^2 - (\mathbf{v}'\mathbf{u})^2$, which by Cauchy–Schwarz inequality is nonnegative (see property 6 of Result 1.2.1). This proves property 3. ∎

We next present a result on "leave-one-out" predictions, which is necessary for a discussion on residuals.

Result 8.3.3. For $i = 1, \ldots, N$, let $\boldsymbol{\beta}^0_{(i)}$ be the LS fit using the data with the ith observation omitted. If $\mathbf{x}_i \in \mathcal{R}(\mathbf{X}_{(i)})$, then

$$\mathbf{X}\boldsymbol{\beta}^0_{(i)} = \mathbf{X}\boldsymbol{\beta}^0 - \frac{\mathbf{X}(\mathbf{X}'\mathbf{X})^{-}\mathbf{x}_i\widehat{\varepsilon}_i}{1-p_{ii}} = \mathbf{y} - \widehat{\boldsymbol{\varepsilon}} - \frac{\mathbf{X}(\mathbf{X}'\mathbf{X})^{-}\mathbf{x}_i\widehat{\varepsilon}_i}{1-p_{ii}}. \tag{8.3.4}$$

Proof. Let $\mathbf{G} = (\mathbf{X}'\mathbf{X})^{-}$. Since $\boldsymbol{\beta}^0_{(i)} = (\mathbf{X}'\mathbf{X} - \mathbf{x}_i\mathbf{x}'_i)^{-}\mathbf{X}'_{(i)}\mathbf{y}_{(i)}$, from (3.1.9), $\mathbf{x}'_i(\mathbf{X}'\mathbf{X})^{-}\mathbf{x}_i = p_{ii}$, and $\mathbf{X}'_{(i)}\mathbf{y}_{(i)} = \mathbf{X}'\mathbf{y} - \mathbf{x}_iY_i$, so

$$\begin{aligned}
\mathbf{X}\boldsymbol{\beta}^0_{(i)} &= \mathbf{X}\left(\mathbf{G} + \frac{\mathbf{G}\mathbf{x}_i\mathbf{x}'_i\mathbf{G}}{1-p_{ii}}\right)(\mathbf{X}'\mathbf{y} - \mathbf{x}_iY_i) \\
&= \mathbf{X}\mathbf{G}\mathbf{X}'\mathbf{y} - \mathbf{X}\mathbf{G}\mathbf{x}_iY_i + \frac{\mathbf{X}\mathbf{G}\mathbf{x}_i\mathbf{x}'_i\mathbf{G}\mathbf{X}'\mathbf{y}}{1-p_{ii}} - \frac{\mathbf{X}\mathbf{G}\mathbf{x}_i\mathbf{x}'_i\mathbf{G}\mathbf{x}_iY_i}{1-p_{ii}} \\
&= \widehat{\mathbf{y}} - \mathbf{X}\mathbf{G}\mathbf{x}_iY_i + \frac{\mathbf{X}\mathbf{G}\mathbf{x}_i\widehat{Y}_i}{1-p_{ii}} - \frac{\mathbf{X}\mathbf{G}\mathbf{x}_ip_{ii}Y_i}{1-p_{ii}} = \widehat{\mathbf{y}} - \frac{\mathbf{X}\mathbf{G}\mathbf{x}_i(Y_i - \widehat{Y}_i)}{1-p_{ii}},
\end{aligned}$$

proving (8.3.4). ∎

Note that when $\mathbf{X}_{(i)}$ has full column rank, then $\mathbf{X}$ has full column rank, and (9.1.13) immediately follows from (8.3.4).

8.3.2 Types of residuals

In Chapter 4, we defined the LS residuals $\widehat{\varepsilon}_i$, $i = 1, \cdots, N$. We have seen that the vector $\widehat{\boldsymbol{\varepsilon}} = (\widehat{\varepsilon}_1, \cdots, \widehat{\varepsilon}_N)'$ has mean $\mathbf{0}$, while $\mathrm{E}(\widehat{\boldsymbol{\varepsilon}}\widehat{\boldsymbol{\varepsilon}}') = \sigma^2(\mathbf{I}-\mathbf{P})$. We now introduce four transformations of the LS residuals, i.e., normalized residuals, standardized residuals, internally Studentized residuals and externally Studentized residuals. In general, the ith transformed residual is defined by

$$\widehat{\varepsilon}_i^* = \widehat{\varepsilon}_i/\,\mathrm{s.d.}(\widehat{\varepsilon}_i), \quad i = 1, \cdots, N, \tag{8.3.5}$$

where $\mathrm{s.d.}(\widehat{\varepsilon}_i)$ denotes the standard deviation of the ith residual $\widehat{\varepsilon}_i$ (the square-root of the ith diagonal element of $\sigma^2(\mathbf{I}-\mathbf{P})$). By using different estimates for the unknown parameter $\mathrm{s.d.}(\widehat{\varepsilon}_i)$ in (8.3.5), we obtain the four definitions shown below.

Definition 8.3.1. Normalized residuals. The ith normalized residual is defined by

$$a_i = \frac{\widehat{\varepsilon}_i}{(\widehat{\varepsilon}'\widehat{\varepsilon})^{1/2}}, \quad i = 1, \cdots, N. \tag{8.3.6}$$

Definition 8.3.2. Standardized residuals. The ith standardized residual is defined by

$$b_i = \frac{\widehat{\varepsilon}_i}{\widehat{\sigma}}, \quad i = 1, \cdots, N, \tag{8.3.7}$$

where $\widehat{\sigma}^2 = MSE = \widehat{\varepsilon}'\widehat{\varepsilon}/(N - \text{r}(\mathbf{X}))$.

Definition 8.3.3. Internally Studentized residuals. The ith internally Studentized residual is

$$r_i = \frac{\widehat{\varepsilon}_i}{\widehat{\sigma}(1 - p_{ii})^{1/2}}, \quad i = 1, \cdots, N. \tag{8.3.8}$$

Definition 8.3.4. Externally Studentized residuals. The ith externally Studentized residual is

$$r_i^* = \frac{\widehat{\varepsilon}_i}{\widehat{\sigma}_{(i)}(1 - p_{ii})^{1/2}}, \quad i = 1, \cdots, N, \tag{8.3.9}$$

where $\widehat{\sigma}^2_{(i)} = MSE_{(i)}$, the residual mean square when the ith observation is omitted from the fit.

Result 8.3.4. The following relationships between the residuals hold:

1. $b_i = a_i(N - \text{r}(\mathbf{X}))^{1/2}$.
2. $r_i = b_i/(1 - p_{ii})^{1/2} = a_i(N - \text{r}(\mathbf{X}))^{1/2}/(1 - p_{ii})^{1/2}$.
3. If $\mathbf{x}_i \in \mathcal{R}(\mathbf{X}_{(i)})$, then $r_i^* = \dfrac{a_i(N - \text{r}(\mathbf{X}) - 1)^{1/2}}{(1 - p_{ii} - a_i^2)^{1/2}} = \dfrac{r_i(N - \text{r}(\mathbf{X}) - 1)^{1/2}}{(N - \text{r}(\mathbf{X}) - r_i^2)^{1/2}}$. (From Result 8.3.2, if $\mathbf{x}_i \notin \mathcal{R}(\mathbf{X}_{(i)})$, then $p_{ii} = 1$, so r_i^* is not defined.)

Proof. The proof of the first two relations follows directly from their definitions. To prove property 3, from Result 8.3.3, the residuals for the LS fit of $\mathbf{y}_{(i)} = \mathbf{X}_{(i)}\boldsymbol{\beta} + \boldsymbol{\varepsilon}_{(i)}$ are $\widehat{\varepsilon}_j + p_{ij}\widehat{\varepsilon}_i/(1 - p_{ii})$, $j \neq i$, while the error of the prediction of Y_i based on $\mathbf{X}_{(i)}$ and $\mathbf{y}_{(i)}$ is $\widehat{\varepsilon}_i/(1 - p_{ii})$. Then

$$\begin{aligned} SSE_{(i)} &= \sum_{j=1}^{N} \left(\widehat{\varepsilon}_j + \frac{p_{ij}\widehat{\varepsilon}_i}{1 - p_{ii}}\right)^2 - \frac{\widehat{\varepsilon}_i^2}{(1 - p_{ii})^2} \\ &= \sum_{j=1}^{N} \widehat{\varepsilon}_j^2 + \frac{2\widehat{\varepsilon}_j}{1 - p_{ii}} \sum_{j=1}^{N} p_{ij}\widehat{\varepsilon}_j + \left(\sum_{j=1}^{N} p_{ij}^2 - 1\right) \frac{\widehat{\varepsilon}_i^2}{(1 - p_{ii})^2}. \end{aligned}$$

From $\sum_{j=1}^N p_{ij}\widehat{\varepsilon}_j = (\mathbf{P}\widehat{\boldsymbol{\varepsilon}})_i = 0$ and (8.3.1), $\sum_{j=1}^N p_{ij}^2 - 1 = p_{ii} - 1$. Then,

$$SSE_{(i)} = \widehat{\varepsilon}'\widehat{\varepsilon} - \widehat{\varepsilon}_i^2/(1 - p_{ii}) = \widehat{\varepsilon}_i^2/a_i^2 - \widehat{\varepsilon}_i^2/(1 - p_{ii}).$$

Note that $r(\mathbf{X}_{(i)}) = r(\mathbf{X})$. Then,

$$\widehat{\sigma}^2_{(i)} = \frac{SSE_{(i)}}{N-1-r(\mathbf{X})} = \frac{(1-p_{ii}-a_i^2)\widehat{\varepsilon}_i^2}{(N-1-r(\mathbf{X}))a_i^2(1-p_{ii})},$$

so $r^{*2} = (N-1-r(\mathbf{X}))a_i^2/(1-p_{ii}-a_i^2)$. Since r^* and a_i have the same sign, the first equality in property 3 follows. The second equality follows directly from the definitions of the related quantities. ■

From property 3, r_i^* is a monotonic transformation of r_i, which itself is a monotonic transformation of a_i. As $r_i^2 \to (N-p)$, $r_i^{*2} \to \infty$, and therefore the latter reflects large deviations more dramatically. Both being constant multiples of $\widehat{\varepsilon}_i$, the normalized residuals a_i and the standardized residuals b_i are equivalent as diagnostic measures. A disadvantage may be that neither residual takes into account $\text{Var}(\widehat{\varepsilon}_i)$, i.e., the diagonal elements of the projection matrix. Although this might not be critically important in some situations, it would be preferable to use the r_i when the p_{ii} vary substantially. We state two distributional properties (for proof, see Chatterjee and Hadi (1988), section 4.2). Result 8.3.5, which is due to Ellenberg (1973), implies that $|r_i|$ cannot exceed $(N-p)^{1/2}$, while Result 8.3.6, which was proved by Beckman and Trussell (1974), shows that the square of the externally Studentized residual follows an F-distribution.

Result 8.3.5. Suppose $r(\mathbf{X}_{(i)}) = r(\mathbf{X})$ for all $1 \le i \le N$. Then for all $i \ne j$, $r_i^2/(N-r(\mathbf{X})-1) \sim \text{Beta}(1/2, (N-r(\mathbf{X})-1)/2)$, and the covariance between r_i and r_j is $\text{Cov}(r_i, r_j) = -(1-p_{ii})^{-1/2}(1-p_{jj})^{-1/2}p_{ij}$.

Result 8.3.6. Provided $r(\mathbf{X}_{(i)}) = r(\mathbf{X})$ for all $1 \le i \le N$, r_i^* are identically distributed as Student t-variables with $N - r(\mathbf{X}) - 1$ d.f.

The externally Studentized residuals r_i^* may be preferred since $\widehat{\sigma}_{(i)}$ is robust to gross errors in the ith observation and the r_i^* itself has a t-distribution for which critical values are easily available. Note that the Studentized residuals stabilize the variance in the ordinary residuals $\widehat{\varepsilon}_i$, but leave the correlation pattern unchanged. Other residuals also play a useful role in the analysis of linear models, such as predicted residuals that are defined below.

Definition 8.3.5. Predicted residuals. The ith predicted residual is defined by

$$\widehat{\varepsilon}_{i(i)} = Y_i - \mathbf{x}_i'\boldsymbol{\beta}^0_{(i)}, \quad i = 1, \cdots, N,$$

where $\boldsymbol{\beta}^0_{(i)}$ denotes the least squares estimate of $\boldsymbol{\beta}$ with the ith case excluded; see Result 8.3.3.

Unlike the ordinary and Studentized residuals, the ith predicted residual vector $\widehat{\boldsymbol{\varepsilon}}_{(i)}$ is based on a fit to the data with the ith case excluded. We may think of $\widehat{\varepsilon}_{i(i)}$ as a prediction error, since we exclude the ith case while obtaining the fit. In Result 9.1.3, we show that if the errors are normally distributed, $\widehat{\varepsilon}_{i(i)}$ will have the same correlation structure as the $\widehat{\varepsilon}_i$, with zero means and variances equal to $\sigma^2/(1-p_{ii})$. It is preferable to use Studentized versions of $\widehat{\varepsilon}_{i(i)}$, which gives back the internally and externally Studentized residuals r_i and r_i^*, as the reader may verify. In Section 9.1.1, we describe partial residuals and discuss their use in assessment of variables to be included in the linear model.

8.3.3 Outliers and high leverage observations

Outliers

Frequently, the data may contain outliers, i.e., anomalous observations that do not reasonably fit the assumed model. In a linear model, the ith observation is an outlier if the magnitude of r_i or r_i^* is large in comparison with the rest of the observations in the data set. The presence of outliers may seriously bias parameter estimation and inference. Graphical displays such as boxplots of the residuals described in the previous section are useful in outlier detection.

Leverage

From (8.3.2) and (8.3.3), we may interpret p_{ij} as the amount of *leverage* each Y_j has on determining $\widehat{Y}_i$, irrespective of its actual value (Hoaglin and Welsch, 1978). The leverage of the ith case is p_{ii}, which is the ith diagonal element of the projection matrix $\mathbf{P}$, while the reciprocal $1/p_{ii}$ is the effective or equivalent number of observations that determine $\widehat{Y}_i$ (Huber, 1981). When $p_{ii} = 1/2$, an equivalent of two observations determine the fitted value $\widehat{Y}_i$; when $p_{ii} = 1$, Y_i solely determines $\widehat{Y}_i$; while if $p_{ii} = 0$, Y_i has no influence on $\widehat{Y}_i$. Huber (1981) suggested that if $p_{ii} > 0.2$, then the ith case is a high-leverage point. Since $\sum_{i=1}^{N} p_{ii} = p$, Hoaglin and Welsch (1978) suggested that cases with $p_{ii} > 2p/N$ may be classified as high-leverage points. A wide variation in the values of p_{ii} indicate nonhomogeneous spacing of the rows of $\mathbf{P}$.

Geometrically, suppose $\mathbf{X}$ contains a constant column $\mathbf{1}_N$, or suppose that the columns of $\mathbf{X}$ are centered at their respective averages. In either case, suppose that $\mathbf{X}$ has full column rank. For a p-dimensional vector $\mathbf{u}$, and a constant c, the quadratic form $\mathbf{u}'(\mathbf{X}'\mathbf{X})^{-1}\mathbf{u} = c$ determines p-dimensional elliptical contours centered at $\overline{\mathbf{x}}$. The smallest convex set which contains the scatter of N sample values of $\mathbf{X}$ lies within ellipsoids of radius c, where $c \leq \max(p_{ii})$. The implication is that a large value of p_{ii} indicates that $\mathbf{x}_i$ is far removed from the center $\overline{\mathbf{x}}$, i.e., it is an outlier in the X space. We define the following quantities that enable us to measure the distance of $\mathbf{x}_i$ from the rest of the cases. Assume that we have a linear model with intercept, and recall that $\mathbf{X} = (\mathbf{1}_N, \widetilde{\mathbf{X}})$.

Definition 8.3.6. The *Mahalanobis distance* is defined by (see Section 6.3)

$$M_i = (N-2)(\widetilde{\mathbf{x}}_i - \overline{\widetilde{\mathbf{X}}}_{(i)})'\{\widetilde{\mathbf{X}}'_{(i)}(\mathbf{I} - (N-1)^{-1}\mathbf{1}\mathbf{1}')\widetilde{\mathbf{X}}_{(i)}\}^{-1}(\widetilde{\mathbf{x}}_i - \overline{\widetilde{\mathbf{X}}}_{(i)}), \tag{8.3.10}$$

for $i = 1, \cdots, N$, where, $\overline{\widetilde{\mathbf{X}}}_{(i)}$ is the average of $\widetilde{\mathbf{X}}_{(i)}$. It is easily shown that

$$M_i = \frac{N(N-2)(p_{ii} - 1/N)}{(N-1)(1-p_{ii})}, \quad i = 1, \cdots, N.$$

Definition 8.3.7. The *weighted squared standardized distance* (WSSD) is defined as the weighted sum of squared distances of X_{ij} from the mean of X_j, the weights being $\widehat{\beta}_j$ (Daniel and Wood, 1971):

$$W_i^* = \sum_{j=1}^{p} c_{ij}^2 / S_Y^2, \quad i = 1, \cdots, N, \tag{8.3.11}$$

where $c_{ij} = \widehat{\beta}_j(X_{ij} - \overline{X}_j)$, $i = 1, \cdots, N$, $j = 1, \cdots, p$ denotes the effect of X_j on $\widehat{Y}_i$, and $\sum_{j=1}^{p} c_{ij} = \widehat{Y}_i - \overline{Y}$. If X_{ij} is far removed from $\overline{X}$, or if $\widehat{\beta}_j$ is large, or both, then W_i^* will be large, and the ith case is influential on the distance between $\widehat{Y}_i$ and $\overline{Y}$.

L–R plot

An L–R plot combines information about leverages and residuals into a single graphical display, and enables us to distinguish between high-leverage points and outliers. It is a scatterplot of leverages p_{ii} versus the squared normalized residuals a_i^2. The scatter of points must lie within the triangle defined by these conditions: (i) $0 \le p_{ii} \le 1$, (ii) $0 \le a_i^2 \le 1$, and (iii) $p_{ii} + a_i^2 \le 1$. Points that lie in the lower right corner of the L–R plot are outliers, while points that lie in the upper left corner have high leverage.

8.3.4 Diagnostic measures based on influence functions

Neither outliers nor high-leverage points are necessarily influential. Also, an observation may be influential in the linear model fit either because it is an outlying response, or it is a high-leverage point, or both. In this section, we discuss measures based on the influence function. The study of *influence* is the study of the dependence of conclusions and inferences on various aspects of a statistical problem formulation. This is implemented via a perturbation scheme in which data are modified by deletion of cases, either singly, or in groups (Hampel, 1974).

In this section, we present results describing the influence of the ith case on the LS estimates, $\widehat{\boldsymbol{\beta}}$, $\widehat{\sigma}^2$, as well as on $\text{Cov}(\widehat{\boldsymbol{\beta}})$, $\widehat{\mathbf{y}}$, and $\text{Cov}(\widehat{\mathbf{y}})$. We use measures based on the theoretical influence function (Hampel, 1974; Huber, 1981); see item 8 in Appendix C.

Result 8.3.7. Suppose $\mathbf{x}$ and Y have joint c.d.f. F and $\text{E}(\|\mathbf{x}\|^2 + Y^2) < \infty$. If $\boldsymbol{\Sigma}_{\mathbf{xx}}(F) = \text{E}(\mathbf{xx}')$ is p.d., $\boldsymbol{\Sigma}_{\mathbf{x}Y}(F) = \text{E}(\mathbf{x}Y)$, and $\sigma_{YY}(F) = \text{E}(Y^2)$, then $\Psi(\boldsymbol{\beta}) = \text{E}[(Y - \mathbf{x}'\boldsymbol{\beta})^2]$ has a unique minimizer given by

$$\widehat{\boldsymbol{\beta}}(F) = \boldsymbol{\Sigma}_{\mathbf{xx}}^{-1}(F)\boldsymbol{\Sigma}_{\mathbf{x}Y}(F), \tag{8.3.12}$$

and

$$\widehat{\sigma}^2(F) = \min_{\boldsymbol{\beta}} \Psi(\boldsymbol{\beta}) = \sigma_{YY}(F) - \boldsymbol{\Sigma}_{\mathbf{x}Y}'(F)\widehat{\boldsymbol{\beta}}(F). \tag{8.3.13}$$

Proof. Since $\Psi(\boldsymbol{\beta}) = \sigma_{YY}(F) - 2\boldsymbol{\Sigma}_{\mathbf{x}Y}'(F)\boldsymbol{\beta} + \boldsymbol{\beta}'\boldsymbol{\Sigma}_{\mathbf{xx}}(F)\boldsymbol{\beta}$ and $\boldsymbol{\Sigma}_{\mathbf{xx}}$ is p.d., $\Psi(\boldsymbol{\beta})$ is strictly convex. Solving $\nabla\Psi(\boldsymbol{\beta}) = 2\boldsymbol{\Sigma}_{\mathbf{xx}}(F)\boldsymbol{\beta} - 2\boldsymbol{\Sigma}_{\mathbf{x}Y}(F) = \mathbf{0}$ then yields (8.3.12). From $\widehat{\sigma}^2(F) = \Psi(\widehat{\boldsymbol{\beta}}(F))$, (8.3.13) follows. ■

When F and T are fixed, $\psi(\mathbf{z}, F, T)$ as a function of $\mathbf{z}$ is called an influence curve for T, denoted by $\text{IC}_{T,F}(\mathbf{z})$.

Result 8.3.8. The influence curves for $\widehat{\boldsymbol{\beta}}$ and $\widehat{\sigma}^2$ are respectively

$$\text{IC}_{\widehat{\boldsymbol{\beta}},F}(\mathbf{x}', y) = \boldsymbol{\Sigma}_{\mathbf{xx}}^{-1}(F)\mathbf{x}[y - \mathbf{x}'\widehat{\boldsymbol{\beta}}(F)], \tag{8.3.14}$$

$$\text{IC}_{\widehat{\sigma}^2,F}(\mathbf{x}', y) = [y - \mathbf{x}'\widehat{\boldsymbol{\beta}}(F)]^2 - \sigma_{YY}(F) + \boldsymbol{\Sigma}_{\mathbf{x}Y}'(F)\widehat{\boldsymbol{\beta}}(F). \tag{8.3.15}$$

Proof. Let $F_\varepsilon = (1-\varepsilon)F + \varepsilon\delta_{(\mathbf{x}',Y)}$. Then

$$\begin{aligned}
\boldsymbol{\Sigma}_{\mathbf{xx}}(F_\varepsilon) &= \boldsymbol{\Sigma}_{\mathbf{xx}}(F) - \varepsilon\mathbf{D},\\
\boldsymbol{\Sigma}_{\mathbf{x}Y}(F_\varepsilon) &= \boldsymbol{\Sigma}_{\mathbf{x}Y}(F) - \varepsilon\mathbf{d},\\
\sigma_{YY}(F_\varepsilon) &= \sigma_{YY}(F) - \varepsilon d_0,
\end{aligned}$$

where $\mathbf{D} = \boldsymbol{\Sigma}_{\mathbf{xx}}(F) - \mathbf{xx}'$, $\mathbf{d} = \boldsymbol{\Sigma}_{\mathbf{x}Y}(F) - \mathbf{x}y$, and $d_0 = \sigma_{YY}(F) - y^2$. From the Sherman–Morrison–Woodbury formula in property 5 of Result 1.3.8,

$$\boldsymbol{\Sigma}_{\mathbf{xx}}^{-1}(F_\varepsilon) = \boldsymbol{\Sigma}_{\mathbf{xx}}^{-1}(F) + \varepsilon\mathbf{M}_\varepsilon,$$

where $\mathbf{M}_\varepsilon = \boldsymbol{\Sigma}_{\mathbf{xx}}^{-1}(F)\mathbf{D}(\mathbf{I} - \varepsilon\boldsymbol{\Sigma}_{\mathbf{xx}}^{-1}(F)\mathbf{D})^{-1}\boldsymbol{\Sigma}_{\mathbf{xx}}^{-1}(F)$, provided all the inverses exist. Then, from (8.3.12), if $|\varepsilon|$ is small enough,

$$\begin{aligned}\widehat{\boldsymbol{\beta}}(F_\varepsilon) - \widehat{\boldsymbol{\beta}}(F) &= (\boldsymbol{\Sigma}_{\mathbf{xx}}^{-1}(F) + \varepsilon\mathbf{M}_\varepsilon)(\boldsymbol{\Sigma}_{\mathbf{x}Y}(F) - \varepsilon\mathbf{d}) - \boldsymbol{\Sigma}_{\mathbf{xx}}^{-1}(F)\boldsymbol{\Sigma}_{\mathbf{x}Y}(F)\\ &= \varepsilon[\mathbf{M}_\varepsilon\boldsymbol{\Sigma}_{\mathbf{x}Y}(F) - \boldsymbol{\Sigma}_{\mathbf{xx}}^{-1}(F)\mathbf{d}] + O(\varepsilon^2).\end{aligned}$$

Divide both sides by ε and let $\varepsilon \to 0$. Since $\mathbf{M}_\varepsilon \to \boldsymbol{\Sigma}_{\mathbf{xx}}^{-1}(F)\mathbf{D}\boldsymbol{\Sigma}_{\mathbf{xx}}^{-1}(F)$,

$$\begin{aligned}\mathrm{IC}_{\widehat{\boldsymbol{\beta}},F}(\mathbf{x}', y) &= \boldsymbol{\Sigma}_{\mathbf{xx}}^{-1}(F)\mathbf{D}\boldsymbol{\Sigma}_{\mathbf{xx}}^{-1}(F)\boldsymbol{\Sigma}_{\mathbf{x}Y}(F) - \boldsymbol{\Sigma}^{-1}(F)\mathbf{d}\\ &= \boldsymbol{\Sigma}_{\mathbf{xx}}^{-1}(F)[\boldsymbol{\Sigma}_{\mathbf{x}Y}(F) - \mathbf{xx}'\widehat{\boldsymbol{\beta}}(F)] - \boldsymbol{\Sigma}_{\mathbf{xx}}^{-1}(F)[\boldsymbol{\Sigma}_{\mathbf{x}Y}(F) - \mathbf{x}y],\end{aligned}$$

yielding (8.3.14). Similarly, from (8.3.13)

$$\begin{aligned}&\widehat{\sigma}^2(F_\varepsilon) - \widehat{\sigma}^2(F)\\ &= [\sigma_{YY}(F) - \varepsilon d_0] - [\boldsymbol{\Sigma}_{\mathbf{x}Y}(F) - \varepsilon\mathbf{d}]'\widehat{\boldsymbol{\beta}}(F_\varepsilon) - \sigma_{YY}(F) + \boldsymbol{\Sigma}_{\mathbf{x}Y}'(F)\widehat{\boldsymbol{\beta}}(F)\\ &= \varepsilon[-d_0 + \mathbf{d}'\widehat{\boldsymbol{\beta}}(F)] - \boldsymbol{\Sigma}_{\mathbf{x}Y}'(F)[\widehat{\boldsymbol{\beta}}(F_\varepsilon) - \widehat{\boldsymbol{\beta}}(F)].\end{aligned}$$

Divide both sides by ε and let $\varepsilon \to 0$. From (8.3.14),

$$\mathrm{IC}_{\widehat{\sigma}^2,F}(\mathbf{x}', y) = -d_0 + \mathbf{d}'\widehat{\boldsymbol{\beta}}(F) - \boldsymbol{\Sigma}_{\mathbf{x}Y}'(F)\boldsymbol{\Sigma}_{\mathbf{xx}}^{-1}(F)\mathbf{x}[y - \mathbf{x}'\widehat{\boldsymbol{\beta}}(F)],$$

yielding (8.3.15). ■

Note that since $y - \mathbf{x}'\widehat{\boldsymbol{\beta}}(F)$ is unbounded, $\mathrm{IC}_{\widehat{\boldsymbol{\beta}},F}(\mathbf{x}', y)$ and $\mathrm{IC}_{\widehat{\sigma}^2,F}(\mathbf{x}', Y)$ are unbounded, so the LS estimates $\widehat{\boldsymbol{\beta}}(F)$ and $\widehat{\sigma}^2(F)$ are not robust estimators.

The theoretical functions are intended to measure the influence on $\widehat{\boldsymbol{\beta}}$ and $\widehat{\sigma}^2$ due to adding one observation $(\mathbf{x}', Y)$ to a very large sample. We do not, however, always have very large samples in practice, and therefore need finite sample approximations of these influence functions. Four approximations to $\mathrm{IC}_{\boldsymbol{\beta},F}(\mathbf{x}', Y)$ are shown below. For more details, see Cook and Weisberg (1982) or Chatterjee and Hadi (1988). Henceforth, assume that $\mathbf{X}_{(i)}$ is of full column rank, $i = 1, \ldots, N$.

Result 8.3.9. Four approximations to the theoretical influence function $\mathrm{IC}_{\boldsymbol{\beta},F}(\mathbf{x}', Y)$ are given below.

1. The *empirical influence function* (EIC) based on N observations is

$$\mathbf{EIC}_i = N(\mathbf{X}'\mathbf{X})^{-1}\mathbf{x}_i\widehat{\varepsilon}_i, \quad i = 1, \cdots, N. \tag{8.3.16}$$

2. The *empirical influence function based on* $(N-1)$ *observations*, ($EIC_{(i)}$) has the form

$$\mathbf{EIC}_{(i)} = (N-1)(\mathbf{X}'\mathbf{X})^{-1}\mathbf{x}_i\frac{\widehat{\varepsilon}_i}{(1-p_{ii})^2}, \quad i = 1, \cdots, N. \tag{8.3.17}$$

3. The *sample influence function* (SIC) based on N observations is

$$\mathbf{SIC}_i = (N-1)(\widehat{\boldsymbol{\beta}} - \widehat{\boldsymbol{\beta}}_{(i)}) = (N-1)(\mathbf{X}'\mathbf{X})^{-1}\mathbf{x}_i\frac{\widehat{\varepsilon}_i}{1-p_{ii}}, \quad i = 1, \cdots, N. \tag{8.3.18}$$

4. The *sensitivity curve* (SC) based on N observations is

$$\mathbf{SC}_i = N(\mathbf{X}'\mathbf{X})^{-1}\mathbf{x}_i\frac{\widehat{\varepsilon}_i}{(1-p_{ii})}, \quad i = 1, \cdots, N. \tag{8.3.19}$$

Proof. The function $\mathbf{EIC}_i$ is obtained by setting $(\mathbf{x}', Y) = (\mathbf{x}_i', Y_i)$, $F = \widehat{F}_N$, $\boldsymbol{\beta}(F) = \widehat{\boldsymbol{\beta}}$ and $\boldsymbol{\Sigma}_{\mathbf{xx}}^{-1}(F) = N(\mathbf{X}'\mathbf{X})^{-1}$ in (8.3.14). The function $\mathbf{EIC}_{(i)}$ is obtained by setting $(\mathbf{x}', Y) = (\mathbf{x}_i', Y_i)$, $F = \widehat{F}_{N(i)}$, $\boldsymbol{\beta}(F) = \widehat{\boldsymbol{\beta}}_{(i)}$ and $\boldsymbol{\Sigma}_{\mathbf{xx}}^{-1}(F) = (N-1)(\mathbf{X}_{(i)}'\mathbf{X}_{(i)})^{-1}$ in (8.3.14), which gives, for $i = 1, \cdots, N$,

$$\mathbf{EIC}_{(i)} = (N-1)(\mathbf{X}_{(i)}'\mathbf{X}_{(i)})^{-1}\mathbf{x}_i(Y_i - \mathbf{x}_i'\widehat{\boldsymbol{\beta}}_{(i)}).$$

Using (2.1.12) and (9.1.12), property 2 follows. Substitute $(\mathbf{x}', Y)$ for z, and $\boldsymbol{\beta}(F)$ from (8.3.12) for the functional T into (C.8) to get

$$\psi\{(\mathbf{x}', Y), F, \boldsymbol{\beta}(F)\} = \lim_{\varepsilon \to 0} \frac{1}{\varepsilon}[\boldsymbol{\beta}\{(1-\varepsilon)F + \varepsilon\delta_{(\mathbf{x}',Y)}\} - \boldsymbol{\beta}\{F\}]. \tag{8.3.20}$$

The sample influence curve $\mathbf{SIC}_i$ is obtained by setting $(\mathbf{x}', Y) = (\mathbf{x}_i', Y_i)$, $F = \widehat{F}_N$, and $\varepsilon = -1/(N-1)$ in (8.3.20), omitting the limit and simplifying. The proof is left as an exercise (Exercise 8.8). To obtain $\mathbf{SC}_i$, set $(\mathbf{x}', Y) = (\mathbf{x}_i', Y_i)$, $F = \widehat{F}_{N(i)}$, and $\varepsilon = 1/N$ in (8.3.20), omit the limit and simplify. ■

The main difference between these approximate influence functions is in the power of $(1 - p_{ii})$. We see that $\mathbf{EIC}_i$ is least sensitive to high leverage points, while $\mathbf{EIC}_{(i)}$ is the most sensitive. Note that $\mathbf{SIC}_i$ and $\mathbf{SC}_i$ are equivalent and are proportional to the distance $(\widehat{\boldsymbol{\beta}} - \widehat{\boldsymbol{\beta}}_{(i)})$. Note that each of these approximate influence curves for $\widehat{\boldsymbol{\beta}}$ is a p-dimensional vector, which is unwieldy. In practice, it would be useful to obtain an ordering of the N observations based on a scalar summary measure of influence. The quantity

$$D_i(\mathbf{M}, c) = [\psi'\{\mathbf{x}_i', Y_i, F, \boldsymbol{\beta}(F)\}\mathbf{M}\psi\{\mathbf{x}_i', Y_i, F, \boldsymbol{\beta}(F)\}]/c$$

for appropriate choices of $\mathbf{M}$ and c is useful. If $D_i(\mathbf{M}, c)$ is large, then the ith observation has strong influence on $\widehat{\boldsymbol{\beta}}$ relative to $\mathbf{M}$ and c. Four different choices of $\mathbf{M}$ and c lead to the following measures that are popular diagnostics measures: (a) Cook's distance (C_i), (b) Modified Cook's distance (MC_i), (c) $DFFITS$ or Welsch–Kuh's distance (WK_i), and (d) Welsch's distance (W_i). The essential difference between these is in the choice of scale. Further, C_i only measures the influence of the ith observation on $\widehat{\boldsymbol{\beta}}$, whereas the other three statistics measure the influence on both $\widehat{\boldsymbol{\beta}}$ and $\widehat{\sigma}^2$.

Definition 8.3.8. Cook's distance. Cook's distance is defined by Cook and Weisberg (1982) and Atkinson (1985) as

$$\begin{aligned} C_i &= D_i(\mathbf{X}'\mathbf{X}, p\widehat{\sigma}^2/(N-1)^2) \\ &= \frac{(\widehat{\boldsymbol{\beta}} - \widehat{\boldsymbol{\beta}}_{(i)})'(\mathbf{X}'\mathbf{X})(\widehat{\boldsymbol{\beta}} - \widehat{\boldsymbol{\beta}}_{(i)})}{p\widehat{\sigma}^2}, \quad i = 1, \cdots, N. \end{aligned} \tag{8.3.21}$$

The $100(1-\alpha)\%$ joint ellipsoidal confidence region for $\boldsymbol{\beta}$ given in (7.4.2) is centered at $\widehat{\boldsymbol{\beta}}$. The quantity C_i measures the change in the center of this ellipsoid when the ith observation is omitted, and thereby assesses its influence. C_i may be interpreted as the scaled distance between $\widehat{\boldsymbol{\beta}}$ and $\widehat{\boldsymbol{\beta}}_{(i)}$, or alternately, as the scaled distance between $\widehat{\mathbf{y}}$ and $\widehat{\mathbf{y}}_{(i)}$. The following result summarizes alternate forms for Cook's distance.

Result 8.3.10. For $i = 1, \cdots, N$,

$$C_i = \frac{(\widehat{\mathbf{y}} - \widehat{\mathbf{y}}_{(i)})'(\widehat{\mathbf{y}} - \widehat{\mathbf{y}}_{(i)})}{p\widehat{\sigma}^2}, \quad \text{or} \tag{8.3.22}$$

$$C_i = \frac{\mathbf{x}_i'(\mathbf{X}'\mathbf{X})^{-1}\mathbf{x}_i}{p(1-p_{ii})} \frac{\widehat{\varepsilon}_i^2}{\widehat{\sigma}^2(1-p_{ii})} = \frac{1}{p}\frac{p_{ii}}{(1-p_{ii})} r_i^2. \tag{8.3.23}$$

Proof. Formula (8.3.22) follows directly from the definition of C_i in (8.3.21). The second one follows by combining (8.3.21) and (9.1.13). ■

It is clear from the previous result that we certainly need not run $(N+1)$ model fits (one with all N observations, and N models where we successively omit observations one at a time). The quantity C_i will be large if p_{ii} is large, or if r_i^2 is large, or both. Although it has been suggested that each C_i be compared to percentiles of the $F_{p,N-p}$ distribution to see if it is large, we can also use a Box-plot or stem-and-leaf plot or index plot of C_i to answer the question, "how large is large?". An *index plot* is a plot of Cook's distance against observation number. Points that are above some threshold value such as the 50th percentile of an $F_{p,N-p}$ distribution are regarded as influential observations. Several modifications of Cook's distance are available for normal linear models (Chatterjee and Hadi (1988), section 4.2, and section 5.4), while Pregibon (1981) used a one-step approximation to extend this statistic to binary response models as well (see Chapter 12). Atkinson (1981) suggested a modification of C_i in order to (a) give more emphasis to extreme points, and (b) be more suitable for graphical displays such as the normal probability plots. The modification consists of replacing $\widehat{\sigma}^2$ by $\widehat{\sigma}^2_{(i)}$, taking the square root of C_i, and adjusting C_i for sample size.

Definition 8.3.9. Modified Cook's distance. We define the modified Cook's distance by

$$MC_i = |r_i^*| \left[\frac{p_{ii}}{(1-p_{ii})} \frac{(N-p)}{p} \right]^{1/2}. \tag{8.3.24}$$

In the case where $p_{ii} = p/N$, $i = 1, \cdots, N$, the plot of MC_i versus i is identical to the plot of $|r_i^*|$.

Figures 8.3.1 (a) and (b) give graphical interpretations of Cook's distance and the modified Cook's distance for a 2-dimensional vector $\boldsymbol{\beta}$. In (a), the ellipsoid is centered at $\widehat{\boldsymbol{\beta}}$. Cases i and l are equally influential on $\widehat{\boldsymbol{\beta}}$ in (a), while case m is more influential since $\widehat{\boldsymbol{\beta}}_{(m)}$ lies on an outer contour corresponding to a larger value of Cook's distance. The modified Cook's distance in (b) measures the distance from $\widehat{\boldsymbol{\beta}}_{(i)}$ to $\widehat{\boldsymbol{\beta}}$ relative to the ellipsoids constructed using all the observations, but with a scale $\widehat{\sigma}^2_{(i)}$ specific to the ith case. Since MC_i does not compare cases relative to a fixed metric, the ellipsoids in (b) can have different shapes. Two other measures are related to the ellipsoidal confidence region for $\boldsymbol{\beta}$ and are defined next.

Definition 8.3.10. Andrews–Pregibon statistic. The Andrews–Pregibon statistic is defined as (Andrews and Pregibon, 1978)

$$AP_i = 1 - \frac{SSE_{(i)}|\mathbf{X}'_{(i)}\mathbf{X}_{(i)}|}{SSE|\mathbf{X}'\mathbf{X}|}, \quad i = 1, \cdots, N, \tag{8.3.25}$$

which measures the influence of the ith observation on the volume of the confidence ellipsoids for $\boldsymbol{\beta}$, with and without the ith observation.

We leave it to the reader to verify the relationships between AP_i and the values of p_{ii} and r_i (Exercise 8.14). In particular, since $1-AP_i = (1-p_{ii})\{1-r_i^2/(N-p)\}$, AP_i combines information about high-leverage and outliers, and is therefore potentially less informative than p_{ii} and r_i (Draper and John, 1981).

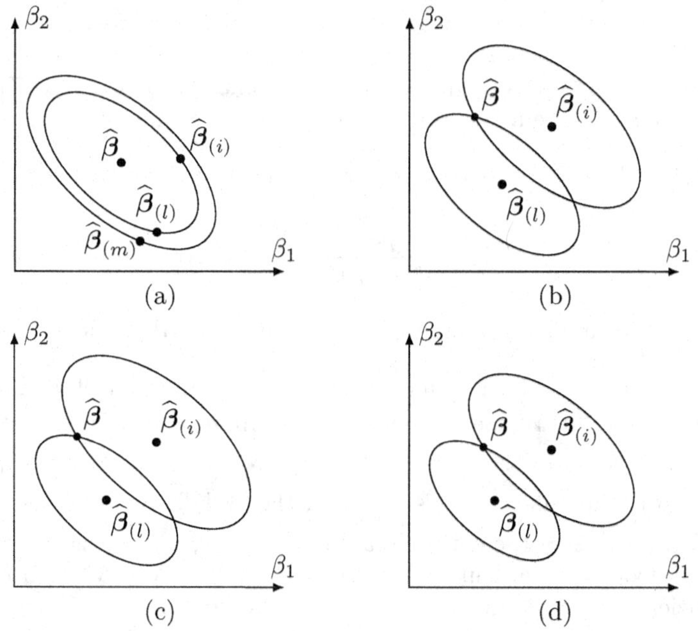

FIGURE 8.3.1. Graphical interpretation of distance measures.

Definition 8.3.11. Cook–Weisberg statistic. The Cook–Weisberg statistic (Cook and Weisberg, 1980) is defined by

$$CW_i = \log\left\{\left(\frac{|\mathbf{X}'_{(i)}\mathbf{X}_{(i)}|}{|\mathbf{X}'\mathbf{X}|}\right)^{1/2}\frac{\widehat{\sigma}^p}{\widehat{\sigma}^p_{(i)}}\left(\frac{F_{p,N-p,\alpha}}{F_{p,N-p-1,\alpha}}\right)^{p/2}\right\} \tag{8.3.26}$$

Definition 8.3.12. Welsch–Kuh distance or DFFITS. The Welsch–Kuh distance or $DFFITS_i$ is a popular statistic which measures the influence of the ith observation on $\widehat{Y}_i$ by the change in the prediction at $\mathbf{x}_i$ by omitting the ith observation relative to s. e.$(\widehat{Y}_i)$ as

$$\begin{aligned} DFFITS_i = WK_i &= \frac{|\mathbf{x}'_i(\widehat{\boldsymbol{\beta}} - \widehat{\boldsymbol{\beta}}_{(i)})|}{\widehat{\sigma}_{(i)}\sqrt{p_{ii}}} \\ &= \frac{|[\widehat{\varepsilon}_i/(1-p_{ii})]\mathbf{x}'_i(\mathbf{X}'\mathbf{X})^{-1}\mathbf{x}_i|}{\widehat{\sigma}_{(i)}\sqrt{p_{ii}}} = |r_i^*|\sqrt{\frac{p_{ii}}{1-p_{ii}}}. \end{aligned} \tag{8.3.27}$$

Note that the relation between MC_i and WK_i is

$$MC_i = WK_i\sqrt{(N-p)/p}.$$

WK_i is also called $DFFITS_i$ (Belsley et al., 1980) because it is the scaled difference between $\widehat{Y}_i$ and $\widehat{Y}_{i(i)}$. It provides a measure of the influence of the ith observation on the prediction at $\mathbf{x}_i$. Large values of $DFFITS_i$ indicate that the ith observation is influential on the fitted model. Again, how large is large? Velleman and Welsch (1981) recommended that "values greater than 1 or 2 seem reasonable to nominate for special attention". Based on the t_{N-p-1} distribution for r_i^*, a cut-off point for WK_i of $t_{N-p-1,\alpha/2}[p/(N-p)]^{1/2}$ could be used.

Alternately, we could replace $t_{N-p-1,\alpha/2}$ by 2 to get a cut-off point. Similarly, the influence of the ith observation on the prediction at $\mathbf{x}_h, h \neq i$ is given by $|\mathbf{x}_h'(\widehat{\boldsymbol{\beta}}-\widehat{\boldsymbol{\beta}}_{(i)})|/(\sigma\sqrt{p_{hh}})$, which can be shown to be at most equal to WK_i. The implication is that if the ith observation is not seen to be influential on the prediction at $\mathbf{x}_i$, as indicated by a small WK_i value, then the ith observation cannot be influential on the prediction at any other $\mathbf{x}_h$, $h \neq i$.

Definition 8.3.13. Weslch distance. The Weslch distance is defined by

$$W_i = \left[(N-1)r_i^{*2}\frac{p_{ii}}{(1-p_{ii})^2}\right]^{1/2} = WK_i\sqrt{\frac{N-1}{1-p_{ii}}}, \tag{8.3.28}$$

and clearly gives more emphasis to high leverage points than WK_i does.

Figures 8.3.1 (c) and (d) give graphical representations of the Welsch distance and the Welsch–Kuh distance. While W_i measures the distance from $\widehat{\boldsymbol{\beta}}_{(i)}$ to $\widehat{\boldsymbol{\beta}}$ relative to the ellipsoid centered at $\widehat{\boldsymbol{\beta}}_{(i)}$, W_j measures the distance from $\widehat{\boldsymbol{\beta}}_{(j)}$ to $\widehat{\boldsymbol{\beta}}$ relative to the ellipsoid centered at $\widehat{\boldsymbol{\beta}}_{(j)}$. Similar to the modified Cook's distance, the Welsch–Kuh distance measures the distance from $\widehat{\boldsymbol{\beta}}_{(i)}$ to $\widehat{\boldsymbol{\beta}}$ using the entire data, but with a different scale for each case.

Until now, we looked at measures that study the influence of the ith observation on the entire coefficient vector $\boldsymbol{\beta}$. In some cases, we may be especially interested in certain components of $\boldsymbol{\beta}$. It might happen that an observation is influential only on one dimension (predictor variable). Further, an observation which has moderate influence on all the β_j's may be judged to be more influential than an observation which has large influence on just one coefficient, and negligible influence on the other coefficients. The following result describes the influence of an observation on a single β_j. Denote the columns of $\mathbf{X}$ by $\mathbf{u}_1, \cdots, \mathbf{u}_p$ and let $\mathbf{w}_j = (\mathbf{I}-\mathbf{P}_j)\mathbf{u}_j$, where $\mathbf{P}_j$ is the projection matrix on the space spanned by the columns other than $\mathbf{u}_j$.

Result 8.3.11. Let $\mathbf{X}$ be of full column rank and $\mathbf{x}_i \in \mathcal{R}(\mathbf{X}_{(i)})$ for each $i = 1, \cdots, N$. Then for any $i, j = 1, \cdots, N$,

$$\widehat{\beta}_j - \widehat{\beta}_{j(i)} = \frac{\widehat{\varepsilon}_i}{(1-p_{ii})}\frac{w_{ij}}{\mathbf{w}_j'\mathbf{w}_j}. \tag{8.3.29}$$

Proof. Denote the columns of the symmetric matrix $(\mathbf{X}'\mathbf{X})^{-1}$ by $\mathbf{h}_1, \cdots, \mathbf{h}_p$. From (9.1.13), for any i and j, $\widehat{\beta}_j - \widehat{\beta}_{j(i)} = \mathbf{h}_j'\mathbf{x}_i\widehat{\varepsilon}_i/(1-p_{ii})$. Suppose $j = 1$. Let $\mathbf{U} = (\mathbf{u}_2, \cdots, \mathbf{u}_p)$. Applying Result 2.1.4 to

$$(\mathbf{X}'\mathbf{X})^{-1} = \begin{pmatrix} \mathbf{u}_1'\mathbf{u}_1 & \mathbf{u}_1'\mathbf{U} \\ \mathbf{U}'\mathbf{u}_1 & \mathbf{U}'\mathbf{U} \end{pmatrix}^{-1}$$

we see that

$$\mathbf{h}_1' = \left(\frac{1}{\mathbf{u}_1'\mathbf{u}_1 - \mathbf{u}_1'\mathbf{U}(\mathbf{U}'\mathbf{U})^{-1}\mathbf{U}'\mathbf{u}_1}, -\frac{\mathbf{u}_1'\mathbf{U}(\mathbf{U}'\mathbf{U})^{-1}}{\mathbf{u}_1'\mathbf{u}_1 - \mathbf{u}_1'\mathbf{U}(\mathbf{U}'\mathbf{U})^{-1}\mathbf{U}'\mathbf{u}_1}\right).$$

From $\mathbf{U}(\mathbf{U}'\mathbf{U})^{-1}\mathbf{U} = \mathbf{P}_1$, it follows that $\mathbf{u}_1'\mathbf{u}_1 - \mathbf{u}_1'\mathbf{U}(\mathbf{U}'\mathbf{U})^{-1}\mathbf{U}'\mathbf{u}_1 = \mathbf{w}_1'\mathbf{w}_1$ and

$$\mathbf{h}_1'\mathbf{X}' = \left(\frac{1}{\mathbf{w}_1'\mathbf{w}_1}, -\frac{\mathbf{u}_1'\mathbf{U}(\mathbf{U}'\mathbf{U})^{-1}}{\mathbf{w}_1'\mathbf{w}_1}\right)\begin{pmatrix}\mathbf{u}_1' \\ \mathbf{U}'\end{pmatrix} = \frac{\mathbf{u}_1' - \mathbf{u}_1'\mathbf{P}_1}{\mathbf{w}_1'\mathbf{w}_1} = \frac{\mathbf{w}_1'}{\mathbf{w}_1'\mathbf{w}_1}.$$

Since $\mathbf{h}_1'\mathbf{x}_i$ is the ith component of the row vector $\mathbf{h}_1'\mathbf{X}'$, it is equal to $w_{1i}/\mathbf{w}_1'\mathbf{w}_1$, which then leads to (8.3.29) for $j = 1$. The proof for other values of j is the same. ■

In addition to the four measures which quantify the influence of the ith observation on the entire vector $\widehat{\boldsymbol{\beta}}$, it is possible to measure its influence on a single coefficient $\widehat{\beta}_j$ using the following statistics.

Definition 8.3.14. DFBETAS.

$$
\begin{aligned}
DFBETAS_{ij} &= r_i \frac{w_{ij}}{\sqrt{\mathbf{w}_j'\mathbf{w}_j}} \frac{1}{\sqrt{1-p_{ii}}}, \quad \text{using estimate } \widehat{\sigma}, \text{ or} \\
&= r_i^* \frac{w_{ij}}{\sqrt{\mathbf{w}_j'\mathbf{w}_j}} \frac{1}{\sqrt{1-p_{ii}}}, \quad \text{using estimate } \widehat{\sigma}_{(i)}. \qquad (8.3.30)
\end{aligned}
$$

Belsley et al. (1980) suggested that values of $|DFBETAS_{ij}|$ exceeding $2/N$ are influential on $\widehat{\beta}_j$.

Definition 8.3.15. Covratio. A diagnostic measure Belsley et al. (1980) which assesses the influence of the ith observation by comparing the estimated variance of $\widehat{\boldsymbol{\beta}}$ and $\widehat{\boldsymbol{\beta}}_{(i)}$ is called the Covratio, and has the form

$$
CR_i = \frac{|\widehat{\sigma}^2_{(i)}(\mathbf{X}_{(i)}'\mathbf{X}_{(i)})^{-1}|}{|\widehat{\sigma}^2(\mathbf{X}'\mathbf{X})^{-1}|}, \quad i = 1, \cdots, N. \qquad (8.3.31)
$$

Now, CR_i will be approximately equal to one when all the N observations have equal influence on $\text{Cov}(\widehat{\boldsymbol{\beta}})$, so that deviation of CR_i from unity suggests that the ith observation is influential. Exercise 8.14 shows a relationship between CR_i and the values of p_{ii} and r_i, and the resulting calibration points for CR_i. The Cook–Weisberg statistic is equivalent to CR_i since

$$
CW_i = -\log(CR_i)/2 + p\log(F_{p,N-p,\alpha}/F_{p,N-p-1,\alpha})/2.
$$

Numerical Example 8.3. Regression diagnostics. The following analysis pertains to a study of production waste and land use. There are $N = 40$ observations on a response variable Y and five predictor variables. The response Y is solid waste (in millions of tons), while X_1 is industrial land (acres), X_2 is fabricated metals (acres), X_3 denotes trucking and wholesale trade (acres), X_4 denotes retail trade (acres), and X_5 is number of restaurants and hotels. The data set "waste" is available in the R package *RobStatTM*. Graphical summaries of the Studentized residuals and some influence diagnostics, as well as a summary of least squares fit, are shown below. $R^2 = 0.85$, the F-statistic is 38.39 and $\widehat{\sigma}^2 = 0.023$.

```
library(RobStatTM)
data(waste)
model <-lm(SolidWaste ~ Land + Metals + Trucking +
           Retail + Restaurants, data = waste)
summary(model)$coefficients
```

```
                 Estimate   Std. Error    t value     Pr(>|t|)
(Intercept)  1.242807e-01 3.159531e-02  3.9335164 3.916380e-04
Land        -5.248585e-05 1.786442e-05 -2.9380107 5.895087e-03
Metals       4.146423e-05 1.532860e-04  0.2705024 7.884091e-01
Trucking     2.503577e-04 8.831333e-05  2.8348799 7.662709e-03
Retail      -8.616159e-04 3.753535e-04 -2.2954787 2.799948e-02
Restaurants  1.335293e-02 2.275630e-03  5.8677924 1.278723e-06
```

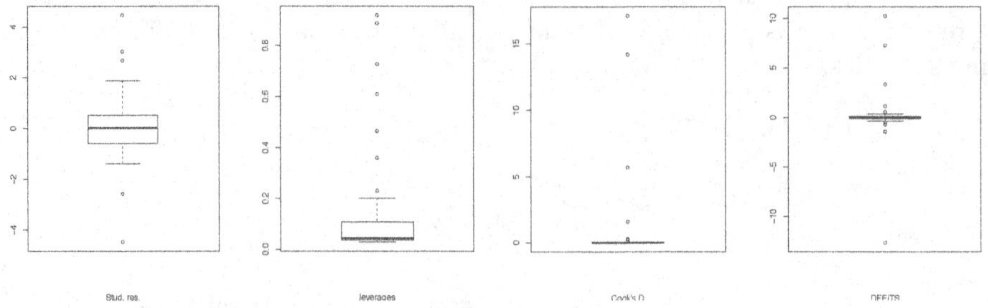

FIGURE 8.3.2. Boxplots of regression diagnostics for the waste data.

```
summary(model)$r.squared
 0.8490727

summary(model)$fstatistic
   value     numdf     dendf
38.25481  5.00000 34.00000

summary(model)$sigma^2
 0.02251635

# Boxplots of various diagnostics
i <- influence(model)
par(mfrow = c(2, 2))
boxplot(rstudent(model), sub = "Stud. res.")
boxplot(i$hat, sub = "leverages")
boxplot(cooks.distance(model), sub = "Cook's D")
boxplot(dffits(model), sub = "DFFITS")
```

The boxplots of the Studentized residuals, leverages, Cook's distance and DFFITS are displayed in Figure 8.3.2 and show anomalous cases in different ways. The *L–R* plot (not shown) can be constructed using the following code and enables us to distinguish outliers from high leverage points (see Section 8.3.3):

```
plot(rstandard(model)^2,i$hat,pch = 19,
  xlab = "res**2",ylab = "leverage")
i0 <- which(i$hat > 0.5)
text(rstandard(model)[i0]^2,i$hat[i0],
  labels = i0,cex = 0.9,font = 2,pos = 1)
```

▲

8.4 Prediction intervals and calibration

We first discuss prediction intervals. Example 8.4.1 illustrates this for a simple linear regression model.

Example 8.4.1. Suppose $X_{0,1}$ is a specified new value of X_1 in an SLR model. The predicted mean response corresponding to this value of the explanatory variable is

$$\widehat{Y}_0 = \widehat{\beta}_0 + \widehat{\beta}_1 X_{0,1} = \overline{Y} + \widehat{\beta}_1(X_{0,1} - \overline{X}_1). \tag{8.4.1}$$

Let $\mathbf{x}_0' = (1, X_{0,1})$. Then we can write $\widehat{Y}_0$ as a linear combination of $\widehat{\boldsymbol{\beta}} = (\widehat{\beta}_0, \widehat{\beta}_1)'$:

$$\widehat{Y}_0 = (1, X_{0,1}) \begin{pmatrix} \widehat{\beta}_0 \\ \widehat{\beta}_1 \end{pmatrix} = \mathbf{x}_0' \widehat{\boldsymbol{\beta}}.$$

Suppose the error $\varepsilon \sim N(0, \sigma^2)$. Then $\widehat{Y}_0$ has a normal distribution with mean $\mathrm{E}(\widehat{Y}_0) = \beta_0 + \beta_1 X_{0,1}$ and variance

$$\begin{aligned}\mathrm{Var}(\widehat{Y}_0) &= \mathrm{Var}(\overline{Y}) + (X_{0,1} - \overline{X}_1)^2 \,\mathrm{Var}(\widehat{\beta}_1) \\ &= \frac{\sigma^2}{N} + \frac{(X_{0,1} - \overline{X}_1)^2 \sigma^2}{\sum_{i=1}^N (X_{i,1} - \overline{X}_1)^2}.\end{aligned}$$

Alternately, we can compute this variance using

$$\mathrm{Var}(\widehat{Y}_0) = \mathrm{Var}(\widehat{\beta}_0) + X_{0,1}^2 \,\mathrm{Var}(\widehat{\beta}_1) + 2X_{0,1} \,\mathrm{Cov}(\widehat{\beta}_0, \widehat{\beta}_1).$$

The estimated standard deviation of $\widehat{Y}_0$ is then

$$\widehat{\mathrm{s.e.}}(\widehat{Y}_0) = \widehat{\sigma} \left\{ \frac{1}{N} + \frac{(X_{0,1} - \overline{X}_1)^2}{\sum_{i=1}^N (X_{i,1} - \overline{X}_1)^2} \right\}^{1/2},$$

which attains a minimum value when $X_{0,1}$ coincides with $\overline{X}_1$ and increases as the distance between the two values increases. Intuitively, this tells us that we expect inaccurate predictions for $X_{0,1}$ values that are outside the observed range of X values. We may now construct a $100(1-\alpha)\%$ prediction interval for the mean value of the distribution of Y corresponding to a given $X_{0,1}$ value:

$$\widehat{Y}_0 \pm t_{N-2,\alpha/2} \times \widehat{\mathrm{s.e.}}(\widehat{Y}_0),$$

where $t_{N-2,\alpha/2}$ corresponds to the upper $(\alpha/2)$th critical point from the t_{N-2}-distribution.

The actual (unobserved) value of Y varies about the true mean value with variance σ^2; the predicted value of an individual response corresponding to $X_{0,1}$ is still $\widehat{Y}_0$, with estimated variance

$$\widehat{\mathrm{Var}}(\widehat{Y}_0) = \widehat{\sigma}^2 \left\{ 1 + \frac{1}{N} + \frac{(X_{0,1} - \overline{X}_1)^2}{\sum_{i=1}^N (X_{i,1} - \overline{X}_1)^2} \right\}.$$

Then,

$$\widehat{Y}_0 \pm t_{N-2,\alpha/2} \widehat{\sigma} \left\{ 1 + \frac{1}{N} + \frac{(X_{0,1} - \overline{X}_1)^2}{\sum_{i=1}^N (X_{i,1} - \overline{X}_1)^2} \right\}^{1/2}$$

is the $100(1-\alpha)\%$ prediction interval for an individual unknown response.

An extension of the marginal prediction intervals approach enables us to predict the average of L unknown observations at the new value $X_{0,1}$ which we denote by $\overline{Y}_0$ (when $L = 1$, we get the case we discussed above). We can verify that $\overline{Y}_0$ has a normal distribution with mean $\beta_0 + \beta_1 X_{0,1}$, and variance σ^2/L, which is independent of the distribution of $\widehat{Y}_0$. Hence,

$$\overline{Y}_0 - \widehat{Y}_0 \sim N(0, \sigma^2/L + \text{Var}(\widehat{Y}_0));$$

replacing σ^2 by its estimate $\widehat{\sigma}^2$, we derive the $100(1-\alpha)\%$ confidence interval for $\overline{Y}_0$ as

$$\widehat{Y}_0 \pm t_{N-2,\alpha/2}\widehat{\sigma}\left\{\frac{1}{L} + \frac{1}{N} + \frac{(X_{0,1} - \overline{X}_1)^2}{\sum_{i=1}^{N}(X_{i,1} - \overline{X}_1)^2}\right\}^{1/2}.$$

These limits are wider than those for the mean response of Y corresponding to $X_{0,1}$, which is to be expected. □

This procedure is useful for constructing point predictions and corresponding confidence intervals for unknown true mean responses or unknown individual responses corresponding to a set of different X values, say, $X_{0,1}, \cdots, X_{0,L}$. The confidence intervals for the responses $Y_{0,1}, \cdots, Y_{0,L}$ are called *marginal* intervals; the jth interval contains $Y_{0,j}$ with probability $1-\alpha$. However, the joint probability that all the intervals simultaneously contain $Y_{0,j}$, $j = 1, \cdots, L$ is usually less than $1-\alpha$. We can alternatively construct simultaneous prediction intervals, as well as simultaneous confidence curves for the whole regression function over its entire range. The latter are *confidence bands* which involve a critical value from the F-distribution.

Example 8.4.2. We now show some results for the multiple linear regression model $\mathbf{y} = \mathbf{X}\boldsymbol{\beta} + \boldsymbol{\varepsilon}$, where $\mathbf{X} = (\mathbf{x}_1, \cdots, \mathbf{x}_N)'$ is of full column rank with each $\mathbf{x}_i = (1, X_{i1}, \cdots, X_{ik})'$, and $\boldsymbol{\varepsilon}$ consists of i.i.d. $N(0, \sigma^2)$ errors. Given specified new values of the predictors, viz., $\mathbf{x}_0' = (1, X_{0,1}, \cdots, X_{0,k})$, the predicted response is $\widehat{Y}_0 = \mathbf{x}_0'\widehat{\boldsymbol{\beta}} = \widehat{\beta}_0 + \widehat{\beta}_1 X_{0,1} + \cdots + \widehat{\beta}_k X_{0,k}$. From Corollary 4.2.1, $\text{E}(\widehat{Y}_0) = \mathbf{x}_0'\boldsymbol{\beta}$ and $\text{Var}(\widehat{Y}_0) = \mathbf{x}_0' \text{Cov}(\widehat{\boldsymbol{\beta}})\mathbf{x}_0 = \sigma^2 \mathbf{x}_0'(\mathbf{X}'\mathbf{X})^{-1}\mathbf{x}_0$. Apart from σ^2, the prediction variance depends on the term $\mathbf{x}_0'(\mathbf{X}'\mathbf{X})^{-1}\mathbf{x}_0$, which we may denote by p_{00}. Clearly, if we let $\widehat{Y}_i$ denote $\mathbf{x}_i'\widehat{\boldsymbol{\beta}}$, then $\text{Var}(\widehat{Y}_i) = \sigma^2 p_{ii}$, where p_{ii} is the ith diagonal element of $\mathbf{P}$. The property $1/N \le p_{ii} \le 1$ implies that $1/N \le \text{Var}(\widehat{Y}_i)/\sigma^2 \le 1$. Apart from σ^2, the sum of $\text{Var}(\widehat{Y}_i)$ over the N data locations is equal to the number of regression parameters, i.e., $\sum_{i=1}^{N} \text{Var}(\widehat{Y}_i)/\sigma^2 = \text{tr}(\mathbf{P}) = k+1$. This suggests the advantage of parsimonious models from the point of view of reducing the overall prediction variance. The $100(1-\alpha)\%$ confidence interval for the true mean value of Y at $\mathbf{x}_0'$ is

$$\widehat{Y}_0 \pm t_{N-k-1,\alpha/2}\,\widehat{\sigma}\{\mathbf{x}_0'(\mathbf{X}'\mathbf{X})^{-1}\mathbf{x}_0\}^{1/2}. \tag{8.4.2}$$

On the other hand, if we wish to construct a prediction estimate of an individual Y response given $\mathbf{x}_0'$, the point estimate is still $\widehat{Y}_0$, while the $100(1-\alpha)\%$ confidence interval is

$$\widehat{Y}_0 \pm t_{N-k-1,\alpha/2}\,\widehat{\sigma}\{1 + \mathbf{x}_0'(\mathbf{X}'\mathbf{X})^{-1}\mathbf{x}_0\}^{1/2} \quad \square \tag{8.4.3}$$

Calibration is the problem of constructing confidence intervals for an unknown X_0 given Y_0. Consider fitting a straight line regression model to the pairs of observations (X_i, Y_i), $i = 1, \cdots, N$, for which the fitted line is $\widehat{Y} = \widehat{\beta}_0 + \widehat{\beta}_1 X$. Our interest is in obtaining point and interval estimates of the predictor variable X_0 corresponding to an observed Y_0. From the fitted least squares model, we may write $Y_0 = \widehat{\beta}_0 + \widehat{\beta}_1 \widehat{X}_0$, so that $\widehat{X}_0 = (Y_0 - \widehat{\beta}_0)/\widehat{\beta}_1$. This estimator $\widehat{X}_0$, which is also the MLE of X_0, is in general, biased. We consider two situations under which the confidence interval for X_0 is determined. First, let Y_0 denote the

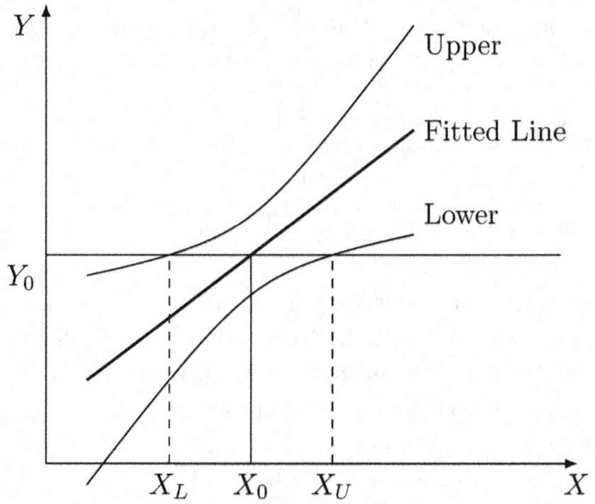

FIGURE 8.4.1. Calibration.

true mean value of the underlying response distribution. We solve for X_L and X_U as points of intersection of the line

$$Y = Y_0 = \widehat{\beta}_0 + \widehat{\beta}_1 \widehat{X}_0$$

and the curves (see Figure 8.4.1)

$$Y = Y_L - t_{N-2,\alpha/2} \left\{ \frac{\widehat{\sigma}^2}{N} + \frac{\widehat{\sigma}^2 (X_L - \overline{X})^2}{s_{XX}} \right\}^{1/2},$$

$$Y = Y_U + t_{N-2,\alpha/2} \left\{ \frac{\widehat{\sigma}^2}{N} + \frac{\widehat{\sigma}^2 (X_U - \overline{X})^2}{s_{XX}} \right\}^{1/2},$$

where $s_{XX} = \sum_{i=1}^{N} (X_i - \overline{X})^2$, $Y_L = \widehat{\beta}_0 + \widehat{\beta}_1 X_L$, and $Y_U = \widehat{\beta}_0 + \widehat{\beta}_1 X_U$. That is, we draw a horizontal line parallel to the x-axis at a height Y_0. From the point of intersection of the fitted line and this horizontal line, we drop a perpendicular to the x-axis, which gives the inverse point estimate $\widehat{X}_0$.

From the points where this line cuts the confidence interval curves, we drop perpendiculars onto the x-axis to give the lower and upper $100(1-\alpha)\%$ inverse confidence limits X_L and X_U. Williams (1959) referred to these as "fiducial limits". Algebraically, we set

$$Y_0 = \widehat{\beta}_0 + \widehat{\beta}_1 X_L + t\{\widehat{\sigma}^2/N + \widehat{\sigma}^2 (X_L - \overline{X})^2 / s_{XX}\}^{1/2}$$

and

$$Y_0 = \widehat{\beta}_0 + \widehat{\beta}_1 X_R - t\{\widehat{\sigma}^2/N + \widehat{\sigma}^2 (X_R - \overline{X})^2 / s_{XX}\}^{1/2},$$

where we denote $t_{N-2,\alpha/2}$ simply by t. By subtracting $\overline{Y} = \widehat{\beta}_0 + \widehat{\beta}_1 \overline{X}$, moving terms around, and finally taking squares on both sides, both $X_L - \overline{X}$ and $X_U - \overline{X}$ solve the following quadratic equation in z:

$$(Y_0 - \overline{Y} - \widehat{\beta}_1 z)^2 = t^2 \{\widehat{\sigma}^2/N + \widehat{\sigma}^2 z^2 / s_{XX}\}.$$

The equation has at most two roots. When both roots are real, the smaller one must be $X_L - \overline{X}$ and the bigger one $X_R - \overline{X}$. By elementary algebra,

$$X_* = \overline{X} + \frac{\widehat{\beta}_1(Y_0 - \overline{Y}) \pm t\widehat{\sigma}\{(Y_0 - \overline{Y})^2/s_{XX} + N^{-1}(\widehat{\beta}_1^2 - t^2\widehat{\sigma}^2/s_{XX})\}^{1/2}}{\widehat{\beta}_1^2 - t^2\widehat{\sigma}^2/s_{XX}},$$

where $X_* = X_L$ (resp. X_R) when the sign in the numerator on the right side is minus (resp. plus).

In general, inverse estimation is not very informative unless the regression of Y on X is significant, i.e., β_1 is significantly different from zero. In such cases, it might happen that (a) the two roots of the quadratic equation in X^* are complex, or (b) the roots X_L and X_U are real-valued, but the interval (X_L, X_U) does not contain $\widehat{X}_0$ because the endpoints lie on the same side of the regression line.

We next assume that Y_0 denotes an individual value from the response distribution. In this case, we solve for X_L and X_U as points of intersection of the line $Y = Y_0 = \widehat{\beta}_0 + \widehat{\beta}_1\widehat{X}_0$ and the curves

$$Y = Y_L - t_{N-2,\alpha/2}\left\{\widehat{\sigma}^2 + \frac{\widehat{\sigma}^2}{N} + \frac{\widehat{\sigma}^2(X_L - \overline{X})^2}{s_{XX}}\right\}^{1/2},$$

$$Y = Y_U + t_{N-2,\alpha/2}\left\{\widehat{\sigma}^2 + \frac{\widehat{\sigma}^2}{N} + \frac{\widehat{\sigma}^2(X_U - \overline{X})^2}{s_{XX}}\right\}^{1/2}.$$

By an argument similar to the previous case, both $X_L - \overline{X}$ and $X_R - \overline{X}$ solve the following quadratic equation in z:

$$(Y_0 - \overline{Y} - \widehat{\beta}_1 z)^2 = t^2\{\widehat{\sigma}^2 + \widehat{\sigma}^2/N + \widehat{\sigma}^2 z^2/s_{XX}\},$$

whose solution again is standard.

Numerical Example 8.4. Inverse simple linear regression. The data "steamuse" in the R package *robustbase* consists of observations taken at intervals from a steam plant which is part of a large industry. The data is a portion of a larger data set used to fit a multiple linear regression model (Draper and Smith, 1998). The response and predictor variables are respectively Y, the monthly use of steam (pounds), and X, the average atmospheric pressure (degrees F). Given a true mean response value of $Y_0 = 10$, we construct a point estimate and an interval estimate of the predictor X_0 corresponding to Y_0.

```
#Inverse linear regression
data("steamUse", package = "robustbase")
attach(steamUse)
# Calibrate
library(EnvStats)
calibrate <- calibrate(Steam ~ temperature, data = steamUse)
summary(calibrate)
anova(calibrate)
newdata <-
  data.frame(temperature = seq(min(temperature),
             max(temperature), length.out = 100))
#predicted values
pred <-
  predict(calibrate, newdata = newdata, se.fit = TRUE)
pointwise <- pointwise(pred, coverage = 0.99, individual = TRUE)
```

```
inversePredictCalibrate(
  calibrate,
  obs.y = 10,
  intervals = TRUE,
  coverage = 0.95,
  individual = TRUE
)

   obs.y    pred.x     lpl.x     upl.x
[1,]    10 45.38455 45.38455 45.38455
```

When the true mean response is $Y_0 = 10$, $\widehat{X}_0 = 45.384$, while $(X_L, X_U) = (40.576, 48.881)$. If, on the other hand, Y_0 is an *individual* value, the value of $\widehat{X}_0$ remains the same, while $(X_L, X_U) = (20.356, 69.102)$. ▲

Exercises

8.1. Suppose $\varepsilon_1, \cdots, \varepsilon_N$ are i.i.d. $N(0,1)$ random variables. Suppose $Y_0 = 0$, and $Y_i = \theta Y_{i-1} + \varepsilon_i$, $i = 1, \cdots, N$, and $|\theta| < 1$. Find the maximum likelihood estimate of θ.

8.2. Consider a two-way ANOVA model with no interaction and $n > 1$ replications per cell:

$$Y_{ijk} = \mu + \tau_i + \theta_j + \varepsilon_{ijk},$$

$i = 1, \cdots, a$, $j = 1, \cdots, b$, and $k = 1, \cdots, n$, where ε_{ijk} are i.i.d. $N(0, \sigma^2)$ random variables.

(a) Show that the MVUE of σ^2 is

$$\widehat{\sigma}^2 = (T_1 + T_2)/N^*,$$

where

$$T_1 = \sum_{i=1}^{a} \sum_{j=1}^{b} n(\overline{Y}_{ij\cdot} - \overline{Y}_{i\cdot\cdot} - \overline{Y}_{\cdot j\cdot} + \overline{Y}_{\cdots})^2,$$

$$T_2 = \sum_{i=1}^{a} \sum_{j=1}^{b} \sum_{k=1}^{n} (Y_{ijk} - \overline{Y}_{ij\cdot})^2,$$

and $N^* = nab - a - b + 1$.

(b) Let $Z_1 = \sum_{i=1}^{a} c_i \tau_i^0$, and $Z_2 = \sum_{j=1}^{b} d_j \theta_j^0$. Show that Z_1, Z_2, and $U = N^* \widehat{\sigma}^2/\sigma^2$ are jointly independent.

8.3. Let $Y_1, \cdots, Y_N$ be independent normal random variables with $E(Y_i) = \beta_1 X_{i1} + \beta_2 X_{i2}$, and $\text{Var}(Y_i) = \sigma^2 X_{i1} X_{i2}$. Assume that X_{ij}'s are positive and the regression matrix $\mathbf{X}$ has full rank. Obtain the MVUE's $\widetilde{\beta}_1$ and $\widetilde{\beta}_2$ of β_1 and β_2. If $\widehat{\beta}_1$ and $\widehat{\beta}_2$ denote the OLS estimates of these parameters, obtain the efficiency of $\widetilde{\beta}_1$ relative to $\widehat{\beta}_1$, i.e., obtain $\text{Var}(\widetilde{\beta}_1)/\text{Var}(\widehat{\beta}_1)$.

8.4. Consider the model

$$\mathbf{y}_i = \mathbf{X}_i\boldsymbol{\beta} + \boldsymbol{\varepsilon}_i, \quad i = 1, \cdots, a,$$

where $\mathbf{y}_i$ and $\boldsymbol{\varepsilon}_i$ are n_i-dimensional vectors, $\mathbf{X}_i$ is an $n_i \times p$ matrix, and $\boldsymbol{\varepsilon}_i$ are normally distributed with $\mathrm{E}(\boldsymbol{\varepsilon}_i\boldsymbol{\varepsilon}_i') = \sigma_i^2\mathbf{I}$ and $\mathrm{E}(\boldsymbol{\varepsilon}_i\boldsymbol{\varepsilon}_j') = \mathbf{O}$, $i \neq j$. Derive the LRT statistic for H_0: $\sigma_1^2 = \cdots = \sigma_a^2$.

8.5. Let $\widehat{\rho}$ be the estimated serial correlation defined in (8.2.5).

(a) Ignoring end effects, show that $DW \backsimeq 2(1 - \widehat{\rho})$.

(b) Show that, if the true serial correlation coefficient $\rho = 0$, then

$$\begin{aligned}\mathrm{E}(DW) &= \mathbf{Q}/(N-p), \text{ and} \\ \mathrm{Var}(DW) &= 2\{\mathbf{R} - \mathbf{Q}\,\mathrm{E}(DW)\}/\{(N-p)(N-p+2)\},\end{aligned}$$

where, for $\mathbf{A}$ defined in (8.2.8),

$$\begin{aligned}\mathbf{Q} &= \mathrm{tr}(\mathbf{A}) - tr[\mathbf{X}'\mathbf{A}\mathbf{X}(\mathbf{X}'\mathbf{X})^-], \text{ and} \\ \mathbf{R} &= \mathrm{tr}(\mathbf{A}^2) - 2tr[\mathbf{X}'\mathbf{A}^2\mathbf{X}(\mathbf{X}'\mathbf{X})^-] + tr[\{\mathbf{X}'\mathbf{A}\mathbf{X}(\mathbf{X}'\mathbf{X})^-\}^2].\end{aligned}$$

[*Hint*: start with (8.2.10) and $tr[\{\mathbf{X}'\mathbf{A}\mathbf{X}(\mathbf{X}'\mathbf{X})^-\}^k] = \mathrm{tr}((\mathbf{AP})^k)$, $k \geq 1$, where $\mathbf{P}$ is the projection matrix onto $\mathcal{C}(\mathbf{X})$.]

8.6. In the regression model (4.1.2), show that the MSE after deleting the ith case is given by

$$MSE_{(i)} = [(N - p - r_i^2)/(N - p - 1)]\ MSE.$$

8.7. Chatterjee and Hadi (1988). Let p_{zii} be the diagonal elements of the matrix $P_Z = \mathbf{Z}(\mathbf{Z}'\mathbf{Z})^{-1}\mathbf{Z}'$ where $\mathbf{Z} = (\mathbf{X}, \mathbf{y})$. Show that

$$p_{zii} = p_{ii} + \widehat{\varepsilon}_i^2/\widehat{\boldsymbol{\varepsilon}}'\widehat{\boldsymbol{\varepsilon}}.$$

8.8. Derive the form of the sample influence function $\mathbf{SIC}_i$ in (8.3.18).

8.9. Consider the model $\mathbf{y} = \mathbf{X}\boldsymbol{\beta} + \boldsymbol{\varepsilon}$, where $\mathbf{X}$ is $N \times p$ and $\mathrm{Cov}(\boldsymbol{\varepsilon}) = \sigma^2\mathbf{I}_N$. Partition $\mathbf{X}$ as $\begin{pmatrix}\mathbf{X}_1 \\ \mathbf{X}_2\end{pmatrix}$, where $\mathbf{X}_i$ is $n_i \times p$, $i = 1, 2$, and $n_1 + n_2 = N$. Partition $\mathbf{y}$ conformably into $\mathbf{y}_1$ and $\mathbf{y}_2$, and $\boldsymbol{\varepsilon}$ into $\boldsymbol{\varepsilon}_1$ and $\boldsymbol{\varepsilon}_2$. Let $\boldsymbol{\beta}^0$ be the LS solution vector for the model, so that $\mathbf{X}_1\boldsymbol{\beta}^0$ is the LS fit for $\mathbf{y}_1$ under the model. On the other hand, suppose one only has $\mathbf{y}_1$ and $\mathbf{X}_1$ available and a LS fit for $\mathbf{y}_1 = \mathbf{X}_1\boldsymbol{\beta} + \boldsymbol{\varepsilon}_1$ yields a solution vector $\widetilde{\boldsymbol{\beta}}$. Show that $\mathrm{Cov}(\mathbf{X}_1\widetilde{\boldsymbol{\beta}}) - \mathrm{Cov}(\mathbf{X}_1\boldsymbol{\beta}^0)$ is n.n.d., and if $\mathcal{R}(\mathbf{X}_1) \cap \mathcal{R}(\mathbf{X}_2) = \{\mathbf{0}\}$, then the two covariances are equal.

8.10. For the SLR model $Y_i = \beta_0 + \beta_1 X_i + \varepsilon_i$, $i = 1, \cdots, N$, show that the weighted squared standardized distance in (8.3.11) is $W_i^* = \frac{(N-1)}{N}(Np_{ii} - 1)\rho_{X,Y}^2$, where $\rho_{X,Y}$ is the simple correlation between X and Y.

8.11. Show that

(a) $a_i^2 \leq (1 - p_{ii})$.

(b) As $a_i^2 \to 1 - p_{ii}$, show that $r_i^2 \to N - p$, and $r_i^{*2} \to \infty$.

8.12. Show that Cook's distance can be expressed as

$$C_i = \{\widehat{\varepsilon}_i^2/(1-p_{ii})^2\}\{p_{ii}/p\widehat{\sigma}^2\}.$$

8.13. (Belsley et al., 1980) Show that

$$CR_i = \{(N-p-r_i^2)/(N-p-1)\}^p/(1-p_{ii}).$$

Show that when $|r_i| \geq 2$, and $p_{ii} = 1/N$, then $CR_i \leq 1 - \{3p/(N-p)\}$ approximately. Also show that when $r_i = 0$, and $p_{ii} \geq 2p/N$, $CR_i \geq 1 + \{3p/(N-p)\}$ approximately. Use these to arrive at the approximate calibration bounds $|CR_i - 1| \geq 3p/N$.

8.14. Show that

$$AP_i = p_{ii} + \frac{\widehat{\varepsilon}_i^2}{\widehat{\varepsilon}'\widehat{\varepsilon}} = p_{ii} + (1-p_{ii})\frac{r_i^2}{N-p}.$$

8.15. Show that for inference on the location parameter, the influence curve for the median functional is

$$\text{IC}_{T,F}(x) = \text{sign}[x - T(F)]/2f[T(F)],$$

where $f(.)$ is the p.d.f. corresponding to the c.d.f. $F(.)$ and the sign(.) function was defined in (5.5.7).

9

Multiple Linear Regression Models

In this chapter, we consider topics specific to multiple linear regression models, which are full-rank general linear models. In Section 9.1, we describe procedures for selecting variables. Section 9.2 discusses orthogonal predictors and the opposite problem of multicollinearity in MLR models, while Section 9.3 describes regression with categorical predictors, often referred to as dummy variable regression (Draper and Smith, 1998).

9.1 Variable selection in regression

We discuss procedures useful for selecting variables from a set of possible predictors $X_1, \cdots, X_k$. Following some graphical methods, we next describe selection of the best regression equation via (a) all possible regressions, using criteria such as R^2, adjusted R^2, $\widehat{\sigma}^2$, Mallows C_p statistic and the $PRESS$ statistic; (b) best subset regression using these criteria; (c) forward selection; (d) backward elimination; and (e) stepwise regression.

9.1.1 Graphical assessment of variables

This Section describes the well-known added-variable plots and partial residual plots which are useful visual aids in variable selection.

Added-variable plots

Added-variable plots, which are also called *partial regression plots*, are useful for understanding the role of a single predictor variable in a multiple linear regression model (Cook and Weisberg, 1982).

Suppose we partition $\mathbf{X} = (\mathbf{X}_1, \mathbf{x}_*)$, where $\mathbf{x}_*$ is the vector of observations of one of the predictor variables. We denote this particular predictor variable by X_*. Let $\mathbf{P}_1$ be the projection matrix onto the column space of $\mathbf{X}_1$. An added-variable plot shows the contribution made by X_* to the variability in Y in the model

$$\mathbf{y} = \mathbf{X}_1\boldsymbol{\beta}_1 + \mathbf{x}_*\beta_* + \boldsymbol{\varepsilon}, \tag{9.1.1}$$

over and beyond the portion explained by $\mathbf{X}_1$ alone. Let $\widehat{\boldsymbol{\varepsilon}}$ denote the vector of residuals from fitting the model (9.1.1).

Result 9.1.1. Let $\widehat{\boldsymbol{\varepsilon}}(\mathbf{y} \mid \mathbf{X}_1) = (\mathbf{I} - \mathbf{P}_1)\mathbf{y}$ and $\widehat{\boldsymbol{\varepsilon}}(\mathbf{x}_* \mid \mathbf{X}_1) = (\mathbf{I} - \mathbf{P}_1)\mathbf{x}_*$ respectively denote the residuals from a LS fit of $\mathbf{y}$ on $\mathbf{X}_1$ and that of $\mathbf{x}_*$ on $\mathbf{X}_1$. Then,

$$\mathrm{E}[\widehat{\boldsymbol{\varepsilon}}(\mathbf{y} \mid \mathbf{X}_1)] = \beta_*\widehat{\boldsymbol{\varepsilon}}(\mathbf{x}_* \mid \mathbf{X}_1). \tag{9.1.2}$$

DOI: 10.1201/9781315156651-9

Further, the LS estimate of β_* in the model (9.1.1) is

$$\widehat{\beta}_* = \frac{\widehat{\varepsilon}'(\mathbf{x}_* \mid \mathbf{X}_1)\widehat{\varepsilon}(\mathbf{y} \mid \mathbf{X}_1)}{\widehat{\varepsilon}'(\mathbf{x}_* \mid \mathbf{X}_1)\widehat{\varepsilon}(\mathbf{x}_* \mid \mathbf{X}_1)}. \tag{9.1.3}$$

Proof. Pre-multiplying both sides of (9.1.1) by $\mathbf{I} - \mathbf{P}_1$, we see that

$$(\mathbf{I} - \mathbf{P}_1)\mathbf{y} = \boldsymbol{\beta}_*(\mathbf{I} - \mathbf{P}_1)\mathbf{x}_* + (\mathbf{I} - \mathbf{P}_1)\boldsymbol{\varepsilon}.$$

Taking expectations on both sides of the equality,

$$\mathrm{E}[\widehat{\varepsilon}(\mathbf{y} \mid \mathbf{X}_1)] = \beta_*(\mathbf{I} - \mathbf{P}_1)\mathbf{x}_* = \beta_*\widehat{\varepsilon}(\mathbf{x}_* \mid \mathbf{X}_1),$$

proving (9.1.2). Using results on partitioned matrices from Section 2.1, and the idempotency of the matrix $(\mathbf{I} - \mathbf{P}_1)$, it is easily verified that the OLS estimate of β_* in the model (9.1.1) is

$$\widehat{\beta}_* = \{\mathbf{x}_*'(\mathbf{I} - \mathbf{P}_1)\mathbf{y}\}/\{\mathbf{x}_*'(\mathbf{I} - \mathbf{P}_1)\mathbf{x}_*\},$$

which leads directly to (9.1.3). ■

Result 9.1.1 implies that a plot of $\widehat{\varepsilon}(\mathbf{y} \mid \mathbf{X}_1)$ versus $\widehat{\varepsilon}(\mathbf{x}_* \mid \mathbf{X}_1)$ is expected to be a straight line through the origin, with estimated slope $\widehat{\beta}_*$, which incidentally, is also the LS estimate in (9.1.1). The added-variable plot is a visual summary of the t-statistic (or extra sum of squares F-statistic) for testing H_0: $\beta_* = 0$. If all the points in the added-variable plot lie exactly on a straight line with slope β_*, $0 < \beta_* < \infty$, i.e., the residuals from the LS fit of $\widehat{\varepsilon}(\mathbf{y} \mid \mathbf{X}_1)$ versus $\widehat{\varepsilon}(\mathbf{x}_* \mid \mathbf{X}_1)$ are all zero, then X_* is a useful addition to the model. If, on the other hand, the points lie exactly on a horizontal line, the regression $\widehat{\mathbf{y}} = \mathbf{X}_1\widehat{\boldsymbol{\beta}}_1$ explains all the variation in Y, and there is no need to include X_* as a predictor. If the points in the added-variable plot lie on a vertical line, then $\widehat{\varepsilon}(\mathbf{x}_* \mid \mathbf{X}_1) = \mathbf{0}$, i.e., X_* is an exact linear combination of the components of $\mathbf{X}_1$, and is therefore superfluous to the model. This is a situation that we recognize as collinearity and discuss in Section 9.2.

Partial residual plots

Partial residual plots, also called residual plus component plots (Larsen and McCleary, 1972; Wood, 1973), are widely used in practice, because they are computationally more convenient than added-variable plots. Again, let $\mathbf{X} = (\mathbf{X}_1, \mathbf{x}_*)$. Since the columns of $\mathbf{X}$ are LIN, in particular, $\mathbf{x}_* \notin \mathcal{C}(\mathbf{X}_1)$, we can define the vector $\boldsymbol{\varepsilon}_*$ of partial residuals corresponding to X_*. Let $\widehat{\boldsymbol{\beta}}_1$ denote the LS estimate of $\boldsymbol{\beta}_1$ in the model $\mathbf{y} = \mathbf{X}_1\boldsymbol{\beta}_1 + \boldsymbol{\varepsilon}$, omitting the variable X_*.

Definition 9.1.1. Partial residuals. The vector of partial residuals corresponding to the predictor X_* is defined by

$$\boldsymbol{\varepsilon}_* = \mathbf{y} - \mathbf{X}_1\widehat{\boldsymbol{\beta}}_1 = \widehat{\boldsymbol{\varepsilon}} + \mathbf{x}_*\widehat{\beta}_*. \tag{9.1.4}$$

In other words, these partial residuals are residuals that have not been adjusted for the predictor variable X_*. A plot of $\boldsymbol{\varepsilon}_*$ versus X_* has estimated slope $\widehat{\beta}_*$ and is called a *partial residual plot.* In general, X_* could be any predictor variable X_j, $1 \le j \le k$; the corresponding partial residuals are residuals from a regression that includes all other variables except X_j in the model. Thus, partial residuals contain the remnant variability in the response Y, after accounting for relationships between Y and X_l, $l \neq j$, $l = 1, \cdots, k$, and therefore constitute the portion of the data used for estimating β_j. A plot of these partial residuals versus X_j will show the partial relationship between Y and X_j. Although they look very different, it may be verified that the partial residual plot and the added-variable plot for X_j have the same slope. Cook and Weisberg (1984) devoted an entire text to the area of regression graphics.

9.1.2 Criteria-based variable selection

We begin with a description of some basic selection criteria.

(a) **Coefficient of determination R^2.** Although R^2 is a traditional criterion for regression model selection, it is not prediction performance oriented, and hence should always be used in conjunction with other criteria for choosing the best predictive model from a set of candidate models. Since the inclusion of a new explanatory variable into a model can never decrease SSR, and consequently the value of R^2, there might be a tendency to overparametrize in order to achieve a large R^2 value.

(b) **Adjusted R^2.** The adjusted R^2 ensures parsimony by imposing a penalty for including marginally important explanatory variables at the cost of error degrees of freedom.

(c) **MSE.** The use of $\widehat{\sigma}^2$ as a model comparison criterion would entail choosing the model which has the smallest $\widehat{\sigma}^2$ value. Let $\widehat{\sigma}^2(k_1)$ and $\widehat{\sigma}^2(k_2)$ denote the residual mean squares from two fitted regression models with k_1 and k_2 predictors, respectively, where $k_1 < k_2$. It could happen that $\widehat{\sigma}^2(k_1) > \widehat{\sigma}^2(k_2)$, which would perhaps imply that the reduction in residual mean squares by fitting the model with k_1 parameters did not counterbalance the loss in residual degrees of freedom.

We next define another criterion for model selection which compares the standardized total mean squared error of prediction for the observed data (Mallows, 1973, 1995).

Definition 9.1.2. Mallows C_p. Let $\mathbf{y} = \mathbf{X}\boldsymbol{\beta} + \boldsymbol{\varepsilon}$ denote the "true" regression model, where $\boldsymbol{\beta} \in \mathcal{R}^q$. Let $\mathbf{y} = \mathbf{X}_1\boldsymbol{\beta}_1 + \boldsymbol{\varepsilon}$ denote the fitted model, where $\boldsymbol{\beta}_1 \in \mathcal{R}^p$. Let $\widehat{\mathbf{y}}_1 = \mathbf{X}_1\widehat{\boldsymbol{\beta}}_1$. Then,

$$\mathrm{E}(\widehat{\mathbf{y}}_1) = \mathbf{X}_1(\mathbf{X}_1'\mathbf{X}_1)^{-1}\mathbf{X}_1'\,\mathrm{E}(\mathbf{X}\boldsymbol{\beta} + \boldsymbol{\varepsilon}) = \mathbf{X}_1(\mathbf{X}_1'\mathbf{X}_1)^{-1}\mathbf{X}_1'\mathbf{X}\boldsymbol{\beta} = \boldsymbol{\eta}_p, \quad \text{say.}$$

The C_p statistic which is useful for comparing various fitted models with p parameters to the full q parameter model is

$$C_p = \frac{SSE_p}{\widehat{\sigma}^2} - (N - 2p) = \frac{\mathbf{y}'(\mathbf{I} - \mathbf{P}_1)\mathbf{y}}{\widehat{\sigma}^2} - (N - 2p), \tag{9.1.5}$$

where SSE_p is the residual sum of squares from a model with p regression coefficients (including the intercept), and $\widehat{\sigma}^2$ is the residual mean square from the "full-model" containing all the predictors and is presumed to be a reliable estimator of σ^2.

Note that C_q is a fixed quantity given by

$$C_q = SSE_q/\widehat{\sigma}^2 - (N - 2q) = (N - q)\widehat{\sigma}^2/\widehat{\sigma}^2 - (N - 2q) = q.$$

On the other hand, for a model with $p < q$ regression coefficients, C_p is a random variable. As the following result shows, it is an approximately unbiased estimator of the expected standardized total mean square of the predicted values.

Result 9.1.2. Let

$$J_p^* = \mathrm{E}(\widehat{\mathbf{y}}_1 - \mathbf{X}\boldsymbol{\beta})'(\widehat{\mathbf{y}}_1 - \mathbf{X}\boldsymbol{\beta})/\sigma^2.$$

Then, $\mathrm{E}(C_p) \approx J_p^*$.

Proof. Since $\mathbf{I} - \mathbf{P}_1$ is symmetric and idempotent,

$$\begin{aligned}\mathrm{E}\{\mathbf{y}'(\mathbf{I} - \mathbf{P}_1)\mathbf{y}\} &= \mathrm{tr}[(\mathbf{I} - \mathbf{P}_1)\sigma^2\mathbf{I}_N] + \boldsymbol{\beta}'\mathbf{X}'(\mathbf{I} - \mathbf{P}_1)\mathbf{X}\boldsymbol{\beta} \\ &= \sigma^2(N-p) + \boldsymbol{\beta}'\mathbf{X}'(\mathbf{I} - \mathbf{P}_1)(\mathbf{I} - \mathbf{P}_1)\mathbf{X}\boldsymbol{\beta} \\ &= \sigma^2(N-p) + (\mathbf{X}\boldsymbol{\beta} - \mathbf{P}_1\mathbf{X}\boldsymbol{\beta})'(\mathbf{X}\boldsymbol{\beta} - \mathbf{P}_1\mathbf{X}\boldsymbol{\beta}) \\ &= \sigma^2(N-p) + (\mathbf{X}\boldsymbol{\beta} - \boldsymbol{\eta}_p)'(\mathbf{X}\boldsymbol{\beta} - \boldsymbol{\eta}_p). \end{aligned} \quad (9.1.6)$$

Also,

$$\begin{aligned}\mathrm{E}(\widehat{\mathbf{y}}_1 - \mathbf{X}\boldsymbol{\beta})'(\widehat{\mathbf{y}}_1 - \mathbf{X}\boldsymbol{\beta}) &= \mathrm{E}(\widehat{\mathbf{y}}_1 - \boldsymbol{\eta}_p)'(\widehat{\mathbf{y}}_1 - \boldsymbol{\eta}_p) + (\mathbf{X}\boldsymbol{\beta} - \boldsymbol{\eta}_p)'(\mathbf{X}\boldsymbol{\beta} - \boldsymbol{\eta}_p) \\ &= \mathrm{tr}\{\mathrm{Cov}(\widehat{\mathbf{y}}_1)\} + (\mathbf{X}\boldsymbol{\beta} - \boldsymbol{\eta}_p)'(\mathbf{X}\boldsymbol{\beta} - \boldsymbol{\eta}_p) \\ &= p\sigma^2 + (\mathbf{X}\boldsymbol{\beta} - \boldsymbol{\eta}_p)'(\mathbf{X}\boldsymbol{\beta} - \boldsymbol{\eta}_p). \end{aligned} \quad (9.1.7)$$

Subtracting (9.1.7) from (9.1.6), we get

$$\mathrm{E}(\widehat{\mathbf{y}}_1 - \mathbf{X}\boldsymbol{\beta})'(\widehat{\mathbf{y}}_1 - \mathbf{X}\boldsymbol{\beta}) - \mathrm{E}\{\mathbf{y}'(\mathbf{I} - \mathbf{P}_1)\mathbf{y}\} = (2p - N)\sigma^2. \quad (9.1.8)$$

The proof follows by dividing both sides of (9.1.8) by $\widehat{\sigma}^2$. ■

Computed values of C_p for fitted models with different subsets of variables are useful for model selection. We make a few useful observations.

1. If the fitted model is unbiased, we have $\boldsymbol{\eta}_p = \mathbf{X}\boldsymbol{\beta}$, so that $\mathrm{E}(C_p) = p$. Adequate models with C_p values close to p will be indicated in a plot of C_p versus p.

2. Biased models with considerable lack of fit will be indicated by points that are substantially above the line $C_p = p$, although because of randomness, adequate models may sometimes fall below the line.

3. Since the actual value of C_p is an estimate of the expected standardized total mean square of the predicted values J_p^*, the height of each plotted point is also important. As predictors are included in the model, SSE_p decreases, and C_p usually increases. The "best" regression model is one that gives the lowest possible value of C_p for which $C_p \approx p$.

4. Note that there are multiple C_p values corresponding to a given p, i.e., corresponding to different subsets of p (out of q) predictor variables.

5. In many cases, when the choice of model is not obvious from the C_p plot, personal judgment may be employed, or alternative approaches may be used.

We next define the *PRESS* or Prediction Sum of Squares criterion. Let $\widehat{Y}_{i,(i)} = \mathbf{x}_i'\widehat{\boldsymbol{\beta}}_{(i)}$ denote the prediction for the ith response from fitting the regression model (4.1.2) with the ith case excluded. That is, we fit a regression model to observations $(Y_l, \mathbf{x}_l')$, $l \neq i$, $l = 1, \cdots, N$. If $\mathbf{X}_{(i)}$ has full column rank, the estimated $\boldsymbol{\beta}$ vector with the ith case excluded is

$$\begin{aligned}\widehat{\boldsymbol{\beta}}_{(i)} &= (\mathbf{X}_{(i)}'\mathbf{X}_{(i)})^{-1}\mathbf{X}_{(i)}'\mathbf{y}_{(i)} \\ &= \left[(\mathbf{X}'\mathbf{X})^{-1} + \frac{(\mathbf{X}'\mathbf{X})^{-1}\mathbf{x}_i\mathbf{x}_i'(\mathbf{X}'\mathbf{X})^{-1}}{1 - p_{ii}}\right]\mathbf{X}_{(i)}'\mathbf{y}_{(i)}. \end{aligned} \quad (9.1.9)$$

The last equation on the right follows from observing that $(\mathbf{X}'_{(i)}\mathbf{X}_{(i)})^{-1} = (\mathbf{X}'\mathbf{X} - \mathbf{x}_i\mathbf{x}'_i)^{-1}$ and property 5 of Result 1.3.8. Let

$$\widehat{\varepsilon}_{i(i)} = Y_i - \widehat{Y}_{i(i)}, \quad i = 1, \cdots, N \tag{9.1.10}$$

denote the corresponding prediction residuals or *PRESS* residuals (we saw this earlier in Definition 8.3.5). Corresponding to each candidate model that we fit, there will be N *PRESS* residuals.

Definition 9.1.3. *PRESS* statistic. Define

$$PRESS = \sum_{i=1}^{N}(Y_i - \widehat{Y}_{i(i)})^2 = \sum_{i=1}^{N}\widehat{\varepsilon}^2_{i(i)}. \tag{9.1.11}$$

This statistic condenses information in the form of N validations, each with a fitting sample of size $(N-1)$. The model which has the smallest computed *PRESS* statistic value is preferred.

Although it seems at first glance that in order to construct the *PRESS* statistic, we must run N separate regressions, this cumbersome computation is not necessary since the *PRESS* residuals are related to the ordinary residuals, as the next result shows.

Result 9.1.3. The predicted residual is related to the ordinary residual by

$$\widehat{\varepsilon}_{i(i)} = \frac{\widehat{\varepsilon}_i}{1 - p_{ii}}. \tag{9.1.12}$$

Proof. It can be shown that (see Section 8.3.1)

$$\widehat{\boldsymbol{\beta}}_{(i)} = \widehat{\boldsymbol{\beta}} - \frac{\widehat{\varepsilon}_i(\mathbf{X}'\mathbf{X})^{-1}\mathbf{x}_i}{1 - p_{ii}}. \tag{9.1.13}$$

Substituting (9.1.13) into (9.1.10), the result follows. ■

If the regression errors are normally distributed, $\widehat{\varepsilon}_{i(i)}$ have the same correlation structure as the $\widehat{\varepsilon}_i$, and $\text{E}(\widehat{\varepsilon}_{i(i)}) = 0$ and $\text{Var}(\widehat{\varepsilon}_{i(i)}) = \sigma^2/(1 - p_{ii})$. Use of $\widehat{\varepsilon}_{i(i)}$ will tend to emphasize cases with large p_{ii}.

Using Result 9.1.3, the *PRESS* statistic is easily computed as

$$PRESS = \sum_{i=1}^{N}\left(\frac{\widehat{\varepsilon}_i}{1 - p_{ii}}\right)^2. \tag{9.1.14}$$

A related statistic based on *PRESS* residuals is

$$R^2_{pred} = 1 - \frac{PRESS}{\sum_{i=1}^{N}(Y_i - \overline{Y})^2}. \tag{9.1.15}$$

All Possible Regressions is a cumbersome procedure that requires the fitting of every possible regression equation that always includes an intercept and may include any of the variables $X_1, \cdots, X_k$. Since each of the k explanatory variables can either be included or not in the regression model, we must fit 2^k possible regressions; even when $k = 10$, this requires 1024 regression fits. Each fitted regression is generally assessed on the basis of three criteria, viz., R^2, $\widehat{\sigma}^2$, and Mallows C_p statistic. In general, when there are several predictors, an analysis of all possible regressions is quite unwarranted in terms of effort and

Source	d.f.	SS	MS
Regression	p	$SSR = \widehat{\mathbf{y}}'\widehat{\mathbf{y}}$	MSR
Residual	$N-p$	$SSE = \widehat{\boldsymbol{\varepsilon}}'\widehat{\boldsymbol{\varepsilon}} = \mathbf{y}'\mathbf{y} - \widehat{\mathbf{y}}'\widehat{\mathbf{y}}$	MSE
Total	N	$SST = \mathbf{y}'\mathbf{y}$	

TABLE 9.1.1. ANOVA table for the multiple linear regression model.

computer time. With some initial thought, a majority of the models that are considered in such an analysis could be avoided and a suitable selection procedure that compares only a subset of all possible regressions, such as *best subset regression*, is usually preferred. These are discussed below, while in Section 14.3, we discuss regularized regression via approaches such as Lasso or elastic net.

9.1.3 Variable selection based on significance tests

9.1.3.1 Sequential and partial F-tests

In the multiple linear regression model, the ANOVA identity represents a partition of the total variation in Y into two orthogonal components, one due to the fitted regression, and the other due to unexplained error (see Section 4.2). It is written as $\mathbf{y}'\mathbf{y} = \widehat{\boldsymbol{\beta}}'\mathbf{X}'\mathbf{y} + \widehat{\boldsymbol{\varepsilon}}'\widehat{\boldsymbol{\varepsilon}}$, or, by subtracting out the effect due to the intercept from the total and regression sum of squares, as $\mathbf{y}'\mathbf{y} - N\overline{Y}^2 = (\widehat{\boldsymbol{\beta}}'\mathbf{X}'\mathbf{y} - N\overline{Y}^2) + \widehat{\boldsymbol{\varepsilon}}'\widehat{\boldsymbol{\varepsilon}}$. The ANOVA table corresponding to each of these forms is shown in Table 9.1.1. In each case, the mean squares (in column 4) are obtained by dividing the sums of squares (in column 3) by the corresponding degrees of freedom (in column 2).

The F-statistic for the joint hypothesis H_0: $\beta_1 = \cdots = \beta_k = 0$ is given by

$$F_0 = \frac{SSR/p}{SSE/(N-p)} = \frac{MSR}{MSE} = \frac{(N-p)R^2}{p(1-R^2)}, \tag{9.1.16}$$

which has an $F_{p,N-p}$ distribution under H_0. We reject the null hypothesis at level of significance α if $F_0 > F_{p,N-p,\alpha}$.

The ANOVA table corresponding to separating the intercept from the remaining k predictor variables is shown in Table 9.1.2, where SSM denotes the sum of squares due to the intercept term (or mean) and $SSR_c = SSR - SSM$ denotes the corrected sum of squares due to regression. The ANOVA decomposition in Table 9.1.3 commonly appears in many standard statistical software. The F-statistic for the joint hypothesis H_0: $\beta_1 = \cdots = \beta_k = 0$ is given by

$$F_0 = \frac{SSR_c/k}{SSE/(N-k-1)} = \frac{MSR_c}{MSE} = \frac{(N-k-1)R^2}{k(1-R^2)}, \tag{9.1.17}$$

which has an $F_{k,N-k-1}$ distribution under this null hypothesis. We reject the null hypothesis at level of significance α if $F_0 > F_{k,N-k-1,\alpha}$.

The regression sum of squares SSR or SSR_c can be partitioned into meaningful components that enable us to assess the effect of a *single* explanatory variable. The partition of $SSR_c = Q(\beta_1, \cdots, \beta_k \mid \beta_0)$ into *sequential* regression sums of squares is given by

$$SSR_c = Q(\beta_1 \mid \beta_0) + Q(\beta_2 \mid \beta_1, \beta_0) + \cdots + Q(\beta_k \mid \beta_{k-1}, \cdots, \beta_1, \beta_0), \tag{9.1.18}$$

where $Q(\theta \mid \nu)$ refers to the "regression explained by θ in the presence of ν". For example,

Source	d.f.	SS	MS
Intercept β_0	1	$SSM = N\overline{Y}^2$	MSM
Regression $\mid \beta_0$	k	$SSR_c = \widehat{\mathbf{y}}'\widehat{\mathbf{y}} - N\overline{Y}^2$	MSR_c
Residual	$N-k-1$	$SSE = \widehat{\boldsymbol{\varepsilon}}'\widehat{\boldsymbol{\varepsilon}} = \mathbf{y}'\mathbf{y} - \widehat{\mathbf{y}}'\widehat{\mathbf{y}}$	MSE
Total	N	$SST = \mathbf{y}'\mathbf{y}$	

TABLE 9.1.2. ANOVA table for multiple linear regression with intercept separated out.

Source	d.f.	SS	MS
Regression $\mid \beta_0$	k	$SSR_c = \widehat{\mathbf{y}}'\widehat{\mathbf{y}} - N\overline{Y}^2$	MSR_c
Residual	$N-k-1$	$SSE = \widehat{\boldsymbol{\varepsilon}}'\widehat{\boldsymbol{\varepsilon}} = \mathbf{y}'\mathbf{y} - \widehat{\mathbf{y}}'\widehat{\mathbf{y}}$	MSE
Total	$N-1$	$SST_c = \mathbf{y}'\mathbf{y} - N\overline{Y}^2$	

TABLE 9.1.3. Mean corrected ANOVA table for multiple linear regression.

$Q(\beta_3 \mid \beta_2, \beta_1, \beta_0)$ denotes the increase in the regression sum of squares when X_3 is included in a model that has the intercept, X_1 and X_2. Equation (9.1.18) represents a partition of SSR_c into single degree of freedom contributions from explanatory variables that are added sequentially one at a time to a model with just an intercept. Each component represents the incremental increase in the variability in Y explained by a particular explanatory variable included into the model. Alternately, we can view $Q(\beta_3 \mid \beta_2, \beta_1, \beta_0)$ as the reduction in the residual sum of squares by the inclusion of X_3 into a model that had the intercept, X_1 and X_2. This sequential sum of squares partitioning enables us to assess the contribution of each explanatory variable individually.

We can also implement a *subset* partitioning of SSR_c. Suppose we subdivide $\mathbf{X} = (\mathbf{1}_N, \mathbf{X}_1, \mathbf{X}_2)$, where $\mathbf{X}_1$ is $N \times k_1$, $\mathbf{X}_2$ is $N \times k_2$, such that $k = k_1 + k_2$; and we correspondingly subdivide $\boldsymbol{\beta} = (\beta_0, \boldsymbol{\beta}_1', \boldsymbol{\beta}_2')'$. The linear regression model in (4.1.2) can be written as

$$\mathbf{y} = \beta_0 + \mathbf{X}_1\boldsymbol{\beta}_1 + \mathbf{X}_2\boldsymbol{\beta}_2 + \boldsymbol{\varepsilon},$$

and $SSR_c = Q(\boldsymbol{\beta}_1, \boldsymbol{\beta}_2 \mid \beta_0)$. We partition

$$SSR_c = Q(\boldsymbol{\beta}_1 \mid \beta_0) + Q(\boldsymbol{\beta}_2 \mid \boldsymbol{\beta}_1, \beta_0),$$

where $Q(\boldsymbol{\beta}_2 \mid \boldsymbol{\beta}_1, \beta_0)$ is an *extra sum of squares* and represents the increase in the regression sum of squares by adding $\mathbf{X}_2$ to a model that already has the intercept and $\mathbf{X}_1$. Such partitions enable us to test whether any subset of regression coefficients is zero.

Suppose we wish to test H_0: $\boldsymbol{\beta}_2 = \mathbf{0}$ versus H_1: $\boldsymbol{\beta}_2 \neq \mathbf{0}$. Under H_0, $R(\boldsymbol{\beta}_2 \mid \boldsymbol{\beta}_1, \beta_0)/\sigma^2 \sim \chi^2_{k_2}$, so that the partial F-statistic

$$F^* = \frac{Q(\boldsymbol{\beta}_2 \mid \boldsymbol{\beta}_1, \beta_0)/k_2}{MSE} \tag{9.1.19}$$

has an $F_{k_2, N-k-1}$ distribution under H_0. If $F^* > F_{k_2, N-k-1, \alpha}$, we reject the null hypothesis at level of significance α, since the extra variation explained by including $\mathbf{X}_2$ in the model in the presence of an intercept and $\mathbf{X}_1$ is greater than what we would attribute to chance. Some authors refer to F^* as a partial F-statistic only when a single regression coefficient is tested, i.e., when $k_2 = 1$ and the resulting statistic will have an $F_{1,N-k-1}$ distribution under H_0. In general, the quantities $Q(\beta_1 \mid \beta_0, \beta_2, \beta_3, \cdots, \beta_k)$, $Q(\beta_2 \mid \beta_0, \beta_1, \beta_3, \cdots, \beta_k)$, $\cdots$,

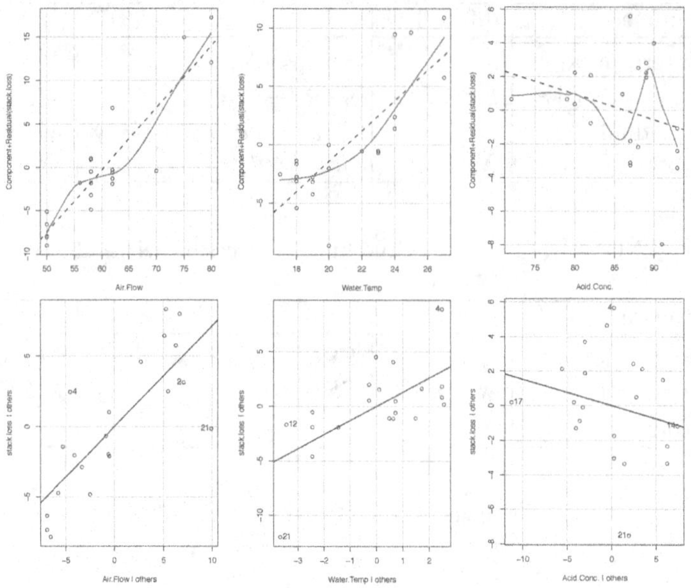

FIGURE 9.1.1. Partial residual plots for the stackloss data (top) and added variable plots for the stackloss data (bottom).

$Q(\beta_k \mid \beta_0, \beta_1, \beta_2, \cdots, \beta_{k-1})$ are partial regression sums of squares. However, these sums of squares need not add up to SSR_c. Hence, these sums of squares and the resulting test statistics are not independent.

Note however that (9.1.18) corresponds to a complete partitioning of SSR_c and the sums of squares on the right side are independent. The resulting F-test statistics are called *sequential F-statistics.* Sequential F-test statistics enable us to test the significance of the contribution of an explanatory variable in a model containing the preceding variables. Clearly, the order of entry of the variables into the model will affect the results. For example, $Q(\beta_3 \mid \beta_0, \beta_1, \beta_2) \neq Q(\beta_3 \mid \beta_0, \beta_1)$ and the contribution of X_3 to SSR_c will depend on which other variables were included in the model previously. Therefore, an objective and complete variable screening cannot be accomplished using only sequential F-tests, unless the selection is implemented in several stages. Many software packages contain both sequential and partial F-tests.

Numerical Example 9.1. We revisit the "stackloss" dataset from the R package *datasets* which we saw in Numerical Example 4.2. We use the `crPlots()` function and the `avPlots()` function in the *cars* package to construct the partial residual plots and the added variable plots, respectively, as shown in Figure 9.1.1.

```
car::crPlots(mod)
car::avPlots(mod)
```

Sequential F-tests are obtained as follows.

```
anova(mod)
```

```
Analysis of Variance Table
Response: y
          Df  Sum Sq Mean Sq  F value    Pr(>F)
x1         1 1450.08 1450.08 242.3679 2.888e-07 ***
x2         1 1207.78 1207.78 201.8705 5.863e-07 ***
x3         1    9.79    9.79   1.6370    0.2366
x4         1    0.25    0.25   0.0413    0.8441
Residuals  8   47.86    5.98
---
Signif. codes:  0 '***' 0.001 '**' 0.01 '*' 0.05 '.' 0.1 ' ' 1
```

The partial F-tests are shown below.

```
car::Anova(mod)
```

```
Anova Table (Type II tests)
Response: stack.loss
            Sum Sq Df F value    Pr(>F)
Air.Flow   296.228  1 28.1601 5.799e-05 ***
Water.Temp 130.308  1 12.3874   0.00263 **
Acid.Conc.   9.965  1  0.9473   0.34405
Residuals  178.830 17
---
Signif. codes:  0 '***' 0.001 '**' 0.01 '*' 0.05 '.' 0.1 ' ' 1
```

▲

We next describe stepwise regression, which combines forward selection and backward elimination.

9.1.3.2 Stepwise regression and variants

Stepwise regression.

This is a widely used approach for variable selection and is available in most standard regression packages. The procedure uses t-statistics (or corresponding p-values) to determine the significance of the predictor variables. At the outset, we choose values of α_{enter} and α_{stay}, say 0.05 each. Here, α_{enter} is the probability of a Type I error related to including a predictor variable into an existing regression model, while α_{stay} is the probability of a Type I error related to retaining in a model a predictor variable that was previously entered.

1. In the first step, we consider k regression models of the form

$$Y = \beta_0 + \beta_1 X_j + \varepsilon,$$

which includes only the jth predictor, $j = 1, \cdots, k$. For each model, we compute the t-statistic (and p-value) for testing H_0: $\beta_1 = 0$ versus H_1: $\beta_1 \neq 0$. Let $X_{[1]}$ denote the variable corresponding to the largest absolute value of the t-statistic (i.e., the smallest p-value) and suppose the corresponding regression model is

$$Y = \beta_0 + \beta_1 X_{[1]} + \varepsilon.$$

If H_0 is not rejected (because the absolute value of the t-statistic is smaller than the $t_{N-2,\alpha/2}$ critical value, with $\alpha = \alpha_{\text{enter}}$, i.e., if the p-value is greater than α_{enter}), then the stepwise procedure terminates and the chosen model is

$$Y = \beta_0 + \varepsilon.$$

If, however, the absolute value of the t-statistic is greater than $t_{N-2,\alpha/2}$, then $X_{[1]}$ is retained, since we see it as being significant at the α_{enter} level.

2. In the second step, we consider $k-1$ possible regressions, each with two predictor variables

$$Y = \beta_0 + \beta_1 X_{[1]} + \beta_2 X_j + \varepsilon,$$

which includes the predictor $X_{[1]}$ chosen in Step 1, and one of the other $k-1$ predictors. For each model, the t-statistic (and p-value) associated with testing H_0: $\beta_2 = 0$ versus H_1: $\beta_2 \neq 0$ is computed. Let $X_{[2]}$ denote the variable corresponding to the largest absolute value of the t-statistic (i.e., the smallest p-value) and suppose the corresponding regression model is

$$Y = \beta_0 + \beta_1 X_{[1]} + \beta_2 X_{[2]} + \varepsilon.$$

The variable $X_{[2]}$ is retained if the t-statistic indicates that it is significant at the α_{enter} level, and the stepwise procedure also checks to see whether or not $X_{[1]}$ should continue to remain in the model. If the p-value corresponding to H_0: $\beta_1 = 0$ versus H_1: $\beta_1 \neq 0$ is smaller than α_{stay}, then $X_{[1]}$ is significant and is retained; otherwise, it is dropped from the model. If $X_{[1]}$ is retained in the model, we have a two-variable model and we proceed to the next step. If, on the other hand, $X_{[1]}$ is dropped, the current one-variable model is

$$Y = \beta_0 + \beta_2 X_{[2]} + \varepsilon.$$

We now are back to the position at the start of the second step, and we must search for another predictor variable that is significant and will be included in the model.

3. We continue with this procedure of adding predictor variables into the model, one at a time. At each step, a variable is added to the model only if it has the largest t-statistic among all variables not in the model, and further, it is significant at the α_{enter} level. After adding a variable, the procedure checks all the variables in the model and removes any variable that is not significant at the α_{stay} level. All necessary removals are made before the procedure attempts to add a "new" variable. The procedure terminates when all the predictor variables not in the model are insignificant at the α_{enter} level, or when the variable to be added in the model is the one that was just removed.

The choice of values for α_{enter} and α_{stay} is arbitrary and has been discussed in the literature. In general, it is recommended that α_{stay} is greater than α_{enter}, since this will preclude the subsequent "easy" inclusion of the same variable which was previously excluded. Draper and Smith (1998) suggest that α_{enter} and α_{stay} are equal, and each is either .05 or .10. If α_{enter} and α_{stay} are set higher than 0.10, more independent variables are likely to be included in the model. In general, it is likely that some important variables (such as higher order terms or interaction terms) may be omitted, while some unimportant variables may be included in the model.

Forward selection.

This is a sequential variable selection procedure that systematically adds explanatory variables to the existing regression model on the basis of partial F-tests (see Section 9.1.3.1). The idea behind the procedure is similar to that of stepwise regression, the difference being that once a predictor variable is included in the model, it is never removed. The initial model contains only the intercept term. In the next step, the explanatory variable which corresponds to the largest partial F-value enters the model, provided it is significant at the α_{enter} level. Let us denote this variable by $X_{[1]}$. In the next step, $X_{[2]}$, the predictor which has the largest partial F-value among the predictors outside the model is included, provided it is significant at the α_{enter} level. This process is continued until we have included all significant predictors into the model. Forward selection is usually considered to be less effective than stepwise regression.

Backward elimination.

This is a sequential procedure that systematically removes a predictor from an existing model based on partial F-test statistics. We first consider a model which includes all k potential predictor variables and an intercept. We pick the independent variable having the smallest partial F-statistic. If this statistic is significant at the α_{stay} level, then the final model contains all k variables, and the procedure is terminated. If, however, this variable is not significant at the α_{stay} level, it is excluded from the model. At the next step, we run a regression with the remaining $(k-1)$ predictors, and repeat the first step. The backward elimination procedure continues by excluding, if possible, one variable at a time from the regression, and terminates when no predictor variable in the model can be removed.

If a modeler prefers to start the model fitting by including all possible variables, then the backward elimination procedure is a reasonable approach. Quite often, it results in the same final model that is given by the stepwise regression procedure.

Numerical Example 9.2. The data "cement" in the R package *MASS* relates to an engineering application concerned with the effect of the composition of cement on heat evolved during hardening and consists of four predictor variables, X_1 (amount of tricalcium aluminate), X_2 (amount of tricalcium silicate), X_3 (amount of tetracalcium alumino ferrite), and X_4 (amount of dicalcium silicate). The response variable Y is the heat evolved per gram of cement (in calories). Results from the different selection procedures are summarized below. All variables left in the model are significant at the 0.10 level.

```
library(MASS)
data(cement)
full.model <- lm(y ~ x1 + x2 + x3 + x4, data = cement)
summary(full.model)

Call:
lm(formula = y ~ x1 + x2 + x3 + x4, data = cement)

Residuals:
   Min     1Q Median    3Q   Max
-3.175 -1.671  0.251 1.378 3.925

Coefficients:
            Estimate Std. Error t value Pr(>|t|)
(Intercept)   62.405     70.071    0.89    0.399
x1             1.551      0.745    2.08    0.071 .
x2             0.510      0.724    0.70    0.501
x3             0.102      0.755    0.14    0.896
x4            -0.144      0.709   -0.20    0.844
---
Signif. codes:  0 '***' 0.001 '**' 0.01 '*' 0.05 '.' 0.1 ' ' 1
```

```
Residual standard error: 2.45 on 8 degrees of freedom
Multiple R-squared:  0.982,     Adjusted R-squared:  0.974
F-statistic:  111 on 4 and 8 DF,  p-value: 4.76e-07
```

We discuss the various variable selection approaches below, using the R package *olsrr*.

```
library(olsrr)
# Rsq and Cp based selection
ols_step_all_possible(full.model)

   Index N  Predictors R-Square Adj. R-Square Mallow's Cp
4      1 1          x4    0.675         0.645      138.73
2      2 1          x2    0.666         0.636      142.49
1      3 1          x1    0.534         0.492      202.55
3      4 1          x3    0.286         0.221      315.15
5      5 2       x1 x2    0.979         0.974        2.68
7      6 2       x1 x4    0.972         0.967        5.50
10     7 2       x3 x4    0.935         0.922       22.37
8      8 2       x2 x3    0.847         0.816       62.44
9      9 2       x2 x4    0.680         0.616      138.23
6     10 2       x1 x3    0.548         0.458      198.09
12    11 3    x1 x2 x4    0.982         0.976        3.02
11    12 3    x1 x2 x3    0.982         0.976        3.04
13    13 3    x1 x3 x4    0.981         0.975        3.50
14    14 3    x2 x3 x4    0.973         0.964        7.34
15    15 4 x1 x2 x3 x4    0.982         0.974        5.00

# Forward selection
ols_step_forward_p(full.model, prem = 0.10)

                          Selection Summary
-----------------------------------------------------------------
        Variable                  Adj.
Step    Entered     R-Square    R-Square      C(p)        AIC        RMSE
-----------------------------------------------------------------
   1    x4            0.6745      0.6450    138.7308    97.7440    8.9639
   2    x1            0.9725      0.9670      5.4959    67.6341    2.7343
   3    x2            0.9823      0.9764      3.0182    63.8663    2.3087
-----------------------------------------------------------------

summary(forward.model <- lm(y ~ x1 + x2 + x4, data = cement))

Call:
lm(formula = y ~ x1 + x2 + x4, data = cement)

Residuals:
   Min     1Q Median     3Q    Max
-3.092 -1.802  0.256  1.282  3.898
```

```
Coefficients:
            Estimate Std. Error t value Pr(>|t|)
(Intercept)   71.648     14.142    5.07  0.00068 ***
x1             1.452      0.117   12.41  5.8e-07 ***
x2             0.416      0.186    2.24  0.05169 .
x4            -0.237      0.173   -1.37  0.20540
---
Signif. codes:  0 ‘***’ 0.001 ‘**’ 0.01 ‘*’ 0.05 ‘.’ 0.1 ‘ ’ 1

Residual standard error: 2.31 on 9 degrees of freedom
Multiple R-squared:  0.982,    Adjusted R-squared:  0.976
F-statistic:  167 on 3 and 9 DF,  p-value: 3.32e-08

# Backward elimination
ols_step_backward_p(full.model,prem=0.10)

                       Elimination Summary
-----------------------------------------------------------------
       Variable                  Adj.
Step   Removed     R-Square   R-Square     C(p)       AIC       RMSE
-----------------------------------------------------------------
   1   x3             0.982      0.976    3.0182   63.8663    2.3087
   2   x4             0.979      0.974    2.6782   64.3124    2.4063
------------------------------------------------------------------

summary(backward.model <- lm(y~x3+x4,data=cement))
Call:
lm(formula = y ~ x3 + x4, data = cement)

Residuals:
   Min     1Q Median     3Q    Max
-4.271 -2.892 -0.644  1.511  8.257

Coefficients:
            Estimate Std. Error t value Pr(>|t|)
(Intercept) 131.2824     3.2748   40.09  2.2e-12 ***
x3           -1.1999     0.1890   -6.35  8.4e-05 ***
x4           -0.7246     0.0723  -10.02  1.6e-06 ***
---
Signif. codes:  0 ‘***’ 0.001 ‘**’ 0.01 ‘*’ 0.05 ‘.’ 0.1 ‘ ’ 1

Residual standard error: 4.19 on 10 degrees of freedom
Multiple R-squared:  0.935,    Adjusted R-squared:  0.922
F-statistic: 72.3 on 2 and 10 DF,  p-value: 1.13e-06
```

```
# Stepwise selection
ols_step_both_p(full.model,prem=0.10)

                    Stepwise Selection Summary
-----------------------------------------------------------------
             Added/                 Adj.
Step  Var.   Removed    R-Square  R-Square     C(p)       AIC      RMSE
-----------------------------------------------------------------
   1   x4    addition    0.675     0.645    138.7310   97.7440   8.9639
   2   x1    addition    0.972     0.967      5.4960   67.6341   2.7343
   3   x2    addition    0.982     0.976      3.0180   63.8663   2.3087
   4   x4    removal     0.979     0.974      2.6780   64.3124   2.4063
-----------------------------------------------------------------

summary(stepwise.model<-lm(y~x1+x2,data=cement))

Call:
lm(formula = y ~ x1 + x2, data = cement)

Residuals:
   Min     1Q Median     3Q    Max
 -2.89  -1.57  -1.30   1.36   4.05

Coefficients:
            Estimate Std. Error t value Pr(>|t|)
(Intercept)  52.5773     2.2862    23.0  5.5e-10 ***
x1            1.4683     0.1213    12.1  2.7e-07 ***
x2            0.6623     0.0459    14.4  5.0e-08 ***
---
Signif. codes:  0 '***' 0.001 '**' 0.01 '*' 0.05 '.' 0.1 ' ' 1

Residual standard error: 2.41 on 10 degrees of freedom
Multiple R-squared:  0.979,     Adjusted R-squared:  0.974
F-statistic:  230 on 2 and 10 DF,  p-value: 4.41e-09
```

▲

9.2 Orthogonal and collinear predictors

Suppose that in the model $\mathbf{y} = \mathbf{X}\boldsymbol{\beta} + \boldsymbol{\varepsilon}$ with $\mathrm{r}(\mathbf{X}) = p$, we partition the $N \times p$ matrix $\mathbf{X}$ into $(r+1)$ sets of columns denoted in matrix form by $\mathbf{X} = (\mathbf{X}_0, \mathbf{X}_1, \cdots, \mathbf{X}_r)$ and that the p-dimensional vector $\boldsymbol{\beta}$ is partitioned conformably so that $\boldsymbol{\beta} = (\boldsymbol{\beta}_0', \boldsymbol{\beta}_1', \cdots, \boldsymbol{\beta}_r')'$. Then the linear regression model can be written as

$$\mathbf{y} = \mathbf{X}_0\boldsymbol{\beta}_0 + \mathbf{X}_1\boldsymbol{\beta}_1 + \cdots + \mathbf{X}_r\boldsymbol{\beta}_r + \boldsymbol{\varepsilon}.$$

9.2.1 Orthogonality in regression

Orthogonality among the predictors comes up occasionally in regression problems. We take a brief look in this section. Figure 9.2.1 corresponds to a model with $p = 2$ and $\mathbf{1}_N$ and $\mathbf{x}$

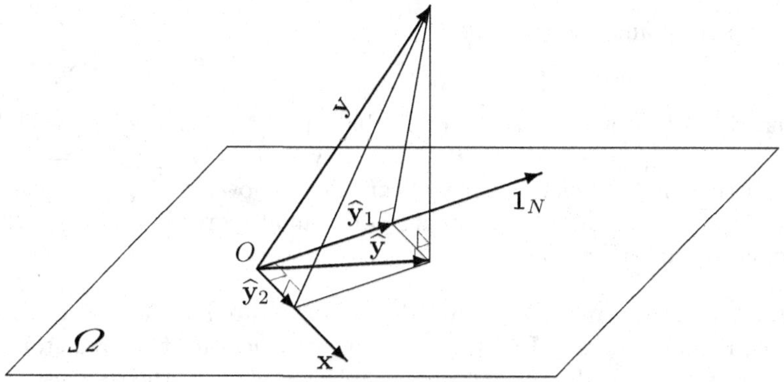

FIGURE 9.2.1. Orthogonal predictors, $p = 2$.

are orthogonal vectors that span Ω. Let $\widehat{\mathbf{y}}$, $\widehat{\mathbf{y}}_1$, and $\widehat{\mathbf{y}}_2$ be the orthogonal projections of $\mathbf{y}$ onto Ω, the space spanned by $\mathbf{1}_N$, and the space spanned by $\mathbf{x}$, respectively. Then $\widehat{\mathbf{y}}_1 \perp \widehat{\mathbf{y}}_2$, $\widehat{\mathbf{y}} = \widehat{\mathbf{y}}_1 + \widehat{\mathbf{y}}_2$, and as a result, $\|\widehat{\mathbf{y}}\|^2 = \|\widehat{\mathbf{y}}_1\|^2 + \|\widehat{\mathbf{y}}_2\|^2$.

Result 9.2.1. Suppose that the columns of $\mathbf{X}_h$ are orthogonal to the columns of $\mathbf{X}_j$ for all $h \neq j$, i.e., $\mathbf{X}_h'\mathbf{X}_j = \mathbf{O}$. The LS solution of $\boldsymbol{\beta}_j$ in the full model $\mathbf{y} = \mathbf{X}\boldsymbol{\beta} + \boldsymbol{\varepsilon}$ is unchanged if any of the other $\boldsymbol{\beta}_h$'s are set equal to zero, i.e., if the corresponding sets of predictors $\mathbf{X}_h$'s are omitted from the model.

Proof. From the assumption, $\mathcal{C}(\mathbf{X}) = \oplus_{j=0}^{r}\mathcal{C}(\mathbf{X}_j)$ and $\mathcal{C}(\mathbf{X}_h) \perp \mathcal{C}(\mathbf{X}_j)$ for $h \neq j$. By repeatedly using property 3 of Result 2.6.6, $\mathbf{P}_{\mathcal{C}(\mathbf{X})} = \sum_{j=0}^{r}\mathbf{P}_{\mathcal{C}(\mathbf{X}_j)}$. Post-multiplying both sides by $\mathbf{y}$, $\mathbf{X}\widehat{\boldsymbol{\beta}} = \sum_{j=0}^{r}\mathbf{P}_{\mathcal{C}(\mathbf{X}_j)}\mathbf{y}$. Since $\mathbf{X}\widehat{\boldsymbol{\beta}}$ consists of the subvectors $\mathbf{X}\widehat{\boldsymbol{\beta}}_j \in \mathcal{C}(\mathbf{X}_j)$, $0 \leq j \leq r$, then $\mathbf{X}\widehat{\boldsymbol{\beta}}_j = \mathbf{P}_{\mathcal{C}(\mathbf{X}_j)}\mathbf{y}$. But the right side is exactly the LS fit of $\mathbf{y}$ versus the predictors in $\mathbf{X}_j$ only, leading to a "geometric" proof of the result.

Note that the matrix $\mathbf{X}'\mathbf{X}$ in its partitioned form has a block-diagonal structure, i.e.,

$$\mathbf{X}'\mathbf{X} = \begin{pmatrix} \mathbf{X}_0'\mathbf{X}_0 & \mathbf{O} & \cdots & \mathbf{O} \\ \mathbf{O} & \mathbf{X}_1'\mathbf{X}_1 & \cdots & \mathbf{O} \\ \vdots & \vdots & \vdots & \vdots \\ \mathbf{O} & \mathbf{O} & \cdots & \mathbf{X}_r'\mathbf{X}_r \end{pmatrix},$$

and $\mathbf{X}'\mathbf{y} = (\mathbf{X}_0'\mathbf{y}, \mathbf{X}_1'\mathbf{y}, \cdots, \mathbf{X}_r'\mathbf{y})'$. The least squares estimator of $\boldsymbol{\beta}$ is then $\widehat{\boldsymbol{\beta}}' = (\widehat{\boldsymbol{\beta}}_0', \widehat{\boldsymbol{\beta}}_1', \cdots, \widehat{\boldsymbol{\beta}}_r')'$, where each $\widehat{\boldsymbol{\beta}}_j = (\mathbf{X}_j'\mathbf{X}_j)^{-1}\mathbf{X}_j'\mathbf{y}$ has the form of the LS estimate of $\boldsymbol{\beta}_j$ in the model $\mathbf{y} = \mathbf{X}_j\boldsymbol{\beta}_j + \boldsymbol{\varepsilon}$. Due to the block-diagonal structure of $(\mathbf{X}'\mathbf{X})^{-1}$, the LS estimate of $\boldsymbol{\beta}_j$ in the full model $\mathbf{y} = \mathbf{X}\boldsymbol{\beta} + \boldsymbol{\varepsilon}$ is unchanged if any of the other $\boldsymbol{\beta}_h$'s are set equal to zero, i.e., if the corresponding sets of regressors $\mathbf{X}_h$'s are omitted from the regression. This is a property unique to orthogonal regressors. In this case the regression sum of squares is

$$\widehat{\boldsymbol{\beta}}'\mathbf{X}'\mathbf{y} = \sum_{j=0}^{r}\widehat{\boldsymbol{\beta}}_j'\mathbf{X}_j'\mathbf{y},$$

so that if we omit the regressor $\mathbf{X}_h$ from the model, the residual sum of squares is increased by $\widehat{\boldsymbol{\beta}}_h'\mathbf{X}_h'\mathbf{y}$. ■

If the model in (4.1.2) has full-rank, then by Gram–Schmidt orthogonalization, the model can be reparametrized and expressed as

$$Y_i = \mathbf{z}_i'\gamma + \varepsilon_i = \gamma_0 + \gamma_1 Z_{i1} + \cdots + \gamma_k Z_{ik} + \varepsilon_i, \quad 1 \le i \le N,$$

where the matrix $\mathbf{Z}$ has orthogonal columns that span the same space as $\mathcal{C}(\mathbf{X})$, and for $r = 0, \cdots, k$, $\gamma_r = \gamma_{r+1} = \cdots = \gamma_k = 0$ if and only if $\beta_r = \beta_{r+1} = \cdots = \beta_k = 0$ (Seber, 1977, p. 60). Orthogonal polynomials have been used for overcoming problems, such as ill-conditioning, that are encountered in fitting polynomial regression models (Hayes, 1974), as the following examples describe.

Example 9.2.1. Orthogonal polynomials in curvilinear regression. In the simple linear regression model (4.1.6), $\text{E}(Y \mid X)$ is a linear function of a predictor X. In some examples, however, there might be a curvilinear relation between the response and predictor variables, which may be adequately modeled by a polynomial model of degree m:

$$Y_i = \beta_0 + \beta_1 X_i + \beta_2 X_i^2 + \cdots + \beta_m X_i^m + \varepsilon_i, \quad i = 1, \cdots, N. \tag{9.2.1}$$

Written in the form $\mathbf{y} = \mathbf{X}\boldsymbol{\beta} + \boldsymbol{\varepsilon}$, with $\{\mathbf{X}\}_{i1} = 1$, and $\{\mathbf{X}\}_{ij} = X_i^j$, $j = 1, \cdots, m$, $i = 1, \cdots, N$, the properties of Corollary 7.1.1 hold for this mth order polynomial model. In practice, the degree m is unknown, and is determined by using suitable tests of hypotheses (see Section 7.3), which could be computationally intensive. Prior to the extensive availability of powerful computing, orthogonal polynomials in X were widely used.

Let $\phi_r(X_i)$ be an rth degree polynomial in X_i, $r = 0, 1, \cdots, m$, and suppose the polynomials are orthogonal over $X_1, \cdots, X_N$, i.e.,

$$\sum_{i=1}^{N} \phi_r(X_i)\phi_s(X_i) = 0, \quad r \neq s. \tag{9.2.2}$$

The orthogonal polynomial regression model

$$Y_i = \sum_{r=0}^{m} \beta_r \phi_r(X_i) + \varepsilon_i \tag{9.2.3}$$

can be written in the form (4.1.1) with $\mathbf{X}$, now defined as

$$\mathbf{X} = \begin{pmatrix} \phi_0(X_1) & \phi_1(X_1) & \cdots & \phi_m(X_1) \\ \phi_0(X_2) & \phi_1(X_2) & \cdots & \phi_m(X_2) \\ \vdots & \vdots & \vdots & \vdots \\ \phi_0(X_N) & \phi_1(X_N) & \cdots & \phi_m(X_N) \end{pmatrix},$$

having mutually orthogonal columns, so that $\mathbf{X}'\mathbf{X}$ is a diagonal matrix with rth diagonal element $\sum_{i=1}^N \phi_r^2(X_i)$, $r = 0, \cdots, m$. From (4.2.3), the least squares estimate of $\boldsymbol{\beta}$ is $\widehat{\boldsymbol{\beta}} = (\widehat{\beta}_0, \widehat{\beta}_1, \cdots, \widehat{\beta}_m)'$, where, for all m,

$$\widehat{\beta}_r = \frac{\sum_{i=1}^N \phi_r(X_i) Y_i}{\sum_{i=1}^N \phi_r^2(X_i)}, \quad r = 0, 1, \cdots, m.$$

If we set $\phi_0(X_i) = 1$, we get $\widehat{\beta}_0 = \overline{Y}$, and

$$\begin{aligned} SSE &= \mathbf{y}'\mathbf{y} - \widehat{\mathbf{y}}'\widehat{\mathbf{y}} = \mathbf{y}'\mathbf{y} - \widehat{\boldsymbol{\beta}}'\mathbf{X}'\mathbf{X}\widehat{\boldsymbol{\beta}} \\ &= \sum_{i=1}^N Y_i^2 - \sum_{r=0}^m \widehat{\beta}_r^2 \sum_{i=1}^N \phi_r^2(X_i) \\ &= \sum_{i=1}^N (Y_i - \overline{Y})^2 - \sum_{r=1}^m \widehat{\beta}_r \sum_{i=1}^N \phi_r^2(X_i). \end{aligned}$$

Suppose we wish to test H: $\beta_m = 0$. By orthogonality,

$$SSE_H = \sum_{i=1}^{N}(Y_i - \overline{Y})^2 - \sum_{r=1}^{m-1} \widehat{\beta}_r^2 \sum_{i=1}^{N} \phi_r^2(X_i)$$
$$= SSE + \widehat{\beta}_m^2 \sum_{i=1}^{N} \phi_m^2(X_i),$$

and it follows from Result 7.2.1 that the test statistic

$$F(H) = \frac{\sum_{i=1}^{N} \phi_m^2(X_i)\widehat{\beta}_m^2}{SSE/(N-m-1)}$$

follows $F_{1,N-m-1}$ under H. □

The advantage of this approach is that the model of polynomial degree m may be easily enlarged to a model of polynomial degree $m+1$ by simply adding one more term $\beta_{m+1}\phi_{m+1}(X_i)$ to (9.2.3), with $\phi_{m+1}(X_i)$ satisfying (9.2.2). It may be verified that the resulting computations are simplified by the assumption of orthogonality.

Example 9.2.2. Response surfaces. Consider the model

$$Y_i = \phi(X_{i1}, \cdots, X_{ik}) + \varepsilon_i, \quad i = 1, \cdots, N, \tag{9.2.4}$$

where $\phi(X_{i1}, X_{i2}, \cdots, X_{ik})$ is a polynomial of degree m in $X_1, \cdots, X_k$ (usually, $m = 2$, or $m = 3$). When $m = 2$, we can write for $i = 1, \cdots, N$,

$$Y_i = \beta_0 + \sum_{j=1}^{k} \beta_j X_{ij} + \sum_{j=1}^{k} \beta_{jj} X_{ij}^2 + \sum_{j=1}^{k}\sum_{l=1}^{k} \beta_{jl} X_{ij} X_{il} + \varepsilon_i. \tag{9.2.5}$$

The function $\phi(X_1, \cdots, X_k)$ is a *response surface*, and the model (9.2.5) can be written in the form (4.1.1) with $\mathbf{X} = (\mathbf{1}_N, \mathbf{X}_1, \mathbf{X}_2)$, where for $i = 1, \cdots, N$, $\{\mathbf{X}_1\}_{i,j} = X_{i,j}$, and $\{\mathbf{X}_2\}_{i,j}$ corresponds to values of X_{ij}^2 and $X_{ij}X_{il}$, $j, l = 1, \cdots, k$, $j \neq l$. Note that $\mathbf{X}_2$ is automatically determined once the variables in $\mathbf{X}_1$ are chosen. For details on optimum choice of $\mathbf{X}_1$ to achieve an optimum response Y, see Myers (1971). The results from Chapter 7 are useful for obtaining the b.l.u.e.'s of the parameters and carrying out inference. □

9.2.2 Multicollinearity

A multiple linear regression model fit is useful if the response variable is highly correlated with the set of explanatory variables. However, it is necessary that the explanatory variables are not highly correlated among themselves. A situation where this occurs is referred to as multicollinearity. In this section, we assume that the predictors are scaled to unit length by dividing each column vector by its own length, but are not centered (Belsley, 1984). The resulting column-equilibrated predictor matrix $\mathbf{X}_E$ (see Example 4.1.2) is used for detecting multicollinearity. Collinearity (or multicollinearity) exists when there is "near-dependency" between the columns of $\mathbf{X}_E$, i.e., when $\mathbf{X}_E'\mathbf{X}_E$ is nearly singular. In such cases, the data/model pair is said to be ill-conditioned, and the resulting least squares estimates tend to be unstable with large variances and covariances. The presence of a high degree of multicollinearity tends to cause the following problems:

1. The standard errors of the regression coefficients may be very large, resulting in small associated t-statistics, thereby leading to the conclusion that truly useful explanatory variables are insignificant in explaining the regression.

2. The sign of regression coefficients may be the opposite of what a mechanistic understanding of the problem would suggest.

3. Deleting a column of the predictor matrix may cause large changes in the coefficient estimates corresponding to the other variables in a model based on the remaining data.

Multicollinearity does not, however, greatly affect the predicted values. Several approaches have been suggested in the literature for the detection of multicollinearity. Detection consists of two aspects: is multicollinearity present, and if it is, how severe is it? The following measures usually indicate severity of multicollinearity:

1. The simple correlation between a pair of predictors is close to 1.0.

2. The multiple correlation coefficient between the explanatory variables is large, and some of the partial correlations are high too (which may help in identifying problem variables).

3. The value of the overall F-statistic is large, but values of (some) t-statistics for individual regression coefficients are small.

4. For $j = 1, \cdots, k$, let R_j^2 denote the coefficient of determination of a regression of the explanatory variable X_j on the remaining $(k-1)$ explanatory variables. A large value of R_j^2 also indicates multicollinearity.

Variance inflation factors formalize the fourth measure in the list, and are defined below.

Definition 9.2.1. Variance inflation factor. For $j = 1, \cdots, k$, the variance inflation factor for X_j, a name due to Marquardt, is defined to be

$$VIF_j = 1/(1 - R_j^2).$$

In the ideal case when X_j is orthogonal to the other predictors, $R_j^2 = 0$, so that $VIF_j = 1$. As R_j^2 increases from zero, VIF_j increases as well. For example, if $R_j^2 = 0.8$, $VIF_j = 5.0$, while if $R_j^2 = 0.99$, $VIF_j = 100$. It has been suggested in the literature that any VIF_j greater than 10 indicates multicollinearity. Alternatively, we may compute the average of the variance inflation factors, i.e., $\overline{VIF} = \sum_{j=1}^{k} VIF_j/k$. If $\overline{VIF}$ substantially exceeds 1, multicollinearity is indicated.

The column-equilibrated matrix $\mathbf{X}_E$ is useful in detecting multicollinearity (Belsley, 1991). We find the singular value decomposition of $\mathbf{X}_E$ (see Result 2.2.7), i.e., $\mathbf{X}_E = \mathbf{UDV}'$, where $\mathbf{U}$ is $N \times p$, $\mathbf{D}$ is $p \times p$, $\mathbf{V}$ is $p \times p$, and $\mathbf{U}'\mathbf{U} = \mathbf{V}'\mathbf{V} = \mathbf{VV}' = \mathbf{I}_p$. Let $\mathbf{D} = \text{diag}(d_1, \cdots, d_p)$, where the nonnegative d_j's are the singular values of $\mathbf{X}_E$. Now

$$\mathbf{X}_E'\mathbf{X}_E = \mathbf{VDU}'\mathbf{UDV}' = \mathbf{VD}^2\mathbf{V}'$$

gives the spectral decomposition of the symmetric matrix $\mathbf{X}_E'\mathbf{X}_E$, so that $\mathbf{D}^2 = \text{diag}(c_1, \cdots, c_p)$, where $c_j = d_j^2$ are the eigenvalues of $\mathbf{X}_E'\mathbf{X}_E$, while $\mathbf{V} = \{v_{ij}\}$ is the corresponding orthogonal eigenvector matrix. We obtain the p condition indices in terms of d_j's and $d_{\max} = \max_{1 \le j \le p} d_j$.

Definition 9.2.2. Condition index. The jth condition index of $\mathbf{X}_E$ is defined by

$$\eta_j = d_{\max}/d_j, \quad j = 1, \cdots, p.$$

A value of d_j which is relatively close to zero will be associated with a large condition index. If d_j is exactly equal to zero, there is an exact linear relationship among the columns of $\mathbf{X}$.

Definition 9.2.3. Condition number. The condition number of $\mathbf{X}_E$ is defined to be

$$C = \frac{d_{\max}}{d_{\min}},$$

where $d_{\min} = \min_{1 \le j \le p} d_j$. The condition number C always exceeds 1.

A large condition number (say, $C > 15$) indicates evidence of strong multicollinearity, and empirical evidence suggests the need for corrective action when $C > 30$.

Note that

$$\text{Cov}(\widehat{\boldsymbol{\beta}})/\sigma^2 = (\mathbf{X}_E'\mathbf{X}_E)^{-1} = \mathbf{V}\mathbf{D}^{-2}\mathbf{V}',$$

where $\mathbf{D}^{-2} = \text{diag}(1/d_1^2, \cdots, 1/d_p^2)$. Corresponding to the predictors $X_j, j = 1, 2, \cdots, k$,

$$\text{Var}(\widehat{\beta}_j)/\sigma^2 = \sum_{h=1}^{k} \frac{v_{jh}^2}{d_h^2}$$

gives a decomposition of the variance of $\widehat{\beta}_j$. Let

$$q_{jh} = \frac{v_{jh}^2/d_h^2}{\sum_{h=1}^{k} v_{jh}^2/d_h^2}$$

denote the proportions of $\text{Var}(\widehat{\beta}_j)/\sigma^2$ corresponding to d_h^2 for $h = 1, \cdots, k$, and let it be the jth column of a $k \times k$ matrix $\mathbf{Q} = \{q_{jh}\}$. Each of the k columns of $\mathbf{Q}$ sums to 1, i.e., $\sum_{h=1}^{k} q_{jh} = 1$ for $j = 1, \cdots, k$. Belsley et al. (1980) suggested that if two or more of the proportions in row h of $\mathbf{Q}$ (corresponding to d_h^2) are relatively large (say, > 0.5), and further, the condition index η_h is also large, then we may assume that the LS fit is affected by multicollinearity.

Once multicollinearity is detected, an obvious remedy is to drop from the model variables that are highly correlated with others. The disadvantage of this is that if the dropped variable is potentially valuable in an understanding of the response, we get absolutely no information about it, and moreover, it may not always be clear as to how the omission of a variable will affect the estimates of the remaining model parameters. Other statistical procedures for dealing with the problem of multicollinearity include ridge regression and principal components regression, which we discuss next.

9.2.3 Ridge regression

Hoerl (1962) first introduced the ridge trace procedure as a solution to certain multicollinearity problems in regression (see also Hoerl and Kennard, 1970a,b).

Result 9.2.2. Consider the centered and scaled form of the multiple linear regression model $\mathbf{y}^* = \mathbf{X}^*\boldsymbol{\beta}^* + \boldsymbol{\varepsilon}$ as described in Example 4.1.2. Suppose $\boldsymbol{\varepsilon}$ satisfies $\text{E}(\boldsymbol{\varepsilon}) = \mathbf{0}$ and $\text{Cov}(\boldsymbol{\varepsilon}) = \sigma^2\mathbf{I}_k$. Let $\theta > 0$ be a fixed constant.

1. The function $\mathcal{S}_\theta(\boldsymbol{\beta}^*) = \|\mathbf{y}^* - \mathbf{X}^*\boldsymbol{\beta}^*\|^2 + \theta\|\boldsymbol{\beta}^*\|^2$ is strictly convex and has a unique minimizer

$$\widehat{\boldsymbol{\beta}}^*(\theta) = \arg\min_{\boldsymbol{\beta}^*} \mathcal{S}_\theta(\boldsymbol{\beta}^*), \tag{9.2.6}$$

known as the ridge regression estimate of $\boldsymbol{\beta}^*$.

2. The ridge regression estimate is given by

$$\widehat{\boldsymbol{\beta}}^*(\theta) = (\mathbf{X}^{*\prime}\mathbf{X}^* + \theta\mathbf{I}_k)^{-1}\mathbf{X}^{*\prime}\mathbf{y}^*. \tag{9.2.7}$$

The corresponding ridge regression estimate of $\boldsymbol{\beta} = (\beta_0, \beta_1, \cdots, \beta_k)'$ in the original model (4.1.2) is given by

$$\widehat{\beta}_j(\theta) = \widehat{\beta}_j^*(\theta) s_Y / s_{X_j}, \quad 1 \le j \le k, \quad \widehat{\beta}_0(\theta) = \overline{Y} - \sum_{j=1}^{k} \widehat{\beta}_j(\theta)\overline{X}_j.$$

3. The expectation and covariance of $\widehat{\boldsymbol{\beta}}^*(\theta)$ are

$$\begin{aligned} \mathrm{E}\{\widehat{\boldsymbol{\beta}}^*(\theta)\} &= (\mathbf{X}^{*\prime}\mathbf{X}^* + \theta\mathbf{I}_k)^{-1}\mathbf{X}^{*\prime}\mathbf{X}^*\boldsymbol{\beta}^*, \quad \text{and} \\ \mathrm{Cov}\{\widehat{\boldsymbol{\beta}}^*(\theta)\} &= (\mathbf{X}^{*\prime}\mathbf{X} + \theta\mathbf{I}_k)^{-1}\mathbf{X}^{*\prime}\mathbf{X}^*(\mathbf{X}^{*\prime}\mathbf{X}^* + \theta\mathbf{I}_k)^{-1}\sigma^2. \end{aligned}$$

4. The total mean square error of the ridge regression estimator is

$$\mathrm{E}\{\|\widehat{\boldsymbol{\beta}}^*(\theta) - \boldsymbol{\beta}^*\|^2\} = \sigma^2 \sum_{j=1}^{k} \lambda_j(\lambda_j + \theta)^{-2} + \theta^2 \boldsymbol{\beta}^{*\prime}(\mathbf{X}^{*\prime}\mathbf{X}^* + \theta\mathbf{I}_k)^{-2}\boldsymbol{\beta}^*,$$

where $\lambda_1, \cdots, \lambda_k$ are the eigenvalues of $\mathbf{X}^{*\prime}\mathbf{X}^*$.

Proof. The Hessian matrix of $\mathcal{S}_\theta(\boldsymbol{\beta}^*)$ is $2(\mathbf{X}^{*\prime}\mathbf{X}^* + \theta\mathbf{I}_k)$, which is p.d. Since $\mathcal{S}_\theta(\boldsymbol{\beta}^*) \ge \theta\|\boldsymbol{\beta}^*\|^2 \to \infty$ as $\|\boldsymbol{\beta}^*\| \to \infty$, property 1 is proved. By solving $\nabla \mathcal{S}_\theta(\boldsymbol{\beta}^*) = \mathbf{0}$, (9.2.7) follows. The transformation to the estimates in terms of the original coefficient vector $\boldsymbol{\beta}$ follows directly from the reasoning in Example 4.1.2. Then property 2 follows. Since $\widehat{\boldsymbol{\beta}}^*(\theta)$ is a linear function of $\mathbf{y}^*$, property 3 follows directly. Recall that for any random vector $\mathbf{z}$,

$$\mathrm{E}(\|\mathbf{z}\|^2) = \|\mathrm{E}(\mathbf{z})\|^2 + \mathrm{tr}(\mathrm{Cov}(\mathbf{z})).$$

Let $\mathbf{z} = \boldsymbol{\beta}^*(\theta) - \boldsymbol{\beta}^*$. From property 3,

$$\mathrm{E}(\mathbf{z}) = -\theta(\mathbf{X}^{*\prime}\mathbf{X}^* + \theta\mathbf{I}_k)^{-1}\boldsymbol{\beta}^*.$$

We see that by Result 2.3.4, $\mathbf{X}^{*\prime}\mathbf{X}^* = \mathbf{QDQ}'$ for an orthogonal matrix $\mathbf{Q}$ and $\mathbf{D} = \mathrm{diag}(\lambda_1, \cdots, \lambda_k)$. Then from property 3,

$$\begin{aligned} \mathrm{Cov}(\widehat{\boldsymbol{\beta}}^*(\theta)) &= \sigma^2\,\mathrm{tr}[(\mathbf{QDQ}' + \theta\mathbf{I})^{-1}(\mathbf{QDQ}')(\mathbf{QDQ}' + \theta\mathbf{I})^{-1}] \\ &= \sigma^2(\mathbf{D} + \theta\mathbf{I}_k)^{-1}\mathbf{D}(\mathbf{D} + \theta\mathbf{I}_k)^{-1}, \end{aligned}$$

which is a diagonal matrix with diagonal elements $\sigma^2\lambda_j(\lambda_j + \theta)^{-2}$. The above three displays then yield the proof of property 4. ■

When $\theta = 0$, the ridge regression estimates reduce to the usual LS estimates. A plot of $\widehat{\beta}_j^*(\theta)$ versus θ or of $\widehat{\beta}_j(\theta)$ versus θ for $j = 1, \cdots, k$, is known as a ridge trace and can be used to select a suitable value θ^* of θ according to the following suggestions by Hoerl and Kennard (1970b), p. 65:

1. The estimates stabilize at a value of θ, with the general characteristics of an orthogonal system.

2. Estimated coefficients do not have unreasonable values.

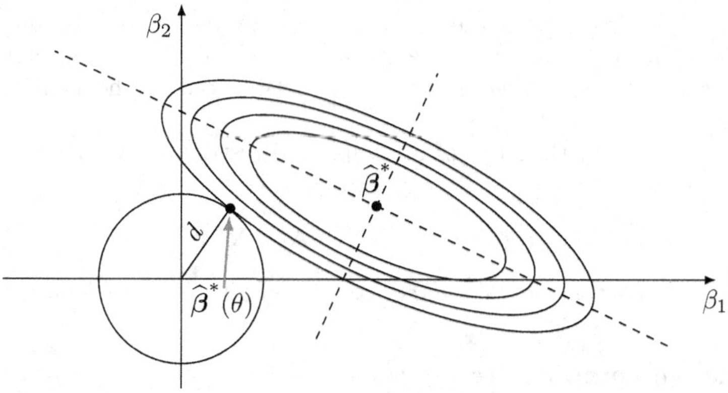

FIGURE 9.2.2. Ridge regression.

3. Estimated values that had apparently incorrect signs at $\theta = 0$ have changed to the proper signs.

4. The SSE does not have an unreasonable value.

Alternatively, there is suggestion in the literature for an automatic choice of θ^* as the value

$$\theta^* = \frac{k\widehat{\sigma}^2}{\{\widehat{\beta}^*(0)\}'\widehat{\beta}^*(0)},$$

where $\widehat{\sigma}^2$ is the MSE from the usual LS fit and

$$\widehat{\beta}^*(0) = (\widehat{\beta}_1^*(0), \cdots, \widehat{\beta}_k^*(0))' = (\widehat{\beta}_1(0)s_{X_1}/s_Y, \cdots, \widehat{\beta}_k(0)s_{X_k}/s_Y)'.$$

The corresponding values $\widehat{\beta}^*(\theta^*)$ are taken to be the final estimated coefficients, and can be used for prediction.

Ridge regression was characterized as a regularized LS problem in (9.2.6). It has an equivalent characterization as a restricted LS problem. Consider LS estimation in the centered and scaled multiple linear regression model $\mathbf{y}^* = \mathbf{X}^*\beta^* + \varepsilon$ subject to the spherical restriction

$$\|\beta^*\| \leq d$$

for a given value $d > 0$. Using the method of Lagrangian multipliers (see section 4.6.1) to minimize

$$S^* = (\mathbf{y}^* - \mathbf{X}^*\beta^*)'(\mathbf{y}^* - \mathbf{X}^*\beta^*) + \theta(\beta^{*\prime}\beta^* - d^2),$$

yields the equations

$$(\mathbf{X}^{*\prime}\mathbf{X}^* + \theta\mathbf{I})\beta^* = \mathbf{X}^{*\prime}\mathbf{y}^*, \quad \beta^{*\prime}\beta^* = d^2,$$

whose solution leads to the ridge estimate of β^* in (9.2.7), and a solution for θ in terms of d. Figure 9.2.2 is a geometric illustration for the case $k = 2$. In the figure, each elliptical contour centered at the LS estimate $\widehat{\beta}^*$ consists of β^* values with $\|\mathbf{y}^* - \mathbf{X}^{*\prime}\beta^*\| = c$ for some $c > 0$, with outer contours determined by larger values of c. Given $d > 0$, $\widehat{\beta}^*(\theta)$

is the first contact point between the elliptical contours as they expand outward and the circle centered at the origin with radius d. As we reduce d down to 0, we obtain the entire sequence of ridge solutions; the *ridge trace* is the path traced by the solution point as the radius of the circle is increased from zero.

Note that if $\mathbf{A}$ is p.d., then use of an ellipsoidal restriction $\boldsymbol{\beta}^{*\prime}\mathbf{A}\boldsymbol{\beta}^{*} \le d^2$ leads to a solution of the form

$$\widehat{\boldsymbol{\beta}}^{*}_{\mathbf{A}}(\theta) = (\mathbf{X}^{*\prime}\mathbf{X}^{*} + \theta\mathbf{A})^{-1}\mathbf{X}^{*\prime}\mathbf{y}. \tag{9.2.8}$$

The choice of the restriction can be viewed as a model selection problem.

9.2.4 Principal components regression

This is a more unified way to handle multicollinearity, but it requires computations beyond standard regression computations. The procedure is based on the observation that every linear regression model can be restated in terms of a set of orthogonal predictor variables, which are constructed as linear combinations of the original variables. The new orthogonal variables are called the principal components of the original variables (Johnson and Wichern, 2007). Principal components regression is an approach that inspects the sample data $(\mathbf{y}, \mathbf{X})$ for directions of variability and uses this information to reduce the dimensionality of the estimation problem. Suppose that $\mathbf{y}$ and $\mathbf{X}$ are already centered so we have the deviation form of the multiple regression model that has no intercept; see Example 4.1.2. Let $\mathbf{X}'\mathbf{X} = \mathbf{QDQ}'$ denote the spectral decomposition of $\mathbf{X}'\mathbf{X}$, where $\mathbf{D} = \text{diag}(\lambda_1, \cdots, \lambda_k)$ is a diagonal matrix consisting of the (real) eigenvalues of $\mathbf{X}'\mathbf{X}$, with $\lambda_1 \ge \cdots \ge \lambda_k \ge 0$ and $\mathbf{Q} = (\mathbf{q}_1, \cdots, \mathbf{q}_k)$ denotes the matrix whose columns are the orthogonal eigenvectors of $\mathbf{X}'\mathbf{X}$ corresponding to the ordered eigenvalues.

Consider the transformation

$$\mathbf{y} = \mathbf{XQQ}'\boldsymbol{\beta} + \boldsymbol{\varepsilon} = \mathbf{Z}\boldsymbol{\theta} + \boldsymbol{\varepsilon},$$

where $\mathbf{Z} = \mathbf{XQ}$, and $\boldsymbol{\theta} = \mathbf{Q}'\boldsymbol{\beta}$. Using this spectral decomposition, every regression model can be expressed in terms of orthogonal predictors, which are linear combinations of the original predictors $X_1, \cdots, X_k$. The matrix $\mathbf{Z} = (\mathbf{z}_1, \cdots, \mathbf{z}_k)$ is called the matrix of principal components of $\mathbf{X}$, while $\mathbf{z}_j = \mathbf{X}\mathbf{q}_j$ is the jth principal component of $\mathbf{X}$. Note that $\mathbf{z}_j'\mathbf{z}_j = \lambda_j$, the jth largest eigenvalue of $\mathbf{X}'\mathbf{X}$, and $\mathbf{z}_i'\mathbf{z}_j = 0$ for $i \ne j$. The elements of $\mathbf{q}_j$ are known as the principal component loadings, while the elements of $\mathbf{z}_j$ are the principal component scores. The elements of $\boldsymbol{\theta}$ are known as the regression parameters of the principal components. The principal components enable us to assess the presence of multicollinearity in a regression problem and provide an alternate estimation approach. Notice however that the principal components lack a simple interpretation, each being some linear combination of the original predictors.

Principal component regression consists of deleting one or more of the variables $\mathbf{z}_j$ corresponding to small values of λ_j, and using LS estimation on the resulting reduced regression model. Suppose we partition $\mathbf{Z} = (\mathbf{Z}_1, \mathbf{Z}_2)$, where $\mathbf{Z}_1 = (\mathbf{z}_1, \cdots, \mathbf{z}_m)$, $\mathbf{Z}_2 = (\mathbf{z}_{m+1}, \cdots, \mathbf{z}_k)$, and $\boldsymbol{\theta}$ conformably, so that we can write the regression model as

$$\mathbf{y} = \mathbf{Z}_1\boldsymbol{\theta}_1 + \mathbf{Z}_2\boldsymbol{\theta}_2 + \boldsymbol{\varepsilon}.$$

Note that $\boldsymbol{\beta} = \mathbf{Q}\boldsymbol{\theta} = \mathbf{Q}_1\boldsymbol{\theta}_1 + \mathbf{Q}_1\boldsymbol{\theta}_2$. If $\mathbf{Z}_1$ corresponds to the transformed predictor variables that will be retained in the model, while $\mathbf{Z}_2\boldsymbol{\theta}_2$ will be discarded from the model, then the LS estimator of $\boldsymbol{\theta}_1$ in the reduced model is

$$\widehat{\boldsymbol{\theta}}_1 = (\mathbf{Z}_1'\mathbf{Z}_1)^{-1}\mathbf{Z}_1'\mathbf{y} = (\lambda_1^{-1}\mathbf{z}_1'\mathbf{y}, \cdots, \lambda_m^{-1}\mathbf{z}_m'\mathbf{y})'.$$

Clearly, $E(\widehat{\boldsymbol{\theta}}_1) = \boldsymbol{\theta}_1$, and $\text{Var}(\widehat{\boldsymbol{\theta}}_1) = \sigma^2(\mathbf{Z}_1'\mathbf{Z}_1)^{-1}$. The principal component estimator of $\boldsymbol{\beta}$ is

$$\widehat{\boldsymbol{\beta}}_P = \mathbf{Q}_1\widehat{\boldsymbol{\theta}}_1 = \sum_{j=1}^{m} \frac{\mathbf{z}_j'\mathbf{y}}{\lambda_j}\mathbf{q}_j.$$

The principal component regression estimator $\widehat{\boldsymbol{\beta}}_P$ is biased but in general has smaller variance than $\widehat{\boldsymbol{\beta}}$.

Numerical Example 9.3. The dataset "Boston" in the R package *MASS* contains information collected by the U.S. Census Bureau concerning housing in the area of Boston, Massachusetts. The variables in the dataset are: crim (per capita crime rate by town), zn (proportion of residential land zoned for lots over 25,000 sq.ft.), indus (proportion of non retail business acres per town), chas (Charles River dummy variable: 1 if tract bounds river; 0 otherwise), nox (nitric oxide concentration in parts per 10 million), rm (average number of rooms per dwelling), age (proportion of owner-occupied units built prior to 1940), dis (weighted distances to five Boston employment centers), rad (index of accessibility to radial highways), tax (full-value property-tax rate per \$10,000), ptratio (pupil-teacher ratio by town), black ($1000(B - 0.63)^2$, where B is the proportion of blacks by town), lstat (lower status of the population), and medv (median value of owner-occupied homes in \$1000's). We discuss multicollinearity diagnostics followed by three remedies, i.e., dropping a variable, ridge regression, and principal components regression. Figure 9.2.3 shows the ridge traces for all the regression coefficients.

```
library(MASS)
data("Boston")
library(olsrr)
model <- lm(medv ~ ., data = Boston)
car::vif(model)   # VIF
crim       zn    indus     chas      nox       rm      age      dis
 1.7922  2.2988  3.9916  1.0740  4.3937  1.9337  3.1008  3.9559
rad      tax   ptratio    black   lstat
7.4845  9.0086  1.7991   1.3485  2.9415
# Refit a model excluding the tax variable
model2 <- lm(medv ~ . - tax, data = Boston)
car::vif(model2)    #all the VIF's < 5
# Ridge regression
x <- Boston[, 1:13]
y <- Boston[, 14]
xs <- scale(x, T, T)
ys <- scale(y, T, T)
plot(lm.ridge(ys ~ xs - 1, lambda = seq(0, 100, 0.1)))
lambda.est <- select(lm.ridge(ys ~ xs - 1, lambda = seq(0, 100, 0.1)))
fit <- lm.ridge(ys ~ xs - 1, lambda = 4.3)
coef(fit)
xscrim      xszn   xsindus    xschas     xsnox      xsrm     xsage
 -0.09741  0.11112  0.00538  0.07563  -0.21150  0.29472  -0.00061
 xsdis     xsrad     xstax xsptratio   xsblack   xslstat
  -0.32569   0.25949  -0.19804  -0.22060   0.09223  -0.40157
# Principal components regression
summary(pcr_model <- pls::pcr(medv ~ ., data = Boston, scale = T))
```

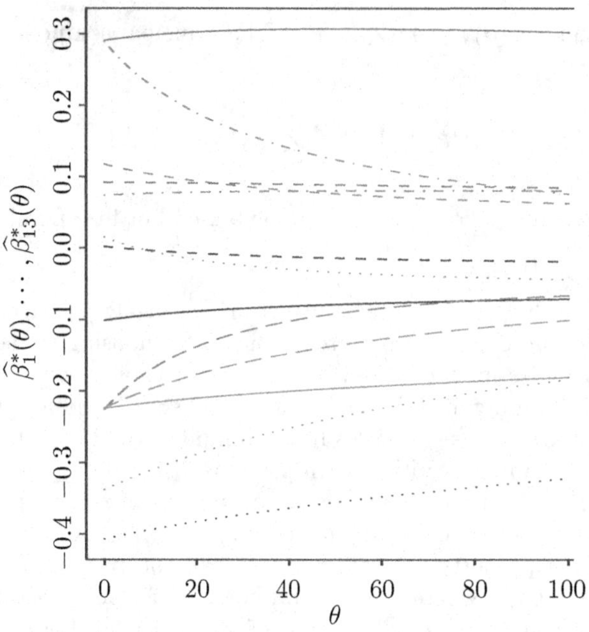

FIGURE 9.2.3. Ridge traces for the Boston data.

```
Method: svdpc
Number of components considered: 13
TRAINING: percent variance explained
        1 comps  2 comps  3 comps  4 comps
X         47.13    58.15    67.71    74.31
medv      37.42    45.59    63.59    64.78
        5 comps  6 comps  7 comps  8 comps
X         80.73    85.79    89.91    92.95
medv      69.70    70.05    70.05    70.56
        9 comps  10 comps  11 comps  12 comps
X         95.08     96.78     98.21     99.51
medv      70.57     70.89     71.30     73.21
        13 comps
X         100.00
medv       74.06
```

▲

9.3 Dummy variables in regression

The explanatory variables in a regression model may be quantitative variables or qualitative variables. Quantitative variables assume continuous values over some interval on the real line. The explanatory variables are qualitative or categorical, when they assume one of a few discrete values. A dummy variable, or indicator variable is any variable in a regression equation that assumes one of a finite number of discrete values and identifies different categories of a nominal variable. The values assumed by such variables do not represent any meaningful measurement; they are values such as 0, 1, or -1, and indicate the category into which the subject falls. If the nominal predictor has L categories, we can create L dummy variables to index these categories. We must include exactly $L-1$ of these indicator variables in a regression model which includes an intercept term, whereas if the regression model does not include an intercept, then we may include all L indicator variables. An example of a dummy variable is gender, i.e., $W_1 = 1$ if the subject is a female, and $W_1 = 0$ if the subject is a male. Suppose a qualitative variable represents the location of a factory in one of three geographic regions of the U.S.: South, West, or Midwest. Two dummy variables suffice: let $W_1 = 1$ if the factory is in the South, and zero otherwise; let $W_2 = 1$ if the factory is in the West, and zero otherwise. In practice, a response variable Y may be explained by L dummy variables and k quantitative predictors. This situation leads to a set of L parallel planes, each representing a regression equation between Y and the k quantitative predictors for a particular level of the dummy variable predictor. When $k = 1$, the model corresponds to sets of L regression lines (see Example 7.2.4) given by

$$\mathcal{S}_{H_0} = \{Y_{l,i} = \beta_{l,0} + \beta_{l,1}X_{l,i} + \varepsilon_{l,i},\ \ i = 1, \cdots, n_l,\ \ l = 1, \cdots, L\}.$$

We derived the least squares estimates both under the assumption of different slopes, and the assumption of the same slope, and also derived a test for parallelism. We continue with inference for sets of regression lines, and derive tests that enable us to compare these lines.

Result 9.3.1. Test for concurrence. The null hypothesis of concurrence of L regression lines at a point on the y-axis (i.e., when $X = 0$) is H_1: $\beta_{1,0} = \beta_{2,0} = \cdots = \beta_{L,0} = \beta_0$, say. The test statistic has an $F_{L-1,N-2L}$ distribution under the null hypothesis.

Proof. We can write

$$\mathcal{S}_{H_1} = \{Y_{l,i} = \beta_0 + \beta_{l,1}X_{l,i} + \varepsilon_{l,i},\ \ i = 1, \cdots, n_l,\ \ l = 1, \cdots, L\}.$$

The least squares estimates of the common intercept and L slopes under H_1 may be derived using (4.6.4) as

$$\begin{aligned}
\widetilde{\beta}_0 &= \frac{Y_{\cdot\cdot} - \{\sum_{l=1}^{L} X_{l\cdot} \sum_{i=1}^{n_l} Y_{l,i}X_{l,i} / \sum_{i=1}^{n_l} X_{l,i}^2\}}{N - \{\sum_{l=1}^{L} X_{l\cdot}^2 / \sum_{i=1}^{n_l} X_{l,i}^2\}}, \\
\widetilde{\beta}_{l,1} &= \frac{\sum_{i=1}^{n_l} (Y_{l,i} - \widetilde{\beta}_0) X_{l,i}}{\sum_{i=1}^{n_l} X_{l,i}}, \quad l = 1, \cdots, L.
\end{aligned}$$

It follows that

$$\begin{aligned}
SSE_{H_1} &= \sum_{l=1}^{L}\sum_{i=1}^{n_l} (Y_{l,i} - \widetilde{\beta}_0 - \widetilde{\beta}_{l,1}X_{l,i})^2, \text{ and} \\
Q(H_0 \,|\, H_1) &= SSE_{H_1} - SSE_{H_0},
\end{aligned}$$

so that under H_1,

$$F(H_0 \,|\, H_1) = \frac{Q(H_0 \,|\, H_1)/(L-1)}{SSE/(N-2L)} \sim F_{L-1,N-2L}. \qquad \blacksquare$$

Result 9.3.2. Test for coincidence. The hypothesis of coincidence of L regression lines is given by H_1: $\beta_{1,0} = \beta_{2,0} = \cdots = \beta_{L,0} = \beta_0$ and $\beta_{1,1} = \beta_{2,1} = \cdots = \beta_{L,1} = \beta_1$. The test statistic has an $F_{2(L-1),N-2L}$ distribution under the null hypothesis.

Proof. The least squares estimates of the common intercept and common slope under H_1 may be derived using (4.6.4) as

$$\widetilde{\beta}_0 = \overline{Y}.., \quad \widetilde{\beta}_{l,1} = \frac{\sum_{l=1}^{L}\sum_{i=1}^{n_l}(Y_{l,i} - \overline{Y}..)(X_{l,i} - \overline{X}..)}{\sum_{l=1}^{L}\sum_{i=1}^{n_l}(X_{l,i} - \overline{X}..)^2}. \tag{9.3.1}$$

We can show that

$$SSE_{H_1} = \sum_{l=1}^{L}\sum_{i=1}^{n_l}(Y_{l,i} - \overline{Y}..)^2 - \widetilde{\beta}^2 \sum_{l=1}^{L}\sum_{i=1}^{n_l}(X_{l,i} - \overline{X}..)^2$$

and under H_1,

$$F(H_0 \,|\, H_1) = \frac{(SSE_{H_1} - SSE_{H_0})/(2L-2)}{SSE/(N-2L)} \sim F_{2(L-1),N-2L}. \quad \blacksquare$$

Although the notation gets cumbersome, this approach can be easily extended to compare L regression hyperplanes as well. With k predictors, the number of model parameters when there is no restriction is $p = (k+1)L$, while the total number of observations is $N = \sum_{l=1}^{L} n_l$.

Numerical Example 9.4. Dummy variables in regression. The dataset "Salaries" in the *carData* package contains 2008-09 nine-month academic salary for Assistant Professors, Associate Professors and Professors in a college in the U.S. The data were collected as part of the ongoing effort of the college's administration to monitor salary differences between male and female faculty members. The data frame has 6 variables: rank (a factor with levels AssocProf, AsstProf, Prof), discipline (a factor with levels A ("theoretical" departments) or B ("applied" departments)), yrs.since.phd, yrs.service, sex (Female, Male), and salary (nine-month salary, in dollars).

```
library(tidyverse)
data("Salaries", package = "carData")
```

We fit a model for salary difference between males and females. Male is included in the model.

```
model <- lm(salary ~ sex, data = Salaries)
summary(model)$coef

             Estimate Std. Error   t value     Pr(>|t|)
(Intercept) 101002.41   4809.386 21.001103 2.683482e-66
sexMale      14088.01   5064.579  2.781674 5.667107e-03
```

If we wish to see a coefficient for Female in the output, we can use the function relevel() to set the baseline category to Male. The coefficient for sex in the output below pertains to Female.

```
Salaries <- Salaries %>% mutate(sex = relevel(sex, ref = "Male"))
model <- lm(salary ~ sex, data = Salaries)
summary(model)$coef

             Estimate Std. Error   t value      Pr(>|t|)
(Intercept) 115090.42   1587.378 72.503463 2.459122e-230
sexFemale   -14088.01   5064.579 -2.781674  5.667107e-03
```

To understand an effect due to a categorical variable with more than two levels, we look at a regression of salary on rank, which has three levels, and has been dummy coded here into two levels. We fit an MLR model using four variables in which three variables are dummy coded (not all output is shown).

```
res <- model.matrix( ~ rank, data = Salaries)
head(res[,-1])
summary(model2 <-
   lm(salary ~ yrs.service + rank + discipline + sex,
      data = Salaries))

Call:
lm(formula = salary ~ yrs.service + rank + discipline + sex,
    data = Salaries)

Residuals:
   Min      1Q Median      3Q     Max
-64202 -14255  -1533  10571  99163

Coefficients:
              Estimate Std. Error t value Pr(>|t|)
(Intercept)   73122.92    3245.27  22.532  < 2e-16 ***
yrs.service     -88.78     111.64  -0.795 0.426958
rankAssocProf 14560.40    4098.32   3.553 0.000428 ***
rankProf      49159.64    3834.49  12.820  < 2e-16 ***
disciplineB   13473.38    2315.50   5.819 1.24e-08 ***
sexFemale     -4771.25    3878.00  -1.230 0.219311
---
Signif. codes:  0 ‘***’ 0.001 ‘**’ 0.01 ‘*’ 0.05 ‘.’ 0.1 ‘ ’ 1

Residual standard error: 22650 on 391 degrees of freedom
Multiple R-squared:  0.4478,Adjusted R-squared:  0.4407
F-statistic: 63.41 on 5 and 391 DF,  p-value: < 2.2e-16


car::Anova(model2)

Anova Table (Type II tests)

Response: salary
                Sum Sq  Df  F value    Pr(>F)
yrs.service 3.2448e+08   1   0.6324    0.4270
rank        1.0288e+11   2 100.2572 < 2.2e-16 ***
discipline  1.7373e+10   1  33.8582 1.235e-08 ***
sex         7.7669e+08   1   1.5137    0.2193
Residuals   2.0062e+11 391
---
Signif. codes:  0 ‘***’ 0.001 ‘**’ 0.01 ‘*’ 0.05 ‘.’ 0.1 ‘ ’ 1

vcov(model2)
```

The last line above enables us to obtain the variance-covariance matrix of the estimated coefficients (result not shown). ▲

Exercises

9.1. Suppose $Y_i = \beta_0 + \beta_1(X_{i1} - \overline{X}_1) + \beta_2(X_{i2} - \overline{X}_2) + \varepsilon_i$, $i = 1, \cdots, N$, where $\overline{X}_j = \sum_{i=1}^N X_{ij}/N$, $j = 1, 2$, and ε_i are i.i.d. normal random variables with mean 0 and variance σ^2. Quantify the width of the marginal 95% confidence interval for β_1 in terms of r_{12}, the sample correlation between X_1 and X_2.

9.2. In the normal linear regression model $Y_i = \beta_0 + \beta_1 X_{i,1} + \cdots + \beta_k X_{i,k} + \varepsilon_i$, $i = 1, \cdots, N$, obtain a test statistic for H: $\beta_j = d$ where $0 \le j \le k$ and d is a constant.

9.3. In the full-rank regression model $\mathbf{y} = \mathbf{X}\boldsymbol{\beta} + \boldsymbol{\varepsilon}$ with $\mathrm{r}(\mathbf{X}) = p$ and $\boldsymbol{\varepsilon} \sim N(\mathbf{0}, \sigma^2\mathbf{I})$, show that $\mathbf{y}'\mathbf{A}_1\mathbf{y}$ with $\mathbf{A}_1 = (\mathbf{I}_N - \mathbf{P})/(N - p + 2)$ has a smaller MSE as an estimator of σ^2 than the LS estimator of σ^2.

9.4. Based on N observations (Y_i, X_{ij}), $j = 1, \cdots, p$, $i = 1, \cdots, N$ following the MLR model (4.1.1) with $\varepsilon_i \sim N(0, \sigma^2)$, suppose we wish to predict the value of a new observation, Y_0 at a new value $\mathbf{x}_0$. Let $\widehat{\boldsymbol{\beta}}$ and $\widehat{\sigma}^2$ denote the least squares estimates of $\boldsymbol{\beta}$ and σ^2, respectively.

(a) Show that the mean of Y_0 and $\mathrm{Var}(Y_0 - \widehat{Y}_0)$ are respectively $\mathbf{x}_0'\boldsymbol{\beta}$ and $\sigma_{Y_0}^2 = \sigma^2\{1 + \mathbf{x}_0'(\mathbf{X}'\mathbf{X})^{-1}\mathbf{x}_0\}$. Find the distribution of $(Y_0 - \mathbf{x}_0'\widehat{\boldsymbol{\beta}})/\sigma\{1 + \mathbf{x}_0'(\mathbf{X}'\mathbf{X})^{-1}\mathbf{x}_0\}^{1/2}$.

(b) Find a 95% symmetric two-sided prediction interval for Y_0.

(c) Let $\eta \in (0, 0.5]$, and suppose the 100ηth percentile of the distribution of Y_0 is $\gamma(\eta) = \mathbf{x}_0'\boldsymbol{\beta} + z_\eta \sigma_{Y_0}$. Find a $100(1 - \alpha)\%$ lower confidence bound for $\gamma(\eta)$.

9.5. Consider the regression lines $Y_{l,i} = \beta_{l,0} + \beta_{l,1}X_{l,i} + \varepsilon_{l,i}$, $l = 1, 2$, $\varepsilon_{l,i}$ being i.i.d. $N(0, \sigma^2)$ variables. Given observations $(X_{l,i}, Y_{l,i})$, $i = 1, \cdots, n_l$, $l = 1, 2$, $N = n_1 + n_2$, carry out the following tests:

(a) H: $\beta_{1,1} = \beta_{2,1}$, i.e., a test for parallelism;

(b) H: $\beta_{1,0} = \beta_{2,0}$, i.e., a test for concurrence;

(c) H: $\beta_{1,0} = \beta_{2,0}$ and $\beta_{1,1} = \beta_{2,1}$, i.e., a test for coincidence.

9.6. Consider the independent regressions

$$Y_{l,i} = \alpha_l + \beta(X_{l,i} - \overline{X}_l) + \varepsilon_{l,i}, \quad i = 1, \cdots, n_l, l = 1, 2,$$

where $\varepsilon_{l,i}$ are i.i.d. $N(0, \sigma^2)$ variables, and $\overline{X}_l = \sum_{i=1}^{n_l} X_{l,i}/n_l$.

(a) Estimate α_1, α_2 and β, and thus the vertical distance between the two lines measured parallel to the y-axis, defined as $D = (\alpha_1 - \alpha_2) + \beta(\overline{X}_1 - \overline{X}_2)$.

(b) Construct a 95% C.I. for this distance.

9.7. Consider the centered and scaled form of the multiple linear regression model in Result 9.2.2. Show that for $\theta > 0$, $\|\widehat{\boldsymbol{\beta}}^*(\theta)\| < \|\widehat{\boldsymbol{\beta}}^*\|$, where $\widehat{\boldsymbol{\beta}}^*$ is the LS estimate of $\boldsymbol{\beta}^*$ and $\widehat{\boldsymbol{\beta}}^*(\theta)$ is the ridge regression estimate given in (9.2.7).

9.8. Consider the MLR model $\mathbf{y} = \mathbf{X}_1\boldsymbol{\beta}_1 + \mathbf{X}_2\boldsymbol{\beta}_2 + \boldsymbol{\varepsilon}$. Suppose the regression matrix $\mathbf{X} = (\mathbf{X}_1, \mathbf{X}_2)$ where $\mathbf{X}_j \in \mathcal{R}^{N\times p_j}$, $j = 1, 2$ and $p_1 + p_2 = p$. Let $SSE_1 = \widehat{\boldsymbol{\varepsilon}}'\widehat{\boldsymbol{\varepsilon}}$, where $\widehat{\boldsymbol{\varepsilon}} = \mathbf{y} - \widehat{\mathbf{y}}$ and $\widehat{\mathbf{y}} = \mathbf{X}_1\widehat{\boldsymbol{\beta}}_1$.

(a) Obtain the LS estimate of $\boldsymbol{\beta}_2$ as a function of $\mathbf{X}_1$, $\mathbf{X}_2$ and $\mathbf{y}$.

(b) Show that $\widehat{\boldsymbol{\beta}}_2$ and SSE_1 are dependent.

9.9. Suppose the true model is $\mathbf{y} = \mathbf{X}_1\boldsymbol{\beta}_1 + \boldsymbol{\varepsilon}$, where $\mathbf{X}_1$ has full column rank and $\text{Cov}(\boldsymbol{\varepsilon}) = \sigma^2\mathbf{I}_N$. Let $\widehat{\boldsymbol{\beta}}_1$ be the LS estimate under the model. Suppose the data is instead fitted by the model $\mathbf{y} = \mathbf{X}_1\boldsymbol{\beta}_1 + \mathbf{X}_2\boldsymbol{\beta}_2 + \boldsymbol{\varepsilon}$, where $\mathcal{C}(\mathbf{X}_1) \cap \mathcal{C}(\mathbf{X}_2) = \{\mathbf{0}\}$ (i.e., the columns of $\mathbf{X}_1$ are LIN of the columns of $\mathbf{X}_2$), and $\mathbf{X}_2$ need not be of full column rank.

(a) Show that $\boldsymbol{\beta}_1$ is still estimable under the expanded model.

(b) Let $\widetilde{\boldsymbol{\beta}}_1$ be the LS estimate of $\boldsymbol{\beta}_1$ under the expanded model. Show that $\text{Cov}(\widehat{\boldsymbol{\beta}}_1) \leq \text{Cov}(\widetilde{\boldsymbol{\beta}}_1)$, with equality holding if and only if all the columns of $\mathbf{X}_2$ are orthogonal to those of $\mathbf{X}_1$.

9.10. Discuss the effect of including an additional regressor X_{k+1} to an MLR model $Y_i = \beta_0 + \sum_{j=1}^k \beta_j X_{ij} + \varepsilon_i$, $i = 1, \cdots, N$, on the variance of the future prediction $\widehat{Y}^*$ at $(X_1^*, \cdots, X_k^*, X_{k+1}^*)$.

9.11. Consider the MLR model $Y_i = \beta_0 + \sum_{j=1}^k \beta_j X_{ij} + \varepsilon_i$, $i = 1, \cdots, N$. Let t_j denote the t-statistic for testing H_0: $\beta_j = 0$ for some $j \leq k$. Let C_k and C_{k-1} be the Mallow's C_p statistic for the regression with k and $k-1$ predictors, respectively. Show that

$$C_{k-1} = \frac{t_j^2 \, SSE_k}{\widehat{\sigma}^2(N-k)} + C_k - 2.$$

10

Fixed-Effects Linear Models

Fixed-effects linear models are general linear models where the response is modeled as a linear function of categorical predictors (factors) with fixed levels. Inference for such models usually follows inference for the less-than-full rank GLM. In Chapter 4, we introduced the least squares approach for balanced fixed-effects models, and discussed the ANOVA decomposition and F-test under normality in Chapter 7. In Section 10.1, we describe parametric inference for unbalanced models, including one-way models and higher-order cross-classified models and nested models. This is followed by a description of two nonparametric procedures in Section 10.2. Section 10.3 describes analysis of covariance procedures. In Section 10.4, we discuss concepts and procedures of multiple testing.

10.1 Inference for unbalanced ANOVA models

The fixed-effects one-way analysis of variance (ANOVA) model that we discussed in Chapters 4 and 7 has the form

$$Y_{ij} = \mu + \tau_i + \varepsilon_{ij}, \quad j = 1, \cdots, n_i, \; i = 1, \cdots, a,$$

where Y_{ij} is the jth observation from the ith level of treatment, μ is the overall mean, τ_i is the ith treatment effect, and the random error components ε_{ij} are i.i.d. $N(0, \sigma^2)$ variables. If $n_1 = \cdots = n_a = n$, we have a balanced model; otherwise, the model is said to be unbalanced. For a detailed study of unbalanced ANOVA models, see Searle (1971, 1987), for a description of "overparametrized" models (of which Example 4.1.3 was a special case) and the *cell means model*. In this section, we discuss these approaches, with slightly more emphasis on the overparametrized model, which seems to be the approach taken in most statistical software.

Unbalanced data can result either due to planned design, or due to unplanned missingness. We focus on the former. *All-cells-filled* data refers to the situation where there is at least one observation in each cell. By contrast, *some-cells-empty* data correspond to some possibly empty cells. This distinction is of importance in determining whether inference on certain interaction effects is possible. Further, with some-cells-empty data, the extent and nature of sparsity determines which effects are estimable.

Definition 10.1.1. Connected data. For every main effects factor in the model, if all differences between levels of a factor are estimable, the data is said to be connected. Otherwise, the data is disconnected.

Example 10.1.1. Consider the two-factor additive model (with no interaction)

$$Y_{ijk} = \mu + \tau_i + \theta_j + \varepsilon_{ijk}, \quad k = 1, \cdots, n_{ij}, i = 1, \cdots, a, \; j = 1, \cdots, b,$$

where the effects μ, τ_i, and θ_j have been defined earlier, and ε_{ijk} are i.i.d. $N(0, \sigma^2)$ variables.

DOI: 10.1201/9781315156651-10

$$\begin{pmatrix} \text{x} & \text{x} & & \\ & \text{x} & \text{x} & \text{x} \\ & & \text{x} & \text{x} \end{pmatrix} \qquad \begin{pmatrix} \text{x} & \text{x} & & \\ & & \text{x} & \text{x} \\ & & \text{x} & \text{x} \end{pmatrix}$$

TABLE 10.1.1. Connected data (left) and disconnected data (right).

Table 10.1.1 shows two sets of data configurations for $a = 3$, and $b = 4$, so that there are $ab = 12$ cells. A symbol "x" in a cell indicates the presence of data, while an empty cell indicates no data.

If the data are disconnected, some differences of the form $\tau_i - \tau_{i'}$, $i \neq i'$, or $\theta_j - \theta_{j'}$, $j \neq j'$ are not estimable. If every such difference can be estimated, then the cell means $\mu_{ij} = \mu + \tau_i + \theta_j$ are estimable for all i, j, irrespective of whether the cell is filled or empty. We leave it to the reader to verify that in Table 10.1.1, the data on the left is connected, while the data on the right is not connected. □

Definition 10.1.2. g-connectedness. In a two-way crossed classification, data are said to be geometrically connected, or g-connected when the filled cells in the $a \times b$ grid can be joined by a continuous line consisting only of horizontal and vertical segments, and which only changes direction in filled cells. All-cells-filled data are trivially g-connected. The data on the left in Table 10.1.1 are g-connected, while the data on the right are not.

Sometimes, reordering rows or columns of the $a \times b$ grid enables us to show g-connectedness. Weeks and Williams (1964) gave an algebraic characterization to verify connectedness in an m-factor model; when $m = 2$, the method reduces to g-connectedness. This gives a sufficient (though not necessary) condition for estimability of differences in the main effects of the factors. Suppose the model has L main effects. Let the array of L subscripts on the sub-most cell mean be $[i_1, \cdots, i_L]$. Two such i-arrays are called "nearly identical" if they differ in only a single element. Connected data are defined as data for which the i-array of every filled sub-most cell is nearly identical to that of at least one other filled sub-most cell. If data is disconnected, separate analyses are recommended for each of the separate sets of data which are connected within themselves.

In the following subsections, we describe inference for one-factor and two-factor unbalanced models. The cell means models are a natural way for carrying out inference on fixed-effects unbalanced ANOVA models. A discussion of this approach for the one-way model is given in Section 10.1.1. In the rest of this section, we describe higher-order *overparametrized* models, and defer the reader to Searle (1987) for a study of the cell means models for these situations.

10.1.1 One-way cell means model

In Chapter 4 and Chapter 7, we described modeling and inference for a one-factor experiment with unequal replication corresponding to the factor levels. The model that we used there is often referred to as the *overparametrized model* because there are $(a+1)$ parameters, but only a cell means $\overline{Y}_{i\cdot}$. The alternative cell means model for the one-way fixed-effects classification is

$$Y_{ij} = \mu_i + \varepsilon_{ij}, \quad j = 1, \cdots, n_i, \; i = 1, \cdots, a, \tag{10.1.1}$$

where μ_i is the population mean of the ith level, i.e., $\mathrm{E}(Y_{ij}) = \mu_i$, and ε_{ij} are assumed to be i.i.d. $N(0, \sigma^2)$ variables. Let $N = \sum_{i=1}^{a} n_i$. The least squares estimates of the unknown cell means are obtained by minimizing with respect to μ_i's the function

$$S(\mu_1, \cdots, \mu_a) = \sum_{i=1}^{a}\sum_{j=1}^{n_i}(Y_{ij} - \mu_i)^2.$$

Solving the resulting set of a equations

$$\sum_{j=1}^{n_i} \widehat{\mu}_i = \sum_{j=1}^{n_i} Y_{ij},$$

we obtain $\widehat{\mu}_i = \overline{Y}_{i\cdot}$, with $\text{Var}(\widehat{\mu}_i) = \sigma^2/n_i$, $i = 1, \cdots, a$. It follows that the fitted values and the residuals are respectively $\widehat{Y}_{ij} = \widehat{\mu}_i = \overline{Y}_{i\cdot}$ and $\widehat{\varepsilon}_{ij} = Y_{ij} - \overline{Y}_{i\cdot}$, $j = 1, \cdots, n_i$, $i = 1, \cdots, a$, while the forms of SSE and $\widehat{\sigma}^2$ coincide with (4.2.29) and (4.2.28). The ANOVA table coincides with Table 7.2.3 and is useful for testing the hypothesis H_0: $\mu_1 = \cdots = \mu_a$.

Result 10.1.1. Let $\omega = \sum_{i=1}^{a} c_i\mu_i$ denote a linear function of the cell means, where c_i's are constants, and let d be a constant.

1. The b.l.u.e. of ω is $\widehat{\omega} = \sum_{i=1}^{a} c_i\overline{Y}_{i\cdot}$ with $\text{Var}(\widehat{\omega}) = \sigma^2\sum_{i=1}^{a} c_i^2/n_i$.

2. A symmetric $100(1-\alpha)\%$ confidence interval for ω is

$$\widehat{\omega} \pm t_{N-a,\alpha/2}[\widehat{\text{Var}}(\widehat{\omega})]^{1/2}.$$

3. The statistic for testing H: $\sum_{i=1}^{a} c_i\mu_i = d$ is

$$F = \frac{(\sum_{i=1}^{a} c_i\overline{Y}_{i\cdot} - d)^2}{\widehat{\sigma}^2\sum_{i=1}^{a} c_i^2/n_i}, \tag{10.1.2}$$

which has an $F_{1,N-a}$ distribution under H.

4. The statistic for testing H: $\sum_{i=1}^{a} c_{1,i}\mu_i = d_1$, $\sum_{i=1}^{a} c_{2,i}\mu_i = d_2$ is

$$F = \frac{(g_2f_1^2 + g_1f_2^2 - 2gf_1f_2)}{2\widehat{\sigma}^2(g_1g_2 - g^2)},$$

and has an $F_{2,N-a}$ distribution under H, where

$$f_1 = \sum_{i=1}^{a} c_{1,i}\overline{Y}_{i\cdot} - d_1, \quad f_2 = \sum_{i=1}^{a} c_{2,i}\overline{Y}_{i\cdot} - d_2,$$

$$g_1 = \sum_{i=1}^{a} c_{1,i}^2/n_i, \quad g_2 = \sum_{i=1}^{a} c_{2,i}^2/n_i, \quad g = \sum_{i=1}^{a} c_{1,i}c_{2,i}/n_i.$$

Proof. The proof of property 1 is a direct consequence of the Gauss–Markov theorem. Property 2 follows immediately (see Section 7.4.1). To prove property 3, we see that the restricted least squares estimates $\widehat{\mu}_{i,H}$ of μ_i, $i = 1, \cdots, a$ are obtained by minimizing

$$\sum_{i=1}^{a}\sum_{j=1}^{n_i}(Y_{ij} - \mu_i)^2 + 2\lambda\left(\sum_{i=1}^{a} c_i\mu_i - d\right)$$

with respect to $\mu_1, \cdots, \mu_a$ and the Lagrange multiplier λ as

$$\widehat{\mu}_{i,H} = \overline{Y}_{i\cdot} - \widehat{\lambda}c_i/n_i, \quad \widehat{\lambda} = \frac{\sum_{i=1}^a (c_i \overline{Y}_{i\cdot} - d)}{\sum_{i=1}^a (c_i^2/n_i)}.$$

It is easy to verify that

$$SSE_H = SSE + \frac{(\sum_{i=1}^a c_i \overline{Y}_{i\cdot} - d)^2}{\sum_{i=1}^a c_i^2/n_i},$$

and (10.1.2) follows. Property 4 corresponds to a *two-part* hypothesis, and its proof is left as Exercise 10.1. ■

The proof of the following result is straightforward from Definition 4.3.3 and is left as Exercise 10.8.

Result 10.1.2. Under the one-way fixed-effects model (4.1.7), two contrasts $\sum_{i=1}^a c_{1,i}\mu_i$ and $\sum_{i=1}^a c_{2,i}\mu_i$ are orthogonal if and only if $\sum_{i=1}^a c_{1,i}c_{2,i}/n_i = 0$. Their b.l.u.e.'s are respectively $\sum_{i=1}^a c_{1,i}\overline{Y}_{i\cdot}$ and $\sum_{i=1}^a c_{2,i}\overline{Y}_{i\cdot}$, while the covariance between them is given by $\sigma^2 \sum_{i=1}^a c_{1,i}c_{2,i}/n_i = 0$.

Numerical Example 10.1. The dataset "schizophrenia" from the R package *HSAUR* analyzes gender differences in the age of onset of schizophrenia. The data frame consists of two variables: age (at the time of diagnosis) and gender (a factor with levels, female or male). We fit a one-way ANOVA model using the `lm()` function.

```
library(HSAUR)
data(schizophrenia)
summary(fit <- lm(age ~ gender, data = schizophrenia))

Call:
lm(formula = age ~ gender, data = schizophrenia)

Residuals:
    Min      1Q  Median      3Q     Max
-24.475  -5.914  -1.914   5.086  34.086

Coefficients:
            Estimate Std. Error t value Pr(>|t|)
(Intercept)  30.4747     0.9989  30.507  < 2e-16 ***
gendermale   -6.5603     1.2837  -5.111  6.4e-07 ***
---
Signif. codes:
0 '***' 0.001 '**' 0.01 '*' 0.05 '.' 0.1 ' ' 1

Residual standard error: 9.939 on 249 degrees of freedom
Multiple R-squared:  0.09493,Adjusted R-squared:  0.0913
F-statistic: 26.12 on 1 and 249 DF,  p-value: 6.401e-07

car::Anova(fit, type = "III") # Type III sums of squares

Anova Table (Type III tests)
```

```
Response: age
            Sum Sq  Df F value    Pr(>F)
(Intercept)  91942   1 930.689 < 2.2e-16 ***
gender        2580   1  26.118 6.401e-07 ***
Residuals    24599 249
---
Signif. codes:
0 '***' 0.001 '**' 0.01 '*' 0.05 '.' 0.1 ' ' 1
```

▲

10.1.2 Higher-order overparametrized models

We present inference for additive cross-classified models as well as models with interaction, and for nested or hierarchical models, all for the unbalanced case. Instead of the cell means models (Searle, 1987), we describe *overparametrized* models.

10.1.2.1 Two-factor additive models

Suppose an experiment involves Factor A with a levels and Factor B with b levels. A model for unbalanced data is

$$Y_{ijk} = \mu + \tau_i + \theta_j + \varepsilon_{ijk}, \quad 1 \le k \le n_{ij},\ 1 \le i \le a,\ 1 \le j \le b, \tag{10.1.3}$$

where μ is the overall mean effect, τ_i is the effect due to the ith level of Factor A, θ_j is the effect due to the jth level of Factor B, and ε_{ijk} are i.i.d. $N(0, \sigma^2)$ variables. Let $p = 1+a+b$ denote the number of model parameters,

$$n_{i\cdot} = \sum_{j=1}^{b} n_{ij}, \quad n_{\cdot j} = \sum_{i=1}^{a} n_{ij}, \quad N = \sum_{i=1}^{a}\sum_{j=1}^{b} n_{ij},$$
$$Y_{i\cdot\cdot} = \sum_{j=1}^{b}\sum_{k=1}^{n_{ij}} Y_{ijk}, \quad Y_{\cdot j\cdot} = \sum_{i=1}^{a}\sum_{k=1}^{n_{ij}} Y_{ijk}, \quad Y_{\cdot\cdot\cdot} = \sum_{i=1}^{a}\sum_{j=1}^{b}\sum_{k=1}^{n_{ij}} Y_{ijk},$$
$$\mathbf{y}_a = (Y_{1\cdot\cdot}, \cdots, Y_{a\cdot\cdot})', \quad \mathbf{y}_b = (Y_{\cdot 1\cdot}, \cdots, Y_{\cdot b\cdot})'.$$

Let $\mathbf{N} = \{n_{ij}\}$ denote the $a \times b$ whose elements are the number of cell replications, and let

$$\mathbf{r}_a = (n_{1\cdot}, \cdots, n_{a\cdot})' = \mathbf{N}\mathbf{1}_b, \quad \mathbf{D}_a = \text{diag}(\mathbf{r}_a),$$
$$\mathbf{r}_b = (n_{\cdot 1}, \cdots, n_{\cdot b})' = \mathbf{N}'\mathbf{1}_a, \quad \mathbf{D}_b = \text{diag}(\mathbf{r}_b),$$
$$\mathbf{U} = \mathbf{D}_a - \mathbf{N}\mathbf{D}_b^{-1}\mathbf{N}', \quad \mathbf{V} = \mathbf{D}_b - \mathbf{N}'\mathbf{D}_a^{-1}\mathbf{N}.$$

For obvious reasons, assume that each level of Factor A has at least one observation and so does each level of Factor B, i.e., $n_{i\cdot} \ge 1$ for all i and $n_{\cdot j} \ge 1$ for all j. The normal equations obtained by minimizing $\mathcal{S}(\mu, \boldsymbol{\tau}, \boldsymbol{\theta}) = \sum_{i=1}^{a}\sum_{j=1}^{b}\sum_{k=1}^{n_{ij}}(Y_{ijk} - \mu - \tau_i - \theta_j)^2$ with respect to μ, $\boldsymbol{\tau}$ and $\boldsymbol{\theta}$ are

$$Y_{\cdot\cdot\cdot} = N\mu^0 + \mathbf{r}_a'\boldsymbol{\tau}^0 + \mathbf{r}_b'\boldsymbol{\theta}^0, \ \text{ i.e., } Y_{\cdot\cdot\cdot} = N\mu^0 + \sum_i n_{i\cdot}\tau_i^0 + \sum_j n_{\cdot j}\theta_j^0, \tag{10.1.4}$$

$$\mathbf{y}_a = \mathbf{r}_a\mu^0 + \mathbf{D}_a\boldsymbol{\tau}^0 + \mathbf{N}\boldsymbol{\theta}^0, \text{ i.e., } Y_{i\cdot\cdot} = n_{i\cdot}(\mu^0 + \tau_i^0) + \sum_j n_{ij}\theta_j^0 \quad \forall i, \tag{10.1.5}$$

$$\mathbf{y}_b = \mathbf{r}_b\mu^0 + \mathbf{N}'\boldsymbol{\tau}^0 + \mathbf{D}_b\boldsymbol{\theta}^0, \text{ i.e., } Y_{\cdot j\cdot} = n_{\cdot j}(\mu^0 + \theta_j^0) + \sum_i n_{ij}\tau_i^0 \quad \forall j. \tag{10.1.6}$$

Since the set of equations in (10.1.5) add up to (10.1.4), as do the set of equations in (10.1.6), the number of LIN normal equations is at most $r = a+b-1$, which is the rank of the matrix $\mathbf{X}'\mathbf{X}$ with the form

$$\mathbf{X}'\mathbf{X} = \begin{pmatrix} N & \mathbf{r}_a' & \mathbf{r}_b' \\ \mathbf{r}_a & \mathbf{D}_a & \mathbf{N} \\ \mathbf{r}_b & \mathbf{N}' & \mathbf{D}_b \end{pmatrix}.$$

Result 10.1.3. The following rank conditions hold.

1. $\mathrm{r}(\mathbf{X}'\mathbf{X}) = \mathrm{r}\begin{pmatrix} \mathbf{D}_a & \mathbf{N} \\ \mathbf{N}' & \mathbf{D}_b \end{pmatrix}.$

2. $\mathrm{r}(\mathbf{X}'\mathbf{X}) = \mathrm{r}\begin{pmatrix} \mathbf{D}_a & \mathbf{N} \\ \mathbf{0} & \mathbf{V} \end{pmatrix} = \mathrm{r}(\mathbf{V}) + a.$

3. $\mathrm{r}(\mathbf{X}'\mathbf{X}) = \mathrm{r}\begin{pmatrix} \mathbf{U} & \mathbf{N} \\ \mathbf{0} & \mathbf{D}_b \end{pmatrix} = \mathrm{r}(\mathbf{U}) + b.$

Proof. As we mentioned earlier, the first column of $\mathbf{X}'\mathbf{X}$ is equal to the sum of the next a columns, as well as to the sum of the last b columns. This proves property 1, since deleting the first row and column of the matrix does not alter its rank. Properties 2 and 3 follow because $\mathbf{D}_a$ and $\mathbf{D}_b$ are diagonal,

$$\mathrm{r}(\mathbf{X}'\mathbf{X}) = \mathrm{r}\begin{pmatrix} \mathbf{I}_a & \mathbf{0} \\ -\mathbf{N}'\mathbf{D}_a^{-1} & \mathbf{I}_b \end{pmatrix}\begin{pmatrix} \mathbf{D}_a & \mathbf{N} \\ \mathbf{N}' & \mathbf{D}_b \end{pmatrix}, \text{ and}$$

$$\mathrm{r}(\mathbf{X}'\mathbf{X}) = \mathrm{r}\begin{pmatrix} \mathbf{D}_a & \mathbf{N} \\ \mathbf{N}' & \mathbf{D}_b \end{pmatrix}\begin{pmatrix} \mathbf{I}_a & \mathbf{0} \\ -\mathbf{D}_b^{-1}\mathbf{N}' & \mathbf{I}_b \end{pmatrix}. \quad \blacksquare$$

Result 10.1.4. Let $\mathbf{w} = (W_1, \cdots, W_a)' = \mathbf{y}_a - \mathbf{N}\mathbf{D}_b^{-1}\mathbf{y}_b$, i.e., $W_i = Y_{i\cdot\cdot} - \sum_{j=1}^{b}\{n_{ij}Y_{\cdot j\cdot}/n_{\cdot j}\}$, $1 \le i \le a$. Then a set of LS solutions to (10.1.4)–(10.1.6) are

$$\mu^0 = 0, \quad \boldsymbol{\tau}^0 = \mathbf{U}^-\mathbf{w}, \quad \boldsymbol{\theta}^0 = \mathbf{D}_b^{-1}\mathbf{y}_b - \mathbf{D}_b^{-1}\mathbf{N}'\boldsymbol{\tau}^0. \tag{10.1.7}$$

Likewise, let $\mathbf{z} = (Z_1, \cdots, Z_b)' = \mathbf{y}_b - \mathbf{N}'\mathbf{D}_a^{-1}\mathbf{y}_a$, another set of LS solutions are

$$\mu^0 = 0, \quad \boldsymbol{\theta}^0 = \mathbf{V}^-\mathbf{z}, \quad \boldsymbol{\tau}^0 = \mathbf{D}_a^{-1}\mathbf{y}_a - \mathbf{D}_a^{-1}\mathbf{N}\boldsymbol{\theta}^0. \tag{10.1.8}$$

Proof. To start with, from the projection perspective, the minimum of the convex function $\mathcal{S}(\mu, \boldsymbol{\tau}, \boldsymbol{\theta})$ can be attained at some finite values of the parameters, so the consistency of the equations (10.1.4)–(10.1.6) is guaranteed. Then, as noted earlier, the equations (10.1.5) and (10.1.6) are consistent. Pre-multiply (10.1.6) by $\mathbf{N}\mathbf{D}_b^{-1}$ and subtract it from (10.1.5) to get

$$\begin{aligned} \mathbf{w} &= \mathbf{y}_a - \mathbf{N}\mathbf{D}_b^{-1}\mathbf{y}_b \\ &= (\mathbf{r}_a - \mathbf{N}\mathbf{D}_b^{-1}\mathbf{r}_b)\mu^0 + (\mathbf{D}_a - \mathbf{N}\mathbf{D}_b^{-1}\mathbf{N}')\boldsymbol{\tau}^0 = \mathbf{U}\boldsymbol{\tau}^0, \end{aligned} \tag{10.1.9}$$

which is a consistent system of equations in $\tau_1^0, \cdots, \tau_a^0$, and has a solution $\boldsymbol{\tau}^0 = \mathbf{U}^-\mathbf{w}$. Letting $\mu^0 = 0$, $\boldsymbol{\theta}^0$ immediately follows from (10.1.6). Then (10.1.7) follows. The solutions (10.1.8) can be obtained similarly. $\blacksquare$

Result 10.1.5. The following properties hold for the model (10.1.3):

1. A linear function $\sum_{i=1}^{a} c_i\tau_i$ is estimable if and only if the vector $\mathbf{c} = (c_1, \cdots, c_a)'$ is a linear combination of the row vectors of the matrix $\mathbf{U}$.

2. $r(\mathbf{U}) = a - 1$ is a sufficient and necessary condition for the data to be connected, i.e., all contrasts in τ's and all contrasts in θ's are estimable.

3. The b.l.u.e. of a contrast $\mathbf{c}'\boldsymbol{\tau}$ is $\mathbf{c}'\mathbf{U}^{-}\mathbf{w}$, with variance $\mathbf{c}'\mathbf{U}^{-}\mathbf{c}\sigma^2$.

4. The b.l.u.e. of an estimable contrast $\mathbf{d}'\boldsymbol{\theta}$ is $\mathbf{d}'\mathbf{D}_b^{-1}(\mathbf{y}_b - \mathbf{N}'\mathbf{U}^{-}\mathbf{w})$, with variance $\mathbf{d}'(\mathbf{D}_b^{-1} + \mathbf{D}_b^{-1}\mathbf{N}'\mathbf{U}^{-}\mathbf{N}\mathbf{D}_b^{-1})\mathbf{d}\sigma^2$.

Proof. The proof of property 1 is left as an exercise. It is straightforward to show that $\mathbf{U}\mathbf{1}_a = \mathbf{0}$, so that $\mathcal{R}(\mathbf{U}) \subset \text{Span}(\mathbf{1}_a)^{\perp}$. If $r(\mathbf{U}) = a - 1$, then $\mathcal{R}(\mathbf{U}) = \text{Span}(\mathbf{1}_a)^{\perp}$ so it contains all vectors $\mathbf{c}$ with $\mathbf{c}'\mathbf{1}_a = 0$. From property 1, every contrast in the τ's is estimable. Further, from properties 2 and 3 of Result 10.1.3, $r(\mathbf{V}) = b-1$. Then, by a completely similar argument, every contrast in the θ's is estimable. Hence the data is connected. Conversely, if the data is connected, then $\mathcal{R}(\mathbf{U})$ contains every $\tau_1 - \tau_i = (\mathbf{e}_1 - \mathbf{e}_i)'\boldsymbol{\tau}$, $2 \le i \le a$. Since $\mathbf{e}_1 - \mathbf{e}_i$ are LIN, then $r(\mathbf{U}) = a - 1$, thus proving property 2. The form of the b.l.u.e. of $\mathbf{c}'\boldsymbol{\tau}$ follows from (10.1.7) and the Gauss–Markov theorem (see Section 4.4). We see that $\text{Cov}(\mathbf{y}_a) = \mathbf{D}_a$, $\text{Cov}(\mathbf{y}_b) = \mathbf{D}_b$, and $\text{Cov}(\mathbf{y}_a, \mathbf{y}_b) = \mathbf{N}$, so that $\text{Cov}(\mathbf{w}) = \text{Cov}(\mathbf{y}_a - \mathbf{N}\mathbf{D}_b^{-1}\mathbf{y}_b) = \sigma^2\mathbf{U}$. If $\mathbf{c}'\boldsymbol{\tau}$ is an estimable contrast, then from property 1, $\mathbf{c}' = \mathbf{u}'\mathbf{U}$ with $\mathbf{u}' = \mathbf{c}'\mathbf{U}^{-}$, and from (10.1.9),

$$\begin{aligned}
\text{Var}(\mathbf{c}'\boldsymbol{\tau}^0) &= \text{Var}(\mathbf{u}'\mathbf{U}\boldsymbol{\tau}^0) = \text{Var}(\mathbf{u}'\mathbf{w}) = \mathbf{u}'\mathbf{U}\mathbf{u}\sigma^2 \\
&= \mathbf{c}'\mathbf{u}\sigma^2 = \mathbf{c}'\mathbf{U}^{-}\mathbf{c}\sigma^2,
\end{aligned}$$

proving property 3. The proof of property 4 is similar. ■

We next discuss sums of squares that enable tests of hypotheses. From $\widehat{Y}_{ijk} = \mu^0 + \tau_i^0 + \theta_j^0$ and the LS solutions given in Result 10.1.4, the model SS is

$$\begin{aligned}
SS &= \sum_{i=1}^{a}\sum_{j=1}^{b}\sum_{k=1}^{n_{ij}} Y_{ijk}(\tau_i^0 + \theta_j^0) \\
&= \sum_{i=1}^{a} Y_{i\cdot\cdot}\tau_i^0 + \sum_{j=1}^{b} Y_{\cdot j\cdot}\left(\frac{Y_{\cdot j\cdot}}{n_{\cdot j}} - \sum_{i=1}^{a}\frac{n_{i'j}\tau_i^0}{n_{\cdot j}}\right) \\
&= \sum_{j=1}^{b}\frac{Y_{\cdot j\cdot}^2}{n_{\cdot j}} + \sum_{i=1}^{a}\tau_i^0 W_i,
\end{aligned} \tag{10.1.10}$$

with $a + b - 1$ d.f. The quantity $\sum_{i=1}^{a}\tau_i^0 W_i$ is called the *SS* due to Factor A, adjusted for Factor B. Also, *SSE* is

$$\begin{aligned}
SSE &= SST_c - SS_A(\text{adj.}) - SS_B(\text{unadj.}) \\
&= \left(\sum_{i=1}^{a}\sum_{j=1}^{b}\sum_{k=1}^{n_{ij}} Y_{ijk}^2 - \frac{Y_{\cdots}^2}{N}\right) - \left(\sum_{j=1}^{b}\frac{Y_{\cdot j\cdot}^2}{n_{\cdot j}} - \frac{Y_{\cdots}^2}{N}\right) - \sum_{i=1}^{a}\tau_i^0 W_i.
\end{aligned}$$

The corresponding sums of squares are shown in Table 10.1.2.

Note that the sum of squares due to Factor B is *unadjusted.* In order to test τ effects, we use Table 10.1.2. However, the same table cannot be used to test hypotheses on θ effects. This situation is peculiar to the unbalanced case. Unlike the balanced case, the unbalanced model is said to be non-orthogonal, i.e., the b.l.u.e.'s of contrasts of τ's and θ's are not orthogonal. Table 10.1.3 shows the form of SS_B(adj.), and two tests of hypotheses are shown in Example 10.1.2 for connected data.

Source	d.f.	SS
Factor A (adj.)	$a-1$	$\sum_{i=1}^a W_i\tau_i^0$
Factor B (unadj.)	$b-1$	$\sum_{j=1}^b Y_{\cdot j\cdot}^2/n_{\cdot j} - Y_{\cdots}/N$
Error	$N-a-b+1$	SSE
Corrected Total	$N-1$	SST_c

TABLE 10.1.2. ANOVA table after eliminating θ effects.

Source	d.f.	SS
Factor A (unadj.)	$a-1$	$\sum_{i=1}^a Y_{i\cdot\cdot}^2/n_{i\cdot} - Y_{\cdots}^2/N$
Factor B (adj.)	$b-1$	$(\sum_{j=1}^b Y_{\cdot j\cdot}/n_{\cdot j} - Y_{\cdots}/N) + \sum_{i=1}^a W_i\tau_i^0$ $-(\sum_{i=1}^a Y_{i\cdot\cdot}^2/n_{i\cdot} - Y_{\cdots}^2/N)$
Error	$N-a-b+1$	SSE
Corrected Total	$N-1$	SST_c

TABLE 10.1.3. ANOVA table after eliminating τ effects.

Example 10.1.2. When the data is connected, the test statistic for the testable hypothesis H_A: $\tau_1 = \cdots = \tau_a$ is

$$F = \frac{\sum_{i=1}^a \tau_i^0 W_i/(a-1)}{SSE/(N-a-b+1)},$$

which has an $F_{a-1,N-a-b+1}$ distribution under H_A. Consider a test of H_B: $\theta_1 = \cdots = \theta_b$. The numerator sum of squares in the F-statistic is

$$\begin{aligned} SS_B(\text{adj.}) &= SST_c - SS_A(\text{unadj.}) - SSE \\ &= \left(\sum_{j=1}^b \frac{Y_{\cdot j\cdot}}{n_{\cdot j}} - \frac{Y_{\cdots}}{N}\right) + \sum_{i=1}^a W_i\tau_i^0 - \left(\sum_{i=1}^a \frac{Y_{i\cdot\cdot}^2}{n_{i\cdot}} - \frac{Y_{\cdots}^2}{N}\right) \end{aligned}$$

with $(b-1)$ d.f. The test statistic is

$$F = \frac{SS_B(\text{adj.})/(b-1)}{SSE/(N-a-b+1)},$$

which follows an $F_{b-1,N-a-b+1}$ distribution under H_B. □

10.1.2.2 Two-factor models with interaction

Consider the model

$$Y_{ijk} = \mu + \tau_i + \theta_j + (\tau\theta)_{ij} + \varepsilon_{ijk}, \quad i = 1, \cdots, a, \ j = 1, \cdots, b, \tag{10.1.11}$$

with $n_{ij} \geq 1$ observations in each of f filled cells ($f \leq ab$); μ is the overall mean effect, τ_i is the effect due to the ith level of Factor A, θ_j is the effect due to the jth level of Factor B, $(\tau\theta)_{ij}$ is the effect due to interaction between the ith level of Factor A and the jth level of Factor B, and ε_{ijk} are i.i.d. $N(0,\sigma^2)$ variables. The number of model parameters is $p = 1+a+b+f$ (the $(\tau\theta)_{ij}$'s correspond to the f filled cells), which we attempt to estimate based on f observed cell means $\overline{Y}_{ij\cdot}$.

We use the matrix notation introduced in Chapter 4 and Chapter 7. It may be verified that $r(\mathbf{X}'\mathbf{X})$ for this model is f. In order to solve the p normal equations in the least squares approach, we set $p-f=1+a+b$ elements of the overall parameter vector equal to zero. It is simplest to set μ^0, τ_i^0, $i=1,\cdots,a$, and θ_j^0, $j=1,\cdots,b$ equal to zero to obtain the reduced normal equations $n_{ij}(\tau\theta)_{ij}^0 = Y_{ij\cdot}$, with solution $(\tau\theta)_{ij}^0 = \overline{Y}_{ij\cdot}$, for the f filled cells. The g-inverse corresponding to this solution is a block diagonal matrix $\mathbf{G} = \text{diag}(\mathbf{O}, \mathbf{D})$, where $\mathbf{D} \in \mathcal{R}^{f\times f}$ consisting of elements $1/n_{ij}$. The proof of the next result is left as Exercise 10.5.

Result 10.1.6. Consider a model with some-cells-empty data. The following functions are estimable:

1. A cell mean $\mu_{ij} = \mu + \tau_i + \theta_j + (\tau\theta)_{ij}$ corresponding to a filled cell, i.e., for $n_{ij} \neq 0$. Its b.l.u.e. is $\widehat{\mu}_{ij} = (\tau\theta)_{ij}^0 = \overline{Y}_{ij\cdot}$, with $\text{Var}(\widehat{\mu}_{ij}) = \sigma^2/n_{ij}$. Further, $\text{Cov}(\widehat{\mu}_{ij}, \widehat{\mu}_{i'j'}) = 0$ if $i \neq i'$ or $j \neq j'$.

2. Any linear function of estimable μ_{ij}'s is estimable. For example, if the $(1,2)$th cell and $(2,2)$th cell are filled, μ_{12} and μ_{22} are estimable, and so is $\mu_{12}-\mu_{22} = \tau_1-\tau_2+((\tau\theta)_{12} - (\tau\theta)_{22})$.

3. For $i \neq i'$, the function $\tau_i - \tau_{i'} + \sum_{j=1}^b c_{ij}(\theta_j + (\tau\theta)_{ij}) - \sum_{j=1}^b c_{i'j}(\theta_j + (\tau\theta)_{i'j})$ is estimable provided $\sum_{j=1}^b c_{ij} = \sum_{j=1}^b c_{i'j} = 1$, and $c_{ij} = 0$ when $n_{ij} = 0$ and $c_{i'j} = 0$ when $n_{i'j} = 0$. Its b.l.u.e. is $\sum_{j=1}^b c_{ij}\overline{Y}_{ij\cdot} - \sum_{j=1}^b c_{i'j}\overline{Y}_{i'j\cdot}$ with variance $\sum_{j=1}^b (c_{ij}^2/n_{ij} + c_{i'j}^2/n_{i'j})\sigma^2$.

4. For $j \neq j'$, the function $\theta_j - \theta_{j'} + \sum_{i=1}^a d_{ij}(\tau_i + (\tau\theta)_{ij}) - \sum_{i=1}^a d_{ij'}(\tau_i + (\tau\theta)_{ij'})$ is estimable provided $\sum_{i=1}^a d_{ij} = \sum_{i=1}^a d_{ij'} = 1$, and $d_{ij} = 0$ when $n_{ij} = 0$ and $d_{ij'} = 0$ when $n_{ij'} = 0$. Its b.l.u.e. is $\sum_{i=1}^a d_{ij}\overline{Y}_{ij\cdot} - \sum_{i=1}^a d_{ij'}\overline{Y}_{ij'\cdot}$ with variance $\sum_{i=1}^a (d_{ij}^2/n_{ij} + d_{ij'}^2/n_{ij'})\sigma^2$.

It is clear that with some-cells-empty data, the pattern of nonzero n_{ij}'s determines which functions are estimable (see Exercise 10.6). Also, differences such as $\tau_i - \tau_{i'}$, $i \neq i'$, or $\theta_j - \theta_{j'}$, $j \neq j'$ are not estimable. For example, $\mu_{ij} - \mu_{i'j} = (\mu+\tau_i+\theta_j+(\tau\theta)_{ij}) - (\mu+\tau_{i'}+\theta_j+(\tau\theta)_{i'j}) = \tau_i - \tau_{i'} + (\tau\theta)_{ij} - (\tau\theta)_{i'j}$, and we cannot eliminate the $(\tau\theta)$'s from the final expression. In other words, no estimable function of μ_{ij}'s is possible which involves only differences such as $\tau_i - \tau_{i'}$ or $\theta_j - \theta_{j'}$, without the $(\tau\theta)_{ij}$'s. Differences between levels of Factor A can be estimated only in the presence of average effects due to Factor B and the interaction. Any testable hypothesis will involve linear functions of μ_{ij}'s.

Result 10.1.7. Consider a model with all-cells-filled data.

1. For all $i \neq i'$, the test statistic for

$$H\colon \tau_i - \tau_{i'} + \frac{1}{b}\left(\sum_{j=1}^b (\tau\theta)_{ij} - \sum_{j=1}^b (\tau\theta)_{i'j}\right) = 0 \tag{10.1.12}$$

is

$$F = \frac{\{\sum_{j=1}^b (\overline{Y}_{ij\cdot} - \overline{Y}_{i'j\cdot})\}^2}{\sum_{j=1}^b (1/n_{ij} + 1/n_{i'j})\widehat{\sigma}^2}, \tag{10.1.13}$$

which has an $F_{1,f-a-b+1}$ distribution under H.

2. The test statistic for

$$H\colon \tau_i + \frac{1}{b}\sum_{j=1}^b (\tau\theta)_{ij} \text{ all equal}, \quad i = 1, \cdots, a \tag{10.1.14}$$

is

$$F = \frac{1}{(a-1)\widehat{\sigma}^2}\left[\sum_{i=1}^{a}(U_i/h_i) - \frac{(\sum_{i=1}^{a} V_i/h_i)^2}{\sum_{i=1}^{a} 1/h_i}\right], \qquad (10.1.15)$$

where $U_i = (\sum_{j=1}^{b} \overline{Y}_{ij\cdot})^2$, $h_i = \sum_{j=1}^{b} 1/n_{ij}$, and $V_i = \sum_{j=1}^{b} \overline{Y}_{ij\cdot}$.

Proof. Property 1 is a special case of property 3 in Result 10.1.6 with $c_{ij} = c_{i'j} = 1/b$. Its b.l.u.e. is $b^{-1}\sum_{j=1}^{b}(\overline{Y}_{ij\cdot} - \overline{Y}_{i'j\cdot})$ with variance $\sigma^2(1/n_{ij} - 1/n_{i'j})/b$, from which (10.1.13) follows directly. The proof of property 2 is left as Exercise 10.9. ■

Observe that (10.1.13) is also the test statistic for H: $\tau_i - \tau_{i'} = 0$, to which the hypothesis (10.1.12) reduces under the restriction $\sum_{j=1}^{b}(\tau\theta)_{ij} = 0$ for all $i = 1, \cdots, a$ in the model. Likewise, the hypothesis (10.1.14) reduces to H: $\tau_1 = \cdots = \tau_a$ under this restriction, and is tested by (10.1.15).

Numerical Example 10.2. The dataset "foster" is available in the *HSAUR* package and describes a foster feeding experiment with rat mothers and litters of four different genotypes. The response is the litter weight after a trial feeding period. The two factor variables are litgen (genotype of the litter), with four levels (A, B, I, and J) and motgen (genotype of the mother), with four levels (A, B, I, and J).

```
library(HSAUR)
data(foster)
fit <- lm(weight ~ litgen * motgen, foster)
fit$coefficients
```

```
      (Intercept)           litgenB           litgenI
        63.680000        -11.355000        -16.580000
          litgenJ           motgenB           motgenI
        -9.330000        -11.280000         -9.555000
          motgenJ litgenB:motgenB litgenI:motgenB
       -14.720000         19.595000         28.546667
litgenJ:motgenB litgenB:motgenI litgenI:motgenI
        13.030000         11.155000         14.055000
litgenJ:motgenI litgenB:motgenJ litgenI:motgenJ
         9.738333          8.295000         17.053333
litgenJ:motgenJ
         9.430000
```

```
car::Anova(fit, type = "III") # Type III SS
```

```
Anova Table (Type III tests)

Response: weight
               Sum Sq Df  F value  Pr(>F)
(Intercept)   20275.7  1 373.8122 < 2e-16 ***
litgen          591.7  3   3.6362 0.01968 *
motgen          582.3  3   3.5782 0.02099 *
litgen:motgen   824.1  9   1.6881 0.12005
Residuals      2440.8 45
---
```

```
Signif. codes:
0 '***' 0.001 '**' 0.01 '*' 0.05 '.' 0.1 ' ' 1
```

We leave it to the reader to fit an additive model to this data and compare the two models in order to appreciate the differences between them. ▲

10.1.2.3 Nested or hierarchical models

We introduced this model in Example 4.2.7, while in Example 7.3.6, we described inference in the balanced case. With unbalanced data, the nested model is written as

$$Y_{ijk} = \mu + \tau_i + \theta_{j(i)} + \varepsilon_{ijk}, \quad 1 \le k \le n_{ij}, 1 \le j \le b_i, 1 \le i \le a. \tag{10.1.16}$$

The total number of parameters is $p = 1 + a + \sum_{i=1}^{a} b_i$. The normal equations are derived by minimizing $\sum_{i=1}^{a}\sum_{j=1}^{b_i}\sum_{k=1}^{n_{ij}}(Y_{ijk} - \mu^0 - \tau_i^0 - \theta_{j(i)}^0)^2$ with respect to the parameters as

$$Y_{\cdots} = N\mu^0 + \sum_{i=1}^{a} n_{i\cdot}\tau_i^0 + \sum_{i=1}^{a}\sum_{j=1}^{b_i} n_{ij}\theta_{j(i)}^0$$

$$Y_{i\cdot\cdot} = n_{i\cdot}\mu^0 + n_{i\cdot}\tau_i^0 + \sum_{j=1}^{b_i} n_{ij}\theta_{j(i)}^0, \quad i = 1, \cdots, a$$

$$Y_{ij\cdot} = n_{ij}(\mu^0 + \tau_i^0\theta_{j(i)}^0), \quad j = 1, \cdots, b_i, i = 1, \cdots, a,$$

where $n_{i\cdot} = \sum_{j=1}^{b_i} n_{ij}$. By adding the last set of equations over j, we get the second set, and by adding them over i and j, we get the first equation. Only $\sum_{i=1}^{a} b_i$ of the p normal equations are LIN. By specifying any values of μ^0 and $\tau_1^0, \cdots, \tau_a^0$, we get one solution. Thus, one set of solutions to the normal equations subject to these constraints are

$$\mu^0 = 0, \ \tau_i^0 = 0, \ \theta_{j(i)}^0 = Y_{ij\cdot}/n_{ij} = \overline{Y}_{ij\cdot}, \quad j = 1, \cdots, b_i, \ i = 1, \cdots, a.$$

The corresponding $p \times p$ g-inverse is

$$\mathbf{G} = \begin{pmatrix} \mathbf{O} & \mathbf{O} \\ \mathbf{O} & \mathbf{D} \end{pmatrix},$$

where $\mathbf{D} = \text{diag}(n_{11}, \cdots, n_{1b_1}, \cdots, n_{a1}, \cdots, n_{ab_a})$. It is routine to verify that $\mu + \tau_i + \theta_{j(i)}$ is estimable with b.l.u.e. $\overline{Y}_{ij\cdot}$. Functions of the form $\theta_{j(i)} - \theta_{j'(i)}$, $j \ne j'$ are also estimable, with b.l.u.e. $\overline{Y}_{ij\cdot} - \overline{Y}_{ij'\cdot}$. In the framework of a nested sequence of hypotheses, we write

$$\begin{aligned} \mathcal{S}_{H_0} = \{Y_{ijk} = \mu + \tau_i + \theta_{j(i)} + \varepsilon_{ijk}\} &\supset \mathcal{S}_{H_1} = \{Y_{ijk} = \mu + \tau_i + \varepsilon_{ijk}\} \\ &\supset \mathcal{S}_{H_2} = \{Y_{ijk} = \mu + \varepsilon_{ijk}\} \\ &\supset \mathcal{S}_{H_3} = \{Y_{ijk} = \varepsilon_{ijk}\}. \end{aligned}$$

Note that each model may be expressed in different ways, with or without model constraints. Here, $\dim(\mathcal{S}_{H_0}) = \sum_{i=1}^{a} b_i$, $\dim(\mathcal{S}_{H_1}) = a$, $\dim(\mathcal{S}_{H_2}) = 1$, and $\dim(\mathcal{S}_{H_3}) = 0$. The fitted values are $\widehat{Y}_{ijk} = \overline{Y}_{ij\cdot}$, and $SSR(H_0) = \widehat{\mathbf{y}}_{H_0}'\widehat{\mathbf{y}}_{H_0} = \sum_i\sum_j Y_{ij\cdot}^2/n_{ij}$, $SST = \sum_i\sum_j\sum_k(Y_{ijk} - \overline{Y}_{\cdots})^2$. Let $SSM = N\overline{Y}_{\cdots}^2$. Table 10.1.4 summarizes the sums of squares.

10.2 Nonparametric procedures

In this section, we give a brief description of two nonparametric procedures for carrying out inference in one-way and two-way ANOVA models. The Kruskal–Wallis (KW) method and

Source	d.f.	SS
Model	$\sum_{i=1}^{a} b_i$	$SSR(H_0) - SSM$
Error	$N - b.$	SSE
Corrected Total	$N - 1$	SST_c

TABLE 10.1.4. ANOVA table for the nested effects model.

Friedman's procedure employ the classical tests with rank transformations, which consists of replacing all the observations in a classical test statistic by their ranks in the entire dataset (Conover and Iman, 1981).

Kruskal–Wallis procedure

In situations where the normality assumption is not justified, the Kruskal–Wallis test (Kruskal and Wallis, 1952) is used as an alternative to the F-test in the one-factor model. Suppose independent random samples of sizes $n_1, \cdots, n_a$ are drawn from a univariate populations with unknown c.d.f.'s F_i, $i = 1, \cdots, a$. We describe a test of the null hypothesis H: $F_1 = \cdots = F_a$ versus alternatives of the form $F_i(x) = F(x - \theta_i)$, for all x, $i = 1, \cdots, a$, with all θ_i's not equal.

The observations Y_{ij}, $j = 1, \cdots, n_i$, $i = 1, \cdots, a$ are first ranked in ascending order, and Y_{ij} is replaced by its rank R_{ij} in the overall sample. If there are ties, i.e., two or more observations are equal, each of the tied observations is given the average of the ranks for which it is tied. Suppose there are G groups of ties. Let t_g be the number of tied observations in group g, $g = 1, \cdots, G$. Let $R_{i\cdot} = \sum_{j=1}^{n_i} R_{ij}$ denote the sum of the ranks in the ith group, and $\overline{R}_{i\cdot} = R_{i\cdot}/n_i$. Clearly, $\sum_{i=1}^{a} R_{i\cdot} = N(N+1)/2$.

Result 10.2.1. The Kruskal–Wallis test statistic is

$$\text{KW} = \frac{1}{A}\left[\frac{12}{N(N+1)}\sum_{i=1}^{a}\frac{R_{i\cdot}^2}{n_i} - 3(N+1)\right], \tag{10.2.1}$$

where $N = \sum_{i=1}^{a} n_i$, and

$$A = 1 - \frac{1}{N(N^2-1)}\sum_{g=1}^{G}(t_g - 1)t_g(t_g + 1). \tag{10.2.2}$$

If the a populations are identical, and n_i's are not too small, KW approximately has a χ^2_{a-1} distribution. With no ties, we set $A = 1$ in (10.2.1). If KW $> \chi^2_{a-1,\alpha}$, we reject the null hypothesis at level of significance α.

Proof. The N-dimensional vector $(R_{11}, \cdots, R_{1n_1}, \cdots, R_{a1}, \cdots, R_{an_a})$ takes as values the $N!$ permutations of $(1, \cdots, N)$ with equal probability under H_0. The test statistic corresponding to the standard analysis of variance (see Example 7.2.7) based on the ranks R_{ij}'s has the form

$$\sum_{i=1}^{a} n_i[\overline{R}_{i\cdot} - (N+1)/2]^2 \Big/ \sum_{i=1}^{a}\sum_{j=1}^{n_i}(R_{ij} - \overline{R}_{i\cdot})^2,$$

and is a monotonically increasing function of

$$\sum_{i=1}^{a} n_i[\overline{R}_{i\cdot} - (N+1)/2]^2 \Big/ \sum_{i=1}^{a}\sum_{j=1}^{n_i}(R_{ij} - (N+1)/2)^2. \tag{10.2.3}$$

The denominator in (10.2.3) is a constant because of the use of ranks, and let B denote its numerator. First, we divide B by $(N^2 - 1)/12$, which is the variance of the uniform distribution on the integers $\{1, 2, \cdots, N\}$, and we multiply the resultant expression by the factor $(N - 1)/N$ (to yield $a - 1$ in expectation under H). An algebraic simplification yields the expression in the numerator of (10.2.1). The term A in the denominator is an adjustment for ties. For a proof of the approximate chi-square distribution of KW, see Kruskal (1952). ■

Conover (1998) showed that the F-test applied to ranks R_{ij} yields the statistic

$$F_{\text{KW}} = \frac{\text{KW}/(a-1)}{(N-1-\text{KW})/(N-a)}, \tag{10.2.4}$$

which is a monotonic function of KW, leading to the statement that the Kruskal–Wallis test is equivalent to applying the classical ANOVA procedure to ranks.

Numerical Example 10.3. The dataset "airquality" in the R package *datasets* has 153 observations on 6 variables to describe the daily air quality measurements in New York, May to September 1973. The data were obtained from the New York State Department of Conservation (ozone data) and the National Weather Service (meteorological data). We model Ozone (ppb) as a function of the factor variable Month (with 12 levels).

```
data(airquality)
kruskal.test(Ozone ~ Month, data = airquality)

Kruskal-Wallis rank sum test

data:  Ozone by Month
Kruskal-Wallis chi-squared = 29.267, df = 4, p-value= 6.901e-06
```

The p-value is very small, indicating that there is significant differences in the effect of the different months of the year on the ozone level. ▲

Friedman's procedure

This procedure relates to the two-way classification. Consider the two-factor fixed-effects additive model $Y_{ij} = \mu + \tau_i + \theta_j + \varepsilon_{ij}$, $i = 1, \cdots, a$, $j = 1, \cdots, b$. That is, suppose we have balanced data with one observation in each of the ab cells, and $N = ab$. Assume that the errors ε_{ij} are independently distributed random variables from some continuous population. Suppose we wish to test H: $\tau_1 = \cdots = \tau_a$. Within each block, rank the a observations in ascending order. Let $R_{ij} = \text{r}(Y_{ij})$. Set $R_i = \sum_{j=1}^{n} R_{ij}$, $R_{i\cdot} = R_i/n$, and $R_{\cdot\cdot} = (a+1)/2$. The test statistic is

$$S = \frac{12n}{a(a+1)}\sum_{i=1}^{a}(R_{i\cdot} - R_{\cdot\cdot})^2 \tag{10.2.5}$$

$$= \frac{12}{na(a+1)}\sum_{i=1}^{a} R_i^2 - 3n(a+1). \tag{10.2.6}$$

As $n \to \infty$, we can show that the statistic S has a limiting χ^2_{a-1} distribution, so that critical values from this distribution determine the rejection region. This test arises naturally if we apply the usual F-statistic in the two-factor additive model to ranks instead of the actual responses. For details, see Conover (1998) and Hollander et al. (2014).

We do not discuss nonparametric tests for other designs here. The reader is referred to Akritas and Arnold (1994) and references therein for a discussion of fully nonparametric hypotheses for general factorial designs.

10.3 Analysis of covariance

Consider an example where we study the effect of three different diets on the weights of subjects in an obesity study. The experiment consists of assigning subjects randomly to the diets and maintaining them on the same diet for a period of three months. The response variable is the weight of a subject at the end of the three-month period. A one-way ANOVA model enables us to study the effect of the treatment (diet) on the weight. Note that it is also reasonable to suppose that the weight of a subject after being on a diet for three months is correlated with the subject's initial weight before the diet started. An analysis of covariance (ANACOVA) model is a combination of ANOVA and regression, which enables us to study the effect of diet on the final weight of the subjects, adjusting for their initial weight. Supposing that the initial weight is linearly related to the final weight, the elimination of this linear effect should result in a smaller MSE of fit. We call the initial weight a covariate or a concomitant variable. Note that as an alternative, we could have blocked the subjects into different groups based on their initial weight and used a two-way ANOVA model, thereby converting a continuous variable (weight) into a class variable. However, use of the ANACOVA model improves the precision of treatment comparisons. In general, we may use k continuous covariates $Z_1, \cdots, Z_k$. Analysis of covariance tests for differences in treatment effects, assuming a constant regression relation among groups. Of course, we must first test whether or not the regression coefficients are similar in the different groups. Regression lines will be parallel when there is a single covariate Z, and there is no interaction between Z and the factor levels. Dissimilar coefficients could reflect the presence of an interaction between treatment groups and covariates.

A general formulation of the ANACOVA model is

$$\mathbf{y} = \mathbf{X}\boldsymbol{\tau} + \mathbf{Z}\boldsymbol{\beta} + \boldsymbol{\varepsilon}, \tag{10.3.1}$$

where $\mathbf{y} \in \mathcal{R}^N$, $\mathbf{X} \in \mathcal{R}^{N\times p}$ is a design matrix with $\mathrm{r}(\mathbf{X}) = r < p$, $\boldsymbol{\tau} \in \mathcal{R}^p$ is a vector of fixed-effects parameters, $\mathbf{Z} \in \mathcal{R}^{N\times q}$ is a regression matrix with $\mathrm{r}(\mathbf{Z}) = q$, $\boldsymbol{\beta} \in \mathcal{R}^q$ is a vector of regression parameters, the columns of $\mathbf{X}$ are linearly independent of the columns of $\mathbf{Z}$, and $\boldsymbol{\varepsilon}$ has an N-variate normal distribution with mean vector $\mathbf{0}$ and covariance matrix $\sigma^2\mathbf{I}_N$. We can rewrite the model in (10.3.1) as

$$\mathbf{y} = \mathbf{W}\boldsymbol{\gamma} + \boldsymbol{\varepsilon}, \quad \text{where } \mathbf{W} = (\mathbf{X},\ \mathbf{Z}) \text{ and } \boldsymbol{\gamma} = \begin{pmatrix}\boldsymbol{\tau}\\ \boldsymbol{\beta}\end{pmatrix}. \tag{10.3.2}$$

Result 10.3.1. Let $\mathbf{P}$ be the matrix of orthogonal projection onto $\mathcal{C}(\mathbf{X})$. The LS solutions for $\boldsymbol{\beta}$ and $\boldsymbol{\tau}$ are

$$\widehat{\boldsymbol{\beta}} = [\mathbf{Z}'(\mathbf{I}-\mathbf{P})\mathbf{Z}]^{-1}\mathbf{Z}'(\mathbf{I}-\mathbf{P})\mathbf{y} \quad \text{and} \tag{10.3.3}$$

$$\boldsymbol{\tau}^0 = (\mathbf{X}'\mathbf{X})^{-}\mathbf{X}'\mathbf{y} - (\mathbf{X}'\mathbf{X})^{-}\mathbf{X}'\mathbf{Z}\widehat{\boldsymbol{\beta}}. \tag{10.3.4}$$

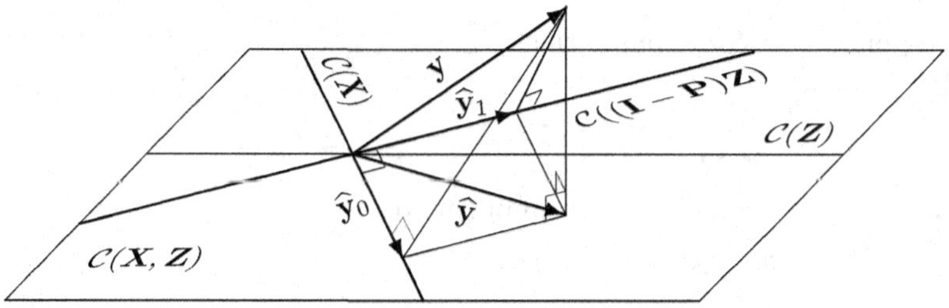

FIGURE 10.3.1. Geometry of ANACOVA.

Proof. Although the result follows immediately from Result 4.2.5, we show some details. The normal equations have the form

$$\begin{pmatrix} \mathbf{X}'\mathbf{X} & \mathbf{X}'\mathbf{Z} \\ \mathbf{Z}'\mathbf{X} & \mathbf{Z}'\mathbf{Z} \end{pmatrix} \begin{pmatrix} \boldsymbol{\tau}^0 \\ \boldsymbol{\beta}^0 \end{pmatrix} = \begin{pmatrix} \mathbf{X}'\mathbf{y} \\ \mathbf{Z}'\mathbf{y} \end{pmatrix}. \tag{10.3.5}$$

Write the normal equations as

$$\mathbf{X}'\mathbf{X}\boldsymbol{\tau}^0 + \mathbf{X}'\mathbf{Z}\boldsymbol{\beta}^0 = \mathbf{X}'\mathbf{y} \tag{10.3.6}$$

$$\mathbf{Z}'\mathbf{X}\boldsymbol{\tau}^0 + \mathbf{Z}'\mathbf{Z}\boldsymbol{\beta}^0 = \mathbf{Z}'\mathbf{y}. \tag{10.3.7}$$

Let $\mathbf{Z}_0 = \mathbf{P}\mathbf{Z}$ and $\mathbf{Z}_1 = (\mathbf{I}-\mathbf{P})\mathbf{Z}$. Then, $\mathbf{Z}_1'\mathbf{Z}_0 = \mathbf{O}$ and $\mathbf{Z}_1'\mathbf{X} = \mathbf{O}$. Writing $\mathbf{Z} = \mathbf{Z}_0 + \mathbf{Z}_1$ in (10.3.6) and (10.3.7) and moving terms around,

$$\begin{aligned} &\mathbf{X}'(\mathbf{X}\boldsymbol{\tau}^0 + \mathbf{Z}_0\boldsymbol{\beta}^0 - \mathbf{y}) = \mathbf{0}, \\ &\mathbf{Z}_0'(\mathbf{X}\boldsymbol{\tau}^0 + \mathbf{Z}_0\boldsymbol{\beta}^0 - \mathbf{y}) = \mathbf{Z}_1'(\mathbf{y} - \mathbf{Z}_1\boldsymbol{\beta}^0). \end{aligned} \tag{10.3.8}$$

Since $\mathbf{Z}_0 = \mathbf{X}\mathbf{T}$ for some $\mathbf{T}$, the two equations lead to $\mathbf{Z}_1'\mathbf{Z}_1\boldsymbol{\beta}^0 = \mathbf{Z}_1'\mathbf{y}$. Because $\mathbf{Z}$ has full column rank and $\mathcal{C}(\mathbf{Z}) \cap \mathcal{C}(\mathbf{X}) = \{\mathbf{0}\}$, $\mathbf{Z}_1$ has full column rank. Then the LS solution for $\boldsymbol{\beta}$ is unique, so it is written as $\widehat{\boldsymbol{\beta}}$ by convention; (10.3.3) follows. Next, from the projection perspective, the system of equations (10.3.5) is consistent. Then, plugging (10.3.3) into (10.3.6), the resulting system of equations in $\boldsymbol{\tau}^0$, which is $\mathbf{X}'\mathbf{X}\boldsymbol{\tau}^0 = \mathbf{X}'(\mathbf{y}-\mathbf{Z}\widehat{\boldsymbol{\beta}})$, is consistent, which yields (10.3.4). ■

Note that although $(\mathbf{X}, \mathbf{Z})$ is not of full rank, we still have a unique estimator $\widehat{\boldsymbol{\beta}}$ of $\boldsymbol{\beta}$. In contrast, there are infinitely many LS solutions $\boldsymbol{\tau}^0$. Geometrically, the projection of $\mathbf{y}$ onto $\mathcal{C}(\mathbf{X}, \mathbf{Z})$ is the sum of its projections onto two orthogonal spaces, one being $\mathcal{C}(\mathbf{X})$, and the other, the orthogonal complement of $\mathcal{C}(\mathbf{X})$ within $\mathcal{C}(\mathbf{X}, \mathbf{Z})$, i.e., $\mathcal{C}((\mathbf{I}-\mathbf{P})\mathbf{Z})$, where $\mathbf{P}$ is the projection onto $\mathcal{C}(\mathbf{X})$ (see Exercise 10.17). The projections are displayed as $\widehat{\mathbf{y}}$, $\widehat{\mathbf{y}}_0$, and $\widehat{\mathbf{y}}_1$ in Figure 10.3.1. Then, from $\widehat{\mathbf{y}} = \mathbf{X}\boldsymbol{\tau}^0 + \mathbf{Z}\widehat{\boldsymbol{\beta}}$, it follows that $\widehat{\mathbf{y}}_0 = \mathbf{P}\widehat{\mathbf{y}} = \mathbf{X}\boldsymbol{\tau}^0 + \mathbf{P}\mathbf{Z}\widehat{\boldsymbol{\beta}}$ and $\widehat{\mathbf{y}}_1 = (\mathbf{I}-\mathbf{P})\widehat{\mathbf{y}} = (\mathbf{I}-\mathbf{P})\mathbf{Z}\widehat{\boldsymbol{\beta}}$.

We next describe inference using the framework of a nested sequence of hypotheses (see Section 7.3.2). Write the ANACOVA model (10.3.1) coordinate-wise as

$$Y_i = \mu + \beta_1 Z_{i1} + \cdots + \beta_k Z_{ik} + \tau_1 X_{i1} + \cdots + \tau_p X_{ip} + \varepsilon_i, \tag{10.3.9}$$

so that $q = k+1$, $\boldsymbol{\beta} = (\mu, \beta_1, \cdots, \beta_k)'$, and $\mathbf{z}_i = (1, Z_{i1}, \cdots, Z_{ik})'$. We can see two ways to look at nested sequence of hypotheses.

Include $\boldsymbol{\beta}$ effects before $\boldsymbol{\tau}$ effects in a model with $\boldsymbol{\mu}$. Starting with the full model

(10.3.9) corresponding to $\mathcal{S}_{H_0}$, we see that

$$\begin{aligned}\mathcal{S}_{H_0} \supset \mathcal{S}_{H_1} &= \{Y_i = \mu + \beta_1 Z_{i1} + \cdots + \beta_k Z_{ik} + \varepsilon_i\} \\ \supset \mathcal{S}_{H_2} &= \{Y_i = \mu + \varepsilon_i\} \supset \mathcal{S}_{H_3} = \{Y_i = \varepsilon_i\},\end{aligned}$$

we get the decomposition of the total sum of squares as

$$SST = SS(\mu) + SS(\boldsymbol{\beta} \mid \mu) + SS(\boldsymbol{\tau} \mid \boldsymbol{\beta}, \mu) + SSE, \tag{10.3.10}$$

where

$SST = \mathbf{y}'\mathbf{y}$ with N d.f.

$SS(\mu) = Q(H_2 \mid H_3) = N\overline{Y}^2$ with 1 d.f.

$SS(\boldsymbol{\beta} \mid \mu) = Q(H_1 \mid H_2)$ with $k = q - 1$ d.f.

$SS(\boldsymbol{\tau} \mid \boldsymbol{\beta}, \mu) = Q(H_0 \mid H_1)$ with r d.f.

$SSE = \mathbf{y}'\mathbf{y} - SS(\mu) - SS(\boldsymbol{\beta} \mid \mu) - SS(\boldsymbol{\tau} \mid \boldsymbol{\beta}, \mu)$, with $N - r - q$ d.f.

Include $\boldsymbol{\tau}$ effects before $\boldsymbol{\beta}$ effects in a model with $\boldsymbol{\mu}$. Starting again with the full model (10.3.9),

$$\begin{aligned}\mathcal{S}_{H_0} \supset \mathcal{S}_{H_1} &= \{Y_i = \mu + \tau_1 X_{i1} + \cdots + \tau_p X_{ip} + \varepsilon_i\} \\ \supset \mathcal{S}_{H_2} &= \{Y_i = \mu + \varepsilon_i\} \supset \mathcal{S}_{H_3} = \{Y_i = \varepsilon_i\},\end{aligned}$$

we get the decomposition of the total sum of squares as

$$SST = SS(\mu) + SS(\boldsymbol{\tau} \mid \mu) + SS(\boldsymbol{\beta} \mid \boldsymbol{\tau}, \mu) + SSE, \tag{10.3.11}$$

where

$SST = \mathbf{y}'\mathbf{y}$ with N d.f.

$SS(\mu) = Q(H_2 \mid H_3) = N\overline{Y}^2$ with 1 d.f.

$SS(\boldsymbol{\tau} \mid \mu) = Q(H_1 \mid H_2)$ with r d.f.

$SS(\boldsymbol{\beta} \mid \boldsymbol{\tau}, \mu) = Q(H_0 \mid H_1)$ with $k = q - 1$ d.f.

$SSE = \mathbf{y}'\mathbf{y} - SS(\mu) - SS(\boldsymbol{\tau} \mid \mu) - SS(\boldsymbol{\beta} \mid \boldsymbol{\tau}, \mu)$, with $N - r - q$ d.f.

Since this decomposition accounts for the effect of the X on Y after adjusting for that of the Z's in the full model (10.3.9) , it may be termed as the "ANACOVA decomposition".

In the next result on ANACOVA, while the null is only about $\boldsymbol{\tau}$, the test statistic has been adjusted for the continuous covariates.

Result 10.3.2. ANACOVA. Suppose ε_i are i.i.d. $\sim N(0, \sigma^2)$. For the model (10.3.1), provided the null hypothesis H_0: $\mathbf{K}'\boldsymbol{\tau} = \mathbf{m}$ is testable, the F-statistic is

$$F = \frac{Q/\,\mathrm{r}(\mathbf{K})}{SSE/(N - r - q)},$$

which has an $F_{\mathrm{r}(\mathbf{K}), N-r-q}$ distribution under H_0, where

$$Q = (\mathbf{K}'\boldsymbol{\tau}^0 - \mathbf{m})'(\mathbf{K}'\mathbf{G}_{11}\mathbf{K})^{-1}(\mathbf{K}'\boldsymbol{\tau}^0 - \mathbf{m})$$

with $\mathbf{G}_{11}$ being the submatrix of

$$\mathbf{G} = \begin{pmatrix} \mathbf{X}'\mathbf{X} & \mathbf{X}'\mathbf{Z} \\ \mathbf{Z}'\mathbf{X} & \mathbf{Z}'\mathbf{Z} \end{pmatrix}^{-} = \begin{pmatrix} \mathbf{G}_{11} & \mathbf{G}_{12} \\ \mathbf{G}_{21} & \mathbf{G}_{22} \end{pmatrix}.$$

Proof. The null hypothesis H_0 can be written as $\mathbf{C}'\boldsymbol{\gamma}$, where $\boldsymbol{\gamma}$ was shown in (10.3.2), and $\mathbf{C} = (\mathbf{K}', \mathbf{O})'$. The proof is a direct consequence of Result 7.2.1. ■

We also remark that to test H_0: $\boldsymbol{\beta} = \mathbf{0}$, we use the test statistic

$$F = \frac{SS(\boldsymbol{\beta} \,|\, \boldsymbol{\tau}, \mu)/q}{SSE/(N - r - q)},$$

where $SS(\boldsymbol{\beta} \,|\, \boldsymbol{\tau}, \mu)$ is the SS accounted for by $\mathbf{Z}$ after adjusting for the effect of $\mathbf{X}$. From Result 10.3.1, $SS(\boldsymbol{\beta} \,|\, \boldsymbol{\tau}, \mu) = \|(\mathbf{I} - \mathbf{P})\mathbf{Z}\widehat{\boldsymbol{\beta}}\|^2$. In Figure 10.3.1, it is illustrated as $\|\widehat{\mathbf{y}}_1\|^2$. Under H_0, F has an $F_{q,N-r-q}$ distribution.

Example 10.3.1. One-factor model with one covariate. We discuss the statistical analysis for an ANACOVA model involving one factor, Factor A, and one continuous covariate Z. Given data on the response and covariate, (Y_{ij}, Z_{ij}), an experimenter may first do a preliminary analysis to verify *equality of the slopes* of the regression of Y on Z in the a groups. If this is indeed so, the ANACOVA model with equal slopes in the a groups is

$$Y_{ij} = \mu + \tau_i + \beta(Z_{ij} - \overline{Z}_{\cdot\cdot}) + \varepsilon_{ij}, \quad j = 1, \cdots, n, i = 1, \cdots, a, \tag{10.3.12}$$

where β is the common slope parameter and ε_{ij} are i.i.d. $N(0, \sigma^2)$ variables.

Let τ_i correspond to fixed effects, with $\sum_{i=1}^{a} \tau_i = 0$. Assume that Z_{ij} are unaffected by the treatments. Our main interest is in testing for equality of the effects due to the levels of Factor A (i.e., $\tau_1 = \cdots, \tau_a$), after eliminating the effect of the continuous covariate Z on the response. Before we do this, we would like to test whether $\beta = 0$.

(i) When $\beta = 0$, the reduction in MSE due to the effect of Z on Y is likely to be very small, and in this case, the data is better analyzed as the simpler ANOVA model

$$Y_{ij} = \mu + \tau_i + \varepsilon_{ij}, \quad j = 1, \cdots, n, i = 1, \cdots, a.$$

(ii) If β is significantly different from zero, we would use each of the a regression lines given by (10.3.12) to estimate the mean value of Y for a given treatment level and for a given value of Z and also to estimate differences among the τ_i for any value of the covariate.

The following notation on sums of squares is useful.

$$S_{YY} = \sum_{i=1}^{a}\sum_{j=1}^{n}(Y_{ij} - \overline{Y}_{\cdot\cdot})^2, \quad S_{ZZ} = \sum_{i=1}^{a}\sum_{j=1}^{n}(Z_{ij} - \overline{Z}_{\cdot\cdot})^2,$$

$$S_{ZY} = \sum_{i=1}^{a}\sum_{j=1}^{n}(Z_{ij} - \overline{Z}_{\cdot\cdot})(Y_{ij} - \overline{Y}_{\cdot\cdot}),$$

$$T_{YY} = \sum_{i=1}^{a}(\overline{Y}_{i\cdot} - \overline{Y}_{\cdot\cdot})^2, \quad T_{ZZ} = \sum_{i=1}^{a}(\overline{Z}_{i\cdot} - \overline{Z}_{\cdot\cdot})^2,$$

$$T_{ZY} = \sum_{i=1}^{a}(\overline{Z}_{i\cdot} - \overline{Z}_{\cdot\cdot})(\overline{Y}_{i\cdot} - \overline{Y}_{\cdot\cdot}),$$

$$E_{YY} = S_{YY} - T_{YY} = \sum_{i=1}^{a}\sum_{j=1}^{n}(Y_{ij} - \overline{Y}_{i\cdot})^2,$$

$$E_{ZZ} = S_{ZZ} - T_{ZZ} = \sum_{i=1}^{a}\sum_{j=1}^{n}(Z_{ij} - \overline{Z}_{i\cdot})^2, \text{ and}$$

$$E_{ZY} = S_{ZY} - T_{ZY} = \sum_{i=1}^{a}\sum_{j=1}^{n}(Z_{ij} - \overline{Z}_{i\cdot})(Y_{ij} - \overline{Y}_{i\cdot}).$$

In the absence of the covariate, we would have $S_{ZY} = S_{ZZ} = T_{ZY} = T_{ZZ} = E_{ZY} = E_{ZZ} = 0$. In this case, the total sum of squares, treatment sum of squares and error sum of squares would respectively be S_{YY}, T_{YY}, and E_{YY}. In the presence of a covariate, however, we must adjust these quantities for the regression of Y on Z.

The least squares solutions of the parameters in the *full model* (10.3.12) are

$$\mu^0 = \overline{Y}_{\cdot\cdot}, \quad \tau_i^0 = (\overline{Y}_{i\cdot} - \overline{Y}_{\cdot\cdot}) - \widehat{\beta}(\overline{Z}_{i\cdot} - \overline{Z}_{\cdot\cdot}), \quad i = 1, \cdots, a,$$

$$\widehat{\beta} = \frac{E_{ZY}}{E_{ZZ}}, \quad \text{and} \quad \widehat{\sigma}^2 = MSE = \frac{SSE}{a(n-1)-1},$$

where

$$SSE = E_{YY} - (E_{ZY})^2/E_{ZZ}$$

is the "adjusted" error sum of squares with $a(n-1)-1$ d.f. We can reduce the full model in two different ways as we discuss below.

Case 1. In the full model (10.3.12), if there were no effects due to the levels of Factor A, i.e., we reject the null hypothesis H: $\tau_1 = \cdots = \tau_a$, then the *reduced model* is

$$Y_{ij} = \mu + \beta(Z_{ij} - \overline{Z}_{\cdot\cdot}) + \varepsilon_{ij}, \quad i = 1, \cdots, a, \ j = 1, \cdots, n,$$

and the least squares estimates of the parameters in this *reduced model* are

$$\widehat{\mu} = \overline{Y}_{\cdot\cdot}, \text{ and } \widehat{\beta} = S_{ZY}/S_{ZZ}.$$

Under this reduced model, the a parallel regression lines coincide. The error sum of squares in this reduced model is

$$SSE' = S_{YY} - (S_{ZY})^2/S_{ZZ} \text{ with } (an-2) \text{ d.f.}$$

Note that $(S_{ZY})^2/S_{ZZ}$ is the reduction in S_{YY} through the regression of Y on Z, and is the regression sum of squares with 1 d.f. The difference $SSE' - SSE$ is a reduction in sum of squares due to τ effects, and has $(a-1)$ d.f.

Case 2. Another way to reduce the full model model (10.3.12) is

$$Y_{ij} = \mu + \tau_i + \varepsilon_{ij}, \quad i = 1, \cdots, a, \ j = 1, \cdots, n.$$

The least squares estimates of the "adjusted treatment means" $\mu + \tau_i$ under this model are

$$\overline{Y}_{i\cdot}(\text{adj.}) = \overline{Y}_{i\cdot} - \widehat{\beta}(\overline{Z}_{i\cdot} - \overline{Z}_{\cdot\cdot}), \quad i = 1, \cdots, a,$$

where $\widehat{\beta} = E_{ZY}/E_{ZZ}$. The standard error of this adjusted treatment mean is

$$\text{s.e.}(\overline{Y}_{i\cdot}(\text{adj.})) = \left\{MSE\left[\frac{1}{n} + \frac{(\overline{Z}_{i\cdot} - \overline{Z}_{\cdot\cdot})^2}{E_{ZZ}}\right]\right\}^{1/2}.$$

The test statistic for H_0: τ_i's equal, $i = 1, \cdots, a$ is

$$F_0 = \frac{(SSE' - SSE)/(a-1)}{SSE/[a(n-1)-1]} \sim F_{(a-1),a(n-1)-1}$$

under H_0. We can represent the ANACOVA as an *adjusted ANOVA*, as shown in Table 10.3.1. The F-statistic is

$$F = \frac{(SSE' - SSE)/(a-1)}{MSE}. \qquad \square$$

Source	d.f.	SS	MS
Regression	1	$(S_{ZY})^2/S_{ZZ}$	$(S_{ZY})^2/S_{ZZ}$
Treatment	$a-1$	$SSE' - SSE$	$[SSE' - SSE]/(a-1)$
Error	$a(n-1)-1$	SSE	$SSE/[a(n-1)-1]$
Total	$an-1$	S_{YY}	

TABLE 10.3.1. Adjusted ANOVA table for an ANACOVA model.

Numerical Example 10.4. Analysis of covariance. The dataset "anxiety" in the R package *datarium* provides a response t_3, an anxiety score for groups of individuals practicing physical exercises at three different levels (grp1: basal, grp2: moderate, and grp3: high). A pretest measure t_1 is also measured.

```
library(datarium)
data("anxiety", package = "datarium")
anxiety$group <-
  factor(anxiety$group, levels = c("grp1", "grp2", "grp3"))
ancova <-
  lm(t3 ~ t1 + group,
     data = anxiety,
     contrasts = list(group = "contr.SAS"))
summary(ancova)$coefficients   # Least Squares Coefficients

             Estimate Std. Error    t value     Pr(>|t|)
(Intercept) -3.227771 0.84409252 -3.823954 4.390914e-04
t1           0.986742 0.04907462 20.106971 7.546396e-23
groupgrp1    2.874306 0.17551282 16.376613 1.363352e-19
groupgrp2    2.328472 0.17639610 13.200247 2.362207e-16
```

Two ANOVA tables are shown below. The first corresponds to the unadjusted sums of squares, followed by the adjusted sums of squares which corresponds to the Type III sums of squares from the Anova() function in the *car* package.

```
anova(ancova)  # unadjusted sum of squares

Analysis of Variance Table

Response: t3
          Df Sum Sq Mean Sq F value    Pr(>F)
t1         1 91.056  91.056  394.29 < 2.2e-16 ***
group      2 69.865  34.933  151.26 < 2.2e-16 ***
Residuals 41  9.468   0.231
---
Signif. codes:  0 '***' 0.001 '**' 0.01 '*' 0.05 '.' 0.1 ' ' 1
```

Note that the Type III sums of squares shown below are the "adjusted sum of squares", and must be used in an ANACOVA model.

```
car::Anova(ancova, type = "III")    # Adjusted sum of squares

Anova Table (Type III tests)
```

```
Response: t3
              Sum Sq Df F value    Pr(>F)
(Intercept)  3.377  1  14.623 0.0004391 ***
t1          93.366  1 404.290 < 2.2e-16 ***
group       69.865  2 151.264 < 2.2e-16 ***
Residuals    9.468 41
---
Signif. codes:  0 '***' 0.001 '**' 0.01 '*' 0.05 '.' 0.1 ' ' 1
```

We can recover the R^2 statistic and $\widehat{\sigma}$ from the ANACOVA model as shown below:

```
summary(ancova)$r.squared
0.9444305

summary(ancova)$sigma
0.4805606
```

▲

10.4 Multiple hypothesis testing

We discuss multiple testing for estimable functions, including several classical multiple comparison procedures. Multiple hypothesis testing, or simply multiple testing, refers to simultaneously testing a set of hypotheses while controlling the overall error. For example, after the *single* null H_0: $\mathbf{c}_1'\boldsymbol{\beta} = 0, \dots, \mathbf{c}_s'\boldsymbol{\beta} = 0$ in an ANOVA model is rejected, one may wish to find out which $\mathbf{c}_i'\boldsymbol{\beta}$ are nonzero by simultaneously testing H_{0i}: $\mathbf{c}_i'\boldsymbol{\beta} = 0$ versus H_{1i}: $\mathbf{c}_i'\boldsymbol{\beta} \neq 0$, $i = 1, \dots, s$. Many issues that arise in multiple testing are due to the increasing difficulty of controlling the Type I error as more hypotheses are tested. By definition, a Type I error, also known as a false positive (FP), is the (incorrect) rejection of a null that is in fact true.

As an example, suppose we conduct independent tests on $m = 2s$ nulls each at the same significance level α. Suppose at least half of the nulls are true. Then the probability of at least one FP is

$$\begin{aligned} \text{P(at least one FP)} &= 1 - \text{P(no FP in the tests on the true nulls)} \\ &\geq 1 - \text{P(no FP in the test on the first true null)}^s \\ &= 1 - (1-\alpha)^s. \end{aligned} \tag{10.4.1}$$

When $m = 200$, at $\alpha = 0.05$, the probability of at least one FP is 99.4%. If the probability is to be controlled at a level no more than 5%, then α has to be less than 5.13×10^{-4}. However, such a small significance level makes it difficult to reject any false nulls, resulting in lower power of testing. Thus, even in this toy example, balancing Type I error control and power requires more than adjusting a single significance level α. In domains such as genomics or high-dimensional testing in many industries, it is routine to perform hundreds or thousands or even more statistical tests. In such high-throughput environments, as a vast majority of tested nulls are expected to be true, multiple testing is like finding needles in a haystack and it becomes critical to get a good tradeoff between Type I error control and power.

We can summarize possible outcomes from m tests using four random variables U, V, T, and S as in Table 10.4.1, where U is the number of true negatives, V is the number of false positives (Type I errors, also called false discoveries), T is the number of false negatives (Type II errors), and S is the number of true positives (true discoveries). Furthermore, $R = V + S$ is the number of rejected null hypotheses (discoveries) whether true or false,

$m_0 = U + V$ is the number of true null hypotheses, and $m - m_0 = T + S$ is the number of true alternative hypotheses. Note that R is observable, while S, T, U, and V are not. Note that $P(R = 0) > 0$.

	Not Significant	Significant	Total
H_0 True	U	V	m_0
H_1 True	T	S	$m - m_0$
	$m - R$	R	m

TABLE 10.4.1. Summary of possible outcomes from multiple tests

10.4.1 Error rates

To conduct multiple testing, we must specify what error rate to control. While the error rate with a single hypothesis test is the Type I error rate, the notion of error rate in multiple testing can be complicated. Several error rates are described below.

Definition 10.4.1. The Per Comparison Error Rate (PCER) is the expectation of the ratio of FPs to the total number of hypotheses:

$$\text{PCER} = \text{E}(V/m) = \text{E}(V)/m. \tag{10.4.2}$$

Definition 10.4.2. The Familywise Error Rate (FWER) is the probability of getting at least one Type I error or FP:

$$\text{FWER} = \text{P}(V \geq 1) = 1 - \text{P}(V = 0). \tag{10.4.3}$$

Hence, ensuring that FWER $\leq \alpha$ controls the probability of making one or more FPs at level α, whereas ensuring that PCER $\leq \alpha$ controls the expected ratio of the number of FPs to m at level α.

Note that the FWER may be computed either conditional on the complete set of nulls, FWER$= \text{P}(V \geq 1 \,|\, H_{0i}, i = 1, \ldots, m)$, or only on a subset of them. We say that

(i) a test procedure has "weak control of the FWER" or "controls the FWER in the weak sense" if the FWER is less than or equal to the level of the test α, only when all null hypotheses are true, i.e., $m_0 = m$;

(ii) a test procedure has "strong control of the FWER" or "controls the FWER in the strong sense" if the FWER control at level α is guaranteed for any configuration of true nulls m_0 and false nulls $m - m_0$.

When the number of nulls m is large while the proportion of false nulls $(m - m_0)/m$ is small, the PCER may not provide sufficient safeguard against FPs. This is because the number of rejections R must be small in order to avoid excessive number of FPs V. Then, $\text{E}(V)/m \leq \text{E}(R)/m$ is small, even if all the rejections are FPs, i.e., $V = R$ or $S = 0$. In contrast, the FWER helps guard against *any* FPs. While such a strong safeguard against FPs is necessary in confirmatory investigations, this is not always the case in exploratory studies. A more flexible trade-off between Type I error (FP) control and power may be achieved by controlling other error rates.

Definition 10.4.3. The False Discovery Rate (FDR) is the expected proportion of Type I errors (FPs or falsely rejected nulls) among the set of R rejected nulls:

$$\text{FDR} = \text{E}[V/(R \vee 1)], \tag{10.4.4}$$

where $R \vee 1 = \max(R, 1)$.

Note that since $R = V + S$,

$$\frac{V}{R \vee 1} = \begin{cases} \frac{V}{V+S} = \frac{V}{R} & \text{if } R > 0 \\ 0 & \text{if } R = 0, \end{cases} \tag{10.4.5}$$

and then

$$\begin{aligned} \text{FDR} &= \text{E}(V/R \,|\, R > 0)\,\text{P}(R > 0) + \text{E}(V/R \,|\, R = 0)\,\text{P}(R = 0) \\ &= \text{E}(V/R \,|\, R > 0)\,\text{P}(R > 0). \end{aligned} \tag{10.4.6}$$

For example, in a microarray experiment, FDR is the expected proportion of genes that are "incorrectly" declared significant, among the R genes that are declared significant. If, say 1000 hypotheses are rejected using FDR at the 5% level, then the conclusion is that on average 50 of them are falsely rejected. While 50 is a lot with Type-1 error control, it is acceptable under FDR. For instance, with genes, it may happen that all 1000 discoveries will be further investigated using a more expensive experiment, and the cost due to the falsely rejected 50 will be negligible, overall. However, if we control the error rate, we may discover very few genes, and miss most of the interesting 950 ones, which FDR allows us to detect.

Definition 10.4.4. An error rate closely related to the FDR is the positive FDR, defined as

$$\text{pFDR} = \text{E}(V/R \,|\, R > 0). \tag{10.4.7}$$

From (10.4.6) and (10.4.7), it is clear that FDR = pFDR $\times$ P$(R > 0)$. However, unlike the FDR, the pFDR may not be controlled at a specified level. For example, if all the nulls are true, i.e., $m_0 = m$, then the pFDR is 1 (see Table 10.4.1). For more details on the pFDR, and the q-value which controls pFDR, see Storey (2002).

Result 10.4.1. Some Properties of Error Rates. For any multiple testing procedure with m hypotheses, the following properties hold:

1. For $i = 1, \ldots, m$, let $\delta_i = I\{H_{0i} \text{ is rejected}\}$ and $\gamma_i = I\{H_{0i} \text{ is false}\}$. Then $R = \sum_{i=1}^m \delta_i$ and $V = \sum_{i=1}^m \delta_i(1 - \gamma_i)$.

2. PCER $\le$ FDR $\le$ FWER.

3. When all the nulls are true, i.e., $m_0 = m$, FDR = FWER.

Proof. Property 1 directly follows from the definition of R and V. Since $R \vee 1 \le m$, $V/m \le V/(R \vee 1)$. Taking expectation on both sides yields PCER $\le$ FDR. Since $V \le R$, it is easy to verify that $V/(R \vee 1) \le I\{V > 0\}$. Taking expectation on both sides yields FDR $\le$ FWER, showing property 2. To show property 3, when all the nulls are true, $V = R$. Then $V/(R \vee 1) = V/(V \vee 1) = I\{V > 0\}$. Taking expectation on both sides yields FDR = FWER. ■

10.4.2 Procedures for Controlling Type I Errors

In multiple testing, most often, the first step is to calculate p-values for individual nulls as though each null is a stand-alone hypothesis, and then a procedure makes rejections based on the entire set of the p-values. Such "marginal" p-values are used when the joint distributions of the test statistics under the nulls are intractable. On the other hand, when the joint distribution is tractable, multiple testing procedures can be developed that critically depend on it; see Section 10.4.3.

We will only consider procedures based on p-values. Let $p_1, \cdots, p_m$ be the p-values for testing $H_{01}, \ldots, H_{0m}$, and α be the control level for an error rate, such as the FWER or FDR. Due to multiplicity, decisions of rejection based on whether $p_i \le \alpha$ or $p_i > \alpha$ will fail to control the error rate at α. The general approach is to adjust each p_i to some $\tilde{p}_i \ge p_i$ which is then compared to α, or equivalently, to adjust α to some smaller $\tilde{\alpha}_i$ which is then compared to p_i. This is known as "adjustment" or "correction" for multiplicity, and its results are known as adjusted p-values or adjusted levels. The adjustment for each null depends not only on the number of tests m, but may also depend on the p-values of the other nulls.

10.4.2.1 FWER control

Let α be the FWER control level. The *Bonferroni procedure* rejects H_{0i} if and only if $p_i \le \tilde{\alpha}$, where $\tilde{\alpha} = \alpha/m$, or equivalently, $\tilde{p}_i \le \alpha$, where $\tilde{p}_i = mp_i$. The procedure is a *single-step* procedure because it makes a single adjustment to the level, or equivalently, the same adjustment to each p-value.

The *Šidák procedure* is another single-step FWER control procedure, which sets $\tilde{\alpha} = 1 - (1-\alpha)^{1/m}$. This procedure is more powerful than the Bonferroni procedure, although the gain is small for large m. Futhermore, if the p-values are dependent, it may fail to control the FWER. Single-step FWER control procedures are conservative and often lack power.

Instead of single-step procedures, one may instead use adaptive procedures. Two standard types of adaptive procedures are *step-up* and *step-down* procedures. Both types specify a sequence of adjusted levels $\alpha_1 \le \alpha_2 \le \ldots \le \alpha_m$ and sort the p-values into $p_{(1)} \le p_{(2)} \le \cdots \le p_{(m)}$. Let $H_{(1)}, H_{(2)}, \ldots, H_{(m)}$ be the nulls corresponding to the sorted p-values.

Step-up procedure. If $p_{(i)} > \alpha_i$ for all i, then accept all the nulls. Otherwise, find $R = \max\{i\colon p_{(i)} \le \alpha_i\}$ and reject all $H_{(i)}$ with $i \le R$.

Step-down procedure. If $p_{(i)} \le \alpha_i$ for all i, then reject all the nulls. Otherwise, find $R = \min\{i\colon p_{(i)} > \alpha_i\} - 1$ and reject all $H_{(i)}$ with $i < R$.

The terms *step-up* and *step-down* originate from multiple tests using upper-tailed p-values of t-tests; see Dunnett and Tamhane (1992). Starting from the smallest t-value (which has the largest p-value), a step-up procedure accepts every null it meets as it goes through the t-values in ascending order and stops at the first one which is greater than the corresponding cut-off, which is the α_i%-th upper percentile of the t-distribution. On the other hand, starting from the largest t-value, a step-down procedure rejects every null it meets as it goes through the t-values in descending order and stops at the first one that is smaller than the corresponding cut-off; the corresponding null, as well as the remaining nulls, are then accepted.

The procedures of Holm (1979) and Hochberg (1988) are two well-known adaptive procedures. They use the same adjusted levels $\alpha_i = \alpha/(m+1-i)$, where α is the FWER control level. However, the *Holm procedure* is a step-down procedure, while the *Hochberg procedure* is a step-up procedure. As a result, the Hochberg procedure is more powerful. In turn, the Holm procedure is more powerful than the Bonferroni procedure. On the other hand, the Holm procedure has strong control of the FWER while the Hochberg procedure only has weak control. For further results on FWER control, see Dmitrienko et al. (2009).

10.4.2.2 FDR control

The most well-known FDR control procedure is the *Benjamini–Hochberg (BH) procedure*. It is a step-up procedure with adjusted levels

$$\alpha_i = i\alpha/m,$$

where α is the FDR control level. The same procedure was proposed by Simes (1986) as a FWER controlling procedure.

Benjamini–Hochberg (BH) procedure. Sort the p-values as $p_{(1)} \le p_{(2)} \le \cdots \le p_{(m)}$. Let $H_{(1)}, H_{(2)}, \ldots, H_{(m)}$ be the nulls corresponding to the sorted p-values. If $p_{(i)} > i\alpha/m$ for all i, then accept all the nulls. Otherwise, find $R = \max\{i\colon p_{(i)} \le i\alpha/m\}$ and reject all $H_{(i)}$ with $i \le R$.

Result 10.4.2. FDR control by the BH procedure. For independent test statistics, if the p-value under each true null is uniformly distributed on $(0, 1)$, then for any configuration of false nulls, the BH procedure controls the FDR at level α, i.e., $\mathrm{E}[V/(R \vee 1)] \le \alpha$.

For details of the proof, see Benjamini and Hochberg (1995). They pointed out that independence of the test statistics corresponding to the false nulls is actually not needed. To allow for dependence of the test statistics, Benjamini and Yekutieli (2001) modified the BH procedure with $\alpha_i = \tilde{\alpha} i/m$, where

$$\tilde{\alpha} = \frac{\alpha}{\sum_{i=1}^{m} 1/i}.$$

The authors proved that this procedure controls FDR when the test statistics have positive regression dependency on each of the test statistics corresponding to the true null hypotheses. The condition for positive dependency is general enough to cover many problems of practical interest, including the comparisons of many treatments with a single control, multivariate normal test statistics with positive correlation matrix and multivariate t. They also showed that the test statistics may be discrete, and composite hypothesis testing can be accommodated. Also, even under other forms of dependency, a simple conservative modification of their procedure controls FDR.

Example 10.4.1. Let $m = 10$ and $p_{(1)} \le p_{(2)} \cdots \le p_{(m)}$ be: 0.001, 0.012, 0.014, 0.122, 0.245, 0.320, 0.550, 0.776, 0.840, 0.995. Let $\alpha = 0.05$. The values of $i\alpha/m$ are: 0.005, 0.010, 0.015, 0.020, 0.025, 0.030, 0.035, 0.040, 0.045, 0.050. Then $R = \max\{i\colon p_{(i)} \le i\alpha/m\} = \max\{1, 3\} = 3$ and the BH procedure rejects nulls associated with the first three p-values, although the second hypothesis had $0.012 > 0.010$. □

How can we explicitly write the BH adjusted p-values using i, m and $p_{(i)}$? Let

$$p^*_{(i),\text{adj}} = \frac{m}{i} p_{(i)}.$$

However, these adjusted p-values may not be strictly increasing. To fix this, set

$$p_{(i),\text{adj}} = \min_{j \ge i} p^*_{(i),\text{adj}}.$$

10.4.2.3 Plug-in approach to estimate the FDR

First, approximate $\mathrm{E}[V/(R \vee 1)]$ by $\mathrm{E}(V)/\mathrm{E}(R)$. Suppose there are m tests and there is a rule to decide significance in each. Let

$$\widehat{\mathrm{E}(R)} = R_{\text{obs}} = \#\{\text{tests declared as significant}\}.$$

Create K permutations of the observed data. For $j = 1, \ldots, K$, let $W_j = \#\{\text{tests declared as significant for the } j\text{-th set of data}\}$. Let $\widehat{EV} = W = \frac{1}{K}\sum_{j=1}^{K} W_j$. Note that W estimates $(m/m_0)\,\mathrm{E}(V)$, since the permutation distribution is based not on m_0 null hypotheses, but on all m. Estimate FDR by

$$\widehat{\mathrm{FDR}} = W/R_{\text{obs}}. \tag{10.4.8}$$

Unfortunately, $\widehat{\mathrm{FDR}}$ overestimates FDR, i.e., it is conservative. Of course, if we can estimate m_0 then, we can better estimate FDR as

$$\widehat{\mathrm{FDR}} = \frac{\widehat{m_0}}{m}\frac{W}{R_{\text{obs}}}.$$

For example, consider a two-sided t-test with $|t_i| > t_{df,\alpha/2}$ or $p_i < \alpha$. Let $R_{\text{obs}} = \sum_{i=1}^{m} I\{|t_i| > t_{d.f.,\alpha/2}\}$. Then, $W_j = \sum_{i=1}^{m} I\{|t_i^j| > t_{d.f.,\alpha/2}\}$, and $W = \frac{1}{K}\sum_{j=1}^{K}\sum_{i=1}^{m} I\{|t_i^j| > t_{d.f.,\alpha/2}\}$.

10.4.3 Multiple comparison procedures

We describe several classical multiple comparison procedures which are widely used for comparing fixed effect means in ANOVA procedures. In the one-way fixed-effects ANOVA model with a treatments, suppose the overall F-test rejects the null hypothesis that all the treatment effects are equal, so there is evidence that some of the $a(a-1)/2$ pairs of means are significantly different; however, the test cannot tell which pairs are significantly different. Finding such pairs requires testing equality for each pair. This type of multiple testing is known as a multiple comparison. Since it is done only *after* an overall ANOVA analysis has indicated the potential existence of significantly different pairs, it is called a *post hoc* analysis. However, if the goal from the beginning is to identify significantly different pairs, then multiple comparisons can be done regardless of other statistical tests on the data, and thus, need not be *post hoc*.

For *post hoc* multiple comparisons, the effects to be compared must be planned in advance. To illustrate, suppose we had used an F-test to reject the hypothesis H_0: $\mu_1 = \mu_2 = \mu_3 = \mu_4 = \mu_5$. Assume that during the analysis, we had observed that the first treatment average, $\overline{Y}_{1\cdot}$ was the largest among the five treatment averages and $\overline{Y}_{3\cdot}$ was the smallest and *then* decided to test for the difference between μ_1 and μ_3. Suppose the test statistic for the pairwise comparison was a Studentized difference $(\overline{Y}_{1\cdot} - \overline{Y}_{3\cdot})/s$, where s^2 was an estimate of $\mathrm{Var}(\overline{Y}_{1\cdot} - \overline{Y}_{3\cdot})$. Denote the observed value of the Studentized difference by t_{obs}. If the pair $(1,3)$ had been predetermined, the p-value would be $\mathrm{P}(|\overline{Y}_{1\cdot} - \overline{Y}_{3\cdot}|/s > t_{obs} \,|\, \mu_1 = \mu_3) = \mathrm{P}(T > t_{obs})$, where T is a t-distributed random variable. However, since the pair (1, 3) was selected because $Y_{1\cdot} = Y_{\max} = \max_{1 \le i \le 5} \overline{Y}_{i\cdot}$ and $Y_{3\cdot} = Y_{\min} = \min_{1 \le i \le 5} \overline{Y}_{i\cdot}$, the p-value was $\mathrm{P}(|\overline{Y}_{1\cdot} - \overline{Y}_{3\cdot}|/s > t_{obs} \,|\, \mu_1 = \mu_3, \overline{Y}_{1\cdot} = \overline{Y}_{\max}, \overline{Y}_{3\cdot} = \overline{Y}_{\min}) = \mathrm{P}(|\overline{Y}_{\max} - \overline{Y}_{\min}|/s > t_{obs} \,|\, \mu_1 = \mu_3, \overline{Y}_{1\cdot} = \overline{Y}_{\max}, \overline{Y}_{3\cdot} = \overline{Y}_{\min})$, a very different one from the p-value if $(1,3)$ had been predetermined.

Ignoring the difference will be an act of *data dredging* or *data snooping*. These are the alternate names given to the practice of first looking at the sample and then choosing to analyze those comparisons that appear to be interesting without carefully accounting for the conditional probability due to the selection.

On the other hand, it is perfectly acceptable to carry out a predetermined multiple comparison procedure with a given confidence coefficient after observing the data. Indeed, given such a procedure, denote by V the number of pairs $\mu_i = \mu_j$ incorrectly identified as different (see Table 10.4.1). If the comparison procedure was carried out only after H_0 had been rejected, then the probability of having at least one misidentified pair would be $\mathrm{P}(V \geq 1 \,|\, H_0 \text{ is rejected})\,\mathrm{P}(H_0 \text{ is rejected}) \leq \mathrm{P}(V \geq 1)$. As long as the multiple comparison procedure controls the FWER when used as a stand-alone procedure, it controls the FWER if used only after H_0 had been rejected; the test on H_0 served as a pre-screening in some situations.

The solution to the multiple comparison problem for one-way fixed-effects ANOVA model is offered by a variety of multiple comparison procedures, ranging from simple and elegant graphical techniques, to sophisticated significance tests, which were pioneered by Tukey and Scheffé; see Miller (1981), or Hochberg and Tamhane (1987). We will describe several methods for the pairwise comparison of treatment means. All the procedures assume that the observations follow the model

$$Y_{ij} = \mu_i + \varepsilon_{ij}, \quad j = 1, \cdots, n_i, \, i = 1, \cdots, a, \tag{10.4.9}$$

where ε_{ij} are i.i.d. $N(0, \sigma^2)$ random variables with unknown variance σ^2. From Example 4.2.5, the least squares estimates are $\widehat{\mu}_i = \overline{Y}_{i\cdot}, \quad i = 1, \cdots, a$, and $\widehat{\sigma}^2 = MSE = (N - a)^{-1} \sum_{i=1}^{a} \sum_{j=1}^{n_i} (Y_{ij} - \overline{Y}_{i\cdot})^2$, where $\overline{Y}_{i\cdot} = (Y_{i1} + \cdots + Y_{in_i})/n_i$ are the treatment means, and $N = \sum_{i=1}^{a} n_i$. Then $\widehat{\mu}_1, \cdots, \widehat{\mu}_a$ are independent of $\widehat{\sigma}^2$, with $\widehat{\mu}_i \sim N(\mu_i, \sigma^2/n_i)$ and $(N-a)\widehat{\sigma}^2/\sigma^2 \sim \chi^2_{N-a}$.

LSD procedure

Fisher's *least significance difference* (LSD) procedure is useful for making pairwise comparisons among a set of a treatment means. We define the least significance difference to be the observed difference between two treatment means that is required in order to decide that the corresponding population means are distinct. Note that for any pair $i \neq l$, $(\widehat{\mu}_i - \widehat{\mu}_l)/(\widehat{\sigma}\sqrt{1/n_i + 1/n_l}) \sim t_{N-a}$. For a given level α, the LSD for comparing μ_i and μ_l in a one-factor fixed-effects ANOVA model is

$$LSD_{il} = t_{N-a,\alpha/2}\widehat{\sigma}\sqrt{1/n_i + 1/n_l},$$

and μ_i and μ_l are declared significantly different if

$$|\widehat{\mu}_i - \widehat{\mu}_l| > LSD_{il} \quad \text{or} \quad \frac{|\widehat{\mu}_i - \widehat{\mu}_l|}{\widehat{\sigma}\sqrt{1/n_i + 1/n_l}} > t_{N-a,\alpha/2}.$$

All the $a(a-1)/2$ pairs of treatment means are compared. The LSD procedure is similar to a two-sample test for any two population means, the difference being that we use the estimate $\widehat{\sigma}^2$ instead of the pooled variances from the ith and lth samples only.

Since the significance level is fixed at α in the calculation of the LSD for each pair of treatment means, the LSD procedure is not adjusted for multiplicity. In general, the procedure is criticized for inflating the Type I error rate. That is, the overall probability of incorrectly declaring some pair of treatment means significantly different, when they are in fact equal, is substantially higher than the specified α value. In practice, however, the procedure is perhaps unjudiciously used often, because of its simplicity and ease of implementation, and, therefore, it is recommended that the LSD procedure be used only after the overall F-test for treatments is seen to be significant. This revised procedure is sometimes called Fisher's protected LSD, and some simulation studies show that the error rate for the protected LSD procedure is controlled on an experiment-wise basis at a level that is approximately equal to α.

Tukey's procedure

Tukey's procedure, also known as Tukey's range test or Tukey's honestly significant difference (HSD) test, is based on the distribution of the Studentized range statistic (see Appendix B).

First, consider the balanced design in which $n_i = n$, $i = 1, \cdots, a$. Since ε_{ij} are i.i.d. $N(0, \sigma^2)$ variables, $\widehat{\mu}_i - \mu_i$ are i.i.d. $N(0, \sigma^2/n)$ variables. Recall that Fisher's LSD procedure in this case is equivalent to comparing $|\widehat{\mu}_i - \widehat{\mu}_l|/(\widehat{\sigma}/\sqrt{n})$ with a single critical value $\sqrt{2}t_{N-a,\alpha/2}$. Tukey's procedure also compares $|\widehat{\mu}_i - \widehat{\mu}_l|/(\widehat{\sigma}/\sqrt{n})$ with a single critical value. To control the FWER at α, the critical value, say C_α, must satisfy

$$\mathrm{P}\{|\widehat{\mu}_i - \widehat{\mu}_l|/(\widehat{\sigma}/\sqrt{n}) > C_\alpha \text{ for some } i \neq l \,|\, \text{all } \mu_i \text{ are equal}\} \leq \alpha.$$

The probability of Type I error is

$$\mathrm{P}\{(\widehat{\mu}_{(a)} - \widehat{\mu}_{(1)})/(\widehat{\sigma}/\sqrt{n}) > C_\alpha \,|\, \text{all } \mu_i \text{ are equal}\},$$

where $\widehat{\mu}_{(1)} \leq \cdots \leq \widehat{\mu}_{(a)}$ are the order statistics of $\widehat{\mu}_i$; $(\widehat{\mu}_{(a)} - \widehat{\mu}_{(1)})/(\widehat{\sigma}\sqrt{n})$ is known as the Studentized range.

Let $Z_i = (\widehat{\mu}_i - \mu_i)/(\sigma/\sqrt{n})$ and $S^2 = \widehat{\sigma}^2/\sigma^2$. Then $Z_1, \cdots, Z_a$ are independent of S with $Z_i \sim N(0,1)$ and $S^2 \sim \chi^2_{N-a}$. Under the null hypothesis that all μ_i are equal, $(\widehat{\mu}_{(a)} - \widehat{\mu}_{(1)})/(\widehat{\sigma}/\sqrt{n})$ is the same as $(Z_{(a)} - Z_{(1)})/S$. Tukey derived the distribution of this random variable, known as the Studentized range distribution, with a means and error d.f. $N - a$. Let $q_\alpha(a, N-a)$ denote the upper 100α percentage point of this distribution. Then μ_i and μ_l are declared significantly different if

$$|\widehat{\mu}_i - \widehat{\mu}_l| > q_\alpha(a, N-a)\widehat{\sigma}/\sqrt{n}.$$

Clearly, if one pair is declared significantly different, then the Studentized range $(\widehat{\mu}_{(a)} - \widehat{\mu}_{(1)})/(\widehat{\sigma}/\sqrt{n})$ is greater than $q_\alpha(a, N-a)$; we also say the range of $\widehat{\mu}_1, \cdots, \widehat{\mu}_a$ is significant according to an α-level Tukey's range test. Tukey's $100(1-\alpha)\%$ simultaneous confidence intervals of all the pairwise differences are

$$(\widehat{\mu}_i - \widehat{\mu}_l) \pm q_\alpha(a, N-a)\widehat{\sigma}/\sqrt{n}.$$

These intervals are narrower than the Scheffé simultaneous confidence intervals because the latter achieves probability $1-\alpha$ of simultaneous coverage of all possible linear combinations of the pairwise differences, i.e., all possible contrasts.

In the unbalanced case where $n_1, \cdots, n_a$ are not all equal, Tukey's procedure is modified so that μ_i and μ_l are declared significantly different if

$$|\widehat{\mu}_i - \widehat{\mu}_l| > q_\alpha(a, N-a)\widehat{\sigma}\sqrt{(n_i^{-1} + n_l^{-1})/2}, \tag{10.4.10}$$

and correspondingly the approximate simultaneous confidence intervals of all the pairwise differences are modified to

$$(\widehat{\mu}_i - \widehat{\mu}_l) \pm q_\alpha(a, N-a)\widehat{\sigma}\sqrt{(n_i^{-1} + n_l^{-1})/2}. \tag{10.4.11}$$

These modifications are known as Tukey–Kramer approximations.

Note that the Studentized range distribution was obtained when all H_{il}: $\mu_i = \mu_l$ are true. Since Tukey–Kramer's procedure critically relies on the distribution, one question is whether it can control the FWER when some of the nulls are false. From Definition 10.4.2, if a procedure achieves FWER control only when all the nulls are true, then it only has weak FWER control. On the other hand, if the procedure achieves FWER control regardless of which and how many nulls are true, then it has strong FWER control. It turns out that the latter is the case for Tukey–Kramer's procedure. The proof of this result is beyond the scope of this book and can be found in Hayter (1984). We only prove the result for the balanced design.

Result 10.4.3.

1. The Tukey–Kramer test (10.4.10) has strong FWER control.
2. Tukey–Kramer simultaneous confidence intervals (10.4.11) are conservative, i.e., the probability of simultaneous coverage is at least $1 - \alpha$ with equality if and only if the design is balanced.

Proof. To prove property 1, we need to show that for any fixed $\mu_1, \cdots, \mu_a$, the FWER of Tukey's test is no greater than α. Partition the indices $1, \cdots, a$ into $I_1, \cdots, I_K$, such that $\mu_i = \mu_l$ if i and l belong to the same I_k, and $\mu_i \neq \mu_l$ otherwise. Then the FWER is the probability of the event $\cup_{k=1}^K \{|\widehat{\mu}_i - \widehat{\mu}_l| > q_\alpha(a, N-a)\widehat{\sigma}/\sqrt{n}$ for some $i, l \in I_k\})$, which is the same as

$$\begin{aligned}
&\mathrm{P}\left(\bigcup_{k=1}^{K}\left\{\frac{\max_{i\in I_k}\widehat{\mu}_i - \min_{i\in I_k}\widehat{\mu}_i}{\widehat{\sigma}/\sqrt{n}} > q_\alpha(a, N-a)\right\}\right)\\
&= \mathrm{P}\left(\bigcup_{k=1}^{K}\left\{\frac{\max_{i\in I_k} Z_i - \min_{i\in I_k} Z_i}{S} > q_\alpha(a, N-a)\right\}\right)\\
&\le \mathrm{P}\left(\frac{Z_{(a)} - Z_{(1)}}{S} > q_\alpha(a, N-a)\right) = \alpha.
\end{aligned}$$

For property 2, using $\widehat{\mu}_i = \mu_i + \sigma Z_i/\sqrt{n}$ and $\widehat{\sigma} = \sigma S$,

$$\begin{aligned}
&\mathrm{P}\left(\mu_i - \mu_l \in (\widehat{\mu}_i - \widehat{\mu}_l) \pm q_\alpha(a, N-a)\widehat{\sigma}/\sqrt{n} \text{ for all } i \text{ and } l\right)\\
&= \mathrm{P}\left(|Z_i - Z_l| \le q_\alpha(a, N-a)S \text{ for all } i \text{ and } l\right)\\
&= \mathrm{P}\left(Z_{(a)} - Z_{(1)} \le q_\alpha(a, N-a)S\right) = 1 - \alpha.
\end{aligned}$$

■

Newman–Keuls procedure

Newman (1939) first derived this test, which got its name due to the revival of the method by Keuls (1954). The Newman-Keuls test is a multiple range test. This type of test declares a pair significantly different if and only if every range of treatment means that contains the pair is significant. To be specific, let $C_2 \le \ldots \le C_a$ be critical values for comparing set of $2, \cdots, a$ means. Let $\widehat{\mu}_{(1)} \le \widehat{\mu}_{(2)} \le \cdots \le \widehat{\mu}_{(a)}$ be the order statistics of $\widehat{\mu}_i$ and $n_{(1)}, n_{(2)}, \cdots, n_{(a)}$ be the corresponding sample sizes. Each pair $\widehat{\mu}_{(i)}$ and $\widehat{\mu}_{(l)}$ in the sorted means is called a $|i - l|$ range. Then

$$\begin{aligned}
&\text{for any } i < l,\ \mu_{(i)} \text{ and } \mu_{(l)} \text{ are significantly different } \Longleftrightarrow\\
&\widehat{\mu}_{(k)} - \widehat{\mu}_{(h)} > C_{k-h+1}\widehat{\sigma}\sqrt{(n_{(h)}^{-1} + n_{(k)}^{-1})/2} \quad \text{for all } h \le i < l \le k. \qquad (10.4.12)
\end{aligned}$$

To test all the pairs, we may maintain a list S of significantly different pairs and a list T of pairs (i, l), $i < l$, to be tested. In T, pairs are ordered by range in descending order, while among pairs with the same range, no particular ordering is required. Before the start, S is empty and T reads $(1, a), (1, a-1), (2, a), (1, a-2), (2, a-1), (3, a), \cdots, (1, 2), (2, 3), \cdots, (a-1, a)$. Then, at each step of the multiple range test, if T is not empty, then remove its first pair (h, k). If $(\widehat{\mu}_{(h)}, \widehat{\mu}_{(k)})$ satisfies (10.4.12), then add it to S, otherwise remove all pairs (i, l) with $h \le i, l \le k$ from T. Continue until T is empty; the list S at the end contains all the significantly different pairs.

Different multiple range tests employ different critical values. The Newman–Keuls procedure at level α uses

$$C_i = q_\alpha(i, N-a), \quad i = 2, \cdots, a.$$

Therefore, at each range, the Newman–Keuls procedure employs a Tukey's test at level α corresponding to that range. Since $q_\alpha(i, N-a)$ is monotonically increasing in i and the Tukey's test can be thought of as a multiple range test with $C_i \equiv q_\alpha(a, N-a)$, the Newman–Keuls procedure is less conservative than Tukey's test. However, a criticism of the procedure is that, unlike Tukey's test, it does not guarantee FWER control.

As an example, suppose $a = 10$, $M \gg \sigma$ (i.e., we choose a value M which is much greater than σ), and we consider the hypotheses: $\mu_{2i-1} = \mu_{2i} = iM$, $i = 1, \ldots, 5$. Then, with high probability which can be arbitrarily close to one by making M/σ large, for any $i < l$, we have $\widehat{\mu}_h \ll \widehat{\mu}_k$, for $h = 2i-1, 2i$, and $k = 2l-1, 2l$, and the pair is declared significantly different. Thus, all pairs of range 1 will be tested; in particular, each of the five pairs $\widehat{\mu}_{2i-1}$ and $\widehat{\mu}_{2i}$ are tested by a Tukey's procedure with $a = 2$ at significance level α. However, the probability that at least one pair is declared significantly different is greater than α, resulting in FWER greater than α.

Duncan's multiple range procedure

Duncan's multiple range procedure is widely used for comparing all pairs of means. At level α, it uses

$$C_i = q_{\alpha_i}(i, N-a), \quad i = 2, \cdots, a,$$

where

$$\alpha_i = 1 - (1-\alpha)^{i-1}.$$

Therefore, unlike the Newman–Keuls procedure, it tests at a level higher than α for every range except for the smallest one. Therefore, it is less conservative than the Newman–Keuls procedure. Generally speaking, Duncan's test is powerful, i.e., it is effective in detecting real differences between treatment means, and is therefore widely used. On the other hand, it provides even less guarantee of FWER control than the Newman–Keuls procedure.

Scheffé's procedure

Scheffé simultaneous confidence intervals in Section 7.4.2 provide a conservative multiple comparison procedure, which we now describe. The one-way fixed-effects ANOVA model (10.4.9) can be written as $\mathbf{y} = \mathbf{X}\boldsymbol{\beta} + \boldsymbol{\varepsilon}$, where $\mathbf{y} = (Y_{11}, \cdots, Y_{1n_1}, \cdots, Y_{a1}, \cdots, Y_{an_a})'$, $\mathbf{X} = \mathbf{1}_{n_1} \oplus \cdots \oplus \mathbf{1}_{n_a}$ (see Definition 2.8.3), and $\boldsymbol{\beta} = (\mu_1, \cdots, \mu_a)'$. Then $\mathrm{r}(\mathbf{X}) = a$ and $\mathbf{G} = \mathrm{diag}(1/n_1, \cdots, 1/n_a)$. On the other hand, the linear space of estimable functions spanned by the pairwise differences $\mu_i - \mu_l$, $1 \le i \ne l \le a$, is the entire set of contrasts of $\boldsymbol{\beta}$, which has dimension $a-1$. From Result 7.4.2, the intervals $I_{\mathbf{c}}$ with end points

$$\mathbf{c}'\widehat{\boldsymbol{\beta}} \pm \widehat{\sigma}\left\{(a-1)F_{a-1,N-a,\alpha}\sum_{i=1}^{a} c_i^2/n_i\right\}^{1/2}$$

are $100(1-\alpha)\%$ simultaneous confidence intervals for all contrasts $\mathbf{c}'\boldsymbol{\beta}$.

In principle, Scheffé's procedure can be applied to test all contrasts, with the rule that $\mathbf{c}'\beta$ is declared nonzero if and only if $0 \notin I_{\mathbf{c}}$. When applied to all-pairwise comparisons only, Scheffé's procedure declares $\mu_i \neq \mu_l$ if and only if

$$|\widehat{\mu}_i - \widehat{\mu}_l| > S_\alpha(a, N-a)\widehat{\sigma}\{(n_i^{-1} + n_l^{-1})/2\}^{1/2},$$

where $S_\alpha(a, f) = \{(a-1)F_{a-1,f,\alpha}\}^{1/2}$. The procedure has strong FWER control, because a Type I error implies $\mathbf{c}'\widehat{\boldsymbol{\beta}} \notin I_{\mathbf{c}}$ for some contrast $\mathbf{c}'\boldsymbol{\beta}$, an event with probability α.

Dunnett's procedure

In many experiments, one treatment is the control, and there is interest in comparing each of the other $a-1$ treatments with this control using $a-1$ comparisons. Suppose that treatment a is the control, and we wish to test the hypothesis H_{i0}: $\mu_i = \mu_a$ versus H_{i1}: $\mu_i \neq \mu_a$ for $i = 1, \cdots, a-1$. For each of the $a-1$ hypotheses, compute observed differences in the sample means $|\widehat{\mu}_i - \widehat{\mu}_a|$, $i = 1, \cdots, a-1$. At the overall significance level α, Dunnett (1964)'s procedure rejects H_{i0} if and only if

$$|\widehat{\mu}_i - \widehat{\mu}_a| > d_\alpha(a-1, N-a)\widehat{\sigma}(1/n_i + 1/n_a)^{1/2},$$

with $d_\alpha(a-1, N-a)$ equal to the upper 100α percentage point of

$$\max\{Z_1, \cdots, Z_{a-1}\}/S,$$

where $(Z_1, \cdots, Z_{a-1})'$ is normally distributed and is independent of S, such that $Z_i \sim N(0,1)$, $\text{Cov}(Z_i, Z_l) = \{(n_a/n_i + 1)(n_a/n_l + 1)\}^{1/2}$, $1 \le i \neq l \le a-1$, and $S \sim \chi^2_{N-a}$. The value of $d_\alpha(a-1, N-a)$ depends on the values $n_1, \cdots, n_a$. It is usually recommended that more observations be used for the control than for other treatments. A rule of thumb is to choose these numbers such that $n_a/n = \sqrt{a}$, where n_a is the number of observations for the control and n denotes the (same) number of observations for each of the other $a-1$ treatments.

Choice of procedure

Which multiple comparison procedure should one use? Unfortunately, there is no precise answer to this question, and there are varied opinions about the relative merits of these procedures. There have been some simulation studies that compare these procedures for different situations. The LSD and Duncan's multiple range test appear to be most powerful for detecting true differences in treatment means. The Newman–Keuls test is more conservative than Duncan's test in that the Type I error rate is smaller, and the test is less powerful. Hence, it is more difficult to declare that two treatment means are significantly different using the Newman–Keuls test than it is using Duncan's test. Scheffé's procedure is used for all possible comparisons and may therefore be extremely conservative when applied to a finite set of treatment differences. Certainly, Dunnett's procedure, which is a modification of the usual t-test is useful when treatment means are compared to a control mean.

The following numerical example, which continues Numerical Example 4.3, is used to illustrate these procedures.

Numerical Example 10.5. Recall the dataset "PlantGrowth" with response Y (Weight) and factor X (Treatment) with three levels, 'ctrl', 'trt1', and 'trt2'. In Numerical Example 4.3, we showed results from the following model one-way ANOVA model fit.

```
data(PlantGrowth)
fit <- aov(weight ~ group, data = PlantGrowth)
```

The following code shows multiple comparison procedures for this data. For Tukey's procedure and Dunnett's procedure, we use the R package *multcomp*.

```
library(multcomp)
# Tukey's procedure
summary(glht(fit, linfct = mcp(group = "Tukey")))

Simultaneous Tests for General Linear Hypotheses
Multiple Comparisons of Means: Tukey Contrasts
Fit: aov(formula = weight ~ group, data = PlantGrowth)
Linear Hypotheses:
                 Estimate Std. Error t value Pr(>|t|)
trt1 - ctrl == 0  -0.3710     0.2788  -1.331    0.391
trt2 - ctrl == 0   0.4940     0.2788   1.772    0.198
trt2 - trt1 == 0   0.8650     0.2788   3.103    0.012 *
---
Signif. codes:  0 '***' 0.001 '**' 0.01 '*' 0.05 '.' 0.1 ' ' 1
(Adjusted p values reported -- single-step method)

# Dunnett's procedure
summary(glht(fit, linfct = mcp(group = "Dunnett")))

Simultaneous Tests for General Linear Hypotheses
Multiple Comparisons of Means: Dunnett Contrasts
Fit: aov(formula = weight ~ group, data = PlantGrowth)
Linear Hypotheses:
                 Estimate Std. Error t value Pr(>|t|)
trt1 - ctrl == 0  -0.3710     0.2788  -1.331    0.323
trt2 - ctrl == 0   0.4940     0.2788   1.772    0.153
(Adjusted p values reported -- single-step method)
```

We can use the R package *agricolae* to perform the procedures shown below.

```
library(agricolae)
#LSD procedure
LSD.test(fit, "group", p.adj = "bonferroni", console = T)
#Duncan multiple range procedure
duncan.test(fit, "group", console = TRUE)
#Newman-Keuls procedure
SNK.test(fit, "group", console = TRUE)
```

▲

10.5 Generalized Gauss–Markov theorem

In Section 4.5, we assumed that the matrix $\mathbf{V}$ is p.d. In some models, however, this assumption is untenable (see Section 11.1). For situations where $\mathbf{V}$ is a known symmetric n.n.d. matrix, the following result, called the generalized Gauss–Markov theorem (Zyskind and Martin, 1969), describes optimal estimation of linear functions of $\boldsymbol{\beta}$.

Result 10.5.1. Let $\mathbf{N}$ be any matrix such that $\mathcal{C}(\mathbf{N}) = \mathcal{C}(\mathbf{X})^{\perp}$. Let $\Omega_{\mathbf{V}}$ denote the class of g-inverses $\mathbf{V}^\sharp$ of $\mathbf{V}$ that satisfy

$$\mathrm{r}(\mathbf{X}'\mathbf{V}^\sharp\mathbf{X}) = \mathrm{r}(\mathbf{X}), \quad \text{and} \tag{10.5.1}$$

$$\mathbf{X}'\mathbf{V}^\sharp\mathbf{V}\mathbf{N} = \mathbf{O}. \tag{10.5.2}$$

Then $\Omega_{\mathbf{V}}$ is nonempty, and for any $\mathbf{V}^\sharp \in \Omega_{\mathbf{V}}$, the linear system

$$\mathbf{X}'\mathbf{V}^\sharp\mathbf{X}\boldsymbol{\beta}^0 = \mathbf{X}'\mathbf{V}^\sharp\mathbf{y} \tag{10.5.3}$$

is consistent and $\mathbf{X}\boldsymbol{\beta}^0$ is invariant to the choice of the solution $\boldsymbol{\beta}^0$. Furthermore, if $\mathbf{c}'\boldsymbol{\beta}$ is estimable, then $\mathbf{c}'\boldsymbol{\beta}^0$ does not depend on the choice of $\boldsymbol{\beta}^0$ and is the unique b.l.u.e. of $\mathbf{c}'\boldsymbol{\beta}$, in the sense that if $d_0 + \mathbf{d}'\mathbf{y}$ is any linear unbiased estimator of $\mathbf{c}'\boldsymbol{\beta}$, then $\mathrm{Var}(\mathbf{c}'\boldsymbol{\beta}^0) \le \mathrm{Var}(d_0 + \mathbf{d}'\mathbf{y})$ with equality holding if and only if $d_0 + \mathbf{d}'\mathbf{y} = \mathbf{c}'\boldsymbol{\beta}^0$. Finally, $\mathbf{X}\boldsymbol{\beta}^0$ is invariant to the choices of $\mathbf{V}^\sharp$ and $\boldsymbol{\beta}^0$.

Proof. We first show that $\Omega_{\mathbf{V}}$ is nonempty. Let $\mathrm{r}(\mathbf{X}) = r$. Then $\mathbf{V} = \mathbf{P}\,\mathrm{diag}(\mathbf{I}_r, \mathbf{0})\,\mathbf{P}'$ for a nonsingular $\mathbf{P}$. Partition $\mathbf{N}'\mathbf{P} = (\mathbf{A}, \mathbf{B})$, where $\mathbf{A}$ has r columns. Let $\mathbf{Q}$ be the orthogonal projection matrix onto $\mathcal{C}(\mathbf{A})$. Then $\mathbf{QA} = \mathbf{A}$ and $\mathbf{QB} = \mathbf{AD}$ for some $\mathbf{D}$. If

$$\mathbf{V}^\sharp = \mathbf{P}^{-1\prime} \begin{pmatrix} \mathbf{I}_r & \mathbf{D} \\ \mathbf{D}' & \mathbf{D}'\mathbf{D} + \mathbf{I} \end{pmatrix} \mathbf{P}^{-1},$$

then $\mathbf{V}^\sharp$ is a symmetric, p.d. g-inverse of $\mathbf{V}$. Letting $\mathbf{L} = (\mathbf{V}^\sharp)^{1/2}$ (see Result 2.4.5), $\mathrm{r}(\mathbf{X}'\mathbf{V}^\sharp\mathbf{X}) = \mathrm{r}((\mathbf{LX})'(\mathbf{LX})) = \mathrm{r}(\mathbf{LX}) = \mathrm{r}(\mathbf{X})$, so that (10.5.1) holds. Next,

$$\begin{aligned}
\mathbf{N}'\mathbf{V}\mathbf{V}^\sharp &= \mathbf{N}'\mathbf{P} \begin{pmatrix} \mathbf{I}_r & \mathbf{O} \\ \mathbf{O} & \mathbf{O} \end{pmatrix} \mathbf{P}'\mathbf{P}^{-1\prime} \begin{pmatrix} \mathbf{I}_r & \mathbf{D} \\ \mathbf{D}' & \mathbf{D}'\mathbf{D} + \mathbf{I} \end{pmatrix} \mathbf{P}^{-1} \\
&= (\mathbf{A}, \mathbf{B}) \begin{pmatrix} \mathbf{I}_r & \mathbf{D} \\ \mathbf{O} & \mathbf{O} \end{pmatrix} \mathbf{P}^{-1} = (\mathbf{A}, \mathbf{AD})\mathbf{P}^{-1} \\
&= (\mathbf{QA}, \mathbf{QB})\mathbf{P}^{-1} = \mathbf{Q}(\mathbf{N}'\mathbf{P})\mathbf{P}^{-1} = \mathbf{QN}'.
\end{aligned}$$

Since both $\mathbf{V}$ and $\mathbf{V}^\sharp$ are symmetric and $\mathbf{X}'\mathbf{N} = \mathbf{O}$, then $\mathbf{X}'\mathbf{V}^\sharp\mathbf{V}\mathbf{N} = \mathbf{X}'\mathbf{N}\mathbf{Q}' = \mathbf{O}$, yielding (10.5.2). Thus, $\Omega_{\mathbf{V}}$ contains $\mathbf{V}^\sharp$, so it is nonempty.

Fix $\mathbf{V}^\sharp \in \Omega_{\mathbf{V}}$. Then $\mathcal{C}(\mathbf{X}'\mathbf{V}^\sharp\mathbf{X}) \subset \mathcal{C}(\mathbf{X}')$. However, from (10.5.1), the two spaces have the same dimension, so they must be equal. Then $\mathcal{C}(\mathbf{X}'\mathbf{V}^\sharp\mathbf{X}) = \mathcal{C}(\mathbf{X}') \supset \mathcal{C}(\mathbf{X}'\mathbf{V}^\sharp)$, so (10.5.3) has a solution. For any two solutions to (10.5.3), let $\mathbf{d}$ be their difference. Then $\mathbf{X}'\mathbf{V}^\sharp\mathbf{X}\mathbf{d} = \mathbf{0}$, i.e., $\mathbf{d} \perp \mathcal{R}(\mathbf{X}'\mathbf{V}^\sharp\mathbf{X})$. Since we can also show that $\mathcal{R}(\mathbf{X}'\mathbf{V}^\sharp\mathbf{X}) = \mathcal{R}(\mathbf{X})$, it follows that $\mathbf{d} \perp \mathcal{R}(\mathbf{X})$, i.e., $\mathbf{Xd} = \mathbf{0}$. Hence $\mathbf{X}\boldsymbol{\beta}^0$ is the same as in the above display regardless of the choice of $\boldsymbol{\beta}^0$.

To find $\mathbf{X}\boldsymbol{\beta}^0$, define the symmetric and n.n.d. matrix

$$\mathbf{S} = [\mathbf{I} - \mathbf{VN}(\mathbf{N}'\mathbf{VN})^-\mathbf{N}']\mathbf{V} = \boldsymbol{\Gamma}'(\mathbf{I} - \mathbf{P}_{\mathcal{C}(\boldsymbol{\Gamma}\mathbf{N})})\boldsymbol{\Gamma},$$

with $\boldsymbol{\Gamma} = \mathbf{V}^{1/2}$. We have $\boldsymbol{\varepsilon} \in \mathcal{C}(\mathbf{V})$, because for any $\mathbf{v} \in \mathcal{C}(\mathbf{V})^\perp$, $\mathrm{Var}(\mathbf{v}'\boldsymbol{\varepsilon}) = \mathbf{v}'\mathbf{V}\mathbf{v} = 0$, giving $\mathbf{v}'\boldsymbol{\varepsilon} = 0$. Let $\boldsymbol{\varepsilon} = \mathbf{V}\boldsymbol{\eta}$. By (10.5.2),

$$\mathbf{X}\mathbf{V}^\sharp\mathbf{y} = \mathbf{X}'\mathbf{V}^\sharp(\mathbf{X}\boldsymbol{\beta} + \boldsymbol{\varepsilon}) = \mathbf{X}'\mathbf{V}^\sharp(\mathbf{X}\boldsymbol{\beta} + \mathbf{S}\boldsymbol{\eta}).$$

On the other hand, $\mathbf{SN} = \boldsymbol{\Gamma}'(\mathbf{I} - \mathbf{P}_{\mathcal{C}(\boldsymbol{\Gamma}\mathbf{N})})\boldsymbol{\Gamma}\mathbf{N} = \mathbf{O}$ and hence $\mathcal{C}(\mathbf{S}) = \mathcal{R}(\mathbf{S}) \subset \mathcal{C}(\mathbf{N})^\perp = \mathcal{C}(\mathbf{X})$. This gives $\mathbf{X}\boldsymbol{\beta} + \mathbf{S}\boldsymbol{\eta} \in \mathcal{C}(\mathbf{X})$, so from the invariance of $\mathbf{X}\boldsymbol{\beta}^0$ to the solution to (10.5.3), we get

$$\mathbf{X}\boldsymbol{\beta}^0 = \mathbf{X}\boldsymbol{\beta} + \mathbf{S}\boldsymbol{\eta} = \mathbf{X}\boldsymbol{\beta} + [\mathbf{I} - \mathbf{VN}(\mathbf{N}'\mathbf{VN})^-\mathbf{N}']\boldsymbol{\varepsilon}.$$

We now can show $\mathbf{c}'\boldsymbol{\beta}^0$ is the unique b.l.u.e. of $\mathbf{c}'\boldsymbol{\beta}$. From Result 4.3.1, $\mathbf{c}' = \mathbf{t}'\mathbf{X}$ for some $\mathbf{t}$. From Section 4.3, $\mathrm{E}(\mathbf{c}'\boldsymbol{\beta}^0) = \mathrm{E}(\mathbf{t}'\mathbf{X}\boldsymbol{\beta}^0) = \mathbf{t}'\mathbf{X}\boldsymbol{\beta} = \mathbf{c}'\boldsymbol{\beta}$, so $\mathbf{c}'\boldsymbol{\beta}^0$ is unbiased. Let $d_0 + \mathbf{d}'\mathbf{y}$ be an unbiased estimator of $\mathbf{c}'\boldsymbol{\beta}$. Then, as in the proof of Result 4.4.1, $d_0 = 0$, and $\mathbf{d}'\mathbf{y} = \mathbf{d}'\widehat{\mathbf{y}} + \mathbf{d}'\widehat{\boldsymbol{\varepsilon}} = \mathbf{c}'\boldsymbol{\beta}^0 + \mathbf{d}'\widehat{\boldsymbol{\varepsilon}}$, where

$$\begin{aligned} \widehat{\mathbf{y}} &= \mathbf{X}\boldsymbol{\beta}^0 = \mathbf{X}\boldsymbol{\beta} + [\mathbf{I} - \mathbf{VN}(\mathbf{N}'\mathbf{VN})^-\mathbf{N}']\boldsymbol{\varepsilon}, \\ \widehat{\boldsymbol{\varepsilon}} &= \mathbf{y} - \widehat{\mathbf{y}} = \mathbf{X}\boldsymbol{\beta} + \boldsymbol{\varepsilon} - \widehat{\mathbf{y}} = \mathbf{VN}(\mathbf{N}'\mathbf{VN})^-\mathbf{N}'\boldsymbol{\varepsilon}. \end{aligned} \tag{10.5.4}$$

Since $\mathbf{SN} = \mathbf{O}$,

$$\begin{aligned} \mathrm{Cov}(\widehat{\mathbf{y}}, \widehat{\boldsymbol{\varepsilon}}) &= [\mathbf{I} - \mathbf{VN}(\mathbf{N}'\mathbf{VN})^-\mathbf{N}]\mathbf{VN}(\mathbf{N}'\mathbf{VN})^{-\prime}\mathbf{N}'\mathbf{V} \\ &= \mathbf{SN}(\mathbf{N}'\mathbf{VN})^{-\prime}\mathbf{N}'\mathbf{V} = \mathbf{O}, \end{aligned}$$

so $\mathrm{Var}(\mathbf{d}'\mathbf{y}) = \mathrm{Var}(\mathbf{d}'\widehat{\mathbf{y}}) + \mathrm{Var}(\mathbf{d}'\widehat{\boldsymbol{\varepsilon}}) = \mathrm{Var}(\mathbf{c}'\boldsymbol{\beta}^0) + \mathrm{Var}(\mathbf{d}'\widehat{\boldsymbol{\varepsilon}}) \geq \mathrm{Var}(\mathbf{c}'\boldsymbol{\beta}^0)$ with equality holding if and only if $\mathrm{Var}(\mathbf{d}'\widehat{\boldsymbol{\varepsilon}}) = 0$. Since $\mathrm{E}(\mathbf{d}'\widehat{\boldsymbol{\varepsilon}}) = \mathbf{0}$, $\mathrm{Var}(\mathbf{d}'\widehat{\boldsymbol{\varepsilon}}) = 0$ if and only if $\mathbf{d}'\widehat{\boldsymbol{\varepsilon}} = 0$, i.e., $\mathbf{d}'\mathbf{y} = \mathbf{c}'\boldsymbol{\beta}^0$. Hence $\mathbf{c}'\boldsymbol{\beta}^0$ is the unique b.l.u.e. Finally, for any row $\mathbf{x}_i'$ of $\mathbf{X}$, $\mathbf{x}_i'\boldsymbol{\beta}$ is estimable. Given $\mathbf{V}_1^\sharp$, $\mathbf{V}_2^\sharp \in \Omega_\mathbf{V}$, let $\boldsymbol{\beta}_1^0$ and $\boldsymbol{\beta}_2^0$ be the corresponding solutions to (10.5.3). Then $\mathbf{x}_i'\boldsymbol{\beta}_1^0$ and $\mathbf{x}_i'\boldsymbol{\beta}_2^0$ each is the unique b.l.u.e. of $\mathbf{x}_i'\boldsymbol{\beta}$, so they are equal. As a result, $\mathbf{X}\boldsymbol{\beta}_1^0 = \mathbf{X}\boldsymbol{\beta}_2^0$, showing that $\mathbf{X}\boldsymbol{\beta}^0$ is invariant to the choice of $\mathbf{V}^\sharp$ and $\boldsymbol{\beta}^0$. ■

When the errors are normally distributed, the following strong result on $\mathbf{X}\boldsymbol{\beta}^0$ holds. Since the result is only used in Section 11.2.1, we omit its detailed proof and assign a sketch of the proof to Exercise 10.22. It suffices to mention that the result is a consequence Result 10.5.2 and the classical theory on complete, sufficient statistics (see Lehmann and Casella, 1998, Chapter 2).

Result 10.5.2. Under the same setup as Result 10.5.1, let $\boldsymbol{\varepsilon} \sim N(\mathbf{0}, \mathbf{V})$. If $\mathbf{c}'\boldsymbol{\beta}$ is estimable, then $\mathbf{c}'\boldsymbol{\beta}^0$ is its unique best unbiased estimator (b.u.e.) in the sense that if $t(\mathbf{y})$ is any unbiased estimator of $\mathbf{c}'\boldsymbol{\beta}$, then $\mathrm{Var}(\mathbf{c}'\boldsymbol{\beta}^0) \leq \mathrm{Var}(t(\mathbf{y}))$, with equality holding if and only if $t(\mathbf{y}) = \mathbf{c}'\boldsymbol{\beta}^0$.

Exercises

10.1. Prove property 4 of Result 10.1.1.

10.2. In the model (10.1.3), show that $\mathbf{U1}_a = \mathbf{0}$, where $\mathbf{U} = \mathbf{D}_a - \mathbf{ND}_b^{-1}\mathbf{N}'$. Hence argue that $\mathrm{r}(\mathbf{U}) \leq a - 1$.

10.3. (a) Prove property 1 of Result 10.1.5.

(b) In the model (10.1.3), show that a linear function $\sum_{j=1}^b d_i\theta_j$ is estimable if and only if the vector $\mathbf{d} = (d_1, \cdots, d_a)'$ is a linear combination of the row vectors of the matrix $\mathbf{V}$. Hence, show that if the data is connected, a necessary and sufficient condition for a linear function of θ's to be estimable is that it is a contrast in θ's.

10.4. In model (10.1.3), show the following:

(a) $\mathrm{Var}(\mathbf{y}_a) = \sigma^2\mathbf{D}_a$, $\mathrm{Var}(\mathbf{y}_b) = \sigma^2\mathbf{D}_b$, and $\mathrm{Cov}(\mathbf{y}_a, \mathbf{y}_b) = \sigma^2\mathbf{N}$.

(b) $\mathrm{Var}(\mathbf{w}) = \sigma^2\mathbf{U}$, $\mathrm{Var}(\mathbf{z}) = \sigma^2\mathbf{V}$, and $\mathrm{Cov}(\mathbf{w}, \mathbf{z}) = -\sigma^2\mathbf{UD}_a^{-1}\mathbf{N}$.

(c) $\mathrm{Cov}(\mathbf{w}, \mathbf{y}_b) = \mathbf{O} = \mathrm{Cov}(\mathbf{z}, \mathbf{y}_a)$.

10.5. Prove Result 10.1.6.

10.6. In (10.1.11), verify whether the following functions are estimable:

(a) $[n_{i\cdot} - \sum_{j=1}^b n_{ij}^2/n_{\cdot j}]\tau_i - \sum_{i \neq i'}^a [\sum_{j=1}^b n_{ij}n_{i'j}/n_{\cdot j}]\tau_{i'}$
$+ \sum_{j=1}^b [n_{ij} - n_{ij}^2/n_{\cdot j}](\tau\theta)_{ij} - \sum_{i\neq i'}^a [\sum_{j=1}^b n_{ij}n_{i'j}/n_{\cdot j}](\tau\theta)_{i'j}$, $i = 1, \cdots, a$.

(b) $[n_{\cdot j} - \sum_{i=1}^a n_{ij}^2/n_{i\cdot}]\theta_j - \sum_{j\neq j'}^b [\sum_{i=1}^a n_{ij}n_{ij'}/n_{i\cdot}]\theta_{j'}$
$+ \sum_{i=1}^a [n_{ij} - n_{ij}^2/n_{i\cdot}](\tau\theta)_{ij} - \sum_{j\neq j'}^b [\sum_{i=1}^a n_{ij}n_{ij'}/n_{i\cdot}](\tau\theta)_{ij'}$, $j = 1, \cdots, b$.

10.7. Suppose $a = 3$ and $b = 4$ in (10.1.11), and suppose cells $(1, 2)$, $(2, 3)$, $(2, 4)$ and $(3, 1)$ are empty. Verify whether the following functions are estimable:

(a) $\tau_2 - \tau_3 + (\tau\theta)_{22} - (\tau\theta)_{32}$,

(b) $\tau_1 - \tau_3 + (\tau\theta)_{12} - (\tau\theta)_{32}$.

10.8. Prove Result 10.1.2.

10.9. Prove property 2 of Result 10.1.7.

10.10. In the model $Y_{ijk} = \mu + \tau_i + \theta_j + (\tau\theta)_{ij} + \varepsilon_{ijk}$, $k = 1, \cdots, n_{ij}$, $i = 1, \cdots, a$, $j = 1, \cdots, b$, suppose $n_{ij} = n_{i\cdot}n_{\cdot j}/N$. Given that ε_{ijk} are i.i.d. $N(0, \sigma^2)$ variables, obtain a statistic for testing H: $(\tau\theta)_{ij} = 0 \; \forall \; i, j$.

10.11. In the two-way unbalanced nested model, suppose we impose the restrictions $\mu^0 = 0$ and $\tau_i^0 = 0$, $i = 1, \cdots, a$. Obtain the least squares estimates of $\theta_{j(i)}$ and write down the corresponding g-inverse of $\mathbf{X}'\mathbf{X}$.

10.12. In the two-way unbalanced nested model, show that the function

$$\mu + \tau_i + \sum_{j=1}^{b_i} w_{ij}\theta_{j(i)}, \quad \sum_j w_{ij} = 1$$

is estimable, with b.l.u.e. given by $\sum_j w_{ij}\overline{Y}_{ij\cdot}$.

10.13. Let $Y_{ij} = \mu_i + \gamma_1 Z_{ij} + \gamma_2 W_{ij} + \varepsilon_{ij}$, $i = 1, \cdots, a$, $j = 1, \cdots, b$, and let ε_{ij} be i.i.d. $N(0, \sigma^2)$ variables.

(a) Derive the least squares estimate of γ_1, and show that it is unbiased.

(b) Find the variance-covariance matrix of the least squares estimate of $\boldsymbol{\gamma} = (\gamma_1, \gamma_2)'$.

(c) Under what conditions are $\widehat{\gamma_1}$ and $\widehat{\gamma_2}$ statistically independent?

10.14. Consider the model $Y_{ij} = \mu_i + \gamma_i Z_{ij} + \varepsilon_{ij}$, $j = 1, \cdots, b$, $i = 1, 2$, where ε_{ij} are i.i.d. $N(0, \sigma^2)$ variables. Set this up as an ANACOVA model, and derive the F-statistic for testing the hypothesis $\gamma_1 = \gamma_2$. Relate this statistic to the usual t-statistic for testing whether two straight lines are parallel.

10.15. Consider the model $Y_{ij} = \mu + \tau_i + \varepsilon_{ij}$, $j = 1, \cdots, n_i$, $i = 1, \cdots, a$, where ε_{ij} are i.i.d. $N(0, \sigma^2)$ variables. Suppose we observe a continuous variate Z_i, $i = 1, \cdots, a$ (for each $j = 1, \cdots, n_i$).

(a) Find the sum of squares due to regression, viz., $SS(\beta_0, \beta_1)$ in the model $Y_{ij} = \beta_0 + \beta_1(Z_i - \overline{Z}) + u_{ij}$, where we define $\overline{Z} = (\sum_{i=1}^a n_i Z_i)/(\sum_{i=1}^a n_i)$.

(b) Compare $SS(\beta_0, \beta_1)$ with $SS(\mu, \tau_1, \cdots, \tau_a)$, which is the "usual" model sum of squares in the one-way fixed-effects ANOVA model.

10.16. Consider the model $Y_{ijk} = \mu_{ij} + \gamma_{ij} Z_{ijk} + \varepsilon_{ijk}$, $i = 1, \cdots, a$, $j = 1, \cdots, b$, and $k = 1, \cdots, c$, where ε_{ijk} are i.i.d. $N(0, \sigma^2)$ variables. Obtain a test statistic for H: $\gamma_{ij} = \gamma$, $i = 1, \cdots, a$, $j = 1, \cdots, b$.

10.17. In the ANACOVA setup, show that $\mathcal{C}((\mathbf{I} - \mathbf{P})\mathbf{Z}) = \mathcal{C}^{\perp}(\mathbf{X}) \cap \mathcal{C}(\mathbf{X}, \mathbf{Z})$, where $\mathbf{P}$ is the projection onto $\mathcal{C}(\mathbf{X})$.

10.18. Write down the steps involved in hypothesis testing for the ANACOVA model in the framework of a nested sequence of hypotheses (see Section 7.3.2).

10.19. Let $Y_{ij} = \mu + \tau_i + \varepsilon_{ij}$, and $\varepsilon_{ij} \sim N(0, \sigma^2)$, $j = 1, \cdots, n$, $i = 1, \cdots, a$. Let $\boldsymbol{\beta} = (\mu, \tau_1, \cdots, \tau_a)'$. For $l = 1, \cdots, (a-1)$, define

$$\mathbf{c}_l'\boldsymbol{\beta} = \frac{1}{\sqrt{l(l+1)}} \left(\sum_{i=1}^{l} \tau_i - l\tau_{l+1} \right).$$

Verify that $\mathbf{c}_l'\boldsymbol{\beta}$ is estimable. Construct marginal and Scheffé simultaneous confidence intervals for these functions of $\boldsymbol{\beta}$.

10.20. Show that if under the null hypothesis, a univariate test statistic has a continuous distribution, then its one-sided p-value follows a uniform distribution.

10.21. [Hsu (1984)] Explaining the assumptions, set up and carry out Hsu's procedure for multiple comparsions with the best in the one-way fixed-effects ANOVA model.

10.22. Let $\mathbf{y} = \mathbf{X}\boldsymbol{\beta} + \boldsymbol{\varepsilon}$, where $\boldsymbol{\varepsilon} \sim N(\mathbf{0}, \mathbf{V})$ and $\mathbf{c}'\boldsymbol{\beta}$ is estimable. Follow the steps below to prove Result 10.5.2:

(a) Suppose $\mathbf{V} = \mathbf{I}$. Show that $\mathbf{X}\boldsymbol{\beta}^0 = \mathbf{P}\mathbf{y}$ is a complete and sufficient statistic of $\mathbf{X}\boldsymbol{\beta}$, where $\mathbf{P} = \mathbf{P}_{\mathcal{C}(\mathbf{X})}$.

(b) Show that if $\mathbf{V}$ is p.d.f., then $\boldsymbol{\beta}^0$ in Result 10.5.1 is the same as $\boldsymbol{\beta}^0_{GLS}$ in Section 4.5 and $\mathbf{X}\boldsymbol{\beta}^0_{GLS}$ is a complete and sufficient statistic of $\mathbf{X}\boldsymbol{\beta}$.

(c) Suppose $r(\mathbf{V}) = n < N$ and $\mathbf{L}$ is a nonsingular matrix such that $\mathbf{L}\mathbf{V}\mathbf{L}' = \text{diag}(\mathbf{I}_n, \mathbf{O})$. Show that the model can be written as

$$\mathbf{z}_1 = \mathbf{M}_1\boldsymbol{\beta} + \boldsymbol{\eta}, \quad \mathbf{z}_2 = \mathbf{M}_2\boldsymbol{\beta}, \tag{10.5.5}$$

where $\mathbf{z}_1$ consists of the first n coordinates of $\mathbf{L}\mathbf{y}$, $\mathbf{M}_1$ consists of the first n rows of $\mathbf{L}\mathbf{X}$, and $\boldsymbol{\eta} \sim N(\mathbf{0}, \mathbf{I}_n)$. Moreover, show that

$$\mathbf{V}^{\sharp} = \mathbf{L}' \begin{pmatrix} \mathbf{I}_n & \mathbf{A} \\ \mathbf{A}' & \mathbf{I}_n + \mathbf{A}'\mathbf{A} \end{pmatrix} \mathbf{L} \in \Omega_{\mathbf{V}},$$

where $\mathbf{A} = -\mathbf{M}_1\mathbf{M}_2'(\mathbf{M}_2\mathbf{M}_2')^{-}$.

(d) Show that if $\boldsymbol{\beta}^0$ minimizes $\|\mathbf{z}_1 - \mathbf{M}_1\boldsymbol{\beta}\|^2$ subject to $\mathbf{z}_2 = \mathbf{M}_2\boldsymbol{\beta}$, then $\boldsymbol{\beta}^0$ is a solution to (10.5.3) with $\mathbf{V}^{\sharp}$ defined as above.

(e) Show that (10.5.5) can be written as $\mathbf{w} = \mathbf{B}\boldsymbol{\gamma} + \boldsymbol{\eta}$, where $\mathbf{w} = \mathbf{z}_1 - \mathbf{M}_1\boldsymbol{\beta}^*$ with $\boldsymbol{\beta}^*$ a solution to $\mathbf{z}_2 = \mathbf{M}_2\boldsymbol{\beta}$, and $\mathbf{B} = \mathbf{M}_1(\mathbf{I} - \mathbf{M}_2^{-}\mathbf{M}_2)$. Furthermore, $\boldsymbol{\beta}^0 = \boldsymbol{\beta}^* + \mathbf{B}\boldsymbol{\gamma}^0$, with $\boldsymbol{\gamma}^0$ minimizes $\|\mathbf{w} - \mathbf{B}\boldsymbol{\gamma}\|^2$.

(f) Combine (a), (c)–(e) to show that $\mathbf{X}\boldsymbol{\beta}^0$ under the setup in Result 10.5.1 is a complete and sufficient statistic of $\mathbf{X}\boldsymbol{\beta}$.

(g) Show that $\mathbf{c}'\boldsymbol{\beta}^0$ is the unique b.u.e. of $\mathbf{c}'\boldsymbol{\beta}$.

11

Random- and Mixed-Effects Models

In Chapter 10, we discussed fixed-effects linear models. In many experiments, the levels of the factor of interest are assumed to be randomly drawn from a large population of levels. We then assume that the effects due to this factor are random. In many practical situations, we find the need to incorporate both fixed effects as well as random effects in order to explain the response variable Y. Such situations are modeled by mixed-effects models, i.e., an additive model involving both fixed effects and random effects. Section 11.1 gives a general description of inference for mixed-effects models, of which the random-effects model is a special case (i.e., there are no fixed effects).

The simplest model corresponds to an experiment with a single random factor, Factor A, with a levels. Inference for this model is described in detail in Section 11.3. Different estimation procedures are described and illustrated on one-factor and two-factor models.

11.1 Setup and examples of mixed-effects linear models

The linear mixed-effects model underlies analysis of a large number of statistical problems, and has the general form

$$\mathbf{y} = \mathbf{X}\boldsymbol{\beta} + \mathbf{Z}\boldsymbol{\gamma} + \boldsymbol{\varepsilon}, \tag{11.1.1}$$

where $\mathbf{y} \in \mathcal{R}^N$, $\mathbf{X}$ and $\mathbf{Z}$ are respectively known $N \times p$ and $N \times q$ matrices of covariates, $\boldsymbol{\beta} \in \mathcal{R}^p$ is a vector of constants representing fixed effects, $\boldsymbol{\gamma} \in \mathcal{R}^q$ is a random vector representing random effects, and $\boldsymbol{\varepsilon} \in \mathcal{R}^N$ is a vector of random errors, such that $\mathrm{E}(\boldsymbol{\gamma}) = \mathbf{0}$, $\mathrm{Cov}(\boldsymbol{\gamma}) = \boldsymbol{\Sigma}_\gamma$, $\mathrm{E}(\boldsymbol{\varepsilon}) = \mathbf{0}$, $\mathrm{Cov}(\boldsymbol{\varepsilon}) = \boldsymbol{\Sigma}_\varepsilon$, and $\mathrm{Cov}(\boldsymbol{\gamma}, \boldsymbol{\varepsilon}) = \mathbf{O}$. Under the model, $\mathrm{E}(\mathbf{y}) = \mathbf{X}\boldsymbol{\beta}$ and $\mathrm{Cov}(\mathbf{y}) = \mathbf{V} = \boldsymbol{\Sigma}_\varepsilon + \mathbf{Z}\boldsymbol{\Sigma}_\gamma\mathbf{Z}'$. In general, $\mathbf{V}$ need not be p.d. In this chapter, we let $\boldsymbol{\nu} = (\nu_1, \cdots, \nu_s)'$ denote the unknown parameter vector which represents the variance components in $\boldsymbol{\Sigma}_\varepsilon$ and $\boldsymbol{\Sigma}_\gamma$. If the model has no fixed effects and only has random effects, then the term $\mathbf{X}\boldsymbol{\beta} = \mathbf{0}$ in (11.1.1) and by convention, the model is referred to as a random-effects model. Also, the mixed-effects model (11.1.1) can be written without explicitly adding the random errors $\boldsymbol{\varepsilon}$, because it can be written as $\mathbf{y} = \mathbf{X}\boldsymbol{\beta} + \widetilde{\mathbf{Z}}\widetilde{\boldsymbol{\gamma}}$ with $\widetilde{\mathbf{Z}} = (\mathbf{Z}, \mathbf{I}_N)$ and $\widetilde{\boldsymbol{\gamma}} = (\boldsymbol{\gamma}', \boldsymbol{\varepsilon}')'$. Indeed, early work on mixed-effects models used this form without separating out the random errors (Harville, 1976); however, their results are entirely transferable to model (11.1.1). In this book, we will follow current convention and use (11.1.1) to denote a mixed-effects linear model. We show a few examples below.

Example 11.1.1. Random-effects one-factor model. In many situations, an experimenter is interested in a factor which has a large number of possible levels. In this case, a of these population levels may be chosen at random for investigation; we then refer to the factor as a random factor. Inference is made not about these randomly selected a levels, but rather about the entire infinite population of factor levels. For example, consider an experiment designed to study the maternal ability of mice using litter weights of ten-day old

DOI: 10.1201/9781315156651-11

litters. There are four mothers, each of which has six litters. A single laboratory technician manages the entire experiment. Let Y_{ij} denote the weight of the jth litter corresponding to the ith mouse; a possible model is the one given in (11.1.2), where μ is an overall mean effect, and γ_i denotes the effect due to the ith mouse, $i = 1, \cdots, a$. Since maternal ability is certainly variable across parents, it is unlikely that the experimenter is interested in these four specific female mice. Rather, these mice may be considered to be a random sample from a very large population of mice, and γ_i is a random effect. As another example, suppose that a textile company weaves a fabric with a large number of looms. The process engineer suspects the existence of significant variation between looms in addition to the usual variation in strength of fabric from the same loom. Four looms are selected at random, and four fabric sample strengths are obtained from each. Once again, the loom effect γ_i is treated as a random effect.

The random-effects model corresponding to a single random Factor A with a levels, and with n_i replications in level i, is given by

$$Y_{ij} = \mu + \gamma_i + \varepsilon_{ij}, \quad j = 1, \cdots, n_i, \quad i = 1, \cdots, a, \tag{11.1.2}$$

where both γ_i and ε_{ij} are random variables, while μ is a constant. We assume that $\gamma_i \sim N(0, \sigma_\gamma^2)$, and are independent of ε_{ij}, which are assumed to be $N(0, \sigma_\varepsilon^2)$ variables. The quantities $\nu_1 = \sigma_\varepsilon^2$ and $\nu_2 = \sigma_\gamma^2$ are called variance components. These are unknown parameters that must be estimated along with the overall mean μ, which is a fixed effect. Since $\text{Cov}(\gamma_i, \gamma_k) = 0$, $i \neq k$, and $\text{Cov}(\gamma_i, \varepsilon_{i'j'}) = 0$, we see that

$$\text{Cov}(Y_{ij}, Y_{i'j'}) = \begin{cases} \sigma_\gamma^2 + \sigma_\varepsilon^2, & \text{if } i = i', j = j' \\ \sigma_\gamma^2, & \text{if } i = i', j \neq j' \\ 0 & \text{if } i \neq i'. \end{cases}$$

The quantity $\rho = \sigma_\gamma^2 / (\sigma_\gamma^2 + \sigma_\varepsilon^2)$ is called the intra-class correlation, and is the proportion of variance of an observation which is due to the differences between treatments.

Let $N = n_1 + \cdots + n_a$. When $n_1 = \cdots = n_a$, the model is called balanced; otherwise, it is unbalanced. Equation (11.1.2) can be written in matrix form as

$$\mathbf{y} = \mathbf{1}_N \mu + \mathbf{Z}\boldsymbol{\gamma} + \boldsymbol{\varepsilon}, \tag{11.1.3}$$

where $\boldsymbol{\gamma} = (\gamma_1, \cdots, \gamma_a)'$ and $\mathbf{Z} = \text{diag}(\mathbf{1}_{n_1}, \cdots, \mathbf{1}_{n_a})$. Then $\text{Cov}(\boldsymbol{\gamma}) = \sigma_\gamma^2 \mathbf{I}_a$ and

$$\text{Cov}(\mathbf{y}) = \sigma_\gamma^2 \mathbf{Z}\mathbf{Z}' + \sigma_\varepsilon^2 \mathbf{I}_N = \mathbf{V} = \begin{pmatrix} \sigma_\gamma^2 \mathbf{J}_{n_1} + \sigma_\varepsilon^2 \mathbf{I}_{n_1} & & \\ & \ddots & \\ & & \sigma_\gamma^2 \mathbf{J}_{n_a} + \sigma_\varepsilon^2 \mathbf{I}_{n_a} \end{pmatrix}, \tag{11.1.4}$$

or in a more compact form, $\mathbf{V} = \text{diag}(\mathbf{V}_1, \cdots, \mathbf{V}_a)$ with $\mathbf{V}_i = \sigma_\gamma^2 \mathbf{J}_{n_i} + \sigma_\varepsilon^2 \mathbf{I}_{n_i}$. □

Example 11.1.2. Random-effects two-factor model. Recall the two-way cross-classified model *with interaction* between the two factors (balanced case)

$$Y_{ijk} = \mu + \gamma_i + \theta_j + (\gamma\theta)_{ij} + \varepsilon_{ijk}, \tag{11.1.5}$$

for $k = 1, \cdots, n$, $i = 1, \cdots, a$, and $j = 1, \cdots, b$. Suppose that the γ, θ, and $(\gamma\theta)$ effects are all random and independent, such that $\gamma_i \sim N(0, \sigma_\gamma^2)$, $\theta_j \sim N(0, \sigma_\theta^2)$, $(\gamma\theta) \sim N(0, \sigma_{\gamma\theta}^2)$, and all are independent from the i.i.d. errors $\varepsilon_{ijk} \sim N(0, \sigma_\varepsilon^2)$. The variance components are $\nu_1 = \sigma_\varepsilon^2$, $\nu_2 = \sigma_\gamma^2$, $\nu_3 = \sigma_\theta^2$, and $\nu_4 = \sigma_{\gamma\theta}^2$. □

Example 11.1.3. Mixed-effects model!two-factor model. Suppose an experiment involves a fixed factor, Factor A, with a levels, and a random factor, Factor B, with b levels. Suppose there are n replications of each combination of the levels of Factor A and Factor B. In the mixed-effects model with interactions,

$$Y_{ijk} = \mu + \tau_i + \theta_j + (\tau\theta)_{ij} + \varepsilon_{ijk}, \quad 1 \le i \le a,\ 1 \le j \le b,\ 1 \le k \le n,$$

where μ, $\tau_1, \cdots, \tau_a$ are the fixed-effects, $\theta_1, \cdots, \theta_b$ are the random main effects, and $(\tau\theta)_{ij}$ are the random interactions. Let $N = abn$. Denote the N-dimensional response vector by $\mathbf{y} = (\mathbf{y}'_{11}, \cdots, \mathbf{y}'_{1b}, \cdots, \mathbf{y}'_{a1}, \cdots, \mathbf{y}'_{ab})'$ with $\mathbf{y}_{ij} = (Y_{ij1}, \cdots, Y_{ijk})'$, and denote the error vector by $\boldsymbol{\varepsilon}$ in a similar way. Let $\boldsymbol{\beta} = (\mu, \tau_1, \cdots, \tau_a)'$, and $\boldsymbol{\gamma} = (\theta_1, \cdots, \theta_b, (\tau\theta)_{11}, \cdots, (\tau\theta)_{ab})'$. In the matrix form of the model $\mathbf{y} = \mathbf{X}\boldsymbol{\beta} + \mathbf{Z}\boldsymbol{\gamma} + \boldsymbol{\varepsilon}$, the design matrices are

$$\mathbf{X} = (\mathbf{1}_N, \mathbf{I}_a \otimes \mathbf{1}_{bn}) \quad \text{and} \quad \mathbf{Z} = (\mathbf{1}_a \otimes \mathbf{I}_b \otimes \mathbf{1}_n, \mathbf{I}_{ab} \otimes \mathbf{1}_n).$$

Suppose $\text{Var}(\theta_j) = \sigma^2_\theta$, $\text{Var}[(\tau\theta)_{jk}] = \sigma^2_{\tau\theta}$, and $\text{Var}(\varepsilon_i) = \sigma^2_\varepsilon$, then

$$\text{Var}(\boldsymbol{\gamma}) = \begin{pmatrix} \sigma^2_\theta \mathbf{I}_b & \mathbf{0} \\ \mathbf{0} & \sigma^2_{\tau\theta}\mathbf{I}_{ab} \end{pmatrix} \quad \text{and} \quad \text{Var}(\boldsymbol{\varepsilon}) = \sigma^2_\varepsilon \mathbf{I}_N.$$

The variance components are $\nu_1 = \sigma^2_\varepsilon$, $\nu_2 = \sigma^2_\theta$, and $\nu_3 = \sigma^2_{\tau\theta}$. For Kronecker product algorithms that are useful in constructing sums of squares and covariance matrices in balanced designs of this type, see Moser and Sawyer (1998). □

11.2 Inference for mixed-effects linear models

In this section, we describe inference for mixed-effects linear models. In Section 11.2.1, we discuss an important optimality result called the extended Gauss–Markov theorem. It is assumed that $\mathbf{V}$ is known. In practice, when $\mathbf{V}$ is unknown, these results may be applied with a suitable estimator of $\mathbf{V}$. In Section 11.2.2, we discuss GLS estimation for the fixed-effects coefficients in (11.1.1). Section 11.2.3 describes the well-known ANOVA method for estimating variance components associated with the random effects $\boldsymbol{\gamma}$. In addition to the ANOVA method, there are alternative estimation approaches for the mixed-effects models. Section 11.2.4 describes maximum likelihood estimation under normality, while Section 11.2.5 discusses the method of restricted maximum likelihood (REML). We end with a discussion of the minimum norm quadratic unbiased estimation (MINQUE) procedure in Section 11.2.6.

11.2.1 Extended Gauss–Markov theorem

In Sections 4.4 and 10.5, optimal linear unbiased estimators were obtained for estimable linear functions of fixed effects. In this section, we extend the results to linear combinations of fixed and random effects under the mixed-effects model (11.1.1), i.e., functions of the form $\mathbf{c}'_1\boldsymbol{\beta} + \mathbf{c}'_2\boldsymbol{\gamma}$. In the following, let $\mathbf{c}'_1\boldsymbol{\beta}$ be estimable, i.e., there is a constant vector $\mathbf{t} \in \mathcal{R}^N$, such that $\text{E}(\mathbf{t}'\mathbf{y}) = \mathbf{c}'_1\boldsymbol{\beta}$. Equivalently, let $\mathbf{c}_1 \in \mathcal{R}(\mathbf{X})$. On the other hand, let $\mathbf{c}_2$ be an arbitrary vector of constants with the same dimension as $\boldsymbol{\gamma}$.

Definition 11.2.1. A function $t(\mathbf{y})$ is called a linear unbiased estimator of $\mathbf{c}'_1\boldsymbol{\beta} + \mathbf{c}'_2\boldsymbol{\gamma}$ if $t(\mathbf{y}) = d_0 + \mathbf{d}'\mathbf{y}$ for some constant $d_0 \in \mathcal{R}$ and $\mathbf{d} \in \mathcal{R}^N$, and if $\text{E}\{t(\mathbf{y})\} = \mathbf{c}'_1\boldsymbol{\beta}$ for any $\boldsymbol{\beta}$ under the model (11.1.1). The mean squared error of the estimator is defined to be $MSE(t(\mathbf{y})) = \text{E}\{[t(\mathbf{y}) - \mathbf{c}'_1\boldsymbol{\beta} - \mathbf{c}'_2\boldsymbol{\gamma}]^2\}$.

Definition 11.2.2. Essentially-unique b.l.u.e. A linear unbiased estimator $t(\mathbf{y})$ of $\mathbf{c}_1'\boldsymbol{\beta} + \mathbf{c}_2'\boldsymbol{\gamma}$ is called an essentially-unique b.l.u.e. if for any linear unbiased estimator $d_0 + \mathbf{d}'\mathbf{y}$ of $\mathbf{c}_1'\boldsymbol{\beta} + \mathbf{c}_2'\boldsymbol{\gamma}$, $MSE(t(\mathbf{y})) \le MSE(d_0 + \mathbf{d}'\mathbf{y})$, with equality holding if and only if $d_0 + \mathbf{dy} = t(\mathbf{y})$ with probability 1.

The following extended Gauss–Markov theorem of Harville (1976) states an important optimality result for mixed-effects linear models.

Result 11.2.1. Let $\boldsymbol{\beta}^0$ denote any solution to the general normal equations

$$(\mathbf{X}'\mathbf{V}^\sharp\mathbf{X})\boldsymbol{\beta}^0 = \mathbf{X}'\mathbf{V}^\sharp\mathbf{y}, \tag{11.2.1}$$

where $\mathbf{V}^\sharp \in \Omega_{\mathbf{V}}$, the same class of g-inverses of $\mathbf{V}$ defined in Result 10.5.1. Let $\widehat{\boldsymbol{\gamma}} = \boldsymbol{\Sigma}_\gamma\mathbf{Z}'\mathbf{V}^-(\mathbf{y} - \mathbf{X}\boldsymbol{\beta}^0)$, where $\mathbf{V}^-$ is any g-inverse of $\mathbf{V}$, not necessarily the same as $\mathbf{V}^\sharp$. The following properties hold.

1. $\mathbf{c}_1'\boldsymbol{\beta}^0 + \mathbf{c}_2'\widehat{\boldsymbol{\gamma}}$ is an essentially-unique b.l.u.e. of $\mathbf{c}_1'\boldsymbol{\beta} + \mathbf{c}_2'\boldsymbol{\gamma}$.

2. If $\boldsymbol{\gamma}$ and $\boldsymbol{\varepsilon}$ are jointly normally distributed, $\mathbf{c}_1'\boldsymbol{\beta}^0 + \mathbf{c}_2'\widehat{\boldsymbol{\gamma}}$ is an essentially-unique b.u.e. of $\mathbf{c}_1'\boldsymbol{\beta} + \mathbf{c}_2'\boldsymbol{\gamma}$; see Result 10.5.2 for the definition of b.u.e.

3. $\mathbf{X}\boldsymbol{\beta}^0$ and $\widehat{\boldsymbol{\gamma}}$ are unique; in particular, they are invariant to the choice of $\mathbf{V}^\sharp$, $\mathbf{V}^-$, and $\boldsymbol{\beta}^0$. (This explains why we use $\widehat{\boldsymbol{\gamma}}$ instead of $\boldsymbol{\gamma}^0$.)

Proof. Let $\xi = \mathbf{c}_1'\boldsymbol{\beta} + \mathbf{c}_2'\boldsymbol{\gamma}$. Then $\mathrm{E}(\xi) = \mathbf{c}_1'\boldsymbol{\beta}$, $\mathrm{Var}(\xi) = \mathbf{c}_2'\boldsymbol{\Sigma}_\gamma\mathbf{c}_2$, and $\mathrm{Cov}(\xi, \mathbf{y}) = \mathbf{c}_2'\,\mathrm{Cov}(\boldsymbol{\gamma}, \mathbf{Z}\boldsymbol{\gamma} + \boldsymbol{\varepsilon}) = \mathbf{c}_2'\boldsymbol{\Sigma}_\gamma\mathbf{Z}'$. Since $\mathrm{E}(\mathbf{y}) = \mathbf{X}\boldsymbol{\beta}$ and $\mathrm{Cov}(\mathbf{y}) = \mathbf{V}$, using Result 5.2.14, we see that the b.l.u.p. of ξ based on $\mathbf{y}$ *if* $\boldsymbol{\beta}$ is known, is

$$\widehat{\xi}(\mathbf{y}, \boldsymbol{\beta}) = \mathbf{c}_1'\boldsymbol{\beta} + \mathbf{c}_2'\boldsymbol{\Sigma}_\gamma\mathbf{Z}'\mathbf{V}^-(\mathbf{y} - \mathbf{X}\boldsymbol{\beta}) = \mathbf{c}_2'\boldsymbol{\Sigma}_\gamma\mathbf{Z}'\mathbf{V}^-\mathbf{y} + \mathbf{c}'\boldsymbol{\beta},$$

where $\mathbf{c}' = \mathbf{c}_1' - \mathbf{c}_2'\boldsymbol{\Sigma}_\gamma\mathbf{Z}'\mathbf{V}^-\mathbf{X}$, and $\xi - \widehat{\xi}(\mathbf{y}, \boldsymbol{\beta})$ is uncorrelated with $\mathbf{y}$ and has mean 0 and variance $\sigma_*^2 = \mathbf{c}_2'\boldsymbol{\Sigma}_\gamma\mathbf{c}_2 - \mathbf{c}_2'\boldsymbol{\Sigma}_\gamma\mathbf{Z}'\mathbf{V}^-\mathbf{Z}\boldsymbol{\Sigma}_\gamma\mathbf{c}_2$, which is invariant to the choice of $\mathbf{V}^-$. Now, fix $\mathbf{V}^-$. For any estimator $t(\mathbf{y})$, let $\tilde{t}(\mathbf{y}) = t(\mathbf{y}) - \mathbf{c}_2'\boldsymbol{\Sigma}_\gamma\mathbf{Z}'\mathbf{V}^-\mathbf{y}$. Then $t(\mathbf{y}) - \xi = t(\mathbf{y}) - \widehat{\xi}(\mathbf{y}, \boldsymbol{\beta}) - [\xi - \widehat{\xi}(\mathbf{y}, \boldsymbol{\beta})] = \tilde{t}(\mathbf{y}) - \mathbf{c}'\boldsymbol{\beta} - [\xi - \widehat{\xi}(\mathbf{y}, \boldsymbol{\beta})]$. If $t(\mathbf{y})$ is a linear function of $\mathbf{y}$, then

$$\mathrm{E}[t(\mathbf{y}) - \xi] = \mathrm{E}[\tilde{t}(\mathbf{y})] - \mathbf{c}'\boldsymbol{\beta} \tag{11.2.2}$$

and since $\xi - \widehat{\xi}(\mathbf{y}, \boldsymbol{\beta})$ is uncorrelated with $\tilde{t}(\mathbf{y})$,

$$\mathrm{E}[t(\mathbf{y}) - \xi]^2 = \mathrm{E}[\tilde{t}(\mathbf{y}) - \mathbf{c}'\boldsymbol{\beta}]^2 + \sigma_*^2. \tag{11.2.3}$$

Since σ_*^2 is a constant independent of $\boldsymbol{\beta}$, it follows that $t(\mathbf{y})$ is an unbiased linear estimator of ξ and has the least MSE among all such estimators if and only if $\tilde{t}(\mathbf{y})$ is the b.l.u.e. of $\mathbf{c}'\boldsymbol{\beta}$. We see that $\mathbf{c} \in \mathcal{R}(\mathbf{X})$ and the mixed-effects model can be written as $\mathbf{y} = \mathbf{X}\boldsymbol{\beta} + \boldsymbol{\eta}$, where $\boldsymbol{\eta} = \mathbf{Z}\boldsymbol{\gamma} + \boldsymbol{\varepsilon}$ has mean zero and covariance $\mathbf{V}$. Then, from Result 10.5.1, $\tilde{t}(\mathbf{y})$ must be $\mathbf{c}'\boldsymbol{\beta}^0$, and so $\mathbf{c}'\boldsymbol{\beta}^0 + \mathbf{c}_2'\boldsymbol{\Sigma}_\gamma\mathbf{Z}'\mathbf{V}^-\mathbf{y} = \mathbf{c}_1'\boldsymbol{\beta}^0 + \mathbf{c}_2'\widehat{\boldsymbol{\gamma}}$ is the only one statistic with the desired property. At this point, note that we cannot rule out possible dependence of $\widehat{\boldsymbol{\gamma}}$ on the choice of $\mathbf{V}^-$. However, since we fixed $\mathbf{V}^-$, $\widehat{\boldsymbol{\gamma}}$ is uniquely defined. Hence, we have shown not only the existence of the b.l.u.e. of ξ, but also its essential-uniqueness, so property 1 is proved.

Next, if $\boldsymbol{\gamma}$ and $\boldsymbol{\varepsilon}$ are normally distributed, then $\xi - \widehat{\xi}(\mathbf{y}, \boldsymbol{\beta})$ and $\mathbf{y}$ are jointly normal, and since they are uncorrelated, they are independent. As a result, the equalities in (11.2.2) and (11.2.3) hold for any estimator $t(\mathbf{y})$ with finite variance, so that $t(\mathbf{y})$ is an essentially-unique b.u.e. of ξ if and only if $\tilde{t}(\mathbf{y})$ is the b.u.e. of $\mathbf{c}'\boldsymbol{\beta}$. From Result 10.5.2, $\tilde{t}(\mathbf{y}) = \mathbf{c}'\boldsymbol{\beta}^0$, showing property 2.

Finally, the uniqueness of $\mathbf{X}\boldsymbol{\beta}^0$ is part of Result 10.5.1. Then $\widehat{\boldsymbol{\gamma}}$ formally only depends on the choice of $\mathbf{V}^-$. Let $\widehat{\boldsymbol{\gamma}}_i$, $i = 1, 2$, be constructed using two different choices of $\mathbf{V}^-$. Then from property 2, for any $\mathbf{c}_2$, both $\mathbf{c}_2'\widehat{\boldsymbol{\gamma}}_i$ are essentially-unique b.l.u.e.'s of $\mathbf{c}_2'\boldsymbol{\gamma}$, so $\mathbf{c}_2'\widehat{\boldsymbol{\gamma}}_1 = \mathbf{c}_2'\widehat{\boldsymbol{\gamma}}_2$ with probability 1. Since $\mathbf{c}_2$ is arbitrary, then $\widehat{\boldsymbol{\gamma}}_1 = \widehat{\boldsymbol{\gamma}}_2$. Alternatively, from (10.5.4), $\mathbf{y} - \mathbf{X}\boldsymbol{\beta}^0 = \mathbf{V}\mathbf{z}$, where $\mathbf{z}$ is a linear function of $\boldsymbol{\eta}$ that does not depend on the choice of $\mathbf{V}^-$. Then $\widehat{\boldsymbol{\gamma}} = \boldsymbol{\Sigma}_\gamma \mathbf{Z}'\mathbf{V}^-\mathbf{V}\mathbf{z}$. Since $\mathbf{V} = \mathbf{Z}\boldsymbol{\Sigma}_\gamma\mathbf{Z}' + \boldsymbol{\Sigma}_\varepsilon$, from Result 3.1.10, $\mathbf{Z}\boldsymbol{\Sigma}_\gamma\mathbf{Z}'\mathbf{V}^-\mathbf{V}\mathbf{z} = \mathbf{Z}\boldsymbol{\Sigma}_\gamma\mathbf{z}$. Letting $\mathbf{L} = \boldsymbol{\Sigma}_\gamma^{1/2}$, $(\mathbf{Z}\mathbf{L})(\mathbf{Z}\mathbf{L})'\mathbf{V}^-\mathbf{V}\mathbf{z} = (\mathbf{Z}\mathbf{L})(\mathbf{Z}\mathbf{L})'\mathbf{z}$, so $(\mathbf{Z}\mathbf{L})'\mathbf{V}^-\mathbf{V}\mathbf{z} = (\mathbf{Z}\mathbf{L})'\mathbf{z}$. Premultiplying both sides of the equation by $\mathbf{L}$, we see that $\widehat{\boldsymbol{\gamma}} = \boldsymbol{\Sigma}_\gamma\mathbf{Z}'\mathbf{z}$, which is invariant to the choice $\mathbf{V}^-$. Property 3 follows. ∎

The dimension of $\mathbf{V}$ is N, which may result in computational burden for large N. In many applications, however, we can assume some structure for $\mathbf{V}$, which reduces the computational effort (see Harville, 1976, section 3). The next result gives the covariance of the essentially-unique b.l.u.e.'s.

Result 11.2.2. Consider $\mathbf{C}_1'\boldsymbol{\beta} + \mathbf{C}_2'\boldsymbol{\gamma}$, where each column of $\mathbf{C}_1$ belongs to $\mathcal{R}(\mathbf{X})$ and $\mathbf{C}_2$ is arbitrary. Let $\boldsymbol{\beta}^0$ and $\widehat{\boldsymbol{\gamma}}$ be as in Result 11.2.1. Then,

$$\begin{aligned}
\operatorname{Cov}(\mathbf{C}_1'\boldsymbol{\beta}^0) &= \mathbf{C}_1'(\mathbf{X}'\mathbf{V}^\sharp\mathbf{X})^-\mathbf{X}'\mathbf{V}^\sharp\mathbf{V}(\mathbf{V}^\sharp)'\mathbf{X}\{(\mathbf{X}'\mathbf{V}^\sharp\mathbf{X})^-\}'\mathbf{C}_1,\\
\operatorname{Cov}(\widehat{\boldsymbol{\gamma}} - \boldsymbol{\gamma}) &= \boldsymbol{\Sigma}_\gamma - \boldsymbol{\Sigma}_\gamma\mathbf{Z}'\mathbf{V}^-\mathbf{Z}\boldsymbol{\Sigma}_\gamma\\
&\quad + \boldsymbol{\Sigma}_\gamma\mathbf{Z}'\mathbf{V}^-\mathbf{X}(\mathbf{X}'\mathbf{V}^\sharp\mathbf{X})^-\mathbf{X}'\mathbf{V}^\sharp\mathbf{V}(\mathbf{V}^-)'\mathbf{Z}\boldsymbol{\Sigma}_\gamma,\\
\operatorname{Cov}(\mathbf{C}_1'\boldsymbol{\beta}^0, \widehat{\boldsymbol{\gamma}} - \boldsymbol{\gamma}) &= -\operatorname{Cov}(\mathbf{C}_1'\boldsymbol{\beta}^0, \boldsymbol{\gamma})\\
&= -\mathbf{C}_1'(\mathbf{X}'\mathbf{V}^\sharp\mathbf{V})^-\mathbf{X}'\mathbf{V}^\sharp\mathbf{V}(\mathbf{V}^-)'\mathbf{Z}\boldsymbol{\Sigma}_\gamma.
\end{aligned}$$

If $\mathcal{C}(\mathbf{X}) \subset \mathcal{C}(\mathbf{V})$, then the covariance matrices simplify to

$$\begin{aligned}
\operatorname{Cov}(\mathbf{C}_1'\boldsymbol{\beta}^0) &= \mathbf{C}_1'(\mathbf{X}'\mathbf{V}^\sharp\mathbf{X})^-\mathbf{C}_1,\\
\operatorname{Cov}(\widehat{\boldsymbol{\gamma}} - \boldsymbol{\gamma}) &= \boldsymbol{\Sigma}_\gamma - \boldsymbol{\Sigma}_\gamma\mathbf{Z}'\mathbf{V}^-\mathbf{Z}\boldsymbol{\Sigma}_\gamma\\
&\quad + \boldsymbol{\Sigma}_\gamma\mathbf{Z}'\mathbf{V}^-\mathbf{X}(\mathbf{X}'\mathbf{V}^\sharp\mathbf{X})^-\mathbf{X}'(\mathbf{V}^-)'\mathbf{Z}\boldsymbol{\Sigma}_\gamma,\\
\operatorname{Cov}(\mathbf{C}_1'\boldsymbol{\beta}^0, \widehat{\boldsymbol{\gamma}} - \boldsymbol{\gamma}) &= -\mathbf{C}_1'(\mathbf{X}'\mathbf{V}^\sharp\mathbf{V})^-\mathbf{X}'(\mathbf{V}^-)'\mathbf{Z}\boldsymbol{\Sigma}_\gamma.
\end{aligned}$$

Note that from Result 11.2.1, the covariance matrices are invariant to the choice of $\boldsymbol{\beta}^0$.

Proof. From (11.2.1), given choices of $\mathbf{V}^\sharp$ and $(\mathbf{X}'\mathbf{V}^\sharp\mathbf{X})^-$, there is a unique solution vector $\boldsymbol{\beta}^0 = (\mathbf{X}'\mathbf{V}^\sharp\mathbf{X})^-\mathbf{X}'\mathbf{V}^\sharp\mathbf{y}$, so that

$$\operatorname{Cov}(\boldsymbol{\beta}^0) = (\mathbf{X}'\mathbf{V}^\sharp\mathbf{X})^-\mathbf{X}'\mathbf{V}^\sharp\mathbf{V}\{(\mathbf{X}'\mathbf{V}^\sharp\mathbf{X})^-\mathbf{X}'\mathbf{V}^\sharp\}'. \tag{11.2.4}$$

The formula for $\operatorname{Cov}(\mathbf{C}_1'\boldsymbol{\beta}^0)$ follows. Since $\widehat{\boldsymbol{\gamma}}$ is a linear function of $\mathbf{y} - \mathbf{X}\boldsymbol{\beta}^0$, from Result 10.5.1, it is uncorrelated with $\mathbf{X}\boldsymbol{\beta}^0$, and so is uncorrelated with $\mathbf{C}_1'\boldsymbol{\beta}^0$, giving $\operatorname{Cov}(\mathbf{C}_1'\boldsymbol{\beta}^0, \widehat{\boldsymbol{\gamma}} - \boldsymbol{\gamma}) = -\operatorname{Cov}(\mathbf{C}_1'\boldsymbol{\beta}^0, \boldsymbol{\gamma})$. To prove the remaining identities, using the argument in the proof of Result 11.2.1, we can get the following matrix version of (11.2.3):

$$\operatorname{Cov}(\mathbf{C}_1'\boldsymbol{\beta}^0 + \mathbf{C}_2'(\widehat{\boldsymbol{\gamma}} - \boldsymbol{\gamma})) = \operatorname{Cov}(\mathbf{C}'\boldsymbol{\beta}^0) + \boldsymbol{\Sigma}_*,$$

where $\mathbf{C}' = \mathbf{C}_1' - \mathbf{C}_2'\boldsymbol{\Sigma}_\gamma\mathbf{Z}'\mathbf{V}^-\mathbf{X}$ and $\boldsymbol{\Sigma}_* = \mathbf{C}_2'\boldsymbol{\Sigma}_\gamma\mathbf{C}_2 - \mathbf{C}_2'\boldsymbol{\Sigma}_\gamma\mathbf{Z}'\mathbf{V}^-\mathbf{Z}\boldsymbol{\Sigma}_\gamma\mathbf{C}_2$. Then, given choices of the g-inverses involved,

$$\begin{aligned}
&\mathrm{Cov}(\mathbf{C}_1'\boldsymbol{\beta}^0) + \mathbf{C}_2'\,\mathrm{Cov}(\widehat{\boldsymbol{\gamma}}-\boldsymbol{\gamma})\mathbf{C}_2 + \mathrm{Cov}(\mathbf{C}_1'\boldsymbol{\beta}^0, \widehat{\boldsymbol{\gamma}}-\boldsymbol{\gamma})\mathbf{C}_2 \\
&+ \mathbf{C}_2'\,\mathrm{Cov}(\widehat{\boldsymbol{\gamma}}-\boldsymbol{\gamma}, \mathbf{C}_1'\boldsymbol{\beta}^0) \\
&= [\mathbf{C}_1' - \mathbf{C}_2'\boldsymbol{\Sigma}_\gamma\mathbf{Z}'\mathbf{V}^-\mathbf{X}]\ \mathrm{Cov}(\boldsymbol{\beta}^0)[\mathbf{C}_1 - \mathbf{X}'(\mathbf{V}^-)'\mathbf{Z}\boldsymbol{\Sigma}_\gamma\mathbf{C}_2] \\
&\quad + \mathbf{C}_2'\boldsymbol{\Sigma}_\gamma\mathbf{C}_2 - \mathbf{C}_2'\boldsymbol{\Sigma}_\gamma\mathbf{Z}'\mathbf{V}^-\mathbf{Z}\boldsymbol{\Sigma}_\gamma\mathbf{C}_2.
\end{aligned}$$

Since this holds for arbitrary $\mathbf{C}_2'$, by comparing the respective quadratic terms and linear terms in $\mathbf{C}_2'$ on both sides,

$$\begin{aligned}
\mathrm{Cov}(\widehat{\boldsymbol{\gamma}}-\boldsymbol{\gamma}) &= \boldsymbol{\Sigma}_\gamma - \boldsymbol{\Sigma}_\gamma\mathbf{Z}'\mathbf{V}^-\mathbf{Z}\boldsymbol{\Sigma}_\gamma + \boldsymbol{\Sigma}_\gamma\mathbf{Z}'\mathbf{V}^-\mathbf{X}\mathbf{D}\mathbf{Z}\boldsymbol{\Sigma}_\gamma, \\
\mathrm{Cov}(\mathbf{C}_1'\boldsymbol{\beta}^0, \widehat{\boldsymbol{\gamma}}-\boldsymbol{\gamma}) &= -\mathbf{C}_1'\mathbf{D}\mathbf{Z}\boldsymbol{\Sigma}_r,
\end{aligned}$$

where $\mathbf{D} = \mathrm{Cov}(\boldsymbol{\beta}^0)\mathbf{X}'(\mathbf{V}^-)'$. From (11.2.4),

$$\mathbf{D} = (\mathbf{X}'\mathbf{V}^\sharp\mathbf{X})^-(\mathbf{X}'\mathbf{V}^\sharp\mathbf{V})\{\mathbf{V}^-\mathbf{X}(\mathbf{X}'\mathbf{V}^\sharp\mathbf{X})^-\mathbf{X}'\mathbf{V}^\sharp\}'.$$

From (10.5.2), all the rows of $\mathbf{X}'\mathbf{V}^\sharp\mathbf{V}$ belong to $\mathcal{C}(\mathbf{N})^\perp = \mathcal{C}(\mathbf{X})$, so $\mathbf{X}'\mathbf{V}^\sharp\mathbf{V} = \mathbf{L}\mathbf{X}'$ for some matrix $\mathbf{L}$. Then

$$\begin{aligned}
\mathbf{D} &= (\mathbf{X}'\mathbf{V}^\sharp\mathbf{X})^-\mathbf{L}\mathbf{X}'\{\mathbf{V}^-\mathbf{X}(\mathbf{X}'\mathbf{V}^\sharp\mathbf{X})^-\mathbf{X}'\mathbf{V}^\sharp\}' \\
&= (\mathbf{X}'\mathbf{V}^\sharp\mathbf{X})^-\mathbf{L}\{\mathbf{V}^-\mathbf{X}(\mathbf{X}'\mathbf{V}^\sharp\mathbf{X})^-\mathbf{X}'\mathbf{V}^\sharp\mathbf{X}\}'.
\end{aligned}$$

Since by (10.5.1), $\mathcal{R}(\mathbf{X}) = \mathcal{R}(\mathbf{X}'\mathbf{V}^\sharp\mathbf{X})$, from Result 3.1.11,

$$\mathbf{D} = (\mathbf{X}'\mathbf{V}^\sharp\mathbf{X})^-\mathbf{L}(\mathbf{V}^-\mathbf{X})' = (\mathbf{X}'\mathbf{V}^\sharp\mathbf{X})^-\mathbf{L}\mathbf{X}'(\mathbf{V}^-)'.$$

Replace $\mathbf{L}\mathbf{X}'$ with $\mathbf{X}'\mathbf{V}^\sharp\mathbf{V}$ and plug the resulting expression for $\mathbf{D}$ into the formulas for the covariance matrices. Then the identities for $\mathrm{Cov}(\widehat{\boldsymbol{\gamma}}-\boldsymbol{\gamma})$ and $\mathrm{Cov}(\mathbf{C}_1'\boldsymbol{\beta}^0, \widehat{\boldsymbol{\gamma}}-\boldsymbol{\gamma})$ follow.

Let $\mathbf{S} = \mathbf{X}'\mathbf{V}^\sharp\mathbf{X}$ and notice that $(\mathbf{V}^\sharp)'$ is a g-inverse of the symmetric matrix $\mathbf{V}$ as well. If $\mathcal{C}(\mathbf{X}) \subset \mathcal{C}(\mathbf{V})$, then from Result 3.1.11, $\mathbf{X}'\mathbf{V}^\sharp\mathbf{V}(\mathbf{V}^\sharp)'\mathbf{X} = \mathbf{X}'\mathbf{V}^\sharp\mathbf{X} = \mathbf{S}$, so that $\mathrm{Cov}(\mathbf{C}_1'\boldsymbol{\beta}^0) = \mathbf{C}_1'\mathbf{S}^-\mathbf{S}\mathbf{S}^{-\prime}\mathbf{C}_1$. From (10.5.1), $\mathcal{R}(\mathbf{S}) = \mathcal{R}(\mathbf{X})$, so $\mathcal{R}(\mathbf{C}_1') \subset \mathcal{R}(\mathbf{S})$. Again by Result 3.1.11, $\mathrm{Cov}(\mathbf{C}_1'\boldsymbol{\beta}^0) = \mathbf{C}_1'\mathbf{S}^{-\prime}\mathbf{C}_1 = \mathbf{C}_1'\mathbf{S}^-\mathbf{C}_1$ due to the symmetry of the covariance matrix. Then the identity for $\mathrm{Cov}(\mathbf{C}_1')\boldsymbol{\beta}^0$ follows, provided $\mathcal{C}(\mathbf{X}) \subset \mathcal{C}(\mathbf{V})$. The other two simplified identities follow from $\mathbf{X}'\mathbf{V}^\sharp\mathbf{V} = \mathbf{X}'$, again using Result 3.1.11. ■

11.2.2 GLS estimation of fixed effects

For the mixed-effects linear model in (11.1.1), let $\boldsymbol{\eta} = \mathbf{Z}\boldsymbol{\gamma} + \boldsymbol{\varepsilon}$ and regard it as an error vector. Since $\mathrm{E}(\boldsymbol{\eta}) = \mathbf{0}$, the LS fit $\widehat{\mathbf{y}} = \mathbf{X}\boldsymbol{\beta}^0$ for the regression $\mathbf{y} = \mathbf{X}\boldsymbol{\beta} + \boldsymbol{\eta}$ is unbiased. However, in general it is not the b.l.u.e. of $\mathbf{X}\boldsymbol{\beta}$, and so it is inefficient. Suppose $\mathbf{V}$ is positive definite. Then from Section 4.5, the b.l.u.e. of $\mathbf{X}\boldsymbol{\beta}$ is $\mathbf{X}\boldsymbol{\beta}^0_{GLS}$ with

$$\boldsymbol{\beta}^0_{GLS} = (\mathbf{X}'\mathbf{V}^{-1}\mathbf{X})^-\mathbf{X}'\mathbf{V}^{-1}\mathbf{y}. \tag{11.2.5}$$

The b.l.u.e. of $\mathbf{X}\boldsymbol{\beta}$ is therefore $\mathbf{X}(\mathbf{X}'\mathbf{V}^{-1}\mathbf{X})^-\mathbf{X}'\mathbf{V}^{-1}\mathbf{y}$. This can also be derived from Section 10.5.

Note that $\boldsymbol{\Sigma}_\varepsilon$ and $\boldsymbol{\Sigma}_\gamma$ are unknown. If they can be estimated, the b.l.u.e. of $\mathbf{X}\boldsymbol{\beta}$ can be approximated by plugging in the estimated $\mathbf{V} = \boldsymbol{\Sigma}_\varepsilon + \mathbf{Z}\boldsymbol{\Sigma}_\gamma\mathbf{Z}'$. Henceforth, let $\boldsymbol{\Sigma}_\varepsilon = \boldsymbol{\Sigma}_\varepsilon(\boldsymbol{\nu})$ and $\boldsymbol{\Sigma}_\gamma = \boldsymbol{\Sigma}_\gamma(\boldsymbol{\nu})$, where $\boldsymbol{\nu}$ is the vector of unknown variance components as we defined earlier.

11.2.3 ANOVA method for estimation

The ANOVA method is a method of moments type approach. We have discussed the ANOVA approach in detail in earlier chapters. The advantage of the ANOVA method is that estimation does not require any distributional assumption on $\mathbf{y}$, although assumption of normality leads to the usual F-tests of hypotheses. The ANOVA estimation of the variance components proceeds as follows. First, partition $\mathbf{Z}$ into $(\mathbf{Z}_1, \cdots, \mathbf{Z}_s)$, $s \le q$, where each $\mathbf{Z}_i$ is a submatrix of $\mathbf{Z}$. Partition $\boldsymbol{\gamma}$ into $\boldsymbol{\gamma}_1, \cdots, \boldsymbol{\gamma}_s$ correspondingly. Regard $\boldsymbol{\gamma}$ as a vector of *fixed effects*. For $i = 0, \ldots, q$, fit a LS regression for the "fixed-effects model"

$$\mathbf{y} = \mathbf{X}\boldsymbol{\beta} + \mathbf{Z}_1\boldsymbol{\gamma}_1 + \cdots + \mathbf{Z}_s\boldsymbol{\gamma}_s + \boldsymbol{\varepsilon}.$$

This yields the SSE as well as a sequence of SSR's

$$SSR(\mathbf{X}),\ SSR(\mathbf{Z}_i \,|\, \mathbf{X}, \mathbf{Z}_1, \cdots, \mathbf{Z}_{i-1}),\ i = 1, \cdots, s.$$

Express the expected values of the SSR's and SSE in terms of elements of $\boldsymbol{\Sigma}_\gamma$ and $\boldsymbol{\Sigma}_\varepsilon$, i.e., as functions of components of $\boldsymbol{\nu}$:

$$\begin{aligned} &\mathrm{E}[SSR(\mathbf{X})] = h(\boldsymbol{\nu}), \quad \mathrm{E}(SSE) = g(\boldsymbol{\nu}), \\ &\mathrm{E}[SSR(\mathbf{Z}_i \,|\, \mathbf{X}, \mathbf{Z}_1, \cdots, \mathbf{Z}_{i-1})] = f_i(\boldsymbol{\nu}), \end{aligned}$$

say. The ANOVA estimate of $\boldsymbol{\nu}$ is a solution to the equations

$$\begin{aligned} &SSR(\mathbf{X}) = h(\widehat{\boldsymbol{\nu}}), \quad SSE = g(\widehat{\boldsymbol{\nu}}), \\ &SSE(\mathbf{Z}_i \,|\, \mathbf{X}, \mathbf{Z}_1, \cdots, \mathbf{Z}_{i-1}) = f_i(\widehat{\boldsymbol{\nu}}). \end{aligned}$$

However, the equations may not have a solution, may have multiple solutions, or may have a solution that yields nonpositive matrices as estimates of $\boldsymbol{\Sigma}_\gamma$. This is a drawback of the ANOVA method in particular, and of the method of moments in general.

This procedure is called the ANOVA method for estimating variance components, because it utilizes the components in the usual ANOVA table that we write down in the fixed-effects model. In Section 11.3, we discuss the ANOVA method for the one-factor random-effects model, while in Section 11.4, we give examples of the ANOVA method for two-factor random-effects and mixed-effects models.

11.2.4 Method of maximum likelihood

We write the mixed-effects model (11.1.1) as

$$\mathbf{y} = \mathbf{X}\boldsymbol{\beta} + \mathbf{Z}_1\boldsymbol{\gamma}_1 + \cdots + \mathbf{Z}_m\boldsymbol{\gamma}_m + \boldsymbol{\varepsilon}, \tag{11.2.6}$$

where $\boldsymbol{\gamma}_j$ is a q_j-dimensional vector of random effects, and $\mathbf{Z}_j$ is the corresponding $N \times q_j$ design matrix, $j = 1, \cdots, m$, $\boldsymbol{\varepsilon}$ is the N-dimensional error vector, $\mathbf{X}$ is an $N \times p$ matrix of known regressors, and $\boldsymbol{\beta}$ is a p-dimensional vector of fixed effects. We assume that $\mathrm{E}(\boldsymbol{\gamma}_j) = \mathbf{0}$, $\mathrm{Cov}(\boldsymbol{\gamma}_j) = \sigma_j^2\mathbf{I}_{q_j}$, $\mathrm{Cov}(\boldsymbol{\gamma}_j, \boldsymbol{\gamma}_l) = \mathbf{O}$, $j \ne l$, $\mathrm{Cov}(\boldsymbol{\varepsilon}) = \sigma_\varepsilon^2\mathbf{I}_N$, and $\mathrm{Cov}(\boldsymbol{\gamma}_j, \boldsymbol{\varepsilon}) = \mathbf{O}$. Then, $\mathrm{E}(\mathbf{y}) = \mathbf{X}\boldsymbol{\beta}$ and

$$\mathbf{V} = \mathrm{Var}(\mathbf{y}) = \sum_{j=1}^{m} \sigma_j^2 \mathbf{Z}_j\mathbf{Z}_j' + \sigma_\varepsilon^2\mathbf{I}_N. \tag{11.2.7}$$

For ease of notation, we set $\boldsymbol{\varepsilon} \equiv \boldsymbol{\gamma}_0$, $N \equiv q_0$, and $\mathbf{Z}_0 \equiv \mathbf{I}_N$. The likelihood function has the form

$$L(\boldsymbol{\beta}, \mathbf{V}; \mathbf{y}, \mathbf{X}) = (2\pi)^{-N/2}|\mathbf{V}|^{-1/2}\exp\left\{-\frac{1}{2}(\mathbf{y} - \mathbf{X}\boldsymbol{\beta})'\mathbf{V}^{-1}(\mathbf{y} - \mathbf{X}\boldsymbol{\beta})\right\}. \tag{11.2.8}$$

Differentiate $\log L(\boldsymbol{\beta}, \mathbf{V}; \mathbf{y}, \mathbf{X})$ with respect to $\boldsymbol{\beta}$ and σ_j^2, to yield the likelihood equations

$$\frac{\partial \log L}{\partial \boldsymbol{\beta}} = \mathbf{X}'\mathbf{V}^{-1}\mathbf{y} - \mathbf{X}'\mathbf{V}^{-1}\mathbf{X}\boldsymbol{\beta} = \mathbf{0}, \tag{11.2.9}$$

and

$$\frac{\partial \log L}{\partial \sigma_j^2} = -\operatorname{tr}(\mathbf{V}^{-1}\mathbf{Z}_j\mathbf{Z}_j')/2 + (\mathbf{y} - \mathbf{X}\boldsymbol{\beta})'\mathbf{V}^{-1}\mathbf{Z}_j\mathbf{Z}_j'\mathbf{V}^{-1}(\mathbf{y} - \mathbf{X}\boldsymbol{\beta}) = 0,$$

$$j = 0, \cdots, m. \tag{11.2.10}$$

Denote by $\widetilde{\boldsymbol{\beta}}$ and $\widetilde{\sigma}_j^2$ the solutions to (11.2.9) and (11.2.10), the latter being nonlinear functions of the variance components. Usually, closed form solutions do not exist; moreover, the solutions of the variance components may be negative. Provided $\widetilde{\sigma}_0^2 > 0$ and $\widetilde{\sigma}_j^2 \geq 0$, $j = 1, \cdots, m$, they are the MLEs of the variance components, which are otherwise set to zero. Then, by the invariance principle, the MLE of $\mathbf{V}$ is obtained by substituting σ^2's by their MLEs in (11.2.7), and the MLE of $\boldsymbol{\beta}$ is obtained by substituting $\mathbf{V}$ by its MLE in (11.2.9). Let us denote the MLEs by $\widehat{\boldsymbol{\beta}}_{ML}$, and $\widehat{\sigma}_{j,ML}^2$, $j = 0, \cdots, m$. For more details, the reader is referred to chapter 6 of Searle et al. (1992). In Exercise 11.5, we ask the reader to derive the asymptotic variances of these estimators.

11.2.5 REML estimation

The method of restricted maximum likelihood (REML) was introduced as a technique for estimating variance components in a random-effects or mixed-effects general linear model (Patterson and Thompson, 1971). Let $\mathbf{y} \sim N(\mathbf{X}\boldsymbol{\beta}, \mathbf{V})$, where $\mathbf{V} = \mathbf{V}(\boldsymbol{\nu})$, for some unknown parameter vector $\boldsymbol{\nu}$. Suppose $\boldsymbol{\nu}$ is the parameter of interest, and $\boldsymbol{\beta}$ is a nuisance parameter. The marginal likelihood approach eliminates $\boldsymbol{\beta}$ from the likelihood function and obtains the REML estimator of $\boldsymbol{\nu}$ as the MLE based on a linear transformation of $\mathbf{y}$ which has zero expectation, and whose distribution is independent of the location parameter $\boldsymbol{\beta}$. More specifically, if $\mathbf{B}$ is a matrix such that $\mathbf{B}'\mathbf{X} = \mathbf{O}$, then $\mathbf{B}'\mathbf{y} \sim N(\mathbf{0}, \mathbf{B}'\mathbf{V}\mathbf{B})$. Provided that $\mathbf{B}'\mathbf{V}\mathbf{B}$ is nonsingular of rank m, the log-likelihood function of $\boldsymbol{\nu}$ based on $\mathbf{B}'\mathbf{y}$ is

$$\ell(\boldsymbol{\nu}; \mathbf{B}'\mathbf{y}) = -\frac{m}{2}\log(2\pi) - \frac{1}{2}\log|\mathbf{B}'\mathbf{V}(\boldsymbol{\nu})\mathbf{B}| - \frac{1}{2}\mathbf{y}'\mathbf{B}[\mathbf{B}'\mathbf{V}(\boldsymbol{\nu})\mathbf{B}]^{-1}\mathbf{B}'\mathbf{y}, \tag{11.2.11}$$

and the corresponding MLE is $\arg\min_{\boldsymbol{\nu}} \log|\mathbf{B}'\mathbf{V}(\boldsymbol{\nu})\mathbf{B}| + \mathbf{y}'\mathbf{B}[\mathbf{B}'\mathbf{V}(\boldsymbol{\nu})\mathbf{B}]^{-1}\mathbf{B}'\mathbf{y}$. Clearly, the MLE depends on the choice of $\mathbf{B}$. However, if $\mathbf{V}(\boldsymbol{\nu})$ is nonsingular, then as long as $\mathbf{B}$ has full column rank $N - r$ such that $\mathbf{B}'\mathbf{X} = \mathbf{O}$, where $r = \mathrm{r}(\mathbf{X})$, the MLE of $\boldsymbol{\nu}$ is independent of any specific choice of $\mathbf{B}$. Indeed, for any two such matrices $\mathbf{B}_1$ and $\mathbf{B}_2$, since $\mathcal{C}(\mathbf{B}_1) = \mathcal{C}(\mathbf{B}_2) = \mathcal{C}(\mathbf{X})^{\perp}$, $\mathbf{B}_2 = \mathbf{B}_1\mathbf{T}$ for some nonsingular $(N - r) \times (N - r)$ matrix $\mathbf{T}$. This implies that the MLE of $\boldsymbol{\nu}$ based on $\mathbf{B}_2'\mathbf{y}$ is identical to the one based on $\mathbf{B}_1'\mathbf{y}$. The REML specifically refers to estimation using such a matrix $\mathbf{B}$.

Result 11.2.3. Suppose $\mathbf{V}(\boldsymbol{\nu})$ is nonsingular for all $\boldsymbol{\nu}$. The REML estimator $\widehat{\boldsymbol{\nu}}_{RM}$ maximizes the function

$$\begin{aligned}\ell_{RM}(\boldsymbol{\nu}) = & -\frac{1}{2}\log|\mathbf{V}(\boldsymbol{\nu})| - \frac{1}{2}\log|\mathbf{X}_1'\mathbf{V}^{-1}(\boldsymbol{\nu})\mathbf{X}_1| \\ & -\frac{1}{2}(\mathbf{y} - \mathbf{X}\boldsymbol{\beta}_{GLS}^0)'\mathbf{V}^{-1}(\boldsymbol{\nu})(\mathbf{y} - \mathbf{X}\boldsymbol{\beta}_{GLS}^0),\end{aligned} \tag{11.2.12}$$

where $\mathbf{X}_1$ is any submatrix of $\mathbf{X}$ consisting of r LIN column vectors and $\boldsymbol{\beta}_{GLS}^0 = \boldsymbol{\beta}_{GLS}^0(\boldsymbol{\nu}, \mathbf{y}) = [\mathbf{X}'\mathbf{V}^{-1}(\boldsymbol{\nu})\mathbf{X}]^{-}\mathbf{X}'\mathbf{V}^{-1}(\boldsymbol{\nu})\mathbf{y}$ is a GLS solution.

Proof. For ease of notation, we use $\mathbf{V}$ to denote $\mathbf{V}(\boldsymbol{\nu})$. We must show that the maximizer of $\ell_{RM}(\boldsymbol{\nu})$ is identical to that of the log-likelihood function in (11.2.11), where as mentioned earlier, the matrix $\mathbf{B}$ has full column rank $N-r$, and $\mathbf{B}'\mathbf{X} = \mathbf{O}$. Comparing (11.2.12) with (11.2.11), we see that it suffices to show

$$\mathbf{y}'\mathbf{B}(\mathbf{B}'\mathbf{V}\mathbf{B})^{-1}\mathbf{B}'\mathbf{y} = (\mathbf{y} - \mathbf{X}\boldsymbol{\beta}^0_{GLS})'\mathbf{V}^{-1}(\mathbf{y} - \mathbf{X}\boldsymbol{\beta}^0_{GLS}) \tag{11.2.13}$$

and

$$|\mathbf{B}'\mathbf{V}\mathbf{B}| = c|\mathbf{V}||\mathbf{X}_1'\mathbf{V}^{-1}\mathbf{X}_1|, \tag{11.2.14}$$

for some constant $c > 0$ that only depends on $\mathbf{B}$ and $\mathbf{X}$. First, from Section 4.5, if $\mathbf{K}$ is any nonsingular matrix with $\mathbf{V}^{-1} = \mathbf{K}'\mathbf{K}$, then

$$(\mathbf{y} - \mathbf{X}\boldsymbol{\beta}^0_{GLS})'\mathbf{V}^{-1}(\mathbf{y} - \mathbf{X}\boldsymbol{\beta}^0_{GLS}) = \|(\mathbf{I} - \mathbf{P}_{\mathcal{C}(\mathbf{KX})})\mathbf{K}\mathbf{y}\|^2.$$

Let $\mathbf{L} = \mathbf{K}^{-1\prime}$. Then, $(\mathbf{LB})'(\mathbf{KX}) = \mathbf{O}$ and $\mathbf{LB}$ has full column rank $N-r$. It follows that $\mathcal{C}(\mathbf{LB}) = \mathcal{C}(\mathbf{KX})^{\perp}$, so $\mathbf{I} - \mathbf{P}_{\mathcal{C}(\mathbf{KX})} = \mathbf{P}_{\mathcal{C}(\mathbf{LB})}$. The sum of squares shown above is equal to $\|\mathbf{P}_{\mathcal{C}(\mathbf{LB})}\mathbf{K}\mathbf{y}\|^2 = \mathbf{y}'\mathbf{B}(\mathbf{B}\mathbf{V}\mathbf{B}')^{-1}\mathbf{B}\mathbf{y}$, showing (11.2.13).

Next, let $\mathbf{Q}_1$ be an $N \times r$ matrix whose column vectors form an orthonormal basis of $\mathcal{C}(\mathbf{X})$, and let $\mathbf{Q}_2$ be an $N \times (N-r)$ matrix whose column vectors form an orthonormal basis of $\mathcal{C}(\mathbf{B})$. Then $\mathbf{Q} = (\mathbf{Q}_1, \mathbf{Q}_2)$ is an orthogonal matrix, and $\mathbf{X}_1 = \mathbf{Q}_1\mathbf{R}_1$ and $\mathbf{B} = \mathbf{Q}_2\mathbf{R}_2$ for some nonsingular matrices $\mathbf{R}_1$ and $\mathbf{R}_2$. Let

$$\mathbf{U} = \mathbf{Q}'\mathbf{V}\mathbf{Q} = \begin{pmatrix} \mathbf{U}_{11} & \mathbf{U}_{12} \\ \mathbf{U}_{21} & \mathbf{U}_{22} \end{pmatrix},$$

where $\mathbf{U}_{11} \in \mathcal{R}^{r\times r}$ and $\mathbf{U}_{22} \in \mathcal{R}^{(N-r)\times(N-r)}$. Likewise, define $(\mathbf{U}^{-1})_{ij}$, $i, j = 1, 2$. Then by $\mathbf{Q}_1'\mathbf{Q} = (\mathbf{I}_r, \mathbf{O})$,

$$\begin{aligned} |\mathbf{X}_1'\mathbf{V}^{-1}\mathbf{X}_1| &= |\mathbf{R}_1'\mathbf{Q}_1'(\mathbf{Q}\mathbf{U}\mathbf{Q}')^{-1}\mathbf{Q}_1\mathbf{R}_1| \\ &= |\mathbf{R}_1|^2|\mathbf{Q}_1'\mathbf{Q}\mathbf{U}^{-1}\mathbf{Q}'\mathbf{Q}_1| = |\mathbf{R}_1|^2|(\mathbf{U}^{-1})_{11}|, \end{aligned}$$

and likewise $|\mathbf{B}'\mathbf{V}\mathbf{B}| = |\mathbf{R}_2|^2|\mathbf{U}_{22}|$. From Results 2.1.3 and 2.1.4, $(\mathbf{U}^{-1})_{11} = (\mathbf{U}_{11.2})^{-1}$ and $|\mathbf{U}| = |\mathbf{U}_{11.2}||\mathbf{U}_{22}|$, where $\mathbf{U}_{11.2} = \mathbf{U}_{11} - \mathbf{U}_{12}\mathbf{U}_{22}^{-1}\mathbf{U}_{21}$. Then,

$$|\mathbf{V}||\mathbf{X}_1'\mathbf{V}^{-1}\mathbf{X}_1| = |\mathbf{U}| \cdot |\mathbf{R}_1|^2|(\mathbf{U}_{11.2})^{-1} = |\mathbf{R}_1|^2|\mathbf{U}_{11}|.$$

Thus, (11.2.14) follows by letting $c = |\mathbf{R}_2|^2/|\mathbf{R}_1|^2$. ■

In estimating random effects, maximum likelihood estimation does not account for the degrees of freedom that are involved in estimating fixed effects. REML (sometimes called residual maximum likelihood) corrects this defect, and obtains estimates of variance components in a general mixed-effects linear model based on residuals from an OLS fit to the fixed effects portion of the model. Notice that REML is a variation of the method of maximum likelihood in which we maximize just that part of the likelihood function that is *location invariant.* For a balanced model, the REML estimator coincides with the ANOVA estimator. We show some examples in Sections 11.3 and 11.4.

11.2.6 MINQUE estimation

This section gives a discussion of minimum norm quadratic unbiased estimation of variance components. We first show a result on optimal estimation of the error variance in the model (4.1.1), which we denote by $\sigma_\varepsilon^2 \geq 0$. Rao (1952) proposed the class $\mathcal{A}$ of nonnegative quadratic unbiased estimates $\mathbf{y}'\mathbf{A}\mathbf{y}$ of $(N-r)\sigma_\varepsilon^2$, where $\mathbf{A}$ is symmetric and n.n.d.

Result 11.2.4. Optimal estimator of σ_ε^2. Consider the linear model (4.1.1). Suppose the errors are i.i.d. and their second, third, and fourth central moments exist and are respectively equal to σ_ε^2, μ_3, and μ_4. Then $(N-r)\widehat{\sigma}_\varepsilon^2$ is the unique nonnegative quadratic unbiased estimator of $(N-r)\sigma_\varepsilon^2$, with minimum variance when $\mu_4 = 3\sigma_\varepsilon^4$, or when the nonzero diagonal elements of the orthogonal projection matrix $\mathbf{P}$ are equal (Atiqullah, 1962).

Proof. For a symmetric and n.n.d. $N \times N$ matrix $\mathbf{A}$, let $\mathbf{a}$ denote the vector of its diagonal elements. We will state sufficient conditions for $\mathbf{y}'(\mathbf{I}-\mathbf{P})\mathbf{y}$ to have minimum variance in $\mathcal{A}$. If $\mathbf{y}'\mathbf{A}\mathbf{y} \in \mathcal{A}$, by Result 4.2.3,

$$(N-r)\sigma_\varepsilon^2 \equiv \mathrm{E}(\mathbf{y}'\mathbf{A}\mathbf{y}) = \sigma_\varepsilon^2 \,\mathrm{tr}(\mathbf{A}) + \boldsymbol{\beta}'\mathbf{X}'\mathbf{A}\mathbf{X}\boldsymbol{\beta}$$

for all $\boldsymbol{\beta}$. Then $\mathrm{tr}(\mathbf{A}) = (N-r)$ and $\mathbf{X}'\mathbf{A}\mathbf{X} = \mathbf{O}$, which in turn implies that $\mathbf{A}\mathbf{X} = \mathbf{O}$ (see Exercise 2.28). Then $\mathbf{y}'\mathbf{A}\mathbf{y} = (\boldsymbol{\beta}'\mathbf{X}' + \boldsymbol{\varepsilon}')\mathbf{A}(\mathbf{X}\boldsymbol{\beta} + \boldsymbol{\varepsilon}) = \boldsymbol{\varepsilon}'\mathbf{A}\boldsymbol{\varepsilon} = \sum_i a_{ii}\varepsilon_i^2 + 2\sum_{i<j} a_{ij}\varepsilon_i\varepsilon_j$, where $\mathbf{A} = \{a_{ij}\}$. It can be shown that

$$\mathrm{Var}(\mathbf{y}'\mathbf{A}\mathbf{y}) = (\mu_4 - 3\sigma_\varepsilon^4)\mathbf{a}'\mathbf{a} + 2\sigma_\varepsilon^4 \,\mathrm{tr}(\mathbf{A}^2). \tag{11.2.15}$$

The proof of (11.2.15) is left as Exercise 11.9. From (4.2.26), $(N-r)\widehat{\sigma}_\varepsilon^2 = \mathbf{y}'(\mathbf{I}-\mathbf{P})\mathbf{y} \in \mathcal{A}$, and by letting $\mathbf{A} = \mathbf{I} - \mathbf{P}$ in (11.2.15), where $\mathbf{P}$ is the projection matrix onto $\mathcal{C}(\mathbf{X})$,

$$\mathrm{Var}[\mathbf{y}'(\mathbf{I}-\mathbf{P})\mathbf{y}] = (\mu_4 - 3\sigma_\varepsilon^4)\mathbf{d}'\mathbf{d} + 2\sigma_\varepsilon^4(N-r), \tag{11.2.16}$$

where $\mathbf{d}$ is the vector of diagonal elements of $\mathbf{I}-\mathbf{P}$. Suppose $\mathbf{A} = (\mathbf{I}-\mathbf{P}) + \mathbf{R}$, and similarly, $\mathbf{a} = \mathbf{d} + \mathbf{r}$. It follows that $\mathbf{R}$ is symmetric, with $\mathrm{tr}(\mathbf{R}) = \mathrm{tr}(\mathbf{A}) - \mathrm{tr}(\mathbf{I}-\mathbf{P}) = 0$. Since $\mathbf{O} = \mathbf{A}\mathbf{P} = \mathbf{R}\mathbf{P}$, we get $\mathbf{R}(\mathbf{I}-\mathbf{P}) = \mathbf{R}$; it follows that $\mathrm{tr}(\mathbf{A}^2) = \mathrm{tr}(\mathbf{I}-\mathbf{P}) + \mathrm{tr}(\mathbf{R}^2) + 2\,\mathrm{tr}(\mathbf{R}) = (N-r) + \mathrm{tr}(\mathbf{R}^2)$. Substitute $\mathbf{a} = \mathbf{d} + \mathbf{r}$ in (11.2.15) and use (11.2.16) to get

$$\begin{aligned}
&\mathrm{Var}(\mathbf{y}'\mathbf{A}\mathbf{y})\\
&= (\mu_4 - 3\sigma_\varepsilon^4)\mathbf{a}'\mathbf{a} + 2\sigma_\varepsilon^4[(N-r) + \mathrm{tr}(\mathbf{R}^2)]\\
&= (\mu_4 - 3\sigma_\varepsilon^4)(\mathbf{d}'\mathbf{d} + 2\mathbf{d}'\mathbf{r} + \mathbf{r}'\mathbf{r}) + 2\sigma_\varepsilon^4[(N-r) + \mathrm{tr}(\mathbf{R}^2)]\\
&= (\mu_4 - 3\sigma_\varepsilon^4)\mathbf{d}'\mathbf{d} + 2\sigma_\varepsilon^4(N-r) + 2\left[(\mu_4 - 3\sigma_\varepsilon^4)\left(\mathbf{d}'\mathbf{r} + \frac{1}{2}\mathbf{r}'\mathbf{r}\right) + \sigma_\varepsilon^4 \,\mathrm{tr}(\mathbf{R}^2)\right]\\
&- \mathrm{Var}(\mathbf{y}'(\mathbf{I}-\mathbf{P})\mathbf{y}) + 2(\mu_4 - 3\sigma_\varepsilon^4)\left(\sum_{i=1}^N d_{ii}r_{ii} + \frac{1}{2}\sum_{i=1}^N r_{ii}^2\right) + 2\sigma_\varepsilon^4 \sum_{i=1}^N\sum_{j=1}^N r_{ij}^2.
\end{aligned}$$

Although minimization of $\mathrm{Var}(\mathbf{y}'\mathbf{A}\mathbf{y})$ subject to $\mathrm{tr}(\mathbf{R}) = 0$ and $\mathbf{R}(\mathbf{I}-\mathbf{P}) = \mathbf{R}$ is difficult in general, the two special cases in the result are easily treatable. First, when $\mu_4 = 3\sigma_\varepsilon^4$, $\mathrm{Var}(\mathbf{y}'\mathbf{A}\mathbf{y}) = \mathrm{Var}(\mathbf{y}'(\mathbf{I}-\mathbf{P})\mathbf{y}) + 2\sigma_\varepsilon^4\sum_{i=1}^N\sum_{j=1}^N r_{ij}^2$, which is minimum when $r_{ij} = 0$ for $i, j = 1, \cdots, N$, i.e., when $\mathbf{A} = \mathbf{I} - \mathbf{P}$. Second, if all the diagonal elements of $(\mathbf{I}-\mathbf{P})$ are equal, then they are necessarily equal to $(N-r)/N$ since the trace of $\mathbf{I}-\mathbf{P}$ is $N-r$. The proof that $\mathrm{Var}(\mathbf{y}'\mathbf{A}\mathbf{y})$ attains a minimum in this case is left as Exercise 11.8. ■

We now describe MINQUE estimation in the context of the model formulation in (11.2.6), with the same moment assumptions on $\boldsymbol{\gamma}_j$'s and $\boldsymbol{\varepsilon}$ as we had earlier. It follows from Rao (1970, 1971a,b, 1972) that

$$\mathrm{Var}(\mathbf{y}) = \sum_{j=1}^m \sigma_j^2\mathbf{Z}_j\mathbf{Z}_j' + \sigma_\varepsilon^2\mathbf{I}_N = \sigma_\varepsilon^2\mathbf{T}_0 + \sum_{j=1}^m \sigma_j^2\mathbf{T}_j,$$

where $\mathbf{T}_j = \mathbf{Z}_j\mathbf{Z}_j'$, $j = 1, \cdots, m$, $\mathbf{T}_0 = \mathbf{Z}_0\mathbf{Z}_0'$, $\mathbf{Z}_0 = \mathbf{I}_N$. For ease of notation, set $\sigma_\varepsilon^2 = \sigma_0^2$.

There are $s = m + 1$ variance components, i.e., $\nu_1 = \sigma_\varepsilon^2, \nu_2 = \sigma_1^2, \cdots, \nu_{m+1} = \sigma_m^2$. The problem is to estimate a function of the variance components, $F = \sum_{j=0}^m f_j\sigma_j^2$, by some quadratic form $\mathbf{y}'\mathbf{A}\mathbf{y}$. We require $\mathbf{y}'\mathbf{A}\mathbf{y}$ to be

(i) invariant with respect to translation in $\boldsymbol{\beta}$, i.e., if $\boldsymbol{\beta} \to \boldsymbol{\beta}_d = \boldsymbol{\beta} - \boldsymbol{\beta}_0$, we require $\mathbf{y}'\mathbf{A}\mathbf{y} = \mathbf{y}_d'\mathbf{A}\mathbf{y}_d$; a condition for this is $\mathbf{A}\mathbf{x} = \mathbf{O}$, and

(ii) we require $\mathrm{E}(\mathbf{y}'\mathbf{A}\mathbf{y}) = F$, for which we must have

$$\mathrm{E}(\mathbf{y}'\mathbf{A}\mathbf{y}) = \sum_{j=0}^m \mathrm{tr}(\mathbf{A}\mathbf{T}_j)\sigma_j^2 = F,$$

which in turn requires that $\mathrm{tr}(\mathbf{A}\mathbf{T}_j) = f_j$, $j = 0, \cdots, m$.

Now, $\mathrm{Var}(\boldsymbol{\gamma}_j) = \sigma_j^2\mathbf{I}_{q_j}$; this implies that if $\boldsymbol{\gamma}_j$'s are known, a natural estimator of σ_j^2 would be $\boldsymbol{\gamma}_j'\boldsymbol{\gamma}_j/q_j$. If $\boldsymbol{\varepsilon}$ were also known, then $\boldsymbol{\varepsilon}'\boldsymbol{\varepsilon}/N$ would be a natural estimate of σ_0^2. In this case, F would be estimated by

$$\widehat{F} = \sum_{j=1}^m f_j\boldsymbol{\gamma}_j'\boldsymbol{\gamma}_j/q_j + f_0\boldsymbol{\varepsilon}'\boldsymbol{\varepsilon}/N = \boldsymbol{\eta}'\boldsymbol{\Delta}\boldsymbol{\eta},$$

say, where,

$$\boldsymbol{\eta} = (\boldsymbol{\gamma}_1', \cdots, \boldsymbol{\gamma}_m', \boldsymbol{\varepsilon}')', \text{ and}$$
$$\boldsymbol{\Delta} = \mathrm{diag}\{(f_1/q_1)\mathbf{1}_{q_1}, \cdots, (f_m/q_m)\mathbf{1}_{q_m}, (f_0/N)\mathbf{1}_N\}.$$

Since $\mathbf{A}\mathbf{X} = \mathbf{O}$, the proposed estimator is

$$\mathbf{y}'\mathbf{A}\mathbf{y} = \left\{\sum_{j=1}^m \mathbf{Z}_j\boldsymbol{\gamma}_j + \mathbf{Z}_0\boldsymbol{\varepsilon}\right\}' \mathbf{A} \left\{\sum_{j=1}^m \mathbf{Z}_j\boldsymbol{\gamma}_j + \mathbf{Z}_0\boldsymbol{\varepsilon}\right\} = \boldsymbol{\eta}'\widetilde{\mathbf{Z}}'\mathbf{A}\widetilde{\mathbf{Z}}\boldsymbol{\eta},$$

where $\widetilde{\mathbf{Z}} = \{\mathbf{Z}_1, \cdots, \mathbf{Z}_m, \mathbf{Z}_0\}$.

Now, $\widetilde{\mathbf{Z}}'\boldsymbol{\Delta}\boldsymbol{\eta}$ will be closest to $\mathbf{y}'\mathbf{A}\mathbf{y}$ if the Frobenius norm (see Definition 1.3.21) $\|\widetilde{\mathbf{Z}}'\mathbf{A}\widetilde{\mathbf{Z}} - \boldsymbol{\Delta}\|$ is small. The matrix $\mathbf{A}$ is chosen to minimize this norm. Since $\|\widetilde{\mathbf{Z}}'\mathbf{A}\widetilde{\mathbf{Z}} - \boldsymbol{\Delta}\|^2$ is the sum of squares of elements of $\widetilde{\mathbf{Z}}'\mathbf{A}\widetilde{\mathbf{Z}} - \boldsymbol{\Delta}$, we see after some simplification that

$$\begin{aligned}\|\widetilde{\mathbf{Z}}'\mathbf{A}\widetilde{\mathbf{Z}} - \boldsymbol{\Delta}\|^2 &= \mathrm{tr}\{(\widetilde{\mathbf{Z}}'\mathbf{A}\widetilde{\mathbf{Z}} - \boldsymbol{\Delta})'(\widetilde{\mathbf{Z}}'\mathbf{A}\widetilde{\mathbf{Z}} - \boldsymbol{\Delta})\} \\ &= \mathrm{tr}(\mathbf{A}\widetilde{\mathbf{Z}}\widetilde{\mathbf{Z}}')^2 - \sum_{j=1}^m \{f_j^2/q_j + f_0^2/N\}.\end{aligned}$$

Let $\mathbf{T} = \sum_{j=0}^m \mathbf{T}_j = \sum_{j=0}^m \mathbf{Z}_j\mathbf{Z}_j' = \widetilde{\mathbf{Z}}\widetilde{\mathbf{Z}}'$. The MINQUE of F is $\mathbf{y}'\mathbf{A}\mathbf{y}$, where $\mathbf{A}$ is such that $\mathrm{tr}(\mathbf{A}\widetilde{\mathbf{Z}}\widetilde{\mathbf{Z}}')^2$, which is equal to $\mathrm{tr}(\mathbf{A}\mathbf{T})^2$, is minimum subject to $\mathbf{A}\mathbf{X} = \mathbf{O}$, and $\mathrm{tr}(\mathbf{A}\mathbf{T}_j) = f_j$, $j = 0, \cdots, m$. Rao (1971a,b) showed that this minimum occurs when $\mathbf{A} = \sum_{j=0}^m \lambda_j\mathbf{S}\mathbf{T}_j\mathbf{S}$, where,

$$\mathbf{S} = \mathbf{T}^{-1} - \mathbf{T}^{-1}\mathbf{X}(\mathbf{X}'\mathbf{T}^{-1}\mathbf{X})^{-}\mathbf{X}'\mathbf{T}^{-1},$$

and λ_j are solutions of $\sum_{j=0}^m \lambda_j\,\mathrm{tr}(\mathbf{S}\mathbf{T}_j\mathbf{S}\mathbf{T}_l) = f_l$, $l = 0, \cdots, m$. Hence, the MINQUE of F is $\mathbf{y}'\mathbf{A}\mathbf{y} = \sum_{j=0}^m \lambda_j\mathbf{y}'\mathbf{S}\mathbf{T}_j\mathbf{S}\mathbf{y} = \sum_{j=0}^m \lambda_j u_j$, where $u_j = \mathbf{y}'\mathbf{S}\mathbf{T}_j\mathbf{S}\mathbf{y}$, $j = 0, \cdots, m$. Let

$\mathbf{u} = (u_0, u_1, \cdots, u_m)'$. To find $\boldsymbol{\lambda} = (\lambda_0, \cdots, \lambda_m)'$, let $w_{jl} = \text{tr}(\mathbf{ST}_j\mathbf{ST}_l)$, $j, l = 0, \cdots, m$, and suppose $\mathbf{W} = \{w_{jl}\}$ is an $(m+1) \times (m+1)$ matrix with these elements. Then, $\sum_{j=0}^{m} \lambda_j \text{tr}(\mathbf{ST}_j\mathbf{ST}_l) = f_l$, $l = 0, \cdots, m$ if and only if $\mathbf{W}\boldsymbol{\lambda} = \mathbf{f}$, where $\mathbf{f}' = (f_0, \cdots, f_m)$; the solution is $\boldsymbol{\lambda} = \mathbf{W}^-\mathbf{f}$. Hence, the MINQUE of F is

$$\mathbf{y}'\mathbf{A}\mathbf{y} = \sum_{j=0}^{m} \lambda_j u_j = \boldsymbol{\lambda}'\mathbf{u} = (\mathbf{W}^-\mathbf{f})'\mathbf{u} = \mathbf{f}'\mathbf{W}^-\mathbf{u} = \mathbf{f}'\widehat{\boldsymbol{\sigma}} = \sum_{j=0}^{m} f_j \widehat{\sigma}_j^2,$$

where $\widehat{\boldsymbol{\sigma}} = (\widehat{\sigma}_0^2, \cdots, \widehat{\sigma}_m^2)' = \mathbf{W}^-\mathbf{u}$.

11.3 One-factor random-effects model

The random-effects model corresponding to a single random Factor A with a levels, and with n_i replications in level $i = 1, \cdots, a$ was given by (11.1.2) and (11.1.3). Note that γ_i and ε_{ij} are random variables, while μ is a constant. The variance components are $\nu_1 = \sigma_\varepsilon^2$ and $\nu_2 = \sigma_\gamma^2$.

Result 11.3.1. The GLS estimate of μ in the model (11.1.3) is

$$\widehat{\mu}_{GLS} = \sum_{i=1}^{a} \frac{n_i \overline{Y}_{i\cdot}}{\sigma_\varepsilon^2 + n_i\sigma_\gamma^2} \Bigg/ \sum_{i=1}^{a} \frac{n_i}{\sigma_\varepsilon^2 + n_i\sigma_\gamma^2}. \tag{11.3.1}$$

Proof. From Section 4.5 and (11.1.4), the GLS estimate of μ is

$$\widehat{\mu}_{GLS} = \frac{\mathbf{1}_N'\mathbf{V}^{-1}\mathbf{y}}{\mathbf{1}_N'\mathbf{V}^{-1}\mathbf{1}_N}, \tag{11.3.2}$$

where $\mathbf{V}^{-1} = \text{diag}(\mathbf{V}_1^{-1}, \cdots, \mathbf{V}_a^{-1})$, with $\mathbf{V}_i^{-1} = (\sigma_\gamma^2\mathbf{J}_{n_i} + \sigma_\varepsilon^2\mathbf{I}_{n_i})^{-1}$. Since, by Sherman–Morrison–Woodbury in Result 1.3.8,

$$\mathbf{V}_i^{-1} = \frac{1}{\sigma_\varepsilon^2}\left(\mathbf{I}_{n_i} - \frac{\sigma_\gamma^2}{\sigma_\varepsilon^2 + n_i\sigma_\gamma^2}\mathbf{J}_{n_i}\right), \tag{11.3.3}$$

it follows that

$$\begin{aligned}\widehat{\mu}_{GLS} &= \frac{\sum_{i=1}^{a}\{Y_{i\cdot} - n_i\sigma_\gamma^2 Y_{i\cdot}/(\sigma_\varepsilon^2 + n_i\sigma_\gamma^2)\}/\sigma_\varepsilon^2}{\sum_{i=1}^{a}\{n_i - n_i^2\sigma_\gamma^2/(\sigma_\varepsilon^2 + n_i\sigma_\gamma^2)\}/\sigma_\varepsilon^2} \\ &= \frac{\sum_{i=1}^{a} n_i\overline{Y}_{i\cdot}/(\sigma_\varepsilon^2 + n_i\sigma_\gamma^2)}{\sum_{i=1}^{a} n_i/(\sigma_\varepsilon^2 + n_i\sigma_\gamma^2)} = \frac{\sum_{i=1}^{a}\overline{Y}_{i\cdot}/\text{Var}(\overline{Y}_{i\cdot})}{\sum_{i=1}^{a} 1/\text{Var}(\overline{Y}_{i\cdot})},\end{aligned} \tag{11.3.4}$$

where $\text{Var}(\overline{Y}_{i\cdot}) = \sigma_\gamma^2 + \sigma_\varepsilon^2/n_i$. The GLS estimate of μ is the weighted average of the cell means, the weights being the reciprocals of their variances. When $n_i = n$, $i = 1, \cdots, a$, $\widehat{\mu}_{GLS} = \overline{Y}_{\cdot\cdot}$. Note that $\widehat{\mu}_{GLS}$ depends on the unknown variance components which define the variance covariance matrix of $\mathbf{y}$. The final form of the estimate will be obtained by substituting estimates for σ_γ^2 and σ_ε^2 into (11.3.4). ■

In Section 11.1, we discussed the ANOVA method, the method of maximum likelihood and the restricted maximum likelihood (REML) method for estimating the variance components. Recall that the ANOVA method starts from the ANOVA table that we have described in the fixed-effects model, and using the expected mean squares, derives an F-statistic. The likelihood based methods use the assumption that the errors ε_{ij} are normally distributed.

ANOVA method

In the *balanced* random-effects model, we estimate the parameters σ_ε^2 and σ_γ^2 as follows. First, we write down the ANOVA decomposition as if all effects are fixed, i.e.,

$$\sum_{i=1}^{a}\sum_{j=1}^{n}(Y_{ij}-\overline{Y}_{..})^2 = \sum_{i=1}^{a} n(\overline{Y}_{i\cdot}-\overline{Y}_{..})^2 + \sum_{i=1}^{a}\sum_{j=1}^{n}(Y_{ij}-\overline{Y}_{i\cdot})^2, \text{ i.e.,}$$

$$SST_c = SS_A + SSE.$$

Result 11.3.2. The ANOVA estimates of the variance components are

$$\widehat{\sigma}_\varepsilon^2 = MSE, \text{ and } \widehat{\sigma}_\gamma^2 = \frac{1}{n}(MS_A - MSE). \tag{11.3.5}$$

Proof. Let MS_A and MSE be the mean squares corresponding to Factor A and the error respectively. Using results from Section 5.4, it can be shown that MS_A and MSE are distributed independently with $\text{E}(MS_A) = \sigma_\varepsilon^2 + n\sigma_\gamma^2$, and $\text{E}(MSE) = \sigma_\varepsilon^2$. It follows that $\text{E}(SS_A) = (a-1)(n\sigma_\gamma^2+\sigma_\varepsilon^2)$, and $\text{E}(SSE) = a(n-1)\sigma_\varepsilon^2$. The ANOVA method of estimation consists of equating the expected sums of squares to the corresponding observed values, i.e.,

$$SS_A = (a-1)(n\widehat{\sigma}_\gamma^2 + \widehat{\sigma}_\varepsilon^2) \quad \text{and} \quad SSE = a(n-1)\widehat{\sigma}_\varepsilon^2,$$

and solving the resulting equations (which are linear in the variance components) for the expressions in (11.3.5). ■

Recall that $\widehat{\mu} = \overline{Y}_{..}$ in the balanced case. These estimators are unbiased for the true parameters and are computationally simple to obtain. One disadvantage of the ANOVA estimator of σ_γ^2 is that it may be negative, which occurs when $MS_A < MSE$. The method itself offers no protection against a negative estimate for σ_γ^2, which may occur with some data. A negative variance component might be interpreted as evidence that σ_γ^2 is in fact equal to zero, and the negative value is a result of sampling variability. This interpretation seems especially sensible when the unbiased estimator $\widehat{\sigma}_\gamma^2$ has a large negative value. Then, the model is reduced to $Y_{ij} = \mu+\varepsilon_{ij}$, and the ANOVA estimator of σ_ε^2 is $\widehat{\sigma}_\varepsilon^2 = SST_c/(an-1)$. Alternatively, a negative $\widehat{\sigma}_\gamma^2$ could indicate model misspecification, and we might try to fit a more suitable model to the data. Yet again, it might be an indication of "undersampling", so that collecting more data might yield a positive estimate. In general, we would wish to avoid negative estimates of the variance components.

Result 11.3.3. Under the normality assumption of the errors,

$$\begin{aligned} \frac{(a-1)MS_A}{\sigma_\varepsilon^2 + n\sigma_\gamma^2} &\sim \chi^2_{a-1}, \\ \frac{a(n-1)MSE}{\sigma_\varepsilon^2} &\sim \chi^2_{N-a} \\ \frac{MS_A/(n\sigma_\gamma^2+\sigma_\varepsilon^2)}{MSE/\sigma_\varepsilon^2} &\sim F_{a-1,a(n-1)}. \end{aligned} \tag{11.3.6}$$

Proof. That the distributions of SS_A and SSE are chi-square follows directly from the results in Section 5.4. Also, SS_A and SSE are independently distributed. Note that

$$\widehat{\sigma}_\varepsilon^2 = MSE \sim \{\sigma_\varepsilon^2/a(n-1)\}\chi^2_{a(n-1)},$$

i.e., $\widehat{\sigma}_\varepsilon^2$ is distributed as a scaled $\chi^2_{a(n-1)}$ variable with scale factor $\sigma_\varepsilon^2/a(n-1)$. Although

MS_A and MSE are each distributed as a multiple of a χ^2 random variable, and are independently distributed, it does not follow that $(MS_A - MSE)$ has a χ^2 distribution; therefore, neither does $\widehat{\sigma}^2_\gamma$. In fact, this estimator does not have a simple closed form distribution. It follows directly from normal distribution theory that

$$\begin{aligned}
\mathrm{Var}(\widehat{\sigma}^2_\varepsilon) &= \mathrm{Var}(MSE) = 2\sigma^4_\varepsilon/a(n-1),\\
\mathrm{Var}(\widehat{\sigma}^2_\gamma) &= \mathrm{Var}[(MS_A - MSE)/n]\\
&= \frac{2}{n^2}\left\{\frac{(n\sigma^2_\gamma+\sigma^2_\varepsilon)^2}{a-1}+\frac{\sigma^4_\varepsilon}{a(n-1)}\right\}, \text{ and}\\
\mathrm{Cov}(\widehat{\sigma}^2_\varepsilon,\widehat{\sigma}^2_\gamma) &= -\frac{1}{n}\mathrm{Var}(MSE) = \frac{-2\sigma^4_\varepsilon}{an(n-1)}.
\end{aligned} \tag{11.3.7}$$

Unbiased estimates of these quantities are obtained by replacing σ^2_ε and σ^2_γ by $\widehat{\sigma}^2_\varepsilon$ and $\widehat{\sigma}^2_\gamma$ respectively, and by replacing $a(n-1)$ by $a(n-1)+2$ in the denominator (i.e., adding 2 to the denominator degrees of freedom) in each formula. Further,

$$\frac{MS_A/(n\sigma^2_\gamma+\sigma^2_\varepsilon)}{MSE/\sigma^2_\varepsilon} \sim F_{a-1,a(n-1)}, \tag{11.3.8}$$

so that MS_A/MSE has a distribution which is a multiple of a central F-distribution, the multiple reducing to 1 when $\sigma^2_\gamma = 0$. ■

Based on the distribution theory under normality, we can construct *confidence interval* estimates for the variance components. Since

$$\mathrm{P}(\chi^2_{N-a,1-\alpha/2} \le (N-a)MSE/\sigma^2_\varepsilon \le \chi^2_{N-a,\alpha/2}) = 1-\alpha,$$

the exact $100(1-\alpha)\%$ confidence interval for σ^2_ε is

$$\left((N-a)MSE/\chi^2_{N-a,\alpha/2}, (N-a)MSE/\chi^2_{N-a,1-\alpha/2}\right). \tag{11.3.9}$$

There is no exact confidence interval for σ^2_γ. The distribution of $\widehat{\sigma}^2_\gamma$ is a linear combination of two χ^2 random variables, i.e., $c_1\chi^2_{a-1} - c_2\chi^2_{N-a}$, where,

$$c_1 = (\sigma^2_\varepsilon + n\sigma^2_\gamma)/(N-n), \text{ and } c_2 = \sigma^2_\varepsilon/\{n(N-a)\}.$$

Since there is no known closed form expression for this distribution, only an approximate confidence interval for σ^2_γ can be obtained. It is, however, possible to construct exact confidence intervals for functions such as $\rho = \sigma^2_\gamma/(\sigma^2_\varepsilon+\sigma^2_\gamma)$, $\sigma^2_\varepsilon/(\sigma^2_\varepsilon+\sigma^2_\gamma)$, and $\sigma^2_\gamma/\sigma^2_\varepsilon$. For instance, the confidence interval for ρ is derived using the fact that

$$\frac{MS_A/(\sigma^2_\varepsilon+n\sigma^2_\gamma)}{MSE/\sigma^2_\varepsilon} \sim F_{a-1,N-a}.$$

Hence,

$$\mathrm{P}\left(F_{a-1,N-a,1-\alpha/2} \le \frac{MS_A/(\sigma^2_\varepsilon+n\sigma^2_\gamma)}{\{MSE/\sigma^2_\varepsilon} \le F_{a-1,N-a,\alpha/2}\right) = 1-\alpha,$$

and after some rearrangement of terms, we obtain the $100(1-\alpha)\%$ confidence interval for

$\sigma_\gamma^2/\sigma_\varepsilon^2$ as (L, U), where

$$L = \frac{1}{n}\left(\frac{MS_A}{(F_{a-1,N-a,\alpha/2})MSE} - 1\right),$$
$$U = \frac{1}{n}\left(\frac{MS_A}{(F_{a-1,N-a,1-\alpha/2})MSE} - 1\right).$$

Then, the $100(1-\alpha)\%$ confidence interval for ρ is $\left(\frac{L}{1+L}, \frac{U}{1+U}\right)$. Let us now discuss hypothesis tests. In the random-effects model, it is no longer meaningful to test hypotheses comparing treatment effects γ_i.

Result 11.3.4. Consider a test of

$$H_0\colon \sigma_\gamma^2 = 0 \text{ versus } H_1\colon \sigma_\gamma^2 > 0. \tag{11.3.10}$$

Under H_0,

$$F_0 = MS_A/MSE \sim F_{a-1,N-a}.$$

Proof. Under the null hypothesis, all treatments are identical, while treatments are variable under H_1. Under the null hypothesis, we have $SS_A/\sigma_\varepsilon^2 \sim \chi^2_{a-1}$, $SSE/\sigma_\varepsilon^2 \sim \chi^2_{N-a}$, and SS_A and SSE are distributed independently. Note that under H_0, both MS_A and MSE are unbiased estimators of σ_ε^2, while under H_1, we would expect that $\mathrm{E}(MS_A) > \mathrm{E}(MSE)$. The ratio $F_0 = MS_A/MSE$ has an $F_{a-1,N-a}$ distribution under H_0, and we would reject H_0 for large values of the test statistic, i.e., if $F_0 > F_{a-1,N-a,\alpha}$, where α is the chosen level of significance. ■

Under normality, it is possible to compute the probability of a negative estimate for σ_γ^2 as

$$\begin{aligned} p = \mathrm{P}(\widehat{\sigma}_\gamma^2 < 0) &= \mathrm{P}(MS_A < MSE) \\ &= \mathrm{P}\left(F_{a-1,N-a} < \frac{\sigma_\varepsilon^2}{\sigma_\varepsilon^2 + n\sigma_\gamma^2}\right) \end{aligned} \tag{11.3.11}$$

for various choices of a and n. Studies indicate that p decreases as either a or n increases. In any experiment, it is more important to have many classes (large a) than it is to have many observations per class (large n). Also, if $\sigma_\gamma^2 > \sigma_\varepsilon^2$, p is zero, except for small values of a (say, $a < 4$). If $\sigma_\gamma^2 \le \sigma_\varepsilon^2/10$, then p can be appreciably large (see Searle et al. (1992) for more details).

Result 11.3.5. In the *unbalanced* case, the ANOVA decomposition is

$$\sum_{i=1}^{a}\sum_{j=1}^{n_i}(Y_{ij} - \overline{Y}_{..})^2 = \sum_{i=1}^{a} n_i(\overline{Y}_{i\cdot} - \overline{Y}_{..})^2 + \sum_{i=1}^{a}\sum_{j=1}^{n_i}(Y_{ij} - \overline{Y}_{i\cdot})^2, \text{ i.e.,}$$
$$SST_c = SS_A + SSE.$$

In this case,

$$\mathrm{E}(SS_A) = \left(N - \sum_{i=1}^{a} n_i^2/N\right)\sigma_\gamma^2 + (a-1)\sigma_\varepsilon^2, \text{ and}$$
$$\mathrm{E}(SSE) = (N-a)\sigma_\varepsilon^2.$$

Proof. By equating the sums of squares to their expected values, as in the balanced case, we obtain estimates of the variance components as

$$\widehat{\sigma}_\varepsilon^2 = MSE, \text{ and } \widehat{\sigma}_\gamma^2 = (MS_A - MSE)/n^*$$

where

$$n^* = \frac{1}{a-1}\left(\sum_{i=1}^a n_i - \sum_{i=1}^a n_i^2 \Big/ \sum_{i=1}^a n_i\right).$$

When $n_i = n$ for $i = 1, \cdots, a$, these formulas reduce to those we saw in the balanced case. Once again, there exists the possibility of a negative estimate for σ_γ^2. In this case, we can show that $SSE/\sigma_\varepsilon^2 \sim \chi_{N-a}^2$. However, although SS_A and SSE are independently distributed in the unbalanced case, neither SS_A nor a multiple of SS_A has a χ^2 distribution. This is in contrast to the balanced case with the random-effects model, as well as the balanced and unbalanced cases with the fixed-effects model. After considerable algebra, we can show that

$$\begin{aligned}
\text{Var}(\widehat{\sigma}_\varepsilon^2) &= \text{Var}(MSE) = \frac{2\sigma_\varepsilon^4}{N-a}, \\
\text{Var}(\widehat{\sigma}_\gamma^2) &= \frac{2N}{(N^2 - \sum_{i=1}^a n_i^2)} \\
&\quad \times [\frac{N(N-1)(a-1)}{(N-a)(N^2 - \sum_{i=1}^a n_i^2)}\sigma_\varepsilon^4 + 2\sigma_\varepsilon^2\sigma_\gamma^2 \\
&\quad + \frac{N^2\sum_{i=1}^a n_i^2 + (\sum_{i=1}^a n_i^2)^2 - 2N\sum_{i=1}^a n_i^3}{N(N^2 - \sum_{i=1}^a n_i^2)}\sigma_\gamma^4], \\
\text{Cov}(\widehat{\sigma}_\varepsilon^2, \widehat{\sigma}_\gamma^2) &= \frac{-2(a-1)\sigma_\varepsilon^4}{(N-a)(N - \sum_{i=1}^a n_i^2/N)}.
\end{aligned} \tag{11.3.12}$$

In the unbalanced random-effects model, MS_A/MSE does not even have a distribution which is a multiple of an F-distribution when $\sigma_\gamma^2 > 0$. When $\sigma_\gamma^2 = 0$, then $SS_A \sim \sigma_\varepsilon^2 \chi_{a-1}^2$, and $F = MS_A/MSE \sim F_{a-1,N-a}$. We can therefore use the F-test for testing H_0: $\sigma_\gamma^2 = 0$ versus H_1: $\sigma_\gamma^2 > 0$ in the unbalanced case as well. ■

Maximum likelihood estimation

We describe maximum likelihood estimation of the parameters in the unbalanced model.

Result 11.3.6. In the unbalanced one-factor random-effects model, the MLE of μ is

$$\widehat{\mu}_{ML} = \left\{\sum_{i=1}^a [n_i/(\widehat{\sigma}_{\varepsilon,ML}^2 + n_i\widehat{\sigma}_{\gamma,ML}^2)]\right\}^{-1} \sum_{i=1}^a \{n_i\overline{Y}_{i\cdot}/(\widehat{\sigma}_{\varepsilon,ML}^2 + n_i\widehat{\sigma}_{\gamma,ML}^2)\}, \tag{11.3.13}$$

and the MLEs $\widehat{\sigma}_{\varepsilon,ML}^2$ and $\widehat{\sigma}_{\gamma,ML}^2$ are solutions to the equations

$$\begin{aligned}
&-\frac{1}{2}\sum_{i=1}^a n_i/\lambda_i + \sum_{i=1}^a \{n_i^2(\overline{Y}_{i\cdot} - \mu)^2\}/2\lambda_i^2 = 0, \text{ and} \\
&-(N-a)/2\sigma_\varepsilon^2 - \frac{1}{2}\sum_{i=1}^a 1/\lambda_i + SSE/2\sigma_\varepsilon^4 + \sum_{i=1}^a \{n_i(\overline{Y}_{i\cdot} - \mu)^2\}/2\lambda_i^2 = 0,
\end{aligned} \tag{11.3.14}$$

where $\lambda_i = \sigma_\varepsilon^2 + n_i\sigma_\gamma^2$, provided the solution σ_γ^2 in (11.3.14) is positive. However, if this estimate is negative, $\widehat{\sigma}_{\gamma,ML}^2 = 0$, while $\widehat{\mu}_{ML} = \overline{Y}_{\cdot\cdot}$.

Proof. Since $\mathbf{y} \sim N(\mu \mathbf{1}_N, \mathbf{V})$, where $\mathbf{V}$ is given in (11.1.4), the likelihood function is defined by

$$L(\mu, \mathbf{V}; \mathbf{y}) = (2\pi)^{-N/2} |\mathbf{V}|^{-1/2} \exp\left\{-\frac{1}{2}(\mathbf{y} - \mu\mathbf{1}_N)'\mathbf{V}^{-1}(\mathbf{y} - \mu\mathbf{1}_N)\right\},$$

where,

$$|\mathbf{V}| = \prod_{i=1}^{a} \sigma_\varepsilon^{2(n_i-1)}(\sigma_\varepsilon^2 + n_i\sigma_\gamma^2), \quad \text{and} \quad \mathbf{V}^{-1} = \text{diag}(\mathbf{V}_1^{-1}, \cdots, \mathbf{V}_a^{-1}),$$

with the first identity due to property 12 of Result 1.3.6 and $\mathbf{V}_i^{-1}$ given in (11.3.3). We can write the likelihood function as

$$\begin{aligned}
&L(\mu, \sigma_\gamma^2, \sigma_\varepsilon^2; \mathbf{y}) \\
&= (2\pi)^{-N/2} \sigma_\varepsilon^{-2[(N-a)/2]} \prod_{i=1}^{a} (\sigma_\varepsilon^2 + n_i\sigma_\gamma^2)^{-1/2} \\
&\times \exp\left\{-\frac{1}{2\sigma_\varepsilon^2}\left[\sum_{i=1}^{a}\sum_{j=1}^{n_i}(Y_{ij} - \mu)^2 - \sum_{i=1}^{a} \frac{\sigma_\gamma^2}{\sigma_\varepsilon^2 + n_i\sigma_\gamma^2}(Y_{i\cdot} - n_i\mu)^2\right]\right\},
\end{aligned}$$

and its logarithm is

$$\begin{aligned}
&\ell(\mu, \sigma_\gamma^2, \sigma_\varepsilon^2; \mathbf{y}) \\
&= -\frac{N}{2}\log 2\pi - \frac{(N-a)}{2}\log\sigma_\varepsilon^2 - \frac{1}{2}\sum_{i=1}^{a}\log(\sigma_\varepsilon^2 + n_i\sigma_\gamma^2) \\
&\quad - \frac{1}{2\sigma_\varepsilon^2}\sum_{i=1}^{a}\sum_{j=1}^{n_i}(Y_{ij} - \mu)^2 + \frac{1}{2\sigma_\varepsilon^2}\sum_{i=1}^{a}\frac{\sigma_\gamma^2}{\sigma_\varepsilon^2 + n_i\sigma_\gamma^2}(Y_{i\cdot} - n_i\mu)^2 \\
&= -\frac{N}{2}\log 2\pi - \frac{(N-a)}{2}\log\sigma_\varepsilon^2 - \frac{1}{2}\sum_{i=1}^{a}\log\lambda_i - \frac{SSE}{2\sigma_\varepsilon^2} - \sum_{i=1}^{a}\frac{n_i(\overline{Y}_{i\cdot} - \mu)^2}{2\lambda_i},
\end{aligned}$$

where, as in the ANOVA method, the SSE is obtained as if all the effects are fixed. Since $\partial\lambda_i/\partial\sigma_\varepsilon^2 = 1$, and $\partial\lambda_i/\partial\sigma_\gamma^2 = n_i$, setting the partial derivatives of the log-likelihood function with respect to each parameter equal to zero yields the set of maximum likelihood equations

$$\begin{aligned}
\frac{\partial\ell}{\partial\mu} &= \sum_{i=1}^{a}\frac{n_i(\overline{Y}_{i\cdot} - \mu)}{\lambda_i} = 0, \\
\frac{\partial\ell}{\partial\sigma_\gamma^2} &= -\frac{1}{2}\sum_{i=1}^{a}\frac{n_i}{\lambda_i} + \sum_{i=1}^{a}\frac{n_i^2(\overline{Y}_{i\cdot} - \mu)^2}{2\lambda_i^2} = 0, \text{ and} \\
\frac{\partial\ell}{\partial\sigma_\varepsilon^2} &= -\frac{(N-a)}{2\sigma_\varepsilon^2} - \frac{1}{2}\sum_{i=1}^{a}\frac{1}{\lambda_i} + \frac{SSE}{2\sigma_\varepsilon^4} + \sum_{i=1}^{a}\frac{n_i(\overline{Y}_{i\cdot} - \mu)^2}{2\lambda_i^2} = 0.
\end{aligned}$$

Solving $\partial\ell/\partial\mu = 0$, the solution, denoted by $\widetilde{\mu}$, is the estimate $\widehat{\mu}_{GLS}$ in (11.3.13). There are no closed form expressions for the solutions $\widetilde{\sigma}_\varepsilon^2$, and $\widetilde{\sigma}_\gamma^2$; they are obtained as numerical solutions to the nonlinear equations $\partial\ell/\partial\sigma_\gamma^2 = 0$ and $\partial l/\partial\sigma_\varepsilon^2 = 0$. These solutions $\widetilde{\mu}$, $\widetilde{\sigma}_\varepsilon^2$, and $\widetilde{\sigma}_\gamma^2$ are the maximum likelihood estimates of the corresponding parameters only if they lie within the support of these parameters. Since the likelihood function tends to zero as $\sigma_\varepsilon^2 \to 0$, or $\sigma_\varepsilon^2 \to \infty$, it must attain its maximum at a positive value of σ_ε^2, thereby precluding the possibility of a negative $\widehat{\sigma}^2_{\varepsilon,ML}$. However, it is possible that $\widetilde{\sigma}^2_\gamma$ is negative. If $0 \le \widetilde{\sigma}^2_\gamma < \infty$, $\widehat{\sigma}^2_{\gamma,ML} = \widetilde{\sigma}^2_\gamma$, $\widehat{\sigma}^2_{\varepsilon,ML} = \widetilde{\sigma}^2_\varepsilon$, and $\widehat{\mu}_{ML} = \widetilde{\mu}$; if, on the other hand, $\widetilde{\sigma}^2_\gamma < 0$, then, we set $\widehat{\sigma}^2_{\gamma,ML} = 0$, $\widehat{\sigma}^2_{\varepsilon,ML} = SST_m/N$, and $\widehat{\mu}_{ML} = \overline{Y}_{\cdot\cdot}$; these estimators have asymptotic normal distributions. ■

In the *balanced* case, the log-likelihood function has the simple form

$$\ell(\mu, \sigma_\varepsilon^2, \sigma_\gamma^2) = -\frac{N}{2}\log(2\pi) - \frac{1}{2}a(n-1)\log\sigma_\varepsilon^2 \\ -\frac{a}{2}\log\lambda - \frac{SSE}{2\sigma_\varepsilon^2} - \frac{SS_A}{2\lambda} - \frac{an(\overline{Y}_{\cdot\cdot} - \mu)^2}{2\lambda}, \tag{11.3.15}$$

where $\lambda = \sigma_\varepsilon^2 + n\sigma_\gamma^2$. Setting the partial derivatives of the log likelihood with respect to μ, σ_ε^2, and σ_γ^2 equal to zero, and solving the resulting equations yields the solutions

$$\widetilde{\mu} = \overline{Y}_{\cdot\cdot}, \quad \widetilde{\sigma}^2_\varepsilon = MSE, \quad \text{and} \quad \widetilde{\sigma}^2_\gamma = \frac{(1 - 1/a)MS_A - MSE}{n}.$$

Once again, these solutions will be the maximum likelihood estimators if they lie within the support of the parameter space.

REML method

In the one-way random-effects model, REML refers to maximizing the portion of the likelihood function that does not involve μ. The restricted likelihood is also referred to as the marginal likelihood of σ_ε^2 and σ_γ^2. In the balanced case, its logarithm has the form, up to an additive constant,

$$\ell_R(\sigma_\varepsilon^2, \sigma_\gamma^2; SS_A, SSE) = -\frac{1}{2}a(n-1)\log\sigma_\varepsilon^2 - \frac{1}{2}(a-1)\log\lambda - \frac{SSE}{2\sigma_\varepsilon^2} - \frac{SS_A}{2\lambda}, \tag{11.3.16}$$

where $\lambda = \sigma_\varepsilon^2 + n\sigma_\gamma^2$ as in (11.3.15). The proof of (11.3.16) is left as Exercise 11.10. The REML estimators are obtained by maximizing (11.3.16) within the parameter space $\sigma_\gamma^2 \ge 0$, and $\sigma_\varepsilon^2 > 0$. Equating the partial derivatives

$$\frac{\partial \ell_R}{\partial \sigma_\varepsilon^2} = \frac{-a(n-1)}{2\sigma_\varepsilon^2} + \frac{SSE}{2\sigma_\varepsilon^4}, \\ \frac{\partial \ell_R}{\partial \lambda} = \frac{-(a-1)}{2\lambda} + \frac{SS_A}{2\lambda^2}$$

to zero, and solving simultaneously, we get the solutions

$$\widetilde{\sigma}^2_{\varepsilon,R} = \frac{SSE}{a(n-1)} = MSE, \\ \widetilde{\lambda}_R = \frac{SS_A}{(a-1)} = MS_A, \tag{11.3.17}$$

so that

$$\widetilde{\sigma}^2_{\gamma,R} = \frac{1}{n}(MS_A - MSE). \tag{11.3.18}$$

These are the REML estimates provided they are nonnegative. Note that these coincide with the ANOVA estimators in the balanced case. The REML solutions are very complicated for the unbalanced case. Westfall (1987) compared the different estimation procedures for the one-way unbalanced random-effects model. The conclusion seems to be that REML is favored for estimating σ_γ^2, the ANOVA method for σ_ε^2, while for simultaneous estimation of both σ_γ^2 and σ_ε^2, the ML method is favored.

Numerical Example 11.1. Random-effects one-factor model. The dataset "Dyestuff" from the R package *lme4* is based on an investigation to find out how much the variation from batch to batch in the quality of an intermediate product contributes to the variation in the yield of the dyestuff made from it. In the experiment, six samples of the intermediate, representing different batches of works manufacture, were obtained, and five preparations of the dyestuff were made in the laboratory from each sample. The equivalent yield of each preparation as grams of standard colour was determined by dye-trial. The dataset consists of one predictor variable X denoting Batch and a response variable Y denoting Yield.

```
data(Dyestuff, package = "lme4")
```

The ANOVA estimates of the variance components are shown below. We use the `aov()` function in the fixed-effects model using the `lm()` function (as discussed in Chapters 4 and 7).

```
Dyestuff$Batch <- factor(Dyestuff$Batch)
fit <- aov(Yield ~ Batch, Dyestuff)
summary(fit)

  Df Sum Sq Mean Sq F value Pr(>F)
Batch          5  56358    11272       4.6 0.0044 **
Residuals     24  58830     2451
---
Signif.  codes:  0 '***' 0.001 '**' 0.01 '*' 0.05 '.' 0.1 ' ' 1


 coef(fit)
(Intercept)     BatchB      BatchC      BatchD      BatchE      BatchF
       1505         23          59          -7          95         -35
```

From the ANOVA table, we can estimate the variance components as $\widehat{\sigma}_\varepsilon^2 = \text{MSE} = 2451$ and $\widehat{\sigma}_\gamma^2 = (MS_A - MSE)/n = (11272 - 2451)/5 = 1764.2$.

Although we can also recover the fixed-effects coefficients for the five levels of Batch using `coef(fit)`, these are meaningless since we have assumed that Batch is a random effect. The linear model fit for a random-effects model is shown below. REML estimates of the two variance components can be obtained from this fit and random-effects for the five levels can also be recovered using the function `ranef()`, as shown below.

```
r <- lmer(Yield ~ 1 | Batch, data = Dyestuff)
summary(r)

Linear mixed model fit by REML ['lmerMod']
Formula: Yield ~ 1 | Batch
   Data: Dyestuff

REML criterion at convergence: 319.7

Scaled residuals:
```

```
    Min      1Q  Median      3Q     Max
-1.4117 -0.7634  0.1418  0.7792  1.8296

Random effects:
 Groups   Name        Variance Std.Dev.
 Batch    (Intercept) 1764     42.00
 Residual             2451     49.51
Number of obs: 30, groups:  Batch, 6

Fixed effects:
            Estimate Std.  Error t value
(Intercept)  1527.50      19.38    78.8
```

The estimated random effects for the $a = 5$ levels of batch can be recovered.

```
ranef(r)

$Batch
  (Intercept)
A -17.6068514
B   0.3912634
C  28.5622256
D -23.0845385
E  56.7331877
F -44.9952868
```

We also show the ML estimates, which is obtained by setting the option REML=F in the lmer() function.

```
m <- lmer(Yield ~ 1 | Batch, data = Dyestuff, REML = FALSE)
summary(m)

Linear mixed model fit by maximum likelihood [
lmerMod]
Formula: Yield ~ 1 | Batch
   Data: Dyestuff

     AIC      BIC   logLik deviance df.resid
   333.3    337.5   -163.7    327.3       27

Scaled residuals:
    Min      1Q  Median      3Q     Max
-1.4315 -0.7972  0.1480  0.7721  1.8037

Random effects:
 Groups   Name        Variance Std.Dev.
 Batch    (Intercept) 1388     37.26
 Residual             2451     49.51
Number of obs: 30, groups:  Batch, 6

Fixed effects:
            Estimate Std.  Error t value
(Intercept)  1527.50      17.69   86.33
```

The estimated random effects are close to but not exactly the same as the REML estimates

```
ranef(m)

$Batch
  (Intercept)
A -16.6282263
B   0.3695161
C  26.9746782
D -21.8014523
E  53.5798403
F -42.4943561
```

For completeness, we show partial output from the MINQUE estimation using the R package `minque()`.

```
(res = minque::lmm(Yield ~ 1 | Batch, Dyestuff))
res$Var
$Var$Var$Yield

             Est        SE    Chi_sq
V(Batch) 1764.05 1432.7513  1.515934
V(e)     2451.25  707.6149 12.000000
              P_value
V(Batch) 0.1091179474
V(e)     0.0002660028


$FixedEffect$FixedEffect$Yield
      Est         SE  z_value P_value
mu 1527.5 19.38341 78.80449       0

res$RandomEffect$RandomEffect$Yield

                 Pre        SE     z_value
Batch(A) -19.9036217 33.91676 -0.58683728
Batch(B)   0.4423027 33.91676  0.01304083
Batch(C)  32.2880974 33.91676  0.95198048
Batch(D) -26.0958595 33.91676 -0.76940888
Batch(E)  64.1338920 33.91676  1.89092013
Batch(F) -50.8648109 33.91676 -1.49969527
            P_value
Batch(A) 0.55731300
Batch(B) 0.98959522
Batch(C) 0.34110688
Batch(D) 0.44165062
Batch(E) 0.05863501
Batch(F) 0.13369336
```

▲

11.4 Two-factor random- and mixed-effects models

In this section, we show a few examples of two-factor models.

Example 11.4.1. Random-effects two-factor nested model. Consider the two-factor model where Factor A has a levels, the b levels of Factor B are *nested* within each of the levels of Factor A, and both factors are random. The two-way nested model is

$$Y_{ijk} = \mu + \gamma_i + \theta_{j(i)} + \varepsilon_{ijk},$$

for $i = 1, \cdots, a$, $j = 1, \cdots, b$, $k = 1, \cdots, n$. Suppose γ_i are i.i.d. $N(0, \sigma_\gamma^2)$ variables, $\theta_{j(i)}$ are i.i.d. $N(0, \sigma_\theta^2)$ variables, ε_{ijk} are i.i.d. $N(0, \sigma_\varepsilon^2)$ variables which are independent of γ_i and $\theta_{j(i)}$. Similar to the fixed-effects model, the ANOVA decomposition is

$$SST_m = SS_A + SS_{B(A)} + SSE,$$

where the sums of squares were defined in Example 7.3.6. When the effects are random, we can verify that the expected mean squares are

$$\begin{aligned} E(MS_A) &= \sigma_\varepsilon^2 + n\sigma_\theta^2 + nb\sigma_\gamma^2, \\ E(MS_{B(A)}) &= \sigma_\varepsilon^2 + n\sigma_\theta^2, \text{ and} \\ E(MSE) &= \sigma_\varepsilon^2. \end{aligned} \tag{11.4.1}$$

This list of expected mean squares serves as a guide for the construction of test statistics. Under the null H_0: $\sigma_\theta^2 = 0$, $E(MS_{B(A)}) = \sigma_\varepsilon^2$, and we use the statistic $F = MS_{B(A)}/MSE$ to test H_0. This statistic has an $F_{a(b-1),ab(n-1)}$ distribution under H_0. Under the hypothesis H_0: $\sigma_\gamma^2 = 0$, we see that $E(MS_A) = \sigma_\varepsilon^2 + n\sigma_\theta^2$, and we reject the null hypothesis at level of significance α if $F = MS_A/MS_{B(A)} > F_{a-1,a(b-1),\alpha}$. □

Example 11.4.2. Random-effects two-factor model with interaction. Recall the two-way cross-classified model *with interaction* between the two factors (balanced case)

$$Y_{ijk} = \mu + \gamma_i + \theta_j + (\gamma\theta)_{ij} + \varepsilon_{ijk}, \tag{11.4.2}$$

for $k = 1, \cdots, n$, $i = 1, \cdots, a$, and $j = 1, \cdots, b$. Suppose that the γ, θ, and $(\gamma\theta)$ effects are all random. Suppose γ_i are i.i.d. $N(0, \sigma_\gamma^2)$ variables, θ_j are i.i.d. $N(0, \sigma_\theta^2)$ variables, $(\gamma\theta)$ are i.i.d. $N(0, \sigma_{\gamma\theta}^2)$ variables, $\text{Cov}(\gamma_i, \theta_j) = 0 = \text{Cov}(\gamma_i, (\gamma\theta)_{ij}) = \text{Cov}(\gamma_i, \varepsilon_{ijk})$, with similar assumptions about θ's and $(\gamma\theta)$'s.

Based on the ANOVA table we saw under the fixed-effects case, we can compute the expected mean squares when both Factor A and Factor B are random as

$$\begin{aligned} E(MS_A) &= bn\sigma_\gamma^2 + n\sigma_{\gamma\theta}^2 + \sigma_\varepsilon^2, \\ E(MS_B) &= an\sigma_\theta^2 + n\sigma_{\gamma\theta}^2 + \sigma_\varepsilon^2, \\ E(MS_{AB}) &= n\sigma_{\gamma\theta}^2 + \sigma_\varepsilon^2, \text{ and} \\ E(MSE) &= \sigma_\varepsilon^2, \end{aligned}$$

which are linear combinations of the unknown variance components.

The ANOVA method of estimating variance components from balanced data consists of equating the observed mean squares to these expected mean squares, and solving the resulting equations for $\widehat{\sigma}^2_\gamma$, $\widehat{\sigma}^2_\theta$, $\widehat{\sigma}^2_{\gamma\theta}$, and $\widehat{\sigma}^2_\varepsilon$. We obtain

$$\begin{aligned}
\widehat{\sigma}^2_\varepsilon &= MSE, \\
\widehat{\sigma}^2_\gamma &= \frac{1}{bn}(MS_A - MS_{AB}), \\
\widehat{\sigma}^2_\theta &= \frac{1}{an}(MS_B - MS_{AB}), \text{ and} \\
\widehat{\sigma}^2_{\gamma\theta} &= \frac{1}{n}(MS_{AB} - MSE).
\end{aligned} \tag{11.4.3}$$

Note that once again, there exists a positive probability of negative estimates. In a balanced model, the ANOVA estimators of the variance components are minimum variance, quadratic unbiased estimators (Graybill and Hultquist, 1961), even if normality is not assumed. Under normality, Graybill (1954) showed that the ANOVA estimators are unbiased and have minimum variance.

Under normality, we obtain the following distributions of the sums of squares:

$$\frac{(a-1)MS_A}{bn\sigma^2_\gamma + n\sigma^2_{\gamma\theta} + \sigma^2_\varepsilon} \sim \chi^2_{a-1}, \qquad \frac{(a-1)(b-1)MS_{AB}}{n\sigma^2_{\gamma\theta} + \sigma^2_\varepsilon} \sim \chi^2_{(a-1)(b-1)},$$

and these are independently distributed. It follows that the distribution of $\widehat{\sigma}^2_\gamma$ is a linear combination of scaled χ^2 variables, i.e.,

$$\widehat{\sigma}^2_\gamma \sim \frac{bn\sigma^2_\gamma + n\sigma^2_{\gamma\theta} + \sigma^2_\varepsilon}{bn(a-1)}\chi^2_{a-1} - \frac{n\sigma^2_{\gamma\theta} + \sigma^2_\varepsilon}{bn(a-1)(b-1)}\chi^2_{(a-1)(b-1)}.$$

However, $\widehat{\sigma}^2_\varepsilon \sim \{\sigma^2_\varepsilon/ab(n-1)\}\chi^2_{ab(n-1)}$, i.e., MSE always has a scaled χ^2 distribution. The expected mean squares often enable us to determine which mean squares are appropriate in the denominators of test statistics. For instance, in this model, the statistic MS_{AB}/MSE has an $F_{(a-1)(b-1),ab(n-1)}$ distribution under H_0: $\sigma^2_{\gamma\theta} = 0$. To test H_0: $\sigma^2_\gamma = 0$, we use the statistic MS_A/MS_{AB}, which has an $F_{a-1,(a-1)(b-1)}$ distribution under this null hypothesis.

To construct confidence intervals, note that the mean squares are distributed independently as multiples of χ^2 variables. Suppose we denote the kth mean square by M_k with corresponding degrees of freedom f_k. Then, an exact $100(1-\alpha)\%$ confidence interval for $\text{E}(M_k)$ has the form

$$\{f_k M_k\}/\chi^2_{f_k,U} \le \text{E}(M_k) \le \{f_k M_k\}/\chi^2_{f_k,L},$$

where $\chi^2_{f_k,U}$ and $\chi^2_{f_k,L}$ are defined by $\text{P}(\chi^2_{f_k,L} \le \chi^2_{f_k} < \chi^2_{f_k,U}) = 1-\alpha$.

For the two-factor model with interaction, the maximum likelihood estimators do not have a closed form even with balanced data. These must be obtained numerically for each dataset. The ML equations are defined by

$$\begin{aligned}
&\frac{SS_A}{\delta^2_{11}} + \frac{SS_B}{\delta^2_{12}} + \frac{SS_{AB}}{\delta^2_1} + \frac{SSE}{\delta^2_0} \\
&= \frac{1}{\delta_4} + \frac{a-1}{\delta_{11}} + \frac{b-1}{\delta_{12}} + \frac{(a-1)(b-1)}{\delta_1} + \frac{ab(n-1)}{\delta_0}, \\
&\frac{SS_A}{\delta^2_{11}} = \frac{1}{\delta_4} + \frac{a-1}{\delta_{11}}, \quad \frac{SS_B}{\delta^2_{12}} = \frac{1}{\delta_4} + \frac{b-1}{\delta_{12}}, \text{ and} \\
&\frac{SS_A}{\delta^2_{11}} + \frac{SS_B}{\delta^2_{12}} + \frac{SS_{AB}}{\delta^2_1} = \frac{1}{\delta_4} + \frac{a-1}{\delta_{11}} + \frac{b-1}{\delta_{12}} + \frac{(a-1)(b-1)}{\delta_1},
\end{aligned}$$

where $\delta_0 = \sigma^2_\varepsilon$, $\delta_1 = \sigma^2_\varepsilon + n\sigma^2_{\gamma\beta}$, $\delta_{11} = \sigma^2_\varepsilon + n\sigma^2_{\gamma\beta} + bn\sigma^2_\gamma$, $\delta_{12} = \sigma^2_\varepsilon + n\sigma^2_{\gamma\beta} + an\sigma^2_\beta$, and $\delta_4 = \delta_{11} + \delta_{12} - \delta_1$. The MLEs are obtained by solving these equations simultaneously. □

Example 11.4.3. Mixed-effects two-factor model. In the two-factor cross-classified model, suppose Factor A is fixed, Factor B is random, and we consider interaction between the factors:

$$Y_{ijk} = \mu + \tau_i + \theta_j + (\tau\theta)_{ij} + \varepsilon_{ijk}.$$

Assume (restricted model) $\sum_{i=1}^{a} \tau_i = 0$ and $\sum_{i=1}^{a} (\tau\theta)_{ij} = 0$, $j = 1, \ldots, b$. The expected mean squares are

$$\begin{aligned} \mathrm{E}(MS_A) &= \sigma_\varepsilon^2 + n\sigma_{\tau\theta}^2 + nb\sum_{i=1}^{a} \tau_i^2/(a-1) \\ \mathrm{E}(MS_B) &= \sigma_\varepsilon^2 + an\sigma_\theta^2, \\ \mathrm{E}(MS_{AB}) &= \sigma_\varepsilon^2 + n\sigma_{\tau\theta}^2, \text{ and} \\ \mathrm{E}(MSE) &= \sigma_\varepsilon^2. \end{aligned} \tag{11.4.4}$$

The ANOVA estimators (which are also REML estimators) are

$$\begin{aligned} \mu^0 &= \overline{Y}_{\ldots}, \\ \tau_i^0 &= \overline{Y}_{i\cdot\cdot} - \overline{Y}_{\ldots}, \quad i = 1, \cdots, a, \\ \widehat{\sigma}_\theta^2 &= (MS_B - MSE)/an, \\ \widehat{\sigma}_{\tau\theta}^2 &= (MS_{AB} - MSE)/n, \text{ and} \\ \widehat{\sigma}_\varepsilon^2 &= MSE. \end{aligned}$$

Under H_0: $\sigma_{\tau\theta}^2 = 0$, we can show that the test statistic $F_0 = MS_{AB}/MSE$ has an $F_{(a-1)(b-1),ab(n-1)}$ distribution. Under H_0: $\sigma_\theta^2 = 0$, the test statistic $F_0 = MS_B/MSE$ has an $F_{b-1,ab(n-1)}$ distribution. Under H_0: $\tau_1 = \cdots = \tau_a$, the test statistic $F_0 = MS_A/MS_{AB}$ has an $F_{a-1,(a-1)(b-1)}$ distribution. □

Example 11.4.4. Consider the two-factor nested model

$$Y_{ijk} = \mu + \tau_i + \theta_{j(i)} + \varepsilon_{ijk},$$

for $i = 1, \cdots, a$, $j = 1, \cdots, b$ and $k = 1, \cdots, n$. Suppose Factor A is fixed, while Factor B is a random factor whose levels are nested within the levels of Factor A. In this mixed-effects model, we assume that $\sum_{i=1}^{a} \tau_i = 0$, and $\theta_{j(i)}$'s are i.i.d. $N(0, \sigma_\theta^2)$ variables , $j = 1, \cdots, b$, $i = 1, \cdots, a$. In this case, μ^0 and τ_i^0 have the same form as in the fixed-effects model, while the estimates of $\widehat{\sigma}_\varepsilon^2$ and $\widehat{\sigma}_\theta^2$ are given under the random-effects model (see Exercise 11.6). Then,

$$\begin{aligned} \mathrm{E}(MS_A) &= \sigma_\varepsilon^2 + n\sigma_\theta^2 + bn\sum_{i=1}^{a} \tau_i^2/(a-1), \\ \mathrm{E}(MS_{B(A)}) &= \sigma_\varepsilon^2 + n\sigma_\theta^2, \text{ and} \\ \mathrm{E}(MSE) &= \sigma_\varepsilon^2. \end{aligned}$$

The hypothesis H_A: $\tau_1 = \cdots = \tau_a$ is tested by $F_0 = MS_A/MS_{B(A)}$ which has an $F_{a-1,a(b-1)}$ distribution under H_A. The test statistic for H_B: $\sigma_\theta^2 = 0$ is $F_0 = MS_{B(A)}/MSE$, which has an $F_{a(b-1),ab(n-1)}$ distribution under H_B. □

Numerical Example 11.2. The dataset "curdies" available in the R package *GFD* relates to an ecological analysis on the Curdies river in Western Victoria in Australia. The response variable Y is dugesia, the number of flatworms counted on a particular stone (in no./dm^2). The two factor variables are X_1 (season, with two levels, "SUMMER" and "WINTER") and X_2 (site, with six levels) nested within X_1.

```
library(GFD)
data(curdies)
curdies$site <- as.factor(curdies$site)
curdies.aov <- aov(dugesia ~ season + Error(site), data=curdies)
summary(lm(curdies.aov))$coefficients # least square coefficients

              Estimate Std.  Error t value Pr(>|t|)
(Intercept)     0.1942       1.06   0.1825    0.856
seasonWINTER    1.8552       1.51   1.2325    0.227
site2           2.1325       1.51   1.4167    0.167
site3          -1.3712       1.51  -0.9110    0.370
site4           0.2249       1.51   0.1494    0.882
site5           0.0348       1.51   0.0231    0.982
```

▲

Exercises

11.1. Show that

(a) $\mathbf{X}(\mathbf{X}'\mathbf{V}^{\sharp}\mathbf{X})^{-}\mathbf{X}'\mathbf{V}^{\sharp}\mathbf{y}$ is invariant to $\mathbf{V}^{\sharp}$ and $(\mathbf{X}'\mathbf{V}^{\sharp}\mathbf{X})^{-}$, and

(b) $\mathcal{C}(\mathbf{X},\mathbf{V}) = \mathcal{C}(\mathbf{X},\mathbf{VN})$,

where $\mathbf{N}$ appeared in (10.5.2), and the other quantities appear in Result 11.2.1.

11.2. In the one-way random-effects model, show that

$$\mathrm{Var}(\widehat{\mu}_{GLS}) = [\sum_{i=1}^{a} n_i/(n_i\sigma_\gamma^2 + \sigma_\varepsilon^2)]^{-1}.$$

11.3. Consider the balanced one-way random-effects model where ε_{ij} are i.i.d. $N(0,\sigma_\varepsilon^2)$ variables, and γ_i are i.i.d. $N(0,\sigma_\gamma^2)$ variables. Show that the sum of squares due to the Factor A can be written as $SS_A = \mathbf{y}'(\mathbf{I}_a \otimes \mathbf{J}_n/n - \mathbf{J}_N/N)\mathbf{y}$, where $\mathbf{y} = (Y_{11},\cdots,Y_{1n},Y_{21},\cdots,Y_{2n},\cdots,Y_{a1},\cdots,Y_{an})'$.

11.4. In the formulation of Exercise 11.3, show that $\mathrm{E}(MS_A) = n\sigma_\gamma^2 + \sigma_\varepsilon^2$. Derive the distributions of $SS_A/\,\mathrm{E}(MS_A)$ and $SSE/\,\mathrm{E}(MSE)$.

11.5. In the mixed-effects linear model (11.2.6), let $\boldsymbol{\sigma}^2 = (\sigma_\varepsilon^2,\sigma_1^2,\cdots,\sigma_m^2)'$. Let $\mathbf{X}$ have full column rank. Show that, as $N\to\infty$,

(a) $\mathrm{Var}(\widehat{\boldsymbol{\beta}}_{ML}) \approx (\mathbf{X}'\mathbf{V}^{-1}\mathbf{X})^{-1}$,

(b) $\mathrm{Var}(\widehat{\boldsymbol{\sigma}}^2_{ML}) = 2\mathbf{S}^{-1}$, where $\mathbf{S} = \{S_{ij}\}$ is an $(m+1)\times(m+1)$ matrix whose (i,j)th element is $S_{ij} \approx \mathrm{tr}(\mathbf{V}^{-1}\mathbf{Z}_i\mathbf{Z}_i'\mathbf{V}^{-1}\mathbf{Z}_j\mathbf{Z}_j')$, and

(c) $\mathrm{Cov}(\widehat{\boldsymbol{\beta}}_{ML},\widehat{\boldsymbol{\sigma}}^2_{ML}) \to \mathbf{O}$.

11.6. Obtain the ANOVA estimators and maximum likelihood estimators of the variance components in the two-factor (balanced) nested model with random effects, which is described in Example 11.4.1.

11.7. Consider the two-factor additive random-effects model

$$Y_{ijk} = \mu + \gamma_i + \theta_j + \varepsilon_{ijk},$$

for $i = 1, \cdots, a$, $j = 1, \cdots, b$, and $k = 1, \cdots, n$. Suppose γ_i are i.i.d. $N(0, \sigma_\gamma^2)$ variables, θ_j are i.i.d. $N(0, \sigma_\theta^2)$ variables, ε_{ijk} are i.i.d. $N(0, \sigma_\varepsilon^2)$ variables, $\text{Cov}(\gamma_i, \theta_j) = 0$, $\text{Cov}(\gamma_i, \varepsilon_{ij}) = 0$, and $\text{Cov}(\theta_j, \varepsilon_{ij}) = 0$. Let $N = nab$. Derive the ANOVA, ML and REML estimators of the variance components.

11.8. In Result 11.2.4, show that $(N - r)\widehat{\sigma}_\varepsilon^2$ is the unique nonnegative quadratic unbiased estimator of $(N - r)\sigma_\varepsilon^2$, with minimum variance when the nonzero diagonal elements of the orthogonal projection matrix $\mathbf{P}$ are equal.

11.9. Prove (11.2.15).

11.10. Prove (11.3.16).

12

Generalized Linear Models

Although linear models with normal responses are very versatile in many applications, like any other models, they have limits. These models are not suitable when responses clearly follow a non-normal distribution, such as when they are discrete counts or binary-valued responses, or they are strictly positive continuous-valued responses, or they exhibit substantial non-linearity in their relationship with the predictors. A motivation for using generalized linear models (GLIM's) is that they permit more general distributions than the normal for the response (Nelder and Wedderburn (1972) and McCullagh and Nelder (1989)). A GLIM generalizes a linear model by allowing the expectation of the response variable to be related to a parametric linear function of the predictors via a suitable link function. The usual normal linear models can be thought of as a particular type of GLIM's. This chapter describes GLIM's for non-normal responses.

12.1 Components of GLIM

Let $Y_1, \cdots, Y_N$ be independent response variables, each associated with the same number of known predictors $X_{i1}, \cdots, X_{ik}$. Denote $\mathbf{y} = (Y_1, \cdots, Y_N)'$ and $\mathbf{x}_i = (1, X_{i1}, \cdots, X_{ik})'$, $i = 1, \cdots, N$. A GLIM for the data $(Y_i, X_{i1}, \cdots, X_{ik})$, $i = 1, \cdots, N$, is specified by the following three components:

1. **Random component.** Each Y_i has a p.m.f. or p.d.f. from an exponential family of distributions, taking the form

$$f(y_i; \theta_i, \phi) = \exp\left\{\frac{y_i\theta_i - b(\theta_i)}{a_i(\phi)} + c_i(y_i, \phi)\right\}, \tag{12.1.1}$$

   where ϕ is a parameter that may or may not be known but is the same for all Y_i, $a_i(\cdot) > 0$, $b(\cdot)$, and $c_i(\cdot, \cdot)$ are known functions, with $b(\cdot)$ being the same for all Y_i, and θ_i, called the canonical parameter, is different for different Y_i; see (B.22). Let $\mu_i = \text{E}(Y_i)$.

2. **Systematic component.** Consider a vector $\boldsymbol{\beta} = (\beta_0, \beta_1, \cdots, \beta_k)$, where the β_j's are unknown coefficients. Given ϕ, the distribution of Y_i only depends on the linear predictor

$$\eta_i = \mathbf{x}_i'\boldsymbol{\beta}. \tag{12.1.2}$$

3. **Link function.** The parameter θ_i in the distribution of Y_i in (12.1.1) is determined by η_i via a continuous 1-to-1 function $g(\cdot)$, such that

$$g(\mu_i) = \eta_i. \tag{12.1.3}$$

   Note that ϕ is a constant and does not depend on $\boldsymbol{\beta}$ and $\mathbf{x}_1, \cdots, \mathbf{x}_N$.

DOI: 10.1201/9781315156651-12

Remark 1. The dependence of θ_i on η_i can be made more explicit than in the definition. Given the first and second derivatives of $b(\theta_i)$ with respect to θ, from item 15 in Appendix B,

$$E(Y_i) = \mu_i = b'(\theta_i), \text{ and } \operatorname{Var}(Y_i) = a_i(\phi)b''(\theta_i), \tag{12.1.4}$$

so that $b''(\cdot) > 0$. As a result, $b'(\cdot)$ is continuous and strictly increasing in its domain. From (12.1.3),

$$b'(\theta_i) = g^{-1}(\mathbf{x}_i'\boldsymbol{\beta}). \tag{12.1.5}$$

For later use, we define for $\mu = b'(\theta)$ that

$$V(\mu) = b''(\theta) = b''((b')^{-1}(\mu)). \tag{12.1.6}$$

Remark 2. Since μ_i and θ_i have a 1-to-1 correspondence due to (12.1.4), we can write the log-likelihood function in terms of the mean vector $\boldsymbol{\mu} = (\mu_1, \cdots, \mu_N)'$ instead of the canonical parameter $\boldsymbol{\theta} = (\theta_1, \cdots, \theta_N)'$, and denote it by $\ell(\boldsymbol{\mu}, \phi; \mathbf{y})$. We can also write the log-likelihood as a function of $\boldsymbol{\beta}$, i.e., $\ell(\boldsymbol{\beta}, \phi; \mathbf{y})$. When ϕ is known, these are written as $\ell(\boldsymbol{\mu}; \mathbf{y})$ and $\ell(\boldsymbol{\beta}; \mathbf{y})$, respectively.

Remark 3. In (12.1.1), ϕ is called the dispersion parameter and $a_i(\phi)$ typically takes the form $a_i(\phi) = \phi/w_i$, where w_i is a known prior weight for Y_i.

Remark 4. In practice, $b(\cdot)$ is usually explicitly given, while $c_i(\cdot,\cdot)$ is left implicit. This is not a problem as far as the MLE of the parameter vector $\boldsymbol{\beta}$ is concerned. However, without an explicit form for $c_i(\cdot,\cdot)$, likelihood based estimation of ϕ and likelihood based inference on $(\boldsymbol{\beta}, \phi)$ are not available.

Remark 5. Each $c_i(\cdot,\cdot)$ is determined by $b(\cdot)$ and $a_i(\cdot)$. Indeed, since the function in (12.1.1) integrates to 1, given ϕ, $\exp\{b(a_i(\phi)\theta)\}$ is the Laplace transform of $e^{c_i(y,\phi)}$, so $c_i(\cdot,\phi)$ is unique due to the one-to-one correspondence between functions and their Laplace transforms, provided the latter are finite.

If $g(\mu_i) = \mu_i$, we call $g(\cdot)$ the identity link function. The usual normal linear model in (4.1.1) has the identity link function. For non-normal GLIM's, the following special link function is more useful.

Definition 12.1.1. Canonical link. The canonical (or natural) link function is the unique function g such that $g(\mu_i) = \theta_i$. More directly, from (12.1.4), the canonical link function is

$$g = (b')^{-1}. \tag{12.1.7}$$

Result 12.1.1. The following properties of the canonical link hold.

1. If g is the canonical link function, then $V(\mu) = 1/g'(\mu)$.
2. Under the GLIM with the canonical link function g,

$$\theta_i = \eta_i = \mathbf{x}_i'\boldsymbol{\beta}, \text{ and } \operatorname{Var}(Y_i) = a_i(\phi)/g'(\mu_i) = a_i(\phi)V(\mu_i).$$

Proof. Since $b'(\theta) = g^{-1}(\theta)$, then $b''(\theta) = 1/g'(g^{-1}(\theta))$, so that $V(\mu) = b''((b')^{-1}(\mu)) = 1/g'(\mu)$, proving property 1. Under the GLIM with the canonical link, $\eta_i = g(\mu_i) = \theta_i$. On the other hand, from (12.1.4), $\operatorname{Var}(Y_i) = a_i(\phi)b''(\theta_i)$. The formula for $\operatorname{Var}(Y_i)$ follows from (12.1.6) and property 1. ■

Example 12.1.1. Binary response. Suppose that for the ith subject, the response Y_i is binary, i.e., it can assume either the value 0, or the value 1. Suppose $\mathrm{P}(Y_i = 1) = \pi_i = 1 - \mathrm{P}(Y_i = 0)$. The p.m.f. of Y_i is

$$f(y_i) = \pi_i^{y_i}(1-\pi_i)^{1-y_i} = (1-\pi_i)\exp\{y_i \log[\pi_i/(1-\pi_i)]\}.$$

For a GLIM, we let $\theta_i = \log[\pi_i/(1-\pi_i)]$. Since $\mathrm{E}(Y_i) = \pi_i$, from Definition 12.1.1, the canonical link function is the logit function $\log[\pi_i/(1-\pi_i)]$. Under the GLIM with the logit link, we set $\log[\pi_i/(1-\pi_i)] = \eta_i = \mathbf{x}_i'\boldsymbol{\beta}$, so that

$$\pi_i = \frac{e^{\mathbf{x}_i'\boldsymbol{\beta}}}{1+e^{\mathbf{x}_i'\boldsymbol{\beta}}}.$$

We can carry out inference on the vector of proportions $\boldsymbol{\pi} = (\pi_1, \cdots, \pi_N)'$ as a function of the linear predictor $\boldsymbol{\eta} = \mathbf{X}\boldsymbol{\beta}$, where $\mathbf{X} = (\mathbf{x}_1', \cdots, \mathbf{x}_N')'$.

The logit link function maps the unit interval $[0,1]$ onto $\mathcal{R}$, as does each of the following alternative link functions that are commonly used with binary responses:

1. the probit or inverse Gaussian link function

$$\eta = \Phi^{-1}(\pi),$$

   $\Phi(\cdot)$ denoting the c.d.f. of $N(0,1)$;

2. the complementary log-log link function

$$\eta = \log\{-\log(1-\pi)\}; \text{ and}$$

3. the log-log link function

$$\eta = -\log\{-\log(\pi)\}.$$

The logit function is widely used due to its interpretability as the logarithm of the odds ratio $\pi/(1-\pi)$, and it induces a logistic regression model. In particular, notice that the link function satisfies the condition that $g(\pi) = H^{-1}(\pi)$, where H is the c.d.f. of a given distribution. For instance, the logit link corresponds to the c.d.f. $H(\pi) = \exp(\pi)/\{1+\exp(\pi)\}$, whereas the complementary log-log link is obtained by specifying $H(\pi) = \exp\{-\exp(-\pi)\}$. We give a detailed discussion of binary response models in Section 12.4. □

Example 12.1.2. Poisson response. Let Y_i denote counts on the ith subject, $i = 1, \cdots, N$, and suppose that Y_i's are independently distributed as Poisson random variables with mean λ_i. Since Y_i has p.m.f.

$$f(y_i) = e^{-\lambda_i}\lambda_i^{y_i}(y_i!)^{-1} = (y_i!)^{-1}e^{-\lambda_i}e^{y_i \log \lambda_i},$$

to set up a GLIM, let $\theta_i = \log \lambda_i$, $a_i(\phi) = 1$, and $b(\theta_i) = \exp(\theta_i)$. From $\mathrm{E}(Y_i) = \lambda_i$ and Definition 12.1.1, the canonical link function is $\log \lambda_i$. The log-linear GLIM postulates the logarithmic link $\log \lambda_i = \eta_i = \mathbf{x}_i'\boldsymbol{\beta}$, $i = 1, \cdots, N$, so that $\lambda_i = e^{\mathbf{x}_i'\boldsymbol{\beta}}$ and the log-likelihood function is (up to an additive term that does not depend on $\boldsymbol{\beta}_i$),

$$\ell(\boldsymbol{\beta}; \mathbf{y}) = \sum_{i=1}^{N} Y_i \mathbf{x}_i'\boldsymbol{\beta} - \sum_{i=1}^{N} \exp(\mathbf{x}_i'\boldsymbol{\beta}). \tag{12.1.8}$$

Log-linear models for count data are discussed in Section 12.5. □

Example 12.1.3. Gamma response. Let Y_i denote a continuous, positive valued response on the ith subject, $i = 1, \cdots, N$, following a Gamma(α, s_i) distribution. Its p.d.f. is given by (B.10), $\mathrm{E}(Y_i) = \mu_i = \alpha s_i$, and $\mathrm{Var}(Y_i) = \alpha s_i^2$. Define $\theta_i = -1/\mu_i = -1/(\alpha s_i)$ and $\phi = 1/\alpha$. Then the p.d.f. of Y_i is

$$f(y_i; \theta_i, \phi) = \frac{y_i^{\alpha-1} e^{-y_i/s_i}}{\Gamma(\alpha) s_i^{\alpha}} = \frac{y_i^{1/\phi - 1}}{\Gamma(1/\phi)(-\phi/\theta_i)^{1/\phi}} e^{y_i\theta_i/\phi}.$$

The following link functions have been used in model fitting (see Exercise 12.2):

1. Canonical link: since $\theta = -1/\mu$, from Definition 12.1.1, the canonical link function is $g(\mu) = -1/\mu$. Under the GLIM with the canonical link, $\mu_i = -1/(\mathbf{x}_i'\boldsymbol{\beta}) > 0$. Then $\boldsymbol{\beta}$ must satisfy $\mathbf{x}_i'\boldsymbol{\beta} < 0$ for all i and the p.d.f. of Y_i is obtained by plugging $\theta_i = -1/\mu_i = \mathbf{x}_i'\boldsymbol{\beta}$ into the above display.

2. Inverse link: $g(\mu) = 1/\mu$, so that $\mu_i = 1/(\mathbf{x}_i'\boldsymbol{\beta})$. Then, $\boldsymbol{\beta}$ must satisfy $\mathbf{x}_i'\boldsymbol{\beta} > 0$ for all i and the p.d.f. of Y_i is obtained by plugging $\theta_i = -1/\mu_i = -\mathbf{x}_i'\boldsymbol{\beta}$ into the above display.

3. Log link: $g(\mu) = \log\mu$, so that $\mu_i = \exp\{\mathbf{x}_i'\boldsymbol{\beta}\}$. Then $\theta_i = -1/\mu_i = -\exp\{-\mathbf{x}_i'\boldsymbol{\beta}\}$. Note that not only is there no constraint on $\boldsymbol{\beta}$, but $\mathbf{x}_i'\boldsymbol{\beta}$ need not be log-transformed.

4. Identity link: $g(\mu) = \mu$, so that $\mu_i = \mathbf{x}_i'\boldsymbol{\beta}$ and $\theta_i = -1/\mu_i = -1/(\mathbf{x}_i'\boldsymbol{\beta})$. Similar to the inverse link, $\boldsymbol{\beta}$ must satisfy $\mathbf{x}_i'\boldsymbol{\beta} > 0$ for all i. □

Example 12.1.4. Inverse Gaussian GLIM. The p.d.f. of an inverse Gaussian distribution with mean $\mu > 0$ and shape parameter $\lambda > 0$ denoted as IGaus(μ, λ) is given in (B.21). Define $\theta = -1/(2\mu^2)$ and $\phi = 1/\lambda$. Then $V(\mu) = \mu^3$ and the p.d.f. of Y_i can be expressed as

$$f(y_i; \theta_i, \phi) = \exp\left\{\frac{\theta_i y_i + (-2\theta_i)^{1/2}}{\phi} - \frac{1}{2\phi y_i} - \frac{1}{2}\log(2\pi\phi y_i^3)\right\},$$

so that $b(\theta_i) = -(-2\theta_i)^{1/2}$, $c(y_i, \phi) = -(2\phi y_i)^{-1} - (1/2)\log(2\pi\phi y_i^3)$, and by Definition 12.1.1, the canonical link is $g(\mu) = -1/(2\mu^2)$. □

12.2 Estimation approaches

Several approaches are useful for estimating parameters in a GLIM. In this section, we discuss maximum likelihood estimation via Fisher scoring, the method of iteratively reweighted least squares (IRLS), and the quasi-likelihood approach.

12.2.1 Score and Fisher information for GLIM

For simplicity of exposition, first consider the GLIM for a single observation,

$$\ell(\boldsymbol{\beta}, \phi; Y, \mathbf{x}) = \frac{Y\theta - b(\theta)}{a(\phi)} + c(Y, \phi),$$

where $\mathbf{x} = (X_0, X_1, \cdots, X_k)'$, $\boldsymbol{\beta} = (\beta_0, \beta_1, \cdots, \beta_k)'$, and θ is uniquely determined via the functional relations $\mu = b'(\theta)$ and $g(\mu) = \eta = \mathbf{x}'\boldsymbol{\beta}$. Throughout the discussion, assume ϕ is known and fixed.

By definition, the score function of $\boldsymbol{\beta}$ is a random vector of $p = k + 1$ dimensions

$$\mathbf{S}(\boldsymbol{\beta}) = \frac{\partial \ell}{\partial \boldsymbol{\beta}} = \left(\frac{\partial \ell}{\partial \beta_0}, \frac{\partial \ell}{\partial \beta_1}, \cdots, \frac{\partial \ell}{\partial \beta_k} \right)'.$$

Note that the score function can be evaluated at any $\boldsymbol{\beta}$, not necessarily the true parameter value. When $\boldsymbol{\beta}$ is the true parameter value, the mean value of the score function is zero and its covariance matrix is known as the Fisher information, which can be shown to be equal to

$$\mathbf{I}(\boldsymbol{\beta}) = \mathrm{E}\left(\frac{\partial \ell}{\partial \boldsymbol{\beta}} \frac{\partial \ell}{\partial \boldsymbol{\beta}'} \right) = -\mathrm{E}\left(\frac{\partial^2 \ell}{\partial \boldsymbol{\beta} \partial \boldsymbol{\beta}'} \right).$$

Moreover, the $p \times p$ matrix

$$\mathbf{J}(\boldsymbol{\beta}) = -\frac{\partial^2 \ell}{\partial \boldsymbol{\beta} \partial \boldsymbol{\beta}'} = \left\{ -\frac{\partial^2 \ell}{\partial \beta_j \partial \beta_h} \right\}$$

is known as the observed Fisher information.

Use the chain rule of differentiation to write

$$\frac{\partial \ell}{\partial \beta_j} = \frac{\partial \ell}{\partial \theta} \frac{d\theta}{d\mu} \frac{d\mu}{d\eta} \frac{\partial \eta}{\partial \beta_j}. \tag{12.2.1}$$

Recall that $b'(\theta) = \mu = \mathrm{E}(Y)$ and $b''(\theta) = V(\mu) = \mathrm{Var}(Y)/a(\phi)$; see (12.1.4) and (12.1.6). Then,

$$\frac{\partial \ell}{\partial \theta} = \frac{Y - b'(\theta)}{a(\phi)} = \frac{Y - \mu}{a(\phi)}, \quad \frac{d\mu}{d\theta} = b''(\theta) = V(\mu), \quad \frac{\partial \eta}{\partial \beta_j} = X_j.$$

Thus, we get

$$\mathbf{S}(\boldsymbol{\beta}) = \frac{Y - \mu}{\mathrm{Var}(Y)} \frac{d\mu}{d\eta} \mathbf{x} = \frac{Y - \mu}{\mathrm{Var}(Y)} \frac{d\mu}{d\eta} \mathbf{x} \quad \text{and} \tag{12.2.2}$$

$$\mathbf{I}(\boldsymbol{\beta}) = \mathrm{E}\left[\left(\frac{Y - \mu}{\mathrm{Var}(Y)} \right)^2 \left(\frac{d\mu}{d\eta} \right)^2 \mathbf{x}\mathbf{x}' \right] = \frac{1}{\mathrm{Var}(Y)} \left(\frac{d\mu}{d\eta} \right)^2 \mathbf{x}\mathbf{x}'. \tag{12.2.3}$$

Example 12.2.1. Score and information under the canonical link. Under the canonical link, $\eta = \theta$, giving $d\mu/d\eta = b''(\theta) = V(\mu)$. From (12.2.2),

$$\mathbf{S}(\boldsymbol{\beta}) = \frac{Y - \mu}{a(\phi)} \mathbf{x},$$

and the observed Fisher information is

$$\mathbf{J}(\boldsymbol{\beta}) = \frac{1}{a(\phi)} \frac{\partial \mu}{\partial \boldsymbol{\beta}} \mathbf{x}' = \frac{1}{a(\phi)} \frac{d\mu}{d\eta} \frac{\partial \eta}{\partial \boldsymbol{\beta}} \mathbf{x}' = \frac{b''(\theta)}{a(\phi)} \mathbf{x}\mathbf{x}',$$

which is constant, a property of the canonical link. As a result, the Fisher information based on a single observation is $\mathbf{I}(\boldsymbol{\beta}) = \mathbf{J}(\boldsymbol{\beta}) = [b''(\theta)/a(\phi)]\mathbf{x}\mathbf{x}'$. □

Now, suppose we have independent observations, $(Y_i, \mathbf{x}_i)$, $i = 1, \cdots, N$. Then from (12.2.2) and (12.2.3), the score function and Fisher information are

$$\mathbf{S}(\boldsymbol{\beta}) = \sum_{i=1}^{N} \mathbf{S}_i(\boldsymbol{\beta}) = \sum_{i=1}^{N} \frac{Y_i - \mu_i}{\mathrm{Var}(Y_i)} \frac{d\mu_i}{d\eta_i} \mathbf{x}_i, \tag{12.2.4}$$

$$\mathbf{I}(\boldsymbol{\beta}) = \sum_{i=1}^{N} \mathbf{I}_i(\boldsymbol{\beta}) = \sum_{i=1}^{N} \frac{1}{\mathrm{Var}(Y_i)} \left(\frac{d\mu_i}{d\eta_i} \right)^2 \mathbf{x}_i \mathbf{x}_i', \tag{12.2.5}$$

where $\mathbf{S}_i$ and $\mathbf{I}_i$ are the score function and Fisher information based on $(Y_i, \mathbf{x}_i)$, respectively.

12.2.2 Maximum likelihood estimation – Fisher scoring

The Fisher scoring algorithm is a variant of the Newton–Raphson algorithm, which is a general method for finding the roots of a function. To maximize a twice differentiable function $U(\boldsymbol{\beta})$, one may instead search for the roots of $\mathbf{u}(\boldsymbol{\beta}) = \partial U/\partial \boldsymbol{\beta}$. The Newton–Raphson algorithm uses the iteration

$$\boldsymbol{\beta}^{(m+1)} = \boldsymbol{\beta}^{(m)} - \left[\frac{\partial \mathbf{u}(\boldsymbol{\beta})}{\partial \boldsymbol{\beta}'}\right]^{-1} \mathbf{u}(\boldsymbol{\beta}^{(m)}), \quad m = 0, 1, \cdots \tag{12.2.6}$$

to approximate a root of $\mathbf{u}(\boldsymbol{\beta})$. When U is a likelihood function, $\mathbf{u}$ is the score function and $-\partial \mathbf{u}/\partial \boldsymbol{\beta}'$ is the observed Fisher information. In the iteration, if the observed Fisher information is replaced with its expectation, i.e., the Fisher information, the resulting algorithm is known as the Fisher scoring algorithm.

Combining the above description with the likelihood function of the GLIM, we get the following result on the Fisher scoring algorithm.

Result 12.2.1. Let $(Y_i, \mathbf{x}_i)$, $i = 1, \cdots, N$, be independent observations. Let

$$\mathbf{A} = \text{diag}\left(\frac{1}{\text{Var}(Y_1)}\frac{d\mu_1}{d\eta_1}, \cdots, \frac{1}{\text{Var}(Y_N)}\frac{d\mu_N}{d\eta_N}\right) \quad \text{and} \tag{12.2.7}$$

$$\mathbf{W} = \text{diag}\left(\frac{1}{\text{Var}(Y_1)}\left(\frac{d\mu_1}{d\eta_1}\right)^2, \cdots, \frac{1}{\text{Var}(Y_N)}\left(\frac{d\mu_N}{d\eta_N}\right)^2\right). \tag{12.2.8}$$

1. The $(m+1)$th iteration of the Fisher scoring algorithm to obtain the MLE of $\boldsymbol{\beta}$ is

$$\boldsymbol{\beta}^{(m+1)} = \boldsymbol{\beta}^{(m)} + (\mathbf{X}'\mathbf{W}^{(m)}\mathbf{X})^{-1}\mathbf{X}'\mathbf{A}^{(m)}(\mathbf{y} - \boldsymbol{\mu}^{(m)}) \tag{12.2.9}$$

   for $m = 0, 1, \cdots$, where $\mathbf{X} = (\mathbf{x}_1, \cdots, \mathbf{x}_N)'$, and $\mathbf{W}^{(m)}$, $\mathbf{A}^{(m)}$, and $\boldsymbol{\mu}^{(m)}$ are evaluated at $\boldsymbol{\beta}^{(m)}$.

2. Under the canonical link, the Fisher scoring algorithm coincides with the Newton–Raphson algorithm.

Proof. From the above general description on Fisher scoring for the MLE, each iteration of the algorithm has the form

$$\boldsymbol{\beta}^{(m+1)} = \boldsymbol{\beta}^{(m)} + [\mathbf{I}(\boldsymbol{\beta}^{(m)})]^{-1}\mathbf{S}(\boldsymbol{\beta}^{(m)}).$$

From (12.2.4), we see that the score function based on the data is

$$\mathbf{S}(\boldsymbol{\beta}) = \sum_{i=1}^{N} \mathbf{x}_i \left[\frac{1}{\text{Var}(Y_i)}\frac{d\mu_i}{d\eta_i}\right](Y_i - \mu_i) = \mathbf{X}'\mathbf{A}(\mathbf{y} - \boldsymbol{\mu}),$$

while the Fisher information based on the data is

$$\mathbf{I}(\boldsymbol{\beta}) = \sum_{i=1}^{N} \mathbf{x}_i \left[\frac{1}{\text{Var}(Y_i)}\left(\frac{d\mu_i}{d\eta_i}\right)^2\right]\mathbf{x}_i' = \mathbf{X}'\mathbf{W}\mathbf{X}.$$

Combining the above three displays, we obtain (12.2.9), hence proving property 1. On the other hand, the Newton–Raphson algorithm uses the observed Fisher information in place of the Fisher information in each iteration. From Example 12.2.1, under the canonical link, the observed Fisher information is equal to the Fisher information, so that the Newton–Raphson algorithm is the same as the Fisher scoring algorithm, proving property 2. ■

Under certain regularity conditions, i.e.,

(i) the true parameter values lie in the interior of the parameter space,

(ii) the log-likelihood function is three times differentiable, and

(iii) the third derivatives of the log-likelihood function are bounded,

it can be shown that $\boldsymbol{\beta}^{(m)} \to \widehat{\boldsymbol{\beta}}_{ML}$ as $m \to \infty$ (Osborne, 1992). Also, for large N,

$$\widehat{\boldsymbol{\beta}}_{ML} \sim N(\boldsymbol{\beta}, (\mathbf{X}'\mathbf{W}\mathbf{X})^{-1}) \quad \text{approximately,} \tag{12.2.10}$$

where $\mathbf{W}$ is evaluated at $\widehat{\boldsymbol{\beta}}_{ML}$.

12.2.3 Iteratively reweighted least squares

When the Fisher scoring algorithm is written in a slightly different form, it is commonly known as the iteratively reweighted least squares (IRLS) estimation algorithm for $\boldsymbol{\beta}$. Recall that Corollary 4.5.2 presented the weighted least squares (WLS) estimate of $\boldsymbol{\beta}$ for the full-rank GLM as

$$\widehat{\boldsymbol{\beta}}_{WLS} = (\mathbf{X}'\mathbf{W}\mathbf{X})^{-1}\mathbf{X}'\mathbf{W}\mathbf{y}.$$

Let

$$\mathbf{D} = \text{diag}\left(\frac{d\eta_1}{d\mu_1}, \cdots, \frac{d\eta_N}{d\mu_N}\right) = \text{diag}(g'(\mu_1), \cdots, g'(\mu_N)). \tag{12.2.11}$$

Define the adjusted dependent variate $\mathbf{z} \in \mathcal{R}^N$ as

$$\mathbf{z} = \boldsymbol{\eta} + \mathbf{D}(\mathbf{y} - \boldsymbol{\mu}), \tag{12.2.12}$$

so that for each $i = 1, \cdots, N$,

$$Z_i = \eta_i + g'(\mu_i)(Y_i - \mu_i). \tag{12.2.13}$$

Result 12.2.2. IRLS estimator. For $m = 0, 1, 2, \cdots$, the $(m + 1)$th step of the iteratively reweighted least squares (IRLS) algorithm for estimating $\boldsymbol{\beta}$ is

$$\boldsymbol{\beta}_{IRLS}^{(m+1)} = (\mathbf{X}'\mathbf{W}^{(m)}\mathbf{X})^{-1}\mathbf{X}'\mathbf{W}^{(m)}\mathbf{z}^{(m)}, \tag{12.2.14}$$

where $\mathbf{W}^{(m)}$ and $\mathbf{z}^{(m)}$ were defined in (12.2.8) and (12.2.12) and evaluated at $\boldsymbol{\beta}_{IRLS}^{(m)}$. When each $a_i(\phi) = \phi/w_i$, where w_i is the known prior weight of Y_i, the IRLS estimation is independent of the dispersion parameter ϕ.

Proof. To relate Fisher scoring to WLS, we rewrite (12.2.9) as

$$\boldsymbol{\beta}^{(m+1)} = (\mathbf{X}'\mathbf{W}^{(m)}\mathbf{X})^{-1}[\mathbf{X}'\mathbf{W}^{(m)}\mathbf{X}\boldsymbol{\beta}^{(m)} + \mathbf{X}'\mathbf{A}^{(m)}(\mathbf{y} - \boldsymbol{\mu}^{(m)})].$$

Then (12.2.14) follows from $\mathbf{A} = \mathbf{WD}$. Since $\text{Var}(Y_i) = V(\mu_i)a_i(\phi)$, when $a_i(\phi) = \phi/w_i$, we can see that ϕ cancels out on the right side of (12.2.14). ■

Estimation of the dispersion parameter ϕ is discussed in Section 12.3. On the other hand, if $a_i(\phi) = \phi/w_i$, where w_i is the known prior weight of Y_i, then from Result 12.2.2, the IRLS algorithm is free of ϕ. We describe its implementation as a result below.

Result 12.2.3. Suppose $a_i(\phi) = \phi/w_i$ with w_i being known. Then the initial iteration of the IRLS algorithm can be carried out by the following steps:

1. Specify an initial vector for $\boldsymbol{\beta}$, say $\boldsymbol{\beta}^{(0)}$.

2. Evaluate $\boldsymbol{\eta}^{(0)} = \mathbf{X}\boldsymbol{\beta}^{(0)}$.

3. Evaluate $\boldsymbol{\mu}^{(0)} = (\mu_1^{(0)}, \cdots, \mu_N^{(0)})'$ and $\boldsymbol{\theta}^{(0)} = (\theta_1^{(0)}, \cdots, \theta_N^{(0)})'$, with $\mu_i^{(0)} = g^{-1}(\eta_i^{(0)})$ and $\theta_i^{(0)} = (b')^{-1}(\mu_i^{(0)})$, $i = 1, \cdots, N$. In particular, for the canonical link, $\boldsymbol{\theta}^{(0)} = \boldsymbol{\eta}^{(0)}$.

4. Evaluate $d_i^{(0)} = g'(\mu_i^{(0)})$ and $v_i^{(0)} = V(\mu_i^{(0)})$, $i = 1, \cdots, N$. In particular, for the canonical link, $v_i^{(0)} = 1/d_i^{(0)}$; see property 2 of Result 12.1.1.

5. Compute the adjusted dependent variate $Z_i^{(0)} = \eta_i^{(0)} + d_i^{(0)}(Y_i - \mu_i^{(0)})$ for $i = 1, \cdots, N$.

6. Compute weights $W_i^{(0)} = w_i/[v_i^{(0)}(d_i^{(0)})^2]$, recalling we let $a_i(\phi) = \phi/w_i$. In particular, for the canonical link, $W_i^{(0)} = w_i v_i$.

7. Let $\mathbf{z}^{(0)} = (Z_1^{(0)}, \cdots, Z_N^{(0)})'$ and $\mathbf{W}^{(0)} = \text{diag}(W_1^{(0)}, \cdots, W_N^{(0)})$.

8. Fit the WLS regression $\mathbf{z}^{(0)} = \mathbf{X}\boldsymbol{\beta} + \boldsymbol{\varepsilon}^{(0)}$ with $\text{Var}(\boldsymbol{\varepsilon}^{(0)}) = \sigma^2(\mathbf{W}^{(0)})^{-1}$, where σ^2 is unknown. Denote the WLS solution by $\boldsymbol{\beta}^{(1)}$, i.e.,

$$\boldsymbol{\beta}^{(1)} = (\mathbf{X}'\mathbf{W}^{(0)}\mathbf{X})^{-1}\mathbf{X}'\mathbf{W}^{(0)}\mathbf{z}^{(0)}. \tag{12.2.15}$$

We iterate the above procedure, i.e., Steps 1–8 followed by (12.2.15) starting from $\boldsymbol{\beta}^{(1)}$, $\boldsymbol{\beta}^{(2)}$, $\cdots$ until convergence. The converged estimate is $\widehat{\boldsymbol{\beta}}_{ML}$. The variance-covariance matrix of $\widehat{\boldsymbol{\theta}}_{ML}$ is estimated by

$$\widehat{\text{Cov}}(\widehat{\boldsymbol{\theta}}_{ML}) = \phi(\mathbf{X}'\widehat{\mathbf{W}}\mathbf{X})^{-1}, \tag{12.2.16}$$

where ϕ has to be replaced with an estimate if it is unknown.

Proof. The steps are direct consequences of the iteration described in Result 12.2.2. Notice that the matrix $\mathbf{W}^{(j)}$ in the steps does not include ϕ, so it is ϕ^{-1} times the weight matrix in (12.2.12). Then the estimate in (12.2.16) follows from (12.2.10). ■

Example 12.2.2. For the binary response model in Example 12.1.1, the log-likelihood function (apart from a constant term) has the form

$$\ell(\boldsymbol{\pi}; \mathbf{y}) = \sum_{i=1}^{N} Y_i \log[\pi_i/(1-\pi_i)] + \log(1-\pi_i),$$

which reduces under the logit link $\log[\pi_i/(1-\pi_i)] = \eta_i = \mathbf{x}_i'\boldsymbol{\beta}$ to

$$\ell(\boldsymbol{\beta}; \mathbf{y}) = \sum_{i=1}^{N} Y_i \mathbf{x}_i'\boldsymbol{\beta} - \sum_{i=1}^{N} \log[1 + \exp(\mathbf{x}_i'\boldsymbol{\beta})]. \tag{12.2.17}$$

We use (12.2.1) and Result 12.2.2 to obtain the IRLS estimate of $\boldsymbol{\beta}$. Details and a numerical example are shown in Section 12.4. □

12.2.4 Quasi-likelihood estimation

The usefulness of a quasi-likelihood function for situations where there is insufficient information to construct a likelihood function is well documented in the literature. It is closely related to the estimating equations approach, whereby given data $\mathbf{y}$, an estimate of a parameter $\boldsymbol{\beta}$ is a solution to $\boldsymbol{\psi}(\boldsymbol{\beta};\mathbf{y}) = \mathbf{0}$, where the problem specific function $\boldsymbol{\psi}(\boldsymbol{\beta};\mathbf{y})$ is known as an estimating function. By analogy to the score function, if $\boldsymbol{\psi}(\boldsymbol{\beta};\mathbf{y})$ can be written as $\partial K(\boldsymbol{\beta};\mathbf{y})/\partial\boldsymbol{\beta}$, it is also referred to as a quasi-score function and $K(\boldsymbol{\beta};\mathbf{y})$ as a quasi-likelihood function.

We first consider the case where the observations are independent. First, under a GLIM, if the data is $\mathbf{y} = (Y_1, \cdots, Y_N)'$, then from (12.2.4) and since $\eta_i = \mathbf{x}_i'\boldsymbol{\beta}$, $\widehat{\boldsymbol{\beta}}_{ML}$ is a solution to

$$\sum_{i=1}^{N} \frac{Y_i - \mu_i}{\mathrm{Var}(Y_i)} \frac{\partial \mu_i}{\partial \boldsymbol{\beta}} = \mathbf{0} \text{ with } \boldsymbol{\mu} = \mathrm{E}(\mathbf{y}). \tag{12.2.18}$$

Remarkably, while (12.2.18) has the GLIM as its origin, its form is quite general, only consisting of means and variances of observations. As a result, (12.2.18) can be used as an estimating equation even when a likelihood function is not available, as long as the mean and variance are modeled parametrically.

Suppose we have specified a model for the means $\mu_i = \mu_i(\boldsymbol{\beta})$, $i = 1, \cdots, N$. To construct a quasi-likelihood function, further assume that $\mathrm{Var}(Y_i) = \phi V_i(\mu_i)$. Define

$$\begin{aligned} K(\boldsymbol{\mu};\mathbf{y}) &= \sum_{i=1}^{N} K_i(\mu_i; Y_i), \text{ with} \\ K_i(\mu_i; Y_i) &= \int_{Y_i}^{\mu_i} \frac{(Y_i - \xi)}{\phi V_i(\xi)}\, d\xi. \end{aligned} \tag{12.2.19}$$

By the chain rule of differentiation and the fundamental theorem of calculus (the derivative of the integral of a function is the function itself), we see that

$$\frac{\partial K(\boldsymbol{\mu}(\boldsymbol{\beta});\mathbf{y})}{\partial \boldsymbol{\beta}} = \sum_{i=1}^{N} \frac{dK_i(\mu_i;Y_i)}{d\mu_i}\frac{\partial \mu_i}{\partial \boldsymbol{\beta}} = \sum_{i=1}^{N} \frac{(Y_i - \mu_i)}{\phi V_i(\mu_i)}\frac{\partial \mu_i}{\partial \boldsymbol{\beta}},$$

which by the assumption on $\mathrm{Var}(Y_i)$ is the estimating function in (12.2.18). Thus, $K(\boldsymbol{\mu};\mathbf{y})$ is a quasi-likelihood function. In a few examples, the quasi-likelihood function corresponds to the log-likelihood function.

Example 12.2.3. Binary and Poisson models. For the binary response model, $\mu_i = \pi_i$, the variance function is $V_i(\pi_i) = \pi_i(1 - \pi_i)$, and $\phi = 1$. Then

$$\begin{aligned} K_i(\pi_i; Y_i) &= \int_{Y_i}^{\pi_i} \frac{Y_i - \xi}{\xi(1-\xi)}\, d\xi \\ &= \begin{cases} -\displaystyle\int_0^{\pi_i} \frac{1}{1-\xi}\, d\xi = \log(1 - \pi_i) & \text{if } Y_i = 0, \\ -\displaystyle\int_{\pi_i}^{1} \frac{1}{\xi}\, d\xi = \log \pi_i & \text{if } Y_i = 1. \end{cases} \end{aligned}$$

Then $K_i(\pi_i; Y_i) = Y_i \log\{\pi_i/(1 - \pi_i)\} + \log(1 - \pi_i)$, the same as the log-likelihood function of the binary response model.

For the Poisson model, recall that the mean is $\mu_i = \lambda_i$, the variance function is $V_i(\mu_i) = \lambda_i$, $i = 1, \cdots, N$, and $\phi = 1$. Then,

$$K_i(\lambda_i; Y_i) = \int_{Y_i}^{\lambda_i} \frac{Y_i - \xi}{\xi} d\xi = Y_i \log \lambda_i - \lambda_i - Y_i \log Y_i + Y_i,$$

where $0 \log 0 = 0$. The log-likelihood function is $\log(e^{-\lambda_i} \lambda_i^{Y_i} / Y_i!) = Y_i \log \lambda_i - \lambda_i - \log(Y_i!)$. Here, the quasi-likelihood is different from the log-likelihood by a constant that depends on Y_i. □

Result 12.2.4. Let the data $Y_1, \cdots, Y_N$ be independent. Let $\mathbf{B}(\boldsymbol{\beta})$ be the $N \times p$ matrix whose (i,j)th element is $B_{ij}(\boldsymbol{\beta}) = \partial \mu_i / \partial \beta_j$. Let $\mathbf{V}(\boldsymbol{\mu}) = \text{diag}(V_1(\mu_1), \cdots, V_N(\mu_N))$. The Fisher scoring algorithm obtains the quasi-likelihood estimator $\widehat{\boldsymbol{\beta}}_{QL}$ by iterating to convergence the sequence of estimates of the form

$$\boldsymbol{\beta}^{(m+1)} = \boldsymbol{\beta}^{(m)} + \{\mathbf{B}^{(m)\prime}(\mathbf{V}^{(m)})^{-1}\mathbf{B}^{(m)}\}^{-1}\mathbf{B}^{(m)\prime}(\mathbf{V}^{(m)})^{-1}(\mathbf{y} - \boldsymbol{\mu}^{(m)}), \qquad (12.2.20)$$

where $\mathbf{B}^{(m)}$, $\mathbf{V}^{(m)}$, and $\boldsymbol{\mu}^{(m)}$ are evaluated at $\boldsymbol{\beta}^{(m)}$.

Proof. Denote the quasi-score function by $\mathbf{u}(\boldsymbol{\beta}) = (u_1(\boldsymbol{\beta}), \cdots, u_p(\boldsymbol{\beta}))'$. The Fisher scoring algorithm uses the iteration

$$\boldsymbol{\beta}^{(m+1)} = \boldsymbol{\beta}^{(m)} - \left\{ \text{E}\left[\left. \frac{\partial \mathbf{u}(\boldsymbol{\beta})}{\partial \boldsymbol{\beta}'} \right|_{\boldsymbol{\beta}=\boldsymbol{\beta}^{(m)}} \right] \right\}^{-1} \mathbf{u}(\boldsymbol{\beta}^{(m)}), \qquad (12.2.21)$$

where the terms are evaluated at $\boldsymbol{\beta}^{(m)}$. For $j = 1, \cdots, p$,

$$u_j(\boldsymbol{\beta}) = \sum_{i=1}^{N} \frac{Y_i - \mu_i}{\phi V_i(\mu_i)} \frac{\partial \mu_i}{\partial \beta_j} = \frac{1}{\phi} \sum_{i=1}^{N} B_{ij}(\boldsymbol{\beta}) \frac{Y_i - \mu_i}{V_i(\mu_i)},$$

and so

$$\mathbf{u}(\boldsymbol{\beta}) = \frac{1}{\phi} [\mathbf{B}(\boldsymbol{\beta})]' [\mathbf{V}(\boldsymbol{\mu})]^{-1} (\mathbf{y} - \boldsymbol{\mu}). \qquad (12.2.22)$$

We see that for $\ell = 1, \cdots, p$,

$$\frac{\partial u_j(\boldsymbol{\beta})}{\partial \beta_\ell} = -\sum_{i=1}^{N} \frac{\partial \mu_i}{\partial \beta_\ell} \left[\frac{1}{\phi V_i(\mu_i)} \frac{\partial \mu_i}{\partial \beta_j} \right] + \sum_{i=1}^{N} (Y_i - \mu_i) \frac{\partial}{\partial \beta_\ell} \left[\frac{1}{\phi V_i(\mu_i)} \frac{\partial \mu_i}{\partial \beta_j} \right].$$

Since $\text{E}(Y_i) = \mu_i$, provided that $\boldsymbol{\beta}$ is the true parameter, we get

$$\begin{aligned}
\text{E}\left[\left. \frac{\partial u_j(\boldsymbol{\beta})}{\partial \beta_\ell} \right|_{\boldsymbol{\beta}=\boldsymbol{\beta}^{(m)}} \right] &= -\sum_{i=1}^{N} \frac{\partial \mu_i}{\partial \beta_\ell} \left[\frac{1}{\phi V_i(\mu_i)} \frac{\partial \mu_i}{\partial \beta_j} \right] \Bigg|_{\boldsymbol{\beta}=\boldsymbol{\beta}^{(m)}} \\
&= -\sum_{i=1}^{N} B_{i\ell}^{(m)} \frac{1}{\phi V_i(\mu_i)} B_{ij}^{(m)},
\end{aligned}$$

so that

$$\text{E}\left[\left. \frac{\partial \mathbf{u}(\boldsymbol{\beta})}{\partial \boldsymbol{\beta}'} \right|_{\boldsymbol{\beta}=\boldsymbol{\beta}^{(m)}} \right] = \frac{1}{\phi} \mathbf{B}^{(m)\prime} [\mathbf{V}^{(m)}]^{-1} \mathbf{B}^{(m)}. \qquad (12.2.23)$$

Combining (12.2.21)–(12.2.23) then yields (12.2.20), and proves the result. ■

For results on the consistency of the QL estimates and properties of QL estimation when we have dependent observations, we refer the reader to McCullagh (1983) and McCullagh and Nelder (1989).

12.3 Residuals and model checking

The residuals as well as goodness of fit criteria enable us to assess the adequacy of fit of the GLIM. Suppose $\boldsymbol{\theta} = (\theta_1, \cdots, \theta_N)'$ denotes the canonical parameter and $a_i(\phi) = \phi/w_i$, $i = 1, \cdots, N$, where the dispersion parameter ϕ is assumed to be fixed.

12.3.1 GLIM residuals

Three types of residuals are usually defined for GLIMs. Let $\widehat{\mu}_i$ be the mean value of Y_i determined by $g(\widehat{\mu}_i) = \widehat{\eta}_i = \mathbf{x}_i'\widehat{\boldsymbol{\beta}}$, where $\widehat{\boldsymbol{\beta}}$ is an estimate of $\boldsymbol{\beta}$.

Definition 12.3.1. The ith Pearson residual is defined by

$$r_{i,P} = \frac{Y_i - \widehat{\mu}_i}{\sqrt{V(\widehat{\mu}_i)}}. \tag{12.3.1}$$

The distribution of $r_{i,P}$ is usually skewed for non-normal response distributions, and Anscombe residuals may be preferred for this reason.

Definition 12.3.2. The ith Anscombe residual has the form

$$r_{i,A} = \frac{A(Y_i) - A(\widehat{\mu}_i)}{A'(\widehat{\mu}_i)\sqrt{V(\widehat{\mu}_i)}}, \tag{12.3.2}$$

where the function $A(\cdot)$ is given by

$$A(z) = \int \frac{dz}{V^{1/3}(z)}.$$

For example, for the Poisson model,

$$r_{i,A} = \frac{3(Y_i^{2/3} - \widehat{\mu}_i^{2/3})}{2\widehat{\mu}_i^{1/6}}.$$

Definition 12.3.3. The ith deviance residual is defined by

$$r_{i,D} = \text{sign}(Y_i - \widehat{\mu}_i)\sqrt{d_i}, \tag{12.3.3}$$

where we will define d_i in (12.3.4) in Section 12.3.2.

For example, for the Poisson model,

$$r_{i,D} = \text{sign}(Y_i - \widehat{\mu}_i)[2(Y_i \log(Y_i/\widehat{\mu}_i) - Y_i + \widehat{\mu}_i)]^{1/2}.$$

The values of Anscombe residuals and deviance residuals are very similar in most cases (see Pierce and Schafer (1986) for details).

Residual diagnostics similar to those defined in Section 8.3 have been studied for GLIMs; see Pregibon (1981) for details.

12.3.2 Goodness of fit measures

We define the deviance and scaled deviance functions, and the generalized Pearson X^2 statistic. The deviance is a generalization of the notion of residual sum of squares (SSE) in the general linear model. It is useful to think about three model classes.

(a) **Null (or constant) model.** Assume that $\mu_i = \mu_0 \; i = 1, \cdots, N$, so we need to estimate a single parameter μ_0, which does not depend on any predictors. Then, $\widehat{\mu}_i = \widehat{\mu}_0 = \overline{Y}$.

(b) **Fitted model with k predictors.** Suppose $g(\mu_i) = \mathbf{x}_i'\boldsymbol{\beta}$, where $\mathbf{x}_i$ is a $(k+1)$-dimensional vector. Then, $\widehat{\mu}_i = g^{-1}(\mathbf{x}_i'\widehat{\boldsymbol{\beta}}_{ML})$. Let $\widehat{\boldsymbol{\theta}} = \boldsymbol{\theta}(\widehat{\boldsymbol{\mu}})$ denote the estimate of the canonical parameter $\boldsymbol{\theta}$ and let $\ell(\widehat{\boldsymbol{\mu}}, \phi; \mathbf{y})$ denote the maximized log-likelihood function under the fitted model.

(c) **Saturated model.** Assume that $\boldsymbol{\mu}$ has N free parameters and is completely free of constraints. In this case, the log-likelihood is maximized at $\widetilde{\boldsymbol{\mu}} = (\widetilde{\mu}_1, \cdots, \widetilde{\mu}_N)'$, where $\widetilde{\mu}_i$ is such that $\widetilde{\theta}_i = (b')^{-1}(\widetilde{\mu}_i)$ maximizes $Y_i\theta_i - b(\theta_i)$. This implies that $Y_i = b'(\widetilde{\theta}_i)$, so that $\widetilde{\boldsymbol{\mu}} = \mathbf{y}$, the observed response vector. Let $\widetilde{\boldsymbol{\theta}} = \widetilde{\boldsymbol{\theta}}(\mathbf{y})$ denote the estimate of the canonical parameter $\boldsymbol{\theta}$ and let $\ell(\mathbf{y}, \phi; \mathbf{y})$ denote the maximized log-likelihood under the saturated model.

The discrepancy of fit between the saturated model and the model with k predictors is proportional to $2[\ell(\mathbf{y}, \phi; \mathbf{y}) - \ell(\widehat{\boldsymbol{\mu}}, \phi; \mathbf{y})]$. The discrepancy for the ith observation becomes

$$d_i = 2w_i[Y_i(\widetilde{\theta}_i - \widehat{\theta}_i) - b(\widetilde{\theta}_i) + b(\widehat{\theta}_i)], \quad i = 1, \cdots, N. \tag{12.3.4}$$

Definition 12.3.4. Deviance. The deviance under the fitted model is defined as the following function of the data:

$$D(\mathbf{y}; \widehat{\boldsymbol{\mu}}) = \sum_{i=1}^{N} d_i = \sum_{i=1}^{N} 2w_i[Y_i(\widetilde{\theta}_i - \widehat{\theta}_i) - b(\widetilde{\theta}_i) + b(\widehat{\theta}_i)], \tag{12.3.5}$$

where $\widehat{\theta}_i = (b(\widehat{\theta}_i)')^{-1}\widehat{\mu}_i$, $\widetilde{\theta}_i = (b(\widetilde{\theta}_i)')^{-1}Y_i$ and w_i was defined in $a_i(\phi) = \phi/w_i$.

It is straightforward to compute the deviance for different models. For example,

$$D(\mathbf{y}; \widehat{\boldsymbol{\mu}}) = \begin{cases} \sum_{i=1}^{N}(Y_i - \widehat{\mu}_i)^2 \text{ or } SSE & \text{(normal model)} \\ 2\sum_{i=1}^{N}[Y_i \log(Y_i/\widehat{\mu}_i) + (1 - Y_i)\log[(1 - Y_i)/(1 - \widehat{\mu}_i)] & \\ & \text{(binary logit model)} \\ 2\sum_{i=1}^{N} Y_i \log(Y_i/\widehat{\mu}_i) & \text{(Poisson log-linear model)} \end{cases}$$

Note that the deviance is a non-decreasing function of k (the number of predictors included in the model).

Definition 12.3.5. The scaled deviance is

$$D^*(\mathbf{y}; \widehat{\boldsymbol{\mu}}) = 2[\ell(\mathbf{y}, \phi; \mathbf{y}) - \ell(\widehat{\boldsymbol{\mu}}, \phi; \mathbf{y})] = \frac{D(\mathbf{y}; \widehat{\boldsymbol{\mu}})}{\phi}, \tag{12.3.6}$$

and is computed as the deviance divided by the dispersion ϕ. When $\phi = 1$, the scaled deviance is the deviance. The scaled deviance has an asymptotic χ^2_{N-p} distribution.

Another criterion that enables us to assess discrepancy in fit is the generalized Pearson X^2 statistic. Result 12.3.1 shows how it enables us to estimate the dispersion parameter ϕ.

Definition 12.3.6. The generalized Pearson X^2 statistic is defined by

$$X^2 = \sum_{i=1}^{N} \frac{(Y_i - \widehat{\mu}_i)^2}{V(\widehat{\mu}_i)}. \tag{12.3.7}$$

Let $a_i(\phi) = \phi$ for $i = 1, \cdots, N$. The scaled generalized Pearson X^2 statistic is defined as

$$X_s^2 = \frac{X^2}{\phi}. \tag{12.3.8}$$

For the normal model, the X^2 statistic is SSE, while for the binomial and Poisson models, it is the usual Pearson X^2 statistic. For the Poisson model, $\sum_{i=1}^N r_{i,P}^2 = X^2$, Pearson's goodness of fit statistic, which leads to the construction of $r_{i,P}$.

Result 12.3.1. The method of moments estimate of the dispersion parameter ϕ is

$$\widehat{\phi} = \frac{X^2}{N-p}. \tag{12.3.9}$$

Proof. Provided the model is correctly specified, $X_s^2 \sim \chi^2_{N-p}$, so that $\text{E}(X_s^2) = N - p$. We get the result by setting $X_s^2 = \frac{X^2}{\phi} = N - p$ and solving for ϕ. ■

12.3.3 Hypothesis testing and model comparisons

We discuss the well-known Wald and likelihood ratio tests for GLIMs.

12.3.3.1 Wald Test

Recall from (12.2.10) that for large samples, under certain regularity conditions, we have $\widehat{\boldsymbol{\beta}}_{ML} \sim N(\boldsymbol{\beta}, (\mathbf{X}'\mathbf{W}\mathbf{X})^{-1})$. Suppose we wish to test H_0: $\boldsymbol{\beta} = \boldsymbol{\beta}_0$. The Wald statistic is

$$\begin{aligned}\text{Wald} &= (\widehat{\boldsymbol{\beta}}_{ML} - \boldsymbol{\beta}_0)'\mathbf{I}(\boldsymbol{\beta})^{-1}(\widehat{\boldsymbol{\beta}}_{ML} - \boldsymbol{\beta}_0)\\ &= (\widehat{\boldsymbol{\beta}}_{ML} - \boldsymbol{\beta}_0)'(\mathbf{X}'\mathbf{W}\mathbf{X})^{-1}(\widehat{\boldsymbol{\beta}}_{ML} - \boldsymbol{\beta}_0)\\ &\sim \chi^2_p \text{ approx. under } H_0. \end{aligned} \tag{12.3.10}$$

It is easy to construct a similar statistic for testing a hypothesis that imposes linear constraints on the elements of $\boldsymbol{\beta}$, for example, H_0: $\mathbf{C}'\boldsymbol{\beta} = \mathbf{d}$, for a known $p \times q$ matrix $\mathbf{C}$ and a known q-dimensional vector $\mathbf{d}$.

It is also easy to see that for testing a single coefficient β_j, the Wald test is an asymptotic z-test:

$$z = \frac{\widehat{\beta}_{j,ML} - \beta_{j,0}}{SE(\widehat{\beta}_{j,ML})} \sim N(0,1) \text{ approx.} \tag{12.3.11}$$

Hypothesis tests can also help us choose between two models, where one model is a subset of (is nested within) another model. Consider a "full model" with k predictors and a model reduced by a null hypothesis H_0 which sets some of the k partial regression coefficients to zero. Suppose $\widehat{\boldsymbol{\mu}}_F$ and $\widehat{\boldsymbol{\mu}}_R$ respectively denote the fitted mean vectors under the full and reduced models.

12.3.3.2 Likelihood ratio test

Recall that we can use the LRT to compare the maximized likelihoods from two models, $\mathcal{M}_1 \subset \mathcal{M}_2$. For instance, $\mathcal{M}_2$ could be the "full model" with all p β coefficients, while $\mathcal{M}_1$ is a model reduced by a null hypothesis H_0 which tests whether some of the β coefficients are zero. Suppose the maximized likelihood under $\mathcal{M}_1$ is

$$\max_{\boldsymbol{\beta} \in \mathcal{M}_1} L(\boldsymbol{\beta}; \mathbf{y}) = L(\widehat{\boldsymbol{\beta}}_{1,ML}; \mathbf{y})$$

and the maximized likelihood under $\mathcal{M}_2$ is

$$\max_{\boldsymbol{\beta}\in\mathcal{M}_2} L(\boldsymbol{\beta};\mathbf{y}) = L(\widehat{\boldsymbol{\beta}}_{2,ML};\mathbf{y}).$$

The LRT statistic is

$$\Lambda = \frac{L(\widehat{\boldsymbol{\beta}}_{1,ML};\mathbf{y})}{L(\widehat{\boldsymbol{\beta}}_{2,ML};\mathbf{y})}. \tag{12.3.12}$$

Under the usual regularity conditions,

$$-2\log\Lambda \sim \chi^2_\nu \text{ approx. under } H_0,$$

where $\nu = \dim(\mathcal{M}_2) - \dim(\mathcal{M}_1)$.

12.3.3.3 Drop-in-deviance test

We discuss the widely used drop-in-deviance test, also called the deviance difference test, in order to compare nested models using the deviance functions from both models. This is similar to the use of the Extra SS F-test in the normal GLM (see Section 9.1.3.1). Let the "full model" be $\mathcal{M}_2$ with p_2 of the p coefficients in $\boldsymbol{\beta}$. Let $\mathcal{M}_1$ be a reduced model with $p_1 < p_2$ coefficients, so $\mathcal{M}_1 \subset \mathcal{M}_2$. Let $\widehat{\boldsymbol{\mu}}_F$ and $\widehat{\boldsymbol{\mu}}_R$ denote the fitted vectors under the full and reduced models.

The deviance difference statistic for testing H_0 that the data prefers Model $\mathcal{M}_1$ versus the alternative of the "full model" is

$$D(\mathbf{y};\widehat{\boldsymbol{\mu}}_R) - D(\mathbf{y};\widehat{\boldsymbol{\mu}}_F) \sim \chi^2_\nu \text{ approx. under } H_0, \tag{12.3.13}$$

where $\nu = \dim(\mathcal{M}_2) - \dim(\mathcal{M}_1) = p_2 - p_1$. It is easy to see that the deviance difference corresponds to the LRT statistic for testing H_0, i.e.,

$$D(\mathbf{y};\widehat{\boldsymbol{\mu}}_R) - D(\mathbf{y};\widehat{\boldsymbol{\mu}}_F) = 2\ell(\widehat{\boldsymbol{\mu}}_F;\mathbf{y}) - 2\ell(\widehat{\boldsymbol{\mu}}_R;\mathbf{y}).$$

To study the discrepancy of fit, see that the LRT for comparing the fitted and saturated models gives (apart from a constant function of ϕ)

$$-2\log\Lambda = -2[\ell(\widehat{\boldsymbol{\mu}},\phi;\mathbf{y}) - \ell(\mathbf{y},\phi;\mathbf{y})]. \tag{12.3.14}$$

The discrepancy of fit between the fitted model and the saturated model is thus proportional to

$$2[\ell(\mathbf{y},\phi;\mathbf{y}) - \ell(\widehat{\boldsymbol{\mu}},\phi;\mathbf{y})],$$

which is the deviance.

12.4 Binary and binomial response models

Recall that the Bernoulli(π) distribution is a special case of the Binom(m,π) distribution. Suppose $T_i \sim \text{Binom}(m_i,\pi_i)$, $i = 1,\cdots,N$, are independent. Let $Y_i = T_i/m_i$. Then, the p.m.f. of Y_i is

$$f(y_i) = \binom{m_i}{m_i y_i} \exp\left\{\frac{y_i\log[\pi_i/(1-\pi_i)] + \log(1-\pi_i)}{1/m_i}\right\}.$$

To set up a GLIM for Y_i, let

$$\theta_i = \text{logit}(\pi_i) = \log[\pi_i/(1-\pi_i)].$$

Comparing with (12.1.1), the dispersion parameter can be set as $\phi = 11$, while $a_i(\phi) = 1/m_i$ and $c_i(y_i, \phi) = \log \binom{m_i}{m_i y_i}$. In particular, the prior weight of the ith observation is $w_i = m_i$.

Recall that $E(Y_i) = \pi_i$, and the canonical link function is $g(\pi_i) = \text{logit}(\pi_i)$. Under the logit link, $\pi_i = e^{\mathbf{x}_i'\boldsymbol{\beta}}/(1+e^{\mathbf{x}_i'\boldsymbol{\beta}})$, so that the log-likelihood function of $\boldsymbol{\beta}$ is, up to an additive term independent of $\boldsymbol{\beta}$,

$$\ell(\boldsymbol{\beta}; \mathbf{y}) = \sum_{i=1}^{N} m_i\{Y_i \mathbf{x}_i'\boldsymbol{\beta} - \log[1+\exp(\mathbf{x}_i'\boldsymbol{\beta})]\} \tag{12.4.1}$$

and the likelihood equations are easily obtained as

$$\frac{\partial \ell}{\partial \beta_r} = \sum_{i=1}^{N} m_i(Y_i - \pi_i)X_{i,r}.$$

The IRLS algorithm to get $\widehat{\boldsymbol{\beta}}_{ML}$ is particularly simple for the logit link. To carry out the steps in Result 12.2.3, given an initial estimate $\boldsymbol{\beta}^{(0)}$, for each $i = 1, \cdots, N$, $\pi_i^{(0)} = 1/[1+\exp(-\mathbf{x}_i'\boldsymbol{\beta}^{(0)})]$ and $\eta_i^{(0)} = \mathbf{x}_i'\boldsymbol{\beta}^{(0)}$. Since $g'(\pi) = 1/[\pi(1-\pi)]$,

$$d_i^{(0)} = \frac{1}{\pi_i^{(0)}(1-\pi_i^{(0)})}, \quad v_i^{(0)} = \pi_i^{(0)}(1-\pi_i^{(0)}),$$

so that the adjusted dependent response and weight are, respectively,

$$Z_i^{(0)} = \eta_i^{(0)} + \frac{(Y_i - \pi_i^{(0)})}{\pi_i^{(0)}(1-\pi_i^{(0)})} \quad \text{and} \quad W_i^{(0)} = m_i \pi_i^{(0)}(1-\pi_i^{(0)}).$$

Below are two other commonly used links.

1. Probit link: $g(\pi) = \Phi^{-1}(\pi)$, so that $E(Y_i \mid \mathbf{x}_i) = \Phi(\mathbf{x}_i'\boldsymbol{\beta})$, where Φ is the c.d.f. of a standard normal variable, i.e., the p.m.f. of Y_i is

$$f(y_i) = \binom{m_i}{m_i y_i} \Phi(\mathbf{x}_i'\boldsymbol{\beta})^{m_i y_i}[1-\Phi(\mathbf{x}_i'\boldsymbol{\beta})]^{m_i(1-y_i)}$$

   and $\theta_i = \text{logit}(\Phi(\mathbf{x}_i'\boldsymbol{\beta}))$.

2. Complementary log-log link: $g(\mu) = \log(-\log(1-\mu))$, so that $E(Y_i \mid \mathbf{x}_i) = 1 - \exp(-\exp(\mathbf{x}_i'\boldsymbol{\beta}))$. Then the p.m.f. of Y_i is

$$f(y_i) = \binom{m_i}{m_i y_i}(1-\exp\{-e^{\mathbf{x}_i'\boldsymbol{\beta}}\})^{m_i y_i} \exp\{-m_i(1-y_i)e^{\mathbf{x}_i'\boldsymbol{\beta}}\}$$

   and $\theta_i = \log(\exp(\exp(\mathbf{x}_i'\boldsymbol{\beta})) - 1)$.

For more complex link functions, see Stukel (1988) or Chen et al. (1999).

Numerical Example 12.1. The dataset "heart" from the R package *catdata* describes a retrospective sample of males in a heart-disease high-risk region of the Western Cape, South Africa. The data frame has 462 observations on 10 variables. The binary response Y takes the value 1 for coronary heart disease and 0 otherwise. The predictors are sbp (systolic blood pressure), tobacco (cumulative tobacco), ldl (low density lipoprotein cholesterol), adiposity, famhist (family history of heart disease), typea (type-A behavior), obesity, alcohol (current alcohol consumption), and age (age at onset).

```
data(heart, package = "catdata")
full_logit <- glm(
  y ~ sbp + tobacco + ldl + adiposity + famhist + typea
  + obesity + alcohol + age,
  data = data.frame(heart),
  family = binomial(link = "logit")
)
summary(full_logit)

Call:
glm(formula = y ~ sbp + tobacco + ldl + adiposity + famhist +
    typea + obesity + alcohol + age,
    family = binomial(link = "logit"), data = data.frame(heart))

Deviance Residuals:
    Min       1Q   Median       3Q      Max
-1.7781  -0.8213  -0.4387   0.8889   2.5435

Coefficients:
              Estimate Std. Error z value Pr(>|z|)
(Intercept) -6.1507209  1.3082600  -4.701 2.58e-06 ***
sbp          0.0065040  0.0057304   1.135 0.256374
tobacco      0.0793764  0.0266028   2.984 0.002847 **
ldl          0.1739239  0.0596617   2.915 0.003555 **
adiposity    0.0185866  0.0292894   0.635 0.525700
famhist      0.9253704  0.2278940   4.061 4.90e-05 ***
typea        0.0395950  0.0123202   3.214 0.001310 **
obesity     -0.0629099  0.0442477  -1.422 0.155095
alcohol      0.0001217  0.0044832   0.027 0.978350
age          0.0452253  0.0121298   3.728 0.000193 ***
---
Signif. codes:  0 '***' 0.001 '**' 0.01 '*' 0.05 '.' 0.1 ' ' 1

(Dispersion parameter for binomial family taken to be 1)

    Null deviance: 596.11  on 461  degrees of freedom
Residual deviance: 472.14  on 452  degrees of freedom
AIC: 492.14

Number of Fisher Scoring iterations: 5
```

We show the deviance difference (chi-squared) test to compare a reduced model with fewer predictors with the full model with all predictors.

```
red_logit <-  glm(
  y ~ tobacco + ldl + famhist + typea + age ,
  data = data.frame(heart),
  family = binomial(link = "logit")
)
summary(red_logit)

Call:
```

```
glm(formula = y ~ tobacco + ldl + famhist + typea + age,
family = binomial(link = "logit"),
    data = data.frame(heart))

Deviance Residuals:
    Min       1Q   Median       3Q      Max
-1.9165  -0.8054  -0.4430   0.9329   2.6139

Coefficients:
            Estimate Std. Error z value Pr(>|z|)
(Intercept) -6.44644    0.92087  -7.000 2.55e-12 ***
tobacco      0.08038    0.02588   3.106  0.00190 **
ldl          0.16199    0.05497   2.947  0.00321 **
famhist      0.90818    0.22576   4.023 5.75e-05 ***
typea        0.03712    0.01217   3.051  0.00228 **
age          0.05046    0.01021   4.944 7.65e-07 ***
---
Signif. codes:  0 '***' 0.001 '**' 0.01 '*' 0.05 '.' 0.1 ' ' 1

(Dispersion parameter for binomial family taken to be 1)

    Null deviance: 596.11  on 461  degrees of freedom
Residual deviance: 475.69  on 456  degrees of freedom
AIC: 487.69

Number of Fisher Scoring iterations: 5

(a.out <- anova(red_logit, full_logit))

Analysis of Deviance Table
Model 1: y ~ tobacco + ldl + famhist + typea + age
Model 2: y ~ sbp + tobacco + ldl + adiposity + famhist
   + typea + obesity +  alcohol + age
  Resid. Df Resid. Dev Df Deviance
1       456     475.69
2       452     472.14  4   3.5455

aval <- a.out$Deviance[2]
adf <- a.out$DF[2]
1 - pchisq(aval, 4)  #p-value for the test
0.4709938
```

The p-value is $0.471 > 0.05$, so the data prefers the reduced model. We can also fit a probit regression model by replacing link = "logit" by link = "probit" in the glm() function. The results are very similar when we use these two link functions. ▲

12.5 Count Models

Log-linear count models based on the Poisson distribution were discussed in Example 12.1.2. Under the log link $g(\lambda_i) = \log \lambda_i$, the log-likelihood function was given in (12.1.8), so that the likelihood equations become

$$\frac{\partial \ell}{\partial \beta_r} = \sum_{i=1}^{N} (Y_i - \lambda_i) X_{i,r},$$

where $\lambda_i = \exp(\mathbf{x}_i' \boldsymbol{\beta})$. It can be shown that in the IRLS algorithm, at any intermediate estimate $\boldsymbol{\beta}^{(0)}$, the weight matrix and the adjusted dependent responses are respectively

$$\mathbf{W}^{(0)} = \text{diag}(\lambda_1^{(0)}, \cdots, \lambda_N^{(0)}) \quad \text{and} \tag{12.5.1}$$

$$Z_i^{(0)} = \eta_i^{(0)} + (Y_i/\lambda_i^{(0)} - 1), \quad i = 1, \cdots, N, \tag{12.5.2}$$

where $\lambda_i^{(0)} = \exp(\eta_i^{(0)})$ and $\eta_i^{(0)} = \mathbf{x}_i' \boldsymbol{\beta}^{(0)}$. The steps in the IRLS algorithm for iterating to the MLE of $\boldsymbol{\beta}$ are similar to what we showed in Section 12.4 for the binomial responses. The derivation of the details is left as Exercise 12.6. Since $\phi = a_i(\phi) = 1$ for the Poisson GLIM, from Result 12.2.3, at convergence, the estimate of the variance-covariance matrix of $\widehat{\boldsymbol{\beta}}_{ML}$ is

$$\widehat{\text{Cov}}(\widehat{\boldsymbol{\beta}}_{\text{ML}}) = (\mathbf{X}' \widehat{\mathbf{W}} \mathbf{X})^{-1}, \tag{12.5.3}$$

where $\widehat{\mathbf{W}}$ is the weights matrix evaluated at $\widehat{\boldsymbol{\beta}}_{ML}$.

Maximization of the log-likelihood function can again be done using Fisher scoring or the Newton–Raphson iterative procedure. Since the log-likelihood function is globally concave, convergence is guaranteed, and convergent estimates are obtained in a few (often less than 20) iterations.

Overdispersion

In some cases, due to clustering of events, or some contaminating influences, there is variation in the responses that does not coincide with that implied by the Poisson distribution where $\text{Var}(Y) = \text{E}(Y) = \mu$. Consider the situation where the dispersion of the data exceeds that implied by the Poisson model, i.e., $\text{Var}(Y) > \text{E}(Y)$. In the absence of knowledge of the precise mechanism that causes the over- (or under-) dispersion, it is convenient to assume that $\text{Var}(Y) = \phi\lambda$, where ϕ is an unknown constant called the overdispersion parameter. A value of $\phi = 1$ corresponds to the usual Poisson model with nominal dispersion, a value $\phi > 1$ implies overdispersion relative to the Poisson and there is underdispersion if $\phi < 1$.

Suppose we fit the usual Poisson regression model to overdispersed responses, the resulting fit will usually have issues. Although the estimate of $\boldsymbol{\beta}$ will still be approximately unbiased, the standard errors of the coefficients are likely to be smaller than they should be.

There are a few simple ways to check for overdispersion. First, we can compare the sample variances to the sample averages for groups of responses with identical covariate values. We can examine the deviance goodness of fit test after fitting a rich model to the data. We can examine the residuals to see if outliers may lead to a large deviance statistic. We can fit an overdispersed Poisson model and check to see if the parameter ϕ is close to 1, using as an estimate of ϕ the sum of squared deviance residuals divided by the residual d.f. We can also use the quasi-likelihood approach for parameter estimation in the overdispersed Poisson model.

Overdispersed Poisson Model.

In order to account for the extra variation, we include an extra parameter ϕ in the Poisson log-linear model and carry out inference on ϕ as well as $\boldsymbol{\beta}$:

$$\text{E}(Y_i \,|\, \mathbf{x}_i) = \mu_i; \quad \text{Var}(Y_i \,|\, \mathbf{x}_i) = \phi\mu_i; \quad \log \mu_i = \mathbf{x}_i'\boldsymbol{\beta}. \tag{12.5.4}$$

From (12.5.3), note that if we set $\phi = 1$, the standard errors of the ML estimates of β_j, $j = 1, \cdots, p$ will be conservative in the presence of over-dispersion. We may correct the standard errors using an estimate of ϕ, instead of assuming it to be 1. From (12.3.7), we see that

$$X_p^2 = \sum_{i=1}^{N} (Y_i - \widehat{\mu}_i)^2 / V(\widehat{\mu}_i) = \sum_{i=1}^{N} (Y_i - \widehat{\mu}_i)^2 / \phi\mu_i. \tag{12.5.5}$$

If the fitted model has p parameters, we compare X_p^2 to the value $N - p$. Since $X_p^2 > N - p$ shows evidence of overdispersion, equate the X_p^2 statistic to its expectation, and solve for ϕ:

$$\widehat{\phi} = \frac{X_p^2}{N - p}. \tag{12.5.6}$$

Since X_p^2 is usually a goodness of fit measure, treating it as pure error in order to estimate ϕ requires that we are very confident about the systematic part of the GLIM. That is, we are confident that the lack of fit is due to overdispersion, and not due to misspecification in the systematic part.

Negative binomial log-linear model.

We include a multiplicative random effect R in the Poisson log-linear model to represent unobserved heterogeneity, leading to the negative binomial log-linear model. Suppose

$$p(Y \,|\, R) \sim \text{Poisson}(\mu R), \tag{12.5.7}$$

so that $\text{E}(Y \,|\, R) = \mu R = \text{Var}(Y \,|\, R)$.

If R were known, then Y would have a marginal Poisson distribution. Since R is unknown, we usually set $\text{E}(R) = 1$, in which case, μ represents the expected outcome for the average individual given covariates $\mathbf{X}$. Assume that $R \sim \text{Gamma}(\alpha, \beta)$, and let $\alpha = \beta = 1/\gamma$, so that $\text{E}(R) = 1$ and $\text{Var}(R) = \gamma^2$. By integrating out R from the joint distribution $p(Y \,|\, R) \times p(R)$, we get the marginal distribution of Y to be negative binomial with p.m.f.

$$p(y; \alpha, \beta, \mu) = \frac{\Gamma(y + \alpha)}{y!\Gamma(\alpha)} \frac{\beta^\alpha \mu^y}{(\mu + \beta)^{\alpha + y}}, \tag{12.5.8}$$

so that $\text{E}(Y) = \mu$ and $\text{Var}(Y) = \mu(1 + \gamma^2\mu)$. If γ^2 is estimated to be zero, we have the Poisson log-linear model. If γ^2 is estimated to be positive, we have an overdispersed model. Most software can fit these models to count data.

Numerical Example 12.2. Count model. The dataset "aids" from the R package *catdata* has 2376 observations on 8 variables and was based on a survey around 369 men who were infected with HIV. The response Y is cd4 (number of CD4 cells). The predictors are time (years since seroconversion), drugs (recreational drug use, yes=1/no=0), partners (number of sexual partners), packs (packs of cigarettes a day), cesd (mental illness score), age (age centered around 30), and person (ID number).

```
data("aids", package = "catdata")
full_pois <-
  glm(cd4 ~ time + drugs + partners + packs + cesd + age,
      data = data.frame(aids),
      family = poisson)
summary(full_pois)

glm(formula = cd4 ~ time + drugs + partners + packs + cesd +
    age, family = poisson, data = data.frame(aids))

Deviance Residuals:
    Min       1Q   Median       3Q      Max
-37.763   -9.679   -2.005    6.242   60.470

Coefficients:
              Estimate Std. Error  z value Pr(>|z|)
(Intercept)  6.583e+00  1.711e-03 3847.187  < 2e-16 ***
time        -1.159e-01  4.279e-04 -270.938  < 2e-16 ***
drugs        6.707e-02  1.863e-03   36.008  < 2e-16 ***
partners    -6.351e-04  2.171e-04   -2.925  0.00345 **
packs        7.497e-02  5.007e-04  149.741  < 2e-16 ***
cesd        -2.607e-03  7.928e-05  -32.888  < 2e-16 ***
age          1.046e-03  1.011e-04   10.345  < 2e-16 ***
---
Signif. codes:  0 '***' 0.001 '**' 0.01 '*' 0.05 '.' 0.1 ' ' 1

(Dispersion parameter for poisson family taken to be 1)

    Null deviance: 467303  on 2375  degrees of freedom
Residual deviance: 352915  on 2369  degrees of freedom
AIC: 372732

Number of Fisher Scoring iterations: 4
```

We check the dispersion parameter as the Residual Deviance/df, which must be approximately 1 for the Poisson case. Here, the dispersion parameter is estimated as $352915/2369 = 149$, which is considerably greater than 1, indicating overdispersion. We fit a negative binomial model or a quasi poisson regression model as shown below.

```
full_nb <-
  glm.nb(cd4 ~ time + drugs + partners + packs + cesd + age,
  data = data.frame(aids))
summary(full_nb)

glm.nb(formula = cd4 ~ time + drugs + partners + packs + cesd +
    age, data = data.frame(aids), init.theta = 4.649974551,
    link = log)

Deviance Residuals:
    Min       1Q   Median       3Q      Max
-5.2016  -0.8134  -0.1593   0.4665   4.3382
```

```
Coefficients:
               Estimate Std. Error z value Pr(>|z|)
(Intercept)  6.5805718  0.0213683 307.959  < 2e-16 ***
time        -0.1126975  0.0053359 -21.121  < 2e-16 ***
drugs        0.0731919  0.0231311   3.164  0.00156 **
partners    -0.0006407  0.0028137  -0.228  0.81987
packs        0.0725523  0.0067755  10.708  < 2e-16 ***
cesd        -0.0026719  0.0010129  -2.638  0.00834 **
age          0.0004858  0.0012973   0.374  0.70806
---
Signif. codes:  0 ‘***’ 0.001 ‘**’ 0.01 ‘*’ 0.05 ‘.’ 0.1 ‘ ’ 1

(Dispersion parameter for Negative Binomial(4.65) family taken to
be 1)

    Null deviance: 3149.1  on 2375  degrees of freedom
Residual deviance: 2465.8  on 2369  degrees of freedom
AIC: 34173

Number of Fisher Scoring iterations: 1
              Theta:  4.650
          Std. Err.:  0.132

 2 x log-likelihood:  -34156.670
```

We can do a deviance difference test based on the reduced model. The data prefers the reduced model.

```
summary(red_nb <- glm.nb(cd4 ~ time + drugs + packs + cesd,
                 data = data.frame(aids)))
anova(full_nb, red_nb)

Likelihood ratio tests of Negative Binomial Models

Response: cd4
                                       Model    theta Resid. df
1                 time + drugs + packs + cesd 4.649635      2371
2 time + drugs + partners + packs + cesd + age 4.649975      2369
     2 x log-lik.   Test    df  LR stat.   Pr(Chi)
1       -34156.85
2       -34156.67 1 vs 2     2 0.1814178 0.9132835
```

▲

Exercises

12.1. Let $(Y \mid P = p) \sim \text{Binom}(m, p)$, where $P \sim \text{Beta}(\alpha, \beta)$, $0 \leq p \leq 1$. For $\alpha > 0$ and $\beta > 0$, find the marginal distribution of Y, and show that it does not belong to the exponential family. Find $\text{E}(Y)$ and $\text{Var}(Y)$.

12.2. For $i = 1, \cdots, N$, let Y_i follow a gamma distribution with shape parameter $\alpha > 0$, scale parameter $s_i > 0$ and p.d.f. $f(y; \alpha, s)$ given by (B.10).

(a) Let $\theta = -1/\mu = -1/(\alpha s)$ and $\phi = 1/\alpha$. Show that $V(\mu) = \mu^2$ and the p.d.f. can be written as

$$f(y;\theta,\phi) = \exp\left\{\frac{\theta y + \log(-\theta)}{\phi} + \frac{1}{\phi}\log(y/\phi) - \log y - \log\Gamma(1/\phi)\right\},$$

so that $b(\theta) = -\log(-\theta)$, $c(y,\phi) = (1/\phi)\log(y/\phi) - \log y - \log\Gamma(1/\phi)$.

(b) By parametrizing the distribution of Y in terms of its mean μ, show that the canonical link is $g(\mu) = -1/\mu$.

(c) Given $\mathbf{x}_i'\boldsymbol{\beta}$ for $i = 1,\cdots,N$, write the p.d.f. $f(y;\boldsymbol{\beta},\phi)$ under the (i) inverse link, (ii) log link and (iii) identity link.

12.3. Assume that Y_t, the number of fatal accidents in a year t is modeled by a Poisson distribution with mean $\lambda_t = \beta_0 + \beta_1 t$. Given data for 20 years, derive the IRLS estimates for β_0 and β_1.

12.4. Let Y_1 denote the number of cars that cross a check-point in one hour, and let Y_2 denote the number of other vehicles (except cars) that cross the point. Consider two models. Under Model 1, assume that Y_1 and Y_2 have independent Poisson distributions with means λ_1 and λ_2 respectively, both unknown. Under Model 2, assume a Binomial specification for Y_1 with unknown probability p and sample size $Y_1 + Y_2$.

(a) If we define $p = \lambda_1/(\lambda_1 + \lambda_2)$, show that Model 1 and Model 2 have the same likelihood. (Hint: Use Definition 12.1.1.)

(b) What is the relationship between the canonical link functions?

(c) How can we incorporate a covariate X, which is a highway characteristic, into the modeling?

12.5. Consider the binary response model defined in Section 12.4. Write down in detail the steps in the IRLS algorithm for estimating $\boldsymbol{\beta}$ under the probit, and complementary log-log links, showing details on the weight matrix $\mathbf{W}$ and the vector of adjusted dependent responses $\mathbf{z}$.

12.6. Consider the Poisson log-linear model defined in Example 12.1.2 and discussed in Section 12.5. Write down in detail the steps in the IRLS algorithm for estimating $\boldsymbol{\beta}$ and show that the weight matrix $\mathbf{W}$ and the vector of adjusted dependent responses $\mathbf{z}$ are given by (12.5.1) and (12.5.2), respectively.

12.7. Consider the Poisson model. For $i = 1,\cdots,N$, show that

(a) the Anscombe residuals are given by

$$r_{i,A} = \frac{3(Y_i^{2/3} - \widehat{\mu}_i^{2/3})}{2\widehat{\mu}_i^{1/6}}.$$

(b) the deviance residuals are given by

$$r_{i,D} = \text{sign}(Y_i - \widehat{\mu}_i)[2(Y_i\log(Y_i/\widehat{\mu}_i) - Y_i + \widehat{\mu}_i)]^{1/2}.$$

12.8. [Cameron and Trivedi (2013)] Let Y follow a negative binomial distribution with p.d.f.

$$f(y;\mu,\alpha) = \frac{\Gamma(y+\alpha^{-1})}{\Gamma(y+1)\Gamma(\alpha^{-1})}\left(\frac{\alpha^{-1}}{\alpha^{-1}+\mu}\right)^{\alpha^{-1}}\left(\frac{\mu}{\alpha^{-1}+\mu}\right)^{y},$$

for $y = 0,1,2,\ldots$, where $\alpha \geq 0$. Let $\mu_i = \exp(\mathbf{x}_i'\boldsymbol{\beta})$.

(a) Show that

$$\log L(\alpha, \boldsymbol{\beta}) = \sum_{i=1}^{N} \{ (\sum_{j=0}^{y_i - 1} \log(j + \alpha^{-1})) - \log \Gamma(y_i + 1)$$
$$- (y_i + \alpha^{-1}) \log(1 + \alpha \exp(\mathbf{x}_i' \boldsymbol{\beta})) + y_i \log \alpha + y_i \mathbf{x}_i' \boldsymbol{\beta} \}.$$

(b) For fixed α, write down the steps in the IRLS algorithm for estimating $\boldsymbol{\beta}$ under the log link.

12.9. [Baker (1994)] Consider a categorical variable with J mutually exclusive levels. For $i = 1, \cdots, N$, and $j = 1, \cdots, J$, suppose that Y_{ij} are non-negative integers with $\sum_{j=1}^{J} Y_{ij} = Y_{i\cdot}$. Given $Y_{i\cdot}$, suppose that the random vector $\boldsymbol{y}_i = (Y_{i1}, \cdots, Y_{iJ})'$ has a multinomial distribution, Multinom$(Y_{i\cdot}, \pi_{i1}, \cdots, \pi_{iJ})$, with p.m.f.

$$p(\boldsymbol{y}_i; \boldsymbol{\pi}_i) = \frac{y_{i\cdot}!}{\prod_{j=1}^{J} y_{ij}!} \prod_{j=1}^{J} \pi_{ij}^{y_{ij}}$$

where $\boldsymbol{\pi}_i = (\pi_{i1}, \cdots, \pi_{iJ})'$, with $0 \leq \pi_{ij} \leq 1$ and $\sum_{j=1}^{J} \pi_{ij} = 1$. Assume that $\pi_{ij} \propto \exp(\boldsymbol{\beta}_j)$, where $\boldsymbol{\beta}_j$ is a k-dimensional vector of unknown regression coefficients. Discuss the "Poisson trick" for modeling the categorical response data.

13

Special Topics

This chapter describes a few special topics which extend the general linear model in useful directions. Section 13.1 introduces multivariate linear models for vector-valued responses. Longitudinal models are described in Section 13.2. Section 5.5.2 discusses a non-normal alternative to the GLM via an elliptically contoured linear model. Section 13.4 discusses the hierarchical Bayesian linear model framework of Lindley and Smith, while Section 13.5 presents the dynamic linear model estimated by Kalman filtering and smoothing.

13.1 Multivariate general linear models

The theory of linear models in Chapters 4 and 7 can be extended to the situation where each subject produces a fixed number of responses that are typically correlated. This Section provides an introduction to multivariate linear models for such responses. For a more comprehensive discussion of the theory and application of multivariate linear models, we refer the reader to Mardia (1970) and Johnson and Wichern (2007).

13.1.1 Model definition

The multivariate general linear model (MV GLM) has the form

$$\mathbf{Y} = \mathbf{X}\mathbf{B} + \mathbf{E}, \tag{13.1.1}$$

where

$$\mathbf{Y} = \begin{pmatrix} Y_{11} & Y_{12} & \cdots & Y_{1m} \\ Y_{21} & Y_{22} & \cdots & Y_{2m} \\ \vdots & \vdots & \ddots & \vdots \\ Y_{N1} & Y_{N2} & \cdots & Y_{Nm} \end{pmatrix} \quad \text{and} \quad \mathbf{E} = \begin{pmatrix} \varepsilon_{11} & \varepsilon_{12} & \cdots & \varepsilon_{1m} \\ \varepsilon_{21} & \varepsilon_{22} & \cdots & \varepsilon_{2m} \\ \vdots & \vdots & \ddots & \vdots \\ \varepsilon_{N1} & \varepsilon_{N2} & \cdots & \varepsilon_{Nm} \end{pmatrix}$$

are $N \times m$ matrices with $\mathbf{Y}$ consisting of m responses on each of N subjects and $\mathbf{E}$ consisting of corresponding errors,

$$\mathbf{X} = \begin{pmatrix} 1 & X_{11} & \cdots & X_{1k} \\ 1 & X_{21} & \cdots & X_{2k} \\ \vdots & \vdots & \vdots & \vdots \\ 1 & X_{N1} & \cdots & X_{Nk} \end{pmatrix}$$

is an $N \times (k+1)$ matrix where the first column is a vector of 1's (corresponding to an intercept) and the remaining columns consist of observations on k explanatory variables,

DOI: 10.1201/9781315156651-13

and

$$\mathbf{B} = \begin{pmatrix} \beta_{01} & \beta_{02} & \cdots & \beta_{0m} \\ \beta_{11} & \beta_{12} & \cdots & \beta_{1m} \\ & \vdots & & \\ \beta_{k1} & \beta_{k2} & \cdots & \beta_{km} \end{pmatrix}$$

is a $(k+1) \times m$ matrix of regression coefficients. Let $N > p = k+1$ and $r = \mathrm{r}(\mathbf{X})$. This accommodates the full-rank multivariate linear regression model when $\mathrm{r}(\mathbf{X}) = p$, as well as the less than full-rank multivariate ANOVA (MANOVA) model useful for designed experiments when $\mathrm{r}(\mathbf{X}) < p$. We can also consider a model without an intercept, by omitting the first column of $\mathbf{X}$ and the first row of $\mathbf{B}$.

In the rest of the section, we will denote the rows of a matrix by $\mathbf{x}_i'$, $\mathbf{y}_j'$, etc., and the columns by $\mathbf{x}_{(i)}$, $\mathbf{y}_{(j)}$, etc. First,

$$\mathbf{X} = \begin{pmatrix} \mathbf{x}_1' \\ \mathbf{x}_2' \\ \vdots \\ \mathbf{x}_N' \end{pmatrix} = \begin{pmatrix} \mathbf{1}_N & \mathbf{x}_{(1)} & \cdots & \mathbf{x}_{(k)} \end{pmatrix},$$

with $\mathbf{x}_i'$ the vector of observations from the ith subject on all the explanatory variables, and $\mathbf{x}_{(j)}$ the vector of observations from all the subjects on the jth explanatory variable. Next, let

$$\mathbf{Y} = \begin{pmatrix} \mathbf{y}_1' \\ \vdots \\ \mathbf{y}_N' \end{pmatrix} \quad \text{and} \quad \mathbf{y}_i' = \mathbf{x}_i'\mathbf{B} + \boldsymbol{\varepsilon}_i', \quad i = 1, \cdots, N, \tag{13.1.2}$$

with $\mathbf{y}_i$ being the m-dimensional response vector for the ith subject, and $\boldsymbol{\varepsilon}_i$ denoting the m-dimensional vector of unobserved errors for the ith subject. Clearly, if $m = 1$, the model (13.1.1) is reduced to the model (4.1.1). We can also write

$$\mathbf{Y} = \begin{pmatrix} \mathbf{y}_{(1)} & \mathbf{y}_{(2)} & \cdots & \mathbf{y}_{(m)} \end{pmatrix}, \quad \mathbf{E} = \begin{pmatrix} \boldsymbol{\varepsilon}_{(1)} & \boldsymbol{\varepsilon}_{(2)} & \cdots & \boldsymbol{\varepsilon}_{(m)} \end{pmatrix},$$
$$\mathbf{B} = \begin{pmatrix} \boldsymbol{\beta}_{(1)} & \boldsymbol{\beta}_{(2)} & \cdots & \boldsymbol{\beta}_{(m)} \end{pmatrix}.$$

Then the MV GLM can be written as m univariate linear models,

$$\begin{aligned} \mathbf{y}_{(1)} &= \beta_{01}\mathbf{1}_N + \beta_{11}\mathbf{x}_{(1)} + \cdots + \beta_{k1}\mathbf{x}_{(k)} + \boldsymbol{\varepsilon}_{(1)} \\ \mathbf{y}_{(2)} &= \beta_{02}\mathbf{1}_N + \beta_{12}\mathbf{x}_{(1)} + \cdots + \beta_{k2}\mathbf{x}_{(k)} + \boldsymbol{\varepsilon}_{(2)} \\ &\vdots \\ \mathbf{y}_{(m)} &= \beta_{0m}\mathbf{1}_N + \beta_{1m}\mathbf{x}_{(1)} + \cdots + \beta_{km}\mathbf{x}_{(k)} + \boldsymbol{\varepsilon}_{(m)}. \end{aligned}$$

That is, each column of $\mathbf{Y}$ can be written as the GLM in (4.1.1)

$$\mathbf{y}_{(j)} = \mathbf{X}\boldsymbol{\beta}_{(j)} + \boldsymbol{\varepsilon}_{(j)}. \tag{13.1.3}$$

While (13.1.1) can be expressed as m univariate GLMs, the reason for carrying out a multivariate analysis rather than m separate univariate analyses is to accommodate the dependence in the errors, which is important for carrying out hypothesis testing and confidence region construction. To extend the basic assumption (4.1.4) for the GLM, suppose that in (13.1.1), the error vectors of different subjects are uncorrelated, each with zero mean and the same unknown covariance $\boldsymbol{\Sigma}$. That is,

$$\mathrm{E}(\boldsymbol{\varepsilon}_i) = \mathbf{0}, \quad \mathrm{Cov}(\boldsymbol{\varepsilon}_i, \boldsymbol{\varepsilon}_j) = \begin{cases} \boldsymbol{\Sigma}, & \text{if } i = j, \\ \mathbf{O} & \text{otherwise,} \end{cases} \qquad i, j = 1, \ldots, N. \tag{13.1.4}$$

Column-wise, if $\boldsymbol{\Sigma} = \{\sigma_{ij}\}$, this can then be written as

$$\mathrm{E}(\boldsymbol{\varepsilon}_{(i)}) = \mathbf{0}, \quad \mathrm{Cov}(\boldsymbol{\varepsilon}_{(i)}, \boldsymbol{\varepsilon}_{(j)}) = \sigma_{ij}\mathbf{I}_N, \quad i, j = 1, \ldots, m. \tag{13.1.5}$$

Alternately, we can write (13.1.3) and (13.1.5) as

$$\begin{pmatrix} \mathbf{y}_{(1)} \\ \mathbf{y}_{(2)} \\ \vdots \\ \mathbf{y}_{(m)} \end{pmatrix} = \begin{pmatrix} \mathbf{X} & \mathbf{O} & \cdots & \mathbf{O} \\ \mathbf{O} & \mathbf{X} & \cdots & \mathbf{O} \\ \vdots & \vdots & \ddots & \vdots \\ \mathbf{O} & \mathbf{O} & \cdots & \mathbf{X} \end{pmatrix} \begin{pmatrix} \boldsymbol{\beta}_{(1)} \\ \boldsymbol{\beta}_{(2)} \\ \vdots \\ \boldsymbol{\beta}_{(m)} \end{pmatrix} + \begin{pmatrix} \boldsymbol{\varepsilon}_{(1)} \\ \boldsymbol{\varepsilon}_{(2)} \\ \vdots \\ \boldsymbol{\varepsilon}_{(m)} \end{pmatrix} \tag{13.1.6}$$

and

$$\mathrm{Cov}\begin{pmatrix} \boldsymbol{\varepsilon}_{(1)} \\ \boldsymbol{\varepsilon}_{(2)} \\ \vdots \\ \boldsymbol{\varepsilon}_{(m)} \end{pmatrix} = \begin{pmatrix} \sigma_{11}\mathbf{I}_N & \sigma_{12}\mathbf{I}_N & \cdots & \sigma_{1m}\mathbf{I}_N \\ \sigma_{21}\mathbf{I}_N & \sigma_{22}\mathbf{I}_N & \cdots & \sigma_{2m}\mathbf{I}_N \\ \vdots & \vdots & \ddots & \vdots \\ \sigma_{m1}\mathbf{I}_N & \sigma_{m2}\mathbf{I}_N & \cdots & \sigma_{mm}\mathbf{I}_N \end{pmatrix}, \tag{13.1.7}$$

respectively. Using the notation in Section 2.8, the above expressions can be compactly written as

$$\mathrm{vec}(\mathbf{Y}) = (\mathbf{I}_m \otimes \mathbf{X})\,\mathrm{vec}(\mathbf{B}) + \mathrm{vec}(\mathbf{E}), \tag{13.1.8}$$

$$\mathrm{E}(\mathrm{vec}(\mathbf{E})) = \mathbf{0}, \quad \mathrm{Cov}(\mathrm{vec}(\mathbf{E})) = \boldsymbol{\Sigma} \otimes \mathbf{I}_N. \tag{13.1.9}$$

13.1.2 Least squares estimation

Similar to the GLM, least squares (LS) solutions for (13.1.1) are values of $\mathbf{B}$ that minimize

$$S(\mathbf{B}) = \|\mathbf{Y} - \mathbf{XB}\|^2 = \mathrm{tr}((\mathbf{Y} - \mathbf{XB})'(\mathbf{Y} - \mathbf{XB})) = \sum_{i=1}^{N}\sum_{j=1}^{m}(y_{ij} - \mathbf{x}_i'\boldsymbol{\beta}_{(j)})^2,$$

where $\|\cdot\|$ is the Frobenious norm of a matrix (see Section 1.3). Let $\mathbf{G} = (\mathbf{X}'\mathbf{X})^-$.

Result 13.1.1. Multivariate LS solution. Let

$$\mathbf{B}^0 = \mathbf{GX}'\mathbf{Y}, \tag{13.1.10}$$

or equivalently,

$$\boldsymbol{\beta}^0_{(j)} = \mathbf{GX}'\mathbf{y}_{(j)}, \quad j = 1, \cdots, m. \tag{13.1.11}$$

Then $\mathbf{B}^0$ minimizes $S(\mathbf{B})$. We will refer to $\mathbf{B}^0$ as a *LS solution.*

Proof. Minimizing $S(\mathbf{B}) = \sum_{i=1}^{m} S_i(\boldsymbol{\beta}_{(i)})$ is equivalent to minimizing each term $S_i(\boldsymbol{\beta}) = (\mathbf{y}_{(i)} - \mathbf{X}\boldsymbol{\beta})'(\mathbf{y}_{(i)} - \mathbf{X}\boldsymbol{\beta})$. By (4.2.4), $\boldsymbol{\beta}^0_{(i)}$ minimizes $S_i(\boldsymbol{\beta}_{(i)})$. ■

As a passing remark, (13.1.11) is just one of many ways to construct linear functions of $\mathbf{Y}$ to minimize $S(\mathbf{B})$. If $\boldsymbol{\beta}^0_{(j)} = \mathbf{G}_j\mathbf{X}'\mathbf{y}_{(j)}$, $j = 1, \dots, m$, where $\mathbf{G}_j$ denotes a different g-inverse of $\mathbf{X}'\mathbf{X}$, then $\mathbf{B}^0$ minimizes $S(\mathbf{B})$ as well. However, in this book, by an LS solution, we specifically mean $\mathbf{B}^0$ of the form (13.1.10).

Definition 13.1.1. Fitted values and residuals. The fitted value and residual of the ith subject's jth response are $\widehat{y}_{ij} = \mathbf{x}_i'\boldsymbol{\beta}^0_{(j)}$ and $\widehat{\varepsilon}_{ij} = y_{ij} - \widehat{y}_{ij}$, respectively. The matrix of fitted values is

$$\widehat{\mathbf{Y}} = \{\widehat{y}_{ij}\} = \mathbf{X}\mathbf{B}^0, \quad \text{i.e.,} \quad \widehat{\mathbf{y}}_{(j)} = \mathbf{X}\boldsymbol{\beta}^0_{(j)}, \quad j = 1, \dots, m. \tag{13.1.12}$$

The matrix of residuals is

$$\widehat{\mathbf{E}} = \{\widehat{\varepsilon}_{ij}\} = \mathbf{Y} - \widehat{\mathbf{Y}} = \mathbf{Y} - \mathbf{X}\mathbf{B}^0, \quad \text{i.e.,} \quad \widehat{\boldsymbol{\varepsilon}}_{(j)} = \mathbf{y}_{(j)} - \widehat{\mathbf{y}}_{(j)}. \tag{13.1.13}$$

Let $\mathbf{P}$ be the orthogonal projection matrix onto $\mathcal{C}(\mathbf{X})$. As we saw in Chapter 4,

$$\widehat{\mathbf{Y}} = \mathbf{P}\mathbf{Y}, \quad \widehat{\mathbf{E}} = (\mathbf{I}_N - \mathbf{P})\mathbf{Y} = (\mathbf{I}_N - \mathbf{P})\mathbf{E}. \tag{13.1.14}$$

Corresponding to Result 4.2.4 for the GLM, we consider the mean and covariance for the multivariate LS solution. Note that with $\widehat{\mathbf{Y}}$ and $\widehat{\mathbf{E}}$ being matrix-valued random variables, their covariances are characterized by covariances of their vectorized versions, or covariances between their individual columns.

Result 13.1.2.

1. Let $\mathbf{H} = \mathbf{G}\mathbf{X}'\mathbf{X}$. For $\mathbf{B}^0$ in (13.1.10), $\mathrm{E}(\mathbf{B}^0) = \mathbf{H}\mathbf{B}$ and $\mathrm{Cov}(\mathrm{vec}(\mathbf{B}^0)) = \boldsymbol{\Sigma}\otimes(\mathbf{G}\mathbf{X}'\mathbf{X}\mathbf{G}')$, i.e., $\mathrm{Cov}(\boldsymbol{\beta}^0_{(i)}, \boldsymbol{\beta}^0_{(j)}) = \sigma_{ij}\mathbf{G}\mathbf{X}'\mathbf{X}\mathbf{G}'$, $i, j = 1, \cdots, m$.

2. $\mathrm{E}(\widehat{\mathbf{Y}}) = \mathbf{X}\mathbf{B}$, $\mathrm{E}(\widehat{\mathbf{E}}) = \mathbf{O}$, $\mathrm{Cov}(\mathrm{vec}(\widehat{\mathbf{Y}})) = \boldsymbol{\Sigma} \otimes \mathbf{P}$, and $\mathrm{Cov}(\mathrm{vec}(\widehat{\mathbf{E}})) = \boldsymbol{\Sigma} \otimes (\mathbf{I} - \mathbf{P})$, i.e., $\mathrm{Cov}(\widehat{\mathbf{y}}_{(i)}, \widehat{\mathbf{y}}_{(j)}) = \sigma_{ij}\mathbf{P}$ and $\mathrm{Cov}(\widehat{\boldsymbol{\varepsilon}}_{(i)}, \widehat{\boldsymbol{\varepsilon}}_{(j)}) = \sigma_{ij}(\mathbf{I} - \mathbf{P})$, $i, j = 1, \cdots, m$.

3. $\mathbf{B}^0$ and $\widehat{\mathbf{E}}$ are uncorrelated, i.e., $\mathrm{Cov}(\boldsymbol{\beta}^0_{(i)}, \widehat{\boldsymbol{\varepsilon}}_{(j)}) = \mathbf{O}$ for all $i, j = 1, \dots, m$. In particular, $\widehat{\mathbf{Y}} = \mathbf{X}\mathbf{B}^0$ and $\widehat{\mathbf{E}}$ are uncorrelated.

4. Let

$$\widehat{\boldsymbol{\Sigma}} = \frac{1}{N - r}\widehat{\mathbf{E}}'\widehat{\mathbf{E}}. \tag{13.1.15}$$

Then $\widehat{\boldsymbol{\Sigma}}$ is an unbiased estimator of $\boldsymbol{\Sigma}$, i.e., $\mathrm{E}(\widehat{\boldsymbol{\Sigma}}) = \boldsymbol{\Sigma}$. Equivalently,

$$\widehat{\sigma}_{ij} = \frac{1}{N - r}\widehat{\boldsymbol{\varepsilon}}'_{(i)}\widehat{\boldsymbol{\varepsilon}}_{(j)}$$

is an unbiased estimator of σ_{ij}.

Proof. The proof of properties 1–3 closely follows the proof of the corresponding properties in Result 4.2.4; we therefore leave it as Exercise 13.1. To show property 4, it suffices to show that for each $1 \le i \le j \le m$, $\mathrm{E}(\widehat{\boldsymbol{\varepsilon}}'_{(i)}\widehat{\boldsymbol{\varepsilon}}_{(j)}) = (N - r)\sigma_{ij}$. From (13.1.14), $\widehat{\boldsymbol{\varepsilon}}'_{(i)}\widehat{\boldsymbol{\varepsilon}}_{(j)} = \boldsymbol{\varepsilon}'_{(i)}(\mathbf{I}_N - \mathbf{P})\boldsymbol{\varepsilon}_{(j)}$. Then $\mathrm{E}(\widehat{\boldsymbol{\varepsilon}}'_{(i)}\widehat{\boldsymbol{\varepsilon}}_{(j)}) = \mathrm{E}(\boldsymbol{\varepsilon}'_{(i)}(\mathbf{I}_N - \mathbf{P})\boldsymbol{\varepsilon}_{(j)}) = \mathrm{tr}(\mathrm{E}[(\mathbf{I}_N - \mathbf{P})\boldsymbol{\varepsilon}_{(j)}\boldsymbol{\varepsilon}'_{(i)}]) = \mathrm{tr}(\sigma_{ij}(\mathbf{I}_N - \mathbf{P})) = \sigma_{ij}(N - r)$. ■

The following corollary is an immediate consequence of Result 13.1.2.

Corollary 13.1.1. If $\mathrm{r}(\mathbf{X}) = p$, the following properties hold.

1. $(\mathbf{X}'\mathbf{X})^{-1}$ exists and the least squares *estimator* of $\mathbf{B}$ is given by

$$\widehat{\mathbf{B}} = (\mathbf{X}'\mathbf{X})^{-1}\mathbf{X}'\mathbf{Y}. \tag{13.1.16}$$

   Equivalently, the least squares *estimator* of $\boldsymbol{\beta}_{(i)}$ is

$$\widehat{\boldsymbol{\beta}}_{(i)} = (\mathbf{X}'\mathbf{X})^{-1}\mathbf{X}'\mathbf{y}_{(i)},\ i = 1, \cdots, m. \tag{13.1.17}$$

2. $\mathrm{E}(\widehat{\mathbf{B}}) = \mathbf{B}$.

3. $\mathrm{Cov}(\widehat{\boldsymbol{\beta}}_{(i)}, \widehat{\boldsymbol{\beta}}_{(j)}) = \sigma_{ij}(\mathbf{X}'\mathbf{X})^{-1}$ for $i, j = 1, \cdots, m$.

Corresponding to the ANOVA decomposition of sum of squares (SS) for the GLM, the Multivariate ANOVA (MANOVA) decomposition of sum of squares and cross-products (SSCP) can be established for the MV GLM. Define

$$\begin{aligned} \text{Total SSCP} &= \mathbf{y}_1\mathbf{y}_1' + \cdots + \mathbf{y}_N\mathbf{y}_N' = \mathbf{Y}'\mathbf{Y}, \\ \text{Model SSCP} &= \widehat{\mathbf{y}}_1\widehat{\mathbf{y}}_1' + \cdots + \widehat{\mathbf{y}}_N\widehat{\mathbf{y}}_N' = \widehat{\mathbf{Y}}'\widehat{\mathbf{Y}}, \\ \text{Error SSCP} &= \widehat{\boldsymbol{\varepsilon}}_1\widehat{\boldsymbol{\varepsilon}}_1' + \cdots + \widehat{\boldsymbol{\varepsilon}}_N\widehat{\boldsymbol{\varepsilon}}_N' = \widehat{\mathbf{E}}'\widehat{\mathbf{E}}. \end{aligned}$$

From (13.1.14), $\widehat{\mathbf{Y}}'\widehat{\mathbf{E}} = \mathbf{Y}'\mathbf{P}(\mathbf{I}_N - \mathbf{P})\mathbf{E} = \mathbf{O}$. Since $\mathbf{Y} = \widehat{\mathbf{Y}} + \widehat{\mathbf{E}}$, we see that $\mathbf{Y}'\mathbf{Y} = \widehat{\mathbf{Y}}'\widehat{\mathbf{Y}} + \widehat{\mathbf{E}}'\widehat{\mathbf{E}}$, i.e.,

$$\text{Total SSCP} = \text{Model SSCP} + \text{Error SSCP}.$$

The MANOVA decomposition can also be established for the corrected SSCP's. Let $\overline{\mathbf{y}} = N^{-1}\sum_{i=1}^N \mathbf{y}_i$. The (mean-)corrected version of the total SSCP is then

$$\text{Corrected Total SSCP} = \sum_{i=1}^N (\mathbf{y}_i - \overline{\mathbf{y}})(\mathbf{y}_i - \overline{\mathbf{y}})' = \sum_{i=1}^N \mathbf{y}_i\mathbf{y}_i' - N\overline{\mathbf{y}}\,\overline{\mathbf{y}}'.$$

The mean-corrected versions of the Model SSCP and the Error SSCP are presumably $\widehat{\mathbf{Y}}'\widehat{\mathbf{Y}} - N\overline{\widehat{\mathbf{y}}}\,\overline{\widehat{\mathbf{y}}}'$ and $\widehat{\mathbf{E}}'\widehat{\mathbf{E}} - N\overline{\widehat{\boldsymbol{\varepsilon}}}\,\overline{\widehat{\boldsymbol{\varepsilon}}}'$, respectively. However, suppose $\mathbf{1}_N \in \mathcal{C}(\mathbf{X})$, which holds for any model with an intercept, then $\overline{\widehat{\mathbf{y}}} = \widehat{\mathbf{Y}}'\mathbf{1}_N/N = \mathbf{Y}'\mathbf{P}\mathbf{1}_N/N = \mathbf{Y}'\mathbf{1}_N/N = \overline{\mathbf{y}}$ and $\overline{\widehat{\boldsymbol{\varepsilon}}} = \widehat{\mathbf{E}}'\mathbf{1}_N/N = \mathbf{Y}'(\mathbf{I}_N - \mathbf{P})\mathbf{1}_N/N = \mathbf{0}$, so that

$$\text{Corrected Model SSCP} = \widehat{\mathbf{Y}}'\widehat{\mathbf{Y}} - N\overline{\mathbf{y}}\,\overline{\mathbf{y}}',$$

while the Error SSCP needs no correction. Then

$$\text{Corrected Total SSCP} = \text{Corrected Model SSCP} + \text{Error SSCP}.$$

Note that the (i, j)th element of the Corrected Total SSCP matrix is proportional to the sample covariance between the ith and jth response variables, while the diagonal elements are the corrected (univariate) SST for the corresponding univariate response. Similarly, the diagonal elements of the Corrected Model SSCP and Error SSCP are respectively SSR and SSE for the corresponding univariate response.

13.1.3 Estimable functions and Gauss–Markov theorem

Estimable functions of $\mathbf{B}$ in the MV GLM (13.1.1) are functions of the linear form $\sum_{i=1}^{p}\sum_{j=1}^{m} c_{ij}\beta_{ij}$, where c_{ij} are constants. Let $\mathbf{C} = \{c_{ij}\} \in \mathcal{R}^{p\times m}$. Then, the linear form can be written as $\text{tr}(\mathbf{C}'\mathbf{B}) = \sum_{j=1}^{m} \mathbf{c}'_{(j)}\boldsymbol{\beta}_{(j)}$. This function is said to be estimable if it is equal to $\text{E}(\sum_{i=1}^{N}\sum_{j=1}^{m} t_{ij}y_{ij}) = \text{E}(\text{tr}(\mathbf{T}'\mathbf{Y}))$ for some $\mathbf{T} = \{t_{ij}\} \in \mathcal{R}^{N\times m}$.

Result 13.1.3. The function $\text{tr}(\mathbf{C}'\mathbf{B})$ is estimable if and only if any one of the following properties hold.

1. Each $\mathbf{c}_{(j)}$ belongs to $\mathcal{R}(\mathbf{X})$, or equivalently, $\mathbf{C}' = \mathbf{T}'\mathbf{X}$ for some $\mathbf{T} \in \mathcal{R}^{N\times m}$.
2. $\mathbf{C}' = \mathbf{C}'\mathbf{H}$, where $\mathbf{H} = \mathbf{G}\mathbf{X}'\mathbf{X}$.
3. $\text{tr}(\mathbf{C}'\mathbf{B}^0)$ is invariant to the LS solution $\mathbf{B}^0$.

Proof. Since $\text{tr}(\mathbf{C}'\mathbf{B}) = \text{vec}(\mathbf{C})'\,\text{vec}(\mathbf{B})$, by applying Result 4.3.1 to the vectorized expression (13.1.8) of the model, $\text{tr}(\mathbf{C}'\mathbf{B})$ is estimable if and only if $\text{vec}(\mathbf{C}) \in \mathcal{R}(\mathbf{I}_m \otimes \mathbf{X}) = \mathcal{R}(\mathbf{X}) \oplus \cdots \oplus \mathcal{R}(\mathbf{X})$. It follows that $\text{tr}(\mathbf{C}'\mathbf{B})$ is estimable if and only if $\mathbf{c}_{(i)} \in \mathcal{R}(\mathbf{X})$ for every i, proving property 1. That properties 1 and 2 are equivalent follows from property 4 of Result 4.3.1.

To prove property 3, observe that $\text{vec}(\mathbf{B}^0)$ is a LS solution to (13.1.8). Then, from property 5 of Result 4.3.1, if $\text{tr}(\mathbf{C}'\mathbf{B})$ is estimable, then $\text{tr}(\mathbf{C}'\mathbf{B}^0) = \text{vec}(\mathbf{C})'\,\text{vec}(\mathbf{B}^0)$ is invariant to $\mathbf{B}^0$. However, the proof of the converse is somewhat subtle. This is because it follows from the remark below the proof of Result 13.1.1 that solutions of the form $\text{vec}(\mathbf{B}^0)$ do not exhaust the LS solutions for (13.1.8); hence, we cannot prove the converse by directly applying property 5 of Result 4.3.1 to (13.1.8).

Instead, we have $\mathbf{C} = \mathbf{C}_0 + \mathbf{D}$, where $\mathcal{C}(\mathbf{C}_0) \subset \mathcal{R}(\mathbf{X})$ and $\mathcal{C}(\mathbf{D}) \perp \mathcal{R}(\mathbf{X})$. From what was just shown, $\text{tr}(\mathbf{C}_0'\mathbf{B}^0)$ is invariant to $\mathbf{B}^0$. Thus, if $\text{tr}(\mathbf{C}'\mathbf{B}^0)$ is invariant to $\mathbf{B}^0 = \mathbf{G}\mathbf{X}'\mathbf{Y}$, then so is $\text{tr}(\mathbf{D}'\mathbf{B}^0)$, yielding the invariance of $\text{E}(\text{tr}(\mathbf{D}'\mathbf{B}^0)) = \text{tr}(\mathbf{D}'\mathbf{G}\mathbf{X}'\mathbf{X}\mathbf{B})$ to $\mathbf{G}$. Since $\mathbf{X}\mathbf{D} = \mathbf{O}$, for any $p\times m$ matrix $\mathbf{M}$, $\mathbf{G} + \mathbf{D}\mathbf{M}'$ is also a g-inverse of $\mathbf{X}'\mathbf{X}$, so the invariance yields $\text{tr}(\mathbf{D}'\mathbf{D}\mathbf{M}'\mathbf{X}'\mathbf{X}\mathbf{B}) = 0$, which holds for all $\mathbf{B}$. As a result, $\mathbf{D}'\mathbf{D}\mathbf{M}'\mathbf{X}'\mathbf{X} = \mathbf{O}$. Given any $\mathbf{u} \in \mathcal{R}^p$ and $\mathbf{v} \in \mathcal{R}^m$, let $\mathbf{M} = \mathbf{u}\mathbf{v}'$. Then $(\mathbf{D}'\mathbf{D}\mathbf{v})(\mathbf{X}'\mathbf{X}\mathbf{u})' = \mathbf{O}$. Since $\mathbf{X} \neq \mathbf{O}$, there is $\mathbf{u}$ such that $\mathbf{X}'\mathbf{X}\mathbf{u} \neq \mathbf{0}$, which then leads to $\mathbf{D}'\mathbf{D}\mathbf{v} = \mathbf{0}$. Since $\mathbf{v}$ is arbitrary, $\mathbf{D}'\mathbf{D} = \mathbf{O}$, so $\mathbf{D} = \mathbf{O}$. As a result, $\mathbf{C}' = \mathbf{C}_0' = \mathbf{T}'\mathbf{X}$. Then from property 1, $\text{tr}(\mathbf{C}'\mathbf{B})$ is estimable, completing the proof of property 3. ■

We now generalize the Gauss–Markov theorem to the MV GLM. In the multivariate setting, a linear estimator is any estimator of the form $d_0 + \text{tr}(\mathbf{D}'\mathbf{Y})$, where d_0 is a constant, and $\mathbf{D}$ is an $N\times m$ matrix of constants.

Result 13.1.4. Gauss–Markov theorem for the MV GLM. Let the assumptions in (13.1.4) hold for nonsingular $\boldsymbol{\Sigma}$. Let $\text{tr}(\mathbf{C}'\mathbf{B})$ be an estimable function of $\mathbf{B}$ and let $\mathbf{B}^0$ be any LS solution. Then, $\text{tr}(\mathbf{C}'\mathbf{B}^0)$ is the unique b.l.u.e. of $\text{tr}(\mathbf{C}'\mathbf{B})$, with variance $\text{Var}(\text{tr}(\mathbf{C}'\mathbf{B}^0)) = \text{tr}(\boldsymbol{\Sigma}\mathbf{C}'\mathbf{G}\mathbf{C})$. Finally, for two estimable functions $\text{tr}(\mathbf{C}_i\mathbf{B})$, $i = 1, 2$, $\text{Cov}(\text{tr}(\mathbf{C}_1'\mathbf{B}^0), \text{tr}(\mathbf{C}_2'\mathbf{B}^0)) = \text{tr}(\boldsymbol{\Sigma}\mathbf{C}_1'\mathbf{G}\mathbf{C}_2)$.

Proof. From Result 13.1.3, $\mathbf{C}' = \mathbf{T}'\mathbf{X}$ for some matrix $\mathbf{T}$. Then $\text{E}(\text{tr}(\mathbf{C}'\mathbf{B}^0)) = \text{E}(\text{tr}(\mathbf{T}'\mathbf{X}\mathbf{B}^0)) = \text{E}(\text{tr}(\mathbf{T}'\widehat{\mathbf{Y}}))$, which by property 2 is equal to $\text{tr}(\mathbf{T}'\mathbf{X}\mathbf{B}) = \text{tr}(\mathbf{C}'\mathbf{B})$. Therefore $\mathbf{C}'\mathbf{B}^0$ is an unbiased estimator of $\text{tr}(\mathbf{C}'\mathbf{B})$. Let $d_0 + \text{tr}(\mathbf{D}'\mathbf{Y})$ be any unbiased estimator of the function. Taking expectation yields $d_0 + \text{tr}(\mathbf{D}'\mathbf{X}\mathbf{B}) = \text{tr}(\mathbf{C}'\mathbf{B})$. Since the equality holds for all $\mathbf{B}$, $\mathbf{C}' = \mathbf{D}'\mathbf{X}$. Then $\text{tr}(\mathbf{D}'\mathbf{Y}) = \text{tr}(\mathbf{D}'\widehat{\mathbf{Y}}) + \text{tr}(\mathbf{D}'\widehat{\mathbf{E}})$. From property 3, $\text{tr}(\mathbf{D}'\widehat{\mathbf{Y}})$ and $\text{tr}(\mathbf{D}'\widehat{\mathbf{E}})$ are uncorrelated. Meanwhile, $\text{tr}(\mathbf{D}'\widehat{\mathbf{Y}}) = \text{tr}(\mathbf{D}'\mathbf{X}\mathbf{B}^0) = \text{tr}(\mathbf{C}\mathbf{B}^0)$. As a result,

$\text{Var}(\text{tr}(\mathbf{D}'\mathbf{Y})) = \text{Var}(\text{tr}(\mathbf{CB}^0)) + \text{Var}(\text{tr}(\mathbf{D}'\widehat{\mathbf{E}}) \geq \text{Var}(\text{tr}(\mathbf{CB}^0))$ and equality holds if and only if $\text{Var}(\text{tr}(\mathbf{D}'\widehat{\mathbf{E}})) = 0$. Since $\text{E}(\text{tr}(\mathbf{D}'\widehat{\mathbf{E}})) = 0$, the latter is equivalent to $\text{tr}(\mathbf{D}'\widehat{\mathbf{E}}) = 0$, or equivalently, $\text{tr}(\mathbf{D}'\widehat{\mathbf{E}}) = \text{tr}(\mathbf{CB}^0)$, so that $\text{tr}(\mathbf{CB}^0)$ is the unique b.l.u.e.

Next, since $\text{tr}(\mathbf{C}'\mathbf{B}^0)) = \text{vec}(\mathbf{C})' \text{vec}(\mathbf{B}^0)$, we have

$$\text{Var}(\text{tr}(\mathbf{C}'\mathbf{B}^0)) = \text{vec}(\mathbf{C})' \text{Cov}(\text{vec}(\mathbf{B}^0)) \text{vec}(\mathbf{C}).$$

From property 1 of Result 13.1.2, the right side is equal to

$$\text{vec}(\mathbf{C})'(\boldsymbol{\Sigma} \otimes (\mathbf{GX}'\mathbf{XG}')) \text{vec}(\mathbf{C}) = \sum_{i,j=1}^{m} \mathbf{c}'_{(i)}(\sigma_{ij}\mathbf{GX}'\mathbf{XG}')\mathbf{c}_{(j)},$$

which, according to property 2 of Result 13.1.3, is equal to $\sum_{i,j=1}^{m} \sigma_{ij}\mathbf{c}'_{(i)}\mathbf{Gc}_{(j)} = \text{tr}(\boldsymbol{\Sigma}\mathbf{C}'\mathbf{GC})$. Finally, the statement on the covariance of two estimable functions follows by a completely similar argument. ■

The result below holds as an application of the above results.

Result 13.1.5. Let $\mathbf{c} \in \mathcal{R}(\mathbf{X})$.

1. Each component of $\mathbf{B}'\mathbf{c}$ is estimable.
2. $\text{E}(\mathbf{B}^{0\prime}\mathbf{c}) = \mathbf{B}'\mathbf{c}$ and $\text{Cov}(\mathbf{B}^{0\prime}\mathbf{c}) = (\mathbf{c}'\mathbf{Gc})\boldsymbol{\Sigma}$.

Proof. The components of $\mathbf{B}'\mathbf{c}$ are $\mathbf{c}'\boldsymbol{\beta}_{(i)} = \mathbf{c}'\mathbf{Be}_i = \text{tr}(\mathbf{e}_i\mathbf{c}'\mathbf{B})$, $i = 1, \ldots, m$. Let $\mathbf{c}' = \mathbf{t}'\mathbf{X}$. Then $\mathbf{e}_i\mathbf{c}' = \mathbf{T}'\mathbf{X}$ with $\mathbf{T} = \mathbf{te}'_i$, so from Result 13.1.3, $\mathbf{c}'\boldsymbol{\beta}_{(i)}$ is estimable, proving property 1. From Result 13.1.4, the first equality in property 2 is clear, and $\text{Cov}(\mathbf{c}'\boldsymbol{\beta}^0_{(i)}, \mathbf{c}'\boldsymbol{\beta}^0_{(j)}) = \text{Cov}(\mathbf{e}_i\mathbf{c}'\mathbf{B}^0, \mathbf{e}_j\mathbf{c}'\mathbf{B}^0) = \text{tr}(\boldsymbol{\Sigma}\mathbf{e}_i\mathbf{c}'\mathbf{Gce}'_j) = \text{tr}(\mathbf{e}'_j\boldsymbol{\Sigma}\mathbf{e}_i\mathbf{c}'\mathbf{Gc}) = (\mathbf{c}'\mathbf{Gc})\sigma_{ij}$, so the second equality holds as well. ■

Numerical Example 13.1. The data "mtcars" was extracted from the 1974 Motor Trend US magazine, and comprises fuel consumption and 10 aspects of automobile design and performance for 32 automobiles (1973–74 models). The data set consists of 11 variables: mpg (miles/gallon), cyl (number of cylinders), disp (displacement in cu.in.), hp (gross horsepower), drat (rear axle ratio), wt (weight in 1000 lbs), qsec (1/4 mile time), vs (engine: 0 = v-shaped, 1 = straight), am (transmission: 0 = automatic, 1 = manual), gear (number of forward gears), and carb (number of carburetors).

```
data(mtcars)
Y <- as.matrix(mtcars[, c("mpg", "disp", "hp", "wt")])
mvmod <- lm(Y ~ cyl + am + carb, data = mtcars)
summary(mvmod)

Response mpg :

Call:
lm(formula = mpg ~ cyl + am + carb, data = mtcars)

Residuals:
    Min      1Q  Median      3Q     Max
-5.8853 -1.1581  0.2646  1.4885  5.4843

Coefficients:
```

```
            Estimate Std. Error t value Pr(>|t|)
(Intercept)  32.1731     2.4914  12.914 2.59e-13 ***
cyl          -1.7175     0.4298  -3.996 0.000424 ***
am            4.2430     1.3094   3.240 0.003074 **
carb         -1.1304     0.4058  -2.785 0.009481 **
---
Signif. codes:  0 '***' 0.001 '**' 0.01 '*' 0.05 '.' 0.1 ' ' 1

Residual standard error: 2.755 on 28 degrees of freedom
Multiple R-squared:  0.8113,Adjusted R-squared:  0.7911
F-statistic: 40.13 on 3 and 28 DF,  p-value: 2.855e-10

Response disp :

Call:
lm(formula = disp ~ cyl + am + carb, data = mtcars)

Residuals:
    Min      1Q  Median      3Q     Max
-75.781 -34.975  -0.247  23.263 123.853

Coefficients:
            Estimate Std. Error t value Pr(>|t|)
(Intercept) -110.123     47.973  -2.296   0.0294 *
cyl           59.001      8.275   7.130  9.3e-08 ***
am           -35.850     25.213  -1.422   0.1661
carb          -3.434      7.814  -0.440   0.6637
---
Signif. codes:  0 '***' 0.001 '**' 0.01 '*' 0.05 '.' 0.1 ' ' 1

Residual standard error: 53.04 on 28 degrees of freedom
Multiple R-squared:  0.8346,Adjusted R-squared:  0.8168
F-statistic: 47.08 on 3 and 28 DF,  p-value: 4.588e-11

Response hp :

Call:
lm(formula = hp ~ cyl + am + carb, data = mtcars)

Residuals:
    Min      1Q  Median      3Q     Max
-58.504 -15.061  -1.157  23.414  48.516

Coefficients:
            Estimate Std. Error t value Pr(>|t|)
(Intercept)  -63.981     26.816  -2.386 0.024041 *
cyl           25.726      4.626   5.561    6e-06 ***
am            11.604     14.094   0.823 0.417291
carb          16.632      4.368   3.808 0.000702 ***
---
Signif. codes:  0 '***' 0.001 '**' 0.01 '*' 0.05 '.' 0.1 ' ' 1
```

```
Residual standard error: 29.65 on 28 degrees of freedom
Multiple R-squared:  0.8311,Adjusted R-squared:  0.813
F-statistic: 45.92 on 3 and 28 DF,  p-value: 6.129e-11

Response wt :

Call:
lm(formula = wt ~ cyl + am + carb, data = mtcars)

Residuals:
     Min       1Q   Median       3Q      Max
-0.61492 -0.32981 -0.06998  0.12396  1.23908

Coefficients:
            Estimate Std. Error t value Pr(>|t|)
(Intercept)  1.88716    0.45627   4.136 0.000291 ***
cyl          0.21007    0.07871   2.669 0.012512 *
am          -0.99370    0.23980  -4.144 0.000285 ***
carb         0.15429    0.07432   2.076 0.047186 *
---
Signif. codes:  0 '***' 0.001 '**' 0.01 '*' 0.05 '.' 0.1 ' ' 1

Residual standard error: 0.5045 on 28 degrees of freedom
Multiple R-squared:  0.7599,Adjusted R-squared:  0.7341
F-statistic: 29.54 on 3 and 28 DF,  p-value: 8.088e-09
```

The estimated coefficient matrix $\widehat{\mathbf{B}}$ can be pulled out as follows.

```
coef(mvmod)
                  mpg       disp        hp         wt
(Intercept) 32.173065 -110.12333 -63.98118  1.8871637
cyl         -1.717492   59.00099  25.72551  0.2100744
am           4.242978  -35.85045  11.60366 -0.9936961
carb        -1.130370   -3.43438  16.63221  0.1542897
```

▲

13.1.4 Maximum likelihood estimation

We now consider maximum likelihood (ML) estimation for $\mathbf{B}$ and $\boldsymbol{\Sigma}$ and describe the sampling distributions of the estimators under the following normality assumption on the errors:

$$\mathbf{E} = (\boldsymbol{\varepsilon}_1, \cdots, \boldsymbol{\varepsilon}_N)' \text{ with } \boldsymbol{\varepsilon}_i \text{ being i.i.d.} \sim N_m(\mathbf{0}, \boldsymbol{\Sigma}), \ \boldsymbol{\Sigma} \text{ is p.d.} \tag{13.1.18}$$

As before, we write $\mathbf{B} = (\boldsymbol{\beta}_{(1)}, \cdots, \boldsymbol{\beta}_{(m)})$.

Result 13.1.6. Suppose $\mathbf{E}$ satisfies (13.1.18).

1. Let $\mathbf{Q}$ be an $N \times N$ orthogonal projection matrix of rank r. Then, $\mathbf{E}'\mathbf{Q}\mathbf{E} \sim W_m(\boldsymbol{\Sigma}, r)$.

2. Let $\mathbf{M}_1$ and $\mathbf{M}_2$ be two matrices of constants, each having N rows, such that $\mathbf{M}_1'\mathbf{M}_2 = \mathbf{O}$. Then, $\mathbf{M}_1'\mathbf{E}$ and $\mathbf{M}_2'\mathbf{E}$ are independent.

Proof. To prove property 1, $\mathbf{Q} = \mathbf{P}' \operatorname{diag}(\mathbf{1}_r, \mathbf{O})\mathbf{P}$ for some orthogonal matrix $\mathbf{P}$. Then $\mathbf{E}'\mathbf{Q}\mathbf{E} = (\mathbf{P}\mathbf{E})' \operatorname{diag}(\mathbf{1}_r, \mathbf{O})(\mathbf{P}\mathbf{E})$. From the normality assumption (13.1.18), for $1 \le i, j \le m$, $\operatorname{Cov}(\mathbf{P}\boldsymbol{\varepsilon}_{(i)}, \mathbf{P}\boldsymbol{\varepsilon}_{(j)}) = \mathbf{P}\operatorname{Cov}(\boldsymbol{\varepsilon}_{(i)}, \boldsymbol{\varepsilon}_{(j)})\mathbf{P}' = \mathbf{P}(\sigma_{ij}\mathbf{I}_N)\mathbf{P}' = \sigma_{ij}\mathbf{I}_N = \operatorname{Cov}(\boldsymbol{\varepsilon}_{(i)}, \boldsymbol{\varepsilon}_{(j)})$. The joint distribution of the $\mathbf{P}\boldsymbol{\varepsilon}_{(i)}$'s is the same as the $\boldsymbol{\varepsilon}_{(i)}$'s, so $\mathbf{P}\mathbf{E} \sim \mathbf{E}$. Then, $\mathbf{E}'\mathbf{Q}\mathbf{E} = \mathbf{E}'\mathbf{P}\operatorname{diag}(\mathbf{1}_r, \mathbf{O})\mathbf{P}\mathbf{E} \sim \mathbf{E}'\operatorname{diag}(\mathbf{1}_r, \mathbf{O})\mathbf{E} = \mathbf{E}'\operatorname{diag}(\mathbf{1}_r, \mathbf{O})\mathbf{E} = \sum_{i=1}^{r} \boldsymbol{\varepsilon}_i\boldsymbol{\varepsilon}_i' \sim W_m(\boldsymbol{\Sigma}, r)$. The proof of property 2 is left as Exercise 13.2. ■

Result 13.1.7. Suppose $\mathrm{r}(\mathbf{X}) = p$. The MLE of $\mathbf{B}$, denoted by $\widehat{\mathbf{B}}_{ML}$, is equal to its LS estimate $\widehat{\mathbf{B}}$ given in (13.1.16) and the MLE of $\boldsymbol{\Sigma}$ is

$$\widehat{\boldsymbol{\Sigma}}_{ML} = \frac{1}{N}\widehat{\mathbf{E}}'\widehat{\mathbf{E}} = (1 - 1/N)\widehat{\boldsymbol{\Sigma}}. \tag{13.1.19}$$

The maximized likelihood is

$$L(\widehat{\mathbf{B}}_{ML}, \widehat{\boldsymbol{\Sigma}}_{ML}; \mathbf{Y}) = \frac{(2\pi e)^{-Nm/2}}{|\widehat{\boldsymbol{\Sigma}}_{ML}|^{N/2}}. \tag{13.1.20}$$

Proof. We follow the proof of Result 6.1.4. From the assumption, $\mathbf{y}_1, \cdots, \mathbf{y}_N$ are independent with $\mathbf{y}_i \sim N_m(\mathbf{B}'\mathbf{x}_i, \boldsymbol{\Sigma})$. The likelihood function is

$$L(\mathbf{B}, \boldsymbol{\Sigma}; \mathbf{Y}) = \frac{(2\pi)^{-Nm/2}}{|\boldsymbol{\Sigma}|^{N/2}} e^{-V/2},$$

where $V = \sum_{i=1}^{N} (\mathbf{y}_i - \mathbf{B}'\mathbf{x}_i)'\boldsymbol{\Sigma}^{-1}(\mathbf{y}_i - \mathbf{B}'\mathbf{x}_i) = \operatorname{tr}((\mathbf{Y} - \mathbf{X}\mathbf{B})'\boldsymbol{\Sigma}^{-1}(\mathbf{Y} - \mathbf{X}\mathbf{B}))$. From property 3 of Result 1.3.5,

$$\begin{aligned} V &= \operatorname{tr}(\boldsymbol{\Sigma}^{-1}(\mathbf{Y} - \mathbf{X}\mathbf{B})'(\mathbf{Y} - \mathbf{X}\mathbf{B})) \\ &= \operatorname{tr}(\boldsymbol{\Sigma}^{-1}[\widehat{\mathbf{E}} - \mathbf{X}(\mathbf{B} - \widehat{\mathbf{B}})]'[\widehat{\mathbf{E}} - \mathbf{X}(\mathbf{B} - \widehat{\mathbf{B}})]). \end{aligned}$$

Since $\widehat{\mathbf{E}}'\mathbf{X} = \mathbf{E}'(\mathbf{I} - \mathbf{P})\mathbf{X} = \mathbf{O}$,

$$V = \operatorname{tr}(\boldsymbol{\Sigma}^{-1}[\widehat{\mathbf{E}}'\widehat{\mathbf{E}} + (\widehat{\mathbf{B}} - \mathbf{B})'\mathbf{X}'\mathbf{X}(\widehat{\mathbf{B}} - \mathbf{B})]) = \operatorname{tr}(\boldsymbol{\Sigma}^{-1}\widehat{\mathbf{E}}'\widehat{\mathbf{E}}) + \operatorname{tr}(\mathbf{D}),$$

where $\mathbf{D} = \mathbf{X}(\widehat{\mathbf{B}} - \mathbf{B})\boldsymbol{\Sigma}^{-1}(\widehat{\mathbf{B}} - \mathbf{B})'\mathbf{X}')$. Then,

$$L(\mathbf{B}, \boldsymbol{\Sigma}; \mathbf{Y}) = \frac{(2\pi)^{-Nm/2}}{|\boldsymbol{\Sigma}|^{N/2}} \exp\left\{-\frac{1}{2}\operatorname{tr}(\boldsymbol{\Sigma}^{-1}\widehat{\mathbf{E}}'\widehat{\mathbf{E}}) - \frac{1}{2}\operatorname{tr}(\mathbf{D})\right\}.$$

Since $\mathbf{D}$ is n.n.d., $\operatorname{tr}(\mathbf{D}) \ge 0$ with $\operatorname{tr}(\mathbf{D}) = 0$ if and only if $\mathbf{D} = \mathbf{O}$. Since $\boldsymbol{\Sigma}^{-1}$ is p.d. and $\mathbf{X}$ has full column rank, $\mathbf{D} = \mathbf{O}$ if and only if $\mathbf{B} = \widehat{\mathbf{B}}$. Hence $\widehat{\mathbf{B}}_{ML} = \widehat{\mathbf{B}}$ and it only remains to maximize $|\boldsymbol{\Sigma}|^{-N/2}\exp\{-\frac{1}{2}\operatorname{tr}(\boldsymbol{\Sigma}^{-1}\widehat{\mathbf{E}}'\widehat{\mathbf{E}})\}$. From Example 2.4.4, (13.1.19) follows. Let $\mathbf{B} = \widehat{\mathbf{B}}_{ML}$ and $\boldsymbol{\Sigma} = \widehat{\boldsymbol{\Sigma}}_{ML}$ in the above display. With $\operatorname{tr}(\widehat{\boldsymbol{\Sigma}}_{ML}^{-1}\widehat{\mathbf{E}}'\widehat{\mathbf{E}}) = \operatorname{tr}(N\mathbf{I}_m) = Nm$ and $\mathbf{D} = \mathbf{O}$, the maximized likelihood is obtained as in (13.1.20). ■

Result 13.1.8. Under the condition in Result 13.1.7, the following properties hold:

1. $\widehat{\boldsymbol{\beta}}_{(i)} \sim N_p(\boldsymbol{\beta}_{(i)}, \sigma_{ii}(\mathbf{X}'\mathbf{X})^{-1})$ and $\operatorname{Cov}(\widehat{\boldsymbol{\beta}}_{(i)}, \widehat{\boldsymbol{\beta}}_{(\ell)}) = \sigma_{i\ell}(\mathbf{X}'\mathbf{X})^{-1}$. We may also write this as $\operatorname{vec}(\widehat{\mathbf{B}}) \sim N_{pm}(\operatorname{vec}(\mathbf{B}), \boldsymbol{\Sigma} \otimes (\mathbf{X}'\mathbf{X})^{-1}))$.

2. If $r = \mathrm{r}(\mathbf{X})$, then $\widehat{\mathbf{E}}'\widehat{\mathbf{E}} \sim W_m(\boldsymbol{\Sigma}, N - r)$, and $\widehat{\mathbf{B}}$ and $\widehat{\mathbf{E}}$ are independently distributed.

Proof. Property 1 directly follows from property 1 of Result 13.1.2. From (13.1.14), $\widehat{\mathbf{E}}'\widehat{\mathbf{E}} = \mathbf{E}'(\mathbf{I}_N - \mathbf{P})\mathbf{E}$. Since $\mathbf{I}_N - \mathbf{P}$ is symmetric and idempotent with rank $N - r$, its distribution follows from property 1 of Result 13.1.6. Since $\widehat{\mathbf{B}} = \mathbf{G}\mathbf{X}'\mathbf{Y} = \mathbf{H}\mathbf{B} + \mathbf{G}\mathbf{X}'\mathbf{E}$, $\widehat{\mathbf{E}} = (\mathbf{I}_N - \mathbf{P})\mathbf{E}$, and $\mathbf{G}\mathbf{X}'(\mathbf{I}_N - \mathbf{P}) = \mathbf{O}$, from property 2 of Result 13.1.6, $\widehat{\mathbf{B}}$ and $\widehat{\mathbf{E}}$ are independent, completing the proof of property 2. ■

13.1.5 Likelihood ratio tests for linear hypotheses

The likelihood ratio test (LRT) is a standard tool for testing a linear hypothesis on the MV GLM (13.1.1). To illustrate the fundamental idea, we will only consider a particular example. More comprehensive accounts of the LRT for hypotheses on the MV GLM can be found in Mardia (1970), chapter 6.

Suppose that the errors ε_i satisfy (13.1.18) and we partition the $p \times m$ matrix $\mathbf{B}$ as

$$\mathbf{B} = \begin{pmatrix} \mathbf{B}_1 \\ \mathbf{B}_2 \end{pmatrix},$$

where $\mathbf{B}_1 \in \mathcal{R}^{q \times m}$ and $\mathbf{B}_2 \in \mathcal{R}^{(p-q) \times m}$. Partition the $N \times p$ matrix $\mathbf{X}$ as

$$\mathbf{X} = (\mathbf{X}_1, \mathbf{X}_2),$$

where $\mathbf{X}_1 \in \mathcal{R}^{N \times q}$ and $\mathbf{X}_2 \in \mathcal{R}^{N \times (p-q)}$. Then, (13.1.1) can be written as

$$\mathbf{Y} = \mathbf{XB} + \mathbf{E} = \mathbf{X}_1\mathbf{B}_1 + \mathbf{X}_2\mathbf{B}_2 + \mathbf{E}.$$

Suppose $\mathrm{r}(\mathbf{X}) = p$. From Result 13.1.3, $\mathbf{B}_2 = (\mathbf{O}, \mathbf{I}_{p-q})\mathbf{B}$ is estimable. Suppose we wish to test H_0: $\mathbf{B}_2 = \mathbf{O}$, i.e., the responses are not affected by $X_{q+1}, \cdots, X_p$. The reduced model under H_0 is

$$\mathbf{Y} = \mathbf{X}_1\mathbf{B}_1 + \mathbf{E}. \tag{13.1.21}$$

The LRT statistic for H_0 is defined by

$$\Lambda = \frac{\sup_{\mathbf{B}_1, \mathbf{\Sigma}} L(\mathbf{B}_1, \mathbf{\Sigma}; \mathbf{Y})}{\sup_{\mathbf{B}, \mathbf{\Sigma}} L(\mathbf{B}, \mathbf{\Sigma}; \mathbf{Y})},$$

where the likelihood in the denominator is evaluated under the full model and the likelihood in the numerator is evaluated under the reduced model (13.1.21). The statistic $\Lambda^{2/N}$ is known as *Wilks' Lambda*. Closely related to the statistic is the Wilks' Lambda distribution defined below.

Definition 13.1.2. If $\mathbf{W}_i \sim W_m(\mathbf{I}_m, \nu_i)$, $i = 1, 2$, are independent and $\nu_1 \geq m$, denote by $\Lambda(m, \nu_1, \nu_2)$ the distribution of

$$\frac{|\mathbf{W}_1|}{|\mathbf{W}_1 + \mathbf{W}_2|} = |\mathbf{I}_m + \mathbf{W}_1^{-1}\mathbf{W}_2|^{-1},$$

known as Wilks' Lambda distribution, with parameters m, ν_1, and ν_2.

Result 13.1.9.

1. For ν_1, $\nu_2 \geq 1$, the Wilks' Lambda distribution $\Lambda(1, \nu_1, \nu_2)$ is identical to the $\mathrm{Beta}(\nu_1/2, \nu_2/2)$ distribution.

2. Let $\mathbf{\Sigma}$ be a $k \times k$ p.d. matrix. Let $\mathbf{W}_i \sim W_m(\mathbf{\Sigma}, \nu_i)$, $i = 1, 2$, be independent and let $\nu_1 \geq m$. Then $|\mathbf{W}_1|/|\mathbf{W}_1 + \mathbf{W}_2| \sim \Lambda(m, \nu_1, \nu_2)$.

Proof. First, $\Lambda(1, \nu_1, \nu_2)$ is the distribution of $X/(X+Y)$, where $X \sim W_1(1, \nu_1)$ and $Y \sim W_1(1, \nu_2)$ are independent. For any $n \geq 1$, $W_1(1, n)$ is identical to χ^2_n, i.e., $\mathrm{Gamma}(n/2, 2)$. Then $X/(X+Y) \sim \mathrm{Beta}(\nu_1/2, \nu_2/2)$, proving property 1. To prove property 2, let $\mathbf{\Gamma} = \mathbf{\Sigma}^{1/2}$. From Exercise 6.2, $\mathbf{\Gamma}^{-1}\mathbf{W}_i\mathbf{\Gamma}^{-1} \sim W_m(\mathbf{I}_m, \nu_i)$ are independent, giving $|\mathbf{W}_1|/|\mathbf{W}_1 + \mathbf{W}_2| = |\mathbf{\Gamma}^{-1}\mathbf{W}_1\mathbf{\Gamma}^{-1}|/|\mathbf{\Gamma}^{-1}\mathbf{W}_1\mathbf{\Gamma}^{-1} + \mathbf{\Gamma}^{-1}\mathbf{W}_2\mathbf{\Gamma}^{-1}| \sim \Lambda(m, \nu_1, \nu_2)$. ■

Result 13.1.10. Let $\widehat{\boldsymbol{\Sigma}}_{ML}$ be the MLE of $\boldsymbol{\Sigma}$ under the full model (13.1.1), and $\widehat{\boldsymbol{\Sigma}}_{0,ML}$ be the MLE under the reduced model (13.1.21).

1. The LRT statistic for H_0: $\mathbf{B}_2 = \mathbf{O}$ has the form

$$\Lambda = \frac{|\widehat{\boldsymbol{\Sigma}}_{ML}|^{N/2}}{|\widehat{\boldsymbol{\Sigma}}_{0,ML}|^{N/2}}. \tag{13.1.22}$$

2. Under H_0, the distribution of the Wilks' Lambda statistic is

$$\Lambda^{2/N} = \frac{|\widehat{\boldsymbol{\Sigma}}_{ML}|}{|\widehat{\boldsymbol{\Sigma}}_{0,ML}|} \sim \Lambda(m, N-p, p-q). \tag{13.1.23}$$

3. For large samples, the modified LRT statistic

$$-\left(N - \frac{m+p+q+1}{2}\right)\log\frac{|\widehat{\boldsymbol{\Sigma}}_{ML}|}{|\widehat{\boldsymbol{\Sigma}}_{0,ML}|}$$

has an approximate $\chi^2_{m(p-q)}$ distribution under H_0. This is known as Bartlett's chi-square approximation (Bartlett, 1938; Mardia et al., 1979).

Proof. Property 1 directly follows from (13.1.20) in Result 13.1.7, where the formula is applied twice, once under the full model, and once again under the reduced model. From Result 13.1.7 and (13.1.14),

$$\Lambda^{2/N} = \frac{|\mathbf{E}'(\mathbf{I}_N - \mathbf{P})\mathbf{E}|}{|\mathbf{E}'(\mathbf{I}_N - \mathbf{P}_1)\mathbf{E}|} = \frac{|\mathbf{E}'(\mathbf{I}_N - \mathbf{P})\mathbf{E}|}{|\mathbf{E}'(\mathbf{I}_N - \mathbf{P})\mathbf{E} + \mathbf{E}'(\mathbf{P} - \mathbf{P}_1)\mathbf{E}|},$$

where $\mathbf{P}$ and $\mathbf{P}_1$ are respectively the orthogonal projections onto $\mathcal{C}(\mathbf{X})$ and $\mathcal{C}(\mathbf{X}_1)$. Both $\mathbf{I}_N - \mathbf{P}$ and $\mathbf{P} - \mathbf{P}_1$ are orthogonal projection matrices, with ranks $N-p$ and $p-q$, respectively. Also, $(\mathbf{I}_N - \mathbf{P})(\mathbf{P} - \mathbf{P}_1) = \mathbf{O}$. Then, by Result 13.1.6, $\mathbf{E}'(\mathbf{I}_N - \mathbf{P})\mathbf{E} \sim W_m(\boldsymbol{\Sigma}, N-p)$, $\mathbf{E}'(\mathbf{P} - \mathbf{P}_1)\mathbf{E} \sim W_m(\boldsymbol{\Sigma}, p-q)$, and the two random matrices are independent. Property 2 then follows from property 2 of Result 13.1.9. The proof of property 3 is beyond the scope of this book but can be found in Bartlett (1938). ■

13.1.6 Confidence region and Hotelling T^2 distribution

Suppose we wish to estimate the mean response vector $\mathbf{y}_0 = \mathbf{B}'\mathbf{x}_0$ for a vector of predictors $\mathbf{x}_0$. Since we have assumed that $\text{r}(\mathbf{X}) = p$, from Result 13.1.5, $\mathbf{B}'\mathbf{x}_0$ is estimable and the estimate is $\widehat{\mathbf{y}}_0 = \widehat{\mathbf{B}}'\mathbf{x}_0$. Suppose the errors are i.i.d. $N(\mathbf{0}, \boldsymbol{\Sigma})$ random vectors, where $\boldsymbol{\Sigma}$ is p.d.

From property 1 in Result 13.1.8, and property 2 of Result 13.1.5,

$$\widehat{\mathbf{y}}_0 \sim N_m(\mathbf{y}_0, (\mathbf{x}_0'\mathbf{G}\mathbf{x}_0)\boldsymbol{\Sigma}). \tag{13.1.24}$$

From Result 5.4.5, $(\mathbf{x}_0'\mathbf{G}\mathbf{x}_0)^{-1}(\widehat{\mathbf{y}}_0 - \mathbf{y}_0)'\boldsymbol{\Sigma}^{-1}(\widehat{\mathbf{y}}_0 - \mathbf{y}_0) \sim \chi^2_m$. If $\boldsymbol{\Sigma}$ were known, a $100(1-\alpha)\%$ confidence region for $\mathbf{B}'\mathbf{x}_0$ would be

$$\left\{\mathbf{y}_0 \in \mathcal{R}^m\colon (\mathbf{x}_0'\mathbf{G}\mathbf{x}_0)^{-1}(\widehat{\mathbf{y}}_0 - \mathbf{y}_0)'\boldsymbol{\Sigma}^{-1}(\widehat{\mathbf{y}}_0 - \mathbf{y}_0) \le \chi^2_{m,\alpha}\right\},$$

where $\chi^2_{m,\alpha}$ is the upper (100α)th percentile from the χ^2_m distribution. Since $\boldsymbol{\Sigma}$ is usually unknown, a natural idea is to replace it with the unbiased estimate $\widehat{\boldsymbol{\Sigma}}$ given in (13.1.15) and replace $\chi^2_{m,\alpha}$ with the corresponding critical value, as discussed below.

Definition 13.1.3. Let $\mathbf{z} \sim N_m(\mathbf{0}, \mathbf{I}_m)$ and $\mathbf{W} \sim W_m(\mathbf{I}_m, \nu)$ be independent of $\mathbf{z}$. Denote by $T^2(m, \nu)$ the distribution of

$$\mathbf{z}'(\mathbf{W}/\nu)^{-1}\mathbf{z} = \nu\mathbf{z}'\mathbf{W}^{-1}\mathbf{z},$$

known as Hotelling T^2 distribution with parameters m and ν.

Result 13.1.11. Let $\mathbf{z} \sim N_m(\mathbf{0}, \boldsymbol{\Sigma})$ and $\mathbf{W} \sim W_m(\boldsymbol{\Sigma}, \nu)$ be independent of $\mathbf{z}$. Then $\nu\mathbf{z}'\mathbf{W}^{-1}\mathbf{z} \sim T^2(m, \nu)$.

Proof. Let $\mathbf{R} = \boldsymbol{\Sigma}^{1/2}$; see property 2 in Result 2.4.5. Then $\mathbf{y} = \mathbf{R}^{-1}\mathbf{z} \sim N_m(\mathbf{0}, \mathbf{I}_m)$ and $\mathbf{M} = \mathbf{R}^{-1}\mathbf{W}\mathbf{R}^{-1} \sim W_m(\mathbf{I}_m, \nu)$ are independent. As a result, $\nu\mathbf{z}'\mathbf{W}^{-1}\mathbf{z} = \nu(\mathbf{R}\mathbf{y})'(\mathbf{R}\mathbf{M}\mathbf{R})^{-1}(\mathbf{R}\mathbf{y}) = \nu\mathbf{y}'\mathbf{M}^{-1}\mathbf{y} \sim T^2(m, \nu)$, yielding the proof. ■

Result 13.1.12. Recall that $\mathbf{X} \in \mathcal{R}^{N \times p}$ has full column rank.

1. $(\mathbf{x}_0'\mathbf{G}\mathbf{x}_0)^{-1}(\widehat{\mathbf{y}}_0 - \mathbf{y}_0)'\widehat{\boldsymbol{\Sigma}}^{-1}(\widehat{\mathbf{y}}_0 - \mathbf{y}_0) \sim T^2(m, N-p)$.
2. For $\alpha \in (0, 1)$, a $100(1-\alpha)\%$ confidence region for $\mathbf{B}'\mathbf{x}_0$ is

$$\left\{\mathbf{y}_0 \in \mathcal{R}^m \colon (\mathbf{x}_0'\mathbf{G}\mathbf{x}_0)^{-1}(\widehat{\mathbf{y}}_0 - \mathbf{y}_0)'\widehat{\boldsymbol{\Sigma}}^{-1}(\widehat{\mathbf{y}}_0 - \mathbf{y}_0) \le T^2(m, N-p, \alpha)\right\},$$

   where $T^2(m, N-r, \alpha)$ is the upper (100α)th percentile of the $T^2(m, N-p)$ distribution.

Proof. From (13.1.24), $\widehat{\mathbf{y}}_0 - \mathbf{y}_0 \sim N_m(\mathbf{0}, (\mathbf{x}_0'\mathbf{G}\mathbf{x}_0)\boldsymbol{\Sigma})$. From property 2 in Result 13.1.8, $\widehat{\mathbf{y}}_0 - \mathbf{y}_0$ is independent of $(\mathbf{x}_0'\mathbf{G}\mathbf{x}_0)\widehat{\mathbf{E}}'\widehat{\mathbf{E}} \sim W_m((\mathbf{x}_0'\mathbf{G}\mathbf{x}_0)\boldsymbol{\Sigma}, N-p)$. Since $\widehat{\boldsymbol{\Sigma}} = \widehat{\mathbf{E}}'\widehat{\mathbf{E}}/(N-p)$, property 1 follows from Result 13.1.11. Property 2 follows immediately from property 1. ■

13.1.7 One-factor multivariate analysis of variance model

Suppose we wish to compare the mean vectors from $a(>2)$ populations. Suppose we have a independent samples, with the ℓth sample consisting of i.i.d. observations $\mathbf{y}_{\ell,1}, \cdots, \mathbf{y}_{\ell,n_\ell}$ from a $N_m(\boldsymbol{\mu}_\ell, \boldsymbol{\Sigma}_\ell)$ population with sample mean $\overline{\mathbf{y}}_\ell$ and sample variance-covariance matrix $\mathbf{S}_\ell$. Let the overall sample mean of all the observations be $\overline{\mathbf{y}}$ and $N = n_1 + \cdots + n_a$.

The one-factor multivariate analysis of variance (MANOVA) model is

$$\mathbf{y}_{\ell,j} = \boldsymbol{\mu} + \boldsymbol{\tau}_\ell + \boldsymbol{\varepsilon}_{\ell,j}, \quad j = 1, \cdots, n_\ell, \; \ell = 1, \cdots, a, \tag{13.1.25}$$

where $\boldsymbol{\mu}$ is an m-dimensional vector denoting the overall mean, $\boldsymbol{\tau}_\ell$ is an m-dimensional vector denoting the ℓth treatment effect, such that

$$\sum_{\ell=1}^{a} n_\ell \boldsymbol{\tau}_\ell = \mathbf{0}. \tag{13.1.26}$$

Also assume that $\boldsymbol{\varepsilon}_{\ell,j}$ are i.i.d. $\sim N_m(\mathbf{0}, \boldsymbol{\Sigma})$.

Under the restricted one-way MANOVA model, the hypothesis

$$H_0\colon \boldsymbol{\tau}_1 = \cdots = \boldsymbol{\tau}_a = \mathbf{0}$$

is testable. In Section 13.1.5, LRT was applied to test a linear hypothesis for a full-rank model. The LRT applies equally well here, although the model (13.1.25) is not of full-rank.

Source	d.f.	SSCP
Treatment	$a-1$	$\boldsymbol{\Theta} = \sum_{\ell=1}^{a} n_\ell(\overline{\mathbf{y}}_\ell - \overline{\mathbf{y}})(\overline{\mathbf{y}}_\ell - \overline{\mathbf{y}})'$
Residual	$N-a$	$\mathbf{W} = \sum_{\ell=1}^{a}\sum_{j=1}^{n_\ell}(\mathbf{y}_{\ell,j} - \overline{\mathbf{y}}_\ell)(\mathbf{y}_{\ell,j} - \overline{\mathbf{y}}_\ell)'$
Total (corrected)	$N-1$	$\boldsymbol{\Theta} + \mathbf{W}$

TABLE 13.1.1. One-way MANOVA table.

We see that under the restriction (13.1.26), $\widehat{\boldsymbol{\mu}}_{ML} = \overline{\mathbf{y}}$, $\widehat{\boldsymbol{\tau}}_{\ell,ML} = \overline{\mathbf{y}}_\ell - \overline{\mathbf{y}}$, $\ell = 1, \cdots, a$, and $\widehat{\boldsymbol{\Sigma}}_{ML} = \mathbf{W}/N$, where

$$\mathbf{W} = \sum_{\ell=1}^{a}\sum_{j=1}^{n_\ell}(\mathbf{y}_{\ell,j} - \overline{\mathbf{y}}_\ell)(\mathbf{y}_{\ell,j} - \overline{\mathbf{y}}_\ell)' = \sum_{\ell=1}^{a}(n_\ell - 1)\mathbf{S}_\ell.$$

On the other hand, under the model reduced by H_0, $\widehat{\boldsymbol{\mu}}_{0,ML} = \overline{\mathbf{y}}$ and

$$\begin{aligned} N\widehat{\boldsymbol{\Sigma}}_{0,ML} &= \sum_{\ell=1}^{a}\sum_{j=1}^{n_\ell}(\mathbf{y}_{\ell,j} - \overline{\mathbf{y}})(\mathbf{y}_{\ell,j} - \overline{\mathbf{y}})' \\ &= \sum_{\ell=1}^{a}\sum_{j=1}^{n_\ell}(\mathbf{y}_{\ell,j} - \overline{\mathbf{y}}_\ell)(\mathbf{y}_{\ell,j} - \overline{\mathbf{y}}_\ell)' + \sum_{\ell=1}^{a} n_\ell(\overline{\mathbf{y}}_\ell - \overline{\mathbf{y}})(\overline{\mathbf{y}}_\ell - \overline{\mathbf{y}})' \\ &= \mathbf{W} + \boldsymbol{\Theta}. \end{aligned}$$

Note that $\boldsymbol{\Theta}$ denotes the Between Groups SSCP, while $\mathbf{W}$ denotes the Within Groups SSCP. Under H_0, for each $\ell = 1, \cdots, a$, $(n_\ell - 1)\mathbf{S}_\ell \sim W_m(\boldsymbol{\Sigma}, n_\ell - 1)$, and so $\mathbf{W} \sim W_m(\boldsymbol{\Sigma}, N-a)$. Meanwhile, $\boldsymbol{\Theta} \sim W_m(\boldsymbol{\Sigma}, a-1)$ and is independent of $\mathbf{W}$. These results give the MANOVA decomposition which can be expressed as

Total Corrected SSCP = Within Groups SSCP + Between Groups SSCP,

with d.f. $N-1$, $N-a$, and $a-1$, respectively. The MANOVA decomposition is summarized in Table 13.1.1.

Now, as in property 1 of Result 13.1.10, the LRT statistic for H_0 is $\Lambda = |\widehat{\boldsymbol{\Sigma}}_{ML}|^{N/2}/|\widehat{\boldsymbol{\Sigma}}_{0,ML}|^{N/2}$, and the Wilk's Lambda statistic is

$$\Lambda^* = \frac{|\mathbf{W}|}{|\boldsymbol{\Theta} + \mathbf{W}|}.$$

From properties 2–3 of Result 13.1.10, under H_0, $\Lambda^* \sim \Lambda(m, N-a, a-1)$. If N is large, then

$$WL = -\left(N - \frac{m+a+2}{2}\right)\log\Lambda^* \sim \chi^2_{p(a-1)} \quad \text{approx.}$$

Approximately, at level α, a LRT rejects H_0 if $WL > \chi^2_{m(a-1),\alpha}$.

The statistic Λ^* can also be written as

$$\Lambda^* = \prod_{i=1}^{s}\frac{1}{1+\widehat{\lambda}_i},$$

where, $\widehat{\lambda}_1, \cdots, \widehat{\lambda}_s$ are the nonzero eigenvalues of $\mathbf{W}^{-1}\boldsymbol{\Theta}$, with $s = \mathrm{r}(\boldsymbol{\Theta}) = \min(p, a-1)$.

Other statistics that are functions of the $\widehat{\lambda}_i$'s and related eigenvalues are shown below. Let $\widehat{\eta}_1, \cdots, \widehat{\eta}_s$ be the nonzero eigenvalues of $\mathbf{W}(\mathbf{B}+\mathbf{W})^{-1}$, and let

$$r_i^2 = \frac{\widehat{\lambda}_i}{(1+\widehat{\lambda}_i)}$$

be the canonical correlations.

1. The Hotelling–Lawley Trace is defined as

$$HLT = \text{tr}(\mathbf{W}^{-1}\mathbf{B}) = \sum_{i=1}^{s} \widehat{\lambda}_i = \sum_{i=1}^{s} \frac{r_i^2}{1-r_i^2}.$$

2. Pillai's trace is defined as

$$PT = \text{tr}\{\mathbf{B}(\mathbf{B}+\mathbf{W})^{-1}\} = \sum_{i=1}^{s} \frac{\widehat{\lambda}_i}{1+\widehat{\lambda}_i} = \sum_{i=1}^{s} r_i^2.$$

3. Wilk's statistic, which is based on the LRT statistic is defined as

$$WS = \frac{|\mathbf{W}|}{|\mathbf{B}+\mathbf{W}|} = \prod_{i=1}^{s} (1+\widehat{\lambda}_i)^{-1} = \sum_{i=1}^{s} (1-r_i^2).$$

4. Roy's Maximum Root statistic is defined as

$$RMRT = \max_{i=1,\cdots,s} \widehat{\eta}_i.$$

For large samples, these statistics are essentially equivalent. It has been suggested that PT is more robust to non-normality than the others. None of these statistics performs uniformly better than the others for all situations, and the choice of statistic is usually left to the user; see Johnson and Wichern (2007) for a discussion.

Suppose we reject H_0 at level α. Then we can construct Bonferroni intervals for contrasts $\boldsymbol{\tau}_k - \boldsymbol{\tau}_\ell$, for $\ell \neq k = 1, \cdots, a$. We have $\widehat{\boldsymbol{\tau}}_k = \overline{\mathbf{y}}_k - \overline{\mathbf{y}}$, i.e., $\widehat{\tau}_{ki} = \overline{y}_{ki} - \overline{y}_i$. Also, $\widehat{\tau}_{ki} - \widehat{\tau}_{\ell i} = \overline{y}_{ki} - \overline{y}_{\ell i}$ and $\text{Var}(\widehat{\tau}_{ki} - \widehat{\tau}_{\ell i}) = \text{Var}(\overline{y}_{ki} - \overline{y}_{\ell i}) = (1/n_k + 1/n_\ell)\sigma_{ii}$, σ_{ii} being the ith diagonal element of $\boldsymbol{\Sigma}$, and is estimated by $w_{ii}/(N-a)$, where w_{ii} is the ith diagonal element of the Within SSCP matrix and $N = \sum_{\ell=1}^{a} n_\ell$. As a result, the $100(1-\alpha)\%$ Bonferroni intervals for $\tau_{ki} - \tau_{\ell i}$ are given by

$$\overline{y}_{ki} - \overline{y}_{\ell i} \pm t_{N-a,\alpha/ma(a-1)} \left[\frac{(1/n_k + 1/n_\ell)w_{ii}}{N-a} \right]^{1/2}$$

for all $i = 1, \cdots, m$ and for all differences $1 \le \ell < k \le a$.

Numerical Example 13.2. The data set "Plastic" with 20 observations from the R package *heplots* describes an experiment to determine the optimum conditions for extruding plastic film. Three responses were measured tear (tear resistance), gloss (film gloss), and opacity (film opacity). The responses were measured in relation to two factors, rate of extrusion (a factor representing change in the rate of extrusion with levels, Low: −10% or High: 10%), and amount of an additive (a factor with levels, Low: 1.0% or High: 1.5%). We model the response as a function of a single factor, rate using the `lm()` function and the `manova()` function. The `pairs()` function shows interesting scatterplots (the figure is not shown here).

```
library(heplots)
data(Plastic)
model <- lm(cbind(tear, gloss, opacity) ~ rate, data = Plastic)
Anova(model)

Type II MANOVA Tests: Pillai test statistic
     Df test stat approx F num Df den Df    Pr(>F)
rate  1   0.58638    7.561      3     16 0.002273 **
---
Signif. codes:  0 '***' 0.001 '**' 0.01 '*' 0.05 '.' 0.1 ' ' 1

pairs(model)
# manova() function
fit <- manova(cbind(tear, gloss, opacity) ~ rate, data = Plastic)
summary(fit, test = "Pillai")

 Df  Pillai approx F num Df den Df    Pr(>F)
rate       1 0.58638    7.561      3     16 0.002273 **
Residuals 18
---
Signif. codes:  0 '***' 0.001 '**' 0.01 '*' 0.05 '.' 0.1 ' ' 1
```

▲

13.2 Longitudinal models

In longitudinal studies, measurements on subjects are obtained repeatedly over time. Measurements obtained by following the subjects forward in time are repeated measures or longitudinal data and are collected prospectively, unlike extraction of data over time from available records which constitutes a retrospective study. Longitudinal studies enable us to estimate changes in the response variable over time and relate them to changes in predictors over time. In this section, we describe general linear models for longitudinal data on m subjects over several time periods. For example, Diggle et al. (1994) in a study of the effect of ozone pollution on growth of Sitka spruce trees, data for 79 trees over two growing seasons were obtained in normal and ozone-enriched chambers. A total of 54 trees were grown with ozone exposure at 70 ppb, while 25 trees were grown in normal (control) atmospheres. The response variable was $Y = \log(hd^2)$, where h is the height and d, the diameter of a tree.

For each subject i ($i = 1, \cdots, m$), we observe a response Y_{ij} at time j (for $j = 1, \cdots, n_i$). In general, in longitudinal studies, the natural experimental unit is not the individual measurement Y_{ij}, but the sequence $\mathbf{y}_i \in \mathcal{R}^{n_i}$ of repeated measurements over time on the ith subject.

13.2.1 Multivariate model for longitudinal data

Let

$$Y_{ij} = \beta_0 + \beta_1 X_{ij1} + \beta_2 X_{ij2} + \cdots + \beta_k X_{ijk} + \varepsilon_{ij} = \mathbf{x}'_{ij}\boldsymbol{\beta} + \varepsilon_{ij}, \tag{13.2.1}$$

where Y_{ij} denotes the response variable, and $\mathbf{x}'_{ij} = (1, X_{ij1}, \cdots, X_{ijk})$, all observed at times t_{ij}, for $j = 1, \cdots, n_i$, and for subjects $i = 1, \cdots, m$. In (13.2.1), $\boldsymbol{\beta} = (\beta_0, \beta_1, \cdots, \beta_k)'$

is a p-dimensional vector of unknown regression coefficients and ε_{ij} denotes the random error component with $E\varepsilon_{ij} = 0$, and $\text{Cov}(\varepsilon_{ij}, \varepsilon_{ij'}) \neq 0$. In matrix notation, let $\mathbf{y}_i = (Y_{i1}, \cdots, Y_{in_i})'$ denote the n_i-dimensional vector of repeated outcomes on subject i with mean vector $\text{E}(\mathbf{y}_i) = \boldsymbol{\mu}_i = (\mu_{i1}, \cdots, \mu_{in_i})'$ and $\text{Cov}(\mathbf{y}_i) = \sigma^2\mathbf{V}_i = \sigma^2\{v_{ijk}\}$. Let $\mathbf{y} = (\mathbf{y}_1', \cdots, \mathbf{y}_m')'$ denote the N-dimensional response vector, where $N = \sum_{i=1}^m n_i$, so that $\text{Cov}(\mathbf{y}) = \sigma^2\mathbf{V} = \sigma^2 \text{ diag}(\mathbf{V}_1, \cdots, \mathbf{V}_m)$. We can write the model in (13.2.1) as

$$\mathbf{y}_i = \mathbf{X}_i\boldsymbol{\beta} + \boldsymbol{\varepsilon}_i, \tag{13.2.2}$$

where $\mathbf{X}_i$ is an $n_i \times p$ matrix with $\mathbf{x}_{ij}$ in the ith row and $\boldsymbol{\varepsilon}_i = (\varepsilon_{i1}, \cdots, \varepsilon_{in_i})$. Clearly, $\mathbf{y} \sim N(\mathbf{X}\boldsymbol{\beta}, \sigma^2\mathbf{V})$, where $\mathbf{X} = (\mathbf{X}_1', \cdots, \mathbf{X}_m')'$ is an $N \times p$ matrix. Although such *multivariate models* for longitudinal data with a general covariance structure are straightforward to implement with balanced data as shown in Result 13.2.1 (McCulloch et al., 2011), they may be cumbersome when subjects are measured at arbitrary times, or when the dimension of $\mathbf{V}$ is large. We consider the case where $\mathbf{V}_i = \mathbf{V}_0$ for $i = 1, \cdots, m$.

Result 13.2.1. Balanced longitudinal data model. For $i = 1, \cdots, m$, assume that $n_i = n$, $\mathbf{X}_i = \mathbf{X}_0 \in \mathcal{R}^{n \times p}$, and $\mathbf{V}_i = \mathbf{V}_0 \in \mathcal{R}^{n \times n}$. Let $\boldsymbol{\mu}_0 = \mathbf{X}_0\boldsymbol{\beta}$ and $\boldsymbol{\Sigma}_0 = \sigma^2\mathbf{V}_0$.

1. Irrespective of what $\boldsymbol{\Sigma}_0$ is, the MLE of $\boldsymbol{\mu}_0$ is

$$\widehat{\boldsymbol{\mu}}_0 = \frac{1}{m}\sum_{i=1}^m \mathbf{y}_i, \tag{13.2.3}$$

   so that $\widehat{\mu}_{0,j} = \overline{Y}_{\cdot,j}$.

2. The MLE of $\boldsymbol{\Sigma}_0$ is $\widehat{\boldsymbol{\Sigma}}_0 = \{\widehat{\sigma}_{0,j,\ell}\}_{j,\ell=1}^n$ with

$$\widehat{\sigma}_{0,j,\ell} = \frac{1}{m}\sum_{i=1}^m (Y_{ij} - \overline{Y}_{\cdot j})(Y_{i\ell} - \overline{Y}_{\cdot \ell}). \tag{13.2.4}$$

Proof. Under the assumption, $\mathbf{y}_1, \cdots, \mathbf{y}_m$ are i.i.d. $\sim N(\boldsymbol{\mu}_0, \boldsymbol{\Sigma}_0)$. Then the result is an immediate consequence of Result 6.1.4. ■

Remark 1. We see that in Result 13.2.1, $\mathbf{y} \sim N(\boldsymbol{\mu}, \sigma^2\mathbf{V})$, where $\boldsymbol{\mu} = \mathbf{1}_m \otimes \boldsymbol{\mu}_0$ and $\mathbf{V} = \mathbf{I}_m \otimes \mathbf{V}_0$.

Remark 2. In general, if $\mathbf{V}$ is known, then the MLE of $\boldsymbol{\beta}$ is the same as its GLS estimator (see (4.5.5)):

$$\widehat{\boldsymbol{\beta}}(\mathbf{V}) = (\mathbf{X}'\mathbf{V}^{-1}\mathbf{X})^{-1}\mathbf{X}'\mathbf{V}^{-1}\mathbf{y}. \tag{13.2.5}$$

Also,

$$SSE(\mathbf{V}) = \{\mathbf{y} - \mathbf{X}\widehat{\boldsymbol{\beta}}(\mathbf{V})\}'\mathbf{V}^{-1}\{\mathbf{y} - \mathbf{X}\widehat{\boldsymbol{\beta}}(\mathbf{V})\}. \tag{13.2.6}$$

Remark 3. If $\mathbf{V}$ is unknown, but is not a function of $\boldsymbol{\beta}$, the MLE of $\boldsymbol{\beta}$ is obtained by substituting the ML estimator $\widehat{\mathbf{V}}_{ML}$ for $\mathbf{V}$ in (13.2.5).

Remark 4. For a symmetric weight matrix $\mathbf{W}$, it follows from (4.5.5) that the WLS estimator of $\boldsymbol{\beta}$ is

$$\widehat{\boldsymbol{\beta}}_{WLS} = (\mathbf{X}'\mathbf{W}\mathbf{X})^{-1}\mathbf{X}'\mathbf{W}\mathbf{y},$$

Note that $\mathbf{W} = \mathbf{I}$ yields the OLS estimate of $\boldsymbol{\beta}$, while setting $\mathbf{W} = \mathbf{V}^{-1}$ leads to the estimator shown in (13.2.5).

Remark 5. It may be possible to consider a structured $\mathbf{V}$ instead of letting it be a general symmetric p.d. matrix. We show examples with two possible assumptions on the block-diagonal structure of $\sigma^2\mathbf{V}$, each using only two parameters.

Example 13.2.1. Uniform correlation model. For each subject $i = 1, \cdots, m$, we assume that $\mathbf{V}_i = \mathbf{V}_0 = (1-\rho)\mathbf{I} + \rho\mathbf{J}$, i.e., we assume a positive correlation ρ between any two measurements on the same subject. This is often called the *compound symmetry* assumption. One interpretation of this model is that we introduce a random subject effect U_i, which are mutually independent variables with variance ν^2 between subjects, so that

$$Y_{ij} = \mu_{ij} + U_i + Z_{ij}, \quad j = 1, \cdots, n, \; i = 1, \cdots, m,$$

where U_i are $N(0, \nu^2)$ variables, and Z_{ij} are mutually independent $N(0, \tau^2)$ variables, which are independent of U_i. In this case, $\rho = \nu^2/(\nu^2 + \tau^2)$, and $\sigma^2 = \nu^2 + \tau^2$. □

Example 13.2.2. Exponential correlation model. We assume that the correlation between any two measurements on the same subject decays towards zero as the time separation between the measurements increases, i.e., the (j,k)th element of $\mathbf{V}_0$ is

$$\text{Cov}(Y_{ij}, Y_{ik}) = \sigma^2 \exp\{-\phi|t_{ij} - t_{ik}|\}.$$

A special case corresponds to equal spacing between observations times; i.e., if $t_{ij} - t_{ik} = d$, for $k = j-1$, and letting $\rho = \exp(-\phi d)$ denote the correlation between successive responses, we can write $\text{Cov}(Y_{ij}, Y_{ik}) = \sigma^2\rho|j-k|$. One justification for this model is to write

$$Y_{ij} = \mu_{ij} + W_{ij}, \quad j = 1, \cdots, n, \; i = 1, \cdots, m,$$

where $W_{ij} = \rho W_{i,j-1} + Z_{ij}$, Z_{ij} are independently distributed as $N(0, \sigma^2\{1-\rho^2\})$ variables. This is often referred to as an AR(1) correlation specification. □

Mixed-effects modeling (see Chapter 11) is useful for longitudinal or repeated measures designs which are employed in experiments or surveys where more than one measurement of the same response variable is obtained on each subject, and we wish to examine changes (over time) in measurements taken on each subject (called within-subjects effects). In such cases, all the subjects may either belong to a single homogeneous population, or to a populations which we then compare.

Example 13.2.3. Repeated measures design. Consider the balanced mixed-effects model

$$Y_{ijk} = \mu + \tau_i + \theta_j + (\tau\theta)_{ij} + \gamma_{k(i)} + \varepsilon_{ijk}, \tag{13.2.7}$$

where $i = 1, \cdots, a$, $j = 1, \cdots, b$ and $k = 1, \cdots, c$, the parameter μ denotes an overall mean effect, τ_i is the (fixed) effect due to the ith level of a treatment (Factor A), θ_j denotes the mean (fixed) effect due to the jth time, $(\tau\theta)_{ij}$ denotes the (fixed) interaction effect between the ith treatment level and the jth time, $\gamma_{k(i)}$ denotes a random effect of the kth subject (replicate) nested in the ith treatment level. We assume that $\sum_{i=1}^{a}\tau_i = 0$, $\sum_{j=1}^{b}\theta_j = 0$, $\sum_{i=1}^{a}(\tau\theta)_{ij} = 0$, $\sum_{j=1}^{b}(\tau\theta)_{ij} = 0$, and $\gamma_{k(i)} \sim N(0, \sigma_\gamma^2)$, distributed independently of $\varepsilon_{ijk} \sim N(0, \sigma_\varepsilon^2)$. The variance structure in the responses follow.

$$\begin{aligned}
\text{Var}(Y_{ijk}) &= \sigma_\varepsilon^2 + \sigma_\gamma^2, \\
\text{Cov}(Y_{ijk}, Y_{ij'k}) &= \sigma_\gamma^2, \text{ so that} \\
\text{Corr}(Y_{ijk}, Y_{ij'k}) &= \rho = \frac{\sigma_\gamma^2}{\sigma_\varepsilon^2 + \sigma_\gamma^2}.
\end{aligned}$$

Source	d.f.	SS
Treatment	$a-1$	$SS_A = bc\sum_{i=1}^{a}(\overline{Y}_{i\cdot\cdot} - \overline{Y}_{\cdots})^2$
Time	$b-1$	$SS_B = ac\sum_{j=1}^{b}(\overline{Y}_{\cdot j\cdot} - \overline{Y}_{\cdots})^2$
Treat.×Time	$(a-1)(b-1)$	$SS_{AB} - c\sum_{i=1}^{a}\sum_{j=1}^{b}\overline{R}_{ij\cdot}^2$ (see *)
Subj. w/i treat.	$a(c-1)$	$SS_C = c\sum_{i=1}^{a}\sum_{k=1}^{c}(\overline{Y}_{i\cdot k} - \overline{Y}_{i\cdot\cdot})^2$
Residual	$a(b-1)(c-1)$	$SSE = \sum_{i=1}^{a}\sum_{j=1}^{b}\sum_{k=1}^{c}R_{ijk}^2$ (see *)
Total	$abc-1$	$SST_c = \sum_{i=1}^{a}\sum_{j=1}^{b}\sum_{k=1}^{c}(Y_{ijk} - \overline{Y}_{\cdots})^2$

* $\overline{R}_{ij\cdot} = \overline{Y}_{ij\cdot} - \overline{Y}_{i\cdot\cdot} - \overline{Y}_{\cdot j\cdot} + \overline{Y}_{\cdots}$, $R_{ijk} = Y_{ijk} - \overline{Y}_{ij\cdot} - \overline{Y}_{i\cdot k} + \overline{Y}_{i\cdot\cdot}$

TABLE 13.2.1. ANOVA table for Example 13.2.3.

An ANOVA table consists of the following sums of squares with corresponding d.f.

Following details in Chapter 11, we test H_0: τ_i's equal (or test H_0: $\sigma_\tau^2 = 0$ if Factor A is a random factor) using $F = MS_A/MS_C$ with $(a-1, a(c-1))$ d.f. We test H_0: θ_j's equal using $F = MS_B/MSE$ with $[b-1, a(b-1)(c-1)]$ d.f. We can also test H_0: $(\tau\theta)_{ij} = 0$ (or $\sigma_{\tau\beta}^2 = 0$) using $F = MS_{AB}/MSE$ with $((a-1)(b-1), a(b-1)(c-1))$ d.f. □

Remark 1. The assumption of *compound symmetry* in a repeated measures design implies an assumption of equal correlation between responses pertaining to the same subject over different time periods. This assumption may not be reasonable in all examples, and we usually test for the validity of the assumption.

Remark 2. Greenhouse and Geisser (1959) discussed situations with possibly unequal correlations among the responses at different time periods (lack of compound symmetry), and suggested use of the same F-statistics as in Example 13.2.3, but with reduced d.f., given by $f(a-1)$ and $f(a-1)(b-1)$ respectively for SS_B and SSE (see Table 13.2.1), where the factor f depends on the covariances among the Y_{ij}'s. The value of the factor f is not known in practice. One approach is to choose the smallest possible value of $f = 1/(a-1)$, such that the resulting F-statistic has $(1, b-1)$ d.f. The actual significance level of the test in this case is smaller than the stated level α, and the test is conservative. However, in cases where b is small, this procedure makes it extremely difficult to reject the null hypothesis. Most software use an F-statistic with $\widehat{f}(a-1)$ and $\widehat{f}(a-1)(b-1)$ d.f., where $\widehat{f}$ is obtained by an approach described in Geisser and Greenhouse (1958).

13.2.2 Two-stage random-effects models

Laird and Ware (1982) described two-stage random-effects models for longitudinal data, which are based on explicit identification of population and individual characteristics. The probability distributions for the response vectors of different subjects are assumed to belong to a single family, but some random-effects parameters vary across subjects, with a distribution specified at the second stage. The two-stage random-effects model for longitudinal data is given below (see (11.1.1)).

Stage 1. For each individual $i = 1, \cdots, m$,

$$\mathbf{y}_i = \mathbf{X}_i\boldsymbol{\tau} + \mathbf{Z}_i\boldsymbol{\gamma}_i + \boldsymbol{\varepsilon}_i, \quad i = 1, \cdots, m, \tag{13.2.8}$$

where $\boldsymbol{\tau}$ is a p-dimensional vector of unknown population parameters, $\mathbf{X}_i$ is an $n_i \times p$ known design matrix, $\boldsymbol{\gamma}_i$ is a q-dimensional vector of unknown individual effects, and $\mathbf{Z}_i$

is a known $n_i \times q$ design matrix. The errors $\boldsymbol{\varepsilon}_i$ are independently distributed as $N(\mathbf{0}, \mathbf{R}_i)$ vectors, while $\boldsymbol{\tau}$ are fixed parameter vectors.

Stage 2. The $\boldsymbol{\gamma}_i$ are i.i.d. $\sim N(\mathbf{0}, \mathbf{D}_\gamma)$, and $\text{Cov}(\boldsymbol{\gamma}_i, \boldsymbol{\varepsilon}_i) = \mathbf{0}$. The population parameters $\boldsymbol{\tau}$ are fixed effects.

For $i = 1, \cdots, m$, $\mathbf{y}_i$ are independent $N(\mathbf{X}_i\boldsymbol{\tau}, \mathbf{V}_i)$, where $\mathbf{V}_i = \mathbf{R}_i + \mathbf{Z}_i\mathbf{D}_\gamma\mathbf{Z}_i'$. Let $\mathbf{W}_i = \mathbf{V}_i^{-1}$. Let $\boldsymbol{\theta}$ denote an vector of parameters determining $\mathbf{R}_i$, $i = 1, \cdots, m$, and $\mathbf{D}_\gamma$. Inference can be carried out using least squares, maximum likelihood or REML approaches.

Estimation of mean effects

The classical approach to inference derives estimates of $\boldsymbol{\tau}$ and $\boldsymbol{\theta}$ based on the marginal distribution of $\mathbf{y}' = (\mathbf{y}_1', \cdots, \mathbf{y}_m')$, while an estimate for $\boldsymbol{\gamma}$ is obtained by use of Harville's extended Gauss–Markov theorem for mixed-effects models (Harville, 1976) (see Result 11.2.1). Let $\boldsymbol{\gamma} = (\boldsymbol{\gamma}_1', \cdots, \boldsymbol{\gamma}_m')'$. Result 13.2.2 discusses estimation for two cases, when $\boldsymbol{\theta}$ is known and when it is unknown.

Result 13.2.2.

1. Suppose $\boldsymbol{\theta}$ is known. Assuming that the necessary matrix inversions exist, let

$$\widehat{\boldsymbol{\tau}} = \left(\sum_{i=1}^m \mathbf{X}_i'\mathbf{W}_i\mathbf{X}_i\right)^{-1} \sum_{i=1}^m \mathbf{X}_i'\mathbf{W}_i\mathbf{y}_i, \text{ and} \tag{13.2.9}$$

$$\widehat{\boldsymbol{\gamma}}_i = \mathbf{D}_\gamma\mathbf{Z}_i'\mathbf{W}_i(\mathbf{y}_i - \mathbf{X}_i\widehat{\boldsymbol{\tau}}). \tag{13.2.10}$$

Then $\widehat{\boldsymbol{\tau}}$ is the MLE of $\boldsymbol{\tau}$, for any vectors of constants $\mathbf{c}_0 \in \mathcal{R}^p$ and $\mathbf{c}_1, \cdots, \mathbf{c}_m \in \mathcal{R}^q$, $\mathbf{c}_0'\widehat{\boldsymbol{\tau}} + \sum_{i=1}^m \mathbf{c}_i'\widehat{\boldsymbol{\gamma}}_i$ is an essentially-unique b.l.u.e. of $\mathbf{c}_0'\boldsymbol{\tau} + \sum_{i=1}^m \mathbf{c}_i'\boldsymbol{\gamma}_i$, and

$$\text{Var}(\widehat{\boldsymbol{\tau}}) = \left(\sum_{i=1}^m \mathbf{X}_i'\mathbf{W}_i\mathbf{X}_i\right)^{-1}, \text{ and} \tag{13.2.11}$$

$$\text{Var}(\widehat{\boldsymbol{\gamma}}_i) = \mathbf{D}_\gamma\mathbf{Z}_i'\left\{\mathbf{W}_i - \mathbf{W}_i\mathbf{X}_i\left(\sum_{i=1}^m \mathbf{X}_i'\mathbf{W}_i\mathbf{X}_i\right)^{-1}\mathbf{X}_i'\mathbf{W}_i\right\}\mathbf{Z}_i\mathbf{D}_\gamma. \tag{13.2.12}$$

2. Suppose, as is usually the case in practice, $\boldsymbol{\theta}$ is unknown. Let $(\widehat{\boldsymbol{\theta}}, \widehat{\boldsymbol{\tau}})$ denote the MLEs of $(\boldsymbol{\theta}, \boldsymbol{\tau})$, respectively. Then,

$$\widehat{\boldsymbol{\tau}} = \left(\sum_{i=1}^m \mathbf{X}_i'\widehat{\mathbf{W}}_i\mathbf{X}_i\right)^{-1} \sum_{i=1}^m \mathbf{X}_i'\widehat{\mathbf{W}}_i\mathbf{y}_i, \tag{13.2.13}$$

where $\widehat{\mathbf{W}}_i = \mathbf{W}_i(\widehat{\boldsymbol{\theta}})$, i.e., the value of $\mathbf{W}_i$ under $\widehat{\boldsymbol{\theta}}$. The MLE of $\text{Cov}(\widehat{\boldsymbol{\tau}})$ follows by replacing $\mathbf{W}_i$ by $\widehat{\mathbf{W}}_i$ in (13.2.11).

Proof. Recall that we estimate $\boldsymbol{\tau}$ and $\boldsymbol{\theta}$ simultaneously by maximizing their joint likelihood based on the marginal distribution of $\mathbf{y}$. The proof is left as an exercise. ■

A better assessment of the error of estimation is provided by

$$\begin{aligned}\text{Var}(\widehat{\boldsymbol{\gamma}}_i - \boldsymbol{\gamma}_i) &= \mathbf{D}_\gamma - \mathbf{D}_\gamma\mathbf{Z}_i'\mathbf{W}_i\mathbf{Z}_i\mathbf{D}_\gamma \\ &\quad + \mathbf{D}_\gamma\mathbf{Z}_i'\mathbf{W}_i\mathbf{X}_i\left(\sum_{i=1}^m \mathbf{X}_i'\mathbf{W}_i\mathbf{X}_i\right)^{-1}\mathbf{X}_i'\mathbf{W}_i\mathbf{Z}_i\mathbf{D}_\gamma,\end{aligned} \tag{13.2.14}$$

which incorporates the variation in γ_i. Under normality, the estimator $\widehat{\boldsymbol{\tau}}$ is the MVUE, and $\widehat{\gamma}$ is the essentially-unique MVUE of γ (see Result 11.2.1). The estimate $\widehat{\boldsymbol{\tau}}$ also maximizes the likelihood based on the marginal normal distribution of $\mathbf{y}$. We can write

$$\widehat{\gamma} = \mathrm{E}(\gamma \,|\, \mathbf{y}, \widehat{\boldsymbol{\tau}}, \boldsymbol{\theta}),$$

so that $\widehat{\gamma}_i$ is the empirical Bayes estimator of γ_i.

Estimation of covariance structure

Popular approaches for estimating $\boldsymbol{\theta}$ are the method of maximum likelihood and the method of restricted maximum likelihood (REML). In balanced ANOVA models, MLE's of variance components fail to account for the degrees of freedom in estimating the fixed effects $\boldsymbol{\tau}$, and are therefore negatively biased. The REML estimates on the other hand are unbiased; the REML estimate of $\boldsymbol{\theta}$ is obtained by maximizing the likelihood of $\boldsymbol{\theta}$ based on any full rank set of error contrasts, $\mathbf{z} = \mathbf{B}'\mathbf{y}$.

Numerical Example 13.3. Repeated measures analysis. The dataset "emotion" in the R package *psycho* consists of emotional ratings of neutral and negative pictures by healthy participants and consists of 912 rows on 11 variables.

```
library(psycho); library(tidyverse)
df <- psycho::emotion %>% select(Participant_ID, Participant_Sex,
Emotion_Condition, Subjective_Valence, Recall)
summary(aov(Subjective_Valence ~ Emotion_Condition +
Error(Participant_ID/Emotion_Condition), data=df))

Error: Participant_ID
          Df Sum Sq Mean Sq F value Pr(>F)
Residuals 18 115474    6415

Error: Participant_ID:Emotion_Condition
                  Df  Sum Sq Mean Sq F value   Pr(>F)
Emotion_Condition  1 1278417 1278417   245.9 6.11e-12 ***
Residuals         18   93573    5198
---
Signif. codes:  0 '***' 0.001 '**' 0.01 '*' 0.05 '.' 0.1 ' ' 1

Error: Within
           Df Sum Sq Mean Sq F value Pr(>F)
Residuals 874 935646    1070
```

The ANOVA output from a linear mixed model (LMM) is shown below.

```
library(lmerTest)
fit <- lmer(Subjective_Valence ~ Emotion_Condition +
 (1|Participant_ID), data=df)
anova(fit)

Type III Analysis of Variance Table with Satterthwaite's method
                   Sum Sq Mean Sq NumDF DenDF F value    Pr(>F)
Emotion_Condition 1278417 1278417     1   892    1108 <2.2e-16***
---
Signif. codes:  0 '***' 0.001 '**' 0.01 '*' 0.05 '.' 0.1 ' ' 1
```

▲

Diggle et al. (1994) describe generalized linear models for longitudinal data (binary responses, count responses, etc.) using (a) marginal models, (b) random effects models, and (c) transition models.

13.3 Elliptically contoured linear model

We start this Section with results on distributions of quadratic forms in elliptical random vectors, which we defined in Section 5.5.2.3.

Result 13.3.1. Let $\mathbf{x} \sim E_k(\mathbf{0}, \mathbf{V}, h)$ with p.d.f. $f(\mathbf{x}) = c_k|\mathbf{V}|^{-1/2}h(\mathbf{x}'\mathbf{V}^{-1}\mathbf{x})$, and let $Q = \mathbf{x}'\mathbf{V}^{-1}\mathbf{x}$. Then

$$f_Q(w) = \frac{\pi^{k/2}}{\Gamma(k/2)} c_k w^{k/2-1} h(w). \tag{13.3.1}$$

Proof. Let $\mathbf{y} = \mathbf{V}^{-1/2}\mathbf{x}$, and $f(\mathbf{y}) = c_k h(\mathbf{y}'\mathbf{y})$. The distributions of Q and $\mathbf{y}'\mathbf{y}$ coincide. ■

For $q < k$,

$$g_q(w) = \frac{\pi^{q/2}}{\Gamma(q/2)} c_q w^{q/2-1} h_q(w) \tag{13.3.2}$$

is the p.d.f. of $\sum_{i=1}^{q} Y_i^2$, for $\mathbf{y} = (Y_1, \cdots, Y_k)'$.

Result 13.3.2. Let $\mathbf{x} \sim E_k(\mathbf{0}, \mathbf{V}, h)$ with finite fourth moment, and let $\mathbf{A}$ be a symmetric matrix of rank r. Then $\mathbf{x}'\mathbf{A}\mathbf{x} \sim g_r(.)$ if and only if

1. $r + \text{r}(\mathbf{V}^{-1} - \mathbf{A}) = k$, or
2. $\mathbf{A} = \mathbf{A}\mathbf{V}\mathbf{A}$.

Proof. To prove property 1, assume first that $r + \text{r}(\mathbf{V}^{-1} - \mathbf{A}) = k$. By Result 2.4.5, there exists a nonsingular matrix $\mathbf{P}$ such that $\mathbf{V} = \mathbf{P}\mathbf{P}'$. Let $\mathbf{y} = \mathbf{P}^{-1}\mathbf{x}$; $f(\mathbf{y}) = c_k h(\mathbf{y}'\mathbf{y})$. Now, $\mathbf{x}'\mathbf{V}^{-1}\mathbf{x} = \mathbf{x}'\mathbf{A}\mathbf{x} + \mathbf{x}'(\mathbf{V}^{-1} - \mathbf{A})\mathbf{x}$, i.e., $\mathbf{y}'\mathbf{y} = \mathbf{y}'\mathbf{P}'\mathbf{A}\mathbf{P}\mathbf{y} + \mathbf{y}'(\mathbf{I} - \mathbf{P}'\mathbf{A}\mathbf{P})\mathbf{y}$. The rank condition ensures the existence of an orthogonal transformation $\mathbf{z} = \mathbf{B}\mathbf{y}$ such that $f(\mathbf{z}) = f(Z_1, \cdots, Z_k) = c_k h(\mathbf{z}'\mathbf{z})$. Now, $\mathbf{x}'\mathbf{A}\mathbf{x} = \mathbf{y}'\mathbf{P}'\mathbf{A}\mathbf{P}\mathbf{y} = \sum_{i=1}^{r} Z_i^2 \sim g_r(w)$. To prove the converse, suppose $\mathbf{x}'\mathbf{A}\mathbf{x} \sim g_r(w)$, where $r = \text{r}(\mathbf{A})$. There exists a nonsingular matrix $\mathbf{P}$ such that $\mathbf{V} = \mathbf{P}\mathbf{P}'$, and an orthogonal matrix $\mathbf{Q}$ such that $\mathbf{Q}'\mathbf{A}\mathbf{Q} = \mathbf{D} = \text{diag}(\lambda_1, \cdots, \lambda_r, \mathbf{0})$. If $\mathbf{x} = \mathbf{Q}\mathbf{P}\mathbf{z}$, it follows that $\mathbf{x}'\mathbf{A}\mathbf{x} = \mathbf{z}'\mathbf{P}'\mathbf{Q}'\mathbf{A}\mathbf{Q}\mathbf{P}\mathbf{z} = \sum_{i=1}^{r} \lambda_i Z_i^2$. Using the second and fourth moments of $\mathbf{z}$, it can be shown that $\sum_{i=1}^{r} \lambda_i = \sum_{i=1}^{r} \lambda_i^2 = r$, so that $\lambda_i = 1$, $i = 1, \cdots, r$. We can therefore write $\mathbf{x}'\mathbf{V}^{-1}\mathbf{x} = \mathbf{x}'\mathbf{A}\mathbf{x} + \mathbf{x}'(\mathbf{V}^{-1} - \mathbf{A})\mathbf{x}$ as $\mathbf{z}'\mathbf{z} = \sum_{i=1}^{r} Z_i^2 + \sum_{i=r+1}^{k} Z_i^2$. This implies that $\text{r}(\mathbf{A}) + \text{r}(\mathbf{V}^{-1} - \mathbf{A}) = k$, proving property 1. The proof of property 2 is similar to the proof of property 2 of Result 5.4.5. ■

Anderson and Fang (1987) discuss Cochran's theorem for elliptical distributions. Matrix variate elliptically contoured distributions extend the elliptical distributions from the vector to the matrix case and an excellent discussion can be seen in Gupta and Varga (1993).

Consider the elliptical linear model

$$\mathbf{y} = \mathbf{X}\boldsymbol{\beta} + \boldsymbol{\varepsilon}, \quad \boldsymbol{\varepsilon} \sim E_N(\mathbf{0}, \sigma^2\mathbf{I}_N, h), \tag{13.3.3}$$

where $\mathrm{r}(\mathbf{X}) = p$, and $h(u)$ is a continuous and differentiable function of $u \geq 0$ which is strictly decreasing on $\{u: f(u) > 0\}$. Then, $\mathbf{y} \sim E_N(\mathbf{X}\boldsymbol{\beta}, \sigma^2\mathbf{I}_N, h)$.

Result 13.3.3. Let $\widehat{\boldsymbol{\beta}}_h$ and $\widehat{\sigma}_h^2$ denote the ML estimators of $\boldsymbol{\beta}$ and σ^2 under the elliptical linear model (13.3.3).

1. Then,

$$\widehat{\boldsymbol{\beta}}_h = \widehat{\boldsymbol{\beta}}_{ML} = (\mathbf{X}'\mathbf{X})^{-1}\mathbf{X}'\mathbf{y}, \tag{13.3.4}$$

$$\widehat{\sigma}_h^2 = \frac{1}{u_0}(\mathbf{y} - \mathbf{X}\widehat{\boldsymbol{\beta}}_h)'(\mathbf{y} - \mathbf{X}\widehat{\boldsymbol{\beta}}_h), \tag{13.3.5}$$

where $u_0 = \arg\sup_{u \geq 0} u^{N/2}h(u)$ and may be obtained as a solution to $h'(u) + (N/2u)h(u) = 0$.

2. The sampling distribution of $\widehat{\boldsymbol{\beta}}_h$ is

$$\widehat{\boldsymbol{\beta}}_h \sim E_p(\boldsymbol{\beta}, \sigma^2(\mathbf{X}'\mathbf{X})^{-1}, h). \tag{13.3.6}$$

3. An unbiased estimator of σ^2 is

$$\widehat{\sigma}_*^2 = \frac{u_0\widehat{\sigma}_h^2}{N - p}. \tag{13.3.7}$$

Proof. Property 1 follows from maximizing the logarithm of the likelihood function

$$L(\boldsymbol{\beta}, \sigma^2; \mathbf{y}) \propto (\sigma^2)^{-N/2}h\{\sigma^{-2}(\mathbf{y} - \mathbf{X}\boldsymbol{\beta})'(\mathbf{y} - \mathbf{X}\boldsymbol{\beta})\}. \tag{13.3.8}$$

To show property 2, we derive the MGF of $\widehat{\boldsymbol{\beta}}_h$ using (5.5.15) as

$$\begin{aligned} M(\mathbf{X}(\mathbf{X}'\mathbf{X})^{-1}\mathbf{X}'\mathbf{t}) &= \psi(\mathbf{t}'\mathbf{X}(\mathbf{X}'\mathbf{X})^{-1}\mathbf{X}'(\sigma^2\mathbf{I}_N)(\mathbf{X}'\mathbf{X})^{-1}\mathbf{X}'\mathbf{X}'\mathbf{t}) \\ &\quad \times \exp(\mathbf{t}'\mathbf{X}(\mathbf{X}'\mathbf{X})^{-1}\mathbf{X}'\mathbf{X}\boldsymbol{\beta}) \\ &= \psi(\mathbf{t}'\mathbf{X}(\sigma^2(\mathbf{X}'\mathbf{X})^{-1})\mathbf{X}'\mathbf{t})\exp(\mathbf{t}'\mathbf{X}\boldsymbol{\beta}) \\ &= \psi(\mathbf{u}'\,\mathrm{Cov}(\widehat{\boldsymbol{\beta}}_h)\mathbf{u})\exp(\mathbf{u}'\boldsymbol{\beta}), \text{ where } \mathbf{u} = \mathbf{X}'\mathbf{t}, \end{aligned}$$

from which it follows that $\widehat{\boldsymbol{\beta}}_h \sim E_p(\boldsymbol{\beta}, \sigma^2(\mathbf{X}'\mathbf{X})^{-1}, h)$. We leave the reader to prove property 3. ■

The estimator $\widehat{\boldsymbol{\beta}}_h$ also coincides with the LS estimate of $\boldsymbol{\beta}$ and is the same for any density generator $h(.)$ in the elliptical family. The fitted and residual vectors are also the same and have the forms shown in Section 4.2 for the full-rank model. Note that the distribution of $\widehat{\boldsymbol{\beta}}_h$ is the same for all $h(.)$ within the elliptical family. However, the distribution of $\widehat{\sigma}_h^2$ is affected by departures from normality within the family of elliptical distributions. When $h(.)$ corresponds to the normal or t-distributions, $u_0 = N$, while for other elliptical distributions, u_0 must be solved numerically. It is straightforward to extend Result 13.3.3 to the case where $\mathbf{y} \sim E_N(\mathbf{X}\boldsymbol{\beta}, \sigma^2\mathbf{V}, h)$, $\mathbf{V}$ being a known symmetric, p.d. matrix. The details are left as Exercise 13.9.

The next result describes testing a linear hypothesis H: $\mathbf{C}'\boldsymbol{\beta} = \mathbf{d}$ on $\boldsymbol{\beta}$ under the model (13.3.3). Similar to Section 7.2 for the normal GLM, we assume that $\mathbf{C}$ is a $p \times s$ matrix with known coefficients, $\mathrm{r}(\mathbf{C}) = s$ and $\mathbf{d} = (d_1, \cdots, d_s)'$ is a vector of known constants.

Result 13.3.4.

1. Irrespective of $h(.)$, the LRT statistic for testing H: $\mathbf{C}'\boldsymbol{\beta} = \mathbf{d}$ is

$$\Lambda = \left(1 + \frac{Q}{SSE}\right)^{-N/2},$$

where SSE is from the LS fit and Q is given in Result 7.2.1.

2. Under H,

$$F(H) = \frac{Q/s}{SSE/(N-p)} \sim F_{s,N-p}$$

Proof. Let $\widehat{\boldsymbol{\varepsilon}}$ denote the vector of residuals of the LS fit and $\widehat{\boldsymbol{\varepsilon}}_H$ the one subject to the constraint by H. Then from (13.3.8),

$$\begin{aligned}
\Lambda &= \frac{\sup_{\sigma^2,\boldsymbol{\beta}:\, \mathbf{C}'\boldsymbol{\beta}=\mathbf{d}} (\sigma^2)^{-N/2} h\{\sigma^{-2}\|\mathbf{y} - \mathbf{X}\boldsymbol{\beta}\|^2\}}{\sup_{\sigma^2,\boldsymbol{\beta}} (\sigma^2)^{-N/2} h\{\sigma^{-2}\|\mathbf{y} - \mathbf{X}\boldsymbol{\beta}\|^2\}} \\
&= \frac{\sup_{\sigma^2} (\sigma^2)^{-N/2} h\{\sigma^{-2} \inf_{\boldsymbol{\beta}:\, \mathbf{C}'\boldsymbol{\beta}=\mathbf{d}} \|\mathbf{y} - \mathbf{X}\boldsymbol{\beta}\|^2\}}{\sup_{\sigma^2} (\sigma^2)^{-N/2} h\{\sigma^{-2}\| \inf_{\boldsymbol{\beta}} \mathbf{y} - \mathbf{X}\boldsymbol{\beta}\|^2\}} \\
&= \frac{\sup_{u\geq 0} u^{N/2} h(u \cdot \widehat{\boldsymbol{\varepsilon}}_H' \widehat{\boldsymbol{\varepsilon}}_H)}{\sup_{u\geq 0} u^{N/2} h(u \cdot \widehat{\boldsymbol{\varepsilon}}' \widehat{\boldsymbol{\varepsilon}})}.
\end{aligned}$$

Since for any $a > 0$, $\sup_{u\geq 0} u^{N/2} h(ua) = a^{-N/2} u_0^{N/2} h(u_0)$, with u_0 as in property 1 of Result 13.3.3, then $\Lambda = (\widehat{\boldsymbol{\varepsilon}}_H' \widehat{\boldsymbol{\varepsilon}}_H / \widehat{\boldsymbol{\varepsilon}}' \widehat{\boldsymbol{\varepsilon}})^{-N/2}$, so that property 1 follows from Result 7.2.2.

Let $\mathbf{C}' = \mathbf{T}'\mathbf{X}$. Denote by $\mathbf{P}$ the projection onto $\mathcal{C}(\mathbf{X})$. Then from Result 7.2.1, we see that under H, $Q = \boldsymbol{\varepsilon}'\mathbf{M}\boldsymbol{\varepsilon}$, where $\mathbf{M} = \mathbf{PT}[\mathbf{C}'(\mathbf{X}'\mathbf{X})^{-1}\mathbf{C}]^{-1}\mathbf{T}'\mathbf{P}$ is symmetric and n.n.d. On the other hand, $SSE = \boldsymbol{\varepsilon}'\mathbf{P}\boldsymbol{\varepsilon}$. Then

$$F(H) = \frac{\boldsymbol{\varepsilon}'\mathbf{M}\boldsymbol{\varepsilon}/s}{\boldsymbol{\varepsilon}'\mathbf{P}\boldsymbol{\varepsilon}/(N-p)} = \frac{\mathbf{u}'\mathbf{M}\mathbf{u}/s}{\mathbf{u}'\mathbf{P}\mathbf{u}/(N-p)},$$

where $\mathbf{u} = \boldsymbol{\varepsilon}/\|\boldsymbol{\varepsilon}\|$. Note that the above identity holds for any distribution of $\boldsymbol{\varepsilon}$ as long as $\mathrm{P}(\boldsymbol{\varepsilon} \neq \mathbf{0}) = 1$. Since the distribution of $F(H)$ only depends on the distribution of $\boldsymbol{\varepsilon}/\|\boldsymbol{\varepsilon}\|$, it is the same for all spherical distributions with zero mass at the origin. Since $N(\mathbf{0}, \mathbf{I}_p)$ is one such spherical distribution, then from Result 7.2.1, property 2 follows. ■

By using the relationship between the likelihood ratio test statistic and the F-statistic, it can be shown that

$$F_0 = \frac{(N-p)}{s} \frac{(\mathbf{C}'\widehat{\boldsymbol{\beta}} - \mathbf{d})'\{\mathbf{C}'(\mathbf{X}'\mathbf{X})^{-1}\mathbf{C}\}^{-1}(\mathbf{C}'\widehat{\boldsymbol{\beta}} - \mathbf{d})}{(N-p)\widehat{\sigma}^2} \tag{13.3.9}$$

has an $F_{s,N-p}$ distribution under H (see Result 8.1.2).

Result 13.3.5. The $100(1-\alpha)\%$ joint confidence region for $\boldsymbol{\beta}$ under the model (13.3.3) is

$$\{\boldsymbol{\beta} \in \mathcal{R}^p\colon (\widehat{\boldsymbol{\beta}}_h - \boldsymbol{\beta})'\mathbf{X}'\mathbf{X}(\widehat{\boldsymbol{\beta}}_h - \boldsymbol{\beta}) \leq p\widehat{\sigma}_*^2 F_{p,N-p,1-\alpha}\}. \tag{13.3.10}$$

The proof of the result follows by inverting the F-statistic in (13.3.9).

These distributional results facilitate inference in linear models with elliptical errors. For more details, the reader is referred to Fang and Anderson (1990) and references given there.

We define the LS (or ML) residuals as well as internally and externally Studentized residuals from fitting the model (13.3.3).

Definition 13.3.1. Residuals. 1. The usual LS (or ML) residuals are $\widehat{\boldsymbol{\varepsilon}} = \mathbf{y} - \widehat{\mathbf{y}} = (\mathbf{I} - \mathbf{P})\mathbf{y}$, where $\mathbf{P} = \mathbf{X}(\mathbf{X}'\mathbf{X})^{-1}\mathbf{X}'$, and

$$\widehat{\boldsymbol{\varepsilon}} \sim E_N(\mathbf{0}, \sigma^2(\mathbf{I} - \mathbf{P}), h),$$
$$\widehat{\varepsilon}_i \sim E(0, \sigma^2(1 - p_{ii}), h), \ i = 1, \ldots, N. \tag{13.3.11}$$

2. The ith internally Studentized residual is

$$r_i = \frac{\widehat{\varepsilon}_i}{\widehat{\sigma}_* \sqrt{1 - p_{ii}}}, \tag{13.3.12}$$

where $\widehat{\sigma}_*^2$ was defined in (13.3.7).

3. The ith externally Studentized residual is

$$r_i^* = \frac{\widehat{\varepsilon}_i}{\widehat{\sigma}_{*(i)} \sqrt{1 - p_{ii}}}, \text{ where,} \tag{13.3.13}$$
$$\widehat{\sigma}_{*(i)}^2 = \frac{u_0^* \widehat{\sigma}_{(i)}^2}{N - 1 - p}, \tag{13.3.14}$$

where

$$\widehat{\sigma}_{(i)}^2 = (\mathbf{y}_{(i)} - \mathbf{X}_{(i)}\widehat{\boldsymbol{\beta}}_{h(i)})'(\mathbf{y}_{(i)} - \mathbf{X}_{(i)}\widehat{\boldsymbol{\beta}}_{h(i)})$$

and $u_0^* = \max(u^{(N-1)/2}h(u), \widehat{\sigma}_{(i)}^2)$.

4. Independent of $h(.)$,

$$\frac{r_i^2}{N - p} \sim \text{Beta}\left(\frac{1}{2}, \frac{N - 1 - p}{2}\right),$$
$$r_i^* \sim t_{N-1-p},$$

similar to their distributions under the normal GLM.

Properties of these residuals and details on other case deletion diagnostics including likelihood displacement (Cook and Weisberg, 1982) have been discussed in Galea et al. (2000). We show an example next.

Numerical Example 13.4. The dataset "luzdat" from the R package *gwer* with 150 observations on four variables is part of a study development by the nutritional department of São Paulo University.

```
library(gwer)
data(luzdat); attach(luzdat)
z1 <- C(factor(x1),treatment)
z2 <- x2
z3 <- x2^2
luz <- data.frame(y,z1,z2,z3)
# Normal fit
elliptical.fitn <-
  elliptical(y ~ z1 + z2 + z3, family = Normal(), data = luz)
summary(elliptical.fitn)
```

```
family(elliptical.fitn)
# Student-t fit
elliptical.fitt <-
  elliptical(y ~ z1 + z2 + z3, family = Student(df = 5),
  data = luz)
summary(elliptical.fitt)
family(elliptical.fitt)
#
elliptical.fitpe <- elliptical(y ~ z1+z2+z3, family =
   Powerexp(k=0.5), data=luz)
summary(elliptical.fitpe)
family(elliptical.fitpe)
```

The output from the power exponential model fit with $k = 0.5$ is shown below. Results for the normal and Student-t fits can be obtained similarly. Other families that are available include Cauchy, LogisI, LogisII, Glogis, Gstudent, and Cnormal.

```
Call: elliptical(formula = y ~ z1 + z2 + z3, family =
    Powerexp(k = 0.5), data = luz)

Coefficients:
                  Value Std. Error    z-value     p-value
(Intercept) 78.8947193 0.41683362 189.271485 0.00000e+00
z12         -0.2923446 0.36084438  -0.810168 4.17844e-01
z13         -0.6090463 0.36084438  -1.687837 9.14426e-02
z14         -1.4096325 0.36084438  -3.906483 9.36491e-05
z15         -0.9022161 0.36084438  -2.500291 1.24091e-02
z2          -1.4567704 0.08099231 -17.986527 2.48449e-72
z3           0.0547718 0.00392593  13.951274 3.09037e-44

Scale parameter for Powerexp : 0.908031  ( 0.128415 )

Degrees of Freedom: 150 Total; 143 Residual
-2*Log-Likelihood 549.111

Number  Iterations: 22

Correlation of Coefficients:
    (Intercept) z12      z13      z14      z15      z2
z12 -0.432840
z13 -0.432840   0.500000
z14 -0.432840   0.500000 0.500000
z15 -0.432840   0.500000 0.500000 0.500000
z2  -0.728639   0.000000 0.000000 0.000000 0.000000
z3   0.631037   0.000000 0.000000 0.000000 0.000000 -0.969458
```

▲

13.4 Bayesian linear models

We begin this Section with the following matrix results which are useful in combining quadratic forms, and find application in a Bayesian treatment of linear models (Box and Tiao, 1973).

Result 13.4.1. Let $\mathbf{x}$, $\mathbf{a}$, and $\mathbf{b}$ denote k-dimensional vectors, and let $\mathbf{A}$ and $\mathbf{B}$ be $k \times k$ symmetric matrices such that $(\mathbf{A}+\mathbf{B})^{-1}$ exists. Then,

$$\begin{aligned}&(\mathbf{x}-\mathbf{a})'\mathbf{A}(\mathbf{x}-\mathbf{a}) + (\mathbf{x}-\mathbf{b})'\mathbf{B}(\mathbf{x}-\mathbf{b})\\ &= (\mathbf{x}-\mathbf{c})'(\mathbf{A}+\mathbf{B})(\mathbf{x}-\mathbf{c}) + (\mathbf{a}-\mathbf{b})'\mathbf{A}(\mathbf{A}+\mathbf{B})^{-1}\mathbf{B}(\mathbf{a}-\mathbf{b}),\end{aligned}$$

where $\mathbf{c} = (\mathbf{A}+\mathbf{B})^{-1}(\mathbf{A}\mathbf{a}+\mathbf{B}\mathbf{b})$.

Proof. Clearly,

$$\begin{aligned}&(\mathbf{x}-\mathbf{a})'\mathbf{A}(\mathbf{x}-\mathbf{a}) + (\mathbf{x}-\mathbf{b})'\mathbf{B}(\mathbf{x}-\mathbf{b})\\ &= \mathbf{x}'(\mathbf{A}+\mathbf{B})\mathbf{x} - 2\mathbf{x}'(\mathbf{A}\mathbf{a}+\mathbf{B}\mathbf{b}) + \mathbf{a}'\mathbf{A}\mathbf{a} + \mathbf{b}'\mathbf{B}\mathbf{b}\\ &= \mathbf{x}'(\mathbf{A}+\mathbf{B})\mathbf{x} - 2\mathbf{x}'(\mathbf{A}+\mathbf{B})\mathbf{c} + \mathbf{c}'(\mathbf{A}+\mathbf{B})\mathbf{c} + d\\ &= (\mathbf{x}-\mathbf{c})'(\mathbf{A}+\mathbf{B})(\mathbf{x}-\mathbf{c}) + d,\end{aligned}$$

where $d = \mathbf{a}'\mathbf{A}\mathbf{a} + \mathbf{b}'\mathbf{B}\mathbf{b} - \mathbf{c}'(\mathbf{A}+\mathbf{B})\mathbf{c}$. The right side is

$$\begin{aligned}\mathbf{c}'(\mathbf{A}+\mathbf{B})\mathbf{c} &= (\mathbf{A}\mathbf{a}+\mathbf{B}\mathbf{b})'(\mathbf{A}+\mathbf{B})^{-1}(\mathbf{A}\mathbf{a}+\mathbf{B}\mathbf{b})\\ &= [\mathbf{A}(\mathbf{a}-\mathbf{b}) + (\mathbf{A}+\mathbf{B})\mathbf{b}]'(\mathbf{A}+\mathbf{B})^{-1}\\ &\quad \times [(\mathbf{A}+\mathbf{B})\mathbf{a} - \mathbf{B}(\mathbf{a}-\mathbf{b})]\\ &= -(\mathbf{a}-\mathbf{b})'\mathbf{A}(\mathbf{A}+\mathbf{B})^{-1}\mathbf{B}(\mathbf{a}-\mathbf{b}) + \mathbf{a}'\mathbf{A}\mathbf{a} + \mathbf{b}'\mathbf{B}\mathbf{b},\end{aligned}$$

and the result follows immediately. ■

Result 13.4.2. Let $\mathbf{x}$, $\mathbf{a}$, and $\mathbf{b}$ denote k-dimensional vectors, and let $\mathbf{A}$ and $\mathbf{B}$ be $k \times k$ p.s.d. symmetric matrices such that $\mathrm{r}(\mathbf{A}+\mathbf{B}) = q < k$. Then, subject to the constraints $\mathbf{G}\mathbf{x} = \mathbf{0}$,

$$\begin{aligned}&(\mathbf{x}-\mathbf{a})'\mathbf{A}(\mathbf{x}-\mathbf{a}) + (\mathbf{x}-\mathbf{b})'\mathbf{B}(\mathbf{x}-\mathbf{b})\\ &= (\mathbf{x}-\mathbf{c}^*)'(\mathbf{A}+\mathbf{B}+\mathbf{M})(\mathbf{x}-\mathbf{c}^*) + (\mathbf{a}-\mathbf{b})'\mathbf{A}(\mathbf{A}+\mathbf{B}+\mathbf{M})^{-1}\mathbf{B}(\mathbf{a}-\mathbf{b}),\end{aligned}$$

where $\mathbf{G}$ is any $(k-q) \times k$ matrix of rank $(k-q)$ such that the rows of $\mathbf{G}$ are LIN of the rows of the matrix $\mathbf{A}+\mathbf{B}$, $\mathbf{M} = \mathbf{G}'\mathbf{G}$, and

$$\mathbf{c}^* = (\mathbf{A}+\mathbf{B}+\mathbf{M})^{-1}(\mathbf{A}\mathbf{a}+\mathbf{B}\mathbf{b}). \tag{13.4.1}$$

Proof. Denote $\mathbf{S} = \mathbf{A}+\mathbf{B}$. Let $\mathbf{K}_1$ and $\mathbf{K}_2$ be any matrices of k columns such that $\mathbf{K}_1'\mathbf{K}_1 = \mathbf{A}$ and $\mathbf{K}_2'\mathbf{K}_2 = \mathbf{B}$. Let $\mathbf{K} = (\mathbf{K}_1', \mathbf{K}_2')'$. Then $\mathbf{S} = \mathbf{K}'\mathbf{K}$. On the other hand, let $\mathbf{L}$ be a $q \times k$ matrix of full row rank, such that $\mathbf{S} = \mathbf{L}'\mathbf{L}$. From property 4 of Result 1.3.10, $\mathcal{R}(\mathbf{K}) = \mathcal{R}(\mathbf{L}) = \mathcal{R}(\mathbf{S})$. It follows that $\mathcal{R}(\mathbf{K}_i) \subset \mathcal{R}(\mathbf{L})$, so $\mathbf{K}_i' = \mathbf{L}'\mathbf{D}_i$ for some matrices $\mathbf{D}_i$. By the assumption, $\mathcal{R}^k = \mathcal{R}(\mathbf{S}) \oplus \mathcal{R}(\mathbf{G})$, so $\mathcal{R}^k = \mathcal{R}(\mathbf{L}) \oplus \mathcal{R}(\mathbf{G})$. Then $\mathbf{H} = (\mathbf{L}', \mathbf{G}')'$ is $k \times k$ nonsingular, so $\mathbf{H}(\mathbf{H}'\mathbf{H})^{-1}\mathbf{H}' = \mathbf{I}_k$. Since $\mathbf{H}'\mathbf{H} = \mathbf{L}'\mathbf{L} + \mathbf{G}'\mathbf{G} = \mathbf{S}+\mathbf{M}$, then

$$\begin{pmatrix}\mathbf{L}\\ \mathbf{G}\end{pmatrix}(\mathbf{S}+\mathbf{M})^{-1}(\mathbf{L}', \mathbf{G}') = \begin{pmatrix}\mathbf{I}_q & \mathbf{O}\\ \mathbf{O} & \mathbf{I}_{k-q}\end{pmatrix}.$$

First, we get $\mathbf{G}(\mathbf{S}+\mathbf{M})^{-1}\mathbf{L}' = \mathbf{O}$. Pre-multiplying the equation by $\mathbf{G}'$ and post-multiplying it by $\mathbf{D}_i\mathbf{K}_i$, we get $\mathbf{M}(\mathbf{S}+\mathbf{M})^{-1}\mathbf{A} = \mathbf{M}(\mathbf{S}+\mathbf{M})^{-1}\mathbf{B} = \mathbf{O}$, and hence $\mathbf{Mc}^* = \mathbf{0}$. Second, $\mathbf{L}(\mathbf{S}+\mathbf{M})^{-1}\mathbf{L}' = \mathbf{I}_q$. Pre-multiplying the equation by $\mathbf{L}'$ and post-multiplying it by $\mathbf{D}_i\mathbf{K}_i$, we get $\mathbf{S}(\mathbf{S}+\mathbf{M})^{-1}\mathbf{A} = \mathbf{A}$ and $\mathbf{S}(\mathbf{S}+\mathbf{M})^{-1}\mathbf{B} = \mathbf{B}$, and hence $\mathbf{Sc}^* = \mathbf{Aa}+\mathbf{Bb}$. Since $\mathbf{Mx} = \mathbf{0}$,

$$
\begin{aligned}
&(\mathbf{x}-\mathbf{a})'\mathbf{A}(\mathbf{x}-\mathbf{a}) + (\mathbf{x}-\mathbf{b})'\mathbf{B}(\mathbf{x}-\mathbf{b}) \\
&= \mathbf{x}'(\mathbf{S}+\mathbf{M})\mathbf{x} - 2\mathbf{x}'(\mathbf{S}+\mathbf{M})\mathbf{c}^* + \mathbf{c}^{*\prime}(\mathbf{S}+\mathbf{M})\mathbf{c}^* + d_1 \\
&= (\mathbf{x}-\mathbf{c}^*)'(\mathbf{S}+\mathbf{M})(\mathbf{x}-\mathbf{c}^*) + d_1,
\end{aligned}
$$

where $d_1 = \mathbf{a}'\mathbf{Aa} + \mathbf{b}'\mathbf{Bb} - \mathbf{c}^{*\prime}(\mathbf{S}+\mathbf{M})\mathbf{c}^*$. We can now show that

$$
\mathbf{c}^{*\prime}(\mathbf{S}+\mathbf{M})\mathbf{c}^* = -(\mathbf{a}-\mathbf{b})'\mathbf{A}(\mathbf{S}+\mathbf{M})^{-1}\mathbf{B}(\mathbf{a}-\mathbf{b}) + \mathbf{a}'\mathbf{Aa} + \mathbf{b}'\mathbf{ABb}
$$

so that $d_1 = (\mathbf{a}-\mathbf{b})'\mathbf{A}(\mathbf{S}+\mathbf{M})^{-1}\mathbf{B}(\mathbf{a}-\mathbf{b})$, which proves the result. ■

13.4.1 Bayesian normal linear model

Consider the normal linear model in (4.1.1),

$$
\mathbf{y} = \mathbf{X}\boldsymbol{\beta} + \boldsymbol{\varepsilon},
$$

where $\boldsymbol{\varepsilon} \sim N(\mathbf{0}, \sigma^2\mathbf{I}_N)$, and $\boldsymbol{\beta} \in \mathcal{R}^p$ and $\sigma^2 > 0$ are assumed to be random variables. Let $\mathrm{r}(\mathbf{X}) = p$. In the Bayesian framework, we will specify priors for $\boldsymbol{\beta}$ and σ^2. These priors can either be proper priors or improper priors.

Result 13.4.3 shows details for the normal linear model under a conjugate prior specification for $\boldsymbol{\beta}$ and σ^2, called the normal or Gaussian inverse-gamma (GIG) prior. In Exercise 13.11, we ask the reader to show that under the prior, $\boldsymbol{\beta}$ has a marginal multivariate t-distribution (see Example 5.5.3).

Result 13.4.3. Let $\pi(\boldsymbol{\beta}\,|\,\sigma^2) = N(\boldsymbol{\mu}, \sigma^2\boldsymbol{\Omega})$, where $\boldsymbol{\mu} \in \mathcal{R}^p$ and $\boldsymbol{\Omega} \in \mathcal{R}^{p\times p}$. Let $\pi(\sigma^2) = \mathrm{IG}(a,b)$, where $a, b > 0$ (see (B.25) in Appendix B). Then,

1. The joint prior distribution is denoted by $\mathrm{GIG}(\boldsymbol{\mu}, \boldsymbol{\Omega}, a, b)$ and has the form

$$
\pi(\boldsymbol{\beta}, \sigma^2) = \frac{(\sigma^2)^{-(a+1+\frac{p}{2})} b^a \exp(-\frac{b}{\sigma^2})}{(2\pi)^{\frac{p}{2}}|\boldsymbol{\Omega}|^{\frac{1}{2}}\Gamma(a)} \exp\left\{-\frac{1}{2\sigma^2}(\boldsymbol{\beta}-\boldsymbol{\mu})'\boldsymbol{\Omega}^{-1}(\boldsymbol{\beta}-\boldsymbol{\mu})\right\}. \quad (13.4.2)
$$

2. The posterior distribution of $(\boldsymbol{\beta}, \sigma^2)$ given the data is a $\mathrm{GIG}(\boldsymbol{\mu}^*, \boldsymbol{\Omega}^*, a^*, b^*)$ distribution where

$$
\begin{aligned}
\boldsymbol{\Omega}^* &= (\boldsymbol{\Omega}^{-1} + \mathbf{X}'\mathbf{X})^{-1}, \\
\boldsymbol{\mu}^* &= (\boldsymbol{\Omega}^{-1} + \mathbf{X}'\mathbf{X})^{-1}(\boldsymbol{\Omega}^{-1}\boldsymbol{\mu} + \mathbf{X}'\mathbf{y}) = \boldsymbol{\Omega}^*(\boldsymbol{\Omega}^{-1}\boldsymbol{\mu} + \mathbf{X}'\mathbf{y}), \\
a^* &= a + \frac{N}{2}, \\
b^* &= b + \frac{1}{2}[\mathbf{y}'\mathbf{y} + \boldsymbol{\mu}'\boldsymbol{\Omega}^{-1}\boldsymbol{\mu} - \boldsymbol{\mu}^{*\prime}\boldsymbol{\Omega}^{*-1}\boldsymbol{\mu}^*]. \quad (13.4.3)
\end{aligned}
$$

Proof. It is easy to show that

$$
\begin{aligned}
\pi(\boldsymbol{\beta}, \sigma^2) &= \pi(\boldsymbol{\beta}\,|\,\sigma^2)\pi(\sigma^2) \\
&= \frac{1}{(2\pi\sigma^2)^{\frac{p}{2}}|\boldsymbol{\Omega}|^{\frac{1}{2}}} \exp\left\{-\frac{1}{2\sigma^2}(\boldsymbol{\beta}-\boldsymbol{\mu})'\boldsymbol{\Omega}^{-1}(\boldsymbol{\beta}-\boldsymbol{\mu})\right\} \times \frac{b^a(\sigma^2)^{-(a+1)}}{\Gamma(a)} \exp\left(-\frac{b}{\sigma^2}\right),
\end{aligned}
$$

from which property 1 follows. By Bayes rule, the posterior p.d.f. of $(\boldsymbol{\beta}, \sigma^2)$ is

$$\begin{aligned}
\pi(\boldsymbol{\beta}, \sigma^2 \,|\, \mathbf{y}) &\propto f(\mathbf{y}; \boldsymbol{\beta}, \sigma^2) \times \pi(\boldsymbol{\beta}, \sigma^2) \\
&= \frac{1}{(2\pi\sigma^2)^{\frac{N}{2}}} \exp\left\{-\frac{1}{2\sigma^2}(\mathbf{y} - \mathbf{X}\boldsymbol{\beta})'(\mathbf{y} - \mathbf{X}\boldsymbol{\beta})\right\} \\
&\quad \times \frac{(\sigma^2)^{-(a+1+\frac{p}{2})} b^a \exp(-\frac{b}{\sigma^2})}{(2\pi)^{\frac{p}{2}} |\boldsymbol{\Omega}|^{\frac{1}{2}} \Gamma(a)} \exp\left\{-\frac{1}{2\sigma^2}(\boldsymbol{\beta} - \boldsymbol{\mu})'\boldsymbol{\Omega}^{-1}(\boldsymbol{\beta} - \boldsymbol{\mu})\right\} \\
&\propto (\sigma^2)^{-(a+\frac{N}{2}+1+\frac{p}{2})} \exp\left\{-\frac{b}{\sigma^2} - \frac{Q(\boldsymbol{\beta})}{2\sigma^2}\right\},
\end{aligned}$$

where $Q(\boldsymbol{\beta}) = \boldsymbol{\beta}'(\boldsymbol{\Omega}^{-1} + \mathbf{X}'\mathbf{X})\boldsymbol{\beta} - 2(\boldsymbol{\Omega}^{-1}\boldsymbol{\mu} + \mathbf{X}'\mathbf{y})'\boldsymbol{\beta} + (\boldsymbol{\mu}'\boldsymbol{\Omega}^{-1}\boldsymbol{\mu} + \mathbf{y}'\mathbf{y})$. The result follows by completing the square (see item 9 in Appendix C) and simplifying. ■

Result 13.4.4 shows that, under noninformative priors for $\boldsymbol{\beta}$ and σ^2, results from the Bayesian framework coincide with results under ML estimation from Chapter 8. The proof of the result is left as Exercise 13.13.

Result 13.4.4. Assume flat (non-informative) priors $\pi(\boldsymbol{\beta}) \propto 1$ and $\pi(\sigma^2) \propto 1/\sigma^2$, and assume independence between them, so that $\pi(\boldsymbol{\beta}, \sigma^2) \propto 1/\sigma^2$. We suppress conditioning on the fixed matrix $\mathbf{X}$. Let $\widehat{\boldsymbol{\beta}}$ and $\widehat{\sigma}^2$ denote the LS estimates of $\boldsymbol{\beta}$ and σ^2, respectively.

1. The conditional posterior distribution of $\boldsymbol{\beta}$ given $\sigma^2, \mathbf{y}$ is

$$\boldsymbol{\beta} \,|\, \sigma^2, \mathbf{y} \sim N(\widehat{\boldsymbol{\beta}}, \sigma^2(\mathbf{X}'\mathbf{X})^{-1}) = N((\mathbf{X}'\mathbf{X})^{-1}\mathbf{X}'\mathbf{y}, \sigma^2(\mathbf{X}'\mathbf{X})^{-1}). \tag{13.4.4}$$

2. The posterior distribution of σ^2 given $\mathbf{y}$ is

$$\sigma^2 \,|\, \mathbf{y} \sim \text{IG}\left(\frac{N-p}{2}, \frac{(N-p)\widehat{\sigma}^2}{2}\right). \tag{13.4.5}$$

3. The marginal posterior of $\boldsymbol{\beta}$ given $\mathbf{y}$ is obtained by integrating out σ^2 in (13.4.4) to get

$$\begin{aligned}
\pi(\boldsymbol{\beta} \,|\, \mathbf{y}) &= \int \pi(\boldsymbol{\beta} \,|\, \sigma^2, \mathbf{y})\pi(\sigma^2 \,|\, \mathbf{y})\, d\sigma^2 \\
&= c(N, p, \mathbf{X}, \widehat{\sigma}^2) \left[1 + \frac{(\boldsymbol{\beta} - \widehat{\boldsymbol{\beta}})'(\mathbf{X}'\mathbf{X})(\boldsymbol{\beta} - \widehat{\boldsymbol{\beta}})}{(N-p)\widehat{\sigma}^2}\right]^{-\frac{N}{2}},
\end{aligned} \tag{13.4.6}$$

which is a p-variate t-distribution with $N - p$ d.f., location $\widehat{\boldsymbol{\beta}}$, and scale matrix $\widehat{\sigma}^2(\mathbf{X}'\mathbf{X})^{-1}$, where

$$c(N, p, \mathbf{X}, \widehat{\sigma}^2) = \frac{\Gamma(\frac{N}{2})}{[\pi(N-p)]^{\frac{p}{2}}\Gamma(\frac{N-p}{2})|\widehat{\sigma}^2(\mathbf{X}'\mathbf{X})^{-1}|}.$$

4. For $j = 1, \ldots, p$, the marginal posterior of $(\beta_j - \widehat{\beta}_j)/\sqrt{\widehat{\sigma}^2 c_i}$ is a univariate t_{N-p} distribution, where c_i is the ith diagonal elements of $(\mathbf{X}'\mathbf{X})^{-1}$.

5. Let $\mathbf{X}^*$ denote data from a new set of predictors. The posterior predictive distribution for the corresponding response $\mathbf{y}^*$ is

$$f(\mathbf{y}^* \,|\, \mathbf{y}) = \int f(\mathbf{y}^*; \boldsymbol{\beta}, \sigma^2)\pi(\boldsymbol{\beta}, \sigma^2 \,|\, \mathbf{y})\, d\boldsymbol{\beta}\, d\sigma^2, \tag{13.4.7}$$

which has a multivariate t-distribution with d.f. $\nu = N - p$, location $\mathbf{X}^*\widehat{\boldsymbol{\beta}}$, and scale matrix $\widehat{\sigma}^2(\mathbf{I} + \mathbf{X}^*(\mathbf{X}'\mathbf{X})^{-1}\mathbf{X}^{*\prime})$.

Example 13.4.1. In the simple linear regression model (4.1.6), suppose a prior on $\boldsymbol{\beta} = (\beta_0, \beta_1)'$ is the noninformative prior $\pi(\boldsymbol{\beta}) = 1$. The OLS estimator $\widehat{\boldsymbol{\beta}}$ is a sufficient statistic for $\boldsymbol{\beta}$. The posterior distribution is $\pi(\boldsymbol{\beta} \,|\, \mathbf{y}) = \pi(\boldsymbol{\beta} \,|\, \widehat{\boldsymbol{\beta}})$, which is a $N_2(\widehat{\boldsymbol{\beta}}, \sigma^2 \boldsymbol{\Sigma})$ distribution, where

$$\boldsymbol{\Sigma} = \frac{1}{SST_c} \begin{pmatrix} \sum_{i=1}^N X_i^2 & -\overline{X} \\ -\overline{X} & 1 \end{pmatrix}$$

and SST_c was defined in Table 7.2.2. Also, the joint density of $(\mathbf{y}, \boldsymbol{\beta})$ given $\mathbf{x}$ is normal. The predictive distribution $p(\mathbf{y} \,|\, X^*)$ of $\mathbf{y}$ given a specific predictor value X^* is also normal with mean $\widehat{\beta}_0 + \widehat{\beta}_1 X^*$ and variance $\sigma^2[1 + 1/N + (\overline{X} - X^*)^2/SST_c]$. □

13.4.2 Hierarchical normal linear model

A general formulation of the hierarchical normal linear model setup from Lindley and Smith (1972) is described below.

Result 13.4.5. Hierarchical Linear Model. Let $\mathbf{y}$ be an N-dimensional vector, let $\boldsymbol{\theta}_1$ be a k_1-dimensional vector, and suppose that

$$\mathbf{y} \,|\, \boldsymbol{\theta}_1 \sim N(\mathbf{A}_1 \boldsymbol{\theta}_1, \mathbf{C}_1). \tag{13.4.8}$$

Also, assume that given a k_2-dimensional vector of hyperparameters $\boldsymbol{\theta}_2$,

$$\boldsymbol{\theta}_1 \,|\, \boldsymbol{\theta}_2 \sim N(\mathbf{A}_2 \boldsymbol{\theta}_2, \mathbf{C}_2). \tag{13.4.9}$$

Assume that $\mathbf{C}_1$, and $\mathbf{C}_2$ are known p.d. matrices of appropriate dimensions (Lindley and Smith, 1972).

1. The marginal distribution of $\mathbf{y}$ is given by

$$\mathbf{y} \sim N(\mathbf{A}_1 \mathbf{A}_2 \boldsymbol{\theta}_2, \mathbf{C}_1 + \mathbf{A}_1 \mathbf{C}_2 \mathbf{A}_1'). \tag{13.4.10}$$

2. The conditional distribution of $\boldsymbol{\theta}_1$ given $\mathbf{y}$ is $N(\mathbf{Bb}, \mathbf{B})$ where

$$\mathbf{B}^{-1} = \mathbf{A}_1' \mathbf{C}_1^{-1} \mathbf{A}_1 + \mathbf{C}_2^{-1} \tag{13.4.11}$$

and

$$\mathbf{b} = \mathbf{A}_1' \mathbf{C}_1^{-1} \mathbf{y} + \mathbf{C}_2^{-1} \mathbf{A}_2 \boldsymbol{\theta}_2. \tag{13.4.12}$$

Proof. From (13.4.8), we can write $\mathbf{y} = \mathbf{A}_1 \boldsymbol{\theta}_1 + \mathbf{u}$, where $\mathbf{u} \sim N(\mathbf{0}, \mathbf{C}_1)$. We can also write (13.4.9) as $\boldsymbol{\theta}_1 = \mathbf{A}_2 \boldsymbol{\theta}_2 + \mathbf{v}$, where $\mathbf{v} \sim N(\mathbf{0}, \mathbf{C}_2)$ and $\mathbf{v}$ is independent of $\mathbf{u}$. From these observations, it follows that

$$\mathbf{y} = \mathbf{A}_1 \mathbf{A}_2 \boldsymbol{\theta}_2 + \mathbf{A}_1 \mathbf{v} + \mathbf{u}.$$

By the independence of $\mathbf{u}$ and $\mathbf{v}$, and Result 5.2.5, $\mathbf{A}_1 \mathbf{v} + \mathbf{u} \sim N(\mathbf{0}, \mathbf{C}_1 + \mathbf{A}_1 \mathbf{C}_2 \mathbf{A}_1')$, which leads directly to the proof of property 1. To prove property 2, from Result 5.2.12, the conditional distribution of $\boldsymbol{\theta}_1$ given $\mathbf{y}$ is a normal distribution whose mean and covariance are explicitly available. However, to get (13.4.11) and (13.4.12) more directly, we see from Bayes' theorem that

$$\begin{aligned} \pi(\boldsymbol{\theta}_1 \,|\, \mathbf{y}) &\propto L(\boldsymbol{\theta}_1; \mathbf{y}) \pi(\boldsymbol{\theta}_1) \\ &\propto \exp\left\{ -\frac{1}{2} (\mathbf{y} - \mathbf{A}_1 \boldsymbol{\theta}_1)' \mathbf{C}_1^{-1} (\mathbf{y} - \mathbf{A}_1 \boldsymbol{\theta}_1) \right\} \\ &\quad \times \exp\left\{ -\frac{1}{2} (\boldsymbol{\theta}_1 - \mathbf{A}_2 \boldsymbol{\theta}_2)' \mathbf{C}_2^{-1} (\boldsymbol{\theta}_1 - \mathbf{A}_2 \boldsymbol{\theta}_2) \right\} = \exp(-Q/2), \end{aligned}$$

where $Q = (\boldsymbol{\theta}_1 - \mathbf{Bb})'\mathbf{B}^{-1}(\boldsymbol{\theta}_1 - \mathbf{Bb}) + [\mathbf{y}'\mathbf{C}_1^{-1}\mathbf{y} + \boldsymbol{\theta}_2'\mathbf{A}_2'\mathbf{C}_2^{-1}\mathbf{A}_2\boldsymbol{\theta}_2 - \mathbf{b}'\mathbf{Bb}]$, from which property 2 follows. ■

We give a more general result (Lindley and Smith, 1972) that involves more levels in the hierarchy. The proof is left as Exercise 13.14.

Result 13.4.6. Suppose

$$\mathbf{y}\,|\,\boldsymbol{\theta}_1 \sim N(\mathbf{A}_1\boldsymbol{\theta}_1, \mathbf{C}_1), \quad \boldsymbol{\theta}_1\,|\,\boldsymbol{\theta}_2 \sim N(\mathbf{A}_2\boldsymbol{\theta}_2, \mathbf{C}_2), \quad \boldsymbol{\theta}_2\,|\,\boldsymbol{\theta}_3 \sim N(\mathbf{A}_3\boldsymbol{\theta}_3, \mathbf{C}_3), \tag{13.4.13}$$

where $\boldsymbol{\theta}_3$ is a known k_3-dimensional vector, and $\mathbf{C}_1$, $\mathbf{C}_2$, and $\mathbf{C}_3$ are known p.d. matrices of appropriate dimensions. The posterior distribution of $\boldsymbol{\theta}_1$ given $\mathbf{y}$ is $N(\mathbf{Dd}, \mathbf{D})$ where

$$\mathbf{D}^{-1} = \mathbf{A}_1'\mathbf{C}_1^{-1}\mathbf{A}_1 + [\mathbf{C}_2 + \mathbf{A}_2\mathbf{C}_3\mathbf{A}_2']^{-1} \tag{13.4.14}$$

and

$$\mathbf{d} = \mathbf{A}_1'\mathbf{C}_1^{-1}\mathbf{y} + [\mathbf{C}_2 + \mathbf{A}_2\mathbf{C}_3\mathbf{A}_2']^{-1}\mathbf{A}_2\mathbf{A}_3\boldsymbol{\theta}_3. \tag{13.4.15}$$

The mean of the posterior distribution is seen to be a weighted average of the least squares estimate $(\mathbf{A}_1'\mathbf{C}_1^{-1}\mathbf{A}_1)^{-1}\mathbf{A}_1'\mathbf{C}_1^{-1}\mathbf{y}$ of $\boldsymbol{\theta}_1$ and its prior mean $\mathbf{A}_2\mathbf{A}_3\boldsymbol{\theta}_3$, and is a point estimate of $\boldsymbol{\theta}_1$. The marginal density (13.4.10) is also called the *predictive* density of $\mathbf{y}$. The three-stage hierarchy can be extended to several stages.

Example 13.4.2. Consider the multiple regression model $\mathbf{y} = \mathbf{X}\boldsymbol{\beta} + \boldsymbol{\varepsilon}$, where $\mathbf{X}$ is a standardized matrix, $\boldsymbol{\varepsilon} \sim N(\mathbf{0}, \sigma^2\mathbf{I}_N)$, and the components of $\boldsymbol{\beta} = (\beta_1, \cdots, \beta_k)'$ are assumed to be exchangeable. Suppose we assume the prior distribution $\beta_j \sim N(\psi, \sigma_\beta^2)$. Consider two cases, one where we assume that $\psi = 0$ and that σ^2 and σ_β^2 are known, and the second and more realistic case in which $\psi \neq 0$ and both σ^2 and σ_β^2 are unknown nuisance parameters. In the first case, the posterior mode of $\boldsymbol{\beta}$ given the data is

$$\boldsymbol{\beta}^* = \{\mathbf{I}_k + c(\mathbf{X}'\mathbf{X})^{-1}\}^{-1}\widehat{\boldsymbol{\beta}},$$

where $c = \sigma^2/\sigma_\beta^2$ and $\widehat{\boldsymbol{\beta}}$ is the OLS estimate. Note the similarity of this estimate to the ridge regression estimate (see Section 9.2.3). In the second case, assume inverse-χ^2 prior distributions for σ^2 and σ_β^2 given by

$$\nu\lambda/\sigma^2 \sim \chi_\nu^2, \quad \nu_\beta\lambda_\beta/\sigma_\beta^2 \sim \chi_{\nu_\beta}^2.$$

Then,

$$\begin{aligned}
\pi(\boldsymbol{\beta}, \sigma^2, \sigma_\beta^2\,|\,\mathbf{y}) &\propto (\sigma^2)^{-(N+\nu+2)/2} \exp\left\{-\frac{1}{2\sigma^2}[\nu\lambda + (\mathbf{y} - \mathbf{X}\boldsymbol{\beta})'(\mathbf{y} - \mathbf{X}\boldsymbol{\beta})]\right\} \\
&\quad \times (\sigma_\beta^2)^{-(k+\nu_\beta+1)/2} \exp\left\{-\frac{1}{2\sigma^2}\left[\nu_\beta\lambda_\beta + \sum_{j=1}^{k}(\beta_j - \widetilde{\beta})^2\right]\right\},
\end{aligned}$$

where $\widetilde{\beta} = \sum_{j=1}^k \beta_j/k$. The marginal posterior distributions of $\boldsymbol{\beta}$, σ^2 and σ_β^2 may be obtained in closed form, and the posterior modes are obtained as

$$\begin{aligned}
\boldsymbol{\beta}^* &= [\mathbf{I}_k + c^*(\mathbf{X}'\mathbf{X})^{-1}(\mathbf{I}_k - k^{-1}\mathbf{J}_k)]^{-1}\widehat{\boldsymbol{\beta}}, \\
\sigma^{2*} &= [\nu\lambda + (\mathbf{y} - \mathbf{X}\boldsymbol{\beta}^*)'(\mathbf{y} - \mathbf{X}\boldsymbol{\beta}^*)]/(N + \nu + 2), \text{ and} \\
\sigma_\beta^{2*} &= [\nu_\beta\lambda_\beta + \textstyle\sum_{j=1}^k(\beta_j^* - \widetilde{\beta}^*)^2]/(k + \nu_\beta + 1),
\end{aligned}$$

where $c^* = \sigma^{2*}/\sigma_\beta^{2*}$ and $\widetilde{\beta}^* = \sum_{j=1}^k \beta_j^*/k$. For more details, see Lindley and Smith (1972). □

Example 13.4.3. Two-way additive random-effects model. Let

$$Y_{ij} = \mu + \tau_i + \theta_j + \varepsilon_{ij},\ i = 1, \cdots, a,\ j = 1, \cdots, b,$$

We can write the model in the form (4.1.1) with $N = ab$, $p = a + b + 1$. Now,

$$\mathbf{A}_1 = \begin{pmatrix} \mathbf{1}_b & \mathbf{1}_b & \mathbf{0} & \cdots & \mathbf{0} & \mathbf{I}_b \\ \mathbf{1}_b & \mathbf{0} & \mathbf{1}_b & \cdots & \mathbf{0} & \mathbf{I}_b \\ \vdots & \vdots & \vdots & \ddots & \vdots & \mathbf{I}_b \\ \mathbf{1}_b & \mathbf{0} & \mathbf{0} & \cdots & \mathbf{1}_b & \mathbf{I}_b \end{pmatrix},$$

with $r(\mathbf{X}) = a + b - 1$. Let $\tau_i \sim N(0, \sigma_\tau^2)$, $\theta_j \sim N(0, \sigma_\theta^2)$, $\varepsilon_{ij} \sim N(0, \sigma_\varepsilon^2)$, and $\mu \sim (\omega, \sigma_\mu^2)$. Let $\mathbf{C} = \text{diag}(1/\sigma^\mu, 1/\sigma_\tau^2 \mathbf{1}_a, 1/\sigma_\theta^2 \mathbf{1}_b)$. From Result 13.4.6, the posterior mean of $\boldsymbol{\beta}$ given the data is (Lindley and Smith, 1972)

$$\boldsymbol{\beta}^* = (\mathbf{A}_1'\mathbf{A}_1 + \sigma^2 \mathbf{C}_2^{-1})^{-1}\mathbf{A}_1'\mathbf{y}. \qquad \square$$

13.4.3 Bayesian model assessment and selection

We describe criteria that enable model assessment and selection in the Bayesian framework. This is a huge and expanding topic, see Vehtari and Ojanen (2012) which presents a survey of Bayesian predictive methods for model assessment, selection and comparison. Here, we only present a few criteria.

Bayes Factors

The Bayes Factor (BF) for a model M_0 against a model M_1 is the ratio of the marginal likelihoods of the data $\mathbf{y}$, see item 14 of Appendix C. Suppose M_0 and M_1 denote two normal linear models written in the form (Smith and Spiegelhalter, 1980)

$$\mathbf{y} \,|\, \boldsymbol{\beta}_\ell, \sigma^2 \sim N(\mathbf{X}_\ell \boldsymbol{\beta}_\ell, \sigma^2 \mathbf{I}_N),\ \ell = 0, 1, \qquad (13.4.16)$$

where, for $\ell = 0, 1$, $\mathbf{X}_\ell$ are known matrices with full column ranks p_ℓ and $\boldsymbol{\beta}_\ell \in \mathcal{R}^{p_\ell}$.

Result 13.4.7. Consider the model setup in Result 13.4.3. Consider a GIG$(\boldsymbol{\mu}_\ell, \boldsymbol{\Omega}_\ell, a, b)$ prior for $(\boldsymbol{\beta}_\ell, \sigma^2)$, $\ell = 0, 1$; let $\nu = 2a$ and $\boldsymbol{\Sigma}_\ell = \mathbf{I} + \mathbf{X}_\ell \boldsymbol{\Omega}_\ell \mathbf{X}_\ell'$. Then B_{01} is given by

$$B_{01} = \left(\frac{\nu\pi b}{a}\right)^{\frac{p_1 - p_0}{2}} \frac{\Gamma((\nu + p_0)/2)}{\Gamma((\nu + p_1)/2)} \frac{|\boldsymbol{\Sigma}_1|^{1/2}}{|\boldsymbol{\Sigma}_0|^{1/2}}$$
$$\times \frac{[1 + \frac{a}{b\nu}(\mathbf{y} - \mathbf{X}_0\boldsymbol{\mu}_0)'\boldsymbol{\Sigma}_0^{-1}(\mathbf{y} - \mathbf{X}_0\boldsymbol{\mu}_0)]^{-\frac{\nu + p_0}{2}}}{[1 + \frac{a}{b\nu}(\mathbf{y} - \mathbf{X}_1\boldsymbol{\mu}_1)'\boldsymbol{\Sigma}_1^{-1}(\mathbf{y} - \mathbf{X}_1\boldsymbol{\mu}_1)]^{-\frac{\nu + p_1}{2}}}. \qquad (13.4.17)$$

Proof. The proof follows directly from the definition of B_{01} (see item 14 of Appendix C) as

$$B_{01} = p(\mathbf{y} \,|\, M_0)/p(\mathbf{y} \,|\, M_1),$$

and Exercise 13.12, from which $p(\mathbf{y} \,|\, M_\ell)$ is the p.d.f. of a multivariate t-distribution with d.f. $\nu = 2a$, location $\mathbf{X}_\ell \boldsymbol{\mu}_\ell$, and scale matrix $(b/a)\boldsymbol{\Sigma}_\ell$. ■

We assign to Exercise 13.19 the derivation of B_{01} for the normal linear model under the prior structure for $(\boldsymbol{\beta}_\ell, \sigma^2)$, $\ell = 0, 1$ as shown in Result 13.4.4.

Deviance information criterion

We describe the Deviance Information Criterion (DIC) which was developed from a Bayesian perspective by Spiegelhalter et al. (2002) to motivate a complexity measure for the effective number of parameters in a model indicating a Bayesian measure of adequacy of a fitted model. The optimal model among a set of competing models is then chosen by selecting one that minimizes the value of DIC. Let $L(\boldsymbol{\beta}; \mathbf{y})$ denote the data likelihood under a statistical model parametrized by $\boldsymbol{\beta} \in \mathcal{R}^p$. Let $\pi(\boldsymbol{\beta} \,|\, \mathbf{y})$ denote the posterior distribution of $\boldsymbol{\beta}$ given the observed data. Let $\overline{\boldsymbol{\beta}}$ be the posterior mean of $\boldsymbol{\beta}$ given $\mathbf{y}$.

The *effective number of parameters* with respect to a model is defined as

$$p_D = -2\,\mathrm{E}[\log L(\boldsymbol{\beta}; \mathbf{y})] + 2 \log L(\mathrm{E}\,\boldsymbol{\beta}), \tag{13.4.18}$$

where the expectation is with respect to $\pi(\boldsymbol{\beta} \,|\, \mathbf{y})$. The *Bayesian deviance* $D(\boldsymbol{\beta})$ is defined as

$$D(\boldsymbol{\beta}) = -2 \log L(\boldsymbol{\beta}; \mathbf{y}) + 2 \log m(\mathbf{y}) \tag{13.4.19}$$

where $m(\mathbf{y})$ is a fully specified standardizing term that is a function of the data $\mathbf{y}$ alone. Let $\overline{D(\boldsymbol{\beta})}$ be the posterior mean of $D(\boldsymbol{\beta})$ given $\mathbf{y}$. DIC is based on the principle of *goodness of fit* plus *complexity.*

Definition 13.4.1. Deviance Information Criterion. DIC is defined as

$$DIC = \overline{D(\boldsymbol{\beta})} + p_D. \tag{13.4.20}$$

The following result expresses DIC in a form that is similar to the information criteria for model selection that we have seen in Chapter 8.

Result 13.4.8. For the normal linear model, DIC has the form

$$DIC = -2 \log L(\overline{\boldsymbol{\beta}}; \mathbf{y}) + 2p_D. \tag{13.4.21}$$

Proof. From (13.4.18) and (13.4.19), we can see that

$$p_D = \overline{D(\boldsymbol{\beta})} - D(\overline{\boldsymbol{\beta}}), \tag{13.4.22}$$

and the result follows directly. ■

While the DIC has become a widely used criterion for model selection, one criticism that p_D and hence DIC is not invariant to reparameterization, so that we may get different values when we parametrize by σ instead of σ^2, say, even under mathematically equivalent prior setups. Another criticism is that it has weak theoretical justification. We refer the reader to Spiegelhalter et al. (2014) and references therein for details.

L measure

We next define the L measure, which is another Bayesian model selection criterion, and was defined by Ibrahim et al. (2001) as follows. Let $\mathbf{y} \in \mathcal{R}^N \sim f(\mathbf{y} \,|\, \boldsymbol{\theta})$ be the observed data, while $\mathbf{y}^* \in \mathcal{R}^N \sim f(\mathbf{y} \,|\, \boldsymbol{\theta})$ denotes an unobserved vector that comes from the same distribution and is independent of $\mathbf{y}$. Let $\mathbf{b} \in \mathcal{R}^N$ be an arbitrary location vector. Let $\pi(\boldsymbol{\theta} \,|\, \mathbf{y})$ denote the posterior distribution of $\boldsymbol{\theta}$ given $\mathbf{y}$ and $f(\mathbf{y}^* \,|\, \mathbf{y})$ be the posterior predictive distribution of $\mathbf{y}^*$ given $\mathbf{y}$. Define

$$L_1(\mathbf{y}, \mathbf{b}, k) = \mathrm{E}[(\mathbf{y}^* - \mathbf{b})'(\mathbf{y}^* - \mathbf{b})] + k(\mathbf{y} - \mathbf{b})'(\mathbf{y} - \mathbf{b}), \tag{13.4.23}$$

where the expectation in (13.4.23) is taken with respect to $f(\mathbf{y}^* \mid \mathbf{y})$ and $k \geq 0$ is a scalar which weights the discrepancy based on the unobserved "future" values $\mathbf{y}^*$ relative to the observed data $\mathbf{y}$. Let $\boldsymbol{\mu}^* = \mathrm{E}(\mathbf{y}^* \mid \mathbf{y})$, and $\nu = k/(k+1)$. Note that $\nu = 0$ if $k = 0$ and $\nu \to 1$ as $k \to \infty$. It can be shown that

$$\arg\min_{\mathbf{b}} L_1(\mathbf{y}, \mathbf{b}, k) = (1-\nu)\boldsymbol{\mu}^* + \nu\mathbf{y},$$

and (13.4.23) can be simplified and expressed as

$$L_2(\mathbf{y}, \nu) = \sum_{i=1}^{N} \mathrm{Var}(Y_i^* \mid \mathbf{y}) + \nu(\boldsymbol{\mu}^* - \mathbf{y})'(\boldsymbol{\mu}^* - \mathbf{y}), \quad 0 \leq \nu \leq 1. \tag{13.4.24}$$

Ibrahim et al. (2001) showed that the L measure $L_2(\mathbf{y}, \nu)$ can be obtained in closed form for the full-rank normal LM. Similar to Result 13.4.3, they assumed that $\pi(\boldsymbol{\beta} \mid \sigma^2)$ is $N(\boldsymbol{\mu}, \sigma^2\boldsymbol{\Omega})$, where $\boldsymbol{\mu} \in \mathcal{R}^p$ and $\boldsymbol{\Omega} \in \mathcal{R}^{p\times p}$ are specified hyperparameters, and for simplicity, σ^2 is assumed known. Let $\widehat{\boldsymbol{\beta}}$ be the LS estimate of $\boldsymbol{\beta}$ and $\mathbf{P}$ be the projection matrix onto $\mathcal{C}(\mathbf{X})$.

Result 13.4.9. For the full-rank normal linear model, the L measure is given by

$$\begin{aligned} L_2(\mathbf{y}, \nu) &= N\sigma^2 + \sigma^2 \mathrm{tr}(\mathbf{I}_p - \boldsymbol{\Lambda}) + \nu(\mathbf{B}\mathbf{y} - \mathbf{X}\boldsymbol{\Lambda}\boldsymbol{\mu})'(\mathbf{B}\mathbf{y} - \mathbf{X}\boldsymbol{\Lambda}\boldsymbol{\mu}), \text{ where} \\ \boldsymbol{\Lambda} &= (\boldsymbol{\Omega}\mathbf{X}'\mathbf{X} + \mathbf{I}_p)^{-1} \text{ and} \\ \mathbf{B} &= \mathbf{I}_N - \mathbf{P} + \mathbf{X}\boldsymbol{\Lambda}(\mathbf{X}'\mathbf{X})^{-1}\mathbf{X}'. \end{aligned} \tag{13.4.25}$$

Proof. Note that (see Exercise 13.20)

$$\begin{aligned} \boldsymbol{\beta} \mid \mathbf{y} &\sim N_p(\boldsymbol{\Lambda}\boldsymbol{\mu} + (\mathbf{I}_p - \boldsymbol{\Lambda})\widehat{\boldsymbol{\beta}}, \sigma^2(\mathbf{X}'\mathbf{X} + \boldsymbol{\Omega}^{-1})^{-1}) \text{ and} \\ \mathbf{y}^* \mid \mathbf{y} &\sim N(\mathbf{X}(\boldsymbol{\Lambda}\boldsymbol{\mu} + (\mathbf{I}_p - \boldsymbol{\Lambda})\widehat{\boldsymbol{\beta}}), \sigma^2(\mathbf{I}_N + \mathbf{X}(\boldsymbol{\Omega}^{-1} + \mathbf{X}'\mathbf{X})^{-1}\mathbf{X}')). \end{aligned} \tag{13.4.26}$$

The result in (13.4.25) follows from substituting these in (13.4.24) and simplifying. ■

Ibrahim et al. (2001) showed that for the linear model, certain values of ν yield highly desirable properties of the L measure and the calibration distribution relative to other values, and that $\nu = 1/2$ is a desirable and justifiable choice for model selection.

13.5 Dynamic linear models

The general linear model in (4.1.1) describes the relationship between independent responses on N subjects and a set of covariates. In contrast to such cross-sectional data, we frequently encounter situations where the responses and covariates are observed sequentially over time. It is of interest to develop inference for such problems in the context of a dynamic linear model.

Let $\mathbf{y}_1, \cdots, \mathbf{y}_T$ denote p-dimensional random variables which are available at times $1, \cdots, T$. Suppose $\mathbf{y}_t$ depends on an unknown q-dimensional *state* vector $\boldsymbol{\theta}_t$ (which may again be scalar or vector-valued) via the *observation equation*

$$\mathbf{y}_t = \mathbf{F}_t\boldsymbol{\theta}_t + \mathbf{v}_t \tag{13.5.1}$$

where $\mathbf{F}_t$ is a known $p \times q$ matrix, and we assume that the observation error $\mathbf{v}_t \sim N(\mathbf{0}, \mathbf{V}_t)$, with known $\mathbf{V}_t$. The dynamic change in $\boldsymbol{\theta}_t$ is represented by the *state equation*

$$\boldsymbol{\theta}_t = \mathbf{G}_t\boldsymbol{\theta}_{t-1} + \mathbf{w}_t \tag{13.5.2}$$

where $\mathbf{G}_t$ is a known $q \times q$ state transition matrix, and the state error $\mathbf{w}_t \sim N(\mathbf{0}, \mathbf{W}_t)$, with known $\mathbf{W}_t$. In addition, we suppose that $\mathbf{v}_t$ and $\mathbf{w}_t$ are independently distributed. Note that $\boldsymbol{\theta}_t$ is a random vector; let

$$\boldsymbol{\theta}_t \,|\, \mathbf{y}_t \sim N(\widehat{\boldsymbol{\theta}}_t, \boldsymbol{\Sigma}_t) \tag{13.5.3}$$

represent the posterior distribution of $\boldsymbol{\theta}_t$. The next two subsections describe the well-known Kalman filter recursions and Kalman smoother recursions that enable estimation of the state vector.

13.5.1 Kalman filter equations

The *Kalman filter* is a recursive procedure for determining the posterior distribution of $\boldsymbol{\theta}_t$, and thereby predicting $\mathbf{y}_t$. Let $\mathbf{Y}_{[t]} = (\mathbf{y}_t', \cdots, \mathbf{y}_1')'$ denote all the observations up to time t and let $\widehat{\boldsymbol{\theta}}_0$ and $\boldsymbol{\Sigma}_0$ denote the initial guess about the mean and variance of the distribution of $\boldsymbol{\theta}$. We assume that $(\boldsymbol{\theta}_{t-1} \,|\, \mathbf{Y}_{[t-1]}) \sim N(\widehat{\boldsymbol{\theta}}_{t-1}, \boldsymbol{\Sigma}_{t-1})$. The filter employs Bayes' theorem which describes the state of knowledge at time t about $\boldsymbol{\theta}_t$:

$$\mathrm{P}(\boldsymbol{\theta}_t \,|\, \mathbf{Y}_{[t]}) = \frac{\mathrm{P}(\mathbf{y}_t \,|\, \boldsymbol{\theta}_t, \mathbf{Y}_{[t-1]})\,\mathrm{P}(\boldsymbol{\theta}_t \,|\, \mathbf{Y}_{[t-1]})}{\int_{\boldsymbol{\theta}_t} \mathrm{P}(\mathbf{y}_t, \boldsymbol{\theta}_t \,|\, \mathbf{Y}_{[t-1]})d\boldsymbol{\theta}_t} \tag{13.5.4}$$

(Meinhold and Singpurwalla, 1983). Equation (13.5.4) represents the posterior distribution for $\boldsymbol{\theta}$ at time t as the product of the likelihood (first term on the right side) and the prior distribution for $\boldsymbol{\theta}$ (second term on the right). We show how this posterior distribution is derived using the notions of Bayes' Theorem and the results on multivariate normality from Section 5.2.

Result 13.5.1. At time t, the prior distribution of $\boldsymbol{\theta}$ is

$$\boldsymbol{\theta}_t \,|\, \mathbf{Y}_{[t-1]} \sim N(\mathbf{G}_t\widehat{\boldsymbol{\theta}}_{t-1}, \mathbf{R}_t), \tag{13.5.5}$$

where

$$\mathbf{R}_t = \mathbf{G}_t\boldsymbol{\Sigma}_{t-1}\mathbf{G}_t' + \mathbf{W}_t. \tag{13.5.6}$$

Proof. Prior to observing $\mathbf{y}_t$, the best guess for $\boldsymbol{\theta}_t$ is based on (13.5.2) and is $\boldsymbol{\theta}_t = \mathbf{G}_t\boldsymbol{\theta}_{t-1} + \mathbf{w}_t$. Since $(\boldsymbol{\theta}_{t-1} \,|\, \mathbf{Y}_{[t-1]}) \sim N(\widehat{\boldsymbol{\theta}}_{t-1}, \boldsymbol{\Sigma}_{t-1})$, we see using Result 5.2.5 that $(\boldsymbol{\theta}_t \,|\, \mathbf{Y}_{[t-1]})$ is normal with mean $\mathbf{G}_t\widehat{\boldsymbol{\theta}}_{t-1}$ and covariance $\mathbf{R}_t = \mathbf{G}_t\boldsymbol{\Sigma}_{t-1}\mathbf{G}_t' + \mathbf{W}_t$. ■

Result 13.5.2. The posterior distribution of $\boldsymbol{\theta}$ at time t after observing $\mathbf{y}_t$ is normal with mean

$$\widehat{\boldsymbol{\theta}}_t = \mathbf{G}_t\widehat{\boldsymbol{\theta}}_{t-1} + \mathbf{R}_t\mathbf{F}_t'(\mathbf{V}_t + \mathbf{F}_t\mathbf{R}_t\mathbf{F}_t')^{-1}\varepsilon_t \tag{13.5.7}$$

and covariance matrix

$$\boldsymbol{\Sigma}_t = \mathbf{R}_t - \mathbf{R}_t\mathbf{F}_t'(\mathbf{V}_t + \mathbf{F}_t\mathbf{R}_t\mathbf{F}_t')^{-1}\mathbf{F}_t\mathbf{R}_t \tag{13.5.8}$$

where

$$\varepsilon_t = \mathbf{y}_t - \widehat{\mathbf{y}}_t = \mathbf{y}_t - \mathbf{F}_t\mathbf{G}_t\widehat{\boldsymbol{\theta}}_{t-1}. \tag{13.5.9}$$

Proof. Let $\widehat{\mathbf{y}}_t$ denote the prediction of $\mathbf{y}_t$ based on $\mathbf{Y}_{[t-1]}$, i.e.,

$$\widehat{\mathbf{y}}_t = \mathrm{E}(\mathbf{y}_t \mid \mathbf{Y}_{[t-1]}) = \mathrm{E}[(\mathbf{F}_t\boldsymbol{\theta}_t + \mathbf{v}_t) \mid \mathbf{Y}_{[t-1]}] = \mathbf{F}_t\mathbf{G}_t\widehat{\boldsymbol{\theta}}_{t-1}. \tag{13.5.10}$$

We denote the prediction error by $\boldsymbol{\varepsilon}_t = \mathbf{y}_t - \widehat{\mathbf{y}}_t$, which is equal to $\mathbf{y}_t - \mathbf{F}_t\mathbf{G}_t\widehat{\boldsymbol{\theta}}_{t-1}$. We can write (13.5.4) as

$$\begin{aligned}\mathrm{P}(\boldsymbol{\theta}_t \mid \mathbf{y}_t, \mathbf{Y}_{[t-1]}) &= \mathrm{P}(\boldsymbol{\theta}_t \mid \boldsymbol{\varepsilon}_t, \mathbf{Y}_{[t-1]}) \\ &\propto \mathrm{P}(\boldsymbol{\varepsilon}_t \mid \boldsymbol{\theta}_t, \mathbf{Y}_{[t-1]})\,\mathrm{P}(\boldsymbol{\theta}_t \mid \mathbf{Y}_{[t-1]}),\end{aligned}$$

where $\mathrm{P}(\boldsymbol{\varepsilon}_t \mid \boldsymbol{\theta}_t, \mathbf{Y}_{[t-1]})$ is equivalent to the likelihood function $L(\boldsymbol{\theta}_t \mid \mathbf{y}_t)$ by virtue of the fact that observing $\mathbf{y}_t$ is the same as observing $\boldsymbol{\varepsilon}_t$, when $\mathbf{F}_t$, $\mathbf{G}_t$, and $\widehat{\boldsymbol{\theta}}_{t-1}$ are known. Using (13.5.1) we can write (13.5.9) as $\boldsymbol{\varepsilon}_t = \mathbf{F}_t(\boldsymbol{\theta}_t - \mathbf{G}_t\boldsymbol{\theta}_{t-1}) + \mathbf{v}_t$, and it follows that

$$\mathrm{E}(\boldsymbol{\varepsilon}_t \mid \boldsymbol{\theta}_t, \mathbf{Y}_{[t-1]}) = \mathbf{F}_t(\boldsymbol{\theta}_t - \mathbf{G}_t\widehat{\boldsymbol{\theta}}_{t-1}), \quad \text{and} \quad \mathrm{Var}(\boldsymbol{\varepsilon}_t \mid \boldsymbol{\theta}_t, \mathbf{Y}_{[t-1]}) = \mathbf{V}_t.$$

Again, it follows from Result 5.2.5 that

$$(\boldsymbol{\varepsilon}_t \mid \boldsymbol{\theta}_t, \mathbf{Y}_{[t-1]}) \sim N(\mathbf{F}_t(\boldsymbol{\theta}_t - \mathbf{G}_t\widehat{\boldsymbol{\theta}}_{t-1}), \mathbf{V}_t). \tag{13.5.11}$$

An application of (13.5.4) and the equivalence of probabilistic information in $\mathbf{Y}_t$ and $\boldsymbol{\varepsilon}_t$ implies

$$\mathrm{P}(\boldsymbol{\theta}_t \mid \mathbf{Y}_{[t]}) = \mathrm{P}(\boldsymbol{\varepsilon}_t \mid \boldsymbol{\theta}_t, \mathbf{Y}_{[t-1]})\,\mathrm{P}(\boldsymbol{\theta}_t \mid \mathbf{Y}_{[t-1]}) \Big/ \int_{\text{all } \boldsymbol{\theta}_t} \mathrm{P}(\boldsymbol{\varepsilon}_t, \boldsymbol{\theta}_t \mid \mathbf{Y}_{[t-1]}) d\boldsymbol{\theta}_t\,.$$

It is possible to evaluate $(\boldsymbol{\theta}_t \mid \mathbf{Y}_{[t]})$ by using properties of the multivariate normal distribution in Section 5.2. Use Result 5.2.12, and let $\mathbf{x}_1$ correspond to $\boldsymbol{\varepsilon}_t \mid \mathbf{Y}_{[t-1]}$, and $\mathbf{x}_2$ correspond to $\boldsymbol{\theta}_t \mid \mathbf{Y}_{[t-1]}$. From (13.5.5), $\boldsymbol{\mu}_2$ corresponds to $\mathbf{G}_t\widehat{\boldsymbol{\theta}}_{t-1}$, while $\boldsymbol{\Sigma}_{22}$ corresponds to $\mathbf{R}_t$. From (13.5.11), it follows that $\boldsymbol{\mu}_{1.2} = \boldsymbol{\mu}_1 + \boldsymbol{\Sigma}_{12}\mathbf{R}_t^{-1}(\boldsymbol{\theta}_t - \mathbf{G}_t\widehat{\boldsymbol{\theta}}_{t-1})$ corresponds to $\mathbf{F}_t(\boldsymbol{\theta}_t - \mathbf{G}_t\widehat{\boldsymbol{\theta}}_{t-1})$, so that $\boldsymbol{\mu}_1$ corresponds to $\mathbf{0}$, and $\boldsymbol{\Sigma}_{12}$ to $\mathbf{F}_t\mathbf{R}_t$. Likewise, $\boldsymbol{\Sigma}_{11.2} = \boldsymbol{\Sigma}_{11} - \boldsymbol{\Sigma}_{12}\boldsymbol{\Sigma}_{22}^{-1}\boldsymbol{\Sigma}_{21} = \boldsymbol{\Sigma}_{11} - \mathbf{F}_t\mathbf{R}_t\mathbf{F}_t'$ corresponds to $\mathbf{V}_t$, from which we see that $\boldsymbol{\Sigma}_{11}$ corresponds to $\mathbf{V}_t + \mathbf{F}_t\mathbf{R}_t\mathbf{F}_t'$. Using the converse relationship discussed following Result 5.2.17, we see that

$$\left[\begin{pmatrix}\boldsymbol{\theta}_t \\ \boldsymbol{\varepsilon}_t\end{pmatrix} \middle| \mathbf{Y}_{[t-1]}\right] \sim N\left[\begin{pmatrix}\mathbf{G}_t\widehat{\boldsymbol{\theta}}_{t-1} \\ \mathbf{0}\end{pmatrix}, \begin{pmatrix}\mathbf{R}_t & \mathbf{R}_t\mathbf{F}_t' \\ \mathbf{F}_t\mathbf{R}_t & \mathbf{V}_t + \mathbf{F}_t\mathbf{R}_t\mathbf{F}_t'\end{pmatrix}\right].$$

Now, let the conditioning variable $\mathbf{x}_2$ correspond to $\boldsymbol{\varepsilon}_t \mid \mathbf{Y}_{[t-1]}$, with corresponding mean and covariance equal to $\mathbf{0}$ and $\mathbf{V}_t + \mathbf{F}_t\mathbf{R}_t\mathbf{F}_t'$, we see from Result 5.2.17 that $(\boldsymbol{\theta}_t \mid \boldsymbol{\varepsilon}_t, \mathbf{Y}_{[t-1]})$ has a normal distribution with mean $\widehat{\boldsymbol{\theta}}_t = \mathbf{G}_t\widehat{\boldsymbol{\theta}}_{t-1} + \mathbf{R}_t\mathbf{F}_t'(\mathbf{V}_t + \mathbf{F}_t\mathbf{R}_t\mathbf{F}_t')^{-1}\boldsymbol{\varepsilon}_t$ and covariance $\boldsymbol{\Sigma}_t = \mathbf{R}_t - \mathbf{R}_t\mathbf{F}_t'(\mathbf{V}_t + \mathbf{F}_t\mathbf{R}_t\mathbf{F}_t')^{-1}\mathbf{F}_t\mathbf{R}_t$, which are the expressions for the posterior mean and covariance of $\boldsymbol{\theta}_t \mid \mathbf{Y}_{[t]}$ given in (13.5.7) and (13.5.8). ■

Example 13.5.1. Suppose an observed univariate quarterly time series $\{Y_t\}$ is expressed as

$$Y_t = T_t + S_t + v_t,$$

where T_t denotes trend, S_t denotes the seasonal component, and v_t, the error in this structural model (Shumway and Stoffer, 2017). Suppose we set

$$T_t = \phi T_{t-1} + w_{t1}, \quad S_t + S_{t-1} + S_{t-2} + S_{t-3} = w_{t2},$$

where $\phi > 1$, to characterize an exponentially increasing trend, and a seasonal component that is expected to sum to zero over 4 quarters. To write this model in the form (13.5.1) and (13.5.2), set the state vector to be $\boldsymbol{\theta}_t = (T_t, S_t, S_{t-1}, S_{t-2})'$. The observation and state equations are

$$Y_t = (1 \quad 1 \quad 0 \quad 0)\begin{pmatrix} T_t \\ S_t \\ S_{t-1} \\ S_{t-2} \end{pmatrix} + v_t,$$

$$\begin{pmatrix} T_t \\ S_t \\ S_{t-1} \\ S_{t-2} \end{pmatrix} = \begin{pmatrix} \phi & 0 & 0 & 0 \\ 0 & -1 & -1 & -1 \\ 0 & 1 & 0 & 0 \\ 0 & 0 & 1 & 0 \end{pmatrix}\begin{pmatrix} T_{t-1} \\ S_{t-1} \\ S_{t-2} \\ S_{t-3} \end{pmatrix} + \begin{pmatrix} w_{t1} \\ w_{t2} \\ 0 \\ 0 \end{pmatrix},$$

where the observation error variance is V_{11}, and the state error variance is $\text{diag}(W_{11}, W_{22}, 0, 0)$. The filter equations follow directly from Result 13.5.1. □

For convenience, we adopt the following notation (Shumway and Stoffer, 2017). Let $\boldsymbol{\theta}_t^s = \text{E}(\boldsymbol{\theta}_t \,|\, \mathbf{Y}_{[s]})$, $\boldsymbol{\Sigma}_{t_1,t_2}^s = \text{E}[(\boldsymbol{\theta}_{t_1} - \boldsymbol{\theta}_{t_1}^s)(\boldsymbol{\theta}_{t_2} - \boldsymbol{\theta}_{t_2}^s)']$, and condense $\boldsymbol{\Sigma}_{t,t}^s$ to $\boldsymbol{\Sigma}_t^s$. Kalman filter estimates are obtained when $s < t$. The derivations in the next Section employ this notation.

13.5.2 Kalman smoothing equations

Estimators of the state vector $\boldsymbol{\theta}_t$ based on the entire data $\mathbf{y}_1, \cdots, \mathbf{y}_T$, where $t \le T$ are called *smoothers* and are denoted by $\boldsymbol{\theta}_t^T$, $t = 1, \cdots, T$. The proof of the next result is along the lines of Rauch et al. (1965).

Result 13.5.3. With initial conditions $\boldsymbol{\theta}_T^T$ and $\boldsymbol{\Sigma}_T^T$ obtained from the filter recursions,

$$\boldsymbol{\theta}_{t-1}^T = \boldsymbol{\theta}_{t-1}^{t-1} + J_{t-1}(\boldsymbol{\theta}_t^T - \boldsymbol{\theta}_t^{t-1}), \text{ and} \tag{13.5.12}$$

$$\boldsymbol{\Sigma}_{t-1}^T = \boldsymbol{\Sigma}_{t-1}^{t-1} + \mathbf{J}_{t-1}(\boldsymbol{\Sigma}_t^T - \boldsymbol{\Sigma}_t^{t-1})\mathbf{J}_{t-1}', \tag{13.5.13}$$

where $\mathbf{J}_{t-1} = \boldsymbol{\Sigma}_{t-1}^{t-1}\mathbf{G}_t'(\boldsymbol{\Sigma}_t^{t-1})^{-1}$.

Proof. Now,

$$\begin{aligned} &\text{P}(\boldsymbol{\theta}_{t-1}, \boldsymbol{\theta}_t \,|\, \mathbf{Y}_{[T]}) \propto \text{P}(\boldsymbol{\theta}_{t-1}, \boldsymbol{\theta}_t, \mathbf{Y}_{[T]}) \\ &\quad = \text{P}(\boldsymbol{\theta}_{t-1}, \boldsymbol{\theta}_t, \mathbf{Y}_{[t-1]}, \mathbf{y}_t, \cdots, \mathbf{y}_T) \\ &\quad = \text{P}(\mathbf{Y}_{[t-1]})\,\text{P}(\boldsymbol{\theta}_{t-1}, \boldsymbol{\theta}_t \,|\, \mathbf{Y}_{[t-1]})\,\text{P}(\mathbf{y}_t, \cdots, \mathbf{y}_T \,|\, \boldsymbol{\theta}_{t-1}, \boldsymbol{\theta}_t, \mathbf{Y}_{[t-1]}), \end{aligned}$$

which can be simplified to

$$\text{P}(\boldsymbol{\theta}_{t-1}, \boldsymbol{\theta}_t \,|\, \mathbf{Y}_{[t-1]}) = \delta_1(\boldsymbol{\theta}_t)\,\text{P}(\boldsymbol{\theta}_{t-1} \,|\, \mathbf{Y}_{[t-1]})\,\text{P}(\boldsymbol{\theta}_t \,|\, \boldsymbol{\theta}_{t-1}),$$

where $\delta_1(\boldsymbol{\theta}_t)$ does not depend on $\boldsymbol{\theta}_{t-1}$. Clearly, $\boldsymbol{\theta}_{t-1} \,|\, \mathbf{Y}_{[t-1]} \sim N(\boldsymbol{\theta}_{t-1}^{t-1}, \boldsymbol{\Sigma}_{t-1}^{t-1})$ and $\boldsymbol{\theta}_{t-1} \,|\, \boldsymbol{\theta}_{t-1} \sim N(\mathbf{G}_t\boldsymbol{\theta}_{t-1}, \mathbf{W}_t)$. The smoothers $\boldsymbol{\theta}_t^T$ and $\boldsymbol{\theta}_{t-1}^T$ are obtained by minimizing

$$\begin{aligned} &-2\log \text{P}(\boldsymbol{\theta}_{t-1}, \boldsymbol{\theta}_t \,|\, \mathbf{Y}_{[t-1]}) \\ &\propto (\boldsymbol{\theta}_{t-1} - \boldsymbol{\theta}_{t-1}^{t-1})\{\boldsymbol{\Sigma}_{t-1}^{t-1}\}^{-1}(\boldsymbol{\theta}_{t-1} - \boldsymbol{\theta}_{t-1}^{t-1})' \\ &\quad + (\boldsymbol{\theta}_t - \mathbf{G}_t\boldsymbol{\theta}_{t-1})\mathbf{W}_t^{-1}(\boldsymbol{\theta}_t - \mathbf{G}_t\boldsymbol{\theta}_{t-1})' + \delta_2(\boldsymbol{\theta}_t), \end{aligned} \tag{13.5.14}$$

where $\delta_2(\boldsymbol{\theta}_t)$ is independent of $\boldsymbol{\theta}_{t-1}$. Substitute the available $\boldsymbol{\theta}_t^T$ for $\boldsymbol{\theta}_t$ in (13.5.14), and minimize the resulting expression with respect to $\boldsymbol{\theta}_{t-1}$ to obtain

$$\boldsymbol{\theta}_{t-1}^T = [\{\boldsymbol{\Sigma}_{t-1}^{t-1}\}^{-1} + \mathbf{G}_t'\mathbf{W}_t^{-1}\mathbf{G}_t]^{-1}[\{\boldsymbol{\Sigma}_{t-1}^{t-1}\}^{-1}\boldsymbol{\theta}_{t-1}^{t-1} + \mathbf{G}_t'\mathbf{W}_t^{-1}\boldsymbol{\theta}_t^T],$$

which yields (13.5.12), on using Exercise 2.30 with $\mathbf{A} = \boldsymbol{\Sigma}_{t-1}^{t-1}$, $\mathbf{B} = \mathbf{W}_t$, and $\mathbf{C} = \mathbf{G}_t$. To derive (13.5.13), we see that from (13.5.12),

$$(\boldsymbol{\theta}_{t-1} - \boldsymbol{\theta}_{t-1}^T) + \mathbf{J}_{t-1}\boldsymbol{\theta}_t^T = (\boldsymbol{\theta}_{t-1} - \boldsymbol{\theta}_{t-1}^{t-1}) + \mathbf{J}_{t-1}\mathbf{G}_t\boldsymbol{\theta}_{t-1}^{t-1},$$

and

$$\boldsymbol{\Sigma}_{t-1}^T + \mathbf{J}_{t-1}\,\mathrm{E}(\boldsymbol{\theta}_t^T\boldsymbol{\theta}_t^{T\prime})\mathbf{J}_{t-1}' = \boldsymbol{\Sigma}_{t-1}^{t-1} + \mathbf{J}_{t-1}\mathbf{G}_t\,\mathrm{E}(\boldsymbol{\theta}_{t-1}^{t-1}\boldsymbol{\theta}_{t-1}^{t-1\prime})\mathbf{G}_t'\mathbf{J}_{t-1}'.$$

Since

$$\mathrm{E}(\boldsymbol{\theta}_t^T\boldsymbol{\theta}_t^{T\prime}) = \mathrm{E}(\boldsymbol{\theta}_t\boldsymbol{\theta}_t') - \boldsymbol{\Sigma}_t^T = \mathbf{G}_t\,\mathrm{E}(\boldsymbol{\theta}_{t-1}\boldsymbol{\theta}_{t-1}')\mathbf{G}_t' + \mathbf{W}_t - \boldsymbol{\Sigma}_t^T,$$

and

$$\mathrm{E}(\boldsymbol{\theta}_{t-1}^{t-1}\boldsymbol{\theta}_{t-1}^{t-1\prime}) = \mathrm{E}(\boldsymbol{\theta}_{t-1}\boldsymbol{\theta}_{t-1}') - \boldsymbol{\Sigma}_{t-1}^{t-1},$$

(13.5.13) follows. ■

In practice, the parameters in the model specification, such as elements of $\mathbf{G}_t$, $\mathbf{V}_t$, and $\mathbf{W}_t$ may be unknown, and must be estimated. We refer the reader to Shumway and Stoffer (2017) for details on parameter estimation via maximizing the innovations form of the likelihood or via a variant of the EM algorithm which we described in Section 14.4. Diagnostics for the dynamic linear model is described in Harrison and West (1991), while West and Harrison (1997) is a good reference for the study of non-normal dynamic linear models.

Numerical Example 13.5. The dataset "jj" in the R package *astsa* shows the quarterly earnings per share of Johnson & Johnson for $T = 84$ quarters (21 years) from the first quarter of 1960 to the last quarter of 1980 and is discussed in Shumway and Stoffer (2017), whose code template we follow here.

```
library(astsa); data(jj)
# set up as a DLM
num = length(jj); A = cbind(1,1,0,0)
# Function to calculate the innovations likelihood
# kf: filter estimates of the state
Linn=function(para){
 Phi = diag(0,4); Phi[1,1] = para[1]
 Phi[2,]=c(0,-1,-1,-1); Phi[3,]=c(0,1,0,0); Phi[4,]=c(0,0,1,0)
 cQ1 = para[2]; cQ2 = para[3]      # sqrt q11 and sqrt q22
 cQ=diag(0,4); cQ[1,1]=cQ1; cQ[2,2]=cQ2
 cR = para[4]        # sqrt r11
 kf = Kfilter0(num,jj,A,mu0,Sigma0,Phi,cQ,cR)
 return(kf$like)
 }
# Initial parameters
mu0 = c(.7,0,0,0); Sigma0 = diag(.04,4)
init.par = c(1.03,.1,.1,.5)  # Phi[1,1], the 2 Qs and R
```

```
# Numerical MLEs
est = optim(init.par, Linn, NULL, method="BFGS", hessian=TRUE,
            control=list(trace=1,REPORT=1))
SE = sqrt(diag(solve(est$hessian)))
u = cbind(estimate=est$par,SE); rownames(u)=c("Phi11","sigw1",
          "sigw2","sigv"); u
```

The approximate MLEs of the hyperparameters together with their SEs are shown below.

```
          estimate         SE
Phi11 1.0350847657 0.00253645
sigw1 0.1397255477 0.02155155
sigw2 0.2208782663 0.02376430
sigv  0.0004655672 0.24174702
```

We can obtain the smoothed estimates of the state vector.

```
Phi = diag(0,4); Phi[1,1] = est$par[1]
Phi[2,]=c(0,-1,-1,-1); Phi[3,]=c(0,1,0,0); Phi[4,]=c(0,0,1,0)
cQ1 = est$par[2]; cQ2 = est$par[3]; cQ = diag(1,4); cQ[1,1]=cQ1;
cQ[2,2]=cQ2
cR = est$par[4]
ks = Ksmooth0(num,jj,A,mu0,Sigma0,Phi,cQ,cR)
```

We can plot the filter estimates and the smoothed estimates together with the observed time series.

```
Tsm = ts(ks$xs[1,,], start=1960, freq=4)
Ssm = ts(ks$xs[2,,], start=1960, freq=4)
p1 = 2*sqrt(ks$Ps[1,1,]); p2 = 2*sqrt(ks$Ps[2,2,])
par(mfrow=c(3,1))
plot(Tsm, main="Trend Component", ylab="Trend")
     lines(Tsm+p1, lty=2, col=4); lines(Tsm-p1,lty=2, col=4)
plot(Ssm, main="Seasonal Component", ylim=c(-5,4), ylab="Season")
lines(Ssm+p2,lty=2, col=4); lines(Ssm-p2,lty=2, col=4)
plot(jj, type="p", main="Data (points) and Trend+Season (line)")
lines(Tsm+Ssm)
```

We forecast 12 steps ahead (3 years) into the future and can plot them together with their upper and lower confident limits.

```
n.ahead=12;
y = ts(append(jj, rep(0,n.ahead)), start=1960, freq=4)
rmspe = rep(0,n.ahead); x00 = ks$xf[,,num]; P00 = ks$Pf[,,num]
Q=t(cQ)%*%cQ;  R=t(cR)%*%(cR)
for (m in 1:n.ahead){
 xp = Phi%*%x00; Pp = Phi%*%P00%*%t(Phi)+Q
 sig = A%*%Pp%*%t(A)+R; K = Pp%*%t(A)%*%(1/sig)
 x00 = xp; P00 = Pp-K%*%A%*%Pp
 y[num+m] = A%*%xp; rmspe[m] = sqrt(sig)  }
plot(y, type="o", main="", ylab="", ylim=c(5,30),
     xlim=c(1975,1984))
upp = ts(y[(num+1):(num+n.ahead)]+2*rmspe, start=1981, freq=4)
low = ts(y[(num+1):(num+n.ahead)]-2*rmspe, start=1981, freq=4)
lines(upp, lty=2);  lines(low, lty=2);  abline(v=1980.75, lty=3)
```

▲

Exercises

13.1. Prove properties 1–3 of Result 13.1.2.

13.2. Prove property 2 of Result 13.1.6.

13.3. Consider the balanced two-factor MANOVA model

$$\mathbf{y}_{\ell kr} = \boldsymbol{\mu} + \boldsymbol{\tau}_\ell + \boldsymbol{\beta}_k + \boldsymbol{\gamma}_{\ell k} + \boldsymbol{\varepsilon}_{\ell kr},$$
$$r = 1, \cdots, n, \ \ell = 1, \cdots, g, \ k = 1, \cdots, b,$$

where $\boldsymbol{\mu}$ is a p-dimensional overall mean vector, $\boldsymbol{\tau}_\ell$ is a p-dimensional effect due to level ℓ of Factor A, $\boldsymbol{\beta}_k$ is a p-dimensional effect due to level k of Factor B, and $\boldsymbol{\gamma}_{\ell k}$ is a p-dimensional effect due to the interaction between level ℓ of Factor A and level k of Factor B; these satisfy the constraints $\sum_{\ell=1}^{g} \boldsymbol{\tau}_\ell = \mathbf{0}$, $\sum_{k=1}^{b} \boldsymbol{\beta}_k = \mathbf{0}$, $\sum_{\ell=1}^{g} \boldsymbol{\gamma}_{\ell k} = \mathbf{0}$, and $\sum_{k=1}^{b} \boldsymbol{\gamma}_{\ell k} = \mathbf{0}$. Assume that $\boldsymbol{\varepsilon}_{\ell kr}$ are i.i.d. $\sim N_m(\mathbf{0}, \boldsymbol{\Sigma})$.

(a) Show details of the MANOVA decomposition.

(b) Derive a test for no interaction effect.

(c) Derive tests for the main effects due to Factor A and Factor B.

13.4. Consider the Result 13.2.1.

(a) What is the distribution of the MLE $\widehat{\boldsymbol{\mu}}_0$ of $\boldsymbol{\mu}_0$?

(b) What is the distribution of the MLE $\widehat{\boldsymbol{\Sigma}}_0$ of $\boldsymbol{\Sigma}_0$?

13.5. Prove Result 13.2.2.

13.6. Consider the "conditional-independence model", i.e., the two-stage random-effects model for longitudinal data in (13.2.8), with $\mathbf{R}_i = \sigma^2\mathbf{I}$. Obtain the maximum likelihood estimates of the parameters.

13.7. If a random vector $\mathbf{y}$ has p.d.f. $f(\mathbf{y}) = c_k h(\mathbf{y}'\mathbf{y})$, show that the p.d.f. of $\|\mathbf{y}\|$ is

$$f_{\|\mathbf{y}\|}(w) = \frac{2c_k \pi^{k/2}}{\Gamma(k/2)} w^{k-1} h(w^2) \tag{13.5.15}$$

13.8. Prove property 3 of Result 13.3.3.

13.9. Extend Result 13.3.3 to the case where $\mathbf{y} \sim E_N(\mathbf{X}\boldsymbol{\beta}, \sigma^2\mathbf{V}, h)$, $\mathbf{V}$ being a known symmetric, p.d. matrix.

13.10. Show that in the model (13.3.3), Cook's distance is invariant to $h(.)$ and has the same form given in Section 8.3.4.

13.11. Starting from the GIG prior in (13.4.2), show that $\pi(\boldsymbol{\beta})$ has a p-variate t-distribution with d.f. $\nu = 2a$, location $\boldsymbol{\mu}$, and scale matrix $\boldsymbol{\Sigma} = (b/a)\boldsymbol{\Omega}$ and identify the parameters of this distribution (see Example 5.5.3).

13.12. In the linear model with GIG prior considered in Result 13.4.3, derive expressions for $p(\mathbf{y} \,|\, \sigma^2)$ and the marginal $p(\mathbf{y})$.

13.13. Prove Result 13.4.4.

13.14. Prove Result 13.4.6.

13.15. Show that b^* which was shown in (13.4.3) can be expressed as

$$b^* = b + \frac{1}{2}(\mathbf{y} - \mathbf{X}\boldsymbol{\mu})'(\mathbf{I}_N + \mathbf{X}\boldsymbol{\Omega}\mathbf{X}')^{-1}(\mathbf{y} - \mathbf{X}\boldsymbol{\mu}).$$

Hint: Use the Sherman–Morrison–Woodbury formula.

13.16. Suppose $\mathbf{y}_j \sim N(\mu_j \mathbf{1}_{N_j}, \sigma_j^2 \mathbf{I}_{N_j})$, $j = 1, 2$. Also, suppose $\pi(\mu_j, \sigma_j^2) \propto 1/\sigma_j^2$, $j = 1, 2$ and $\pi(\mu_1, \sigma_1^2)$ and $\pi(\mu_2, \sigma_2^2)$ are independent. What is the posterior distribution of $\{S_1^2/S_2^2\}/\{\sigma_1^2/\sigma_2^2\}$?

13.17. Let $\mathbf{y} \sim N(\mu \mathbf{1}_N, \sigma^2 \mathbf{I}_N)$.

(a) Suppose (μ, σ^2) has a normal-inverse χ^2 prior. Show that the posterior distribution $\pi(\mu, \sigma^2 \mid \mathbf{y})$ also has the normal-inverse χ^2 form and derive its parameters.

(b) Suppose we use a noninformative prior, i.e., $\pi(\mu, \sigma^2) \propto 1/\sigma^2$. Derive the form of the posterior distribution $\pi(\mu, \sigma^2 \mid \mathbf{y})$. What is the form of the marginal posterior distribution of $\sqrt{N}(\mu - \overline{Y})/S$?

13.18. Let $F(.)$ denote the c.d.f. of $Y \mid \theta$, and suppose $\pi_j(\theta)$, $j = 1, \cdots, L$ is the conjugate prior p.d.f. for θ. Consider the class $\pi(\theta) = \sum_{j=1}^{L} w_j \pi_j(\theta)$ of finite mixture prior densities, where the weights satisfy $w_j \geq 0$, $j = 1, \cdots, L$ and $\sum_{j=1}^{L} w_j = 1$. Is this also a conjugate class?

13.19. Derive the form of the Bayes factor B_{01} for the normal linear model under the prior structure for $(\boldsymbol{\beta}_i, \sigma^2)$, $i = 1, 2$ discussed in Result 13.4.4.

13.20. Verify (13.4.26).

13.21. Write a univariate AR(1) model $Y_t = \rho Y_{t-1} + v_t$ as a dynamic linear model and derive the Kalman filter equations.

13.22. Let Z_t be an observed time series at time t, $t = 1, \cdots, T$, which measures a signal with error. Suppose the signal and error terms are additive, and the signal satisfies an AR(1) model. Assuming independent Gaussian errors for the observation and signal processes, derive the Kalman filter recursions.

14

Miscellaneous Topics

This last chapter gives short descriptions of a few miscellaneous topics that are closely aligned with linear model theory. In Section 14.1, we discuss ideas from robust regression analysis such as least absolute deviation (LAD) regression and M-regression. This is followed by a description of a few nonparametric regression procedures in Section 14.2. Section 14.3 looks at regularized regression approaches. Section 14.4 gives a brief look at missing data analysis and the expectation-maximization (EM) algorithm. The reader is encouraged to follow the references given in each Section to learn more on these topics.

14.1 Robust regression

Least squares estimates have several optimality properties within the class of normal linear models, which in addition to their computational simplicity, have made this procedure popular. However, the method of least squares can be extremely sensitive to (a) a departure of the error distribution from normality, and (b) the presence of outliers, even under normality. In general, the class of L_d-estimators are obtained by minimizing the L_d-norm $\{\sum_{i=1}^N |\varepsilon_i|^d\}^{1/d}$, $d \geq 1$. When $d = 2$, we have the Euclidean metric, or L_2-norm, which leads to the least squares estimator. The case $d = 1$ corresponds to the *absolute* metric or L_1-norm, and yields the L_1-estimator or *LAD* estimator, which we discuss below. Note that $d = \infty$ leads to the minimax method (Berger, 1980).

Many robust and resistant methods have been developed since the 1960's with the purpose of obtaining statistical procedures that are less sensitive to outliers or anomalous data values, and are reasonably efficient with data from the ideal Gaussian distribution or a range of alternative distributions. Robust methods are designed to have high efficiency in a neighborhood of an assumed model (Huber, 1964). In the following sections, we describe *LAD* or L_1-regression and M-regression (see Birkes and Dodge, 1993).

14.1.1 Least absolute deviations regression

Least absolute deviation (LAD) regression or L_1-regression is a natural generalization of the median to a regression problem and consists of finding the estimator $\widehat{\boldsymbol{\beta}}_{LAD}$ as any solution to the minimization problem

$$\min_{\boldsymbol{\beta}} \sum_{i=1}^{N} |Y_i - \mathbf{x}_i'\boldsymbol{\beta}|, \tag{14.1.1}$$

where each Y_i is an observed response, $\mathbf{x}_i = (1, X_{i1}, \cdots, X_{ik})'$ is the associated p-dimensional vector of known predictors with $p = k + 1$, and $\boldsymbol{\beta}$ is a p-dimensional vector of unknown parameters. Note that in this chapter we use $\widehat{\boldsymbol{\beta}}$ instead of $\boldsymbol{\beta}^0$ to denote any solution to an optimization problem treated as estimation, even if the solution is non-unique.

DOI: 10.1201/9781315156651-14

As usual, denote $\mathbf{y} = (Y_1, \cdots, Y_N)'$ and $\mathbf{X} = (\mathbf{x}_1', \cdots, \mathbf{x}_N')'$. For $\mathbf{z} = (z_1, \cdots, z_N)'$, let $\|\mathbf{z}\|_1 = \sum_{i=1}^N |z_i|$, known as the L_1-norm of $\mathbf{z}$. Then

$$\|\mathbf{y} - \mathbf{X}\widehat{\boldsymbol{\beta}}_{LAD}\|_1 = \min_{\boldsymbol{\beta}} \|\mathbf{y} - \mathbf{X}\boldsymbol{\beta}\|_1 = \min_{\widehat{\mathbf{y}} \in \mathcal{C}(\mathbf{X})} \|\mathbf{y} - \widehat{\mathbf{y}}\|_1. \tag{14.1.2}$$

LAD regression applies to the following statistical model

$$Y_i = \mathbf{x}_i'\boldsymbol{\beta} + \varepsilon_i, \quad i = 1, \cdots, N,$$

where ε_i are i.i.d. with mean zero and c.d.f. $F(x/\sigma)$ with the scale $\sigma > 0$ being unknown. Unlike the GLM, the errors need not have finite variance.

Example 14.1.1. Suppose the errors ε_i are random samples from a double exponential distribution,

$$f(\varepsilon_i) = (2\sigma)^{-1} \exp(-|\varepsilon_i|/\sigma), \quad -\infty < \varepsilon_i < \infty$$

(see (B.15)). Assuming that σ is fixed, it is clear that maximum likelihood estimation of $\boldsymbol{\beta}$ involves minimization of the expression in (14.1.1). □

In the literature, LAD regression has alternately been referred to as minimum absolute deviations (MAD) regression, minimum absolute errors (MAE) regression, least absolute residuals (LAR) regression, least absolute values (LAV) regression, or simply L_1-regression. It is a special case of L_p-regression where we minimize $\sum_{i=1}^N |Y_i - \mathbf{x}_i'\boldsymbol{\beta}|^p$.

LAD regression is substantially different from LS regression The result below may provide some idea on this. Recall that a simplex in $\mathcal{R}^N$ is a set $S = \{\sum_{j=0}^k a_j \mathbf{y}_j \colon \sum_{j=0}^k a_j = 1,\ a_j \geq 0\}$, such that $\mathbf{y}_1 - \mathbf{y}_0, \cdots, \mathbf{y}_k - \mathbf{y}_0$ are LIN. The set $\partial S = \{\sum_{j=0}^k a_j \mathbf{y}_j \in S \colon a_j = 0$ for some $j\}$ is called the boundary of S and $S \setminus \partial S$ is the interior of S. By the definition, the interior of a 0-simplex, i.e., a single point, is the point itself. A simplex with all its vertices also being vertices of S is called a sub-simplex of S. Any point in ∂S is in the interior of a unique sub-simplex of S.

Result 14.1.1. Let $\mathrm{r}(\mathbf{X}) = p$. Then there is a solution $\widehat{\boldsymbol{\beta}}_{LAD}$ to (14.1.2) such that $Y_i = \mathbf{x}_i'\widehat{\boldsymbol{\beta}}_{LAD}$ for at least p different $i = 1, \cdots, N$.

Proof. Let $r = \min_{\boldsymbol{\beta}} \|\mathbf{y} - \mathbf{X}\boldsymbol{\beta}\|_1$. Denote $B_r(\mathbf{y}) = \{\mathbf{z} \in \mathcal{R}^N \colon \|\mathbf{z} - \mathbf{y}\|_1 \leq r\}$. The boundary of $B_r(\mathbf{y})$, i.e., $\partial B_r(\mathbf{y}) = \{\mathbf{z} \in \mathcal{R}^N \colon \|\mathbf{y} - \mathbf{z}\|_1 = r\}$ consists of 2^N facets, each being a simplex. We see that $K = \mathcal{C}(\mathbf{X}) \cap B_r(\mathbf{y}) = \{\mathbf{X}\widehat{\boldsymbol{\beta}}_{LAD}\}$ is a convex polytope entirely in a facet of $B_r(\mathbf{y})$, say S_0. Let $\widehat{\mathbf{y}} = \mathbf{X}\boldsymbol{\beta}^*$ be any vertex of K. We show that $\boldsymbol{\beta}^*$ has the property asserted in Result 14.1.1.

Let S be the unique sub-simplex of S_0 that contains $\widehat{\mathbf{y}}$ in its interior and $\mathcal{V}$ be the linear space spanned by $\{\mathbf{z} - \widehat{\mathbf{y}} \colon \mathbf{z} \in S\}$. Given $\mathbf{v} \in \mathcal{V} \setminus \{\mathbf{0}\}$, there is $a > 0$, such that $\widehat{\mathbf{y}} + s\mathbf{v} \in S$ for all $|s| < a$. If $\mathbf{v} \in \mathcal{C}(\mathbf{X})$, then $\widehat{\mathbf{y}} + s\mathbf{v} \in \mathcal{C}(\mathbf{X}) \cap S \subset K$ for all $|s| < a$, contradicting $\widehat{\mathbf{y}}$ being a vertex of K. Hence $\mathbf{v} \notin \mathcal{C}(\mathbf{X})$, giving $\mathcal{C}(\mathbf{X}) \cap \mathcal{V} = \{\mathbf{0}\}$. Next, noticing $\mathbf{y} - \widehat{\mathbf{y}} \notin \mathcal{V}$, let $\mathcal{U} = \mathrm{Span}(\mathbf{y} - \widehat{\mathbf{y}}) \oplus \mathcal{V}$. If $\mathcal{U} \cap \mathcal{C}(\mathbf{X})$ has a nonzero $\mathbf{u}$, then $\mathbf{u} = a(\mathbf{y} - \widehat{\mathbf{y}}) + \mathbf{v}$ for some $a \neq 0$ and $\mathbf{v} \in \mathcal{V}$. Then for any $c > 0$, $\mathbf{u}_c = c(\mathbf{y} - \widehat{\mathbf{y}}) + (c/a)\mathbf{v} \in \mathcal{U} \cap \mathcal{C}(\mathbf{X})$. If $0 < c < 1$ is small enough, then $\mathbf{v}_c = \widehat{\mathbf{y}} + \frac{c}{a(1-c)}\mathbf{v} \in S$. Then $\mathbf{u}_c + \widehat{\mathbf{y}} = c\mathbf{y} + (1 - c)\mathbf{v}_c$ is in the interior of $B_r(\mathbf{y})$, i.e., $\|\mathbf{y} - (\mathbf{u}_c + \widehat{\mathbf{y}})\| < r$. However, $\mathbf{u}_c + \widehat{\mathbf{y}} \in \mathcal{C}(\mathbf{X})$, which contradicts $\mathcal{C}(\mathbf{X})$ only intersecting with $B_r(\mathbf{y})$ on its boundary. Hence $\mathcal{U} \cap \mathcal{C}(\mathbf{X}) = \{\mathbf{0}\}$. Now suppose S has h vertices. We see that $\dim(\mathcal{U}) = h \leq N - \dim(\mathcal{C}(\mathbf{X})) = N - p$. Write the vertices of S as $\mathbf{y} + c_1\mathbf{e}_{i_1}, \cdots, \mathbf{y} + c_h\mathbf{e}_{i_h}$, where $c_i = \pm 1$ and $i_1 < \cdots < i_h$. Then $\widehat{\mathbf{y}} = \mathbf{y} + \sum_{j=1}^h a_j c_j \mathbf{e}_{i_j}$, where $a_1 + \cdots + a_h = 1$ and $a_j > 0$. It follows that $\widehat{Y}_i = Y_i$ for all $i \notin \{i_1, \cdots, i_h\}$, i.e., $\mathbf{x}_i'\boldsymbol{\beta}^* = Y_i$ for at least p different i's. ■

Although there is a solution that fits (14.1.1) exactly at p observations, this solution is not necessarily unique. If we define $\beta_j^+ = \max(\beta_i, 0)$ and $\beta_j^- = \max(-\beta_j, 0)$ for $j = 0, 1, \cdots, k$, and $u_i = |Y_i - \mathbf{x}_i'\boldsymbol{\beta}|$ for $i = 1, \cdots, N$, then given $Y_1, \cdots, Y_N$ and $\mathbf{x}_1, \cdots, \mathbf{x}_N$, (14.1.1) can be written as

$$\begin{aligned}
&\text{minimize} \sum_{i=1}^{N} u_i \text{ over } u_1, \cdots, u_N, \beta_0^+, \beta_1^+, \cdots, \beta_k^+, \beta_0^-, \beta_1^-, \cdots, \beta_k^-,\\
&\text{subject to } - u_i - \sum_{j=1}^{p} X_{ij}(\beta_j^+ - \beta_j^-) \leq -Y_i, \quad i = 1, \cdots, N,\\
&\qquad\qquad - u_i + \sum_{j=1}^{p} X_{ij}(\beta_j^+ - \beta_j^-) \leq Y_i, \quad i = 1, \cdots, N,\\
&\qquad\qquad \text{and all } u_i \geq 0,\ \beta_j^+ \geq 0,\ \beta_j^- \geq 0.
\end{aligned}$$

This form is known as a canonical form of a linear program, so that the computation of the solution to (14.1.1) may be carried out by linear programming (Charnes and Cooper, 1955; Barrodale and Roberts, 1973; Dantzig, 1963).

In some cases (Dodge and Jurečková, 2000), LAD regression is faster than LS regression for sufficiently large N and moderate p. Provided the c.d.f. of ε_i has a continuous and positive derivative in a neighborhood of the population median, it is true that as $N \to \infty$, $\widehat{\boldsymbol{\beta}}_{LAD} \sim N(\boldsymbol{\beta}, \sigma^2(\mathbf{X}'\mathbf{X})^{-1})$. In practice, σ^2 is unknown, and is estimated as follows based on the LAD residuals $\widehat{\varepsilon}_{i,LAD} = Y_i - \mathbf{x}_i'\widehat{\boldsymbol{\beta}}_{LAD}$, $i = 1, \cdots, N$. Arrange the $M = N - p$ nonzero residuals in ascending order, and denote these by $\widehat{\varepsilon}_{(1),LAD} \leq \cdots \leq \widehat{\varepsilon}_{(M),LAD}$. Let $K_1 = [(M+1)/2 - \sqrt{M}]$ and $K_2 = [(M+1)/2 + \sqrt{M}]$. Then

$$\widehat{\sigma}^2_{LAD} = \frac{\sqrt{M}\{\widehat{\varepsilon}_{(K_2),LAD} - \widehat{\varepsilon}_{(K_1),LAD}\}}{2z_{\alpha/2}}, \tag{14.1.3}$$

where $z_{\alpha/2}$ is the upper $(\alpha/2)$th critical value from a $N(0,1)$ distribution.

Example 14.1.2. Consider the simple linear regression model $Y_i = \beta_0 + \beta_1 X_i + \varepsilon_i$. The LAD estimates of β_0 and β_1, as well as σ^2 are obtained as described above. It is clear that the p-value of the test H_0: $\beta_1 = 0$ is $\mathrm{P}(|T| > |t|)$, where,

$$|t| = \frac{|\widehat{\beta}_{1,LAD}|\{\sum_{i=1}^{N}(X_i - \overline{X})^2\}^{1/2}}{\widehat{\sigma}_{LAD}},$$

and $T \sim t_{N-2}$. It has been proved that as $N \to \infty$, $\sqrt{N}(\widehat{\boldsymbol{\beta}}_{LAD} - \boldsymbol{\beta})$ converges in distribution to the $N(\mathbf{0}, \{2f(0)\}^{-2}\mathbf{Q}^{-1})$ distribution, where $f(0)$ is the p.d.f. at the median, and $\mathbf{Q} = \lim_{N\to\infty}(\mathbf{X}'\mathbf{X}/N)$. □

A more general formulation given by Bassett and Koenker (1978) is useful. Suppose as before that the errors ε_i in the linear model are i.i.d. with c.d.f. F, which is symmetric about 0. The θth regression quantile $(0 < \theta < 1)$ is defined as any solution to the minimization problem

$$\min_{\boldsymbol{\beta}} \left[\sum_{i:\, Y_i \geq \mathbf{x}_i'\boldsymbol{\beta}} \theta|Y_i - \mathbf{x}_i'\boldsymbol{\beta}| + \sum_{i:\, Y_i < \mathbf{x}_i'\boldsymbol{\beta}} (1-\theta)|Y_i - \mathbf{x}_i'\boldsymbol{\beta}| \right]. \tag{14.1.4}$$

When $\theta = 1/2$, this is equivalent to minimizing $\sum_{i=1}^{N} |Y_i - \mathbf{x}_i'\boldsymbol{\beta}|$, and the resulting estimator coincides with $\widehat{\boldsymbol{\beta}}_{LAD}$. Let $\widehat{\boldsymbol{\beta}}^*(\theta)$ denote the solution to (14.1.4). For a general discussion of asymptotic properties of this estimator, and for a discussion of the trimmed least squares estimator, see Bassett and Koenker (1978) and Ruppert and Carroll (1980).

Example 14.1.3. Censored regression. We describe a *limited dependent variable* model in which the response variable, which is observed over time, is censored. The tobit model is, for $t = 1, \cdots, N$, $Y_t = \mathbf{x}_t'\boldsymbol{\beta} + \varepsilon_t$, while the observation is Y_t^+, where we assume that the ε_t's are i.i.d., whose c.d.f. F has positive derivative $f(0)$ at zero, and has median zero. We also assume that the true regression parameter belongs to a bounded open subset of $\mathcal{R}^p$. Let I denote the indicator function of a set of values; i.e., $I(u) = 1$ if u belongs to that set, otherwise $I(u) = 0$.

Powell (1984) studied the properties of the LAD estimator of $\boldsymbol{\beta}$ which is obtained by minimizing

$$\sum_{i=1}^{N} |Y_t^+ - (\mathbf{x}_t'\widehat{\boldsymbol{\beta}}_{LAD})^+|. \tag{14.1.5}$$

Under some regularity conditions, Powell (1984) and Rao and Zhao (1993) proved the asymptotic normality of this estimator.

Suppose we wish to test H_0: $\mathbf{C}'\boldsymbol{\beta} = \mathbf{d}$ versus the alternative H_1: $\mathbf{C}'\boldsymbol{\beta} \neq \mathbf{d}$, where $\mathbf{C}$ is a $p \times s$ matrix of rank s, and $\mathbf{d}$ is a constant vector. Let $\widehat{\boldsymbol{\beta}}_{LAD,H_0}$ denote the solution to (14.1.5) subject to H_0. The likelihood ratio test statistic is

$$LRT = \sum_{i=1}^{N} |Y_t^+ - (\mathbf{x}_t'\widehat{\boldsymbol{\beta}}_{LAD,H_0})^+| - \sum_{i=1}^{N} |Y_t^+ - (\mathbf{x}_t'\widehat{\boldsymbol{\beta}}_{LAD})^+|. \tag{14.1.6}$$

In practice, $f(0)$ is unknown and is estimated by

$$\begin{aligned}\widehat{f}_N(0) = h &\left[\sum_{i=1}^{N} I(\mathbf{x}_t'\widehat{\boldsymbol{\beta}}_{LAD} > 0)^{-1}\right] \\ &\times \left[\sum_{i=1}^{N} I(\mathbf{x}_t'\widehat{\boldsymbol{\beta}}_{LAD} > 0) I(\mathbf{x}_t'\widehat{\boldsymbol{\beta}}_{LAD} < Y_t^+ \leq \mathbf{x}_t'\widehat{\boldsymbol{\beta}}_{LAD} + h)\right],\end{aligned}$$

where $h > 0$ is a fixed parameter. Under certain conditions (Rao and Zhao, 1993), $4\widehat{f}(0)LRT$ has a limiting χ_s^2 distribution under H_0. □

14.1.2 *M*-regression

The M-estimator, which was introduced by Huber (1964), is a robust alternative to the LS estimator in the linear model. It is attractive because it is relatively simple to compute, and it offers good performance and flexibility. In the model (4.1.2), we assume that ε_i are i.i.d. random variables with c.d.f. F, which is symmetric about 0. The distribution of ε_i can be fairly general; in fact we need not assume existence of the mean and variance of the distribution. Recall from Section 8.1.1 that the MLE of $\boldsymbol{\beta}$ is obtained as a solution to the likelihood equations

$$\sum_{i=1}^{N} \mathbf{x}_i \{f'(Y_i - \mathbf{x}_i'\boldsymbol{\beta})/f(Y_i - \mathbf{x}_i'\boldsymbol{\beta})\} = \mathbf{0}, \tag{14.1.7}$$

where f denotes the p.d.f. of ε_i, and f' denotes its first derivative. Suppose (14.1.7) cannot be solved explicitly; by replacing f'/f by a suitable function ψ, we obtain a "pseudo maximum likelihood" estimator of $\boldsymbol{\beta}$, which is called its M-estimator. The function ψ is generally chosen such that it leads to an estimator $\widehat{\boldsymbol{\beta}}_M$ which is robust to alternative specifications of f. Suppose ρ is a convex, continuously differentiable function on $\mathcal{R}$, and suppose $\psi = \rho'$. The M-estimator $\widehat{\boldsymbol{\beta}}_M$ of $\boldsymbol{\beta}$ in the model (4.1.2) is obtained by solving the minimization problem

$$\min_{\boldsymbol{\beta}} \sum_{i=1}^{N} \rho(Y_i - \mathbf{x}_i'\boldsymbol{\beta}), \tag{14.1.8}$$

or, equivalently, by solving the set of p equations

$$\sum_{i=1}^{N} \mathbf{x}_i \psi(Y_i - \mathbf{x}_i'\boldsymbol{\beta}) = \mathbf{0}. \tag{14.1.9}$$

When (a) the function ρ is not convex, (b) the first derivative of ρ is not continuous, or (c) the first derivative exists everywhere except at a finite or countably infinite number of points, we consider $\widehat{\boldsymbol{\beta}}_M$ to be a solution of (14.1.8), since (14.1.2) might not have a solution, or might lead to an incorrect solution. By a suitable choice of ρ and ψ, we can obtain an estimator which is robust to possible heavy-tailed behavior in f.

To ensure that $\widehat{\boldsymbol{\beta}}_M$ is scale-invariant, we generalize (14.1.8) and (14.1.2) respectively to

$$\min_{\boldsymbol{\beta}} \sum_{i=1}^{N} \rho\{(Y_i - \mathbf{x}_i'\boldsymbol{\beta})/\sigma\}, \text{ and} \tag{14.1.10}$$

$$\sum_{i=1}^{N} \mathbf{x}_i \psi\{(Y_i - \mathbf{x}_i'\boldsymbol{\beta})/\sigma\} = \mathbf{0}. \tag{14.1.11}$$

In practice, σ is unknown, and is replaced by a robust estimate. One such estimate is the median of the absolute residuals from LAD regression.

As we have seen, M-regression generalizes LS estimation by allowing a choice of objective functions. LS regression corresponds to $\rho(t) = t^2$. In general, we have considerable flexibility in choosing the function ρ, and this choice in turn determines the properties of $\widehat{\beta}_M$. The choice involves a balance between efficiency and robustness.

Example 14.1.4. Suppose the true regression function is the conditional median of $\mathbf{y}$ given $\mathbf{X}$. The choice of $\rho(t)$ is $|t|$, and the corresponding problem is called median regression. □

Example 14.1.5. W-regression is an alternative form of M-regression; $\widehat{\beta}_W$ is a solution to the p simultaneous equations

$$\sum_{i=1}^{N} \left(\frac{Y_i - \mathbf{x}_i'\beta}{\sigma}\right) w_i \mathbf{x}_i' = \mathbf{0}, \tag{14.1.12}$$

where $w_i = w\{(Y_i - \mathbf{x}_i'\beta)/\sigma\}$. The equations in (14.1.12) are obtained by replacing $\psi(t)$ by $tw(t)$ in (14.1.8); $w(t)$ is called a weight function. □

Numerical Example 14.1. The dataset is "stackloss" in the R package *MASS*. M-regression using Huber, Hampel and bisquare ψ functions are shown below.

```
data(stackloss, package = MASS)
summary(rlm(stack.loss ~ ., stackloss))

rlm(stack.loss ~ ., stackloss, psi = psi.huber, init = "lts")
Call:
rlm(formula = stack.loss ~ ., data = stackloss, psi = psi.huber,
    init = "lts")
Converged in 12 iterations

Coefficients:
(Intercept)    Air.Flow  Water.Temp  Acid.Conc.
-41.0263481   0.8293999   0.9259955  -0.1278426

Degrees of freedom: 21 total; 17 residual
Scale estimate: 2.44

rlm(stack.loss ~ ., stackloss, psi = psi.hampel, init = "lts")
Call:
rlm(formula = stack.loss ~ ., data = stackloss, psi = psi.hampel,
    init = "lts")
Converged in 9 iterations

Coefficients:
(Intercept)    Air.Flow  Water.Temp  Acid.Conc.
-40.4747671   0.7410846   1.2250749  -0.1455246

Degrees of freedom: 21 total; 17 residual
Scale estimate: 3.09

rlm(stack.loss ~ ., stackloss, psi = psi.bisquare)
Call:
rlm(formula = stack.loss ~ ., data = stackloss,
    psi = psi.bisquare)
Converged in 11 iterations

Coefficients:
(Intercept)    Air.Flow  Water.Temp  Acid.Conc.
-42.2852537   0.9275471   0.6507322  -0.1123310

Degrees of freedom: 21 total; 17 residual
Scale estimate: 2.28
```

▲

14.2 Nonparametric regression methods

Parametric regression models, such as GLMs and GLIMs, assume specific forms of the regression function $\mathrm{E}(Y \mid \mathbf{x})$ that can be completely determined by a finite number of parameters, so that model inference reduces to estimation and inference pertaining to the parameters (under some distributional assumption on the errors). In contrast, nonparametric methods only make a few general assumptions about the regression function (Hastie and Tibshirani, 1990; Hastie et al., 2009; Thisted, 1988). In this section, we give a brief introduction to computer-intensive regression methods that are useful for nonparametric fitting, when the predictor $\mathbf{x}$ is multidimensional, or when a parametric linear model does not satisfactorily explain the relationship between Y and $\mathbf{x}$.

Additive models and projection pursuit regression are described in the next two subsections. The additive model is a generalization of the multiple linear regression (MLR) model, where the usual linear function of observed predictors is replaced by a sum of unspecified smooth functions of these predictors. Projection pursuit (Friedman and Tukey, 1974) is a general approach that seeks to unravel structure within a high dimensional predictor by finding interesting projections of the data onto a low dimensional linear subspace, such as a line or a plane. We also discuss multivariate adaptive regression splines (MARS) and neural nets regression.

14.2.1 Regression splines

The MLR model can provide an inadequate fit when a linear function is unsuitable to explain the dependence of Y on the explanatory variables. One remedy is to add polynomial terms in some or all of the explanatory variables, resulting in a *polynomial regression model.* These models may however have limited scope, because, it is often difficult to guess what the appropriate order of the polynomials should be, and also because of the global nature of its fit. In many cases, a *piecewise polynomial model* is more suitable. Consider a functional relation between a response Y and a single predictor X:

$$Y = f(X) + \varepsilon \quad \text{with } \mathrm{E}(\varepsilon) = 0,\ \mathrm{Var}(\varepsilon) = \sigma^2. \tag{14.2.1}$$

Definition 14.2.1. Piecewise polynomial regression. Fix $q \geq 1$. An order-q piecewise polynomial with fixed *knots* $\xi_1 < \cdots < \xi_L$ is a function such that, within each of the intervals $I_0 = (-\infty, \xi_1)$, $I_\ell = [\xi_\ell, \xi_{\ell+1})$, $\ell = 1, \ldots, L-1$, and $I_L = [\xi_L, \infty)$, it is a polynomial of degree at most $q-1$ with real-valued coefficients

$$f_{\ell,q}(x) = \sum_{j=1}^{q} c_{\ell,j} x^{j-1}, \quad x \in I_\ell,\ \ell = 0, 1, \cdots, L. \tag{14.2.2}$$

Piecewise polynomial regression consists of fitting the data by a regression model (14.2.1), with f assumed to be a piecewise polynomial with no constraints on its coefficients.

Although piecewise polynomial functions are more flexible than polynomial functions, one disadvantage is that they are not necessarily smooth functions. In many cases, they can be discontinuous at the knots. Regression splines retain the flexibility of piecewise polynomials while incorporating a degree of smoothness.

Definition 14.2.2. Polynomial spline. An order-q polynomial spline with fixed knots $\xi_1 < \cdots < \xi_L$ is a piecewise polynomial f with those knots and with continuous derivatives up to order $q-2$, that is

$$f^{(k)}(\xi_\ell -) = f^{(k)}(\xi_\ell +), \quad k = 0, \cdots, q-2 \tag{14.2.3}$$

for each $\ell = 1, \cdots, L$.

Polynomial splines in the context of the regression model (14.2.1) are known as regression splines. They facilitate flexible fitting of the data with a degree of smoothness. Usually,

$q = 1, 2$, or 4 are used. When $q = 4$, we call these cubic splines. If an order-q polynomial spline with L knots is expressed as in (14.2.2), there are $(L + 1)q$ coefficients in total. However, from (14.2.3) $f^{(k)}_{\ell-1,q}(\xi_\ell) = f^{(k)}_{\ell,q}(\xi_\ell)$, $0 \le k \le q - 2$, so there are $q - 1$ linear constraints on the coefficients at each of the L knots. As a result, for $q > 1$, direct estimation of the coefficients is a constrained optimization problem. In practice, since the dimension of the space of order-q splines with L knots is $(L + 1)q - L(q - 1) = L + q$, regression spline fitting is carried out by unconstrained optimization using a set of basis functions $h_1, \cdots, h_{L+q}$ of the space, so that the data is fit with

$$f(x) = \sum_{\ell=1}^{L+q} a_\ell h_\ell(x), \tag{14.2.4}$$

with $a_1, \cdots, a_{L+q}$ being unconstrained coefficients.

Example 14.2.1. Truncated power basis. The space of order-q regression splines with knots $\xi_1 < \cdots < \xi_L$ has a basis consisting of the functions $h_j(x) = x^{j-1}$, $j = 1, \cdots, q$, and $h_{q+\ell}(x) = (x - \xi_\ell)^{q-1}_+$, $\ell = 1, \cdots, L$, where as before, $t_+ = \max(0, t)$. □

A widely used class of spline basis is the class of B-splines. Suppose all the X_i's are in $[a, b]$ and $a < \xi_1 < \cdots < \xi_L < b$. Fix $m \ge 1$. Define an augmented sequence of knots $\tau_1 \le \cdots \le \tau_{L+2m}$, such that

$$\tau_\ell = \begin{cases} a & \text{if } \ell \le m \\ \xi_{\ell-m} & \text{if } \ell = m + 1, \cdots, m + L \\ b & \text{if } \ell > m + L. \end{cases}$$

Denote by $B_{i,j}(x)$ the ith B-spline basis function of order j for the augmented knot sequence, where $i = 1, \cdots, L + 2m - j$ and $j = 1, \cdots, q$. To start with,

$$B_{i,1}(x) = \begin{cases} 1 & \text{if } \tau_i \le x < \tau_{i+1} \\ 0 & \text{otherwise.} \end{cases}$$

For $j = 2, \ldots, q$, recursively define

$$B_{i,j}(x) = \frac{x - \tau_i}{\tau_{i+j-1} - \tau_i} B_{i,j-1}(x) + \frac{\tau_{i+j} - x}{\tau_{i+j} - \tau_{i+1}} B_{i,j}(x),$$

where the convention $0/0 = 0$ is used. Then, $B_{i,j}$, $i = m - j + 1, \cdots, m + L$, constitutes a basis of splines of order j with the original knots $\xi_1, \cdots, \xi_L$.

Definition 14.2.3. Smoothing spline. Given data (Y_i, X_i), for $i = 1, \cdots, N$, a smoothing spline minimizes the following expression, as a compromise between the accuracy of fit and the degree of smoothness:

$$\sum_{i=1}^{N} [Y_i - f(X_i)]^2 + \lambda \int \{f''(x)\}^2 dx$$

over all twice-continuously differentiable functions f, where $\lambda \ge 0$ is a fixed smoothing parameter.

It is a remarkable fact that given $\lambda > 0$, a minimizer of the expression above not only exists, but is also a unique *natural cubic spline* with knots at the unique values of

$X_1, \cdots, X_N$, i.e., it is linear on $(-\infty, \min X_i)$ and $(\max X_i, \infty)$, respectively. This is the essence of the definition. For details, see Green and Silverman (1994). For larger λ, the spline has less curvature, while for smaller λ, it becomes rougher, yielding more of an interpolation to the data.

Regression spline fitting is a very rich area. For more details, see Eubank (1999). We show an example to illustrate cubic spline and smoothing spline fitting.

Numerical Example 14.2. We use the dataset "Wage" from the R package *ISLR*, which consists of data on 3000 male workers in the Mid-Atlantic region of the U.S. The response variable is wage (worker's raw income) which we model as a function of the age of the worker. Note that age ranges from A scatterplot of wage versus age (figure not shown here) suggests age values of 25, 35 and 65 as possible knots. Figure 14.2.1 shows the cubic splines and smoothing spline fits to the data.

```
library(ISLR)
data(Wage)
library(splines)
(agelims <- range(age))
(age.grid <- seq(from = agelims[1], to = agelims[2]))
# Cubic spline with 3 knots/cutpoints at ages 30,45,65
cuts = c(30, 45, 65)
summary(fit <- lm(wage ~ bs(age, knots = cuts), data = Wage))

Coefficients:
                        Estimate Std. Error t value Pr(>|t|)
(Intercept)               58.989      7.791   7.571 4.90e-14 ***
bs(age, knots = cuts)1    13.916     11.368   1.224   0.2210
bs(age, knots = cuts)2    58.694      7.720   7.603 3.86e-14 ***
bs(age, knots = cuts)3    59.147      9.483   6.237 5.09e-10 ***
bs(age, knots = cuts)4    63.320      9.347   6.774 1.50e-11 ***
bs(age, knots = cuts)5    35.770     14.556   2.457   0.0141 *
bs(age, knots = cuts)6    27.350     19.800   1.381   0.1673
---
Signif. codes:  0 '***' 0.001 '**' 0.01 '*' 0.05 '.' 0.1 ' ' 1

Residual standard error: 39.92 on 2993 degrees of freedom
Multiple R-squared:  0.0867,Adjusted R-squared:  0.08487
F-statistic: 47.36 on 6 and 2993 DF,  p-value: < 2.2e-16

# Smoothing spline
fit1 <- smooth.spline(age, wage, df = 16)

smooth.spline(x = age, y = wage, df = 16)
Smoothing Parameter spar= 0.4732071  lambda= 0.0006537868 (13
iterations)
Equivalent Degrees of Freedom (Df): 16.00237
Penalized Criterion (RSS): 61597.01
GCV: 1599.69

# Plot cubic and smoothing splines
plot(age, wage, xlab = "age", ylab = "wage")
points(age.grid, predict(fit, newdata = list(age = age.grid)),
```

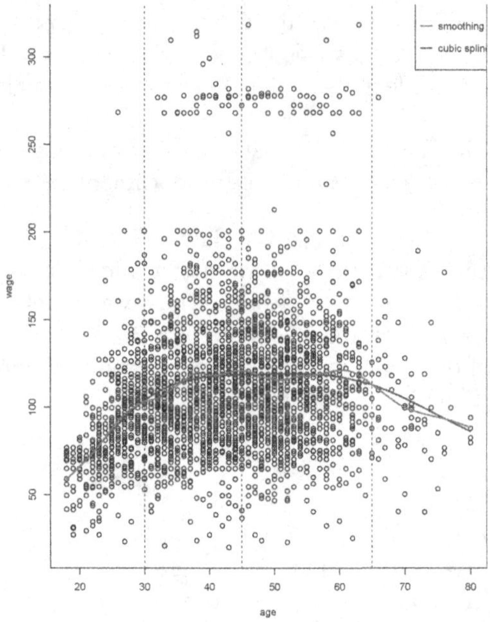

FIGURE 14.2.1. Cubic and smoothing splines for the Wage data.

```
  col = "blue", lwd = 2, type = "l")
abline(v = cuts, lty = 2, col = "blue") #add cutpoints
lines(fit1, col = "red", lwd = 2)
legend("topright", c("smoothing spline", "cubic spline"),
  col = c("red", "blue"), lwd = 2)
```

▲

14.2.2 Additive and generalized additive models

Let $\mathbf{x} = (X_1, \cdots, X_k)'$. An additive model is defined by

$$Y_i = \tau + \sum_{j=1}^{k} g_j(X_{ij}) + \varepsilon_i, \quad \mathrm{E}(\varepsilon_i) = 0, \;\; \mathrm{Var}(\varepsilon_i) = \sigma^2, \tag{14.2.5}$$

where the g_j's are unspecified functions, one for each predictor. A simple assumption may be that each of these component functions is univariate and smooth; however, the smoothness assumption may be relaxed, and it is possible to have a function that is continuous and multi-dimensional, or even categorical. One attractive feature of the additive model is that the variation of the response function obtained by holding all but one predictor fixed does not depend on the values of the fixed predictors, which makes it possible to plot the j component functions separately in order to assess their usefulness in modeling Y after fitting the model to data. Hence, apart from their use in predicting the response variable, additive models enable us to explore the appropriate shape of each predictor effect on the response without a rigid parametric assumption about that effect, such as linearity or an exponential form. Hastie and Tibshirani (1990) present a detailed description of this for data analysis.

In an additive model, no parametric form is imposed on $g_1, \cdots, g_k$. The most general approach for fitting the model to data consists of estimating these functions iteratively using *scatterplot smoothers*. Generally speaking, a smoother is any nonparametric tool for estimating the trend of a response as a function of a set of predictors. The estimated trend is less variable than the response itself, which leads to the name "smoother". The scatterplot smoother is a special case where there is a single predictor. Regression splines and smoothing splines in Section 14.2.1 as well as locally-weighted running lines and kernel smoothers are all examples of scatterplot smoothers.

The backfitting algorithm is a general algorithm to carry out the iterative fitting. For each $j = 1, \cdots, k$, select a scatterplot smoother S_j that fits its input as a function of X_{ij}, $i = 1, \cdots, N$. The algorithm consists of the following steps:

(i) Initialization: Set $\tau = \overline{Y}$, and set $g_j = g_j^0$, $j = 1, \cdots, k$.

(ii) Update: For $j = 1, \cdots, k$, set $g_j = S_j(\{Z_{ij}, i = 1, \cdots, N\})$, where $Z_{ij} = Y_i - \tau - \sum_{l \neq j} g_l(X_{il})$.

(iii) Continue (ii) until the individual functions do not change.

When the observations are assigned with weights, the algorithm simply sets τ equal to the weighted average of Y_i at initialization and uses weighted scatterplot smoothers to update g_j, with each Z_{ij} assigned the same weight as $(Y_i, \mathbf{x}_i)$.

The backfitting algorithm is motivated by the fact that, if the additive model is correct, then for any $j = 1, \cdots, k$,

$$\mathrm{E}\left\{ Y - \tau - \sum_{l \neq j} g_l(X_l) \middle| X_j \right\} = g_j(X_j). \tag{14.2.6}$$

The usual linear regression estimates may be used for the initialization. By design, the algorithm is only given as a modular skeleton; details for a particular case would depend on the choice of the smoothers and the data analytic context in which the model is employed. In particular, a model fitting criterion must be specified, e.g., the penalized sum of squares

$$\sum_{i=1}^{N} \left\{ Y_i - \tau - \sum_{j=1}^{J} g_j(X_{ij}) \right\}^2 + \sum_{j=1}^{J} \lambda_j \int \{g_j''\}^2,$$

so that the smoothers S_j are employed for its optimization. It has been pointed out that when the smoothers are linear operators, the backfitting algorithm is a Gauss–Seidel algorithm for solving a certain set of estimating equations, and convergence in many practical situations has been proven. When all the smoothers S_j are projection operators, the entire backfitting algorithm may be replaced by a global projection that involves no iteration and whose convergence is guaranteed. Although smoothers such as smoothing splines are not projections, they do possess properties of projections that guarantee that the solution converges.

Numerical Example 14.3. Additive model. We illustrate additive model fitting using simulated data consisting of 100 observations on six variables discussed in the R package *gam*. The variable y denotes the response, x is a numeric predictor, z is a numeric noise predictor, f denotes the true function, $probf$ denotes a probability function, and $ybin$ is a binary response.

```
library(gam)
data(gam.data)
gam(y ~ s(x) + z, data = gam.data)

Call:
gam(formula = y ~ s(x) + z, data = gam.data)

Degrees of Freedom: 99 total; 93.99988 Residual
Residual Deviance: 7.90768
```

▲

Generalized additive models (GAMs) are an extension of additive models to GLIMs that were described in Chapter 12. We assume that the response variable Y_i associated with $\mathbf{x}_i = (X_{i1}, \cdots, X_{ik})'$ has an exponential family density as in (12.1.1) with the mean $\mu_i = \mathrm{E}(Y_i \,|\, \mathbf{x}_i)$ linked to the predictors by the functional form

$$g(\mu_i) = \eta_i = \tau + \sum_{j=1}^{k} f_j(X_{ij}). \tag{14.2.7}$$

We estimate $\tau, f_1, \cdots, f_k$ by an algorithm called the *local scoring procedure*, which is a generalization (using local averaging) of Fisher's scoring procedure that we discussed under the GLIM setup in Section 12.2 and which consists of the following steps (for details, see Hastie and Tibshirani, 1990).

(i) Initialization: Set $\tau = g(\overline{Y})$, and $f_j^0 \equiv 0$, $j = 1, \cdots, k$. Set $\boldsymbol{\eta}^0 = (\eta_1^0, \cdots, \eta_N^0)'$ and $\boldsymbol{\mu}^0 = (\mu_1^0, \cdots, \mu_N^0)'$ according to (14.2.7) with $f_j = f_j^0$.

(ii) Update: For each $i = 1, \cdots, N$, construct an adjusted dependent variable

$$\begin{aligned} Z_i &= \eta_i^0 + (Y_i - \mu_i^0) g'(\mu_i^0), \text{ where} \\ \eta_i^0 &= \tau + \sum_{j=1}^{k} f_j^0(X_{ij}) \text{ and } \mu_i^0 = g^{-1}(\eta_i^0) \end{aligned} \tag{14.2.8}$$

and assign Z_i the weight

$$W_i = \frac{[g'(\mu_i^0)]^2}{V_i^0}, \tag{14.2.9}$$

where $V_i^0 = a_i(\phi)/g'(\mu_i^0)$ is the variance of Y_i at μ_i^0; see Result 12.1.1. Use the backfitting algorithm to fit a weighted additive model $Z_i = \tau + \sum_{j=1}^{k} f_j(X_{ij})$, $i = 1, \cdots, N$, to obtain τ^1, $f_1^1, \cdots, f_k^1$. Set $\boldsymbol{\eta}^1$ and $\boldsymbol{\mu}^1$ according to (14.2.7) with $f_j = f_j^1$. Compute the convergence criterion

$$\Delta = \sum_{j=1}^{k} \|f_j^1 - f_j^0\| \Big/ \sum_{j=1}^{k} \|f_j^0\|, \tag{14.2.10}$$

where, $\|f\|$ may be computed as the length of the vector of evaluations of f at the N sample points.

(iii) Continue (ii), replacing $\boldsymbol{\eta}^0$ by $\boldsymbol{\eta}^1$, until the convergence criterion Δ is smaller than a threshold.

Numerical Example 14.4. Data on the results of a spinal operation (laminectomy) on children, to correct for a condition called *kyphosis*, is available in the dataset "kyphosis" in the R package *gam*. The data frame has 81 observations on four variables. The response Y is a binary variable kyphosis with levels absent or present. The predictors are Age (months), Number (vertebra involved in the operation) and Start (level of the operation). A GAM can be fit to this data using the following code.

```
library(gam)
data(kyphosis)
Mod <- gam(Kyphosis ~ poly(Age, 2) + s(Start), data = kyphosis,
      family = binomial)
summary(Mod)

Call: gam(formula = Kyphosis ~ poly(Age, 2) + s(Start),
    family = binomial, data = kyphosis)
Deviance Residuals:
     Min        1Q    Median        3Q       Max
-1.68915 -0.42271 -0.18871 -0.02362  2.12393

(Dispersion Parameter for binomial family taken to be 1)

    Null Deviance: 83.2345 on 80 degrees of freedom
Residual Deviance: 49.0815 on 73.9998 degrees of freedom
AIC: 63.082

Number of Local Scoring Iterations: 10

Anova for Parametric Effects
             Df Sum Sq Mean Sq F value   Pr(>F)
poly(Age, 2)  2  5.794  2.8970  4.4428 0.015061 *
s(Start)      1  5.328  5.3277  8.1705 0.005527 **
Residuals    74 48.253  0.6521
---
Signif. codes:  0 '***' 0.001 '**' 0.01 '*' 0.05 '.' 0.1 ' ' 1

Anova for Nonparametric Effects
             Npar Df Npar Chisq  P(Chi)
(Intercept)
poly(Age, 2)
s(Start)           3     7.3817 0.06069 .
---
Signif. codes:  0 '***' 0.001 '**' 0.01 '*' 0.05 '.' 0.1 ' ' 1
```
▲

14.2.3 Projection pursuit regression

In the linear regression model (4.1.2), we assumed that the response function $\mathrm{E}(Y \mid \mathbf{x})$ has a fixed linear form. Friedman and Stuetzle (1981) proposed projection pursuit regression which assumes

$$\mathrm{E}(Y \mid \mathbf{x}) = \sum_{j=1}^{k} S_j(\mathbf{c}_j' \mathbf{x}), \tag{14.2.11}$$

where $\mathbf{c}_1, \cdots, \mathbf{c}_k$ are unknown unit vectors, and $S_1, \cdots, S_k$ are unknown functions that must be smooth but are otherwise arbitrary. Each $S_j(\mathbf{c}_j'\mathbf{x})$ is called a ridge function of $\mathbf{x}$. Hence, projection pursuit regression employs an additive model that only depends on several selected projections of the predictors. It can be viewed as a low dimensional expansion approach. The univariate arguments of S_j are not prespecified, but are adjusted to best fit the data. As shown by Diaconis and Shahshahani (1984), the model (14.2.11) can approximate arbitrary continuous functions of $\mathbf{x}$ when k is sufficiently large.

Given data $(Y_i, \mathbf{x}_i)$, $i = 1, \cdots, N$, the model (14.2.11) can be fit by an iterative algorithm (Friedman and Stuetzle, 1981) with the following steps. Let $\mathcal{S}$ denote a selected scatterplot smoother whose output is a smooth function.

(i) Set the counter of terms $k = 0$, and set initial vector of residuals $\widehat{\boldsymbol{\varepsilon}}^{(0)} = \mathbf{y}$.

(ii) Search for the next term in the model. For a vector $\mathbf{c}$, define

$$I(\mathbf{c}) = \sum_{i=1}^{N} \left[\widehat{\varepsilon}_i^{(k)} - S_{\mathbf{c}}(\mathbf{c}'\mathbf{x}_i)\right]^2$$

where $S_{\mathbf{c}}$ is the output of $\mathcal{S}$ by fitting $\widehat{\varepsilon}_i^{(k)}$ with a smooth function of $\mathbf{c}'\mathbf{x}_i$. Find $\widehat{\mathbf{c}}$ to minimize $I(\mathbf{c})$

(iii) If $1 - I(\widehat{\mathbf{c}})/\|\widehat{\boldsymbol{\varepsilon}}^{(k)}\|^2$ is smaller than a preselected threshold, then stop. Otherwise, add $S_{k+1} = S_{\widehat{\mathbf{c}}}$ to (14.2.11), update the residuals to $\widehat{\varepsilon}_i^{(k+1)} = \widehat{\varepsilon}_i^{(k)} - S_j(\widehat{\mathbf{c}}'\mathbf{x}_i)$, $i = 1, \cdots, N$, increment the counter to $k+1$, and repeat from Step (ii).

The scatterplot smoother used by Friedman and Stuetzle (1981) requires several passes of fitting. The minimization of $I(\mathbf{c})$ in Step (ii) can be carried out iteratively by a Gauss–Raphson type of algorithms (Hastic ct al., 2009). Unlike additive models, projection pursuit regression models can include interactions between the X_j's in the model function, and they are therefore useful for representing general regression functions.

Numerical Example 14.5. The dataset "rock" in the R package *datasets* consists of measurements on 48 rock samples from a petroleum reservoir. Twelve core samples from petroleum reservoirs were sampled by 4 cross-sections. Each core sample was measured for permeability, and each cross-Section has total area of pores, total perimeter of pores, and shape. The data consists of four variables: area (of pores space, in pixels out of 256 by 256), peri (perimeter in pixels), shape (perimeter/sqrt(area)), and perm (permeability in milli-Darcies).

```
data(rock)
attach(rock)
area1 <- area / 10000
peri1 <- peri / 10000
rock.ppr <- ppr(log(perm) ~ area1 + peri1 + shape,
    data = rock, nterms = 2, max.terms = 5)
summary(rock.ppr)

Call:
ppr(formula = log(perm) ~ area1 + peri1 + shape, data = rock,
    nterms = 2, max.terms = 5)

Goodness of fit:
```

```
 2 terms  3 terms  4 terms  5 terms
8.737806 5.289517 4.745799 4.490378

Projection direction vectors ('alpha'):
      term 1      term 2
area1  0.34357179  0.37071027
peri1 -0.93781471 -0.61923542
shape  0.04961846  0.69218595

Coefficients of ridge terms ('beta'):
   term 1    term 2
1.6079271 0.5460971
```

▲

14.2.4 Multivariate adaptive regression splines

We looked at regression splines in Section 14.2.1. We now discuss the multivariate adaptive regression splines (MARS) algorithm (Friedman, 1991), which can be viewed as a generalization of regression splines to high-dimensional settings. MARS is a completely data-driven procedure. It produces continuous models and, unlike additive models, it allows interactions between predictors up to some order that may be user specified.

MARS uses expansion in product spline basis functions. From Section 14.2.1, given t, $(x-t)_+$ and $(t-x)_+$ are splines in x of order 2 with knot t. They will be referred to as *hinge functions*. MARS assumes $\mathrm{E}(Y \mid \mathbf{x})$ to take the form

$$f(\mathbf{x}) = a_0 + \sum_{m=1}^{M} a_m B_m(\mathbf{x}), \quad \mathbf{x} = (X_1, \cdots, X_k)', \tag{14.2.12}$$

where each a_m is a constant coefficient, and each $B_m(\mathbf{x})$ is either a hinge function in one of the predictors X_j, or a product of two or more such functions, but each with a different X_j. MARS selects the knot of each hinge function of X_j from its observed values. As a result, given data $(Y_i, \mathbf{x}_i)$, $i = 1, \cdots, N$, the set of available hinge functions for (14.2.12) is

$$\mathcal{C} = \bigcup_{j=1}^{k} \{(X_j - t)_+, (t - X_j)_-, t \in \{X_{1j}, \cdots, X_{Nj}\}\}.$$

From the identity $(x-t)_+ - (t-x)_+ + t = x$, we see that (14.2.12) allows linear functions of $\mathbf{x}$ to be included.

The MARS algorithm consists of a forward stepwise procedure followed by a backward deletion procedure. The underlying principle of the forward procedure is the same as that of the forward variable selection for linear regression (see Section 9.1). It has the following steps:

(i) Set the collection of $B_m(\mathbf{x})$ in (14.2.12) equal to $\mathcal{M} = \{B_0(\mathbf{x})\}$, where $B_0(\mathbf{x}) \equiv 1$.

(ii) For each $B_\ell(\mathbf{x}) \in \mathcal{M}$ and each pair $(X_j - t)_+$, $(t - X_j)_+$ in $\mathcal{C}$, such that both X_j and t have *not* appeared in the expression of $B_\ell(\mathbf{x})$, denote by $SSE(\ell, j, t)$ the SSE of the LS fit of the data by the model

$$\begin{aligned} Y = &\sum_{B_m \in \mathcal{M}} a_m B_m(\mathbf{x}) + a_{M+1} B_\ell(\mathbf{x})(X_j - t)_+ \\ &+ a_{M+2} B_\ell(\mathbf{x})(t - X_j)_+ + \varepsilon \end{aligned}$$

Find (ℓ, j, t) that minimizes $SSE(\ell, j, t)$. Then add $B_\ell(\mathbf{x})(X_j - t)_+$ and $B_\ell(\mathbf{x})(t - X_j)_+$ to $\mathcal{M}$.

(iii) Repeat Step (ii) until a pre-determined maximum number of terms are added.

The above steps can also be carried out with an upper limit on the number of hinge functions in each term.

The backward deletion procedure is like the backward variable elimination for linear regression (see Section 9.1). At each step, a term is removed using the criterion of minimal increase in the SSE. MARS estimates the optimum number of terms, λ, by minimizing a generalized cross-validation criterion

$$GCV(\lambda) = \frac{\sum_{i=1}^{N}(Y_i - f_\lambda(\mathbf{x}_i))^2}{(1 - M(\lambda)/N)^2}$$

where $f_\lambda(\mathbf{x})$ is the model with λ terms produced by the backward deletion procedure, and $M(\lambda)$ is effective number of parameters. If a total of r linearly independent hinge functions and a total of K knots appear in the λ functions $B_m(\mathbf{x})$ of the model, then $M(\lambda) = r + cK$, where $c = 3$ if some $B_m(\mathbf{x})$ are products of two or more hinge functions, and $c = 2$ if every $B_m(\mathbf{x})$ is a hinge function.

Numerical Example 14.6. We use the dataset "trees" from the R package *mda*. The data provides measurements of the diameter (inches) measured at 4 ft 6 in above the ground, height (feet), and volume of timber (cubic feet) in 31 felled black cherry trees. The source is Ryan et al. (1976). The response Y is the volume and the two predictors are X_1 (diameter) and X_2 (height). The `mars()` function is used to fit the model $Y = f(X_1, X_2) + \varepsilon$.

```
library(mda)
data(trees)
diameter <- trees[, 1]
height <- trees[, 2]
volume <- trees[, 3]
fit.tr <- mars(cbind(diameter, height), volume)
```

Partial output from the results are shown, including coefficients from the fit.

```
fit.tr$coefficients

          [,1]
[1,] 26.2609773
[2,]  6.1128688
[3,] -3.1671107
[4,]  0.4976885


fit.tr$cuts
showcuts <- function(obj)
{tmp <- obj$cuts[obj$sel, ]
dimnames(tmp) <- list(NULL, names(trees)[-3])
tmp}
showcuts(fit.tr)

Girth Height
[1,]   0.0      0
[2,]  13.8      0
```

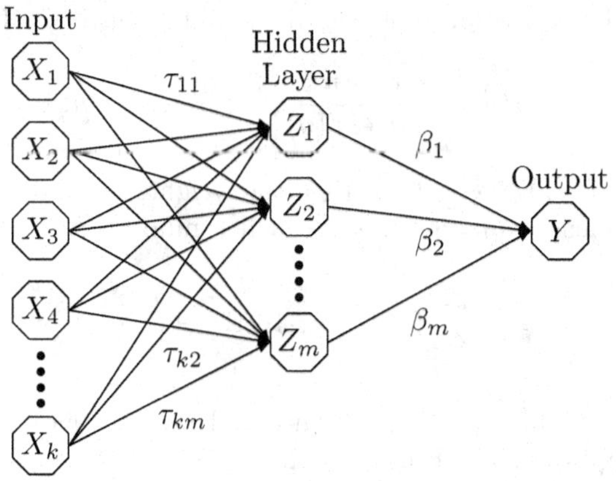

FIGURE 14.2.2. A feed-forward neural network with one hidden layer.

```
[3,]  13.8     0
[4,]   0.0    72
```

▲

14.2.5 Neural networks regression

Neural networks are mathematical objects representing a system of interconnected computational units (Cheng and Titterington, 1994; Hertz et al., 1991). That is, they constitute a collection of a possibly large number of simple computational units which are interlinked by a system of possibly intricate connections. In the neural networks framework, specification of the general linear model and associated parametric inference are handled by constructing a network of units and links from which the linear model can be written down.

A flexible approach to regression or classification uses feed-forward neural networks with a single hidden layer. For regression, there is often only one output unit. A diagram of such a neural network is shown in Figure 14.2.2. The input units $X_1, \cdots, X_k$ distribute the inputs to the *hidden* units $Z_1, \cdots, Z_m$ in the second layer. The hidden units accumulate their inputs together with a constant term (the bias) as a weighted sum. A known activation function ϕ_ℓ is applied to the weighted sum, giving

$$Z_\ell = \phi_\ell(\mathbf{x}'\boldsymbol{\tau}_\ell) = \phi_\ell\left(\tau_{0\ell} + \sum_{j=1}^{k} X_j\tau_{j\ell}\right), \quad \ell = 1, \cdots, m;$$

where $\mathbf{x} = (1, X_1, \cdots, X_k)'$ and $\boldsymbol{\tau}_\ell = (\tau_{0\ell}, \tau_{1\ell}, \cdots, \tau_{k\ell})'$. The single output unit has the same form, except with output function ψ. This gives

$$Y = \psi(\mathbf{z}'\boldsymbol{\beta}) = \psi\left(\beta_0 + \sum_{\ell=1}^{m} Z_\ell\beta_\ell\right),$$

where $\mathbf{z} = (1, Z_1, \cdots, Z_m)'$. Denote $\mathbf{w} = (\boldsymbol{\tau}_1', \cdots, \boldsymbol{\tau}_m', \boldsymbol{\beta}')'$. Then,

$$Y = f(\mathbf{x}; \mathbf{w}) = \psi\left(\beta_0 + \sum_{\ell=1}^{m} \beta_\ell\phi_\ell(\mathbf{x}'\boldsymbol{\tau}_\ell)\right). \tag{14.2.13}$$

The usual choice for the activation function of the hidden layer is the logistic function

$$\phi_\ell(u) = \frac{\exp(u)}{1+\exp(u)}, \quad \ell = 1, \cdots, m.$$

Given observations $(\mathbf{x}_1, Y_1), \cdots, (\mathbf{x}_N, Y_N)$, the weights $\tau_{j\ell}$ and β_ℓ are estimated numerically according to the least squares criterion, which minimizes

$$R(\mathbf{w}) = \frac{1}{2}\sum_{i=1}^{N}\{Y_i - f(\mathbf{x}_i, \mathbf{w})\}^2 = \frac{1}{2}\sum_{i=1}^{N}[\widehat{\varepsilon}_i(\mathbf{w})]^2 \tag{14.2.14}$$

(Rumelhart et al., 1986; Thisted, 1988; Warner and Misra, 1996). Gradient descent is the standard approach employed, which uses the iteration

$$\mathbf{w}^{(n+1)} = \mathbf{w}^{(n)} - \delta_n \sum_{i=1}^{N} \widehat{\varepsilon}_i(\mathbf{w}^{(n)}) \nabla \widehat{\varepsilon}_i(\mathbf{w}^{(n)}), \quad n = 1, 2, 3, \cdots$$

where $\delta_n > 0$ is the learning rate. The procedure is commonly known as *back-propagation*, referring to the way in which the errors at the output unit, i.e., $\widehat{\varepsilon}_i(\mathbf{w}^{(n)})$, are used to update the weights backward. That is, by the chain rule of differentiation, the errors are first used to update the weights between the hidden layer and the output unit, i.e., $\boldsymbol{\beta}$, and then they are used to update the weights between the input layer and the hidden layer, i.e., $\boldsymbol{\tau}_1, \cdots, \boldsymbol{\tau}_m$ (see Hastie et al., 2009).

Numerical Example 14.7. The dataset "marketing" from the package *datarium* is useful for studying the impact of three advertising media options (youtube, facebook and newspaper) on sales. The data frame consists of the advertising budget (in 1000's of USD) along with the sales. The advertising experiment was repeated 200 times.

```
library(datarium)
data(marketing)
# scale features to be in (0,1)
scale01 <- function(x) {
  (x - min(x)) / (max(x) - min(x))}
marketing <- marketing %>% mutate_all(scale01)
# split into training and test
marketing_train <- sample_frac(tbl = marketing,
              replace = FALSE, size = 0.80)
marketing_test <- anti_join(marketing, marketing_train)
```

We fit a simple 1-hidden layer neural net with 1 neuron (plot is not shown).

```
library(neuralnet)
set.seed(123457)
nn.1 <- neuralnet(sales ~ youtube + facebook + newspaper,
            data = marketing_train)
plot(nn.1, rep = "best")
```

It is useful to see what the SSE of the test data is.

```
test_nn.1_out <- compute(nn.1,marketing_test[, 1:3])$net.result
nn.1_test_SSE <- sum((test_nn.1_out - marketing_test[, 4])^2)/2
nn.1_test_SSE
```

▲

14.3 Regularized regression

While the least squares estimator has the smallest MSE of all unbiased linear estimators in the general linear model (4.1.1), we have seen that it is possible to achieve a smaller MSE or better prediction accuracy by considering *biased estimation*, by trading a little bias for a large reduction in variance. An example is the ridge regression approach described in Section 9.2.3. It can be viewed as a *regularized least squares method*, using the square of the L_2-norm of the coefficients, i.e., $\boldsymbol{\beta}'\boldsymbol{\beta}$, as *penalty*, and resulting in the (ridge) estimator which is a linear function of $\mathbf{y}$.

Definition 14.3.1. The regularized or penalized least squares estimation consists of the optimization

$$\min_{\boldsymbol{\beta}\in\mathcal{R}^k} \left\{\|\mathbf{y}-\mathbf{X}\boldsymbol{\beta}\|_2^2 + \lambda J(\boldsymbol{\beta})\right\}, \tag{14.3.1}$$

where $J(\boldsymbol{\beta})$ is a penalty function and $\lambda \geq 0$ is a regularization parameter that typically depends on the number of observations N. Note that the square of the L_2-norm $\|\mathbf{y}-\mathbf{X}\boldsymbol{\beta}\|_2^2$ is the same as $\|\mathbf{y}-\mathbf{X}\boldsymbol{\beta}\|^2$ that we defined in Chapter 4. The constrained least square estimation consists of the optimization

$$\min_{\boldsymbol{\beta}\in\mathcal{R}^k} \|\mathbf{y}-\mathbf{X}\boldsymbol{\beta}\|_2^2 \quad \text{subject to } \boldsymbol{\beta}\in C, \tag{14.3.2}$$

where C is a set, usually convex.

A few commonly used penalty functions are $J(\boldsymbol{\beta}) = \|\boldsymbol{\beta}\|_p$ with $p = 0, 1, 2$, where

$$\|\boldsymbol{\beta}\|_p = \begin{cases} \sum_{j=1}^k I\{\beta_j \neq 0\} & p = 0 \\ \sum_{j=1}^k |\beta_j|^p & p \in (0,1) \\ (\sum_{j=1}^k |\beta_j|^p)^{1/p} & p \geq 1 \end{cases}$$

are referred to as the L_p-norm; $\|\boldsymbol{\beta}\|_0$ is the number of non-zero β's. Note that for $p < 1$, $\|\boldsymbol{\beta}\|_p$ are not vector norms, because for $c \in \mathcal{R}$, $\|c\boldsymbol{\beta}\|_p = |c|^p\|\boldsymbol{\beta}\|_p \not\equiv |c|\|\boldsymbol{\beta}\|_p$ (see Definition 1.3.20). When $\|\boldsymbol{\beta}\|_p$ is used as a penalty, the constrained set in (14.3.2) usually takes the form $C = \{\boldsymbol{\beta}\colon \|\boldsymbol{\beta}\|_p \leq c\}$ for some $c > 0$. The set is convex when $p \geq 1$. However, C is not convex when $p < 1$, making the computation and analysis of the constrained optimization more difficult. Below are some comments on the regularized and constrained LS estimations based on these L_p-norms.

1. The L_0-regularized LS estimation and its constrained version respectively take the forms

$$\min_{\boldsymbol{\beta}\in\mathcal{R}^k} \left\{\frac{1}{2}\|\mathbf{y}-\mathbf{X}\boldsymbol{\beta}\|_2^2 + \lambda\|\boldsymbol{\beta}\|_0\right\}, \quad \text{and} \tag{14.3.3}$$

$$\min_{\boldsymbol{\beta}\in\mathcal{R}^k} \|\mathbf{y}-\mathbf{X}\boldsymbol{\beta}\|_2^2 \quad \text{subject to } \|\boldsymbol{\beta}\|_0 \leq u, \tag{14.3.4}$$

   where $\lambda \geq 0$ and $u \in \{0, 1, 2, \cdots\}$. The estimation (14.3.4) leads to best subset selection (see Section 9.1).

2. The L_1-regularized LS estimation and its constrained version respectively take the forms

$$\min_{\boldsymbol{\beta}\in\mathcal{R}^k} \left\{\frac{1}{2}\|\mathbf{y}-\mathbf{X}\boldsymbol{\beta}\|_2^2 + \lambda\|\boldsymbol{\beta}\|_1\right\}, \quad \text{and} \tag{14.3.5}$$

$$\min_{\boldsymbol{\beta}\in\mathcal{R}^k} \|\mathbf{y}-\mathbf{X}\boldsymbol{\beta}\|_2^2 \quad \text{subject to } \|\boldsymbol{\beta}\|_1 \leq u, \tag{14.3.6}$$

where $\lambda \geq 0$ and $u \geq 0$. The regularized regression (14.3.5) is also known as Lasso (Tibshirani, 1996). Lasso has become a popular approach to regression in the high-dimensional setting. Unlike the classical setting, which has fixed dimensionality k and large sample size N, in a high-dimensional setting, k can grow polynomially with the sample size N, i.e., $k = O(N^\alpha)$ for some $\alpha > 0$, or even faster, and may be bigger than N.

3. The L_2-regularized LS estimation and its constrained version respectively take the forms

$$\min_{\beta \in \mathcal{R}^k} \left\{ \frac{1}{2} \|\mathbf{y} - \mathbf{X}\beta\|_2^2 + \lambda \|\beta\|_2^2 \right\}, \quad \text{and} \tag{14.3.7}$$

$$\min_{\beta \in \mathcal{R}^k} \|\mathbf{y} - \mathbf{X}\beta\|_2^2 \text{ subject to } \|\beta\|_2^2 \leq u, \tag{14.3.8}$$

where $\lambda \geq 0$ and $u \geq 0$. This is also known as the ridge regression.

It can be shown that the regularized Lasso (14.3.5) and its constrained form (14.3.6) are equivalent. That is, for any $\lambda \geq 0$ and solution $\widehat{\beta}$ to (14.3.5), there is a $u \geq 0$ such that $\widehat{\beta}$ solves (14.3.6), and vice versa. Likewise, (14.3.7) and (14.3.8) are equivalent. Note that LS regression with L_2 penalty may also mean that

$$\min_{\beta \in \mathcal{R}^k} \frac{1}{2} \|\mathbf{y} - \mathbf{X}\beta\|_2^2 + \lambda \|\beta\|_2,$$

which has $\|\beta\|_2$ as the penalty, instead of $\|\beta\|_2^2$ as in (14.3.7). Despite the difference, the regression is equivalent to ridge regression (14.3.7) but not as computationally convenient to use. For both L_1-regularized and L_2-regularized regressions, there is another often used constrained version that takes the form

$$\min_{\beta \in \mathcal{R}^k} \|\beta\|_p \quad \text{subject to } \|\mathbf{y} - \mathbf{X}\beta\|_2^2 \leq R, \quad p = 1, 2, \tag{14.3.9}$$

where $R \geq 0$. When $p = 1$, (14.3.9) is referred to as relaxed basis pursuit by Chen et al. (1998). This constrained form is equivalent to the above two forms.

On the other hand, (14.3.3) and (14.3.4) are not equivalent. For every $\lambda \geq 0$ and solution $\widehat{\beta}$ in (14.3.3), $\widehat{\beta}$ solves (14.3.4) with $u = \|\widehat{\beta}\|_0$. However, if for some u, (14.3.4) has two solutions $\widehat{\beta}_1$ and $\widehat{\beta}_2$ with $\|\widehat{\beta}_1\|_0 < \|\widehat{\beta}_2\|_0 \leq u$, then it is not possible for $\widehat{\beta}_2$ to be solution to (14.3.3) for any $\lambda \geq 0$. Ridge regression was discussed in Section 9.2.3. In the following sections, we describe L_0- and L_1-regularizations in linear models.

14.3.1 L_0-regularization

While the most immediate goal of the L_0-regularized LS estimation is to determine the parameter values of a given linear model in order to best fit the data, it can also be understood from the perspective of model selection via penalized empirical risk minimization, with model dimension as penalty. For a comprehensive theory on risk bounds for this approach to model selection, see Barron et al. (1999).

Suppose $Y_i = f(\mathbf{x}_i) + \varepsilon_i$, $i = 1, \cdots, N$, where $\mathbf{x}_1, \cdots, \mathbf{x}_N$ are known while the function f is unknown, and $\varepsilon_1, \cdots, \varepsilon_N$ are unobservable i.i.d. $\sim N(0, \sigma^2)$. Here $\mathbf{x}_i$ are elements of a vector space $\mathcal{X}$ that may or may not be finite dimensional. Suppose we wish to use a linear model to approximate the unknown function f. Consider a family of models $\{S_m, m \in \mathcal{M}\}$ with each model S_m being a finite dimensional space of linear functions. If $\mathcal{X} = \mathcal{R}^k$, then each S_m can be written as $\{\mathbf{x}'\beta\colon \beta \in E_m\}$, where E_m is a linear subspace of $\mathcal{R}^k$. Then

$\dim(S_m) = \dim(E_m) \le k$. On the other hand, if $\mathcal{X}$ is infinite dimensional, then $\dim(S_m)$ may be unbounded. For example, for $\mathcal{X} = \{(x_1, x_2, \cdots): \sum_i x_i^2 < \infty\}$, S_m could consist of all linear functions $\beta_1 x_1 + \cdots + \beta_m x_m$, so that $\dim(S_m) = m$.

For each m, let $\widehat{f}_m$ be the LS estimator using model S_m, so that

$$\widehat{f}_m = \arg\min_{g \in S_m} \frac{1}{N\sigma^2} \sum_{i=1}^{N} [Y_i - g(\mathbf{x}_i)]^2.$$

The penalized model selection then selects model $S_{\widehat{m}}$ so that

$$\widehat{m} = \arg\min_m \left[\frac{1}{N\sigma^2} \sum_{i=1}^{N} [Y_i - \widehat{f}_m(\mathbf{x}_i)]^2 + \text{pen}(m) \right],$$

where $\text{pen}(m)$ is a penalty term assigned to model S_m that takes the form

$$\text{pen}(m) = \kappa w_m D_m / n,$$

where κ is some positive constant, $D_m = \dim(S_m)$, and $w_m \ge 1$ is a weight that satisfies a condition of the type

$$\sum_m \exp(-w_m D_m) \le A < \infty.$$

The penalty term takes into account both the difficulty of estimating within the model S_m (the role of D_m) and the additional noise due to the size of the list of models (the role of w_m) and derives from exponential probability bounds for the empirical risk. Then by choosing a suitable κ,

$$\begin{aligned} &\mathrm{E}\left[\frac{1}{N\sigma^2} \sum_{i=1}^{N} [f(\mathbf{x}_i) - \widehat{f}_{\widehat{m}}(\mathbf{x}_i)]^2 \right] \\ &\le \kappa_1 \inf_m \left[\frac{1}{N\sigma^2} \inf_{g \in S_m} \sum_{i=1}^{N} [f(\mathbf{x}_i) - g(\mathbf{x}_i)]^2 + \text{pen}(m) \right] + \frac{\kappa_2 A}{N}, \end{aligned}$$

where κ_1 and κ_2 are numerical constants. Barron et al. (1999) showed that if $\kappa = 24$, then one can set $\kappa_1 = 3$ and $\kappa_2 = 32$.

Now suppose the general linear model (4.1.1) holds, so that $\mathbf{y} = \mathbf{X}\boldsymbol{\beta} + \boldsymbol{\varepsilon}$, where $\mathbf{X} = (\mathbf{x}_1, \cdots, \mathbf{x}_N)'$ with each $\mathbf{x}_i \in \mathcal{R}^k$. For complete variable selection, the family of models is $\{S_m, m \in \mathcal{M}\}$, so that m enumerates all the subsets of the predictors. From the above general result, one can set $w_m = 1 + \log k$ for all m so that the complete variable selection is exactly the L_0-regularized regression in (14.3.3) with $\lambda = \kappa(1 + \log k)/N$. As a result, the solution $\widehat{\boldsymbol{\beta}}_{L_0}$ to (14.3.3) satisfies

$$\begin{aligned} &\mathrm{E}\left\{ \frac{1}{N} \|\mathbf{X}\widehat{\boldsymbol{\beta}}_{L_0} - \mathbf{X}\boldsymbol{\beta}\|_2^2 \right\} \\ &\le \kappa_1 \inf_{E \subset \{1, \cdots, k\}} \left\{ \frac{1}{N} \|(\mathbf{I} - \mathbf{P}_E)\mathbf{X}\boldsymbol{\beta}\|_2^2 + \frac{\sigma^2 \kappa (1 + \log k)|E|}{N} \right\}, \end{aligned}$$

where κ_1 is some positive constant and $\mathbf{P}_E$ is the projection matrix onto the space spanned by the columns of $\mathbf{X}$ with indices in E. This upper bound on the prediction risk shows the trade-off between the approximation error (model bias) and the price we pay in the search

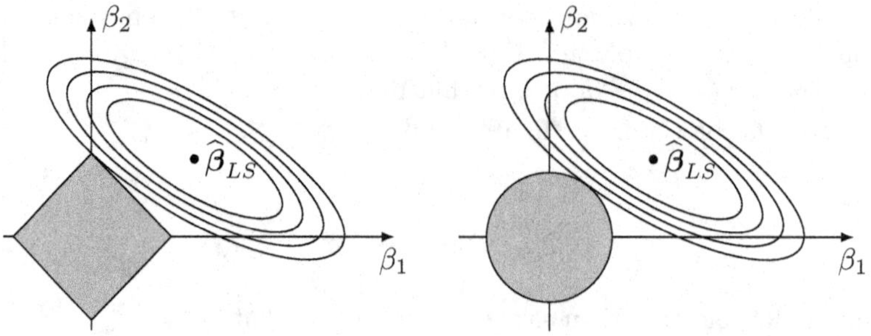

FIGURE 14.3.1. Geometry of Lasso and ridge regression.

over a large family of models. Note that the result is for the prediction loss $N^{-1}\|\mathbf{X}\widehat{\boldsymbol{\beta}}_{L_0} - \mathbf{X}\boldsymbol{\beta}\|_2^2$. Without further assumptions on $\mathbf{X}$, it is challenging to derive results for other losses such as the L_p-loss $\|\widehat{\boldsymbol{\beta}}_{L_0} - \boldsymbol{\beta}\|_p$, $p > 0$.

Although L_0-regularization methods have appealing risk properties, the computation for a global minimizer may be infeasible in high-dimensional situations because of the combinatorial optimization involved in (14.3.3) and (14.3.4). Unfortunately it may be infeasible to implement even in moderate dimensions, with k between, say 50 and 100. The computational difficulty stems from the discontinuity and nonconvexity of the L_0-norm. While it is possible to investigate non-convex methods and the behavior of local minimizers for variable selection Fan and Lv (2010), a natural idea is to replace the L_0-norm with some continuous or convex penalty function, as discussed in the following sections.

14.3.2 L_1-regularization and Lasso

Also known as L_1-regularized LS estimation, Lasso is an acronym for "least absolute shrinkage and selection operator". Three equivalent definitions of Lasso are given in (14.3.5), (14.3.6), and (14.3.9), respectively. Due to the nature of an L_1 penalty or constraint, Lasso is able to estimate some coefficients as exactly 0 and hence performs variable selection. Lasso enjoys some favorable properties of subset selection (interpretable models) and ridge regression (stability).

14.3.2.1 Geometry of Lasso

To understand some of the properties of Lasso, it is useful to take a look at its geometry. Consider the constrained form of Lasso (14.3.6). The constraint $\|\boldsymbol{\beta}\|_1 \leq u$ means that the search region for solutions is the L_1-ball of "radius" u, i.e., the region $\{|\beta_1|+|\beta_2|+\cdots+|\beta_k| \leq u\}$. When $k = 2$, this is a square centered at the origin with 45° rotation on the β_1–β_2 plane, as shown in the left panel of Figure 14.3.1. On the other hand, provided that $\mathbf{X}$ has full column rank, the isolines of each $N^{-1}\|\mathbf{y} - \mathbf{X}\boldsymbol{\beta}\|_2^2$ are elliptical contours with the same orientation and center; we see that the center is the LS estimate $\widehat{\boldsymbol{\beta}}_{LS}$. The rotated square and the "smallest" elliptical contour that "hits" it have a point in common, which is the solution $\widehat{\boldsymbol{\beta}}_{Lasso}$ to (14.3.6). Note that $\widehat{\boldsymbol{\beta}}_{Lasso}$ is necessarily closer in L_1-norm to the origin than $\widehat{\boldsymbol{\beta}}_{LS}$ is, unless $\widehat{\boldsymbol{\beta}}_{LS}$ is within the rotated square, in which case the constraint is too loose and the two estimates are equal. Therefore, Lasso performs shrinkage estimation. Furthermore, depending on the center and orientation of the elliptical contours, $\widehat{\boldsymbol{\beta}}_{Lasso}$ can be one of the corners of the rotated square, so that one of the coefficients is zero. As k

increases, the L_1-ball $\|\boldsymbol{\beta}\|_1 \le u$ not only has an increasing number of corners, i.e., vertices, but also an increasing number of low dimensional simplices on its surface, and so it is likely that some coefficients will be set equal to zero. This amounts to variable selection by identifying the active set of $\widehat{\boldsymbol{\beta}}_{Lasso}$ defined as follows.

Definition 14.3.2. Active set or support and Active signs. The active set or support of a solution vector $\widehat{\boldsymbol{\beta}}$ is

$$A = \text{supp}(\widehat{\boldsymbol{\beta}}) = \{j\colon \widehat{\beta}_j \neq 0\}.$$

Active signs are defined by

$$s_A \in \{-1, +1\}^{|A|},$$

indicating whether $\widehat{\beta}_j > 0$ or $\widehat{\beta}_j < 0$ for each $j \in A$.

Hence, Lasso performs shrinkage and (effectively) variable selection. Under suitable conditions on $\mathbf{X}$ and $\boldsymbol{\varepsilon}$, if the active set of the true value of $\boldsymbol{\beta}$ is "sparse", i.e., small compared to k, then with high-probability, the active set of $\widehat{\boldsymbol{\beta}}_{Lasso}$ is also sparse and closely matches the active set of $\boldsymbol{\beta}$, resulting in approximate variable selection consistency. In contrast, for ridge regression (see (14.3.8)), the constraint corresponds to a sphere around the origin, so it has no corners to hit; see the right panel of Figure 14.3.1. As a result, while ridge regression also performs shrinkage estimation, its estimate rarely has zero coefficients, and so even when the true value of $\boldsymbol{\beta}$ has the sparsity property, the ridge estimate rarely does, making it unsuitable for variable selection.

The Lasso solution $\widehat{\boldsymbol{\beta}}_{Lasso}$ depends on the regularization parameter λ. As $\lambda \to \infty$, $\widehat{\boldsymbol{\beta}}_{Lasso}$ moves continuously to zero. A useful graphical method to examine the path of $\widehat{\boldsymbol{\beta}}_{Lasso}$ is to plot each coordinate of $\widehat{\boldsymbol{\beta}}_{Lasso}$ versus λ. In practice, we need to determine a value of λ, say, λ^* (obtained using some cross-validation techniques, say) and use the corresponding $\widehat{\boldsymbol{\beta}}_{Lasso}$ as the final estimator. This requires fast algorithms for computing $\widehat{\boldsymbol{\beta}}_{Lasso}$ for a grid of λ values.

We discuss two well-known computational algorithms for Lasso: the Fast Forward Stagewise (FS) Regression algorithms (Tibshirani, 2015) and Least angle regression (LARS) discussed in Efron et al. (2004).

14.3.2.2 Fast forward stagewise selection

This algorithm follows a very simple strategy for constructing a sequence of sparse regression estimates: it starts with all $\beta_j = 0$ and iteratively updates the coefficient (by a small amount) of that X variable which has the largest absolute inner product with the current residual. Assume standardized covariates and centered response (mean of Y is 0). Fix the step size or learning rate $\delta > 0$ (say at 0.01). Let $\boldsymbol{\beta}^{(0)} = \mathbf{0}$. For steps $\ell = 1, 2, \cdots$, repeat

$$\iota \in \arg\max_{j=1,\cdots,k} |\mathbf{x}_j'(\mathbf{y} - \mathbf{X}\boldsymbol{\beta}^{(\ell-1)})|, \text{ and} \tag{14.3.10}$$

$$\boldsymbol{\beta}^{(\ell)} = \boldsymbol{\beta}^{(\ell-1)} + \delta \ \text{sign}(\mathbf{x}_\iota'(\mathbf{y} - \mathbf{X}\boldsymbol{\beta}^{(\ell-1)}))\mathbf{e}_\iota \tag{14.3.11}$$

where $\mathbf{e}_\iota$ is a k-dimensional vector which has 1 in the ιth position and 0 elsewhere, and $\mathbf{x}_j$ is an N-dimensional vector. At each iteration, the procedure greedily selects the variable indexed by ι that has the largest absolute inner product (or correlation, for standardized variables) with the residual, and adds $s_\iota\delta$ to its coefficient, where s_ι is the sign of this inner product.

This procedure thus "slows down" the learning process. It may require 1000s of iterations to reach a model with < 100 predictors. The slow learning is the difference between the forward stagewise (FS) procedure and the forward/stepwise selection procedure (see Section 9.1). Recall that at each iteration $\ell = 1, 2, \cdots$, the forward selection procedure chooses a variable similar to (14.3.10). Suppose A denotes the set of variables selected at the end of iteration $\ell - 1$. At iteration ℓ, the forward selection procedure chooses X_ι such that when Y is regressed onto the variables in $A \cup \{\iota\}$, the SSE is smallest, i.e., it chooses ι such that

$$|\tilde{\mathbf{x}}_\iota'(\mathbf{y} - \mathbf{X}\boldsymbol{\beta}^{(\ell-1)})|$$

is largest, where $\tilde{\mathbf{x}}_\iota'$ is the residual from regressing X_ι on all variables in A. Both are greedy algorithms.

14.3.2.3 Least angle regression

This is a unified approach that motivates Lasso and FS regression. Least angle regression (LARS) provides fast implementation and a fast way to choose the tuning parameter λ. LARS is a stylized version of the FS procedure. Only k steps are required for the full set of solutions, instead of the 1000's of steps taken by FS. The steps are as follows:

1. Standardize the predictors to have mean zero and unit norm. Center the response. Start with $\widehat{\boldsymbol{\beta}} = \mathbf{0}$ (similar to FS).

2. Find the predictor most correlated with Y, say X_{j_1}.

3. Take the largest step possible in the direction of this predictor X_{j_1} until some other predictor, say X_{j_2}, has as much correlation with the current residual.

4. At this point, LARS parts company with FS. Instead of continuing along X_{j_1}, LARS proceeds in a direction equiangular between the two predictors until a third variable X_{j_3} earns its way into the "most correlated" set.

5. Then, LARS proceeds equiangularly between $X_{j_1}, X_{j_2}, X_{j_3}$, i.e., along the "least angle direction", until a fourth variable enters. This procedure is continued.

6. LARS builds up estimates $\widehat{\boldsymbol{\mu}} = \mathbf{X}\widehat{\boldsymbol{\beta}}$, in successive steps, each step adding one covariate to the model, so that after k steps, k of the $\widehat{\boldsymbol{\beta}}_j$'s are non-zero.

It is interesting to see the LARS algorithm in "Lasso mode". Set an iteration counter $\ell = 0$. Let $\lambda_0 = \infty$, active set $A = \emptyset$, and active signs $s_A = \emptyset$; see Definition 14.3.2. Let $\mathbf{X}_A$ denote the matrix that only consists of the column vectors of $\mathbf{X}$ with indices in A. While $\lambda_\ell > 0$, iterate the following steps for $\ell = 1, 2, \cdots$. Let $\max^+$ denote the maximum of the arguments that are smaller than λ_ℓ.

(i) As λ decreases from a knot λ_ℓ, compute the lasso solution:

$$\widehat{\boldsymbol{\beta}}_A(\lambda) = (\mathbf{X}_A'\mathbf{X}_A)^{-1}(\mathbf{X}_A'\mathbf{y} - \lambda s_A), \quad \widehat{\boldsymbol{\beta}}_{-A}(\lambda) = \mathbf{0}. \tag{14.3.12}$$

(ii) Compute the next "hitting time" as

$$\lambda_{\ell+1}^{\text{hit}} = \max^+_{j \notin A, s_j \in \{-1,1\}} \frac{\mathbf{x}_j'(\mathbf{I} - \mathbf{X}_A(\mathbf{X}_A'\mathbf{X}_A)^{-1}\mathbf{X}_A')\mathbf{y}}{s_j - \mathbf{x}_j'(\mathbf{I} - \mathbf{X}_A(\mathbf{X}_A'\mathbf{X}_A)^{-1} s_A}. \tag{14.3.13}$$

(iii) Compute the next "crossing time" as

$$\lambda_{\ell+1}^{\text{cross}} = \max_{j\in A}{}^{+} \frac{[(\mathbf{X}_A'\mathbf{X}_A)^{-1}\mathbf{X}_A'\mathbf{y}]_j}{[(\mathbf{X}_A'\mathbf{X}_A)^{-1}s_A]_j}. \tag{14.3.14}$$

(iv) Decrease λ until you reach $\lambda_{\ell+1}$ defined as

$$\lambda_{\ell+1} = \max\{\lambda_{\ell+1}^{\text{hit}}, \lambda_{\ell+1}^{\text{cross}}\}. \tag{14.3.15}$$

(v) If $\lambda_{\ell+1}^{\text{hit}} > \lambda_{\ell+1}^{\text{cross}}$, we add the hitting predictor variable to the active set A and its sign to the active signs s_A. Else, if $\lambda_{\ell+1}^{\text{hit}} \le \lambda_{\ell+1}^{\text{cross}}$, remove the crossing variable from A and remove its sign from s_A. Update ℓ to $\ell+1$.

In Step (i), when we decrease λ from a knot λ_ℓ, we can rewrite the Lasso update as

$$\widehat{\boldsymbol{\beta}}_A(\lambda) = \widehat{\boldsymbol{\beta}}_A(\lambda_\ell) + (\lambda_\ell - \lambda)(\mathbf{X}_A'\mathbf{X}_A)^{-1}s_A, \quad \widehat{\boldsymbol{\beta}}_{-A}(\lambda) = \mathbf{0} \tag{14.3.16}$$

For decreasing λ, the LARS algorithm moves the active $\boldsymbol{\beta}$ coefficients in the direction

$$(\lambda_\ell - \lambda)(\mathbf{X}_A'\mathbf{X}_A)^{-1}s_A \tag{14.3.17}$$

and corresponding fitted values proceed as

$$\mathbf{X}\widehat{\boldsymbol{\beta}}_A(\lambda) = \mathbf{X}\widehat{\boldsymbol{\beta}}_A(\lambda_\ell) + (\lambda_\ell - \lambda)\mathbf{X}_A(\mathbf{X}_A'\mathbf{X}_A)^{-1}s_A \tag{14.3.18}$$

Efron et al. (2004) called $\mathbf{X}_A(\mathbf{X}_A'\mathbf{X}_A)^{-1}s_A$ the equiangular direction.

Although it is not as elegant as LARS, the coordinate descent algorithm (Fu, 1998; Tseng, 2001) is a simple, stable and efficient algorithm for a variety of high-dimensional models. Coordinate descent algorithms optimize a target function with respect to a single parameter at a time, iteratively cycling through all parameters until convergence is reached. They are ideal for problems that have a simple closed form solution in a single dimension but lack this in higher dimensions. See Friedman et al. (2007) for properties of this algorithm as well as some extensions beyond Lasso.

14.3.3 Elastic net

Similar to ridge regression (but unlike the best subset selection that we have seen in Chapter 9), Lasso pertains to a convex optimization problem. However, unlike ridge regression, Lasso is not always strictly convex, and therefore, it need not always have a unique solution. Zou and Hastie (2005) defined "elastic net", which is always strictly convex, and combines the predictive properties of ridge regression with the sparsity properties of Lasso.

Let $\mathbf{y}$ denote a mean centered response vector and $\mathbf{X}$ denote a standardized matrix of predictors. The elastic net criterion is

$$E(\boldsymbol{\beta}) = \|\mathbf{y} - \mathbf{X}\boldsymbol{\beta}\|_2^2 + \lambda\{\alpha\|\boldsymbol{\beta}\|_1 + (1-\alpha)\|\boldsymbol{\beta}\|_2^2\},$$

where $\lambda > 0$ and $\alpha \in (0,1)$ are unknown parameters. The elastic net penalty

$$P_\alpha(\boldsymbol{\beta}) = \alpha\|\boldsymbol{\beta}\|_1 + (1-\alpha)\|\boldsymbol{\beta}\|_2^2 \tag{14.3.19}$$

is a convex combination of the Lasso penalty and the ridge penalty. The naive elastic net estimator $\widehat{\boldsymbol{\beta}}_{NEN}$ is then obtained as

$$\widehat{\boldsymbol{\beta}}_{NEN} = \arg\min_{\boldsymbol{\beta}} E(\boldsymbol{\beta}). \tag{14.3.20}$$

Since the elastic net criterion can be written as

$$E(\boldsymbol{\beta}) = \|\tilde{\mathbf{y}} - \tilde{\mathbf{X}}\boldsymbol{\beta}\|_2^2 + \lambda_1\|\boldsymbol{\beta}\|_1,$$

where

$$\tilde{\mathbf{y}} = \begin{pmatrix} \mathbf{y} \\ \mathbf{0}_k \end{pmatrix}, \quad \tilde{\mathbf{X}} = \begin{pmatrix} \mathbf{X} \\ \sqrt{\lambda_2}\mathbf{I}_k \end{pmatrix}$$

with $\lambda_1 = \lambda\alpha$ and $\lambda_2 = \lambda(1-\alpha)$, the optimization problem for the elastic net can be transformed into one for the Lasso. The elastic net estimator is obtained from $\widehat{\boldsymbol{\beta}}_{NEN}$ via a scaling correction:

$$\widehat{\boldsymbol{\beta}}_{EN} = (1+\lambda_2)\widehat{\boldsymbol{\beta}}_{NEN}.$$

It can be shown that

$$\widehat{\boldsymbol{\beta}}_{EN} = \arg\min_{\boldsymbol{\beta}} \left\{ \boldsymbol{\beta}' \left(\frac{\mathbf{X}'\mathbf{X} + \lambda_2\mathbf{I}}{1+\lambda_2} \right) \boldsymbol{\beta} - 2\mathbf{y}'\mathbf{X}\boldsymbol{\beta} + \lambda_1\|\boldsymbol{\beta}\|_1. \right\}. \tag{14.3.21}$$

Since

$$\widehat{\boldsymbol{\beta}}_{Lasso} = \arg\min_{\boldsymbol{\beta}}\{\boldsymbol{\beta}'\mathbf{X}'\mathbf{X}\boldsymbol{\beta} - 2\mathbf{y}'\mathbf{X}\boldsymbol{\beta} + \lambda_1\|\boldsymbol{\beta}\|_1\},$$

$\widehat{\boldsymbol{\beta}}_{EN}$ can be viewed as a stabilized version of $\widehat{\boldsymbol{\beta}}_{Lasso}$.

Numerical Example 14.8. We illustrate regularized regression Lasso and elastic net using the "Hitters" dataset from the R package *ISLR*, which deals with baseball related data. The data frame has 322 rows for major league players on 20 variables on their offensive and defensive statistics as well as their salary. After removing rows with missing data, we analyze a complete data frame with 263 rows on 20 variables.

```
library(ISLR)
data(Hitters, package = "ISLR")
Hitters <- na.omit(Hitters)
X <- model.matrix(Salary ~ ., Hitters)[, -1]
y <- Hitters$Salary
```

In the code below, the LS fit is first obtained, and then the R package *glmnet* is used to perform ridge estimation ($\alpha = 0$), Lasso estimation ($\alpha = 1$), and elastic net estimation ($\alpha = 0.25$ and 0.75). The plots in Figure 14.3.2 show the traces of coefficients versus $\log \lambda$; the numbers on the top of each plot shows the number of predictors included in the model.

```
#LS fit
fit <- lm(Salary ~ ., Hitters)
c.ls <- coef(fit)
library(glmnet)
ridge    <- glmnet(X, y, alpha = 0.0)
lasso    <- glmnet(X, y, alpha = 1.0)
enet.1 <- glmnet(X, y, alpha = 0.25) #close to ridge
enet.2 <- glmnet(X, y, alpha = 0.75) #close to Lasso
par(mfrow = c(2, 2))
plot(ridge, xvar = "lambda", main = "Ridge (alpha = 0)")
plot(lasso, xvar = "lambda", main = "Lasso (alpha = 1)")
plot(enet.1, xvar = "lambda", main = "Elastic Net (alpha = .25)")
plot(enet.2, xvar = "lambda", main = "Elastic Net (alpha = .75)")
```

Cross-validation is used to select the parameter λ using the `cv.glmnet()` function with the (default) 10 folds, and the penalty term using the minimum λ value is shown in each case for Lasso and two elastic net fits. We then show the matrix of estimated coefficients from each fit. Clearly, Lasso penalizes the most, followed by elastic net with $\alpha = 0.75$.

```
#Ridge fit
fit.ridge.cv <- cv.glmnet(X, y, alpha = 0)
fit.ridge.cv
c.ridge <- coef(fit.ridge.cv)
sum(coef(fit.ridge.cv, s = "lambda.min")[-1] ^ 2)
#Lasso fit
fit.lasso.cv <- cv.glmnet(X, y, alpha = 1)
fit.lasso.cv
c.lasso <- coef(fit.lasso.cv, s = "lambda.min")
sum(coef(fit.lasso.cv, s = "lambda.min")[-1] ^ 2)
# Elastic net fit, alpha = 0.25
fit.enet.1.cv <- cv.glmnet(X, y, alpha = 0.25)
fit.enet.1.cv
c.enet.1 <- coef(fit.enet.1.cv, s = "lambda.min")
# Elastic net fit, alpha = 0.75
fit.enet.2.cv <- cv.glmnet(X, y, alpha = 0.75)
fit.enet.2.cv
c.enet.2 <- coef(fit.enet.2.cv, s = "lambda.min")
(c <- round(cbind(c.ls, c.ridge, c.lasso, c.enet.1, c.enet.2), 3))
```

```
20 x 5 sparse Matrix of class "dgCMatrix"
                c.ls       1        1        1        1
(Intercept)  163.104 226.844  129.416  144.700  133.580
AtBat         -1.980   0.087   -1.613   -1.666   -1.636
Hits           7.501   0.353    5.806    5.779    5.824
HmRun          4.331   1.144    .        0.486    0.060
Runs          -2.376   0.569    .       -0.106    .
RBI           -1.045   0.570    .        0.010    .
Walks          6.231   0.735    4.847    5.140    4.923
Years         -3.489   2.397   -9.972   -9.881  -10.370
CAtBat        -0.171   0.007    .       -0.042    .
CHits          0.134   0.028    .        0.130    .
CHmRun        -0.173   0.208    0.537    0.635    0.553
CRuns          1.454   0.056    0.681    0.729    0.693
CRBI           0.808   0.058    0.390    0.392    0.394
CWalks        -0.812   0.057   -0.556   -0.608   -0.575
LeagueN       62.599   2.850   32.465   50.200   33.335
DivisionW   -116.849 -20.329 -119.348 -120.927 -119.942
PutOuts        0.282   0.049    0.274    0.279    0.276
Assists        0.371   0.007    0.186    0.265    0.201
Errors        -3.361  -0.128   -2.165   -3.215   -2.397
NewLeagueN   -24.762   2.654    .      -15.960    .
```

▲

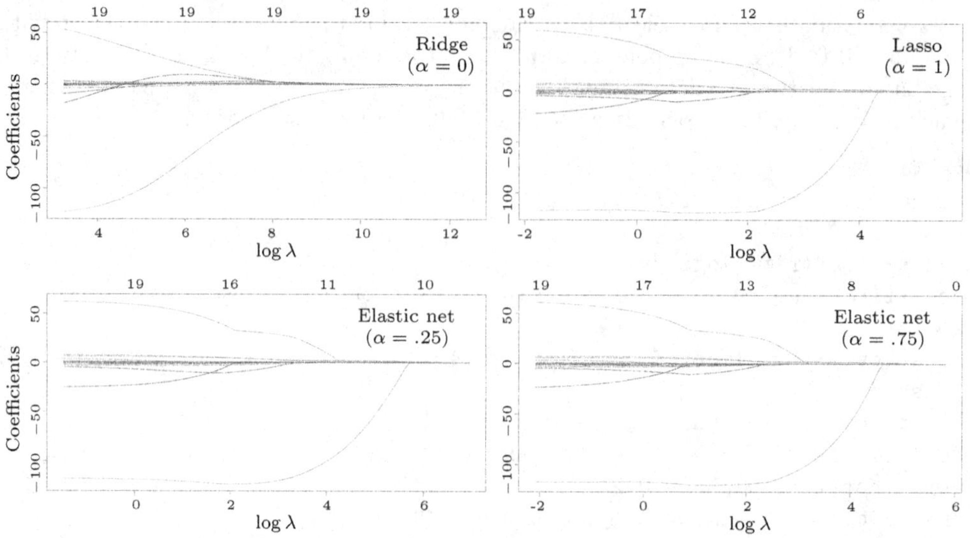

FIGURE 14.3.2. Traces of regularized estimates for the Hitters data.

14.4 Missing data analysis

In many situations, a portion of the data might be missing; these missing observations may have a significant impact on the statistical analysis of the non-missing data. Suppose $\mathbf{y}_{tot} = (\mathbf{y}_{obs}, \mathbf{y}_{mis}) \in \mathcal{R}^N$ denotes the *complete data,* where $\mathbf{y}_{obs}$ and $\mathbf{y}_{mis}$ respectively denote the N_1- and N_2-dimensional vectors of *observed data* and the *missing data,* with $N_1 + N_2 = N$. The missing values are said to be *missing at random* (MAR) if the observed units are a random subsample of the sampled units; in this case, the missing data mechanism is ignorable. On the other hand, if the probability that $Y_{tot,i}$ is observed depends on the value of $Y_{tot,i}$, then the missing values are not MAR, and the missing data mechanism is nonignorable (Little and Rubin, 1987). Suppose we partition the $N \times p$ predictor matrix as $\mathbf{X} = (\mathbf{X}_{obs}, \mathbf{X}_{mis})$, where $\mathbf{X}_{obs}$ is an $N_1 \times p$ matrix and $\mathbf{X}_{mis}$ is an $N_2 \times p$ matrix. In general, there are three approaches for handling missing data: (i) complete case analysis, (ii) imputation for missing values, and (iii) model based techniques via the likelihood function. We take a brief look at (i) and (ii), before discussing (iii) in more detail.

As the name suggests, *complete case analysis* consists of doing the usual statistical analyses using only the cases where all observations are available. Standard software may be used, which is attractive. The disadvantage is that if the percentage of missingness (which is $100N_2/N$) is large, and if the data is not MAR, such analyses tend to be inefficient. An alternative approach is *imputation of missing data,* usually occuring in the $\mathbf{X}$ matrix. Standard procedures are then used to analyze the data with $\widehat{\mathbf{X}}^* = (\mathbf{X}_{obs}, \widehat{\mathbf{X}}^*_{mis})$. The following methods of imputation are commonly used in practice. Hot deck imputation consists of selecting imputed values from the sample distribution. In cold deck imputation, we replace a missing value by a constant, usually obtained extraneously. Mean imputation consists of replacing missing values by the averages of available observations. Sometimes, we may use model based imputation. For instance, regression (or correlation) imputation consists of fitting a regression model of the missing data on observed values, and substituting the resulting predicted values for the missing values. We next discuss model based techniques which use a factorization of the *complete data likelihood.*

Assume that the data are MAR. Let $f(\mathbf{y}_{\text{tot}};\boldsymbol{\theta}) = f(\mathbf{y}_{\text{obs}},\mathbf{y}_{\text{mis}};\boldsymbol{\theta})$ denote the joint p.d.f. of $\mathbf{y}_{\text{obs}}$ and $\mathbf{y}_{\text{mis}}$, from which we obtain the marginal p.d.f. of $\mathbf{y}_{\text{obs}}$ as

$$f(\mathbf{y}_{\text{obs}};\boldsymbol{\theta}) = \int f(\mathbf{y}_{\text{obs}},\mathbf{y}_{\text{mis}};\boldsymbol{\theta})\, d\mathbf{y}_{\text{mis}}.$$

The likelihood function for $\boldsymbol{\theta}$ which ignores the missing-data mechanism, and is based only on $f(\mathbf{y}_{\text{obs}};\boldsymbol{\theta})$ is called the *incomplete or observed data likelihood* and is

$$L(\boldsymbol{\theta};\mathbf{y}_{\text{obs}}) \propto f(\mathbf{y}_{\text{obs}};\boldsymbol{\theta}). \tag{14.4.1}$$

The objective is to maximize $L(\boldsymbol{\theta};\mathbf{y}_{\text{obs}})$ with respect to $\boldsymbol{\theta}$. If $L(\boldsymbol{\theta};\mathbf{y}_{\text{obs}})$ is unimodal and differentiable, the MLE of $\boldsymbol{\theta}$ is obtained either as the closed-form solution to the ML equations

$$\partial \log L(\boldsymbol{\theta};\mathbf{y}_{\text{obs}})/\partial\boldsymbol{\theta} = \mathbf{0},$$

or by using iterative methods such as the Newton–Raphson algorithm. In many cases, it may be cumbersome to maximize the incomplete data likelihood $L(\boldsymbol{\theta};\mathbf{y}_{\text{obs}})$. The EM algorithm is useful for such situations.

14.4.1 EM algorithm

Suppose we denote the complete data by $\mathbf{z}$, with unknown p.d.f. $f(\mathbf{z};\boldsymbol{\theta})$. In numerous applications, $\mathbf{z}$ can only be partially observed as $\mathbf{y} = M(\mathbf{z})$, where M is a many-to-one mapping. Hence, observing $\mathbf{y}$ does not fully recover $\mathbf{z}$. For example, in missing data analysis, $M(\mathbf{y}_{\text{tot}}) = \mathbf{y}_{\text{obs}}$. In other applications, $\mathbf{y}$ can denote observed data, $\mathbf{u}$ can denote unobserved (latent) information, and $\mathbf{z} = (\mathbf{y}',\mathbf{u}')'$ denote complete data.

In principle, one could calculate the p.d.f. of $\mathbf{y}$ by integrating $f(\mathbf{z};\boldsymbol{\theta})$ over $\mathbf{z}$ with $M(\mathbf{z}) = \mathbf{y}$. Then $\widehat{\boldsymbol{\theta}}_{ML}$ can be found by solving the equation

$$\partial\ell(\boldsymbol{\theta};\mathbf{y})/\partial\boldsymbol{\theta} = \mathbf{0}, \tag{14.4.2}$$

where $\ell(\boldsymbol{\theta};\mathbf{y}) = \log f(\mathbf{y};\boldsymbol{\theta})$ is the observed-data log-likelihood function based on $\mathbf{y}$. There are, however, some issues in the actual numerical implementation of this approach. The integration of $f(\mathbf{z};\boldsymbol{\theta})$ to get the p.d.f. of $\mathbf{y}$ is often complicated and lacks tractable forms, and as a result, standard optimization algorithms, such as Newton–Raphson algorithm or gradient descent, can be messy to implement when applied to numerical integrals of $f(\mathbf{z};\boldsymbol{\theta})$. An alternative, iterative approach is the expectation-maximization (EM) algorithm (Dempster et al., 1977). Although the convergence of the EM algorithm is typically slower than the Newton–Raphson algorithm, it is widely used due to simplicity of implementation and stable ascent of likelihood; see Result 14.4.1 below.

Denote by $f(\mathbf{z}\,|\,\mathbf{y};\boldsymbol{\theta})$ the conditional p.d.f. of $\mathbf{z}$ given $M(\mathbf{z}) = \mathbf{y}$ if the complete data $\mathbf{z}$ has p.d.f. $f(\mathbf{z};\boldsymbol{\theta})$, and let $\mathrm{E}(\cdot\,|\,\mathbf{y};\boldsymbol{\theta})$ denote the corresponding conditional expectation. Given $\mathbf{y}$, define

$$\begin{aligned} Q(\boldsymbol{\theta},\boldsymbol{\theta}^*) = Q(\boldsymbol{\theta},\boldsymbol{\theta}^*;\mathbf{y}) &= \mathrm{E}[\ell(\boldsymbol{\theta};\mathbf{z})\,|\,\mathbf{y};\boldsymbol{\theta}^*] \\ &= \int \ell(\boldsymbol{\theta};\mathbf{z}) f(\mathbf{z}\,|\,\mathbf{y};\boldsymbol{\theta}^*)\, d\mathbf{z}. \end{aligned}$$

The EM algorithm starts with an initial estimate $\boldsymbol{\theta}^{(0)}$ and at the $(m+1)$st iteration, $m = 0, 1, \cdots$, it updates the estimate to

$$\boldsymbol{\theta}^{(m+1)} = \arg\max_{\boldsymbol{\theta}} Q(\boldsymbol{\theta}, \boldsymbol{\theta}^{(m)}),$$

or, in other words,

$$Q(\boldsymbol{\theta}^{(m+1)}, \boldsymbol{\theta}^{(m)}) = \max_{\boldsymbol{\theta}} Q(\boldsymbol{\theta}, \boldsymbol{\theta}^{(m)}).$$

In the iteration, the E step refers to the calculation of $Q(\boldsymbol{\theta}, \boldsymbol{\theta}^{(m)})$, and the M step to the maximization of $Q(\boldsymbol{\theta}, \boldsymbol{\theta}^{(m)})$. The iteration continues until a stopping criterion is met, for example, $|\boldsymbol{\theta}^{(m+1)} - \boldsymbol{\theta}^{(m)}|$ or $|Q(\boldsymbol{\theta}^{(m+1)}, \boldsymbol{\theta}^{(m)}) - Q(\boldsymbol{\theta}^{(m)}, \boldsymbol{\theta}^{(m)})|$ is less than a pre-specified value. If instead of maximizing $Q(\boldsymbol{\theta}, \boldsymbol{\theta}^{(m)})$ with respect to $\boldsymbol{\theta}$, each iteration simply finds an update $\boldsymbol{\theta}^{(m+1)}$ so that

$$Q(\boldsymbol{\theta}^{(m+1)}, \boldsymbol{\theta}^{(m)}) > Q(\boldsymbol{\theta}^{(m)}, \boldsymbol{\theta}^{(m)}),$$

then the algorithm is known as a generalized EM (GEM) algorithm.

Comparing to (14.4.2), which relies on the marginal p.d.f. of $\mathbf{y}$, a GEM algorithm relies on integrals with respect to the conditional p.d.f. of $\mathbf{z}$ given $M(\mathbf{z}) = \mathbf{y}$, which in many cases are much more tractable. Another attractive feature of a GEM algorithm is that each iteration increases the likelihood. However, there is no guarantee that the limit is a global maximum of $\ell(\boldsymbol{\theta}; \mathbf{y})$ or even a local maximum of $\ell(\boldsymbol{\theta}; \mathbf{y})$. More precisely, the following holds.

Result 14.4.1. In a GEM algorithm, for any $\boldsymbol{\theta}^{(m+1)}$ with $Q(\boldsymbol{\theta}^{(m+1)}, \boldsymbol{\theta}^{(m)}) > Q(\boldsymbol{\theta}^{(m)}, \boldsymbol{\theta}^{(m)})$, we have $\ell(\boldsymbol{\theta}^{(m+1)}; \mathbf{y}) > \ell(\boldsymbol{\theta}^{(m)}; \mathbf{y})$.

Proof. Since $f(\mathbf{z}; \boldsymbol{\theta}) = f(\mathbf{y}; \boldsymbol{\theta}) f(\mathbf{z} \,|\, \mathbf{y}; \boldsymbol{\theta})$, then

$$\ell(\boldsymbol{\theta}; \mathbf{z}) = \ell(\boldsymbol{\theta}; \mathbf{y}) + \log f(\mathbf{z} \,|\, \mathbf{y}; \boldsymbol{\theta}).$$

Multiply both sides by $f(\mathbf{z} \,|\, \mathbf{y}; \boldsymbol{\theta}^{(m)})$ and integrate over $\mathbf{z}$. It follows that

$$Q(\boldsymbol{\theta}, \boldsymbol{\theta}^{(m)}) = \ell(\boldsymbol{\theta}; \mathbf{y}) + H(\boldsymbol{\theta}, \boldsymbol{\theta}^{(m)}),$$

where

$$H(\boldsymbol{\theta}, \boldsymbol{\theta}^{(m)}) = \int \{\log f(\mathbf{z} \,|\, \mathbf{y}; \boldsymbol{\theta})\} f(\mathbf{z} \,|\, \mathbf{y}; \boldsymbol{\theta}^{(m)})\, d\mathbf{z}.$$

By Jensen's inequality (see Appendix C)

$$\begin{aligned} H(\boldsymbol{\theta}, \boldsymbol{\theta}^{(m)}) - H(\boldsymbol{\theta}^{(m)}, \boldsymbol{\theta}^{(m)}) &= \int \left\{ \log \frac{f(\mathbf{z} \,|\, \mathbf{y}; \boldsymbol{\theta})}{f(\mathbf{z} \,|\, \mathbf{y}; \boldsymbol{\theta}^{(m)})} \right\} f(\mathbf{z} \,|\, \mathbf{y}; \boldsymbol{\theta}^{(m)})\, d\mathbf{z} \\ &\le \log \int \left\{ \frac{f(\mathbf{z} \,|\, \mathbf{y}; \boldsymbol{\theta})}{f(\mathbf{z} \,|\, \mathbf{y}; \boldsymbol{\theta}^{(m)})} \right\} f(\mathbf{z} \,|\, \mathbf{y}; \boldsymbol{\theta}^{(m)})\, d\mathbf{z} = 0. \end{aligned}$$

Then

$$\begin{aligned} &\ell(\boldsymbol{\theta}^{(m+1)}; \mathbf{y}) - \ell(\boldsymbol{\theta}^{(m)}; \mathbf{y}) \\ &= [Q(\boldsymbol{\theta}^{(m+1)}, \boldsymbol{\theta}^{(m)}) - Q(\boldsymbol{\theta}^{(m)}, \boldsymbol{\theta}^{(m)})] + [H(\boldsymbol{\theta}^{(m)}, \boldsymbol{\theta}^{(m)}) - H(\boldsymbol{\theta}^{(m+1)}, \boldsymbol{\theta}^{(m)})] \\ &\ge Q(\boldsymbol{\theta}^{(m+1)}, \boldsymbol{\theta}^{(m)}) - Q(\boldsymbol{\theta}^{(m)}, \boldsymbol{\theta}^{(m)}) > 0. \end{aligned}$$

■

Example 14.4.1. For the missing data problem,

$$Q(\boldsymbol{\theta}, \boldsymbol{\theta}^*) = \int \ell(\boldsymbol{\theta}; \mathbf{y}_{\text{tot}}) f(\mathbf{y}_{\text{mis}} \,|\, \mathbf{y}_{\text{obs}}; \boldsymbol{\theta}^*)\, d\mathbf{y}_{\text{mis}}.$$

Dempster et al. (1977) showed that if $l(\boldsymbol{\theta} \,|\, \mathbf{y}_{\text{obs}})$ is bounded, then $l(\boldsymbol{\theta}^{(m)} \,|\, \mathbf{y}_{\text{obs}})$ converges to some value ℓ^*, while if $f(Y_{\text{tot}} \,|\, \boldsymbol{\theta})$ is a regular exponential family (see Appendix B), and $\ell(\boldsymbol{\theta} \,|\, \mathbf{y}_{\text{obs}})$ is bounded, then $\boldsymbol{\theta}^{(m)}$ converges to some stationary point $\boldsymbol{\theta}^*$. In general, the rate of convergence is inversely proportional to the proportion of missing information. When the distribution of the complete data $\mathbf{y}_{\text{tot}}$ belongs to the exponential family, the EM algorithm is especially simple. Then, the E step consists of estimating the complete data sufficient statistics $\mathbf{T}(\mathbf{y}_{\text{tot}})$ by

$$\mathbf{T}^{(m+1)} = \mathrm{E}\{\mathbf{T}(\mathbf{y}_{\text{tot}}) \,|\, \mathbf{y}_{\text{obs}}, \boldsymbol{\theta}^{(m)}\},$$

while the M step obtains the new iterate $\boldsymbol{\theta}^{(m+1)}$ as the solution to the likelihood equations

$$\mathrm{E}\{\mathbf{T}(\mathbf{y}_{\text{tot}}) \,|\, \boldsymbol{\theta}\} = \mathbf{T}^{(m+1)}.$$ □

Example 14.4.2. Positron emission tomography (PET) is a technology to study the metabolic activity in organs. Since its introduction (Shepp and Vardi, 1982), the EM algorithm and its variants have been a popular statistical methods for PET image reconstruction. To generate a PET scan, some radioactive material is administered into an organ within a body. Then the gamma-emissions resulting from the decay of the radioactive material are recorded using a PET scanner, which consists of an array of detectors surrounding the body.

The statistical model used by the EM algorithm for the PET data is as follows. The region of interest is divided into small volume, or voxels. The number of photons Z_{ij} received by detector i from voxel j is assumed to follow a Poisson distribution

$$Z_{ij} \sim \text{Poisson}(a_{ij}\theta_j),$$

where the coefficient a_{ij} is known and determined beforehand based on the biological and physiological properties of the tissue in voxel j, and θ_j is the emission density in voxel j. Since θ_j can be used to quantify the metabolic activity within the region of interest, the goal is to estimate $\boldsymbol{\theta} = (\theta_j)$. The estimates are then transformed into a visually more interpretable color image.

Since each detector i receives photons from all the voxels, only the sum

$$Y_i = \sum_j Z_{ij}$$

is observable. Thus, one has to use $\mathbf{y} = (Y_i)$ instead of $\mathbf{z} = (Z_{ij})$ to estimate $\boldsymbol{\theta}$. It is reasonable to assume that all Z_{ij} are independent. Then Y_i are independent and

$$Y_i \sim \text{Poisson}(\mathbf{a}_i'\boldsymbol{\theta}),$$

where $\mathbf{a}_i' = (a_{i1}, a_{i2}, \cdots)$. Since

$$\ell(\boldsymbol{\theta}; \mathbf{y}) = \log \prod_i [(\mathbf{a}_i'\boldsymbol{\theta})^{Y_i} e^{-\mathbf{a}_i'\boldsymbol{\theta}} / Y_i!] = \sum_i [Y_i \log(\mathbf{a}_i'\boldsymbol{\theta}) - \mathbf{a}_i'\boldsymbol{\theta} - \log(Y_i!)],$$

then

$$\widehat{\boldsymbol{\theta}}_{ML} = \arg\max_{\boldsymbol{\theta}} \sum_i [Y_i \log(\mathbf{a}_i'\boldsymbol{\theta}) - \mathbf{a}_i'\boldsymbol{\theta}].$$

It follows that $\widehat{\boldsymbol{\theta}}_{ML}$ is a root of

$$\nabla \ell(\boldsymbol{\theta};\mathbf{y}) = \sum_i \left(\frac{Y_i}{\mathbf{a}_i'\boldsymbol{\theta}} - 1\right)\mathbf{a}_i = \mathbf{0}.$$

For the PET data, the dimension of $\boldsymbol{\theta}$ is very large, equal to the number of voxels in the region. Thus, a direct solution to the equation is difficult and an iterative approach is needed.

According to the EM algorithm,

$$\begin{aligned}
&\boldsymbol{\theta}^{(m+1)} \\
&= \arg\max_{\boldsymbol{\theta}} \mathrm{E}\left[\ell(\boldsymbol{\theta};\mathbf{z})\,\middle|\,\sum_j Z_{ij} = Y_i\ \forall i; \boldsymbol{\theta}^{(m)}\right] \\
&= \arg\max_{\boldsymbol{\theta}} \mathrm{E}\left[\sum_{ij}[Z_{ij}\log(a_{ij}\theta_j) - a_{ij}\theta_j - \log(Z_{ij}!)]\,\middle|\,\sum_j Z_{ij} = Y_i\ \forall i; \boldsymbol{\theta}^{(m)}\right] \\
&= \arg\max_{\boldsymbol{\theta}} \left\{\sum_j \log\theta_j \sum_i \mathrm{E}\left[Z_{ij}\,\middle|\,\sum_j Z_{ij} = Y_i\ ; \boldsymbol{\theta}^{(m)}\right] - \sum_j\left(\sum_i a_{ij}\right)\theta_j\right\}.
\end{aligned}$$

Recall that for independent Poisson random variables with parameters $\lambda_1, \cdots, \lambda_s$, conditional on their sum being equal to y, their expectations are $\lambda_1 y/\lambda, \cdots \lambda_s y/\lambda$, where $\lambda = \lambda_1 + \cdots + \lambda_s$. Since $Z_{ij} \sim \text{Poisson}(a_{ij}\theta_j^{(m)})$ are independent under $f(\mathbf{z};\boldsymbol{\theta}^{(m)})$,

$$\boldsymbol{\theta}^{(m+1)} = \arg\max_{\boldsymbol{\theta}} \left\{\sum_j \theta_j^{(m)} \log\theta_j \sum_i \frac{a_{ij}Y_i}{\mathbf{a}_i'\boldsymbol{\theta}^{(m)}} - \sum_j\left(\sum_i a_{ij}\right)\theta_j\right\}.$$

It follows that each $\theta_j^{(m+1)}$ must maximize

$$\theta_j^{(m)} \log\theta_j \sum_i \frac{a_{ij}Y_i}{\mathbf{a}_i'\boldsymbol{\theta}^{(m)}} - \left(\sum_i a_{ij}\right)\theta_j$$

with respect to θ_j. Therefore,

$$\theta_j^{(m+1)} = \frac{\theta_j^{(m)}}{\sum_i a_{ij}} \sum_i \frac{a_{ij}Y_i}{\mathbf{a}_i'\boldsymbol{\theta}^{(m)}}.$$

This is the basic EM iteration for the PET reconstruction. □

A

Multivariate Probability Distributions

1. **Random vector.** A multivariate random variable, or random vector, is a random variable taking values in $\mathcal{R}^n$ with $n \geq 1$. Denote by $X_1, \cdots, X_n$ the coordinates of the random vector.

2. **Joint distribution.** The cumulative distribution function (c.d.f.) of $X_1, \cdots, X_n$ is

$$F(\mathbf{x}) = F(x_1, \cdots, x_n) = \mathrm{P}(X_1 \leq x_1, \cdots, X_n \leq x_n) \tag{A.1}$$

for any fixed $\mathbf{x} = (x_1, \cdots, x_n)' \in \mathcal{R}^n$, and its probability density function (p.d.f.), if it exists, is a function $f(\mathbf{x}) \geq 0$ such that for any (Borel measurable) set $\mathcal{A} \subset \mathcal{R}^n$,

$$\mathrm{P}\{(X_1, \cdots, X_n)' \in \mathcal{A}\} = \int_{\mathcal{A}} f(\mathbf{x})\, d\mathbf{x}.$$

In particular, if the c.d.f. F is n times differentiable everywhere, then

$$f(\mathbf{x}) = \frac{\partial^n F(\mathbf{x})}{\partial x_1 \cdots \partial x_n}. \tag{A.2}$$

Note that a p.d.f. f must satisfy

$$\int_{\mathcal{R}^n} f(\mathbf{x})\, d\mathbf{x} = \int_{-\infty}^{\infty} \cdots \int_{-\infty}^{\infty} f(x_1, \cdots, x_n)\, dx_1 \cdots dx_n = 1.$$

We also call F and f respectively as the joint c.d.f. and joint p.d.f. of $X_1, \cdots, X_n$.

3. **Marginal distributions.** For any subvector $(X_{i_1}, \cdots, X_{i_k})'$, where $1 \leq i_1 < \cdots < i_k \leq n$ and $k < n$, its c.d.f. and p.d.f. are called marginal when considered in relation to those of $(X_1, \cdots, X_n)'$. The marginal p.d.f. can be obtained as

$$f_{i_1, \cdots, i_k}(x_{i_1}, \cdots, x_{i_k}) = \int_{-\infty}^{\infty} \cdots \int_{-\infty}^{\infty} f(x_1, \cdots, x_n) \prod_{j \neq i_1, \cdots, i_k} dx_j, \tag{A.3}$$

by integrating out variates not in the subvector. Denote $\mathbf{I} = (i_1, \cdots, i_k)$ and $\mathbf{J} = (j_1, \cdots, j_{n-k})$, where $j_1 < \cdots < j_{n-k}$ are all numbers in $\{1, \cdots, n\} \setminus \{i_1, \cdots, i_k\}$. Then the marginal p.d.f. can be written as

$$f_{\mathbf{I}}(\mathbf{x}_{\mathbf{I}}) = \int_{\mathcal{R}^{n-k}} f(\mathbf{x})\, d\mathbf{x}_{\mathbf{J}}.$$

4. **Conditional distributions.** The conditional p.d.f. of $(X_{i_1}, \cdots, X_{i_k})'$ given $X_j = x_j$, $j \neq i_1, \cdots, i_k$, is defined as

$$g_{\mathbf{I}|\mathbf{J}}(\mathbf{x}_{\mathbf{I}} \,|\, \mathbf{x}_{\mathbf{J}}) = \frac{f(\mathbf{x})}{f_{\mathbf{J}}(\mathbf{x}_{\mathbf{J}})}. \tag{A.4}$$

For example,

$$g_{1, \cdots, k|k+1, \cdots, n}(x_1, \cdots, x_k \,|\, x_{k+1}, \cdots, x_n) = \frac{f(x_1, \cdots, x_k, x_{k+1}, \cdots, x_n)}{f_{k+1, \cdots, n}(x_{k+1}, \cdots, x_n)}.$$

DOI: 10.1201/9781315156651-A

5. **Moment generating function.** The moment generating function (m.g.f.) of $(X_1, \cdots, X_n)'$ is

$$M_{X_1,\cdots,X_n}(\mathbf{t}) = \mathrm{E}[\exp(\textstyle\sum_{i=1}^n t_i X_i)] \tag{A.5}$$

for any fixed $\mathbf{t} = (t_1, \cdots, t_n)' \in \mathcal{R}^n$. Since $\exp(\sum_{i=1}^n t_i X_i)$ is positive, the expectation is always well defined but may be infinity. Define $\mathcal{D} = \{\mathbf{t} \in \mathcal{R}^n : M_{X_1,\cdots,X_n}(\mathbf{t}) < \infty\}$. It can be shown that $\mathcal{D}$ is convex. Clearly, $M_{X_1,\cdots,X_n}(\mathbf{0}) = 1$, so $\mathbf{0} \in \mathcal{D}$. However, it may well happen that $\mathcal{D}$ only contains $\mathbf{0}$. If random vectors $(X_1, \cdots, X_n)$ and $(Y_1, \cdots, Y_n)$ have the same distribution, then evidently their m.g.f.s are identical. Importantly, if $M_{X_1,\cdots,X_n}(\mathbf{t}) = M_{Y_1,\cdots,Y_n}(\mathbf{t}) < \infty$ on an nonempty open set, then the two random vectors have the same distribution. This provides a way to characterize a distribution and leads to the well-known moment generating function technique for determining the distribution of a "new" random variable, which we use in various places throughout this book.

6. **Cumulant generating function.** The cumulant generating function (c.g.f.) of $(X_1, \cdots, X_n)'$ is defined as

$$\mathcal{K}_{X_1,\cdots,X_n}(\mathbf{t}) = \log[M_{X_1,\cdots,X_n}(\mathbf{t})].$$

7. **Moments of random vectors.** If the moment generating function $M_{X_1,\cdots,X_n}(\mathbf{t})$ of the random vector $(X_1, \cdots, X_n)'$ exists in an open set containing the origin, then moments of all orders of the random vector exist. A momemt of order r of the random vector $(X_1, \cdots, X_n)'$ has the form

$$\mu_{i_1,\cdots,i_k}^{(r_1,\cdots,r_k)} = \mathrm{E}(X_{i_1}^{r_1} \cdots X_{i_k}^{r_k}),$$

where $1 \le i_1 < \cdots < i_k \le n$ and $r_i > 0$ are integers with $r_1 + \cdots + r_k = r$. The rth moment of X_i whose marginal p.d.f. is $f_i(x_i)$ is given by

$$\mu_i^{(r)} = \int_{-\infty}^{\infty} x_i^r f_i(x_i)\, dx_i = \int_{-\infty}^{\infty} \cdots \int_{-\infty}^{\infty} x_i^r f(x_1, \cdots, x_n)\, dx_1 \cdots dx_n. \tag{A.6}$$

The mean of X_i is the first moment,

$$\mu_i = \mu_i^{(1)} = \mathrm{E}(X_i). \tag{A.7}$$

The variance of X_i is the second central moment

$$\sigma_i^2 = \mu_i^{(2)} - \mu_i^2 = \mathrm{Var}(X_i) \tag{A.8}$$

and the standard deviation of X_i is the positive square root of the variance and is denoted by σ_i. The covariance $\mathrm{Cov}(X_i, X_j)$ between X_i and X_j, for $i \ne j$, is

$$\begin{aligned} \sigma_{ij} &= \mathrm{E}[(X_i - \mu_i)(X_j - \mu_j)] \\ &= \int_{-\infty}^{\infty} \int_{-\infty}^{\infty} (x_i - \mu_i)(x_j - \mu_j) f_{ij}(x_i, x_j)\, dx_i\, dx_j \\ &= \int_{-\infty}^{\infty} \cdots \int_{-\infty}^{\infty} (x_i - \mu_i)(x_j - \mu_j) f(x_1, \cdots, x_n)\, dx_1 \cdots dx_n. \end{aligned} \tag{A.9}$$

We denoted by $\mathbf{x}$ the random vector $(X_1, \cdots, X_n)'$. Its mean $\mathrm{E}(\mathbf{x})$ is defined to be $\boldsymbol{\mu} = (\mu_1, \cdots, \mu_n)'$. The variance-covariance matrix of $\mathbf{x} = (X_1, \cdots, X_n)'$ is defined by

$\boldsymbol{\Sigma} = \{\sigma_{ij}\}$. The ith diagonal element of $\boldsymbol{\Sigma}$ is $\mathrm{Var}(X_i)$, and the (i,j)th off-diagonal element is $\mathrm{Cov}(X_i, X_j)$. From $\mathrm{Cov}(X_i, X_j) = \mathrm{Cov}(X_j, X_i)$, $\boldsymbol{\Sigma}$ is symmetric. It is also non-negative definite (n.n.d.), because for nonzero $\mathbf{t}$, $\mathbf{t}'\boldsymbol{\Sigma}\mathbf{t} = \sum_{i=1}^n \sum_{j=1}^n t_i t_j \,\mathrm{Cov}(X_i, X_j) = \mathrm{Var}(\sum_{i=1}^n t_i X_i) \geq 0$ with equality if and only if $\sum_{i=1}^n t_i X_i$ is a constant.

For any $m \times n$ matrix of jointly distributed random variables $\mathbf{Y} = \{Y_{ij}\}$, define $\mathrm{E}(\mathbf{Y})$ to be the $m \times n$ matrix of $\{\mathrm{E}(Y_{ij})\}$. We see that $(\mathbf{x} - \boldsymbol{\mu})(\mathbf{x} - \boldsymbol{\mu})'$ is an $n \times n$ matrix and

$$\boldsymbol{\Sigma} = \mathrm{Cov}(\mathbf{x}) = \mathrm{E}[(\mathbf{x} - \boldsymbol{\mu})(\mathbf{x} - \boldsymbol{\mu})'] = \mathrm{E}(\mathbf{x}\mathbf{x}') - \mathrm{E}(\mathbf{x})\,\mathrm{E}(\mathbf{x}'). \tag{A.10}$$

Now, let $\mathbf{y}$ be an m-dimensional random vector jointly distributed with $\mathbf{x}$. The covariance matrix of $\mathbf{x}$ and $\mathbf{y}$ is defined as

$$\mathrm{Cov}(\mathbf{x}, \mathbf{y}) = \mathrm{E}\{[\mathbf{x} - \mathrm{E}(\mathbf{x})][\mathbf{y} - \mathrm{E}(\mathbf{y})]'\} = \{\mathrm{Cov}(\mathbf{y}, \mathbf{x})\}'. \tag{A.11}$$

We say $\mathbf{x}$ and $\mathbf{y}$ are uncorrelated if and only if $\mathrm{Cov}(\mathbf{x}, \mathbf{y}) = \mathbf{O}$. Let $\mathbf{A}$ and $\mathbf{B}$ be $p \times n$ and $q \times m$ matrices, respectively. Then,

$$\begin{aligned}\mathrm{Cov}(\mathbf{Ax}, \mathbf{By}) &= \mathrm{E}\{[\mathbf{Ax} - \mathrm{E}(\mathbf{Ax})][\mathbf{By} - \mathrm{E}(\mathbf{By})]'\} \\ &= \mathbf{A}\,\mathrm{E}\{[\mathbf{x} - \mathrm{E}(\mathbf{x})][\mathbf{y} - \mathrm{E}(\mathbf{y})]'\}\mathbf{B}' \\ &= \mathbf{A}\,\mathrm{Cov}(\mathbf{x}, \mathbf{y})\mathbf{B}'. \end{aligned} \tag{A.12}$$

It is easy to see that $\boldsymbol{\Sigma} = \mathbf{O}$ if and only if $\mathbf{x}$ is a constant. Furthermore, we have the following.

Result A.1. Suppose $\mathbf{x}$ has mean $\boldsymbol{\mu}$ and variance-covariance matrix $\boldsymbol{\Sigma}$. Then the support of $\mathbf{x}$ is contained in the affine space $\boldsymbol{\mu} + \mathcal{C}(\boldsymbol{\Sigma})$, but not in any affine space of dimension less than $\mathrm{r}(\boldsymbol{\Sigma})$.

Proof. Let $\mathbf{P}$ be the orthogonal projection matrix onto $\mathcal{C}(\boldsymbol{\Sigma})$. Then by (A.12), $\mathrm{Cov}((\mathbf{I}-\mathbf{P})(\mathbf{x}-\boldsymbol{\mu})) = (\mathbf{I}-\mathbf{P})\boldsymbol{\Sigma}(\mathbf{I}-\mathbf{P}) = (\boldsymbol{\Sigma}-\mathbf{P}\boldsymbol{\Sigma})(\mathbf{I}-\mathbf{P}) = \mathbf{O}$. Therefore, $(\mathbf{I}-\mathbf{P})(\mathbf{x}-\boldsymbol{\mu})$ is a constant, and since its mean is $\mathbf{0}$, it is constant $\mathbf{0}$. Then $\mathbf{x} = \boldsymbol{\mu}+\mathbf{P}(\mathbf{x}-\boldsymbol{\mu}) \in \boldsymbol{\mu}+\mathcal{C}(\boldsymbol{\Sigma})$. Assume that the support of $\mathbf{x}$ is in an affine space of dimension less than $\mathrm{r}(\boldsymbol{\Sigma})$. Then $\mathbf{x} - \boldsymbol{\mu}$ is in a vector subspace $\mathcal{V}$ with $\dim(\mathcal{V}) < \mathrm{r}(\boldsymbol{\Sigma})$. For any $\mathbf{u} \perp \mathcal{V}$, $\mathbf{u}'(\mathbf{x} - \boldsymbol{\mu}) = 0$, so $\mathrm{Var}(\mathbf{u}'(\mathbf{x} - \boldsymbol{\mu})) = 0$. Then from (A.12), $\mathbf{u}'\boldsymbol{\Sigma}\mathbf{u} = 0$. Since $\boldsymbol{\Sigma}$ is non-negative definite (n.n.d.), then $\boldsymbol{\Sigma}\mathbf{u} = \mathbf{0}$ (see Exercise 2.28). It follows that $\mathcal{V}^{\perp} \subset \mathcal{N}(\boldsymbol{\Sigma})$. Then by property 1 of Result 1.3.10, $\dim(\boldsymbol{\Sigma}) \leq \dim(\mathcal{V})$. As this contradicts the assumption on $\mathcal{V}$, the proof is then complete. ■

For further details and examples on variance-covariance matrices, the reader is referred to Casella and Berger (1990) or Mukhopadhyay (2000).

8. **Moments of a partitioned random vector.** It is sometimes useful to consider a partitioned random vector, such as in the case of conditional p.d.f. Let $\mathbf{x}$ be partitioned as $\mathbf{x}' = (\mathbf{x}_1', \cdots, \mathbf{x}_m')$, where $\mathbf{x}_i$ is q_i-dimensional, $q_i > 0$, and $\sum_{i=1}^m q_i = n$. The mean vector $\boldsymbol{\mu}$ and the variance-covariance matrix $\boldsymbol{\Sigma}$ of $\mathbf{x}$ then should be partitioned conformably as $(\boldsymbol{\mu}_1', \cdots, \boldsymbol{\mu}_m')'$ and $\{\boldsymbol{\Sigma}_{ij}\}_{1 \leq i,j \leq n}$, respectively, so that $\boldsymbol{\mu} = \mathrm{E}(\mathbf{x}_i)$ and $\boldsymbol{\Sigma}_{ij} = \mathrm{Cov}(\mathbf{x}_i, \mathbf{x}_j)$, which is a $q_i \times q_j$ matrix.

   If $\mathbf{x}_1, \cdots, \mathbf{x}_n$ are independent k-dimensional random vectors with a common variance-covariance matrix $\boldsymbol{\Sigma}$, then the "rolled out" vector of the $\mathbf{x}_i$'s, viz., $\mathbf{y} = (\mathbf{x}_1', \cdots, \mathbf{x}_n')' =$

vec($\mathbf{X}$) with $\mathbf{X} = (\mathbf{x}_1, \cdots, \mathbf{x}_n)$ (see Section 2.8) has

$$\text{Cov}(\mathbf{y}) = \begin{pmatrix} \boldsymbol{\Sigma} & \mathbf{O} & \cdots & \mathbf{O} \\ \mathbf{O} & \boldsymbol{\Sigma} & \cdots & \mathbf{O} \\ \vdots & \vdots & \vdots & \vdots \\ \mathbf{O} & \mathbf{O} & \cdots & \boldsymbol{\Sigma} \end{pmatrix}.$$

This vectorized version is useful in a discussion of multivariate linear model theory, including multivariate time series.

9. **Mutual independence of random vectors.** We say that $X_1, \cdots, X_n$ are mutually independent if and only if

$$F(\mathbf{x}) = \prod_{i=1}^{n} F_i(x_i) \tag{A.13}$$

for all fixed $\mathbf{x} \in \mathcal{R}^n$, where $F_i(x_i)$ denotes the marginal c.d.f. of X_i, or equivalently, if

$$M_{X_1, \cdots, X_n}(\mathbf{t}) = \prod_{i=1}^{n} M_{X_i}(t_i) \tag{A.14}$$

for all $\mathbf{t}$ in a nonempty open set in $\mathcal{R}^n$. We similarly define mutual independence between random vectors by replacing each X_i with a random vector. Note that functions of independent random vectors are themselves independent.

10. **Transformation of random vectors.** We next consider the p.d.f. of a random vector under transformation. Recall that a domain in $\mathcal{R}^n$ is a connected open set.

Result A.2. Transformation. Let $(X_1, \cdots, X_n)'$ be a random vector that takes values in a domain $\mathcal{D} \subset \mathcal{R}^n$ with probability one. Suppose $\mathbf{g} = (g_1, \cdots, g_n)' : \mathcal{D} \to \mathcal{R}^n$ is a one-to-one differentiable transformation and its inverse $\mathbf{g}^* = (g_1^*, \cdots, g_n^*)' : \mathcal{E} \to \mathcal{D}$ is also differentiable, where $\mathcal{E} = \mathbf{g}(\mathcal{D})$. Let

$$Y_i = g_i(X_1, \cdots, X_n), \quad i = 1, \cdots, n.$$

If $(X_1, \cdots, X_n)'$ has a p.d.f. $f(\mathbf{x})$, then $(Y_1, \cdots, Y_n)'$ also has a p.d.f. $h(\mathbf{y})$ which is given by

$$h(\mathbf{y}) = \begin{cases} f(\mathbf{g}^*(\mathbf{y}))J(\mathbf{y}), & \text{if } \mathbf{y} \in \mathcal{E} \\ 0 & \text{else} \end{cases}$$

where $J(\mathbf{y}) = |\det(\mathbf{J}(\mathbf{y}))|$ denotes the absolute value of the Jacobian of the transformation, with

$$\mathbf{J}(\mathbf{y}) = \{J_{ij}(\mathbf{y})\} \text{ with } J_{ij}(\mathbf{y}) = \frac{\partial}{\partial y_j} g_i^*(\mathbf{y}), \quad i, j = 1, \ldots, n.$$

and $\det(\cdot)$ denotes the determinant of a matrix.

Proof. Since $(X_1, \cdots, X_n)' \in \mathcal{D}$ with probability one, then $(Y_1, \cdots, Y_n)' \in \mathcal{E}$ with probability one, so we can always define the p.d.f. of $(Y_1, \cdots, Y_n)'$ outside $\mathcal{E}$ to be zero. It suffices to show that if $\mathcal{B}$ is a measurable subset of $\mathcal{E}$, then

$$\text{P}((Y_1, \cdots, Y_n)' \in \mathcal{B}) = \int_{\mathcal{B}} f(\mathbf{g}^*(\mathbf{y}))J(\mathbf{y})\, d\mathbf{y}. \tag{A.15}$$

Let $\mathcal{A} = \mathbf{g}^*(\mathcal{B})$. Then

$$\mathrm{P}((Y_1, \cdots, Y_n)' \in \mathcal{B}) = \mathrm{P}((X_1, \cdots, X_n)' \in \mathcal{A}) = \int_{\mathcal{A}} f(\mathbf{x})\, d\mathbf{x}.$$

Using change of variable for multiple integrals, we see that

$$\int_{\mathcal{A}} f(\mathbf{x})\, d\mathbf{x} = \int_{\mathcal{B}} f(g^*(\mathbf{y}))J(\mathbf{y})\, d\mathbf{y}.$$

Combining the above two displays, (A.15) follows. ■

11. **Nonsingular linear transformations.** The class of nonsingular linear transformations, which is a simple case of Result A.2, plays a special role in the theory of linear models. Suppose $\mathbf{g}(\mathbf{x}) = \mathbf{Tx}$, where $\mathbf{T}$ is a nonsingular $n \times n$ matrix. Then $\mathbf{g}^*(\mathbf{y}) = \mathbf{T}^{-1}\mathbf{y}$, and $\mathbf{J}(\mathbf{y}) = \mathbf{T}^{-1}$, giving $J(\mathbf{y}) = 1/|\det(\mathbf{T})|$. Let

$$\begin{pmatrix} Y_1 \\ \vdots \\ Y_n \end{pmatrix} = \mathbf{T} \begin{pmatrix} X_1 \\ \vdots \\ X_n \end{pmatrix}. \tag{A.16}$$

Then the p.d.f. of $(Y_1, \cdots, Y_n)'$ is

$$h(\mathbf{y}) = \frac{f(\mathbf{T}^{-1}\mathbf{y})}{|\det(\mathbf{T})|}. \tag{A.17}$$

B

Common Families of Distributions

1. **Normal distribution**. A normal distribution with mean $-\infty < \mu < \infty$ and variance $\sigma^2 > 0$, denoted $N(\mu, \sigma^2)$, has p.d.f.

$$f(x; \mu, \sigma^2) = \frac{1}{\sqrt{2\pi\sigma^2}} \exp\left\{\frac{-(x-\mu)^2}{2\sigma^2}\right\}, \quad -\infty < x < \infty. \tag{B.1}$$

If $X \sim N(\mu, \sigma^2)$, then $\mathrm{E}(X) = \mu$ and $\mathrm{Var}(X) = \sigma^2$. The random variable $Z = (X-\mu)/\sigma$, called the standard normal variable, has $\mathrm{E}(Z) = 0$, $\mathrm{Var}(Z) = 1$, and p.d.f.

$$f(z) = \frac{e^{-z^2/2}}{\sqrt{2\pi}}, \quad -\infty < z < \infty. \tag{B.2}$$

The moment generating functions of X and Z are respectively

$$M_X(t) = \mathrm{E}[e^{tX}] = \exp\{\mu t + \sigma^2 t^2/2\} \tag{B.3}$$

and

$$M_Z(t) = \mathrm{E}[e^{tZ}] = \exp\{t^2/2\}. \tag{B.4}$$

2. **Chi-square distribution**. A chi-square distribution with k degrees of freedom (d.f.), denoted χ^2_k, has p.d.f.

$$f(u) = \frac{1}{2^{k/2}\Gamma(k/2)} u^{(k/2)-1} e^{-u/2}, \quad 0 < u < \infty. \tag{B.5}$$

If $U \sim \chi^2_k$, then $\mathrm{E}(U) = k$, $\mathrm{Var}(U) = 2k$, and its m.g.f. is

$$M_U(t) = (1-2t)^{-k/2}, \quad t < 1/2. \tag{B.6}$$

Let $X_1, \cdots, X_{k+1}$ be i.i.d. $\sim N(\mu, \sigma^2)$. Then $kS^2/\sigma^2 \sim \chi^2_k$, where

$$S^2 = \frac{1}{k}\sum_{i=1}^{k+1}(X_i - \overline{X})^2$$

is the sample variance and

$$\overline{X} = \frac{1}{k+1}\sum_{i=1}^{k+1} X_i.$$

is the sample mean.

DOI: 10.1201/9781315156651-B

3. **Student's *t*-distribution**. A Student's *t*-distribution with k d.f., denoted t_k, has p.d.f.

$$f(t) = \frac{1}{\sqrt{k\pi}} \frac{\Gamma((k+1)/2)}{\Gamma(k/2)} \left(1 + \frac{t^2}{k}\right)^{(k+1)/2}, \quad -\infty < t < \infty. \tag{B.7}$$

Let $T \sim t_k$. If $k = 1$, then the mean of T does not exist, while if $k > 1$, then $\mathrm{E}(T) = 0$. If $k \leq 2$, then the variance of T does not exist, while if $k > 2$, then $\mathrm{Var}(T) = k/(k-2)$. The m.g.f. of T does not exist. Let $X_1, \cdots, X_{k+1}$ be i.i.d. $\sim N(\mu, \sigma^2)$. Then $\sqrt{k+1}(\overline{X} - \mu)/S \sim t_k$.

4. **Snedecor's *F*-distribution**. A Snedecor's F-distribution with numerator d.f. p and denominator d.f. q, denoted $F_{p,q}$, has p.d.f.

$$f(u) = \frac{\Gamma((p+q)/2)}{\Gamma(p/2)\Gamma(q/2)} \left(\frac{p}{q}\right)^{p/2} \frac{u^{p/2-1}}{[1 + (p/q)u]^{(p+q)/2}}, \quad 0 < u < \infty. \tag{B.8}$$

If $F \sim F_{p,q}$, then

$$\mathrm{E}(F) = q/(q-2), \quad q > 2$$

$$\mathrm{Var}(F) = 2\left(\frac{q}{q-2}\right)^2 \frac{p+q-2}{p(q-4)}, \quad q > 4$$

and its rth raw moment is

$$\mathrm{E}(F^r) = \frac{\Gamma((p+2r)/2)\Gamma((q-2r)/2)}{\Gamma(p/2)\Gamma(q/2)} \left(\frac{q}{p}\right)^r, \quad r < q/2.$$

The m.g.f. of F does not exist. It can be shown that

(i) if $U \sim F_{p,q}$, then $1/U \sim F_{q,p}$;
(ii) if $U \sim t_q$, then $U^2 \sim F_{1,q}$;
(iii) if $U \sim F_{p,q}$, then $\frac{(p/q)U}{[1+(p/q)U]} \sim \mathrm{Beta}(p/2, q/2)$;
(iv) if $U_1 \sim \chi^2_p$ and $U_2 \sim \chi^2_q$ are independent, then $(U_1/p)/(U_2/q) \sim F_{p,q}$.

Let $X_1, \cdots, X_{n_1}$ i.i.d. $\sim N(\mu_x, \sigma_x^2)$ and $Y_1, \cdots, Y_{n_2}$ i.i.d. $\sim N(\mu_y, \sigma_y^2)$ be two independent samples with sample variances S_x^2 and S_y^2, respectively. Then

$$\frac{(n_1-1)S_x^2/\sigma_x^2}{(n_2-1)S_y^2/\sigma_y^2} \sim F_{n_1-1, n_2-1}. \tag{B.9}$$

5. **Gamma distribution**. A gamma distribution with shape parameter $\alpha > 0$ and scale parameter $\beta > 0$, denoted Gamma(α, β), has p.d.f.

$$f(x; \alpha, \beta) = \frac{1}{\Gamma(\alpha)\beta^\alpha} x^{\alpha-1} \exp\{-x/\beta\}, \quad x > 0, \tag{B.10}$$

where

$$\Gamma(\alpha) = \int_0^\infty u^{\alpha-1} \exp\{-u\}\, du \tag{B.11}$$

is known as the Gamma function. If $X \sim \mathrm{Gamma}(\alpha, \beta)$, then

$$\mathrm{E}(X) = \alpha\beta, \quad \mathrm{Var}(X) = \alpha\beta^2,$$

and its m.g.f. is

$$M_X(t) = (1 - \beta t)^{-\alpha}, \quad t < 1/\beta.$$

6. **Beta distribution.** A beta distribution with parameters $\alpha > 0$ and $\beta > 0$, denoted $\text{Beta}(\alpha, \beta)$ and also known as a beta distribution of the *first kind*, has p.d.f.

$$f(x) = \frac{x^{\alpha-1}(1-x)^{\beta-1}}{B(\alpha,\beta)}, \quad 0 \le x < 1, \tag{B.12}$$

where $B(\alpha,\beta) = \Gamma(\alpha)\Gamma(\beta)/\Gamma(\alpha+\beta)$ is called the Beta function. If $X \sim \text{Beta}(\alpha,\beta)$, then

$$\text{E}(X) = \frac{\alpha}{\alpha+\beta}, \quad \text{Var}(X) = \frac{\alpha\beta}{(\alpha+\beta)^2(\alpha+\beta+1)},$$

and its m.g.f. is

$$M_X(t) = 1 + \sum_{k=1}^{\infty}\left(\prod_{j=0}^{k-1}\frac{\alpha+j}{\alpha+\beta+j}\right)\frac{t^k}{k!},$$

which cannot be simplified further. If $V \sim F_{p,q}$, then

$$\frac{(p/q)V}{1+(p/q)V} \sim \text{Beta}(p/2, q/2). \tag{B.13}$$

A beta prime distribution with parameters $\alpha > 0$ and $\beta > 0$, denoted $\text{Beta}'(\alpha,\beta)$ and also known as a beta distribution of the *second kind*, has p.d.f.

$$f(y) = \frac{1}{B(\alpha,\beta)}y^{\alpha-1}(1+y)^{-(m+n)}, \quad 0 < y < \infty.$$

If $Y \sim \text{Beta}'(\alpha,\beta)$, then $Y/(1+Y) \sim \text{Beta}(\alpha,\beta)$.

7. **Cauchy distribution.** A Cauchy distribution has p.d.f.

$$f(x;\mu,\sigma) = \frac{1}{\pi\sigma[1+(x-\mu)/\sigma]^2}, \quad -\infty < x < \infty, \tag{B.14}$$

where $-\infty < \mu < \infty$ is the location parameter and $\sigma > 0$ is the scale parameter. The mean and variance of this distribution do not exist and neither does the m.g.f. If U and V are i.i.d. $\sim N(0,1)$, then U/V has a Cauchy distribution with location 0 and unit scale, which is also a Student's t-distribution with 1 d.f.

8. **Double exponential (or Laplace) distribution.** A double exponential (or Laplace) distribution with location parameter $-\infty < \mu < \infty$ and scale parameter $\sigma > 0$, denoted $\text{Laplace}(\mu,\sigma)$, has p.d.f.

$$f(x;\mu,\sigma) = \frac{1}{2\sigma}\exp\left\{-\frac{|x-\mu|}{\sigma}\right\}, \quad -\infty < x < \infty. \tag{B.15}$$

If $X \sim \text{Laplace}(\mu,\sigma)$, then

$$\text{E}(X) = \mu, \quad \text{Var}(X) = 2\sigma^2,$$

and its m.g.f. is

$$M_X(t) = \frac{e^{\mu t}}{1-\sigma^2 t^2}, \quad t < 1/\sigma.$$

The standard form of the distribution is obtained by setting $\mu = 0$ and $\sigma = 1$, and has p.d.f.

$$f(z) = \frac{1}{2}\exp\{-|z|\}, \quad -\infty < z < \infty.$$

9. **Finite mixture distribution.** Let X be a random variable with p.d.f.

$$f(x) = p_1 f_1(x) + \cdots + p_L f_L(x), \tag{B.16}$$

where $p_j > 0$, $j = 1, \cdots, L$, $\sum_{j=1}^{L} p_j = 1$, and $f_j(x)$ are themselves valid p.d.f.'s. Then X is said to have a finite mixture distribution with L mixands, and with mixing proportions (or mixing weights) $p_1, \cdots, p_L$. The p.d.f.'s $f_j(x)$ are known as the components of the mixture.

10. **Mixture of normals distribution.** Also referred to as normal mixtures in the literature, these are among the most widely studied finite mixture distributions. The jth component of a finite normal mixture is a $N(\mu_j, \sigma_j^2)$ distribution. If the mixing proportions are $p_1, \cdots, p_L$, then the p.d.f. of the mixture is given by

$$f(x) = \frac{1}{\sqrt{2\pi}} \sum_{j=1}^{L} \frac{p_j}{\sigma_j} \exp\left\{-\frac{1}{2}\left(\frac{x-\mu_j}{\sigma_j}\right)^2\right\}, \quad -\infty < x < \infty. \tag{B.17}$$

This p.d.f. is determined by $3L-1$ parameters, consisting of $\mu_j \in (-\infty, \infty)$, $\sigma_j^2 > 0$, and $p_j > 0$, $j = 1, \cdots, L$, subject to the constraint $\sum_{j=1}^{L} p_j = 1$. A normal mixture affords a great deal of flexibility for modeling by allowing for multimodality. The expectation and variance of a mixture random variable can be obtained from the means and variances of its components:

$$\begin{aligned} \mathrm{E}(X) &= \overline{\mu} = \sum_{j=1}^{L} p_j \mu_j, \\ \mathrm{Var}(X) &= \sum_{j=1}^{L} p_j \sigma_j^2 + \sum_{j=1}^{L} p_j (\mu_j - \overline{\mu})^2. \end{aligned} \tag{B.18}$$

11. **Scale mixture of normals distribution.** The distribution of $Z\Lambda$ is said to be a scale mixture of normals (SMN) distribution if $Z \sim N(0,1)$ and Λ is a positive random variable independent of Z. The p.d.f. of the distribution is

$$f(x;\Lambda) = \int_0^\infty \frac{1}{\sqrt{2\pi}\lambda} \exp\left\{-\frac{x^2}{2\lambda^2}\right\} dF_\Lambda(\lambda), \tag{B.19}$$

where $F_\Lambda(\lambda)$ is the c.d.f. of Λ and can be discrete or continuous (Andrews and Mallows, 1974).

12. **Stable distribution.** A random variable X has a four-parameter stable distribution $S_\alpha(\sigma, \beta, \delta)$ if its characteristic function has the form

$$\mathrm{E}[\exp(i\theta X)] = \begin{cases} \exp\{-|\sigma\theta|^\alpha(1 - i\beta\,\mathrm{sign}(\theta)\tan(\pi\alpha/2) + i\delta\theta\} & \text{if } \alpha \neq 1 \\ \exp\{-|\sigma\theta|(1 + \frac{2}{\pi} i\beta \log|\theta|\,\mathrm{sign}(\theta) + i\delta\theta\} & \text{if } \alpha = 1, \end{cases} \tag{B.20}$$

where θ is a real number, and

$$\mathrm{sign}(\theta) = \begin{cases} 1, & \text{if } \theta > 0 \\ 0, & \text{if } \theta = 0 \\ -1, & \text{if } \theta < 0 \end{cases}$$

(Samorodnitsky and Taqqu, 1994). The stability parameter α lies in the range $(0, 2]$, and measures the heaviness of the tails of the p.d.f. When $\alpha = 2$, the stable distribution reduces to a normal distribution $N(\delta, 2\sigma^2)$. The skewness parameter β, which lies in the range $[-1, 1]$, measures the departure of the distribution from symmetry. The distribution is symmetric when $\beta = 0$, is skewed to the right when $\beta > 0$, and is skewed to the left when $\beta < 0$. The location parameter δ lies in the range $(-\infty, \infty)$, and shifts the distribution to the right or the left. The scale parameter σ lies in the range $(0, \infty)$, and is the parameter in proportion to which the distribution of X around δ is compressed or extended. When $\beta = 1$, $0 < \alpha < 1$, and $\delta = 0$, the distribution is totally skewed to the right, and is referred to as the positive stable distribution. The p.d.f. of this distribution is not available in closed form.

13. **Inverse Gaussian distribution**. The inverse Gaussian distribution with mean $\mu > 0$ and shape parameter $\lambda > 0$, denoted IGaus(μ, λ), has p.d.f.

$$f(x; \mu, \lambda) = \left(\frac{\lambda}{2\pi x^3}\right)^{1/2} \exp\left\{-\frac{\lambda}{2\mu^2}\frac{(x-\mu)^2}{y}\right\}, \quad x > 0. \tag{B.21}$$

If $X \sim$ IGaus(μ, λ), then $\mathrm{E}(X) = \mu$ and $\mathrm{Var}(X) = \mu^3/\lambda$.

14. **Studentized range distribution**. Let $X_1, \cdots, X_n$ be i.i.d. $\sim N(\mu, \sigma^2)$. Let $R = \max_i X_i - \min_i X_i$ denote the range of the n variables. Let S^2 be independent of $X_1, \cdots, X_n$ such that $\nu S^2/\sigma^2 \sim \chi^2_\nu$. The distribution of R/S is called the Studentized range distribution with (n, ν) d.f.

15. **Exponential family**. The p.d.f. or p.m.f. from an exponential family has the form

$$f(x; \theta, \phi) = \exp\{[x\theta - b(\theta)]/a(\phi) + c(x, \phi)\}, \tag{B.22}$$

where θ and ϕ are respectively the location and scale parameters, while $a(\phi), b(\theta)$ and $c(x, \phi)$ are known functions. In many cases, $a(\phi)$ has the form

$$a(\phi) = \frac{\phi}{w}$$

for a known prior weight w, which is usually set to 1. If the dispersion parameter ϕ is known, (B.22) is an exponential family with canonical parameter θ. If $\phi = 1$, we have the unit dispersion exponential family with

$$f(x; \theta) = \exp\{x\theta - b(\theta) + c(x)\}, \tag{B.23}$$

For example, the normal distribution is in the exponential family, because each $N(\mu, \sigma^2)$ has p.d.f.

$$f(x; \mu, \sigma^2) = \exp\left\{\frac{x\mu - \mu^2/2}{\sigma^2} - \frac{x}{2\sigma^2} - \frac{1}{2}\log(2\pi\sigma^2)\right\},$$

so that $\theta = \mu$, $a(\phi) = \phi = \sigma^2$, $b(\theta) = \mu^2/2$, and $c(x, \phi) = -[x^2/\sigma^2 + \log(2\pi\sigma^2)]/2$.

Let $\ell(\theta, \phi; x) = \log f(x; \theta, \phi)$ denote the logarithm of the likelihood function based on x, and let $b'(\theta)$ and $b''(\theta)$ be the first and second derivatives of $b(\theta)$. Since

$$\partial\ell/\partial\theta = \{x - b'(\theta)\}/a(\phi), \quad \partial^2\ell/\partial\theta^2 = -b''(\theta)/a(\phi),$$

and

$$\mathrm{E}(\partial \ell/\partial \theta) = 0, \quad \mathrm{E}(\partial^2 \ell/\partial \theta^2) + \mathrm{E}[(\partial \ell/\partial \theta)^2] = 0,$$

it follows that $\mathrm{E}(X) = \mu = b'(\theta)$, and $\mathrm{Var}(X) = b''(\theta)a(\phi)$. The function $b''(\theta)$ depends on the canonical parameter, and hence on μ, and is called the variance function, denoted by $V(\mu)$.

The p.d.f. or p.m.f. from a k-parameter exponential family has the form

$$f(x; \boldsymbol{\theta}, \phi) = \exp\left[\frac{\sum_{j=1}^{k} \xi_j(\boldsymbol{\theta}) t_j(x) - b(\boldsymbol{\theta})}{a(\phi)} + c(x, \phi)\right]. \tag{B.24}$$

Here, $\xi_1(\boldsymbol{\theta}), \cdots, \xi_k(\boldsymbol{\theta})$ are known real-valued functions of the k-dimensional parameter $\boldsymbol{\theta}$ (they do not depend on x), $t_1(x), \cdots, t_k(x)$ are known real-valued functions of x (they do not depend on $\boldsymbol{\theta}$), $b(\boldsymbol{\theta})$ is a known cumulant function and $a(\phi) = \phi/w$ is a known function of the dispersion parameter ϕ and known weight w.

16. **Inverse gamma distribution.** A random variable X is said to follow an inverse gamma distribution with parameters $\alpha, \beta > 0$, denoted $X \sim \mathrm{IG}(a, b)$, if $1/X \sim \mathrm{Gamma}(\alpha, 1/\beta)$. The p.d.f. of X is

$$f(x; \alpha, \beta) = \frac{\beta^\alpha}{\Gamma(\alpha)} x^{-(\alpha+1)} \exp\left(-\frac{\beta}{x}\right), \quad x > 0. \tag{B.25}$$

We can verify that $\mathrm{E}(X) = \beta/(\alpha - 1)$ if $\alpha > 1$, and $\mathrm{Var}(X) = \beta^2/\{(\alpha-1)^2(\alpha-2)\}$ if $\alpha > 2$.

C

Some Useful Statistical Notions

1. **Kullback–Leibler divergence**. The Kullback–Leibler (KL) divergence, which is also called relative entropy, is a measure of how one probability distribution is different from a second probability distribution, usually a reference distribution. The KL-divergence of a p.d.f. $g(x)$ from another p.d.f. $f(x)$ is defined as

$$\mathrm{KL}(g \,\|\, f) = \int g(x) \log \left[\frac{g(x)}{f(x)}\right] dx. \tag{C.1}$$

If f and g are p.m.f.'s, we replace the integral by a sum.

2. **Jensen's inequality**. Let X be a random variable and let $g(.)$ be a convex function. Then,

$$g(\mathrm{E}(X)) \leq \mathrm{E}(g(X)). \tag{C.2}$$

The difference between the two sides of the inequality is called the Jensen gap. For concave functions, the direction of the equality is opposite.

3. **Markov's inequality**. Markov's inequality gives an upper bound for the probability that a nonnegative random variable is greater than or equal to some positive constant. Let X be a nonnegative random variable and $a > 0$. Then,

$$\mathrm{P}(X \geq a) \leq \frac{\mathrm{E}(X)}{a}. \tag{C.3}$$

4. **Laplace approximation**. Consider the ratio of integrals given by

$$\frac{\int w(\theta)\, e^{\ell(\theta)}\, d\theta}{\int \pi(\theta)\, e^{\ell(\theta)}\, d\theta}, \tag{C.4}$$

where θ is an $m \times 1$ vector of parameters and $\ell(\theta)$ is the log-likelihood function of θ based on n observations $\mathbf{y} = (y_1, \cdots, y_n)$ coming from the probability model $p(y \,|\, \theta)$, that is,

$$\ell(\theta) = \sum_{i=1}^{n} \log p(y_i \,|\, \theta).$$

The quantities $w(\theta)$ and $\pi(\theta)$ are functions of θ which may need to satisfy certain conditions depending on the context.

In Bayesian formulation $\pi(\theta)$ is the prior and $w(\theta) = u(\theta)\pi(\theta)$ where $u(\theta)$ is some function of θ which is of interest. Thus, the ratio represents the posterior expectation of $u(\theta)$, that is, $\mathrm{E}[u(\theta) \,|\, \mathbf{y}]$. For example, if $u(\theta) = \theta$ then the ratio in (C.4) gives us

DOI: 10.1201/9781315156651-C

the posterior mean of θ. Similarly, if $u(\theta) = p(y \mid \theta)$, it yields the posterior predictive distribution value at y. Alternatively, we can write the above ratio as

$$\frac{\int u(\theta)e^{\Lambda(\theta)}d\theta}{\int e^{\Lambda(\theta)}d\theta}, \tag{C.5}$$

where $\Lambda(\theta) = \ell(\theta) + \log \pi(\theta)$.

Lindley (1980) develops asymptotic expansions for the above ratio of integrals in (C.5) as the sample size n gets large. The idea is to obtain a Taylor series expansion of all the above functions of θ about $\widehat{\theta}$, the posterior mode. Lindley's approximation to $\mathrm{E}[u(\theta) \mid \mathbf{y}]$ is given by:

$$\mathrm{E}[u(\theta) \mid \mathbf{y}] \approx u(\widehat{\theta}) + \frac{1}{2}\left(\sum_{i,j=1}^{m} u_{i,j}\sigma_{i,j} + \sum_{i,j,k,l=1}^{m} \Lambda_{i,j,k}\, u_l\, \sigma_{i,j}\sigma_{k,l}\right), \tag{C.6}$$

where

$$u_i \equiv \left.\frac{\partial u(\theta)}{\partial \theta_i}\right|_{\theta=\hat{\theta}}, \quad u_{i,j} \equiv \left.\frac{\partial^2 u(\theta)}{\partial \theta_i \partial \theta_j}\right|_{\theta=\hat{\theta}}, \quad \Lambda_{i,j,k} \equiv \left.\frac{\partial^3 \Lambda(\theta)}{\partial \theta_i \partial \theta_j \partial \theta_k}\right|_{\theta=\widehat{\theta}},$$

and $\sigma_{i,j}$ are the elements in the negative inverse Hessian of Λ at $\widehat{\theta}$.

Lindley's approximation (C.6) involves third order differentiation and therefore is computationally cumbersome in highly parameterized cases. Tierney and Kadane (1986) propose an alternative approximation which involves only the first and second order derivatives. This is achieved by using the mode of the product $u(\theta)e^{\Lambda(\theta)}$ rather than the mode of the posterior $e^{\Lambda(\theta)}$ and evaluating the second derivatives at this mode. Tierney–Kadane approximation approximates $\mathrm{E}[u(\theta) \mid \mathbf{y}]$ by

$$\mathrm{E}[u(\theta) \mid \mathbf{y}] \approx \left(\frac{|\Sigma^*(\widehat{\theta})|}{|\Sigma(\widehat{\theta})|}\right)^{1/2} \exp\{n[\Lambda^*(\widehat{\theta}) - \Lambda(\widehat{\theta})]\}, \tag{C.7}$$

where $\Lambda^*(\theta) = \log u(\theta) + \Lambda(\theta)$ and $\Sigma^*(\widehat{\theta})$ and $\Sigma(\widehat{\theta})$ are the corresponding negative inverse Hessians of Λ^* and Λ evaluated at $\widehat{\theta}$.

5. **Stirling's formula**. As $n \to \infty$,

$$n! = (n/e)^n \sqrt{2\pi n}[1 + o(1)].$$

6. **Newton–Raphson algorithm in optimization**. The Newton–Raphson algorithm is used on a twice-differentiable function $f(x) : \mathcal{R} \to \mathcal{R}$ in order to find solutions to $f'(x) = 0$, which are stationary points of $f(x)$. These solutions may be minima, maxima, or saddle points of $f(x)$. Let $x_0 \in \mathcal{R}$ be an initial value and let $x_k \in \mathcal{R}$ be the value at the kth iteration. Using a sequence of second-order Taylor series approximations of $f(x)$ around the current iterate x_k, $k = 0, 1, 2, \cdots$, the Newton–Raphson method updates x_k to x_{k+1} by

$$x_{k+1} = x_k - \frac{f'(x_k)}{f''(x_k)}.$$

7. **Convergence in probability**. If the sequence of estimators $\hat{\theta}_N$ based on N observations converges in probability to a constant θ, we say that θ is the limit in probability of the sequence $\widehat{\theta}_N$, and write $\mathrm{plim}_{N\to\infty}\, \widehat{\theta}_N = \theta$.

8. **Hampel's influence function.** Suppose $(z_1, \cdots, z_N)$ denotes a large random sample from a population with c.d.f. F. Let $\widehat{F}_N = F_N(z_1, \cdots, z_N)$ denote the empirical c.d.f. and let $T_N = T(z_1, \cdots, z_N)$ be a (scalar or vector-valued) statistic of interest. The study of influence consists of assessing the change in T_N when some specific aspect of the problem is slightly changed. The first step is to find a statistical functional T which maps (a subset of) the set of all c.d.f.'s onto $\mathcal{R}^p$, so that $T(\widehat{F}_N) = T_N$. We assume that F_N converges to F and that T_N converges to T. For example, when $T_N = \overline{Z}$, the corresponding functional is $T(F) = \int z\,dF(z)$, and $T(\widehat{F}_N) = \int z\,d\widehat{F}_N = \overline{Z}$. The influence of an estimator T_N is said to be unbounded if it is sensitive to extreme observations, in which case, T_N is said to be nonrobust. To assess this, one more observation z is added to the *large* sample, and we monitor the change in T_N, and the conclusions based on T_N.

   Let z denote one observation that is added to the large sample $(z_1, \cdots, z_N)$ drawn from a population with c.d.f. F, and let T denote a functional of interest. The influence function is defined by

$$\psi(z, F, T) = \lim_{\varepsilon \to 0} \frac{1}{\varepsilon}[T\{(1-\varepsilon)F + \varepsilon\delta_z\} - T\{F\}], \tag{C.8}$$

   provided the limit exists for every $z \in \mathcal{R}$, and where $\delta_z = 1$ at z and zero otherwise. The influence curve is the ordinary right-hand derivative, evaluated at $\varepsilon = 0$, of the unction $T[(1-\varepsilon)F + \varepsilon\delta_z]$ with respect to ε.

   The influence curve is useful for studying asymptotic properties of an estimator as well as for comparing estimators. For example, if $T = \mu = \int z\,dF$, then

$$\psi[z, F, T] = \lim_{\varepsilon \to 0}\{[(1-\varepsilon)\mu + \varepsilon z] - \mu\}/\varepsilon = z - \mu, \tag{C.9}$$

   which is "unbounded", so that $T_N = \overline{Z}$ is nonrobust. To use the influence function in the regression context, we must first construct appropriate functionals corresponding to $\boldsymbol{\beta}$ and σ^2.

9. **Multivariate completion of square.** Let $\mathbf{x} \in \mathcal{R}^p$, $\mathbf{a} \in \mathcal{R}^p$, and $\mathbf{A} \in \mathcal{R}^{p \times p}$ be a symmetric, p.d. matrix. Then,

$$\mathbf{x}'\mathbf{A}\mathbf{x} - 2\mathbf{a}'\mathbf{x} = (\mathbf{x} - \mathbf{A}^{-1}\mathbf{a})'\mathbf{A}(\mathbf{x} - \mathbf{A}^{-1}\mathbf{a}) - \mathbf{a}'\mathbf{A}^{-1}\mathbf{a}. \tag{C.10}$$

   This is useful in deriving expressions in a Bayesian framework.

10. **Bayes' theorem.** Bayes' theorem, is a time-honored result dating back to the late 18th century. Let $\mathcal{B} = (B_1, \cdots, B_k)$ denote a partition of a sample space $\mathcal{S}$, and let A denote an event with $\mathrm{P}(A) > 0$. By the definition of conditional probability (Casella and Berger, 1990), we have

$$\mathrm{P}(B_j \mid A) = \mathrm{P}(B_j \cap A)/\,\mathrm{P}(A) = \mathrm{P}(A \mid B_j)\,\mathrm{P}(B_j)/\,\mathrm{P}(A).$$

   By substituting for $\mathrm{P}(A)$ from the law of total probability, i.e.,

$$\mathrm{P}(A) = \sum_{i=1}^{k} \mathrm{P}(A \mid B_i)\,\mathrm{P}(B_i),$$

   it follows that for any event B_j in $\mathcal{B}$,

$$\mathrm{P}(B_j \mid A) = \mathrm{P}(A \mid B_j)\,\mathrm{P}(B_j)\Big/\sum_{i=1}^{k} \mathrm{P}(A \mid B_i)\,\mathrm{P}(B_i).$$

When the partition $\mathcal{B}$ represents all possible mutually exclusive states of nature or hypotheses, we refer to $\mathrm{P}(B_j)$ as the prior probability of an event B_j. An event A is then observed, and this modifies the probabilities of the events in $\mathcal{B}$. We call $\mathrm{P}(B_j \mid A)$ the posterior probability of B_j.

Now consider the setup in terms of continuous random vectors. Let $\mathbf{x}$ be a k-dimensional random vector with joint pdf $f(\mathbf{x}; \boldsymbol{\theta})$, where $\boldsymbol{\theta}$ is a q-dimensional parameter vector. We assume that $\boldsymbol{\theta}$ is also a continuous random vector with p.d.f. $\pi(\boldsymbol{\theta})$, which we refer to as the *prior* density of $\boldsymbol{\theta}$. Given the likelihood function is $L(\boldsymbol{\theta}; \mathbf{x}) = f(\mathbf{x}; \boldsymbol{\theta})$, an application of Bayes' theorem gives the posterior density of $\boldsymbol{\theta}$ as

$$\pi(\boldsymbol{\theta} \mid \mathbf{x}) = L(\boldsymbol{\theta}; \mathbf{x})\pi(\boldsymbol{\theta}) \Big/ \int L(\boldsymbol{\theta}; \mathbf{x})\pi(\boldsymbol{\theta})\, d\boldsymbol{\theta},$$

where the term in the denominator is called the marginal distribution or likelihood of $\mathbf{x}$ and is usually denoted by $m(\mathbf{x})$.

11. **Posterior summaries.** Let $\pi(\boldsymbol{\theta} \mid \mathbf{x})$ denote the posterior distribution of $\boldsymbol{\theta}$ given data $\mathbf{x}$. The expectations below are taken with respect to $\pi(\boldsymbol{\theta} \mid \mathbf{x})$.

    (a) The posterior mean is the Bayes estimator (and also the minimum mean squared estimator, MMSE) of $\boldsymbol{\theta}$ and is defined as

    $$\mathrm{E}(\boldsymbol{\theta} \mid \mathbf{x}) = \int_{\Theta} \boldsymbol{\theta}\pi(\boldsymbol{\theta} \mid \mathbf{x}) d\boldsymbol{\theta}.$$

    (b) The posterior variance matrix is defined as

    $$\mathrm{Var}(\boldsymbol{\theta} \mid \mathbf{x}) = \mathrm{E}[(\boldsymbol{\theta} - \mathrm{E}(\boldsymbol{\theta} \mid \mathbf{x}))(\boldsymbol{\theta} - \mathrm{E}(\boldsymbol{\theta} \mid \mathbf{x})'].$$

    (c) The largest posterior mode or the generalized MLE of $\boldsymbol{\theta}$ and is

    $$\pi(\widehat{\boldsymbol{\theta}} \mid \mathbf{x}) = \sup_{\boldsymbol{\theta} \in \Theta} \pi(\boldsymbol{\theta} \mid \mathbf{x}).$$

12. **Credible set and interval.** A $100(1-\alpha)\%$ credible set for $\boldsymbol{\theta}$ is a subset $C \subset \Theta$ of the parameter space satisfying the condition

    $$1 - \alpha \le \mathrm{P}(C \mid \mathbf{x}) = \int_C \pi(\boldsymbol{\theta} \mid \mathbf{x}) d\boldsymbol{\theta}. \tag{C.11}$$

    If the posterior distribution of a scalar θ is continuous, symmetric, and unimodal, the credible interval can be written as $(\theta^{(L)}, \theta^{(U)})$, where

    $$\int_{-\infty}^{\theta^{(L)}} \pi(\theta \mid \mathbf{x}) = \int_{\theta^{(U)}}^{\infty} \pi(\theta \mid \mathbf{x}) = \alpha/2. \tag{C.12}$$

13. **HPD Interval.** The $100(1-\alpha)\%$ highest posterior density (HPD) credible set for $\boldsymbol{\theta}$ is a subset $C \subset \Theta$ given by

    $$C = C(k(\alpha)) = \{\boldsymbol{\theta} \in \Theta : \pi(\boldsymbol{\theta} \mid \mathbf{x}) \ge k(\alpha)\}, \tag{C.13}$$

    where $k(\alpha)$ is the largest constant satisfying

    $$\mathrm{P}(C(k(\alpha) \mid \mathbf{x})) \ge 1 - \alpha. \tag{C.14}$$

14. **Bayes Factor**. Consider a model selection problem in which we must choose between two models, M_1 and M_2, parametrized by $\boldsymbol{\theta}_1$ and $\boldsymbol{\theta}_2$ respectively, based on observed data $\mathbf{x}$. The Bayes factor is defined as the ratio of the marginal likelihoods

$$BF_{12} = p(\mathbf{x} \,|\, M_1)/p(\mathbf{x} \,|\, M_2) \\ = \frac{p(M_1 \,|\, \mathbf{x})}{p(M_2 \,|\, \mathbf{x})} \times \frac{p(M_2)}{p(M_1)}, \tag{C.15}$$

which can also be expressed as the ratio of the posterior odds of M_1 to M_2 divided by their priors odds.

D

Solutions to Selected Exercises

Chapter 1

1.1 Verify that $|\mathbf{a} \bullet \mathbf{b}| \le \|\mathbf{a}\| \cdot \|\mathbf{b}\|$.

1.3 $a = \pm 1/\sqrt{2}$ and $b = \pm 1/\sqrt{2}$.

1.7 Solve the system $\begin{pmatrix} 2 \\ 3 \end{pmatrix} = c_1 \begin{pmatrix} 1 \\ 2 \end{pmatrix} + c_2 \begin{pmatrix} 3 \\ 5 \end{pmatrix}$ to get $c_1 = -1$ and $c_2 = 1$, showing that $\mathbf{u}$ is in Span$\{\mathbf{v}_1, \mathbf{v}_2\}$.

1.12 Computing the product on both sides, and equate them to get the conditions.

1.14 Show that $\mathbf{C} = \mathbf{A}^{k-1} + \mathbf{A}^{k-2}\mathbf{B} + \cdots + \mathbf{A}\mathbf{B}^{k-2} + \mathbf{B}^{k-1}$.

1.17 Use Result 1.3.5

1.22 $\Delta_n = (1 + a^2 + a^4 + \cdots + a^{2n}) = [1 - a^{2(n+1)}]/(1 - a^4)$ if $a \neq 1$, and $\Delta_n = n + 1$ if $a = 1$.

1.30 For (a), use the definition of orthogonality and Result 1.3.6. For (b), use Definition 1.2.9 and Definition 1.3.11.

1.33 $\mathrm{r}(\mathbf{A}) = 2$.

1.35 Use Definition 1.2.8 and proof by contradiction.

1.39 The eigenvector corresponding to $\lambda = 3$ is $\mathbf{v} = t(1, 1, 1)'$, for arbitrary $t \neq 0$.

Chapter 2

2.3 Use Result 2.1.3 to write the determinant of $\mathbf{A}$ as $|1| \cdot |\mathbf{P} - \mathbf{x}\mathbf{x}'| = |\mathbf{P}| \cdot |1 - \mathbf{x}'\mathbf{P}^{-1}\mathbf{x}|$. Simplify.

2.5 Use Example 2.1.3 and note that when $\mathbf{A}_2$ is an $n \times 1$ matrix, the expression $\mathbf{A}_2'(\mathbf{I} - \mathbf{P}_1)\mathbf{A}_2$ is a scalar.

2.7 Show that $\mathbf{A}\mathbf{a} = \mathbf{a}\mathbf{a}'\mathbf{a} = (\mathbf{a}'\mathbf{a})\mathbf{a}$, and use Definition 1.3.16.

2.12 For (a), use Result 2.3.4 to show that $\mathrm{r}(\mathbf{A}) = \mathrm{r}(\mathbf{D})$, which is equal to the number of nonzero elements of the diagonal matrix $\mathbf{D}$. To show (b), $\|\mathbf{A}\|^2 = \mathrm{tr}(\mathbf{A}'\mathbf{A}) = \mathrm{tr}(\mathbf{A}^2) = \mathrm{tr}(\mathbf{A}\mathbf{P}\mathbf{P}'\mathbf{A}\mathbf{P}\mathbf{P}') = \mathrm{tr}(\mathbf{P}'\mathbf{A}\mathbf{P}\mathbf{P}'\mathbf{A}\mathbf{P}) = \mathrm{tr}(\mathbf{D}^2)$.

2.16 Construct an $n \times (n-k)$ matrix $\mathbf{V}$ such that the columns of $(\mathbf{U}, \mathbf{V})$ form an orthogonal basis for $\mathcal{R}^n$.

2.21 Given that $\mathbf{Q}\mathbf{A}\mathbf{Q}^{-1} = \mathbf{D}$, invert both sides.

2.23 Suppose on the contrary, that $\mathrm{r}(\mathbf{C}) < q$, and use Exercise 2.10 to contradict this assumption.

DOI: 10.1201/9781315156651-D

2.26 $-1/(n-1) < a < 1$

2.30 For (a), use property 5 of Result 1.3.8 and simplify. Substitute for $(\mathbf{A}^{-1} + \mathbf{C}'\mathbf{B}^{-1}\mathbf{C})^{-1}$ from (a) into the LHS of (b) and simplify.

2.34 If possible, let $\mathbf{P}_1$ and $\mathbf{P}_2$ be two such matrices. Since $\mathbf{u}$ is unique, $(\mathbf{P}_1 - \mathbf{P}_2)\mathbf{y} = \mathbf{0}$ for all $\mathbf{y} \in \mathcal{R}^n$, so that $\mathbf{P}_1$ must equal $\mathbf{P}_2$.

Chapter 3

3.2 $\mathbf{G}$ is a g-inverse of $\mathbf{A}$ if and only if it has the form $\begin{pmatrix} \mathbf{u}' & a \\ \mathbf{I}_{n-1} & \mathbf{v} \end{pmatrix}$, where $\mathbf{u}$, $\mathbf{v}$ are any vectors in $\mathcal{R}^{n-1}$ and $a \in \mathcal{R}$.

3.5 For (a), use properties 1 and 2 of Result 3.1.9. For (b), transpose both sides of (a). Use (3.1.7) to obtain (c). For (d), use property 3 of Result 3.1.9. (e) follows directly from (d).

3.8 Since $\mathbf{G}$ is a g-inverse of $\mathbf{A}$, $\mathrm{r}(\mathbf{A}) \le \mathrm{r}(\mathbf{G})$. If $\mathbf{A}$ is a g-inverse of $\mathbf{G}$, then likewise $\mathrm{r}(\mathbf{G}) \le \mathrm{r}(\mathbf{A})$, so $\mathrm{r}(\mathbf{A}) = \mathrm{r}(\mathbf{G})$. Conversely, from Result 3.1.3, if $\mathbf{A} = \mathbf{BDC}$, where $\mathbf{B}$ and $\mathbf{C}$ are nonsingular and $\mathbf{D} = \mathrm{diag}(\mathbf{I}_r, \mathbf{0})$ with $r = \mathrm{r}(\mathbf{A})$, then $\mathbf{G} = \mathbf{C}^{-1}\mathbf{E}\mathbf{B}^{-1}$, where $\mathbf{E}$ has the form $\begin{pmatrix} \mathbf{I}_r & \mathbf{K} \\ \mathbf{L} & \mathbf{M} \end{pmatrix}$. Use Result 2.1.3.

3.14 Use Result 2.6.1 and property 2 of Result 1.3.10, followed by Result 3.1.13.

3.17 (a) Let $\mathbf{G} = \mathbf{H}\mathbf{A}^{-1}$. If $\mathbf{H} = \mathbf{B}^-$, then $\mathbf{ABGAB} = \mathbf{ABHA}^{-1}\mathbf{AB} = \mathbf{ABHB} = \mathbf{AB}$. Conversely, let $\mathbf{ABGAB} = \mathbf{AB}$. Then $\mathbf{BGAB} = \mathbf{B}$, i.e., $\mathbf{BHB} = \mathbf{B}$. The solution for (b) is similar.

3.20 The unique solution is $(-1, -4, 2, -1)'$.

3.22 $\mathcal{C}(\mathbf{c}) \subset \mathcal{C}(\mathbf{A})$ and $\mathcal{R}(\mathbf{b}) \subset \mathcal{R}(\mathbf{A})$; use Result 3.2.8.

Chapter4

4.1 Use property 8 of Result 1.3.11 to show that $\mathrm{r}(\mathbf{X}'\mathbf{X}, \mathbf{X}'\mathbf{y}) \ge \mathrm{r}(\mathbf{X}'\mathbf{X})$. Use properties 4 and 6 of Result 1.3.11 to show that $\mathrm{r}(\mathbf{X}'\mathbf{X}, \mathbf{X}'\mathbf{y}) = \mathrm{r}(\mathbf{X}'\mathbf{X})$.

4.6 $n_1 = 2n_2$.

4.9 (i) $\widehat{\beta}_1 = (Y_2 + Y_4 + Y_6 - Y_1 - Y_3 - Y_5)/6$, with variance $\sigma^2/6$. (ii) $\widehat{\beta}_1 = \{5(Y_2 - Y_5) + 8(Y_4 - Y_3) + 11(Y_6 - Y_1)\}/48$, with variance $420\sigma^2/(48)^2$. The ratio of variances is 32/35.

4.19 For (a), equate (4.5.5) with (4.2.3) and set $\mathbf{y} = \mathbf{Xz}$. Part (b) follows since $\mathbf{V}^{-1} = (1-\rho)^{-1}\mathbf{I} - \rho(1-\rho)^{-1}[1 + (N-1)\rho]^{-1}\mathbf{J}$, and $\mathbf{1}'\mathbf{y} = 0$.

4.23 From (4.5.7), $\mathbf{WX} = \mathbf{KP}_{\mathcal{C}(\mathbf{LX})}\mathbf{LX} = \mathbf{KLX} = \mathbf{X}$ as $\mathbf{L} = \mathbf{K}^{-1}$. From (4.5.7), $\mathcal{C}(\mathbf{W}) = \mathcal{C}(\mathbf{X}(\mathbf{X}'\mathbf{V}^{-1}\mathbf{X})^-\mathbf{X}'\mathbf{V}^{-1}) \subset \mathcal{C}(\mathbf{X})$. Finally, $\mathbf{W}^2 = \mathbf{KP}_{\mathcal{C}(\mathbf{LX})}\mathbf{LKP}_{\mathcal{C}(\mathbf{LX})} = \mathbf{KP}_{\mathcal{C}(\mathbf{LX})}\mathbf{P}_{\mathcal{C}(\mathbf{LX})} = \mathbf{W}$.

4.24 $\widehat{\theta}_i = Y_i - \overline{Y} + 60°$, $i = 1, 2, 3$.

4.26 $\widehat{\boldsymbol{\beta}} = \mathbf{y}$, and $\widehat{\boldsymbol{\beta}}_r = [\mathbf{I}_N - \mathbf{VC}(\mathbf{C}'\mathbf{VC})^{-1}\mathbf{C}']\mathbf{y}$.

4.29

$$\begin{aligned}\mathrm{E}(SSE_r) &= \sigma^2(N-r) + \sigma^2\,\mathrm{tr}(\mathbf{I}_q) + +(\mathbf{A}'\boldsymbol{\beta} - \mathbf{b})'(\mathbf{A}'\mathbf{G}\mathbf{A})^{-1}(\mathbf{A}\boldsymbol{\beta} - \mathbf{b}) \\ &= \sigma^2(N-r+q) + (\mathbf{A}'\boldsymbol{\beta} - \mathbf{b})'(\mathbf{A}'\mathbf{G}\mathbf{A})^{-1}(\mathbf{A}\boldsymbol{\beta} - \mathbf{b}).\end{aligned}$$

Chapter 5

5.2 (a) Use Result 5.1.1. (b) $\pi/\sqrt{3}$.

5.3 $a = 3.056$.

5.5 $\boldsymbol{\mu} = \begin{pmatrix} 4 \\ -2 \\ 1 \end{pmatrix}$ and $\boldsymbol{\Sigma} = \begin{pmatrix} 5 & -1 & 2 \\ -1 & 3 & 1 \\ 2 & 1 & 6 \end{pmatrix}$.

5.7 1/3.

5.10 Use the singular value decomposition.

5.13 See Result 5.2.14; show that the b.l.u.p. of $\mathbf{x}_1$ based on $\mathbf{x}_2$ is exactly $\mathbf{x}_1$ if and only if $\mathbf{x}_1 = \mathbf{A}\mathbf{x}_2 + \mathbf{a}$ for some matrix $\mathbf{A}$ and vector $\mathbf{a}$ of constants.

5.14 For (a), $X_2 \sim N(\mu_2, \sigma^2)$ and $X_3 \sim N(\mu_3, \sigma^2)$. For (b), $(X_1 \mid X_2, X_3)$ is normal with mean $\mu_1 + \rho(x_2 - \mu_2)/(1-\rho^2) - \rho^2(x_3 - \mu_3)/(1-\rho^2)$ and variance $\sigma^2(1-2\rho^2)/(1-\rho^2)$; it reduces to the marginal distribution of X_1 when $\rho = 0$. For (c), $\rho = -1/2$.

5.20 Show that the product of the m.g.f.'s of the two random variables on the right side gives the m.g.f. of the random variable on the left side.

5.23 For (a), by Result 5.4.5 and Result 5.3.4, $\mathrm{E}(U) = k + 2\lambda$ and $\mathrm{Var}(U) = 2(k+4\lambda)$. For (b), using Result 5.4.2, $U \sim \chi^2(k, \lambda)$ with $\lambda = \boldsymbol{\mu}'\boldsymbol{\Sigma}^{-1}\boldsymbol{\mu}$. For (c), $\mathbf{x}'\mathbf{A}\mathbf{x} \sim \chi^2(k-1, \boldsymbol{\mu}'\mathbf{a}\boldsymbol{\mu}/2)$.

5.35 Verify that $Q_1 = \mathbf{x}'\mathbf{A}_1\mathbf{x}$, and $Q_2 = \mathbf{x}'\mathbf{A}_2\mathbf{x}$, with $\mathbf{A}_1 = \begin{pmatrix} 1 & -1 \\ -1 & 1 \end{pmatrix}$ and $\mathbf{A}_2 = \begin{pmatrix} 1 & 1 \\ 1 & 1 \end{pmatrix}$. Use idempotency of $\mathbf{A}_1\boldsymbol{\Sigma}$ and that $\mathbf{A}_1\boldsymbol{\Sigma}\mathbf{A}_2 = \mathbf{O}$.

5.37 Use independence of $\mathbf{x}'\mathbf{A}_j\mathbf{x}$ to show necessity of Result 5.4.9. From Result 5.4.7, for $i \neq j$, $\mathbf{A}_i'\mathbf{A}_j = \mathbf{O}$, so $\mathcal{C}(\mathbf{A}_i) \perp \mathcal{C}(\mathbf{A}_j)$ and it is easy to show the result.

5.39 We show property 3. See that

$$f(\mathbf{x}_2) = c_{k-q}|\mathbf{V}_{22}|^{-1/2}h_{k-q}(\mathbf{x}_2'\mathbf{V}_{22}^{-1}\mathbf{x}_2),$$

and the conditional covariance is

$$\int (\mathbf{x}_1 - \mathbf{V}_{12}\mathbf{V}_{22}^{-1}\mathbf{x}_2)(\mathbf{x}_1 - \mathbf{V}_{12}\mathbf{V}_{22}^{-1}\mathbf{x}_2)'\, dF_{\mathbf{x}_1|\mathbf{x}_2}(\mathbf{x}_1).$$

Let $\mathbf{V}_{11.2} = \mathbf{D}'\mathbf{D}$, and let $\mathbf{y} = (Y_1, \cdots, Y_q)' = \mathbf{D}^{-1}(\mathbf{x}_1 - \mathbf{V}_{12}\mathbf{V}_{22}^{-1}\mathbf{x}_2)$. The covariance matrix is

$$\int \mathbf{y}\mathbf{y}'dF_{\mathbf{y},\mathbf{x}_2}(\mathbf{y},\mathbf{x}_2)/f(\mathbf{x}_2) = \mathbf{C} = \{c_{ij}\}, \text{ say},$$

and we are given that $\mathbf{C}$ does not depend on $\mathbf{x}_2$.

$$c_{11} = \int_{-\infty}^{\infty} Y_1^2\, dF_{\mathbf{y},\mathbf{x}_2}(\mathbf{y},\mathbf{x}_2)/f(\mathbf{x}_2),$$

and we can write

$$c_{11}c_{k-q}|\mathbf{V}_{22}|^{-1/2}h_{k-q}(\mathbf{x}_2'\mathbf{V}_{22}^{-1}\mathbf{x}_2)$$
$$= c_{k-q+1}|\mathbf{V}_{22}|^{-1/2}\int_{-\infty}^{\infty} Y_1^2 h_{k-q+1}(Y_1^2+\mathbf{x}_2'\mathbf{V}_{22}^{-1}\mathbf{x}_2)\,d\mathbf{y}_1.$$

Set $Z = \mathbf{x}_2'\mathbf{V}_{22}^{-1}\mathbf{x}_2$, and $b = 2c_{k-q+1}/c_{k-q}$.

Chapter 6

6.2 $\mathbf{B}'\mathbf{W}\mathbf{B} \sim \sum_{j=1}^m \mathbf{B}'\mathbf{x}_j(\mathbf{B}'\mathbf{x}_j)'$. Since $\mathbf{B}'\mathbf{x}_1,\cdots,\mathbf{B}'\mathbf{x}_m$ are i.i.d. $\sim N_q(\mathbf{0},\mathbf{B}'\mathbf{\Sigma}\mathbf{B})$, then $\mathbf{B}'\mathbf{W}\mathbf{B} \sim W_q(\mathbf{B}'\mathbf{\Sigma}\mathbf{B}, m)$.

6.4 Since $\widehat{\mathbf{\Sigma}}_{ML} = \{(N-1)/N\}\mathbf{S}_N$, $|\widehat{\mathbf{\Sigma}}_{ML}|^{N/2} = (1-\frac{1}{N})^{kN/2}|\mathbf{S}_N|^{N/2}$; substituting this into the expression (6.1.9),

$$L(\widehat{\boldsymbol{\mu}}_{ML}, \widehat{\mathbf{\Sigma}}_{ML}) = \frac{\exp(-\frac{Nk}{2})}{(2\pi)^{Nk/2}(1-\frac{1}{N})^{kN/2}|\mathbf{S}_N|^{N/2}}.$$

6.5 The first identity is immediate from the definition of the sample mean, and some algebra. To show the second identity, write

$$\begin{aligned}\mathbf{S}_{m+1} &= \sum_{i=1}^{m+1}(\mathbf{x}_i-\overline{\mathbf{x}}_{m+1})(\mathbf{x}_i-\overline{\mathbf{x}}_{m+1})' \\ &= \sum_{i=1}^{m}(\mathbf{x}_i-\overline{\mathbf{x}}_m+\overline{\mathbf{x}}_m-\overline{\mathbf{x}}_{m+1})(\mathbf{x}_i-\overline{\mathbf{x}}_m+\overline{\mathbf{x}}_m-\overline{\mathbf{x}}_{m+1})' \\ &\quad + (\mathbf{x}_{m+1}-\overline{\mathbf{x}}_{m+1})(\mathbf{x}_{m+1}-\overline{\mathbf{x}}_{m+1})' \\ &= \mathbf{S}_m + m(\overline{\mathbf{x}}_m-\overline{\mathbf{x}}_{m+1})(\overline{\mathbf{x}}_m-\overline{\mathbf{x}}_{m+1})' + (\mathbf{x}_{m+1}-\overline{\mathbf{x}}_{m+1})(\mathbf{x}_{m+1}-\overline{\mathbf{x}}_{m+1})',\end{aligned}$$

use the first identity for $\overline{\mathbf{x}}_{m+1}$ and simplify.

6.9 Use the proof of Result 6.2.3 to show that $\mathbf{S}_{11.2}$ is independent of $\mathbf{Z}$, hence independent of $\mathbf{S}_{22} = \mathbf{Z}'\mathbf{A}\mathbf{Z}$. Show the conditional independence of $\mathbf{S}_{11.2}$ and $\mathbf{S}_{21}$ given $\mathbf{Z}$. Argue that the conditional distribution of $\mathbf{S}_{11.2}$ is the same as its unconditional distribution, from which the (unconditional) independence follows.

6.12 Beta$(k/2,(N-k-1)/2)$ distribution, which follows from (6.2.11) and (B.13).

6.14 For (a), note that $\sum_{j=1}^N Z_{(j)} = \sum_{j=1}^N Z_j = N\overline{Z}_N$. Also, $Z_{(i)} - \overline{Z}_N$ is a function of $(Z_1-\overline{Z}_N,\cdots,Z_N-\overline{Z}_N)$, which is independent of $\overline{Z}_N$. (b) follows directly.

Chapter 7

7.1 Property 1 of Corollary 7.1.1 is a direct consequence of Result 5.2.5, while property 2 follows from Result 5.2.4. Property 3 follows from the orthogonality of $\mathbf{X}$ and $\mathbf{I}-\mathbf{P}$.

7.2 (c) The test statistic $F_0 = (N-3)Q/SSE \sim F_{1,9}$ under H_0, where $Q = \sum\widehat{\beta}_1X_i(2\widehat{\beta}_0 + \widehat{\beta}_1X_i + 2\widehat{\beta}_2X_i^2)$ and $SSE = \sum(Y_i-\widehat{\beta}_0-\widehat{\beta}_1X_i-\widehat{\beta}_2X_i^2)$.

7.4 Let $N = n_1 + n_2 + n_3$. $SSE_H = \sum_{i=1}^{3}\sum_{j=1}^{n_i} Y_{ij}^2 - \sum_{i=1}^{3}(n_i\overline{Y}_{i\cdot})^2/N$ and $SSE = \sum_{i=1}^{3}\sum_{j=1}^{n_i} Y_{ij}^2 - \sum_{i=1}^{3} Y_{i\cdot}^2/n_i$, so that $F(H) = (N-3)(SSE_H - SSE)/(2SSE) \sim F_{2,N-3}$ under H.

7.8 (a) Let $\boldsymbol{\beta}' = (\lambda_1, \lambda_2)$. Then, $\widehat{\lambda}_1 = [(1+c^2)Y_k - c^3Y_{k+1} - c^2Y_{k+2}]/(1+c^2+c^4)$ and $\widehat{\lambda}_2 = [cY_k + Y_{k+1} - c(1+c^2)Y_{k+2}]/(1+c^2+c^4)$, with

$$\text{Var}(\widehat{\boldsymbol{\beta}}) = \frac{\sigma^2}{(1+c^2+c^4)}\begin{pmatrix} 1+c^2 & c \\ c & 1+c^2 \end{pmatrix}.$$

(b) Under the restriction $\lambda_1 = -\lambda_2 = \lambda$, say, using results from section 4.6.1, we get $\widehat{\lambda} = [Y_k - (1+c)Y_{k+1} + cY_{k+2}]/2(1+c+c^2)$, with $\text{Var}(\widehat{\lambda}) = \sigma^2/2(1+c+c^2)$.

7.10 $\text{E}(SSE_H) - \text{E}(SSE) = \sigma^2(r - r_2) + \boldsymbol{\beta}_1'\mathbf{X}_1'(\mathbf{I} - \mathbf{P}_2)\mathbf{X}_1\boldsymbol{\beta}_1$.

7.14 H_0: $X^* = 0$ implies H_0: $\beta_1 = 0$. The test statistic $F_0 = (N-3)Q/SSE \sim F_{1,9}$ under H_0, where $Q = \sum \widehat{\beta}_1 X_i(2\widehat{\beta}_0 + \widehat{\beta}_1 X_i + 2\widehat{\beta}_2 X_i^2)$ and $SSE = \sum(Y_i - \widehat{\beta}_0 - \widehat{\beta}_1 X_i - \widehat{\beta}_2 X_i^2)$.

7.16 Due to orthogonality, the least squares estimates of β_0 and β_1 are unchanged, and are $\widehat{\beta}_0 = (Y_1 + Y_2 + Y_3)/3$, and $\widehat{\beta}_1 = (Y_3 - Y_1)/2$

7.18 The hypothesis H can be writen as $\mathbf{C}'\boldsymbol{\beta} = \mathbf{d}$, where $\mathbf{C}'$ is an $(a-1)\times a$ matrix

$$\begin{pmatrix} 1 & -2 & 0 & 0 & \cdots & 0 & 0 \\ 0 & 2 & -3 & 0 & \cdots & 0 & 0 \\ 0 & 0 & 3 & -4 & \cdots & 0 & 0 \\ & & & \vdots & & & \\ 0 & 0 & 0 & 0 & \cdots & a-1 & -a \end{pmatrix},$$

$\boldsymbol{\beta} = (\mu_1, \cdots, \mu_a)'$, and $\mathbf{d}$ is an $(a-1)$-dimensional vector of zeroes. The resulting F-statistic is obtained from (7.2.9) and has an $F_{a-1,a(n-1)}$ distribution under H.

7.19 $F(H) = (3n-3)Q/SSE \sim F_{1,3n-3}$ under H, where $Q = \frac{2n}{3}[\overline{Y}_{2\cdot} - \frac{1}{2}(\overline{Y}_{1\cdot} + \overline{Y}_{1\cdot})]^2$, and $SSE = \sum_{i=1}^{3}\sum_{j=1}^{n}(Y_{ij} - \overline{Y}_{i\cdot})^2$.

7.23 The 95% C.I. for β_1 is $\widehat{\beta}_1 \pm 4.73\sqrt{11/4}$. For β_2, it is $\widehat{\beta}_2 \pm 4.73\sqrt{3}$. For β_3, it is $\widehat{\beta}_3 \pm 4.73$. For $\beta_1 - \beta_2$, it is $\widehat{\beta}_1 - \widehat{\beta}_2 \pm 4.73\sqrt{59/4}$. For $\beta_1 + \beta_3$, it is $\widehat{\beta}_1 + \widehat{\beta}_3 \pm 4.73\sqrt{27/4}$.

7.25 Scheffé's simultaneous confidence set has the form

$$\{\boldsymbol{\beta}: |\mathbf{c}'\boldsymbol{\beta}_{GLS}^0 - \mathbf{c}'\boldsymbol{\beta}| \le \widehat{\sigma}_{GLS}(dF_{d,N-r,\alpha})^{1/2}[\mathbf{c}(\mathbf{X}'\boldsymbol{\Sigma}^{-1}\mathbf{X})^{-}\mathbf{c}]^{1/2}\forall\ \mathbf{c} \in \mathcal{L}\}.$$

Chapter 8

8.1 The MLE is

$$\widehat{\theta} = \frac{\sum_{t=1}^{N} Y_t Y_{t-1}}{\sum_{t=1}^{N} Y_t^2}.$$

8.3 It can be verified that

$$\text{Var}(\widetilde{\beta}_1) = \sigma^2 \left\{ \sum_{i=1}^{N} \frac{X_{i1}}{X_{i2}} - \frac{N^2}{\sum_{i=1}^{N} \frac{X_{i2}}{X_{i1}}} \right\}^{-1}, \text{ and}$$

$$\text{Var}(\widetilde{\beta}_2) = \sigma^2 \left\{ \sum_{i=1}^{N} \frac{X_{i2}}{X_{i1}} - \frac{N^2}{\sum_{i=1}^{N} \frac{X_{i1}}{X_{i2}}} \right\}^{-1}.$$

These may be compared with $\text{Var}(\widehat{\beta}_1)$ and $\text{Var}(\widehat{\beta}_2)$ respectively.

8.7 By Example 2.1.3, $\mathbf{P}_Z = \mathbf{P}_X + \{(\mathbf{I}-\mathbf{P}_X)\mathbf{y}\mathbf{y}'(\mathbf{I}-\mathbf{P}_X)\}/\mathbf{y}'(\mathbf{I}-\mathbf{P}_X)\mathbf{y} = \mathbf{P}_X + \widehat{\varepsilon}\widehat{\varepsilon}'/\widehat{\varepsilon}'\widehat{\varepsilon}$.

8.8 $SIC_i = (N-1)\{T(\widehat{F}_N) - T(\widehat{F}_{(i)})\} = (N-1)(\widehat{\beta} - \widehat{\beta}_{(i)})$. Simplify to get the result.

8.11 (a) Since

$$\mathbf{P}_Z = \mathbf{P}_X + \frac{\widehat{\varepsilon}\widehat{\varepsilon}'}{\widehat{\varepsilon}'\widehat{\varepsilon}},$$

$p_{Zii} \leq 1$ for all i, implying that $p_{ii} + \frac{\widehat{\varepsilon}_i^2}{\widehat{\varepsilon}'\widehat{\varepsilon}}$. (b) These follow directly from the relations in properties 2 and 3 of Result 8.3.4.

8.14 Use Exercise 8.7. The second result follows directly.

Chapter 9

9.1 $\text{Var}(\widehat{\beta}_1) = \sigma^2/\sum_{i=1}^{N}(X_{i1} - \overline{X}_1)^2(1 - r_{12}^2)$. The width of the 95% confidence interval for β_1 is $2t_{N-3,.025}\,\text{s.e.}(\widehat{\beta}_1)$.

9.3 $MSE(\widehat{\sigma}^2) = 2\sigma^4/(N-p)$, while $MSE(\mathbf{y}'\mathbf{A}_1\mathbf{y}) = 2\sigma^4/(N-p+2)$.

9.6 (a) $\widehat{\alpha}_i = \overline{Y}_i$, $i = 1, 2$, $\widehat{\beta} = [S_{XY}^{(1)} + S_{XY}^{(2)}]/[S_{XX}^{(1)} + S_{XX}^{(2)}]$. The vertical distance between the lines is $D = (\alpha_1 - \alpha_2) + \beta(\overline{X}_1 - \overline{X}_2)$, with $\widehat{D} = (\overline{Y}_1 - \overline{Y}_2) - \widehat{\beta}(\overline{X}_1 - \overline{X}_2)$. Then, $\text{E}(\widehat{D}) = D$. (b) A 95% symmetric C.I. for D is $\widehat{D} \pm \widehat{\sigma}[F_{1,n_1+n_2-3,.05}]^{1/2}\,\text{s.e.}(\widehat{D})$.

9.9 (a) From Exercise 4.14, $\mathbf{X}_1\boldsymbol{\beta}_1$ is estimable under the expanded model. Since $\mathbf{X}_1$ has full column rank, $\boldsymbol{\beta}_1$ is estimable. (b) Use (4.2.36) to show that

$$\text{Cov}(\widetilde{\boldsymbol{\beta}}_1) = \sigma^2[\mathbf{X}_1'(\mathbf{I} - \mathbf{P}_2)\mathbf{X}_1]^{-1}.$$

On the other hand, $\text{Cov}(\widehat{\boldsymbol{\beta}}_1) = \sigma^2(\mathbf{X}_1'\mathbf{X}_1)^{-1}$. Use the Sherman–Morrison–Woodbury formula.

Chapter 10

10.2 $\mathbf{U}\mathbf{1}_a = \sum_{j=1}^{a} U_{ij} = U_{ii} + \sum_{j\neq i}^{a} U_{ij} = N_{i\cdot} - \sum_{l=1}^{b}\sum_{j=1}^{a} n_{ij}n_{lj}/N_{\cdot j} = N_{i\cdot} - N_{i\cdot} = 0$, $i = 1, \cdots, a$, implying $\text{r}(\mathbf{U}) \leq a - 1$.

10.3 (b) In order that $\text{E}(\sum_{i=1}^{a}\sum_{j=1}^{b} c_{ij}Y_{ij}) = \mu\sum_{i=1}^{a}\sum_{j=1}^{b} c_{ij} + \sum_{i=1}^{a}(\sum_{j=1}^{b} c_{ij})\tau_i + \sum_{j=1}^{b}(\sum_{i=1}^{a} c_{ij})\beta_j$ is to be function only of β's, we must have $\sum_{i=1}^{a}\sum_{j=1}^{b} c_{ij} = 0$, $\sum_{j=1}^{b} c_{ij} = 0$, $i = 1, \cdots, a$. Then, $RHS = \sum_{j=1}^{b}(\sum_{i=1}^{a} c_{ij})\beta_j = \sum_{j=1}^{b} d_j\beta_j$, where $\sum_{j=1}^{b} d_j = \sum_{j=1}^{b}\sum_{i=1}^{a} c_{ij} = 0$.

10.6 (a) For $i = 1, \cdots, n$, write the function as $\sum_{j=1}^{b}\{n_{ij}(\mu+\tau_i+\theta_j+\gamma_{ij}) - \sum_{k=1}^{a} \frac{n_{ij}n_{kj}}{n_{\cdot j}}(\mu+\tau_k+\theta_j+\gamma_{kj})\}$, which shows it is estimable. The proof of (b) is similar.

10.7 (a) Yes, (b) No.

10.11 $\theta^0_{j(i)} = Y_{ij\cdot}/n_{ij} = \overline{Y}_{ij\cdot}$, $j = 1, \cdots, b_i$; $i = 1, \cdots, a$. The corresponding g-inverse is $\mathbf{G} = \begin{pmatrix} 0 & \mathbf{0} \\ \mathbf{0} & \mathbf{D} \end{pmatrix}$, where $\mathbf{D} = \text{diag}(1/n_{11}, \cdots, 1/n_{aa})$.

10.15 (a) $SS(\beta_0, \beta_1) = [Y_{\cdot\cdot}^2/N] + \widehat{\beta}_1^2 \sum_i n_i(Z_i - \overline{Z})^2$. (b) $SS(\mu, \tau_1, \cdots, \tau_a) = \sum_{i=1}^{a}(Y_{i\cdot}^2/n_i) - Y_{\cdot\cdot}^2/N$. Hence, $SS(\beta_0, \beta_1) - SS(\mu, \tau_1, \cdots, \tau_a) = A/B$, where $A = [\sum_{i=1}^{a} c_i^2 \sum_{i=1}^{a} d_i^2 - (\sum_{i=1}^{a} c_i d_i)^2]$ and $B = [\sum_{i=1}^{a} n_i(Z_i - \overline{Z})^2]$, $c_i = \sqrt{n_i}(\overline{Y}_{i\cdot} - \overline{Y}_{\cdot\cdot})$ and $d_i = \sqrt{n_i}(Z_i - \overline{Z})$. The difference in SS is always nonnegative (by Cauchy–Schwarz inequality), with equality only when $c_i \propto d_i$, $i = 1, \cdots, a$, i.e., only when $\overline{Y}_{i\cdot} - \overline{Y}_{\cdot\cdot} = w(Z_i - \overline{Z})$, where w is a constant.

10.19 Write $\mathbf{c}_l' = \frac{1}{\sqrt{l(l+1)}}(0, 1, \cdots, 1, -l, 0, \cdots, 0)$. For $l = 1, \cdots, a-1$, the 95% marginal C.I. for $\mathbf{c}_l'\boldsymbol{\beta}$ is

$$\mathbf{c}_l'\boldsymbol{\beta}^0 \pm \frac{\widehat{\sigma}}{n} t_{a(n-1),.025}$$

and the 95% Scheffé intervals are

$$\mathbf{c}_l'\boldsymbol{\beta}^0 \pm \sqrt{MSE\frac{(a-1)}{n}F_{a-1,N-a,0.05}}.$$

Chapter 11

11.3 Let the $N \times (a+1)$ matrix $\mathbf{X} = (\mathbf{1}_N, \mathbf{x}_1, \cdots, \mathbf{x}_a)$, where the N-dimensional vector $\mathbf{x}_i = \mathbf{e}_i \otimes \mathbf{1}_a$, with $\mathbf{e}_i$ the ith standard basis vector. Then, $SSA = \sum_{i=1}^{a} Y_{i\cdot}^2/n - N\overline{Y}_{\cdot\cdot}^2 = \mathbf{y}'\{\frac{1}{n} \oplus_{i=1}^{a} \mathbf{x}_i\mathbf{x}_i' - \frac{1}{N}\mathbf{1}_N\mathbf{1}_N'\}\mathbf{y}$, which gives the result.

11.4 Suppose $SSA = \mathbf{y}'\mathbf{M}\mathbf{y}$, then, $\text{E}(SSA) = \text{tr}(\mathbf{MV})$, where $\mathbf{V}$ denotes $\text{Cov}(\mathbf{y})$; simplify.

11.5 Find the first and second partial derivatives of the log-likelihood function with respect to $\boldsymbol{\tau}$ and σ_j^2. Use the results $\text{E}(\mathbf{y} - \mathbf{X}\boldsymbol{\tau}) = \mathbf{0}$ and $\text{E}(\mathbf{y} - \mathbf{X}\boldsymbol{\tau})'\mathbf{A}(\mathbf{y} - \mathbf{X}\boldsymbol{\tau}) = \text{tr}(\mathbf{AV})$.

11.6 Let $SSA = \sum_i nb(\overline{Y}_{i\cdot\cdot} - \overline{Y}_{\cdot\cdot\cdot})^2$, $SSB(A) = \sum_{i,j} n(\overline{Y}_{ij\cdot} - \overline{Y}_{i\cdot\cdot})^2$, $SSE = \sum_{i,j,k}(Y_{ijk} - \overline{Y}_{ij\cdot})^2$ with respective d.f. $(a-1)$, $a(b-1)$, and $ab(n-1)$. The ANOVA estimators are $\widehat{\sigma}_\tau^2 = \{MSA - MSB(A)\}/nb$, $\widehat{\sigma}_\beta^2 = \{MSB(A) - MSE\}/n$, and $\widehat{\sigma}_\varepsilon^2 = MSE$. The ML solutions are $\widetilde{\sigma}_\tau^2 = \{(1 - 1/a)MSA - MSB(A)\}/nb$, while $\widetilde{\sigma}_\beta^2$ and $\widetilde{\sigma}_\varepsilon^2$ coincide with their ANOVA estimators.

11.9 Since $\mathbf{AX} = \mathbf{O}$, $\mathbf{y}'\mathbf{A}\mathbf{y} = \boldsymbol{\varepsilon}'\mathbf{A}\boldsymbol{\varepsilon} = \sum_i a_{ii}\varepsilon_i^2 + 2\sum_{i<j} a_{ij}\varepsilon_i\varepsilon_j$, and

$$\text{E}(\mathbf{y}'\mathbf{A}\mathbf{y}) = \text{E}(\boldsymbol{\varepsilon}'\mathbf{A}\boldsymbol{\varepsilon}) = \sigma_\varepsilon^2 \sum_i a_{ii}.$$

It is enough to show that

$$\begin{aligned}\text{E}[(\mathbf{y}'\mathbf{A}\mathbf{y})^2] &= \sum_{i \neq k} \sigma_\varepsilon^4 a_{ik}^2 + \sum_{i \neq j} \sigma_\varepsilon^4 a_{ij}^2 + \sum_{i \neq j} \sigma_\varepsilon^4 a_{ii}a_{jj} + \sum_i \mu_4 a_{ii}^2 \\ &= 2\sigma_\varepsilon^4 \,\text{tr}(\mathbf{A}^2) + \sigma_\varepsilon^4 \left(\sum_i a_{ii}\right)^2 + (\mu_4 - 3\sigma_\varepsilon^2)\mathbf{a}'\mathbf{a}.\end{aligned}$$

11.9 We see from the proof of Result 11.2.4 that $\mathbf{y}'\mathbf{A}\mathbf{y} = \boldsymbol{\varepsilon}'\mathbf{A}\boldsymbol{\varepsilon} = \sum_i a_{ii}\varepsilon_i^2 + 2\sum_{i<j} a_{ij}\varepsilon_i\varepsilon_j$. Then

$$\mathrm{E}(\mathbf{y}'\mathbf{A}\mathbf{y}) = \mathrm{E}(\boldsymbol{\varepsilon}'\mathbf{A}\boldsymbol{\varepsilon}) = \sigma_\varepsilon^2 \sum_i a_{ii}. \tag{D.1}$$

Next

$$\begin{aligned}\mathrm{E}[(\mathbf{y}'\mathbf{A}\mathbf{y})^2] &= \mathrm{E}(\boldsymbol{\varepsilon}'\mathbf{A}\boldsymbol{\varepsilon}\boldsymbol{\varepsilon}'\mathbf{A}\boldsymbol{\varepsilon}) = \mathrm{E}[\mathrm{tr}(\boldsymbol{\varepsilon}\boldsymbol{\varepsilon}'\mathbf{A}\boldsymbol{\varepsilon}\boldsymbol{\varepsilon}'\mathbf{A})] \\ &= \mathrm{E}\left[\sum_{i,j,k,l} \varepsilon_i\varepsilon_j a_{jk}\varepsilon_k\varepsilon_l a_{li}\right] = \sum_{i,j,k,l} \mathrm{E}[\varepsilon_i\varepsilon_j a_{jk}\varepsilon_k\varepsilon_l a_{li}].\end{aligned}$$

Since the ε_i are i.i.d. with mean 0, $\mathrm{E}[\varepsilon_i\varepsilon_j a_{jk}\varepsilon_k\varepsilon_l a_{li}] = 0$ if one of i, j, k, l is not equal to any of the others. Hence the expectation is nonzero only in the following four cases, i) $i = j \neq k = l$, with value $\sigma_\varepsilon^4 a_{ik}^2$, ii) $i = k \neq j = l$, with value $\sigma^2 a_{ij}^2$, iii) $i = l \neq j = k$, with value $\sigma_\varepsilon^4 a_{ii}a_{jj}$, and iv) $i = j = k = l$, with value $\mu_4 a_{ii}^2$. Therefore, show that

$$\begin{aligned}\mathrm{E}[(\mathbf{y}'\mathbf{A}\mathbf{y})^2] &= \sum_{i\neq k} \sigma_\varepsilon^4 a_{ik}^2 + \sum_{i\neq j} \sigma_\varepsilon^4 a_{ij}^2 + \sum_{i\neq j} \sigma_\varepsilon^4 a_{ii}a_{jj} + \sum_i \mu_4 a_{ii}^2 \\ &= 2\sigma_\varepsilon^4\,\mathrm{tr}(\mathbf{A}^2) + \sigma_\varepsilon^4\left(\sum_i a_{ii}\right)^2 + (\mu_4 - 3\sigma_\varepsilon^2)\mathbf{a}'\mathbf{a}.\end{aligned}$$

This combined with (D.1) then yields (11.2.15).

Chapter 12

12.1 Y has a Beta-Binomial distribution with $\mathrm{E}(Y) = m\alpha/(\alpha+\beta)$ and $\mathrm{Var}(Y) = m\alpha\beta/(\alpha+\beta)^2$.

$$\begin{aligned}\mathrm{E}(Y) &= \mathrm{E}_p\{\mathrm{E}(Y\,|\,p)\} = m\,\mathrm{E}(p) = m\alpha/(\alpha+\beta), \quad \text{and} \\ \mathrm{Var}(Y) &= \mathrm{E}_p\{\mathrm{Var}(Y\,|\,p)\} + \mathrm{Var}_p\{\mathrm{E}(Y\,|\,p)\} \\ &= \frac{m\alpha\beta}{(\alpha+\beta)(\alpha+\beta+1)} + \frac{m^2\alpha\beta}{(\alpha+\beta)^2(\alpha+\beta+1)} = \frac{m\alpha\beta(\alpha+\beta+m)}{(\alpha+\beta)^2(\alpha+\beta+1)}.\end{aligned}$$

12.2 For (a),

$$Y_i\,|\,\mathbf{x}_i, \boldsymbol{\beta} \sim \exp\left\{\frac{-\mathbf{x}_i'\boldsymbol{\beta}\cdot Y_i + \log(\mathbf{x}_i'\boldsymbol{\beta})}{\phi_i} + \frac{1}{\phi_i}\log(Y_i/\phi_i) - \log Y_i - \log\Gamma(1/\phi_i)\right\}.$$

The other cases are similar.

12.4 (a) Under Model 1 and Model 2, the likelihood is

$$L = \exp\{-(\lambda_1+\lambda_2)\}\frac{\lambda_1^{y_1}\lambda_2^{y_2}}{y_1!y_2!}.$$

(c) In order to incorporate a covariate X, set $g(\lambda_i) = \log\lambda_i = \alpha + \beta X_i$, say.

12.8 (a) For integer y, it can be seen that

$$\frac{\Gamma(y+c)}{\Gamma(c)} = \prod_{j=0}^{y-1}(j+c).$$

Apply this in the p.m.f. $f(y;\mu,\alpha)$ and use the log-link which gives $\mu_i = \exp(\mathbf{x}_i'\boldsymbol{\beta})$. This gives the required likelihood. Part(b) follows the IRLS steps shown in the text.

Chapter 13

13.2 For any i, j, $\text{Cov}(\mathbf{M}_1'\boldsymbol{\varepsilon}_{(i)}, \mathbf{M}_2'\boldsymbol{\varepsilon}_{(j)}) = \mathbf{O}$, so $\mathbf{M}_1'\boldsymbol{\varepsilon}_{(i)}$ and $\mathbf{m}_2'\boldsymbol{\varepsilon}_{(j)}$ are uncorrelated, and hence by joint normality, are independent.

13.3 The MANOVA decomposition is

$$\mathbf{y}_{\ell kr} = \overline{\mathbf{y}} + (\overline{\mathbf{y}}_{\ell\cdot} - \overline{\mathbf{y}}) + (\overline{\mathbf{y}}_{\cdot k} - \overline{\mathbf{y}}) + (\overline{\mathbf{y}}_{\ell k} - \overline{\mathbf{y}}_{\ell\cdot} - \overline{\mathbf{y}}_{\cdot k} + \overline{\mathbf{y}}) + (\mathbf{y}_{\ell kr} - \overline{\mathbf{y}}_{\ell k}).$$

Summing over $\ell = 1, \cdots, g; j = 1, \cdots, b; r = 1, \cdots, n$ gives the breakup of the Corrected Total SSCP as

$$\begin{aligned}\text{Total SSCP}_c &= \text{SSCP(Factor A)} + \text{SSCP(Factor B)}\\ &+ \text{SSCP(Interaction)} + \text{SSCP(Error)}\end{aligned}$$

from which the MANOVA table follows and the rest can be based on this.

13.9 The GLS estimator of $\boldsymbol{\beta}$ has the form in (4.5.5), i.e., $\widehat{\boldsymbol{\beta}} = (\mathbf{X}'\mathbf{V}^{-1}\mathbf{X})^{-1}\mathbf{X}'\mathbf{V}^{-1}\mathbf{y}$, while the GLS estimator of σ^2 is $\widehat{\sigma}^2 = \frac{1}{N-p}SSE_{GLS}$, where SSE_{GLS} was shown in (4.5.9).

13.16 F_{N_1-1,N_2-1}.

13.21 The state equation is $x_{t+1} = \rho x_t + w_t$, and the observation equation is $Y_t = x_t$.

13.22 The autoregressive signal-plus-noise model is defined by

$$\begin{aligned}Z_t &= Y_t + v_t,\\ Y_t &= \rho Y_{t-1} + w_t,\end{aligned}$$

v_t and w_t are independent error processes, $v_t \sim N(0, \sigma_v^2)$ and $w_t \sim N(0, \sigma_w^2)$. Express this in the form of (13.5.1) and (13.5.2), and the Kalman filter recursions are obtained from (13.5.7)-(13.5.9).

Bibliography

Akaike, H. (1974), "A new look at the statistical model identification," *IEEE Transactions on Automatic Control*, 19, 716–723.

Akritas, M. G. and Arnold, S. F. (1994), "Fully nonparametric hypotheses for factorial designs I: Multivariate repeated measures designs," *Journal of the American Statistical Association*, 89, 336–343.

Anderson, T. W. (2003), *An Introduction to Multivariate Statistical Analysis*, Wiley Series in Probability and Statistics, Wiley-Interscience [John Wiley & Sons]: Hoboken, NJ, 3rd ed.

Anderson, T. W. and Fang, K.-T. (1987), "Cochran's theorem for elliptically contoured distributions," *Sankhyā,: The Indian Journal of Statistics, Series A*, 305–315.

Andrews, D. F., Gnanadesikan, R., and Warner, J. L. (1971), "Transformations of multivariate data," *Biometrics*, 825–840.

Andrews, D. F. and Mallows, C. L. (1974), "Scale mixtures of normal distributions," *Journal of the Royal Statistical Society Series B*, 36, 99–102.

Andrews, D. F. and Pregibon, D. (1978), "Finding the outliers that matter," *Journal of the Royal Statistical Society Series B*, 40, 85–93.

Atiqullah, M. (1962), "The estimation of residual variance in quadratically balanced least-squares problems and the robustness of the F-test," *Biometrika*, 49, 83–91.

Atkinson, A. C. (1981), "Two graphical displays for outlying and influential observations in regression," *Biometrika*, 68, 13–20.

— (1985), *Plots, Transformations and Regression: An Introduction to Graphical Methods of Diagnostic Regression Analysis*, Oxford University Press: UK.

Bai, Z. and Silverstein, J. W. (2010), *Spectral Analysis of Large Dimensional Random Matrices*, Springer-Verlag: New York, 2nd ed.

Baker, S. G. (1994), "The multinomial-Poisson transformation," *Journal of the Royal Statistical Society Series D*, 43, 495–504.

Barrodale, I. and Roberts, F. D. K. (1973), "An improved algorithm for discrete l_1 linear approximation," *SIAM J. Numer. Anal.*, 10, 839–848.

Barron, A., Birgé, L., and Massart, P. (1999), "Risk bounds for model selection via penalization," *Probability Theory and Related Fields*, 113, 301–413.

Bartlett, M. S. (1937), "Properties of sufficiency and statistical tests," *Proceedings of the Royal Society of London Series A*, 160, 268–282.

— (1938), "Further aspects of the theory of multiple regression," *Proceedings of the Cambridge Philosophical Society*, 34, 33–40.

Bassett, G. and Koenker, R. (1978), "Asymptotic theory of least absolute error regression," *Journal of the American Statistical Association*, 73, 618–622.

Basu, D. (1964), "Recovery of ancillary information," *Sankhyā,: The Indian Journal of Statistics, Series A*, 26, 3–16.

Beckman, R. J. and Trussell, H. J. (1974), "The distribution of an arbitrary Studentized residual and the effects of updating in multiple regression," *Journal of the American Statistical Association*, 69, 199–201.

Belsley, D. A. (1984), "Demeaning conditioning diagnostics through centering," *The American Statistician*, 38, 73–77.

— (1991), *Conditioning Diagnostics: Collinearity and Weak Data in Regression*, John Wiley & Sons, Inc.: New York.

Belsley, D. A., Kuh, E., and Welsch, R. E. (1980), *Regression Diagnostics: Identifying Influential Data and Sources of Collinearity*, John Wiley & Sons: New York-Chichester-Brisbane.

Benjamini, Y. and Hochberg, Y. (1995), "Controlling the false discovery rate: A practical and powerful approach to multiple testing," *Journal of the Royal Statistical Society Series B*, 57, 289–300.

Benjamini, Y. and Yekutieli, D. (2001), "The control of the false discovery rate in multiple testing under dependency," *The Annals of Statistics*, 1165–1188.

Berger, J. O. (1980), *Statistical Decision Theory: Foundations, Concepts, and Methods*, Springer-Verlag, New York-Heidelberg.

Bickel, P. J. and Doksum, K. A. (1981), "An analysis of transformations revisited," *Journal of the American Statistical Association*, 76, 296–311.

Birkes, D. and Dodge, Y. (1993), *Alternative Methods of Regression*, John Wiley & Sons, Inc., New York.

Box, G. E. P. and Cox, D. R. (1964), "An analysis of transformations," *Journal of the Royal Statistical Society Series B*, 26, 211–243.

Box, G. E. P. and Tiao, G. C. (1973), *Bayesian Inference in Statistical Analysis*, Addison-Wesley Publishing Co., Reading, Mass.-London-Don Mills, Ont.

Breusch, T. S. and Pagan, A. R. (1979), "A simple test for heteroscedasticity and random coefficient variation," *Econometrica*, 1287–1294.

Brockwell, P. J. and Davis, R. A. (2006), *Time Series: Theory and Methods*, Springer, New York, reprint of the second (1991) edition.

Cameron, A. C. and Trivedi, P. K. (2013), *Regression Analysis of Count Data*, vol. 53, Cambridge University Press, Cambridge, 2nd ed.

Carroll, R. J. and Ruppert, D. (1988), *Transformation and Weighting in Regression*, Chapman & Hall, New York.

Casella, G. and Berger, R. L. (1990), *Statistical Inference*, Wadsworth & Brooks/Cole Advanced Books & Software, Pacific Grove, CA.

Charnes, A. and Cooper, W. W. (1955), "Optimal estimation of executive compensation by linear programming," *Management Sci.*, 1, 138–151.

Chatterjee, S. and Hadi, A. S. (1988), *Sensitivity Analysis in Linear Regression*, John Wiley & Sons, Inc., New York.

Chen, M.-H., Dey, D. K., and Shao, Q.-M. (1999), "A new skewed link model for dichotomous quantal response data," *Journal of the American Statistical Association*, 94, 1172–1186.

Chen, S. S., Donoho, D. L., and Saunders, M. A. (1998), "Atomic decomposition by basis pursuit," *SIAM Journal on Scientific Computing*, 20, 33–61.

Cheng, B. and Titterington, D. M. (1994), "Neural networks: A review from a statistical perspective," *Statistical Science*, 9, 2–30.

Chmielewski, M. A. (1981), "Elliptically symmetric distributions: A review and bibliography," *International Statistical Review*, 49, 67–74.

Cochran, W. G. (1934), "The distribution of quadratic forms in a normal system, with applications to the analysis of covariance," *Mathematical Proceedings of the Cambridge Philosophical Society*, 30, 178–191.

Conover, W. J. (1998), *Practical Nonparametric Statistics*, John Wiley & Sons: New York, 3rd ed.

Conover, W. J. and Iman, R. L. (1981), "Rank transformations as a bridge between parametric and nonparametric statistics," *The American Statistician*, 35, 124–129.

Cook, R. D. and Weisberg, S. (1980), "Characterizations of an empirical influence function for detecting influential cases in regression," *Technometrics*, 22, 495–508.

— (1982), *Residuals and Influence in Regression*, Chapman & Hall, London.

— (1984), *An Introduction to Regression Graphics*, John Wiley & Sons: New York.

Craig, A. T. (1943), "Note on the independence of certain quadratic forms," *The Annals of Mathematical Statistics*, 14, 195–197.

Daniel, C. and Wood, F. S. (1971), *Fitting Equations to Data: Computer Analysis of Multifactor Data*, John Wiley & Sons: New York.

Dantzig, G. B. (1963), *Linear Programming and Extensions*, Princeton University Press, Princeton, N.J.

Dempster, A. P., Laird, N. M., and Rubin, D. B. (1977), "Maximum likelihood from incomplete data via the EM algorithm," *Journal of the Royal Statistical Society Series B*, 39, 1–22.

Devlin, S. J., Gnanadesikan, R., and Kettenring, J. R. (1976), "Some multivariate applications of elliptical distributions," in *Essays in Probability and Statistics*, ed. Ikeda, S., pp. 365–395.

Diaconis, P. and Shahshahani, M. (1984), "On nonlinear functions of linear combinations," *SIAM Journal on Scientific and Statistical Computing*, 5, 175–191.

Diggle, P. J., Liang, K.-Y., and Zeger, S. L. (1994), *Analysis of Longitudinal Data*, Oxford University Press: UK.

Dixon, L. C. W. (1972), *Nonlinear Optimisation*, English University Press: London.

Dmitrienko, A., Bretz, F., Westfall, P. H., Troendle, J., Wiens, B. L., Tamhane, A. C., and Hsu, J. C. (2009), "Multiple testing methodology," in *Multiple Testing Problems in Pharmaceutical Statistics*, Chapman and Hall/CRC, pp. 53–116.

Dodge, Y. and Jurečková, J. (2000), *Adaptive Regression*, Springer-Verlag, New York.

Draper, N. R. and John, J. A. (1981), "Influential observations and outliers in regression," *Technometrics*, 23, 21–26.

Draper, N. R. and Smith, H. (1998), *Applied Regression Analysis*, John Wiley & Sons, Inc., New York, 3rd ed.

Driscoll, M. F. (1999), "An improved result relating quadratic forms and chi-square distributions," *The American Statistician*, 53, 273–275.

Driscoll, M. F. and Krasnicka, B. (1995), "An accessible proof of Craig's theorem in the general case," *The American Statistician*, 49, 59–62.

Dunnett, C. W. (1964), "New tables for multiple comparisons with a control," *Biometrics*, 20, 482–491.

Dunnett, C. W. and Tamhane, A. C. (1992), "A step-up multiple test procedure," *Journal of the American Statistical Association*, 87, 162–170.

Durbin, J. and Watson, G. S. (1950), "Testing for serial correlation in least squares regression: I," *Biometrika*, 37, 409–428.

— (1951), "Testing for serial correlation in least squares regression: II," *Biometrika*, 38, 159–177.

Efron, B., Hastie, T., Johnstone, I., and Tibshirani, R. (2004), "Least angle regression," *The Annals of Statistics*, 32, 407–499.

Ellenberg, J. H. (1973), "The joint distribution of the standardized least squares residuals from a general linear regression," *Journal of the American Statistical Association*, 68, 941–943.

Englefield, M. J. (1966), "The commuting inverses of a square matrix," *Mathematical Proceedings of the Cambridge Philosophical Society*, 62, 667–671.

Eubank, R. L. (1999), *Nonparametric Regression and Spline Smoothing*, Marcel Dekker, Inc., New York, 2nd ed.

Fan, J. and Lv, J. (2010), "A selective overview of variable selection in high dimensional feature space," *Statistica Sinica*, 20, 101–148.

Fang, K. T. and Anderson, T. W. (1990), *Statistical Inference in Elliptically Contoured and Related Distributions*, Allerton Press Inc.: New York.

Fang, K. T., Kotz, S., and Ng, K. (1990), *Symmetric Multivariate and Related Distributions*, Chapman & Hall: London.

Fisher, R. A. (1921), "On the "probable error" of a coefficient of correlation deduced from a small sample," *Metron*, 1, 3–32.

Friedman, J., Hastie, T., Höfling, H., and Tibshirani, R. (2007), "Pathwise coordinate optimization," *The Annals of Applied Statistics*, 1, 302–332.

Friedman, J. H. (1991), "Multivariate adaptive regression splines," *The Annals of Statistics*, 1–67.

Friedman, J. H. and Stuetzle, W. (1981), "Projection pursuit regression," *Journal of the American Statistical Association*, 76, 817–823.

Friedman, J. H. and Tukey, J. W. (1974), "A projection pursuit algorithm for exploratory data analysis," *IEEE Transactions on Computers*, 100, 881–890.

Fu, W. J. (1998), "Penalized regressions: the bridge versus the lasso," *Journal of Computational and Graphical Statistics*, 7, 397–416.

Galea, M., Riquelme, M., and Paula, G. A. (2000), "Diagnostic methods in elliptical linear regression models," *Brazilian Journal of Probability and Statistics*, 14, 167–184.

Geisser, S. and Greenhouse, S. W. (1958), "An extension of Box's results on the use of the F distribution in multivariate analysis," *Annals of Mathematical Statistics*, 29, 885–891.

Ghosh, M. (1996), "Wishart distribution via induction," *The American Statistician*, 50, 243–246.

Gnanadesikan, R. (1997), *Methods for Statistical Data Analysis of Multivariate Observations*, John Wiley & Sons, Inc., New York, 2nd ed.

Goldfeld, S. M. and Quandt, R. E. (1965), "Some tests for homoscedasticity," *Journal of the American Statistical Association*, 60, 539–547.

Golub, G. H. and Styan, G. P. H. (1973), "Numerical computations for univariate linear models," *Journal of Statistical Computation and Simulation*, 2, 253–274.

Golub, G. H. and Van Loan, C. F. (1989), *Matrix Computations*, Johns Hopkins University Press, Baltimore, 2nd ed.

Graybill, F. A. (1954), "On quadratic estimates of variance components," *The Annals of Mathematical Statistics*, 25, 367–372.

— (1961), *An Introduction to Linear Statistical Models*, McGraw- Hill: New York.

— (1983), *Matrices with Applications in Statistics*, Wadsworth Advanced Books and Software, Belmont, Calif., 2nd ed.

Graybill, F. A. and Hultquist, R. A. (1961), "Theorems concerning Eisenhart's model II," *The Annals of Mathematical Statistics*, 32, 261–269.

Green, P. J. and Silverman, B. W. (1994), *Nonparametric Regression and Generalized Linear Models: A Roughness Penalty Approach*, vol. 58, Chapman & Hall, London.

Greenhouse, S. W. and Geisser, S. (1959), "On methods in the analysis of profile data," *Psychometrika*, 24, 95–112.

Guenther, W. C. (1964), "Another derivation of the non-central chi-square distribution," *Journal of the American Statistical Association*, 59, 957–960.

Gupta, A. K. and Varga, T. (1993), *Elliptically Contoured Models in Statistics*, Kluwer Academic Publishers: Netherlands.

Halmos, P. R. and Savage, L. J. (1949), "Application of the Radon–Nikodym theorem to the theory of sufficient statistics," *The Annals of Mathematical Statistics*, 20, 225–241.

Hampel, F. R. (1974), "The influence curve and its role in robust estimation," *Journal of the American Statistical Association*, 69, 383–393.

Harrison, J. and West, M. (1991), "Dynamic linear model diagnostics," *Biometrika*, 78, 797–808.

Harville, D. A. (1976), "Extension of the Gauss–Markov theorem to include the estimation of random effects," *The Annals of Statistics*, 4, 384–395.

— (1997), *Matrix Algebra from a Statistician's Perspective*, Springer-Verlag: New York.

Hastie, T. J. and Tibshirani, R. J. (1990), *Generalized Additive Models*, Chapman & Hall: London.

Hastie, T. J., Tibshirani, R. J., and Friedman, J. (2009), *The Elements of Statistical Learning: Data Mining, Inference, and Prediction*, Spring-Verlag: New York, 2nd ed.

Hayes, J. G. (1974), "Numerical methods for curve and surface fitting (by polynomials and splines)," *Journal of Inst. Math. Appl.*, 10, 144–152.

Hayes, K. and Haslett, J. (1999), "Simplifying general least squares," *The American Statistician*, 53, 376–381.

Hayter, A. J. (1984), "A proof of the conjecture that the Tukey–Kramer multiple comparisons procedure is conservative," *The Annals of Statistics*, 61–75.

Hertz, J., Krogh, A., and Palmer, R. G. (1991), *Introduction to the Theory of Neural Computation*, Addison-Wesley Publishing Company, Advanced Book Program, Redwood City, CA.

Hinkley, D. V. (1975), "On power transformations to symmetry," *Biometrika*, 62, 101–111.

Hoaglin, D. C. and Welsch, R. E. (1978), "The hat matrix in regression and ANOVA," *The American Statistician*, 32, 17–22.

Hochberg, Y. (1988), "A sharper Bonferroni procedure for multiple tests of significance," *Biometrika*, 75, 800–802.

Hochberg, Y. and Tamhane, A. C. (1987), *Multiple Comparison Procedures*, John Wiley & Sons, Inc., New York.

Hoerl, A. E. (1962), "Applications of ridge analysis to regression problems," *Chem. Eng. Progress.*, 58, 54–59.

Hoerl, A. E. and Kennard, R. W. (1970a), "Ridge regression: Applications to nonorthogonal problems," *Technometrics*, 12, 69–82.

— (1970b), "Ridge regression: Biased estimation for nonorthogonal problems," *Technometrics*, 12, 55–67.

Hollander, M., Wolfe, D. A., and Chicken, E. (2014), *Nonparametric Statistical Methods*, John Wiley & Sons: New York, 3rd ed.

Holm, S. (1979), "A simple sequentially rejective multiple test procedure," *Scandinavian Journal of Statistics*, 6, 65–70.

Horn, R. A. and Johnson, C. R. (2013), *Matrix Analysis*, Cambridge University Press, Cambridge, 2nd ed.

Hsu, J. C. (1984), "Constrained simultaneous confidence intervals for multiple comparisons with the best," *The Annals of Statistics*, 12, 1136–1144.

Huber, P. J. (1964), "Robust estimation of a location parameter," *The Annals of Mathematical Statistics*, 35, 73–101.

— (1981), *Robust Statistics*, John Wiley & Sons, Inc., New York.

Ibrahim, J. G., Chen, M.-H., and Sinha, D. (2001), "Criterion-based methods for Bayesian model assessment," *Statistica Sinica*, 11, 419–443.

James, G. S. (1952), "Notes on a theorem of Cochran," *Proc. Camb. Phil. Soc*, 48, 443–446.

John, J. A. and Draper, N. R. (1980), "An alternative family of transformations," *Journal of the Royal Statistical Society Series C*, 29, 190–197.

Johnson, R. A. and Wichern, D. W. (2007), *Applied Multivariate Statistical Analysis*, Pearson Prentice Hall, Upper Saddle River, NJ, sixth ed.

Judge, G. G., Griffiths, W. E., Hill, R. C., Lutkepohl, H., and Lee, T.-C. (1985), *The Theory and Practice of Econometrics*, John Wiley & Sons: New York, 2nd ed.

Kelker, D. (1970), "Distribution theory of spherical distributions and a location-scale parameter generalization," *Sankhyā,: The Indian Journal of Statistics, Series A*, 32, 419–430.

Kendall, M. G. and Stuart, A. (1958), *The Advanced Theory of Statistics, Vol. 1: Distribution Theory*, Hafner Publishing: New York.

Keuls, M. (1954), "Testing differences between means in an analysis of variance," *Biometrics*, 10, 167–168.

Khuri, A. I. (1999), "A necessary condition for a quadratic form to have a chi-squared distribution: an accessible proof," *International Journal of Mathematical Education in Science and Technology*, 30, 335–339.

Kotz, S., Balakrishnan, N., and Johnson, N. L. (2000), *Continuous Multivariate Distributions. Vol. 1*, John Wiley & Sons: New York, 2nd ed.

Kruskal, W. H. (1952), "A nonparametric test for the several sample problem," *The Annals of Mathematical Statistics*, 23, 525–540.

Kruskal, W. H. and Wallis, W. A. (1952), "Use of ranks in one-criterion variance analysis," *Journal of the American Statistical Association*, 47, 583–621.

Laha, R. G. (1956), "On the stochastic independence of two second-degree polynomial statistics in normally distributed variates," *The Annals of Mathematical Statistics*, 27, 790–796.

Laird, N. M. and Ware, J. H. (1982), "Random-effects models for longitudinal data," *Biometrics*, 38, 963–974.

Larsen, W. A. and McCleary, S. J. (1972), "The use of partial residual plots in regression analysis," *Technometrics*, 14, 781–790.

Lee Rodgers, J. and Nicewander, W. A. (1988), "Thirteen ways to look at the correlation coefficient," *The American Statistician*, 42, 59–66.

Lehmann, E. L. and Casella, G. (1998), *Theory of Point Estimation*, Springer-Verlag, New York, 2nd ed.

Levene, H. (1960), "Robust tests for equality of variances," in *Contributions to Probability and Statistics. Essays in Honor of Harold Hotelling*, Stanford University Press, pp. 278–292.

Lindley, D. V. (1980), "Approximate Bayesian methods," *Trabajos de estadística y de investigación operativa*, 31, 223–245.

Lindley, D. V. and Smith, A. F. M. (1972), "Bayes estimates for the linear model," *Journal of the Royal Statistical Society Series B*, 34, 1–18.

Little, R. J. A. and Rubin, D. B. (1987), *Statistical Analysis with Missing Data*, John Wiley & Sons, Inc., New York.

Magnus, J. and Neudecker, H. (1988), *Matrix Differential Calculus with Applications in Statistics and Econometrics*, John Wiley & Sons: New York.

Mallows, C. L. (1973), "Some comments on C_p," *Technometrics*, 15, 661–675.

— (1995), "Some comments on C_p," *Technometrics*, 37, 362–372.

Manly, B. F. J. (1976), "Exponential data transformations," *Journal of the Royal Statistical Society Series D*, 25, 37–42.

Mardia, K. V. (1970), "Measures of multivariate skewness and kurtosis with applications," *Biometrika*, 57, 519–530.

— (1975), "Assessment of multinormality and the robustness of Hotelling's T^2 test," *Journal of the Royal Statistical Society Series C*, 24, 163–171.

Mardia, K. V., Kent, J. T., and Bibby, J. M. (1979), *Multivariate Analysis*, Academic Press [Harcourt Brace Jovanovich, Publishers], London-New York-Toronto, Ont.

McCullagh, P. (1983), "Quasi-likelihood functions," *The Annals of Statistics*, 11, 59–67.

McCullagh, P. and Nelder, J. A. (1989), *Generalized Linear Models*, Chapman & Hall, London, 2nd ed.

McCulloch, C. E., Searle, S. R., and Neuhaus, J. M. (2011), *Generalized, Linear, and Mixed Models*, John Wiley & Sons: New York, 2nd ed.

McElroy, F. W. (1967), "A necessary and sufficient condition that ordinary least-squares estimators be best linear unbiased," *Journal of the American Statistical Association*, 62, 1302–1304.

Meinhold, R. J. and Singpurwalla, N. D. (1983), "Understanding the Kalman filter," *The American Statistician*, 37, 123–127.

Miller, Jr., R. G. (1981), *Simultaneous Statistical Inference*, Springer-Verlag, New York-Berlin, 2nd ed.

Moser, B. K. and Sawyer, J. K. (1998), "Algorithms for sums of squares and covariance matrices using Kronecker products," *The American Statistician*, 52, 54–57.

Muirhead, R. J. (1982), *Aspects of Multivariate Statistical Theory*, John Wiley & Sons, Inc., New York.

Mukhopadhyay, N. (2000), *Probability and Statistical Inference*, Marcel Dekker: New York.

Myers, R. H. (1971), *Response Surface Methodology*, Allyn & Bacon: Boston.

Nelder, J. A. and Wedderburn, R. W. M. (1972), "Generalized linear models," *Journal of the Royal Statistical Society Series A*, 135, 370–384.

Newman, D. (1939), "The distribution of range in samples from a normal population, expressed in terms of an independent estimate of standard deviation," *Biometrika*, 31, 20–30.

Ogasawara, T. and Takahashi, M. (1951), "Independence of quadratic quantities in a normal system," *Journal of Science of the Hiroshima University, Series A (Mathematics, Physics, Chemistry)*, 15, 1–9.

Ogawa, J. (1950), "On the independence of quadratic forms in a non-central normal system," *Osaka Mathematical Journal*, 2, 151–159.

— (1993), "A history of the development of Craig–Sakamoto's theorem viewed from Japanese standpoint," *Proc. Ann. Inst. Statist. Math*, 41, 47–59.

Osborne, M. R. (1992), "An effective method for computing regression quantiles," *IMA Journal of Numerical Analysis*, 12, 151–166.

Patterson, H. D. and Thompson, R. (1971), "Recovery of inter-block information when block sizes are unequal," *Biometrika*, 58, 545–554.

Pierce, D. A. and Schafer, D. W. (1986), "Residuals in generalized linear models," *Journal of the American Statistical Association*, 81, 977–986.

Powell, J. L. (1984), "Least absolute deviations estimation for the censored regression model," *Journal of Econometrics*, 25, 303–325.

Pregibon, D. (1981), "Logistic regression diagnostics," *The Annals of Statistics*, 9, 705–724.

Pringle, R. M. and Rayner, A. A. (1971), *Generalized Inverse Matrices with Applications to Statistics*, Hafner Publishing Co., New York.

R Core Team (2018), *R: A Language and Environment for Statistical Computing*, R Foundation for Statistical Computing, Vienna, Austria.

Rao, C. R. (1952), "Some theorems on minimum variance estimation," *Sankhyā,: The Indian Journal of Statistics, Series A*, 12, 27–42.

— (1970), "Estimation of heteroscedastic variances in linear models," *Journal of the American Statistical Association*, 65, 161–172.

— (1971a), "Estimation of variance and covariance components – MINQUE theory," *Journal of Multivariate Analysis*, 1, 257–275.

— (1971b), "Minimum variance quadratic unbiased estimation of variance components," *Journal of Multivariate Analysis*, 1, 445–456.

— (1972), "Estimation of variance and covariance components in linear models," *Journal of the American Statistical Association*, 67, 112–115.

— (1973), *Linear Statistical Inference and Its Applications*, John Wiley & Sons: New York, 2nd ed.

Rao, C. R. and Mitra, S. K. (1971), *Generalized Inverse of Matrices and Its Applications*, John Wiley & Sons, Inc., New York-London-Sydney.

Rao, C. R. and Toutenberg, H. (1995), *Linear Models: Least Squares and Alternatives*, Springer-Verlag: New York, 2nd ed.

Rao, C. R. and Zhao, L. C. (1993), "Asymptotic normality of LAD estimator in censored regression models," *Mathematical Methods of Statistics*, 2, 228–239.

Rauch, H. E., Tung, F., and Striebel, C. T. (1965), "Maximum likelihood estimates of linear dynamic systems," *AIAA journal*, 3, 1445–1450.

Reid, J. G. and Driscoll, M. F. (1988), "An accessible proof of Craig's theorem in the noncentral case," *The American Statistician*, 42, 139–142.

Rumelhart, D. E., Hinton, G. E., and Williams, R. J. (1986), "Learning representations by back-propagating errors," *Nature*, 323, 533–536.

Ruppert, D. and Carroll, R. J. (1980), "Trimmed least squares estimation in the linear model," *Journal of the American Statistical Association*, 75, 828–838.

Ryan, T. A., Joiner, B. L., and Ryan, B. F. (1976), *The Minitab Student Handbook*, Duxbury Press: New York.

Samorodnitsky, G. and Taqqu, M. S. (1994), *Stable Non-Gaussian Random Processes: Stochastic Models with Infinite Variance*, Chapman & Hall: New York.

Saxena, K. M. L. and Alam, K. (1982), "Estimation of the non-centrality parameter of a chi-squared distribution," *The Annals of Statistics*, 1012–1016.

Scheffé, H. (1953), "A method for judging all contrasts in the analysis of variance," *Biometrika*, 40, 87–110.

— (1959), *The Analysis of Variance*, John Wiley & Sons, Inc., New York; Chapman & Hall, Ltd., London.

Schwarz, G. (1978), "Estimating the dimension of a model," *The Annals of Statistics*, 6, 461–464.

Searle, S. R. (1971), *Linear Models*, John Wiley & Sons, Inc., New York-London-Sydney.

— (1982), *Matrix Algebra Useful for Statistics*, John Wiley & Sons, Ltd., Chichester.

— (1987), *Linear Models for Unbalanced Data*, John Wiley & Sons, Inc., New York.

Searle, S. R., Casella, G., and McCulloch, C. E. (1992), *Variance Components*, John Wiley & Sons, Inc., New York, a Wiley-Interscience Publication.

Seber, G. A. F. (1977), *Linear Regression Analysis*, John Wiley & Sons, New York-London-Sydney.

— (1984), *Multivariate Observations*, John Wiley & Sons, Inc., New York.

Seely, J. F., Birkes, D., and Lee, Y. (1997), "Characterizing sums of squares by their distributions," *The American Statistician*, 51, 55–58.

Shanbhag, D. N. (1966), "On the independence of quadratic forms," *Journal of the Royal Statistical Society Series B*, 28, 582–583.

Shapiro, S. S. and Wilk, M. B. (1965), "An analysis of variance test for normality (complete samples)," *Biometrika*, 52, 591–611.

Shepp, L. A. and Vardi, Y. (1982), "Maximum likelihood reconstruction for emission tomography," *IEEE Transactions on Medical Imaging*, 1, 113–122.

Shumway, R. H. and Stoffer, D. S. (2017), *Time Series Analysis and Its Applications*, Springer, Cham, 4th ed.

Simes, R. J. (1986), "An improved Bonferroni procedure for multiple tests of significance," *Biometrika*, 73, 751–754.

Smith, A. F. M. and Spiegelhalter, D. J. (1980), "Bayes factors and choice criteria for linear models," *Journal of the Royal Statistical Society Series B*, 42, 213–220.

Solomon, P. J. (1985), "Transformations for components of variance and covariance," *Biometrika*, 72, 233–239.

Spiegelhalter, D. J., Best, N. G., Carlin, B. P., and Van Der Linde, A. (2002), "Bayesian measures of model complexity and fit," *Journal of the Royal Statistical Society Series B*, 64, 583–639.

Spiegelhalter, D. J., Best, N. G., Carlin, B. P., and Van der Linde, A. (2014), "The deviance information criterion: 12 years on," *Journal of the Royal Statistical Society Series B*, 76, 485–493.

Stewart, G. W. (1973), *Introduction to Matrix Computations*, Academic Press [A subsidiary of Harcourt Brace Jovanovich, Publishers], New York-London.

Storey, J. D. (2002), "A direct approach to false discovery rates," *Journal of the Royal Statistical Society Series B*, 64, 479–498.

Stukel, T. A. (1988), "Generalized logistic models," *Journal of the American Statistical Association*, 83, 426–431.

Thisted, R. A. (1988), *Elements of Statistical Computing*, Chapman & Hall, New York.

Tibshirani, R. (1996), "Regression shrinkage and selection via the lasso," *Journal of the Royal Statistical Society Series B*, 58, 267–288.

Tibshirani, R. J. (2015), "A general framework for fast stagewise algorithms," *The Journal of Machine Learning Research*, 16, 2543–2588.

Tierney, L. and Kadane, J. B. (1986), "Accurate approximations for posterior moments and marginal densities," *Journal of the American Statistical Association*, 81, 82–86.

Tseng, P. (2001), "Convergence of a block coordinate descent method for nondifferentiable minimization," *Journal of optimization theory and applications*, 109, 475–494.

Tukey, J. W. (1957), "On the comparative anatomy of transformations," *The Annals of Mathematical Statistics*, 28, 602–632.

Vehtari, A. and Ojanen, J. (2012), "A survey of Bayesian predictive methods for model assessment, selection and comparison," *Statistics Surveys*, 6, 142–228.

Velleman, P. F. and Welsch, R. E. (1981), "Efficient computing of regression diagnostics," *The American Statistician*, 35, 234–242.

Warner, B. and Misra, M. (1996), "Understanding neural networks as statistical tools," *The American Statistician*, 50, 284–293.

Weeks, D. L. and Williams, D. R. (1964), "A note on the determination of connectedness in an N-way cross classification," *Technometrics*, 6, 319–324.

West, M. and Harrison, J. (1997), *Bayesian Forecasting and Dynamic Models*, Springer-Verlag, New York, 2nd ed.

Westfall, P. H. (1987), "A comparison of variance component estimates for arbitrary underlying distributions," *Journal of the American Statistical Association*, 82, 866–874.

White, H. (1980), "A heteroskedasticity-consistent covariance matrix estimator and a direct test for heteroskedasticity," *Econometrica*, 48, 817–838.

Williams, E. J. (1959), *Regression Analysis*, John Wiley & Sons, Inc., New York.

Wood, F. S. (1973), "The use of individual effects and residuals in fitting equations to data," *Technometrics*, 15, 677–695.

Wu, L. S.-Y., Hosking, J. R. M., and Ravishanker, N. (1993), "Reallocation outliers in time series," *Journal of the Royal Statistical Society Series C*, 42, 301–313.

Zou, H. and Hastie, T. (2005), "Regularization and variable selection via the elastic net," *Journal of the Royal Statistical Society Series B*, 67, 301–320.

Zyskind, G. and Martin, F. B. (1969), "On best linear estimation and general Gauss–Markov theorem in linear models with arbitrary nonnegative covariance structure," *SIAM Journal on Applied Mathematics*, 17, 1190–1202.

Author Index

Subject Index

Printed in the United States
by Baker & Taylor Publisher Services